# STATISTICAL FLUID MECHANICS: Mechanics of Turbulence

Volume II

A. S. MONIN and A. M. YAGLOM

*English edition updated, augmented and revised by the authors*

Edited by John L. Lumley

DOVER PUBLICATIONS, INC.
Mineola, New York

### *Copyright*



### *Bibliographical Note*

This Dover edition, first published in 2007, is an unabridged republication of the work published in 1975 by The MIT Press, Cambridge, Massachusetts. That work was originally published in 1965 by Nauka Press, Moscow, under the title *Statisicheskaya gridomekhanika-Mekhanika Turbulenosti,* by Andrey S. Monin and Akiva M. Yaglom. Translated from the Russian by Scripta Graphica, Inc. This Dover edition is published by special arrangement with The MIT Press, 55 Hayward Street, Cambridge, MA 02142.

*Library of Congress Cataloging-in-Publication Data*

Monin, A. S.
Statistical fluid mechanics : mechanics of turbulence / A.S. Monin an A.M. Yaglom ; edited by John Lumley.
p. cm.
Originally published: Cambridge, Mass. : MIT Press, c1971–1975.
Includes bibliographical references and index.
**ISBN-13: 978-0-486-45891-5 (pbk.)**
**ISBN-10: 0-486-45891-1 (pbk.)**
1. Hydrodynamics. 2. Fluid mechanics. 3. Turbulence. I. Title.

QA911 .M6313 2007
532'.0527—dc22

2006053473

Manufactured in the United States by LSC Communications
45891106 2022
**www.doverpublications.com**

# AUTHORS' PREFACE TO THE ENGLISH EDITION

We are very happy that this book, the preparation of which has taken so many years of our lives, is now available to English readers in its entirety. We have been very encouraged by the warm reception accorded to Volume 1 (judging from the reviews that we have seen), although it is not easy for readers to form an opinion on the basis of the first half, when the two volumes were prepared for use as a whole. As in Volume 1, we have done our best to make the book up-to-date by adding new material and figures (particularly in Chapter 8) amounting to nearly ten percent of the total, and more than 250 references to works which appeared after the publication of the Russian edition. We have also added supplementary remarks and errata to Volume 1. We understand that the incorporation of the additions in the main body of the book has made the work of our friendly editor, J. L. Lumley, much more difficult, and we are grateful to him for his attempts to make the book better.

**A. S. Monin**
**A. M. Yaglom**

# EDITOR'S PREFACE TO THE ENGLISH EDITION

As with Volume 1, the translated manuscript was sent to A. M. Yaglom, who revised both content and style extensively. The manuscript was then returned to me for editing. Afterthoughts, however, continued to arrive as postscripts to letters until at last one manuscript page had nearly doubled in length by five successive accretions. A halt was finally declared at the end of May, 1973. I have found the seven years' interaction necessary to prepare these volumes for publication in English a thoroughly enjoyable and rewarding experience.

**J. L. Lumley**

# CONTENTS

Authors' Preface to the English Edition .................... iii
Editor's Preface to the English Edition .................... v

Chapter 6 Mathematical Description of Turbulence. Spectral Functions .................... 1

11. Spectral Representations of Stationary Processes and Homogeneous Fields .................... 1
- 11.1 Spectral Representation of Stationary Processes .......... 1
- 11.2 Spectral Representation of Homogeneous Fields .......... 16
- 11.3 Partial Derivatives of Homogeneous Fields. Divergence and Curl of a Vector Field .......... 23

12. Isotropic Random Fields .................... 29
- 12.1 Correlation Functions and Spectra of Scalar Isotropic Fields .......... 29
- 12.2 Correlation Functions and Spectra of Isotropic Fields .......... 35
- 12.3 Solenoidal and Potential Isotropic Vector Fields .......... 49
- 12.4 One-Point and Two-Point Higher-Order Moments of Isotropic Fields .......... 58
- 12.5 Three-Point Moments of Isotropic Fields .......... 75

13. Locally Homogeneous and Locally Isotropic Random Fields .................... 80
- 13.1 Processes with Stationary Increments .......... 80
- 13.2 Locally Homogeneous Fields .......... 93
- 13.3 Locally Isotropic Fields .......... 98

Chapter 7 Isotropic Turbulence .................... 113

14. Equations for the Correlation and Spectral Functions of Isotropic Turbulence .................... 113
- 14.1 Definition of Isotropic Turbulence and the Possibilities of its Experimental Realization .......... 113
- 14.2 Equations for the Velocity Correlations .......... 117
- 14.3 Equations for the Velocity Spectra .......... 123
- 14.4 Correlations and Spectra Containing Pressure .......... 130
- 14.5 Correlations and Spectra Containing the Temperature .......... 136

15. **The Simplest Consequences of the Correlation and Spectral Equations** . . . . . . . . . . . . . . . . . . . . . . . . . . 141
15.1 Balance Equations for Energy, Vorticity, and Temperature-Fluctuation Intensity. . . . . . . . . . . . . . . 141
15.2 The Loitsyanskii and Corrsin Integrals . . . . . . . . . . . 146
15.3 Final Period of Decay of Isotropic Turbulence . . . . . . . 152
15.4 Experimental Data on the Final Period of Decay. The Decay of Homogeneous Turbulence . . . . . . . . . . . 162
15.5 Asymptotic Behavior of the Correlations and Spectra of Homogeneous Turbulence in the Range of Large Length Scales (or Small Wave Numbers) . . . . . . . . . . . 169
15.6 The Influence of the Spectrum Singularity on the Final Period Decay . . . . . . . . . . . . . . . . . . . . . . . . . . 174
16. **Self-Preservation Hypotheses** . . . . . . . . . . . . . . . . . . . . . . . 177
16.1 The von Kármán Hypothesis on the Self-Preservation of the Velocity Correlation Functions . . . . . . . . . . . 177
16.2 Less Stringent Forms of the von Kármán Hypothesis . . . 181
16.3 Spectral Formulation of the Self-Preservation Hypotheses . . . . . . . . . . . . . . . . . . . . . . . . . . . . . . 185
16.4 Experimental Verification of the Self-Preservation Hypotheses . . . . . . . . . . . . . . . . . . . . . . . . . . . . . . 189
16.5 The Kolmogorov Hypotheses on Small-Scale Self-Preservation at High Enough Reynolds Numbers . . . 197
16.6 Conditions for the Existence of Kolmogorov Self-Preservation in Grid Turbulence . . . . . . . . . . . . . 204
16.7 The Meso-Scale Quasi-Equilibrium Hypothesis. Self-Preservation of Temperature Fluctuations . . . . . . . 210
17. **Spectral Energy-Transfer Hypotheses** . . . . . . . . . . . . . . . . . . 212
17.1 Approximate Formulas for the Spectral Energy Transfer . . . . . . . . . . . . . . . . . . . . . . . . . . . . . . . . 212
17.2 Application of the Energy Transfer Hypotheses to the Study of the Shape of the Spectrum in the Quasi-Equilibrium Range . . . . . . . . . . . . . . . . . . . . . 225
17.3 Application of the Energy-Transfer Hypotheses to Decaying Turbulence behind a Grid . . . . . . . . . . . . . 235
17.4 Self-Preserving Solutions of the Approximate Equations for the Energy Spectrum . . . . . . . . . . . . . 237
18. **The Millionshchikov Zero-Fourth-Cumulant Hypothesis and its Application to the Investigation of Pressure and Acceleration Fluctuations** . . . . . . . . . . . . . . . . . . . . . . . 241
18.1 The Zero-Fourth-Cumulant Hypothesis and the Data on Velocity Probability Distributions . . . . . . . . . 241
18.2 Calculation of the Pressure Correlation and Spectra . . . 250
18.3 Estimation of the Turbulent Acceleration Fluctuations . . 256

19. **Dynamic Equations for the Higher-Order Moments and the Closure Problem** . . . . . 260
19.1 Equations for the Third-Order Moments of Flow Variables . . . . . 260
19.2 Closure of the Moment Equations by the Moment Discard Assumption . . . . . 267
19.3 Closure of the Second- and Third-Order Moment Equations Using the Millionshchikov Zero-Fourth-Cumulant Hypothesis . . . . . 271
19.4 Zero-Fourth-Cumulant Approximation for Temperature Fluctuations in Isotropic Turbulence . . . 286
19.5 Space-Time Correlation Functions. The Case of Stationary Isotropic Turbulence . . . . . 290
19.6 Application of Perturbation Theory and the Diagram Technique . . . . . 295
19.7 Equations for the Finite-Dimensional Probability Distributions of Velocities . . . . . 310
20. **Turbulence in Compressible Fluids** . . . . . 317
20.1 Invariants of Isotropic Compressible Turbulence . . . . . 317
20.2 Linear Theory; Final Period of Decay of Compressible Turbulence . . . . . 321
20.3 Quadratic Effects; Generation of Sound by Turbulence . . . . . 328

Chapter 8 **Locally Isotropic Turbulence** . . . . . 337
21. **General Description of the Small-Scale Structure of Turbulence at Large Reynolds Numbers** . . . . . 337
21.1 A Qualitative Scheme for Developed Turbulence . . . . . 337
21.2 Definition of Locally Isotropic Turbulence . . . . . 341
21.3 The Kolmogorov Similarity Hypotheses . . . . . 345
21.4 Local Structure of the Velocity Fluctuations . . . . . 351
21.5 Statistical Characteristics of Acceleration, Vorticity, and Pressure Fields . . . . . 368
21.6 Local Structure of the Temperature Field for High Reynolds and Peclet Numbers . . . . . 377
21.7 Local Characteristics of Turbulence in the Presence of Buoyancy Forces and Chemical Reactions. Effect of Thermal Stratification . . . . . 387
22. **Dynamic Theory of the Local Structure of Developed Turbulence** . . . . . 395
22.1 Equations for the Structure and Spectral Functions of Velocity and Temperature . . . . . 395
22.2 Closure of the Dynamic Equations . . . . . 403

22.3 Behavior of the Turbulent Energy Spectrum in the Far Dissipation Range . . . . . . . . . . . . . . . . . . . . . . . . . 421
22.4 Behavior of the Temperature Spectrum at Very Large Wave Numbers . . . . . . . . . . . . . . . . . . . . . 433
**23. Experimental Data on the Fine Scale Structure of Developed Turbulence** . . . . . . . . . . . . . . . . . . . . . . . . . . 449
23.1 Methods of Measurement; Application of Taylor's Frozen-Turbulence Hypothesis . . . . . . . . . . . . . . . . . . 449
23.2 Verification of the Local Isotropy Assumption. . . . . . . 453
23.3 Verification of the Second Kolmogorov Similarity Hypothesis for the Velocity Fluctuations . . . . . . . . . 461
23.4 Verification of the First Kolmogorov Similarity Hypothesis for the Velocity Field . . . . . . . . . . . . . . 486
23.5 Data on the Local Structure of the Temperature and other Scalar Fields Mixed by Turbulence . . . . . . . . . 494
23.6 Data on Turbulence Spectra in the Atmosphere beyond the Low-Frequency Limit of the Inertial Subrange . . . . . . . . . . . . . . . . . . . . . . . . . . . . 517
**24. Diffusion in an Isotropic Turbulence** . . . . . . . . . . . . . . . . . . 527
24.1 Diffusion in an Isotropic Turbulence. Statistical Characteristics of the Motion of a Fluid Particle . . . . . 527
24.2 Statistical Characteristics of the Motion of a Pair of Fluid Particles . . . . . . . . . . . . . . . . . . . . . . . . . . 536
24.3 Relative Diffusion and Richardson's Four-Thirds Law . . . . . . . . . . . . . . . . . . . . . . . . . . . . . . . . . . 551
24.4 Hypotheses on the Probability Distributions of Local Diffusion Characteristics . . . . . . . . . . . . . . . . 567
24.5 Material Line and Surface Stretching in Turbulent Flows . . . . . . . . . . . . . . . . . . . . . . . . . . . . . . . . 578
**25. Refined Treatment of the Local Structure of Turbulence, Taking into Account Fluctuations in Dissipation Rate** . . . . . . . 584
25.1 General Considerations and Model Examples . . . . . . . 584
25.2 Refined Similarity Hypothesis . . . . . . . . . . . . . . . . . 590
25.3 Statistical Characteristics of the Dissipation. . . . . . . . . 594
25.4 Refined Expressions for the Statistical Characteristics of Small-Scale Turbulence . . . . . . . . . . . . . . . . . . . 640
25.5 More General Form of the Refined Similarity Hypothesis . . . . . . . . . . . . . . . . . . . . . . . . . . . . . 650

Chapter 9 Wave Propagation Through Turbulence . . . . . . . . . 653
**26. Propagation of Electromagnetic and Sound Waves in a Turbulent Medium** . . . . . . . . . . . . . . . . . . . . . . . . . . . . . 653
26.1 Foundations of the Theory of Electromagnetic Wave Propagation in a Turbulent Medium . . . . . . . . . 653

26.2 Sound Propagation in a Turbulent Atmosphere. . . . . . . 668
26.3 Turbulent Scattering of Electromagnetic and Sound Waves . . . . . . . . . . . . . . . . . . . . . . . . . . . 674
26.4 Fluctuations in the Amplitude and Phase of Electromagnetic and Sound Waves in a Turbulent Atmosphere . . . . . . . . . . . . . . . . . . . . . . . . . . . 685
26.5 Strong Fluctuations of Wave Amplitude . . . . . . . . . . . 704

**27. Stellar Scintillation.** . . . . . . . . . . . . . . . . . . . . . . . . . . . 721
27.1 Fluctuations in the Amplitude and Phase of Star Light Observed on the Earth's Surface . . . . . . . . . . . 721
27.2 The Effect of Telescope Averaging and Scintillation of Stellar and Planetary Images . . . . . . . . . . . . . . . . . 729
27.3 Time Spectra of Fluctuations in the Intensity of Stellar Images in Telescopes . . . . . . . . . . . . . . . . . . . . 733
27.4 Chromatic Stellar Scintillation . . . . . . . . . . . . . . . . . . 737

**Chapter 10 Functional Formulation of the Turbulence Problem** . . . . . . . . . . . . . . . . . . . . . . . . . . . . 743

**28. Equations for the Characteristic Functional.** . . . . . . . . . . . . . . 743
28.1 Equations for the Spatial Characteristic Functional of the Velocity Field . . . . . . . . . . . . . . . . . . . . . . . . 743
28.2 Spectral Form of the Equations for the Spatial Characteristic Functional . . . . . . . . . . . . . . . . . . . . . . 751
28.3 Equations for the Space-Time Characteristic Functional . . . . . . . . . . . . . . . . . . . . . . . . . . . . . . . 760
28.4 Equations for the Characteristic Functional in the Presence of External Forces . . . . . . . . . . . . . . . . . . . 763

**29. Methods of Solving the Equations for the Characteristic Functional** . . . . . . . . . . . . . . . . . . . . . . . . . . . . . . . . 773
29.1 Use of a Functional Power Series . . . . . . . . . . . . . . 773
29.2 Zero-Order Approximation in the Reynolds Number . . . 783
29.3 Expansion in Powers of the Reynolds Number . . . . . . . 791
29.4 Other Expansion Schemes . . . . . . . . . . . . . . . . . . . . . 798
29.5 Use of Functional Integrals . . . . . . . . . . . . . . . . . . . 802

**Bibliography.** . . . . . . . . . . . . . . . . . . . . . . . . . . . . . . . . . . . 813
**Supplementary Remarks to Volume 1** . . . . . . . . . . . . . . . . . . . . 853
**References** . . . . . . . . . . . . . . . . . . . . . . . . . . . . . . . . . . . . 854
**Errata to Volume 1** . . . . . . . . . . . . . . . . . . . . . . . . . . . . . . . 855
**Author Index** . . . . . . . . . . . . . . . . . . . . . . . . . . . . . . . . . . . 863
**Subject Index** . . . . . . . . . . . . . . . . . . . . . . . . . . . . . . . . . . 871

# 6 MATHEMATICAL DESCRIPTION OF TURBULENCE. SPECTRAL FUNCTIONS

## 11. SPECTRAL REPRESENTATIONS OF STATIONARY PROCESSES AND HOMOGENEOUS FIELDS

### 11.1 Spectral Representation of Stationary Processes

The general concept of the random field was discussed in Chapt. 2 (see Vol. 1) where the main statistical characteristics of such fields, i.e., the mean values and the correlation functions of various orders, were introduced. This was sufficient for discussion of the simplest properties of turbulent flows in Chapts. 3–5. However, when we consider the finer properties of turbulence, we find that this requires a number of new mathematical ideas, and these will in fact be developed in the present chapter.

This chapter will be mainly concerned with the application of *harmonic analysis* to random functions of one or more variables, i.e., random processes and random fields. Harmonic analysis, i.e., the representation of functions by Fourier series or integrals, is very widely used in mathematical physics. It must be remembered, however, that representation by Fourier series is possible only for periodic functions, while representation by Fourier integrals is possible only for functions which vanish sufficiently rapidly at infinity. In applications, however, we frequently encounter nonperiodic functions which do not vanish at infinity and which, strictly speaking, cannot be represented by Fourier series or integrals. From this point of view, random functions turn out to have certain advantages as compared with ordinary, i.e., nonrandom, functions. The point is that a Fourier expansion (or, in other words, *spectral representation*) of a special form, and with a clear physical interpretation, is possible for any *stationary* random process and *homogeneous* random fields which, by definition, do not vanish at infinity.[1]

Let us begin with a consideration of the case of a stationary process $u(t)$. We shall suppose that $\overline{u(t)}=0$ [since, otherwise, we need only consider the process $u'(t)=u(t)-\overline{u(t)}$) instead of $u(t)$]. The idea of a spectral expansion of the process $u(t)$ is most readily understood by considering the following example. Let $u(t)$ be a random function (in general complex)[2] of the form

$$u(t)=\sum_{k=1}^{n} Z_k e^{i\omega_k t}, \tag{11.1}$$

where $\omega_1, \ldots, \omega_n$ are given numbers and $Z_1, \ldots, Z_n$ are complex random variables with the following properties:

$$\overline{Z_k}=0, \quad \overline{Z_k^* Z_l}=0 \quad \text{when } k \neq l \tag{11.2}$$

[1] The spectral representation of stationary random processes and homogeneous random fields by the superposition of harmonic oscillations or plane waves is a special case of the representation of a random function by the superposition of components of a given functional form with random and mutually uncorrelated coefficients, which is possible under very general conditions (see, for example, Yaglom, 1962, 1963, and Lumley, 1967, 1970). For random functions defined on a finite interval, or a finite region of space, or even given on an unbounded region, but such that the mean value of the integral of the square of the function, taken over the entire region, is finite, this "generalized spectral representation" takes the form of a series expansion in terms of a denumerable set of special orthogonal functions. For functions, defined in an unbounded region, and having infinite "energy", it is written in the form of an integral expansion in terms of a continuous set of functions which coincides with the set of one-dimensional or multidimensional harmonics only in the case of stationary processes and homogeneous fields.

[2] It is known that the complex form of the Fourier expansion is the most convenient in harmonic analysis (i.e., expansion in terms of the functions $e^{i\omega t}$). It will therefore be convenient to take the functions $u(t)$ as complex to start with, and only later consider the limitations which follow from the condition that these functions are, in fact, real.

(here and henceforth an asterisk will denote the complex conjugate). The correlation function of a complex process $u(t)$ is conveniently defined by

$$B(t_1, t_2) = \overline{u^*(t_1)\, u(t_2)} \tag{11.3}$$

(for real processes, this definition is of course the same as that given in Chapt. 2). From (11.1) and (11.2) we then have

$$B(t_1, t_2) = \sum_{k=1}^{n} F_k e^{i\omega_k (t_2 - t_1)}, \qquad F_k = \overline{|Z_k|^2} \geqslant 0. \tag{11.4}$$

We see that the correlation function depends only on $t_2 - t_1$, as should be the case for a stationary random process. Under certain additional conditions imposed on the variables $Z_k$ (which are automatically fulfilled when the multidimensional probability distributions for the variables $\mathrm{Re}\, Z_k$ and $\mathrm{Im}\, Z_k$ are Gaussian), all the higher-order moments and finite-dimensional probability distributions of the values of $u(t)$ will also depend only on the time differences, i.e., the process $u(t)$ will be stationary. Equation (11.1) will, in this case, define the spectral representation of this stationary process.[3]

The process defined by Eq. (11.1) will be real if, and only if, the number $n = 2m$ is even, and when the $2m$ terms on the right-hand side of Eq. (11.1) split into $m$ complex conjugate pairs of the form $(Z_k e^{i\omega_k t}, Z_k^* e^{-i\omega_k t})$. When this condition is satisfied, Eq. (11.1) may be rewritten in the form

$$u(t) = \sum_{k=1}^{m} \left(Z_k^{(1)} \cos \omega_k t + Z_k^{(2)} \sin \omega_k t\right) = \sum_{k=1}^{m} W_k \cos(\omega_k t - \varphi_k), \tag{11.5}$$

where

$$Z_k^{(1)} = Z_k + Z_k^*, \quad Z_k^{(2)} = i(Z_k - Z_k^*), \quad W_k = 2\,|Z_k|, \quad \varphi_k = \operatorname{arctg}\left(Z_k^{(2)}/Z_k^{(1)}\right)$$

[3] We note, moreover, that all the facts relating to spectral representations are valid, not only for ordinary stationary processes, but also for a wider class of processes which satisfy only the conditions $\overline{u(t)} =$ const and $B(t_1, t_2) = B(t_2 - t_1)$. These are the so-called "wide sense stationary" processes. A similar situation occurs in the case of the spectral representation of homogeneous random fields which will be discussed in the next subsection.

and consequently

$$\overline{Z_k^{(i)} Z_l^{(j)}} = \delta_{ij}\delta_{kl}E_k, \qquad E_k = \frac{1}{2}\overline{W_k^2} > 0. \tag{11.6}$$

Hence it is clear that a real process of the form Eq. (11.1) is a superposition of uncorrelated harmonic oscillations with random amplitudes and phases. The correlation function for the process Eq. (11.5) is given by

$$B(\tau) = \sum_{k=1}^{m} E_k \cos\omega_k\tau; \tag{11.7}$$

It depends on the mean squares of the amplitudes $W_k$, but is independent of the statistical characteristics of the phases $\varphi_k$.

It can be shown that an arbitrary stationary random process $u(t)$ has a spectral representation which is a direct generalization of Eq. (11.1). However, in general, we must take the limiting form of Eq. (11.1) as $n \to \infty$ by assuming that the frequencies $\omega_1, \ldots, \omega_n$ can approach each other without limit, and their number in a given interval along the $\omega$ axis can be infinite (although the sum of the complex amplitudes $Z_k$ is finite). Let the sum of the amplitudes $Z_k$ corresponding to frequencies $\omega_k < \omega$ be denoted by $Z(\omega)$. The function $Z(\omega)$ will be a random complex function such that $\overline{Z(\omega)} \equiv 0$ if $\overline{u(t)} = 0$, and

$$\overline{[Z^*(\omega^{(2)}) - Z^*(\omega^{(1)})][Z(\omega^{(4)}) - Z(\omega^{(3)})]} = 0 \tag{11.8}$$

for $\omega^{(1)} < \omega^{(2)} \leqslant \omega^{(3)} < \omega^{(4)}$ [because of the properties of the amplitudes $Z_k$ defined by Eq. (11.2)]. Equation (11.8) can also be formally written down in the more convenient differential form

$$\overline{dZ^*(\omega)\, dZ(\omega_1)} = 0 \text{ when } \omega \neq \omega_1. \tag{11.8'}$$

The fact that the process $u(t)$ is obtained from Eq. (11.1) as $n \to \infty$ and $\sum_{\omega_k < \omega} Z_k \to Z(\omega)$ means that

$$u(t) = \lim_{\Omega\to\infty} \left\{ \lim_{\omega_{k+1}-\omega_k \to 0} \sum_{k=0}^{n-1} [Z(\omega_{k+1}) - Z(\omega_k)]\, e^{i\omega_k' t} \right\}, \tag{11.9}$$

where $-\Omega = \omega_0 < \omega_1 < \ldots < \omega_n = \Omega$ and $\omega_k \leqslant \omega_k' \leqslant \omega_{k+1}$ (the limits of sequences of random variables are understood as the mean square limits: $W = \lim_{n\to\infty} W_n$ if $\lim \overline{|W - W_n|^2} = 0$). However, the right-hand side of Eq. (11.9) is equal to the improper Stieltjes integral (evaluated between $-\infty$ and $\infty$), so that we can write Eq. (11.9) in the symbolic form

$$u(t) = \int_{-\infty}^{\infty} e^{i\omega t}\, dZ(\omega). \tag{11.10}$$

This is, in fact, the general spectral representation of a stationary process $u(t)$ first derived by Kolmogorov (1940a) and Cramer (1942) [see also Doob (1953), Yaglom (1962), and Rozanov (1967)]. In view of Eq. (11.9), the significance of this expansion lies in the possibility that any stationary random process $u(t)$ can be replaced, to any required approximation, by the sum of uncorrelated harmonic oscillations with random amplitudes and phases.[4]

From Eq. (11.10) we can obtain the following "inversion formula" which enables us to express $Z(\omega)$ in terms of $u(t)$:

$$Z(\omega) = \lim_{T\to\infty} \frac{1}{2\pi} \int_{-T}^{T} \frac{e^{-i\omega t} - 1}{-it}\, u(t)\, dt + \text{const}, \tag{11.11}$$

so that

[4] Strictly speaking, this means that for any stationary process $u(t)$ and any $\epsilon > 0$, and $T > 0$, it is possible to choose a number $n$, numbers $\omega_1, \omega_2, \ldots, \omega_n$, and random variables $Z_1, Z_2, \ldots, Z_n$, satisfying Eq. (11.2), so that for each $t$ in the range $-T < t < T$ we have

$$\overline{\left| u(t) - \sum_k Z_k e^{i\omega_k t} \right|^2} < \varepsilon.$$

Hence, it follows that, for any $\epsilon > 0$, $\eta > 0$, and $T > 0$, it is possible to find such $n$, $\omega_1, \ldots, \omega_n$, and $Z_1, \ldots, Z_n$ that

$$P\left\{ \left| u(t) - \sum_k Z_k e^{i\omega_k t} \right| > \varepsilon \right\} < \eta$$

for $|t| < T$ where $P\{\ldots\}$ represents the probability that the relationship written in braces will be satisfied. This is more readily understood by considering the transition from equation (4.68) to (4.70) in Vol. 1. However, the individual realizations of the process $u(t)$ may not necessarily be capable of expansion into Fourier-Stieltjes integrals of the form given by Eq. (11.10).

$$Z(\omega_2) - Z(\omega_1) = Z([\omega_1, \omega_2]) = \lim_{T\to\infty} \frac{1}{2\pi} \int_{-T}^{T} \frac{e^{-i\omega_2 t} - e^{-i\omega_1 t}}{-it} u(t)\, dt. \tag{11.11'}$$

It is clear from Eq. (11.11) that in the case of a Gaussian process $u(t)$, the probability distributions of the values of $Z(\omega)$ will also be Gaussian. If the process $u(t)$ is real (henceforth we shall consider only real processes), then it is clear that $Z(-\omega_1)-Z(-\omega_2)=Z^*(\omega_2)-Z^*(\omega_1)$ for any $\omega_2 > \omega_1 \geqslant 0$, i.e., $dZ(-\omega)=dZ^*(\omega)$. The spectral representation (11.10) can then be rewritten in a real form similar to Eq. (11.5).

The real physical significance of the spectral representation Eq. (11.10) can be seen from the fact that the spectral components of the process, corresponding in this representation to different parts of the frequency spectrum, can be isolated experimentally by means of suitably chosen *filters*. Filters are devices which transmit harmonic oscillations in a given frequency range, but reject oscillations corresponding to other frequencies. When a real stationary process described by Eq. (11.10) is acted upon by a filter with a bandwidth $\Delta\omega = [\omega_1, \omega_2]$, the signal at the output of the filter is given by

$$u(\Delta\omega, t) = \int_{\omega_1}^{\omega_2} e^{i\omega t}\, dZ(\omega) + \int_{-\omega_2}^{-\omega_1} e^{i\omega t}\, dZ(\omega) = 2\,\mathrm{Re} \int_{\Delta\omega} e^{i\omega t}\, dZ(\omega), \tag{11.12}$$

where Re denotes the real part. This expression gives the *spectral component of the process $u(t)$ corresponding to the frequency interval* $\Delta\omega$. The condition given by Eq. (11.8) [or Eq. (11.8')] shows that spectral components corresponding to nonoverlapping frequency intervals are uncorrelated with each other. It is readily shown that, for a sufficiently narrow frequency range $\Delta\omega$, and a moderate time interval, the spectral component $u(\Delta\omega, t)$ can be replaced with a high degree of accuracy by the harmonic oscillation $2\,\mathrm{Re}\, Z(\Delta\omega)\, e^{i\omega t}$, although in the general case of a continuous spectrum, i.e., a process which cannot be represented by a finite or infinite sum of the form (11.1), the components $u(\Delta\omega, t)$ will not be strictly periodic for any $\Delta\omega$.

The fact that the spectral components $u(\Delta\omega, t)$ can be isolated experimentally gives a real significance to the *frequency distribution of the (mean) energy of the process $u(t)$*. In physical applications, the energy of a process $u(t)$ is usually proportional to $[u(t)]^2$ (for example, if $u(t)$ is the velocity, then $[u(t)]^2$ is proportional to the kinetic energy). It follows that, for stationary random functions $u(t)$, the quantity $\overline{[u(t)]^2} = B(0)$ plays the role of the mean energy. Using Eqs. (11.12) and (11.8) we can readily show that the mean energy of a spectral component $u(\Delta\omega, t)$ i.e., the mean energy corresponding to the harmonic oscillations with frequencies in the range $\Delta\omega = [\omega_1, \omega_2]$), is given by

$$\overline{[u(\Delta\omega, t)]^2} = \overline{|Z(\Delta\omega)|^2} + \overline{|Z(-\Delta\omega)|^2} = 2\overline{|Z(\Delta\omega)|^2}, \quad (11.13)$$

where

$$Z(\Delta\omega) = Z([\omega_1, \omega_2]) = Z(\omega_2) - Z(\omega_1).$$

We thus see that the numerical (nonrandom) nonnegative function $\overline{|Z(\Delta\omega)|^2}$ describes the distribution of the energy of the process $u(t)$ over the frequency spectrum in the range $-\infty < \omega < \infty$, while the function $2\overline{|Z(\Delta\omega)|^2}$ describes the energy distribution over the spectrum range $0 \leqslant \omega < \infty$.

The energy distribution over the spectrum for a turbulent flow is always continuous.[5] The quantity $\overline{|Z(\Delta\omega)|^2}$ is therefore represented by the integral of the *spectral density function* (or the *spectrum*) $F(\omega)$ of the process $u(t)$:

$$\overline{|Z(\Delta\omega)|^2} = \int_{\Delta\omega} F(\omega)\, d\omega. \quad (11.14)$$

Instead of the spectrum $F(\omega)$ defined for $-\infty < \omega < \infty$ (and even with respect to $\omega$), it is frequently more convenient to consider the spectrum $E(\omega) = 2F(\omega)$ where $0 \leqslant \omega < \infty$. According to Eq. (11.14) we have

$$\overline{|dZ(\omega)|^2} = F(\omega)\, d\omega = \frac{1}{2} E(\omega)\, d\omega. \quad (11.15)$$

[5] It can be shown that in a turbulent flow there is energy transfer along the spectrum, which is due to nonlinear terms in the fluid dynamic equations; this transfer rapidly redistributes the energy at any discrete point on the spectrum over a continuous spectral region (in this connection see, for example, Fig. 42 in Sect. 19). Mathematically, the continuity of the energy distribution is reflected in the fact that the correlation function $B(\tau)$ decreases rapidly at infinity, and can therefore be expanded into a Fourier integral. In the presence of a discrete spectrum, the function $B(\tau)$ would contain undamped periodic terms [see Eq. (11.4)], and therefore could be represented only by a Fourier-Stieltjes integral similar to that given by Eq. (11.10).

In view of Eqs. (11.15) and (11.8′), the evaluation of the mean values of double integrals with respect to $Z^*(\omega)$ and $Z(\omega_1)$ yields the correct result if we use the symbolic equation

$$\overline{dZ^*(\omega)\,dZ(\omega_1)} = \delta(\omega-\omega_1)\,F(\omega)\,d\omega\,d\omega_1 =$$

$$= \frac{1}{2}\,\delta(\omega-\omega_1)\,E(\omega)\,d\omega\,d\omega_1, \qquad (11.16)$$

where $\delta(\omega)$ is the improper Dirac $\delta$-function. This symbolic equation is equivalent to the rigorous relationship $\overline{dZ^*(\omega)dZ(\omega_1)} = \chi(\omega-\omega_1)F(\omega)d\omega$, where $\chi(x) = 1$ when $x = 0$, and $\chi(x) = 0$ when $x \neq 0$.

We can readily show from Eqs. (11.10) and (11.16) that the correlation function $B(\tau) = \overline{u^*(t)\,u(t+\tau)}$ is the Fourier transform of the corresponding spectral density:

$$B(\tau) = \int_{-\infty}^{\infty} e^{i\omega\tau} F(\omega)\,d\omega = \int_{0}^{\infty} \cos\omega\tau\,E(\omega)\,d\omega, \qquad (11.17)$$

so that

$$F(\omega) = \frac{1}{2\pi}\int_{-\infty}^{\infty} e^{-i\omega\tau}\,B(\tau)\,d\tau, \quad E(\omega) = \frac{2}{\pi}\int_{0}^{\infty} \cos\omega\tau\,B(\tau)\,d\tau. \qquad (11.18)$$

The special case of Eq. (11.17) corresponding to $\tau = 0$, i.e.,

$$B(0) = \int_{-\infty}^{\infty} F(\omega)\,d\omega = \int_{0}^{\infty} E(\omega)\,d\omega \qquad (11.19)$$

has a particularly simple physical interpretation. It shows that the total energy of the process $u(t)$ is the sum of the energies of the individual spectral components.

It is clear from Eqs. (11.14) and (11.18) that the Fourier transform of the correlation function of a stationary process should be a nonnegative function. This constitutes Khinchin's theorem on the Fourier expansion of correlation functions [Khinchin (1934)]. Khinchin also showed that each function which has a nonnegative Fourier transform is the correlation function of some stationary random process. Therefore, to verify whether a given function is the correlation function of a stationary random process, we must find its

Fourier transform and establish whether or not it is always nonnegative. For example, it follows that the functions

$$B(\tau) = Ce^{-\alpha|\tau|}, \tag{11.20}$$
$$B(\tau) = Ce^{-\alpha\tau^2}, \tag{11.20'}$$
$$B(\tau) = \begin{cases} C(1-\alpha|\tau|) & \text{when } |\tau| \leqslant 1/\alpha, \\ 0 & \text{when } |\tau| > 1/\alpha, \end{cases} \tag{11.20''}$$
$$B(\tau) = Ce^{-\alpha\tau}\cos\beta\tau, \tag{11.20'''}$$
$$B(\tau) = C(\alpha\tau)^{\nu} K_{\nu}(\alpha\tau) \tag{11.20^{IV}}$$

which correspond to the Fourier transforms

$$E(\omega) = \frac{2C\alpha}{\pi(\alpha^2+\omega^2)}, \tag{11.21}$$
$$E(\omega) = \frac{C}{\sqrt{\alpha\pi}} e^{-\omega^2/4\alpha}, \tag{11.21'}$$
$$E(\omega) = \frac{4C\alpha}{\pi} \frac{\sin^2(\omega/2\alpha)}{\omega^2}, \tag{11.21''}$$
$$E(\omega) = \frac{2C\alpha(\omega^2+\alpha^2+\beta^2)}{\pi[(\omega^2-\alpha^2-\beta^2)^2+4\alpha^2\omega^2]}, \tag{11.21'''}$$
$$E(\omega) = \frac{2^{\nu-1}\Gamma(\nu+1/2)\alpha^{2\nu}C}{\sqrt{\pi}(\alpha^2+\omega^2)^{\nu+1/2}} \tag{11.21^{IV}}$$

can all be correlation functions. In Eq. (11.20$^{IV}$) $K_\nu$ is the so-called Basset's (or Macdonald's) function (modified Bessel function of the third kind), and $\Gamma$ is the $\Gamma$-function. At the same time the function equal to $C(1-\alpha^2\tau^2)$ for $|\tau| \leqslant 1/\alpha$ and 0 for $|\tau| > 1/\alpha$ cannot be a correlation function for any stationary random process, since it can be readily verified that its Fourier transform is not nonnegative for all $\omega$.

The equations (11.17) and (11.18) can be verified directly by experiment, since the function $B(\tau)$ can usually be determined from a single measured realization of the process by means of time averaging (see Vol. 1, Sect. 4.7), and the function $E(\omega)$ can be measured independently with the aid of a set of narrow-band filters with different bandwidths [using Eqs. (11.13) and (11.15)]. Suppose, for example, that the process $u(t)$ is realized in the form of a fluctuating voltage [if $u(t)$ is a velocity or temperature fluctuation at a point in a turbulent flow, their transformation into voltage fluctuations is usually achieved automatically by the measuring instruments; see Vol. 1, Sect. 8.3]. Let us apply the signal $u(t)$ to a filter which will only transmit oscillations with frequency less than $\omega_0$, and let us determine the power at the output of the filter with the aid of a watt meter. The pointer of this instrument will indicate the value of the integral $\int_0^{\omega_0} E(\omega)\, d\omega$ (the time averaging in the formulas given by Eqs. (11.13)–(11.15) is, as a rule, performed directly by the watt meter which has a certain inertia). By varying the value of $\omega_0$ and differentiating the resulting empirical curve, we can find the function $E(\omega)$ which is necessary for the verification of the formulas given by Eqs. (11.17) and (11.18).

This type of verification of Eqs. (11.17) and (11.18) was first carried out by Taylor (1938b).[6] For the function $u(t)$, Taylor used the longitudinal velocity fluctuations at a fixed point in grid-generated turbulence in a wind tunnel. The corresponding spectral density $E(\omega)$ was measured by Simmons and Salter (1938) with the aid of a special set of electric filters (Fig. 1). Since no direct data were available at the time on the function $B(\tau)$, Taylor used measurements of the space correlation function $B(x)$, i.e., the mean

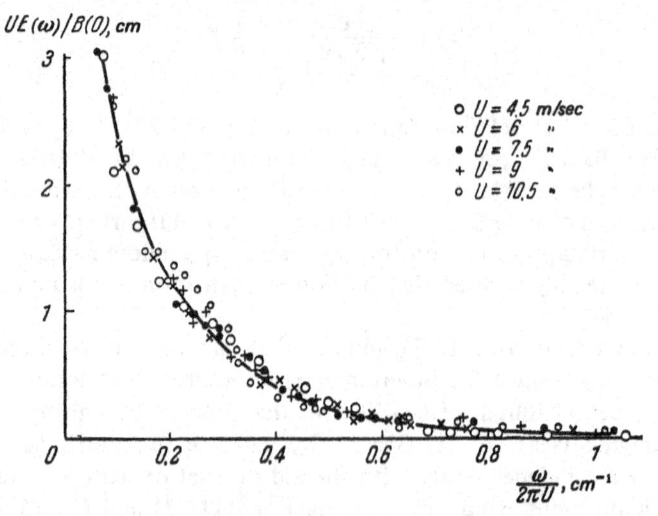


**FIG. 1** Spectral density of the longitudinal velocity component behind a grid in a wind tunnel.

[6] At the time, Taylor was apparently unaware of the earlier work by Khinchin (1934); therefore, equations (11.17) and (11.18) were derived by Taylor from a mathematical theorem of Wiener's by a method that was not completely rigorous.

product of the simultaneous values of the longitudinal velocity components at two points separated by a distance $x$ (in the direction of the wind-tunnel axis). Since the mean velocity $\bar{u} = U$ in the wind-tunnel was much greater than the fluctuating velocity $u' = u - U$, Taylor assumed that the disturbances (eddies) were transported with the mean velocity without appreciable distortion. Hence, it follows that that

$$B(x) = B(U\tau) \tag{11.22}$$

The assumption that this equation is valid has since been known as the *Taylor hypothesis*. In Fig. 2, which is taken from Taylor's paper (1938b), the full points represent direct measurements of the function $B(x)$, while crosses show the values of this function calculated from Eqs. (11.17) and (11.22) using the data of Fig. 1. As can be seen, the agreement is completely satisfactory.

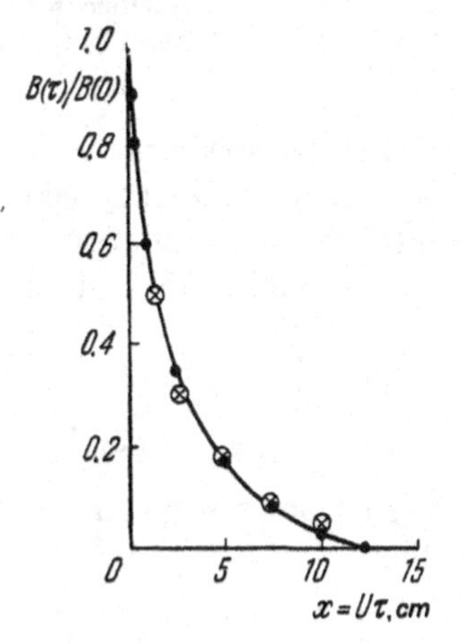


**FIG. 2** Comparison of measured values of the velocity correlation function (solid points) with calculations based on Eqs. (11.17) and (11.22) (circles with crosses).

Similar results have been frequently reported by many subsequent workers. Recent developments in electronics have permitted the direct determination of the time correlation functions $B(\tau)$[7] without the use of Taylor's hypothesis. Such measurements can, of course, be used to verify directly the Taylor hypothesis [see, for example Favre et al. (1962), Frenkiel and Klebanoff (1966) or Comte-Bellot and Corrsin (1971)]. As an example, Fig. 3, which is taken from the paper by Favre et al. (1954) shows the results of filter measurements of the spectral density $E(\omega)$ of longitudinal velocity fluctuations in a wind tunnel, and also the values of $E(\omega)$ calculated from Eq. (11.18) using experimental data on the function $B(\tau)$. Here again the agreement is completely satisfactory.

[7] This is achieved by the use of a "delay line" so that the values of $u(t)$ and $u(t + \tau)$ can be simultaneously applied to a multiplying device; many designs of such "delay lines" have been described in the literature.

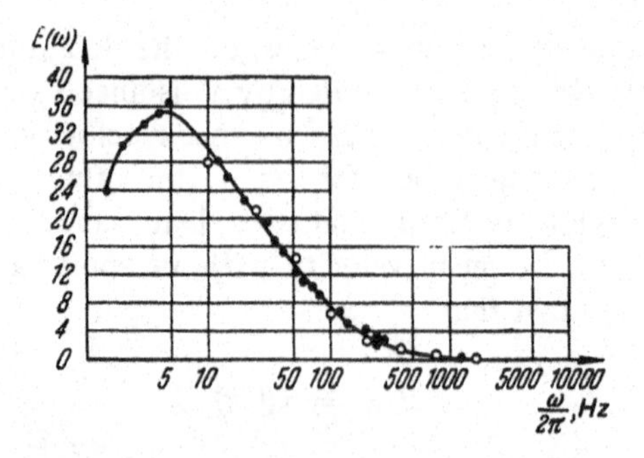


**FIG. 3** Comparison of the measured time velocity spectrum (solid points) with values calculated from Eq. (11.18) (open circles).

Let us now briefly consider the spectral representation of the derivatives of a stationary process $u(t)$. If we define the derivative of $u'(t)$ of a random function as the mean square limit of the corresponding finite differences, i.e.,

$$\lim_{h\to 0} \overline{\left| u'(t) - \frac{u(t+h) - u(t)}{h} \right|^2} = 0, \tag{11.23}$$

then from the spectral representation (11.10) we can readily show that

$$u'(t) = \int_{-\infty}^{\infty} e^{i\omega t} i\omega \, dZ(\omega). \tag{11.24}$$

We note, however, that since the left-hand side of Eq. (11.23) contains the statistical mean, the derivative given by Eq. (11.24) characterizes the entire statistical ensemble of the realizations of the process $u(t)$, and not each realization individually. It follows that, even if all the individual realizations of our process were smooth differentiable functions, but the statistical variance of the realization derivative at the point $t$ were infinite, the derivative defined by Eq. (11.23) would not exist. On the other hand, if only the "mean square derivative" (11.24) of the stationary process $u(t)$ is finite, it can be shown that this implies that all the realizations of this process will be differentiable, and the derivative of the realization can then be regarded as coinciding with the realization of the random process (11.24) [see, for example, Doob (1953), Chapt. 11, Sect. 9, where a more rigorous formulation of this statement can be found]. Hence, processes for which the derivative (11.24) is finite can be referred to as *differentiable processes.*

Equation (11.24) defines the spectral representation of the derivative $u'(t)$. It shows that the spectrum of the derivative is equal to $\omega^2 E(\omega)$ [see Eq. (11.16)]. Of course, this spectrum is possible only if the function $E(\omega)$ vanishes at infinity sufficiently rapidly to ensure that

$$\int_0^\infty \omega^2 E(\omega)\, d\omega < \infty. \tag{11.25}$$

If this is not so, the process $u(t)$ will not be differentiable in the above sense. In view of Eq. (11.17), the left-hand side of Eq. (11.25) is equal to $B''(0)$, and the condition (11.25) itself is equivalent to the condition for the existence of a finite second derivative of the function $B(\tau)$. The correlation function for the process (11.24) is given by

$$\overline{u'(t)\, u'(t+\tau)} = \int_0^\infty \cos \omega\tau\, \omega^2 E(\omega)\, d\omega = -B''(\tau). \tag{11.26}$$

Similarly, for the $n$th derivative, we have

$$u^{(n)}(t) = \int_{-\infty}^{\infty} e^{i\omega t} (i\omega)^n\, dZ(\omega) \tag{11.27}$$

and

$$\overline{u^{(n)}(t)\, u^{(n)}(t+\tau)} = \int_0^\infty \cos \omega\tau\, \omega^{2n} E(\omega)\, d\omega = (-1)^n B^{(2n)}(\tau). \tag{11.28}$$

These equations are meaningful provided

$$\int_0^\infty \omega^{2n} E(\omega)\, d\omega < \infty \tag{11.29}$$

or provided the function $B(\tau)$ is $2n$-fold differentiable at the point $\tau=0$. In particular, the process $u(t)$ will have derivatives of all orders if the integral in Eq. (11.29) converges for any $n$, i.e., the spectral density should fall to zero more rapidly than any finite negative power of $|\omega|$ as $|\omega|\to\infty$ (and the function $B(\tau)$ should have derivatives of all orders at the point $\tau=0$).

Since the correlation function and the spectrum are mutual Fourier transforms, all the formal relationships between these two functions can be inverted. In particular, the second equation in Eq. (11.28) corresponds to

$$E^{(2n)}(\omega)=\frac{2(-1)^n}{\pi}\int_0^\infty \cos\omega\tau\, \tau^{2n}B(\tau)\,d\tau, \tag{11.30}$$

which shows that the function $E(\omega)$ will have at least $2n$ derivatives if the correlation function $B(\tau)$ decreases at infinity more rapidly than $|\tau|^{-2n-1}$, so that

$$\int_0^\infty \tau^{2n}B(\tau)\,d\tau<\infty. \tag{11.31}$$

Hence, it is clear that the spectrum $E(\omega)$ will be infinitely differentiable provided only the correlation function $B(\tau)$ decreases at infinity more rapidly than any negative power of $|\tau|$.

Suppose now that $u(t)=\{u_1(t), \ldots, u_n(t)\}$ is a multidimensional (vector) real random stationary process such that $\overline{u_k(t)}=0$, $k=1, \ldots, n$. We shall assume that the correlation functions $B_{kl}(\tau)=\overline{u_k(t)\,u_l(t+\tau)}$ fall to zero at infinity sufficiently rapidly, so that they can be represented by the Fourier integrals

$$B_{kl}(\tau)=\int_{-\infty}^{\infty} e^{i\omega\tau}F_{kl}(\omega)\,d\omega, \qquad k, l=1, \ldots, n, \tag{11.32}$$

where

$$F_{kl}(\omega)=\frac{1}{2\pi}\int_{-\infty}^{\infty} e^{-i\omega\tau}B_{kl}(\tau)\,d\tau. \tag{11.33}$$

It is clear that the functions $F_{kk}(\omega)$, $k=1, \ldots, n$ are the ordinary nonnegative and even (with respect to $\omega$) spectra of the one-dimensional processes $u_k(t)$, $k=1, \ldots, n$. As regards the functions $F_{kl}(\omega)$ with $k \neq l$, i.e., the *cross-spectral densities* (or simply cross-spectra) of the processes $u_k(t)$ and $u_l(t)$, these functions can in general be complex.[8] However, since the functions $B_{kl}(\tau)$ are real, we always have

$$F_{kl}(\omega) = F^*_{kl}(-\omega). \tag{11.34}$$

Next, the obvious relationship

$$B_{kl}(\tau) = B_{lk}(-\tau) \tag{11.35}$$

shows that

$$F_{kl}(\omega) = F^*_{lk}(\omega), \tag{11.36}$$

so that the matrix $\| F_{kl}(\omega) \|$ is Hermitian for all $\omega$. Moreover, for any complex $c_1, \ldots, c_n$ and any $\omega$ we have

$$\sum_{k, l=1}^{n} F_{kl}(\omega)\, c^*_k c_l \geqslant 0, \tag{11.37}$$

so that the matrix $\| F_{kl}(\omega) \|$ is nonnegative-definite for any $\omega$.[9] Cramer (1940) and Kolmogorov (1941e) have also shown that any matrix $\| F_{kl}(\omega) \|$ which satisfies the above conditions will be the spectral matrix, i.e., the matrix of the cross spectra, of a multidimensional stationary process.

The spectral representation of a multidimensional stationary process $\boldsymbol{u}(t)$ is of the form

$$\boldsymbol{u}(t) = \int_{-\infty}^{\infty} e^{i\omega t}\, d\boldsymbol{Z}(\omega), \tag{11.38}$$

[8] The real part of the function $F_{kl}(\omega)$ [or the doubled real part $E_{kl}(\omega) = 2\mathrm{Re}F_{kl}(\omega)$], is called the *co-spectrum* (or the *co-phase spectrum*) of the processes $u_k(t)$ and $u_l(t)$, while the imaginary part of the same function is called the *quadrature spectrum*.

[9] To prove this we need only note that the left-hand side of Eq. (11.37) coincides with the spectrum of the complex process $c_1u_1(t) + \ldots + c_nu_n(t)$.

where $\boldsymbol{Z}(\omega) = \{Z_1(\omega), \ldots, Z_n(\omega)\}$ and the functions $Z_1(\omega), \ldots, Z_n(\omega)$ are such that

$$\overline{dZ_k^*(\omega)\, dZ_l(\omega_1)} = \delta(\omega - \omega_1)\, F_{kl}(\omega)\, d\omega\, d\omega_1 \tag{11.39}$$

[when $k = l$, Eq. (11.39) is obviously identical with Eq. (11.16)]. It is is clear that Eq. (11.39) leads directly to Eqs. (11.36) and (11.37).

## 11.2 Spectral Representation of Homogeneous Fields

The above discussion of the spectral representation of stationary random processes $u(t)$ can be extended to homogeneous random fields $u(\boldsymbol{x})$ (except for the experimental verification with the aid of filters). In this case, the harmonic oscillations $e^{i\omega t}$ are replaced by the plane waves $e^{i\boldsymbol{k}\boldsymbol{x}}$, and the random function $Z(\Delta\omega) = Z([\omega_1, \omega_2]) = Z(\omega_2) - Z(\omega_1)$ of the frequency interval $\Delta\omega$ is replaced by the random function $Z(\Delta\boldsymbol{k})$ of the multidimensional interval $\Delta\boldsymbol{k} = [\boldsymbol{k}', \boldsymbol{k}'']$ in the space of the wave-number vectors $\boldsymbol{k}$. Let $Z(d\boldsymbol{k}) = Z([\boldsymbol{k}, \boldsymbol{k} + d\boldsymbol{k}]) = dZ(\boldsymbol{k})$; then we can write down the spectral representation of a homogeneous random field $u(\boldsymbol{x})$ in the form

$$u(\boldsymbol{x}) = \int e^{i\boldsymbol{k}\boldsymbol{x}} dZ(\boldsymbol{k}), \tag{11.40}$$

where the integral is evaluated over the entire wave-number vector space and has the same meaning as the integral in Eq. (11.10). We shall suppose that the field $u(\boldsymbol{x})$ is real, has a zero mean, and is such that

$$\int |B(\boldsymbol{r})|\, d\boldsymbol{r} < \infty, \tag{11.41}$$

where $B(\boldsymbol{r}) = \overline{u(\boldsymbol{x})\, u(\boldsymbol{x} + \boldsymbol{r})}$ (the last condition ensures the continuity of the energy distribution of the field $u(\boldsymbol{x})$ in the wave-number vector space). The random amplitudes $dZ(\boldsymbol{k})$ of the representation (11.40) will then have the following properties:

$$dZ(-\boldsymbol{k}) = dZ^*(\boldsymbol{k}), \tag{11.42}$$

$$\overline{dZ(\boldsymbol{k})} = 0, \tag{11.43}$$

$$\overline{dZ^*(\boldsymbol{k})\, dZ(\boldsymbol{k}_1)} = \delta(\boldsymbol{k} - \boldsymbol{k}_1)\, F(\boldsymbol{k})\, d\boldsymbol{k}\, d\boldsymbol{k}_1, \tag{11.44}$$

where $\delta(\boldsymbol{k})$ is the multidimensional $\delta$-function. When $\boldsymbol{k}=\boldsymbol{k}_1$, the equation given by Eq. (11.44) should be understood as meaning that $F(\boldsymbol{k})\,d\boldsymbol{k}=\overline{|dZ(\boldsymbol{k})|^2}$ and hence it is clear that $F(-\boldsymbol{k})=F(\boldsymbol{k})$. The function $F(\boldsymbol{k})$ will be called the *three-dimensional spectral density* (or simply the *three-dimensional spectrum*[10]) of the field $u(\boldsymbol{x})$.[11] The formulas given by Eqs. (11.40) and (11.44) show that each homogeneous field can be approximated as closely as desired [in the sense explained in the discussion following Eq. (11.10)] by a finite sum of uncorrelated plane waves with different wavelengths and orientations and random amplitudes and phases. The complex amplitudes $dZ(\boldsymbol{k})$ of these plane waves can be found from the field $u(\boldsymbol{x})$ with the aid of the "inversion formulas"

$$Z([\boldsymbol{k}',\boldsymbol{k}''])=\frac{1}{8\pi^3}\int\!\!\int_{-\infty}^{\infty}\!\!\int \frac{e^{-ik_1''x_1}-e^{-ik_1'x_1}}{-ix_1}\,\frac{e^{-ik_2''x_2}-e^{-ik_2'x_2}}{-ix_2}\times$$
$$\times\frac{e^{-ik_3''x_3}-e^{-ik_3'x_3}}{-ix_3}\,u(\boldsymbol{x})\,d\boldsymbol{x}, \qquad (11.45)$$

where the integrals between infinite limits are to be understood as the limits (mean square limits, in fact) of the integrals between $-A$ and $+A$ as $A\to\infty$ [see Eq. (11.11')].

It follows from Eqs. (11.40) and (11.44) that

$$B(\boldsymbol{r})=\int e^{i\boldsymbol{k}\boldsymbol{r}}F(\boldsymbol{k})\,d\boldsymbol{k}. \qquad (11.46)$$

We note that the possibility of representation of the function $B(\boldsymbol{r})$ by a Fourier integral of the form (11.46), follows from the condition (11.41), and the spectral expansion (11.40) is only required to show that the function $F(\boldsymbol{k})=\lim_{\Delta\boldsymbol{k}\to\infty}\overline{|Z(\Delta\boldsymbol{k})|^2}/\Delta\boldsymbol{k}$ must be nonnegative. The converse of this is also true: Any function $B(\boldsymbol{r})$ which has a

[10] Below, we shall be exclusively concerned with fields $u(\boldsymbol{x})$ in three-dimensional space, and will not consider the obvious modifications which arise in the case of random fields on a plane. Moreover, we shall take the quantity $\overline{[u(\boldsymbol{x})]^2}$ as being equal to the energy, although in physical applications its meaning will depend on which particular field $u(\boldsymbol{x})$ is considered.

[11] Note that this usage differs from that of other writers; what is called here the "three-dimensional spectrum" would be the "spectrum"; what is called here the "spectrum" would be the "three-dimensional spectrum" (editor's note).

nonnegative Fourier transform, is the correlation function of some homogeneous random field [see, for example, Kampé de Fériet (1953) or Yaglom (1957, 1962)].

It follows from Eq. (11.46) that

$$F(\boldsymbol{k}) = \frac{1}{8\pi^3} \int e^{-i\boldsymbol{k}\boldsymbol{r}} B(\boldsymbol{r})\, d\boldsymbol{r}. \tag{11.47}$$

Hence, by specifying the three-dimensional spectrum $F(\boldsymbol{k})$ we simultaneously specify the correlation function $B(\boldsymbol{r})$. Nevertheless, some workers use, instead of $F(\boldsymbol{k})$, the less complete statistical characteristic

$$E(k) = \iint\limits_{|\boldsymbol{k}|=k} F(\boldsymbol{k})\, dS(\boldsymbol{k}) \tag{11.48}$$

where $dS(\boldsymbol{k})$ is an area element on the sphere $|\boldsymbol{k}|=k$. This function depends on the single argument $k$ (instead of the three arguments $k_1, k_2, k_3$). The function $E(k)$ will be referred to simply as the *spectrum* (without the adjective three-dimensional) of the field $u(\boldsymbol{x})$. The spectrum $E(k)$ does not uniquely define the function $B(\boldsymbol{r})$ since the energy $E(k)\,dk$ of plane waves with wave-number vectors in the range $(k, k+dk)$ can be distributed in different ways between the waves of different orientation. However, the total energy of the field, $\overline{[u(\boldsymbol{x})]^2} = B(0)$, is simply expressed in terms of $E(k)$:

$$B(0) = \int_0^\infty E(k)\, dk. \tag{11.49}$$

For any fixed value $k = k_0$ we can split $B(0)$ into two parts, namely,

$$B(0) = \int_0^{k_0} E(k)\, dk + \int_{k_0}^\infty E(k)\, dk. \tag{11.50}$$

This corresponds to the division of the field $u(\boldsymbol{x})$ into two uncorrelated parts, namely, the *macrocomponent* $\overline{u}(k_0, \boldsymbol{x})$ (set of disturbances with wavelengths greater than $2\pi/k_0$), and the *microcomponent* $\underline{u}(k_0, x)$ (set of disturbances with wavelengths smaller than $2\pi/k_0$):

$$u(x) = \int_{|k|<k_0} e^{ikx}\, dZ(k) + \int_{|k|>k_0} e^{ikx}\, dZ(k) = \bar{u}(k_0, x) + \underline{u}(k_0, x). \tag{11.51}$$

This is the analog of the usual division of an arbitrary homogeneous field into the mean value $\overline{u(x)}$ and a fluctuation $u(x) - \overline{u(x)} = u'(x)$. The latter division can be regarded as a special case of the general relation (11.51) corresponding to $k_0 = 0$.

In the case of a multidimensional (vector) homogeneous random field $u(x) = \{u_1(x), \ldots, u_n(x)\}$ with $\overline{u_j(x)} = 0$, $j = 1, \ldots, n$, the formula (11.40) will be valid for each component $u_j(x)$, so that

$$u_j(x) = \int e^{ikx}\, dZ_j(k). \tag{11.52}$$

The quantities $dZ_j(k)$, $j = 1, \ldots, n$, will then satisfy Eqs. (11.42) and (11.43), and the condition (11.44) in the case of a field with sufficiently rapidly decreasing correlation functions $B_{jl}(r) = \overline{u_j(x)\, u_l(x + r)}$ is replaced by the following more general condition:

$$\overline{dZ_j^*(k)\, dZ_l(k_1)} = \delta(k - k_1)\, F_{jl}(k)\, dk\, dk_1. \tag{11.53}$$

From equations (11.52) and (11.53) it follows that

$$B_{jl}(r) = \int e^{ikr} F_{jl}(k)\, dk, \tag{11.54}$$

and consequently

$$F_{jl}(k) = \frac{1}{8\pi^3} \int e^{-ikr} B_{jl}(r)\, dr. \tag{11.55}$$

Since $B_{jl}(r)$ are real functions, and $B_{jl}(r) = B_{lj}(-r)$, the *three-dimensional cross spectra* $F_{jl}(k)$ of the fields $u_j(x)$ and $u_l(x)$ satisfy the conditions

$$F_{jl}(k) = F_{jl}^*(-k) = F_{lj}^*(k). \tag{11.56}$$

Hence it follows that the spectral matrix

$$\| F_{jl}(\boldsymbol{k}) \| \tag{11.57}$$

is Hermitian for any $\boldsymbol{k}$. Moreover, it follows from equation (11.53) that the matrix (11.57) is nonnegative-definite for any $\boldsymbol{k}$, i.e., it is such that

$$\sum_{j,\, l} F_{jl}(\boldsymbol{k})\, c_j^* c_l \geqslant 0$$

for any $c_1, \ldots, c_n$. The converse statement is also true: Each matrix $\| F_{jl}(\boldsymbol{k}) \|$ which satisfies Eq. (11.56) and is nonnegative-definite for all $\boldsymbol{k}$, is the spectral matrix of a multidimensional homogeneous random field [see, for example, Kampé de Fériet (1953) or Yaglom (1957, 1962)].

In the important special case when $\boldsymbol{u}(\boldsymbol{x})$ is the velocity field of a fluid flow, the quantity

$$\frac{1}{2}\overline{\boldsymbol{u}^2} = \frac{1}{2}\sum \overline{u_j^2(\boldsymbol{x})} = \frac{1}{2}B_{jj}(0)$$

is the mean kinetic energy per unit mass of the fluid. Therefore, in addition to the spectral densities $F_{jl}(\boldsymbol{k})$, it is convenient to introduce the scalar spectral densities

$$F(\boldsymbol{k}) = \frac{1}{2}F_{jj}(\boldsymbol{k}), \quad E(k) = \iint\limits_{|\boldsymbol{k}|=k} F(\boldsymbol{k})\, dS(\boldsymbol{k}), \tag{11.58}$$

which are such that

$$\frac{1}{2}\overline{\boldsymbol{u}^2(\boldsymbol{x})} = \iiint\limits_{-\infty}^{\infty} F(\boldsymbol{k})\, d\boldsymbol{k} = \int\limits_0^{\infty} E(k)\, dk. \tag{11.59}$$

If we represent each component of the vector $\boldsymbol{u}(\boldsymbol{x})$ in the form of the sum (11.51), the vector field $\boldsymbol{u}(\boldsymbol{x})$ will be divided into a macrocomponent $\bar{\boldsymbol{u}}(k_0, \boldsymbol{x})$, with energy equal to

$$\int\limits_0^{k_0} E(k)\, dk$$

and a microcomponent $\underline{u}(k_0, x)$, with energy equal to

$$\int_{k_0}^{\infty} E(k)\, dk$$

Let us now go on to consider the higher-order moments of homogeneous random fields. We note that, in all real cases, certain special combinations of these moments, which are referred to as the cumulants, rapidly tend to zero when any of their arguments tends to infinity, and consequently they can be represented by Fourier integrals (see Vol. 1, Sect. 4.2). However, in the case of cumulants of order higher than 2, the precise conditions which must be satisfied by their Fourier transforms are not known. We cannot, therefore, say just which functions can be higher order cumulants (or moments) of a homogeneous field and which cannot. Nevertheless, we shall write down the representations in the form of Fourier integrals for some simple combinations of two- and three-point moments of the four-dimensional homogeneous field $\{u_1(x), u_2(x), u_3(x), \vartheta(x)\} = \{u(x), \vartheta(x)\}$ with $\overline{u_i(x)} = \overline{\vartheta(x)} = 0$, which will be useful in our subsequent discussion:

$$B_{ij,l}(r) = \overline{u_i(x)\,u_j(x)\,u_l(x+r)} = \int e^{ikr} F_{ij,l}(k)\, dk, \quad (11.60)$$

$$B_{j\vartheta,\vartheta}(r) = \overline{u_j(x)\vartheta(x)\,\vartheta(x+r)} = \int e^{ikr} F_{j\vartheta,\vartheta}(k)\, dk, \quad (11.60')$$

$$B_{ijl}(r, r') = \overline{u_i(x)\,u_j(x+r)\,u_l(x+r')} = \iint e^{i(kr+k'r')} F_{ijl}(k, k')\, dk\, dk', \quad (11.60'')$$

$$B_{j\vartheta\vartheta}(r, r') = \overline{u_j(x)\vartheta(x+r)\,\vartheta(x+r')} = \iint e^{i(kr+k'r')} F_{j\vartheta\vartheta}(k, k')\, dk\, dk', \quad (11.60''')$$

$$B^{(0)}_{ij,l,m}(r, r') = B_{ij,l,m}(r, r') - B_{ij}(0)\,B_{lm}(r'-r) = \overline{u_i(x)\,u_j(x)\,u_l(x+r)\,u_m(x+r')} - \overline{u_i(x)\,u_j(x)} \times \overline{u_l(x+r)\,u_m(x+r')} = \iint e^{i(kr+k'r')} F_{ij,l,m}(k, k')\, dk\, dk', \quad (11.60^{IV})$$

$$B^{(0)}_{ij,\vartheta,\vartheta}(r, r') = B_{ij,\vartheta,\vartheta}(r, r') - B_{ij}(0)\,B_{\vartheta\vartheta}(r'-r) = \overline{u_i(x)\,u_j(x)\vartheta(x+r)\,\vartheta(x+r')} - \overline{u_i(x)\,u_j(x)}\,\overline{\vartheta(x+r)\,\vartheta(x+r')} = \iint e^{i(kr+k'r')} F_{ij,\vartheta,\vartheta}(k, k')\, dk\, dk', \quad (11.60^{V})$$

$$B^{(0)}_{i\vartheta,\,j,\,\vartheta}(\boldsymbol{r},\,\boldsymbol{r}')=B_{i\vartheta,\,j,\,\vartheta}(\boldsymbol{r},\,\boldsymbol{r}')-B_{i\vartheta}(0)\,B_{j\vartheta}(\boldsymbol{r}'-\boldsymbol{r})=$$
$$=\overline{u_i(\boldsymbol{x})\,\vartheta(\boldsymbol{x})\,u_j(\boldsymbol{x}+\boldsymbol{r})\,\vartheta(\boldsymbol{x}+\boldsymbol{r}')}-\overline{u_i(\boldsymbol{x})\,\vartheta(\boldsymbol{x})}\times$$
$$\times\overline{u_j(\boldsymbol{x}+\boldsymbol{r})\,\vartheta(\boldsymbol{x}+\boldsymbol{r}')}=\int\int e^{i(\boldsymbol{kr}+\boldsymbol{k}'\boldsymbol{r}')}F_{i\vartheta,\,j,\,\vartheta}(\boldsymbol{k},\,\boldsymbol{k}')\,d\boldsymbol{k}\,d\boldsymbol{k}'. \quad (11.60^{\mathrm{VI}})$$

The possibility of these representations follows from the fact that the third-order moments are equal to the corresponding cumulants for zero mean values, while combinations of the moments $B^{(0)}_{i,\,j,\,l,\,m}$, $B^{(0)}_{i,\,j,\,\vartheta,\,\vartheta}$, and $B^{(0)}_{i\vartheta,\,j,\,\vartheta}$ differ from the corresponding fourth-order cumulants by terms which also tend to zero when $|\boldsymbol{r}|\to\infty$ or $|\boldsymbol{r}'|\to\infty$ [for example, the terms $B_{il}(\boldsymbol{r})\,B_{jm}(\boldsymbol{r}')+B_{im}(\boldsymbol{r}')\,B_{jl}(\boldsymbol{r})$ in the case of the function $B^{(0)}_{ij,\,l,\,m}(\boldsymbol{r},\,\boldsymbol{r}')$]. It is clear that the functions $F_{ij,\,l}(\boldsymbol{k})$, $F_{j,\,\vartheta\vartheta}(\boldsymbol{k})$, $F_{ijl}(\boldsymbol{k},\,\boldsymbol{k}')$ etc., on the right-hand sides of Eqs. (11.60)–(11.60$^{\mathrm{VI}}$) uniquely determine the corresponding moments (or combinations of moments) of which they are the Fourier transforms. We shall refer to them as the higher-order spectra (or spectral tensors) of the homogeneous fields $\boldsymbol{u}(\boldsymbol{x})$ or $\{\boldsymbol{u}(\boldsymbol{x}),\,\vartheta(\boldsymbol{x})\}$. Since $B_{ij,\,l}(\boldsymbol{r})=B_{ijl}(0,\,\boldsymbol{r})$ and $B_{j\vartheta,\,\vartheta}(\boldsymbol{r})=B_{j\vartheta\vartheta}(0,\,\boldsymbol{r})=B_{j\vartheta\vartheta}(\boldsymbol{r},\,0)$, the first four higher-order spectra satisfy the following relations:

$$F_{ij,\,l}(\boldsymbol{k})=\int F_{ijl}(\boldsymbol{k}',\,\boldsymbol{k})\,d\boldsymbol{k}', \quad (11.61)$$

$$F_{j\vartheta,\,\vartheta}(\boldsymbol{k})=\int F_{j\vartheta\vartheta}(\boldsymbol{k}',\,\boldsymbol{k})\,d\boldsymbol{k}'=\int F_{j\vartheta\vartheta}(\boldsymbol{k},\,\boldsymbol{k}')\,d\boldsymbol{k}'. \quad (11.61')$$

Moreover, it is clear that the tensors $F_{ij}$, $F_{ij,\,l,\,m}$, and $F_{ij,\,\vartheta,\,\vartheta}$ are symmetric in subscripts $i$, $j$, and that all the higher-order spectra transform to the complex-conjugate quantities when the signs of their arguments are reversed (the moments are real). They also satisfy the following simple relationships:

$$F_{ijl}(\boldsymbol{k},\,\boldsymbol{k}')=F_{ilj}(\boldsymbol{k}',\,\boldsymbol{k})=F_{jli}(\boldsymbol{k}',\,-\boldsymbol{k}-\boldsymbol{k}');\ \ F_{j\vartheta\vartheta}(\boldsymbol{k},\,\boldsymbol{k}')=F_{j\vartheta\vartheta}(\boldsymbol{k}',\,\boldsymbol{k});$$
$$F_{ij,\,l,\,m}(\boldsymbol{k},\,\boldsymbol{k}')=F_{ij,\,m,\,l}(\boldsymbol{k}',\,\boldsymbol{k})\ \text{etc.}$$

since

$$B_{ijl}(\boldsymbol{r},\,\boldsymbol{r}')=B_{ilj}\boldsymbol{r}',\,\boldsymbol{r})=B_{jli}(\boldsymbol{r}'-\boldsymbol{r},\,-\boldsymbol{r});$$
$$B_{j\vartheta\vartheta}(\boldsymbol{r},\,\boldsymbol{r}')=B_{j\vartheta\vartheta}(\boldsymbol{r}',\,\boldsymbol{r});\ \ B^{(0)}_{ij,\,l,\,m}(\boldsymbol{r},\,\boldsymbol{r}')=B^{(0)}_{ij,\,m,\,l}(\boldsymbol{r}',\,\boldsymbol{r})\ \text{etc.}$$

### 11.3 Partial Derivatives of Homogeneous Fields. Divergence and Curl of A Vector Field

As in the case of stationary random processes, the number of times one can differentiate the spectral density of a homogeneous random field $F(\boldsymbol{k})$, increases with the rate at which the correlation function $B(\boldsymbol{r})$ of this field falls to zero at infinity. Conversely, the rate at which the spectrum $F(\boldsymbol{k})$ decreases to zero at infinity governs the smoothness of the correlation function $B(\boldsymbol{r})$. Both these statements are simple consequences of the fact that the functions $B(\boldsymbol{r})$ and $F(\boldsymbol{k})$ are mutual Fourier transforms, so that

$$\frac{\partial^{m_1+m_2+m_3}B(\boldsymbol{r})}{\partial r_1^{m_1}\partial r_2^{m_2}\partial r_3^{m_3}}=i^{m_1+m_2+m_3}\int e^{i\boldsymbol{kr}}k_1^{m_1}k_2^{m_2}k_3^{m_3}F(\boldsymbol{k})\,d\boldsymbol{k} \tag{11.62}$$

and

$$\frac{\partial^{m_1+m_2+m_3}F(\boldsymbol{k})}{\partial k_1^{m_1}\partial k_2^{m_2}\partial k_3^{m_3}}=\frac{(-i)^{m_1+m_2+m_3}}{8\pi^3}\int e^{-i\boldsymbol{kr}}r_1^{m_1}r_2^{m_2}r_3^{m_3}B(\boldsymbol{r})\,d\boldsymbol{r}. \tag{11.63}$$

It follows from the last formula that if for $|\boldsymbol{r}|\to\infty$ the function $B(\boldsymbol{r})$ falls to zero as $r^{-N}$ where $N$ is an integer [or, in other words, if $B(\boldsymbol{r})=O(r^{-N})$], then the spectrum $F(\boldsymbol{k})$ will be $(N-4)$-fold differentiable with respect to its arguments for all values of $\boldsymbol{k}$, but its partial derivatives of order $N-3$ will exist only for $\boldsymbol{k}\neq 0$ but not at $\boldsymbol{k}=0$ [the function $B(\boldsymbol{r})$ behaves in an analogous fashion if $F(\boldsymbol{k})=O(k^{-N})$ as $|\boldsymbol{k}|=k\to\infty$)]. Since the integral on the right-hand side of Eq. (11.63) diverges logarithmically when $B(\boldsymbol{r})=O(r^{-N})$, $m_1+m_2+m_3=N-3$ and $\boldsymbol{k}=0$, and the function $e^{-i\boldsymbol{kr}}$ can be replaced by unity when $|\boldsymbol{r}|<C/|\boldsymbol{k}|$ (where $C$ is a small constant), it is clear that, if $B(\boldsymbol{r})=O(r^{-N})$, the partial derivates of order $N-3$ of the function $F(\boldsymbol{k})$ (which are continuous functions of $\boldsymbol{k}$ for $\boldsymbol{k}\neq 0$) will in general increase logarithmically as $k\to 0$. The situation is precisely the same for the more general functions $B_{jl}(\boldsymbol{r})$ and $F_{jl}(\boldsymbol{k})$, which correspond to the multidimensional homogeneous field $\boldsymbol{u}(\boldsymbol{x})=\{u_1(\boldsymbol{x}), \ldots, u_n(\boldsymbol{x})\}$. Therefore, for example, the general Taylor expansion of the function $F_{jl}(\boldsymbol{k})$ in the neighborhood of $\boldsymbol{k}=0$, which is of the form

$$F_{jl}(\boldsymbol{k})=f_{jl}+f_{jl,\,l}k_l+f_{jl,\,lm}k_lk_m+f_{jl,\,lmn}k_lk_mk_n+\ldots, \tag{11.64}$$

will for $B(r) = O(r^{-N})$ contain only terms up to order $N-4$ inclusive, and these will be followed by a residual term of the order of $k^{N-4}\ln k$ (when $j=l$, all the odd terms in Eq. (11.64) will, of course, be equal to zero). On the other hand, to insure that the spectrum $F_{jl}(\boldsymbol{k})$ will be infinitely differentiable, the correlation function $B_{jl}(\boldsymbol{k})$ should decrease to zero at infinity more rapidly than any finite negative power of $|\boldsymbol{r}|$. Similarly, to ensure infinite differentiability of the function $B_{jl}(\boldsymbol{r})$, we must have the appropriate rate of decrease to zero at infinity for the spectrum $F_{jl}(\boldsymbol{k})$.

Let us suppose that

$$\int k_1^{2m_1} k_2^{2m_2} k_3^{2m_3} F(\boldsymbol{k})\, d\boldsymbol{k} < \infty. \tag{11.65}$$

When this is so, the partial derivative

$$\frac{\partial^{m_1+m_2+m_3} u(\boldsymbol{x})}{\partial x_1^{m_1}\, \partial x_2^{m_2}\, \partial x_3^{m_3}} = i^{m_1+m_2+m_3} \int e^{i\boldsymbol{k}\boldsymbol{x}} k_1^{m_1} k_2^{m_2} k_3^{m_3}\, dZ(\boldsymbol{k}) \tag{11.66}$$

will exist in the sense analogous to Eq. (11.23), and so will all the lower-order partial derivatives. Moreover, in view of Eq. (11.66),

$$\overline{\frac{\partial^{m_1+m_2+m_3} u(\boldsymbol{x})}{\partial x_1^{m_1}\, \partial x_2^{m_2}\, \partial x_3^{m_3}} \frac{\partial^{n_1+n_2+n_3} u(\boldsymbol{x}')}{\partial x_1^{n_1}\, \partial x_2^{n_2}\, \partial x_3^{n_3}}} =$$

$$= (-i)^{m_1+m_2+m_3} i^{n_1+n_2+n_3} \int e^{i\boldsymbol{k}(\boldsymbol{x}'-\boldsymbol{x})} k_1^{m_1+n_1} k_2^{m_2+n_2} k_3^{m_3+n_3} F(\boldsymbol{k})\, d\boldsymbol{k} =$$

$$= (-1)^{m_1+m_2+m_3} \left. \frac{\partial^{m_1+m_2+m_3+n_1+n_2+n_3} B(\boldsymbol{r})}{\partial r_1^{m_1+n_1}\, \partial r_2^{m_2+n_2}\, \partial r_3^{m_3+n_3}} \right|_{\boldsymbol{r}=\boldsymbol{x}'-\boldsymbol{x}}. \tag{11.67}$$

Similarly, if

$$\int k_1^{2m_1} k_2^{2m_2} k_3^{2m_3} F_{jj}(\boldsymbol{k})\, d\boldsymbol{k} < \infty, \quad \int k_1^{2n_1} k_2^{2n_2} k_3^{2n_3} F_{ll}(\boldsymbol{k})\, d\boldsymbol{k} < \infty \tag{11.68}$$

(no summation over $j$ and $l$), the derivatives

$$\frac{\partial^{m_1+m_2+m_3} u_j(\boldsymbol{x})}{\partial x_1^{m_1}\, \partial x_2^{m_2}\, \partial x_3^{m_3}}, \quad \frac{\partial^{n_1+n_2+n_3} u_l(\boldsymbol{x})}{\partial x_1^{n_1}\, \partial x_2^{n_2}\, \partial x_3^{n_3}},$$

will exist, and

$$\overline{\frac{\partial^{m_1+m_2+m_3}u_j(\boldsymbol{x})}{\partial x_1^{m_1}\,\partial x_2^{m_2}\,\partial x_3^{m_3}}\;\frac{\partial^{n_1+n_2+n_3}u_l(\boldsymbol{x}')}{\partial x_1'^{n_1}\,\partial x_2'^{n_2}\,\partial x_3'^{n_3}}}=$$

$$=(-i)^{m_1+m_2+m_3}\,i^{n_1+n_2+n_3}\int e^{i\boldsymbol{k}(\boldsymbol{x}'-\boldsymbol{x})}k_1^{m_1+n_1}k_2^{m_2+n_2}k_3^{m_3+n_3}F_{jl}(\boldsymbol{k})\,d\boldsymbol{k}=$$

$$=(-1)^{m_1+m_2+m_3}\left.\frac{\partial^{m_1+m_2+m_3+n_1+n_2+n_3}B_{jl}(\boldsymbol{r})}{\partial r_1^{m_1+n_1}\,\partial r_2^{m_2+n_2}\,\partial r_3^{m_3+n_3}}\right|_{\boldsymbol{r}=\boldsymbol{x}'-\boldsymbol{x}}. \tag{11.69}$$

Of all the multidimensional random fields, the most important case for the theory of turbulence is the three-dimensional velocity field $\boldsymbol{u}(\boldsymbol{x})=\{u_1(\boldsymbol{x}), u_2(\boldsymbol{x}), u_3(\boldsymbol{x})\}$. In view of Eq. (11.66),

$$\frac{\partial u_j}{\partial x_m}=i\int e^{i\boldsymbol{k}\boldsymbol{x}}k_m\,dZ_j(\boldsymbol{k}) \tag{11.70}$$

and, hence,

$$\operatorname{div}\boldsymbol{u}(\boldsymbol{x})=\frac{\partial u_j(\boldsymbol{x})}{\partial x_j}=i\int e^{i\boldsymbol{k}\boldsymbol{x}}k_j\,dZ_j(\boldsymbol{k}), \tag{11.71}$$

$$\omega_j(\boldsymbol{x})=\varepsilon_{jlm}\frac{\partial u_m(\boldsymbol{x})}{\partial x_l}=i\varepsilon_{jlm}\int e^{i\boldsymbol{k}\boldsymbol{x}}k_l\,dZ_m(\boldsymbol{k}), \tag{11.72}$$

where $\omega(\boldsymbol{x})=\operatorname{curl}\boldsymbol{u}(\boldsymbol{x})=\nabla\times\boldsymbol{u}(\boldsymbol{x})$ and $\varepsilon_{jlm}$ is a unit tensor of rank 3 which is antisymmetric in all its indices. Consequently,

$$\overline{\operatorname{div}\boldsymbol{u}(\boldsymbol{x})\operatorname{div}\boldsymbol{u}(\boldsymbol{x}+\boldsymbol{r})}=\int e^{i\boldsymbol{k}\boldsymbol{r}}k_jk_lF_{jl}(\boldsymbol{k})d\boldsymbol{k}=-\frac{\partial^2B_{jl}(\boldsymbol{r})}{\partial r_j\,\partial r_l} \tag{11.73}$$

$$\overline{\omega_j(\boldsymbol{x})\,\omega_i(\boldsymbol{x}+\boldsymbol{r})}=\varepsilon_{jlm}\varepsilon_{ipq}\int e^{i\boldsymbol{k}\boldsymbol{r}}k_lk_pF_{mq}(\boldsymbol{k})d\boldsymbol{k}=-\varepsilon_{jlm}\varepsilon_{ipq}\frac{\partial^2B_{mq}(\boldsymbol{r})}{\partial r_l\,\partial r_p}. \tag{11.74}$$

It follows that the correlation function of the divergence of the field $\boldsymbol{u}(\boldsymbol{x})$ is equal to $-\frac{\partial^2B_{jl}(\boldsymbol{r})}{\partial r_j\,\partial r_l}$, and the spectrum is $k_jk_lF_{jl}(\boldsymbol{k})$. Similarly, the correlation tensor of the vorticity of the field $\boldsymbol{u}(\boldsymbol{x})$ is equal to $-\varepsilon_{jlm}\varepsilon_{ipq}\frac{\partial^2B_{mq}(\boldsymbol{r})}{\partial r_l\,\partial r_p}$, and the spectral tensor of the vorticity field is $\varepsilon_{jlm}\varepsilon_{ipq}k_lk_pF_{mq}(\boldsymbol{k})$. The formulas for the correlation and spectral tensors of the vorticity field $\boldsymbol{\omega}(\boldsymbol{x}) =$ curl $\boldsymbol{u}(\boldsymbol{x})$ can be transformed with the aid of the following identity

$$\varepsilon_{jlm}\varepsilon_{ipq} = \delta_{ij}\delta_{lp}\delta_{mq} + \delta_{jp}\delta_{lq}\delta_{mi} + \delta_{jq}\delta_{li}\delta_{mp} - \\ - \delta_{ij}\delta_{lq}\delta_{mp} - \delta_{jp}\delta_{li}\delta_{mq} - \delta_{jq}\delta_{lp}\delta_{mi}. \quad (11.75)$$

which is readily verified. Moreover, it follows from Eq. (11.69) that

$$\overline{\left(\frac{\partial u_j}{\partial x_m}\right)^2} = \int k_m^2 F_{jj}(\boldsymbol{k})\, d\boldsymbol{k} \quad (11.76)$$

(no summation over $j$), and

$$\bar{\varepsilon} = \frac{\nu}{2} \sum_{j,\, m} \overline{\left(\frac{\partial u_j}{\partial x_m} + \frac{\partial u_m}{\partial x_j}\right)^2} = \nu \int [k^2 F_{jj}(\boldsymbol{k}) + k_j k_m F_{jm}(\boldsymbol{k})]\, d\boldsymbol{k}. \quad (11.77)$$

We shall now restrict our attention to the case where the vector field $\boldsymbol{u}(\boldsymbol{x})$ is *solenoidal*, i.e., $\partial u_j(\boldsymbol{x})/\partial x_j = 0$. We then have

$$\overline{\frac{\partial u_j(\boldsymbol{x})}{\partial x_j} u_l(\boldsymbol{x}+\boldsymbol{r})} = \overline{u_j(\boldsymbol{x}+\boldsymbol{r}) \frac{\partial u_l(\boldsymbol{x})}{\partial x_l}} = 0. \quad (11.78)$$

In view of Eq. (11.70) this means that

$$k_j F_{jl}(\boldsymbol{k}) = k_l F_{jl}(\boldsymbol{k}) = 0 \quad (11.79)$$

$$\frac{\partial B_{jl}(\boldsymbol{r})}{\partial r_j} = \frac{\partial B_{jl}(\boldsymbol{r})}{\partial r_l} = 0. \quad (11.80)$$

The four relationships in Eqs. (11.79) and (11.80) are not, of course, independent: From any one of them the other three follow automatically. Instead of Eqs. (11.79) and (11.80) we can also use the equivalent relationship $k_j\, dZ_j(\boldsymbol{k}) = 0$.

Substituting the identity given by Eq. (11.75) into Eq. (11.74), and using Eqs. (11.80) and (11.79), we can verify that for a solenoidal field $\boldsymbol{u}(\boldsymbol{x})$ the correlation and spectral tensors of the vorticity field are given by

$$B_{\omega_j\omega_l}(\boldsymbol{r}) = \overline{\omega_j(\boldsymbol{x})\,\omega_l(\boldsymbol{x}+\boldsymbol{r})} =$$

$$= -\delta_{jl}\,\Delta B_{ll}(\boldsymbol{r}) + \frac{\partial^2 B_{ll}(\boldsymbol{r})}{\partial r_j \partial r_l} + \Delta B_{lj}(\boldsymbol{r}) \quad (11.81)$$

and

$$F_{\omega_j\omega_l}(\boldsymbol{k}) = (\delta_{jl}k^2 - k_j k_l)\,F_{ll}(\boldsymbol{k}) - k^2 F_{lj}(\boldsymbol{k}). \quad (11.82)$$

Particularly simple results are obtained for the traces of the tensors $B_{\omega_j\omega_l}(\boldsymbol{r})$ and $F_{\omega_j\omega_l}(\boldsymbol{k})$:

$$B_{\omega_j\omega_j}(\boldsymbol{r}) = -\Delta B_{jj}(\boldsymbol{r}), \quad F_{\omega_j\omega_j}(\boldsymbol{k}) = k^2 F_{jj}(\boldsymbol{k}). \quad (11.83)$$

The relationships given by Eq. (11.79) enable us to determine the general form of the tensor $F_{jl}(\boldsymbol{k})$ [see Kampé de Fériet (1948)]. In view of Eq. (11.79), the vector $\boldsymbol{k}$ is the eigenvector of the Hermitian matrix $\| F_{jl}(\boldsymbol{k}) \|$ corresponding to its zero eigenvalue. Moreover, this matrix should have two further (in general, complex) eigenvectors, $\boldsymbol{a}^{(1)}$ and $\boldsymbol{b}^{(1)}$, which are orthogonal to each other and to the vector $\boldsymbol{k}$ and correspond to nonnegative eigenvalues $\lambda_1(\boldsymbol{k})$ and $\lambda_2(\boldsymbol{k})$. Let us normalize the vectors $\boldsymbol{a}^{(1)}$ and $\boldsymbol{b}^{(1)}$ by the conditions $a_l^{(1)*}a_l^{(1)} = b_l^{(1)*}b_l^{(1)} = 1$, and expand $dZ(\boldsymbol{k})$ in terms of the vectors $\boldsymbol{a}^{(1)}, \boldsymbol{b}^{(1)}$, and $\boldsymbol{k}$. Since $\boldsymbol{k}\,d\boldsymbol{Z}(\boldsymbol{k}) = 0$, we have

$$d\boldsymbol{Z}(\boldsymbol{k}) = dZ_a(\boldsymbol{k})\,\boldsymbol{a}^{(1)}(\boldsymbol{k}) + dZ_b(\boldsymbol{k})\,\boldsymbol{b}^{(1)}(\boldsymbol{k}), \quad (11.84)$$

where $dZ_a(\boldsymbol{k}) = dZ_j(\boldsymbol{k})\,a_j^{(1)*}$, $dZ_b(\boldsymbol{k}) = dZ_j(\boldsymbol{k})\,b_j^{(1)*}$, and hence

$$\overline{|dZ_a(\boldsymbol{k})|^2} = \lambda_1(\boldsymbol{k})\,d\boldsymbol{k}, \quad \overline{|dZ_b(\boldsymbol{k})|^2} = \lambda_2(\boldsymbol{k})\,d\boldsymbol{k}, \quad (11.85)$$
$$\overline{dZ_a^*(\boldsymbol{k})\,dZ_b(\boldsymbol{k})} = 0$$

[since $F_{jl}(\boldsymbol{k})\,a_j^{(1)} = \lambda_1 a_l^{(1)}$ and $F_{jl}(\boldsymbol{k})\,b_j^{(1)} = \lambda_2 b_l^{(1)}$ by definition of $\boldsymbol{a}^{(1)}$ and $\boldsymbol{b}^{(1)}$]. Using the notations $\boldsymbol{a}(\boldsymbol{k}) = \sqrt{\lambda_1(\boldsymbol{k})}\,\boldsymbol{a}^{(1)}(\boldsymbol{k})$ and $\boldsymbol{b}(\boldsymbol{k}) = \sqrt{\lambda_2(\boldsymbol{k})}\,\boldsymbol{b}^{(1)}(\boldsymbol{k})$ and recalling that $F_{jl}(\boldsymbol{k}) = \lim \overline{dZ_j^*(\boldsymbol{k})\,dZ_l(\boldsymbol{k})}/d\boldsymbol{k}$, we have from Eqs. (11.84) and (11.85)

$$F_{jl}(\boldsymbol{k}) = a_j^*(\boldsymbol{k})\,a_l(\boldsymbol{k}) + b_j^*(\boldsymbol{k})\,b_l(\boldsymbol{k}). \quad (11.86)$$

This is, in fact, the general form of the tensor $F_{jl}(\boldsymbol{k})$ [since the tensor (11.86) will satisfy all the requirements imposed on the spectral tensor for $a_j(-\boldsymbol{k}) = a_j^*(\boldsymbol{k})$, $b_j(-\boldsymbol{k}) = b_j^*(\boldsymbol{k})$]. This tensor depends on six complex functions of the argument $\boldsymbol{k}$, which are related by the three conditions $k_j a_j = k_j b_j = a_j^* b_j = 0$. Since, moreover, $\boldsymbol{a}/a$, $\boldsymbol{b}/b$, $\boldsymbol{k}/k$ [where $a = |\boldsymbol{a}|$, $b = |\boldsymbol{b}|$] is an orthonormal triad of vectors, and consequently

$$\frac{k_j k_l}{k^2} + \frac{a_j^* a_l}{a^2} + \frac{b_j^* b_l}{b^2} = \delta_{jl},$$

we can eliminate $b_j^* b_l$ from Eq. (11.86) and rewrite the expression for $F_{jl}(\boldsymbol{k})$ in the form

$$F_{jl}(\boldsymbol{k}) = b^2\left(\delta_{jl} - \frac{k_j k_l}{k^2}\right) + a_j^* a_l\left(1 - \frac{b^2}{a^2}\right), \quad (11.87)$$

The functions $b^2(\boldsymbol{k})$, $a^2(\boldsymbol{k})$, $a_j(\boldsymbol{k})$ are here related by

$$k_j a_j = 0, \quad a_j^* a_j = a^2 \geqslant 0, \quad b^2 \geqslant 0.$$

Using the conditions given by Eq. (11.79), we can establish another general representation of the tensor $F_{jl}(\boldsymbol{k})$ which is occasionally quite convenient. Namely, let us consider the tensor

$$\Delta_{jm}(\boldsymbol{k}) = \delta_{jm} - \frac{k_j k_m}{k^2}. \tag{11.88}$$

This tensor is symmetric and orthogonal to the vector $\boldsymbol{k}$, i.e., it satisfies the condition $k_j \Delta_{jm}(\boldsymbol{k}) = 0$. Moreover, if $k_m F_m(\boldsymbol{k}) = 0$ then, clearly, $\Delta_{jm}(\boldsymbol{k}) F_m(\boldsymbol{k}) = F_j(\boldsymbol{k})$. On the other hand, if $\boldsymbol{F}(\boldsymbol{k}) = \{F_1(\boldsymbol{k}), F_2(\boldsymbol{k}), F_3(\boldsymbol{k})\}$ is an arbitrary vector field, then the field $F'_j(\boldsymbol{k}) = \Delta_{jm}(\boldsymbol{k}) F_m(\boldsymbol{k})$ will satisfy the condition $k_j F'_j(\boldsymbol{k}) = 0$. Therefore, the transformation with the matrix $\| \Delta_{jm}(\boldsymbol{k}) \|$ isolates the part of the vector $\boldsymbol{F}(\boldsymbol{k})$ which is orthogonal to the vector $\boldsymbol{k}$. Therefore, for example, the tensor $F_{jl}(\boldsymbol{k})$ which satisfies Eq. (11.79) can be always written in the form

$$F_{jl}(\boldsymbol{k}) = \Delta_{jm}(\boldsymbol{k}) \Delta_{ln}(\boldsymbol{k}) \Phi_{mn}(\boldsymbol{k}), \tag{11.89}$$

where $\Phi_{mn}(\boldsymbol{k})$ is a symmetric tensor. We can take the original tensor $F_{mn}(\boldsymbol{k})$ for $\Phi_{mn}(\boldsymbol{k})$. It is possible, however, to choose $\Phi_{mn}(\boldsymbol{k})$ to be an arbitrary Hermitian symmetric tensor with a nonnegative-definite matrix, in which case the tensor (11.89) will still satisfy all the requirements imposed on the spectral tensor of a solenoidal vector field.

If $\boldsymbol{u}(\boldsymbol{x})$ is a solenoidal homogeneous vector field, while $\vartheta(\boldsymbol{x})$ is a homogeneous scalar field which is homogeneously correlated with $\boldsymbol{u}(\boldsymbol{x})$, then

$$\overline{\frac{\partial u_j(\boldsymbol{x})}{\partial x_j} \vartheta(\boldsymbol{x}+\boldsymbol{r})} = \overline{\vartheta(\boldsymbol{x}-\boldsymbol{r}) \frac{\partial u_j(\boldsymbol{x})}{\partial x_j}} = 0,$$

i.e.,

$$\frac{\partial B_{j\vartheta}(\boldsymbol{r})}{\partial r_j} = \frac{\partial B_{\vartheta j}(\boldsymbol{r})}{\partial r_j} = 0 \tag{11.90}$$

and

$$k_j F_{j\vartheta}(\boldsymbol{k}) = k_j F_{\vartheta j}(\boldsymbol{k}) = 0. \tag{11.91}$$

According to Eq. (11.91), the vector $F_{j\vartheta}(\boldsymbol{k})$ is orthogonal to $\boldsymbol{k}$, and consequently it can be represented in the form

$$F_{j\vartheta}(\boldsymbol{k}) = \Delta_{jm}(\boldsymbol{k}) \Phi_m(\boldsymbol{k}). \tag{11.92}$$

The higher-order spectral tensors (or vectors) of a solenoidal field, [for example, $F_{lj,l}(\boldsymbol{k})$, $F_{ljl}(\boldsymbol{k}, \boldsymbol{k}')$, or $F_{j\vartheta\vartheta}(\boldsymbol{k}, \boldsymbol{k}')$] can be represented in a similar way. In fact, if $\partial u_j / \partial x_j = 0$, then

$$\overline{u_l(\boldsymbol{x}) u_j(\boldsymbol{x}) \frac{\partial u_l(\boldsymbol{x}')}{\partial x'_l}} = 0,$$

$$\overline{u_l(\boldsymbol{x}) u_j(\boldsymbol{x}') \frac{\partial u_l(\boldsymbol{x}'')}{\partial x''_l}} = \overline{u_l(\boldsymbol{x}) \frac{\partial u_j(\boldsymbol{x}')}{\partial x'_j} u_l(\boldsymbol{x}'')} = \overline{\frac{\partial u_l(\boldsymbol{x})}{\partial x_l} u_j(\boldsymbol{x}') u_l(\boldsymbol{x}'')} = 0$$

$$\overline{\frac{\partial u_j(\boldsymbol{x})}{\partial x_j} \vartheta(\boldsymbol{x}') \vartheta(\boldsymbol{x}'')} = 0$$

so that

$$k_l F_{ij,l}(\boldsymbol{k}) = 0, \quad k'_l F_{ijl}(\boldsymbol{k}, \boldsymbol{k}') = k_j F_{ijl}(\boldsymbol{k}, \boldsymbol{k}') = \\ = (-k'_l - k_l) F_{ijl}(\boldsymbol{k}, \boldsymbol{k}') = 0, \qquad (-k_j - k'_j) F_{j\vartheta\vartheta}(\boldsymbol{k}, \boldsymbol{k}') = 0. \tag{11.93}$$

Consequently,

$$F_{ij,l}(\boldsymbol{k}) = \Delta_{lm}(\boldsymbol{k}) \Phi_{ij,m}(\boldsymbol{k}), \tag{11.94}$$

$$F_{ijl}(\boldsymbol{k}, \boldsymbol{k}') = \Delta_{im}(\boldsymbol{k}'') \Delta_{jn}(\boldsymbol{k}) \Delta_{lp}(\boldsymbol{k}') \Phi_{mnp}(\boldsymbol{k}, \boldsymbol{k}'), \quad \boldsymbol{k}'' = -\boldsymbol{k}' - \boldsymbol{k}, \tag{11.94'}$$

$$F_{j\vartheta\vartheta}(\boldsymbol{k}, \boldsymbol{k}') = \Delta_{jm}(\boldsymbol{k}'') \Phi_m(\boldsymbol{k}, \boldsymbol{k}'). \tag{11.94''}$$

The above results lead us to certain conclusions with regard to the behavior of the spectra of the field $\boldsymbol{u}(\boldsymbol{x})$ in the neighborhood of the origin of the wave vector space. Let us consider, for example, the tensor $F_{jl}(\boldsymbol{k})$ and suppose that it can be expanded in the Taylor series Eq. (11.64) in the neighborhood of the point $\boldsymbol{k} = 0$. In that case, the functions $a_j(\boldsymbol{k})$ and $b_j(\boldsymbol{k})$ in Eq. (11.86) can also be expanded in a Taylor series. Since, however, the vectors $a(\boldsymbol{k})$ and $b(\boldsymbol{k})$ should be orthogonal to all the vectors $\boldsymbol{k}/|\boldsymbol{k}|$ for $\boldsymbol{k} = 0$ (i.e., they should be equal to zero), it is clear that the expansion of $F_{jl}(\boldsymbol{k})$ in a Taylor series should begin with second order terms:

$$F_{jl}(\boldsymbol{k}) = f_{jl,mn} k_m k_n + \dots \tag{11.95}$$

Similarly, it can be shown that the Taylor series for the vector $F_{j\vartheta}(\boldsymbol{k})$ begins with first order terms:

$$F_{j\vartheta}(\boldsymbol{k}) = f_{j\vartheta,m} k_m + \dots \tag{11.96}$$

If we pass from the spectra to the correlation matrices, we have from Eq. (11.95),

$$\int B_{jl}(\boldsymbol{r})\, d\boldsymbol{r} = \int B_{jl}(\boldsymbol{r})\, r_m\, d\boldsymbol{r} = 0, \tag{11.97}$$

and from Eq. (11.96) we have

$$\int B_{j\vartheta}(\boldsymbol{r})\, d\boldsymbol{r} = 0. \tag{11.98}$$

## 12. ISOTROPIC RANDOM FIELDS

### 12.1 Correlation Functions and Spectra of Scalar Isotropic Fields

The spectra of homogeneous random fields will, in general, depend on the three variables $k_1, k_2, k_3$ (while the higher-order spectra will depend on a still greater number of variables). In the subsequent analysis we shall be particularly concerned with special *homogeneous and isotropic fields* for which the main statistical characteristics depend only on one variable. Such homogeneous and isotropic fields will be discussed in this section.

A scalar random field $u(\boldsymbol{x})$ is called isotropic if all the finite-dimensional probability densities $p_{\boldsymbol{x}_1\boldsymbol{x}_2\ldots\boldsymbol{x}_N}(u_1, u_2, \ldots, u_N)$ corresponding to it are unaffected by all possible rotations of the set of points $\boldsymbol{x}_1, \boldsymbol{x}_2, \ldots, \boldsymbol{x}_N$ about axes passing through the origin, and by mirror reflections of this set of points in any plane passing through the origin. We shall confine our attention to fields which are both homogeneous and isotropic, and for the sake of brevity we shall refer to such fields simply as *isotropic*. In other words, by isotropic fields we shall understand random fields $u(\boldsymbol{x})$ for which the probability densities $p_{\boldsymbol{x}_1\boldsymbol{x}_2\ldots\boldsymbol{x}_N}(u_1, u_2, \ldots, u_N)$ are unaffected by any translations (i.e., shifts), rotations, and mirror reflections (i.e., space symmetries) of the set of points $\boldsymbol{x}_1, \boldsymbol{x}_2, \ldots, \boldsymbol{x}_N$.

Since the field $u(\boldsymbol{x})$ is homogeneous, its mean value $\overline{u(\boldsymbol{x})}$ should be a constant. Without loss of generality we can assume that $\overline{u(\boldsymbol{x})} = 0$, having replaced, if necessary, the original field $u(\boldsymbol{x})$ by the field $u'(\boldsymbol{x}) = u(\boldsymbol{x}) - \overline{u(\boldsymbol{x})}$. We shall adopt this procedure henceforth. The correlation function $B(\boldsymbol{x}, \boldsymbol{x}') = \overline{u(\boldsymbol{x})u(\boldsymbol{x}')}$ of an isotropic field $u(\boldsymbol{x})$ must, of course, assume equal values for any two pairs of points $(\boldsymbol{x}, \boldsymbol{x}')$ and $(\boldsymbol{x}_1, \boldsymbol{x}_1')$ which can be made to coincide as a result of a rigid motion (translation or rotation) of the pair of points $(\boldsymbol{x}_1, \boldsymbol{x}_1')$. In other words, if the distance between the points $\boldsymbol{x}$ and $\boldsymbol{x}'$ is equal to the distance between $\boldsymbol{x}_1$ and $\boldsymbol{x}_1'$ then $B(\boldsymbol{x}, \boldsymbol{x}') = B(\boldsymbol{x}_1, \boldsymbol{x}_1')$ so that the function $B(\boldsymbol{x}, \boldsymbol{x}')$ can depend only on the distance $r = |\boldsymbol{x}' - \boldsymbol{x}|$ between the points $\boldsymbol{x}$ and $\boldsymbol{x}' = \boldsymbol{x} + \boldsymbol{r}$:

$$\overline{u(\boldsymbol{x})u(\boldsymbol{x}+\boldsymbol{r})} = B(r), \qquad r = |\boldsymbol{r}|. \tag{12.1}$$

We thus see that in the case of an isotropic field $u(\boldsymbol{x})$, the correlation function does, in fact, depend on the single variable $r$.

If we substitute Eq. (12.1) into the general formula (11.47) which defines the three-dimensional spectrum $F(\boldsymbol{k})$ of a homogeneous field in terms of $B(\boldsymbol{r})$ and transform to spherical coordinates, we find that, in the isotropic case, the spectrum will depend only on $k = |\boldsymbol{k}|$:

$$F(\boldsymbol{k}) = \frac{1}{8\pi^3} \int\int\int_{-\infty}^{\infty} e^{-i\boldsymbol{k}\boldsymbol{r}} B(r)\, d\boldsymbol{r} =$$

$$= \frac{1}{8\pi^3} \int_0^{\infty}\int_{-\pi}^{\pi}\int_0^{\pi} e^{-ikr\cos\theta} B(r)\, r^2 \sin\theta\, d\theta\, d\varphi\, dr =$$

$$= \frac{1}{2\pi^2} \int_0^{\infty} \frac{\sin kr}{kr} B(r)\, r^2\, dr = F(k). \tag{12.2}$$

Conversely, if

$$F(\boldsymbol{k}) = F(k), \quad \overline{|dZ(\boldsymbol{k})|^2} = F(k)\,d\boldsymbol{k}, \tag{12.3}$$

then the correlation function $B(\boldsymbol{r})$ will depend only on $r = |\boldsymbol{r}|$, and in view of Eq. (11.46) we have in this case

$$B(r) = 4\pi \int_0^\infty \frac{\sin kr}{kr} F(k)\,k^2\,dk. \tag{12.4}$$

The function $F(k)$ in equations (12.2) and (12.4) must be nonnegative. The set of correlation functions for all the possible isotropic random fields is therefore determined by the condition that the integral (12.2) should be nonnegative for all $k \geqslant 0$. Hence, it follows that the functions

$$B(r) = Ce^{-\alpha r}, \tag{12.5}$$

$$B(r) = Ce^{-\alpha r^2}, \tag{12.5'}$$

$$B(r) = C(\alpha r)^\nu K_\nu(\alpha r), \quad \nu > 0, \tag{12.5''}$$

are the correlation functions of isotropic fields. In fact, it is readily verified that the three-dimensional spectra which correspond to these functions are, respectively,

$$F(k) = \frac{C\alpha}{\pi^2(\alpha^2 + k^2)^2}, \tag{12.6}$$

$$F(k) = \frac{C}{8(\alpha\pi)^{3/2}} e^{-k^2/4\alpha}, \tag{12.6'}$$

$$F(k) = \frac{2^{\nu-1}\Gamma(\nu + 3/2)\,\alpha^{2\nu} C}{\pi^{3/2}(\alpha^2 + k^2)^{\nu+3/2}}, \tag{12.6''}$$

i.e., they are nonnegative.

It is clear that, if we replace the variable $r$ by $|\tau|$ in the functions (12.5), (12.5′), and (12.5″), we obtain examples of correlation functions of stationary processes. This is explained by the fact that the values of an isotropic field $u(\boldsymbol{x})$ at points on the straight line $x_2 = x_3 = 0$ form a homogeneous field on the straight line (i.e., a stationary process in the variable $x_1$). The one-dimensional Fourier transform of the function $B(r)$ which is defined for $r < 0$ with the aid of the condition $B(-r) = B(r)$ must be nonnegative throughout:

$$F_1(k_1) = \frac{1}{2\pi} \int_{-\infty}^{\infty} e^{-ik_1r} B(r)\, dr = \frac{1}{\pi} \int_0^{\infty} \cos k_1 r\, B(r)\, dr \geqslant 0. \quad (12.7)$$

The function $F_1(k_1)$ defined by Eq. (12.7) or, what amounts to the same thing, by the inverse formulas

$$B(r) = \int_{-\infty}^{\infty} e^{ik_1r} F_1(k_1)\, dk_1 = 2 \int_0^{\infty} \cos k_1 r\, F_1(k_1)\, dk_1, \quad (12.8)$$

will be called the *one-dimensional spectral density* (or the *one-dimensional spectrum*) of the field $u(\boldsymbol{x})$. Comparison of Eq. (12.8) with the special case of Eq. (11.46) obtained for $r_2 = r_3 = 0$, will readily show that

$$F_1(k_1) = \iint_{-\infty}^{\infty} F\left(\sqrt{k_1^2 + k_2^2 + k_3^2}\right) dk_2\, dk_3 =$$

$$= 2\pi \int_0^{\infty} F\left(\sqrt{k_1^2 + \varkappa^2}\right) \varkappa\, d\varkappa = 2\pi \int_{k_1}^{\infty} F(k)\, k\, dk. \quad (12.9)$$

Hence it also follows that for $F(k) \geqslant 0$, the function $F_1(k_1)$ will always be nonnegative. On the other hand, it is readily seen that, from the fact that the function $F_1(k_1)$ of Eq. (12.9) is nonnegative, it does not immediately follow that the three-dimensional spectrum

$$F(k) = -\frac{1}{2\pi k} \frac{dF_1(k)}{dk}. \quad (12.10)$$

is nonnegative. In fact, if we substitute, for example, the function given by Eq. (11.21″), or the function given by Eq. (11.21‴) with $\alpha < \sqrt{3}\,\beta$, for $2F_1(k)$ (with the argument $\omega$ replaced by $k$), we obtain a function $F(k)$ which will also assume negative values.[12] We thus

[12] We note that according to Eq. (12.10) the function $F(k)$ will be nonnegative if the function $F_1(k) = E_1(k)/2$ is nonincreasing along the entire half-axis $0 < k < \infty$. It is readily seen that this condition is not satisfied in the case of the function (11.21″) or (11.21‴) with $\alpha < \sqrt{3}\,\beta$.

conclude the functions $B(r) = C \max\{1 - \alpha r, 0\}$ or $B(r) = Ce^{-\alpha r}\cos\beta r$, $\alpha < \sqrt{3}\,\beta$, can be the correlation functions of homogeneous fields on a straight line (stationary processes), but they cannot be correlation functions of isotropic fields in three-dimensional space.

Instead of the three-dimensional spectrum $F(k)$ we can also use the spectrum

$$E(k) = \int\limits_{|\mathbf{k}|=k} F(\mathbf{k})\, d\mathbf{k} = 4\pi k^2 F(k). \tag{12.11}$$

Similarly, the function $F_1(k_1)$ can be replaced by $E_1(k_1) = 2F_1(k_1)$ which we shall again call the one-dimensional spectrum (or one-dimensional spectral density). When applied to the functions $E(k)$ and $E_1(k_1)$, the formulas given by Eqs. (12.2), (12.4), (12.9), and (12.10) assume the form

$$E(k) = \frac{2}{\pi}\int\limits_0^\infty kr \sin kr B(r)\, dr, \quad B(r) = \int\limits_0^\infty \frac{\sin kr}{kr} E(k)\, dk \tag{12.12}$$

and

$$E_1(k_1) = \int\limits_{k_1}^\infty \frac{E(k)}{k}\, dk, \quad E(k) = -k\frac{dE_1(k)}{dk}. \tag{12.13}$$

As already indicated, the rate at which the correlation function $B(r)$ decreases to zero at infinity, determines the smoothness of the spectra $F(k)$ and $E(k)$, while the rate at which the spectrum $F(k)$ or $E(k)$ decreases, governs the degree of smoothness of the correlation function $B(r)$. Moreover, the coefficients of the Taylor expansion for $F(k)$ or $E(k)$ (if the expansion is possible here) can be simply expressed in terms of the moments of the function $B(r)$, while the Taylor expansion coefficients for $B(r)$ can be expressed in terms of the moments of the function $F(k)$ or $E(k)$. In fact, by expanding, for example, the function $\sin kr/kr$ in Eqs. (12.2) and (12.4) in a Taylor series, we can verify that the Taylor expansion for the functions $F(k)$, $E(k) = 4\pi k^2 F(k)$ and $B(r)$ at the point $k = 0$, and hence $r = 0$, are of the form

$$F(k) = f_0 + f_2 k^2 + f_4 k^4 + \ldots,$$
$$E(k) = 4\pi f_0 k^2 + 4\pi f_2 k^4 + 4\pi f_4 k^6 + \ldots, \tag{12.14}$$
$$B(r) = b_0 + b_2 r^2 + b_4 r^4 + \ldots, \tag{12.15}$$

where

$$f_{2n} = \frac{1}{(2n)!}\left(\frac{d^{2n}F(k)}{dk^{2n}}\right)_{k=0} = \frac{(-1)^n}{2\pi^2(2n+1)!}\int_0^\infty r^{2n+2}B(r)\,dr, \tag{12.16}$$

$$b_{2n} = \frac{1}{(2n)!}\left(\frac{d^{2n}B(r)}{dr^{2n}}\right)_{r=0} = \frac{(-1)^n 4\pi}{(2n+1)!}\int_0^\infty k^{2n+2}F(k)\,dk =$$

$$= \frac{(-1)^n}{(2n+1)!}\int_0^\infty k^{2n}E(k)\,dk. \tag{12.17}$$

It is clear from Eq. (12.17) that, in particular, the derivatives of $B(r)$ of order $4n$ at the point $r=0$ are nonnegative, while the derivatives of order $4n+2$ at this point are nonpositive. Next, if $F(k) \sim k^4$, i.e., $E(k) \sim k^6$), as $k \to 0$ then in view of Eqs. (12.14) and (12.16), we have

$$\int_0^\infty r^2 B(r)\,dr = \int_0^\infty r^4 B(r)\,dr = 0. \tag{12.18}$$

We shall use this result below.

From the formulas given by Eqs. (12.4) and (12.2) or (12.12) we also have

$$\int_0^\infty B(r)\,dr = 2\pi^2\int_0^\infty kF(k)\,dk = \frac{\pi}{2}\int_0^\infty \frac{E(k)}{k}\,dk, \tag{12.19}$$

$$\int_0^\infty rB(r)\,dr = 4\pi\int_0^\infty F(k)\,dk = \int_0^\infty \frac{E(k)}{k^2}\,dk. \tag{12.20}$$

Therefore,

$$L = \frac{1}{B(0)}\int_0^\infty B(r)\,dr = \frac{\pi}{2}\frac{\int_0^\infty kF(k)\,dk}{\int_0^\infty k^2F(k)\,dk} = \frac{\pi}{2}\frac{\int_0^\infty k^{-1}E(k)\,dk}{\int_0^\infty E(k)\,dk}, \tag{12.21}$$

where $L$ is a characteristic length scale of the field $u(x)$, which is called the *integral length scale* and is of the order of the distance over which there is an appreciable correlation between the values of the field at two points. Another and usually much smaller characteristic length scale of the field $u(x)$ is the *differential length scale* which is given by the expression

$$\lambda=\left\{-\frac{B(0)}{2B''(0)}\right\}^{1/2}=$$
$$=\left\{\frac{3\int\limits_0^\infty E(k)\,dk}{2\int\limits_0^\infty k^2E(k)\,dk}\right\}^{1/2}. \qquad (12.22)$$

This length scale is equal to the length of the segment cut on the horizontal axis by the parabola touching the graph of the correlation function $B(r)$ at its apex (Fig. 4).

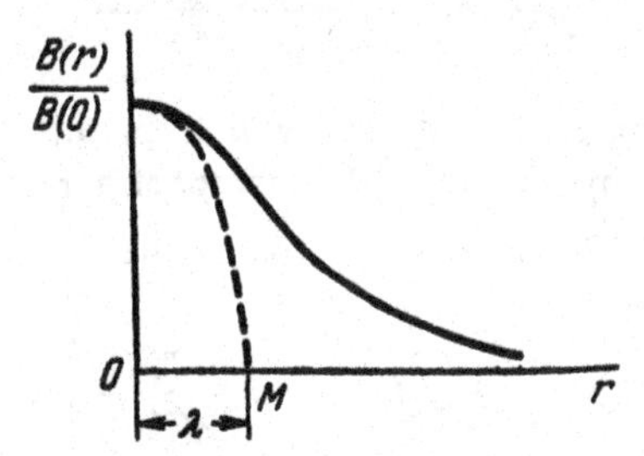


FIG. 4 Definition of the length scale $\lambda$.

## 12.2 Correlation Functions and Spectra of Isotropic Vector Fields

We shall now consider multidimensional isotropic random fields. At first sight it would appear natural to define the multidimensional field $u(x)=\{u_1(x), \ldots, u_n(x)\}$ as isotropic if all the probability densities for the values of any components of this field at a given set of points in space do not change under translations, rotations, and reflections of this set of points. We shall see later that another definition of the multidimensional isotropic field is more important for the theory of turbulence, but we shall begin our discussion with multidimensional fields which are isotropic in the sense indicated above. A multidimensional field of this kind is characterized by the correlation matrix

$$\| B_{jl}(r) \| = \| \overline{u_j(\boldsymbol{x}) u_l(\boldsymbol{x}+\boldsymbol{r})} \|, \tag{12.23}$$

in which all the elements depend only on $r = |\boldsymbol{r}|$. By analogy with the derivation of Eqs. (12.2) and (12.4), it can be shown that all the elements of this matrix can be written in the form

$$B_{jl}(r) = 4\pi \int_0^\infty \frac{\sin kr}{kr} F_{jl}(k) k^2 \, dk, \quad F_{jl}(k) = \frac{1}{2\pi^2} \int_0^\infty \frac{\sin kr}{kr} B_{jl}(r) r^2 \, dr, \tag{12.24}$$

where

$$F_{jl}(k) = \lim_{dk \to 0} \frac{\overline{dZ_j^*(\boldsymbol{k}) \, dZ_l(\boldsymbol{k})}}{dk}$$

are the corresponding cross spectra which, in this case, depend only on $k = |\boldsymbol{k}|$. Since by Eq. (12.24) the spectra $F_{jl}(k)$ are real, they satisfy the conditions

$$F_{jl}(k) = F_{lj}(k), \quad F_{jj}(k) \geqslant 0 \text{ and } \sum_{j,l=1}^{n} F_{jl}(k) c_j c_l > 0 \tag{12.25}$$

for any $k$ and any real $c_1, \ldots, c_n$. Conversely, any functions which can be written in the form (12.24), where $F_{jl}(k)$ satisfy the condition given by Eq. (12.25), are the elements of the spectral matrix of an isotropic random field.

Examples of multidimensional random fields in the theory of turbulence are the velocity field $\boldsymbol{u}(\boldsymbol{x}) = \{u_1(\boldsymbol{x}), u_2(\boldsymbol{x}), u_3(\boldsymbol{x})\}$, a combination of the velocity field and some scalar fields (for example, the five-dimensional field $\{\boldsymbol{u}(\boldsymbol{x}), p(\boldsymbol{x}), \vartheta(\boldsymbol{x})\}$ where $p(\boldsymbol{x})$ is the pressure and $\vartheta(\boldsymbol{x})$ is the temperature) and, finally, a combination of scalar fields (for example, the two-dimensional field $\{p(\boldsymbol{x}), \vartheta(\boldsymbol{x})\}$). However, the above definition of isotropic multidimensional fields is completely satisfactory only for fields of the last type. The reason is that when the points $\boldsymbol{x}$ are rotated, the components of the velocity vector $\boldsymbol{u}(\boldsymbol{x})$ undergo a linear transformation (in accordance with the general transformation rule for the components of a vector). Hence, when applied to a vector field $\boldsymbol{u}(\boldsymbol{x}) = \{u_1(\boldsymbol{x}), u_2(\boldsymbol{x}), u_3(\boldsymbol{x})\}$, the isotropy condition must be formulated as follows: A random vector field $\boldsymbol{u}(\boldsymbol{x})$ is called *isotropic* if the probability densities for the values of its components at an arbitrary set of points $\boldsymbol{x}_1, \boldsymbol{x}_2, \ldots, \boldsymbol{x}_N$ are unaffected by any translations, rotations and/or reflections accompanied by simultaneous rotation and/or reflection of the coordinate system with respect to which the components of the vector are determined. Thus, for example, the two-dimensional probability density for the values of $u_1(\boldsymbol{x}_1)$ and $u_2(\boldsymbol{x}_2)$ (see Fig. 5) should coincide with the probability density for the values of $u_1'(\boldsymbol{x}_1')$ and $u_2'(\boldsymbol{x}_2')$, where $u_i'(\boldsymbol{x}_i')$ is the projection of the vector $\boldsymbol{u}(\boldsymbol{x}_i')$ along the $Ox_i'$ axis (i.e., the $i$th component of $\boldsymbol{u}(\boldsymbol{x}_i')$ relative to the "rotated" set of coordinates). For a field of the form $\{\boldsymbol{u}(\boldsymbol{x}), p(\boldsymbol{x}), \vartheta(\boldsymbol{x})\}$, which in addition to the vector $\boldsymbol{u}(\boldsymbol{x})$ contains some scalar components, the isotropy condition requires that the corresponding probability densities are unaffected by translations, rotations, and reflections, which are accompanied by a linear transformation of the components of the vector $\boldsymbol{u}(\boldsymbol{x})$.

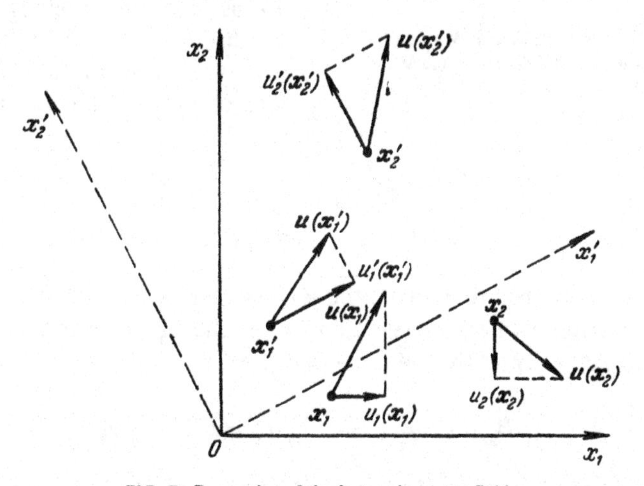


FIG. 5 Properties of the isotropic vector field.

Let us suppose now that $\boldsymbol{u}(\boldsymbol{x})$ is an isotropic random vector field. Since the field is homogeneous, its mean value $\overline{\boldsymbol{u}(\boldsymbol{x})} = \boldsymbol{U}$ is a constant vector. However, since the vector is isotropic, it must be unaffected by rotations. Consequently, it must be a zero vector. It follows that the mean value of an isotropic random vector field is necessarily equal to zero.

The situation is much more complicated in the case of the correlation tensor of an isotropic vector field

$$\overline{u_j(M)\,u_l(M')} = \overline{u_j(x)\,u_l(x+r)} = B_{jl}(r). \tag{12.26}$$

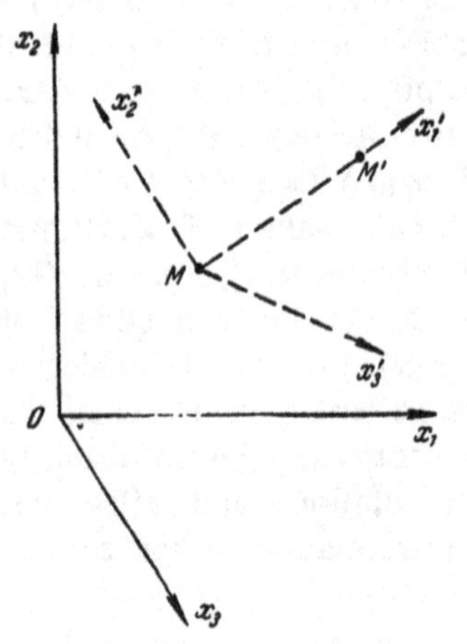


**FIG. 6** Coordinate system $Mx_1'x_2'x_3'$ associated with the points $M$ and $M'$.

When we investigate this tensor it is convenient to transform to a special set of coordinates $Mx_1'x_2'x_3'$, the first axis of which lies along the vector $r$, while the two other axes, $Mx_2'$ and $Mx_3'$, are perpendicular to this vector as shown in Fig. 6. Let the values of the components of the tensor $B_{jl}(r)$ in this new set of coordinates be denoted by $B'_{jl}(r)$ [the functions $B'_{jl}(r)$ will, clearly, depend only on $r=|r|$, since the set of coordinates $Mx_1'x_2'x_3'$ will rotate together with the points $(M,\ M')$ as they are rotated]. Since rotation through 180° about the $Mx_1'$ axis results in the reversal of the directions of the $Mx_2'$ and $Mx_3'$ axes, we must have

$$B'_{12}(r) = -B'_{12}(r) = 0,$$
$$B'_{13}(r) = B'_{21}(r) = B'_{31}(r) = 0.$$

Moreover, by rotating about the $Mx_1'$ axis we can transform the axis $Mx_2'$ into $Mx_3'$, whereas reflection in the $Mx_1'x_3'$ plane results in the replacement of $Mx_2'$ by $-Mx_2'$, so that

$$B'_{22}(r) = B'_{33}(r), \quad B'_{23}(r) = B'_{32}(r) = 0.$$

Thus it is clear that the tensor $B'_{jl}(r)$ [and, consequently, $B_{jl}(r)$ also] is symmetric, and of the six different components $B'_{jl}(r)$, three components are zero and the two remaining components are equal. Therefore, there remain only two nonequal components which we shall denote as follows:

$$B'_{11}(r) = B_{LL}(r) = \overline{u_L(M)\,u_L(M')},$$
$$B_{22}(r) = B_{33}(r) = B_{NN}(r) = \overline{u_N(M)\,u_N(M')} \tag{12.27}$$

Here, and henceforth, $u_L$ and $u_N$ will denote the projections of the vector $\boldsymbol{u}$ on to the direction of $\boldsymbol{r}$ and, correspondingly, on to any perpendicular direction. The functions $B_{LL}(r)$ and $B_{NN}(r)$ are called the longitudinal and lateral correlation functions of the field $\boldsymbol{u}(\boldsymbol{x})$, respectively (see Fig. 7). When $r = 0$, these functions must be equal:

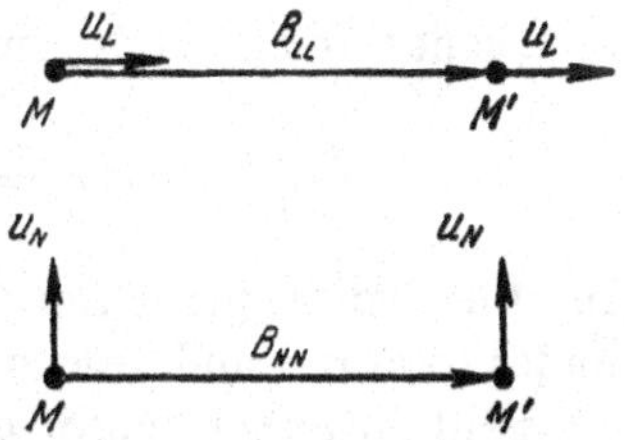

$$B_{LL}(0) = B_{NN}(0) = \frac{1}{3}\overline{[\boldsymbol{u}(\boldsymbol{x})]^2}. \tag{12.28}$$

FIG. 7 The correlation functions $B_{LL}$ and $B_{NN}$.

Next, if we resolve the unit vectors of the old coordinate system $Ox_1x_2x_3$ along the axes of the system $Mx'_1x'_2x'_3$, and use equations (12.26) and (12.27), we obtain

$$B_{jl}(\boldsymbol{r}) = [B_{LL}(r) - B_{NN}(r)]\frac{r_j r_l}{r^2} + B_{NN}(r)\,\delta_{jl}, \tag{12.29}$$

where $\delta_{jl} = 1$ when $j = l$ and $\delta_{jl} = 0$ when $j \neq l$, where $r_j$, $j = 1, 2, 3$ are the three components of the vector $\boldsymbol{r}$ in the system $Ox_1x_2x_3$. We have thus expressed all the components of the correlation tensor of an isotropic vector field in terms of the two scalar functions $B_{LL}(r)$ and $B_{NN}(r)$.

In view of the importance of Eq. (12.29) we shall now give another derivation based on an important general idea (Robertson, 1940). We shall use the fact that $B_{jl}(\boldsymbol{r})$ is the tensor function of the vector $\boldsymbol{r}$ which is invariant under all rotations and reflections. Therefore, if $\boldsymbol{a}$ and $\boldsymbol{b}$ are arbitrary unit vectors, then the quadratic form

$$B(\boldsymbol{r}, \boldsymbol{a}, \boldsymbol{b}) = B_{jl}(\boldsymbol{r})\,a_j b_l$$

will be a scalar depending on the three vectors $\boldsymbol{r}$, $\boldsymbol{a}$, and $\boldsymbol{b}$ (the last two have a unit length), and its dependence on $\boldsymbol{a}$ and on $\boldsymbol{b}$ will be linear. We know from the theory of invariants under rotation and reflection groups (see Weyl, 1939) that the scalar $B$ should be expressible in terms of the principal invariants of the above three vectors, namely, the length $r = (r_j r_j)^{1/2}$ and the scalar products

$\boldsymbol{ra} = r_j a_j$, $\boldsymbol{rb} = r_j b_j$ and $\boldsymbol{ab} = a_j b_j$ (we recall that $a_j a_j = b_j b_j = 1$). The general form of the function $B(\boldsymbol{r}, \boldsymbol{a}, \boldsymbol{b})$ which depends linearly on $\boldsymbol{a}$ and $\boldsymbol{b}$ is therefore given by

$$B(\boldsymbol{r}, \boldsymbol{a}, \boldsymbol{b}) = A_1(r)\, r_j a_j r_l b_l + A_2(r)\, a_j b_j$$

and, hence, the general form of the tensor $B_{jl}(\boldsymbol{r})$ is

$$B_{jl}(\boldsymbol{r}) = A_1(r)\, r_j r_l + A_2(r)\, \delta_{jl}. \tag{12.30}$$

In other words, the tensor function of the vector $\boldsymbol{r}$ which is invariant under rotations and reflections should be a linear combination of the constant invariant tensor $\delta_{jl}$ and the tensor $r_j r_l$, with coefficients depending on a unique invariant which can be made up of the components of $\boldsymbol{r}$, i.e., the length $r = |\boldsymbol{r}|$.[13] If we again take the set of coordinates $Mx_1'x_2'x_3'$ (in which $r_1 = r$, $r_2 = r_3 = 0$), and consider all the possible combinations of the subscripts $j$ and $l$ in Eq. (12.30), we find that $B'_{12} = B'_{13} = B'_{21} = B'_{23} = B'_{31} = B'_{32} = 0$ and

$$B_{LL}(r) = B'_{11}(r) = A_1(r)\, r^2 + A_2(r), \quad B_{NN}(r) = B'_{22}(r) = B'_{33}(r) = A_2(r).$$

Hence, it is clear that the formula given by Eq. (12.30) is identical with Eq. (12.29).

In order to settle finally the question as to what is the acceptable form of the tensor $B_{jl}(\boldsymbol{r})$, we need only consider the possible forms

[13] We note two further examples of the above idea. An additional tensor (other than the unit tensor $\delta_{jl}$) which is invariant under all rotations (but not reflections) is the constant tensor $\varepsilon_{ljl}$ which is antisymmetric in all its subscripts. The general form of $B_{jl}(\boldsymbol{r})$ which is invariant under rotations only is, therefore, the following

$$B_{jl}(\boldsymbol{r}) = A_1(r)\, r_j r_l + A_2(r)\, \delta_{jl} + A_3(r)\, \varepsilon_{ljl} r_l$$

This shows that the tensor $B_{jl}$ can, in fact, be nonsymmetric. Similarly, the general form of an axisymmetric tensor $B_{jl}(\boldsymbol{r})$ of rank 2, which is invariant under rotations about axes lying along the unit vector $\boldsymbol{\lambda}$, and under reflections in the planes containing $\boldsymbol{\lambda}$ or perpendicular to $\boldsymbol{\lambda}$, is given by

$$B_{jl}(\boldsymbol{r}) = A_1 r_j r_l + A_2 \lambda_j \lambda_l + A_3 \delta_{jl} + A_4 r_j \lambda_l + A_5 r_l \lambda_j,$$

where $A_1, \ldots, A_5$ are arbitrary functions of the two variables $r = (\boldsymbol{r}^2)^{1/2}$ and $\boldsymbol{r\lambda}$. Papers devoted to axisymmetric random fields, whose statistical characteristics are invariant under the above transformation, are those by Batchelor (1946) and Chandrasekhar (1950). Here, we shall restrict our attention to isotropic random fields, although fields which are invariant under translations but not under reflections, and axisymmetric random fields, are also used in the theory of turbulence.

of the functions $B_{LL}(r)$ and $B_{NN}(r)$. With this aim, let us consider the representation (11.54) of the tensor $B_{jl}(\boldsymbol{r})$ in the form of the Fourier transform of the spectral tensor $F_{jl}(\boldsymbol{k})$. For an isotropic field $\boldsymbol{u}(\boldsymbol{x})$, the tensor $F_{jl}(\boldsymbol{k})$ will be an isotropic (i.e., invariant under rotations and reflections) tensor function of the vector $\boldsymbol{k}$. In view of the discussion given above, we therefore have

$$F_{jl}(\boldsymbol{k}) = [F_{LL}(k) - F_{NN}(k)] \frac{k_j k_l}{k^2} + F_{NN}(k)\,\delta_{jl}, \tag{12.31}$$

where

$$F_{LL}(k) = \lim_{dk \to 0} \frac{\overline{|dZ_L(\boldsymbol{k})|^2}}{dk}, \quad F_{NN}(k) = \lim_{dk \to 0} \frac{\overline{|dZ_N(\boldsymbol{k})|^2}}{dk} \tag{12.32}$$

are functions of the single variable $k = |\boldsymbol{k}|$. The functions are called *the longitudinal and lateral three-dimensional spectra* of the field $\boldsymbol{u}(\boldsymbol{x})$ (the subscripts $L$ and $N$ now represent projections along the direction of $\boldsymbol{k}$ and the direction perpendicular to $\boldsymbol{k}$). The condition $F_{jl}(\boldsymbol{k}) = F^*_{lj}(-\boldsymbol{k})$ shows that both functions $F_{LL}(k)$ and $F_{NN}(k)$ must be real. Moreover, the matrix $\| F_{jl}(\boldsymbol{k}) \|$ should be nonnegative for all $\boldsymbol{k}$, and this is readily shown to be equivalent to the condition that both functions $F_{LL}(k)$ and $F_{NN}(k)$ should be nonnegative [this also follows from their definition given by Eq. (12.32)]. To insure that the tensor $B_{jl}(\boldsymbol{r})$ should be the correlation tensor of an isotropic random vector field, its Fourier transform should be representable in the form given by Eq. (12.31), where

$$F_{LL}(k) \geqslant 0, \quad F_{NN}(k) \geqslant 0. \tag{12.33}$$

If we substitute Eq. (12.31) into the general formula (11.54), and complete the integration with respect to the angular variables, which is facilitated by the identity

$$\int e^{i\boldsymbol{k}\boldsymbol{r}} k_j k_l F(\boldsymbol{k})\, d\boldsymbol{k} = -\frac{\partial^2}{\partial r_j \partial r_l} \left( \int e^{i\boldsymbol{k}\boldsymbol{r}} F(\boldsymbol{k})\, d\boldsymbol{k} \right),$$

we obtain the formulas which express $B_{LL}(r)$ and $B_{NN}(r)$ directly in terms of $F_{LL}(k)$ and $F_{NN}(k)$. These formulas are of the form

$$B_{LL}(r)=4\pi\int_0^\infty\left\{\frac{\sin kr}{kr}+2\frac{\cos kr}{(kr)^2}-2\frac{\sin kr}{(kr)^3}\right\}F_{LL}(k)\,k^2\,dk+$$
$$+4\pi\int_0^\infty\left\{-2\frac{\cos kr}{(kr)^2}+2\frac{\sin kr}{(kr)^3}\right\}F_{NN}(k)\,k^2\,dk,$$
$$B_{NN}(r)=4\pi\int_0^\infty\left\{-\frac{\cos kr}{(kr)^2}+\frac{\sin kr}{(kr)^3}\right\}F_{LL}(k)\,k^2\,dk+ \tag{12.34}$$
$$+4\pi\int_0^\infty\left\{\frac{\sin kr}{kr}+\frac{\cos kr}{(kr)^2}-\frac{\sin kr}{(kr)^3}\right\}F_{NN}(k)\,k^2\,dk$$

[see Yaglom (1948)]. Substituting Eq. (12.29) into Eq. (11.55), which expresses $F_{jl}(\boldsymbol{k})$ in terms of $B_{jl}(r)$, and proceeding as above, we obtain

$$F_{LL}(k)=\frac{1}{2\pi^2}\int_0^\infty\left\{\frac{\sin kr}{kr}+2\frac{\cos kr}{(kr)^2}-2\frac{\sin kr}{(kr)^3}\right\}B_{LL}(r)\,r^2\,dr+$$
$$+\frac{1}{2\pi^2}\int_0^\infty\left\{-2\frac{\cos kr}{(kr)^2}+2\frac{\sin kr}{(kr)^3}\right\}B_{NN}(r)\,r^2\,dr$$
$$F_{NN}(k)=\frac{1}{2\pi^2}\int_0^\infty\left\{-\frac{\cos kr}{(kr)^2}+\frac{\sin kr}{(kr)^3}\right\}B_{LL}(r)\,r^2\,dr+ \tag{12.35}$$
$$+\frac{1}{2\pi^2}\int_0^\infty\left\{\frac{\sin kr}{kr}+\frac{\cos kr}{(kr)^2}-\frac{\sin kr}{(kr)^3}\right\}B_{NN}(r)\,r^2\,dr.$$

Thus, to verify whether the given functions $B_{LL}(r)$ and $B_{NN}(r)$ are the longitudinal and lateral correlation functions of an isotropic vector field, we must substitute these functions into Eq. (12.35) and determine whether the corresponding functions $F_{LL}(k)$ and $F_{NN}(k)$ are nonnegative or not (see examples in Sect. 12.3 below).

If $\boldsymbol{u}(\boldsymbol{x})=\boldsymbol{u}(x_1, x_2, x_3)$ is an isotropic random vector field, then the variables $u_1(x_1, 0, 0)$ and $u_2(x_1, 0, 0)$ will obviously represent homogeneous fields on the straight line $x_2=x_3=0$ (i.e., stationary processes of the variable $x_1$), while the functions $B_{LL}(r)$ and $B_{NN}(r)$ will be the correlation functions for these fields on the straight line. Hence, it follows that the one-dimensional Fourier transforms of the functions $B_{LL}(r)$ and $B_{NN}(r)$, namely,

$$F_1(k_1) = \frac{1}{2\pi} \int_{-\infty}^{\infty} e^{-ik_1 r} B_{LL}(r)\, dr = \frac{1}{\pi} \int_0^{\infty} \cos k_1 r\, B_{LL}(r)\, dr \qquad (12.36)$$

and

$$F_2(k_1) = \frac{1}{2\pi} \int_{-\infty}^{\infty} e^{-ik_1 r} B_{NN}(r)\, dr = \frac{1}{\pi} \int_0^{\infty} \cos k_1 r\, B_{NN}(r)\, dr, \qquad (12.37)$$

will always be nonnegative. These transforms will be called the *longitudinal and lateral one-dimensional spectra of the field* $\boldsymbol{u}(\boldsymbol{x})$. It is readily seen that

$$\begin{aligned} F_1(k_1) &= \iint_{-\infty}^{\infty} F_{11}(k_1, k_2, k_3)\, dk_2\, dk_3, \\ F_2(k_1) &= \iint_{-\infty}^{\infty} F_{22}(k_1, k_2, k_3)\, dk_2\, dk_3 \end{aligned} \qquad (12.38)$$

[see the analogous formula (12.9) for the scalar field $u(\boldsymbol{x})$]. Substituting Eq. (12.31) into these expressions, and transforming to polar coordinates in the $(k_1, k_2)$ plane, we have after integration with respect to the angular variable:

$$\begin{aligned} F_1(k_1) &= 2\pi \int_{k_1}^{\infty} \left[F_{LL}(k)\, k_1^2 + F_{NN}(k)(k^2 - k_1^2)\right] k^{-1}\, dk, \\ F_2(k_1) &= \pi \int_{k_1}^{\infty} \left[F_{LL}(k)(k^2 - k_1^2) + F_{NN}(k)(k^2 + k_1^2)\right] k^{-1}\, dk. \end{aligned} \qquad (12.39)$$

These formulas show that if $F_{LL}(k) \geqslant 0$ and $F_{NN}(k) \geqslant 0$, we must also have $F_1(k) \geqslant 0$ and $F_2(k) \geqslant 0$. However, the converse may not be true.

In the isotropic case, the distribution of the kinetic energy $\overline{\boldsymbol{u}^2(\boldsymbol{x})}/2$ among the disturbances with various wave numbers is given by the function

$$E(k) = 4\pi k^2 \frac{F_{jj}(k)}{2} = 2\pi k^2 [F_{LL}(k) + 2F_{NN}(k)]. \qquad (12.40)$$

However, knowledge of the spectrum $E(k)$ is not sufficient for an unambiguous determination of the spectral tensor $F_{jl}(\boldsymbol{k})$.

If we replace the trigonometric functions in Eqs. (12.34) and (12.35) by the corresponding power series, we find that the Taylor expansions for the functions $B_{LL}(r)$, $B_{NN}(r)$, $F_{LL}(k)$ and $F_{NN}(k)$ contain only the even powers of the independent variable:

$$\begin{aligned} B_{LL}(r) &= b_0^{(LL)} + b_2^{(LL)} r^2 + b_4^{(LL)} r^4 + \dots, \\ B_{NN}(r) &= b_0^{(NN)} + b_2^{(NN)} r^2 + b_4^{(NN)} r^4 + \dots \end{aligned} \tag{12.41}$$

and

$$\begin{aligned} F_{LL}(k) &= f_0^{(LL)} + f_2^{(LL)} k^2 + f_4^{(LL)} k^4 + \dots, \\ F_{NN}(k) &= f_0^{(NN)} + f_2^{(NN)} k^2 + f_4^{(NN)} k^4 + \dots. \end{aligned} \tag{12.42}$$

For the coefficients on the right-hand side of Eqs. (12.41) and (12.42) we then have the following relationships:

$$\begin{aligned} b_{2n}^{(LL)} &= \frac{1}{(2n)!}\left(\frac{d^{2n}B_{LL}(r)}{dr^{2n}}\right)_{r=0} = \\ &= \frac{(-1)^n \cdot 4\pi}{(2n)!\,(2n+1)\,(2n+3)}\left[(2n+1)\int_0^\infty F_{LL}(k)\,k^{2n+2}\,dk + \right. \\ &\qquad \left. + 2\int_0^\infty F_{NN}(k)\,k^{2n+2}\,dk\right], \\ b_{2n}^{(NN)} &= \frac{1}{(2n)!}\left(\frac{d^{2n}B_{NN}(r)}{dr^{2n}}\right)_{r=0} = \\ &= \frac{(-1)^n \cdot 4\pi}{(2n)!\,(2n+1)\,(2n+3)}\left[\int_0^\infty F_{LL}(k)\,k^{2n+2}\,dk + \right. \\ &\qquad \left. + 2(n+1)\int_0^\infty F_{NN}(k)\,k^{2n+2}\,dk\right] \end{aligned} \tag{12.43}$$

from which it follows that

$$B_{LL}^{(4n)}(0) \geqslant 0, \quad B_{NN}^{(4n)}(0) \geqslant 0, \quad B_{LL}^{(4n+2)}(0) \leqslant 0, \quad B_{NN}^{(4n+2)}(0) \leqslant 0$$

and

$$f_{2n}^{(LL)}=\frac{1}{(2n)!}\left(\frac{d^{2n}F_{LL}(k)}{dk^{2n}}\right)_{k=0}=$$
$$=\frac{(-1)^n}{2\pi^2(2n)!\,(2n+1)\,(2n+3)}\left[(2n+1)\int_0^\infty B_{LL}(r)\,r^{2n+2}\,dr+2\int_0^\infty B_{NN}(r)\,r^{2n+2}\,dr\right],\quad (12.44)$$
$$f_{2n}^{(NN)}=\frac{1}{(2n)!}\left(\frac{d^{2n}F_{NN}(k)}{dk^{2n}}\right)_{k=0}=$$
$$=\frac{(-1)^n}{2\pi^2(2n)!\,(2n+1)\,(2n+3)}\left[\int_0^\infty B_{LL}(r)\,r^{2n+2}\,dr+2(n+1)\int_0^\infty B_{NN}(r)\,r^{2n+2}\,dr\right].$$

These relations show the rate at which the functions $F_{LL}(k)$ and $F_{NN}(k)$ [or $B_{LL}(r)$ and $B_{NN}(r)$] should decrease at infinity in order that the functions $B_{LL}(r)$ and $B_{NN}(r)$ [and, correspondingly, $F_{LL}(k)$ and $F_{NN}(k)$] should be at least $2n$-fold differentiable. We note also that if the function $B_{jj}(r)=B_{LL}(r)+2B_{NN}(r)$ falls to zero at infinity more rapidly than $r^{-3}$, then in view of Eq. (12.44),

$$F_{LL}(0)=F_{NN}(0)=\frac{1}{6\pi^2}\int_0^\infty [B_{LL}(r)+2B_{NN}(r)]\,r^2\,dr. \quad (12.45)$$

This last result is connected with the fact that if $F_{LL}(0)\neq F_{NN}(0)$, then the tensor $F_{jl}(\boldsymbol{k})$ will have a singularity at the point $\boldsymbol{k}=0$. Hence it also follows that the functions $B_{jl}(\boldsymbol{r})$ should in this case decrease sufficiently slowly at infinity.

In the special case when $F_{LL}(0)=F_{NN}(0)=0$ [i.e., when $F_{jl}(\boldsymbol{k})\sim c_0k^2\delta_{jl}+c_1k_jk_l$ as $\boldsymbol{k}\to 0$], it follows from Eq. (12.45) that

$$\int_0^\infty [B_{LL}(r)+2B_{NN}(r)]\,r^2\,dr=0. \quad (12.46)$$

When $n=0$, the formulas given by Eq. (12.43) lead to the following expression:

$$B_{LL}(0)=B_{NN}(0)=\frac{4\pi}{3}\int_0^\infty [F_{LL}(k)+2F_{NN}(k)]\,k^2\,dk, \quad (12.47)$$

which is similar to Eq. (12.45) and equivalent to Eq. (12.40). The integral on the right-hand side is, of course, always positive.

It is readily shown from Eqs. (12.34) and (12.35) that

$$\begin{aligned}\int_0^\infty B_{LL}(r)\,dr &= 2\pi^2\int_0^\infty F_{NN}(k)\,k\,dk,\\ \int_0^\infty B_{NN}(r)\,dr &= \pi^2\int_0^\infty [F_{LL}(k)+F_{NN}(k)]\,k\,dk\end{aligned} \quad (12.48)$$

and, correspondingly,

$$\begin{aligned}\int_0^\infty F_{LL}(k)\,dk &= \frac{1}{4\pi}\int_0^\infty B_{NN}(r)\,r\,dr,\\ \int_0^\infty F_{NN}(k)\,dk &= \frac{1}{8\pi}\int_0^\infty [B_{LL}(r)+B_{NN}(r)]\,r\,dr.\end{aligned} \quad (12.49)$$

Using Eqs. (12.48) and (12.47), we can readily express the longitudinal and lateral integral length scales $L_1$ and $L_2$ of the field $\boldsymbol{u}(\boldsymbol{x})$:

$$L_1=\frac{1}{B_{LL}(0)}\int_0^\infty B_{LL}(r)\,dr, \quad L_2=\frac{1}{B_{NN}(0)}\int_0^\infty B_{NN}(r)\,dr. \quad (12.50)$$

in terms of the spectral densities $F_{LL}(k)$ and $F_{NN}(k)$. Similarly, the longitudinal and lateral differential scales $\lambda_1$ and $\lambda_2$, which are given by

$$\lambda_1=\left\{-\frac{B_{LL}(0)}{2B''_{LL}(0)}\right\}^{1/2}, \quad \lambda_2=\left\{-\frac{B_{NN}(0)}{2B''_{NN}(0)}\right\}^{1/2}, \quad (12.51)$$

can readily be expressed in terms of $F_{LL}(k)$ and $F_{NN}(k)$ with the aid of Eqs. (12.43) and (12.47) (cf. the end of Sect. 12.1).

Suppose now that in addition to the isotropic vector field $\boldsymbol{u}(\boldsymbol{x})$ we also have an isotropic scalar field $\vartheta(\boldsymbol{x})$ with $\overline{\vartheta(\boldsymbol{x})}=0$). Moreover, the four-dimensional field $\{\boldsymbol{u}(\boldsymbol{x}), \vartheta(\boldsymbol{x})\}$ is also isotropic. This four-dimensional field is characterized by the correlation tensor $B_{jl}(\boldsymbol{r})$, which can be written down in the form of Eq. (12.29), a scalar correlation function $B_{\vartheta\vartheta}(\boldsymbol{r})$, and the cross correlation vectors

$$B_{j\vartheta}(\boldsymbol{r})=\overline{u_j(\boldsymbol{x})\,\vartheta(\boldsymbol{x}+\boldsymbol{r})},\qquad B_{\vartheta j}(\boldsymbol{r})=\overline{\vartheta(\boldsymbol{x})\,u(\boldsymbol{x}+\boldsymbol{r})}=B_{j\vartheta}(-\boldsymbol{r}),$$
$$j=1,\,2,\,3. \qquad (12.52)$$

Similarly to the derivation of Eq. (12.29), it can be shown that

$$B_{j\vartheta}(\boldsymbol{r})=B_{L\vartheta}(r)\frac{r_j}{r},\quad B_{\vartheta j}(\boldsymbol{r})=B_{\vartheta L}(r)\frac{r_j}{r}=-B_{L\vartheta}(r)\frac{r_j}{r}, \qquad (12.53)$$

where $B_{L\vartheta}(r)=\overline{u_L(\boldsymbol{x})\,\vartheta(\boldsymbol{x}+\boldsymbol{r})}$ (and $\overline{u_N(\boldsymbol{x})\,\vartheta(\boldsymbol{x}+\boldsymbol{r})}=0$). Consider the spectral densities $F_{j\vartheta}(\boldsymbol{k})$ which are defined by

$$B_{j\vartheta}(\boldsymbol{r})=\int e^{i\boldsymbol{k}\boldsymbol{r}}F_{j\vartheta}(\boldsymbol{k})\,d\boldsymbol{k},\quad F_{j\vartheta}(\boldsymbol{k})=\frac{1}{8\pi^3}\int e^{-i\boldsymbol{k}\boldsymbol{r}}B_{j\vartheta}(\boldsymbol{r})\,d\boldsymbol{r} \qquad (12.54)$$

and, similarly, for $F_{\vartheta j}(\boldsymbol{k})$. We then have formulas similar to Eq. (12.53):

$$F_{j\vartheta}(\boldsymbol{k})=iF_{L\vartheta}(k)\frac{k_j}{k},\qquad F_{\vartheta j}(\boldsymbol{k})=iF_{\vartheta L}(k)\frac{k_j}{k} \qquad (12.55)$$

where the factor $i$ is added for convenience. Moreover,

$$iF_{L\vartheta}(\boldsymbol{k})\,d\boldsymbol{k}=\overline{dZ_L^*(\boldsymbol{k})\,dZ_\vartheta(\boldsymbol{k})},\qquad \overline{dZ_N^*(\boldsymbol{k})\,dZ_\vartheta(\boldsymbol{k})}=0, \qquad (12.56)$$

where $dZ_\vartheta(\boldsymbol{k})$ are the complex Fourier amplitudes of the scalar field $\vartheta(\boldsymbol{x})$, and $dZ_L(\boldsymbol{k})$ and $dZ_N(\boldsymbol{k})$ are the projections of the amplitudes $d\boldsymbol{Z}(\boldsymbol{k})$ of the vector field $\boldsymbol{u}(\boldsymbol{x})$ along the direction of $\boldsymbol{k}$ and any perpendicular direction. Since $F_{j\vartheta}(\boldsymbol{k})=F^*_{j\vartheta}(-\boldsymbol{k})=F^*_{\vartheta j}(\boldsymbol{k})$ [see Eq. (11.56)], the functions $F_{L\vartheta}(k)$ and $F_{\vartheta L}(k)$ will be real and, moreover, $F_{L\vartheta}(k)=-F_{\vartheta L}(k)$. It is readily verified that the condition that the matrix $\|F_{jl}(\boldsymbol{k})\|$, $j,\ l=1,\,2,\,3,\ \vartheta$ should be nonnegative-definite can be reduced to the conditions $F_{LL}(k)\geqslant 0, F_{NN}(k)\geqslant 0, F_{\vartheta\vartheta}(k)\geqslant 0$, and

$$F_{\vartheta\vartheta}(k)\,F_{LL}(k)-[F_{L\vartheta}(k)]^2\geqslant 0. \qquad (12.57)$$

Similarly, in the case of the five-dimensional isotropic field $\{u(x), P(x), S(x)\}$, where $P(x)$ and $S(x)$ are scalar fields (we shall encounter such a five-dimensional field below), the corresponding spectral densities $F_{LL}(k)$, $F_{NN}(k)$, $F_{LP}(k)$, $F_{LS}(k)$, $F_{PP}(k)$, $F_{SS}(k)$, and $F_{PS}(k)$ should satisfy the following conditions which guarantee that the matrix $\| F_{jl}(k) \|$, $j, l = 1, 2, 3, P, S$ is nonnegative:

$$\begin{gathered} F_{LL}(k) \geqslant 0, \quad F_{NN}(k) \geqslant 0, \quad F_{PP}(k) \geqslant 0, \quad F_{SS}(k) \geqslant 0, \\ F_{PP}(k) F_{LL}(k) - F_{LP}^2(k) \geqslant 0, \quad F_{SS}(k) F_{LL}(k) - F_{LS}^2(k) \geqslant 0, \\ F_{PP}(k) F_{SS}(k) - F_{PS}^2(k) \geqslant 0, \\ F_{LL}(k) \left[F_{PP}(k) F_{SS}(k) - F_{PS}^2(k)\right] - F_{LP}^2(k) F_{SS}(k) - \\ - F_{LS}^2(k) F_{PP}(k) + 2F_{LP}(k) F_{LS}(k) F_{PS}(k) \geqslant 0 \end{gathered} \tag{12.58}$$

[see Yaglom (1948)].

Substituting Eqs. (12.53) and (12.55) into Eq. (12.54), and carrying out the integration with respect to the angular variables, we obtain the following equations relating $B_{L\theta}(r)$ and $F_{L\theta}(k)$:

$$B_{L\theta}(r) = 4\pi \int_0^\infty \left\{ -\frac{\cos kr}{kr} + \frac{\sin kr}{(kr)^2} \right\} F_{L\theta}(k) k^2 \, dk, \tag{12.59}$$

$$F_{L\theta}(k) = \frac{1}{2\pi^2} \int_0^\infty \left\{ -\frac{\cos kr}{kr} + \frac{\sin kr}{(kr)^2} \right\} B_{L\theta}(r) r^2 \, dr. \tag{12.60}$$

Hence, it is clear that $B_{L\theta}(0) = 0$, $F_{L\theta}(0) = 0$ (this was expected in view of the fact that $B_{N\theta}(r) \equiv 0$, $F_{N\theta}(k) \equiv 0$, and also the fact that the subscripts $L$ and $N$ are equivalent at the origin), and that the Taylor expansion for the functions $B_{L\theta}(r)$ and $F_{L\theta}(k)$ are of the form

$$B_{L\theta}(r) = b_1^{(L\theta)} r + b_3^{(L\theta)} r^3 + b_5^{(L\theta)} r^5 + \ldots, \tag{12.61}$$

$$F_{L\theta}(k) = f_1^{(L\theta)} k + f_3^{(L\theta)} k^3 + f_5^{(L\theta)} k^5 + \ldots, \tag{12.62}$$

where

$$b_{2n+1}^{(L\theta)} = \frac{1}{(2n+1)!} \left( \frac{d^{2n+1} B_{L\theta}(r)}{dr^{2n+1}} \right)_{r=0} = \frac{(-1)^n \cdot 4\pi}{(2n+1)!\,(2n+3)} \int_0^\infty F_{L\theta}(k) k^{2n+3} \, dk, \tag{12.63}$$

$$f_{2n+1}^{(L\vartheta)} = \frac{1}{(2n+1)!}\left(\frac{d^{2n+1}F_{L\vartheta}(k)}{dk^{2n+1}}\right)_{k=0} =$$
$$= \frac{(-1)^n}{2\pi^2(2n+1)!(2n+3)}\int_0^\infty B_{L\vartheta}(r)\,r^{2n+3}\,dr. \quad (12.64)$$

Moreover, it can be shown from Eqs. (12.59) and (12.60) that

$$\int_0^\infty B_{L\vartheta}(r)\,dr = 4\pi\int_0^\infty F_{L\vartheta}(k)\,k\,dk, \quad (12.65)$$

$$\int_0^\infty F_{L\vartheta}(k)\,dk = \frac{1}{2\pi^2}\int_0^\infty B_{L\vartheta}(r)\,r\,dr \quad (12.66)$$

[cf. equations (12.19), (12.20), and (12.48), and (12.49) above].

## 12.3 Solenoidal and Potential Isotropic Vector Fields

The above results on the correlation functions and spectra of an isotropic vector field $\boldsymbol{u}(\boldsymbol{x})$ are considerably simplified when it is known in advance that this field is solenoidal (zero divergence) or potential (zero curl). In fact, if for example the field $\boldsymbol{u}(\boldsymbol{x})$ is solenoidal, then its correlation tensor $B_{jl}(\boldsymbol{r})$ must satisfy the conditions given by Eq. (11.80), while its spectral tensor $F_{jl}(\boldsymbol{k})$ must satisfy the conditions given by Eq. (11.79). However, in the isotropic case, we can use the fact that the tensor $B_{jl}(\boldsymbol{r})$ must be given by Eq. (12.29), while the tensor $F_{jl}(\boldsymbol{k})$ must be of the form (12.31). At the same time, it is readily verified that the conditions given by Eq. (11.80) and Eq. (11.79) now assume the form

$$B_{NN}(r) = B_{LL}(r) + \frac{r}{2}\frac{d}{dr}B_{LL}(r) \quad (12.67)$$

so that

$$B_{LL}(r) = \frac{2}{r^2}\int_0^r B_{NN}(x)\,x\,dx$$

and, consequently,

$$F_{LL}(k) = 0. \quad (12.68)$$

The conditions given by Eqs. (12.67) and (12.68) [like the original conditions (11.79) and (11.80)] are equivalent to each other. The first of them was found by von Kármán (1937) and has since been named after him. The second was established much later [see Yaglom (1948) and Batchelor (1949a)].

Similarly, in the case of a homogeneous potential vector field

$$\overline{\left[\frac{\partial u_j(x)}{\partial x_l}-\frac{\partial u_l(x)}{\partial x_j}\right]u_i(x+r)}=0$$

and hence

$$\frac{\partial B_{jl}(r)}{\partial r_l}-\frac{\partial B_{ll}(r)}{\partial r_j}=0, \quad k_l F_{jl}(k)-k_j F_{ll}(k)=0. \qquad (12.69)$$

Substituting the general expressions given by Eqs. (12.29) and (12.31) into Eq. (12.69), we can verify that, in the isotropic case, the condition for the field $u(x)$ to be potential can be written in one of the following two equivalent forms:

$$B_{LL}(r)=B_{NN}(r)+r\frac{dB_{NN}(r)}{dr} \qquad (12.70)$$

so that

$$B_{NN}(r)=\frac{1}{r}\int_0^r B_{LL}(x)\,dx$$

or

$$F_{NN}(k)=0 \qquad (12.71)$$

[see Obukhov (1954) and Obukhov and Yaglom (1951)]. The equations given by (12.67), (12.68), (12.70), and (12.71) show that, in the case of a solenoidal or potential isotropic vector field $u(x)$, the correlation and spectral tensors are determined by a single scalar function rather than by two such functions. In particular, the spectral tensor is uniquely expressed in terms of the spectrum $E(k)$ of Eq. (12.40), which is related to the spectra $F_{LL}(k)$ and $F_{NN}(k)$ by obvious relations

$$E(k)=\begin{cases}4\pi k^2F_{NN}(k) & \text{for solenoidal field,}\\ 2\pi k^2F_{LL}(k) & \text{for potential field.}\end{cases} \tag{12.72}$$

The spectral tensor of a solenoidal isotropic field can thus be written

$$F_{jl}(\boldsymbol{k})=\frac{E(k)}{4\pi k^2}\left(\delta_{jl}-\frac{k_jk_l}{k^2}\right), \tag{12.73}$$

while the spectral tensor of a potential isotropic field is given by

$$F_{jl}(\boldsymbol{k})=\frac{E(k)\,k_jk_l}{2\pi k^4}. \tag{12.74}$$

In view of Eqs. (12.34), (12.35), and (12.67), (12.68), in the case of a solenoidal field we have

$$B_{LL}(r)=2\int_0^\infty\left[-\frac{\cos kr}{(kr)^2}+\frac{\sin kr}{(kr)^3}\right]E(k)\,dk, \tag{12.75}$$

$$B_{NN}(r)=\int_0^\infty\left[\frac{\sin kr}{kr}+\frac{\cos kr}{(kr)^2}-\frac{\sin kr}{(kr)^3}\right]E(k)\,dk \tag{12.76}$$

and (assuming that $r^2B_{LL}(r)\to 0$ as $r\to\infty$)

$$E(k)=\frac{1}{\pi}\int_0^\infty kr\sin kr\,[3B_{LL}(r)+rB'_{LL}(r)]\,dr=$$

$$=\frac{1}{\pi}\int_0^\infty (kr\sin kr-k^2r^2\cos kr)\,B_{LL}(r)\,dr. \tag{12.77}$$

As in the case of a potential field, we need only retain the first term the right-hand side of Eq. (12.34), and replace $F_{LL}(k)$ by $E(k)/2\pi k^2$. Equation (12.35) can then be rewritten in the form

$$E(k)=\frac{1}{\pi}\int_0^\infty kr\sin kr\,[3B_{NN}(r)+rB'_{NN}(r)]\,dr=$$

$$=\frac{1}{\pi}\int_0^\infty (kr\sin kr-k^2r^2\cos kr)\,B_{NN}(r)\,dr. \tag{12.78}$$

In the case of a solenoidal field $u(x)$, the formulas given by Eqs. (12.43) and (12.44) assume the form

$$b_{2n}^{(LL)} = \frac{2(-1)^n}{(2n)!\,(2n+1)\,(2n+3)} \int_0^\infty E(k)\,k^{2n}\,dk = \frac{1}{n+1}\,b_{2n}^{(NN)}, \quad (12.79)$$

$$4\pi f_{2n-2}^{(NN)} = \frac{1}{(2n)!}\left(\frac{d^{2n}E(k)}{dk^{2n}}\right)_{k=0} =$$

$$= \frac{(-1)^n}{\pi\,(2n-3)!\,(2n-1)} \int_0^\infty B_{LL}(r)\,r^{2n}\,dr, \qquad n = 1, 2, \ldots \quad (12.80)$$

The formulas transform in an analogous way when the field $u(x)$ is a potential field

Henceforth, when we discuss a solenoidal isotropic field, we shall simply write $F(k)$ for $F_{NN}(k)$ [while in the case of a potential function we shall write $F(k)$ for $F_{LL}(k)$]. The formulas given by equations (12.67) and (12.70) show that if the sum $B_{LL}(r) + 2B_{NN}(r)$ falls to zero at infinity more rapidly than $r^{-3}$, then

$$\int_0^\infty [B_{LL}(r) + 2B_{NN}(r)]\,r^2\,dr = 0 \quad (12.81)$$

for both solenoidal and potential fields. In view of Eq. (12.45), this means that in both cases $F(0) = 0$, i.e., $E(k) \sim k^4$, for small $k$ (the last condition guarantees the absence of a singularity in the tensor $F_{ij}(\boldsymbol{k})$ at the point $\boldsymbol{k} = 0$). Finally, the formulas given by Eqs. (12.48) and (12.49) show that for solenoidal fields

$$\int_0^\infty B_{LL}(r)\,dr = 2\int_0^\infty B_{NN}(r)\,dr = \frac{\pi}{2}\int_0^\infty \frac{E(k)}{k}\,dk,$$
$$\int_0^\infty rB_{NN}(r)\,dr = 0, \quad \int_0^\infty rB_{LL}(r)\,dr = 2\int_0^\infty \frac{E(k)}{k^2}\,dk, \quad (12.82)$$

while for potential fields

$$\int_0^\infty B_{LL}(r)\,dr = 0, \quad \int_0^\infty B_{NN}(r)\,dr = \frac{\pi}{2}\int_0^\infty \frac{E(k)}{k}\,dk,$$
$$\int_0^\infty rB_{NN}(r)\,dr = -\int_0^\infty rB_{LL}(r)\,dr = 2\int_0^\infty \frac{E(k)}{k^2}\,dk. \tag{12.83}$$

From the second equation in (12.82) and the first equation in (12.83) it follows in particular that, in the case of a solenoidal field, the function $B_{NN}(r)$ must also assume negative values, while in the case of a potential field, the function $B_{LL}(r)$ must change sign. The same conclusions are reached by starting with Eq. (12.67) and (12.70), which show that, in the case of a solenoidal isotropic field $\boldsymbol{u}(\boldsymbol{x})$, we have

$$\int_0^\infty \left[B_{NN}(r) + \frac{m-1}{2} B_{LL}(r)\right] r^m\,dr = 0, \tag{12.84}$$

if $r^{m+1}B_{LL}(r) \to 0$ as $r \to \infty$, while in the case of a potential field $\boldsymbol{u}(\boldsymbol{x})$, we have

$$\int_0^\infty [B_{LL}(r) + mB_{NN}(r)]\, r^m\,dr = 0. \tag{12.85}$$

if $r^{m+1}B_{NN}(r) \to 0$ as $r \to \infty$.

Comparison of Eqs. (12.77) and (12.12) will show that, to ensure that a given function $B_{LL}(r)$ will be the longitudinal correlation function of a solenoidal vector field, the function $3B_{LL} + rB'_{LL}$ must be the correlation function of a scalar isotropic field. Similarly, it follows from Eq. (12.78) that the function $B_{NN}$ can be the lateral correlation function of a potential field, provided only the function $3B_{NN} + rB'_{NN}$ is the correlation function of an isotropic scalar field. Hence, it is clear that *the set of longitudinal correlation functions of all the possible solenoidal fields coincides with the set of lateral correlation functions of all the possible potential fields* [this result is valid for isotropic fields in a space of any number of dimensions; see Yaglom, 1957]. The last result can also be shown to follow from Eq. (12.39), according to which in the case of a solenoidal field

$$E_1(k_1) = \int_{k_1}^{\infty} \left(1 - \frac{k_1^2}{k^2}\right) \frac{E(k)}{k}\, dk, \tag{12.86}$$

$$E_2(k_1) = \frac{1}{2} \int_{k_1}^{\infty} \left(1 + \frac{k_1^2}{k^2}\right) \frac{E(k)}{k}\, dk = \frac{1}{2} E_1(k_1) - \frac{1}{2} k_1 \frac{dE_1(k_1)}{dk_1}, \tag{12.87}$$

so that

$$E(k) = \frac{1}{2}\left(k^2 \frac{d^2}{dk^2} - k\frac{d}{dk}\right) E_1(k) = \frac{1}{2} k^3 \frac{d}{dk}\left(\frac{1}{k}\frac{d}{dk} E_1(k)\right), \tag{12.88}$$

while in the case of a potential field

$$E_2(k_1) = \int_{k_1}^{\infty} \left(1 - \frac{k_1^2}{k^2}\right) \frac{E(k)}{k}\, dk \tag{12.89}$$

and

$$E(k) = \frac{1}{2}\left(k^2 \frac{d^2}{dk^2} - k\frac{d}{dk}\right) E_2(k) = \frac{1}{2} k^3 \frac{d}{dk}\left(\frac{1}{k}\frac{d}{dk} E_2(k)\right), \tag{12.90}$$

where $E_1(k) = 2F_1(k)$ and $E_2(k) = 2F_2(k)$ are the longitudinal and lateral one-dimensional spectra respectively. It follows to ensure that the function $B_{LL}(r)$ is the longitudinal correlation function of a solenoidal field, and that the function $B_{NN}(r)$ is the lateral correlation function of a potential field, it is only necessary that the application of operator

$$k^2 \frac{d^2}{dk^2} - k\frac{d}{dk} = k^3 \frac{d}{dk}\frac{1}{k}\frac{d}{dk}$$

to the one-dimensional Fourier transform of the function under consideration should lead to a nonnegative function. From this, or directly from equations (12.77) and (12.78), it can be readily verified that all three functions (12.5)–(12.5″) which were given as examples of correlation functions of scalar isotropic fields, can also be the longitudinal correlation functions of solenoidal vector fields, or the lateral correlation functions of potential vector fields. In both cases, these functions will correspond to the spectra $E(k)$, respectively given by

$$E(k)=\frac{8C\alpha}{\pi}\frac{k^4}{(k^2+\alpha^2)^3},$$

$$E(k)=\frac{C}{8\pi^{1/2}\alpha^{5/2}}k^4e^{-k^2/4\alpha},$$

$$E(k)=\frac{2^{\nu+1}\Gamma(\nu+5/2)\,\alpha^{2\nu}C}{\pi^{1/2}}\frac{k^4}{(\alpha^2+k^2)^{\nu+5/2}}.$$

Examples of lateral correlation functions of a solenoidal field, and longitudinal correlation functions of a potential field, can be obtained from those given in Eqs. (12.5)–(12.5″) with the aid of Eqs. (12.67) and (12.70).

The integral and differential length scales $L_1$, $L_2$ and $\lambda_1$, $\lambda_2$ of equations (12.50) and (12.51) will, in the case of a solenoidal field, be given by

$$L_1=\frac{1}{2}L_2=\frac{3\pi\int_0^\infty E(k)\,k^{-1}\,dk}{4\int_0^\infty E(k)\,dk}=\frac{3\pi\int_0^\infty F(k)\,k\,dk}{4\int_0^\infty F(k)\,k^2\,dk}, \tag{12.91}$$

$$\lambda_1=\sqrt{2}\,\lambda_2=\left\{\frac{5\int_0^\infty E(k)\,dk}{\int_0^\infty E(k)\,k^2\,dk}\right\}^{1/2}=\left\{\frac{5\int_0^\infty F(k)\,k^2\,dk}{\int_0^\infty F(k)\,k^4\,dk}\right\}^{1/2}. \tag{12.92}$$

Similarly, for a potential field

$$L_1=0,\quad L_2=\frac{3\pi\int_0^\infty E(k)\,k^{-1}\,dk}{4\int_0^\infty E(k)\,dk},\quad \lambda_1=\frac{\lambda_2}{\sqrt{3}}=\left\{\frac{5\int_0^\infty E(k)\,dk}{3\int_0^\infty E(k)\,k^2\,dk}\right\}^{1/2}. \tag{12.93}$$

The rate of energy dissipation $\bar{\varepsilon}$ of a solenoidal isotropic field $\boldsymbol{u}(\boldsymbol{x})$

is, in view of equation (11.77), (12.88), and (12.87), given by

$$\bar{\varepsilon} = 2\nu \int_0^\infty k^2 E(k)\, dk = 15\nu \int_0^\infty k^2 E_1(k)\, dk = \frac{15}{2} \nu \int_0^\infty k^2 E_2(k)\, dk, \qquad (12.94)$$

while the vorticity field $\boldsymbol{\omega}(\boldsymbol{x})$ of $\boldsymbol{u}(\boldsymbol{x})$ will be solenoidal with a three-dimensional spectrum $k^2 F(k)$ [and a spectrum $k^2 E(k)$].

In the case of a pair of isotropic vector fields $\boldsymbol{u}(\boldsymbol{x})$ and $\boldsymbol{v}(\boldsymbol{x})$, such that the six-dimensional field $\{\boldsymbol{u}(\boldsymbol{x}), \boldsymbol{v}(\boldsymbol{x})\}$ is also isotropic, the cross correlation and spectral tensors $B_{jl}^{(uv)}(\boldsymbol{r}) = \overline{u_j(\boldsymbol{x}) v_l(\boldsymbol{x}+\boldsymbol{r})}$ and $F_{jl}^{(uv)}(\boldsymbol{k})$ will have many of the properties of the tensors $B_{jl}(\boldsymbol{r})$ and $F_{jl}(\boldsymbol{k})$. In particular, the representations given by equations (12.29), (12.31), (12.34), and (12.35) will be valid for them (except that the functions $F_{LL}^{(uv)}(k)$ and $F_{NN}^{(uv)}(k)$ will no longer be necessarily nonnegative). If at least one of the fields $\boldsymbol{u}(\boldsymbol{x})$, $\boldsymbol{v}(\boldsymbol{x})$ is solenoidal, then the functions $B_{LL}^{(uv)}(r)$ and $B_{NN}^{(uv)}(r)$ will satisfy the von Kármán condition (12.67), and the equation $F_{LL}^{(uv)}(k) = 0$ will be satisfied. In exactly the same way, if at least one of the fields $\boldsymbol{u}(\boldsymbol{x})$, $\boldsymbol{v}(\boldsymbol{x})$ is potential, then the functions $B_{LL}^{(uv)}(r)$, $B_{NN}^{(uv)}(r)$ will satisfy the Obukhov condition (12.70), and the equation $F_{NN}^{(uv)}(k) = 0$ will be satisfied. Hence, it follows that if one of the fields $\boldsymbol{u}(\boldsymbol{x})$, $\boldsymbol{v}(\boldsymbol{x})$ is solenoidal, while the other is potential, then $F_{LL}^{(uv)}(k) = F_{NN}^{(uv)}(k) = 0$ and consequently $B_{jl}^{(uv)}(\boldsymbol{r}) = 0$. Therefore, two isotropic vector fields of which one is solenoidal and the other potential are necessarily uncorrelated with each other.

On the other hand, following Obukhov (1954), it is readily shown that any isotropic random vector field $\boldsymbol{u}(\boldsymbol{x})$ can be represented as the sum of two uncorrelated isotropic fields, one of which is solenoidal and the other potential. In fact, the condition that the field $\boldsymbol{u}(\boldsymbol{x})$ is solenoidal means that

$$k_j\, dZ_j(\boldsymbol{k}) = 0, \quad \text{i.e.,} \quad dZ_L(\boldsymbol{k}) \equiv 0; \qquad (12.95)$$

while the condition that the field is potential means that

$$\varepsilon_{ijl} k_j\, dZ_l(\boldsymbol{k}) = 0, \quad \text{i.e.,} \quad dZ_N(\boldsymbol{k}) = 0. \qquad (12.96)$$

Therefore, to expand an arbitrary uniform field $\boldsymbol{u}(\boldsymbol{x})$ into a solenoidal and a potential component, we must expand the vector function $d\boldsymbol{Z}(\boldsymbol{k})$ corresponding to this field into a "longitudinal component" $d\boldsymbol{Z}_L(\boldsymbol{k})$ and a "lateral component" $d\boldsymbol{Z}_N(\boldsymbol{k})$, using the formulas

$$d\boldsymbol{Z}_L(\boldsymbol{k}) = \frac{\boldsymbol{k}}{k^2} k_j \, dZ_j(\boldsymbol{k}), \quad d\boldsymbol{Z}_N(\boldsymbol{k}) = d\boldsymbol{Z}(\boldsymbol{k}) - \frac{\boldsymbol{k}}{k^2} k_j \, dZ_j(\boldsymbol{k}) \quad (12.97)$$

and then set up the expressions

$$\boldsymbol{u}_s(\boldsymbol{x}) = \int e^{i\boldsymbol{k}\boldsymbol{x}} \, d\boldsymbol{Z}_N(\boldsymbol{k}), \quad \boldsymbol{u}_p(\boldsymbol{x}) = \int e^{i\boldsymbol{k}\boldsymbol{x}} \, d\boldsymbol{Z}_L(\boldsymbol{k}). \quad (12.98)$$

In that case, $\boldsymbol{u}_s(\boldsymbol{x})$ will be the solenoidal component of the field $\boldsymbol{u}(\boldsymbol{x})$, and $\boldsymbol{u}_p(\boldsymbol{x})$ will be its potential component. The fields $\boldsymbol{u}_s(\boldsymbol{x})$ and $\boldsymbol{u}_p(\boldsymbol{x})$ will be uncorrelated, and

$$\boldsymbol{u}(\boldsymbol{x}) = \boldsymbol{u}_s(\boldsymbol{x}) + \boldsymbol{u}_p(\boldsymbol{x}). \quad (12.99)$$

The spectral tensor $F_{jl}(\boldsymbol{k})$ of the original field $\boldsymbol{u}(\boldsymbol{x})$ will then split into the components $F_{NN}(k)(\delta_{jl} - k^{-2}k_jk_l)$ and $F_{LL}(k)\,k^{-2}k_jk_l$, which correspond to the components $\boldsymbol{u}_s(\boldsymbol{x})$ and $\boldsymbol{u}_p(\boldsymbol{x})$, while the correlation tensor $B_{jl}(\boldsymbol{r})$ will split into the components $B_{jl}^{(s)}(\boldsymbol{r})$ and $B_{jl}^{(p)}(\boldsymbol{r})$, the first of which satisfies the condition

$$\frac{\partial B_{jl}^{(s)}}{\partial r_l} = 0,$$

and the second the condition

$$\frac{\partial B_{jl}^{(p)}}{\partial r_l} - \frac{\partial B_{ll}^{(p)}}{\partial r_j} = 0.$$

We shall now show that a scalar isotropic field $\vartheta(\boldsymbol{x})$ cannot be correlated with a solenoidal isotropic vector field $\boldsymbol{u}(\boldsymbol{x})$. We shall use the fact that, in view of Eq. (12.56), the complex Fourier amplitude $dZ_\vartheta(\boldsymbol{k})$ of an isotropic scalar field $\vartheta(\boldsymbol{x})$ can be correlated only with $d\boldsymbol{Z}_L(\boldsymbol{k})$, but not with $d\boldsymbol{Z}_N(\boldsymbol{k})$. Hence, it immediately follows that the field $\vartheta(\boldsymbol{x})$ itself can be correlated only with the potential component of the isotropic vector field $\boldsymbol{u}(\boldsymbol{x})$, but not with its solenoidal component. The same result is obtained without using the spectral representations of random fields in the form of Fourier-Stieltjes integrals. In fact, if $\boldsymbol{u}(\boldsymbol{x})$ is a solenoidal homogeneous vector field, and $\vartheta(\boldsymbol{x})$ is a homogeneous scalar field homogeneously correlated with it, then obviously

$$\frac{\partial B_{j\vartheta}(\boldsymbol{r})}{\partial r_j}=0, \qquad \text{i.e.,} \quad k_j F_{j\vartheta}(\boldsymbol{k})=0. \tag{12.100}$$

Using the fact that $F_{j\vartheta}(\boldsymbol{k})=iF_{L\vartheta}(k)\,k^{-1}k_j$, we obtain

$$F_{L\vartheta}(k)=0, \qquad \text{i.e.,} \quad F_{j\vartheta}(\boldsymbol{k})\equiv 0 \tag{12.101}$$

[see also inequality (12.57)]. Following von Kármán and Howarth (1938), we can also use the fact that the first equation in Eq. (12.100) takes the form

$$\frac{d}{dr}B_{L\vartheta}(r)+\frac{2}{r}B_{L\vartheta}(r)=0, \tag{12.102}$$

when $B_{j\vartheta}(\boldsymbol{r})=B_{L\vartheta}(r)\,r^{-1}r_j$, and the solution $B_{L\vartheta}(r)=Cr^{-2}$ of this equation becomes infinite at $r=0$ for $C\neq 0$. Consequently, $B_{L\vartheta}(r)=0$.

## 12.4 One-Point and Two-Point Higher-Order Moments of Isotropic Fields

Consider, to begin with, the one-point higher-order moments of the derivatives $\partial u_i(\boldsymbol{x})/\partial x_j$ of a solenoidal isotropic vector field $\boldsymbol{u}(\boldsymbol{x})$. These moments are independent of $\boldsymbol{x}$, i.e., they are constants. It is clear that they satisfy the usual inequalities for moments which were discussed in the end of Sect. 4.1 of Vol. 1. It turns out, however, that if the field $\boldsymbol{u}(\boldsymbol{x})$ is isotropic and solenoidal, there are additional inequalities for the moments of derivatives $\partial u_i/\partial x_j$ which are more restrictive than those which are satisfied by the moments of arbitrary random variables. Following Betchov (1956), we shall derive such an additional inequality which relates the dimensionless quantities

$$s=\overline{\left(\frac{\partial u_1}{\partial x_1}\right)^3}\Big/\left[\overline{\left(\frac{\partial u_1}{\partial x_1}\right)^2}\right]^{3/2} \text{ and } \delta=\overline{\left(\frac{\partial u_1}{\partial x_1}\right)^4}\Big/\left[\overline{\left(\frac{\partial u_1}{\partial x_1}\right)^2}\right]^2. \tag{12.103}$$

The quantities $s$ and $\delta$ should, in any case, satisfy the inequality $|s|\leqslant\delta^{1/2}$ (page 225, Vol. 1). Let $\partial u_i/\partial x_i=a_{i,\,i}$ (no summation over $i$), and let us use the fact that since the field $\boldsymbol{u}(\boldsymbol{x})$ is isotropic and solenoidal, the quantities $a_{1,1}, a_{2,2}, a_{3,3}$ have the same probability distributions, and are such that $a_{1,1}+a_{2,2}+a_{3,3}=0$. It is readily shown that

$$\min_{a+b+c=0} \frac{(a^2+b^2+c^2)(a^4+b^4+c^4)}{(a^3+b^3+c^3)^2}=3$$

(the minimum is reached for $a=b=-c/2$). Consequently,

$$|a_{1,1}^3+a_{2,2}^3+a_{3,3}^3| \leqslant \\ \leqslant \frac{1}{\sqrt{3}}(a_{1,1}^2+a_{2,2}^2+a_{3,3}^2)^{1/2}(a_{1,1}^4+a_{2,2}^4+a_{3,3}^4)^{1/2}. \quad (12.104)$$

Since,

$$\overline{|a_{1,1}^3+a_{2,2}^3+a_{3,3}^3|} \geqslant |\overline{a_{1,1}^3+a_{2,2}^3+a_{3,3}^3}| = 3\,|\overline{a_{1,1}^3}|$$

and

$$\overline{(a_{1,1}^2+a_{2,2}^2+a_{3,3}^2)^{1/2}(a_{1,1}^4+a_{2,2}^4+a_{3,3}^4)^{1/2}} \leqslant \\ \leqslant [\overline{(a_{1,1}^2+a_{2,2}^2+a_{3,3}^2)}\,\overline{(a_{1,1}^4+a_{2,2}^4+a_{3,3}^4)}]^{1/2} = 3\,(\overline{a_{1,1}^2})^{1/2}(\overline{a_{1,1}^4})^{1/2},$$

it follows that, by taking the average of both sides of Eq. (12.104), we obtain

$$|\overline{a_{1,1}^3}| \leqslant \frac{1}{\sqrt{3}}(\overline{a_{1,1}^2})^{1/2}(\overline{a_{1,1}^4})^{1/2}, \quad \text{i.e.,} \quad |s| \leqslant \frac{1}{\sqrt{3}}\delta^{1/2} \approx 0.58\delta^{1/2}. \quad (12.105)$$

The last inequality is stronger than the more general inequality $|s| \leqslant \delta^{1/2}$. However, even this result can be strengthened if we use the fact that $\partial u_1/\partial x_1$, $\partial u_2/\partial x_2$ and $\partial u_3/\partial x_3$ are the diagonal elements of the symmetric tensor

$$a_{i,j}=\frac{1}{2}\left(\frac{\partial u_i}{\partial x_j}+\frac{\partial u_j}{\partial x_i}\right)$$

with zero trace [when the inequality (12.105) was derived, the tensorial character of the quantities $a_{i,j}$ was not taken into account]. Let the eigenvalues of the matrix $\|a_{i,j}\|$ be denoted by $a_1$, $a_2$, and $a_3$, respectively. In that case, we have $a_1+a_2+a_3=0$ and hence

$$|a_1^3 + a_2^3 + a_3^3| \leqslant \frac{1}{\sqrt{3}} (a_1^2 + a_2^2 + a_3^2)^{1/2} (a_1^4 + a_2^4 + a_3^4)^{1/2} \quad (12.106)$$

[compare this with Eq. (12.104)]. Let us now transform to the set of coordinates $OX_1X_2X_3$, the axes of which are the principal directions of the tensor $a_{i,j}$. In this set of coordinates, the above tensor will have the elements $a_i\delta_{ij}$, i.e., it will correspond to a diagonal matrix with elements $a_i$ along the diagonal. Let $\theta$ be the angle between the $x_1$ and $X_1$ axes (the "latitude" of $Ox_1$ with respect to $OX_1X_2X_3$), and let $\varphi$ be the angle between the $x_1X_1$ plane and the $X_2$ axis (the "longitude" of $Ox_1$). Transforming the tensor $a_i\delta_{ij}$ to the set of coordinates $Ox_1x_2x_3$ in accordance with the general transformation rule for tensors, we obtain

$$\partial u_1/\partial x_1 = a_{1,1} = a_1 \cos^2\theta + a_2 \sin^2\theta \cos^2\varphi + a_3 \sin^2\theta \sin^2\varphi. \quad (12.107)$$

In the last formula, both the eigenvalues $a_1, a_2, a_3$ and the angles $\theta$, $\varphi$ are random variables. The variables $a_1$, $a_2$, and $a_3$ do not then depend on the orientation of the coordinate axes, while the angles $\theta$ and $\varphi$ assume all the possible values as the coordinate system is rotated. Since in the case of an isotropic field $\boldsymbol{u}(\boldsymbol{x})$ all the probability distributions should be independent of the orientation of the axes, the probability distribution for the angles $\theta$ and $\varphi$ should be invariant under all possible rotations, i.e., the directions of the $x_1$ axis in the $X_1X_2X_3$ system should be uniformly distributed for all $a_1$, $a_2$, $a_3$. However, this means that the probability distribution for the angles $\theta$ and $\varphi$ is independent of the distribution of $a_1$, $a_2$, $a_3$, and is specified by the probability density $p(\theta, \varphi) = (4\pi)^{-1} \sin\theta$. It follows that when we average any function of $\partial u_1/\partial x_1$ we must first integrate the corresponding function of $a_1$, $a_2$, $a_3$, $\theta$, and $\varphi$ with respect to the angles $0 \leqslant \varphi \leqslant 2\pi$ and $0 \leqslant \theta \leqslant \pi$ with the weight $(4\pi)^{-1} \sin\theta$, and then take the average of the result with respect to all the possible values of $a_1$, $a_2$, and $a_3$. Since, $a_1 + a_2 + a_3 = 0$, and consequently

$$-2(a_1a_2 + a_1a_3 + a_2a_3) = a_1^2 + a_2^2 + a_3^2,$$

$$-(a_1^2a_2 + a_1^2a_3 + a_2^2a_1 + a_2^2a_3 + a_3^2a_1 + a_3^2a_2) = 3a_1a_2a_3 = \\ = a_1^3 + a_2^3 + a_3^3,$$

$$\frac{1}{2}(a_1^2 + a_2^2 + a_3^2)^2 = a_1^4 + a_2^4 + a_3^4$$

and so on, we can readily show that

$$\overline{\left(\frac{\partial u_1}{\partial x_1}\right)^2} = \frac{2}{15}\left(\overline{a_1^2} + \overline{a_2^2} + \overline{a_3^2}\right),$$
$$\overline{\left(\frac{\partial u_1}{\partial x_1}\right)^3} = \frac{8}{105}\left(\overline{a_1^3} + \overline{a_2^3} + \overline{a_3^3}\right), \qquad (12.108)$$
$$\overline{\left(\frac{\partial u_1}{\partial x_1}\right)^4} = \frac{8}{105}\left(\overline{a_1^4} + \overline{a_2^4} + \overline{a_3^4}\right).$$

Substituting these formulas into the averaged inequality (12.106), we obtain the Betchov inequality

$$\left|\overline{\left(\frac{\partial u_1}{\partial x_1}\right)^3}\right| \leqslant \frac{2}{\sqrt{21}}\left[\overline{\left(\frac{\partial u_1}{\partial x_1}\right)^2}\right]^{1/2}\left[\overline{\left(\frac{\partial u_1}{\partial x_1}\right)^4}\right]^{1/2},$$

i.e.,

$$|s| \leqslant \frac{2}{\sqrt{21}}\,\delta^{1/2} \approx 0.44\delta^{1/2}, \qquad (12.109)$$

This is clearly stronger than Eq. (12.105).

Similar inequalities can be obtained for the higher-order moments of $\partial u_1/\partial x_1$ or joint moments of it and other random variables. Let us consider briefly the inequalities for the joint moment

$$\overline{\left(\frac{\partial \vartheta}{\partial x_1}\right)^2 \frac{\partial u_1}{\partial x_1}}$$

of the four-dimensional isotropic field $\{\boldsymbol{u}(\boldsymbol{x}),\ \vartheta(\boldsymbol{x})\}$, where $\vartheta$ is a scalar and $\boldsymbol{u}$ a solenoidal vector. Consider the dimensionless parameters

$$s_\vartheta = \overline{\left(\frac{\partial \vartheta}{\partial x_1}\right)^2 \frac{\partial u_1}{\partial x_1}} \Big/ \overline{\left(\frac{\partial \vartheta}{\partial x_1}\right)^2}\left[\overline{\left(\frac{\partial u_1}{\partial x_1}\right)^2}\right]^{1/2},$$
$$\delta_\vartheta = \overline{\left(\frac{\partial \vartheta}{\partial x_1}\right)^4} \Big/ \left[\overline{\left(\frac{\partial \vartheta}{\partial x_1}\right)^2}\right]^{1/2}. \qquad (12.110)$$

From the general inequality $|\overline{u_1^2 u_2}| \leqslant (\overline{u_1^4})^{1/2}(\overline{u_2^2})^{1/2}$, which is valid for any $u_1$, $u_2$, it follows that

$$\left|\overline{\left(\frac{\partial\vartheta}{\partial x_1}\right)^2\frac{\partial u_1}{\partial x_1}}\right| \leqslant \left[\overline{\left(\frac{\partial\vartheta}{\partial x_1}\right)^4}\right]^{1/2}\left[\overline{\left(\frac{\partial u_1}{\partial x_1}\right)^2}\right]^{1/2}, \quad \text{i.e.,} \quad |s_\vartheta| \leqslant \delta_\vartheta^{1/2}. \tag{12.111}$$

We now note that

$$\min_{a+b+c=0} \frac{(a^2+b^2+c^2)(\alpha^4+\beta^4+\gamma^4)}{(a\alpha^2+b\beta^2+c\gamma^2)^2} = \frac{3}{2}$$

(the minimum is reached for $a=b=-c/2$, $\alpha=\beta=0$). Consequently

$$|a\alpha^2+b\beta^2+c\gamma^2| \leqslant \sqrt{\frac{2}{3}}\,(a^2+b^2+c^2)^{1/2}(\alpha^4+\beta^4+\gamma^4)^{1/2}. \tag{12.112}$$

Substituting $a=\partial u_1/\partial x_1$, $b=\partial u_2/\partial x_2$, $c=\partial u_3/\partial x_3$, $\alpha=\partial\vartheta/\partial x_1$, $\beta=\partial\vartheta/\partial x_2$, $\gamma=\partial\vartheta/\partial x_3$, we have, after taking the average,

$$\left|\overline{\left(\frac{\partial\vartheta}{\partial x_1}\right)^2\frac{\partial u_1}{\partial x_1}}\right| \leqslant \sqrt{\frac{2}{3}}\left[\overline{\left(\frac{\partial\vartheta}{\partial x_1}\right)^4}\right]^{1/2}\left[\overline{\left(\frac{\partial u_1}{\partial x_1}\right)^2}\right]^{1/2},$$

i.e.,

$$|s_\vartheta| \leqslant \sqrt{\frac{2}{3}}\,\delta_\vartheta^{1/2} \approx 0{,}82\delta_\vartheta^{1/2} \tag{12.113}$$

[compare this with the derivation of Eq. (12.105)]. The last result is not, however, final. It can be made stronger by substituting

$a=3a_1+a_2+a_3$, $b=a_1+3a_2+a_3$, $c=a_1+a_2+3a_3$, $\alpha=\dfrac{\partial\vartheta}{\partial X_1}$, $\beta=\dfrac{\partial\vartheta}{\partial X_2}$, $\gamma=\dfrac{\partial\vartheta}{\partial X_3}$,

where, as above, $a_1$, $a_2$, $a_3$ are the eigenvalues of the tensor

$$a_{i,j} = \frac{1}{2}\left(\frac{\partial u_i}{\partial x_j}+\frac{\partial u_j}{\partial x_i}\right)$$

and $X_1X_2X_3$ is the set of coordinates defined by the principal directions of this tensor. Since, under these conditions,

$$\frac{\partial\vartheta}{\partial x_1}=\alpha\cos\theta+\beta\sin\theta\cos\varphi+\gamma\sin\theta\sin\varphi, \qquad (12.114)$$

we can verify, using Eq. (12.107), that

$$\overline{\left(\frac{\partial\vartheta}{\partial x_1}\right)^2\frac{\partial u_1}{\partial x_1}}=\frac{1}{15}[\overline{\alpha^2(3a_1+a_2+a_3)}+\overline{\beta^2(a_1+3a_2+a_3)}+ \\ +\overline{\gamma^2(a_1+a_2+3a_3)}]=\frac{1}{15}\overline{(a\alpha^2+b\beta^2+c\gamma^2)}. \qquad (12.115)$$

Similarly, it can be shown that

$$\overline{\left(\frac{\partial\vartheta}{\partial x_1}\right)^4}=\frac{1}{5}\overline{(\alpha^2+\beta^2+\gamma^2)^2}\geqslant\frac{1}{5}\overline{(\alpha^4+\beta^4+\gamma^4)}. \qquad (12.116)$$

$$\overline{\left(\frac{\partial u_1}{\partial x_1}\right)^2}=\frac{1}{30}[\overline{(3a_1+a_2+a_3)^2+(a_1+3a_2+a_3)^2+(a_1+a_2+3a_3)^2}]= \\ =\frac{1}{30}\overline{(a^2+b^2+c^2)} \qquad (12.117)$$

[the last equation is equivalent to the first equation in (12.108)]. Taking the average of both sides of equation (12.112), and using Eqs. (12.115)–(12.117), we obtain

$$\overline{\left(\frac{\partial\vartheta}{\partial x_1}\right)^2\frac{\partial u_1}{\partial x_1}}\leqslant\frac{2}{3}\left[\overline{\left(\frac{\partial\vartheta}{\partial x_1}\right)^4}\right]^{1/2}\left[\overline{\left(\frac{\partial u_1}{\partial x_1}\right)^2}\right]^{1/2},$$

i.e.,

$$|s_\vartheta|\leqslant\frac{2}{3}\delta_\vartheta^{1/2}\approx 0{,}67\delta_\vartheta^{1/2} \qquad (12.118)$$

This result is due to Yaglom (1967a).

Let us now consider the two-point third and fourth moments of isotropic fields. We shall restrict our attention to the elementary consequences of the isotropy condition which is concerned with the general functional form of such moments. We shall begin with the tensor of two-point third moments of an isotropic vector field $\boldsymbol{u}(\boldsymbol{x})$:

$$B_{ij,l}(\boldsymbol{r})=\overline{u_i(\boldsymbol{x})\,u_j(\boldsymbol{x})\,u_l(\boldsymbol{x}+\boldsymbol{r})}. \qquad (12.119)$$

Since the invariant tensor $\delta_{ij}$ and the components of the vector $\boldsymbol{r}$ can be used to make up only three tensors of rank 3 symmetric in the indices $i$, $j$, namely, $r_i r_j r_l$, $r_i \delta_{jl} + r_j \delta_{il}$, and $r_l \delta_{ij}$, the general form of the tensor $B_{ij,l}(\boldsymbol{r})$ in the isotropic case should be

$$B_{ij,l}(\boldsymbol{r}) = B_1(r) r_i r_j r_l + \\ + B_2(r)(\delta_{jl} r_i + \delta_{il} r_j) + B_3(r) \delta_{ij} r_l, \qquad (12.120)$$

where $B_1(r)$, $B_2(r)$, and $B_3(r)$ are scalar functions of $r = |\boldsymbol{r}|$ [cf. the derivation of Eq. (12.30)]. The significance of the functions $B_1(r)$, $B_2(r)$ can be understood by transforming to the coordinates $x_1'$, $x_2'$, $x_3'$, where the $x_1'$ axis lies along the vector $\boldsymbol{r}$, while the $x_2'$ and $x_3'$ axes are perpendicular to $\boldsymbol{r}$ (see Fig. 6). It is readily seen that in this system of coordinates the only nonzero components of the tensor $B_{ij,l}(\boldsymbol{r})$ are

$$\begin{aligned} B_{LL,L}(r) &= B'_{11,1}(r) = \overline{u_L^2(\boldsymbol{x})\, u_L(\boldsymbol{x}+\boldsymbol{r})}, \\ B_{NN,L}(r) &= B'_{22,1}(r) = B'_{33,1}(r) = \overline{u_N^2(\boldsymbol{x})\, u_L(\boldsymbol{x}+\boldsymbol{r})}, \qquad (12.121) \\ B_{LN,N}(r) &= B'_{12,2}(r) = B'_{13,3}(r) = \overline{u_L(\boldsymbol{x})\, u_N(\boldsymbol{x})\, u_N(\boldsymbol{x}+\boldsymbol{r})} \end{aligned}$$

(see Fig. 8). Since in the new set of coordinates $r_1 = r$, $r_2 = r_3 = 0$, we have from Eq. (12.120)

$$B_{LL,L}(r) = B_1(r) r^3 + [2B_2(r) + B_3(r)]\, r,$$
$$B_{NN,L}(r) = B_3(r)\, r, \quad B_{LN,N}(r) = B_2(r)\, r.$$

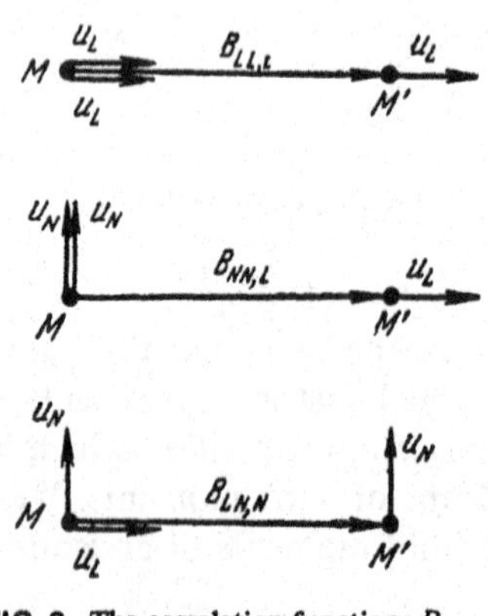

FIG. 8 The correlation functions $B_{LL,L}$, $B_{NN,L}$, and $B_{LN,N}$.

The functions $B_1(r)$, $B_2(r)$, and $B_3(r)$ are thus expressed in terms of the functions $B_{LL,L}(r)$, $B_{NN,L}(r)$, and $B_{LN,N}(r)$, with the result that the formula given by Eq. (12.120) is transformed to the von Kármán-Howarth formula

$$B_{ij,l}(\boldsymbol{r}) = \frac{B_{LL,L}(r) - B_{NN,L}(r) - 2B_{LN,N}(r)}{r^3} r_i r_j r_l + \\ + \frac{B_{LN,N}(r)}{r}(\delta_{jl} r_i + \delta_{il} r_j) + \frac{B_{NN,L}(r)}{r}\delta_{ij} r_l. \quad (12.122)$$

It is clear from equation (12.121) that $B_{LL,L}(0) = B_{NN,L}(0) = B_{LN,N}(0) = 0$. Moreover, in view of the homogeneity of the field $\boldsymbol{u}(\boldsymbol{x})$ we have

$$\left.\frac{dB_{LL,L}(r)}{dr}\right|_{r=0} = \frac{\partial}{\partial r_1}\overline{u_1^2(\boldsymbol{x})\, u_1(\boldsymbol{x}+\boldsymbol{r})}\Big|_{r=0} = \\ = \overline{u_1^2(\boldsymbol{x})\frac{\partial u_1(\boldsymbol{x})}{\partial x_1}} = \frac{1}{3}\frac{\partial}{\partial x_1}\overline{u_1^3(\boldsymbol{x})} = 0$$

and since $\overline{u_1^2\, \partial^2 u_1/\partial x_1^2}$ changes sign when the $x_1$ axis is replaced by the $-x_1$ axis, we have

$$\left.\frac{d^2 B_{LL,L}(r)}{dr^2}\right|_{r=0} = \frac{\partial^2}{\partial r_1^2}\overline{u_1^2(\boldsymbol{x})\, u_1(\boldsymbol{x}+\boldsymbol{r})}\Big|_{r=0} = \overline{u_1^2(\boldsymbol{x})\frac{\partial^2 u_1(\boldsymbol{x})}{\partial x_1^2}} = 0$$

(for the same reason $B^{(2n)}_{LL,L}(0) = 0$ for all integral $n$. Consequently

$$B_{LL,L}(r) = b_3^{(LL,L)} r^3 + b_5^{(LL,L)} r^5 + \ldots, \quad (12.123)$$

where

$$b_3^{(LL,L)} = \frac{1}{6} B'''_{LL,L}(0) = \frac{1}{6}\overline{u_1^2\frac{\partial^3 u_1}{\partial x_1^3}} = \frac{1}{6}\overline{\left(\frac{\partial u_1}{\partial x_1}\right)^3} \quad (12.123')$$

(here again, we have used the fact that the field $\boldsymbol{u}(\boldsymbol{x})$ is homogeneous). The functions $B_{NN,L}(r)$ and $B_{LN,N}(r)$ are also odd, but their Taylor expansions will in general begin with the terms $b_1^{(NN,L)} r + \ldots$ and $b_1^{(LN,N)} r + \ldots$, where

$$b_1^{(NN,L)} + 2b_1^{(LN,N)} = \overline{u_2^2\frac{\partial u_1}{\partial x_1}} + \overline{2u_1 u_2\frac{\partial u_2}{\partial x_1}} = \frac{\partial}{\partial x_1}\overline{u_1 u_2^2} = 0$$

The last condition together with Eq. (12.123) insures that the tensor (12.122) is regular at the origin.

The tensor $F_{ij,l}(\boldsymbol{k})$, is the Fourier transform of the tensor $B_{ij,l}(\boldsymbol{r})$ and can be represented in a form similar to Eq. (12.122):

$$F_{ij,l}(\boldsymbol{k})=i\left\{\frac{F_{LL,L}(k)-F_{NN,L}(k)-2F_{LN,N}(k)}{k^3}k_ik_jk_l+\right.$$
$$\left.+\frac{F_{LN,N}(k)}{k}(\delta_{jl}k_i+\delta_{il}k_j)+\frac{F_{NN,L}(k)}{k}\delta_{ij}k_l\right\} \quad (12.124)$$

The factor $i$ is added to ensure that the functions $F_{LL,L}(k)$, $F_{NN,L}(k)$, and $F_{LN,N}(k)$ are real. From Eq. (12.124) it is clear that the functions $F_{ij,l}(\boldsymbol{k})$ will be regular at the point $\boldsymbol{k}=0$ (to ensure this, it is sufficient that the functions $B_{ij,l}(\boldsymbol{r})$ fall to zero sufficiently rapidly at infinity) provided only $F_{LL,L}(0)=F_{NN,L}(0)= F_{LN,N}(0)=0$ and $f_1^{(LL,L)}-f_1^{(NN,L)}-2f_1^{(LN,L)}=0$ where $f_1^{(LL,L)}$ etc., are the coefficients of the first power of $k$ in the Taylor expansions of the functions $F_{LL,L}(k)$ etc.

In the case of the mixed two-point third moment $B_{j\vartheta,\vartheta}(\boldsymbol{r})=\overline{u_j(\boldsymbol{x})\vartheta(\boldsymbol{x})\vartheta(\boldsymbol{x}+\boldsymbol{r})}$ of the isotropic field $\{\boldsymbol{u}(\boldsymbol{x}),\ \vartheta(\boldsymbol{x})\}$, the formulas analogous to Eqs. (12.122) and (12.124) have a much simpler form. In this case we have

$$B_{j\vartheta,\vartheta}(\boldsymbol{r})=B_{L\vartheta,\vartheta}(r)\frac{r_j}{r},\quad B_{L\vartheta,\vartheta}(r)=\overline{u_L(\boldsymbol{x})\vartheta(\boldsymbol{x})\vartheta(\boldsymbol{x}+\boldsymbol{r})} \quad (12.125)$$

and

$$F_{j\vartheta,\vartheta}(\boldsymbol{k})=iF_{L\vartheta,\vartheta}(k)\frac{k_j}{k}, \quad (12.125')$$

where $B_{L\vartheta,\vartheta}(r)$ and $F_{L\vartheta,\vartheta}(k)$ are real functions of one variable such that $B_{L\vartheta,\vartheta}(0)=F_{L\vartheta,\vartheta}(0)=0$. Substituting Eqs. (12.125) and (12.125′) into Eq. (11.60′) and in the corresponding "inversion formula" we obtain

$$B_{L\vartheta,\vartheta}(r)=4\pi\int_0^\infty\left\{-\frac{\cos kr}{kr}+\frac{\sin kr}{(kr)^2}\right\}F_{L\vartheta,\vartheta}(k)\,k^2\,dk, \quad (12.126)$$

$$F_{L\vartheta,\vartheta}(k)=\frac{1}{2\pi^2}\int_0^\infty\left\{-\frac{\cos kr}{kr}+\frac{\sin kr}{(kr)^2}\right\}B_{L\vartheta,\vartheta}(r)\,r^2\,dr \quad (12.126')$$

[compare this with equations (12.59) and (12.60)]. Hence it follows that relations of the form (12.61)–(12.66) are satisfied also for the functions $B_{L\vartheta,\vartheta}(r)$ and $F_{L\vartheta,\vartheta}(k)$ (provided the subscripts $L\vartheta$ are replaced by $L\vartheta,\ \vartheta$ throughout). We note that the first coefficient

$$b_1^{(L\vartheta,\,\vartheta)} = \frac{4\pi}{3}\int_0^\infty F_{L\vartheta,\vartheta}(k)\,k^3\,dk$$

in the Taylor expansion of the function $B_{L\vartheta,\vartheta}(r)$ is given by

$$b_1^{(L\vartheta,\,\vartheta)} = \frac{\partial\overline{u_1(x)\,\vartheta(x)\,\vartheta(x+r)}}{\partial r_1}\Bigg|_{r=0} = \overline{u_1(x)\,\vartheta(x)\,\frac{\partial\vartheta(x)}{\partial x_1}} = -\frac{1}{2}\,\overline{\frac{\partial u_1(x)}{\partial x_1}\,\vartheta^2(x)} \qquad (12.126'')$$

since

$$\frac{\partial}{\partial x_1}\,\overline{u_1\vartheta^2} = 0$$

in view of the homogeneity of the field $\{u(x),\ \vartheta(x)\}$.

Consider now the two-point fourth moments of the isotropic vector field $u(x)$. There are two types of such moment:

$$B_{ij,\,kl}(r) = \overline{u_i(x)\,u_j(x)\,u_k(x+r)\,u_l(x+r)},$$
$$B_{ijk,\,l}(r) = \overline{u_i(x)\,u_j(x)\,u_k(x)\,u_l(x+r)}. \qquad (12.127)$$

The tensor $B_{ij,\,kl}(r)$ is isotropic, symmetric in the indices $i,\ j$ and $k,\ l$, and is unaffected by the replacement of $i,\ j$ by $k,\ l$ and vice versa. We therefore, have a formula analogous to Eq. (12.120):

$$B_{ij,\,kl}(r) = C_1(r)\,r_ir_jr_kr_l + C_2(r)\,(r_ir_j\delta_{kl} + r_kr_l\delta_{ij}) + C_3(r)\,(r_ir_k\delta_{jl} + r_ir_l\delta_{jk} + r_jr_k\delta_{il} + r_jr_l\delta_{ik}) + C_4(r)\,(\delta_{ik}\delta_{jl} + \delta_{il}\delta_{jk}) + C_5(r)\,\delta_{ij}\delta_{kl}. \qquad (12.128)$$

Similarly, it can be shown that

$$B_{ijk,\,l}(r) = D_1(r)\,r_ir_jr_kr_l + D_2(r)\,(r_ir_j\delta_{kl} + r_ir_k\delta_{jl} + r_jr_k\delta_{il}) + D_3(r)\,r_l\,(r_i\delta_{jk} + r_j\delta_{ik} + r_k\delta_{ij}) + D_4(r)\,(\delta_{ij}\delta_{kl} + \delta_{ik}\delta_{jl} + \delta_{jk}\delta_{il}). \qquad (12.129)$$

The scalar functions $C_1(r), \ldots, D_4(r)$ in Eq. (12.128) and Eq. (12.129) can be expressed in terms of the components of the corresponding tensors in the special set of coordinates $Mx_1'x_2'x_3'$, which has a clear statistical significance. It is readily seen that in this set of coordinates the only nonzero components of the tensor $B_{ij,\,kl}(r)$ will be

$$B_{LL,\,LL}(r),\ B_{LL,\,NN}(r),\ B_{LN,\,LN}(r),\ B_{NN,\,NN}(r),\qquad\qquad (12.130)$$
$$B_{NN,\,MM}(r),\ B_{NM,\,NM}(r)$$

where the subscript $M$ represents the direction orthogonal to $L$ (i.e., $\boldsymbol{r}$) and $N$ (see Fig. 9a). The pair of indices before and after the comma can be interchanged since $B_{ij,\,kl}(r)=B_{kl,\,ij}(r)$. Since the six functions in Eq. (12.130) can be expressed in terms of $C_1(r), \ldots, C_5(r)$, it is clear that they cannot all be independent. In fact, if we express $B_{LL,\,LL}(r), \ldots, B_{NM,\,NM}(r)$ in terms of $C_1(r), \ldots, C_5(r)$ with the aid of Eq. (12.128), we obtain

$$B_{NN,\,NN}(r)=2B_{NM,\,NM}(r)+B_{NN,\,MM}(r) \qquad (12.131)$$

This identity was first found by Millionshchikov (1941a, b). The representation given by Eq. (12.128) can be rewritten in a form containing only the functions (12.130). The result is

$$\begin{aligned}B_{ij,\,kl}(r)=&\frac{B_{LL,\,LL}(r)-2B_{LL,\,NN}(r)-4B_{LN,\,LN}(r)+B_{NN,\,NN}(r)}{r^4}r_ir_jr_kr_l+\\&+\frac{B_{LL,\,NN}(r)-B_{NN,\,MM}(r)}{r^2}(r_ir_j\delta_{kl}+r_kr_l\delta_{ij})+\\&+\frac{B_{LN,\,LN}(r)-B_{NM,\,NM}(r)}{r^2}(r_ir_k\delta_{jl}+r_ir_l\delta_{jk}+r_jr_k\delta_{il}+r_jr_l\delta_{ik})+\\&+B_{NM,\,NM}(r)(\delta_{ik}\delta_{jl}+\delta_{il}\delta_{jk})+B_{NN,\,MM}(r)\delta_{ij}\delta_{kl}.\end{aligned} \qquad (12.132)$$

Similarly, in the set of coordinates $Mx_1'x_2'x_3'$ the only nonzero components of $B_{ijk,\,l}(r)$ are

$$B_{LLL,\,L}(r),\ B_{LLN,\,N}(r),\ B_{LNN,\,L}(r),\ B_{NNN,\,N}(r),\ B_{NNM,\,M}(r) \qquad (12.133)$$

(see Fig. 9b). These functions can be expressed in terms of $D_1(r), \ldots, D_4(r)$ with the aid of Eq. (12.129). It turns out that

$$B_{NNN,\,N}(r)=3B_{NNM,\,M}(r). \qquad (12.134)$$

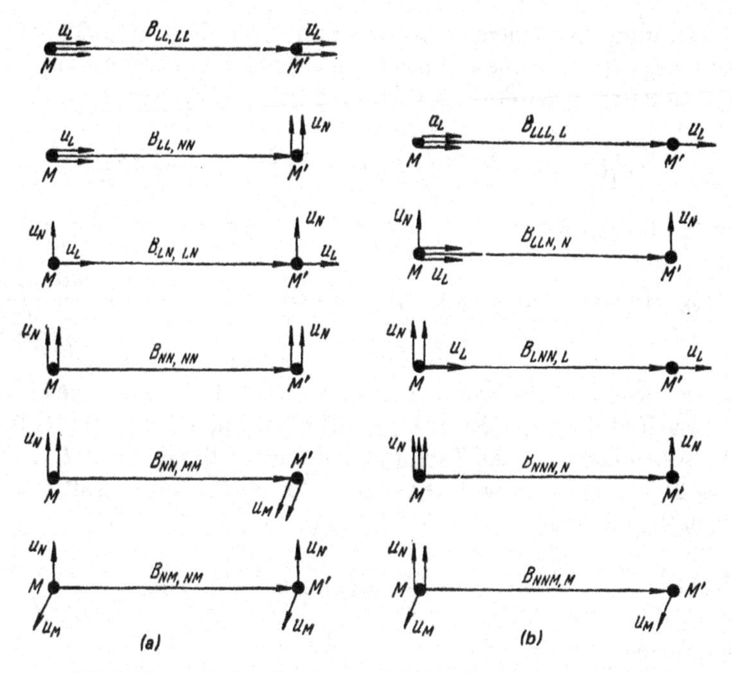


**FIG. 9** Two-point fourth moments of the first (a) and second (b) type.

In terms of the functions (12.133), the formula (12.129) assumes the form

$$\begin{aligned}B_{ijk,l}(r)=&\frac{B_{LLL,L}(r)-3B_{LLN,N}(r)-3B_{LNN,L}(r)+B_{NNN,N}(r)}{r^4}r_ir_jr_kr_l+\\&+\frac{3B_{LLN,N}(r)-B_{NNN,N}(r)}{3r^2}(r_ir_j\delta_{kl}+r_ir_k\delta_{jl}+r_jr_k\delta_{il})+\\&+\frac{3B_{LNN,L}(r)-B_{NNN,N}(r)}{3r^2}r_l(r_i\delta_{jk}+r_j\delta_{ik}+r_k\delta_{ij})+\\&+\frac{1}{3}B_{NNN,N}(r)(\delta_{ij}\delta_{kl}+\delta_{ik}\delta_{jl}+\delta_{jk}\delta_{il}).\end{aligned} \tag{12.135}$$

The Fourier transforms $F_{ij,kl}(\boldsymbol{k})$ and $F_{ijk,l}(\boldsymbol{k})$ of the tensors $B^{(0)}_{ij,kl}(\boldsymbol{r})=B_{ij,kl}(\boldsymbol{r})-B_{ij}(0)\,B_{kl}(0)$ and $B_{ijk,l}(\boldsymbol{r})$ can be represented by analogy with Eqs. (12.132) and (12.135).

Let us now suppose that the isotropic vector field $\boldsymbol{u}(\boldsymbol{x})$ is solenoidal. It is clear that, in this case,

$$\frac{\partial B_{ij,l}(\boldsymbol{r})}{\partial r_l}=0 \quad\text{and}\quad k_lF_{ij,l}(\boldsymbol{k})=0. \tag{12.136}$$

Substituting the general representation (12.120) or (12.122) of the tensor $B_{ij,l}(r)$ into the left-hand side of the first equation (12.136), and equating to zero the coefficients of $r_i r_j$ and $\delta_{ij}$, we obtain

$$B_1(r) = \frac{1}{r}\frac{dB_3(r)}{dr}, \quad B_2(r) = -\frac{3}{2}B_3(r) - \frac{r}{2}\frac{dB_3(r)}{dr}$$

or, in other words,

$$B_{NN,L}(r) = -\frac{1}{2}B_{LL,L}(r), \quad B_{LN,N}(r) = \frac{1}{2}B_{LL,L}(r) + \frac{r}{4}\frac{dB_{LL,L}(r)}{dr} \tag{12.137}$$

These relationships were first established by von Kármán and Howarth (1938). In view of Eq. (12.123) it follows from this that, in the solenoidal case, the Taylor expansions of the functions $B_{NN,L}(r)$ and $B_{LN,N}(r)$ begin with terms of order $r^3$ (the same result follows from the fact that

$$\frac{\partial}{\partial x_2}\overline{u_2^3(\boldsymbol{x})} = 0$$

and hence

$$b_1^{(NN,L)} = \overline{u_2^2\frac{\partial u_1}{\partial x_1}} = \frac{1}{2}\overline{u_2^2\frac{\partial u_l}{\partial x_l}} = 0\Big).$$

The formula (12.122) for the solenoidal field $\boldsymbol{u}(\boldsymbol{x})$ assumes the form

$$B_{ij,l}(r) = \frac{B_{LL,L}(r) - rB'_{LL,L}(r)}{2r^3}r_i r_j r_l + \\ + \frac{2B_{LL,L}(r) + rB'_{LL,L}(r)}{4r}(r_i\delta_{jl} + r_j\delta_{il}) - \frac{B_{LL,L}(r)}{2r}r_l\delta_{ij}. \tag{12.138}$$

Substituting the representation (12.124) of the tensor $F_{ij,l}(\boldsymbol{k})$ into the second equation (12.136), we obtain[14]

[14]This result is easily explained by the fact that, in the solenoidal case, we have $dZ_L(\boldsymbol{k}) = 0$ where $dZ_L$ is the component of $d\boldsymbol{Z}$ parallel to $\boldsymbol{k}$. In fact, from Eq. (11.52) we have

$$F_{ij,l}(\boldsymbol{k})\,d\boldsymbol{k} = \int \overline{dZ_i^*(\boldsymbol{k}_1)\,dZ_j^*(\boldsymbol{k}-\boldsymbol{k}_1)\,dZ_l(\boldsymbol{k})}$$

(integration with respect to $\boldsymbol{k}_1$, with $\boldsymbol{k}$ and $d\boldsymbol{k}$ fixed). Hence, it is clear that $F_{ij,L}(\boldsymbol{k}) \equiv 0$ for any $i$ and $j$.

$$F_{NN,L}(k)=F_{LL,L}(k)=0. \tag{12.139}$$

We thus see that the spectral tensor $F_{ij,l}(\boldsymbol{k})$ is determined by the single scalar function $F_{LN,N}(k)$. For the sake of brevity, we shall denote this function by $F_3(k)$, and we shall call it simply the third order spectrum of the field $\boldsymbol{u}(\boldsymbol{x})$. Moreover,

$$F_{ij,l}(\boldsymbol{k})=iF_3(k)\left\{\delta_{jl}\frac{k_i}{k}+\delta_{il}\frac{k_j}{k}-\frac{2k_ik_jk_l}{k^3}\right\}. \tag{12.140}$$

It is clear that this spectrum will be regular at the origin provided the Taylor expansion of the function $F_3(k)$ begins with the term of order $k^3$.

Substituting Eq. (12.140) into Eq. (11.60), and transforming the resulting equation to the form (12.120), we can readily verify

$$B_{LL,L}(r)=16\pi\int_0^\infty\left\{\frac{\sin kr}{(kr)^2}+\frac{3\cos kr}{(kr)^3}-\frac{3\sin kr}{(kr)^4}\right\}F_3(k)\,k^2\,dk. \tag{12.141}$$

The converse relationship which expresses $F_3(k)$ in terms of $B_{LL,L}(r)$ can be obtained by substituting Eq. (12.138) into the formula inverse to Eq. (11.60). It has form

$$F_3(k)=\frac{1}{8\pi^2}\int_0^\infty\left\{\sin kr+\frac{3\cos kr}{kr}-\frac{3\sin kr}{(kr)^2}\right\}B_{LL,L}(r)\,r^2\,dr. \tag{12.142}$$

Instead of the relatively complicated formulas (12.138) and (12.140), we can use the fact that the functions

$$B_{ij,i}(r)=\left[2B_{LL,L}(r)+\frac{r}{2}B'_{LL,L}(r)\right]\frac{r_j}{r} \quad \text{and} \quad F_{ij,i}(\boldsymbol{k})=2iF_3(k)\,k_jk^{-1}$$

and the functions

$$K(r)=\frac{\partial B_{ij,i}(r)}{\partial r_j}=\left(r\frac{\partial}{\partial r}+3\right)\left(\frac{1}{2}\frac{\partial}{\partial r}+\frac{2}{r}\right)B_{LL,L}(r)$$

and $\Gamma(k) = ik_j F_{ij,l}(k) = -2kF_3(k)$ are the Fourier transforms of each other. From this we can readily show that

$$2B_{LL,L}(r) + \frac{r}{2}\frac{dB_{LL,L}(r)}{dr} = 8\pi \int_0^\infty \left\{ \frac{\cos kr}{kr} - \frac{\sin kr}{(kr)^2} \right\} F_3(k)\, k^2\, dk, \tag{12.141'}$$

$$F_3(k) = \frac{1}{4\pi^2} \int_0^\infty \left\{ \frac{\cos kr}{kr} - \frac{\sin kr}{(kr)^2} \right\} \left\{ 2B_{LL,L}(r) + \frac{r}{2}\frac{dB_{LL,L}(r)}{dr} \right\} r^2\, dr, \tag{12.142'}$$

$$K(r) = 4\pi \int_0^\infty \frac{\sin kr}{kr} \Gamma(k)\, k^2\, dk, \tag{12.141''}$$

$$\Gamma(k) = \frac{1}{2\pi^2} \int_0^\infty \frac{\sin kr}{kr} K(r)\, r^2\, dr, \tag{12.142''}$$

which in the case of a sufficiently rapidly decreasing function $B_{LL,L}(r)$ are equivalent to Eqs. (12.141) and (12.142).

From Eq. (12.142) [or Eqs. (12.142') or (12.142'')] it is readily shown that all the even-order derivatives of $F_3(k)$ vanish at $k = 0$, and that if $B_{LL,L}(r)$ falls off more rapidly than $r^{-6}$ as $r \to \infty$, then for small $k$,

$$F_3(k) \approx -\frac{k^3}{120\pi^2} \int_0^\infty B_{LL,L}(r)\, r^5\, dr. \tag{12.143}$$

Similarly, it follows from Eq. (12.141) [or Eq. (12.141') or (12.141'')] that, in the neighborhood of the point $r = 0$,

$$B_{LL,L}(r) = -\frac{16\pi}{15} \int_0^\infty F_3(k)\, k^3\, dk \cdot r + \frac{8\pi}{105} \int_0^\infty F_3(k)\, k^5\, dk \cdot r^3 + \dots,$$

so that in view of Eqs. (12.123) and (12.123')

$$\int_0^\infty F_3(k)\, k^3\, dk = 0, \quad \int_0^\infty F_3(k)\, k^5\, dk = \frac{35}{16\pi} \overline{\left(\frac{\partial u_1}{\partial x_1}\right)^3}. \tag{12.144}$$

In the next chapter, instead of the function $F_3(k)$ we shall usually use the functions $\Gamma(k) = -2kF_3(k)$ or $T(k) = -8\pi k^3 F_3(k)$. It is clear that all the formulas containing $F_3(k)$ can be rewritten in a form containing $\Gamma(k)$ or $T(k)$. For example,

$$B_{LL,L}(r) = -2\int_0^\infty \left\{\frac{\sin kr}{(kr)^2} + \frac{3\cos kr}{(kr)^3} - \frac{3\sin kr}{(kr)^4}\right\}\frac{T(k)}{k}\,dk, \tag{12.141'''}$$

$$T(k) = \frac{1}{\pi}\int_0^\infty \{(k^2r^2-3)\,kr\sin kr + 3k^2r^2\cos kr\}\frac{B_{LL,L}(r)}{r}\,dr =$$

$$= \frac{2}{\pi}\int_0^\infty kr\sin kr K(r)\,dr, \tag{12.142'''}$$

$$\int_0^\infty T(k)\,dk = 0, \quad \int_0^\infty T(k)\,k^2\,dk = -\frac{35}{4}\overline{\left(\frac{\partial u_1}{\partial x_1}\right)^3} \tag{12.144'}$$

and so on.

The condition for the field $\boldsymbol{u}(\boldsymbol{x})$ to be solenoidal enables us to simplify and generalize the form of the tensor $B_{ijk,l}(\boldsymbol{r})$. Under this condition, it will depend only on two scalar functions instead of four.[15] The general form of the moments $B_{j\vartheta,\vartheta}(\boldsymbol{r})$ and $B_{ij,kl}(\boldsymbol{r})$ (and, in particular, the number of scalar functions on which they depend) is unaffected by whether the $\boldsymbol{u}(\boldsymbol{x})$ is solenoidal or not. This does not mean, however, that the fact that the field $\boldsymbol{u}(\boldsymbol{x})$ is solenoidal will not, in general, affect the quantities $B_{j\vartheta,\vartheta}(\boldsymbol{r})$ and $B_{ij,kl}(\boldsymbol{r})$. For example, the equation (12.126″) shows that, in general, the function $B_{L\vartheta,\vartheta}(r)$ will be proportional to the first power of $r$ for small $r$. However, in the solenoidal case we have, clearly,

$$b_1^{(L\vartheta,\vartheta)} = -\frac{1}{2}\overline{\frac{\partial u_1}{\partial x_1}\vartheta^2} = -\frac{1}{6}\overline{\frac{\partial u_l}{\partial x_l}\vartheta^2} = 0$$

[15] In particular, if we use the representation of the corresponding spectral tensor $F_{ijk,l}(\boldsymbol{k})$ in the form analogous to Eq. (12.135), then from [illegible] zero-divergence condition, $k_l F_{ijk,l}(\boldsymbol{k}) = 0$, it will follow that $F_{LLL,L}(k) = F_{LNN,L}(k) = 0$, i.e., $F_{ijk,L}(\boldsymbol{k}) = 0$ for any $i, j, k$.

and, consequently,

$$B_{L\vartheta,\vartheta}(r) = b_3^{(L\vartheta,\vartheta)} r^3 + b_5^{(L\vartheta,\vartheta)} r^5 + \ldots, \quad b_3^{(L\vartheta,\vartheta)} = \frac{1}{6} \overline{u_1(\boldsymbol{x})\,\vartheta(\boldsymbol{x})\,\frac{\partial^3\vartheta(\boldsymbol{x})}{\partial x_1^3}}. \tag{12.145}$$

The expression for the coefficient $b_3^{(L\vartheta,\vartheta)}$ can be transformed to a more convenient form as follows. If the solenoidal field $\boldsymbol{u}(\boldsymbol{x})$ is homogeneous, and $\vartheta(\boldsymbol{x})$ is an arbitrary homogeneous scalar field, we have

$$\overline{u_1(\boldsymbol{x})\,\frac{\partial\vartheta(\boldsymbol{x})}{\partial x_1}} = -\overline{\frac{\partial u_1}{\partial x_1}\,\vartheta} = 0.$$

However, since the field $\{\boldsymbol{u}(\boldsymbol{x}), \vartheta^2(\boldsymbol{x})\}$ is assumed to be isotropic, we have

$$\overline{u_l(\boldsymbol{x})\,\frac{\partial^3\vartheta^2(\boldsymbol{x})}{\partial x_l\,\partial x_m\,\partial x_n}} = A\,(\delta_{ll}\delta_{mn} + \delta_{lm}\delta_{ln} + \delta_{ln}\delta_{lm}) =$$
$$= \frac{1}{3}\,\overline{u_1\,\frac{\partial^3\vartheta^2}{\partial x_1^3}}\,(\delta_{ll}\delta_{mn} + \delta_{lm}\delta_{ln} + \delta_{ln}\delta_{lm})$$

Therefore,

$$\overline{u_1\,\frac{\partial\,\Delta\vartheta^2}{\partial x_1}} = \overline{u_1\,\frac{\partial^3\vartheta^2}{\partial x_1\,\partial x_l\,\partial x_l}} = \frac{5}{3}\,\overline{u_1\,\frac{\partial^3\vartheta^2}{\partial x_1^3}} = 0,$$

and hence,

$$\overline{u_1\,\frac{\partial^3\vartheta^2}{\partial x_1^3}} = 2\overline{u_1\,\vartheta\,\frac{\partial^3\vartheta}{\partial x_1^3}} + 6\overline{u_1\,\frac{\partial\vartheta}{\partial x_1}\,\frac{\partial^2\vartheta}{\partial x_1^2}} = 0.$$

Consequently, if the field $\boldsymbol{u}(\boldsymbol{x})$ is solenoidal then

$$b_3^{(L\vartheta,\vartheta)} = -\frac{1}{4}\,\overline{u_1\,\frac{\partial}{\partial x_1}\left(\frac{\partial\vartheta}{\partial x_1}\right)^2} = \frac{1}{4}\,\overline{\frac{\partial u_1}{\partial x_1}\left(\frac{\partial^2\vartheta}{\partial x_1}\right)^2},$$
$$B_{L\vartheta,\vartheta}(r) = \frac{1}{4}\,\overline{\frac{\partial u_1}{\partial x_1}\left(\frac{\partial\vartheta}{\partial x_1}\right)^2}\,r^3 + \ldots \tag{12.146}$$

In view of Eq. (12.126), the equations given by Eqs. (12.145) and

(12.146) also show that if

$$\frac{\partial u_j}{\partial x_j} = 0,$$

then

$$\int_0^\infty F_{L\vartheta,\,\vartheta}(k)\,k^3\,dk = 0, \quad \int_0^\infty F_{L\vartheta,\,\vartheta}(k)\,k^5\,dk = -\frac{15}{8\pi}\,\overline{\frac{\partial u_1}{\partial x_1}\left(\frac{\partial\vartheta}{\partial x_1}\right)^2}. \quad (12.147)$$

When the field $\boldsymbol{u}(\boldsymbol{x})$ is potential, the higher-order correlation and spectral tensors are also appreciably simplified. Thus, for example, in this case we have $F_{LN,\,N}(k) \equiv 0$. However, potential fields are less important for turbulence theory and we shall not pause to consider them here.

## 12.5 Three-Point Moments of Isotropic Fields

The three-point moments of a scalar isotropic field $u(\boldsymbol{x})$ (for example, $B(\boldsymbol{r},\ \boldsymbol{r}') = \overline{u(\boldsymbol{x})\,u(\boldsymbol{x}+\boldsymbol{r})\,u(\boldsymbol{x}+\boldsymbol{r}')}$ or $B_{2,1,1}(\boldsymbol{r},\ \boldsymbol{r}') = \overline{u^2(\boldsymbol{x})\,u(\boldsymbol{x}+\boldsymbol{r})\,u(\boldsymbol{x}+\boldsymbol{r}')}$) depend on three scalar arguments: $r = |\boldsymbol{r}|$, $r' = |\boldsymbol{r}'|$ and $\rho = \boldsymbol{r}\boldsymbol{r}'/rr'$ (instead of $\rho$ we can also use the product $\boldsymbol{r}\boldsymbol{r}'$ or $r'' = |\boldsymbol{r}' - \boldsymbol{r}|$ or any other function of $r$, $r'$, $\boldsymbol{r}\boldsymbol{r}'$). Similarly, the six-dimensional Fourier transform (with respect to $\boldsymbol{r}$ and $\boldsymbol{r}'$) of the three-point moments will depend on three arguments: $k = |\boldsymbol{k}|$, $k' = |\boldsymbol{k}'|$ and $\mu = \boldsymbol{k}\boldsymbol{k}'/kk'$ (or, what is the same, $k$, $k'$ and $\boldsymbol{k}\boldsymbol{k}'$ or $k$, $k'$ and $k'' = |\boldsymbol{k}' + \boldsymbol{k}''|$).

The situation is more complicated in the case of the three-point moments of an isotropic vector field $\boldsymbol{u}(\boldsymbol{x})$ or the isotropic four-dimensional field $\{\boldsymbol{u}(\boldsymbol{x}),\ \vartheta(\boldsymbol{x})\}$ containing a vector and a scalar component. Since we shall require such moments in the subsequent analysis, let us consider them in somewhat greater detail. We shall confine our attention to the most important third order three-point moments.

Let us begin with the mixed moment $B_{j\vartheta\vartheta}(\boldsymbol{r},\ \boldsymbol{r}') = \overline{u_j(\boldsymbol{x})\,\vartheta(\boldsymbol{x}+\boldsymbol{r})\,\vartheta(\boldsymbol{x}+\boldsymbol{r}')}$ which is a vector function of the vectors $\boldsymbol{r}$ and $\boldsymbol{r}'$ invariant under any rotations and reflections. The scalar $B(\boldsymbol{r},\ \boldsymbol{r}',\ \boldsymbol{a}) = B_{j\vartheta\vartheta}(\boldsymbol{r},\ \boldsymbol{r}')\,a_j$ can depend only on the invariants $r$, $r'$, $\boldsymbol{r}\boldsymbol{r}' = rr'\rho$, $\boldsymbol{r}\boldsymbol{a}$, and $\boldsymbol{r}'\boldsymbol{a}$ for any unit vector $\boldsymbol{a}$, and moreover, it is a linear function of the last two invariants. Hence, it follows that

$$B_{j\vartheta\vartheta}(\boldsymbol{r},\ \boldsymbol{r}') = B(r,\ r',\ \rho)\,\frac{r_j}{r} + B_1(r,\ r',\ \rho)\,\frac{r'_j}{r'}, \quad (12.148)$$

where $B$ and $B_1$ are scalar functions of three arguments. However, since $B_{j00}(\boldsymbol{r}, \boldsymbol{r}') = B_{j00}(\boldsymbol{r}', \boldsymbol{r})$ we have $B_1(r, r', \rho) = B(r', r, \rho)$, i.e.,

$$B_{j00}(\boldsymbol{r}, \boldsymbol{r}') = B(r, r', \rho)\frac{r_j}{r} + B(r', r, \rho)\frac{r'_j}{r'}. \quad (12.149)$$

Therefore, the moment $B_{j00}(\boldsymbol{r}, \boldsymbol{r}')$ is determined by the single scalar function $B(r, r', \rho)$. If the vector field $\boldsymbol{u}(\boldsymbol{x})$ is solenoidal, then

$$\left(\frac{\partial}{\partial r_j} + \frac{\partial}{\partial r'_j}\right) B_{j00}(\boldsymbol{r}, \boldsymbol{r}') = 0. \quad (12.150)$$

In this case, the function $B(r, r', \rho)$ satisfies a first order partial differential equation containing $B(r, r', \rho)$ and $B(r', r, \rho)$.

The six-dimensional Fourier transform of the function $B_{j00}(\boldsymbol{r}, \boldsymbol{r}')$ which we have denoted by the symbol $F_{j00}(\boldsymbol{k}, \boldsymbol{k}')$, can by analogy with Eq. (12.149) be written in the form

$$F_{j00}(\boldsymbol{k}, \boldsymbol{k}') = -i\left[F(k, k', \mu)\frac{k_j}{k} + F(k', k, \mu)\frac{k'_j}{k'}\right] \quad (12.151)$$

Since $F_{j00}(-\boldsymbol{k}, -\boldsymbol{k}') = F^*_{j00}(\boldsymbol{k}, \boldsymbol{k}')$, the factor $i$ insures that the function $F(k, k', \mu)$, where $\mu = \boldsymbol{k}\boldsymbol{k}'/kk'$, is real. If the field $\boldsymbol{u}(\boldsymbol{x})$ is solenoidal then $(k_j + k'_j)\, F_{j00}(\boldsymbol{k}, \boldsymbol{k}') = 0$, i.e.,

$$F(k, k', \mu)(k + k'\mu) + F(k', k, \mu)(k\mu + k') = 0. \quad (12.152)$$

Consider the scalar function

$$\Gamma_0(k, k', \mu) = ik_j F_{j00}(\boldsymbol{k}, \boldsymbol{k}'). \quad (12.153)$$

The equation (12.152) then assumes the form

$$\Gamma_0(k, k', \mu) = -\Gamma_0(k', k, \mu). \quad (12.154)$$

On the other hand, it is readily seen that, in this case,

$$F(k, k', \mu) = \frac{k\mu + k'}{kk'(1-\mu^2)}\Gamma_0(k, k', \mu). \quad (12.155)$$

Using Eqs. (12.155) and (12.151), the spectrum $F_{j00}(\boldsymbol{k}, \boldsymbol{k}')$ can be uniquely determined by the real function $\Gamma_0(k, k', \mu)$ which is antisymmetric in $k$ and $k'$. Since $B_{j00}(\boldsymbol{r}, 0) = B_{j0,0}(\boldsymbol{r})$, we have

$$B(r, 0, \mu) = B_{L0,0}(r), \quad B(0, r, \mu) = 0 \qquad (12.156)$$

for any $\mu$ (where $-1 \leqslant \mu \leqslant 1$), and [cf. Eq. (11.61′)]

$$\int ik_j F_{j00}(\boldsymbol{k}, \boldsymbol{k}')\, d\boldsymbol{k}' = ik_j F_{j0,0}(\boldsymbol{k})$$

i.e.,

$$2\pi \int_0^{\infty} \int_{-1}^{1} \Gamma_0(k, k', \mu) k'^2 \, d\mu \, dk' = -kF_{L0,0}(k). \qquad (12.157)$$

Let us now consider the three-point third moments $B_{ijl}(\boldsymbol{r}, \boldsymbol{r}') = \overline{u_i(\boldsymbol{x}) u_j(\boldsymbol{x}+\boldsymbol{r}) u_l(\boldsymbol{x}+\boldsymbol{r}')}$ of the vector field $\boldsymbol{u}(\boldsymbol{x})$. From the components $r_i$, $r'_i$, and the unit tensor $\delta_{ij}$ we can construct fourteen different tensors of rank 3. Therefore, instead of Eq. (12.148) we now obtain a formula containing 14 terms, the coefficients of which depend on the three arguments $r$, $r'$, and $\rho$, or on $r$, $r'$, and $r'' = |\boldsymbol{r}' - \boldsymbol{r}|$ [cf. Eq. (12.160)]. The symmetry conditions

$$B_{ijl}(\boldsymbol{r}, \boldsymbol{r}') = B_{ilj}(\boldsymbol{r}', \boldsymbol{r}) \mp B_{jil}(-\boldsymbol{r}, \boldsymbol{r}' - \boldsymbol{r}) \qquad (12.158)$$

enable us to reduce the number of independent functions on which the tensor $B_{ijl}(\boldsymbol{r}, \boldsymbol{r}')$ depends. Nevertheless, the final result is still very complicated. If the field $\boldsymbol{u}(\boldsymbol{x})$ is solenoidal then the tensor $B_{ijl}(\boldsymbol{r}, \boldsymbol{r}')$ should satisfy the additional conditions

$$\left(\frac{\partial}{\partial r_i} + \frac{\partial}{\partial r'_i}\right) B_{ijl}(\boldsymbol{r}, \boldsymbol{r}') = \frac{\partial}{\partial r_j} B_{ijl}(\boldsymbol{r}, \boldsymbol{r}') = \frac{\partial}{\partial r'_l} B_{ijl}(\boldsymbol{r}, \boldsymbol{r}') = 0. \qquad (12.159)$$

These conditions, however, are difficult to use to simplify the general form of the tensor $B_{ijl}(\boldsymbol{r}, \boldsymbol{r}')$. Following Proudman and Reid (1954) we shall therefore, consider the general form of the spectral tensor $F_{ijl}(\boldsymbol{k}, \boldsymbol{k}')$ (six-dimensional Fourier transform of the tensor $B_{ijl}(\boldsymbol{r}, \boldsymbol{r}')$). The isotropic tensor $F_{ijl}(\boldsymbol{k}, \boldsymbol{k}')$ can be written in the form of a sum of 14 terms, namely,

$$F_{ijl}(\boldsymbol{k}, \boldsymbol{k}') = -i[F_1 k_i k_j k_l + F_2 k_i k_j k'_l + F_3 k_i k'_j k_l + F_4 k'_i k_j k_l + \\ + F_5 k'_i k'_j k'_l + F_6 k'_i k'_j k_l + F_7 k'_i k_j k'_l + F_8 k_i k'_j k'_l + F_9 k_i \delta_{jl} + \\ + F_{10} k_j \delta_{il} + F_{11} k_l \delta_{ij} + F_{12} k'_i \delta_{jl} + F_{13} k'_j \delta_{il} + F_{14} k'_l \delta_{ij}], \tag{12.160}$$

where $F_1, \ldots, F_{14}$ are functions of $k$, $k'$, and $\mu$ (or of $k$, $k'$, and $k'' = |\boldsymbol{k}' + \boldsymbol{k}|$), while the factor $i$ is added to ensure that all these functions are real [since $F_{ijl}(-\boldsymbol{k}, -\boldsymbol{k}') = F^*_{ijl}(\boldsymbol{k}, \boldsymbol{k}')$]. The symmetry conditions (12.158) now assume the form

$$F_{ijl}(\boldsymbol{k}, \boldsymbol{k}') = F_{ilj}(\boldsymbol{k}', \boldsymbol{k}) = F_{jil}(-\boldsymbol{k} - \boldsymbol{k}', \boldsymbol{k}'), \tag{12.161}$$

while the conditions ensuring that the vector is solenoidal, [Eq. (12.159)], show that

$$k_j F_{ijl}(\boldsymbol{k}, \boldsymbol{k}') = k'_l F_{ijl}(\boldsymbol{k}, \boldsymbol{k}') = (k_i + k'_i) F_{ijl}(\boldsymbol{k}, \boldsymbol{k}') = 0. \tag{12.162}$$

The last conditions enable us to reduce considerably the number of functions determining the general form of $F_{ijl}(\boldsymbol{k}, \boldsymbol{k}')$. Let us substitute

$$\Delta_{jm}(\boldsymbol{k}) = \delta_{jm} - \frac{k_j k_m}{k^2}, \quad \boldsymbol{k}'' = -\boldsymbol{k} - \boldsymbol{k}';$$

in which case $k_j \Delta_{jm}(\boldsymbol{k}) = 0$ and $\Delta_{jm}(\boldsymbol{k}) f_m(\boldsymbol{k}) = f_j(\boldsymbol{k})$, if $k_m f_m = 0$. Therefore,

$$\Delta_{im}(\boldsymbol{k}'') \Delta_{jn}(\boldsymbol{k}) \Delta_{lp}(\boldsymbol{k}') F_{mnp}(\boldsymbol{k}, \boldsymbol{k}') = F_{ijl}(\boldsymbol{k}, \boldsymbol{k}') \tag{12.163}$$

Substituting Eq. (12.160) into the left-hand side of Eq. (12.163), we can verify that eight of the fourteen terms in Eq. (12.160) will vanish on contraction with $\Delta_{jn}(\boldsymbol{k})$ or $\Delta_{lp}(\boldsymbol{k}')$. As a result, on the left-hand side of Eq. (12.163) we will have only terms containing the expressions $k_m k'_n k_p, k'_m k'_n k_p, k_m \delta_{np}, k_p \delta_{mn}, k'_m \delta_{np}$, and $k'_n \delta_{mp}$. Moreover, since

$$\Delta_{im}(\boldsymbol{k}'')(k_m k'_n k_p + k'_m k'_n k_p) = 0 \ \text{ and } \ \Delta_{im}(\boldsymbol{k}'')(k_m \delta_{np} + k'_m \delta_{np}) = 0,$$

we can also exclude terms containing $k'_m k'_n k_p$ and $k'_m \delta_{np}$. The final expression thus becomes

$$F_{ijl}(\boldsymbol{k}, \boldsymbol{k}') = -i\Delta_{lm}(\boldsymbol{k}'')\Delta_{jn}(\boldsymbol{k})\Delta_{ip}(\boldsymbol{k}')[\Phi k_m\delta_{np} + \\ + \Phi_1 k'_n\delta_{mp} + \Phi_2 k_p\delta_{nm} + \Psi k_m k'_n k_p]. \quad (12.164)$$

Using the symmetry conditions (12.161), we can readily verify that

$$\begin{aligned} &\Phi(k, k', k'') = -\Phi(k', k, k''), \\ &\Psi(k, k', k'') = -\Psi(k', k, k'') = -\Psi(k, k'', k'), \\ &\Phi_1(k, k', k'') = -\Phi(k'', k', k), \ \Phi_2(k, k', k'') = \Phi(k, k'', k'), \end{aligned} \quad (12.165)$$

so that

$$F_{ijl}(\boldsymbol{k}, \boldsymbol{k}') = -i\Delta_{lm}(\boldsymbol{k}'')\Delta_{jn}(\boldsymbol{k})\Delta_{ip}(\boldsymbol{k}')[\Phi(k, k', k'')k_m\delta_{np} - \\ - \Phi(k'', k', k)k'_n\delta_{mp} + \Phi(k, k'', k')k_p\delta_{nm} + \\ + \Psi(k, k', k'')k_m k'_n k_p]. \quad (12.166)$$

Thus, in the solenoidal case, the tensor $F_{ijl}(\boldsymbol{k}, \boldsymbol{k}')$ is determined by the two functions $\Phi(k, k', k'')$ and $\Psi(k, k', k'')$, the first of which is antisymmetric in its first two arguments and the second in any pair of arguments.

The functions $\Phi$ and $\Psi$ can be replaced by any other two independent scalar functions constructed from $F_{ijl}(\boldsymbol{k}, \boldsymbol{k}')$. In the ensuing analysis we shall also encounter the function

$$\Gamma(k, k', \mu) = ik_l F_{ljj}(\boldsymbol{k}, \boldsymbol{k}'), \quad (12.167)$$

which can readily be expressed in terms of $\Phi$ and $\Psi$ with the aid of Eq. (12.166). In view of Eqs. (12.161) and (12.162), this function is antisymmetric in $k$ and $k'$:

$$\Gamma(k, k', \mu) = -\Gamma(k', k, \mu). \quad (12.168)$$

Moreover, by Eqs. (11.61) and (12.140) it must satisfy the relation

$$2\pi \int_0^\infty \int_{-1}^1 \Gamma(k, k', \mu) k'^2 \, d\mu \, dk' = -2kF_3(k) = \Gamma(k), \quad (12.169)$$

which we shall use in the next chapter.

## 13. LOCALLY HOMOGENEOUS AND LOCALLY ISOTROPIC RANDOM FIELDS

### 13.1 Processes with Stationary Increments

The first section of this chapter was devoted to random functions which were stationary or homogeneous. We shall now consider a generalization of these properties which is important in the theory of turbulence.

We shall begin with random processes which are functions of the single variable $t$. Consider, as an example, the diffusion (molecular or turbulent) of particles suspended in a fluid. Let $u(t)$ be a coordinate of one such particle at time $t$. The process $u(t)$ is, of course nonstationary, since the probability that a particle will depart far from its initial position increases with time. If, however, the properties of the fluid are time-independent, and the flow is steady and everywhere the same, then the probability distribution for a path traversed by the particle in a time $\tau$ (between $t$ and $t+\tau$) will be independent of $t$. Moreover, the multidimensional probability distribution for the random variables $u(t+t_2)-u(t+t_1)$, $u(t+t_4)-u(t+t_3)$, ..., $u(t+t_{2n})-u(t+t_{2n-1})$ for any $n$ and $t$, $t_1$, $t_2$, ..., $t_{2n-1}$, $t_{2n}$ will also be independent of $t$ in this case. In other words, the multidimensional probability densities for the differences

$$u(t_2)-u(t_1),\ u(t_4)-u(t_3),\ \ldots,\ u(t_{2n})-u(t_{2n-1})$$

will be unaffected in this case when the entire group of points $t_1$, $t_2$, ..., $t_{2n}$ is arbitrarily shifted along the time axis. Random processes $u(t)$ satisfying the latter condition are quite frequently encountered in applications (including many problems in the theory of turbulence). Following Kolmogorov (1940a), we shall call them *processes with stationary increments*.

It is clear that any stationary random process is at the same time a process with stationary increments. There are, however, nonstationary processes with stationary increments. They include, in particular, the above time-dependence of the coordinates of a diffusing particle. Another broad class of processes of this kind consists of the indefinite integrals of stationary processes $v(t)$, i.e., the processes

$$u(t)=\int_{t_0}^{t} v(t')\,dt'+c, \tag{13.1}$$

where $t_0$ is a fixed number, and $c$ is a constant random variable [the simplest examples of processes of the form (13.1) are the linear functions $u(t) = vt + c$, where $v$ and $c$ are constants].[16] These examples show that the concept of a process with stationary increments is an important generalization of the concept of a stationary process.

The moments of random variables $\Delta_\tau u(t) = u(t+\tau) - u(t)$ which are functions of $\tau$ only are important statistical characteristics of processes $u(t)$ with stationary increments. The simplest of these is the first moment

$$m(\tau) = \overline{[u(t+\tau) - u(t)]}. \tag{13.2}$$

In view of the obvious equation $u(t+\tau_1+\tau_2) - u(t) = [u(t+\tau_1+\tau_2) - u(t+\tau_2)] + [u(t+\tau_2) - u(t)]$, the function $m(\tau)$ should satisfy the equation

$$m(\tau_1+\tau_2) = m(\tau_1) + m(\tau_2). \tag{13.3}$$

We shall suppose that $m(\tau)$ is a continuous function of $\tau$. From Eq. (13.3) we then find that

$$m(\tau) = c_1\tau, \tag{13.4}$$

where $c_1$ is a constant. Let $\overline{u(0)} = c_0$, i.e., let us suppose that $\overline{u(0)}$ exists.[17] Then if we write $u(t)$ in the form $[u(t) - u(0)] + u(0)$ we

[16] We note, moreover, that if $\overline{v(t)} = 0$, and

$$\int_{-\infty}^{\infty} F_v(\omega)\,\omega^{-2}\,d\omega < \infty$$

where $F_v(\omega)$ is the spectrum of $v(t)$, the constant $c$ can be chosen so that the process defined by Eq. (13.1) is not only a process with stationary increments, but simply a stationary process. However, even in such cases, stationarity will occur only for a particular choice of the random variable $c$, while for all other values of $c$ the process $u(t)$ will be nonstationary.

[17] In general, we can sssume in the theory of processes with stationary increments that for the values of the process under consideration the probability distributions do not even exist, while the one-dimensional and multidimensional distributions exist only for the differences of the values of the process $u(t)$ at two points. Hence it follows in particular that the probability distributions for

$$\int_{-\infty}^{\infty} u(t)\,\theta(t)\,dt$$

(Continued on page 82)

obtain, using Eq. (13.4),

$$\overline{u(t)} = c_0 + c_1 t. \tag{13.5}$$

Let us recall that in the case of a stationary process $u(t)$ the mean $\overline{u}(t)$ should be a constant [see Eq. (4.59) of Vol. 1]. Stationary processes can therefore, be used for the description of turbulence only in the case of steady flows for which all the average characteristics are time-independent. In the case of processes with stationary increments, on the other hand, we have the possibility of being able to describe the turbulence even in nonsteady flows (but only for time intervals over which all the mean characteristics of the flow can be approximately regarded as linear).[18] This is one of the reasons for the considerable interest in processes with stationary increments in the theory of turbulence.

It is clear that in the case of a process $u(t)$ with stationary increments, the values of the difference $u(t+\tau) - u(t) = \Delta_\tau u(t)$ at different $t$ and fixed $\tau$ constitute a stationary process depending on the parameter $\tau$. In the experimental determination of the constant $c_1$, therefore, we can use the time average of the differences $\Delta_\tau u(t)$, based on the usual ergodic theorem for stationary processes (see Sect. 4.7 of Vol. 1). Having determined $c_1$ we can subtract the linear function $c_0 + c_1 t$ from $u(t)$, where $c_0$ is arbitrary, and thus transform from $u(t)$ to the new process $u_1(t) = u(t) - c_1 t - c_0$ which satisfies the condition $\overline{[u_1(t+\tau) - u_1(t)]} = 0$. Therefore, without loss of generality we can consider only such processes $u(t)$ with stationary

exist in this case only provided

$$\int_{-\infty}^{\infty} \theta(t)\, dt = 0$$

The characteristic functional $\Phi[\theta(t)]$ for processes with stationary increments will therefore be meaningful only for such functions $\theta(t)$ for which

$$\int_{-\infty}^{\infty} \theta(t)\, dt = 0$$

[cf. Yaglom 1955, 1957)].

[18] If the assumption of linear dependence of the mean characteristics on time seems to be inadmissible, we can use the more complicated theory of processes with stationary higher-order increments (see Yaglom, 1955, or Gel'fand and Vilenkin, 1961). We shall, however, not linger here on this point.

increments for which $c_1 = 0$. We shall adopt this procedure from now on.

Consider now the general joint second moment of the differences $\Delta_\tau u(t) = u(t+\tau) - u(t)$, which is a function of three variables:

$$\overline{[u(t_1+\tau_1)-u(t_1)][u(t_2+\tau_2)-u(t_2)]} = D(t_2-t_1, \tau_1, \tau_2). \quad (13.6)$$

We shall use the following elementary identity:

$$(a-b)(c-d) = \frac{1}{2}[(a-d)^2+(b-c)^2-(a-c)^2-(b-d)^2]. \quad (13.7)$$

Substituting $a = u(t_2+\tau_2)$, $b = u(t_2)$, $c = u(t_1+\tau_1)$, and $d = u(t_1)$, we find that the function (13.6) can be written in the form

$$D(t, \tau_1, \tau_2) = \frac{1}{2}[D(t+\tau_2)+D(t-\tau_1)-D(t-\tau_1+\tau_2)-D(t)], \quad (13.8)$$

where $D(\tau)$ is a nonnegative function of a single variable, which is defined by the equation

$$D(\tau) = \overline{[u(t+\tau)-u(t)]^2} \quad (13.9)$$

i.e., $D(\tau) = D(0, \tau, \tau)$. We shall call this the *structure function* of the process $u(t)$. According to Eq. (13.8), in the case of a process with stationary increments, this function will unambiguously determine the joint second moment (13.6) of the differences $\Delta_\tau u(t)$. It follows that it is just as important for the theory of such processes as the second moment of $u(t)$, i.e., the correlation function $B(\tau)$, is for the theory of stationary processes.

By its very definition, the structure function is always nonnegative, even, and such that $D(0) = 0$. In contrast to the correlation function $B(\tau)$ which is always bounded [because $|B(\tau)| \leqslant B(0)$], the function $D(\tau)$ can increase without limit as $\tau \to \infty$. Thus, for example, if $u(t) = vt + c$ then $D(\tau) = A\tau^2$, $A = \overline{v^2} > 0$. We shall now show that the increase of the structure function as $\tau \to \infty$ cannot be as rapid as desired, or even more rapid than in the above example. Using the well known inequality

$$\frac{\overline{[u(\tau_2)-u(\tau_1)][u(\tau_4)-u(\tau_3)]}}{\{\overline{[u(\tau_2)-u(\tau_1)]^2}\}^{1/2}\{\overline{[u(\tau_4)-u(\tau_3)]^2}\}^{1/2}} \leqslant 1$$

where the left-hand side is equal to the correlation coefficient for the variables $u(\tau_2) - u(\tau_1)$ and $u(\tau_4) - u(\tau_3)$, which always lies between $-1$ and $+1$. Consequently

$$D(\tau) = \overline{[u(\tau) - u(0)]^2} = \overline{\left\{\sum_{i=1}^{n}\left[u\left(\frac{i\tau}{n}\right) - u\left(\frac{(i-1)\tau}{n}\right)\right]\right\}^2} =$$

$$= \sum_{i=1}^{n}\sum_{j=1}^{n}\overline{\left[u\left(\frac{i\tau}{n}\right) - u\left(\frac{(i-1)\tau}{n}\right)\right]\left[u\left(\frac{j\tau}{n}\right) - u\left(\frac{(j-1)\tau}{n}\right)\right]} \leqslant$$

$$\leqslant \sum_{i=1}^{n}\sum_{j=1}^{n}\left\{\overline{\left[u\left(\frac{i\tau}{n}\right) - u\left(\frac{(i-1)\tau}{n}\right)\right]^2}\right\}^{1/2}\left\{\overline{\left[u\left(\frac{j\tau}{n}\right) - u\left(\frac{(j-1)\tau}{n}\right)\right]^2}\right\}^{1/2} =$$

$$= n^2\left\{D\left(\frac{\tau}{n}\right)\right\}^{1/2}\left\{D\left(\frac{\tau}{n}\right)\right\}^{1/2} = n^2 D\left(\frac{\tau}{n}\right)$$

or, in other words,

$$\frac{D(\tau)}{\tau^2} \leqslant \frac{D(\tau/n)}{(\tau/n)^2}. \tag{13.10}$$

If now we use the symbol $A$ to denote the maximum of the continuous function $D(\tau)/\tau^2 = \varphi(\tau)$ in the interval $1 \leqslant \tau \leqslant 2$, and recall that any $\tau \geqslant 1$ always lies between $n$ and $2n$, where $n$ is an integer (so that $1 \leqslant \tau/n \leqslant 2$), we find directly from Eq. (13.10) that

$$D(\tau) \leqslant A\tau^2, \qquad \tau \geqslant 1. \tag{13.11}$$

In the special case when the function $D(\tau)$ is twice-differentiable at the point $\tau = 0$, we can use the properties of $D(\tau)$ to write this second derivative in the form

$$D''(0) = \lim_{h \to 0}\frac{D(h) - 2D(0) + D(-h)}{h^2} = \lim_{h \to 0}\frac{2D(h)}{h^2}$$

Then, assuming that $n \to \infty$ in Eq. (13.10), we obtain immediately

$$D(\tau) \leqslant \frac{1}{2}D''(0)\tau^2. \tag{13.11'}$$

Consider the further special case where $u(t)$ is not merely a process with stationary increments but a stationary process in the usual sense. As usual, we shall suppose that $\overline{u(t)} = 0$. In that case the function $D(\tau)$ will be simply expressed in terms of the correlation function $B(\tau)$: from Eq. (13.9) we have

$$D(\tau)=2\,[B(0)-B(\tau)]. \tag{13.12}$$

It follows from this equation that, in the special case under consideration, the structure function is definitely bounded: $|D(\tau)|\leqslant 4B(0)$. Next, it is clear from this that, if we know the correlation function for a stationary process, this means that we have a knowledge of the structure function also. The converse statement is not, in general, correct. In the general case, the formula given by Eq. (13.12) will not uniquely determine $B(\tau)$ in terms of the function $D(\tau)$. If, however, it is clear from physical considerations that the correlation between $u(t)$ and $u(t+\tau)$ vanishes as $\tau\to\infty$ and therefore $B(\tau)\to 0$ as $\tau\to\infty$, then by Eq. (13.12) we have

$$\lim_{\tau\to\infty} D(\tau)=D(\infty)=2B(0), \qquad B(\tau)=\frac{1}{2}\,[D(\infty)-D(\tau)]. \tag{13.13}$$

Therefore, for stationary random processes with a correlation function $B(\tau)$ which vanishes at infinity, the statistical characteristics of $B(\tau)$ and $D(\tau)$ are mutually interchangeable: Each can be determined from the other.[19]

In the experimental determination of the structure function $D(\tau)$ the statistical average is usually replaced by the time average, i.e., we take the average of the quantities $\Delta_\tau u(t)=u(t+\tau)-u(t)$ (where $\tau$ is fixed) over a sufficiently long time interval $T$ for a sequence of values of $\tau$. It then frequently turns out that, for a stationary process $u(t)$, the structure function $D(\tau)$ found by averaging over a given time interval $T$ is obtained with a smaller error than in the case of the correlation function $B(\tau)$.[20] Therefore, even in those cases where the

[19] Unfortunately, in applied problems the quantity $D(\infty)=2B(0)$ is only very approximately found from measurements of the function $D(\tau)$ (since the approach of the function $D(\tau)$ to a constant value is often very slow as $\tau\to\infty$). In such cases, the quantity $B(0)=\overline{u^2(t)}$ must be found directly by taking the time average of the square of the process. We note also that when $\overline{u(t)}=$ const $\neq 0$ we can use the formula (13.12) where $B(\tau)$ is now understood to be either $\overline{u(t)\,u(t+\tau)}$ or $\overline{u'(t)\,u'(t+\tau)}=\overline{[u(t)-\bar{u}]\,[u(t+\tau)-\bar{u}]}$ (since in the stationary case the values of $D(\tau)$ are independent of $\bar{u}$). However, the function $\overline{u(t)\,u(t+\tau)}$ which depends on $\bar{u}$ when $\bar{u}\neq 0$ is no longer uniquely determined by $D(\tau)$.

[20] This is due to the fact that, when we consider the differences $u(t+\tau)-u(t)$ instead of $u(t)$, we no longer have to deal with the spectral components of the process $u(t)$ having the longest periods (much greater than $\tau$) which introduce the largest errors for a relatively restricted averaging time.

correlation function $B(\tau)$ is our main interest, it is frequently useful to find $D(\tau)$ and $B(0)=\overline{u^2(t)}$ from the data, and then use the formula given by Eq. (13.13).

Let us now consider the spectral decomposition of a process $u(t)$ with stationary increments and the corresponding structure functions $D(\tau)$. It is convenient to begin with the case of differentiable processes. Here, the derivative $u'(t)=v(t)$ will be a stationary process having a spectral representation of the form (11.10), and $u(t)$ will be expressed in terms of $v(t)$ with the aid of the formula (13.1). Substituting the spectral representation of $v(t)$ into Eq. (13.1) and changing the order of integration with respect to $t$ and $Z(\omega)$, we obtain

$$u(t)=\int_{-\infty}^{\infty}(e^{i\omega t}-1)\frac{dZ_v(\omega)}{i\omega}+u_0, \tag{13.14}$$

where $Z_v(\omega)$ is a random function of the frequency $\omega$ which enters into the spectral representation of $v(t)$, and $u_0$ is a constant random variable [equal to $u(0)$].[21] The formula (13.14) can also be rewritten in the form

$$u(t)=\int_{-\infty}^{\infty}(e^{i\omega t}-1)\,dZ(\omega)+u_0, \tag{13.15}$$

where $Z(\omega)$ is defined for $\omega>0$ and $\omega<0$ by the condition

$$Z(\omega_2)-Z(\omega_1)=\int_{\omega_1}^{\omega_2}\frac{dZ_v(\omega)}{i\omega},\quad 0<\omega_1<\omega_2 \quad\text{or}\quad \omega_1<\omega_2<0. \qquad (13.16)^{22}$$

[21] If the point $\omega=0$ is a jump discontinuity of $Z_v(\omega)$, i.e.,

$$\lim_{\varepsilon\to 0}\,[Z_v(\varepsilon)-Z_v(-\varepsilon)]=u_1\neq 0),$$

then the contribution of this discontinuity to the integral in Eq. (13.14) must be regarded as equal to $u_1\lim\limits_{\omega\to 0}\dfrac{e^{i\omega t}-1}{i\omega}=u_1t$. We shall always assume, however, that the function $Z_v(\omega)$ has no jump discontinuities, since discrete spectra are not encountered in the theory of turbulence.

[22] For intervals $[\omega_1,\omega_2]$ containing the point $\omega=0$, the integral in Eq. (13.16) may

(Continued on page 87)

It is clear that the increments $Z(\Delta\omega) = Z(\omega_2) - Z(\omega_1)$ of the function $Z(\omega)$ in intervals which do not contain the point $\omega = 0$ will have the usual properties defined by Eq. (11.8), and also the property $Z(-\Delta\omega) = Z^*(\Delta\omega)$. If the process $v(t) = u'(t)$ has the spectral density $F_v(\omega) = E_v(\omega)/2$, the equation (13.16) implies

$$\overline{|Z(\Delta\omega)|^2} = \int_{\Delta\omega} F(\omega)\,d\omega, \quad \overline{|dZ(\omega)|^2} = F(\omega)\,d\omega = \frac{1}{2}E(\omega)\,d\omega, \qquad (13.17)$$

where $E(\omega) = 2F(\omega) = 2F_v(\omega)/\omega^2 = E_v(\omega)/\omega^2$ is a nonnegative function which is defined for $\omega > 0$ and satisfies the condition

$$\int_0^\infty \omega^2 E(\omega)\,d\omega < \infty. \qquad (13.18)$$

For the structure function (13.9) we have, from Eqs. (13.15) and (13.17), the following representation

$$D(\tau) = 2\int_0^\infty (1 - \cos\omega\tau)\,E(\omega)\,d\omega. \qquad (13.19)$$

We note that by equation (13.18), the function $E(\omega)$ can rapidly tend to infinity as $\omega$ approaches zero. Nevertheless, the formula (13.19) is always meaningful since the increase of $E(\omega)$ is compensated by the fact that $1 - \cos\omega\tau$ tends to zero [this function is proportional to $\omega^2$ for small $\omega$]. The function $E(\omega)$ [or $F(\omega) = E(\omega)/2$)] is called the *spectral density* or simply the *spectrum* of the process with stationary increments, and the representations given by equations (13.15) and (13.19) are called the *spectral representation* of this process and its structure function, respectively.

It can be shown that a spectral representation of a similar form exists also for any nondifferentiable process $u(t)$ with stationary

not exist. Therefore, the improper integral on the right-hand side of Eq. (13.15) must, strictly speaking, be understood as representing the limit

$$\int_{-\infty}^{\infty} = \lim_{\substack{T\to\infty \\ \varepsilon\to 0}} \left\{ \int_{-T}^{-\varepsilon} + \int_{\varepsilon}^{T} \right\}.$$

increments. The only difference as compared with the differentiable case is that, in the general case, the spectrum $E(\omega)$ determined by Eq. (13.17) may not decrease at infinity rapidly enough to ensure that the integral (13.18) will be finite. Instead, it is only necessary that, for any $\omega_0 > 0$,

$$\int_0^{\omega_0} \omega^2 E(\omega)\, d\omega < \infty, \qquad \int_{\omega_0}^{\infty} E(\omega)\, d\omega < \infty. \tag{13.20}$$

In other words, near zero frequency, the integral $\int_0^{\omega_0} \omega^2 E(\omega)\, d\omega$ must converge, while at infinity the integral $\int_{\omega_0}^{\infty} E(\omega)\, d\omega$ must converge. It is readily seen that this is sufficient to insure that the integral (13.19) is convergent for all $\tau$. It also turns out that any function $D(\tau)$ which can be represented by Eq. (13.19), where $E(\omega)$ is a nonnegative function satisfying Eq. (13.20), is the structure function of some random process with stationary increments. The proof of all these statements was first given by Kolmogorov (1940a), and in a different connection by von Neumann and Schoenberg (1941). It can be found, for example, in Doob's book (1953)[23]

In the special case when not only the conditions (13.20), but also the, more restrictive condition $\int_0^{\infty} E(\omega)\, d\omega < \infty$, are satisfied, the process $u(t)$ will be not only a process with stationary increments but also a stationary process in the usual sense. In fact, when the above condition is satisfied, the integral on the right-hand side of Eq. (13.15) can be written as the difference between two convergent integrals the first of which is the usual spectral representation (11.10) of a stationary process and the second is a constant random variable (independent of $t$). The right-hand side of Eq. (13.19) will then also be representable by the difference between two integrals. In the other special case when Eq. (13.18) is satisfied, the process

[23] Kolmogorov and also Doob consider even the more general case where the frequency distribution of the variables $\overline{|dZ(\omega)|^2}$ may not be continuous and is not defined by the spectral density $F(\omega)$ or $E(\omega) = 2F(\omega)$.

$u(t)$ will be differentiable. The derivative $u'(t)$ will then be a stationary process with the spectrum $\omega^2 E(\omega)$.

Let us note that, when the condition (13.20) is satisfied, the integral $\int_0^\infty E(\omega)\, d\omega$, (which is usually interpreted as the energy in the case of stationary processes) may become infinite. This fact shows that, for processes with stationary increments, this integral will either have a different physical meaning, or it will be infinite only because of the mathematical idealization of the process under consideration which cannot really be used to describe the true behavior of the spectral density at the lowest frequencies (as $\omega \to 0$).

We shall now show that from the spectral representation of processes with stationary increments we can also derive the limitation on the rate of increase of $D(\tau)$ as $\tau \to \infty$ which was found above by another method. In fact, if we use the first of the two elementary inequalities

$$|1 - \cos \omega\tau| < \frac{\omega^2\tau^2}{2}, \quad |1 - \cos \omega\tau| \leqslant 2$$

for $0 < \omega \leqslant \omega_0$, and the second for $\omega > \omega_0$, we have from Eq. (13.19)

$$\frac{D(\tau)}{\tau^2} < \int_0^{\omega_0} \omega^2 E(\omega)\, d\omega + \frac{4}{\tau^2} \int_{\omega_0}^{\infty} E(\omega)\, d\omega. \tag{13.21}$$

By equation (13.20) the right-hand side is bounded, and consequently $D(\tau)$ cannot increase at infinity more rapidly than $A\tau^2$ where $A$ is a constant. Moreover, it follows from Eq. (13.21) that when the spectral density $E(\omega)$ exists, the increase of $D(\tau)$ for $\tau \to \infty$ is always slower than $\tau^2$ so that

$$\lim_{\tau \to \infty} \frac{D(\tau)}{\tau^2} = 0. \tag{13.22}$$

In fact let $\delta$ be an arbitrary positive number as small as desired. Since $\int_0^{\omega_0} \omega^2 E(\omega)\, d\omega$ converges near the origin, $\omega_0$ can be chosen so that the first term on the right-hand side of Eq. (13.21) is less than $\delta/2$. The value of $4\int_{\omega_0}^{\infty} E(\omega)\,d\omega$ will then be fixed, and we can choose $T$ so that the second term on the right-hand side of Eq. (13.21) is less than $\delta/2$ if $\tau > T$. Thus, for sufficiently large $\tau$, the right-hand side of Eq. (13.21) will be less than any positive $\delta$. This is expressed by Eq. (13.22).[24]

An important class of structure functions is defined by

$$D(\tau) = A\tau^{\gamma}, \quad A > 0, \qquad 0 < \gamma < 2 \tag{13.23}$$

[$\gamma > 0$ because $D(0) = 0$, and $\gamma < 2$ in view of Eq. (13.22)].[25] It is readily seen that the functions given by Eq. (13.23) can be written in the form (13.19) if we take $E(\omega)$ in the form

$$E(\omega) = \frac{C}{\omega^{1+\gamma}}, \tag{13.24}$$

where

$$C = \frac{A}{2\int_0^{\infty} \frac{1-\cos x}{x^{\gamma+1}}\,dx} = \frac{A\gamma}{2\cos\frac{\pi\gamma}{2}\,\Gamma(1-\gamma)} = \frac{\Gamma(1+\gamma)\sin\frac{\pi\gamma}{2}}{\pi}A. \tag{13.24'}$$

Since the functions (13.24) are positive, and satisfy the conditions (13.20), it follows that the functions (13.23) can, in fact, be structure functions of processes with stationary increments. We note that as $\tau \to \infty$ all these functions increase without limit.

The structure functions (13.23) have the interesting property that their form is invariant under a similarity transformation group $t \to Tt$, $u \to Uu$. In fact, in this case $D(\tau) = U^2 D(T\tau)$ for $U = T^{-\gamma/2}$ and any $T > 0$. Functions having this property will be referred to as *self-preserving*. They are such that no characteristic scale can be associated with their form. It is readily shown that the correlation

[24] We note also that Eq. (13.22) cannot be proved unless $E(\omega)$ exists, since for $u(t)$ containing a linear component $u_1(t)$ this relation will not be valid.

[25] As a limiting case we can also consider $\gamma = 2$ which corresponds to processes of the form $u(t) = u_0 t + u_1$ where $u_0$ and $u_1$ are constant random variables.

function of a stationary random process cannot be self-preserving, and that functions of the form given by Eq. (13.23) are the only self-preserving structure functions (see Kolomogorov, 1940b). We shall see later that self-preserving structure functions play an important role in the theory of developed turbulence.

Let us now go on to consider multidimensional random processes with stationary increments, i.e., multidimensional processes $\boldsymbol{u}(t) = \{u_1(t), \ldots, u_n(t)\}$ for which the probability density for the differences between any components in an aribtrary set of pairs of points is unaffected by the simultaneous translation of all the points along the time axis. We shall suppose that the mean values of the differences $u_k(t+\tau) - u_k(t)$ are zero for all $k = 1, 2, \ldots, n$ and any real $t$, $\tau$. The spectral representation of the multidimensional process $\boldsymbol{u}(t)$ can then be written in the form

$$\boldsymbol{u}(t) = \int_{-\infty}^{\infty} (e^{i\omega t} - 1)\, d\boldsymbol{Z}(\omega) + \boldsymbol{u}_0, \qquad (13.25)$$

where $\boldsymbol{u}_0$ is a random vector and $\boldsymbol{Z}(\omega) = \{Z_1(\omega), \ldots, Z_k(\omega)\}$ is a random vector function which is such that $\overline{dZ_k^*(\omega)\, dZ_l(\omega_1)} = 0$ for $\omega \neq \omega_1$ and any $k$ and $l$. We shall suppose, moreover, that all processes $u_k(t)$, $k = 1, 2, \ldots, n$ possess spectral densities [in other words, we shall suppose that the correlation coefficient between the increments $u_k(t+\tau) - u_k(t)$ and $u_k(\tau) - u_k(0)$ decrease for all $k = 1, 2, \ldots, n$ as $t \to \infty$ sufficiently rapidly to ensure that it can be expanded in a Fourier integral in $t$]. We then have

$$\overline{dZ_k^*(\omega)\, dZ_l(\omega)} = F_{kl}(\omega)\, d\omega \qquad (13.26)$$

This can be combined with the condition $\overline{dZ_k^*(\omega)\, dZ_l(\omega_1)} = 0$ for $\omega \neq \omega_1$ into the symbolic equation (11.39). The functions $F_{kl}(\omega)$ are complex functions satisfying the conditions (11.34), (11.36), and (11.37), and are such that for any $\omega_0 > 0$

$$\int_0^{\omega_0} \omega^2 F_{kk}(\omega)\, d\omega < \infty, \qquad \int_{\omega_0}^{\infty} F_{kk}(\omega)\, d\omega < \infty, \qquad k = 1, 2, \ldots, n. \qquad (13.27)$$

From Eqs. (13.25) and (11.39) we find that the joint second moment

$$[u_k(t_1+\tau_1)-u_k(t_1)]\,[u_l(t_2+\tau_2)-u_l(t_2)]=D_{kl}(t_2-t_1,\ \tau_1,\ \tau_2), \quad (13.28)$$

is given by the equation

$$D_{kl}(t,\ \tau_1,\ \tau_2)=\int_{-\infty}^{\infty} e^{i\omega t}\,(e^{-i\omega\tau_1}-1)\,(e^{i\omega\tau_2}-1)\,F_{kl}(\omega)\,d\omega. \quad (13.29)$$

In particular, the functions

$$D_{kl}(\tau)=\overline{[u_k(t+\tau)-u_k(t)]\,[u_l(t+\tau)-u_l(t)]}, \quad (13.30)$$

which we shall refer to as the *structure functions of a multi-dimensional process with stationary increments* are given by

$$D_{kl}(\tau)=2\int_{-\infty}^{\infty}(1-\cos\omega\tau)\,F_{kl}(\omega)\,d\omega \quad (13.31)$$

since $D_{kl}(\tau)=D_{kl}(0,\ \tau,\ \tau)$. These structure functions are real by definition (but for $k\neq l$ they are not necessarily nonnegative). They are also even and symmetric in the indices [so that $D_{kl}(\tau)=D_{lk}(\tau)$], and are such that $D_{kl}(0)=0$. Moreover, it is readily seen that

$$\sum_{k,\ l=1}^{n} D_{kl}(\tau)\,c_k c_l \geqslant 0 \quad (13.32)$$

for any real $\tau$ and $c_1, \ldots, c_n$. We note, however, that in the multidimensional case the functions $D_{kl}(t,\ \tau_1,\ \tau_2)$ cannot, in general, be expressed in terms of the functions $D_{kl}(\tau)$: The identity (13.7) is no longer valid. It is more convenient, in this case, to use the following identity:

$$(a_k-b_k)(c_l-d_l)+(a_l-b_l)(c_k-d_k)=(a_k-d_k)(a_l-d_l)+$$
$$+(b_k-c_k)(b_l-c_l)-(a_k-c_k)(a_l-c_l)-(b_k-d_k)(b_l-d_l). \quad (13.33)$$

Substituting $a_j=u_j(t_2+\tau_2), b_j=u_j(t_2), c_j=u_j(t_1+\tau_1)$ and $d_j=u_j(t_1)$ for $j=k$ and $l$, we find that at least the sum $D_{kl}(t,\ \tau_1,\ \tau_2)+D_{lk}(t,\ \tau_1,\ \tau_2)$ can always be expressed in terms of the functions $D_{kl}(\tau)$:

$$D_{kl}(t,\ \tau_1,\ \tau)+D_{lk}(t,\ \tau_1,\ \tau_2)=$$
$$=D_{kl}(t+\tau_2)+D_{kl}(t-\tau_1)-D_{kl}(t-\tau_1+\tau_2)-D_{kl}(t). \quad (13.34)$$

In the special case when

$$D_{kl}(t, \tau_1, \tau_2) = D_{lk}(t, \tau_1, \tau_2), \tag{13.35}$$

it follows from Eq. (13.34) that $D_{kl}(t, \tau_1, \tau_2)$ can also be expressed in terms of the multidimensional structure functions. According to Eq. (13.29) the relationship given by Eq. (13.35) will be satisfied provided $F_{kl}(\omega) = F_{lk}(\omega)$ which, in view of Eq. (11.36), is equivalent to all the spectral densities $F_{kl}(\omega)$ being real.

## 13.2 Locally Homogeneous Fields

For random functions of the space point $\boldsymbol{x}$, the analog of the process with stationary increments is the *locally homogeneous random field* (or, what is the same, a *random field with homogeneous increments*). By this we mean a random field $u(\boldsymbol{x})$ for which all the probability distributions for the field differences in a set of pairs of points are unaffected by any translation of all the points under consideration.[26]

It can easily be shown (similarly as for a process with stationary increments) that the mean value of the increments $\Delta_r u(\boldsymbol{x}) = u(\boldsymbol{x}+\boldsymbol{r}) - u(\boldsymbol{x})$ is a linear function of the vector $\boldsymbol{r}$:

$$\overline{[u(\boldsymbol{x}+\boldsymbol{r}) - u(\boldsymbol{x})]} = m(\boldsymbol{r}) = \boldsymbol{c}_1\boldsymbol{r}, \tag{13.36}$$

where $\boldsymbol{c}_1$ is a constant vector (so that, for example, for fields in three-dimensional space $\boldsymbol{c}_1\boldsymbol{r} = c_{11}r_1 + c_{12}r_2 + c_{13}r_3$). The mean value of the field $u(\boldsymbol{x})$ itself, if it exists, is given by

$$\overline{u(\boldsymbol{x})} = \boldsymbol{c}_1\boldsymbol{x} + c_0, \tag{13.37}$$

where $c_0$ is a constant. When the ergodicity conditions indicated in Sect. 4.7 of Vol. 1 are satisfied, the value of the vector $\boldsymbol{c}_1$ can be determined by taking the space average of the differences $\Delta_r u(\boldsymbol{x}) = u(\boldsymbol{x}+\boldsymbol{r}) - u(\boldsymbol{x})$ with respect to $\boldsymbol{x}$. We can then subtract the linear

[26] The probability distributions for the values of $u(x)$ themselves are not, in general, considered in the theory of locally homogeneous fields, and may even be regarded as nonexistent. Quantities of the form $\int u(\boldsymbol{x})\,\theta(\boldsymbol{x})\,d\boldsymbol{x}$ and the characteristic functional $\Phi[\theta(\boldsymbol{x})]$ are meaningful in this context only if $\int \theta(\boldsymbol{x})\,d\boldsymbol{x} = 0$.

function $c_1 x + c_0$ from the field (where $c_0$ can be chosen arbitrarily), and then assume that $c_1 = 0$. We shall adopt this procedure below.

The spectral theory of locally homogeneous fields is related to the spectral theory of processes with stationary increments and was developed by Yaglom (1957) (see also Gel'fand and Vilenkin, 1961). The main theorem of this theory is that any locally homogeneous field $u(\boldsymbol{x})$ can be written in the form

$$u(\boldsymbol{x}) = \int (e^{i\boldsymbol{k}\boldsymbol{x}} - 1)\, dZ(\boldsymbol{k}) + u_0, \tag{13.38}$$

where the integral is evaluated over the entire space of the wave-number vectors $\boldsymbol{k}$ with the exception of the point $\boldsymbol{k} = 0$, $u_0 = u(0)$ is a constant random variable, and $dZ(\boldsymbol{k}) = Z(d\boldsymbol{k})$ is the value of the function $Z(\Delta\boldsymbol{k})$ of the multidimensional interval $\Delta\boldsymbol{k} = [\boldsymbol{k}', \boldsymbol{k}'']$ (defined for all intervals not containing the origin) which corresponds to the infinitesimal interval $d\boldsymbol{k} = [\boldsymbol{k}, \boldsymbol{k} + d\boldsymbol{k}]$.[27]

The differentials $dZ(\boldsymbol{k})$ have the properties (11.44), and the improper integral on the right-hand side of Eq. (13.38) is interpreted by analogy with the integral on the right-hand side of Eq. (13.15). The representation (13.38) is called the *spectral representation* of the locally homogeneous field $u(\boldsymbol{x})$. The general second moment of the increments $\Delta_{\boldsymbol{r}} u(\boldsymbol{x})$ of the field $u(\boldsymbol{x})$

$$\overline{[u(\boldsymbol{x}_1 + \boldsymbol{r}_1) - u(\boldsymbol{x}_1)]\,[u(\boldsymbol{x}_2 + \boldsymbol{r}_2) - u(\boldsymbol{x}_2)]} = D(\boldsymbol{x}_2 - \boldsymbol{x}_1, \boldsymbol{r}_1, \boldsymbol{r}_2) \tag{13.39}$$

can be expressed through the *structure function* for this field

$$D(\boldsymbol{r}) = \overline{[u(\boldsymbol{x} + \boldsymbol{r}) - u(\boldsymbol{x})]^2} \tag{13.40}$$

with the aid of the equation

[27] As was done when the processes with stationary increments were considered, we are assuming right at the outset that the spectral density exists for the field $u(\boldsymbol{x})$ [for which it is sufficient that the correlation coefficient between the increments $\Delta_{\boldsymbol{r}} u(\boldsymbol{x}_1)$ and $\Delta_{\boldsymbol{r}} u(\boldsymbol{x} + \boldsymbol{x}_1)$ can be expanded in a Fourier integral with respect to $\boldsymbol{x}$, i.e., it must tend to zero sufficiently rapidly as $|\boldsymbol{x}| \to \infty$]. From this assumption it follows, in particular, that the point $\boldsymbol{k} = 0$ in wave-number vector space cannot be a point of the discrete spectrum, i.e., it does not introduce a finite contribution to $u(\boldsymbol{x})$. If this were not so, the right-hand side of Eq. (13.38) would have to be augmented by the term $\boldsymbol{u}_1 \boldsymbol{x}$, where $\boldsymbol{u}_1$ is a constant vector (cf. a footnote after Eq. (13.14).

$$D(\boldsymbol{x}, \boldsymbol{r}_1, \boldsymbol{r}_2) = \frac{1}{2}\{D(\boldsymbol{x} - \boldsymbol{r}_1) + D(\boldsymbol{x} + \boldsymbol{r}_2) - D(\boldsymbol{x} - \boldsymbol{r}_1 + \boldsymbol{r}_2) - D(\boldsymbol{x})\}. \tag{13.41}$$

where we have used the identity (13.7). The three-dimensional spectral density (or the three-dimensional spectrum) $F(\boldsymbol{k})$ of the field $u(\boldsymbol{x})$ is defined by the equation (11.44) or, more precisely, by the equation

$$\overline{|dZ(\boldsymbol{k})|^2} = F(\boldsymbol{k})\,d\boldsymbol{k}. \tag{13.42}$$

where $F(\boldsymbol{k})$ is a nonnegative even function of the wave vector $\boldsymbol{k}$ satisfying the inequalities

$$\int\limits_{|\boldsymbol{k}|<k_0} k^2 F(\boldsymbol{k})\,d\boldsymbol{k} < \infty, \qquad \int\limits_{|\boldsymbol{k}|>k_0} F(\boldsymbol{k})\,d\boldsymbol{k} < \infty. \tag{13.43}$$

for any $k_0 > 0$. The structure function $D(\boldsymbol{r})$ can be expressed in terms of $F(\boldsymbol{k})$ by the equation

$$D(\boldsymbol{r}) = 2\int (1 - \cos \boldsymbol{kr})\,F(\boldsymbol{k})\,d\boldsymbol{k}. \tag{13.44}$$

The converse is also true: Any function which can be written in the form (13.44), where $F(\boldsymbol{k})$ satisfies the above conditions, is the structure function of a locally homogeneous field.

In addition to the three-dimensional spectral density $F(\boldsymbol{k})$, which describes the distribution in the wave-number vector spece, we can also consider the spectral density $E(k)$ along the wave-number radius, which is expressed in terms of $F(\boldsymbol{k})$ through Eq. (11.48). The conditions (13.43) now assume a form analogous to Eq. (13.20):

$$\int_0^{k_0} k^2 E(k)\,dk < \infty, \qquad \int_{k_0}^{\infty} E(k)\,dk < \infty. \tag{13.43'}$$

A natural generalization of the concept of a locally homogeneous field is the *multidimensional locally homogeneous random field*, i.e., a multidimensional field $\boldsymbol{u}(\boldsymbol{x}) = \{u_1(\boldsymbol{x}), \ldots, u_n(\boldsymbol{x})\}$ such that all the probability distributions for the differences between the values of its components in a given set of pairs of points is unaffected by all the possible translations (shifts) of all these points. By Eqs. (13.36) and (13.37), the mean values of the components of such a field are linear functions of the coordinates. As in the case of a one-dimensional field, these averages can be regarded as identically zero without loss of generality. We shall suppose, moreover, that all the correlation coefficients between the increments $u_j(\boldsymbol{x}+\boldsymbol{r}_1)-u_j(\boldsymbol{x})$ and $u_j(\boldsymbol{r}_2)-u_j(0)$ decrease to zero for $\boldsymbol{x}\to\infty$ (and fixed $\boldsymbol{r}_1$ and $\boldsymbol{r}_2$), and do so sufficiently rapidly so that they can be represented by Fourier integrals with respect to $\boldsymbol{x}$. The spectral representation of the field $\boldsymbol{u}(\boldsymbol{x})$ will then be of the form

$$\boldsymbol{u}(\boldsymbol{x})=\int (e^{i\boldsymbol{k}\boldsymbol{x}}-1)\, d\boldsymbol{Z}(\boldsymbol{k})+\boldsymbol{u}_0, \tag{13.45}$$

where $\boldsymbol{u}_0=\boldsymbol{u}(0)$ is a constant $n$-dimensional random vector, and $d\boldsymbol{Z}(\boldsymbol{k})$ is determined by the values $\boldsymbol{Z}(\Delta\boldsymbol{k})=\{Z_1(\Delta\boldsymbol{k}), \ldots, Z_n(\Delta\boldsymbol{k})\}$ of the random vector function of the interval $\Delta\boldsymbol{k}$ in wave-number-vector space (this is meaningful only for intervals which do not contain the origin of this space). The differentials $dZ_j(\boldsymbol{k})$ have the properties (11.53) and the property $dZ_j(-\boldsymbol{k})=dZ_j^*(\boldsymbol{k})$. In view of Eq. (11.53), we have

$$\overline{dZ_j^*(\boldsymbol{k})\, dZ_l(\boldsymbol{k})}=F_{jl}(\boldsymbol{k})\, d\boldsymbol{k}, \tag{13.46}$$

where $F_{jl}(\boldsymbol{k})$ are *three-dimensional spectral densities of the field* $\boldsymbol{u}(\boldsymbol{x})$. These densities have the properties (11.56), and are such that the matrix $\|F_{jl}(\boldsymbol{k})\|$ is nonnegative for any $\boldsymbol{k}\neq 0$, while for any $k_0>0$

$$\int\limits_{|\boldsymbol{k}|<k_0} k^2 F_{ll}(\boldsymbol{k})\, d\boldsymbol{k}<\infty, \qquad \int\limits_{|\boldsymbol{k}|>k_0} F_{ll}(\boldsymbol{k})\, d\boldsymbol{k}<\infty. \tag{13.47}$$

If we introduce the *structure functions* of the field $\boldsymbol{u}(\boldsymbol{x})$ by means of the equation

$$D_{jl}(\boldsymbol{r})=\overline{[u_j(\boldsymbol{x}+\boldsymbol{r})-u_j(\boldsymbol{x})]\,[u_l(\boldsymbol{x}+\boldsymbol{r})-u_l(\boldsymbol{x})]}, \tag{13.48}$$

we find that these functions can be expressed in terms of $F_{jl}(\boldsymbol{k})$in the form

$$D_{jl}(\boldsymbol{r}) = 2 \cdot \int (1 - \cos \boldsymbol{kr}) F_{jl}(\boldsymbol{k}) \, d\boldsymbol{k}. \tag{13.49}$$

The functions $D_{jl}(\boldsymbol{r})$ are even and symmetric in the indices, i.e., $D_{jl}(\boldsymbol{r}) = D_{lj}(\boldsymbol{r})$. Moreover, they are such that $D_{jl}(0) = 0$. Finally, for any vector $\boldsymbol{r}$ and real numbers $c_1, \ldots, c_n$ we have

$$\sum_{j, l=1}^{n} D_{jl}(\boldsymbol{r}) c_j c_l \geqslant 0. \tag{13.50}$$

The identity (13.33) shows that to ensure that the function

$$D_{jl}(\boldsymbol{x}, \boldsymbol{r}_1, \boldsymbol{r}_2) = \overline{[u_j(\boldsymbol{r}_1) - u_j(0)][u_l(\boldsymbol{x} + \boldsymbol{r}_2) - u_l(\boldsymbol{x})]}, \tag{13.51}$$

can be expressed in terms of $D_{jl}(\boldsymbol{r})$ it is only necessary that

$$D_{jl}(\boldsymbol{x}, \boldsymbol{r}_1, \boldsymbol{r}_2) = D_{lj}(\boldsymbol{x}, \boldsymbol{r}_1, \boldsymbol{r}_2). \tag{13.52}$$

It is readily seen that this condition is equivalent to $F_{jl}(\boldsymbol{k}) = F_{lj}(\boldsymbol{k})$, which means that the functions $F_{jl}(\boldsymbol{k})$ are real because the matrix $\| F_{jl}(\boldsymbol{k}) \|$ is Hermitian.

When $\boldsymbol{u}(\boldsymbol{x})$ is a velocity field, it is convenient to consider in addition to the tensor spectral densities $F_{jl}(\boldsymbol{k})$ the scalar spectral density $F(\boldsymbol{k})$, which is equal to one-half of the trace of the tensor $F_{jl}(\boldsymbol{k})$, and the spectral density $E(k)$ along the wave-number radius, which is the integral of $F(\boldsymbol{k})$ over the sphere $|\boldsymbol{k}| = k$ [see Eq. (11.58)]. The conditions (13.47) can then be rewritten in one form of conditions imposed on the density $E(k)$.

Since the first-order partial derivatives of the field $u(\boldsymbol{x})$ [or $\boldsymbol{u}(\boldsymbol{x})$] are defined as the limits of the field differences between two points divided by the corresponding increments of the arguments, it is clear that the probability distributions for the derivatives of a locally homogeneous field at several points will be invariant under any translation of these points. Thus, all the first-order partial derivatives (and derivatives of all higher orders) of any one-dimensional or multidimensional locally homogeneous field, for which the derivatives exist, form homogeneous random fields. We can now use Eq. (13.33) to express the correlation functions for the derivatives of the field $\boldsymbol{u}(\boldsymbol{x})$ satisfying the conditions (13.52) in terms of the corresponding structure functions:

$$\overline{\frac{\partial u_j(\boldsymbol{x})}{\partial x_m} \frac{\partial u_l(\boldsymbol{x}')}{\partial x'_n}} = \frac{1}{2} \frac{\partial^2 D_{jl}(\boldsymbol{r})}{\partial r_m \partial r_n}\bigg|_{\boldsymbol{r}=\boldsymbol{x}'-\boldsymbol{x}} \tag{13.53}$$

[compare this with Eq. (11.69)]. In the one-dimensional case (or when $j = l$), the above formula is deduced from Eq. (13.7) without any additional limitations. The spectral representations (13.38) and (13.45) for locally homogeneous fields show that the spectral representations for the derivatives of such fields are given by equations (11.66) and (11.70), which were derived above for the derivatives of homogeneous fields. The condition for the differentiability of a field can also be written in the form which we already know, i.e., Eq. (11.68) (it is readily shown that this condition can be reformulated as the condition for the existence of a derivative of twice the order of the structure function $D_{jj}(\boldsymbol{r})$ of the field $u_j(\boldsymbol{x})$ at the point $\boldsymbol{r}=0$).[28] Finally, all the remaining results of Sect. 11.3 can be extended to the case of a locally homogeneous field without modification.

### 13.3 Locally Isotropic Fields

Let us now consider locally homogeneous fields which are also locally isotropic, i.e., they have the property that the probability distributions for the field differences in any set of pairs of points are also unaffected by arbitrary rotations and reflections of the entire set of points. Such locally homogeneous and locally isotropic fields (which we shall refer to simply as *locally isotropic fields*) play an important role in the theory of turbulence. We shall, therefore, summarize the most important facts relating to them (for proofs see the paper by Yaglom, 1957).

We shall begin with the case of one (scalar) locally isotropic random field $u(\boldsymbol{x})$. The mean value of the increment $u(\boldsymbol{x}+\boldsymbol{r})-u(\boldsymbol{x})$ should be given by Eq. (13.36). However, since in our case the vector $c_1$ should be unaffected by all space rotations, it can only be the zero vector. The mean value of the increments of any scalar locally isotropic field must therefore be identically zero. If the mean value

[28]It is clear that the condition given by Eq. (11.68) imposes a limitation only on the rate of decrease of the spectral density $F_{jj}(\boldsymbol{k})$ at infinity. In view of Eq. (13.47), the integral given by Eq. (11.68) will converge at the origin for any $m_1$, $m_2$, $m_3$ not all of which are zero.

$\overline{u(\boldsymbol{x})} = U(\boldsymbol{x})$ of the field $u(\boldsymbol{x})$ exists, then it should be a constant.

The structure function $D(\boldsymbol{r})$ of a locally isotropic field $u(\boldsymbol{x})$ can, clearly, depend only on $r = |\boldsymbol{r}|$:

$$\overline{[u(\boldsymbol{x}+\boldsymbol{r}) - u(\boldsymbol{x})]^2} = D(r). \tag{13.54}$$

Hence, it may be concluded that the corresponding three-dimensional spectral density $F(\boldsymbol{k}) = F(k)$ depends only on the length $k$ of the vector $\boldsymbol{k}$, i.e., it is uniquely determined by the spectral density $E(k) = 4\pi k^2 F(k)$. If we now transform in Eq. (13.44) to spherical polar coordinates and integrate with respect to the angular variables, we obtain

$$D(r) = 2\int_0^\infty \left(1 - \frac{\sin kr}{kr}\right) E(k)\, dk, \tag{13.55}$$

where $E(k) = 4\pi k^2 F(k)$ is a nonnegative function satisfying, for any $k_0 > 0$, the conditions

$$\int_0^{k_0} k^2 E(k)\, dk < \infty, \qquad \int_{k_0}^\infty E(k)\, dk < \infty. \tag{13.56}$$

Conversely, any function $D(r)$ which can be represented in the form (13.55), where $E(k)$ is a nonnegative function satisfying Eq. (13.56), is the structure function of a scalar locally isotropic random field.

It is clear that any function $D(r)$ which is the structure function of a locally isotropic random field in three-dimensional space will also be the structure function of a similar field on a straight line, i.e., it will be the structure function of a process with stationary increments. As in Sect. 12.1, it can be shown that the corresponding one-dimensional spectral density $E_1(k_1)$ in the spectral representation

$$D(r) = 2\int_0^\infty (1 - \cos k_1 r)\, E_1(k_1)\, dk_1, \tag{13.57}$$

will be related to the density $E(k)$ by the same relations (12.13) which are valid for isotropic fields.

As an example of the application of the last result, let us verify whether the function

$$D(r) = Ar^{\gamma}, \quad A > 0, \qquad 0 < \gamma < 2. \tag{13.58}$$

can be the structure function of a locally isotropic field $u(\boldsymbol{x})$. We have already seen in Sect. 13.1 that such a function is the structure function of a process with stationary increments with the one-dimensional spectral density $F_1(k_1) = C/k^{1+\gamma}$, where $C = \Gamma(1+\gamma) \times \sin\frac{\pi\gamma}{2} A\pi^{-1}$. Therefore, the spectral density $E(k)$ is now given by

$$E(k) = \frac{C_1}{k^{1+\gamma}}, \qquad C_1 = \frac{\Gamma(2+\gamma)}{\pi} \sin\frac{\pi\gamma}{2} A. \tag{13.59}$$

Since this function is nonnegative and satisfies Eq. (13.56), the function (13.58) can be the structure function of a scalar locally isotropic field in three-dimensional space.

If the field $u(\boldsymbol{x})$ is isotropic in the sense of Sect. 12.1, it will also be locally isotropic. The structure function $D(r)$ in this case can be simply expressed in terms of the correlation function $B(r)$:

$$D(r) = 2[B(0) - B(r)], \tag{13.60}$$

while the spectral density $E(k)$ in Eq. (13.55) will be the same as the density in the expansion (12.12) of $B(r)$. It will be such that

$$\int_0^\infty E(k)\,dk < \infty \tag{13.61}$$

which replaces the more general condition (13.56). If we assume that $B(r) \to 0$ then on passing to the limit as $r \to \infty$ we have

$$\lim_{r\to\infty} D(r) = D(\infty) = 2B(0) = 2\overline{u^2}, \quad B(r) = \frac{1}{2}[D(\infty) - D(r)]. \tag{13.62}$$

Therefore, the functions $B(r)$ and $D(r)$ are here uniquely related.

The odd-order two-point moments of increments of a locally isotropic field $u(\boldsymbol{x})$ are identically zero. In particular,

$$\overline{[u(x+r)-u(x)]^3}=0. \tag{13.63}$$

In fact, any odd moment of the difference $u(x+r)-u(x)$, is a function of the vector $r$ which changes sign when $r$ is replaced by $-r$ (this is equivalent to changing the order of the points $x+r$ and $x$). However, in view of the local isotropy of the field $u(x)$ this moment must depend only on $r=|r|$ and must vanish.

Let us now consider a *multidimensional locally isotropic field* $u(x)=\{u_1(x), \ldots, u_n(x)\}$ which has the property that the probability distributions for the differences between arbitrary components of this field on any set of pairs of points is unaffected by all translations, rotations, and reflections of the set. It is clear that such multidimensional locally isotropic fields are a generalization of the multidimensional isotropic fields defined in the very beginning of Sect. 12.2. Their structure functions

$$D_{jl}(r)=\overline{[u_j(x+r)-u_j(x)][u_l(x+r)-u_l(x)]}, \qquad r=|r|, \tag{13.64}$$

have a representation of the form (13.55), i.e.,

$$D_{jl}(r)=2\int_0^\infty\left(1-\frac{\sin kr}{kr}\right)E_{jl}(k)\,dk. \tag{13.65}$$

where $E_{jl}(k)$ are real functions such that the matrix $\|E_{jl}(k)\|$ is symmetric and nonnegative, while its trace $E(k)=E_{jj}(k)$ satisfies the condition (13.56). Since all the functions $E_{jl}(k)$ are real [this is so because $E_{jl}(k)=E^*_{jl}(-k)$ and the field is isotropic, i.e., $E_{jl}(k)=E_{jl}(k)$, $k=|k|$], this enables us, in view of the discussion given in Sect. 13.2, to express the general structure function

$$D_{jl}(x, r_1, r_2)=\overline{[u_j(y+r_1)-u_j(y)][u_l(x+y+r_2)-u_l(x+y)]} \tag{13.66}$$

in terms of the functions $D_{jl}(r)$ of a single variable:

$$D_{jl}(x, r_1, r_2)=\frac{1}{2}[D_{jl}(|x-r_1|)+D_{jl}(|x+r_2|)-\\ -D_{jl}(|x-r_1+r_2|)-D_{jl}(|x|)]. \tag{13.67}$$

The multidimensional fields which we have just discussed form a set of one-dimensional (scalar) locally isotropic fields which are correlated to each other in a locally isotropic fashion (see Sect. 12.2). From our point of view, a more interesting concept is that of a *locally isotropic vector field* $\boldsymbol{u}(\boldsymbol{x})=\{u_1(\boldsymbol{x}), u_2(\boldsymbol{x}), u_3(\boldsymbol{x})\}$, for which the probability distributions for the field-component differences on a set of pairs of points are unaffected by translations of this set, and by its rotations or reflections, accompanied by the simultaneous rotation or reflection of the coordinate system with respect to which the components are taken (see the analogous definition of an isotropic vector field in Sect. 12.2). The mean value of an increment of such a field is, in general, given by

$$\overline{[\boldsymbol{u}(\boldsymbol{x}+\boldsymbol{r})-\boldsymbol{u}(\boldsymbol{x})]}=c_1\boldsymbol{r}, \tag{13.68}$$

where $c_1$ is an arbitrary constant. However, if $c_1 \neq 0$, this means that the field $\boldsymbol{u}(\boldsymbol{x})$ contains a linear nonrandom component of the form $\boldsymbol{c}_0+c_1\boldsymbol{x}$ which can be eliminated by transforming to the differences $\boldsymbol{u}'(\boldsymbol{x})=\boldsymbol{u}(\boldsymbol{x})-c_1\boldsymbol{x}$ (this component is frequently absent in most applications right from the beginning). We shall, therefore, assume henceforth that $\overline{[\boldsymbol{u}(\boldsymbol{x}+\boldsymbol{r})-\boldsymbol{u}(\boldsymbol{x})]}=0$.

Let us now consider the second moments of a locally isotropic vector field $\boldsymbol{u}(\boldsymbol{x})$. We shall use spectral theory below to show that the general structure functions (13.66) of such a field are always symmetric in the indices $j$, $l$, so that, in this case also, they can be expressed in terms of the simpler functions

$$D_{jl}(\boldsymbol{r})=\overline{[u_j(\boldsymbol{x}+\boldsymbol{r})-u_j(\boldsymbol{x})][u_l(\boldsymbol{x}+\boldsymbol{r})-u_l(\boldsymbol{x})]} \tag{13.64'}$$

with the aid of equation

$$D_{jl}(\boldsymbol{x}, \boldsymbol{r}_1, \boldsymbol{r}_2)=\frac{1}{2}[D_{jl}(\boldsymbol{x}-\boldsymbol{r}_1)+D_{jl}(\boldsymbol{x}+\boldsymbol{r}_2)-$$
$$-D_{jl}(\boldsymbol{x}-\boldsymbol{r}_1+\boldsymbol{r}_2)-D_{jl}(\boldsymbol{x})] \tag{13.67'}$$

The only difference between these two equations and the equations (13.64) and (13.67) is that $D_{jl}(\boldsymbol{r})$ will now depend on the vector $\boldsymbol{r}$ and not simply on its length. The symmetric tensor $D_{jl}(\boldsymbol{r})$ is a tensor function of the vector $\boldsymbol{r}$ which is invariant under rotations and reflections. Therefore,

$$D_{jl}(\boldsymbol{r})=[D_{LL}(r)-D_{NN}(r)]\frac{r_j r_l}{r^2}+D_{NN}(r)\delta_{jl} \tag{13.69}$$

(cf. the derivation of equation (12.29) in Sect. 12.2). In this expression $D_{LL}(r)$ and $D_{NN}(r)$ are functions of one variable, and are called the *longitudinal and lateral structure functions* of the field $\boldsymbol{u}(\boldsymbol{x})$. They are defined by

$$D_{LL}(r)=\overline{[u_L(\boldsymbol{x}+\boldsymbol{r})-u_L(\boldsymbol{x})]^2}, \qquad D_{NN}(r)=\overline{[u_N(\boldsymbol{x}+\boldsymbol{r})-u_N(\boldsymbol{x})]^2}, \tag{13.70}$$

where $u_L$ and $u_N$ have the same significance as in Eq. (12.27). If the field $\boldsymbol{u}(\boldsymbol{x})$ is not only locally isotropic but simply isotropic, it is clear that

$$D_{LL}(r)=2[B(0)-B_{LL}(r)], \qquad D_{NN}(r)=2[B(0)-B_{NN}(r)], \tag{13.71}$$

where

$$B(0)=B_{LL}(0)=B_{NN}(0)=\overline{u^2}/3. \tag{13.72}$$

If, in addition to the field $\boldsymbol{u}(\boldsymbol{x})$ we consider the random scalar field $\vartheta(\boldsymbol{x})$, which is such that the four-dimensional field $\{\boldsymbol{u}(\boldsymbol{x}), \vartheta(\boldsymbol{x})\}$ is also locally isotropic, then the joint second moment

$$D_{j\vartheta}(\boldsymbol{x}, \boldsymbol{r}_1, \boldsymbol{r}_2)= \\ =\overline{[u_j(\boldsymbol{y}+\boldsymbol{r}_1)-u_j(\boldsymbol{y})][\vartheta(\boldsymbol{x}+\boldsymbol{y}+\boldsymbol{r}_1)-\vartheta(\boldsymbol{x}+\boldsymbol{y})]} \tag{13.73}$$

will be an isotropic vector depending on $\boldsymbol{x}$, $\boldsymbol{r}_1$, $\boldsymbol{r}_2$, which cannot be expressed in terms of the simpler moments

$$D_{j\vartheta}(\boldsymbol{r})=\overline{[u_j(\boldsymbol{x}+\boldsymbol{r})-u_j(\boldsymbol{x})][\vartheta(\boldsymbol{x}+\boldsymbol{r})-\vartheta(\boldsymbol{x})]}. \tag{13.74}$$

In fact, we shall see later that $D_{j\vartheta}(\boldsymbol{x}, \boldsymbol{r}_1, \boldsymbol{r}_2)$ is not identical with $D_{\vartheta j}(\boldsymbol{x}, \boldsymbol{r}_1, \boldsymbol{r}_2)$, so that the identity (13.33) cannot be used in this context. It is readily shown that the moments (13.74) are all identically zero. To establish this we need only note that

$$D_{j\vartheta}(\boldsymbol{r})=D_{L\vartheta}(r)\frac{r_j}{r}, \quad D_{L\vartheta}(r)=\overline{[u_L(M')-u_L(M)][\vartheta(M')-\vartheta(M)]}, \tag{13.75}$$

where $M'$ and $M$ are the points $\boldsymbol{x}+\boldsymbol{r}$ and $\boldsymbol{x}$, and $u_L$ is the component of $\boldsymbol{u}$ along $\overrightarrow{MM'}$. But $u_L = -u_{-L}$, where $u_{-L}$ is the component of $\boldsymbol{u}$ along the opposite direction $\overrightarrow{M'M}$. Therefore,

$$\overline{[u_L(M') - u_L(M)][\vartheta(M') - \vartheta(M)]} =$$
$$= -\overline{[u_{-L}(M') - u_{-L}(M)][\vartheta(M') - \vartheta(M)]} =$$
$$= -\overline{[u_{-L}(M) - u_{-L}(M')][\vartheta(M) - \vartheta(M')]}.$$

The right and left sides of the last equation transformed into each other when the points $M$ and $M'$ and the vectors $\boldsymbol{u}(M)$ and $\boldsymbol{u}(M')$ are rotated through 180° about any axis perpendicular to $MM'$ and drawn through its midpoint. It follows that both sides of the last equation are equal and consequently

$$D_{j\vartheta}(r) = 0. \tag{13.76}$$

Similarly it can be shown that

$$D_{LL\vartheta}(r) = \overline{[u_L(x+r) - u_L(x)]^2 [\vartheta(x+r) - \vartheta(x)]} = 0$$

and

$$D_{NN\vartheta}(r) = \overline{[u_N(x+r) - u_N(x)]^2 [\vartheta(x+r) - \vartheta(x)]} = 0;$$

hence

$$D_{jl\vartheta}(r) = \overline{[u_j(x+r) - u_j(x)][u_l(x+r) - u_l(x)][\vartheta(x+r) - \vartheta(x)]} = 0 \tag{13.76'}$$

This conclusion will not, of course, be valid for the third moments

$$D_{j\vartheta\vartheta}(r) = \overline{[u_j(M') - u_j(M)][\vartheta(M') - \vartheta(M)]^2} \tag{13.77}$$

In this case,

$$D_{j\vartheta\vartheta}(r) = D_{L\vartheta\vartheta}(r)\frac{r_j}{r}, \quad D_{L\vartheta\vartheta}(r) = \overline{[u_L(M') - u_L(M)][\vartheta(M') - \vartheta(M)]^2}, \tag{13.78}$$

where the function $D_{L\vartheta\vartheta}(r)$ is not zero. In the special case of isotropic fields we have

$$D_{L\vartheta\vartheta}(r) = 2\,[2B_{L\vartheta,\vartheta}(r) - B_{L,\vartheta\vartheta}(r)], \quad B_{L\vartheta,\vartheta}(r) = \overline{u_L(M)\,\vartheta(M)\,\vartheta(M')},$$
$$B_{L,\vartheta\vartheta}(r) = \overline{u_L(M)\,\vartheta^2(M')} \tag{13.79}$$

The third- and fourth-order two-point moments of the increments of a locally isotropic vector field $\boldsymbol{u}(\boldsymbol{x})$ can be written in the form

$$D_{ijl}(\boldsymbol{r}) = \overline{[u_i(M') - u_i(M)]\,[u_j(M') - u_j(M)]\,[u_l(M') - u_l(M)]} =$$
$$= [D_{LLL}(r) - 3D_{LNN}(r)]\,\frac{r_i r_j r_l}{r^3} +$$
$$+ D_{LNN}(r)\left[\frac{r_i}{r}\,\delta_{jl} + \frac{r_j}{r}\,\delta_{il} + \frac{r_l}{r}\,\delta_{ij}\right], \tag{13.80}$$

$$D_{ijkl}(\boldsymbol{r}) =$$
$$= \overline{[u_i(M') - u_i(M)]\,[u_j(M') - u_j(M)]\,[u_k(M') - u_k(M)]\,[u_l(M') - u_l(M)]} =$$
$$= [D_{LLLL}(r) - 6D_{LLNN}(r) + D_{NNNN}(r)]\,\frac{r_i r_j r_k r_l}{r^4} +$$
$$+ \left[D_{LLNN}(r) - \frac{1}{3}D_{NNNN}(r)\right]\left[\frac{r_i r_j}{r^2}\,\delta_{kl} + \frac{r_i r_k}{r^2}\,\delta_{jl} + \frac{r_i r_l}{r^2}\,\delta_{jk} + \frac{r_j r_k}{r^2}\,\delta_{il} +\right.$$
$$\left. + \frac{r_j r_l}{r^2}\,\delta_{ik} + \frac{r_k r_l}{r^2}\,\delta_{ij}\right] + \frac{1}{3}D_{NNNN}(r)\,[\delta_{ij}\delta_{kl} + \delta_{ik}\delta_{jl} + \delta_{il}\delta_{jk}], \tag{13.81}$$

where

$$D_{LLL}(r) = \overline{[u_L(M') - u_L(M)]^3},$$
$$D_{LNN}(r) = \overline{[u_L(M') - u_L(M)]\,[u_N(M') - u_N(M)]^2} \tag{13.82}$$

and

$$D_{LLLL}(r) = \overline{[u_L(M') - u_L(M)]^4},$$
$$D_{LLNN}(r) = \overline{[u_L(M') - u_L(M)]^2\,[u_N(M') - u_N(M)]^2},$$
$$D_{NNNN}(r) = \overline{[u_N(M') - u_N(M)]^4} =$$
$$= \frac{1}{3}\overline{[u_N(M') - u_N(M)]^2\,[u_M(M') - u_M(M)]^2} \tag{13.83}$$

These formulas are somewhat simpler than Eqs. (12.122), (12.132), and (12.135) because the moments $D_{ijl}(\boldsymbol{r})$ and $D_{ijkl}(\boldsymbol{r})$ are symmetric in all the indices. In the special case of an isotropic field $\boldsymbol{u}(\boldsymbol{x})$, the functions (13.82) and (13.83) can readily be expressed in terms of the moment functions (12.121), (12.130), and (12.133):

$$D_{LLL}(r)=6B_{LL,L}(r),\quad D_{LNN}(r)=2B_{NN,L}(r)+4B_{LN,N}(r),$$
$$D_{LLLL}(r)=2B^{(4)}(0)-8B_{LLL,L}(r)+6B_{LL,LL}(r),$$
$$B^{(4)}(0)=B_{LLLL}(0)=\overline{u_1^4}=\frac{1}{5}\overline{(u^2)^2},$$
$$D_{LLNN}(r)=\frac{2}{3}B^{(4)}(0)+2B_{LL,NN}(r)+4B_{LN,LN}(r)-$$
$$-4B_{LLN,N}(r)-4B_{LNN,L}(r),$$
$$D_{NNNN}(r)=2B^{(4)}(0)-8B_{NNN,N}(r)+6B_{NN,NN}(r). \tag{13.84}$$

If the vector field $u(x)$ has continuous partial derivatives $\partial u_j/\partial x_k$, the structure function $D_{jl}(r)$ will be at least twice differentiable with respect to the components of $r$. We shall now suppose that the field $u(x)$ is solenoidal so that $\partial u_l(x)/\partial x_l=0$. In this case,

$$\overline{[u_j(x+r)-u_j(x)]\frac{\partial u_l(x)}{\partial x_l}}=0,\quad \frac{\partial u_l(x)}{\partial x_l}=\lim_{h\to 0}\frac{u_l(x+h_l)-u_l(x)}{h},$$

where $h_l=hi_l$ and $i_l$ is a unit vector along the $l$th coordinate axis. Hence, using Eq. (13.67′), we can readily show that

$$\frac{\partial D_{jl}(r)}{\partial r_l}=\left.\frac{\partial D_{jl}(r)}{\partial r_l}\right|_{r=0}. \tag{13.85}$$

Since, however, $D_{jl}(r)=D_{jl}(-r)$, the partial derivatives of this function at $r=0$ should all be zero, i.e.,

$$\frac{\partial D_{jl}(r)}{\partial r_l}=0. \tag{13.86}$$

If we now substitute Eq. (13.69) into this expression we have

$$D_{NN}(r)=D_{LL}(r)+\frac{r}{2}\frac{\partial D_{LL}(r)}{\partial r} \tag{13.87}$$

This is in complete correspondence with Eq. (12.67) which relates the longitudinal and lateral correlation functions of a solenoidal isotropic field $u(x)$. In precisely the same way it can be shown that for a locally isotropic potential field $u(x)$

$$\frac{\partial D_{ij}(r)}{\partial r_l} - \frac{\partial D_{il}(r)}{\partial r_j} = 0, \tag{13.88}$$

i.e.,

$$D_{LL}(r) = D_{NN}(r) + r\frac{\partial D_{NN}(r)}{\partial r} \tag{13.89}$$

[compare this with Eq. (12.70)]. From the above equations we can readily conclude that the structure functions of an arbitrary locally isotropic vector field can be uniquely resolved into components corresponding to the solenoidal and potential components of $\boldsymbol{u}(\boldsymbol{x})$ (cf. the similar derivation in Sect. 12.3).

For a solenoidal isotropic field the first two formulas in Eq. (13.84) can be written in the form

$$D_{LNN}(r) = \frac{1}{6}\left[D_{LLL}(r) + r\frac{\partial D_{LLL}(r)}{\partial r}\right]. \tag{13.90}$$

where we have used Eq. (12.137). Consequently, the tensor $D_{ijl}(r)$ can be expressed in terms of the single function $D_{LLL}(r)$:

$$D_{ijl}(r) = \frac{1}{2}\left[D_{LLL}(r) - r\frac{\partial D_{LLL}(r)}{\partial r}\right]\frac{r_i r_j r_l}{r^3} + \\ + \frac{1}{6}\left[D_{LLL}(r) + r\frac{\partial D_{LLL}(r)}{\partial r}\right]\left[\frac{r_i}{r}\delta_{jl} + \frac{r_j}{r}\delta_{il} + \frac{r_l}{r}\delta_{ij}\right]. \tag{13.91}$$

If we use the spectral representation (13.45) of the field $\boldsymbol{u}(\boldsymbol{x})$ we can justify Eq. (13.90) also for any locally isotropic solenoidal field (using the fact that $dZ_L(\boldsymbol{k}) = 0$ for such a field). However, this general proof is considerably more complicated.

The spectral representation (13.45) of an arbitrary locally isotropic field $\boldsymbol{u}(\boldsymbol{x})$, together with formula (13.46) enable us to write the general structure function in the form

$$D_{jl}(\boldsymbol{x}, \boldsymbol{r}_1, \boldsymbol{r}_2) = \int e^{i\boldsymbol{k}\boldsymbol{x}}(e^{-i\boldsymbol{k}\boldsymbol{r}_1} - 1)(e^{i\boldsymbol{k}\boldsymbol{r}_2} - 1)F_{jl}(\boldsymbol{k})\,d\boldsymbol{k}. \tag{13.92}$$

If the field $\boldsymbol{u}(\boldsymbol{x})$ is not only locally homogeneous but also locally isotropic, then $F_{jl}(\boldsymbol{k})$ will be an isotropic tensor function of the vector $\boldsymbol{k}$ and, consequently, it will be expressible in terms of the two scalar functions $F_{LL}(k)$ and $F_{NN}(k)$ in accordance with Eq. (12.31). The conditions that $\|F_{jl}(\boldsymbol{k})\|$ is Hermitian and nonnegative demand that the functions $F_{LL}(k)$ and $F_{NN}(k)$ be real and nonnegative. The same result ensues from Eq. (12.32) which is also valid in the present case. We thus see that the functions $F_{jl}(\boldsymbol{k})$ are real and, therefore, $D_{jl}(\boldsymbol{x}, \boldsymbol{r}_1, \boldsymbol{r}_2) = D_{lj}(\boldsymbol{x}, \boldsymbol{r}_1, \boldsymbol{r}_2)$ so that we can safely use Eq. (13.67′) which was employed earlier without proof. Finally, Eq. (13.47) shows that for any $k_0$

$$\int_0^{k_0} k^4 [F_{LL}(k) + F_{NN}(k)]\, dk < \infty; \quad \int_{k_0}^{\infty} k^2 [F_{LL}(k) + F_{NN}(k)]\, dk < \infty. \tag{13.93}$$

If we substitute Eq. (12.31) into Eq. (13.49), tranform to spherical coordinates, and integrate with respect to the angular variables, we can readily show that

$$\begin{aligned} D_{LL}(r) = 8\pi \int_0^{\infty} \left\{ \frac{1}{3} - \frac{\sin kr}{kr} - 2\frac{\cos kr}{(kr)^2} + 2\frac{\sin kr}{(kr)^3} \right\} F_{LL}(k)\, k^2\, dk + \\ + 8\pi \int_0^{\infty} \left\{ \frac{2}{3} + 2\frac{\cos kr}{(kr)^2} - 2\frac{\sin kr}{(kr)^3} \right\} F_{NN}(k)\, k^2\, dk, \end{aligned} \tag{13.94}$$

$$\begin{aligned} D_{NN}(r) = 8\pi \int_0^{\infty} \left\{ \frac{1}{3} + \frac{\cos kr}{(kr)^2} - \frac{\sin kr}{(kr)^3} \right\} F_{LL}(k)\, k^2\, dk + \\ + 8\pi \int_0^{\infty} \left\{ \frac{2}{3} - \frac{\sin kr}{kr} - \frac{\cos kr}{(kr)^2} + \frac{\sin kr}{(kr)^3} \right\} F_{NN}(k)\, k^2\, dk \end{aligned}$$

In the special case when $4\pi \int_0^{\infty} [F_{LL} + 2F_{NN}]\, k^2\, dk = 3B(0) < \infty$, these formulas turn out to be equivalent to Eq. (12.34), in accordance with Eq. (13.71). Conversely, to insure that the given functions $D_{LL}(r)$ and $D_{NN}(r)$ are the longitudinal and lateral structure functions of a locally isotropic vector field they must be expressible in the form of Eq. (13.94) with $F_{LL}(k) \geqslant 0$ and $F_{NN}(k) \geqslant 0$. The condition given by Eq. (13.93) will then always be satisfied since it is necessary for the convergence of the integrals in Eq. (13.94).

It follows, of course, from Eq. (13.94) that $D_{LL}(0)=D_{NN}(0)=0$. As regards the derivatives of the functions $D_{LL}(r)$ and $D_{NN}(r)$ at $r=0$, the expressions which result for them differ from Eq. (12.43) only by a factor of $-2$.

We note further that the functions $D_{LL}(r)$ and $D_{NN}(r)$ given by Eq. (13.94) are also the structure functions of the locally homogeneous field $u_1(x_1, 0, 0)$ or, correspondingly, $u_2(x_1, 0, 0)$ on the $x_1$ axis (compare this with the analogous discussion for isotropic fields given in Sect. 12.2). The one-dimensional spectra $F_1(k_1)$ and $F_2(k_1)$ corresponding to these "fields on a line" can be expressed in terms of $F_{LL}(k)$ and $F_{NN}(k)$ through the same formulas (12.39) as in the case of isotropic fields.

In the special case when

$$D_{LL}(r)=A_1 r^{\gamma}, \quad D_{NN}(r)=A_2 r^{\gamma}, \qquad 0<\gamma<2, \tag{13.95}$$

the formulas (13.94) [or the formulas of the form (13.19), which relate $D_{LL}(r)$ and $D_{NN}(r)$ with $F_1(k)$ and $F_2(k)$, and the formulas (12.39)] show that the spectra $F_{LL}(k)$ and $F_{NN}(k)$ will be pure power functions:

$$F_{LL}(k)=C_1 k^{-3-\gamma}, \qquad F_{NN}(k)=C_2 k^{-3-\gamma}, \tag{13.96}$$

where it can be shown that

$$\begin{aligned} C_1 &= \frac{\Gamma(2+\gamma)\sin \pi\gamma/2}{4\pi^2\gamma}\left[(\gamma+2)A_1-2A_2\right], \\ C_2 &= \frac{\Gamma(2+\gamma)\sin \pi\gamma/2}{4\pi^2\gamma}\left[(\gamma+1)A_2-A_1\right]. \end{aligned} \tag{13.97}$$

Hence, it is clear that the functions given by Eqs. (13.95) will be the longitudinal and lateral structure functions of a locally isotropic vector field if the constants $A_1>0$ and $A_2>0$ are such that

$$A_2 \leqslant (1+\gamma/2)A_1, \qquad A_1 \leqslant (1+\gamma)A_2. \tag{13.98}$$

It is also readily seen that when $\gamma>2$ the functions (13.95) will not be the structure functions of the field $\boldsymbol{u}(\boldsymbol{x})$, while the case $\gamma=2$ corresponds to a linear field of the form $u_j(\boldsymbol{x})=v_{jl}x_l+\text{const}$, which does not have a spectral density.

The divergence of a locally isotropic vector field will be a scalar isotropic field, the spectral density of which is equal to $k^2F_{LL}(k)$. It follows that a locally isotropic vector field $\boldsymbol{u}(\boldsymbol{x})$ will be solenoidal if and only if $F_{LL}(k)=0$. Similarly, to ensure that the field $\boldsymbol{u}(\boldsymbol{x})$ is potential, we must have $F_{NN}(k)=0$. In the case of a solenoidal field, Eq. (13.94) assumes the form

$$D_{LL}(k)=4\int_0^\infty\left\{\frac{1}{3}+\frac{\cos kr}{(kr)^2}-\frac{\sin kr}{(kr)^3}\right\}E(k)\,dk,$$

$$D_{NN}(k)=2\int_0^\infty\left\{\frac{2}{3}-\frac{\sin kr}{kr}-\frac{\cos kr}{(kr)^2}+\frac{\sin kr}{(kr)^3}\right\}E(k)\,dk, \tag{13.99}$$

where $E(k)=4\pi k^2F_{NN}(k)$ so that the tensor $F_{jl}(k)$ is given by Eq. (12.73). In view of Eq. (13.97), the functions given by Eq. (13.95) will be the longitudinal and lateral structure functions of a solenoidal locally isotropic vector field $\boldsymbol{u}(\boldsymbol{x})$ if and only if, $A_2=(1+\gamma/2)A_1$, which is also in accordance with Eq. (13.87). Moreover, in this case,

$$E(k)=Ck^{-1-\gamma},\quad C=\frac{(3+\gamma)\,\Gamma(2+\gamma)\sin\pi\gamma/2}{2\pi}A_1. \tag{13.100}$$

Similarly, the functions (13.95) will be the structure functions of the potential field $\boldsymbol{u}(\boldsymbol{x})$ if and only if $A_1=(1+\gamma)A_2$, in accordance with Eq. (13.89).

It is readily shown using Eqs. (13.99) and (13.94) that, in the case of a locally isotropic solenoidal field,

$$\frac{3}{2}D_{LL}(r)+\frac{r}{2}D'_{LL}(r)=2\int_0^\infty\left(1-\frac{\sin kr}{kr}\right)E(k)\,dk, \tag{13.101}$$

while in the case of a potential field

$$3D_{NN}(r)+rD'_{NN}(r)=4\int_0^\infty\left(1-\frac{\sin kr}{kr}\right)E(k)\,dk, \tag{13.102}$$

where $E(k)$ is the same as in Eq. (12.72). It is clear that to ensure that the functions $D_{LL}(r)$ and $D_{NN}(r)$ are the longitudinal and transverse structure functions of a locally isotropic solenoidal vector field, it is only necessary that the function $3D_{LL}(r)+rD'_{LL}(r)$ be the structure function of a locally isotropic scalar field, while $D_{NN}(r)$ can be determined from $D_{LL}(r)$ by Eq. (13.87). In the same way, to ensure that $D_{LL}(r)$ and $D_{NN}(r)$ are the structure functions of a potential field, the function $3D_{NN}(r)+rD'_{NN}(r)$ should be the structure function of a scalar field, while $D_{LL}(r)$ should be determined through $D_{NN}(r)$ by Eq. (13.89). Therefore, the classes of functions which can be the longitudinal structure functions of a locally isotropic solenoidal field and the lateral structure functions of a potential field are identical with each other.

For solenoidal locally isotropic fields, the formulas relating the spectrum $E(k)=4\pi k^2 F_{NN}(k)$ and the longitudinal and lateral one-dimensional spectra $E_1(k_1)$ and $E_2(k_1)$ will have the form (12.86)–(12.88), as in the case of solenoidal isotropic fields. Similarly, for potential locally isotropic fields we have the relations (12.89) and (12.90).

For locally isotropic vector fields $\boldsymbol{u}(\boldsymbol{x})$ we have the spectral representation (13.45). In the case of a pair of locally isotropic fields which are related in a locally isotropic fashion, and one of which is the scalar field $\vartheta(\boldsymbol{x})$ and the other the vector field $\boldsymbol{u}(\boldsymbol{x})$, the general joint structure function $D_{j\vartheta}(\boldsymbol{x}, \boldsymbol{r}_1, \boldsymbol{r}_2)$ defined by Eq. (13.73) will have a representation

$$D_{j\vartheta}(\boldsymbol{x}, \boldsymbol{r}_1, \boldsymbol{r}_2)=\int e^{i\boldsymbol{k}\boldsymbol{x}}(e^{-i\boldsymbol{k}\boldsymbol{r}_1}-1)(e^{i\boldsymbol{k}\boldsymbol{r}_2}-1)F_{j\vartheta}(\boldsymbol{k})\,d\boldsymbol{k}, \quad (13.103)$$

where

$$F_{j\vartheta}(\boldsymbol{k})=iF_{L\vartheta}(k)\frac{k_j}{k}, \quad (13.104)$$

and $F_{L\vartheta}(k)$ is a real function. Since the function $F_{j\vartheta}(\boldsymbol{k})$ is purely imaginary (and not real), it follows that $D_{j\vartheta}(\boldsymbol{x},\boldsymbol{r}_1,\boldsymbol{r}_2)\neq D_{\vartheta j}(\boldsymbol{x},\boldsymbol{r}_1,\boldsymbol{r}_2)$; therefore, the function $D_{j\vartheta}(\boldsymbol{x}, \boldsymbol{r}_1, \boldsymbol{r}_2)$ cannot be expressed in terms of the simpler function $D_{j\vartheta}(\boldsymbol{r})$ of Eq. (13.74). Moreover, it follows from the equation $F_{j\vartheta}(-\boldsymbol{k})=-F_{j\vartheta}(\boldsymbol{k})$ that $D_{j\vartheta}(\boldsymbol{r})=D_{j\vartheta}(0, \boldsymbol{r}, \boldsymbol{r})=0$. This result has already been noted above. If, on the other hand, the locally isotropic field $\boldsymbol{u}(\boldsymbol{x})$ is solenoidal, then as for the isotropic case $F_{j\vartheta}(\boldsymbol{k})=0$. Therefore, subject to this condition, $D_{j\vartheta}(\boldsymbol{x},\boldsymbol{r}_1,\boldsymbol{r}_2)=0$ for any $\boldsymbol{x}$, $\boldsymbol{r}_1$, and $\boldsymbol{r}_2$.

# 7 ISOTROPIC TURBULENCE

## 14. EQUATIONS FOR THE CORRELATION AND SPECTRAL FUNCTIONS OF ISOTROPIC TURBULENCE

### 14.1 Definition of Isotropic Turbulence and the Possibilities of its Experimental Realization

Turbulence is said to be *homogeneous* if all the fluid dynamic fields form homogeneous random fields. It is referred to as *isotropic* if all the fluid dynamic fields form homogeneous and isotropic random fields. In this chapter we shall be concerned with isotropic turbulence.

No real turbulent motion can, of course, be exactly isotropic. This is clear from the fact that the only turbulence which can be completely isotropic is that occurring in a fluid occupying an infinite space whereas any real flow must have a boundary. The concept of isotropic turbulence is, therefore, a mathematical idealization which, at best, is convenient only for the approximate description of certain special types of turbulent flow. It is also clear that approximate isotropy can be expected only under certain very special circumstances. We shall see below that the isotropy conditions are apparently satisfactorily fulfilled for a certain class of turbulent flows produced in laboratory wind tunnels. However, such flows are not of great importance in practice.

For a theoretician, however, the case of homogeneous and isotropic turbulence is very attractive. From the mathematical point of view this case is undoubtedly the simplest. It is, therefore, natural to begin with this case and to try to use it to exhibit some of the characteristic properties of turbulent motion in general. Without such a preliminary study there will hardly be any hope of obtaining any concrete theoretical results for the more general cases. In fact, the introduction of the concept of isotropic turbulence by Taylor (1935) has played a major role in the development of the modern statistical theory of turbulence. Kolmogorov (1941 a, d) subsequently used this idea as the basis for the more general concept of locally isotropic turbulence which corresponds more closely to turbulent flows encountered in practice; he showed that many of the ideas and methods of the theory of isotropic turbulence can be successfully transplanted into this more general theory. This last fact is the main reason why we are at present interested in the theory of isotropic turbulence.

Let us first consider how isotropic turbulence could be generated. Theoretically, the simplest method is to produce a homogeneous and isotropic system of randomly distributed local disturbances (eddies) in an initially quiet body of fluid. It is a simple matter to derive mathematical formulas for the initial velocity field corresponding to such a "random system of disturbances." However, this is not enough to enable us to study the dynamics of turbulence. We also need the solutions of the equations of motion corresponding to the above initial conditions. The derivation of these solutions is not at all a simple matter, and it is therefore not surprising that, so far, only certain approximate results have been obtained in this field. Moreover, it has been necessary to simplify the equations of motion so much that the resulting solutions provide only a very idealized description of a real isotropic turbulent flow (Synge and Lin, 1943; Chow Pei-Yuan and Tsai Shu-Tang, 1957).

The experimental excitation of an isotropic set of random disturbances in a fluid initially at rest requires the introduction of some small disturbing devices into the fluid, which could first be simultaneously introduced into it and then instantaneously removed.

We shall not pretend that this type of experiment can, in fact, be performed in a laboratory. We must therefore proceed in a different way. One of the simplest possible methods of introducing into a fluid an almost isotropic set of disturbances consist of the following. Let us imagine that we have introduced into the fluid a moving grid consisting of very thin rods, such that it produces disturbances only in those fluid elements through which it passes. If the grid moves through the fluid very rapidly (in comparison with the characteristic velocity of the disturbances produced by it) then we may assume that all the disturbances are produced simultaneously. The resulting set of eddies will be homogeneous but, in general, it will not be isotropic because the directions of the rods in the grid and the direction of its motion will be special directions. Since, however, this method will generate a set of small-scale disturbances at a very large number of points lying close to each other, one would hope that, after a relatively short time, all these disturbances will mix together and form a homogeneous and isotropic system. In other words, after a short time following the passage of the grid, we should obtain isotropic turbulence in the fluid.

The above method is the basis of an important way of generating turbulence which is nearly isotropic. It is, however, convenient to invert the experimental conditions, i.e., to keep the grid fixed and move the fluid relative to it. The motion of a large mass of a "fluid" (ordinary air) at constant velocity can be produced in a wind tunnel. If we place the grid at the beginning of the working region of the tunnel, then in the set of coordinates moving with the fluid, the turbulence generated in the working section will be the same as that obtained with a stationary fluid and a moving grid.

It is implicitly assumed in the above account that the fluid occupies all space. This is not so in the wind tunnel in which the working section has a finite volume. However, in the case of a wind tunnel with a sufficiently large working section (in comparison with the dimensions of the turbulent disturbances which we shall investigate), we can expect that the effects of the walls will be relatively small in the central region, and the motion of the eddies will not be very different from that in an infinite space. To ensure that the motion is similar in character to isotropic turbulence, we must also eliminate from our analysis that part of the flow which occurs in the immediate neighborhood of the grid, i.e., we must neglect the time interval immediately after the passage of the fluid element through the grid. We must also demand that the mean flow velocity $U$ in the wind tunnel be considerably greater than the fluctuating velocity $u'$, and confine our attention to such differences $x''-x'$ for which the ratio $(x''-x')/U$ is small compared with the characteristic decay time of the turbulence (so that the turbulence at distances $x''$ and $x'$ from the grid is approximately the same). In this expression $x$ represents the distance along the axis of the wind tunnel. By considering different "layers" of the flow with different values of $x$, we shall simultaneously obtain a picture of the time evolution of the turbulence. The time $t$ can then be interpreted as the ratio $x/U$.

The idea of using a grid in a wind tunnel for the experimental verification of theoretical results on isotropic turbulence was first put forward by Taylor (1935) who succeeded in obtaining some very important results. Subsequent very extensive experiments have confirmed that the turbulence behind the grid is very nearly homogeneous and isotropic at distances exceeding the grid bar spacing by a factor of 30–40. This has been a considerable attraction to theoreticians since it provides a means of verifying theoretical results on isotropic turbulence. We shall quote many examples of this below. It is important to emphasize, however, that the turbulence produced behind the grid in the wind tunnel is, in fact, only approximately isotropic; moreover, the question of the degree of its isotropy is still not quite clear. In fact, the measurements of Grant (1958), Grant and Nisbet (1957), Uberoi (1963), Uberoi and Wallis (1966, 1967, 1969), Comte-Bellot and Corrsin (1966), Van Atta and Chen (1968, 1969a) and of some other authors indicate that there are departures from the isotropic state in grid turbulence even at rather great distances from the grid; however, the degree of anisotropy is not the same according to different works. Later Portfors and Keffer (1969) came to the conclusion that all these results are influenced by errors in the usual method of inclined hot-wire calibration; they used the corrected calibration method of Champagne and Sleicher (1967) (cf. also Hill and Sleicher, 1969) and found that grid turbulence is practically isotropic at distances from the grid exceeding 30 mesh lengths. Even if this last result is quite correct, there are nevertheless certain theoretical conclusions referring to isotropic turbulence which cannot in principle be verified by wind tunnel measurements (for example, the asymptotic behavior of the correlation functions for $r \to \infty$). In special cases, a more exact description of real turbulent flow behind the grid in a wind tunnel can be achieved by investigating more general theoretical models involving, for example, axisymmetric (i.e., homogeneous and two-dimensionally isotropic), or general homogeneous but nonisotropic, or stratified inhomogeneous turbulence [see, for example, I. M. Yaglom (1947), Batchelor and Stewart (1950), Tan and Lin (1963), Lee and Tan (1967)]. We shall not, however, be concerned with these generalizations because it is not our aim to investigate in great detail turbulence behind grids. Such experimental data will be quoted only to illustrate the predictions of the theory of isotropic turbulence.

### 14.2 Equations for the Velocity Correlations

Let us now derive the basic dynamic equations for the correlation functions of isotropic turbulence. With the exception of Sect. 20 of this chapter, we shall assume throughout that we are concerned with an incompressible fluid whose motion can be described by the Navier-Stokes equations (1.6) without the external forces $X_i$, and the continuity equation (1.5). We shall confine our attention to special correlation functions referring to a given time $t$, and begin by considering functions containing only the velocity field $\boldsymbol{u}(\boldsymbol{x}, t) = \{u_1(\boldsymbol{x}, t), u_2(\boldsymbol{x}, t), u_3(\boldsymbol{x}, t)\}$.

According to the discussion given in Sect. 12.2, the mean value $\overline{\boldsymbol{u}(\boldsymbol{x}, t)}$ should be identically zero in isotropic turbulence [hence the velocity $\boldsymbol{u}(\boldsymbol{x}, t)$ is the same as the velocity fluctuation], and the correlation tensor $B_{ij}(r, t) = \overline{u_i(\boldsymbol{x}, t)\, u_j(\boldsymbol{x}+\boldsymbol{r}, t)}$ should be of the form

$$B_{ij}(r, t) = [B_{LL}(r, t) - B_{NN}(r, t)] \frac{r_i r_j}{r^2} + B_{NN}(r, t)\delta_{ij}, \quad (14.1)$$

where

$$B_{LL}(r, t) = \overline{u_L(\boldsymbol{x}, t)\, u_L(\boldsymbol{x}+\boldsymbol{r}, t)}, \quad B_{NN}(r, t) = \overline{u_N(\boldsymbol{x}, t)\, u_N(\boldsymbol{x}+\boldsymbol{r}, t)}$$

(see Fig. 7). Similarly, the tensor

$$B_{ij,k}(r, t) = \overline{u_i(\boldsymbol{x}, t)\, u_j(\boldsymbol{x}, t)\, u_k(\boldsymbol{x}+\boldsymbol{r}, t)}$$

can be expressed with the aid of Eq. (12.122) in terms of the three scalar functions

$$B_{ij,k}(r, t) = [B_{LL,L}(r, t) - 2B_{LN,N}(r, t) - B_{NN,L}(r, t)]\frac{r_i r_j r_k}{r^3} + \\ + B_{NN,L}(r, t)\frac{r_k}{r}\delta_{ij} + B_{LN,N}(r, t)\left[\frac{r_i}{r}\delta_{jk} + \frac{r_j}{r}\delta_{ik}\right], \quad (14.2)$$

whose significance is clear from the notation (see Fig. 8). In view of the continuity equation, the functions $B_{LL}(r, t)$ and $B_{NN}(r, t)$ are related by Eq. (12.67):

$$B_{NN}(r, t) = B_{LL}(r, t) + \frac{r}{2}\frac{\partial B_{LL}(r, t)}{\partial r}, \quad (14.3)$$

whereas functions $B_{LL,L}(r, t)$, $B_{LN,N}(r, t)$ and $B_{NN,L}(r, t)$ are related by Eq. (12.137):

$$B_{NN,L}(r, t) = -\frac{1}{2}B_{LL,L}(r, t), \\ B_{LN,N}(r, t) = \frac{1}{2}B_{LL,L}(r, t) + \frac{r}{4}\frac{\partial B_{LL,L}(r, t)}{\partial r}. \quad (14.4)$$

The last two formulas show that each of the tensors $B_{ij}(r, t)$ and $B_{ij,k}(r, t)$ is completely determined by a single scalar function of the two arguments $r$ and $t$. Moreover, as indicated in Sect. 12.3, it also follows from the continuity equation that in isotropic turbulence the velocity field is uncorrelated with the pressure field (or any other scalar flow variable):

$$B_{pi}(r, t) = B_{pL}(r, t)\frac{r_i}{r} = 0. \quad (14.5)$$

The relation (14.3) is the simplest relation of the theory of isotropic turbulence which can be verified experimentally in a wind tunnel flow behind a grid. All that is required is to determine independently the values of the two functions $B_{LL}(r, t)$ and $B_{NN}(r, t)$. Since such measurements are not at all difficult,[1] and can

[1] We note, however, that it is simpler to determine the function $B_{NN}(r)$ since the usual hot-wire anemometer (with the wire normal to the mean flow) reacts to longitudinal (along the mean flow) velocity fluctuations, and when two such anemometers are placed one after another along the flow, the first of them may produce additional disturbances which will affect the readings of the second (this is the so-called "aerodynamic shadow"). It is therefore better to use Eq. (14.3) (after its confirmation) to calculate $B_{LL}$ from the measured values of $B_{NN}$.

be carried out with the aid of a pair of hot-wire anemometers, it is not surprising that the experimental verification of Eq. (4.34) was carried out as far back as 1937 by Taylor. It has since been frequently repeated by other workers. Measurements of this kind have established that Eq. (14.3) is in good agreement with experiment (see, for example, Fig. 10, taken from the paper by MacPhail, 1940). This was the very first considerable success of the new theory and, at the same time, the first satisfactory confirmation of the hypothesis of isotropy as applied to grid generated turbulence. The general form of the correlation functions $B_{LL}$ and $B_{NN}$ shown in Fig. 10 (and confirmed by all other measurements on grid turbulence) is characteristic of the longitudinal and lateral correlation functions of a solenoidal vector field; in particular, $B_{NN}(r)$ has a distinct negative part according to all the data (cf. Sect. 12.3). The situation with regard to the function $B_{LL}(r)$ is less clear in this respect: The majority of the measurements in grid turbulence lead to strictly positive values of this function, but some of the most recent works indicate that $B_{LL}$ apparently has a very small negative minimum (see for example, Van Atta and Chen, 1968 and 1969b).

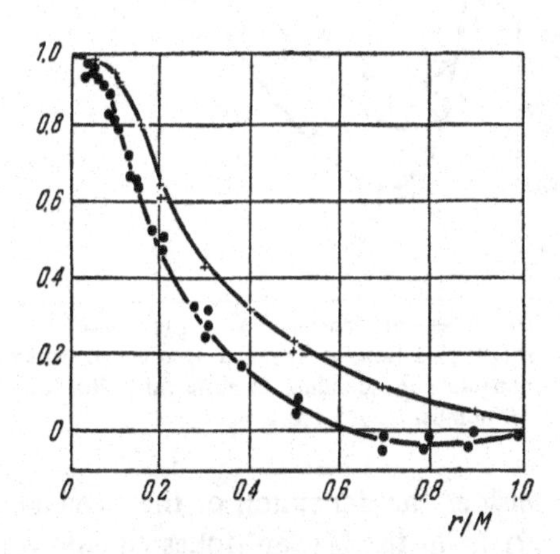


**FIG. 10** Dependence of measured correlation functions $B_{LL}(r)/B_{LL}(0)$ (crosses) and $B_{NN}(r)/B_{NN}(0)$ (points) on $r/M$, where $M$ is the grid mesh size [MacPhail (1940)]. The upper curve is drawn through the experimental points; the lower curve is calculated from this curve using Eq. (14.3).

The relations (14.4) can, in principle, also be used in an analogous experimental verification, but here the measurements are more difficult because third-order correlation functions are relatively difficult to determine, and the resulting data are very much less accurate than in the case of $B_{LL}$ and $B_{NN}$. This is why papers reporting data on the functions (14.4) [see, for example, Townsend (1947), Stewart (1951), Mills, Kistler, O'Brien and Corrsin (1958)] have not discussed the verification of Eq. (14.4) and give usually only one of these functions (in most cases the function $B_{LL,L}(r, t)$ whose typical shape is shown in Fig. 11a taken from Stewart's paper, 1951). The only exception is the work by Van Atta and Chen (1969a, b): These authors have measured both the functions $B_{LL,L}(r)$ and $B_{LN,N}(r)$ [assuming that Taylor's hypothesis (11.22) is valid]. Their data on the general shape of the function $B_{LN,N}(r)$ are shown in Fig. 11b.

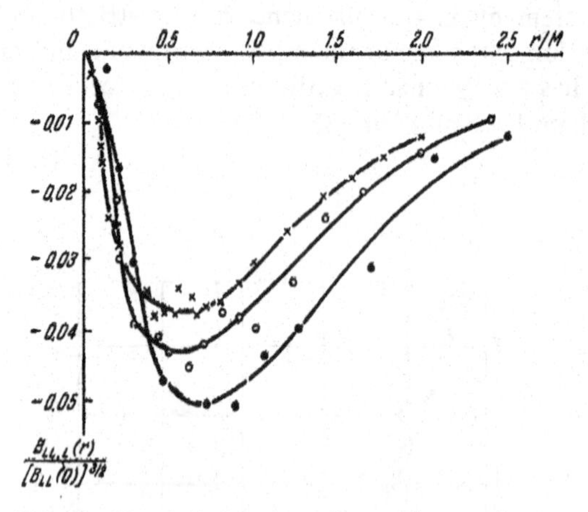


FIG. 11a Measured functions $B_{LL,L}(r)$ according to Stewart (1951). Points, crosses, and open circles represent measurements corresponding to different values of the Reynolds number $Re_M = UM/\nu$.

Let us now consider the derivation of the dynamic equations for the tensor $B_{ij}(\boldsymbol{r}, t)$ from the Navier-Stokes equations (1.6). To find $\frac{\partial}{\partial t} B_{ij}(\boldsymbol{r}, t)$ we must write down equations (1.6) for the $i$th velocity component at the point $\boldsymbol{x}$, and the $j$th velocity component at the point $\boldsymbol{x}+\boldsymbol{r}=\boldsymbol{x}'$ (the values of the flow variables at this point will henceforth be indicated by a prime), and multiply the first of them

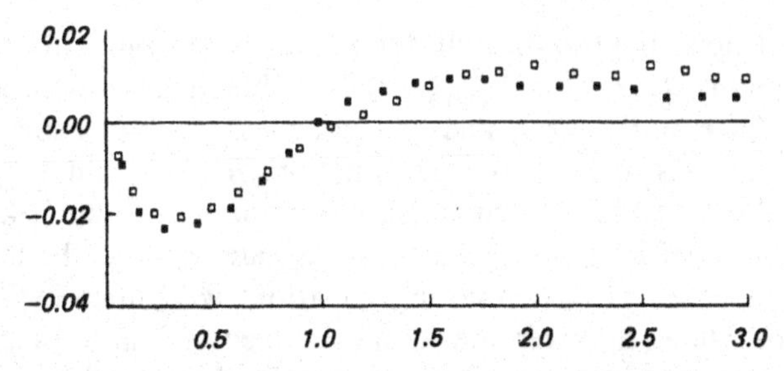


**Fig. 11b** Empirical values of the normalized function $B_{LN,N}(r)$ after Van Atta and Chen (1969a, b). The values of $B_{LN,N}(r)$ shown in the figure were obtained from data measured on the third order time correlation function under the assumption of the truth of Taylor's hypothesis. (Open and closed points correspond to two different measurements under slightly different conditions).

by $u'_j$ and the second by $u_l$. We then add both equations together and take an average [see the general formula (6.2) in Vol. 1]. Having performed all these operations, we obtain the following equation:

$$\frac{\partial\overline{u_l u'_j}}{\partial t}+\frac{\partial\overline{u_l u_k u'_j}}{\partial x_k}+\frac{\partial\overline{u_l u'_j u'_k}}{\partial x'_k}=-\frac{1}{\rho}\left(\frac{\partial\overline{p u'_j}}{\partial x_l}+\frac{\partial\overline{p' u_l}}{\partial x'_j}\right)+ +\nu\left(\frac{\partial^2\overline{u_l u'_j}}{\partial x_k\,\partial x_k}+\frac{\partial^2\overline{u_l u'_j}}{\partial x'_k\,\partial x'_k}\right) \quad (14.6)$$

where we have taken into account the fact that the unprimed variables do not depend on $x'_k$ whereas primed variables do not depend on $x_k$. Since, however, for homogeneous turbulence all the two-point moments are functions of only the vector $\boldsymbol{r}=\boldsymbol{x}'-\boldsymbol{x}$, it follows that $\frac{\partial}{\partial x_k}$ and $\frac{\partial}{\partial x'_k}$ can be replaced by $-\frac{\partial}{\partial r_k}$ and $\frac{\partial}{\partial r_k}$, respectively. This means that Eq. (14.6) now assumes the form

$$\frac{\partial B_{lj}(\boldsymbol{r},t)}{\partial t}=\frac{\partial}{\partial r_k}[B_{lk,\,j}(\boldsymbol{r},t)-B_{l,\,jk}(\boldsymbol{r},t)]+ +\frac{1}{\rho}\left[\frac{\partial B_{pj}(\boldsymbol{r},t)}{\partial r_l}-\frac{\partial B_{lp}(\boldsymbol{r},t)}{\partial r_j}\right]+2\nu\,\frac{\partial^2 B_{lj}(\boldsymbol{r},t)}{\partial r_k\,\partial r_k}. \quad (14.7)$$

which is the basic dynamic equation relating the second and third velocity moments in homogeneous turbulence.

Let us now suppose that the turbulence is isotropic. In that case the functions $B_{pl}(\mathbf{r}, t)$ and $B_{lp}(\mathbf{r}, t)$ must identically vanish and the tensors $B_{lj}(\mathbf{r}, t)$, $B_{lk,j}(\mathbf{r}, t)$ and $B_{l,jk}(\mathbf{r}, t)=B_{jk,l}(-\mathbf{r}, t)$ can be expressed in terms of the two scalar functions $B_{LL}(r, t)$ and $B_{LL,L}(r, t)$. If we substitute the corresponding expressions into Eq. (14.7) and equate the coefficients of the tensors $\delta_{lj}$ and $r_l r_j$ on the left- and right-hand sides of the resulting equation, we obtain two scalar equations which, however, are found to be equivalent to each other. It is therefore sufficient to consider only the equation which is obtained when the coefficients of the tensor $\delta_{lj}$ are equated:

$$\left(1+\frac{r}{2}\frac{\partial}{\partial r}\right)\frac{\partial B_{LL}}{\partial t}=\left(1+\frac{r}{2}\frac{\partial}{\partial r}\right)\left[\left(\frac{\partial}{\partial r}+\frac{4}{r}\right)B_{LL,L}+\right.$$
$$\left.+2\nu\left(\frac{\partial^2}{\partial r^2}+\frac{4}{r}\frac{\partial}{\partial r}\right)B_{LL}\right]. \quad (14.8)$$

The only solution of the equation $f(r)+\frac{r}{2}f'(r)=0$ which has no singularity at $r=0$ vanishes identically. Equation (14.8) is therefore equivalent to

$$\frac{\partial B_{LL}(r, t)}{\partial t}=\left(\frac{\partial}{\partial r}+\frac{4}{r}\right)\left[B_{LL,L}(r, t)+2\nu\frac{\partial B_{LL}(r, t)}{\partial r}\right]. \quad (14.9)$$

This equation was first derived by von Kármán and Howarth (1938). It plays a basic part in all subsequent studies in the theory of isotropic turbulence.

Since the von Kármán-Howarth equation (14.9) is a single relation connecting two unknown functions, it cannot be "solved," i.e., it cannot be used to determine the form of the functions $B_{LL}(r, t)$ and $B_{LL,L}(r, t)$. This is a special case of the general difficulty which we encountered in the case of the Reynolds equation in Chapt. 3 of Vol. 1; it is a consequence of the nonlinearity of the fluid dynamic equations. Nevertheless, Eq. (14.9) is undoubtedly of considerable interest since it restricts substantially the possible values of the functions $B_{LL}$ and $B_{LL,L}$, and leads to a series of consequences which can be verified experimentally. It is also clear that Eq. (14.9) can itself be verified experimentally, since all its terms can be measured directly (to determine the derivative $\frac{\partial B_{LL}}{\partial t}$ it is sufficient to have the values of the functions $B_{LL}(r)$ at two close distances $x$ from the grid). Examples of this type of verification can be found in the papers by Stewart (1951) and Mills, Kistler, O'Brien, and Corrsin (1958). This experimental work has shown that there is good agreement between theory and experiment.

Since the continuity and Navier-Stokes equations express the physical laws of conservation of mass and momentum, it is clear that all the consequences of these equations introduced in this section are also the consequences of these physical laws. Almost immediately after the publication of the very first papers on the theory of isotropic turbulence it was noted by Prandtl that, for example, the von Kármán relation (14.3) can be derived from the integral form of the mass conservation law without transformation to the differential continuity equation (see, for example, Wieghardt, 1941). Subsequently Mattioli (1951) and Hasselmann (1958) showed that an analogous derivation, involving only the integral form of the mass and momentum conservation laws, is also possible for Eqs. (14.4), (14.5), and (14.9).

## 14.3 Equations for the Velocity Spectra

In addition to Eq. (14.9) there is also an equation relating the spectral function $F(k, t)$ [or the spectral energy density $E(k, t) = 4\pi k^2 F(k, t)$] with the third-order spectral function $F_3(k, t)$ [or the functions $\Gamma(k, t) = -2kF_3(k, t)$ or $T(k, t) = 4\pi k^2 \Gamma(k, t)$] which defines the Fourier transform of the tensor $B_{ij,k}(\mathbf{r}, t)$. This equation is, in fact, merely a new form of the von Kármán-Howarth equation (14.9). However, the "spectral form" of Eq. (14.9) is occasionally more convenient. It also has a simple physical interpretation which is important for the understanding of the mechanism of turbulent mixing.

The equation relating the spectral functions $F(k, t)$ and $\Gamma(k, t)$ can be derived directly from the Kármán-Howarth equation (14.9) (see Lin, 1949). To do this, however, Eq. (14.9) must be substantially transformed, since the relation between $B_{LL}(k, t)$ and $F(k, t)$ [and $B_{LL,L}(r, t)$ and $\Gamma(k, t)$] is not at all simple. It is therefore more convenient to start not with Eq. (14.9) itself but with Eq. (14.7), which we have used to derive Eq. (14.9). If we apply the three-dimensional Fourier transformation with respect to $\mathbf{r}$ to all the terms in Eq. (14.7) we obtain the following relation (which is valid for any homogeneous turbulence):

$$\frac{\partial F_{ij}(\boldsymbol{k}, t)}{\partial t} = \Gamma_{ij}(\boldsymbol{k}, t) + \Pi_{ij}(\boldsymbol{k}, t) - 2\nu k^2 F_{ij}(\boldsymbol{k}, t), \quad (14.10)$$

where

$$\Gamma_{ij}(\boldsymbol{k}, t) = ik_l [F_{il,\, j}(\boldsymbol{k}, t) - F_{jl,\, i}(-\boldsymbol{k}, t)], \quad (14.11)$$

$$\Pi_{ij}(\boldsymbol{k}, t) = \frac{i}{\rho} [k_j F_{ip}(\boldsymbol{k}, t) - k_i F_{jp}(-\boldsymbol{k}, t)], \quad (14.12)$$

and $F_{ij}(\boldsymbol{k}, t)$ and $F_{il,\, j}(\boldsymbol{k}, t)$ are given by Eqs. (11.54) and (11.60); $F_{ip}(\boldsymbol{k}, t)$ is the Fourier transform of the function $B_{ip}(\boldsymbol{r}, t)$. Substituting $i = j$ and summing over $j$, we find that

$$\frac{\partial F_{jj}(k, t)}{\partial t} = \Gamma_{jj}(k, t) - 2\nu k^2 F_{jj}(k, t); \; \Gamma_{jj}(k, t) = -2k_l \operatorname{Im} F_{jl,\, j}(\boldsymbol{k}, t) \quad (14.13)$$

[The term $\Pi_{jj}(\boldsymbol{k}, t)$ cancels out in view of Eq. (11.91)]. In the case of isotropic turbulence $\Pi_{ij}(\boldsymbol{k}, t) = 0$ and the tensors $F_{ij}(\boldsymbol{k}, t)$ and $F_{il,\, j}(\boldsymbol{k}, t)$ are given by Eqs. (12.73) and (12.140) in terms of the scalar functions $F(k, t) = E(k, t)/4\pi k^2$ and $F_3(k, t)$. The equations (14.10) and (14.13) are therefore equivalent in the isotropic case to the equation

$$\frac{\partial F(k, t)}{\partial t} = \Gamma(k, t) - 2\nu k^2 F(k, t), \quad \Gamma(k, t) = -2kF_3(k, t), \quad (14.14)$$

or

$$\frac{\partial E(k, t)}{\partial t} = T(k, t) - 2\nu k^2 E(k, t); \; T(k, t) = -8\pi k^3 F_3(k, t), \quad (14.15)$$

This is the required "spectral form" of the von Kármán-Howarth equation.

Equations (14.13)–(14.15) describe the time variation of the wave number distribution of turbulent energy. The last term on the right-hand side of these equations describes the viscous energy dissipation. We thus see that viscosity leads to a decrease in the kinetic energy of disturbances with wave number $k$ which is proportional to the intensity of these disturbances multiplied by $2\nu k^2$. The energy of long-wave disturbances (small values of $k$) is thus found to decrease under the action of viscosity much more slowly than the energy of short-wave disturbances. This was expected because viscous friction is proportional to the velocity gradient. The first term on the right-hand side of Eqs. (14.13)–(14.15), on the other hand, describes the variation of the energy of the "spectral component" of the turbulence with wave number $k$, due to the nonlinear "inertial terms" in the fluid dynamic equations. It is important to note that the variation reduces to an energy redistribution among the individual spectral components, without any change in the energy of the turbulent motion as a whole. In fact, for any $i$ and $j$ we have

$$\int \Gamma_{ij}(\boldsymbol{k}, t)\, d\boldsymbol{k} = 0, \tag{14.16}$$

since in view of Eq. (11.60) and (14.11) the left-hand side of this equation is equal to

$$\overline{\frac{\partial u_l(\boldsymbol{x})\, u_i(\boldsymbol{x})}{\partial x_l} u_j(\boldsymbol{x})} + \overline{u_i(\boldsymbol{x}) \frac{\partial u_l(\boldsymbol{x})\, u_j(\boldsymbol{x})}{\partial x_l}} = \frac{\partial \overline{u_i(\boldsymbol{x})\, u_j(\boldsymbol{x})\, u_l(\boldsymbol{x})}}{\partial x_l}.$$

This expression vanishes since the turbulence is assumed to be homogeneous. In particular, if we set $i = j$, and sum over this index, we have

$$\int \Gamma_{jj}(\boldsymbol{k}, t)\, d\boldsymbol{k} = 0, \tag{14.17}$$

and for isotropic turbulence we have

$$\int_0^\infty k^2 \Gamma(k, t)\, dk = \int_0^\infty T(k, t)\, dk = 0 \tag{14.18}$$

This relation is identical with the first equation in Eq. (12.144); in view of Eq. (12.141) it is also equivalent to the equation $B'_{LL,L}(0) = 0$ which was proved in another way in Sect. 12.4 [below Eq.

(12.122)]. From Eq. (14.18) and (14.15) it follows that the rate of change of the total turbulent energy is due to viscous forces alone and is given by

$$\frac{\partial}{\partial t}\frac{\overline{u_j u_j}}{2}=\frac{\partial}{\partial t}\int_0^\infty E(k,\,t)\,dk=-2\nu\int_0^\infty k^2E(k,\,t)\,dk. \quad (14.19)$$

Figure 12 shows schematically the energy spectrum $E(k)$, the energy dissipation spectrum $2\nu k^2E(k)$, and the function $T(k)$ which governs the energy redistribution over the spectrum (see also Figs. 20–22 and 26). The fact that the function $T(k)$ is negative for small $k$ and positive for large $k$ corresponds to the intuitive expectation that turbulent mixing should lead to the "spliting" of turbulent disturbances, i.e., to the conversion of the energy of large-scale components into the energy of small-scale components which lose their energy directly by overcoming "viscous friction." The fact that viscosity plays an important role only for the small scale components of the motion, which are characterized by high local velocity gradients, is expressed in Fig. 12 by the fact that the maximum of the dissipation spectrum $2\nu k^2E(k)$ lies further to the right along the wave-number axis than the maximum of $E(k)$. The intervals along the $k$ axis within which most of the total energy $\int_0^\infty E(k)\,dk$ and total dissipation $2\nu\int_0^\infty k^2E(k)\,dk$ (say 80–90%) is localized will be called the *energy range* and the *dissipation range*, respectively. Figure 12 shows the disposition of these ranges in the case which we are considering. It is clear that the dissipation range always lies to the right of the energy range along the $k$ axis. In general, however, they can overlap each other.

The terms of the spectral equation (14.15) cannot be measured directly. However, assuming that the turbulence is isotropic, we can calculate the functions $E(k)$ and $T(k)$ from the one-dimensional Fourier transforms of the second- and third-order velocity correlations using the relations of Sects. 12.3 and 12.4. We can evaluate all the terms of the equation (14.15) from the data of velocity measurements at two stations behind the grid in the wind tunnel corresponding to close values of $x = Ut$ (and assuming the

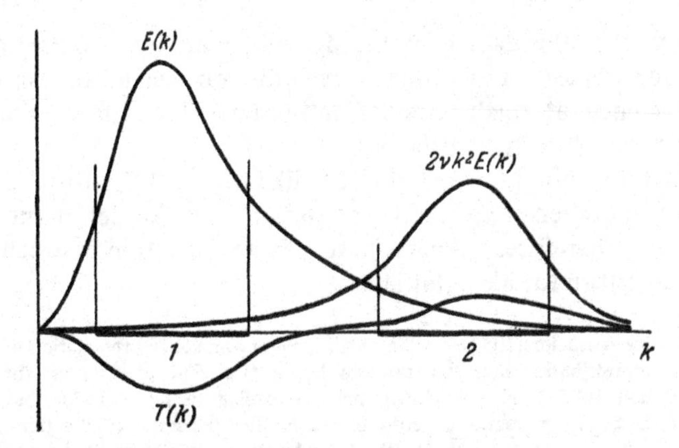


**FIG. 12** Schematic illustration of the energy spectrum $E(k)$, the energy dissipation spectrum $2\nu k^2 E(k)$, and the function $T(K)$. 1) Energy range; 2) dissipation range.

validity of the Taylor hypothesis). Then we can verify the validity of Eq. (14.15) for grid turbulence, i.e., verify whether the turbulence is isotropic or not. Such a verification was made by Van Atta and Chen (1969a) who compared the values of $T(k)$ calculated from the third-order correlations with the values computed from measured spectra with the aid of isotropic dynamic equation (14.15); their result is shown in Fig. 12a [the normalizing parameters $\eta$ and $v_\eta$ are determined in equation (16.35) below]. We see that the

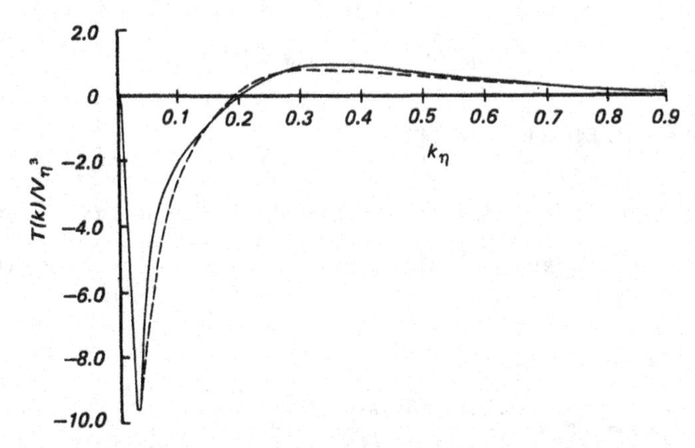


**FIG. 12a** Comparison of the values of $T(k)$ calculated from the measurements of the third order correlation (—) with values obtained from the dynamical equation (14.15) (– – –).

agreement of the data with the dynamic equation is quite satisfactory for values of $k$ which are not too small; the moderate inconsistency at small $k$ is evidently connected with a (not quite reliable) deviation from isotropy.

Equations (14.14) and (14.15), like Eq. (14.9), form a single relation between two unknown functions. To determine these functions, therefore, we must have a further relation between them. We shall return to this point later.

Following Batchelor (1953), we can obtain a more complete representation of the rate of energy redistribution over the spectrum $\Gamma_{jj}(k, t)$ or $T(k, t)$ if we use the spectral representation (11.52) of the velocity field. According to Eq. (14.11), the function $\Gamma_{ij}(\boldsymbol{k}, t)$ is simply expressed in terms of the Fourier transform of the third moment $B_{il,j}(\boldsymbol{r}) = \overline{u_i(\boldsymbol{x})\, u_l(\boldsymbol{x})\, u_j(\boldsymbol{x}+\boldsymbol{r})}$ (the dependence on $t$ will be omitted for the sake of brevity). However, from the spectral representation (11.52) of the field $u(x)$ it follows that

$$u_i(\boldsymbol{x})\, u_l(\boldsymbol{x}) = \int e^{i\boldsymbol{k}\boldsymbol{x}}\, dZ_{il}(\boldsymbol{k}); \quad dZ_{il}(\boldsymbol{k}) = \int dZ_i(\boldsymbol{k}_1)\, dZ_l(\boldsymbol{k}-\boldsymbol{k}_1), \tag{14.20}$$

$$u_i(\boldsymbol{x})\, u_l(\boldsymbol{x})\, u_j(\boldsymbol{x}+\boldsymbol{r}) = \int\int e^{i\boldsymbol{k}\boldsymbol{r}+i(\boldsymbol{k}-\boldsymbol{k}_1)\boldsymbol{x}}\, dZ_j(\boldsymbol{k})\, dZ^*_{il}(\boldsymbol{k}_1). \tag{14.21}$$

It is clear that the function $\overline{u_i(\boldsymbol{x})\, u_l(\boldsymbol{x})\, u_j(\boldsymbol{x}+\boldsymbol{r})}$ will depend only on $\boldsymbol{r}$, and can be represented by a Fourier integral provided

$$\overline{dZ_j(\boldsymbol{k})\, dZ^*_l(\boldsymbol{k}_2)\, dZ^*_i(\boldsymbol{k}_1)} = \delta(\boldsymbol{k}-\boldsymbol{k}_1-\boldsymbol{k}_2)\, Q_{il,j}(\boldsymbol{k}, \boldsymbol{k}_1)\, d\boldsymbol{k}\, d\boldsymbol{k}_1\, d\boldsymbol{k}_2. \tag{14.22}$$

Subject to this condition,

$$F_{il,j}(\boldsymbol{k}) = \int Q_{il,j}(\boldsymbol{k}, \boldsymbol{k}_1)\, d\boldsymbol{k}_1. \tag{14.23}$$

According to Eqs. (14.11) and (14.23),

$$\Gamma_{ij}(\boldsymbol{k}) = ik_l \int [Q_{il,j}(\boldsymbol{k}, \boldsymbol{k}_1) - Q_{jl,i}(-\boldsymbol{k}, -\boldsymbol{k}_1)]\, d\boldsymbol{k}_1, \tag{14.24}$$

$$\Gamma_{jj}(\boldsymbol{k}) = \int Q_{jj}(\boldsymbol{k}, \boldsymbol{k}_1)\, d\boldsymbol{k}_1, \tag{14.25}$$

where, in view of Eq. (14.22),

$$Q_{jj}(\boldsymbol{k}, \boldsymbol{k}_1) = \lim_{\substack{d\boldsymbol{k}\to 0\\ d\boldsymbol{k}_1\to 0}} i \left[ \frac{\overline{dZ_j(\boldsymbol{k})\, dZ^*_j(\boldsymbol{k}_1)\, k_l\, dZ^*_l(\boldsymbol{k}-\boldsymbol{k}_1)}}{d\boldsymbol{k}\, d\boldsymbol{k}_1} - \frac{\overline{dZ^*_j(\boldsymbol{k})\, dZ_j(\boldsymbol{k}_1)\, k_l\, dZ_l(\boldsymbol{k}-\boldsymbol{k}_1)}}{d\boldsymbol{k}\, d\boldsymbol{k}_1} \right]. \tag{14.26}$$

It is evident from this definition that

$$Q_{jj}(\boldsymbol{k}, \boldsymbol{k}_1) = -Q_{jj}(\boldsymbol{k}_1, \boldsymbol{k}) \qquad (14.27)$$

Equation (14.17) follows from this. By Eqs. (14.25) and (14.27) the quantity $Q_{jj}(\boldsymbol{k}, \boldsymbol{k}_1)$ can be interpreted as the specific rate of energy transfer from an element $d\boldsymbol{k}_1$ of the wave-number space into an element $d\boldsymbol{k}$ (per unit volume of wave-number space $\Omega$). In the isotropic case, the function $Q_{jj}(\boldsymbol{k}, \boldsymbol{k}_1)$ will depend only on the lengths $k$ and $k_1$ of the vectors $\boldsymbol{k}$ and $\boldsymbol{k}_1$, and also on $\cos\theta$ where $\theta$ is the angle between $\boldsymbol{k}$ and $\boldsymbol{k}_1$. Therefore, in this case,

$$T(k) = 2\pi k^2 \Gamma_{jj}(k) = 4\pi^2 k^2 \int_0^\infty \int_0^\pi Q_{jj}(k, k_1, \cos\theta)\, k_1^2 \sin\theta \, d\theta \, dk_1. \qquad (14.28)$$

Equation (14.27) remains valid even if we do not imply summation over the index $j$ within it; in Eqs. (14.25) and (14.26) we also need not sum over $j$. Therefore, the inertial terms in the equations of motion lead to an energy redistribution over the wave-number spectrum of each individual component of the velocity field, but they cannot give rise to the conversion of the energy of one of the components into the energy of another. In the case of isotropic turbulence, the energy of each component of the velocity field will be equally reduced by viscosity, so that the turbulence will continue to be isotropic. In the case of arbitrary homogeneous turbulence, the situation is different because of the effect of pressure which leads to the appearance of the term $\Pi_{ij}(\boldsymbol{k}, t)$ in Eq. (14.10). In the absence of isotropy this term cancels out only if we let $i = j$ and sum over $j$. It follows that, in the anisotropic case, pressure forces can give rise to the transfer of energy from one component of the disturbance of the vector velocity field with given $\boldsymbol{k}$ to other components of the same disturbance, but does not give rise to energy transfer between disturbances with different $\boldsymbol{k}$ (see Sect. 6.2 of Vol. 1). The fact that the essentially anisotropic turbulence produced by a grid in the wind tunnel is usually approximately isotropic at a certain distance from the grid shows that, in the absence of an anisotropic energy flux, the pressure forces tend to bring the turbulence to the isotropic state, i.e., there is energy transfer from the components carrying more energy. This last fact is also confirmed by certain special wind tunnel experiments (see, for example, Corrsin, 1959). However, no one has yet succeeded in proving this theoretically from Eq. (14.10).

The complexity of this problem is deepened still further by the fact that the rate at which the isotropic state is approached is different for disturbances with different wave numbers. In the case of disturbances with the longest wavelengths (very small values of $k$) the approach to isotropy is quite slow. For the idealized model of homogeneous turbulence in infinite space (which is the only case in which we can meaningfully speak of very small $k$), it can even be shown theoretically for certain special "regularity conditions" imposed on the velocity field, that the asymptotic form of the spectral tensor $F_{ij}(\boldsymbol{k}, t)$ will remain unaltered as $k \to 0$ during the evolution of the turbulence, so that the initial anisotropy will never vanish, at least in the extreme long wavelength region (see Sect. 15.2). In the energy range of the wave-number spectrum, the approach to the isotropic state definitely occurs, but is relatively slow [it is characterized by the same time scales as the general process of turbulent-energy decay due to viscosity; cf. the papers by Townsend (1954), Uberoi (1957), Mills and Corrsin (1959), Corrsin (1959), Uberoi and Wallis (1967), and Portfors and Keffer (1969), which are not in full agreement with one another].[3] Conversely, in the range of the

[3] The approach of the turbulence behind a grid to the isotropic state can be accelerated by introducing an artificial contraction of the flow near the beginning of the wind tunnel which equalizes the root mean square values of the longitudinal and lateral components of the velocity fluctuations [see, for example, Uberoi and Wallis (1966), Comte-Bellot and Corrsin (1966) where other references can be found].

extremely small-scale disturbances, which are characterized by large values of $k$, the approach to the isotropic state occurs very rapidly. For sufficiently large Reynolds numbers the isotropy of this part of the spectrum is established before the total turbulence energy is observed to decrease appreciably. This is the basis of the extremely important theory of locally isotropic turbulence which will be discussed in detail in the next chapter, where the experimental data confirming the above remarks will be presented.

## 14.4 Correlations and Spectra Containing Pressure

So far we have considered only the statistical characteristics of the vector velocity field. Let us now go on to investigate the statistical characteristics of scalar flow variables, and begin with the case of the pressure field $p(\boldsymbol{x})$.

It is well known that in an incompressible fluid the pressure is related to the space derivatives of the velocity field by the Poisson equation which is a consequence of the Navier-Stokes and the continuity equations [see Eq. (1.9) in Vol. 1]. If we multiply together the left and right hand sides of the two equations (1.9) corresponding to different points $\boldsymbol{x}$ and $\boldsymbol{x}'=\boldsymbol{x}+\boldsymbol{r}$ of the turbulent flow, and take the average of the result, we find that

$$\Delta_{\boldsymbol{x}}\Delta_{\boldsymbol{x}'}\overline{pp'}=\rho^2\frac{\partial^4\overline{u_iu_ju'_mu'_n}}{\partial x_i\,\partial x_j\,\partial x'_m\,\partial x'_n} \tag{14.29}$$

where, as usual, the primes indicate quantities evaluated at the point $\boldsymbol{x}'$. In the case of homogeneous turbulence this equation can be rewritten in the form

$$\Delta^2 B_{p'p'}(\boldsymbol{r},\ t)=\rho^2\frac{\partial^4 B_{ij,mn}(\boldsymbol{r},\ t)}{\partial r_i\,\partial r_j\,\partial r_m\,\partial r_n}, \tag{14.30}$$

where

$$B_{p'p'}(\boldsymbol{r},\ t)=\overline{[p(\boldsymbol{x},\ t)-\bar{p}]\,[p(\boldsymbol{x}+\boldsymbol{r},\ t)-\bar{p}]}=B_{pp}(\boldsymbol{r},\ t)-\bar{p}^2$$

is the centered pressure correlation function. If, on the other hand, the turbulence is not only homogeneous but also isotropic, then $B_{p'p'}(\boldsymbol{r})$ will depend only on $r=|\boldsymbol{r}|$ (the dependence on time will not be indicated for the sake of brevity). Therefore, the right-hand side of Eq. (14.30) will also be a scalar function depending only on $r$:

$$\frac{\partial^4 B_{ij,mn}(\boldsymbol{r})}{\partial r_i\,\partial r_j\,\partial r_m\,\partial r_n}=Q(r). \tag{14.31}$$

The three dimensional Fourier transform of $Q(\boldsymbol{r})$ will, clearly, be equal to $k_i k_j k_m k_n F_{ij,mn}(\boldsymbol{k})$ where $F_{ij,mn}(\boldsymbol{k})$ is the Fourier transform of $B^{(0)}_{ij,mn}(\boldsymbol{r}) = B_{ij,mn}(\boldsymbol{r}) - B_{ij}(0) B_{mn}(0)$, i.e., it is a scalar function of $\boldsymbol{k}$ for which the Taylor series at $\boldsymbol{k}=0$ begins with terms proportional to $k^4$ According to the discussion given in Sect. 12.1 [cf. Eq. 12.18)] it follows from this that

$$\int_0^\infty r^2 Q(r)\,dr = \int_0^\infty r^4 Q(r)\,dr = 0; \tag{14.32}$$

These equations will be used later.

Since for functions depending only on $r$

$$\Delta^2 = \left(\frac{d^2}{dr^2} + \frac{2}{r}\frac{d}{dr}\right)^2 = \frac{d^4}{dr^4} + \frac{4}{r}\frac{d^3}{dr^3}, \tag{14.33}$$

in the case of isotropic turbulence Eq. (14.31) assumes the form

$$\frac{d^4 B_{p'p'}(r)}{dr^4} + \frac{4}{r}\frac{d^3 B_{p'p'}(r)}{dr^3} = \rho^2 Q(r). \tag{14.34}$$

In addition to this equation, the function $B_{p'p'}(r)$ must also satisfy the following boundary conditions:

$$B_{p'p'}(0) < \infty, \quad B_{p'p'}(r) \to 0 \text{ when } r \to \infty. \tag{14.35}$$

Since the homogeneous equation corresponding to the operator (14.33) has four linearly independent solutions, namely 1, $r$, $r^2$ and $r^{-1}$, it is a simple matter to write down Green's function for Eq. (14.34), which corresponds to the solution bounded at the origin and tending to zero at infinity. It is of the form

$$G(r, y) = \begin{cases} -\dfrac{r^2 y}{6} + \dfrac{r y^2}{2} - \dfrac{y^3}{2} & \text{when } r \leqslant y, \\ -\dfrac{y^4}{6r} & \text{when } r \geqslant y. \end{cases} \tag{14.36}$$

Hence it follows that the solution of the inhomogeneous equation (14.34) in which we are interested is given by

$$B_{p'p'}(r) = \rho^2 \int_0^\infty G(r, y) Q(y)\, dy =$$

$$= \rho^2 \left[ -\frac{1}{6r} \int_0^r y^4 Q(y)\, dy - \frac{1}{6} \int_r^\infty (r^2 y - 3ry^2 + 3y^3) Q(y)\, dy \right]. \tag{14.37}$$

Using the fact that $\int_0^\infty y^4 Q(y)\, dy = 0$ according to Eq. (14.32), we can rewrite this formula in a somewhat shorter form:

$$B_{p'p'}(r) = \frac{\rho^2}{6r} \int_r^\infty y(y - r)^3 Q(y)\, dy. \tag{14.38}$$

This last result (first established by Batchelor, 1961) expresses explicitly the function $B_{p'p'}(r)$ in terms of the fourth moments of the velocity field [an expression for the right-hand side of Eq. (14.38) in terms of the scalar functions $B_{LL,LL}(r)$, $B_{LL,NN}(r)$ and so on was given by Uberoi, 1953].[3]

In a similar way we can investigate the pressure-velocity correlations in isotropic turbulence. The cross correlation functions $B_{pl}(r)$ in incompressible isotropic turbulence always vanish identically. However, the tensor of second rank

$$B_{p',mn}(r, t) = \overline{[p(x) - \bar{p}]\, u_m(x')\, u_n(x')} =$$

$$= B_{p,mn}(r, t) - \frac{1}{3} \overline{p\overline{u^2}}\, \delta_{mn}, \quad r = x' - x, \tag{14.39}$$

[3] Let us note that if $Q(r)$ decreases sufficiently rapidly at infinity, then the function $B_{p'p'}(r) = B_{pp}(r) - \bar{p}^2$ will decrease more rapidly than $r^{-1}$ as $r \to \infty$. The boundary conditions (14.35) can therefore be replaced by the single "fourfold condition"

$$rB_{p'p'}(r) \to 0 \quad \text{when} \quad r \to \infty,$$

which corresponds to the Green function

$$G_1(r, y) = \begin{cases} \dfrac{y}{6r}(y - r)^3 & \text{when } r \leqslant y, \\ 0 & \text{when } r \geqslant y. \end{cases}$$

The solution (14.38) of Eq. (14.34) which corresponds to this Green function should automatically satisfy the conditions (14.35), in view of Eq. (14.32). Similar analysis (without using the Green functions) was employed by Batchelor to derive equation (14.38). The Green-function method has been simultaneously used by Obukhov and Yaglom (1951) to obtain a related result.

which also describes the correlation between pressure and velocity fields will in general be nonzero. In the isotropic case

$$B_{p',mn}(\boldsymbol{r}) = P_1(r)\, r_m r_n + P_2(r)\, \delta_{mn}, \tag{14.40}$$

where

$$P_1(r) = \frac{B_{p',LL}(r) - B_{p',NN}(r)}{r^2}, \quad P_2(r) = B_{p',NN}(r) \tag{14.41}$$

[cf. Eq. (12.29)]. Multiplying both sides of Eq. (1.9) by $u'_m u'_n$ and taking the average of the result, we can verify that in homogeneous turbulence

$$\Delta B_{p',mn}(\boldsymbol{r}) = -\rho \frac{\partial^2 B_{lj,mn}(\boldsymbol{r})}{\partial r_l \partial r_j}. \tag{14.42}$$

If, in addition, the turbulence is also isotropic, then the tensor on the left-hand side of Eq. (14.42) will be an isotropic symmetric tensor of second rank, and consequently

$$\frac{\partial^2 B_{lj,mn}(\boldsymbol{r})}{\partial r_l \partial r_j} = Q_1(r)\, r_m r_n + Q_2(r)\, \delta_{mn}. \tag{14.43}$$

The three dimensional Fourier transform of the tensor (14.43) will be of the form $-k_l k_j F_{lj,mn}(\boldsymbol{k})$, i.e., for small $k$ it behaves as $a_0 k^2 \delta_{mn} + a_1 k_m k_n$. According to the analysis given in Sect. 12.2 [cf. Eq. (12.46)], the functions $Q_1(r)$ and $Q_2(r)$ should, therefore, satisfy the condition

$$\int_0^\infty [r^2 Q_1(r) + 3Q_2(r)]\, r^2\, dr = 0. \tag{14.44}$$

Substituting Eqs. (14.40) and (14.43) into Eq. (14.42) and equating the coefficients of $r_m r_n$ and $\delta_{mn}$ on the left- and right-hand sides of the resulting equation, we obtain the following two equations for the unknown functions $P_1(r)$ and $P_2(r)$:

$$\frac{d^2 P_1}{dr^2} + \frac{6}{r}\frac{dP_1}{dr} = -\rho Q_1(r), \tag{14.45}$$

$$\frac{d^2P_2}{dr^2}+\frac{2}{r}\frac{dP_2}{dr}+2P_1=-\rho Q_2(r). \qquad (14.46)$$

Let us replace $P_2(r)$ by the new unknown

$$P_3(r)=r^2P_1(r)+3P_2(r)=B_{p',\,LL}(r)+2B_{p',\,NN}(r)=$$
$$=\overline{[p(\boldsymbol{x})-\bar{p}]\,u^2(\boldsymbol{x}+\boldsymbol{r})}; \qquad (14.47)$$

If we then multiply Eq. (14.45) by $r^2$ and add the result to Eq. (14.46) multiplied by 3, we obtain the following equation for $P_3(r)$:

$$\frac{d^2P_3}{dr^2}+\frac{2}{r}\frac{dP_3}{dr}=-\rho Q_3(r); \quad Q_3(r)=r^2Q_1(r)+3Q_2(r). \qquad (14.48)$$

The linearly independent solutions of the homogeneous equation corresponding to Eq. (14.45) are 1 and $r^{-5}$, whereas those corresponding to Eq. (14.48) are 1 and $r^{-1}$. In view of the boundary conditions

$$P_i(0)<\infty; \quad P_i(r)\to 0 \text{ when } r\to\infty; \; i=1,\,3, \qquad (14.49)$$

we readily find from Eqs. (14.45) and (14.48) that

$$P_1(r)=\frac{\rho}{5}\left[\frac{1}{r^5}\int_0^r y^6Q_1(y)\,dy+\int_r^\infty yQ_1(y)\,dy\right], \qquad (14.50)$$

and

$$P_3(r)=\rho\left[\frac{1}{r}\int_0^r y^2Q_3(y)\,dy+\int_r^\infty yQ_3(y)\,dy\right]. \qquad (14.51)$$

In view of Eq. (14.44), the last of these formulas can also be rewritten in the form

$$P_3(r)=-\rho\int_r^\infty y\left(\frac{y}{r}-1\right)Q_3(y)\,dy. \qquad (14.52)$$

For the functions

$$B_{p',LL}(r)=\frac{1}{3}[2r^2P_1(r)+P_3(r)] \text{ и } B_{p',NN}(r)=-\frac{1}{3}[r^2P_1(r)-P_3(r)]$$

we have from Eqs. (14.50)–(14.52),

$$B_{p',LL}(r)=\frac{\rho}{15}\left\{\frac{2}{r^3}\int_0^r y^6Q_1(y)\,dy-\frac{1}{r}\int_r^\infty [r^2y(5y-7r)Q_1(y)+ + 15y(y-r)Q_2(y)]\,dy\right\}, \quad (14.53)$$

$$B_{p',NN}(r)=-\frac{\rho}{15}\left\{\frac{1}{r^3}\int_0^r y^6Q_1(y)\,dy+ +\frac{1}{r}\int_r^\infty [r^2y(5y-4r)Q_1(y)+15y(y-r)Q_2(y)]\,dy\right\}. \quad (14.54)$$

These results can readily be reformulated in terms of the spectral functions. Let $F_{pp}(\boldsymbol{k})$ be the pressure spectrum, i.e., the three dimensional Fourier transform of the function $B_{p'p'}(\boldsymbol{r})$ (which depends only on $k=|\boldsymbol{k}|$) and let $F_{ij,mn}(\boldsymbol{k})$ be the "fourth-order spectral function" which can be written in a form analogous to Eq. (12.132). In that case, using Eqs. (14.30), (12.132), and (12.131), we have

$$k^4F_{pp}(\boldsymbol{k})=\rho^2k_ik_jk_mk_nF_{ij,mn}(\boldsymbol{k})=\rho^2F_{LL,LL}(\boldsymbol{k})\,k^4,$$

i.e.,

$$F_{pp}(\boldsymbol{k})=\rho^2F_{LL,LL}(\boldsymbol{k}). \quad (14.55)$$

This formula is the "spectral equivalent" of Eqs. (14.34) and (14.38). Similarly, if $F_{p,mn}(\boldsymbol{k})$ is the three dimensional Fourier transform of the tensor $B_{p',mn}(\boldsymbol{r})$, then

$$F_{p,mn}(\boldsymbol{k})=\frac{F_{p,LL}(k)-F_{p,NN}(k)}{k^2}k_mk_n+F_{p,NN}(k)\,\delta_{mn}. \quad (14.56)$$

In view of Eq. (14.42), we have $k^2F_{p,mn}(\boldsymbol{k})=-\rho k_ik_jF_{ij,mn}(\boldsymbol{k})$, and it follows that

$$F_{p,LL}(k)=-\rho F_{LL,LL}(k),\quad F_{p,NN}(k)=-\rho F_{LL,NN}(k). \quad (14.57)$$

These formulas provide a new form of the relations (14.53) and (14.54).

## 14.5 Correlations and Spectra Containing the Temperature

Let us now consider isotropic turbulence in a temperature-inhomogeneous fluid. In this case the temperature fluctuations will also form a homogeneous and isotropic random field. Subject to the usual assumption that the velocity $\boldsymbol{u}(\boldsymbol{x}, t)$ is small in comparison with the sound velocity, and the temperature changes are small in comparison with the mean absolute temperature, we can assume that the density $\rho$, the molecular kinematic viscosity $\nu = \eta/\rho$, and the temperature diffusivity $\chi = \varkappa/c_p\rho$ can be regarded as constants. We shall also adopt the usual assumptions that radiative heat transfer and the heating of the medium due to the kinetic energy dissipation can be neglected. The temperature fluctuations $\vartheta(\boldsymbol{x}, t)$ will then satisfy the usual thermal conduction equation (1.72) which is precisely the same as the diffusion equation for a passive admixture with a molecular diffusion coefficient $\chi$. Below we shall start from Eq. (1.72), so that all the subsequent discussion will be valid for both the temperature and the concentration of a passive admixture. However, since the temperature is more important in applications, and is the most accessible to experimental verification, we shall always refer to $\vartheta$ as the temperature.

Multiplying Eq. (1.72) for the point $\boldsymbol{x}$ by $\vartheta' = \vartheta(\boldsymbol{x}')$, and the same equation for the point $\boldsymbol{x}'$ by $\vartheta(\boldsymbol{x})$, and adding the two term by term, we find after taking the average that in the case of homogeneous turbulence

$$\frac{\partial B_{\vartheta\vartheta}(\boldsymbol{r}, t)}{\partial t} = \frac{\partial}{\partial r_k}\left[B_{k\vartheta,\vartheta}(\boldsymbol{r}, t) - B_{k\vartheta,\vartheta}(-\boldsymbol{r}, t)\right] + 2\chi\frac{\partial^2 B_{\vartheta\vartheta}(\boldsymbol{r}, t)}{\partial r_k \partial r_k} \tag{14.58}$$

(see Yaglom, 1949b). If the turbulence is isotropic, then $B_{\vartheta\vartheta}(\boldsymbol{r}) = B_{\vartheta\vartheta}(r)$ where $r = |\boldsymbol{r}|$ and $B_{k\vartheta,\vartheta}(\boldsymbol{r}) = B_{L\vartheta,\vartheta}(r)\frac{r_k}{r}$.

In this case, Eq. (14.58) assumes the form

$$\frac{\partial B_{\vartheta\vartheta}(r, t)}{\partial t} = 2\left(\frac{\partial}{\partial r} + \frac{2}{r}\right)\left[B_{L\vartheta,\vartheta}(r, t) + \chi\frac{\partial B_{\vartheta\vartheta}(r, t)}{\partial r}\right]. \tag{14.59}$$

The last equation, which plays the role of the von Kármán-Howarth equation for the temperature field, was first established by Corrsin (1951a).

We note also that, in the thermal conduction equation (1.72), we can interpret $\vartheta$ as the deviation of temperature (or concentration) at a given point from the constant mean value $\overline{\vartheta}$, i.e., we can consider that $\overline{\vartheta}=0,\ \vartheta=\vartheta'$ and $B_{\vartheta\vartheta}=B_{\vartheta'\vartheta'}=\overline{[\vartheta(\boldsymbol{x})-\overline{\vartheta}][\vartheta(\boldsymbol{x}')-\overline{\vartheta}]}$. We shall adopt this convention henceforth.

As in the case of the von Kármán-Howarth equation, the Corrsin equation (14.59) relates the two unknown functions $B_{\vartheta\vartheta}(r,\ t)$ and $B_{L\vartheta,\vartheta}(r,\ t)$.

These functions can be measured simultaneously with a resistance thermometer and a hot-wire anemometer. As an example, Figs. 13a and 13b, which are taken from the paper by Mills et al. (1958), show examples of normalized correlation functions $\widetilde{B}_{\vartheta\vartheta}(r)=B_{\vartheta\vartheta}(r)/B_{\vartheta\vartheta}(0)$ and $\widetilde{B}_{L\vartheta,\vartheta}(r)=B_{L\vartheta,\vartheta}(r)/B_{\vartheta\vartheta}(0)\,[B_{LL}(0)]^{1/2}$, measured in a wind tunnel behind an electrically heated grid. If we know the values of the function $B_{\vartheta\vartheta}(r)$ at different distances $x$ from the grid, we can also determine $\frac{\partial}{\partial t}B_{\vartheta\vartheta}(r,\ t)=-\frac{1}{U}\frac{\partial}{\partial x}B_{\vartheta\vartheta}(r,x)$ and verify equation (14.59). This verification was undertaken by Mills et al. who showed that there is good agreement between theory and experiment.

Equation (14.59) can be transformed to a form containing the temperature spectrum $F_{\vartheta\vartheta}(k,\ t)$ [or the function $E_{\vartheta\vartheta}(k,t)=4\pi k^2F_{\vartheta\vartheta}(k,\ t)$ which we shall occasionally denote by the symbol $E^{(\vartheta)}(k,\ t)$] and the function $F_{L\vartheta,\vartheta}(k,\ t)$ given by

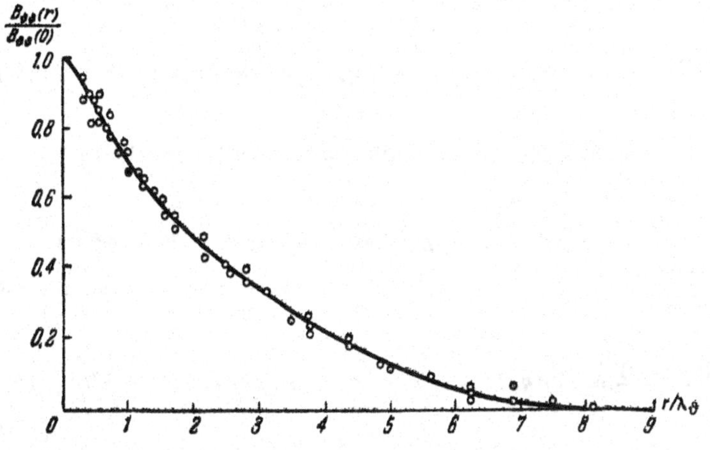


FIG. 13a Measured normalized temperature correlation function $B_{\vartheta\vartheta}(r)/B_{\vartheta\vartheta}(0)$ as a function of $r/\lambda_\vartheta$, where $\lambda_\vartheta^2=-2B_{\vartheta\vartheta}(0)/B''_{\vartheta\vartheta}(0)$.

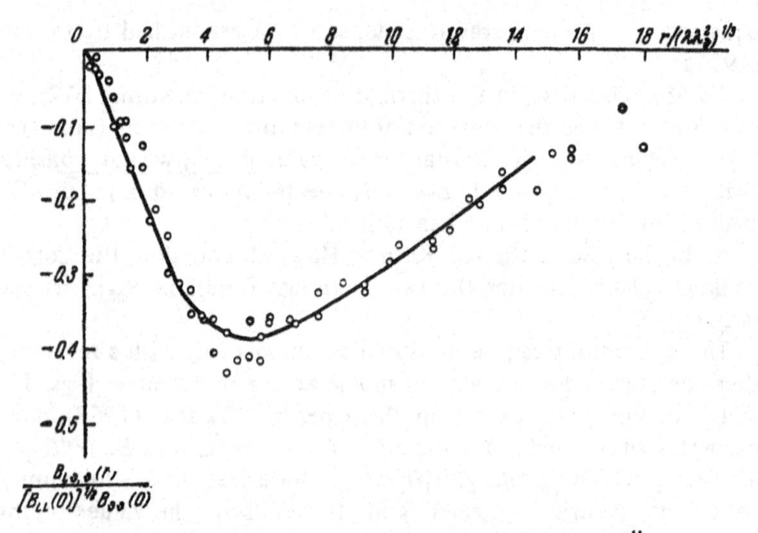

FIG. 13b Measured velocity-temperature correlation $B_{L\vartheta,\,\vartheta}(r)/[B_{LL}(0)]^{1/2}B_{\vartheta\vartheta}(0)$ as a function of $r/(\lambda\lambda_\vartheta^2)^{1/3}$, where $\lambda^2 = -2B_{LL}(0)/B''_{LL}(0)$.

$$F_{j\vartheta,\,\vartheta}(\boldsymbol{k}) = iF_{L\vartheta,\,\vartheta}(k)\frac{k_j}{k}, \tag{14.60}$$

where $F_{j\vartheta,\,\vartheta}(\boldsymbol{k})$ is the Fourier transform of the vector $B_{j\vartheta,\,\vartheta}(\boldsymbol{r})$. In fact, if we take the three dimensional Fourier transform of all the terms in Eq. (14.58) we obtain

$$\frac{\partial F_{\vartheta\vartheta}(\boldsymbol{k},\,t)}{\partial t} = ik_j[F_{j\vartheta,\,\vartheta}(\boldsymbol{k},\,t) - F_{j\vartheta,\,\vartheta}(-\boldsymbol{k},\,t)] - 2\chi k^2 F_{\vartheta\vartheta}(\boldsymbol{k},\,t). \tag{14.61}$$

By Eq. (14.60), the last equation can be rewritten in the form

$$\frac{\partial F_{\vartheta\vartheta}(k,\,t)}{\partial t} = \Gamma_{\vartheta\vartheta}(k,\,t) - 2\chi k^2 F_{\vartheta\vartheta}(k,\,t);\quad \Gamma_{\vartheta\vartheta}(k,\,t) = -2kF_{L\vartheta,\,\vartheta}(k,\,t), \tag{14.62}$$

or

$$\frac{\partial E_{\vartheta\vartheta}(k,\,t)}{\partial t} = T_{\vartheta\vartheta}(k,\,t) - 2\chi k^2 E_{\vartheta\vartheta}(k,\,t);\quad T_{\vartheta\vartheta}(k,\,t) = -8\pi k^3 F_{L\vartheta,\,\vartheta}(k,\,t). \tag{14.63}$$

Equation (14.62) [or Eq. (14.63)] is in fact the required spectral form of Eq. (14.59) (cf. Corrsin, 1951b).

Equation (14.63) describes the time variation of the wave-number distribution of the temperature-fluctuation intensity $\overline{\vartheta'^2} = B_{\vartheta\vartheta}(0)$, which is a natural measure of the inhomogeneity of the temperature field $\vartheta(\boldsymbol{x})$. This "inhomogeneity measure" will, clearly, vary only under the action of thermal conduction which leads to an equalization of the temperature field. Turbulent mixing of the fluid which is produced by the velocity field $\boldsymbol{u}(\boldsymbol{x})$ will then, however, play a very important role: It will lead to random approaches of particles with very different temperatures, i.e., it will produce large temperature gradients resulting in a rapid enhancement of heat transfer due to molecular thermal conduction. If we reformulate these physical ideas in terms of the "spectral language," this will mean that turbulent mixing [described by the term $T_{\vartheta\vartheta}(k, t)$ in Eq. (14.63)] will give rise to a redistribution of the temperature-field disturbances over the wave-number spectrum; namely, it will lead to a conversion of the intensities $E_{\vartheta\vartheta}(k)$ for small values of $k$ into the values of $E_{\vartheta\vartheta}(k)$ for large $k$, without any effect on the "total intensity" $\int_0^\infty E_{\vartheta\vartheta}(k)\,dk = \overline{\vartheta'^2}$.

It is clear that the function $T_{\vartheta\vartheta}(k)$ should satisfy the equation

$$\int_0^\infty T_{\vartheta\vartheta}(k)\,dk = 0, \qquad (14.64)$$

which is analogous to Eq. (14.18). In fact, using Eq. (14.60) we can readily show that the left-hand side of Eq. (14.64) is equal to

$$\overline{u_l(\boldsymbol{x})\frac{\partial\vartheta^2(\boldsymbol{x})}{\partial x_l}} = \frac{\partial\overline{u_l(\boldsymbol{x})\,\vartheta^2(\boldsymbol{x})}}{\partial x_l},$$

which vanishes because the turbulence is homogeneous. Therefore, by integrating Eq. (14.63) with respect to $k$ we obtain

$$\frac{\partial\overline{\vartheta'^2}}{\partial t} = \frac{\partial}{\partial t}\int_0^\infty E_{\vartheta\vartheta}(k)\,dk = -2\chi\int_0^\infty k^2 E_{\vartheta\vartheta}(k)\,dk, \qquad (14.65)$$

as expected.

Using the spectral representation of the velocity and temperature fields, we can readily show that

$$u_l(\boldsymbol{x})\,\vartheta(\boldsymbol{x})\,\vartheta(\boldsymbol{x}+\boldsymbol{r}) = \int\int\int e^{i\boldsymbol{k}\boldsymbol{r}+i(\boldsymbol{k}-\boldsymbol{k}_1)\boldsymbol{x}}\, dZ^*_\vartheta(\boldsymbol{k}_2)\, dZ^*_l(\boldsymbol{k}_1-\boldsymbol{k}_2)\, dZ_\vartheta(\boldsymbol{k}) \tag{14.66}$$

[cf. Eq. (14.21)]. It is clear that the function $\overline{u_l(\boldsymbol{x})\,\vartheta(\boldsymbol{x})\,\vartheta(\boldsymbol{x}+\boldsymbol{r})}$ will be independent of $\boldsymbol{x}$ and can be written in the form of a Fourier integral with respect to $\boldsymbol{r}$ if

$$\overline{dZ_\vartheta(\boldsymbol{k})\, dZ^*_l(\boldsymbol{k}_1)\, dZ^*_\vartheta(\boldsymbol{k}_2)} = \delta(\boldsymbol{k}-\boldsymbol{k}_1-\boldsymbol{k}_2)\, Q_{l\vartheta,\vartheta}(\boldsymbol{k},\boldsymbol{k}_1)\, d\boldsymbol{k}\, d\boldsymbol{k}_1\, d\boldsymbol{k}_2. \tag{14.67}$$

However, in that case, $\Gamma_{\vartheta\vartheta}(\boldsymbol{k})$ will be given by

$$\Gamma_{\vartheta\vartheta}(\boldsymbol{k}) = \int Q_{\vartheta\vartheta}(\boldsymbol{k},\boldsymbol{k}_1)\, d\boldsymbol{k}_1, \tag{14.68}$$

where

$$Q_{\vartheta\vartheta}(\boldsymbol{k},\boldsymbol{k}_1) = -Q_{\vartheta\vartheta}(\boldsymbol{k}_1,\boldsymbol{k}) = \lim_{\substack{d\boldsymbol{k}\to 0\\ d\boldsymbol{k}_1\to 0}} i\left[\frac{\overline{dZ_\vartheta(\boldsymbol{k})\, dZ^*_\vartheta(\boldsymbol{k}_1)\, k_l\, dZ^*_l(\boldsymbol{k}-\boldsymbol{k}_1)}}{d\boldsymbol{k}\, d\boldsymbol{k}_1} - \frac{\overline{dZ^*_\vartheta(\boldsymbol{k})\, dZ_\vartheta(\boldsymbol{k}_1)\, k_l\, dZ_l(\boldsymbol{k}-\boldsymbol{k}_1)}}{d\boldsymbol{k}\, d\boldsymbol{k}_1}\right]. \tag{14.69}$$

The quantity $Q_{\vartheta\vartheta}(\boldsymbol{k},\boldsymbol{k}_1)$ can be interpreted as the specific (per unit volume of the wave-number space $\Omega$) rate of transfer of the inhomogeneity measure of the field $\vartheta$ from the element $d\boldsymbol{k}_1$ of $\Omega$ into the element $d\boldsymbol{k}$. In the isotropic case, $Q_{\vartheta\vartheta}(\boldsymbol{k},\boldsymbol{k}_1)$ will, of course, depend only on $k = |\boldsymbol{k}|, k_1 = |\boldsymbol{k}_1|$, and $\cos\theta = \boldsymbol{k}\cdot\boldsymbol{k}_1/kk_1$, so that here

$$T_{\vartheta\vartheta}(k) = 8\pi^2 k^2 \int_0^\infty \int_0^\pi Q_{\vartheta\vartheta}(k, k_1, \cos\theta)\, k_1^2 \sin\theta\, d\theta\, dk_1. \tag{14.70}$$

We note also that the theory outlined above can readily be generalized to take into account certain additional effects which we have so far neglected. For example, let $\vartheta(\boldsymbol{x})$ be the concentration of a chemically active admixture at the point $\boldsymbol{x}$, which enters into a chemical reaction with some other admixture or with the medium in which it is immersed. We shall suppose, however, that this admixture is passive in the sense that neither its transfer nor the chemical reactions involving it have an appreciable effect on the velocity field $\boldsymbol{u}(\boldsymbol{x}, t)$ (this assumption can be regarded as acceptable if the admixture concentration is low and chemical reactions in the flow occur relatively slowly and quietly). We shall also suppose that the reaction rate depends only on the concentration $\vartheta(\boldsymbol{x})$, i.e., that it is specified by some function $\Phi(\vartheta)$ which is independent of $\boldsymbol{x}$ and $t$ (this will be the case if the concentration $\vartheta$ is much less than the concentration of the material with which the admixture interacts). Under these conditions, the diffusion equation (1.72) must be augmented by a further term:

$$\frac{\partial\vartheta}{\partial t} + u_j\frac{\partial\vartheta}{\partial x_j} = \chi\,\Delta\vartheta - \Phi(\vartheta). \tag{14.71}$$

It follows that the Corrsin equation (14.59) must now be replaced by

$$\frac{\partial B_{\vartheta\vartheta}}{\partial t}=2\left(\frac{\partial}{\partial r}+\frac{2}{r}\right)\left(B_{L\vartheta,\vartheta}+\chi\frac{\partial B_{\vartheta\vartheta}}{\partial r}\right)-2B_{\Phi\vartheta},\tag{14.72}$$

where $B_{\Phi\vartheta}(r)=\overline{\Phi[\vartheta(\boldsymbol{x})]\,\vartheta'(\boldsymbol{x}+\boldsymbol{r})}$ (see Corrsin, 1958a). Since, however, $\overline{\vartheta}=\overline{\vartheta}(t)$ is now a function of time, we must have the further equation

$$\frac{d\overline{\vartheta}}{dt}=-\overline{\Phi(\overline{\vartheta}+\vartheta')}.\tag{14.73}$$

in addition to Eq. (14.72).

The function $\Phi(\vartheta)$ can frequently be approximately replaced by the power function $\Phi(\vartheta)=\mu\vartheta^n$ where $\mu$ = const. In that case, equations (14.72) and (14.73) can be expressed in terms of the moments of the field $\vartheta(\boldsymbol{x})$. In the simplest case, when $n=1$, i.e., $\Phi(\vartheta)=\mu\vartheta$ (the case of "first-order reaction"), we have

$$\frac{d\overline{\vartheta}}{dt}=-\mu\overline{\vartheta},\qquad \overline{\vartheta}=\overline{\vartheta}_0e^{-\mu t}\tag{14.74}$$

and

$$\frac{\partial B_{\vartheta\vartheta}}{\partial t}=2\left(\frac{\partial}{\partial r}+\frac{2}{r}\right)\left(B_{\vartheta L,\vartheta}+\chi\frac{\partial B_{\vartheta\vartheta}}{\partial r}\right)-2\mu B_{\vartheta\vartheta}.\tag{14.75}$$

In the "spectral form," Eq. (14.75) will become

$$\frac{\partial E_{\vartheta\vartheta}}{\partial t}=T_{\vartheta\vartheta}-2\,(\chi k^2+\mu)\,E_{\vartheta\vartheta}.\tag{14.76}$$

However, for any other $n>1$, Eq. (14.73) will contain higher moments of the concentration fluctuations $\vartheta'$, and will not be integrable in an explicit form, while Eq. (14.72) will contain more than two unknown functions (Corrsin, 1958a).

Equations (14.74)–(17.76) can also be used in the case of a decaying radioactive admixture with a half-life $\ln 2/\mu$. When $\vartheta(\boldsymbol{x})$ is the temperature field, we can use the same method to take into account approximately the effect of radiative transfer.

Analogous equations can be written down for the concentration field of product of reaction or of a number of admixtures related through one or more first-order chemical reactions [see, for example, Corrsin (1962b) and Pao (1964)].

# 15. THE SIMPLEST CONSEQUENCES OF THE CORRELATION AND SPECTRAL EQUATIONS

## 15.1 Balance Equations for Energy, Vorticity, and Temperature-Fluctuation Intensity

The equations for the correlation and spectral functions discussed in the last section involve functions of two variables, namely, $r$ and $t$, or $k$ and $t$. These equations lead to a number of predictions about the numerical values of parameters describing the turbulence as a whole, i.e., parameters independent of $r$ or $k$. To obtain these results it is sufficient, for example, to expand somehow the functions of $r$, or of $k$, in the equations of Sect. 14, and then equate the corresponding coefficients on either side of the resulting equation. In particular, if we use the expansion of the correlation functions into a Taylor series in powers of $r$, or the spectra into a Taylor series in powers of $k$, we can obtain relations which have a clear physical interpretation and therefore deserve special consideration.

Following von Kármán and Howarth (1938), let us begin with the equations which are obtained by expanding all the terms in Eq. (14.9) into Taylor series. The zero-order term of this expansion will be obtained by setting $r=0$ in Eq. (14.9):

$$\frac{dB_{LL}(0)}{dt}=10\nu\left(\frac{\partial^2 B_{LL}}{\partial r^2}\right)_{r=0}. \tag{15.1}$$

Equation (15.1) can also be rewritten in the form

$$\frac{d}{dt}\left(\frac{3}{2}\overline{u^2}\right)=15\nu\overline{u^2}f''(0)=-\frac{15\nu\overline{u^2}}{\lambda^2}, \tag{15.2}$$

where $u$ is the velocity component along the $x$ axis and

$$f(r)=\frac{B_{LL}(r)}{B_{LL}(0)}, \qquad \lambda^2=-\frac{1}{f''(0)}. \tag{15.3}$$

Equation (15.2) is the *energy balance equation* for isotropic turbulence. It describes the rate of viscous decrease of the mean kinetic energy of the turbulence. The parameter $\lambda$ in this equation is the differential length scale which is usually referred as the *Taylor microscale* of the turbulence [it was first introduced by Taylor (1935) who was the first to derive Eq. (15.2)], or the *energy dissipation length scale*. In view of the von Kármán relation (14.3), the scale $\lambda$ can also be written in the form

$$\lambda^2=-\frac{2}{g''(0)}, \qquad g(r)=\frac{B_{NN}(r)}{B_{NN}(0)}. \tag{15.4}$$

Hence it follows that this scale can readily be determined from measured values of the function $B_{NN}(r)$ by inserting the parabola into the graph of the normalized correlation function $g(r)$ (see Fig. 4).

Equations (15.3) and (15.4) can, of course, be rewritten in the form

$$\overline{\left(\frac{\partial u}{\partial x}\right)^2} = \frac{\overline{u^2}}{\lambda^2}, \quad \overline{\left(\frac{\partial u}{\partial y}\right)^2} = 2\frac{\overline{u^2}}{\lambda^2}. \tag{15.5}$$

The first of these equations enables us to determine the scale $\lambda^2$ in a relatively simple way. In fact, in grid turbulence $\overline{\left(\frac{\partial u'}{\partial x}\right)^2} = \frac{1}{U^2}\overline{\left(\frac{\partial u'}{\partial t}\right)^2}$ where $U$ is the mean flow velocity and $u' = u - U$. Therefore, to determine $\lambda$ it is sufficient to transform the fluctuations $u'$ into voltage fluctuations, to pass this signal through a differentiating circuit and then to feed it into a watt meter. The scale for grid turbulence is usually of the order of a fraction of a centimeter.

If we have independent measurements of the parameter $\lambda$ and the turbulence intensity $\overline{u^2}$ at different distances $x = Ut$ from the grid, we can directly verify the validity of Eq. (15.2). This verification was first performed by Taylor (1935). It has subsequently been frequently repeated by other workers, and the results were almost always found to be quite satisfactory [see, for example, Batchelor (1953), Batchelor and Townsend (1948a), and Mills et al. (1958).[4] It has also been found that the time dependence of $\overline{u^2}$ is frequently given by the simple power function

$$\overline{u^2} \sim (t - t_0)^{-n}, \tag{15.6}$$

where $t_0$ is the origin of time, and the value of $n$ will be discussed below. Hence, we have the following linear relation for $\lambda^2$:

$$\lambda^2 = \frac{10\nu}{n}(t - t_0). \tag{15.7}$$

Since $B_{LL}(r)$ is an even function of $r$, and $B_{LL,L}(r)$ is an odd function of $r$ (see Sects. 12.2 and 12.4), both sides of Eq. (14.9)

[4] Moderate disagreement was found only by some authors according to whom grid turbulence is appreciably anisotropic [see, for example, Uberoi (1963), Uberoi and Wallis (1969) or Van Atta and Chen (1969a).

contain only even powers of $r$. If we equate the coefficients of $r^2$ in the Taylor series expansions of both sides of this equation, we obtain the equation

$$\frac{1}{2}\frac{d}{dt}B''_{LL}(0) = \frac{7}{6}B'''_{LL,L}(0) + \frac{7}{3}\nu B^{IV}_{LL}(0), \tag{15.8}$$

which, in view of Eq. (11.83), can be interpreted as the equation for the rate of change of the mean square vorticity $\overline{\omega^2}$, i.e., as the *vorticity balance equation*. This equation can also be derived directly from the fluid dynamic equations. In fact, if we multiply all the terms in Eq. (1.7) for the $k$th component of vorticity by $2\omega_k$, sum over $k$, and take the average of the result, we find that for homogeneous turbulence

$$\frac{d\overline{\omega^2}}{dt} = 2\overline{\omega_k\omega_\alpha\frac{\partial u_k}{\partial x_\alpha}} - 2\nu\overline{\frac{\partial\omega_k}{\partial x_\alpha}\frac{\partial\omega_k}{\partial x_\alpha}} \tag{15.9}$$

since in this case

$$\overline{2u_\alpha\omega_k\frac{\partial\omega_k}{\partial x_\alpha}} = \frac{\partial\overline{u_\alpha\omega_k^2}}{\partial x_\alpha} = 0$$

and

$$\overline{\omega_k\frac{\partial^2\omega_k}{\partial x_\alpha^2}} + \overline{\left(\frac{\partial\omega_k}{\partial x_\alpha}\right)^2} = \frac{1}{2}\frac{\partial^2\overline{\omega_k^2}}{\partial x_\alpha^2} = 0.$$

The last term on the right-hand side of Eq. (15.9) describes the decrease in vorticity due to viscous damping. It is always negative, and in the isotropic case it is equal to $-70\nu B^{IV}_{LL}(0)$ (we recall that according to Sect. 12.2, we always have $B^{IV}_{LL}(0) > 0$). As regards the first term on the right-hand side of Eq. (15.9), this describes the effect of inertial terms in the dynamic equations on the variation of the mean square vorticity. This effect has been investigated in detail by Taylor (1938a) who showed (both theoretically and experimentally) that it always leads to a general increase in the vorticity. In the case of isotropic turbulence, it is readily verified that

$$\overline{\omega_l\omega_k\frac{\partial u_l}{\partial x_k}} = -\frac{35}{2}B'''_{LL,L}(0);$$

It follows in particular that $B'''_{LL,L}(0) < 0$. It also follows that in the isotropic case, Eq. (15.9) differs from Eq. (15.8) only by a numerical factor.

In the case of grid turbulence, all the terms in Eq. (15.8) can be measured directly with a hot-wire anemomenter and a differentiating device, since

$$B''_{LL}(0) = -\overline{\left(\frac{\partial u_1}{\partial x_1}\right)^2}, \quad B'''_{LL,L}(0) = \overline{\left(\frac{\partial u_1}{\partial x_1}\right)^3}, \quad B^{IV}_{LL}(0) = \overline{\left(\frac{\partial^2 u_1}{\partial x_1^2}\right)^2}.$$

Such measurements have been described by Batchelor and Townsend (1947). According to this data the agreement between the measured values of the right- and left-hand sides of Eq. (15.8) is completely satisfactory.

Similar consequences can be derived from the Corrsin equation (14.59) for the temperature correlation function. Substituting $r = 0$ into this equation, we obtain

$$\frac{d\overline{\vartheta^2}}{dt} = 6\chi B''_{\vartheta\vartheta}(0) = -\frac{12\chi\overline{\vartheta^2}}{\lambda_\vartheta^2}, \tag{15.10}$$

where

$$\lambda_\vartheta^2 = -\frac{2\overline{\vartheta^2}}{B''_{\vartheta\vartheta}(0)} \tag{15.11}$$

is the temperature microscale introduced by Corrsin (1951a). It corresponds to the distance $OM$ in Fig. 4 if $B(r)$ is interpreted as $B_{\vartheta\vartheta}(r)$ and, as usual, we assume that $\bar{\vartheta} = 0$ where $\vartheta$ is the temperature fluctuation. The length scale $\lambda_\vartheta$ can also be determined from the formula

$$\frac{1}{\lambda_\vartheta^2} = \frac{2}{\overline{\vartheta^2}}\overline{\left(\frac{\partial\vartheta}{\partial x}\right)^2} = \frac{2}{\overline{\vartheta^2}U^2}\overline{\left(\frac{\partial\vartheta}{\partial t}\right)^2}. \tag{15.12}$$

Equation (15.10) is analogous to Eq. (15.2). It describes the rate of decrease of the mean square of temperature fluctuations (the intensity of the temperature fluctuations, or the "measure of the temperature inhomogeneity") due to thermal conductivity.

Verification of this equation for turbulence behind a heated grid in a wind tunnel was carried out by Mills et al. (1958) who determined the values of $\lambda_\vartheta$ for different $x$ with the aid of the construction given in Fig. 4, and compared the results with independent determinations of $\lambda_\vartheta$ obtained from the values of $\frac{d\overline{\vartheta^2}}{dt} = U\frac{d\overline{\vartheta^2}}{dx}$ using Eq. (15.10). The values of $\lambda_\vartheta$ obtained by the two methods were in good agreement and this confirmed the validity of Eq. (15.10) under the conditions of these experiments.

Expanding all the terms in Eq. (14.59) in series in powers of $r^2$, and equating the coefficients of $r^2$ on either side (multiplied by $-3$), we obtain the following equation for the rate of change of the mean square temperature gradient:

$$-\frac{3}{2}\frac{d}{dt}B''_{\vartheta\vartheta}(0) = -5B'''_{L\vartheta,\vartheta}(0) - 5\chi B^{\text{IV}}_{\vartheta\vartheta}(0). \tag{15.13}$$

This equation is similar in form to Eq. (15.8). The last term on the right-hand side is always negative (see Sect. 12.1). It describes the decrease in the mean square temperature gradient due to molecular thermal diffusivity. The first term on the right-hand side of Eq. (15.13) is always positive, since it describes the increase in the temperature gradient due to the inertial approach of fluid particles with very different temperatures.

Similar equations can also be obtained for the concentration of chemically active (or radioactively decaying) admixtures using Eq. (14.75) (see Corrsin, 1958a).

## 15.2 The Loitsyanskii and Corrsin Integrals

The ordinary differential equations, which are obtained by expanding the partial differential equations (14.9) or (14.59) in power series, describe the time variation of local characteristics of isotropic turbulence at a fixed point in the flow. Such equations can be verified with the aid of instruments recording the flow fluctuations at a given point. Equivalent equations can also be obtained by multiplying all the terms of the spectral equations (14.14) or (14.62) by the corresponding power of $k$ and integrating over all values of $k$ [in particular, Eq. (15.2) is equivalent to Eq. (14.19), and Eq. (15.10) is equivalent to Eq. (14.65)]. If, however, we expand all the terms of Eq. (14.14) or (14.62) in a Taylor series in $k$, and equate the corresponding coefficients on the right- and left-hand sides, we obtain equations which have a completely different character. These new equations relate quantities characterizing the behavior of spectral densities near the point $k=0$, i.e., they govern the asymptotic behavior of the longest wavelength components of the flow variables. Such quantities are the integral characteristics of turbulence, and depend on the values of the correlation functions for all values of $r$ between zero and infinity. The corresponding relations cannot, of course, be verified on the basis of measurements behind a grid in a wind tunnel, or any other experimental data. The fact that any real turbulence can be regarded as approximately isotropic only in a finite region, plays a decisive role in this context. Nevertheless, the asymptotic relations governing the behavior of the spectral functions at low wave numbers are interesting from the theoretical viewpoint. We shall see later that they will enable us to derive certain results which have a clear physical interpretation and, in some cases, can be indirectly verified experimentally.

Let us begin with the velocity spectral equation (14.14). We shall suppose that the spectral functions $F(k)$ and $F_3(k)$ are infinitely differentiable at $k=0$, i.e., the correlation functions $B_{LL}(r)$ and $B_{LL,L}(r)$ fall off exponentially as $r\to\infty$ (in fact, more rapidly than any finite power of $r^{-1}$). The validity of this assumption will be discussed later. Substituting the Taylor expansions of $F(k)$ and $F_3(k)$ into Eq. (14.14), and equating the coefficients of equal powers of $k^2$, we obtain the following infinite set of equations:

$$\frac{df_2}{dt}=0, \quad f_2(t)=\text{const}, \tag{15.14}$$

$$\frac{df_4}{dt}=-2g_3-2\nu f_2, \tag{15.15}$$

. . . . . . . . . . . .

where $f_n$ and $g_n$ are the Taylor coefficients of $F(k)$ and $F_3(k)$. Equation (15.14), which has the form of a conservation law, is of particular interest. The coefficient $f_2$ can be expressed in terms of the longitudinal correlation function $B_{LL}(r)$ with the aid of Eq. (12.80), so that Eq. (15.14) can be rewritten in the form

$$\int_0^\infty r^4 B_{LL}(r)\,dr=\Lambda=\text{const}. \tag{15.16}$$

This relation was first obtained in this form by Loitsyanskii (1939), and therefore the quantity $\Lambda$ is usually referred to as the *Loitsyanskii integral* (or *Loitsyanskii invariant*). The spectral significance of this invariant was later established by Kolmogorov (see Yaglom, 1948), Lin (1949) and Batchelor (1949a).

Loitsyanskii (1939) showed that Eq. (15.16) can be derived directly from the von Kármán-Howarth equation (14.9). In fact, if we multiply all the terms in Eq. (14.9) by $r^4$, and integrate between $r=0$ and $r=R$, we obtain

$$\frac{d}{dt}\int_0^R r^4 B_{LL}(r)\,dr = R^4 B_{LL,\,L}(R) + 2\nu R^4 \frac{dB_{LL}(R)}{dR}. \quad (15.17)$$

It is clear that if the integral (15.16) converges, i.e., if $B_{LL}(r)$ tends to zero as $r \to \infty$ more rapidly than $r^{-5}$, then

$$\frac{d}{dt}\int_0^\infty r^4 B_{LL}(r)\,dr = \lim_{R\to\infty} [R^4 B_{LL,\,L}(R)]. \quad (15.18)$$

Thus, Eq. (15.16) will be valid if the integral on the left-hand side of Eq. (15.16) converges and the function $B_{LL,\,L}(r)$ falls off more rapidly than $r^{-4}$. This is equivalent to demanding that the function $F(k)$ is twice differentiable at $k=0$ and $F_3(k)$ tends to zero more rapidly than $k^2$ as $k \to 0$. It is interesting, however, that the case when the right-hand side of Eq. (15.18) is not zero and $\Lambda'$ is finite but not constant has turned out to be of some importance [see Sedov (1951), Proudman and Reid (1954), Batchelor and Proudman (1956); see also Sects. 15.6, 16.2, and 19.3].

Analogous conclusions can be established from the spectral equation (14.62). Assuming that the functions $F_{\vartheta\vartheta}(k)$ and $F_{L\vartheta,\,\vartheta}(k)$ can be expanded in Taylor series in the neighborhood of the point $k=0$, we obtain

$$\frac{df_0^{(\vartheta)}}{dt} = 0, \qquad f_0^{(\vartheta)} = \text{const}, \quad (15.19)$$

$$\frac{df_2}{dt} = -2g_1^{(\vartheta)} - 2\chi f_0^{(\vartheta)}, \quad (15.19')$$

. . . . . . . . . . . . .

where $f_n^{(\vartheta)}$ and $g_n^{(\vartheta)}$ are the coefficients of $k^n$ in the expansions of $F_{\vartheta\vartheta}(k)$ and $F_{L\vartheta,\vartheta}(k)$. The first of these equations has the form of a conservation law and, in view of Eq. (12.16), this conservation law can be rewritten in the form

$$\int_0^\infty r^2 B_{\vartheta\vartheta}(r)\,dr = K = \text{const.} \tag{15.20}$$

This result can be simply obtained from Eq. (14.59). This is in fact, the way in which it was first derived by Corrsin (1951); the quantity $K$ is sometimes called the *Corrsin integral*. To show this, let us multiply all the terms in Eq. (14.59) by $r^2$, integrate the resulting equation between $r=0$ and $r=R$, and then let $R$ tend to infinity. Assuming that the integral (15.20) converges, i.e., $B_{\vartheta\vartheta}(r)$ tends to zero more rapidly than $r^{-3}$ as $r\to\infty$, we obtain

$$\int_0^\infty r^2 B_{\vartheta\vartheta}(r)\,dr = \lim_{R\to\infty}\,[2R^2 B_{L\vartheta,\vartheta}(R)]. \tag{15.21}$$

Hence it is clear that the conservation law (15.20) will be valid provided only that the integral on the left-hand side of this equation converges, and the function $B_{L\vartheta,\vartheta}(r)$ tends to zero more rapidly than $r^{-2}$ as $r\to\infty$.

In the more general case of homogeneous but not necessarily isotropic turbulence, the conservation laws (15.14) and (15.16) degenerate into a family of similar laws. In fact, let us suppose that the functions $F_{ij}(\boldsymbol{k})$, $F_{il,j}(\boldsymbol{k})$, and $F_{pl}(\boldsymbol{k})$ can be expanded in Taylor series, and let us write the Taylor expansions of all the terms of the tensor equation (14.10)

$$\begin{aligned} F_{ij}(\boldsymbol{k}) &= f_{ij,mn}k_m k_n + \dots, \\ \Gamma_{ij}(\boldsymbol{k}) &= \gamma_{ij,mn}k_m k_n + \dots, \\ \Pi_{ij}(\boldsymbol{k}) &= \pi_{ij,mn}k_m k_n + \dots, \end{aligned} \tag{15.22}$$

where

$$\gamma_{ij,mn} = \frac{i}{2}\,[f_{imj,n} + f_{jmi,n} + f_{inj,m} + f_{jni,m}],$$

$$\pi_{ij,mn} = -\frac{i}{2\rho}\,[\delta_{im}f_{pj,n} + \delta_{jm}f_{pi,n} + \delta_{in}f_{pj,m} + \delta_{jn}f_{pi,m}],$$

and $f_{imj,n}$ and $f_{pj,n}$ are the coefficients of $k_n$ in the expansions of the functions $F_{im,j}(\boldsymbol{k})$ and $F_{pj}(\boldsymbol{k})$. Moreover, let us use the facts that the coefficients $f_{imj,n}$ and $f_{pj,n}$ are related by

$$\frac{1}{\rho} f_{pj,\,n}\delta_{lm}k_l k_m k_n = -f_{lmj,\,n}k_l k_m k_n,$$

since $\frac{k^2}{\rho} F_{pl}(\boldsymbol{k}) = -k_j k_l F_{jl,\,l}(\boldsymbol{k})$ [this follows from Eq. (1.9)]. Then it is easy to show that

$$\gamma_{lj,\,mn} + \pi_{lj,\,mn} = 0 \tag{15.23}$$

[see Batchelor (1949a) or Sect. 5.3 of Batchelor (1953)]. Substituting Eq. (15.22) into Eq. (14.10), and using Eq. (15.23), we obtain

$$\frac{d}{dt} f_{lj,\,mn} = 0, \quad \text{i.e.,} \quad f_{lj,\,mn} = \text{const.} \tag{15.24}$$

This is a family of conservation laws which generalizes Eq. (15.14) to the case of arbitrary homogeneous turbulence [in the isotropic case we have, of course,

$$f_{lj,\,mn} = \left[\delta_{mn}\delta_{lj} - \frac{1}{2}(\delta_{lm}\delta_{jn} + \delta_{ln}\delta_{jm})\right] f_2$$

so that Eq. (15.24) is identical with Eq. (15.14)]. In the language of correlation functions, the conservation laws (15.24) are of the form

$$\int r_m r_n B_{lj}(\boldsymbol{r})\, d\boldsymbol{r} = \Lambda_{lj,\,mn} = \text{const.} \tag{15.25}$$

In the isotropic case, the left-hand side of Eq. (15.25) will be either zero or it will differ from Eq. (15.16) only by a numerical factor.[5]

Similarly, it follows from equation (14.61) that, in the general case of arbitrary homogeneous turbulence, there is also a conservation law of the form (15.19). However, $f_0^{(\vartheta)}$ must now be interpreted as the value of the spectral density $F_{\vartheta\vartheta}(\boldsymbol{k}, t)$ at the point $\boldsymbol{k} = 0$. This conservation law can also be rewritten in the form

$$\int B_{\vartheta\vartheta}(\boldsymbol{r})\, d\boldsymbol{r} = K' = \text{const} \tag{15.26}$$

In the isotropic case, the last equation is the same as Eq. (15.20).

Landau and Lifshitz (1963) have noted that the conservation law (15.16) is closely related to the conservation of angular momentum, which is valid for any physical system not subject to external forces [there is a similar situation in the case of the more general conservation laws given by Eq. (15.25)]. Loosely speaking, it can be said that in the case of isotropic turbulent flows with finite and nonzero $\Lambda$, the mean square of the total angular momentum of a large volume $V$ of the fluid will be proportional to this volume, and the proportionality factor will differ from $\Lambda$ multiplied by the square of the density only by a numerical factor:

[5] Analogous invariants, but having an integral rather than local, character can be obtained also for several extended classes of inhomogeneous turbulent flows (Lumley, 1966).

$$\rho^2\Lambda \sim \lim_{V\to\infty} \frac{\overline{[M(V)]^2}}{V} = \lim_{V\to\infty} \frac{1}{V} \sum_{i<j} \overline{\left[\rho \int_V m_{ij}(\boldsymbol{x})\, d\boldsymbol{x}\right]^2}, \tag{15.27}$$

where $M(V)$ is the total angular momentum in the volume $V$, and $m_{ij}(\boldsymbol{x}) = x_i u_j(\boldsymbol{x}) - x_j u_i(\boldsymbol{x})$ is the tensor of the angular momentum density. These statements are not rigorous because as the quantities $m_{ij}(\boldsymbol{x})$ increase without limit as $|\boldsymbol{x}| \to \infty$, the mean square of the angular momentum of a very thin fluid layer adjacent to the boundary of the volume $V$ may make an important contribution to the mean square on the right-hand side of Eq. (15.27) so that, strictly speaking, the limit on the right-hand side of this equation may depend on the shape of the volume $V$. To obtain a rigorous result, it is convenient to follow Kolmogorov's suggestion and replace the right-hand side of Eq. (15.27) by the expression

$$\lim_{\alpha\to 0} \frac{\alpha^{3/2}}{\pi^{3/2}} \overline{[M(\alpha)]^2} = \lim_{\alpha\to 0} \frac{\alpha^{3/2}}{\pi^{3/2}} \rho^2 \sum_{i<j} \overline{\left[\int m_{ij}(\boldsymbol{x})\, e^{-\alpha x^2}\, d\boldsymbol{x}\right]^2}, \tag{15.27'}$$

where the integral on the right is evaluated over all the vector space $\boldsymbol{x}$ [the normalizing factor $\alpha^{3/2}/\pi^{3/2}$ is equal to the reciprocal of the integral $\int e^{-\alpha x^2}\, d\boldsymbol{x}$, i.e., it corresponds to the factor $1/V$ in Eq. (15.27)]. If we rewrite the squares of the integrals in Eq. (15.27') as double integrals with respect to $\boldsymbol{x}$ and $\boldsymbol{x}'$, and then change to the new variables $\boldsymbol{r} = \boldsymbol{x}' - \boldsymbol{x}$ and $\boldsymbol{r}_1 = \boldsymbol{x}' + \boldsymbol{x}$, we can transform the right-hand side of Eq. (15.27') as follows:

$$\begin{aligned}
&\lim_{\alpha\to 0} \frac{\alpha^{3/2}}{\pi^{3/2}} \overline{[M(\alpha)]^2} = \\
&= \lim_{\alpha\to 0} \frac{\rho^2}{2} \frac{\alpha^{3/2}}{\pi^{3/2}} \int\int \overline{(x_i u_j - x_j u_i)(x_i' u_j' - x_j' u_i')}\, e^{-\alpha(x^2 + x'^2)}\, d\boldsymbol{x}\, d\boldsymbol{x}' = \\
&= \lim_{\alpha\to 0} \rho^2 \frac{\alpha^{3/2}}{\pi^{3/2}} \int\int \left[\boldsymbol{x}\boldsymbol{x}' B_{jj}(\boldsymbol{x}' - \boldsymbol{x}) - x_i x_j' B_{ij}(\boldsymbol{x}' - \boldsymbol{x})\right] e^{-\alpha(x^2 + x'^2)}\, d\boldsymbol{x}\, d\boldsymbol{x}' = \\
&= \lim_{\alpha\to 0} \rho^2 \frac{\alpha^{3/2}}{8\pi^{3/2}} \int\int \left[(r_1^2 - r^2) B_{jj}(\boldsymbol{r}) - (r_{1i} r_{1j} - r_i r_j) B_{ij}(\boldsymbol{r})\right] \times \\
&\qquad\qquad \times e^{-\frac{\alpha}{2}(r^2 + r_1^2)}\, d\boldsymbol{r}\, d\boldsymbol{r}_1
\end{aligned}$$

(in the last line we have omitted the zero summands containing integrals of odd functions). For an incompressible isotropic turbulence with a finite integral $\Lambda$ we have from Eqs. (14.1), (14.3), and (15.16)

$$\begin{aligned}
&\int B_{jj}(\boldsymbol{r})\, e^{-\frac{1}{2}\alpha r^2}\, d\boldsymbol{r} = -\frac{\alpha}{2} \int B_{jj}(\boldsymbol{r})\, r^2\, d\boldsymbol{r} + o(\alpha) = 4\pi\alpha\Lambda + o(\alpha), \\
&\int r^2 B_{jj}(\boldsymbol{r})\, e^{-\frac{1}{2}\alpha r^2}\, d\boldsymbol{r} = -8\pi\Lambda + O(\alpha), \\
&\int r_i r_j B_{ij}(\boldsymbol{r})\, e^{-\frac{1}{2}\alpha r^2}\, d\boldsymbol{r} = 4\pi\Lambda + O(\alpha).
\end{aligned}$$

Hence it is readily shown that

$$\lim_{\alpha\to 0} \frac{\alpha^{3/2}}{\pi^{3/2}} \overline{[M(\alpha)]^2} = 5\sqrt{2}\,\pi\rho^2\Lambda. \tag{15.28}$$

This result shows the relation between the conservation law (15.16) and the general law of conservation of angular momentum.

In the homogeneous (but not necessarily isotropic) case, similar analysis shows that the limit

$$\lim_{\alpha\to 0} \frac{\alpha^{3/2}}{\pi^{3/2}} \overline{M_{ij}(\alpha) M_{kl}(\alpha)} = \lim_{\alpha\to 0} \rho^2 \frac{\alpha^{3/2}}{\pi^{3/2}} \int\int \overline{m_{ij}(x)\, m_{kl}(x')}\, e^{-\alpha(x^2+x'^2)}\, dx\, dx' \tag{15.29}$$

can be written in the form of a linear combination of the integrals $\Lambda_{ij,\,mn}$ multiplied by $\rho^2$ and that, conversely, all the $\rho^2\Lambda_{ij,\,mn}$ can be represented as linear combinations of normalized products (15.29) of the components of the angular momentum tensor.

Similarly, we can rewrite Eq. (15.26) in the form

$$\lim_{V\to\infty} \frac{1}{V} \overline{\left[\int_V \vartheta(x)\, dx\right]^2} = K' = \text{const}, \tag{15.30}$$

which shows that Eq. (15.20) expresses the general law of conservation of heat (or mass of passive admixture). In fact, it is readily seen that if the quantity (15.26) [which is equal to $8\pi^3 F_{\vartheta\vartheta}(0)$] is finite then

$$\overline{\left[\int_V \vartheta(x)\, dx\right]^2} = \int_V \int_V B_{\vartheta\vartheta}(x'-x)\, dx\, dx' = K'V + o(V). \tag{15.31}$$

This equation can also be replaced by

$$4\pi K = K' = \lim_{\alpha\to 0} \frac{\alpha^{3/2}}{\pi^{3/2}} \overline{\left[\int \vartheta(x)\, e^{-\alpha x^2}\, dx\right]^2}, \tag{15.32}$$

The form of this is very similar to that of Eq. (15.28) although the proof is considerably simpler.

## 15.3 Final Period of Decay of Isotropic Turbulence

The behavior of the spectral density in the neighborhood of the origin in wave-number space, i.e., in the region of turbulence components with the longest wavelengths, is also important in the investigation of the final period of decay of isotropic turbulence. In fact, we saw in Sect. 14 that the rate of decrease of the velocity fluctuations (or temperature fluctuations) with given wave number $k$ under the action of viscosity (or thermal conduction) is proportional to $2\nu k^2$ (or $2\chi k^2$), i.e., it increases rapidly with increasing $k$. To be specific, we shall consider the mean square *velocity fluctuations*, i.e., we shall be concerned with turbulent energy. The treatment of temperature fluctuations is completely analogous. During the first stage of turbulence decay, the viscous dissipation of the turbulent energy may be compensated by the inflow of energy from other regions of wave-number space due to turbulent mixing. If, however, there is no external energy source then, sooner or later, a stage will be reached when it will no longer be possible to maintain an appreciable energy transfer from one wave-number region to another, which would be comparable in magnitude with the rate of dissipation. From this moment onward, the spectral density will decrease exponentially for all values of $\boldsymbol{k}$ outside a small neighborhood of $\boldsymbol{k}=0$, and the spectrum will vary more slowly only for very small values of $|\boldsymbol{k}|$. It is clear that the asymptotic behavior of the correlation functions for very large values of $t$ should be determined exclusively by the behavior of the initial spectrum in the neighborhood of the point $\boldsymbol{k}=0$.

Let us try to establish this last statement in a more rigorous fashion. Let us define the final period of turbulence decay as the period during which the redistribution of energy and of intensity of temperature fluctuations over the turbulent spectrum can be neglected in comparison with dissipation processes. In other words, the final period of decay will be defined as the period during which the overall reduction in the mean squares of fluctuation of the flow variables due to viscosity and thermal conduction has already ensured that the quadratic inertial terms in the dynamic equations, (which are responsible for the energy redistribution over the spectrum), have become negligible in comparison with the linear disipative terms, i.e., the Reynolds number and the Péclet number have become much less than unity. At this time the flow can no longer be regarded as turbulent in the usual sense of this word, since turbulent mixing has ceased, and there is only a quiet "laminar" decay of individual large-scale disturbances. Further evolution of the velocity and temperature fields in this completely decayed turbulence will be described by linear equations of the form

$$\frac{\partial \boldsymbol{u}}{\partial t}=\nu\Delta\boldsymbol{u}, \qquad (15.33)^{6}$$

[6] It is known that the terms in the Navier-Stokes equations which contain the pressure are of the same order as the terms which are quadratic in the velocities [see Eq. (1.9) of Vol. 1], and therefore if the quadratic terms are assumed to be negligible, then we must also neglect the pressure gradient.

and

$$\frac{\partial \vartheta}{\partial t} = \chi \Delta \vartheta. \tag{15.34}$$

If we use these equations to investigate the final period of decay, then in the correlation and spectral equations we can neglect the third-order correlation functions and their Fourier transforms in comparison with the ordinary second-order correlation and spectra. In other words, these equations now assume the form

$$\frac{\partial B_{LL}(r, t)}{\partial t} = 2\nu \left(\frac{\partial}{\partial r} + \frac{4}{r}\right) \frac{\partial B_{LL}(r, t)}{\partial r}, \tag{15.35}$$

$$\frac{\partial B_{\vartheta\vartheta}(r, t)}{\partial t} = 2\chi \left(\frac{\partial}{\partial r} + \frac{2}{r}\right) \frac{\partial B_{\vartheta\vartheta}(r, t)}{\partial r} \tag{15.36}$$

and correspondingly

$$\frac{\partial F(k, t)}{\partial t} = -2\nu k^2 F(k, t), \tag{15.37}$$

$$\frac{\partial F_{\vartheta\vartheta}(k, t)}{\partial t} = -2\chi k^2 F_{\vartheta\vartheta}(k, t). \tag{15.38}$$

The last four equations are linear and each of them contains one unknown function. Therefore, if $t_0$ is an arbitrary instant of time, contained in the final period of decay, then these equations can be used to determine unambiguously the values of any of the functions $B_{LL}(r, t)$, $B_{\vartheta\vartheta}(r, t)$, $F(k, t)$, and $F_{\vartheta\vartheta}(k, t)$ for $t > t_0$ from their initial values at time $t_0$.

The simplest equations are (15.37) and (15.38). Their solutions are given by

$$F(k, t) = F(k, t_0)\, e^{-2\nu k^2 (t - t_0)}, \tag{15.39}$$

$$F_{\vartheta\vartheta}(k, t) = F_{\vartheta\vartheta}(k, t_0)\, e^{-2\chi k^2 (t - t_0)}. \tag{15.40}$$

According to Eqs. (12.75) and (12.4), the solutions of Eqs. (15.35) and (15.36) can be written in the form

$$B_{LL}(r, t) = -8\pi \int_0^\infty \frac{kr\cos kr - \sin kr}{k^3r^3} e^{-2\nu k^2(t-t_0)} F(k, t_0)\, k^2\, dk \quad (15.41)$$

and

$$B_{\vartheta\vartheta}(r, t) = 4\pi \int_0^\infty \frac{\sin kr}{kr} e^{-2\chi k^2(t-t_0)} F_{\vartheta\vartheta}(k, t_0)\, k^2\, dk, \quad (15.42)$$

where $F(k, t_0)$ and $F_{\vartheta\vartheta}(k, t_0)$ can in turn be expressed in terms of $B_{LL}(r, t_0)$ and $B_{\vartheta\vartheta}(r, t_0)$ with the aid of Eqs. (12.77) and (12.2). If we split the integrals on the right-hand sides of Eqs. (15.41) and (15.42) into a) integrals between zero and some small (but fixed) number $\delta > 0$, and b) integrals between $k = \delta$ and $k = \infty$, then the second of these integrals will decrease exponentially as $t - t_0 \to \infty$ (at least as rapidly as $e^{-2\nu\delta^2(t-t_0)}$ or, correspondingly, $e^{-2\chi\delta^2(t-t_0)}$), and the first integrals will fall more slowly. Hence it follows that the behavior of the functions $B_{LL}(r, t)$ and $B_{\vartheta\vartheta}(r, t)$ for $t - t_0 \to \infty$ is independent of the values of $F(k, t_0)$ and $F_{\vartheta\vartheta}(k, t_0)$ for $k > \delta$, where $\delta$ is an arbitrary fixed positive number.

Substituting Eqs. (12.77) and (12.2) into Eqs. (15.41) and (15.42), and integrating with respect to $k$, we obtain formulas expressing $B_{LL}(r, t)$ and $B_{\vartheta\vartheta}(r, t)$ in terms of $B_{LL}(r, t_0)$ and $B_{\vartheta\vartheta}(r, t_0)$. The same formulas can be obtained directly from Eqs. (15.35) and (15.36) if we use, for example, the fact that the right-hand side of Eq. (15.35) is identical with the radial part of the five-dimensional Laplace operator, and the right-hand side of Eq. (15.36) is identical with the radial part of the usual three-dimensional Laplace operator. Therefore, the general solution of Eq. (15.35) can be written in the form

$$B_{LL}(r, t) = \frac{1}{[8\pi\nu(t-t_0)]^{5/2}} \int\int\int_{-\infty}^{\infty}\int\int B_{LL}(\xi, t_0)\, e^{-\frac{\sum_{i=1}^{5}(r_i-\xi_i)^2}{8\nu(t-t_0)}}\, d\xi_1\, d\xi_2\, d\xi_3\, d\xi_4\, d\xi_5, \quad (15.43)$$

where $\xi_1, \ldots, \xi_5$ are the integration variables, $\xi = \left(\sum_{i=1}^{5} \xi_i^2\right)^{1/2}$, and $r_1, \ldots, r_5$ are arbitrary numbers such that $\left(\sum_{i=1}^{5} r_i^2\right)^{1/2} = r$. If we transform in Eq. (15.43) to spherical coordinates in five-dimensional space, and evaluate the integrals with respect to the angular variables, we obtain

$$B_{LL}(r, t) = \frac{e^{-\frac{r^2}{8\nu(t-t_0)}}}{4\nu(t-t_0)\, r^{3/2}} \int_0^{\infty} B_{LL}(\xi, t_0)\, \xi^{5/2} I_{3/2}\left[\frac{r\xi}{4\nu(t-t_0)}\right] e^{-\frac{\xi^2}{8\nu(t-t_0)}}\, d\xi, \tag{15.44}$$

where $I_{3/2}(x)$ is the Bessel function of order 3/2 of an imaginary argument. It can be expressed in terms of exponential and power functions (see Millionshchikov, 1939a, and Loitsyanskii, 1939). Equation (15.36) on the other hand is identical with the ordinary equation of thermal conduction for a spherically-symmetric temperature distribution. The general solution of this is known to be of the form

$$B_{\vartheta\vartheta}(r, t) = \frac{e^{-\frac{r^2}{8\chi(t-t_0)}}}{[2\pi\chi(t-t_0)]^{1/2}\, r} \int_0^{\infty} B_{\vartheta\vartheta}(\xi, t_0)\, \mathrm{sh}\left[\frac{r\xi}{4\chi(t-t_0)}\right] e^{-\frac{\xi^2}{8\chi(t-t_0)}}\, \xi\, d\xi. \tag{15.45}$$

Let us now consider the asymptotic behavior of the solutions (15.44) and (15.45) as $t - t_0 \to \infty$. We shall begin with the case of a simpler function $B_{\vartheta\vartheta}(r, t)$. It is easily shown from Eq. (15.45) that the asymptotic behavior of this function will be determined only by certain simple integral characteristics of the "initial function" $B_{\vartheta\vartheta}(r, t_0)$. In fact, $\exp\left[-\frac{\xi^2}{8\nu(t-t_0)}\right] \to 1$ as $t - t_0 \to \infty$ for all $\xi$, for which $B_{\vartheta\vartheta}(\xi, t_0)$ differs significantly from zero, and the function $\mathrm{sh}\left[\frac{r\xi}{4\chi(t-t_0)}\right]$, where $r$ is fixed, can be replaced by the first term of its power series in this region. If we do this we can readily verify that, if at $t = t_0$ the Corrsin integral (15.20) is finite and not zero, then for sufficiently large $t - t_0$ we have

$$B_{\vartheta\vartheta}(r, t) \approx \frac{K}{4\sqrt{2\pi}\,[\chi(t-t_0)]^{3/2}}\, e^{-\frac{r^2}{8\chi(t-t_0)}} \tag{15.46}$$

and

$$\overline{\vartheta^2(t)} = B_{\vartheta\vartheta}(0, t) \approx \frac{K}{4\sqrt{2\pi}\,[\chi(t-t_0)]^{3/2}}. \tag{15.47}$$

Thus the asymptotic behavior of $B_{\vartheta\vartheta}(r, t)$ and $\overline{\vartheta^2}(t)$ now depends only on the quantity $K$. The right-hand side of Eq. (15.46) is identical with the Green function of the three-dimensional thermal conduction equation, i.e., the exact solution of this equation corresponding to the case where for $t = t_0$ the heat is localized at the single point $r = 0$, so that the "initial value" $B_{\vartheta\vartheta}(r, t_0)$ is a $\delta$-function. According to Eqs. (15.46) and (15.47), when $0 < K < \infty$ the function $B_{\vartheta\vartheta}(r, t)/B_{\vartheta\vartheta}(0, t)$ will asymptotically approach the density of the Gaussian distribution with a width ("variance") linearly increasing with $t$, and $\overline{\vartheta^2}(t)$ will decrease in proportion to $t^{-3/2}$.

Similarly, it can be shown that if at $t = t_0$ the Corrsin integral $K$ is zero, but the integral

$$K_1 = \int_0^\infty B_{\vartheta\vartheta}(r, t_0)\, r^4\, dr, \tag{15.48}$$

is finite and nonzero, then the asymptotic behavior of $B_{\vartheta\vartheta}(r, t)$ will be described by

$$B_{\vartheta\vartheta}(r, t) \approx -\frac{K_1}{\sqrt{2\pi}\,[4\chi(t - t_0)]^{5/2}}\left[1 - \frac{r^2}{12\chi(t - t_0)}\right] e^{-\frac{r^2}{8\chi(t - t_0)}}. \tag{15.49}$$

This formula will also determine the special solution of Eq. (15.36) corresponding to a "source-type initial value" which is zero for all $r \neq 0$, i.e., the solution corresponding to the initial condition at time $t_0 = 0$ in the form of a set of elementary thermal dipoles with all possible orientations in space located at the point $r = 0$. In general, if for $t = t_0$ the integrals

$$K_p = \int_0^\infty B_{\vartheta\vartheta}(r, t_0)\, r^{2p+2}\, dr \tag{15.50}$$

with $p = 0, 1, \ldots, s-1$ vanish, and the integral $K_s$ is finite and nonzero, then the asymptotic behavior of the solution $B_{\vartheta\vartheta}(r, t)$ as $t - t_0 \to \infty$ [which is readily obtained through the expansion of the functions $\exp\{-\xi^2/8\chi(t - t_0)\}$ and $\mathrm{sh}\{r\xi/4\chi(t - t_0)\}$ in Eq. (15.45) in power series] will be given by

$$\overline{\vartheta^2(t)} = \frac{C_s}{(t-t_0)^{s+\frac{3}{2}}}, \quad C_s = \frac{(-1)^s K_s}{\sqrt{2\pi}\, 2^{3s+2} s!\, \chi^{s+\frac{3}{2}}}, \tag{15.51}$$

and

$$B_{\vartheta\vartheta}(r, t) = \frac{C_s}{(t-t_0)^{s+\frac{3}{2}}} \left\{ 1 - \frac{2s}{3!} \frac{r^2}{4\chi(t-t_0)} + \frac{2^2 s(s-1)}{5!} \frac{r^4}{[4\chi(t-t_0)]^2} - \dots \right.$$
$$\left. \dots + \frac{(-1)^s 2^s s!}{(2s+1)!} \frac{r^{2s}}{[4\chi(t-t_0)]^s} \right\} e^{-\frac{r^2}{8\chi(t-t_0)}}, \tag{15.52}$$

i.e., it will depend only on $K_s$. It is readily verified that the right-hand side of Eq. (15.52) is equal to the $s$th time derivative of the right-hand side of Eq. (15.46) multiplied by a constant factor. It is clear that Eq. (15.52) is also an exact solution of the thermal conduction equation (15.36) corresponding to certain "improper" initial values localized at the point $r = 0$.

Similar analysis can be applied to the equation for the velocity correlation. Since $\exp\left[-\frac{\xi^2}{8\nu(t-t_0)}\right] \to 1$ as $t - t_0 \to \infty$ for all $\xi$ for which $B_{LL}(\xi, t_0)$ differs appreciably from zero, and the function $I_{3/2}\left[\frac{r\xi}{4\nu(t-t_0)}\right]$ can be represented by the first term $\frac{(r\xi)^{3/2}}{12\sqrt{2\pi}\,[\nu(t-t_0)]^{3/2}}$ of the corresponding power series for any fixed $r$ and $t - t_0 \to \infty$, it follows from Eq. (15.44) that, when the Loitsyanskii integral (15.16) is finite and nonzero for $t = t_0$, we have for large values of $t - t_0$

$$B_{LL}(r, t) \approx \frac{\Lambda}{48\sqrt{2\pi}\,[\nu(t-t_0)]^{5/2}} e^{-\frac{r^2}{8\nu(t-t_0)}} \tag{15.53}$$

and

$$\overline{u^2(t)} = B_{LL}(0, t) \approx \frac{\Lambda}{48\sqrt{2\pi}\,[\nu(t-t_0)]^{5/2}} \tag{15.54}$$

(Millionshchikov, 1939a). Therefore, $B_{LL}(r, t)$ and $\overline{u^2(t)}$ will now depend only on the initial value of the quantity $\Lambda$. If, on the other hand, $\Lambda = 0$ at time $t = t_0$, and some integral of the form

$$\Lambda_p = \int_0^\infty B_{LL}(r, t_0) r^{2(p+2)} dr, \qquad p = 1, 2, \ldots, \tag{15.55}$$

is nonzero and finite, then by expanding the exponential and Bessel functions on the right-hand side of Eq. (15.44) in power series we can readily show that as $t - t_0 \to \infty$ we have

$$\overline{u^2(t)} \approx \frac{A_s}{(t-t_0)^{\frac{5}{2}+s}}, \qquad A_s = \frac{(-1)^s \Lambda_s}{3 \cdot 2^{3s+4} \sqrt{2\pi}\, s!\, \nu^{\frac{5}{2}+s}} \tag{15.56}$$

and

$$B_{LL}(r, t) \approx \frac{A_s e^{-\frac{r^2}{8\nu(t-t_0)}}}{(t-t_0)^{\frac{5}{2}+s}} \left\{1 - \frac{2s}{5} \frac{r^2}{8\nu(t-t_0)} + \frac{2^2 s(s-1)}{2!\,5 \cdot 7} \frac{r^4}{[8\nu(t-t_0)]^2} - \ldots \right.$$
$$\left. \ldots + \frac{(-1)^s 2^s}{5 \cdot 7 \ldots (3+2s)} \frac{r^{2s}}{[8\nu(t-t_0)]^s} \right\}, \tag{15.57}$$

where $\Lambda_s$ is the first nonzero moment in Eq. (15.55) [see Sedov (1944, 1951)].

The right-hand sides of Eqs. (15.54) and (15.57) are the special solutions of Eq. (15.35) corresponding to certain "source-type" initial values $B_{LL}(r, t_0)$ which are zero for $r \neq 0$ and infinite (or meaningless) for $r = 0$ (such initial conditions can approximately describe a very intense but very small-scale turbulence). We note also that if the right-hand side of Eq. (15.53) is denoted by $\varphi^{(0)}(r, t)$, then the right-hand side of Eq. (15.57) will differ from $\varphi^{(s)}(r, t) = \frac{d^s}{dt^s} \varphi^{(0)}(r, t)$ only by a constant factor.

Formula (15.52) does not exhaust all the possible types of behavior of the correlation functions $B_{\vartheta\vartheta}(r, t)$ for large values of $t - t_0$. If we assume that at time $t_0$ all the moments given by Eq. (5.50) are either zero or infinite, then we can obtain a large number of new asymptotic formulas. If we further assume that the initial conditions for $t = t_0$ affect the asymptotic behavior of $B_{\vartheta\vartheta}(r, t)$ for $t - t_0 \to \infty$ only through a constant factor $A$, which is an "integral characteristic" of the function $B_{\vartheta\vartheta}(r, t_0)$, then all the asymptotic formulas can be simply investigated by dimensional arguments as was done by Sedov (1944, 1951) for velocity fluctuations. In fact, let the dimensions of the quantity $A$ be $[A] = \Theta^2 L^m T^n$ where $\Theta$ represents the dimension of $\vartheta$ [which appears as a square since $A$ is a linear function of $B_{\vartheta\vartheta}(r, t_0)$]. The ratio $B_{\vartheta\vartheta}(r, t) \chi^{\frac{m}{2}} (t-t_0)^{n+\frac{m}{2}} / A$ will then be a dimensionless function of $r$ and $t-t_0$ which, in view of Eq. (15.36), can depend only on the parameter $\chi$. Consequently, the asymptotic behavior of $B_{\vartheta\vartheta}(r, t)$ will be determined by the solution of Eq. (15.36) which is of the form

$$B_{\vartheta\vartheta}(r, t) = \frac{C}{(t-t_0)^{\alpha}} \varphi\left[\frac{r^2}{\chi(t-t_0)}\right] \tag{15.58}$$

where

$$C = A\chi^{-\frac{m}{2}}, \quad \alpha = n + \frac{m}{2}$$

Substituting Eq. (15.58) into Eq. (15.36), we can readily verify that the function $\varphi(x)$, which is normalized by the condition $\varphi(0) = 1$, can be written in the form

$$\varphi(x) = M\left(\alpha, \frac{3}{2}, -\frac{x}{8}\right), \tag{15.59}$$

where

$$M(\alpha, \gamma, x) = 1 + \frac{\alpha}{\gamma}\frac{x}{1!} + \frac{\alpha(\alpha+1)}{\gamma(\gamma+1)}\frac{x^2}{2!} + \dots$$

is the confluent hypergeometric function [it is also sometimes denoted as $\Phi(\alpha, \gamma; x)$ or as ${}_1F_1(\alpha;\gamma;x)$]. As $r \to \infty$, the solution (15.58)–(15.59) will decrease as $r^{-2\alpha}$, so that when $\alpha < 3/2$ the solution will correspond to an infinite Corrsin integral (15.20), while for $\alpha < 3/2 + s$, where $s$ is an integer, all the integrals $K_p$ with $p \geqslant s$ will be infinite, and the integrals $K_0, K_1, \dots, K_{s-1}$ will be finite [and zero for nonintegral $\alpha - 3/2$ since otherwise we would obtain one of the solutions in Eq. (15.52)]. When $\alpha = 3/2 + s$, where $s$ is an integer, the solution (15.58)–(15.59) will be identical with Eq. (15.52).

In the language of spectral theory, the exact solutions (15.46), (15.52), and (15.58)–(15.59) of Eq. (15.36) correspond to the solutions of Eq. (15.38) of the form

$$F_{\vartheta\vartheta}(k, t) = \frac{K}{2\pi^2} e^{-2\chi k^2 (t-t_0)}, \tag{15.60}$$

$$F_{\vartheta\vartheta}(k, t) = \frac{(-1)^s K_s}{2\pi^2 (2s+1)!} k^{2s} e^{-2\chi k^2 (t-t_0)} \tag{15.61}$$

and

$$F_{\vartheta\vartheta}(k, t) = C_1 k^{2\alpha-3} e^{-2\chi k^2 (t-t_0)}, \quad 0 < \alpha < \infty \tag{15.62}$$

(see the analogous formulas given by Yaglom, 1948). We could use Eqs. (15.60)–(15.62) to devise a simple explanation of many of the properties of Eqs. (15.46), (15.52), and (15.58)–(15.59). Thus, for example, all the facts noted above about the "initial values" $B_{\vartheta\vartheta}(r, t_0)$ of these solutions, and the corresponding values of the integrals (15.20) and (15.50), will follow, in view of Eqs. (12.14) and (12.16), from the power-function form of the corresponding "initial values" $F_{\vartheta\vartheta}(k, t_0)$. It also follows that for the solution (15.52), the integrals $K_0, \dots, K_{s-1}$ will be zero for all $t > t_0$, and $K_s$ will have the same value for all $t > t_0$, i.e., it will be an invariant. Next, since the asymptotic behavior of $B_{\vartheta\vartheta}(r, t)$ as $t - t_0 \to \infty$ depends only on the behavior of the function $F_{\vartheta\vartheta}(k, t_0)$ in the neighborhood of the origin, it follows that in the case of a function $F_{\vartheta\vartheta}(k, t_0)$ admitting a power-function approximation near $k = 0$, this asymptotic behavior can be specified by one of the formulas (15.46), (15.52), and (15.58)–(15.59). In particular, when the function $F_{\vartheta\vartheta}(k, t_0)$ can be expanded in a Taylor series in powers of $k^2$, the asymptotic behavior of $B_{\vartheta\vartheta}(r, t)$ is the same as that obtained when $F_{\vartheta\vartheta}(k, t_0)$ is replaced by the first nonzero term of its Taylor series, i.e., it is given by a formula of the form of Eq. (15.52).

The simple formulas of Eq. (15.53) or (15.57) will similarly, be valid only when at the initial time $t = t_0$ at least one of the integrals (15.16), (15.55) is finite and nonzero. They do not, however, exhaust all possible types of asymptotic behavior of the solutions of Eq. (15.35). Thus, for example, von Kármán and Howarth (1938) noted that Eq. (15.35) has a one-parameter family of linearly independent exact solutions which can be written in the form [according to Sedov (1944, 1951)]

$$B_{LL}(r, t) = \frac{C}{(t - t_0)^{\alpha}} M\left(\alpha, \frac{5}{2}, -\frac{r^2}{8\nu(t - t_0)}\right), \qquad 0 < \alpha < \infty \qquad (15.63)$$

[see also a note by D. A. Lee (1965) where another representation of Eq. (15.63) with $\alpha = 2$ is indicated]. It is clear that the decay of the energy $\overline{u^2(t)}$ during the final period of decay of isotropic turbulence can, in principle, be described by a power function with an arbitrary power index. The solutions (15.63) are similar to Eqs. (15.53) and (15.57) in the sense that they contain the single parameter $C$ which may depend on the initial conditions. In this connection, Sedov (1944, 1951) has noted that, conversely, if we suppose that the initial conditions do affect the asymptotic behavior of the function $B_{LL}(r, t)$ for $t \to \infty$, but only in the form of a constant (dimensional) factor, we necessarily obtain a solution of the form (15.63) [see the derivation of Eqs. (15.58)–(15.59)]. As $r \to \infty$, the solution given by Eq. (15.63) will asymptotically decrease as $r^{-2\alpha}$ hence it is clear that for $\alpha < 5/2$, the solution will correspond to an infinite Loitsyanskii integral (15.16), while for $\alpha < 5/2 + s$, where $s$ is an integer, the integrals in Eq. (15.55) with $p \geqslant s$ will be infinite. It is also readily verified that when $\alpha = 5/2$, the solution (15.63) will become identical with the Loitsyanskii-Millionshchikov solution (15.53), and for $\alpha = 5/2 + s$, where $s$ is an integer, it will become identical with the Sedov solution (15.57).

The properties of the exact solution (15.53), (15.57), and (15.63) of Eq. (15.35) can readily be established from the fact that they correspond to the solutions

$$F(k, t) = \frac{\Lambda}{12\pi^2} k^2 e^{-2\nu k^2 (t - t_0)}, \qquad (15.64)$$

$$F(k, t) = \frac{(-1)^s \Lambda_s}{4(2s+3)(2s+1)!\,\pi^2} k^{2s+2} e^{-2\nu k^2 (t - t_0)} \qquad (15.65)$$

and, correspondingly,

$$F(k, t) = C_1 k^{2\alpha - 3} e^{-2\nu k^2 (t - t_0)}, \qquad 0 < \alpha < \infty, \qquad (15.66)$$

of the spectral equation (15.37) [see Yaglom, 1948]. The fact that the solution (15.57) differs from the $s$th time derivative of Eq. (15.53) only by a constant factor is also an obvious consequence of Eqs. (15.64) and (15.65).

Since the asymptotic behavior of the function $B_{LL}(r, t)$ as $t \to \infty$ is determined by the behavior of the function $F(k, t_0)$ in a small neighborhood of the point $k = 0$, it follows that if the spectral density $F(k, t_0)$ can be expanded in a Taylor series, then for very large $t$ we can use the first nonzero term of this series instead of the exact value of $F(k, t_0)$. Consequently, for such $F(k, t_0)$ the asymptotic behavior of the turbulent energy $\overline{u^2}(t)$ as $t \to \infty$, and of the correlation function $B_{LL}(r, t)$ as $t \to \infty$, will be described by Eqs. (15.54) and (15.53), or (15.56) and (15.57). If, on the other hand, the behavior of $F(k, t_0)$ in the neighborhood of the point $k = 0$ is described by an arbitrary power function, then $B_{LL}(r, t)$ will asymptotically approach Eq. (15.63).

## 15.4 Experimental Data on the Final Period of Decay. The Decay of Homogeneous Turbulence

The general solutions of the differential equations (15.35)–(15.38), which describe the final period of decay of isotropic turbulence, depend on functions of a single variable, namely, the initial value of the corresponding correlation or spectral functions at a time $t_0$ (during the final period). To compare solutions with experimental data we must, therefore, carefully measure the correlation or spectral function for isotropic turbulence at two well-separated instants of time during the final period of decay. Such measurements present considerable difficulties, so that it is not surprising that the general solutions of Eqs. (15.35)–(15.38) [of the form (15.44), (15.45), etc.] have not as yet been verified experimentally (see, however, Figs. 14 and 15).

The asymptotic behavior of the correlation functions for $t - t_0 \to \infty$ is more suitable for experimental verification. We have seen that the asymptotic correlation functions are given by a relatively simple formula under sufficiently general conditions. In

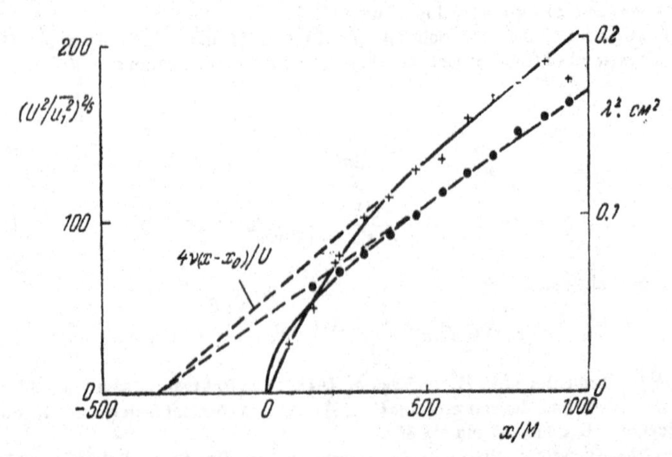


FIG. 14 Verification of the turbulence-decay law for the final period of decay.

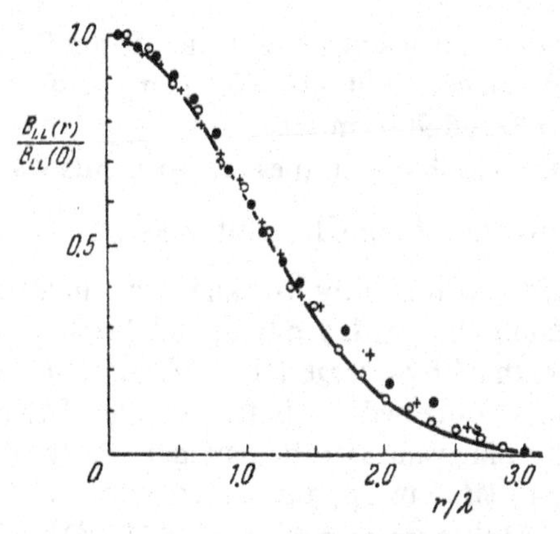


**FIG. 15** Verification of Eq. (15.53′) for the longitudinal velocity correlation function.

particular, the formulas given by Eqs. (15.46), (15.47), and (15.53), (15.54) correspond to the case of spectral densities that are regular at the origin [and at the same time describe the exact solutions of Eqs. (15.35), (15.36) which correspond to very fine-scale initial turbulence with finite Corrsin and Loitsyanskii integrals]. It is natural to try to compare these formulas with experimental data for large values of $t$ [see the papers by Millionshchikov (1939a), Loitsyanskii (1939) and Corrsin (1951b) in which they predict that these formulas should be valid at the end of the decay process].

From the theoretical point of view, the decay of temperature fluctuations is simpler than the decay of velocity fluctuations. However, the experimental study of temperature fluctuations involves a number of technical difficulties which have not really been overcome as yet. Therefore, when we speak of experimental data referring to the final period of decay, we are restricted to velocity-fluctuation data.

The first attempt to compare the theoretical formulas given by Eqs. (15.53) and (15.54) with measurements of velocity fluctuations behind a grid in a wind tunnel is due to Millionshchikov (1939b). However, this attempt was unsuccessful because, in 1939, there were no experimental data on the final period of decay [see the discussion of this attempt by Kolmogorov (1941c) and Frenkiel (1948)]. Greater success was achieved by Batchelor and Townsend (1948b) who carried out special measurements to verify the formulas given by Eqs. (15.53) and (15.54) for grid turbulence during the final period of decay. To achieve sufficiently low values of the Reynolds number, they used a fine-mesh grid ($M = 0.16$ cm) and low free-stream velocities ($U = 620$ and 900 cm/sec).

The use of a fine-mesh grid in these experiments was convenient in two respects. First, the Reynolds number $\mathrm{Re}_M = \frac{UM}{\nu}$ describing the flow near the grid was relatively low and, even more important, the measurements could be carried out up to distances exceeding the characteristic length $M$ by a large factor. Figure 14 shows the values of $(U^2/\overline{u_1^2})^{2/5}$ (points) obtained by Batchelor and Townsend together with the corresponding values of the parameter $\lambda^2$ (crosses). It is clear that, from $x = 450M$ onward, the dependence of $(U^2/\overline{u_1^2})^{2/5}$ on $x$ is practically linear [this corresponds to Eq. (15.54)] while $\lambda^2$ follows the law $\frac{4\nu}{U}(x - x_0)$ [which in view of Eq. (15.7) corresponds to the same law of decay of turbulent energy]. The value of $x_0$ was found to approach $-350M$ (to ensure that $t_0 = x_0/U$ could be taken to be a time during the final period of decay, it would have been necessary to continue the measurements for much larger values of $x$, but this was impossible because of the finite dimensions of the wind tunnel).

Since the decay law (15.54) corresponds to $\lambda^2 = 4\nu(t - t_0)$, we can rewrite equation (15.53) in the form

$$\frac{B_{LL}(r, t)}{\overline{u_1'^2}(t)} = e^{-\frac{r^2}{2\lambda^2}}, \tag{15.53'}$$

which does not contain the time explicitly. This formula was also compared with data by Batchelor and Townsend (1948b). The results of this comparison are shown in Fig. 15 where the solid line represents the "theoretical curve" (15.53'), and the points, circles, and crosses represent values of $B_{LL}(r)$ found experimentally for different values of $x$ during the final period of decay. The agreement is undoubtedly satisfactory.[7]

The data shown in Figs. 14 and 15 suggest that the experimental results confirm the theory given in Sect. 15.3, and give the impression

[7] However, it was pointed out by D. A. Lee (1965), that equally good agreement can be obtained by using Eq. (15.63) with $\alpha = 2$, which is even closer than Eq. (15.53') to the subsequent data of Tan and Ling (1963).

at first glance that the decay of grid turbulence is in very good agreement with the model of idealized homogeneous and isotropic turbulence with a regular spectrum at $k=0$ and a finite nonzero Loitsyanskii integral. In actual fact, however, the situation is not so simple. First, special measurements of the ratio $\overline{u_1^2}/\overline{u_2^2}$ ($u_2$ is one of the components of the turbulent velocity in a direction parallel to the grid) at different distances from the grid, which were carried out by Batchelor and Stewart (1950) [see also Batchelor (1953)], have shown that although at small distances from the grid the turbulence can be regarded as nearly isotropic (i.e., $\overline{u_1^2}/\overline{u_2^2}$ is close to unity), the turbulence at greater distances which correspond to the final period of decay is obviously not isotropic ($\overline{u_1^2}/\overline{u_2^2}$ approaches 1.5). The results shown in Figs. 14 and 15 must therefore be explained in terms of the theory of anisotropic turbulence. Second, we shall see below that the regularity of the turbulence spectrum at the point $k=0$ is also subject to considerable doubt. These two weaknesses of the theory developed above will now be discussed in detail.

Let us begin by investigating the final period of decay of an arbitrary homogeneous (but nonisotropic), turbulence. We have already seen in Sect. 11 [see Eq. (11.95)] that in the "regular case" the spectral tensor $F_{ij}(\boldsymbol{k})$ of such turbulence can be expanded in a Taylor series of the form

$$F_{ij}(\boldsymbol{k})=f_{ij,\,mn}k_m k_n+O(k^3). \tag{15.67}$$

We shall assume that this expansion corresponds to the "initial time" $t_0$ of the final period of decay. When applied to this final period of decay, Eq. (14.2) is written in the form

$$\frac{\partial F_{ij}(\boldsymbol{k},\,t)}{\partial t}=-2\nu k^2 F_{ij}(\boldsymbol{k},\,t). \tag{15.68}$$

Therefore, for $t>t_0$

$$F_{ij}(\boldsymbol{k},\,t)=F_{ij}(\boldsymbol{k},\,t_0)\,e^{-2\nu k^2(t-t_0)}, \tag{15.69}$$

and hence in the regular case

$$F_{ij}(\boldsymbol{k},\,t)=[f_{ij,\,mn}k_m k_n+O(k^3)]\,e^{-2\nu k^2(t-t_0)}. \tag{15.70}$$

If we substitute Eq. (15.70) into the general formula (11.54), we obtain

$$B_{ij}(r, t) = \int e^{ikr} [f_{ij,\,mn} k_m k_n + O(k^3)] e^{-2\nu k^2 (t-t_0)} \, dk. \quad (15.71)$$

Let us divide the range of integration in Eq. (15.71) into two parts, namely, a sphere centered on the origin and having a small radius $\delta > 0$, and an external region $|\boldsymbol{k}| > \delta$. For $t = t_0 \to \infty$, the integral over the external region will fall more rapidly than $e^{-2\nu\delta^2 (t-t_0)}$, while the integral over the region $|\boldsymbol{k}| \leqslant \delta$ will be determined by terms in the brackets which are bilinear in $k_m$. Consequently, to within terms which decay rapidly with time, we have

$$B_{ij}(r, t) = \int f_{ij,\,mn} k_m k_n e^{ikr - 2\nu k^2 (t-t_0)} \, dk \quad (15.72)$$

The region of integration in this expression can now be extended to the entire space of the vectors $\boldsymbol{k}$, since this merely adds an exponentially decaying term. The integral on the right-hand side of Eq. (15.72) can readily be evaluated: It is equal to

$$-f_{ij,\,mn} \frac{\partial^2}{\partial r_m \partial r_n} \int e^{ikr - 2\nu k^2 (t-t_0)} \, dk =$$

$$= -4\pi f_{ij,\,mn} \frac{\partial^2}{\partial r_m \partial r_n} \int_0^\infty \frac{\sin kr}{kr} e^{-2\nu k^2 (t-t_0)} k^2 \, dk =$$

$$= -\frac{\sqrt{2\pi^3} f_{ij,\,mn}}{4 [\nu (t-t_0)]^{3/2}} \frac{\partial^2}{\partial r_m \partial r_n} e^{-\frac{r^2}{8\nu (t-t_0)}},$$

Hence when $t - t_0 \to \infty$ we have

$$B_{ij}(r, t) \approx \frac{\sqrt{2\pi^3}}{16 [\nu (t-t_0)]^{5/2}} \left[ f_{ij,\,mm} - \frac{f_{ij,\,mn} r_m r_n}{4\nu (t-t_0)} \right] e^{-\frac{r^2}{8\nu (t-t_0)}}. \quad (15.73)$$

When we use this formula to determine the longitudinal correlation function $B_{LL}(r, t) = \overline{u_L(\boldsymbol{x}) u_L(\boldsymbol{x} + \boldsymbol{r})}$ we must set $i = j = 1$ and $r_1 = r,\ r_2 = r_3 = 0$ Since, moreover, Eqs. (11.95) and (11.79) imply that $f_{11,\,11} = 0$, we find that when the $x_1$ axis is parallel to $\boldsymbol{r}$ we have

$$B_{LL}(r, t) = \frac{\sqrt{2\pi^3} f_{11,\,mm}}{16 [\nu (t-t_0)]^{5/2}} e^{-\frac{r^2}{8\nu (t-t_0)}}. \quad (15.74)$$

Hence it is clear that formula (15.53) [in which $\lambda^2$ is defined by Eq. (15.3) as in the case of isotropic turbulence] will be valid for the final period of decay of any homogeneous turbulence, so that the results given in Fig. 15 do not require the assumption of isotropy for their explanation. From Eq. (15.74) it follows that the normalized longitudinal correlation function $B_{LL}(\mathbf{r}, t)/B_{LL}(0, t)$ will be the same for all directions of the vector $\mathbf{r}$ (so that $\lambda^2$ will also be independent of the direction), while the unnormalized functions $B_{LL}(\mathbf{r}, t)$ for different directions of $\mathbf{r}$ will differ only by the factors $f_{ii,\,mm}$ (no summation over $i$!) which are proportional to the corresponding "energies" $\overline{u_i^2}$. In the course of time the mean squares of the velocity components will decay by the same 5/2 law:

$$\overline{u_i^2(t)} = \frac{C_i}{(t-t_0)^{5/2}}, \quad i = 1, 2, 3; \quad \overline{u^2(t)} = \frac{C}{(t-t_0)^{5/2}}, \tag{15.75}$$

which is in complete agreement with the data shown in Fig. 14. Moreover, it follows from Eq. (15.75) that the ratios of the mean squares of the individual velocity components will remain constant throughout the final period of decay, i.e.,

$$\frac{\overline{u_i^2(t)}}{\overline{u_j^2(t)}} = \frac{C_i}{C_j} = \text{const.} \tag{15.76}$$

This result is also confirmed by data of grid turbulence measurements [see, for example, Batchelor and Stewart (1950) and Batchelor (1953)].

If we assume that at the initial time $t_0$ of the final period of decay the turbulent spectrum is regular at the point $\mathbf{k} = 0$, then all the results obtained in the wind tunnel experiments for the final period of decay can be explained in terms of the homogeneous (but not isotropic) model of turbulence filling an unbounded space. If, however, we apply this model to the preceding stage of decay, during which we cannot neglect nonlinear terms in the dynamic equations, and try to obtain a description of the spectrum at $t_0$, then we arrive at somewhat unexpected results. The works of Proudman and Reid (1954) and Batchelor and Proudman (1956), which we shall discuss in detail below, have shown that in the case of homogeneous turbulence in an infinite space the pressure forces (which are of the same order as the inertial forces) should, as a rule, lead to a singularity in the spectrum at the point $k = 0$. However, when these results are compared with data for grid turbulence it must be remembered that, in practice, the turbulence always occupies a finite (and not very large) volume, so that long-wave components cannot be present, and the singularity at $k = 0$ may not appear.[8] In such a situation, however, there is no reason to interpret the data of Fig. 15 in terms of homogeneous turbulence in an infinite space and the formula given by Eq. (15.74). It is therefore reasonable to assume at present that, to obtain a sufficiently satisfactory theoretical explanation of the data of Fig. 15, it is necessary to take into account the inhomogeneity of real turbulence behind the grid in the region of sufficiently small wave numbers, which we have so far neglected entirely.

Let us also finally note that the experimental results of Batchelor and Townsend (1948b) are quite unique and were not confirmed by any posterior experiments. It is worth mentioning in this respect the work of Ling and Huang (1970) who measured the parameters of grid turbulence in a low speed water tunnel at very low Reynolds numbers and obtained quite different results (principally that the turbulence was practically homogeneous and isotropic and that the third-order process of spectral energy redistribution plays a significant part even at the smallest value of Re attained).

If, however, we abandon any attempt to explain the data on velocity fluctuations behind the grid in the wind tunnel and consider "pure" turbulence theory, then there is no doubt that the theory developed in this and in the preceding sections is valid for idealized turbulent flows satisfying the above initial conditions at time $t = 0$. Nevertheless, the behavior in the neighborhood of the wave-number origin of unbounded homogeneous turbulence with nonzero third-order correlations is also interesting independently of whether or not there is any correspondence with real grid turbulence. We shall therefore discuss this last problem in detail in the next section.

[8] Let us note that a conservation law, generalizing the conservation law Eq. (15.16) (i.e., the Loitsyanskii invariant) and Eq. (15.25), can be obtained even in the case of arbitrary inhomogeneous turbulence in a closed volume, bounded by a nontorque applying surface. In this case, however, it relates to an integral of the form (15.26) in which, for $B_{ij}(r)$ is understood an integral of $B_{ij}(x,r) = \overline{u_i(x-r/2)u_j(x+r/2)}$ over all values of $x$ (see Lumley, 1966). We will not dwell here on the consequences of this result.

## 15.5 Asymptotic Behavior of the Correlations and Spectra of Homogeneous Turbulence in the Range of Large Length Scales (or Small Wave Numbers)

The theory of Fourier integrals shows that regularity of the spectrum of homogeneous turbulence at the origin is equivalent to a rapid decrease of correlation between velocities at two points as the distance between them increases. In particular, the analyticity of the spectrum at the point $\boldsymbol{k} = 0$ is equivalent to an exponential decrease of all the functions $B_{ij}(\boldsymbol{r})$ as $r \to \infty$. The assumption that the correlations decrease in this way seems natural at first sight, and was in fact assumed for a number of years in most papers on the theory of turbulence. However, after Proudman and Reid (1954) pointed out that this assumption is not valid for one simple statistical model of turbulent flows [see Sect. 19.4, Batchelor and Proudman (1956)] indicated that an incompressible fluid has certain paradoxical physical properties suggesting that the above assumption must be treated with extreme caution. The point is that an incompressible fluid is an idealized model of real continuous media in which disturbances can propagate with an infinite velocity (it is well known that the dynamic equations of an incompressible fluid are obtained from the equations for a compressible fluid by allowing the velocity of sound to tend to infinity). The dynamic equations for an incompressible fluid must therefore include long-range forces which ensure that a velocity disturbance occurring in a finite region is instantaneously propagated to all points within the medium. In that case, however, it is quite possible that the long-range forces will also produce statistical correlations between velocity values, and this will upset the analytical character of the spectrum at $\boldsymbol{k} = 0$. The behavior of the spectral tensor $F_{ij}(\boldsymbol{k})$ in the neighborhood of $\boldsymbol{k} = 0$ must therefore be investigated separately.

Mathematically, the above paradoxical properties of an incompressible fluid are reflected in the fact that the pressure field depends on the velocity in an integral fashion [see Eq. (1.9') of Vol. 1]. As a result of this, any local change in the velocity has an immediate effect on pressures throughout all space, and this immediately affects the acceleration field which governs the velocities at all subsequent times. To investigate the effect of this on the time variation of the velocity correlations, let us suppose that at the initial time $t = 0$ we have produced a homogeneous velocity field $\boldsymbol{u}(\boldsymbol{x},0)$ in an infinite space, such that the tensors $B_{ij}(\boldsymbol{r}, 0)$, $B_{ij,l}(\boldsymbol{r}, 0)$ decay exponentially at infinity together with all the cumulants of any order. We shall now establish the behavior of the correlation tensor $B_{ij}(\boldsymbol{r}, t)$ for times $t > 0$.

We shall start with the general equation (14.7) for $t = 0$:

$$\frac{\partial B_{ij}(\boldsymbol{r}, 0)}{\partial t} = \frac{\partial}{\partial r_k}\left(\overline{u_i u_k u'_j} - \overline{u_i u'_k u'_j}\right) + \frac{1}{\rho}\frac{\partial \overline{p u'_j}}{\partial r_i} - \frac{1}{\rho}\frac{\partial \overline{p' u_i}}{\partial r_j} + 2\nu\, \Delta \overline{u_i u'_j} \qquad (15.77)$$

and the relation

$$\frac{1}{\rho}\overline{p u'_j} = \frac{1}{4\pi}\int \frac{\partial^2 B_{mn,j}(\boldsymbol{r}', 0)}{\partial r'_m \partial r'_n}\frac{d\boldsymbol{r}'}{|\boldsymbol{r} - \boldsymbol{r}'|}, \qquad (15.78)$$

which is a consequence of Eq. (1.9'). If we expand the kernel $1/|\boldsymbol{r}-\boldsymbol{r}'|$ in Eq. (15.78) in a Taylor series in terms of the components of the vector $\boldsymbol{r}'$, we find that

$$\frac{4\pi}{\rho}\overline{pu'_j}=\frac{1}{r}\cdot\int\frac{\partial^2 B_{mn,j}(\mathbf{r}')}{\partial r'_m\,\partial r'_n}\,d\mathbf{r}'-\frac{\partial}{\partial r_p}\left(\frac{1}{r}\right)\cdot\int\frac{\partial^2 B_{mn,j}(\mathbf{r}')}{\partial r'_m\,\partial r'_n}r'_p\,d\mathbf{r}'+$$
$$+\frac{1}{2!}\frac{\partial^2}{\partial r_p\,\partial r_s}\left(\frac{1}{r}\right)\cdot\int\frac{\partial^2 B_{mn,j}(\mathbf{r}')}{\partial r'_m\,\partial r'_n}r'_p r'_s\,d\mathbf{r}'-$$
$$-\frac{1}{3!}\frac{\partial^3}{\partial r_p\,\partial r_s\,\partial r_t}\left(\frac{1}{r}\right)\cdot\int\frac{\partial^2 B_{mn,j}(\mathbf{r}')}{\partial r'_m\,\partial r'_n}r'_p r'_s r'_t\,d\mathbf{r}'+\ldots. \qquad 15.79)$$

Hence it is clear that the function $\overline{pu'_j}$ will decrease at infinity at $t=0$ only as some finite power of $r^{-1}$. This power will be greater than the first or second: The first two terms on the right-hand side of Eq. (15.79) will vanish since the corresponding volume integrals can be reduced to surface integrals and the function $B_{mn,j}(\mathbf{r},0)$ has already been assumed to decay exponentially for $|\mathbf{r}|\to\infty$. Moreover, the third term on the right-hand side of Eq. (15.79) can be integrated by parts giving

$$\frac{\partial^2}{\partial r_m\,\partial r_n}\left(\frac{1}{r}\right)\cdot\int B_{mn,j}(\mathbf{r}')\,d\mathbf{r}',$$

which is again zero because $B_{mn,j}(\mathbf{r}')$ is solenoidal in $j$ for any $m, n$ [see Eq. (11.98)]. If we integrate the fourth term by parts, we find that as $r\to\infty$

$$\frac{1}{\rho}\overline{pu'_j}=-T_{mnjp}\frac{\partial^3}{\partial r_m\,\partial r_n\,\partial r_p}\left(\frac{1}{r}\right)+O(r^{-5}),$$
$$T_{mnjp}=\int B_{mn,j}(\mathbf{r}',0)\frac{r'_p\,d\mathbf{r}'}{4\pi} \qquad (15.80)$$

and consequently

$$\frac{\partial B_{ij}(\mathbf{r},0)}{\partial t}=-T_{mnjp}\frac{\partial^4}{\partial r_i\,\partial r_m\,\partial r_n\,\partial r_p}\left(\frac{1}{r}\right)-$$
$$-T_{mnip}\frac{\partial^4}{\partial r_j\,\partial r_m\,\partial r_n\,\partial r_p}\left(\frac{1}{r}\right)+O(r^{-6}), \qquad (15.81)$$

where the coefficients $T_{mnjp}$ will not in general be zero. Therefore, in the general case $\frac{\partial B_{ij}(\mathbf{r},0)}{\partial t}=O(r^{-5})$, so that the functions $B_{ij}(\mathbf{r},t)$ themselves will decrease as $r^{-5}$ as $r\to\infty$ for $t>0$.

Somewhat more complicated considerations have led Batchelor and Proudman (1956) to the conclusion that $\frac{\partial^2 B_{ij}(\mathbf{r},0)}{\partial t^2}$, $\frac{\partial^3 B_{ij}(\mathbf{r},0)}{\partial t^3}$, and all the subsequent time derivatives of the tensor $B_{ij}(\mathbf{r},t)$ at $t=0$ should also decrease as $r^{-5}$ for $r\to\infty$. Therefore, even if the initial values of $B_{ij}(\mathbf{r},0)$ decrease exponentially as $r\to\infty$, the functions $B_{ij}(\mathbf{r},t)$ will in general decrease as $r^{-5}$ for any $t>0$. It may be concluded that the decay of correlation between the velocities at two points, which is proportional to the reciprocal of the fifth power of the distance between them, does in fact correspond to the long-range effect which appears in an incompressible fluid due to the integral dependence of the pressure forces on the velocity.

To find the explicit form of the term of order $r^{-5}$ in the expression for $B_{ij}(r, t)$ when $t > 0$, we note that the terms of order $r^{-5}$ in the formulas for the time derivatives of $B_{ij}(r, t)$ at $t = 0$ appear, in the final analysis, as a result of the expansion of the integrand in Eq. (1.9') in a Taylor series. All these terms are therefore linear combinations of fourth-order partial derivatives of functions of the order of $r^{-1}$. The entire term of order $r^{-5}$ in the asymptotic expression for $B_{ij}(r, t)$ should therefore be of the same form [it will be seen later that, in fact, this term is a combination of higher-order derivatives; see Eq. (15.89)].

Let us now multiply all the terms in Eq. (15.77) by the operator $-\varepsilon_{kml}\varepsilon_{lnj}\dfrac{\partial^2}{\partial r_m \partial r_n}$.

Terms containing the pressure will then cancel out. Consequently, in the case of an initial velocity field with exponentially decreasing correlations $B_{ij}(r, 0)$ and $B_{ik,j}(r, 0)$ the derivative $\dfrac{\partial}{\partial t}B_{\omega_k\omega_l}(r, 0) = \dfrac{\partial}{\partial t}\overline{\omega_k\omega_l'}$, where $\omega = \text{rot}\, u = \nabla \times u$, will also decrease exponentially as $r \to \infty$. If we transform the second derivative

$$\frac{\partial^2}{\partial t^2}B_{\omega_k\omega_l}(r, 0) = \overline{\frac{\partial^2\omega_k}{\partial t^2}\omega_l'} + 2\overline{\frac{\partial\omega_k}{\partial t}\frac{\partial\omega_l'}{\partial t}} + \overline{\omega_k\frac{\partial^2\omega_l'}{\partial t^2}}$$

by means of the Navier-Stokes equations for the velocity field and the implied equations (1.7) for the vorticity field, we can express this derivative in terms of the initial moments of the flow variables. It then turns out that if all the cumulants (of the two- and three-point type) of the velocity field decrease exponentially at the initial time, then the second derivative will also decrease exponentially as $r \to \infty$. However, the expression for the next time derivative $\dfrac{\partial^3}{\partial t^3}B_{\omega_k\omega_l}(r, t)$ will now contain terms which decrease as reciprocal powers of $r$ for $r \to \infty$. In general, the effect of pressure forces leads to the appearance of terms which decrease as reciprocal powers of $r$ in the time derivatives of sufficiently high order of any velocity cumulants, and in all their space derivatives. However, the highest power of $r^{-1}$ in the asymptotic expression for $\dfrac{\partial^3}{\partial t^3}B_{\omega_k\omega_l}(r, 0)$ is found to be equal to 8 and not 7 as expected from Eq. (11.81) and from the fact that $B_{ij}(r, t) = O(r^{-5})$ (terms of the order of $r^{-5}$ in the expressions for the derivatives of the tensor $B_{ij}(r, 0)$ cancel out when $\dfrac{\partial^3}{\partial t^3}B_{\omega_k\omega_l}(r, 0)$ is evaluated. It can be shown that this will also occur for all the higher derivatives $\dfrac{\partial^n}{\partial t^n}B_{\omega_k\omega_l}(r, t)$. Hence it follows that terms of the order of $r^{-5}$ in the asymptotic expression for $B_{ij}(r, t)$, $t > 0$, will cancel out as a result of the application of the operator $-\varepsilon_{kml}\varepsilon_{lnj}\dfrac{\partial^2}{\partial r_m \partial r_n}$:

$$-\varepsilon_{ikl}\varepsilon_{jmn}\frac{\partial^2 B_{ln}(r, t)}{\partial r_k \partial r_m} = B_{\omega_i\omega_j}(r, t) = O(r^{-8}) \text{ when } t > 0. \tag{15.82}$$

The fact that $B_{ij}(r, t)$ decreases as a reciprocal power of $r$ means that the corresponding spectral tensor $F_{ij}(k, t)$ cannot be analytic for $t > 0$ at $k = 0$, and must have a singularity at this point. Let us now establish the nature of this singularity. Since $B_{ij}(r, t) = O(r^{-5})$, the tensor $F_{ij}(k, t)$ is continuously differentiable with respect to the components of the vector $k$ for all $k$. The second derivatives $\dfrac{\partial^2 F_{ij}(k, t)}{\partial k_m \partial k_n}$ exist and are continuous functions of $k$ for all $k \neq 0$, but as $k \to 0$ these derivatives can, in general, tend logarithmically to infinity (see Sect. 11.3). However, since terms of the order of $r^{-5}$ in the asymptotic expressions for the tensor $B_{ij}(\boldsymbol{r}, t)$ can be written in the form of a linear combination of third-order derivatives of functions of the order $r^{-2}$, integrals of the form

$$\int\limits_{|\boldsymbol{r}|>R} B_{ij}(\boldsymbol{r}, t)\, r_m r_n\, d\boldsymbol{r},$$

where $R$ is sufficiently large, can be reduced after double integration by parts to the sum of a convergent integral [corresponding to the term of the order of $r^{-6}$ or higher in the asymptotic expressions for $B_{ij}(\boldsymbol{r}, t)$] and certain surface integrals which cancel out. Hence it is clear that moments of the form

$$\Lambda_{ij,\, mn} = \int B_{ij}(\boldsymbol{r}, t)\, r_m r_n\, d\boldsymbol{r} \tag{15.83}$$

which are dynamic invariants for an analytic spectrum (see Sect. 15.2), will be convergent integrals (but, in general, not absolutely convergent) for $t > 0$ in the case of homogeneous turbulence with exponentially decaying correlations at $t = 0$. Therefore, the derivatives $\dfrac{\partial^2 F_{ij}(\boldsymbol{k}, t)}{\partial k_m\, \partial k_n}$ will, in fact, be bounded in the neighborhood of the origin, although it cannot be concluded that $-\dfrac{1}{8\pi^3}\Lambda_{ij,\, mn}$ will be equal to the above derivatives at $\boldsymbol{k} = 0$ (we shall see later that $F_{ij}(\boldsymbol{k}, t)$ has no second-order derivatives at this point for $t > 0$). Since in the present case

$$F_{ij}(0, t) = \frac{1}{8\pi^3}\int B_{ij}(\boldsymbol{r}, t)\, d\boldsymbol{r} = 0,$$

$$\left.\frac{\partial F_{ij}(\boldsymbol{k}, t)}{\partial k_m}\right|_{\boldsymbol{k}=0} = -\frac{i}{8\pi^3}\int B_{ij}(\boldsymbol{r}, t)\, r_m\, d\boldsymbol{r} = 0,$$

it follows from the fact that the derivatives $\dfrac{\partial^2 F_{ij}(\boldsymbol{k}, t)}{\partial k_m\, \partial k_n}$ are bounded that

$$F_{ij}(\boldsymbol{k}, t) = O(k^2) \text{ when } \boldsymbol{k} \to 0. \tag{15.84}$$

The singularity of the spectrum must thus be such that the ratio $F_{ij}(\boldsymbol{k}, t)/k^2$ does not become infinite near the point $\boldsymbol{k} = 0$, although $\lim\limits_{\boldsymbol{k}\to 0} F_{ij}(\boldsymbol{k}, t)/k^2$ can, for example, assume different values for $\boldsymbol{k}$ tending to zero along different directions.

We must now take into account the fact that, by Eq. (15.82), the tensor $F_{\omega_i\omega_j}(\boldsymbol{k}, t)$ is at least four-fold continuously differentiable with respect to the components of $\boldsymbol{k}$ for all $\boldsymbol{k}$ and $t > 0$, while for $\boldsymbol{k} \neq 0$ it will also have fifth-order continuous derivatives with respect to $k_l$ which may tend logarithmically to infinity as $\boldsymbol{k} \to 0$ (see Sect. 11.3). Since, by Eqs. (15.84) and (11.82), the tensor $F_{\omega_i\omega_j}(\boldsymbol{k}, t)$ should decrease as $k^4$ for $\boldsymbol{k} \to 0$, we have for small $\boldsymbol{k}$

$$F_{\omega_i\omega_j}(\boldsymbol{k}, t) = C_{ijmnpq} k_m k_n k_p k_q + O(k^5 \log k) \tag{15.85}$$

and, consequently, using Eqs. (11.82) and (11.83),

$$k^4 F_{ij}(\boldsymbol{k}, t) = C_{ijmnpqrs} k_m k_n k_p k_q k_r k_s + O(k^7 \log k). \tag{15.86}$$

However, the tensor $F_{ij}(\boldsymbol{k}, t)$ should be symmetric and should satisfy the orthogonality conditions (11.79), i.e., it must be represented in the form

$$F_{ij}(\boldsymbol{k}, t) = \left(\delta_{lm} - \frac{k_l k_m}{k^2}\right)\left(\delta_{jn} - \frac{k_j k_n}{k^2}\right)\Phi_{mn}(\boldsymbol{k}, t), \tag{15.87}$$

where $\Phi_{mn}$ is a symmetric tensor which depends on $\boldsymbol{k}$ and $t$. To satisfy Eq. (15.86) we must demand that

$$\Phi_{mn}(\boldsymbol{k}, t) = C_{mnpq}(t)\, k_p k_q + O(k^3 \log k),$$

for any $t > 0$, where $C_{mnpq}$ is a tensor symmetric in $m$, $n$ and $p$, $q$. Hence for $t > 0$ we have

$$F_{lj}(\boldsymbol{k}, t) = C_{mnpq}(t)\left(\delta_{lm} - \frac{k_l k_m}{k^2}\right)\left(\delta_{jn} - \frac{k_j k_n}{k^2}\right) k_p k_q + O(k^3 \log k). \tag{15.88}$$

This is the general formula of Batchelor and Proudman for the case of arbitrary homogeneous turbulence with exponentially decreasing initial cumulants. We thus see that the second derivatives of the function $F_{lj}(\boldsymbol{k}, t)$ with respect to the components of the vector $\boldsymbol{k}$ at the point $\boldsymbol{k} = 0$ do not, in fact, exist in this case.

If we apply the Fourier transformation to Eq. (15.88), we find that the asymptotic behavior of the correlation tensor $B_{lj}(\boldsymbol{r}, t)$ for $\boldsymbol{r} \to \infty$ and $t > 0$ is given by

$$B_{lj}(\boldsymbol{r}, t) = \pi^2 C_{mnpq}\left(\delta_{lm}\Delta - \frac{\partial^2}{\partial r_l \partial r_m}\right)\left(\delta_{jn}\Delta - \frac{\partial^2}{\partial r_j \partial r_n}\right)\frac{\partial^2 r}{\partial r_p \partial r_q} + O(r^{-6}), \tag{15.89}$$

where $\Delta = \frac{\partial^2}{\partial r_l \partial r_l}$. The formulas (15.88) and (15.89) determine the principal terms of the spectral tensor $F_{lj}(\boldsymbol{k}, t)$ for $\boldsymbol{k} \to 0$, and of the correlation tensor $B_{lj}(\boldsymbol{r}, t)$ for $\boldsymbol{r} \to \infty$, to within the undetermined coefficients $C_{mnpq}$ [which may depend on the time $t$ and are, in fact, linear combinations of the integrals Eq. (15.83)].

The asymptotic form of the correlation functions $B_{lj,\,l}(\boldsymbol{r}, t)$ and $B_{pl}(\boldsymbol{r}, t)$, for $r \to \infty$ i.e., the behavior of the spectra $F_{lj,\,l}(\boldsymbol{k}, t)$ and $F_{pl}(\boldsymbol{k}, t)$ for $k \to 0$, can be investigated in a similar fashion. It turns out that in homogeneous turbulence with exponentially decreasing initial cumulants for $t > 0$, we have, in general,

$$B_{lj,\,l}(\boldsymbol{r}, t) = 2\pi^2 D_{ljmn}\frac{\partial^3}{\partial r_l \partial r_m \partial r_n}\left(\frac{1}{r}\right) + O(r^{-5}), \tag{15.90}$$

$$B_{pl}(\boldsymbol{r}, t) = \pi^2 D_{ljmn}\left(\delta_{ln}\Delta - \frac{\partial^2}{\partial r_l \partial r_n}\right)\frac{\partial^3 r}{\partial r_l \partial r_j \partial r_m} + O(r^{-5}) \tag{15.91}$$

and, correspondingly,

$$F_{lj,\,l}(\boldsymbol{k}, t) = iD_{ljmn}\left(\delta_{ln} - \frac{k_l k_n}{k^2}\right)k_m + O(k^2 \log k), \tag{15.92}$$

$$F_{pl}(\boldsymbol{k}, t) = -iD_{ljmn}\left(\delta_{ln} - \frac{k_l k_n}{k^2}\right)\frac{k_l k_j k_m}{k^2} + O(k^2 \log k). \tag{15.93}$$

where the $D_{ijmn}$ are new undetermined coefficients which are functions of time and are such that $D_{ijmn} = D_{jimn}$. Since $B_{ij,l}(r)$ falls off as $r^{-4}$, it follows that the $\Lambda_{ij,mn}$ will now be functions of $t$.

The above results are based on the assumption that the functions $B_{ij}(\boldsymbol{r})$ decrease exponentially at $t = 0$ as $r \to \infty$. If this were not so, the behavior of the functions $F_{ij}(\boldsymbol{k}, t)$ near $\boldsymbol{k} = 0$, and of the functions $B_{ij}(\boldsymbol{r}, t)$ as $r \to \infty$, could be quite different than indicated above. For example, if we follow Saffman (1967a) and assume that, at $t = 0$, only the vorticity correlation functions $B_{\omega_i\omega_j}(\boldsymbol{r})$ decrease exponentially at infinity, while the functions $B_{ij}(\boldsymbol{r}, 0)$ can decrease more slowly, then it can be readily shown that the asymptotic form of $B_{ij}(\boldsymbol{r}, t)$ for $r \to \infty$ and $t > 0$ will, in general, be given by equation

$$B_{ij}(\boldsymbol{r}, t) \approx -\pi^2 C_{pq}\left(\delta_{ip}\Delta - \frac{\partial^2}{\partial r_i \partial r_p}\right)\left(\delta_{jq}\Delta - \frac{\partial^2}{\partial r_j \partial r_q}\right) r, \tag{15.89'}$$

where $C_{pq}$ are constants such that in the isotropic case $C_{pq} = C\delta_{pq}$, i.e., $B_{ij}(\boldsymbol{r}) = 2\pi^2 C\,(3r_ir_j/r^5 - \delta_{ij}/r^3)$, $B_{LL}(r) = 4\pi^2 C/r^3)$. Therefore, in the present case the functions $B_{ij}(\boldsymbol{r})$ will decrease as $r^{-3}$ at infinity. The corresponding spectral tensor will be

$$F_{ij}(\boldsymbol{k}, t) \approx C_{pq}\left(\delta_{ip} - \frac{k_ik_p}{k^2}\right)\left(\delta_{jq} - \frac{k_jk_q}{k^2}\right); \tag{15.88'}$$

and in the isotropic case $C_{pq} = C\delta_{pq}$, i.e., $E(k) \approx 4\pi C k^2$. Saffman constructed a simple statistical model of the initial distribution of impulsive forces generating the turbulence, which corresponds to the exponentially decreasing functions $B_{\omega_i\omega_j}(\boldsymbol{r}, 0)$, but to the functions $B_{ij}(\boldsymbol{r}, 0)$ decreasing in accordance with a "minus three power law" where the coefficients $C_{pq}$ are independent of time (i.e., they are dynamical invariants of the motion). However, this model also corresponds to the idealized case of homogeneous turbulence in an infinite space, and cannot be used to describe real turbulent flows such as occur in the turbulence behind the grid in the wind tunnel.

## 15.6 The Influence of the Spectrum Singularity on the Final Period Decay

We shall now consider the final period of decay of homogeneous turbulence for which at $t = 0$ all the cumulants decrease exponentially at infinity. We already know that the spectrum of such turbulence for any $t > 0$ will, in general, be of the form given by Eq. (15.88). In particular, the spectrum at the time $t_0$ will be of this form with certain $C^0_{mnpq} = C_{mnpq}(t_0)$, where $t_0$ is such a time that, for $t > t_0$, the nonlinear terms in the dynamic equations will no longer affect the evolution of the turbulence. Further change of the tensor $F_{ij}(\boldsymbol{k}, t)$ with time will then be determined by the formula (15.69). Since for $t - t_0 \to \infty$ the factor $\exp\{-2\nu k^2(t - t_0)\}$ will be negligible everywhere except for the neighborhood of the point $\boldsymbol{k} = 0$, the asymptotic form of the functions $B_{ij}(\boldsymbol{r}, t)$ for $t - t_0 \to \infty$ will be determined by the leading term of the formula (15.88). Therefore, for large values of $t - t_0$ we can assume that

$$F_{ij}(\boldsymbol{k}, t) = C^0_{mnpq}\left(\delta_{im} - \frac{k_ik_m}{k^2}\right)\left(\delta_{jn} - \frac{k_jk_n}{k^2}\right) k_pk_q e^{-2\nu k^2(t-t_0)}, \tag{15.94}$$

where $C^0_{mnpq}$ are certain constants and, consequently,

$$B_{ij}(\boldsymbol{r}, t) = C^0_{mnpq} \int \left(\delta_{im} - \frac{k_i k_m}{k^2}\right)\left(\delta_{jn} - \frac{k_j k_n}{k^2}\right) k_p k_q e^{i\boldsymbol{k}\boldsymbol{r} - 2\nu k^2 (t-t_0)} d\boldsymbol{k} =$$
$$= C^0_{mnpq}\left(\delta_{im}\Delta - \frac{\partial^2}{\partial r_i \partial r_m}\right)\left(\delta_{jn}\Delta - \frac{\partial^2}{\partial r_j \partial r_n}\right)\frac{\partial^2}{\partial r_p \partial r_q} \times$$
$$\times \int \frac{1 - \cos \boldsymbol{kr}}{k^4} e^{-2\nu k^2 (t-t_0)} d\boldsymbol{k} \quad (15.95)$$

The term $k^{-4}$, which does not contain the variables $r_i$, is added to the first factor under the integral sign on the right-hand side in order to insure convergence. By transforming Eq. (15.95), Batchelor and Proudman showed that in the present case the asymptotic form of $B_{ij}(\boldsymbol{r}, t)$ when $t - t_0 \to \infty$ is given by

$$B_{ij}(\boldsymbol{r}, t) = \frac{\sqrt{\pi^3} C^0_{mnpq}}{[8\nu(t-t_0)]^{5/2}}\left(\delta_{im}\Delta_y - \frac{\partial^2}{\partial y_i \partial y_m}\right) \times$$
$$\times \left(\delta_{jn}\Delta_y - \frac{\partial^2}{\partial y_j \partial y_n}\right)\frac{\partial^2}{\partial y_p \partial y_q}\left\{e^{-y^2} + \sqrt{\pi}\left(y + \frac{1}{2y}\right)\Phi(y)\right\}, \quad (15.96)$$

where

$$y = \frac{r}{\sqrt{8\nu(t-t_0)}}, \quad \Delta_y = \frac{\partial^2}{\partial y_i \partial y_i}, \quad \Phi(y) = \frac{2}{\sqrt{\pi}} \int_0^y e^{-t^2} dt.$$

It is clear from Eq. (15.96) that here again we have a relation of the form of Eq. (15.75):

$$\overline{u^2(t)} = \frac{\text{const}}{(t-t_0)^{5/2}}. \quad (15.97)$$

This was, of course, expected because the spectrum given by Eq. (15.88) [like the spectrum given by Eq. (15.67)] is a quadratic function of the components $k_i$ for small $|\boldsymbol{k}| = k$.[9] However, the general form of the correlation functions is now found to be quite complicated and different from Eq. (15.73). In particular, as $r \to \infty$, the function given by Eq. (15.96) decreases as a power function and not exponentially. It is only in the special case where the constants $C^0_{mnpq}$ are such that the leading term in the asymptotic formula (15.89) is found to vanish, that the integrals in Eq. (15.86) converge absolutely and are equal to the second derivatives of the spectral tensor at $\boldsymbol{k} = 0$. Therefore, for such $C^0_{mnpq}$ the behavior of the tensor $F_{ij}(\boldsymbol{k}, t)$ for small $|\boldsymbol{k}|$ can be described by Eq. (15.67), while the behavior of $B_{ij}(\boldsymbol{r}, t)$ for $t - t_0 \to \infty$ is given by Eq. (15.73). It is readily verified that, with the above limitation on $C^0_{mnpq}$, formula (15.96) does in fact become identical with Eq. (15.73). The experimental results shown in Fig. 15 can be interpreted as an indication of the fact that grid turbulence (at any rate under the conditions under which the data of Fig. 15 were obtained) does in fact correspond to the constants $C^0_{mnpq}$ satisfying the above conditions. On the other hand, we have already noted in Sect. 15.5 that there is considerable doubt as to whether the results on the behavior of the spectrum of homogeneous turbulence for $|\boldsymbol{k}| \to 0$ can be applied to the real turbulence behind a grid.

[9] Since in the case of Eq. (15.88') the dependence of $F_{ij}(\boldsymbol{k})$ on $k_i$ for small $k$ is quite different, it is not surprising that in this case $\overline{u^2(t)} \sim (t-t_0)^{-3/2}$ for $t - t_0 \to \infty$ (Saffman, 1967a).

In conclusion let us briefly consider the case of homogeneous and isotropic turbulence for which all the cumulants decrease exponentially at infinity at the initial instant. All the above discussion will of course apply to the present case except that $C_{mnpq}$ must now be an isotropic tensor which is symmetric in the subscripts $m$, $n$ and $p$, $q$, i.e., it must be of the form

$$C_{mnpq} = A\delta_{mn}\delta_{pq} + B(\delta_{mp}\delta_{nq} + \delta_{mq}\delta_{np}), \tag{15.98}$$

where $A$ and $B$ are scalars (in general, functions of time). However, in that case the leading term in Eq. (15.89) is easily shown to be identically zero, so that in the isotropic case the tensor $B_{ij}(\boldsymbol{r}, t)$ [and, consequently, the scalar function $B_{LL}(r, t)$ which determines it uniquely] will decay at infinity not slower than $r^{-6}$ for $t > 0$. This result is not simply a consequence of the fact that the pressure field in isotropic turbulence is absent from the Kármán-Howarth equation (14.9) for $B_{LL}(r, t)$. In fact, Eq. (14.9) determines only the first time derivative of the tensor $B_{ij}(\boldsymbol{r}, t)$, and the expressions for its higher-order time derivatives do contain the pressure field in the isotropic case. Unfortunately, direct investigation of the derivatives $\dfrac{\partial^2 B_{ij}(\boldsymbol{r}, t)}{\partial t^2}$, $\dfrac{\partial^3 B_{ij}(\boldsymbol{r}, t)}{\partial t^3}$, etc., is very difficult and does not allow us to determine whether these expressions contain terms which decrease as power functions or otherwise. However, the above analysis shows that none of these terms can decrease more slowly than $r^{-6}$.

Since $B_{ij}(\boldsymbol{r}, t)$ decreases at infinity at least as $r^{-6}$, the integrals in Eq. (15.83), which in the isotropic case can be simply expressed in terms of the Loitsyanskii integral $\Lambda$, converge absolutely and coincide with the second derivatives of the spectral tensor at $\boldsymbol{k} = 0$. The asymptotic behavior of the function $F_{ij}(\boldsymbol{k}, t)$ for $|\boldsymbol{k}| \to 0$ can in this case be obtained by substituting Eq. (15.98) in Eq. (15.88):

$$F_{ij}(\boldsymbol{k}, t) = A(\delta_{ij}k^2 - k_i k_j) + O(k^3 \log k). \tag{15.99}$$

We note that the first term on the right-hand side of this formula is regular at $\boldsymbol{k} = 0$. We can use Eq. (15.99) to express the factor $A$ in terms of the Loitsyanskii integral (15.16):

$$A = \frac{1}{3}\Lambda. \tag{15.100}$$

When applied to the final period of decay, the "initial" ($t = t_0$) spectrum of the form given by Eq. (15.99) leads to the usual formulas (15.54) and (15.53) for the decay of turbulent energy and for the form of the longitudinal correlation function. Hence it is clear that in isotropic turbulence (at least during the final decay stage), the quantity $\Lambda$ is an invariant of the motion. Initially, however, when the nonlinear terms of the equations of motion remain appreciable, it is not certain whether $\Lambda$ will remain constant. In general, in the isotropic case, the tensor $D_{ijmn}$ in Eqs. (15.90)–(15.93) will be given by

$$D_{ijmn} = C\delta_{ij}\delta_{mn} + D(\delta_{im}\delta_{jn} + \delta_{in}\delta_{jm}),$$

where $C$ and $D$ are scalar coefficients, so that in this case

$$B_{ij,l}(\boldsymbol{r}, t) = 4\pi^2 D\frac{\partial^3(1/r)}{\partial r_i\,\partial r_j\,\partial r_l} + O(r^{-5}), \quad B_{LL,L}(r, t) = -\frac{24\pi^2 D}{r^4} + O(r^{-5}).$$

However, Eq. (15.18) will now become

$$\frac{d\Lambda}{dt} = -24\pi^2 D. \tag{15.101}$$

Therefore, if $D = D(t)$ is nonzero, the Loitsyanskii integral (15.16) will at first vary with time and will become constant only after the final period of turbulence decay has been reached [a special case of such a turbulence with a finite but variable value of $\Lambda$ was first noted by Sedov (1951); see the very end of Sect. 16.1].

The asymptotic form of the spectrum and of the correlation function for temperature fluctuations can be investigated in an analogous fashion. If all the cumulants decrease exponentially at infinity at $t = 0$, then by Eq. (14.58), $dB_{\vartheta\vartheta}(r, t)/dt$ will also decrease exponentially for $t = 0$. However, the expressions for the subsequent time derivatives of $B_{\vartheta\vartheta}(r, t)$ will contain the pressure field, so that one can expect that, in general, the function $B_{\vartheta\vartheta}(r, t)$ will decrease as a power function for $r \to \infty$ and $t > 0$. Detailed studies of the order of this decrease are not, however, particularly interesting. In fact, it is clear that the effect of the pressure forces definitely does not lead to a decrease of $B_{\vartheta\vartheta}(r, t)$ that is slower than $O(r^{-4})$. Therefore, the integral (15.26) can now be considered as absolutely convergent, and the spectrum $F_{\vartheta\vartheta}(\boldsymbol{k}, t)$ as continuous and continuously differentiable with respect to the components of $\boldsymbol{k}$ throughout the wave-vector space. However, it is clear from this that the validity of the asymptotic formulas given by Eqs. (15.46) and (15.47), which describe the "general case" of the final period of decay of isotropic temperature fluctuations, is subject to no doubt in the present case. In precisely the same way, there is no doubt as to the validity of the conservation law given by Eq. (15.26), since the only requirement used in its derivation was that the function $B_{L\vartheta,\vartheta}(r, t)$ should decrease at least as rapidly as $O(r^{-3})$. In fact, it is readily seen that a slower decrease of the function $B_{L\vartheta,\vartheta}(r, t)$ cannot be due to pressure forces, provided only that, at the initial time, all the cumulants of the turbulence decrease sufficiently rapidly.

# 16. SELF-PRESERVATION HYPOTHESES

## 16.1 The von Kármán Hypothesis on the Self-Preservation of the Velocity Correlation Functions

Simple consequences of the basic equations of the theory of isotropic turbulence, i.e., the von Kármán-Howarth equation (14.9) and the equivalent spectral equation (14.15) were discussed above. However, these equations are not sufficient for the description of the time evolution of isotropic turbulence. Many workers have therefore tried to augment Eqs. (14.9) and (14.15) by special hypotheses containing additional information about the time variation of the correlation and spectral functions. In this section we shall consider a number of such hypotheses—the so-called *self-preservation* (or *self-similarity*) hypotheses—which impose certain restrictions on the general character of the variation of the correlation and spectral functions with time.

The simplest hypothesis on the self-preservation of the correlation functions of isotropic turbulence was put forward by von Kármán (see von Kármán and Howarth, 1938). According to this hypothesis the functions $B_{LL}(r, t)$ and $B_{LL,L}(r, t)$ have similar shapes for different values of $t$, i.e., they differ only by choice of the scales along the coordinate axes. In other words, the von Kármán hypothesis assumes that the functions $B_{LL}(r, t)$ and $B_{LL, L}(r, t)$ can be simultaneously reduced to functions of a single variable through a suitable choice of the length and velocity scales $l = l(t)$ and $v = v(t)$.

$$B_{LL}(r, t) = v^2(t) f\left(\frac{r}{l(t)}\right), \quad B_{LL, L}(r, t) = v^3(t) h\left(\frac{r}{l(t)}\right). \tag{16.1}$$

The scale $v(t)$ can be taken to be the turbulence intensity $\sigma_u = [\overline{u^2}]^{1/2} = [B_{LL}(0)]^{1/2}$ at time $t$ [assuming that $f(0) = 1$], while $l(t)$ can be taken to be, for example, the integral length scale $L_1$ of Eq. (12.50), or the Taylor differential length scale $\lambda$ of Eq. (15.3), or again some other length scale which is uniquely determined by the correlation function $B_{LL}(r, t)$ or $B_{LL, L}(r, t)$ [in view of Eq. (16.1) all these scales can differ only by constant factors]. In particular, if the Loitsyanskii integral (15.16) converges and is nonzero, then

$$\Lambda = v^2 \int_0^\infty f\left(\frac{r}{l}\right) r^4 \, dr = v^2 l^5 \int_0^\infty f(\xi) \xi^4 \, d\xi, \tag{16.2}$$

and hence it is clear that

$$l = a\left(\frac{\Lambda}{v^2}\right)^{1/5}, \tag{16.3}$$

where

$$a = \left[\int_0^\infty f(\xi) \xi^4 \, d\xi\right]^{-1/5}$$

is a numerical constant. In this case we can also assume that $l = (\Lambda/v^2)^{1/5}$, i.e., we can determine $l$ from the normalization condition $\int_0^\infty f(\xi) \xi^4 \, d\xi = 1$. If, moreover, $\lim_{\xi \to \infty} \xi^4 h(\xi) = 0$ so that $\Lambda =$ const, then formula (16.3) determines the relation between the time variations of the scales $v(t)$ and $l(t)$.

Comparisons of the von Kármán hypothesis (16.1) with measurements of grid turbulence will be discussed below. For the moment, let us elucidate its significance. Even the early experiments on grid generated turbulence showed that the disturbances produced by the grid are rapidly mixed and converted into approximately isotropic turbulence. It can then be assumed that the final result of the mixing of these disturbances is that all the disturbances adjust to each other and the initial conditions affect only the characteristic length and velocity scales of the resulting turbulence, but not the general features of its statistical state. It can also be expected that the "universal equilibrium" which is eventually achieved will persist for some time, and only the "turbulence intensity" $v(t)$ (which decreases with time) and the "characteristic length scale" $l(t)$ (which increases with time because small-scale disturbances decay more rapidly than large-scale disturbances) will undergo a change. This assumption does in fact lead directly to the von Kármán hypothesis (16.1). All this leads also to the following natural generalization of the hypothesis (16.1): One would expect that, even if this hypothesis is not valid, then at least a part of the turbulent disturbances behind the grid will at some time vary in a self-preserving fashion. We shall discuss this generalization in detail later. For the moment, let us try to investigate the consequences of Eq. (16.1) in its initial form [we shall follow on the whole the analysis given by Sedov, 1961].

Let us consider first the consequences for the final period of decay of isotropic turbulence. Substituting Eq. (16.1) into Eqs. (15.1) and (15.35), we obtain

$$\frac{dv^2}{dt} = -C\nu\frac{v^2}{l^2}, \qquad C = -\frac{10f''(0)}{f(0)} \tag{16.4}$$

[if we assume that $v(t) = [\overline{u^2(t)}]^{1/2}$ and $l = \lambda$, then $f(0) = -f''(0) = 1$ and $C = 10$] and

$$f''(\xi) + \frac{4}{\xi}f'(\xi) + \frac{C}{2}f(\xi) + \frac{l}{2\nu}\frac{dl}{dt}\xi f'(\xi) = 0. \tag{16.5}$$

To insure that Eq. (16.5) is valid for $f(\xi) \neq \text{const}$ we must have

$$\frac{l}{2\nu}\frac{dl}{dt} = c_1 = \text{const.} \tag{16.6}$$

From Eqs. (16.6) and (16.4) we have $l^2(t) = 4\nu c_1(t - t_0)$ and $v^2(t) \sim (t - t_0)^{-\alpha}$ where $t_0$ is the integration constant and $\alpha = \frac{C}{4c_1}$.

Thus, in the present case

$$B_{LL}(r,\ t)=\frac{A}{(t-t_0)^{\alpha}}\,\varphi\left(\frac{r^2}{\nu(t-t_0)}\right). \tag{16.7}$$

Substituting this formula into Eq. (15.35), or using Eq. (16.5), we obtain the family of solutions (15.63) found by von Kármán and Howarth. If we suppose further that the Loitsyanskii integral $\Lambda$ converges and is nonzero, then $\alpha = 5/2$ in view of Eq. (16.3), i.e., the solution (16.7) now becomes identical with Eq. (15.53). We have thus shown once again that a finite and nonzero Loitsyanskii integral corresponds only to the solution (15.53).

Let us now consider the more important case when the third moments are not negligible. Equation (16.4) remains unaltered, but Eq. (16.5) must now be replaced by

$$\left[f''(\xi)+\frac{4}{\xi}f'(\xi)+\frac{C}{2}f(\xi)\right]+\frac{l}{2\nu}\frac{dl}{dt}\xi f'(\xi)+ \\ +\frac{vl}{2\nu}\left[h'(\xi)+\frac{4}{\xi}h(\xi)\right]=0, \tag{16.8}$$

which is obtained by substituting Eq. (16.1) into the von Kármán-Howarth equation. If we suppose that of the three functions $f''(\xi)+\frac{4}{\xi}f'(\xi)+\frac{C}{2}f(\xi)$, $\xi f'(\xi)$ and $h'(\xi)+\frac{4}{\xi}h(\xi)$ no two are linearly dependent, then it will follow from Eq. (16.8) that

$$\frac{l}{2\nu}\frac{dl}{dt}=c_1=\text{const},\qquad \frac{vl}{2\nu}=c_2=\text{const}, \tag{16.9}$$

i.e.,

$$l^2(t)=4\nu c_1(t-t_0)\sim t-t_0,\qquad v^2(t)=\frac{\nu c_2^2}{c_1(t-t_0)}\sim\frac{1}{t-t_0} \tag{16.10}$$

and consequently $C=4c_1$. We thus see that decay of the turbulent energy is now described by

$$\overline{u^2(t)}=\frac{a}{t-t_0}. \tag{16.11}$$

As regards the functions $f(\xi)$ and $h(\xi)$, one of them, say $f(\xi)$, can be chosen arbitrarily, and the second function will then be determined unambiguously by Eqs. (16.8) and (16.9). We note that none of these self-preserving solutions of the von Kármán-Howarth equation (established by Dryden, 1943) will correspond to turbulence with a finite and nonzero time-invariant Loitsyanskii integral $\Lambda$. In fact, in view of Eqs. (16.3) and (16.6), the decay of the turbulent energy for $0 < \Lambda < \infty$ and $\Lambda =$ const should be of the form $\overline{u^2(t)} \sim (t - t_0)^{-5/2}$, i.e., it will not be of the form given by Eq. (16.11).

There are also solutions of Eq. (16.8) which are such that the functions

$$f''(\xi) + \frac{4}{\xi} f'(\xi) + \frac{C}{2} f(\xi), \ \xi f'(\xi) \text{ and } h'(\xi) + \frac{4}{\xi} h(\xi)$$

are linearly dependent. These special solutions were all investigated by Sedov (1944, 1951) (see also the survey paper by Batchelor, 1948). In addition to the solutions with $h(\xi) = 0$, which we have already considered, and the physically impossible solutions with the spectrum $F(k)$ in the form of a $\delta$-function, there are also solutions for which $dl^2/dt = c_1 v l + c_2 \nu$ while $f(\xi)$ and $h(\xi)$ are uniquely determined by $c_1$, $c_2$, and $C$. In particular, if $c_2 = 1/2$ [which can always be achieved by suitably normalizing the length scale $l(t)$], then $f(\xi) = M(C, 5/2, \xi 2/8)$ and the dependence of $l$ and $v$ on $t$ is given by Eq. (16.4) together with the above formula for $dl^2/dt$ (in which we must assume that $c_1 > 0$). For these solutions $\Lambda = 0$ when $C > 5/2$, and $\Lambda = \infty$ when $C < 5/2$; while $\Lambda$ is finite but not constant for $C = 5/2$.

## 16.2 Less Stringent Forms of the von Kármán Hypothesis

Let us now consider to what extent the conclusions of Sect. 16.1 can remain valid if the von Kármán hypothesis is assumed satisfied but only for a certain range of values of $r$. Let us begin with the case where Eq. (16.1) is satisfied for a finite interval $0 \leqslant r \leqslant R$. Here $v(t)$ can again be identified with $\sigma_u = [\overline{u^2(t)}]^{1/2} = [B_{LL}(0)]^{1/2}$ assuming that $f(0) = 1$, and for $l(t)$ we can take, for example, the differential length scale $\lambda(t)$ which is determined by the behavior of the function $B_{LL}(r)$ near $r = 0$. Equation (16.4) will then remain valid and, moreover, Eq. (16.8) will be satisfied for a certain range of values of $\xi$. This is sufficient to ensure that all the conclusions regarding the functions $l(t)$ and $v(t)$ given in the previous section will remain valid. Moreover, the results for the functions $f(\xi)$ and $h(\xi)$, which are now defined only for not too large values of $\xi$, will also be valid. However, Eqs. (16.2) and (16.3), and all the conclusions regarding the Loitsyanskii integral $\Lambda$ [which depends on the values of $B_{LL}(r)$ for large $r$] are now no longer meaningful.

Further relaxation of the self-preservation hypothesis was put forward by Lin (1948) who considered the case when formulas of the form of Eq. (16.1) are available not for the correlation functions but only for the structure functions

$$D_{LL}(r, t) = \overline{[u_L(x+r, t) - u_L(x, t)]^2} = 2\overline{u^2(t)} - 2B_{LL}(r, t), \tag{16.12}$$
$$D_{LLL}(r, t) = \overline{[u_L(x+r, t) - u_L(x, t)]^3} = 6B_{LL, L}(r, t),$$

Lin assumed that

$$D_{LL}(r, t) = v^2 f_1\left(\frac{r}{l}\right), \quad D_{LLL}(r, t) = v^3 h_1\left(\frac{r}{l}\right), \quad 0 \leqslant r \leqslant R, \tag{16.13}$$

but that the relations (16.1) may not be valid [it is evident that Eq. (16.1) implies Eq. (16.13) but not vice-versa]. It is clear that in this case $f_1(0) = h_1(0) = 0$, and we cannot, therefore, set $v^2(t) = \overline{u^2(t)} = B_{LL}(0)$. A further fact connected with this is that Eq. (16.4) no longer follows from the energy-balance equation (15.1), and instead we have the weaker equation

$$\frac{d\overline{u^2}}{dt} = -C\nu \frac{v^2}{l^2}, \quad C = 5f_1''(0), \tag{16.14}$$

which is obtained by substituting the first formulas in Eqs. (16.12) and (16.13) into the right-hand side of Eq. (15.1). Next, substituting Eqs. (16.12)–(16.14) into Eq. (14.9), we have instead of Eq. (16.8)

$$\left[C - f_1''(\xi) - \frac{4}{\xi} f_1'(\xi)\right] - \frac{l}{2\nu}\frac{dl}{dt}\xi f_1'(\xi) +$$
$$+ \frac{l^2}{2\nu v^2}\frac{dv^2}{dt} f_1(\xi) + \frac{vl}{6\nu}\left[h_1'(\xi) + \frac{4}{\xi} h_1(\xi)\right] = 0. \tag{16.15}$$

This equation is most simply satisfied if we assume that

$$\frac{l}{2\nu}\frac{dl}{dt} = c_1, \quad \frac{l^2}{2\nu v^2}\frac{dv^2}{dt} = c_2, \quad \frac{vl}{6\nu} = c_3. \tag{16.16}$$

From the first and third equations in (16.16) it follows that $l^2(t) = 4\nu c_1(t-t_0)$, $v^2(t) = \frac{9\nu c_3^2}{c_1(t-t_0)}$. The second equation is then also valid, and $c_2 = -2c_1$. Substituting the above values of $l(t)$ and $v(t)$ into Eq. (16.14), we obtain the following energy decay law:

$$\overline{u^2(t)} = \frac{a_1}{t-t_0} + a_2, \tag{16.17}$$

which is somewhat more general than Eq. (16.11). As regards the unknown functions $f_1(\xi)$ and $h_1(\xi)$, the first of them can be chosen arbitrarily in which case $h_1(\xi)$ is uniquely determined by Eqs. (16.15), (16.16) and the boundary condition $h_1(0) = 0$. Further possible relaxation of the von Kármán hypothesis involves the assumption that equations of the form of Eq. (16.1) are satisfied only for $R \leqslant r < \infty$. In this case we again cannot assume that $v(t) = [\overline{u^2(t)}]^{1/2}$ and, moreover, we cannot use the energy-balance equation (15.1), so that $\overline{u^2(t)}$ cannot be related to the scales $v(t)$ and $l(t)$. For $r > R$, the von Kármán-Howarth equation will now differ from Eq. (16.8) only by the fact that the term $\frac{C}{2} f(\xi)$ is replaced by $\frac{l^2}{2\nu v^2} \frac{dv^2}{dt} f(\xi)$. To satisfy it, we must again assume that conditions (16.16) are valid. Consequently, $l^2(t)$ and $v^{-2}(t)$ will now again, in general, be linear functions of $t - t_0$. Next, we can arbitrarily specify values of $f(\xi)$ for $\frac{R}{l} = \xi_0 \leqslant \xi < \infty$, so that $h(\xi)$ is then uniquely determined by the von Kármán-Howarth equation and the condition $h(\infty) = 0$.

The situation is somewhat different when, for sufficiently large $r$, self-preservation occurs for turbulence with very low viscosity, i.e., very large initial Reynolds number $\mathrm{Re}_l = \frac{lv}{\nu}$ Let us rewrite the equation obtained by substituting Eq. (16.1) in Eq. (14.9) in the form:

$$\left[h'(\xi) + \frac{4}{\xi} h(\xi)\right] + \frac{1}{v}\frac{dl}{dt}\xi f'(\xi) - \frac{l}{v^3}\frac{dv^2}{dt} f(\xi) + \\ + \frac{2\nu}{vl}\left[f''(\xi) + \frac{4}{\xi} f'(\xi)\right] = 0. \tag{16.18}$$

The coefficient of $f'' + \frac{4}{\xi} f'$ in this equation is equal to $2/\mathrm{Re}_l$ and therefore we may expect that, for large $\mathrm{Re}_l$, these terms can be included. We must emphasize that the rejection of viscous terms cannot be valid if we consider Eq. (16.18) for all values of $r$. However low the viscosity, it will always ensure that the function $B_{LL}(r)$ will be differentiable at $r=0$, i.e., it will lead to the appearance of a parabolic top on the $y=B_{LL}(r)$ curve. It is natural to suppose, however, that for large Reynolds numbers the parabolic region will extend only to some small value $r=r_0$, while for somewhat larger values of $r$, the form of the function $B_{LL}(r)$ will be independent of the viscosity $\nu$. This means that we can reject the terms containing $\nu$ in the equation for the correlation functions when $r>r_0$. If, moreover, we assume that Eq. (16.1) is valid for $r>r_0$, then for such $r$ we have

$$h'(\xi)+\frac{4}{\xi}h(\xi)+\frac{1}{v}\frac{dl}{dt}\xi f'(\xi)-\frac{l}{v^3}\frac{dv^2}{dt}f(\xi)=0. \qquad (16.19)$$

Next, for very large Re it is natural to suppose that within the small parabolic region, i.e., for $r<r_0$, the function $B_{LL}(r)$ will change only slightly, so that $B_{LL}(r_0)$ will be almost equal to $B_{LL}(0)$. In that case we can again replace $v(t)$ in Eq. (16.1) by $(\overline{u^2})^{1/2}$, and $l(t)$ can be replaced by, say, the turbulence integral length scale $L_1$ or, if the Loitsyanskii integral converges, by the scale $l_0=\left(\frac{\Lambda}{v^2}\right)^{1/5}$ (but not, of course, by the Taylor micro scale $\lambda$).

From Eq. (16.19) we have

$$\frac{1}{v}\frac{dl}{dt}=c_1, \qquad \frac{l}{v^3}\frac{dv^2}{dt}=c_2 \qquad (16.20)$$

[It is readily verified that Eq. (16.19) does not have meaningful "Sedov-type solutions" for which the equations (16.20) would not be satisfied]. We can use Eqs. (16.20) and (16.19) to determine unambiguously the function $h(\xi)$ from the values of $f(\xi)$, which can again be chosen in an arbitrary fashion. Moreover, it follows from Eq. (16.20) that

$$l(t)=A_1(t-t_0)^{\frac{2c_1}{2c_1-c_2}}, \qquad v(t)=A_2(t-t_0)^{\frac{c_2}{2c_1-c_2}}, \qquad (16.21)$$

where $A_1/A_2=c_1-c_2/2$ [from Eq. (16.21) it follows that $c_1/c_2<0$].

The formulas (16.21) were found by von Kármán and Howarth as far back as 1938. These authors were the first to consider the case of self-preserving turbulence at very high Reynolds numbers (see also Sedov, 1951). In view of Eq. (16.3), the Loitsyanskii integral for Eq. (16.21) can be finite and nonzero provided only that $c_2 = -5c_1$, i.e.,

$$v^2(t) \sim (t-t_0)^{-10/7}, \quad l(t) \sim (t-t_0)^{2/7} \tag{16.22}$$

(see Kolmogorov, 1941c).

It may at first appear that in Kolmogorov's paper (1941c), the relations (16.22) are derived under the important additional assumption that the so-called Kolmogorov self-preservation is valid in a certain range of $r$ (this type of self-preservation will be discussed in Sect. 6.5 and Chapt. 8). However, this additional assumption was in fact used by Kolmogorov only for another derivation of the second equation in (16.20). It can be shown on the other hand that if we assume the validity of the "Kolmogorov self-preservation" (or more precisely, if we assume the existence of the "inertial subrange" which follows from it), and that the Loitsyanskii integral is finite and invariant, then relations (16.22) can be justified even without the assumption of the von Kármán self-preservation (16.1) for any $r$ [see Comte-Bellot and Corrsin, 1966].

But the assumptions about the Loitsyanskii integral are decisive for the derivation. If for example we replace these assumptions by Saffman's assumptions (1967a) that $\int_0^\infty r^2 [3B_{LL}(r) + rB'_{LL}(r)]\,dr = C$ is nonzero, finite and invariant (i.e., that $B_{LL}(r) \approx C/r^3$ at large $r$ where $C$ is a dynamical invariant), then Eq. (16.3) will be replaced by $l \sim (C/v^2)^{1/3}$ and we obtain: $c_2 = -3c_1$, i.e.,

$$v^2(t) \sim (t-t_0)^{-6/5}, l(t) \sim (t-t_0)^{2/5} \tag{16.22'}$$

The same result may be established with the replacement of the von Kármán's self-preservation hypothesis by the assumption of the existence of the "inertial subrange" [see Saffman (1967b)].

## 16.3 Spectral Formulation of the Self-Preservation Hypotheses

Let us briefly consider the spectral formulation of the above results. The von Kármán hypothesis (16.1) can also be rewritten in the form

$$E(k, t) = lv^2\varphi(kl), \qquad T(k, t) = v^3\psi(kl). \tag{16.23}$$

If we assume that $v^2(t) = \overline{u^2(t)}$ then the function $\varphi(\xi)$ will be normalized by

$$\int_0^\infty \varphi(\xi)\,d\xi = \frac{3}{2}. \tag{16.24}$$

If, moreover, $\varphi(\xi) \sim \xi^4$ as $\xi \to 0$, i.e., the Loitsyanskii integral (15.16) is finite and nonzero, we have

$$\Lambda = 3\pi \lim_{\xi\to\infty} \frac{\varphi(\xi)}{\xi^4} \tag{16.25}$$

[we can then let $l = (\Lambda/v^2)^{1/5}$, i.e., we can assume that $\lim_{\xi\to 0} \varphi(\xi)/\xi^4 = 1/3\pi$ ].

In view of the first formula in Eq. (16.23), the spectral energy-balance equation (14.19) can be rewritten in the form

$$\frac{d\overline{u^2}}{dt} = -C\nu \frac{v^2}{l^2}, \quad C = \frac{2}{3}\int_0^\infty \xi^2\varphi(\xi)\,d\xi. \tag{16.26}$$

Substituting Eq. (16.23) in the spectral equation (14.15), we obtain

$$\frac{l^2}{2\nu v^2}\frac{dv^2}{dt}\varphi(\xi) + \frac{l}{2\nu}\frac{dl}{dt}[\varphi(\xi) + \xi\varphi'(\xi)] - \frac{vl}{2\nu}\psi(\xi) + \xi^2\varphi(\xi) = 0. \tag{16.27}$$

If we assume that $v^2 = \overline{u^2}$ we can eliminate $dv^2/dt$ with the aid of Eq. (16.26), and this will result in an equation equivalent to Eq. (16.8). This equation is most simply satisfied by adopting the conditions given by Eq. (16.9). For $l(t)$ and $v^2(t) = \overline{u^2(t)}$ we then obtain equations (16.10), and $\varphi(\xi)$ can be taken to be any nonnegative function, while $\psi(\xi)$ can be determined from $\varphi(\xi)$ through Eqs. (16.27) and (16.9). Another (more special) class of solutions corresponding to the final stage of decay) is obtained for $\psi(\xi) = 0$. In that case, Eq. (16.9) becomes identical with Eq. (16.16), and from Eq. (16.27) we have the explicit formula

$$\varphi(\xi) = A\xi^\alpha e^{-\beta\xi^2}, \quad \alpha = \frac{C}{2c_1} - 1, \quad \beta = \frac{1}{2c_1}. \tag{16.28}$$

Moreover, Eq. (16.27) has certain special solutions corresponding to the solutions of Eq. (16.8) found by Sedov (1951). Of all the solutions of Eq. (16.27), only Eq. (16.28) with $\alpha = 4$ will satisfy $\varphi(\xi) \approx a\xi^4$ for $\xi \to 0$ where $a$ = const. In the case of the solutions satisfying Eq. (16.9) and having $\Psi(\xi) \neq 0$ the condition $\varphi(\xi) \approx a\xi^4$ implies that $\Psi(\xi) \approx (3ac_1/c_2)\xi^4$ at small $\xi$; therefore $a$ will not be invariant here.

Let us now suppose that the self-preservation (16.23) occurs only for $k > k_0$. In that case, for $k_0 l = \xi_0 < \xi < \infty$ we again have Eq. (16.27), but the answer to the question as to whether we can assume that $v^2 = \overline{u^2}$ and use Eq. (16.26) will depend on the position of the boundary $k_0$ of the self-preserving range. If $k_0$ lies to the left of the "energy range" of the spectrum, so that the principal energy-containing disturbances correspond to wave numbers greater than $k_0$ and

$$\int_0^{k_0} E(k)\,dk \ll \int_0^{\infty} E(k)\,dk,$$

then the assumption that $v^2 = \overline{u^2}$ will be justified, and so will Eq. (16.26). Therefore, in the case under consideration, all the consequences of the hypothesis remain valid, except for the behavior of $\varphi(\xi)$ as $\xi \to 0$. This case is equivalent to the case of self-preservation of correlation functions for $0 \leqslant r \leqslant R$.

Let us now suppose that the part of the spectrum which is not self-preserving contains an appreciable fraction of the total energy, so that $\int_0^{k_0} E(k)\,dk$ is no longer negligible in comparison with $\int_0^{\infty} E(k)\,dk$. Under this condition, we cannot assume that $v^2 = \overline{u^2}$. Let us now consider the case where the part of the spectrum which is not self-preserving contains an appreciable fraction of the total energy but involves a negligible fraction of the total energy dissipation, so that

$$\int_0^{k_0} k^2 E(k)\,dk \ll \int_0^{\infty} k^2 E(k)\,dk.$$

In this case, Eq. (16.26) where

$$C = \frac{2}{3} \int_{\xi_0}^{\infty} \xi^2 \varphi(\xi)\,d\xi$$

remains valid, and if you adopt Eq. (16.16), the time dependence of the turbulent energy will be given by Eq. (16.17). This was to be expected, since it can be shown that the assumed existence of a restricted self-preservation of the spectrum $E(k)$ leads to the self-preservation of the structure function $D_{LL}(r)$ used in the derivation of Eq. (16.17).

The case where the part of the spectrum which is not self-preserving provides a negligible contribution only to the integrals $\int_0^\infty k^4 E(k)\,dk$ and $\int_0^\infty k^2 T(k)\,dk$, i.e., when

$$\int_0^{k_0} k^4 E(k)\,dk \ll \int_0^\infty k^4 E(k)\,dk$$

and

$$\int_0^{k_0} k^2 T(k)\,dk \ll \int_0^\infty k^2 T(k)\,dk$$

can be considered in a similar fashion. In fact, from the equation

$$\frac{d^2\overline{u^2(t)}}{dt^2} = 4\nu\left\{-\int_0^\infty k^2 T(k,\,t)\,dk + 2\nu\int_0^\infty k^4 E(k,\,t)\,dk\right\},$$

which is a consequence of Eqs. (14.19) and (14.15), it follows that in this case

$$\frac{d^2\overline{u^2}}{dt^2} = C_1\frac{v^3}{l^3} + C_2\frac{v^2}{l^4}, \tag{16.29}$$

so that, in view of Eq. (16.16), we have

$$\overline{u^2(t)} = \frac{a_1}{t-t_0} + a_2 + a_3(t-t_0). \tag{16.30}$$

This formula was indicated by Goldstein (1951). Goldstein also noted a further generalization of this result to the case where the part of the spectrum that is not self-preserving provides a negligible contribution only for the higher-order moments of $E(k)$ and $T(k)$. However, this further generalization can hardly be regarded as useful.

Finally, if $E(k, t)$ and $T(k, t)$ are self-preserving for $k \geqslant k_0$ and the Reynolds number is large (i.e., the viscosity is low), so that the dissipation and energy ranges of the spectrum do not overlap, and if the self-preservation occurs only for $k$ lying to the left of the dissipation range, then the last term in Eq. (16.27) can simply be neglected. To satisfy the "truncated" equation we must then adopt the conditions (16.20), which lead to the von Kármán formulas (16.21) for $v(t)$ and $l(t)$, and to the equation

$$\psi(\xi) = (c_1 + c_2)\varphi(\xi) + c_1 \xi \varphi'(\xi). \tag{16.31}$$

If, at the same time, the self-preserving region of the spectrum contains all the energy range, then $v(t)$ can be taken to be the root mean square velocity $[\overline{u^2(t)}]^{1/2}$. If, moreover, self-preservation is valid for $k$ as small as desired, and $E(k) = ak^4 + o(k^4)$ as $k \to 0$, where $a = \Lambda/3\pi$ is independent of $t$, then, in view of Eq. (16.25), the coefficients $c_1$ and $c_2$ in Eq. (16.21) should be related by $c_1/c_2 = -0.2$, so that we have the Kolmorogov formulas (16.22).

## 16.4 Experimental Verification of the Self-Preservation Hypotheses

Let us first emphasize that, so far, the data suitable for verification of the self-preservation hypotheses for grid turbulence are still incomplete and, what is worse, are frequently contradictory. Therefore, in some cases we shall have to remember that the specific results which we shall be discussing will be in disagreement with the results of other authors.

Experimental data on the final period of decay were discussed in Sect. 15.4 and we shall not review them again here.[10] As regards the case of fully developed grid turbulence, the results obtained by the Cambridge group [see for example, Stewart and Townsend (1951), Stewart (1951), and Proudman (1951)] have shown that the dissipation range and the energy range overlap substantially for $Re_M = UM/\nu$ up to $5 \cdot 10^4$. If this is indeed the case (we shall mention later some data that do not confirm this conclusion), then the "nonviscous self-preservation" of von Kármán and Howarth (1938) which covers the entire energy range of the spectrum, but does not

[10] We shall not discuss also Ling and Huang's data (1970) who found that the second-order moments of very weak decaying grid turbulence are self-preserving; however, third-order moments are not self-preserving with the same scales, but are continuously adapting to the evolution of the second moments.

involve the dissipation range, will definitely not be observable in the wind tunnel at such $Re_M$. The conclusion of von Kármán and Lin (1949, 1951) that the Kolmogorov decay law (16.22) should be valid during the intermediate decay stage (between the "initial period" when either Eq. (16.11) or Eq. (16.17) is valid and the "final period" is accordingly also unjustified.

Let us now briefly review the main results of Stewart and Townsend (1951) who have carried out a series of measurements on the turbulence behind different grids (with $Re_M$ between $2 \cdot 10^3$ and $2 \cdot 10^4$ and occasionally up to $10^5$) in order to verify the self-preservation hypotheses. Let us begin with the data concerned with the self-preservation of the correlation function with respect to the scales $v(t) = [\overline{u^2(t)}]^{1/2}$ and $l(t) = \lambda(t)$. Figures 16 and 17, taken from the paper by Stewart and Townsend (1951), show experimental plots of the functions

$$f(\xi) = \frac{1}{\overline{u^2(t)}} B_{LL}\left(\frac{r}{\lambda}\right) \quad \text{and} \quad h(\xi) = \frac{1}{[\overline{u^2(t)}]^{3/2}} B_{LL,L}\left(\frac{r}{\lambda}\right)$$

for $Re_M = 5300$ at different distances from the grid (the broken portions of the curves in Fig. 17 were regarded by the authors as unreliable). If self-preservation is valid, then curves referring to different $x/M$, i.e., different $t = x/U$, should coincide. We see, however, that this is not so, with the exception of the small region near $r = 0$ where $B_{LL}(r, t) \approx \overline{u^2}\left(1 - \frac{r^2}{2\lambda^2}\right)$. It is clear that self-preservation of the correlation functions does not hold (for all $r$, and

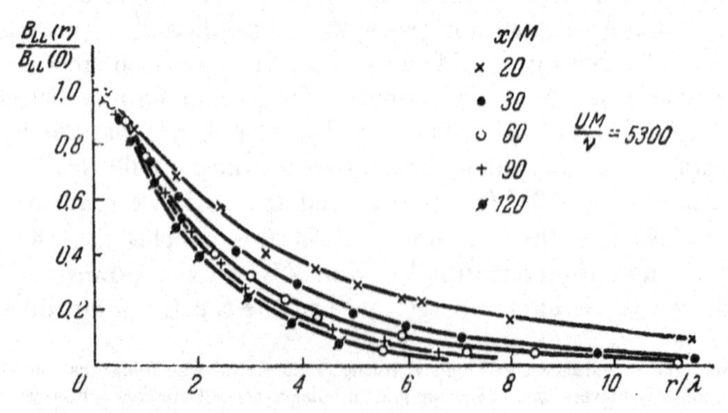


**FIG. 16** Longitudinal velocity correlation functions at different distances from the grid, according to Stewart and Townsend (1951).

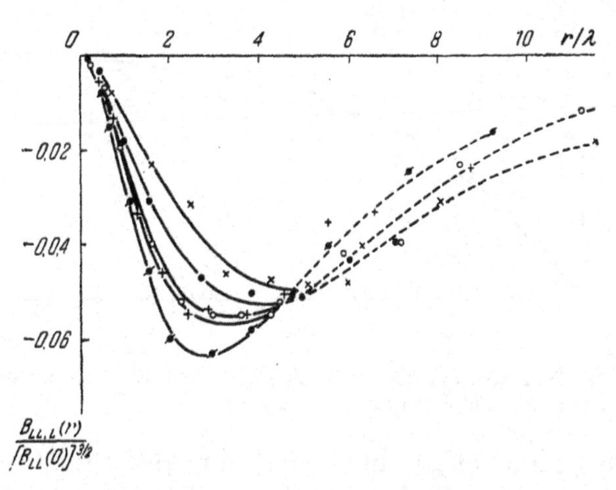


**FIG. 17** **Third-order velocity correlation functions at different distances from the grid [Stewart and Townsend (1951)]. The notation is the same as in Fig. 16.**

even for $0 \leqslant r \leqslant R$, where $R$ exceeds the linear size of the parabolic region near $r = 0$). This is also confirmed by the data of Uberoi (1963) which are quite different in many respects from those of Stewart and Townsend. It may thus be concluded that this situation is typical for turbulent flows behind a grid in a wind tunnel.

We note however that the curves of Fig. 16 have similar shapes although they fall at different rates. This means that after we transform from the correlation functions to the structure functions $D_{LL}(r, t)$ of Eq. (16.12), the latter functions can be made to coincide quite closely if we divide them by suitably chosen factors $v^2(t)$ [which differ from $\overline{u^2(t)}$]. This is shown by Fig. 18 which confirms that, to within the limits of experimental scatter, the functions $D_{LL}(r, t)$ can be regarded as self-preserving. Figure 19 shows the same construction for the third-order structure functions $D_{LLL}(r, t)$. The coincidence of the plots of $h_1(\xi)=\frac{1}{v^3} D_{LLL}\left(\frac{r}{\lambda}\right)$ corresponding to different $x/M$ is found to be less accurate than in Fig. 18. In view of the lower accuracy of the data, it may nevertheless be considered that self-preservation holds for $D_{LLL}(r, t)$ in this case for all $r$ that are not too large.

Therefore, the data of Stewart and Townsend (1951) suggest that the self-preservation hypothesis put forward by Lin (1948) and embodied in Eq. (16.13) may be in relatively good agreement with

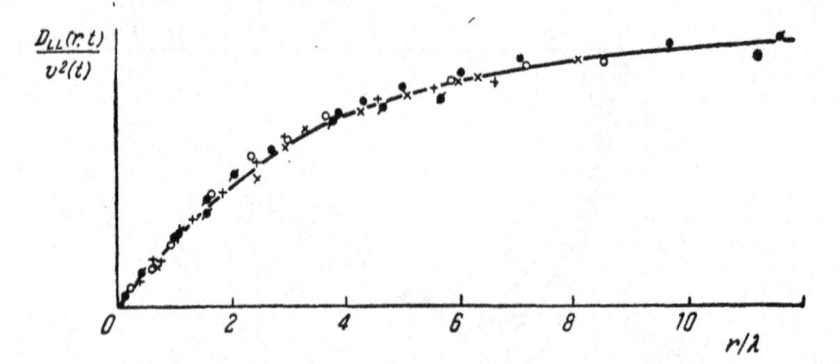


**FIG. 18** Verification of the self-preservation of the longitudinal velocity structure functions measured by Stewart and Townsend (1951).

the initial period of grid turbulence decay [in contrast to the von Kármán hypothesis (16.1) which is definitely incorrect]. A similar conclusion can be reached by considering the spectral data of Stewart and Townsend. Figure 20, which is also taken from their paper, shows the values of the normalized spectral function $\varphi(k\lambda)=\frac{E(k)}{v^2\lambda}$ (at different distances $x$ from the grid) deduced from the curves of Fig. 16, using the formula (12.77). The normalizing factors $v^2=v^2(x)$ are chosen so as to obtain the best possible coincidence of the values of $\varphi(\xi)$ for different $x$ in the short-wave part of the spectrum (large $\xi=k\lambda$). We see that the values of $\varphi(\xi)$ for $k\lambda>1$ can, in fact, be made to coincide but in the region, $k\lambda<1$,

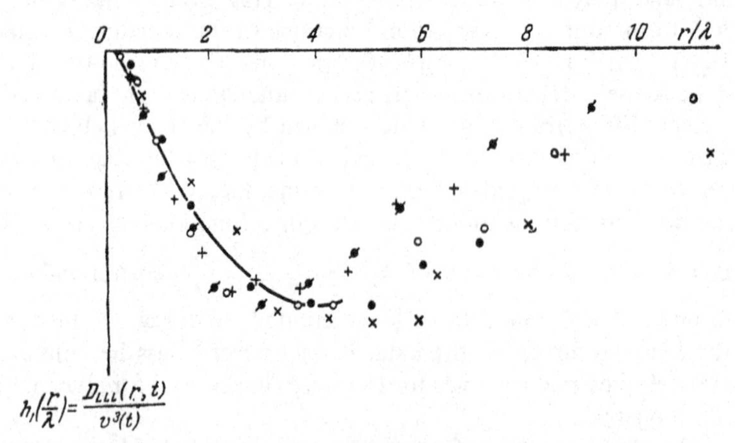


**FIG. 19** Verification of the self-preservation of the longitudinal third-order structure functions measured by Stewart and Townsend (1951).

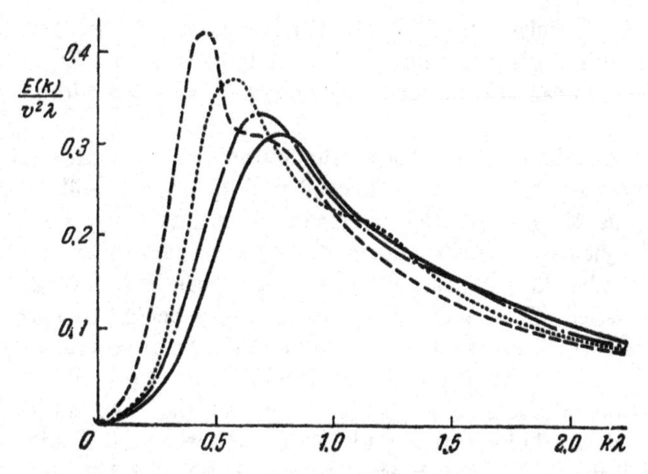


**FIG. 20** Normalized spectral density functions at different distances from the grid [Stewart and Townsend (1951)].

which contains not less than 25% of the total energy, i.e., includes an appreciable part of the energy range, the self-preservation is definitely violated. On the other hand, it is readily verified that the curves $\varphi_1(\xi)=\xi^2\varphi(\xi)$ obtained for different values of $x$ can be regarded as coinciding with sufficient accuracy. It may therefore be concluded that the data of Fig. 20 are also in relatively good agreement with Lin's assumption of self-preservation. Subsequent experiments by Tsuji (1956), who measured the one-dimensional spectra $E_1(k, x)$ at different distances from a grid with mesh size $M =$ 1 cm and mean velocity $U = 5, 10, 15$ m/sec, have also confirmed that the energy spectra do show self-preservation only in the region $k\lambda > 1$, while the functions $k^2E_1(k, x)$ can be regarded as self-preserving for all $k$.

If self-preservation of the above type does in fact occur in grid turbulence, then Eqs. (16.15) and (16.27) should be valid. The general solution of these equations includes the conditions (16.16), which imply that the energy decay law should be of the form (16.17). The data of Batchelor and Townsend (1947, 1948a) and Stewart and Townsend (1951) (see also Batchelor, 1953) suggest that the formula

$$\overline{u^2(t)}=\frac{a_1}{t-t_0}, \tag{16.32}$$

which is a special case of Eq. (16.17), is in good agreement with data for the initial period of decay. According to this data, Eqs. (16.15) and (16.27) can also be relatively accurately satisfied without much difficulty.

The introduction of a further undetermined constant, namely $a_2$, i.e., the use of Eq. (16.17) instead of Eq. (16.32), will of course enable us to improve the agreement with data when Eq. (16.32) already yields moderate agreement (see for example Lin, 1959). A further advantage appears in those cases where Eq. (16.32) is not well satisfied. These include turbulence decay behind two grids, one of which has larger mesh size than the other. Such experiments were carried out by Tsuji and Hama (1953) and Tsuji (1955) at the suggestion of Goldstein (1951). Tsuji and Hama found that the energy decay behind two grids with mesh sizes of 5 and 1 cm, respectively, is in clear disagreement with Eq. (16.32), but can be described with satisfactory accuracy by Eq. (16.17) with $a_2 \neq 0$. Tsuji (1956) has also investigated the form of the turbulence spectra behind the same two grids, and found that, when the separation between the grids is not too small, the spectrum $E_1(k, t)$ can also be regarded as self-preserving for $k\lambda > 1$, while the energy-dissipation spectrum $k^2E_1(k, t)$ is completely self-preserving. Later, Uberoi (1963), Uberoi and Wallis (1967, 1969), and Comte-Bellot and Corrsin (1966) established a number of empirical results for grid turbulence which, in many respects, were contradictory to the results of the Cambridge group. In particular, according to Uberoi (and other American workers), grid turbulence is not completely isotropic, and the decay law for it can be described by the power function

$$\overline{u_1^2} \sim \overline{u_2^2} \sim (t - t_0)^{-n}, \tag{16.32'}$$

where $n$ lies between 1.2 and 1.4 and depends on the geometry of the grid. Hence, it follows that complete self-preservation of the spectrum and of the correlation functions cannot occur in these experiments (since, otherwise, the exponent $n$ should be equal to unity). At the same time, according to the data of Uberoi (1963), which correspond to $n \approx 1.2$, the self-preserving variation of the spectrum in the dissipation range was adequately confirmed by his experiments (we shall return to this point in Sect. 16.6). Hence, it is clear that the decay law given by Eq. (16.17) should in this case be in satisfactory agreement with the data. More recent data of

Portfors and Keffer (1969) show again that the initial period grid turbulence is quite isotropic and the decay law is close to a linear relation (16.32); these data have been already referred to in Sect. 14.1.

Uberoi's data (1963) are in conflict with the data of British workers in one further respect, namely, according to Uberoi the energy and the dissipation ranges do not overlap even for $\mathrm{Re}_M$ = 26,000 (in the sense that it is possible to find a wave number $k_1$ such that the region $k < k_1$ will contain 80% of the turbulent energy while the region $k > k_1$ will contain 80% of energy dissipation). It follows that it is possible in principle to achieve the von Kármán "nonviscous self-preservation" which refers only to the energy range but not to the dissipation range and leads to relations (16.21) with arbitrary $c_1$, $c_2$, and $v^2(t) = \overline{u_1^2}$. To verify the presence of this self-preservation, Uberoi measured the one-dimensional spectrum of the variable $u_1 + \sqrt{2}u_2$ at three distances from the grid, and then used these data to calculate the three-dimensional spectrum $E(k)$ and the function $T(k) = \frac{\partial}{\partial t}E + 2\nu k^2 E$ with the aid of formulas that are valid for isotropic turbulence (the use of the variable $u_1 + \sqrt{2}u_2$, instead of $u_1$ or $u_2$ lead to a simplification of the calculation of $E(k)$ and reduced the effect of the moderate anisotropy of the grid turbulence observed by Uberoi). The results are shown in Figs. 21 and 22.[11]

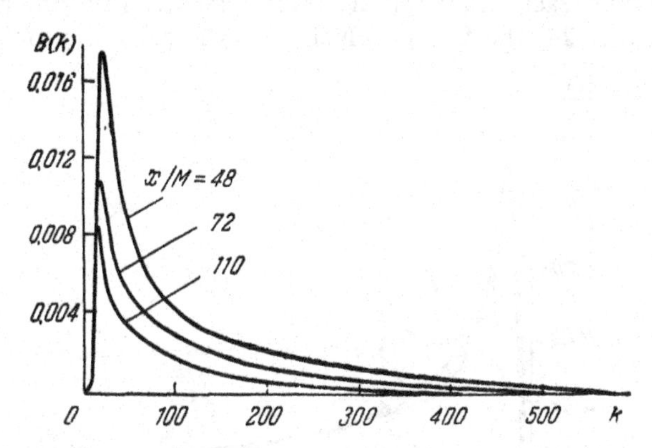


**FIG. 21** Spectral functions $E(k)$ at different distances from the grid [Uberoi (1963)].

[11] These results are, however, apparently not quite reliable (see the critical remarks on them in Van Atta and Chen, 1969a).

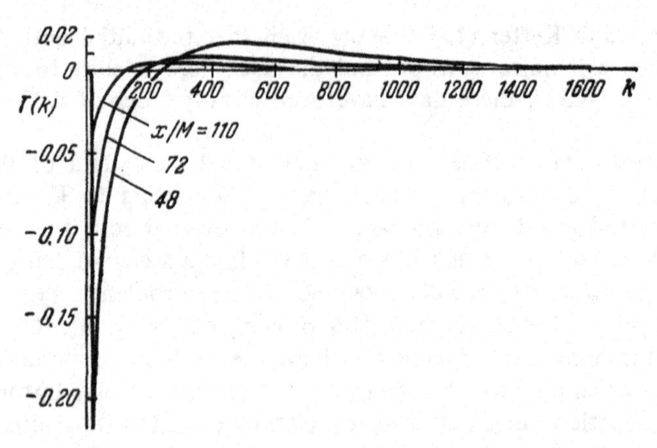


**FIG. 22** The functions $T(k)$ at different distances from the grid. [Uberoi (1963)].

Next, Uberoi tried to choose the length scale $l(t)$ so that Eq. (16.23) was satisfied as closely as possible for the functions $E(k, t)$ and $T(k, t)$ in the energy range [the quantity $v^2(t)$ was replaced by $\overline{u_1^2} + 2\overline{u_2^2}$). It was found that if it is assumed that $l(t) \sim (t - t_0)^{0,4}$, which is in complete agreement with Eqs. (16.21) and (16.32′) with $n = 1.2$, then the functions $\varphi(kl) = \frac{E(k)}{lv^2}$ and $\psi(kl) = \frac{T(k)}{v^3}$ are very close to each other for the three distances from the grid (see Figs. 23 and 24, the first of which gives the data corresponding to

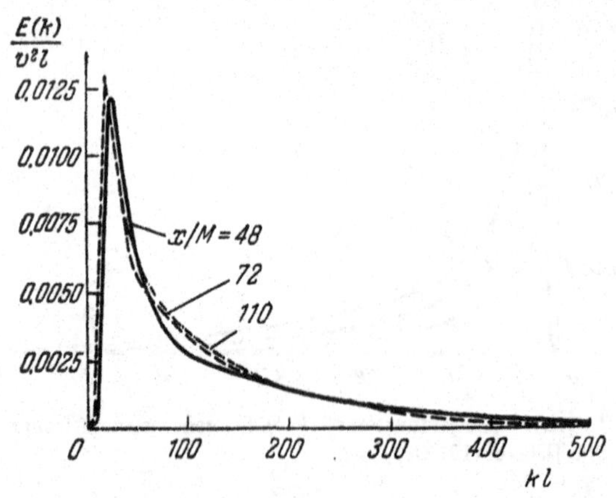


**FIG. 23** Verification of the self-preservation of $E(k)$ [Uberoi (1963)].

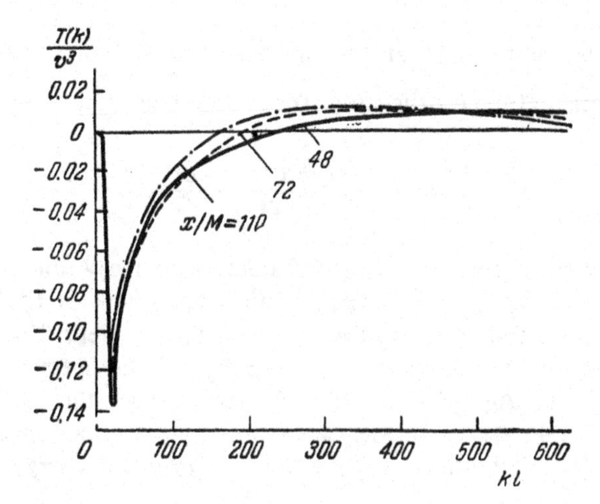


**FIG. 24** Verification of the self-preservation of $T(k)$ [Uberoi (1963).

$x/M = 72$ and $x/M = 110$ which are seen to coincide for the majority of the values of $kl$).

## 16.5 The Kolmogorov Hypotheses on Small-Scale Self-Preservation at High Enough Reynolds Numbers

The data considered above corresponded to moderate Reynolds numbers for which the energy and the dissipation ranges of the spectrum are in contact with each other or may even overlap. However, the case of very high Reynolds numbers is much simpler from the theoretical point of view. Here the two ranges are well separated. For the small-scale turbulence we can then formulate general self-preservation hypotheses based on definite physical representations of the mechanism of turbulent mixing.

Consider a small range $(k, k+\Delta k)$ along the $k$ axis. This will correspond to a definite set of turbulent disturbances (total intensity $E(k)\Delta k$), which will have its own "characteristic length scale" $l_k \sim 1/k$ and also its own "characteristic velocity scale" $U'_k$ as well as a characteristic time scale $T_k = \frac{1}{kU'_k}$. The time scale $T_k$ can be estimated with relatively good accuracy for perturbations in the energy range. In fact, the total turbulent energy $\frac{3}{2}\overline{u'^2} = \frac{3}{2}U'^2$, where $U' = \left(\overline{u'^2}\right)^{1/2}$, will decay with time at the rate $\bar{\varepsilon} = -\frac{3}{2}\frac{dU'^2}{dt}$. Accordingly, the characteristic time for the turbulence decay will be given by

$$T_0 = -U'^2 \Big/ \frac{dU'^2}{dt} \sim \frac{U'^2}{\bar{\varepsilon}}. \qquad (16.33)$$

It is natural to expect that the "characteristic decay time" will be of the same order as the "characteristic time scale" of the principal energy-containing disturbances which can be identified as corresponding to the maximum $E_{max} = E(k_0)$ of the spectral function $E(k)$. If we suppose that for such disturbances the characteristic velocity scale $U'_{k_0}$ is of the same order as the root mean square velocity $U'$ determined from the total turbulent energy, then we obtain

$$-\frac{1}{U'^2}\frac{dU'^2}{dt} \sim U'k_0, \quad \text{i.e.,} \quad \frac{2}{3}\bar{\varepsilon} = -\frac{dU'^2}{dt} = A\frac{U'^3}{L}, \qquad (16.34)$$

where $A$ is a numerical factor of the order of unity and $L = 1/k_0$ is a length which can be regarded as the "characteristic length scale" of the turbulence as a whole [this length is of the same order as the "integral length scales" in Eq. (12.50)]. The formula (16.34) is confirmed by existing data. For example, according to Batchelor and Townsend (1948a), if we identify $L$, with the longitudinal integral length scale $L_1$, then for the turbulence behind a grid, the coefficient $A$ is found to be a very slowly-varying function of $x$ in the range between $20M$ and $180M$, and of $\mathrm{Re}_M$ in the range between 2,800 and 22,000. In these ranges the coefficient $A$ assumes values changing between 0.8 and 1.4 [see Batchelor (1953), Sect. 6.1].

Formula (16.34) suggests that the entire energy decay in fact occurs within a few characteristic time scales of the energy-containing disturbances, which explains the absence of strict self-preservation for such disturbances: They do not have sufficient time to readjust to each other in a universal fashion.[12]

[12] Von Kármán and Lin (1949, 1951) have, nevertheless, considered the case where the spectrum varies in the energy range according to some self-preserving laws and, moreover, definite self-preservation occurs for the extreme large-scale disturbances with $k$ lying to the left of the energy range. However, the problem as to what are the conditions under which this self-preservation is valid remains unresolved.

However, it is natural to suppose (and all existing measurements fully confirm this) that as $k$ increases, the characteristic time scale $T_k = (kU'_k)^{-1}$ will decrease. In other words, as the spatial scale of the disturbances decrease, there is an increase in the frequency of the corresponding fluctuations. It can then be assumed, however, that under certain conditions (in the first instance, for high enough Reynolds numbers), a definite fraction of the disturbances will be characterized by frequencies such that the turbulence decay can then be regarded as very slow. These disturbances must naturally rapidly readjust to each other and pass into a state of statistical quasi-equilibrium which is independent of the conditions under which the turbulence was generated.[13] They subsequently adjust also to the slow variation of the turbulent state, remaining throughout in the same quasi-stationary quasi-equilibrium state, which depends only on certain integral parameters of the turbulence, but does not depend directly on time.

We shall see later that it is not in general necessary for the turbulence to be isotropic in order to ensure that the short-wave disturbances are self-preserving. According to the general Kolmogorov theory, the statistical state of a set of small-scale disturbances is universal for all turbulent flows with sufficiently high Reynolds numbers, and hence it follows that all the statistical parameters of such disturbances are self-preserving. Since the isotropy of turbulence plays no essential role in this case, we shall discuss the Kolmogorov theory in detail in the next chapter. Here, we shall confine our attention to a brief formulation of certain basic assumptions of this theory, which are directly relevant to the study of isotropic turbulence and are important for the remainder of the present chapter.

We must first establish the "external parameters" of the entire flow upon which the statistical characteristics of small-scale turbulence may depend in the case of very high Reynolds numbers. We shall confine our attention to disturbances that are much smaller than those corresponding to the energy maximum (and to a time scale of the order of a typical energy decay time). In that case the statistical state of these disturbances can naturally be assumed to be independent of the total kinetic energy $\mathscr{E}$ of the turbulence and of

[13] It is usual to speak about universal equilibrium, but since there are some important distinctions between this state and other physical equilibrium states, we prefer the term quasi-equilibrium. This question will be discussed in more detail in Chapt. 8.

the root mean square velocity $U' = \left(\frac{2}{3}\mathscr{E}\right)^{1/2}$. The statistical state should also be independent of the parameters characterizing the initial conditions governing the turbulence generation (such as the grid spacing in the case of grid turbulence). On the other hand, if the energy dissipation range does not overlap the energy range (this is necessary for the existence of a universal quasi-equilibrium statistical state), then it is natural to suppose that the state of the disturbances responsible for most of the energy dissipation will be universal, so that the parameter $\bar{\varepsilon} = \frac{d\mathscr{E}}{dt}$ must be included among the external parameters in which we are interested. In addition to $\bar{\varepsilon}$, the universal statistical state of small-scale fluctuations may depend on the fluid parameters, namely, the density $\rho$, the molecular transfer coefficients such as the viscosity $\nu$, and the temperature diffusivity $\chi$. In the case of the velocity fluctuations, viscosity is the only fluid parameter that can play a significant role, since it determines the position of the dissipation range in the spectrum. It is therefore expected that, for sufficiently high Reynolds numbers, all the statistical characteristics of velocity disturbances with wave numbers $k$ exceeding some "sufficiently large" wave number $k^{(0)}$ will depend only on the parameters $\bar{\varepsilon}$ and $\nu$. This is the substance of the general Kolmogorov hypothesis (see Sect. 21.3 for further details).

One consequence of the Kolmogorov hypothesis is that the small-scale statistical characteristics of the velocity fluctuations in different turbulent flows with high Re can differ only by the length and time scales or, what is the same, by the length and velocity scales (which depend on the parameters $\bar{\varepsilon}$ and $\nu$). If, on the other hand, we measure all lengths in units of $\eta$ and all velocities in units of $v_\eta$, where

$$\eta = \left(\frac{\nu^3}{\bar{\varepsilon}}\right)^{1/4}, \qquad v_\eta = (\nu\bar{\varepsilon})^{1/4}, \tag{16.35}$$

then the probability distributions for small-scale velocity fluctuations in different high Re flows will be exactly the same. In particular, the wave-number distribution of the turbulent energy for $k > k^{(0)}$ should be of the form

$$E(k) = \eta v_\eta^2 \varphi(\eta k) = \bar{\varepsilon}^{1/4} \nu^{5/4} \varphi(\nu^{3/4} \bar{\varepsilon}^{-1/4} k), \tag{16.36}$$

where $\varphi(\xi)$ is a universal function. In view of Eq. (12.86), the longitudinal one-dimensional spectrum $E_1(k)$ is of the same form:

$$E_1(k) = \eta v_\eta^2 \varphi_1(\eta k), \quad \varphi_1(\xi) = \int_\xi^\infty \frac{x^2 - \xi^2}{x^3} \varphi(x)\,dx. \qquad (16.36')$$

It is readily seen that it then follows that for $k > k^{(0)}$ the function $T(k)$ can also be written in the form $T(k) = v_\eta^3 \psi(\eta k)$ [see the formula (16.40) below]. Therefore, for $k > k^{(0)}$ we have self-preservation of the form indicated by Eq. (16.23), and the corresponding length and velocity scales can be determined from Eq. (16.35).

The universal functions $\varphi(\xi)$ and $\xi^2\varphi(\xi)$ should, of course, have the same general form as the functions $E(k)$ and $2\nu k^2 E(k)$ shown in Fig. 12. In particular, the dissipation range on the $\xi$ axis will occupy the fixed finite interval $c_1 < \xi < c_2$, and the function $\xi^2\varphi(\xi)$ will reach a maximum at a point $c_0$ within this range. In these expressions $c_1$, $c_2$, and $c_0$ are dimensionless universal constants of the order of unity. In the dimensional form, the limits of the dissipation range will be the wave numbers $c_1/\eta$ and $c_2/\eta$, and the dissipation maximum will occur at $k = c_0/\eta$. It is clear that $k_\eta = 1/\eta$ is the order of magnitude of the wave numbers associated with most of the energy dissipation.

We shall suppose now that the universal statistical state extends well to the left of the point $k_\eta$ on the wave-number axis, so that the function $\varphi(\xi)$ is defined for $\xi \ll 1$ (this requires that the energy and dissipation ranges should be sufficiently well separated along the $k$ axis). The asymptotic form of $\varphi(\xi)$ for $\xi \ll 1$ can then be determined to within a numerical factor as follows. When $\xi \ll 1$, i.e., $k \ll k_\eta$, energy dissipation is negligible, i.e., the viscosity of the fluid has practically no effect. It is therefore natural to suppose that the statistical characteristics of velocity fluctuations in this part of the spectrum should be independent of $\nu$. Hence it follows that when $k \ll k_\eta$, the function $E(k)$ can depend only on $k$ and $\bar{\varepsilon}$. Dimensional considerations then suggest that

$$E(k) = C_1 \bar{\varepsilon}^{2/3} k^{-5/3} \quad \text{for} \quad k \ll k_\eta, \qquad (16.37)$$

where $C_1$ is a dimensionless constant. This formula expresses the so-called *five-thirds law* for the energy spectrum in the inertial subrange, and was first given by Obukhov (1949a, b). If we compare this formula with Eqs. (16.36) and (16.35), we find that

$$\varphi(\xi) = C_1 \xi^{-5/3} \quad \text{for} \quad \xi \ll 1. \tag{16.38}$$

Since the one-dimensional spectrum $E_1(k)$ has the same dimensionality as $E(k)$, we have

$$E_1(k) = C_2 \bar{\varepsilon}^{2/3} k^{-5/3} \quad \text{for} \quad k \ll k_\eta; \quad \varphi_1(\xi) = C_2 \xi^{-5/3} \quad \text{for} \quad \xi \ll 1, \tag{16.37'}$$

where, by Eq. (16.36′), $C_2 = (18/55)C_1$. The subrange $k^{(0)} < k \ll k_\eta$ of the universal equilibrium region $k^{(0)} < k < \infty$, in which Eqs. (16.37), (16.38), and (16.37′) are valid, is called the *inertial subrange*, or the *inertial interval* of the spectrum, since the fluctuations in this subrange are independent of viscous friction and are determined exclusively by inertial forces.

Let us now consider the consequences of the basic spectral equation (14.15) of the theory of isotropic turbulence in the universal quasi-equilibrium range. Since the time dependence of the statistical parameters can be neglected in this range, Eq. (14.15) now assumes the very simple form

$$T(k) = 2\nu k^2 E(k), \tag{16.39}$$

which by Eqs. (16.36) and (16.35) shows that

$$T(k) = v_\eta^3 \psi(k\eta), \qquad \psi(\xi) = 2\xi^2 \varphi(\xi). \tag{16.40}$$

It will be convenient for subsequent applications to integrate Eq. (14.15):

$$\frac{\partial}{\partial t} \int_0^k E(k', t)\, dk' = \int_0^k T(k', t)\, dk' - 2\nu \int_0^k k'^2 E(k', t)\, dk'. \tag{16.41}$$

If $k$ is greater than the upper limit of the energy range, then we may suppose that

$$\int_0^k E(k', t)\, dk' = \frac{3}{2} U'^2(t) \quad \text{and} \quad \frac{\partial}{\partial t} \int_0^k E(k', t)\, dk' = -\bar{\varepsilon}.$$

Next, the second term on the right-hand side of Eq. (16.41) will, in fact, be determined by only the "quasi-stationary" values of $E(k', t)$ with $k' \gtrsim k_\eta > k^{(0)}$, i.e., they will be time-independent. Therefore, the first term, which by Eq. (14.18) can be written in the form $-\int_k^\infty T(k', t)\,dk'$, will also be time independent for $k > k^{(0)}$. It follows that, in the universal equilibrium range, the spectral equation will be of the form

$$W(k) + 2\nu \int_0^k k'^2 E(k')\,dk' = \bar{\varepsilon}. \qquad (16.42)$$

Here $W(k) = -\int_0^k T(k')\,dk' = \int_k^\infty T(k')\,dk'$ is the amount of energy transferred per unit time from disturbances with wave numbers smaller than $k$ to the other disturbances and is called the *spectral energy transfer function*. In exactly the same way, the equation

$$\frac{\partial}{\partial t} \int_k^\infty E(k', t)\,dk' = \int_k^\infty T(k', t)\,dk' - 2\nu \int_k^\infty k'^2 E(k', t)\,dk' \qquad (16.41')$$

which augments Eq. (16.41) for $k$ in the universal quasi-equilibrium range, can be written in the form

$$W(k) = 2\nu \int_k^\infty k'^2 E(k')\,dk'. \qquad (16.42')$$

This equation together with the "normalization condition"

$$2\nu \int_0^\infty k'^2 E(k')\,dk' = \bar{\varepsilon} \qquad (16.43)$$

is precisely equivalent to Eq. (16.42).

## 16.6 Conditions for the Existence of Kolmogorov Self-Preservation in Grid Turbulence

We shall now analyze in greater detail the conditions under which the above Kolmogorov self-preservation can be observed behind a grid in a wind tunnel. This requires above all that the energy range of the spectrum (localized in the wave-number region around $k_0 = 1/L$) and the dissipation range (which in the case of Kolmogorov self-preservation is characterized by wave numbers of the order of $k_\eta = 1/\eta = \bar{\varepsilon}^{1/4}\nu^{-3/4}$) are separated by a sufficiently long interval of wave numbers. Hence, it is clear that Kolmogorov self-preservation cannot occur unless the following condition is satisfied:

$$k_0 \ll k_\eta, \quad \text{i.e.,} \quad L \gg \eta. \tag{16.44}$$

If we use Eqs. (16.35) and (16.34), according to which $\bar{\varepsilon} \sim U'^3/L$, the last condition can be rewritten in the more convenient form

$$\left(\frac{LU'}{\nu}\right)^{3/4} = (\mathrm{Re}_L)^{3/4} \gg 1 \tag{16.45}$$

By substituting Eq. (16.34) in Eq. (15.2), it can readily be shown that $L/\lambda \sim \frac{U'\lambda}{\nu} = \mathrm{Re}_\lambda$, so that Eq. (16.45) can be rewritten in the form $\left(\frac{U'\lambda}{\nu}\right)^{3/2} = (\mathrm{Re}_\lambda)^{3/2} \gg 1$). Therefore, we again conclude that the universal statistical state predicted by Kolmogorov will exist if the Reynolds number is "sufficiently high" (the more precise meaning of this statement can be determined only from experimental data). We note further that comparison of Eq. (15.2) with Eqs. (16.35) and (16.34) shows that $\lambda \sim \eta(\mathrm{Re}_\lambda)^{1/2} \sim L/\mathrm{Re}_\lambda \sim (\eta^2 L)^{1/3}$, i.e., $L \gg \lambda \gg \eta$. In other words, the Taylor microscale $\lambda$ is intermediate between the length scale $L$ of the energy-containing disturbances, and the length scale $\eta$ of disturbances associated with most of the dissipation. It does not determine any characteristic point in the spectrum, and is convenient only in that it can be readily determined experimentally.

The condition (16.45) is not sufficient for the existence of the inertial subrange of the spectrum. The further condition is that there should exist a wave-number interval such that

$$k_0 \ll k \ll k_\eta. \tag{16.46}$$

Assuming that the ratios $k/k_0$ and $k_\eta/k$ should be roughly of the same order of magnitude, Batchelor (1953) concluded that the inertial subrange will exist if the following more stringent inequality is satisfied:

$$(\mathrm{Re}_L)^{3/8} \gg 1 \tag{16.47}$$

or $(\mathrm{Re}_\lambda)^{3/4} \gg 1$.

Batchelor and Townsend (1948) [see also Batchelor (1953), Sect. 6.5], considered the generation of turbulence by a grid consisting of rods having a diameter of $M/5.3$ where $M$ is the mesh size. If we define the Reynolds number as $\mathrm{Re}_{L_1} = \frac{L_1 U'}{\nu}$ where $L_1$ is the integral length scale defined by Eq. (12.50) and is of the same order as the scale $L$ in Eq. (16.34), then according to Batchelor and Townsend the latter does not vary to any considerable extent during the initial period of decay, and can be estimated from the formula $\mathrm{Re}_{L_1} \approx \frac{\mathrm{Re}_M}{134} = \frac{1}{134}\frac{UM}{\nu}$ where $U$ is the mean velocity. Hence, for example, we find that for $\mathrm{Re}_M = 4.2 \cdot 10^4$ we have $\mathrm{Re}_{L_1} \approx 300$, $(\mathrm{Re}_{L_1})^{3/4} \approx 80$, $(\mathrm{Re}_{L_1})^{3/8} \approx 9$.

Stewart (1951) tried to verify the validity of Eq. (16.42) in a flow behind a grid of this kind with $\mathrm{Re}_M = 42{,}000$ and $x = 30M$, using the formulas (12.77) and (12.142) and the measured values of the functions $B_{LL}(r)$ and $B_{LL,L}(r)$. According to his data, this equation is clearly invalid under the above conditions even for very high values of $k$ which definitely lie in the dissipation range. Therefore, the value $\mathrm{Re}_{L_1} \approx 300$, i.e., $(\mathrm{Re}_{L_1})^{3/4} \approx 80$ must be regarded as insufficiently high for the appearance of Kolmogorov self-preservation. According to Batchelor (1953), in these experiments $L_1/\lambda \approx 0.1\, \mathrm{Re}_\lambda$ and, therefore, the values $\mathrm{Re}_M = 4.2 \cdot 10^4$ and $\mathrm{Re}_{L_1} \approx 300$ now correspond to $\mathrm{Re}_\lambda = 55$ (which is again "insufficiently high" according to the results of Stewart).

To ensure that the inertial subrange of the spectrum exists in grid turbulence, the Reynolds number must be very much higher. According to very approximate estimates by Proudman (1951), Stewart and Townsend (1951) and Gibson and Schwarz (1963b), an appreciable inertial subrange can appear in the turbulence spectrum only for $Re_\lambda$ of the order of many hundreds or even a thousand, i.e., for $Re_M$ of the order of one or more millions.[14] Such values of $Re_M$ are very difficult to achieve in existing wind tunnels. On the other hand, wind-tunnel measurements of the function $E_1(k)$ behind a grid for values of $Re_M$ of the order of a few thousand or ten thousand which are reported by Simmons and Salter (1938), Liepmann et al. (1951), Stewart and Townsend (1951), Sato (1951), Favre et al. (1952), Uberoi (1963), and many others, do not confirm the existence of an appreciable range of values of $k$ within which $E_1(k)$ is proportional to $k^{-5/3}$.[15]

Von Kármán (1948b) has tried to approximate the data of Liepmann et al. (1951) on the function $E_1(k)$ (or $Re_M = 10^5$ and $3 \cdot 10^5$ by a function proportional to $k^{-5/3}$ for moderate values of $k$. However, the spread of the experimental points was such that this approximation was highly unsatisfactory. Gibson and Schwarz (1963b) have measured the spectrum $E_1(k)$ in grid flow in a water tunnel for $2 \cdot 10^4 \leqslant Re_M \leqslant 3.8 \cdot 10^4$. At the maximum value of $Re_M$ the spectrum obtained by them (see Fig. 75 below) seems to be proportional to $k^{-5/3}$ over a small range of values of $k$, but here again the accuracy is insufficient to attach too much importance to this conclusion. Finally, Kistler and Vrebalovich (1966) have measured $E_1(k)$ and $E_2(k)$ behind a grid for $1.2 \cdot 10^5 \leqslant Re_M \leqslant 2.4 \cdot 10^6$ in a very large wind tunnel (unfortunately this tunnel was dismounted soon after the first measurements, so that the results could not be subsequently verified). According to their results the measured spectrum $E_1(k)$ was proportional to $k^{-5/3}$ over an appreciable range of

[14] The results of Bradshaw (1969) may seem to contradict this statement. It is found in this work that the value of $C_2$ obtained by drawing a $-5/3$ power law tangent to experimental spectra $E_1(k)$ and using Eq. (16.37′) is approximately constant (and close to a value 0.5 which will be established in Sect. 23.3 below) in grid turbulence and various types of shear flow for values of $Re_\lambda$ down to $Re_\lambda \approx 100$ where the microscale $\lambda$ is defined by $\bar{\epsilon} = 15\nu\overline{u^2}/\lambda^2$. However, Bradshaw does not assert that the condition $Re_\lambda \gg 100$ is sufficient for the existence of a true observable inertial range. He emphasizes in fact that his result only constitutes a useful empirical fact which permits one to evaluate $\bar{\epsilon}$ easily.

[15] Uberoi (1963) has pointed out that a good method of verifying the existence of the inertial subrange is to determine the functions $T(k) = -\frac{dW(k)}{dk}$ [from measured values of $B_{LL,L}(r)$ or $E_1(k, t)$ and $\partial E_1(k, t)/\partial t$]. In the inertial subrange the equation (16.42′) assumes the form $W(k) = \bar{\varepsilon} =$ const. Consequently, throughout this subrange we should have $T(k) = 0$. At the same time, it is clear from Fig. 24 that, for example, the data of Uberoi show no traces of $T(k)$ vanishing over an appreciable range of wave numbers. Consequently, the inertial subrange was absent.

values of $k$. However, the proportionality factor was different from that reported elsewhere (see Sect. 23.3), and the shape of the spectrum $E_2(k)$ for these values of $k$ was different from that expected for the case of isotropic turbulence (which is in direct conflict with the prediction of the Kolmogorov theory). We must conclude, therefore, that when we speak of an empirical verification of the existence of the inertial subrange, the only reliable evidence we have are the data for nonisotropic turbulent flows which will be discussed in Sect. 23.3.

Results of direct verifications of Eq. (16.36) for grid turbulence at moderate values of $\mathrm{Re}_M$ carried out by Townsend (1951), Uberoi (1963), and by some other authors may seem to be in partial conflict with the above discussion on the validity of Kolmogorov self-preservation in grid turbulence. According to Stewart and Townsend (Fig. 25), the empirical dependence of the ratio $E_1(k)/\eta v_\eta^2$ on $k\eta = k/k_\eta$ for different distances $x$ from the grid and different $\mathrm{Re}_M$

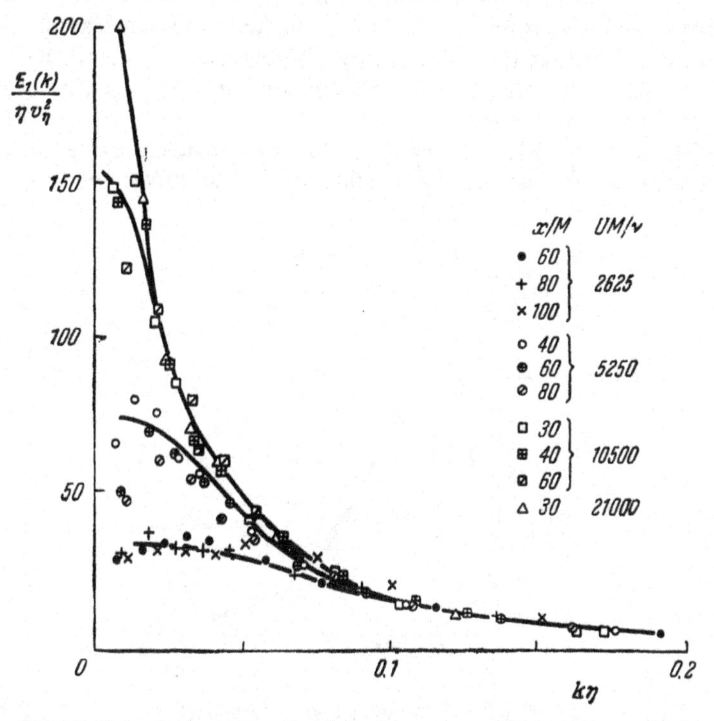


**FIG. 25** Normalized one-dimensional velocity spectra at different distances from the grid and for different Reynolds numbers [Stewart and Townsend (1951)].

(in the range from 2,500 to 20,000) for $k\eta > 0.1$ turns out to be "universal," in complete agreement with Eq. (16.36'). In the region $k\eta < 0.1$, which contains most of the energy, the values of $E_1(k)/\eta v_\eta^2$ for different $\mathrm{Re}_M$ are, however in clear disagreement with each other (there is a smaller discrepancy between spectra for a given flow and different distances from the grid, which is clearly seen in Fig. 20 but is masked in Fig. 25 by the spread of the experimental points). Figure 26 thus suggests that the universal statistical state predicted by Kolmogorov exists in grid turbulence even for moderate values of $\mathrm{Re}_M$ when $k\eta > 0.1$. The same universal form of the one-dimensional energy spectra of grid turbulence at moderate values of $\mathrm{Re}_M$ and small enough values of $k\eta$ was established by the more recent data of Gibson and Schwarz (1963) and Van Atta and Chen (1968, 1969a). An analogous result was obtained by Uberoi (1963) who used the data of Fig. 21 to evaluate the energy-dissipation spectra $k^2E(k)$ at three distances from the grid, and determined the normalized dissipation spectrum $\eta k^2 E(k)/v_\eta^2$ (Fig. 26). We see that this normalized spectrum is now approximately the same at different distances, so that the Kolmogorov self-preservation of the spectrum in the dissipation range is adequately confirmed by the Uberoi data even for $\mathrm{Re}_M \approx 3 \cdot 10^4$.

The data of Figs. 25 and 26 do not provide a quite accurate estimate of the degree of coincidence of the curves for $k\eta > 0.1$

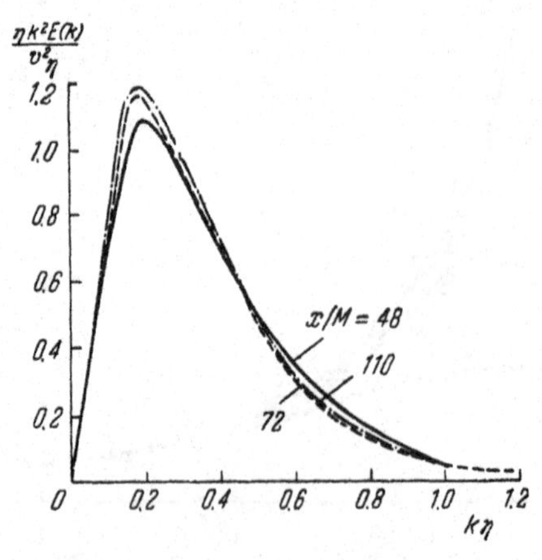


**FIG. 26** Normalized energy dissipation spectra at different distances from the grid [Uberoi (1963)].

where the absolute values of the spectrum are relatively low. To obtain more reliable results, Stewart and Townsend measured both the one-dimensional energy spectrum and the one-dimensional spectrum of the derivatives $\frac{\partial u}{\partial t} = \frac{1}{U}\frac{\partial u}{\partial x}$, $\frac{\partial^2 u}{\partial t^2} = \frac{1}{U^2}\frac{\partial^2 u}{\partial x^2}$ and $\frac{\partial^3 u}{\partial t^3} = \frac{1}{U^3}\frac{\partial^3 u}{\partial x^3}$, which are respectively proportional to the functions $k^2E_1(k)$, $k^4E_1(k)$, and $k^6E_1(k)$. Graphs of the ratios $\frac{\eta^2k^2E_1(k)}{\eta v_\eta^2}$ and $\frac{\eta^4k^4E_1(k)}{\eta v_\eta^2}$ as functions of $k\eta$, deduced from these data, have shown a possibility of a small systematic discrepancy as compared with the corresponding functions deduced from measurements on flows with different $\mathrm{Re}_M$ throughout the range $k\eta < 0.6$, while the values of the same functions corresponding to the same flow but different distances from the grid were found to coincide to within experimental error (in complete agreement with the conclusions drawn from Fig. 20).[16] Moreover, the measured values of the spectrum of $\frac{\partial^3 u_1}{\partial t^3}$ appear to suggest that in the region $k\eta > 0.6$, which lies appreciably to the right of the dissipation range, the dependence of the dimensionless spectrum $\frac{E_1(k)}{\eta v_\eta^2}$ on $k\eta$ for all the $\mathrm{Re}_M$ used in these experiments and all $x > 60M$ (but not for $x = 30M$ or $x = 40M$) is in fact the same, i.e., it is universal. Let us also recall that the data of Gibson and Schwarz (1963), Uberoi (1963) and Van Atta and Chen (1968) suggest that the spectra have a universal shape for an even more extended range of values of $k\eta$. Batchelor (1953) proposed that, even for moderate Reynolds numbers, the behavior of the spectrum for sufficiently large values of $k$ and sufficiently large distances from the grid is determined exclusively by the values of $\bar{\varepsilon}$ and $\nu$, i.e., it is independent of the details of the "initial conditions" and is universal in the sense of the universal statistical state of Kolmogorov for high Reynolds numbers. Subsequently, Gibson and

[16] The fact that the length scales in Figs. 20 and 25 were different is unimportant because during the initial period of decay, when $\overline{u^2} \sim (t - t_0)^{-1}$, we obviously have $\bar{\varepsilon} \sim (t - t_0)^{-2}$, $\lambda \sim (t - t_0)^{1/2}$, $\eta \sim (t - t_0)^{1/2}$, i.e., $\lambda(t)$ and $\eta(t)$ differ only by a constant factor. Transformation from the three-dimensional spectrum $E(k)$ to the one-dimensional spectrum $E_1(k)$ is also unimportant because in view of their connection for $k > k_0$, these spectra may or may not be simultaneously self-preserving.

Schwarz (1963b) noted that instead of the true value of the parameter $\bar{\varepsilon}$ it is probably better to use the "corrected" value $\bar{\varepsilon}_1 \neq \bar{\varepsilon}$ at moderate values of $\mathrm{Re}_M$ (see Sect. 23.4), but that in practice this correction can frequently be neglected because $\eta \sim \bar{\varepsilon}^{-1/4}$ is a very slowly varying function even for large variations of $\bar{\varepsilon}$. So far, all these conclusions must be regarded as purely preliminary.

## 16.7 The Meso-Scale Quasi-Equilibrium Hypothesis. Self-Preservation of Temperature Fluctuations

Since the behavior of the specturm in the extreme short-wave range appears to be determined exclusively by the values of $\bar{\varepsilon}$ and $\nu$, it may be considered that the behavior of the grid turbulence spectrum in the adjacent region of moderate values of $k$ will also depend on a small number of important parameters. This is, in fact, the basis of all the above self-preservation hypotheses in which the dominant parameters were taken to be some characteristic values of the length scale $l$ and the velocity scale $v$. Heisenberg (1948b) and Batchelor (1953) have noted certain general qualitative physical considerations which throw light on the possible origin of the self-preserving quasi-equilibrium in the range of moderate wave numbers. The basic assumption in this reasoning is that in the wave number range in the immediate neighborhood of the "universal range of the spectrum" (determined by the values of $\bar{\varepsilon}$ and $\nu$), the spectrum will depend on only one additional parameter characterizing the decay stage of the turbulence. The simplest assumption of this kind adopted by Batchelor was that the additional parameter can be the time $t-t_0$ where $t_0$ is the virtual "initial time." Hence it follows in particular that the only dimensionless parameter which can be constructed out of $\bar{\varepsilon}$, $\nu$, and $t-t_0$, namely, $R=\bar{\varepsilon}(t-t_0)^2/\nu$, should remain constant throughout the entire period of this meso-scale quasi-equilibrium (since it cannot depend on the dimensional variable $t-t_0$). Consequently, during this period

$$\frac{d\overline{u^2}}{dt} = -\frac{2}{3}\bar{\varepsilon} = -\frac{\frac{2}{3}R\nu}{(t-t_0)^2} \tag{16.48}$$

and hence

$$\overline{u^2} = \frac{\frac{2}{3}R\nu}{t-t_0} + a, \tag{16.49}$$

i.e., we again have the decay law (16.17) which we originally derived, following Lin, from apparently quite different considerations. The quantity $R$ in Eq. (16.49) should depend on the initial conditions for the generation of turbulence. In particular, for grid turbulence, it can depend only on the type of grid and the mean velocity $U$, while for geometrically similar grids it should be a single-valued function of $\mathrm{Re}_M = \frac{UM}{\nu}$. When turbulence decay is described by Eq. (16.49), with zero or very small $a$, the number $R$ turns out to be proportional to $\mathrm{Re}_M$ (for grids with square mesh, as used in the experiments of Townsend and Stewart, Batchelor (1953) found that $R \approx \mathrm{Re}_M/90$).[17] Thus, in the present case, and in contrast to the previous derivation of Eq. (16.17), the coefficient of the term $\sim (t-t_0)^{-1}$ in the decay law, has a definite physical meaning, whereas the significance of the coefficient $a$ even now remains unclear [attempts by von Kármán and Lin (1951) and by Batchelor (1953) to explain the physical meaning of this parameter have not been successful]. For the spectral function $E(k,t)$ in the meso-scale quasi-equilibrium range we have the formula

$$E(k,\ t) = \eta v_\eta^2 \Phi\left(\eta k,\ \frac{\bar{\varepsilon} t^2}{\nu}\right), \tag{16.50}$$

where $t$ is measured from the virtual initial time $t_0$ and $\Phi(\xi, \zeta)$ is a universal function of two variables. This function can be made to agree with the existence of a small-scale quasi-equilibrium range by demanding that, for sufficiently large $\xi$, the function $\Phi(\xi, \zeta)$ must be independent of $\zeta$ (at any rate for $\zeta$ not too small). In the case of turbulence behind geometrically similar grids, the formula (16.50) can be rewritten in the form

$$E(k,\ t) = \eta v_\eta^2 \Phi_1\left(\eta k,\ \frac{UM}{\nu}\right). \tag{16.51}$$

Moreover, since $\bar{\varepsilon} \sim t^{-2}$ and hence $\eta \sim t^{1/2}$, $v_\eta \sim t^{-1/2}$, we can replace $\eta$ by any characteristic length $l(t)$ proportional to $t^{1/2}$, and $v_\eta$ by any velocity $v(t)$ proportional to $t^{-1/2}$. In particular, when $a$ is zero or negligible, we can replace $\eta$ by $\lambda$ and $v_\eta$ by $U' = (\overline{u^2})^{1/2}$, and assume that[18]

$$E(k,\ t) = \lambda U'^2 \Phi_2\left(\lambda k,\ \frac{\lambda U'}{\nu}\right). \tag{16.52}$$

All these facts are in adequate agreement with existing data on grid turbulence.

Instead of the time $t$ (or $t-t_0$) we can introduce the rate of change of energy dissipation $d\bar{\varepsilon}/dt$ as the additional parameter in the meso-scale quasi-equilibrium spectral range (see Goldstein, 1951). This procedure may appear to be physically preferable but it is not difficult to see that it leads to the same results as before.

[17] We note that if $a = 0$ then $t - t_0 = -\overline{u^2} \Big/ \frac{d\overline{u^2}}{dt} = \frac{\lambda^2}{10\nu}$ and, consequently

$$R = \frac{15\nu\overline{u^2}}{\lambda^2}\left(\frac{\lambda^2}{10\nu}\right)^2 \frac{1}{\nu} = 0{,}15\,(\mathrm{Re}_\lambda)^2.$$

[18] The equation (16.52) is a special case of a general formula [Rotta (1950, 1953)]

$$E(k,\ t) = l v^2 \Psi(\lambda k,\ \mathrm{Re}),\quad \mathrm{Re} = \mathrm{Re}\,(t) = \frac{l(t)\, v(t)}{\nu}, \tag{16.52'}$$

for all Re and $k$, where $\Psi$ is a universal function. Since the Reynolds number can now be a function of time, this formula does not assume that the spectrum is self-preserving, but it includes the very strong assumption of a definite similarity between all the existing isotropic turbulent flows throughout their evolution. However, when he determined the specific form of the function $\Psi$, Rotta neglected terms including the derivative $\partial\Psi/\partial\mathrm{Re}$, i.e., he effectively used the usual hypothesis of self-preservation and, moreover, assumed the still more special Heisenberg hypothesis on the spectral energy transfer function which we shall discuss in Sect. 17.

Practically all the results of the present section can be extended without difficulty to the case of self-preserving decay of temperature fluctuations (or the concentration of a passive admixture) $\vartheta(\boldsymbol{x}, t)$ in isotropic turbulence [some of these extensions were formulated, e.g., by Corrsin (1951a, b) and G. W. Sutton (1968)]. All that is required in this case is to introduce three different scales, namely the length scale $l_\vartheta = l_\vartheta(t)$ [which in general may differ from $l(t)$], the temperature scale $\Theta(t)$, and the velocity scale $v(t)$ (which can be taken to be the same as in the hypotheses on the velocity self-preservation). Since we must use the velocity scale $v(t)$, the self-preserving decay of the velocity and temperature fluctuations can conveniently be considered together. The relationships (16.9) or (16.16) must now be augmented by certain new relationships of a similar type which describe the decay of temperature fluctuations and can, in principle, be verified experimentally. In most cases, however, comparison with experiment cannot be performed at present because of the absence of the necessary data. The corresponding theoretical derivations will therefore not be considered here. The question of the "universal quasi-equilibrium statistical state" of small-scale temperature fluctuations at high enough values of Re and Pe is however somewhat more favorable because the existence of this state is physically likely and the resulting analog of the Kolmorogov self-preservation is confirmed by many empirical results. Since, however, practically all the results are concerned not with grid turbulence, but other clearly nonisotropic turbulent flows, we shall postpone the discussion of this problem to the next chapter (see Sect. 23.5).

# 17. SPECTRAL ENERGY-TRANSFER HYPOTHESES

## 17.1 Approximate Formulas for the Spectral Energy Transfer

The self-preservation hypotheses discussed in the last section enable us to reduce the degree of arbitrariness in the choice of the solutions of the basic equations for isotropic turbulence. Nevertheless, they still do not enable us to close these equations. In this section we shall consider certain other hypotheses which will enable us to use the general spectral equations to deduce new equations containing only one unknown function. We note, however, that all the hypotheses discussed below are more or less speculative and lacking sound physical bases. Moreover none of them can be exactly valid since all of them contradict one or another physical requirement. The use of these hypotheses leads only to "model" equations which have certain features in common with the exact spectral equations for isotropic turbulence, and have a number of consequences which are in general qualitative agreement with experimental data on real turbulence.

We shall restrict our attention in this section to the equation for the energy spectrum, and all our hypotheses will be formulated for this particular case. It is of course quite easy to extend the discussion given below to the case of the spectral equation (14.63) for the temperature fluctuations. In fact, with each of the hypotheses concerned with $W(k, t)$, and given below, we can associate an analogous hypothesis for the quantity $W_\vartheta(k, t) = \int_k^\infty T_{\vartheta\vartheta}(k', t)\, dk'$. Since, however, there is very little experimental information on temperature fluctuations in isotropic turbulence, hypotheses of this latter type cannot be verified and must be regarded as unhelpful at present.

We shall start with the integrated spectral equations (16.41) and (16.41′):

$$\frac{\partial}{\partial t} \int_0^k E(k', t)\, dk' = -W(k, t) - 2\nu \int_0^k k'^2 E(k', t)\, dk' \qquad (17.1)$$

and

$$\frac{\partial}{\partial t} \int_k^\infty E(k', t)\, dk' = W(k, t) - 2\nu \int_k^\infty k'^2 E(k', t)\, dk'. \qquad (17.2)$$

where $E(k, t)$ is the energy spectrum,

$$\int_0^k E(k', t)\, dk' \text{ and } \int_k^\infty E(k', t)\, dk'$$

are the energies associated with long-wave and short-wave velocity disturbances with wave numbers $k' < k$ and $k' > k$, respectively, [the macro- and micro-components of the turbulence in the Obukhov terminology (1941a, b)], $2\nu \int_0^k k'^2 E(k', t)\, dk'$ is the rate of energy dissipation of long-wave disturbances (per unit mass of the fluid), $2\nu \int_k^\infty k'^2 E(k', t)\, dk'$ is the rate of energy dissipation of short-wave disturbances and, finally,

$$W(k, t) = \int_k^\infty T(k', t)\, dk'$$

is the energy transfer through the point $k$ of the spectrum, i.e., the amoung of energy transferred per unit time from the macro- to the micro-components.

If the spectral energy transfer $W(k, t)$ could be expressed in terms of the spectral function $E(k, t)$, then both Eqs. (17.1) and (17.2) would be immediately closed. However, the function $W(k, t)$ depends on the third moments of the velocity field, and cannot be determined from $E(k, t)$. All we can do, therefore, is to try to deduce approximations to the unknown functions $W(k, t)$ from the values of $E(k, t)$. This is in fact, the purpose of the hypotheses discussed below.

*The Kovasznay Hypothesis*

To ensure that the expression deduced from $E(k)$ and $k$ can be regarded as "similar" to the energy transfer function $W(k)$, we must first demand that this expression should have the same dimensions as $W(k)$. Since

$$[W(k)] = L^2T^{-3}, \qquad [E(k)] = L^3T^{-2}, \qquad [k] = L^{-1}$$

(where the brackets represent the dimensions, and $L$ and $T$ symbolize the length and time), the simplest assumption is to set

$$W(k) = 2\gamma_K [E(k)]^{3/2} k^{5/2}, \tag{17.3}$$

where $\gamma_K$ is a dimensionless constant (the factor 2 is introduced for convenience in subsequent analysis). This was first proposed by Kovasznay (1948). It is clear that Eq. (17.3) can only be a very rough approximation. In fact, the function $W(k)$ must also depend on the statistical characteristics of all the disturbances with wave numbers $k'$ both larger and smaller than $k$, and therefore it cannot be expressed only in terms of the values of $E(k)$ and $k$. Nevertheless, the simple formula (17.3) leads to apparently acceptable conclusions about the shape of the energy spectrum that are not too different from the conclusions deduced from some of the other hypotheses about $W(k)$ discussed below (see, for example, Fig. 29). This shows that the shape of the spectrum is not very sensitive to changes in the spectral energy transfer. This is precisely why the use of the hypotheses discussed in this chapter is frequently convenient.

The Kovasznay hypothesis takes into account only the dimensions of $W(k)$ and does not involve any physical considerations. The subsequent hypotheses are distinguished by the fact that they involve certain intuitive ideas about the mechanism of energy transfer over the spectrum. Since the process is in fact very complicated, it is not surprising that there are a number of different expressions for $W(k)$, each of which reflects a particular physical aspect of the process.

*The Leith Diffusion Hypothesis*

If we assume that $W(k)$ may depend, for example, on $\partial E(k)/\partial k$ in addition to $k$ and $E(k)$, then the number of possible dimensionally correct expressions for $W(k)$ will increase enormously. One such expression was selected by Leith (1967), who started from the formal similarity between the spectral energy transfer in wavenumber $k$-space and radiation or neutron transport in the usual physical $x$-space. Based on the fruitfulness of the so-called diffusion approximation in radiation and neutron transport theory, Leith suggested that a relationship of the form $W(k) = -D(\partial Q/\partial k)$ is approximately valid where $D$ and $Q$ depend on $k$ and $E(k)$. Requiring also that the nonviscous spectral equation have a stationary solution of the form $E(k) \sim k^2$, $W(k) = 0$ (cf. the discussion of the modified von Kármán hypotheses below), Leith came to a special hypotheses of the form

$$W(k) = -2\gamma_L k^{13/2} \frac{\partial}{\partial k} (k^{-3} [E(k)]^{3/2}) \tag{17.3'}$$

where $\gamma_L$ is another dimensionless constant.

*The Obukhov Hypothesis*

This is the earliest hypothesis, put forward as far back as 1941. It is based on the Reynolds representation (which is quite rigorous) of the energy transfer from the mean motion to the velocity fluctuations by the expression $\overline{u_i' u_j'} \frac{\partial \bar{u}_i}{\partial x_j}$ (see Sect. 6.2 in Vol. 1). However, the decomposition of the isotropic velocity field given by Eq. (11.51) into the macro-velocity $\bar{\boldsymbol{u}}(k)$ and the micro-velocity $\underline{\boldsymbol{u}}(k)$ can be regarded as the analog of the Reynolds decomposition of the velocity $\boldsymbol{u}$ into the mean and the fluctuation components $\bar{\boldsymbol{u}}$ and $\boldsymbol{u}' = \boldsymbol{u} - \bar{\boldsymbol{u}}$. The quantity $W(k)$ should therefore be proportional to the mean value of the product $\overline{\underline{u}_i(k)\underline{u}_j(k)}$ for the micro-velocity (evaluated for fixed values of the macro-velocity) multiplied by the velocity gradient of the macro-velocity. In the first approximation, the mean Reynolds stresses of the micro-component can be regarded as proportional to the energy

$$\frac{1}{2}\overline{\underline{u}_i(k)\,\underline{u}_i(k)} = \int_k^\infty E(k')\,dk',$$

and the root mean square of the derivatives $\frac{\partial \overline{u}_i(k)}{\partial x_j}$ can be taken as proportional to

$$\left[2\int_0^k k'^2 E(k')\,dk'\right]^{1/2}.$$

in view of Eq. (11.76). Obukhov (1941a, b) used these ideas to propose that

$$W(k) = 2\gamma_O\left[\int_0^k k'^2 E(k')\,dk'\right]^{1/2}\int_k^\infty E(k'')\,dk'', \qquad (17.4)$$

where $\gamma_O$ is a new dimensionless constant. A more detailed analysis of the physical assumptions leading to Eq. (17.4) can be found in Tchen (1954).

*Modified Obukhov Hypothesis*

Since some of the consequences of Obukhov's theory are physically unlikely, Ellison (1962) introduced a modification of this theory by assuming that the Reynolds stresses of the micro-component were largely determined by the extreme large-scale disturbances in this component (with wave numbers approaching $k$). In other words, he assumed that the mean Reynolds stresses of the micro-component can be approximately expressed in terms of the values of $E(k)$ and $k$ alone. In that case, dimensional analysis shows that they should be proportional to $kE(k)$, and hence

$$W(k) = 2\gamma_E\, k\, E(k)\left[\int_0^k k'^2 E(k')\,dk'\right]^{1/2}, \qquad (17.5)$$

where $\gamma_E$ is a further dimensionless constant.

The expression (17.5) is, in a sense, intermediate between the Kovasznay and Obukhov formulas: It depends on the behavior of $E(k')$ in the range $k' \leqslant k$, but is independent of the statistical characteristics of small-scale disturbances with $k' > k$. From the physical point of view, the assumption that $W(k)$ is independent of the values of $E(k')$ with $k' > k$ is not very satisfactory. Since, however, the Obukhov formula (17.4) is itself only a rather rough approximation, it can be supposed that the replacement of the factor $\int_k^\infty E(k'')\,dk''$ by $kE(k)$ in this formula should not be of crucial importance. In fact, the spectra $E(k)$ corresponding to Eqs. (17.4) and (17.5) are in many ways similar to each other, and the differences between them are such as to suggest that Eq. (17.5), which at first sight seems more approximate, gives a better result (see Sect. 17.2).

*The Heisenberg Hypothesis*

A few years after the publication of Obukhov's papers (1941a, b), Weizsäcker (1948) and Heisenberg (1948a), who were not familiar with Obukhov's work, put forward a further theory of spectral energy transfer. These were based on the classical Boussinesq idea that the Reynolds stress $\overline{\rho u_i' u_j'}$ can be written as the product of the virtual eddy viscosity coefficient $\rho K$ and the deformation $\frac{\partial \bar{u}_i}{\partial x_j} + \frac{\partial \bar{u}_j}{\partial x_i}$ of the mean velocity field. In that case, however, the expression $W = -\overline{u_\alpha' u_\beta'} \frac{\partial \bar{u}_\alpha}{\partial x_\beta}$, which describes the energy transfer from the mean motion to the fluctuating motion, takes the form

$$W = \frac{1}{2} K \sum_{\alpha,\beta} \left( \frac{\partial \bar{u}_\alpha}{\partial x_\beta} + \frac{\partial \bar{u}_\beta}{\partial x_\alpha} \right)^2, \tag{17.6}$$

which is similar to the Stokes dissipation function [see Eq. (1.69), Vol. 1]. If, as above, we interpret the velocity fluctuation as the velocity of the micro-component of the flow, the mean velocity as the macro-velocity, then if we transform to the spectral form we obtain

$$W(k) = 2K(k) \int_0^k k'^2 E(k')\,dk', \tag{17.7}$$

where $K(k)$ is the eddy viscosity due to all disturbances with wave numbers exceeding $k$. This formula corresponds to the assumption that the effect of the turbulence can be reduced exclusively to an increase in friction in the fluid, i.e., that the transfer of energy from the mean motion (macro-component) to the fluctuating motion (micro-component) has the same character as the conversion of the kinetic energy of a fluid into the thermal energy of molecular motion due to collisions between the molecules. This description is, in general, acceptable although it is only an approximation to the true situation.

If we define the coefficient $K(k)$ by analogy with the kinetic theory of gases (which is the basis of the semiempirical Boussinesq theory), then it must be assumed that the contribution to the eddy viscosity due to velocity disturbances with wavelength $l' = 2\pi/k'$ is proportional to the product of the corresponding "mixing length" $l_{k'}$ (which differs from $1/k'$ only by a numerical factor) and the "characteristic velocity scale" $v_{k'}$ of the disturbances. Using these ideas (in the qualitative form developed by Weizsäcker; see also Chen, 1954) together with dimensional considerations, Heisenberg put forward the following formula

$$K(k) = \gamma_H \int_k^\infty \sqrt{\frac{E(k')}{k'^3}}\, dk', \tag{17.8}$$

where $\gamma_H$ is a dimensionless constant, i.e., he proposed that

$$W(k) = 2\gamma_H \int_k^\infty \sqrt{\frac{E(k')}{k'^3}}\, dk' \int_0^k k''^2 E(k'')\, dk''. \tag{17.9}$$

*Modified Heisenberg Hypotheses*

The assumption that $W(k)$ can be written in the form of Eq. (17.7) is consistent with many other expressions for $K(k)$. For example, Stewart and Townsend (1951) have noted that without coming into conflict with dimensional reasoning, we can replace Eq. (17.8) by the more general expression

$$K(k) = \gamma_{ST} \left\{ \int_k^\infty \frac{[E(k')]^{\frac{1}{2c}}}{(k')^{1+\frac{1}{2c}}}\, dk' \right\}^c. \tag{17.10}$$

where $c$ is an arbitrary positive number and $\gamma_{ST}$ is a dimensionless constant. It can even be replaced by

$$K(k)=\sum_{l}\gamma_{l}\left\{\int_{k}^{\infty}\frac{[E(k')]^{\frac{1}{2c_{l}}}}{(k')^{1+\frac{1}{2c_{l}}}}\,dk'\right\}^{c_{l}}, \qquad (17.11)$$

where $c_l$ and $\gamma_l$ are nonnegative numbers. When $c=1/2$, Eq. (17.10) assumes a particularly simple analytic form. This was why this particular formula for $K(k)$ was used, for example, by Howells (1960) and Monin (1962). On the other hand, Ogura and Miyakoda (1953) proposed the replacement of Eq. (17.8) by

$$K(k)=\frac{\gamma_{OM}}{k}\left[\int_{k}^{\infty}E(k')\,dk'\right]^{\frac{1}{2}}, \qquad (17.12)$$

which corresponds to the assumption that, in the formula $K(k)\sim l_k v_k$, the velocity scale $v_k$ can be replaced by the root mean square velocity of the micro-component and $l_k\sim 1/k$.

*Von Kármán Hypothesis*

So far, we have not used the spectral representation (14.28) of $T(k)$ which is closely connected with $W(k)$. This representation contains the mean values of the products of the Fourier transforms of the three velocity components and cannot, of course, provide an exact expression for $W(k)$ in terms of $E(k)$ and $k$. However, it enables us to explain certain properties of the true function $W(k)$ which can then be compared with the approximate expressions (17.3)–(17.12). Thus, first, it is clear from Eqs. (14.28) and (14.27) that the quantity $W(k)=\int_{k}^{\infty}T(k')\,dk'$ can be written in the form

$$W(k)=\int_{k}^{\infty}\int_{0}^{k}P(k',k'')\,dk'\,dk'', \qquad (17.13)$$

where

$$P(k', k'') = \int\limits_{|\boldsymbol{k}'|=k'} \int\limits_{|\boldsymbol{k}''|=k''} Q_{jj}(\boldsymbol{k}', \boldsymbol{k}'')\, d\boldsymbol{k}'\, d\boldsymbol{k}''. \tag{17.14}$$

According to the end of Sect. 14.3, the quantity $P(k', k'')\,dk'\,dk''$ can be interpreted as the amount of energy transferred per unit time from disturbances with wave numbers in the range $dk' = (k', k'+dk')$ to disturbances in the range $dk'' = (k'', k''+dk'')$. Hence, it is clear that

$$P(k', k'') = -P(k'', k') \tag{17.15}$$

[cf. Eq. (14.27)]. Since by Eq. (14.26) the function $Q_{jj}(\boldsymbol{k}', \boldsymbol{k}'')$ depends on the complex Fourier amplitudes $d\boldsymbol{Z}(\boldsymbol{k})$ of the velocity field at the points $\boldsymbol{k}'$, $\boldsymbol{k}''$, and $\boldsymbol{k}' - \boldsymbol{k}'' = \boldsymbol{k}' + (-\boldsymbol{k}'')$ in the wave-vector space, it seems probable that a good approximation to the true value of $W(k)$ can be obtained by equating $P(k', k'')$ to an expression which depends in some reasonable way on $E(k')$, $E(k'')$, and all the possible values of $E(|\boldsymbol{k}'+\boldsymbol{k}''|)$ where $|\boldsymbol{k}'| = k'$, $|\boldsymbol{k}''| = k''$. None of the above formulas can be reduced to this type of approximation since none of them contains the quantities $E(|\boldsymbol{k}'+\boldsymbol{k}''|)$. However, even if we neglect this fact and confine our attention to the possibility of writing $W(k)$ in the form of the integral (17.13) of the function $P(k', k'')$, which is defined by only the values of $k'$, $k''$, $E(k')$, and $E(k'')$, then the result is still quite unsatisfactory. It is readily seen that, of all the functions $W(k)$ given by Eqs. (17.3)–(17.12), only the function (17.9) can be written in this form (with

$$P(k', k'') = 2\gamma_H (k')^{-3/2} (k'')^2 [E(k')]^{1/2} E(k'')$$

for $k' > k''$).

Let us now demand that the quantity $W(k)$ should be given by Eq. (17.13), and let us suppose that the function $P(k', k'')$ can be expressed, if only very approximately, in terms of $k'$, $k''$, $E(k')$, and $E(k'')$. In that case, Eq. (17.9) can readily be generalized. As noted by von Kármán (1948a, b) we can assume, without coming into conflict with dimensional considerations that for $k' > k''$

$$P(k', k'') = 2\gamma'_K (k')^m (k'')^{\frac{1}{2}-m} [E(k')]^n [E(k'')]^{\frac{3}{2}-n}, \tag{17.16}$$

where $m$, $n$ and $\gamma'_K$ are arbitrary dimensionless constants. According to Eq. (17.16),

$$W(k) = 2\gamma'_K \int_k^\infty (k')^m [E(k')]^n \, dk' \int_0^k (k'')^{\frac{1}{2}-m} [E(k'')]^{\frac{3}{2}-n} \, dk'' \qquad (17.17)$$

When $m = -3/2$, $n = 1/2$ this formula is identical with Eq. (17.9), while for $m = 0$, $n = 1$ it was called "the modified Obukhov formula" by Hinze (1959) and is identical with the formula deduced from completely different considerations by Meksyn (1963) and used particularly by Eschenroeder (1965); the case $m = 0$, $n = 3/2$ was called "the modified Kovasznay formula" by Hinze. The expression (17.17) is the most general expression having the correct dimensions, which is consistent with the assumption that $W(k)$ can be written in the form of the integral (17.13), where for $k' > k''$ the function $P(k', k'')$ is a power function of $k'$, $k''$, $E(k')$, and $E(k'')$. This assumption about the form of the function $P(k', k'')$ ensures that Eq. (17.17) is relatively simple, and makes it particularly convenient in calculations. On the other hand, the quantity $P(k', k'')$ is then again found to be independent of the statistical parameters of disturbances with wave numbers $\mathbf{k}' + \mathbf{k}''$ and, moreover, it changes discontinuously when the pair $(k', k'')$ passes through the point $k' = k''$, which is physically unnatural.

*The Goldstein Hypothesis*

The von Kármán, Obukhov, Heisenberg, and Stewart-Townsend hypotheses given by Eqs. (17.17), (17.4), (17.9), and (17.7), (17.10), respectively, are special cases of the following general formula put forward by Goldstein (1951):

$$W(k) = 2\gamma_G \left[ \int_k^\infty (k')^m [E(k')]^n \, dk' \right]^\lambda \left[ \int_0^k (k'')^{m_1} [E(k'')]^{n_1} \, dk'' \right]^{\lambda_1}, \qquad (17.18)$$

where $\gamma_G$ and $m, n, \lambda, m_1, n_1, \lambda_1$ are real numbers, the last six of which are related by

$$(m+1)\lambda + (m_1+1)\lambda_1 = \frac{5}{2}, \quad n\lambda + n_1\lambda_1 = \frac{3}{2}. \qquad (17.19)$$

which are deduced from dimensional analysis. However, the general formula given by Eq. (17.18) is too complicated and contains too many undetermined constants to be of any real value in practice.

*Modified von Kármán Hypothesis*

The main formula of the von Kármán theory, Eq. (17.16), can be replaced by many other expressions for $P(k', k'')$ with the correct dimensions. Thus, for example, following Kraichnan and Spiegel (1962) we can add an arbitrary numerical function $\varphi(k'/k'')$ to the right hand side of Eq. (17.16), which is such that $\varphi(1)=1$, i.e., we can set

$$P(k', k'') = 2\gamma_\gamma (k')^m (k'')^{\frac{1}{2}-m} [E(k')]^n [E(k'')]^{\frac{3}{2}-n} \varphi\left(\frac{k'}{k''}\right) \quad \text{for } k' > k'' \tag{17.20}$$

(see Yaglom, 1969). Equations (17.20) and (17.13) define the functional set of new expressions for $W(k)$. All these expressions are less convenient in calculations than Eq. (17.17) in which the integrals with respect to $k'$ and $k''$ can be separated. However, physical intuition suggests that the behavior of the function $\varphi(k'/k'')$ which rapidly decays as $k'/k''$ increases (and thus guarantees a reduced energy transfer from the given interval $dk''$ to all sufficiently distant intervals $dk'$) has some real justification (see Sect. 17.2).

The generalized von Kármán formula (17.20), and the original formula (17.16), as well as all the other formulas for $W(k)$ discussed above, describe the energy transfer in wave-number space, which occurs in a definite direction, namely, from small wave numbers to large wave numbers (since the function $P(k', k'')$ with $\gamma$ positive is always nonnegative for $k' > k''$ and, consequently nonpositive for $k' < k''$). However, in point of fact, the energy flows only *mainly* from large-scale to small-scale disturbances. If, however, disturbances with a given $1/k_0$ are specially excited, so that the function $E(k)$ has a sharp peak at $k = k_0$, then the energy of these disturbances will be transferred in both directions (see, for example Fig. 42b below). To take this into account we can assume, following Kraichnan and Spiegel (1962), that disturbances with wave numbers in the range $dk''$ transfer a definite amount of energy $Q(k', k'')\,dk'\,dk''$ into the interval $dk'$ (where $k'$ can be greater or smaller than $k''$), and at the same time receive from this interval a certain other quantity of energy $Q(k'', k')\,dk'\,dk''$. In that case

$$P(k', k'') = Q(k', k'') - Q(k'', k'), \tag{17.21}$$

and it automatically follows that $P(k', k'') = -P(k'', k')$. The function $Q(k', k'')$ can in general be specified by an arbitrary fomula of the form of Eq. (17.20). It is more natural, however, to assume that it depends only on the intensity $E(k'')$ of disturbances transferring the energy, and on the mutual disposition of the intervals $dk''$ and $dk'$ (by analogy with the case of radiative transfer of energy in ordinary space). Since, moreover, it was shown by Hopf and Pitt (1953) that the dynamic equations for an ideal fluid formally allow a stationary solution for which disturbances with all the possible wave numbers are equally excited and hence $F(k) =$ const, $E(k) \sim k^2$ (see Sect. 29.1), Kraichnan and Spiegel demanded that $P(k', k'') \equiv 0$ for $E(k) \sim k^2$. From this and from dimensional considerations they deduced that

$$Q(k', k'') = 2\gamma_{KS} \left[(k'')^{-2} E(k'')\right]^{3/2} (k'k'')^{7/4} \varphi\left(\frac{k'}{k''}\right), \tag{17.22}$$

where $\gamma_{KS}$ is a positive constant, $\varphi(x)$ is an arbitrary numerical function, such that $\varphi(1) = 1$ and

$$\varphi(x) = \varphi\left(\frac{1}{x}\right). \tag{17.23}$$

Equations (17.21)–(17.23) describe the process which could be referred to as "radiative energy transfer in wave-number space." It is easy to see that the Leith equation (17.3′) can be obtained as a special limiting case of those equations.

The function $P(k', k'')$ defined by Eqs. (17.21)–(17.23) passes continuously through zero at $k' = k''$ [in contrast to the original von Kármán function (17.16)]. However, it is again independent of the statistical characteristics of disturbances with wave-number vectors $\mathbf{k}' - \mathbf{k}''$ in contrast with the exact formula (14.26). If, however, we recall Eqs. (14.26) and (17.14), and allow for the fact that $dZ^*(\mathbf{k}) = dZ(-\mathbf{k})$, then we must replace Eq. (17.13) by

$$W(k) = \int\limits_{|\mathbf{k}'| < k} \int\int P(\mathbf{k}', \mathbf{k}'', \mathbf{k}''') \delta(\mathbf{k}' + \mathbf{k}'' + \mathbf{k}''') \, d\mathbf{k}' \, d\mathbf{k}'' \, d\mathbf{k}''', \tag{17.24}$$

where the second and third integrals on the right-hand side are evaluated over all space of the vectors $\boldsymbol{k}''$ and $\boldsymbol{k}'''$ and the function $P(\boldsymbol{k}', \boldsymbol{k}'', \boldsymbol{k}''')$ depends on the velocity disturbances with wave-number vectors $\boldsymbol{k}'$, $\boldsymbol{k}''$, and $\boldsymbol{k}'''$ and satisfies the relation

$$P(\boldsymbol{k}', \boldsymbol{k}'', \boldsymbol{k}''')+P(\boldsymbol{k}'', \boldsymbol{k}''', \boldsymbol{k}')+P(\boldsymbol{k}''', \boldsymbol{k}', \boldsymbol{k}'')=0, \quad (17.25)$$

which replaces Eq. (17.15). Therefore, to obtain a more accurate description of the spectral energy transfer, we can try to express the function $P(\boldsymbol{k}', \boldsymbol{k}'', \boldsymbol{k}''')$ approximately in terms of $E(\boldsymbol{k}')$, $E(\boldsymbol{k}'')$, $E(\boldsymbol{k}''')$, $\boldsymbol{k}'$, $\boldsymbol{k}''$, and $\boldsymbol{k}'''$, and use Eq. (17.24). The model theory then turns out to be very complicated and, since there are no data at present which would enable us to select the best expression for $P(\boldsymbol{k}', \boldsymbol{k}'', \boldsymbol{k}''')$, we shall not consider hypotheses of this kind.[19] Certain additional hypotheses on the spectral energy transfer [including some new generalizations of the von Kármán hypothesis (17.16)] will be given in Sect. 22.2 [see, for example, Eq. (22.27)].

*Hypotheses Formulated in Terms of Correlation Functions*

Among hypotheses of this kind we recall the attempt by Hasselmann (1958) to apply the eddy viscosity hypothesis (17.7) directly to the von Kármán-Howarth equation. By analogy with Eq. (17.7), Hasselmann proposed that the terms in Eq. (14.9) which contain $B_{LL,L}(r, t)$ can be represented in the form

$$\frac{\partial B_{LL,L}}{\partial r}+\frac{4}{r}B_{LL,L}=2K(r)\left[\frac{\partial^2 B_{LL}}{\partial r^2}+\frac{4}{r}\frac{\partial B_{LL}}{\partial r}\right], \quad (17.26)$$

where $K(r)=\gamma v^2(r)\tau(r)$, $\gamma$ is a dimensionless constant of the order of unity, $v(r)$ is the velocity scale of the set of disturbances whose scale does not exceed $r$, and $\tau(r)$ is the time scale (or "mixing time") for these disturbances. Next, he assumed that

[19]We note that one particular choice of the function $P(\boldsymbol{k}', \boldsymbol{k}'', \boldsymbol{k}''')$ which satisfies Eq. (17.25) is provided by the Kraichnan theory (1959, 1961, 1962, 1961a) which is based on the exact equations (14.26) and (14.28) but is augmented by certain special hypotheses of a statistical character whose significance cannot, as yet, be regarded as fully established (see Sect. 19.6). We note, however, that in the Kraichnan theory the function $P(\boldsymbol{k}', \boldsymbol{k}'', \boldsymbol{k}''')$ is an explicit function of $U'=(\overline{u'^2})^{1/2}$, which does have an important effect on the statistical state of disturbances with any value of $k$ (in direct conflict with the Kolmogorov theory on the universal quasi-equilibrium state discussed in Sect. 16.5 and Chapt. 8).

$$v^2(r) = \overline{[u_L(x+r) - u_L(x)]^2} = 2[B_{LL}(0) - B_{LL}(r)]$$

and used for the time scale $\tau(r)$ an expression which was not only a fairly complicated function of $B_{LL}(r)$ but also contained $K(r)$, i.e., it determined $\tau(r)$ only implicitly. The Hasselmann assumptions are equivalent to a fairly complicated relationship between $W(k)$ and $E(k)$. However, the nature of this relationship is such that it is very difficult to verify even whether the corresponding spectrum is always positive. Moreover, the final equation [containing the only unknown function $B_{LL}(r,t)$] turned out to be more complicated and physically less clear than most equations using approximate expressions for $W(k)$. Therefore, although according to Hasselmann the consequences of his equation (and obtained through a numerical integration) are in a better agreement with the data of Stewart and Townsend (1951) then the consequences of the Heisenberg equation (17.9), we shall not consider it any further.

## 17.2 Application of the Energy Transfer Hypotheses to the Study of the Shape of the Spectrum in the Quasi-Equilibrium Range

All the above formulas for $W(k)$ convert Eqs. (17.1) and (17.2) into equations involving one unknown function, namely, $E(k,t)$. The solutions of these equations can of course be of only limited value because the formulas used for $W(k)$ are not rigorous or accurate. Since, however, the development of a rigorous theory of isotropic turbulence is extremely difficult, there is some sense in performing a preliminary comparison between the consequences of these model hypotheses and the existing experimental data (which, in any case, are rather scant at present).

To begin with, it will be useful to apply the approximate formulas for $W(k)$ to the calculation of the spectrum in the universal quasi-equilibrium range in which the Kolmogorov self-preservation is valid. In this interval, $W(k)$ and $E(k)$ should be universal functions depending on the two parameters $\bar{\varepsilon}$ and $\nu$, and therefore the assumption that $W(k)$ can be expressed in a universal way in terms of $E(k)$ may not be completely unjustified.

### *Values of the Spectra $E(k)$ in the Kovasznay, Obukhov and Heisenberg Theories*

The general spectral equations given by (17.1) and (17.2) assume the simple form of Eq. (16.42) and (16.42′) in the universal quasi-equilibrium range. If we take $H(k) = \int_0^k k'^2 E(k')\,dk'$ as the main unknown function, so that $E(k) = \frac{1}{k^2}\frac{dH(k)}{dk}$, and substitute the Kovasznay formula (17.3) into Eq. (16.42), we obtain a simple first-order differential equation for $H(k)$. The solution of this equation which satisfies the boundary conditions $H(0) = 0$ and $H(\infty) = \frac{\bar{\varepsilon}}{2\nu}$ is of the form

$$H(k)=\begin{cases}\dfrac{\bar{\varepsilon}}{2\nu}\left\{1-\left[1-\left(\dfrac{k}{k_1}\right)^{4/3}\right]^3\right\} & k<k_1,\\ \dfrac{\bar{\varepsilon}}{2\nu} & k>k_1,\end{cases} \tag{17.27}$$

where

$$k_1=2^{5/4}\gamma_K^{1/2}\bar{\varepsilon}^{1/4}\nu^{-3/4}=2^{5/4}\gamma_K^{1/2}k_\eta. \tag{17.28}$$

Hence we have

$$E(k)=\begin{cases}(2\gamma_K)^{-2/3}\bar{\varepsilon}^{2/3}k^{-5/3}\left[1-\left(\dfrac{k}{k_1}\right)^{4/3}\right]^2 & k<k_1,\\ 0 & k\geqslant k_1.\end{cases} \tag{17.29}$$

This formula was first found by Kovasznay (1948) (in a form containing an unimportant misprint). It is clear that the solution (17.28)–(17.29) can also be rewritten in the self-preserving form (16.36). It is more convenient, however, to use the formula

$$E(k)=\gamma_K^{-3/2}\bar{\varepsilon}^{1/4}\nu^{5/4}\Phi\left(\frac{k}{k_1}\right),\quad \Phi(x)=\begin{cases}2^{-11/4}x^{-5/3}(1-x^{4/3})^2 & x<1,\\ 0 & x>1,\end{cases} \tag{17.30}$$

which contains the universal function $\Phi(x)$ which is independent of the value of $\gamma_K$ [and differs from the function $\varphi(\xi)$ of Eq. (16.36) only by the choice of the length and velocity scales]. When $k \ll k_1 \sim k_\eta$, the spectral function (17.29) satisfies the five-thirds law (16.37), as expected [with $C_1=(2\gamma_K)^{-2/3}$]. Further increase of $k$ results in a faster decrease of $E(k)$, and the spectrum reaches zero at the point $k=k_1=2^{5/4}\gamma_K^{1/2}k_\eta$ (together with the derivative $\partial E/\partial k$). When $k \geqslant k_1$, the spectrum $E(k)$ is identically zero (see Figs. 27 and 28).

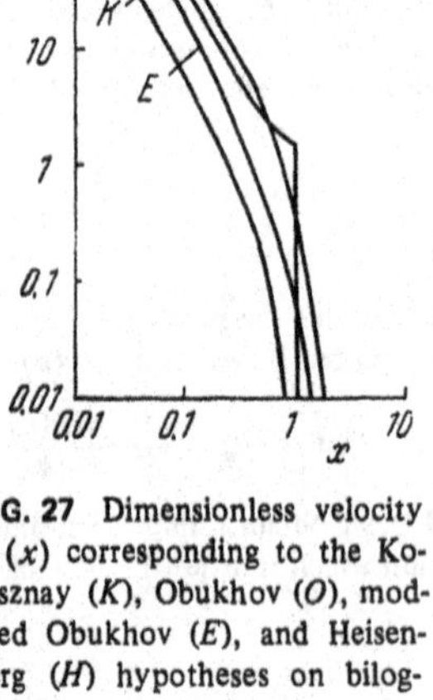

FIG. 27 Dimensionless velocity $\Phi(x)$ corresponding to the Kovasznay ($K$), Obukhov ($O$), modified Obukhov ($E$), and Heisenberg ($H$) hypotheses on bilogarithmic scale.

Let us now substitute the Obukhov formula (17.4) into Eq. (16.42) and again regard $H(k)$ as the main unknown. If we divide all the terms of the resulting equation by $[H(k)]^{1/2}$ and then differentiate with respect to $k$, we obtain

$$-\frac{2\gamma_O}{k^2}\frac{dH(k)}{dk}+\frac{\nu}{[H(k)]^{1/2}}\frac{dH(k)}{dk}=-\frac{\bar{\varepsilon}}{2[H(k)]^{3/2}}\frac{dH(k)}{dk}. \tag{17.31}$$

Hence, it follows that either $H(k)=\text{const}$, i.e., $E(k)=0$, or $H(k)=\dfrac{\bar{\varepsilon}}{2\nu}h\left(\dfrac{k}{k_1}\right)$ where

$$k_1=2^{-1/4}\gamma_O^{1/2}\bar{\varepsilon}^{1/4}\nu^{-3/4}=2^{-1/4}\gamma_O^{1/2}k_\eta. \tag{17.32}$$

and $h(x)$ satisfies the cubic equation $4h^3 = x^4(1+h)^2$ for any $x$. It is readily verified that this equation has only one nonnegative root for any $x$, and this root increases monotonically from zero to unity as $x$ increases from zero to unity. However, $h(1) = 1$ means that $\int_0^{k_1} k^2 E(k)\,dk = \frac{\bar{\varepsilon}}{2\nu}$ and, therefore, in accordance with Eq. (16.43) we must use the trivial solution $H(k) = \frac{\bar{\varepsilon}}{2\nu} = \text{const}$ of Eq. (17.31) when $k > k_1$. Therefore, the spectral function

$$E(k) = \gamma_O^{-3/2}\bar{\varepsilon}^{1/4}\nu^{5/4}\Phi\left(\frac{k}{k_1}\right), \quad \Phi(x) = \begin{cases} 2^{-1/4}\dfrac{h'(x)}{x^2} & \text{of} \quad x < 1, \\ 0 & \text{of} \quad x > 1, \end{cases} \tag{17.33}$$

is now positive only for $k < k_1$, and falls to zero discontinuously at $k = k_1$. When $k \ll k_1$, i.e., $x \ll 1$, it is clear that $h(x) \approx 2^{-2/3}x^{4/3}$ and hence we again have the five-thirds law with $C_1 = \frac{2^{2/3}}{3}\gamma_O^{-2/3}$ for the spectrum $E(k)$. This was in fact the way in which this law was first established by Obukhov (1941a, b). It is readily seen that, moreover, $h(x) \approx 1 - 2(1-x)$ for $1 - x \ll 1$ and, therefore, $\Phi(1) = 2^{3/4}$ and $E(k_1) = 2^{3/4}\gamma_O^{-3/2}\bar{\varepsilon}^{1/4}\nu^{5/4}$. The behavior of the function $\Phi(x)$ at intermediate values of $x$ is shown in Figs. 27 and 28 [the very complicated analytic expression for this function is of little interest; it can be obtained by using the Cardan formula to solve the cubic equations; see Millsaps (1955) and Reid (1956a, 1960)].

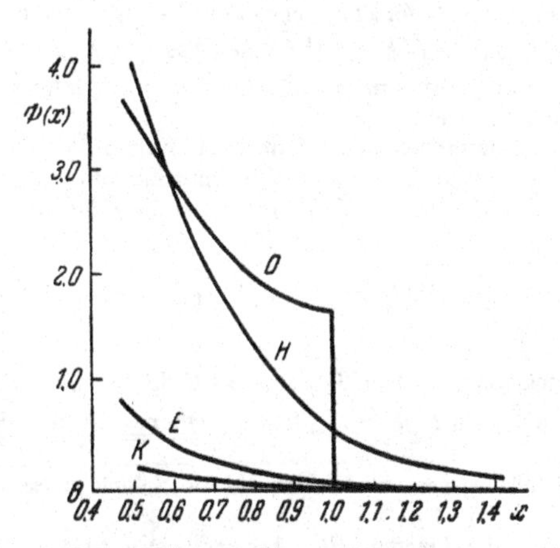


**FIG. 28** Same as Fig. 27 on linear scale.

The discontinuous change in the spectral function at $k = k_1$ is physically quite unlikely. It is therefore interesting to note that the modified form of the Obukhov hypothesis put forward by Ellison and given by Eq. (17.5) leads to a completely different result. In fact, if we substitute Eq. (17.5) into Eq. (16.42) we obtain a first-order differential equation for the function $H(k)$, which together with the boundary condition $H(0) = 0$ turns out to be equivalent to

$$-k^2 = \frac{2\gamma_E}{\nu}\left\{2[H(k)]^{1/2} + 2\sqrt{\frac{\bar{\varepsilon}}{2\nu}} \ln \frac{1 - \left[\frac{2\nu}{\bar{\varepsilon}} H(k)\right]^{1/2}}{\left[1 - \frac{2\nu}{\bar{\varepsilon}} H(k)\right]^{1/2}}\right\}. \quad (17.34)$$

In other words, in this case,

$$E(k) = \gamma_E^{-3/2}\bar{\varepsilon}^{1/4}\nu^{5/4}\Phi\left(\frac{k}{k_1}\right), \quad \Phi(x) = 2^{-7/4}\frac{h'(x)}{x^2}, \quad (17.35)$$

where

$$k_1 = 2^{1/4}\gamma_E^{1/2}\bar{\varepsilon}^{1/4}\nu^{-3/4} = 2^{1/4}\gamma_E^{1/2}k_\eta, \quad (17.36)$$

and the function $h(x) = 2\nu\bar{\varepsilon}^{-1}H(k/k_1)$ satisfies the equation

$$2\sqrt{h} + 2\ln\frac{1-\sqrt{h}}{\sqrt{1-h}} + x^2 = 0. \quad (17.37)$$

It is readily seen that $h(x) \approx (3/2)^{2/3} x^{4/3}$ for $x \ll 1$, and therefore for $k \ll k_1 \sim k_\eta$ we again obtain the five-thirds law, with $C_1 = 3^{-1/3}\gamma_E^{-2/3}$. In this case, however, the function $h(x)$ is less than unity for all $x$, and $h(x) \approx 1 - 4e^{-x^2}$ for $x \gg 1$. Therefore, in this case, $E(k) \approx 2^{3/2}\gamma_E^{-1}\bar{\varepsilon}^{1/2}\nu^{1/2}k^{-1}e^{-(k/k_1)^2}$ for $k \gg k_1 \sim k_\eta$, i.e., the spectrum decays exponentially at infinity. The variation of $\Phi(x)$ with $x$ for intermediate values of this variable is shown in Figs. 27 and 28.

Let us now substitute the Heisenberg formula (17.9) into Eq. (16.42). Dividing all the terms of the resulting equation by $H(k)$, and differentiating with respect to $k$, we obtain the differential equation for the function $H(k)$ which shows that either $H'(k) = 0$, i.e., $E(k) = 0$, or

$$H(k) = \left(3\gamma_H^2\right)^{-1/3}\bar{\varepsilon}^{2/3}k^{4/3}\left(1 + Ck^4\right)^{-1/3}. \quad (17.38)$$

where $C$ is an integration constant. For any real $C$ the function (17.38) tends to zero as $\left(3\gamma_H^2\right)^{-1/3}\bar{\varepsilon}^{2/3}k^{4/3}$ when $k \to 0$ $\left(\text{this corresponds to the five-thirds law with } C_1 = \frac{4}{3^{4/3}}\gamma_H^{-2/3}\right)$ while an increase of $k$ results in a monotonic increase of this function. If, during this increase, the function $H(k)$ becomes equal to $\bar{\varepsilon}/2\nu$ $\left(\text{this requires that } 1 + Cm^4\bar{\varepsilon}\nu^{-3} = \frac{8}{3}\gamma_H^{-2}m^4\right)$ at some finite value of $k$ (say, $mk_\eta$ where $m$ is an arbitrary number), then for $k > mk_\eta$ we must use the trivial solution $H(k) = \frac{\bar{\varepsilon}}{2\nu} = \text{const}$, $E(k) = 0$. We obtain the family of spectral functions corresponding to the Heisenberg hypothesis which are of the form

$$E(k) = \begin{cases} \dfrac{4}{3^{4/3}} \gamma_H^{-2/3} \bar{\varepsilon}^{2/3} k^{-5/3} \left[1 + \left(\dfrac{8}{3\gamma_H^2} - \dfrac{1}{m^4}\right)\left(\dfrac{k}{k_\eta}\right)^4\right]^{-4/3} & k < mk_\eta, \\ 0 & k > mk_\eta, \end{cases} \tag{17.39}$$

All these functions are of the same type as the Obukhov spectrum (17.33), i.e., they fall monotonically from some positive value $E(k_c)$ at $k = k_c = mk_\eta$, and then suddenly become equal to zero. Since this discontinuous change of the spectrum seems unlikely, it is particularly interesting to consider the limiting case of the spectrum (17.39) which is obtained for $m = \infty$, i.e., for $H(k) \rightarrow \dfrac{\bar{\varepsilon}}{2\nu}$ : as $k \rightarrow \infty$:

$$E(k) = \frac{4}{3^{4/3}} \gamma_H^{-2/3} \bar{\varepsilon}^{2/3} k^{-5/3} \left[1 + \frac{8}{3\gamma_H^2}\left(\frac{k}{k_\eta}\right)^4\right]^{-4/3} = \gamma_H^{-3/2} \bar{\varepsilon}^{1/4} \nu^{5/4} \Phi\left(\frac{k}{k_1}\right),$$
$$\Phi(x) = \frac{2^{13/4}}{3^{7/4}} x^{-5/3} (1 + x^4)^{-4/3}, \tag{17.40}$$

where

$$k_1 = \left(\frac{3}{8}\right)^{1/4} \gamma_H^{1/2} \bar{\varepsilon}^{1/4} \nu^{-3/4} = \left(\frac{3}{8}\right)^{1/4} \gamma_H^{1/2} k_\eta. \tag{17.41}$$

This spectrum is positive and falls monotonically for all $k$. When $k \rightarrow \infty$ the spectrum falls as $k^{-7}$. The asymptotic formulas $E(k) \sim k^{-5/3}$ for $k \ll k_\eta$, and $E(k) \sim k^{-7}$ for $\gamma_K$, $\gamma_O$, $\gamma_E$ and $\gamma_H$, which correspond to Eq. (17.9), were obtained by Heisenberg (1948a). The exact formula (17.40), which is suitable for all $k$, was pointed out by Bass (1949) and Chandrasekhar (1949), while the more general solution (17.39) was obtained by Goldstein (1951).

To compare the results obtained under different hypotheses about the energy transfer, Figs. 27 and 28 show also the graphs of $\Phi(x)$. The comparison of the different spectra in Figs. 27 and 28 is somewhat formal, since the variable $x = k/k_1$ is different in the different cases and, moreover, depends on the parameter $\gamma$ which can have an arbitrary value. To obtain results which are more readily interpreted, we note that the spectra (17.30), (17.33), (17.35), and (17.40) will coincide within the inertial subrange, i.e., when the five-thirds law is valid, provided the constants $\gamma_K$, $\gamma_O$, $\gamma_E$ and $\gamma_H$ are in the following ratio:

$$\gamma_K : \gamma_O : \gamma_E : \gamma_H = \frac{9}{16} : \frac{\sqrt{3}}{4} : \frac{3\sqrt{3}}{8} : 1. \tag{17.42}$$

Accordingly, Fig. 29 shows graphs of the dimensionless universal spectrum

$$\Phi_1(k/k_1') = \frac{E(k)}{\gamma_1^{-3/2} \bar{\varepsilon}^{1/4} \nu^{5/4}}, \tag{17.43}$$

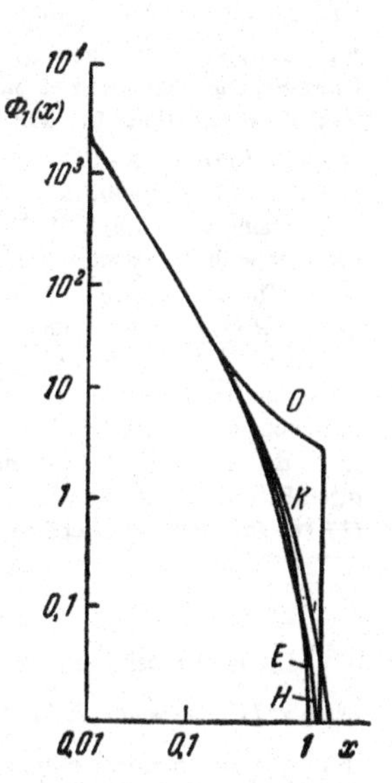


FIG. 29 Dimensionless spectra $\Phi_1(x)$ the four-energy transfer hypotheses.

where $\gamma_1 = \gamma_H$ for the spectrum (17.40), $\gamma_1 = (16/9)\gamma_K$ for the spectrum (17.29), $\gamma_1 = (4/\sqrt{3})\gamma_O$ for the spectrum (17.33), and $\gamma_1 = (8/3\sqrt{3})\gamma_E$ for the spectrum (17.35), while $k_1'$ is given by the formula $k_1' = \left(\frac{3}{8}\right)^{1/4} \gamma_1^{1/2} k_\eta$ in all cases. For $x \ll 1$, i.e., $k \ll k_1'$, the four functions $\Phi_1(x)$ will coincide, while the discrepancies between them for $x \gg 1$ clearly indicate the differences between the spectra which are connected with the particular formula chosen for the energy transfer $W(k)$.

## *Asymptotic Form of Spectra Corresponding to other Energy Transfer Hypotheses*

The first-order differential equation for the function $H(k)$, which enables us to determine the form of the spectrum $E(k)$, can also be obtained under other assumptions about the form of $W(k)$, for example, the assumptions represented by Eqs. (17.7) and (17.10) or (17.12).[20] However, the solution of the above equations is difficult and not particularly interesting. We shall therefore restrict our attention to the asymptotic behavior of the corresponding spectral functions $E(k)$ for very small and very large values of $k$.

For small wave numbers $k \ll k_\eta = (\bar{\varepsilon}\nu^{-3})^{1/4}$, any reasonable theory should, of course, lead to the five-thirds law. Since $2\nu \int_0^k k'^2 E(k')\,dk' \ll \bar{\varepsilon}$ for sufficiently small $k$ in this region of $k$ we can replace Eq. (16.42) by

$$W(k) = \bar{\varepsilon} = \text{const}. \tag{17.44}$$

Let us substitute any of the above approximate formulas for $W(k)$ into Eq. (17.44). This results in an equation for $E(k)$ which does in fact have a solution of the form $E(k) = C_1 \bar{\varepsilon}^{2/3} k^{-5/3}$ where $C_1 \sim \gamma^{-2/3}$, $\gamma$ being the coefficient in the expression for $W(k)$. However, the undetermined parameters or functions which are present in the formula for $W(k)$ must sometimes be additionally restricted by conditions which guarantee that all the integrals converge in the range between 0 and $k$ for $E(k) \sim k^{-5/3}$ [see Yaglom (1969)]. For the case of hypotheses (16.3') the corresponding equation (16.42') was integrated numerically by Leith (1967) who found that the function $E(k)$ vanishes in his theory together with its derivative at some cutoff value $k = k_1$ ($\approx 9\gamma_L^{1/2} k_\eta$) and is equal to zero at $k > k_1$. The general shape of $E(k)$ in the region $k < k_1$ is in this theory rather close to the shape of $E(k)$ obtained from the modified Obukhov hypothesis.

When $k \gg k_\eta$, the integral between 0 and $k$ in the expresion for $W(k)$ can be replaced by its limiting value which is obtained for $k \to \infty$ (when this integral diverges for $k \to \infty$, no consistent results can be obtained). Equation (16.42'), which for large $k$ is more convenient than Eq. (16.42), is then much simpler, and the required asymptotic formula can often be obtained without difficulty. Thus, for example, if we take the Goldstein formula (17.18) for $W(k)$, and replace the integral in it by a constant, the corresponding form of

[20] The only exception is the case of Eq. (17.10) with $c = 1/2$ which reduces to the cubic $h^3 = x^4(1-h)$ where $H(k) = \frac{\bar{\varepsilon}}{2\nu} h\left(\frac{k}{k_1}\right)$, $k_1 = 2^{-1/2}\gamma_{ST}^{1/2} k_\eta$. The positive root $h(x)$ of this equation increases monotonically from zero to unity as $x$ increases from zero to infinity. In particular, $h(x) \sim x^{4/3}$, i.e., $E(k) \sim k^{-5/3}$, for $x \ll 1$, and $1 - h(x) \sim x^{-4}$, i.e., $E(k) \sim k^{-7}$, for $x \gg 1$.

Eq. (16.42′) has a solution of the form

$$E(k) \sim k^{-\frac{3-(m+1)\lambda}{1-n\lambda}} \tag{17.45}$$

[The equation will of course have the further trivial solution $E(k) = 0$ which corresponds to a spectrum with a sharp cutoff; see Yaglom (1969)]. To ensure that the integral between 0 and $k$ in Eq. (17.18) should converge for the spectrum given by Eq. (17.45) as $k \to 0$, while the dissipation $\bar{\varepsilon} = 2\nu \int_0^\infty k^2 E(k)\, dk$ is finite, i.e., to ensure that there is a solution corresponding to a spectrum which does not drop to zero discontinuously at $k = k_1$, the parameters in Eq. (17.18) must satisfy the inequalities

$$\frac{3-(m+1)\lambda}{1-n\lambda} > 3 \quad \text{and} \quad n_1 \frac{3-(m+1)\lambda}{1-n\lambda} > 1 + m_1.$$

In the case of the von Kármán equation (17.17), these two inequalities reduce to the single inequality $\frac{2-m}{1-n} > 3$. These conditions are always satisfied in the case of Eqs. (17.9) and (17.10) (with $c > 0$), while in the case of the Obukhov formula (17.4) they are not. In the special case of the Stewart-Townsend hypothesis given by Eqs. (17.7) and (17.10), the formula of Eq. (17.45) becomes equivalent to $E(k) \sim k^{-7}$, which also corresponds to the Heisenberg spectrum (17.40). The same asymptotic behavior is obtained for $E(k)$ as $k \to 0$ in the case of the hypotheses given by Eqs. (17.11) and (17.12). Therefore, all the theories which employ the concept of and eddy viscosity lead to spectral functions $E(k)$ which behave in the same way, not only for $k \to 0$, but also for $k \to \infty$, and are therefore very similar to each other.

Let us now write the energy transfer $W(k)$ in the form of the double integral (17.13) of the function $P(k', k'')$. Equations (16.42) and (16.42′) can now be rewritten in the form

$$2\nu k^2 E(k) = \int_0^\infty P(k, k')\, dk' = \int_0^k P(k, k')\, dk' - \int_k^\infty P(k', k)\, dk',$$

which shows that the rate of energy dissipation in the interval $dk$ is equal to the difference between the inflow of energy into this interval and the outflow of energy from it. It may be shown that when $k \gg k_\eta$ the second term on the right-hand side of the last equation can be neglected, and this means physically that for $k \gg k_\eta$ most of the energy reaching the interval $dk$ is spent in overcoming viscous friction, and only a very small proportion is transferred to higher wave numbers. When $k \gg k_\eta$ the first term on the right-hand side for $P(k, k')$ given by Eq. (17.20) or (17.21), Eq. (17.22) can be investigated by the usual methods for studying the asymptotic behavior of integrals [see Yaglom (1969b)].[21] In particular, when $\varphi(x) = x^{-c}$, $c > 0$, the formula given by Eq. (17.20) becomes identical with Eq. (17.16) (provided $m$ is replaced by $m - c$). In accordance with Eq. (17.45), in this case

[21] We note, moreover, that if $P(k', k'') = Q(k', k'') - Q(k'', k')$, where the function $Q(k', k'')$ is given by a formula of the form of Eq. (17.20) [Eq. (17.22) is a special case of this], the term $-Q(k'', k')$ which describes the energy transfer to larger-scale disturbances can be neglected as $k \to \infty$. Hence the case when $P(k', k'')$ is given by Eqs. (17.21)–(17.23) need not be investigated separately.

$$E(k) \sim k^{-\frac{c+2-m}{1-n}} \tag{17.45'}$$

where the inequality $\dfrac{c+2-m}{1-n} > 3$ must be satisfied. The relatively slow decrease of the spectrum $E(k)$ is here connected with the relatively slow decrease of the function $P(k', k'')$ as $k'/k''$ increases, i.e., there is a relatively nonlocal transfer of energy in wave-number space. The exponent in (17.45') increases in absolute magnitude as $c$ increases. If, on the other hand, the function $\varphi(x)$ decreases as $x \to \infty$ more rapidly than any power of $x$ (and $n < 1$, since otherwise the spectrum will cut off), then the spectrum $E(k)$ will also decrease at infinity more rapidly than by a power law. Thus, for example, if $\varphi(x) = e^{-ax^s}$, $s > 0$ then the spectrum $E(k)$ will decrease for $k \gg k_\eta$ in accordance with a formula which, in the first approximation, is given by[22]

$$E(k) \sim e^{-e^{B\sqrt{\ln k}}}, \qquad B = \sqrt{2s \ln \frac{3-2n}{2-2n}}, \tag{17.45''}$$

i.e., much more rapidly than $k^{-\beta} \sim e^{-\beta e^{\ln \ln k}}$ for any $\beta$ [a more precise asymptotic formula for this case is given by Yaglom (1969)]. When, however, $\varphi(x) = 1$ for $1 < x < a$, and $\varphi(x) = 0$ for $x > a$ (so that the energy transfer from $dk$ involves only a finite region of the wave-number space), then the spectrum $E(k)$ will decay still more rapidly, namely,

$$E(k) \sim e^{-Ak^B} \sim e^{-e^{B \ln k}}, \qquad B = \frac{\ln \dfrac{3-2n}{2-2n}}{\ln a}. \tag{17.45'''}$$

we see that $B \to \infty$ as $a \to 0$ (i.e., as the degree of energy transfer localization increases without limit. In the special case when the function $P(k', k'')$ is given by the Kraichnan and Spiegel formulas (17.21)–(17.23) we must set $n = 0$ in the above expressions [and $m = 7/4$ in the case of Eq. (17.45')].

Certain additional results concerning the dependence of the rate of decrease of $E(k)$ on the energy-transfer localization in wave-number space are discussed in Sect. 22.2

## *Discussion of the above Results*

To establish the validity or otherwise of the above hypotheses, it is natural to try to compare their consequences with data on real turbulent flows. We indicated above a number of such consequences for the quasi-equilibrium range for which there is a universal statistical state. Unfortunately, we know from Sect. 16.6 that, in the case of grid turbulence, which is the only real model of isotropic turbulence, it is very difficult to establish conditions under which the state of disturbances with sufficiently large wave numbers is in fact universal. Nevertheless, since we have devoted a considerable amount of space to the discussion of the spectra $E(k)$ deduced from the various energy-transfer hypotheses, it will be useful to consider, at least briefly, the possibility of verification of the above results.

This is not a simple problem because the three-dimensional spectrum $E(k)$ cannot be measured directly. When it is determined with the aid of Eq. (12.77), using measured values

[22] Strictly speaking, $k$ in Eqs. (17.45'') and (17.45''') denotes the dimensionless ratio $k/k_\eta$; however, since we are interested here in the relative fall-off rate only, it is more convenient to present these equations in the given form.

of the correlation function $B_{LL}(r)$, the most accurate results for $E(k)$ are found to be those lying in the energy range. On the other hand, the values of $E(k)$ for large values of $k$ (which are the only ones that can be referred to the universal quasi-equilibirum range) are subject to considerable uncertainties. Therefore, when the calculated values of $E(k)$ are compared with experimental data, it is necessary to select some readily measured statistical parameters of the velocity fluctuations, which are unambiguously determined by the behavior of the spectrum $E(k)$ in the universal quasi-equilibrium range, and these values must be used as the basis for comparing the conclusions of the theory with experiment. Among the statistical characteristics that satisfy these conditions there are, for example, the mean squares of the velocity derivatives for $n \geqslant 1$, or the mean cube $\overline{\left(\frac{\partial u_1}{\partial x_1}\right)^3}$ which in the case of universal quasi-equilibrium can be determined from the formula

$$\overline{\left(\frac{\partial u_1}{\partial x_1}\right)^3} = -\frac{4}{35}\nu\int_0^\infty k^4 E(k)\,dk \tag{17.46}$$

which follows from Eqs. (15.8) and (12.79). Other possibilities are the longitudinal and lateral one-dimensional velocity spectra for wave numbers $k$ in the quasi-equilibrium range. We note that the "theoretical values" of some of these readily measured statistical quantities corresponding to a particular hypothesis about $W(k)$ can be found in the literature. Thus, T. D. Lee (1950), Reid (1956a, b), and Reid and Harris (1959) have determined the values of the skewness

$$S = \frac{\overline{\left(\frac{\partial u_1}{\partial x_1}\right)^3}}{\left[\overline{\left(\frac{\partial u_1}{\partial x_1}\right)^2}\right]^{3/2}} = -\frac{3\sqrt{30}}{7}\nu\frac{\int_0^\infty k^4 E(k)\,dk}{\left[\int_0^\infty k^2 E(k)\,dk\right]^{3/2}}, \tag{17.47}$$

for the Heisenberg, Obukhov, and Kovasznay spectra given by Eqs. (17.40), (17.33), and (17.29) respectively. Batchelor and Townsend (1949) have calculated the values of the dimensionless ratio

$$\frac{\overline{\left(\frac{\partial^3 u_1}{\partial x_1^3}\right)^2}\,\overline{\left(\frac{\partial u_1}{\partial x_1}\right)^2}}{\left[\overline{\left(\frac{\partial^2 u_1}{\partial x_1^2}\right)^2}\right]^2} = \frac{35}{27}\frac{\int_0^\infty k^6 E(k)\,dk\cdot\int_0^\infty k^2 E(k)\,dk}{\left[\int_0^\infty k^4 E(k)\,dk\right]^2} \tag{17.48}$$

for the Kovasznay, Obukhov, and Heisenberg-type spectra with cutoffs, given by Eqs. (17.29), (17.33), and (17.39) respectively.[23] Reid (1960) used numerical integration to determine the longitudinal one-dimensional spectra

[23] We note that the spectrum (17.40) which has no cutoff cannot be used here because $\overline{\left(\frac{\partial^3 u_1}{\partial x_1^3}\right)^2}$ is then infinite, and hence Eq. (17.48) is also infinite.

$$\Phi_1\left(\frac{k}{k_1}\right)=\frac{\gamma^{3/2}E_1(k/k_1)}{\bar{\epsilon}^{1/4}\nu^{5/4}}=\int_{k/k_1}^{\infty}\left(1-\frac{k^2}{k_1^2x^2}\right)\frac{\Phi(x)}{x}\,dx, \tag{17.49}$$

corresponding to Eqs. (17.30), (17.33), and (17.40). Leith (1967) found numerically the longitudinal one-dimensional spectrum corresponding to his hypothesis (17.3'), while Walton (1968) made similar calculations for the Kraichnan-Spiegel model of $w(k)$ where it was supposed that $\varphi(x) = 1$ for $0.5 < x < 2$ or for $0.86 < x < 1.16$ and $\varphi(x) = 0$ elsewhere. However, the attempts by Lee, Reid, Batchelor, and Townsend to compare their calculated values with experiment (to verify the theory or to find at least the most suitable values of the coefficients $\gamma_K$, $\gamma_0$, and $\gamma_H$, were concerned with measurements behind grids for relatively low values of $\mathrm{Re}_M$ (of the order of 1,000–10,000) and, therefore, in view of the discussion of Sect. (16.6), the results obtained by them cannot be regarded as satisfactory.

We note further that physical intuition suggests that any spectrum which suddenly vanishes discontinuously at the point $k = k_1$ must be regarded as unnatural and unlikely. The spectra corresponding to the Kovasznay and Leith hypotheses, which fall monotonically to zero from a finite value at $k = k_1$ and then remain identically zero, seem less strange. However, even in the case of this type of spectrum, it must still be assumed that some of the quantities which depend on $k$, for example, the derivative $\frac{d^2E(k)}{dk^2}$, in the case of the Kovasznay hypothesis change discontinuously at $k = k_1$. Moreover, the complete absence of turbulent disturbances with wave numbers greater than $k_1$ is subject to considerable doubt: Owing to the viscosity, the intensity of the disturbances should decrease as their scale decreases, but it is difficult to understand why velocity fluctuations with wavelengths less than $2\pi/k_1$ must be necessarily absent. Many specialists therefore still consider that the Heisenberg hypothesis, which corresponds to the spectrum (17.40) that continues to infinity, is more satisfactory than, for example, the Obukhov and Kovasznay hypotheses [see for example, Lin and Reid (1963)].[24] In fact, however, even the Heisenberg hypothesis provides only a very approximate model, and there are serious objections even to the spectrum given by Eq. (17.40). For example, according to the discussion of Sect. 11.3, this spectrum coresponds to a velocity field which is only doubly differentiable, and the third derivative $\partial^3u_1(x)/\partial^3x_1$ does not exist for it. At the same time, it is natural to expect that, in the case of a finite (though small) viscosity, the velocity field $u(x,t)$ should be differentiable any number of times.[25]

[24]Of course this remark is relevant only when the hypotheses are considered as strict laws valid precisely for all $k \ll 1/L$. However, all the hypotheses may provide a good approximation in some specific limited ranges of $k$ values. In particular there are arguments suggesting that the Kovasznay hypothesis is rather accurate near the small-scale boundary of the inertial subrange, i.e., when viscosity effects are still small (see Lumley, 1967b).

[25]We note that there are theorems in the theory of differential equations which enable us to show quite rigorously that the solutions of the Navier-Stokes equations for any initial conditions at $t = 0$ are in many cases infinitely differentiable for any $t > 0$. This does not, however, help us in any way because it refers only to the differentiability of individual fields $u(x,t)$. At the same time, even if all the individual realizations $u(x,t)$ are infinitely differentiable but, for example, the root mean square of the derivative $\partial^3u_1/\partial x_1^3$ is infinite, then it will follow that $\int_0^\infty k^6E(k)\,dk = \infty$. We must therefore rely on intuition which suggests that, for finite viscosity $\nu$, one would not expect the appearance of the very large values of $\frac{\partial^3u_1}{\partial x_1^3}$ which are necessary to ensure that the quantity $\overline{\left[\frac{\partial^3u_1}{\partial x_1}\right]^2}$ is infinite.

There is, thus, a strong suspicion that the most acceptable spectra from the point of view of physical intuition are the continuous spectra which do not vanish anywhere but decrease more rapidly than any finite negative power of $k$ as $k \to \infty$. In other words, the most probable spectra are those obtained from the modified Obukhov hypothesis (17.5), or the modified von Kármán hypotheses (17.20) [or (17.21)–(17.23)] with a sufficiently rapidly decreasing function $\varphi(x)$.[26] It will be shown in Sect. 22.3 that it is possible to attempt to determine the behavior of $E(k)$ for $k \to \infty$ in terms of certain physical considerations which are better justified than the speculative hypotheses of Sect. 17.1; such attempts lead to an exponentially falling spectrum.

## 17.3 Application of the Energy-Transfer Hypotheses to Decaying Turbulence behind a Grid

We have already noted that all the experimental studies of isotropic turbulence have been based on data obtained behind grids in the wind tunnel. Since the universal statistical state predicted by Kolmogorov does not occur under these conditions, any experimentally verifiable consequences of the hypotheses of Sect. 17.1 can only be deduced from the complete equations (17.1) and (17.2) for the time-dependent spectrum $E(k,t)$. If we substitute any of the expressions for $W(k)$ obtained in Sect. 17.1 into one of these equations, we obtain the integro-differential equation for the function $E(k,t)$. All that remains then is to isolate the solution of the resulting equation, which corresponds to the evolution of the turbulence behind the grid, and compare this with direct measurements.

Since the above integro-differential equation contains only the first-order time derivative, it follows that its solution is uniquely determined by the initial condition $E(k, 0) = E_0(k)$. Therefore, if we know the form of the spectrum at the initial time from experiment, we can determine this spectrum for all subsequent times by numerical integration. This type of calculation has been carried out by Tollmien (1952–1953) and Meetz (1956a, b) for the Heisenberg equations

$$\frac{\partial}{\partial t}\int_0^k E(k', t)\,dk' = -2\left[\nu + \gamma_H \int_k^\infty \sqrt{\frac{E(k'', t)}{k''^3}}\,dk''\right]\int_0^k k'^2 E(k', t)\,dk' \quad (17.50)$$

and the data of Stewart and Townsend (1951). Tollmien took $E_0(k)$ in the form of the data of Fig. 20, which corresponded to $x = 30M$, and let the parameter $\gamma_H$ in Eq. (17.50) assume the value of 0.45, in accordance with Proudman's estimate (1951) which will be discussed below. The resulting function $E(k,t)$ did not display self-preserving behavior in time. Nevertheless, the turbulent energy $\int_0^\infty E(k, t)\,dk = \frac{3}{2}\overline{u^2(t)}$ calculated by Tollmien was found to be in very good agreement with the empirical "minus one law" (16.11) right up to $t = 0.5$, and this law is usually regarded as strong evidence for self-preservation. Next, Tollmien attempted to normalize his solutions in accordance with the data of Fig. 20, so that the values of $E(k,t)$ for large $k$ were as near time-independent as possible, and expressed the resulting values as functions of $k\lambda$. The curves obtained in this way (Fig. 30) were found to be qualitatively similar to the empirical curves of Stewart and Townsend shown in Fig. 20.

[26] The rate of decrease of the spectrum $E(k)$ cannot, of course, be determined on the basis of physical intuition alone. It may be noted, however, that the expected asymptotic behavior of the form of Eq. (17.45") (which is intermediate between a power and an exponential decrease) has never been encountered until now in real physical problems.

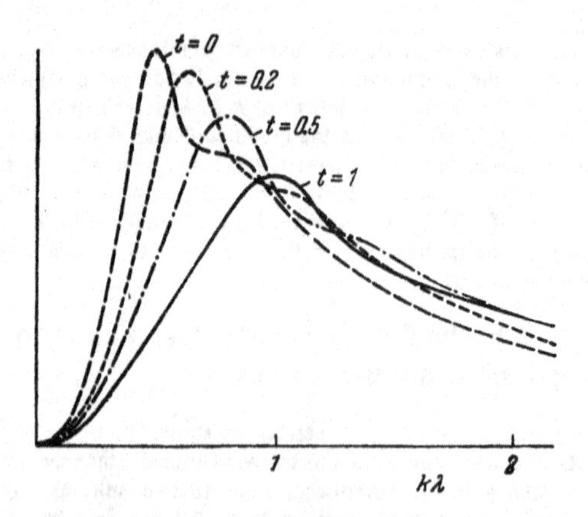


**FIG. 30** Normalized spectra $E(k,t)$ for different $t$, calculated by Tollmien (1952–1953).

Tollmien's results suggest that the general character of energy transfer along the wave-number spectrum is adequately described by Heisenberg's formula (17.9). A similar conclusion can be drawn from results of Meetz (1956a, b), who repeated Tollmien's calculations for a number of values of $\gamma_H$, and carried out a much more careful comparison of his results with the data of Stewart and Townsend. He found that the law given by Eq. (16.11) is well satisfied for values of $x = Ut$ that are not too large, and all the values of $\gamma_H$ which he considered. The data on turbulent-energy decay behind a grid were found to lie between the curve corresponding to $\gamma_H = 0.47$ and $\gamma_H = 0.75$, although they tend to be closer to the $\gamma_H = 0.75$ curve. However, if we use the calculated spectra $E(k,t)$ to determine the correlation functions $B_{LL}(r,t)$ from Eq. (12.75), and $B_{LL,L}(r,t)$ from Eqs. (13.141'''), (17.5) and the formula $T(k) = dW(k)/dk$, we find that the calculated values of $B_{LL}$ are practically the same as the data of Fig. 16 for both $\gamma_H = 0.47$ and $\gamma_H = 0.75$. However, the calculated values of $B_{LL,L}$ for $\gamma_H = 0.47$ are in more or less satisfactory agreement with the data, but for $\gamma_H = 0.75$ there is a clear discrepancy with experiment. The results of Meetz thus seem to confirm Tollmien's conclusion that the Heisenberg formula (17.9) can be used with $\gamma_H$ approaching 0.5, and at the same time show that the value $\gamma_H = 0.75$ is less suitable in some respects.

We shall now compare Meetz's results with those of Tanenbaum and Mintzer (1960), Uberoi (1963) and Van Atta and Chen (1969a). These workers have compared the values of $W(k)$ calculated from the semiempirical hypotheses of Sect. 17.1 [using values of $E(k)$ found with the aid of Eqs. (12.77) and (12.88), from the directly measured functions $B_{LL}(r)$ or $E_1(k)$] with the values of the same functions determined independently from the hypotheses . Namely, they determined $W(k)$ from $dW(k)/dk = -T(k)$ and Eq. (12.142'''), using measured values of $B_{LL,L}(r)$, or with the aid of Eq. (14.5), using the empirical values of $E_1(k)$ and $\partial E_1(k)/\partial t$. In particular, Tanenbaum and Mintzer tried to verify the Kovasznay and von Kármán hypotheses (17.3) and (17.7) for a number of values of $m$ and $n$ [$m = -5/2$, $n = 1/4$; $m = -3/2$, $n = 1/2$ which corresponds to Eq. (17.9); and $m = -5/2$, $n = 1/2$], using the simultaneous measurements of $B_{LL}$ and $B_{LL,L}$ due to Stewart (1951) and Stewart and Townsend (1951). It was found that, in the case of the data for $\mathrm{Re}_M = 21.2 \cdot 10^3$ and $42.4 \cdot 10^3$ (and $x = 30M$), the values of $W(k)$ determined from $B_{LL,L}(r)$ are in clear disagreement with calculations based on both Eqs. (17.3) and (17.17) for all values of $\gamma_k$, $m$, $n$, and $\gamma'_k$. Nevertheless, Tanenbaum and Mintzer believed that this discrepancy was not very decisive because the important values of $E(k)$ in the case of the von Kármán formula are those corresponding to large $k$, which are not very accurately determinable from $B_{LL}(r)$ and, moreover, for $x = 30M$ one would expect a relatively large departure of grid generated turbulence from the isotropic state. From this point of view, the data of Stewart and Townsend obtained for $\text{Re}_M = 5.3 \cdot 10^3$ and a number of distances $x$ from the grid ($20M$, $30M$, $60M$, $90M$, and $120M$) are more convenient because the spectrum $E(k)$ is much narrower and can therefore be determined more precisely from $B_{LL}$, while the higher values of $x/M$ provide a better approximation to isotropy. If, however, we restrict our attention to the data for $\text{Re}_M = 5.3 \cdot 10^3$ and $x/M = 60$, 90, and 120, then it is found that the Kovasznay formula (17.3) still gives very poor agreement between the values of $W(k)$ determined by the two different methods for any $\gamma_K$. The use of the von Kármán formula, on the other hand, with any of the above three values of $m$, $n$ does ensure more or less satisfactory agreement (the agreement appears to be best for $m = -5/2$, $n = 1/4$ and $\gamma_K' \approx 0.6$; this last conclusion, however, cannot be regarded as reliable owing to the insufficient accuracy of the original data).

An analogous verification of Heisenberg's formula (17.9) was carried out by Uberoi (1963) who started with the values of $E_1(k, t)$ and $\frac{\partial E_1(k, t)}{\partial t}$, measured in grid turbulence for $\text{Re}_M = 2.6 \cdot 10^4$ and $x/M = 48, 72$, and 110. According to these data, Eq. (17.9) gives fully acceptable accuracy when $\gamma_H \approx 0.2$. This conclusion cannot, however, be regarded as particularly reliable (we recall that Uberoi also found that the turbulence which he investigated was appreciably anisotropic for all $x/M$).

Finally Van Atta and Chen (1969a) repeated the Tanenbaum and Mintzer calculations based on their own quite reliable measurements of the functions $B_{LL}(r)$ and $B_{LL,L}(r)$ at $\text{Re}_M \approx 25{,}000$ and $x/M = 48$ and using a much better digital method for the calculations of $E(k)$ and $W(k)$. They compared the calculated values of $T(k) = -dW(k)/dk$ with the results obtained from the hypotheses of Kovasznay, Heisenberg and von Kármán (for $m = 0, n = 1$, and for $m = 0, n = 3/2$). They came to the conclusion that none of these hypotheses except Heisenberg's can fit the data well for any value of $\gamma_k$ or $\gamma_k'$, but that Heisenberg's expression fits the high wave-number data fairly well with $\gamma_k = 0.25$.

## 17.4 Self-Preserving Solutions of the Approximate Equations for the Energy Spectrum

Since the determination of the solutions of the integro-differential equations for the isotropic-turbulence spectrum $E(k,t)$ from the empirical initial conditions is complicated and not very fruitful, many workers have tried to solve the simpler problem, namely, that of finding special "self-preserving solutions" for these equations. It is clear that if we introduce any particular hypothesis about the form of the energy-transfer function $W(k)$, the number of possible self-preserving solutions of the von Kármán-Howarth equation, or the equivalent spectral equation, is sharply reduced, and it can be expected that definite expressions capable of simple comparison with experimental data will thus become available for $B_{LL}(r,t)$ and $E(k,t)$. All that needs to be remembered is that the agreement between these expressions and experiment is possible only to the extent to which the experiment does not contradict the general idea of self-preservation. We cannot, therefore, expect too much from this approach in view of the conclusions of Sect. 16.

Heisenberg (1948b) was the first to consider self-preserving solutions of the spectral equation (17.1) with the term $W(k)$ expressed in terms of $E(k,t)$, and applied this to Eq. (17.50). He assumed that the self-preserving state is rapidly established in the turbulence behind a grid, and that the spectral function $E(k,t)$ satisfies the first equation in Eq. (16.23) for all $k$ apart from values in a small range near $k = 0$, which contains a negligible fraction of the total energy. As a result of the decay of the turbulence, the energy range of the spectrum gradually shifts toward longer wavelengths, and eventually includes a part of the region where self-preservation does not obtain. From this moment onward, the self-preserving solution no longer governs the law of energy decay which can, therefore, undergo a change. Subsequently, the experiments of Stewart and Townsend (1957) and Uberoi (1963) showed that the usual situation is in fact different, and an appreciable part of the energy range lies outside the interval of the spectrum in which self-preservation occurs. The results of Heisenberg and other workers, who have investigated completely self-preserving solutions of the spectral equation, can therefore be used to explain the turbulence-decay data for the region behind the grid, but only to a very limited extent.

In view of Eq. (16.10), which relates the scales $l(t)$ and $v(t)$, the general self-preservation condition (16.23) can be rewritten in the form

$$E(k, t) = \frac{\alpha^{3/2}\nu^{3/2}}{\gamma_H^2 \sqrt{t}} \varphi(k\sqrt{\alpha\nu t}) \tag{17.51}$$

where the factors which are functions of $\nu$ are chosen from dimensional considerations, while the dimensionless constants $\gamma$ and $\alpha$ are introduced in a form convenient for subsequent discussion. When Eq. (17.51) is substituted into Eq. (17.50) we obtain the following integro-differential equation for $\varphi(\xi)$:

$$\frac{1}{4}\int_0^\xi [\varphi(\xi') - \xi'\varphi'(\xi')]\, d\xi' = \left\{\int_\xi^\infty \sqrt{\frac{\varphi(\xi'')}{\xi''^3}}\, d\xi'' + \frac{1}{\alpha}\right\} \int_0^\xi \xi'^2 \varphi(\xi')\, d\xi'. \tag{17.52}$$

It follows from this equation that $\varphi(\xi) \sim \xi$ as $\xi \to 0$ [i.e., $E(k)$ should be a linear function when $k \to 0$], since otherwise the right-hand side of Eq. (17.52) would be of a small quantity of a higher order for small $\xi$ than the left-hand side.[27] Heisenberg has shown that when $\nu = 0$, the self-preserving solution of Eq. (17.50) for which $E(k, t) \sim t^{-1/2}\varphi(kt^{1/2})$ is determined practically uniquely and decays as $k^{-5/3}$ when $k \to \infty$, while for $\nu \neq 0$ the solution $\varphi(\xi)$ decreases in proportion to $\xi^{-7}$ as $\xi \to \infty$. In the case of a finite but very small $\nu$, the solution $\varphi(\xi)$ turns out to be close to the function const $\cdot\ \xi^{-5/3}$ for a considerable range of "relatively large" values of $\xi$, and begins to vary as $\xi^{-7}$ only for still larger values of $\xi$. Tollmien (1952–1953) subsequently found that analogous results are also valid for nonself-preserving solutions of Eq. (17.50). Moreover, it is clear from Eq. (17.52) that the function $\varphi(k\sqrt{\alpha\nu t}) \sim \gamma_H^2 t^{1/2} E(k, t)$ is independent of the value of $\gamma_H$, although it is possible that the value of the Reynolds number for isotropic self-preserving turbulence, which corresponds to a given function $\varphi(\xi)$, will also be different for different $\gamma_H$.

The exact solution of the very approximate equation (17.52) is, of course, merely a mathematical exercise, although a relatively extensive literature has been devoted to it. This solution can be found only by numerical methods. It is convenient to transform the variables so that the integro-differential equation (17.52) becomes an ordinary differential equation [Chandrasekhar (1949), Blanch and Ferguson (1959)]. To each value of $\alpha$ there corresponds one solution of Eq. (17.52) describing the energy spectrum of self-preserving isotropic turbulence with a particular Reynolds number. The role of the Reynolds number

[27] We note that this conclusion is independent of the hypothesis given by Eq. (17.9): It follows from the assumption of self-preservation given by Eq. (17.51), in view of the general assumption that disturbances with the longest wavelengths must remain unaltered. This in turn, follows from Eq. (14.15) and the fact that $T(k)$ is a small quantity of higher order than $E(k)$ for small $k$.

is then played by the parameter $\alpha$, since comparison of Eqs. (16.23), (16.9), (16.10), and (17.51) shows that $\alpha = \gamma_H \frac{lv}{\nu} = \gamma_H$ Re [by Eq. (16.10), the number Re $= lv/\nu$ does not vary in this case during the decay process, i.e., it is independent of $t$]. To compare solutions of Eq. (17.52) with data for which $l(t)$ and $v(t)$ are unknown, it is convenient to follow Proudman (1951) and define the Reynolds number corresponding to a given solution $\varphi(\xi)$ by the equation $2\int_0^\xi \xi^2\varphi(\xi)\,d\xi = \gamma_H^2 R$ where $R$ has the same significance as in Sect. 16.7. The number $R$ is readily related to the measured Reynolds numbers $\mathrm{Re}_\lambda = \frac{U'\lambda}{\nu}$ and $\mathrm{Re}_M = \frac{UM}{\nu}$: We have already noted that $R = 0.15\,\mathrm{Re}_\lambda^2 = 1/90\,\mathrm{Re}_M$. The last of these equations is valid only for grids of the type used by Townsend and Stewart. The value of $\gamma_H$ now appears only in the form of the products $\gamma_H \mathrm{Re}_\lambda$ and $\gamma_H^2 \mathrm{Re}_M$ and, therefore, any uncertainty in this number will affect only the value of the Reynolds number corresponding to the given solution.

Chandrasekhar (1949) has carried out a numerical solution of Eq. (17.52) and found six solutions $\varphi(\xi)$ corresponding to values of $\gamma_H^2 R$ in the range between 0.25 and $\infty$ [the last case corresponds to the solution of Eq. (17.52) when $1/\alpha = 0$]. Moreover, additional solutions of Eq. (17.42) have been found by Rotta (1950), Proudman (1951), Meetz (1966a, b), Blanch and Ferguson (1959) and Blanch [see Lin and Reid (1963)]. Figure 31 (taken from Batchelor's book) shows three of the Chandrasekhar solutions corresponding to the largest values of $\gamma_H^2 R$ used by him, and also the Proudman solution. The fact that the maxima of all four curves have nearly the same positions, and also the fact that they all touch the straight line $\varphi = 4\xi$ at the origin, is related to the special choice of scale for $\xi$, which depends on the Reynolds number $\alpha$. We note that the form of the spectrum in Fig. 31 remains practically the same thoughout the energy range as $\gamma_H^2 R$ increases from 21.9 to $\infty$.

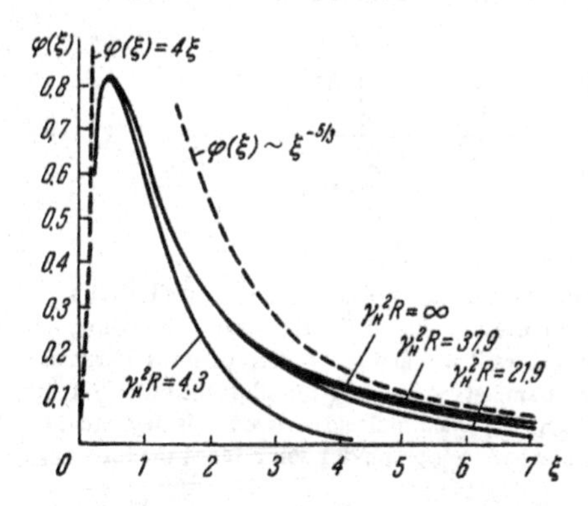


**FIG. 31** Dimensionless spectra obtained by using both the self-preservation hypothesis and the energy-transfer hypothesis.

Proudman has compared the numerical solutions with the data of Stewart and Townsend (1951). He based this on a comparison of the measured correlation functions $B_{LL}(r)$ and $B_{LL,L}(r)$, calculated from the functions $E(k)$ and $T(k) = -W'(k)$. He found that for $\gamma_H$

$= 0.45 \pm 0.05$, the agreement between the calculated and the measured values of $B_{LL}(r)$ and $B_{LL,L}(r)$ was quite satisfactory. This appears to suggest, above all, that the form of the correlation functions is not very sensitive to changes in the form of the spectrum, since the assumption of self-preservation was not satisfied too well in the experiments of Stewart and Townsend.[28]

Analogous calculations have been carried out by Reid and Harris (1959) for the simplest model of spectral energy transfer, which is described by the Kovasznay formula (17.3). In this case, substitution of Eqs. (17.51) and (17.3) into Eq. (14.15), where $T(k) = -dW(k)/dk$, leads to a nonlinear differential equation for the function $\varphi(\xi)$. By changing the variables, this equation can be transformed so that it indicates the explicit form of the general solution, but the resulting expressions are then so complicated that it is more convenient to use numerical methods. Reid and Harris found eleven numerical solutions $\varphi(\xi)$ corresponding to $\gamma_K^2 R$ in the range between 0.35 and $\infty$. Their spectra are very similar to those obtained from the Heisenberg hypothesis. This similarity is so considerable that Reid and Harris were able to use Proudman's results to show that the Kovasznay hypothesis provides a satisfactory explanation of the form of the correlation functions $B_{LL}(r)$ and $B_{LL,L}(r)$ observed by Stewart and Townsend, provided $\gamma_k$ is taken to be close to 0.3.

Some more general self-preserving decaying solutions of the spectral equation corresponding to formula (17.3′) for $W(k)$ were computed numerically by Leith (1967) and compared with the similar results following from the Heisenberg hypothesis. Results of the same type for the Kraichnan-Spiegel model of energy-transfer (with $\varphi(x)$ equal to one on some interval and zero elsewhere) were published by Walton (1968).

In addition to the correlation functions, there are also other directly measurable quantities which can be calculated from the "theoretical" three-dimensional self-preserving spectra deduced from any particular energy-transfer hypothesis, and then compared with experimental data. This includes, for example, the one-dimensional longitudinal spectra $E_1(k)$ which for some self-preserving three-dimensional spectra of the Heisenberg theory were found numerically by Rotta (1950) and Lin and Reid (1963), or the skewness

$$s = \frac{\overline{\left(\frac{\partial u_1}{\partial x_1}\right)^3}}{\left[\overline{\left(\frac{\partial u_1}{\partial x_1}\right)^2}\right]^{3/2}} = -\frac{3\sqrt{30}}{14}\frac{\int_0^\infty k^2 T(k)\,dk}{\left[\int_0^\infty k^2 E(k)\,dk\right]^{3/2}}, \quad T(k) = -\frac{dW(k)}{dk}, \tag{17.53}$$

This last quantity was calculated by Reid (1956a, b) and Reid and Harris (1959) for the self-preserving spectra corresponding to the Heisenberg, Obukhov, and Kovasznay hypotheses in the two limiting cases $R = \infty$ and $R \to 0$. Comparisons of calculations with experiment have shown that the parameters $\gamma_H, \gamma_O, \gamma_K$ are all of the order of unity.

As indicated in Sect. 16, when molecular viscosity is neglected, the dynamic equations for isotropic turbulence have self-preserving solutions of the form (17.51), and also a whole

[28] We note that since the Heisenberg theory is not exact, good agreement with experimental data may demand an adjustment of the value of $\gamma_H$ which may be different in different cases. It is, therefore, not surprising that in the comparisons reported by Heisenberg (1948a), T. D. Lee (1950), Rotta (1950), Proudman (1951), Meetz (1956a, b), Reid (1960), Ellison (1962) and Uberoi (1963), the values obtained for $\gamma_H$ lay in the rather wide range between 0.2 and 0.85 (not all these comparisons were however based on adequate experimental material as far as the conditions assumed in the theoretical predictions were concerned).

family of additional self-preserving solutions determined by the value of $c_1/c_2$ in Eqs. (16.20) and (16.21). If we then assume some particular energy-transfer hypothesis, then the Eq. (17.1) will enable us to obtain an integro-differential equation for the function $\varphi(\xi)$ of Eq. (16.23) for any value of $c_1/c_2$. Examples of analyses and numerical integrations of this equation, corresponding to the Heisenberg and von Kármán hypotheses for $W(k)$, can be found in the papers of Rotta (1950, 1953), Sen (1951), and Ghosh (1954, 1955).

A further type of self-preserving solution was obtained in Sect. 16 for the final period of decay of isotropic turbulence when the third moments of the velocity fluctuations could be neglected in comparison with the second moments. Since this is equivalent to the assumption that $T(k) = 0$, the spectral energy-transfer hypotheses cannot be directly applied to the final period of decay. They can, however, be used to investigate the approach of isotropic turbulence to the final period of decay [see, for example, Reid (1956c) and Reid and Harris (1959)].

Finally, we note that to ensure that a self-preserving solution of the equation for the energy spectrum $E(k)$ is meaningful, it must be stable against small disturbances. In other words, a small departure of $E(k, t)$ from a given solution, must decay more rapidly with time than the solution itself. The usual mathematical methods can be used to investigate the stability of self-preserving solutions corresponding to any of the hypotheses of Sect. 17.1. In particular, the stability problem for the case of $W(k)$ corresponding to the Heisenberg formula (17.49) was discussed briefly by Rotta (1950), Lin (1953b), Lin and Reid (1963), and Sen (1957). According to Lin's results, the self-preserving solutions are stable with respect to disturbances localized in a narrow spectral band, for large wave numbers $k$ (including all values of $k$ for which $E(k) \sim k^{-\frac{5}{3}}$ or $E(k) \sim k^{-7}$), but are unstable for small $k$ for which $E(k) \sim k$. This can be compared with the data of Stewart and Townsend (1951), according to which the grid-generated turbulence varies with time in a self-preserving fashion, but only in the short-wave part of the spectrum which contains about two-thirds of the total energy. Similar studies of the nonviscous self-preserving solution with $c_1/c_2 = -0.2$ and $\overline{u^2(t)} \sim t^{-\frac{10}{7}}$ were carried out by Sen, who showed that this solution is unstable for all $k$. This is in some agreement with the absence of data indicating the existence of self-preservation of this kind in any real turbulent flows. It must be remembered, however, that all these results were obtained only for a model "spectral" equation based on the Heisenberg hypothesis (17.9). The question as to their dependence on the form of $W(k)$ is still an open one.

# 18. THE MILLIONSHCHIKOV ZERO-FOURTH-CUMULANT HYPOTHESIS AND ITS APPLICATION TO THE INVESTIGATION OF PRESSURE AND ACCELERATION FLUCTUATIONS

## 18.1 The Zero-Fourth-Cumulant Hypothesis and the Data on Velocity Probability Distributions

We already know that the fourth-order velocity correlation functions are occasionally encountered in the theory of turbulence. This occurs, for example, in the study of pressure correlations (see Sect. 14.4). In this case, the dynamic correlation equations contain the new unknown functions $B_{LL,LL}(r, t)$, $B_{LL,NN}(r, t)$, etc., which were introduced in Sect. 12.4. Since this complicates the analysis enormously, it is very desirable to be able to express these new unknown functions in terms of the simpler and better known second and third order velocity correlations. However, there are no exact relations enabling us to express $B_{LL,\,LL}(r,\,t)$, and so on in terms of, say, $B_{LL}(r,\,t)$ and $B_{LL,\,L}(r,\,t)$. Therefore, we can only hope to establish some approximate relationships between the fourth- and lower-order moments which are valid with satisfactory accuracy under some particular conditions. An important approximate relationship of this kind is provided by the zero fourth-cumulant hypothesis formulated by Millionshchikov (1941a, b) and subsequently, in a somewhat different form, by Heisenberg (1948a). The present section is devoted to this hypothesis and some of its consequences.

Since the fourth-order correlation terms of the turbulence equations have a less clear physical interpretation than the term $T(k,\,t)$ in the spectral equation (14.15) [or $W(k,\,t)$ in Eqs. (17.1) and (17.2)], the Millionshchikov hypothesis is purely statistical in nature, in contrast to the more physical spectral energy-transfer hypotheses. According to this hypothesis, the fourth-order velocity correlations are approximately related to the second-order correlations by expressions which are valid for normal (Gaussian) probability distributions. In other words, the hypothesis postulates that fourth-order cumulants can be neglected in comparison with the fourth-order correlation functions. For correlation functions of the form $\overline{u_i u_j u'_m u'_n} = B_{ij,\,mn}(r,\,t)$, this means that we can use the equation

$$\overline{u_i u_j u'_m u'_n} = \overline{u_i u_j} \cdot \overline{u'_m u'_n} + \overline{u_i u'_m} \cdot \overline{u_j u'_n} + \overline{u_i u'_n} \cdot \overline{u_j u'_m} \tag{18.1}$$

[see Eq. (4.29) in Sect. 4.3 of Vol. 1], i.e.,

$$B_{ij,\,mn}(r,\,t) = B_{ij}(0,\,t)\,B_{mn}(0,\,t) + B_{im}(r,\,t)\,B_{jn}(r,\,t) + \\ + B_{in}(r,\,t)\,B_{jm}(r,\,t) \tag{18.1'}$$

The term "the Millionshchikov hypothesis" is occasionally used to designate this last equation. On the other hand we shall also use in some cases the same term for the more general hypothesis of the vanishing of the fourth-order velocity cumulants, namely for the hypothesis of the form

$$\overline{u_i u'_j u''_m u'''_n} = \overline{u_i u'_j} \cdot \overline{u''_m u'''_n} + \overline{u_i u''_m} \cdot \overline{u'_j u'''_n} + \overline{u_i u'''_n} \cdot \overline{u'_j u''_m} \tag{18.1''}$$

where all the velocity components are taken at four quite arbitrary space-time points.

The hypothesis (18.1) admits of a "spectral formulation" involving not just the velocity $u_j(\boldsymbol{x})$ but its "random Fourier amplitudes" $dZ_j(\boldsymbol{k})$. If we take the first term on the right-hand side of Eq. (18.1) over to the left-hand side, and apply the Fourier transformation to both sides of the resulting equation, we obtain an expression equivalent to Eq. (18.1), namely,

$$F_{ij,mn}(\boldsymbol{k}) = \int_{-\infty}^{\infty} F_{im}(\boldsymbol{k}-\boldsymbol{k}') F_{jn}(\boldsymbol{k}') d\boldsymbol{k}' + \int_{-\infty}^{\infty} F_{in}(\boldsymbol{k}-\boldsymbol{k}') F_{jm}(\boldsymbol{k}') d\boldsymbol{k}' \tag{18.2}$$

where $F_{ij,mn}(\boldsymbol{k})$ represents, as usual, the three-dimensional Fourier transform of the tensor $B^{(0)}_{ij,mn}(\boldsymbol{r}) = B_{ij,mn}(\boldsymbol{r}) - B_{ij}(0) B_{mn}(0)$. This relation was used by Heisenberg (1948a) who formulated it independently of Millionshchikov. Heisenberg based his derivation on the stronger hypothesis of approximate statistical independence of uncorrelated Fourier amplitudes $dZ_j(\boldsymbol{k})$, and used it to calculate the mean values of the products of four such amplitudes from which he obtained Eq. (18.2). In point of fact, the use of Heisenberg's stronger hypothesis is not strictly necessary.

When the hypothesis (18.1) was first put forward there were absolutely no data on the fourth-order velocity correlations.[29] Later some measurements of the probability distributions and the fourth-order velocity moments in grid turbulence were published; these can be used to test the validity of the hypothesis. The first data on the probability distributions of the velocity fluctuations were concerned with the one-dimensional distributions of the values of $u_i(\boldsymbol{x})$ where $i =$ 1, 2, or 3. These distributions were measured by Townsend (1947) who showed that in the wind tunnel flow behind square-mesh grids they were practically normal (Fig. 32). This conclusion was subsequently confirmed by the careful measurements of Frenkiel and Klebanoff (1965b) and Van Atta and Chen (1968; 1969a, b). According to all the existing data, the one-point fourth-order moments of the velocity fluctuations satisfies the following relations

[29] We note that, just before Millionshchikov's work, Chou Pei Yuan (1940) suggested that the moment $\overline{u_i u_j u'_m u'_n}$ could be approximated with good accuracy by one-half of the right-hand side of Eq. (18.1).

to within the experimental error (which did not exceed 10% of the measured quantities):

$$\overline{u_i^4} = 3(\overline{u_i^2})^2, \overline{u_i^2 u_j^2} = \overline{u_i^2}\,\overline{u_j^2} \tag{18.3}$$

where $i, j = 1, 2, 3; i \neq j$. These relations are special cases of Eq. (18.1).

In addition to the probability distributions and the moments of $u_1$, Townsend (1947) investigated the probability distributions and moments of the derivative $\partial u_1/\partial x_1$. In this case the empirical distributions were clearly non-Gaussian (Fig. 33). In particular, Townsend obtained a value of −0.4 for the skewness $s = \overline{\left(\frac{\partial u_1}{\partial x_1}\right)^3} \Big/ \left[\overline{\left(\frac{\partial u_1}{\partial x_1}\right)^2}\right]^{3/2}$ [subsequently Uberoi (1963) found that $s \approx -0.5$], while for the excess $\tilde{\delta}_1 = \frac{\overline{(\partial u_1/\partial x_1)^4}}{[\overline{(\partial u_1/\partial x_1)^2}]^2} - 3$ Townsend obtained first a value of 0.3; in subsequent experiments carried out under similar, but not exactly the same conditions, this value was found to be closer to 0.9 (see Batchelor and Townsend, 1949). The values of the excess $\tilde{\delta}_n = \frac{\overline{(\partial^n u_1/\partial x_1^n)^4}}{[\overline{(\partial^n u_1/\partial x_1^n)^2}]^2} - 3$ for $\frac{\partial^n u_1}{\partial x_1^n}$ when $n \geqslant 2$ were found to be still greater. According to Batchelor and Townsend (1949) [see also Batchelor (1953) Sect. 8.1] for moderate Reynolds numbers in grid turbulence, $\tilde{\delta}_2 \approx 1.9$ and $\tilde{\delta}_3 \approx 2.9$, on the average, and all the values of $\tilde{\delta}_n$ increase slightly with increasing Reynolds number.

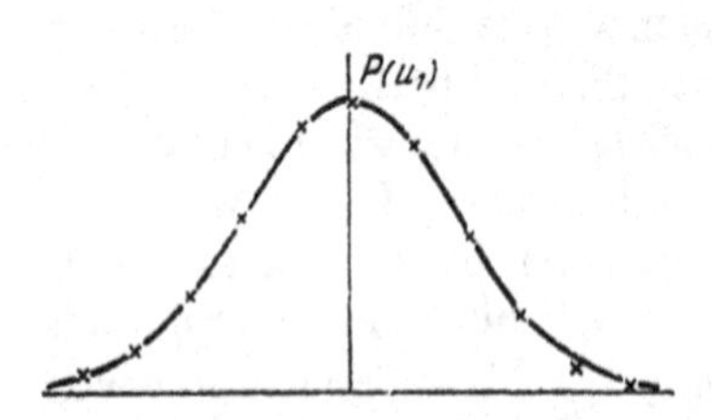


**FIG. 32** Probability density for the longitudinal velocity component behind a grid in a wind tunnel. The curve corresponds to the normal density; crosses represent measured values.

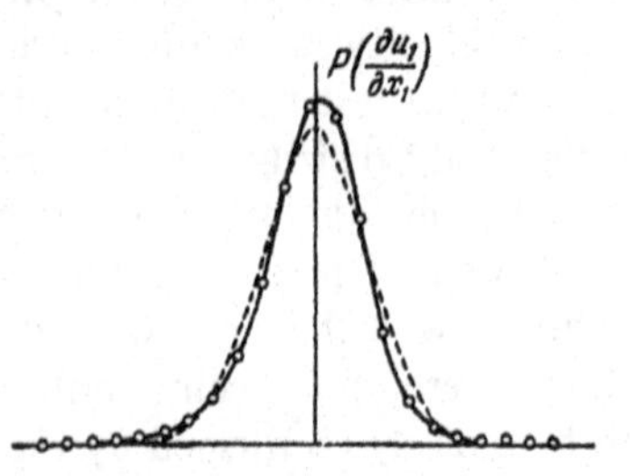


**FIG. 33** Probability density for the derivative $\partial u_1/\partial x_1$. Broken curve shows the normal probability density with variance equal to the measured variance.

More recently Frenkiel and Klebanoff (1971) studied in detail the statistical behavior of velocity derivatives in grid turbulence. They measured the probability density function of $\partial u_1/\partial t = U\partial u_1/\partial x_1$ and the moments and correlations of this variable up to the eighth order. Considerable departures of the measured distributions from the Gaussian distributions were found in this study. Extensive data on probability distributions and statistical moments of the first and second velocity derivatives $\partial u_1/\partial t$ and $\partial^2 u_1/\partial t^2$ and also of filtered velocity signals obtained with bandpass or highpass frequency filters were obtained also by Kuo and Corrsin (1971). These authors performed the measurements in a number of grid-generated and free turbulent flows in the range of Reynolds numbers $\mathrm{Re}_\lambda$ from 12 to 830 and found marked increases in deviations from normality as Reynolds number or/and the order of the derivative (or the wave-number range of the signal) increases. Moreover, the probability distributions of differences $u_1(t) - u_1(t-\tau) \approx u_1(x) - u_1(x - \bar{u}\tau)$ for several values of $\tau$ were measured by Van Atta and Park (1971) in the atmospheric surface layer above the sea; these authors found that the distributions are almost Gaussian for large values of $\bar{u}\tau$, but clearly non-Gaussian for small values of $\bar{u}\tau$.

The possible reason for the great departures from Gaussian behavior of the probability distributions for velocity derivatives and close velocity differences will be discussed in Sect. 25.2 where additional data on probability distributions of the velocity derivatives of high Reynolds number turbulence will also be given. For the moment, let us merely note the following. The one-point values of $u_l$ for which Eq. (18.1) is very accurate are determined in the first instance by the amplitudes $d\boldsymbol{Z}(\boldsymbol{k})$ with $k = |\boldsymbol{k}|$ in the energy range. The values of $\partial u_1/\partial x_1$, on the other hand, are determined by amplitudes $d\boldsymbol{Z}(\boldsymbol{k})$ with $k$ in the dissipation range. The fact that the excess $\tilde{\delta}_1$ is nonzero appears to suggest a departure of the distribution of $d\boldsymbol{Z}(\boldsymbol{k})$ from the Gaussian distribution when $k$ lies in the dissipation range. The higher-order derivatives $\partial^n u_1/\partial x_1^n$ with $n = 2$ and 3 are determined by still smaller-scale velocity disturbances, and therefore the high values of $\tilde{\delta}_2$ and $\tilde{\delta}_3$ should mean that the amplitude distribution for the smallest-scale velocity disturbances is apparently very different from Gaussian. The data on the probability distributions and moments of the derivatives $\partial^n u_1/\partial x_1^n$ thus show that the Millionshchikov hypothesis can be satisfied in grid turbulence, but only with low accuracy, for disturbances in the dissipation range, and even more so for the smaller-scale disturbances. At the same time, data on $u_l(\boldsymbol{x})$ suggest that this hypothesis will apparently be sufficiently accurate when applied to disturbances in the energy range.

Subsequent measurements of various authors who investigated two-, three-, and four-point higher-order correlations in a grid-generated turbulence have confirmed the above conclusion. The first measurements of higher-order two-point moments were made by Stewart (1951) who studied the skewness and excess (or flatness factor) of the longitudinal velocity difference $u_1' - u_1$ at two points along the $x_1$-axis separated by a distance $r$ (and determined largely by disturbances with a scale of order $r$). According to his data shown in Fig. 34 the flatness factor $\delta(r) = \overline{(u_1' - u_1)^4}/[\overline{(u_1' - u_1)^2}]^2$ rapidly approaches its asymptotic value $\delta_\infty \approx 2.9$ as $r$ increases, and this asymptotic value differs only 3–4% from the value $\delta = 3$ which corresponds to the Gaussian probability distribution; at small $r$, however, the deviation of $\delta(r)$ from 3 is appreciably greater. Similar results were later obtained also by Frenkiel and Klebanoff (1967), who measured velocity fourth-order time-correlations and transformed them into space-correlations with the aid of Taylor's hypothesis. According to their data (confirmed also by Van Atta and Chen's measurements, 1968) the values of the velocity-difference flatness-factor are even considerably closer to 3 than 2.9 for all not too small values of $r = Ut$. On the other hand the values of the velocity-difference skewness $S(r) = \overline{(u_1' - u_1)^3}/[\overline{(u_1' - u_1)^2}]^{3/2}$, approach the limiting value $S = 0$ much more slowly according to the data of Stewart, Frenkiel and Klebanoff, and Van Atta and Chen (see for example Fig. 35 taken from the Stewart paper, 1951); the limiting

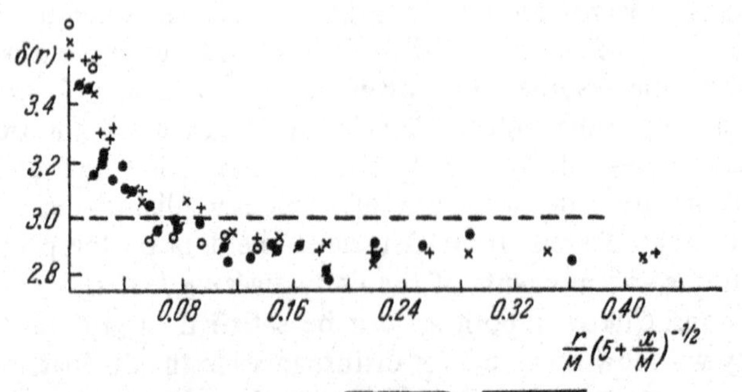


**FIG. 34** The quantity $\delta(r) = \overline{(u_1' - u_1)^4}/[\overline{(u_1' - u_1)^2}]^2$, according to Stewart (1951). The different symbols represent values of $\delta(r)$ for different $x/M$ and $\mathrm{Re}_M$. The scale along the $x$-axis is chosen in order to produce maximum coincidence of data at different $x/M$.

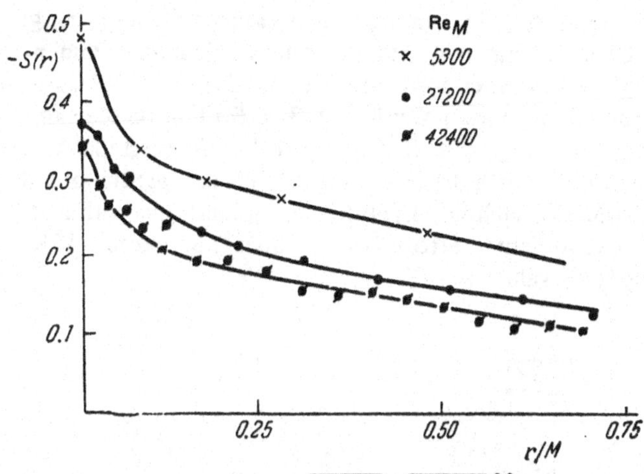


**FIG. 35** The skewness $S(r) = \overline{(u_1'-u_1)^3}/[\overline{(u_1'-u_1)^2}]^{3/2}$, according to Stewart (1951) for $x/M = 30$ and different $Re_M$.

value $S = 0$ here also corresponds to a Gaussian probability distribution. Additional data on the higher odd-order two- and three-point velocity time-correlations in grid turbulence and on the one- and two-point two-dimensional probability distributions for velocity components behind a grid can be found in the papers of Frenkiel and Klebanoff (1965a, b; 1967), Van Atta and Chen (1968; 1969a, b) and Van Atta and Yeh (1969); according to these data, in particular, the one-point two-dimensional probability distribution of the values $u_1$ and $u_2$ is very close to Gaussian, while the two-point two-dimensional probability distribution of the values $u_1$ and $u_1'$ is definitely non-Gaussian. However, these results are of secondary interest to us and we shall not linger on them here.

More important to us are the data on fourth-order velocity correlations which allow direct verification of the Millionshchikov zero-fourth-cumulant hypothesis. The first measurements of the fourth-order correlations were made by Uberoi (1953, 1954). This author measured both the second-order correlations $B_{LL}(r)$ and $B_{NN}(r)$ [which were found to satisfy quite accurately the von Kármán relation (14.3)] and also the functions $B_{LL,LL}(r)$, $B_{NN,NN}(r)$, $B_{NN,MM}(r)$, $B_{LL,NN}(r)$, and two linearly independent combinations of the functions $B_{LL,LL}(r)$, $B_{NN,NN}(r)$, $B_{LL,NN}(r)$ and $B_{LN,LN}(r)$ in grid-generated turbulence (at $x/M = 48$, $Re_\lambda \approx 60$). The measured values of the fourth-order correlation functions were determined largely by the disturbances in the energy range; these values Uberoi compared with the values calculated from measured values of second-order correlations using Eq. (18.1′). Two examples of this comparison are given in Fig. 36a. Both in these examples and in all the remaining cases it was found that the discrepancy between the measured fourth-order moments and the predictions deduced from the Millionshchikov hypothesis lie practically within the limits of the experimental errors (i.e., it does not exceed 15% of the corresponding values).

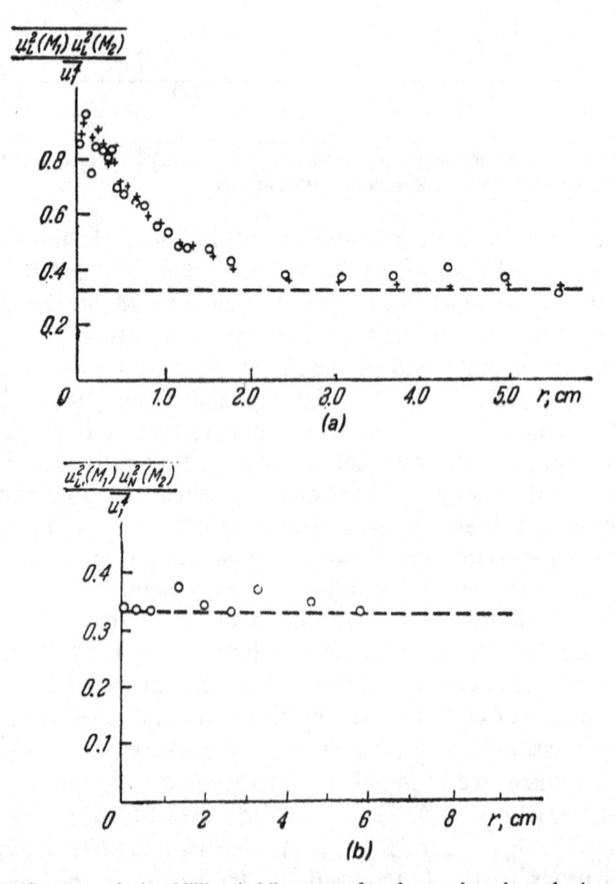


**Fig. 36** Verification of the Millionshchikov zero-fourth-cumulant hypothesis using the Uberoi (1953, 1954) wind-tunnel data. Measurements are represented by the open circles, calculations from measurements of the second moments by the crosses (Fig. a) and the broken curve (Fig. b).

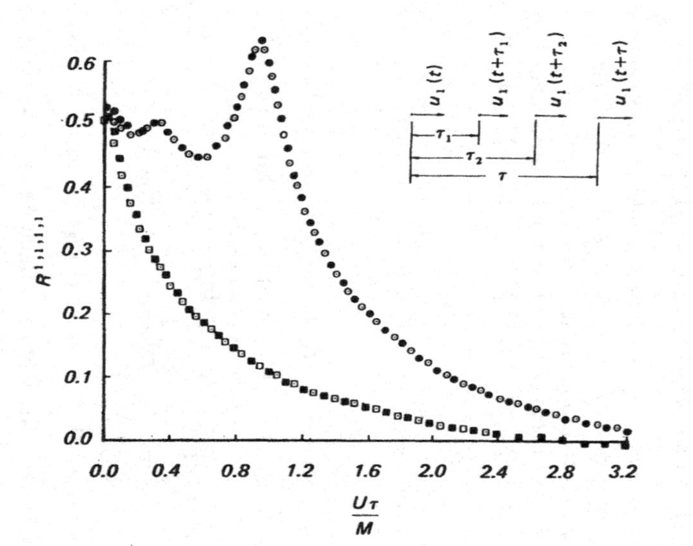


**Fig. 36c** Fourth-order correlations

$$R^{1,1,1,1}(\tau_1,\tau_2,\tau) = \overline{u_1(t)u_1(t+\tau_1)u_1(t+\tau_2)u_1(t+\tau)}/\overline{(u_1^2)}^2$$

for fixed $U\tau_1/M = 0.32388$ and $U\tau_2/M = 0.94464$. Measured correlations: ●, positive $\tau$; ■, negative $\tau$. Calculated from second-order correlations with the aid of Millionshchikov hypothesis: ⊙, positive $\tau$; ⊡, negative $\tau$.

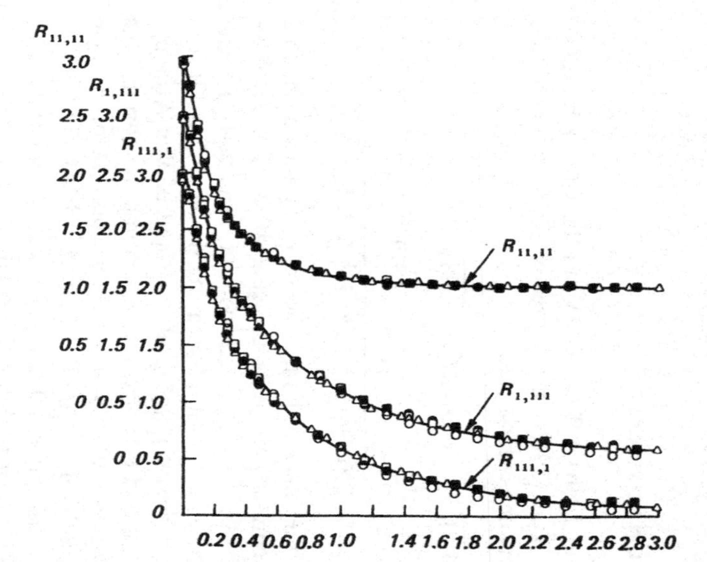


**Fig. 36d** Two-point normalized fourth-order velocity correlations

$$R_{11,11}(\tau) = \overline{u_1^2(t)u_1^2(t+\tau)}/\overline{(u_1^2)}^2, \quad R_{111,1}(\tau) = \overline{u_1^3(t)u_1(t+\tau)}/\overline{(u_1^2)}^2,$$

and

$$R_{1,111}(\tau) = \overline{u_1(t)u_1^3(t+\tau)}/\overline{(u_1^2)}^2$$

in grid turbulence according to Frenkiel and Klebanoff (1967). △, ●, □, ○, four examples of measured fourth-order correlations; ——, calculated from measured second-order correlations using Millionshchikov hypothesis.

Later Frenkiel and Klebanoff (1967), Van atta and Chen (1968) and Van Atta and Yeh (1969) have directly measured two- and four-point four-order time-correlations

$$B_{11,11}(\tau)=\overline{u_1^2(t)u_1^2(t+\tau)},\ B_{111,1}(\tau)=\overline{u_1^3(t)u_1(t+\tau)},$$
$$B_{1,111}(\tau)=\overline{u_1(t)u_1^3(t+\tau)},$$

and

$$B_{1,1,1,1}(\tau_1,\tau_2,\tau)=\overline{u_1(t)u_1(t+\tau_1)u_1(t+\tau_2)u_1(t+\tau)}$$

and also compared the measured values with the values calculated from second-order correlations using the general zero-fourth-order-cumulant hypothesis (18.1″). The results confirm in all cases the high accuracy of this hypothesis; some of them are shown in Figs. 36c, d. Moreover the data of Frenkiel and Klebanoff and of Van Atta and Chen show also that in grid turbulence all the even-order correlations up to eighth-order of the velocity fluctuations (but not of velocity derivatives) can be determined quite accurately from the second-order correlations assuming that the corresponding many-dimensional probability distribution is Gaussian.

## 18.2 Calculation of the Pressure Correlation and Spectra

As the first application of the Millionshchikov hypothesis let us consider the approximate calculation (based on the use of this hypothesis) of the correlation functions $B_{p',LL}(r)$ and $B_{p',NN}(r)$ describing the relation between pressure fluctuations and velocity fluctuations. Substituting Eq. (18.1′) in Eq. (14.43), and using Eq. (11.80), we obtain

$$Q_1(r)r_mr_n+Q_2(r)\delta_{mn}=\frac{\partial^2 B_{ij,mn}(r)}{\partial r_i\,\partial r_j}=2\frac{\partial B_{im}(r)}{\partial r_j}\frac{\partial B_{jn}(r)}{\partial r_i}. \tag{18.4}$$

Since $B_{im}(r)=-B'_{LL}(r)\frac{r_ir_m}{2r}+\left[B_{LL}(r)+\frac{r}{2}B'_{LL}(r)\right]\delta_{im}$ [in view of Eqs. (14.1) and (14.3)], it is readily verified that this is equivalent to

$$\begin{aligned}Q_1(r)&=\frac{6}{r^2}[B'_{LL}(r)]^2+\frac{1}{r}B'_{LL}(r)B''_{LL}(r),\\ Q_2(r)&=-3[B'_{LL}(r)]^2-rB'_{LL}(r)B''_{LL}(r).\end{aligned} \tag{18.5}$$

It is clear that the function $Q_1(r)r^2+3Q_2(r)=Q_3(r)$ is equal to $-\frac{1}{r^2}\frac{d}{dr}\{r^3[B'_{LL}(r)]^2\}$ in the zero-fourth-cumulant approximation [this is in good agreement with the requirements indicated by Eq. (14.44)]. Substituting this value of $Q_3(r)$ into Eq. (14.52) and integrating by parts, we obtain

$$P_3(r) = \overline{[p(\boldsymbol{x}) - \bar{p}]\, u^2(\boldsymbol{x}+\boldsymbol{r})} = -\rho \int_r^\infty [B'_{LL}(y)]^2\, y\, dy. \quad (18.6)$$

This formula and also the subsequent formulas (18.7) and (18.8) were first obtained by Limber (1951). It shows that if the Millionshchikov hypothesis is correct, then in isotropic turbulence the correlation coefficient between pressure fluctuations at a given point and the kinetic energy at some other point is always negative, and its maximum is reached when the two points coincide. Substituting $r = 0$ into Eq. (18.6), and using Eq. (18.3) and the equation (18.11) below, we can readily calculate the correlation coefficient between $p(\boldsymbol{x})$ and $u^2(\boldsymbol{x})$ corresponding to a given function $B_{LL}(r)$. In particular, in the case of the correlation function (15.53), the coefficient turns out to be approximately equal to $-0.2$, while in the case of the function $B_{LL}(r)$ which corresponds to the self-preserving spectrum of Fig. 31 with $R = \infty$ (which is similar in form to the spectra usually obtained behind grids in a wind tunnel), the result is close to $-0.1$.

Substituting Eq. (18.5) in Eq. (14.53) and Eq. (14.54), we can readily show that

$$\frac{1}{\rho} B_{p',LL}(r) = \frac{7}{15r^3} \int_0^r [B'_{LL}(y)]^2\, y^4\, dy + \int_r^\infty \left(\frac{4r^2}{5y} - \frac{y}{3}\right) [B'_{LL}(y)]^2\, dy, \quad (18.7)$$

$$\frac{1}{\rho} B_{p',NN}(r) = -\frac{7}{30r^3} \int_0^r [B'_{LL}(y)]^2\, y^4\, dy - \int_r^\infty \left(\frac{2r^2}{5y} + \frac{y}{3}\right) [B'_{LL}(y)]^2\, dy, \quad (18.8)$$

which enable us to find the functions $B_{p',LL}(r)$ and $B_{p',NN}(r)$ from known values of $B_{LL}(r)$. Figure 37, which is taken from Limber's paper, illustrates the general character of these functions in grid turbulence. It shows the results obtained when $B_{LL}(r)$ is taken to correspond to the spectrum of Fig. 31 with $R = \infty$.

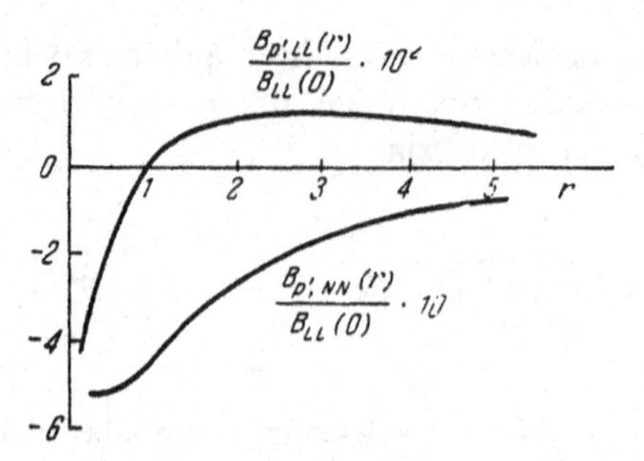


**Fig. 37** Examples of joint pressure-velocity correlation functions, calculated from the Millionshchikov hypothesis.

The function $B_{p'p'}(r)$ can be calculated in a similar way with the aid of the Millionshchikov hypothesis. Substituting Eq. (18.1′) into Eq. (14.31), replacing $B_{ij}(r)$ by Eq. (14.1), taking into account Eq. (14.3), and using some tedious algebra it is possible to reduce the function $Q(r)$ to the form

$$Q(r) = 2\frac{\partial^2 B_{ij}(r)}{\partial r_m \partial r_n}\frac{\partial^2 B_{mn}(r)}{\partial r_i \partial r_j} = 4\left\{\frac{3}{r^2}[B'_{LL}(r)]^2 + \right.$$
$$\left. + \frac{10}{r}B'_{LL}(r)B''_{LL}(r) + 2B'_{LL}(r)B'''_{LL}(r) + 2[B''_{LL}(r)]^2\right\}, \quad (18.9)$$

which obviously satisfies Eq. (14.32). Substituting Eq. (18.9) into Eq. (14.38) and integrating by parts repeatedly, we finally obtain

$$B_{p'p'}(r) = 2\rho^2 \int_r^\infty \left(y - \frac{r^2}{y}\right)[B'_{LL}(y)]^2\,dy. \quad (18.10)$$

It follows from this that

$$B_{p'p'}(0) = \overline{[p(x) - \bar{p}]^2} = 2\rho^2 \int_0^\infty y\,[B'_{LL}(y)]^2\,dy \quad (18.11)$$

and

$$\overline{(\text{grad}\ p)^2} = -3B''_{p'p'}(0) = 12\rho^2 \int_0^\infty [B'_{LL}(y)]^2 \frac{dy}{y}. \quad (18.12)$$

Equations (18.10)–(18.12) are due to Batchelor (1951) [very similar formulas were obtained somewhat earlier by Obukhov (1949b) and Obukhov and Yaglom (1951)]. However, the first application of the Millionshchikov hypothesis to the calculation of turbulent pressure fluctuations was made by Heisenberg (1948) who used Eq. (18.2) to calculate the spectrum $E_{pp}(k)=4\pi k^2 F_{pp}(k)$. Heisenberg paid particular attention to the quantity

$$\overline{(\text{grad } p)^2}=\int_0^\infty k^2 E_{pp}(k)\,dk,$$

for which he obtained the following expression:

$$\overline{(\text{grad } p)^2}=\rho^2\int_0^\infty\int_0^\infty E(k')E(k'')\,k'k''J\left(\frac{k'}{k''}\right)dk'\,dk'', \tag{18.13}$$

where $E(k)$ is the spectral energy density, and

$$J(s)=J\left(\frac{1}{s}\right)=-\frac{1}{8}(s^3+s^{-3})+\frac{11}{24}(s+s^{-1})+ \\ +\frac{1}{16}(s-s^{-1})^4\ln\frac{1+s}{|1-s|}. \tag{18.13'}$$

Later, Batchelor (1951) showed that the expression for $B_{p'p'}(0)=\overline{(p-\overline{p})^2}$, which is the analog of Eq. (18.13), is

$$\overline{(p-\overline{p})^2}=\rho^2\int_0^\infty\int_0^\infty E(k')E(k'')\,I\left(\frac{k'}{k''}\right)dk'\,dk'', \tag{18.14}$$

where

$$I(s)=I\left(\frac{1}{s}\right)=\frac{1}{2}(s^2+s^{-2})-\frac{1}{3}-\frac{1}{4}(s+s^{-1})(s-s^{-1})^2\ln\frac{1+s}{|1-s|}, \tag{18.14'}$$

and reported a more general expression for the function $E_{pp}(k)$ which is even more complicated.

Comparison of Eqs. (18.13), (18.14), with Eqs. (18.12) and (18.11) shows that the method of correlation functions has definite advantages as compared with spectral methods when one is concerned with calculations of the pressure fluctuations. However, in one respect, Eqs. (18.13) and (18.14) are more convenient than Eqs. (18.12) and (18.11): They indicate more clearly which ranges of velocity disturbances are responsible for the main contribution to $\overline{(\text{grad } p)^2}$ and $\overline{(p-\bar{p})^2}$. The functions $J(s)$ and $I(s)$ assume maximum values (of the order of unity) for $s$ of the order of unity, while for higher values of $s$ they are very small, since $J(s) \approx \frac{16}{15s}$ and $I(s) \approx \frac{16}{15s^2}$ for $s \gg 1$. Therefore, the only values of $k'$ and $k''$ that are important in Eqs. (18.13) and (18.14) are those which are of the same order of magnitude. Hence, it follows that the main contribution to $\overline{(p-\bar{p})^2}$ is due to disturbances with wave numbers $k$ for which $E(k)$ is a maximum, i.e., disturbances from the energy range. The value of $\overline{(\text{grad } p)^2}$ is in the first instance determined by disturbances for which $kE(k)$ is near to its maximum value, i.e., by values of $k$ which are intermediate between the energy and dissipation ranges. It is therefore expected that, for real turbulence behind a grid, the accuracy of Eq. (18.13) is lower than the accuracy of Eq. (18.14).[30]

During the final period of turbulence decay the function $B_{LL}(r)$ appears apparently to approach Eq. (15.53′), and therefore it follows from Eqs. (18.10)–(18.12) that during this period

$$B_{p'p'}(r) = \rho^2 (\overline{u^2})^2 e^{-r^2/\lambda^2} = \rho^2 [B_{LL}(r)]^2, \tag{18.15}$$

$$\overline{(p-\bar{p})^2} = B_{p'p'}(0) = \rho^2 (\overline{u^2})^2, \quad \overline{(\text{grad } p)^2} = -3B''_{p'p'}(0) = \frac{6\rho^2 (\overline{u^2})^2}{\lambda^2}. \tag{18.16}$$

The accuracy of these formulas should be very high because the probability distributions during the final period of decay are nearly Gaussian. If we determine the length $\lambda_p$ (which is analogous to $\lambda$) from the equation

$$\frac{1}{\rho^2} \overline{\left(\frac{\partial p}{\partial x}\right)^2} = \frac{(\overline{u^2})^2}{\lambda_p^2}, \tag{18.17}$$

[30] This conclusion can also be obtained from a comparison of the weight functions $y$ and $1/y$ in Eqs. (18.11) and (18.12).

then, in view of Eq. (18.16), we have $\frac{\lambda_p}{\lambda} \to \frac{1}{\sqrt{2}} \approx 0{,}7$ for $\mathrm{Re}_\lambda \to 0$. For moderate values of $\mathrm{Re}_\lambda$, which are typical for real flows in the wind tunnel, the values of $B_{p'p'}(r)$ can be estimated from any approximate formula for $B_{LL}(r)$ or directly from experimental data. Thus, for example Batchelor (1951, 1953) has suggested the use of the correlation function corresponding to the self-preserving spectrum of Fig. 31 for $B_{LL}(r)$ with $R = \infty$ in the calculation of $B_{p'p'}(r)$, while Hinze (1959) assumed that $B_{LL}(r) = B_{LL}(0) \exp(-r/L_1)$, where $L_1$ is the integral length scale of Eq. (12.50). On this basis, Batchelor found that $\overline{(p-\bar{p})^2} \approx 0.34\rho^2(\overline{u^2})^2$ and Hinze that $\overline{(p-\bar{p})^2} \approx 0.5\rho^2(\overline{u^2})^2$. The normalized function $B_{p'p'}(r)/B_{p'p'}(0)$ deduced from these two calculations was found to be approximately the same.

Examples of calculations of the function $B_{p'p'}(r)$ from experimental data on $B_{LL}(r)$ may be found in the papers of Uberoi (1953, 1954) who used his own data and the data of Liepmann et al. (1951) and in the paper by Meetz (1956b). Some of their results are shown in Figs. 38 and 39. In the case of Fig. 39 it must, however, be remembered that the values of $\overline{(\mathrm{grad}\ p)^2}$, and consequently of $\lambda_p/\lambda$ can be estimated only very approximately from measurements of the correlation functions since, in such measurements the values of $B'_{LL}(y)$ for small $y$, which enter with the maximum weight into the righ-hand side of Eq. (18.12), are only very approximately determined [and the formula (18.12) itself is not very accurate].

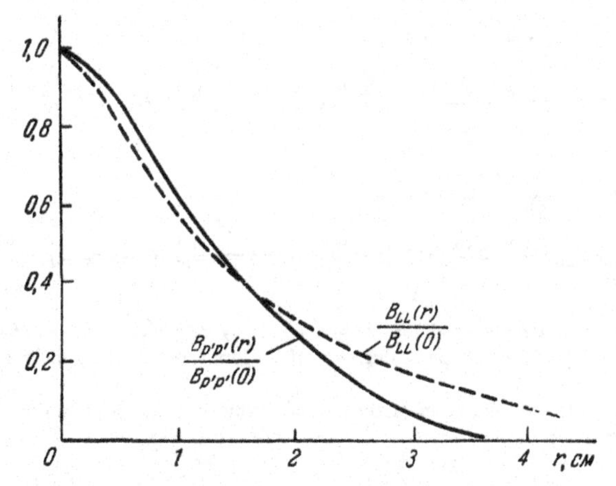


**Fig. 38** Normalized correlation functions for the pressure and velocity [Uberoi (1953, 1954)].

## 18.3 Estimation of the Turbulent Acceleration Fluctuations

Evaluations of the mean square pressure gradient are closely connected with evaluations of the correlation functions of the fluid-particle accelerations in an isotropic turbulence. The acceleration $A(x)$ is given by

$$A_l(x) = \frac{Du_l}{Dt} = \frac{\partial u_l}{\partial t} + u_\alpha \frac{\partial u_l}{\partial x_\alpha} = -\frac{1}{\rho}\frac{\partial p}{\partial x_l} + \nu\,\Delta u_l. \qquad (18.18)$$

Consequently, its correlation tensor is

$$\overline{A_l(x+r)\,A_j(x)} = \overline{A_l A'_j} = -\frac{1}{\rho^2}\frac{\partial^2 \overline{pp'}}{\partial r_l\,\partial r_j} + \frac{2\nu}{\rho}\left(\Delta\frac{\partial \overline{pu'_j}}{\partial r_l} - \Delta\frac{\partial \overline{p'u_l}}{\partial r_j}\right) + \nu^2\Delta^2\overline{u_l u'_j}. \qquad (18.19)$$

In the isotropic case, the terms in the middle of the right-hand side of this equation are equal to zero, and it is very easy to express the extreme terms in terms of $B_{p'p'}(r)$ and $B_{lj}(r)$. This gives

$$\overline{A_l(x+r)\,A_j(x)} = -\frac{1}{\rho^2}\frac{\partial^2 B_{p'p'}(r)}{\partial r_l\,\partial r_j} + \nu^2\Delta^2 B_{lj}(r). \qquad (18.20)$$

If we use the formula

$$\overline{A_l(x+r)\,A_j(x)} = \frac{B_{LL}^{(A)}(r) - B_{NN}^{(A)}(r)}{r^2}\, r_l r_j + B_{NN}^{(A)}(r)\,\delta_{lj}, \qquad (18.21)$$

which is the analog of Eq. (14.1), and if we replace $B_{lj}(r)$ by Eq. (14.1), where $B_{NN}(r)$ is given by Eq. (14.3), we obtain

$$B_{LL}^{(A)}(r) = -\frac{1}{\rho^2}B''_{p'p'}(r) + \nu^2\left[B_{LL}^{IV}(r) + \frac{8}{r}B'''_{LL}(r) + \frac{8}{r^2}B''_{LL}(r) - \frac{8}{r^3}B'_{LL}(r)\right], \qquad (18.22)$$

$$B_{NN}^{(A)}(r) = -\frac{1}{\rho^2 r}B'_{p'p'}(r) + \nu^2\left[\frac{1}{2}rB_{LL}^{V}(r) + 5B_{LL}^{IV}(r) + \frac{8}{r}B'''_{LL}(r) - \frac{4}{r^2}B''_{LL}(r) + \frac{4}{r^3}B'_{LL}(r)\right]. \qquad (18.23)$$

These formulas were reported almost simultaneously by Batchelor (1951) and Obukhov and Yaglom (1951) [see also the preceding note of Yaglom (1949a) where the case $r = 0$ was considered].

An important kinematic characteristic of turbulence is the mean square of the fluid-particle acceleration

$$\overline{A^2} = \overline{A_l A_l} = B_{LL}^{(A)}(0) + 2B_{NN}^{(A)}(0) = 3B_{LL}^{(A)}(0). \qquad (18.24)$$

According to Eqs. (18.22) and (18.23)

$$\overline{A^2} = -\frac{3}{\rho^2} B''_{p'p'}(0) + 35\nu^2 B^{IV}_{LL}(0). \tag{18.25}$$

The relative weight of the two terms on the right-hand side, which represent the pressure acceleration and the viscous acceleration respectively, is different for different Reynolds numbers. For very small Reynolds numbers, which are characteristic for the final period of decay, the ratio of the second term on the right of Eq. (18.25) to the first can readily be calculated from Eqs. (15.53′) and (18.15), and is equal to $35/2\,(\mathrm{Re}_\lambda)^2$, i.e., it is much greater than unity. Consequently, the acceleration of fluid particles is in this case produced mainly by viscous friction. The absolute magnitude of the acceleration can then be relatively high: The data of Fig. 14 can be used to show that in the corresponding experiments the quantity $(\overline{A^2})^{1/2}$ was of the order of 10 cm/sec² [for $(\overline{u^2})^{1/2}$ of the order of 1 cm/sec]. For very high Reynolds numbers, the second term on the right of Eq. (18.25) will amount to only a few percent of the first, so that the acceleration will be largely governed by the pressure gradients (see also Sect. 22.2). In the case of moderate Reynolds numbers, the relative magnitude of the second term can be roughly estimated with the aid of the Millionshchikov hypothesis and the data of Batchelor and Townsend (1947) according to which $\lambda^4 B^{IV}_{LL}(0)/\overline{u^2} \approx 4.3 + 0.2\,\mathrm{Re}_\lambda$ in the turbulence behind a grid. This can then be used to show that, for example, for flows with $\mathrm{Re}_\lambda = 60$, 90, and 150, for which Uberoi (1954) estimated the values of $\lambda_p/\lambda$, the second term on the right of Eq. (18.25) is smaller by a factor of 30 to 40 than the first, i.e., the acceleration is also largely determined by pressure forces.

The quantity $\overline{A^2}$ can be roughly estimated from the data on diffusion in wind tunnels. Consider a linear source of a passive admixture or of heat (the latter is usually employed in a real experiment) in the flow behind the grid. We can then use considerations similar to those employed in the derivation of the classical Taylor formula (9.31) (see Vol. 1) to show that the following equation is valid for the mean square $\overline{Y^2(t)}$ of lateral (i.e., perpendicular to the source and to the mean flow) displacements of diffusing particles at time $t$:

$$\frac{d}{dt}\overline{Y^2(t)} = 2\int_0^t B^{(L)}_{22}(t, \tau)\,d\tau, \tag{18.26}$$

where $B^{(L)}_{22}(t, \tau) = \overline{v(t)\,v(t+\tau)}$ is the lateral Lagragian velocity correlation function. We shall use the equation

$$\frac{1}{2\lambda_\xi^2} = \lim_{\tau\to 0}\left[1 - \frac{B^{(L)}_{22}(t, \tau)}{[\overline{v^2(t)}\,\overline{v^2(t+\tau)}]^{1/2}}\right]\tau^{-2} \tag{18.27}$$

to determine the differential time scale $\lambda_\xi = \lambda_\xi(t)$ which plays the same role in the case of the function $B^{(L)}_{22}(t, \tau)$ as the Taylor length scale $\lambda$ in the case of the function $B_{LL}(r)$. If we expand the functions $B^{(L)}_{22}(t, \tau) = \overline{v(t)\,v(t+\tau)}$ and $\overline{v^2(t+\tau)}$ in Taylor series in powers of $\tau$, we can readily show that the time scale $\lambda_\xi$ is related to the mean square of turbulent acceleration, $\overline{A_2^2} = \overline{\left(\frac{Dv}{Dt}\right)^2}$, along the $y$ axis by

$$\frac{1}{\lambda_\xi^2} = \frac{1}{\overline{v^2}}\left[\overline{\left(\frac{Dv}{Dt}\right)^2} - \left(\frac{d(\overline{v^2})^{1/2}}{dt}\right)^2\right] \tag{18.28}$$

[see for example Uberoi and Corrsin (1953)]. From Eqs. (18.26) and (18.27) it follows that, by measuring the mean square half-width $\overline{Y^2(t)}$ of the turbulent wake behind the source, we can determine the time scale $\lambda_\xi$ by differentiation. This method of finding $\lambda_\xi$ was used by Simmons (see Taylor, 1935), Collis (1948), and especially by Uberoi and Corrsin (1953). If we recall that $\frac{d\,(\overline{v^2})^{1/2}}{dt} = -\frac{5\nu\,(\overline{v^2})^{1/2}}{\lambda^2}$, according to Eq. (15.2), then we can find the mean square acceleration $\overline{\left(\frac{Dv}{Dt}\right)^2} = \overline{A_2^2} = \frac{1}{3}\,\overline{A^2}$ from $\lambda_\xi$ and the known values of $\nu$, $\lambda$, and $\overline{v^2} = \overline{u^2}$. If we then use the estimate given by Batchelor and Townsend (1947) for $B_{LL}^{\mathrm{IV}}(0)$ (see above), we can estimate the ratio $35\rho^2\nu^2 B_{LL}^{\mathrm{IV}}(0)/\overline{(\mathrm{grad}\,p)^2}$ of the two terms on the right-hand side of Eq. (18.25), and also the mean square of the pressure gradient, i.e., the length scale $\lambda_p$ in Eq. (18.17). The results of such calculations carried out by Batchelor (1951) using the data of Simmons and Collis, and by Uberoi (1953, 1954) using Uberoi and Corrsin's data are shown in Fig. 39 together with the values of $\lambda_p/\lambda$ obtained by the method described in the last section [$\lambda_p/\lambda = \frac{1}{\sqrt{2}} \approx 0.7$ corresponds to the correlation function (15.53′), i.e., the case of very small Re].

The Millionshchikov hypothesis enables us also to calculate the correlation functions for the local (Eulerian) accelerations $a_l = \frac{\partial u_l}{\partial t} = -u_j\frac{\partial u_l}{\partial x_j} - \frac{1}{\rho}\frac{\partial p}{\partial x_l} + \nu\,\Delta u_l$ or the inertial accelerations $\alpha_l = u_j\frac{\partial u_l}{\partial x_j}$, and also the cross-correlation functions for these accelerations (see Lin, 1953a). In particular, for the inertial acceleration $\alpha_l$ it follows from Eq. (18.1′) that[31]

$$\overline{\alpha^2} = \overline{\alpha_l\alpha_l} = -\frac{\partial^2}{\partial r_j\,\partial r_k}\left[B_{ll}(r)\,B_{jk}(r) + B_{lj}(r)\,B_{lk}(r)\right]\Big|_{r=0} = \frac{15\,(\overline{u^2})^2}{\lambda^2}. \quad (18.29)$$

Since, on the other hand,

$$\overline{A^2} \approx \frac{1}{\rho^2}\,\overline{|\,\mathrm{grad}\,p\,|^2} \approx \frac{(\overline{u^2})^2}{\lambda^2}\cdot 3\left(\frac{\lambda}{\lambda_p}\right)^2, \quad (18.30)$$

where as a rough approximation we can use the values given in Fig. 39 to estimate $\lambda/\lambda_p$, it is clear from Eq. (18.29) that the inertial acceleration in grid turbulence is much greater than the total acceleration of the fluid particles.

The mean square of the local acceleration $a$ can be calculated in an analogous fashion; for this quantity Lin (1953a) obtained the equation

[31] Uberoi and Corrsin (1953) have pointed out that it is possible to obtain a strict upper bound for the quantity $\overline{\alpha^2}$, which is close in form to Eq. (18.29) (but is much more approximate). This can be deduced from the general Schwarz inequality [according to which the modulus of the correlation coefficient for two random variables $x$ and $y$ cannot exceed unity so that $|\overline{xy}| \leqslant (\overline{x^2}\cdot\overline{y^2})^{1/2}$] and the exact Eqs. (15.5), and does not require the use of the inexact Millionshchikov hypothesis. In fact

$$\overline{\alpha_l\alpha_l} = \overline{u_j\frac{\partial u_l}{\partial x_j}\,u_k\frac{\partial u_l}{\partial x_k}} \leqslant \left[\overline{u_j^2}\,\overline{u_k^2}\,\overline{\left(\frac{\partial u_l}{\partial x_j}\right)^2}\,\overline{\left(\frac{\partial u_l}{\partial x_k}\right)^2}\right]^{1/2} = 3\,(9+4\sqrt{2})\,\frac{(\overline{u^2})^2}{\lambda^2} \approx 44\,\frac{(\overline{u^2})^2}{\lambda^2}.$$

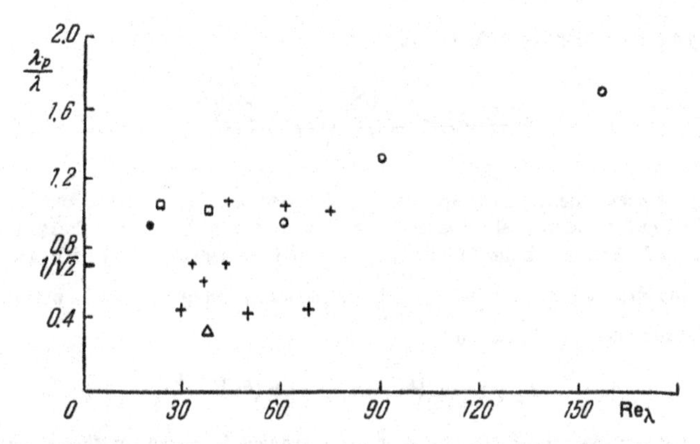


**Fig. 39** Values of $\lambda_p/\lambda$ for different $Re_\lambda$ in grid turbulence, according to the diffusion experiments of Simmons (△), Collis (□), and Uberoi and Corrsin (+), and the calculations of Uberoi (○) and Meetz (•).

$$\overline{a^2} = \overline{a_i a_i} = 35\nu^2 B_{LL}^{\mathrm{IV}}(0) + 140\nu B_{LL,\,L}'''(0) + \overline{\alpha^2} - \frac{1}{\rho^2}\overline{(\operatorname{grad} p)^2}. \tag{18.31}$$

According to existing data on $B_{LL}^{\mathrm{IV}}(0)$ and $B_{LL,\,L}'''(0)$ (see for example Batchelor and Townsend, 1947) and the calculations given above, the dominant term on the right-hand side of Eq. (18.31) in the case of grid turbulence is $\overline{\alpha^2}$, so that the local acceleration $a = \frac{\partial u}{\partial t}$ is also usually much greater than $A = \frac{Du}{Dt}$. Therefore, according to Lin's calculations the substantive acceleration $Du/Dt$ in grid turbulence is a small difference between the two large numbers describing the local and the inertial accelerations. This result can hardly be connected with the approximate character of the Millionshchikov hypothesis. It is probably explained by the fact that the velocity $u(x)$ at a fixed point in isotropic turbulent flow is determined mainly by the large-scale disturbances from the energy range, which are transferred past the point and thus lead to rapid changes in the Eulerian velocity.

Munch and Wheelon (1958) have calculated the correlation tensor $B_{ij}^{(a)}(r) = \overline{\frac{\partial u_i}{\partial t}\frac{\partial u_j'}{\partial t}}$ for the local acceleration field in the case of isotropic stationary turbulence whose statistical characteristics are time-independent. They have also used the Millionshchikov zero fourth cumulant hypothesis and assumed that, in partial agreement with Eq. (17.40), the energy spectrum is given by

$$E(k) = \begin{cases} E_0\left(\frac{k_0}{k}\right)^{5/3}\left[1+\left(\frac{k}{k_1}\right)^4\right]^{-4/3} & \text{for } k \geqslant k_0, \\ 0 & \text{for } k < k_0, \end{cases} \tag{18.32}$$

where $k_0$ is the minimum wave number determined by the length scale $l_0 = 2\pi/k_0$ of the largest disturbances, $E_0 = V_0^2/k_0$, and $V_0$ is the velocity scale of these large-scale disturbances [third-order moments can be eliminated from the stationary expression for $B_{ij}^{(a)}(r)$, so that they do not require any special assumptions]. The tensor $B_{ij}^{(a)}(r)$ is symmetric and solenoidal in both indices. Therefore, its Fourier transform $F_{ij}^{(a)}(\boldsymbol{k})$ can be written in a form analogous to Eq. (12.73):

$$F_{ij}^{(a)}(\boldsymbol{k}) = \frac{E_a(k)}{4\pi k^2}\left(\delta_{ij} - \frac{k_i k_j}{k^2}\right). \qquad (18.33)$$

Munch and Wheelon deduced asymptotic formulas describing the subrange behavior of the function $E_a(k)$ in the inertial subrange $k_0 \ll k \ll k_1$ and in the extreme short-wave range $k \gg k_1$, and determined the function $E_a(k)$ for intermediate values of $k$ and several values of the Reynolds number $\mathrm{Re} = \left(\frac{k_1}{k_0}\right)^{4/3}$ using numerical integration. In particular, for the inertial subrange they found that

$$E_a(k) = E(k)\, E_0 k_0 k^2 = E(k)\, V_0^2 k^2. \qquad (18.34$$

The fact that the spectrum $E_a(k)$ for $k \ll k_0$ is strongly dependent on the velocity $V_0$ of the largest-scale disturbances shows again that the statistical properties of the local accelerations are largely determined by the convective action of the largest-scale velocity fluctuations.

Since, in view of Eqs. (11.53) and (12.40)

$$\frac{E(k)}{4\pi k^2}\, dk = \overline{|dZ(\boldsymbol{k})|^2}, \quad \frac{E_a(k)}{4\pi k^2}\, dk = \overline{\left|\frac{\partial(dZ(\boldsymbol{k}))}{\partial t}\right|^2},$$

we can interpret Eq. (18.34) as an indication of the fact that, in stationary isotropic turbulence, disturbances with wave number $k$ have the characteristic period (relaxation time)

$$\tau_k = \left[\frac{E(k)}{E_a(k)}\right]^{1/2} = \frac{1}{V_0 k}. \qquad (18.35)$$

In this form the result follows from dimensional considerations alone (see Heisenberg, 1948a).

# 19. DYNAMIC EQUATIONS FOR THE HIGHER-ORDER MOMENTS AND THE CLOSURE PROBLEM

## 19.1 Equations for the Third-Order Moments of Flow Variables

Since dynamic equations for the usual correlation functions, i.e., the second-order moments, of various variables are nonlinear, they contain new unknown functions, namely, the third-order moments. We can use the dynamic equations to calculate the time derivatives of these functions. The resulting equations will, however, contain fourth-order moments. By setting up the expressions for the time derivatives of fourth- fifth-, etc., order moments, we obtain further new equations, but the number of unknown functions will increase more rapidly than the number of equations, so that the set of equations will never be closed (see Sect. 6.1, Vol. 1). Therefore, to determine even a single correlation function, we must, strictly speaking, consider an infinite set of coupled equations for the correlation functions of all orders (or the equivalent equation for the characteristic functional, which will be discussed in Chapt. 10).

There are no general methods at present for solving infinite sets of coupled partial differential equations, and therefore the determination of the exact solutions of the equations for the moments of all possible orders is still a hopeless task. However, the equations for the higher-order moments can be used to determine approximately the statistical characteristics of turbulence. This requires the introduction of some additional hypotheses which enable us to close the truncated set of the first few such equations. This approach is a natural generalization of the closure methods discussed above for the single equation relating second- and third-order moments. The present section is devoted to the closure methods for higher order moment equations.[32]

Let us begin with the derivation of the equation for the third-order velocity moments. We must first decide which particular moments we shall consider, i.e., one-point, two-point or three-point. The von Kármán-Howarth equation (14.9) contains the two-point third-order moment $B_{LL,L}(r, t)$, so that it would appear that, in addition to Eq. (14.9), it is natural to consider the equation for the two-point tensor $B_{ij,l}(\boldsymbol{r}, t)$ which is simply expressed in terms of $B_{LL,L}(r, t)$. Let us therefore apply the general formula for the time derivative of a moment [Eq. (6.2) of Vol. 1] to the moment

$$B_{ij,l}(\boldsymbol{r}, t) = \overline{u_i(\boldsymbol{x}, t)\, u_j(\boldsymbol{x}, t)\, u_l(\boldsymbol{x}', t)}.$$

However, the right-hand side of the resulting equation will then contain the two-point third- and fourth-order moments, as well as the joint pressure-velocity third-order moments of the form

[32] All the closing methods of the higher-order moment equations are based on some hypothetical analytical relations between moments of different orders; therefore the methods are sometimes called the *analytical theories of turbulence*. A comparatively detailed presentation of analytical theories can be found in the forthcoming book by Orszag (1975); many related considerations are given in the survey papers of Saffman (1968), Orszag (1969) and Kraichnan (1970a) coutaining extensive bibliographies. We shall not pretend to duplicate (or restate) all the contents of these works; we shall limit ourselves to the consideration, first of the few comparatively old and now quite settled theories, and shall discuss only very briefly some of the more recent and still developing approaches.

$$\overline{\frac{\partial p(\boldsymbol{x})}{\partial x_l} u_j(\boldsymbol{x}) u_l(\boldsymbol{x}')},$$

and also unknown velocity derivative moments of the form $\overline{\frac{\partial u_l(\boldsymbol{x})}{\partial x_m} \frac{\partial u_j(\boldsymbol{x})}{\partial x_m}} u_l(\boldsymbol{x}')$. Moments containing pressure can, of course, be expressed in terms of velocity moments of order greater by one, using the formula $\Delta p = -\rho \frac{\partial^2 u_l u_j}{\partial x_l \partial x_j}$. However, in that case, we have to deal with the three-point as well as the two-point velocity moments. It is, therefore, best to start with the equation for the three-point third-order velocity moments. All the space derivatives in this equation can then be taken from under the averaging sign, and the number of moments will not be increased as a result of the presence of moments containing the derivatives of the main variables. This is, in fact, the method we shall adopt.

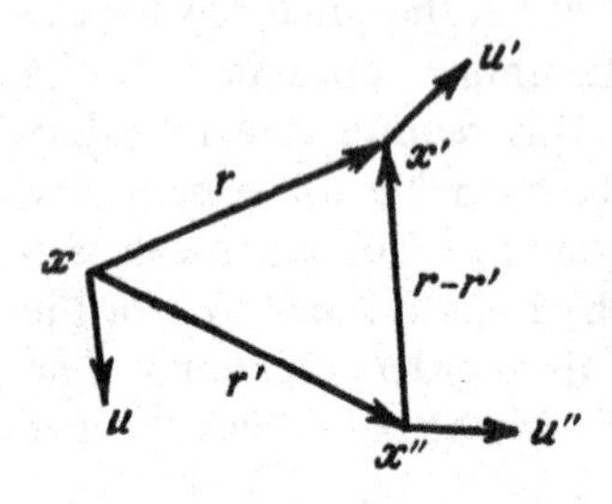


Fig. 40 The three-point moments of the velocity field.

Thus, suppose that $u_l = u_l(\boldsymbol{x}, t)$, $u'_j = u_j(\boldsymbol{x}', t)$, $u''_l = u_l(\boldsymbol{x}'', t)$ (see Fig. 40), and let us write down the Navier-Stokes equations for the velocity components $u_l$, $u'_j$, and $u''_l$ and multiply the first of these by $u'_j u''_l$, the second by $u_l u''_l$, and the third by $u_l u'_j$, and then add the resulting equations. We thus obtain

$$\begin{aligned}\frac{\partial}{\partial t} \overline{u_l u'_j u''_l} = &- \frac{\partial}{\partial x_\alpha} \overline{u_l u_\alpha u'_j u''_l} - \frac{\partial}{\partial x'_\alpha} \overline{u_l u'_j u'_\alpha u''_l} - \\ &- \frac{\partial}{\partial x''_\alpha} \overline{u_l u'_j u''_l u''_\alpha} - \frac{1}{\rho}\left(\frac{\partial \overline{p u'_j u''_l}}{\partial x_l} + \frac{\partial \overline{p' u_l u''_l}}{\partial x'_j} + \frac{\partial \overline{p'' u_l u'_j}}{\partial x''_l}\right) + \\ &+ \nu\left(\frac{\partial^2}{\partial x_\alpha \partial x_\alpha} + \frac{\partial^2}{\partial x'_\alpha \partial x'_\alpha} + \frac{\partial^2}{\partial x''_\alpha \partial x''_\alpha}\right) \overline{u_l u'_j u''_l}. \qquad (19.1)\end{aligned}$$

In the case of homogeneous turbulence, all the functions in this equation will depend on time and on the vectors $\boldsymbol{r} = \boldsymbol{x}' - \boldsymbol{x}$ and $\boldsymbol{r}' = \boldsymbol{x}'' - \boldsymbol{x}$, so that

$$\frac{\partial}{\partial x_m} = -\frac{\partial}{\partial r_m} - \frac{\partial}{\partial r'_m}, \quad \frac{\partial}{\partial x'_m} = \frac{\partial}{\partial r_m}, \quad \frac{\partial}{\partial x''_m} = \frac{\partial}{\partial r'_m}. \qquad (19.2)$$

However, $\overline{p'u_l u''_l} = B_{pll}(-\boldsymbol{r}, \boldsymbol{r}'-\boldsymbol{r})$, $\overline{u_i u'_j u'_\alpha u''_l} = B_{j\alpha, l, l}(-\boldsymbol{r}, \boldsymbol{r}'-\boldsymbol{r})$ and $\overline{p'' u_i u'_j} = B_{plj}(-\boldsymbol{r}', \boldsymbol{r}-\boldsymbol{r}')$, $\overline{u_i u'_j u''_l u''_\alpha} = B_{l\alpha, l, j}(-\boldsymbol{r}', \boldsymbol{r}-\boldsymbol{r}')$, so that Eq. (19.1) can be rewritten in the form

$$\frac{\partial}{\partial t} B_{ijl}(\boldsymbol{r}, \boldsymbol{r}') = \left(\frac{\partial}{\partial r_\alpha} + \frac{\partial}{\partial r'_\alpha}\right) B_{l\alpha, j, l}(\boldsymbol{r}, \boldsymbol{r}') - \\ - \frac{\partial}{\partial r_\alpha} B_{j\alpha, l, l}(-\boldsymbol{r}, \boldsymbol{r}'-\boldsymbol{r}) - \frac{\partial}{\partial r'_\alpha} B_{l\alpha, l, j}(-\boldsymbol{r}', \boldsymbol{r}-\boldsymbol{r}') + \\ + \frac{1}{\rho}\left[\left(\frac{\partial}{\partial r_l} + \frac{\partial}{\partial r'_l}\right) B_{pjl}(\boldsymbol{r}, \boldsymbol{r}') - \frac{\partial}{\partial r_j} B_{pll}(-\boldsymbol{r}, \boldsymbol{r}'-\boldsymbol{r}) - \\ - \frac{\partial}{\partial r'_l} B_{plj}(-\boldsymbol{r}', \boldsymbol{r}-\boldsymbol{r}')\right] + 2\nu\left[\frac{\partial^2}{\partial r_\alpha \partial r_\alpha} + \frac{\partial^2}{\partial r'_\alpha \partial r'_\alpha} + \frac{\partial^2}{\partial r_\alpha \partial r'_\alpha}\right] B_{ijl}(\boldsymbol{r}, \boldsymbol{r}') \tag{19.3}$$

(for the sake of simplicity, we are not indicating the dependence of the moments on $t$).

Let us now apply the Fourier transformation to Eq. (19.3). We must remember that the moments of the form $B_{l\alpha, j, l}(\boldsymbol{r}, \boldsymbol{r}')$ do not tend to zero as $|\boldsymbol{r}| \to \infty$ and $|\boldsymbol{r}'| \to \infty$, but $|\boldsymbol{r}-\boldsymbol{r}'|$ remains bounded. They do not, therefore, have a six-dimensional Fourier transform (see Sect. 11.2). Only the corresponding cumulants $S_{l\alpha, j, l}(\boldsymbol{r}, \boldsymbol{r}')$ or the functions $B^{(0)}_{l\alpha, j, l}(\boldsymbol{r}, \boldsymbol{r}') = B_{l\alpha, j, l}(\boldsymbol{r}, \boldsymbol{r}') - B_{l\alpha}(0) B_{jl}(\boldsymbol{r}'-\boldsymbol{r})$ will have Fourier transforms. Since, however,

$$\left(\frac{\partial}{\partial r_\alpha} + \frac{\partial}{\partial r'_\alpha}\right) B_{jl}(\boldsymbol{r}'-\boldsymbol{r}) = \frac{\partial}{\partial r_\alpha} B_{il}(\boldsymbol{r}') = \frac{\partial}{\partial r'_\alpha} B_{lj}(\boldsymbol{r}) = 0,$$

all the three-point fourth-order moments of the right-hand side of Eq. (19.3) can be replaced by the corresponding functions $B^{(0)}_{rs, t, u}$. Therefore, all the terms in Eq. (19.3) will have six-dimensional Fourier transforms, and the equation itself can be written in the form

$$\frac{\partial}{\partial t} F_{ijl}(\boldsymbol{k}, \boldsymbol{k}') = i\left[(k_\alpha + k'_\alpha) F_{l\alpha, j, l}(\boldsymbol{k}, \boldsymbol{k}') - k_\alpha F_{j\alpha, l, l}(-\boldsymbol{k}-\boldsymbol{k}', \boldsymbol{k}') - \\ - k'_\alpha F_{l\alpha, l, j}(-\boldsymbol{k}-\boldsymbol{k}', \boldsymbol{k})\right] + \\ + \frac{i}{\rho}\left[(k_l + k'_l) F_{pjl}(\boldsymbol{k}, \boldsymbol{k}') - k_j F_{pll}(-\boldsymbol{k}-\boldsymbol{k}', \boldsymbol{k}') - \\ - k_l F_{plj}(-\boldsymbol{k}-\boldsymbol{k}', \boldsymbol{k})\right] - 2\nu\left(k^2 + k'^2 + \boldsymbol{k}\boldsymbol{k}'\right) F_{ijl}(\boldsymbol{k}, \boldsymbol{k}'), \tag{19.4}$$

where $F_{ijl}$, $F_{l\alpha, j, l}$ and $F_{pjl}$ are the Fourier transforms of the functions $B_{ijl}$, $B^{(0)}_{i\alpha,\,j,\,l}$, and $B_{p'jl}$. If we use the symbol $\Gamma_{ijl}(\boldsymbol{k},\boldsymbol{k}')$ to denote the set of the first three terms on the right-hand side of Eq. (19.4), which contain the fourth-order spectral functions, then $\Gamma_{ijl}(\boldsymbol{k},\boldsymbol{k}')$ will satisfy the relation

$$\int \Gamma_{ijl}(\boldsymbol{k},\boldsymbol{k}')\,d\boldsymbol{k}\,d\boldsymbol{k}'=0, \tag{19.5}$$

since its left-hand side is equal to $\frac{\partial}{\partial x_\alpha}\overline{u_i(\boldsymbol{x})\,u_j(\boldsymbol{x})\,u_l(\boldsymbol{x})\,u_\alpha(\boldsymbol{x})}$, which is zero because of homogeneity. This equation is similar to Eq. (14.6). The nonlinear terms in the dynamic equations thus lead only to a change in the spectral composition of the functions $B_{ijl}(\boldsymbol{r},\boldsymbol{r}')$, but produce no change with time in the functions $B_{ijl}(0,0)=\overline{u_i(\boldsymbol{x})\,u_j(\boldsymbol{x})\,u_l(\boldsymbol{x})}$.

Equation (1.9) of volume 1 which relates $\Delta p$ with the velocity derivatives, can be used to express the cross spectral functions $F_{pjl}(\boldsymbol{k},\boldsymbol{k}')$ of the velocity and pressure fields in terms of the spectral functions $F_{lm,\,j,\,l}(\boldsymbol{k},\boldsymbol{k}')$. In fact, if we multiply both sides of Eq. (1.9) by $u'_j u''_l$ and take the average of the result, we obtain

$$\frac{1}{\rho}\frac{\partial^2\overline{pu'_ju''_l}}{\partial x_\alpha\,\partial x_\alpha}=-\frac{\partial^2\overline{u_\alpha u_\beta u'_j u''_l}}{\partial x_\alpha\,\partial x_\beta}=-\frac{\partial^2}{\partial x_\alpha\,\partial x_\beta}\left[\overline{u_\alpha u_\beta u'_j u''_l}-\overline{u_\alpha u_\beta}\cdot\overline{u'_j u''_l}\right]. \tag{19.6}$$

Hence it follows that

$$\frac{1}{\rho}F_{pjl}(\boldsymbol{k},\boldsymbol{k}')=-\frac{(k_\alpha+k'_\alpha)(k_\beta+k'_\beta)}{(\boldsymbol{k}+\boldsymbol{k}')^2}F_{\alpha\beta,\,j,\,l}(\boldsymbol{k},\boldsymbol{k}'). \tag{19.7}$$

Following Proudman and Reid (1954) and Tatsumi (1957a) we can write down the result of the substitution of Eq. (19.7) in Eq. (19.4) in the form

$$\frac{\partial}{\partial t}F_{ijl}(\boldsymbol{k},\boldsymbol{k}')=-i\sum_{\text{cyc}}k''_\alpha\Delta_{i\beta}(\boldsymbol{k}'')\,F_{\beta\alpha,\,j,\,l}(\boldsymbol{k},\boldsymbol{k}')-$$
$$-\nu\left(k^2+k'^2+k''^2\right)F_{ijl}(\boldsymbol{k},\boldsymbol{k}'), \tag{19.8}$$

where as usual

$$\boldsymbol{k}''=-\boldsymbol{k}-\boldsymbol{k}'\ (\text{т. е. } \boldsymbol{k}+\boldsymbol{k}'+\boldsymbol{k}''=0),\ \Delta_{ij}(\boldsymbol{k})=\delta_{ij}-k_ik_j/k^2. \tag{19.9}$$

and $\sum\limits_{cyc}$ represents the sum of the term written down and two further terms obtained from it after the simultaneous application of the same cyclic permutation of the subscripts $(i, j, l)$ and of the vectors $(\boldsymbol{k}, \boldsymbol{k}', \boldsymbol{k}'')$. In the isotropic case, the tensor $F_{ijl}(\boldsymbol{k}, \boldsymbol{k}')$ can be expressed in terms of the two scalar functions $\Phi(k, k', k'')$ and $\Psi(k, k', k'')$ with the aid of Eq. (12.166). The tensor Eq. (19.8) can thus be reduced to two scalar equations containing $\frac{\partial}{\partial t}\Phi(k, k', k'')$ and $\frac{\partial}{\partial t}\Psi(k, k', k'')$ in the left-hand side. Such equations will not, however, be used below.

From the tensor Eq. (19.8) we can derive a series of scalar equations by contracting the indices $i$, $j$, and $l$ among themselves, or with the components of some other vectors constructed from $\boldsymbol{k}$ and $\boldsymbol{k}'$. Let us write down the most important equation, for the scalar function antisymmetric in $k$ and $k'$,

$$\Gamma(k, k', \mu) = ik_l F_{ljj}(\boldsymbol{k}, \boldsymbol{k}'), \quad \mu = \boldsymbol{k}\boldsymbol{k}'/kk'. \tag{19.10}$$

By Eq. (19.8), this equation is of the form

$$\begin{aligned}\frac{\partial\Gamma(k, k', \mu)}{\partial t} &= k_l k''_\alpha \Delta_{l\beta}(\boldsymbol{k}'') F_{\beta\alpha,\, j,\, j}(\boldsymbol{k}, \boldsymbol{k}') + \\ &+ k_l k_\alpha \Delta_{j\beta}(\boldsymbol{k}) F_{\beta\alpha,\, j,\, l}(\boldsymbol{k}', \boldsymbol{k}'') + k_l k'_\alpha \Delta_{j\beta}(\boldsymbol{k}') F_{\beta\alpha,\, l,\, j}(\boldsymbol{k}'', \boldsymbol{k}) - \\ &- \nu(k^2 + k'^2 + k''^2)\Gamma(k, k', \mu).\end{aligned} \tag{19.11}$$

The importance of this equation is connected with the fact that the function $\Gamma(k) = ik_l F_{lj,\, j}(\boldsymbol{k})$ of Eq. (14.14) can be simply expressed in terms of $\Gamma(k, k', \mu)$. In fact, according to Eq. (11.16)

$$F_{lj,\, j}(\boldsymbol{k}) = \int F_{ljj}(\boldsymbol{k}, \boldsymbol{k}')\, d\boldsymbol{k}', \tag{19.12}$$

and hence

$$\Gamma(k) = \int \Gamma(\boldsymbol{k}, \boldsymbol{k}')\, d\boldsymbol{k}' = 2\pi \int_0^\infty \int_{-1}^1 \Gamma(k, k', \mu)\, k'^2\, dk'\, d\mu. \tag{19.13}$$

Equations (19.13) and (19.11) augment the main spectral equation

(14.14) or (14.15) (which is equivalent to the von Kármán-Howarth equation). If the functions $F_{mn,j,l}(\boldsymbol{k}, \boldsymbol{k}')$ could be expressed in terms of spectral functions of lower (second or third) orders, then Eqs. (14.14), (19.13), and (19.11) would form a closed set of equations for $F(k)$, i.e., $E(k)$, and $\Gamma(k)$.

Similarly, the Corrsin equation (14.59) for the correlation function $B_{\vartheta\vartheta}(r)$, or the corresponding spectral equation (14.62), can be augmented by the equation for the three-point joint moment

$$B_{j\vartheta\vartheta}(\boldsymbol{r}, \boldsymbol{r}') = \overline{u_j(\boldsymbol{x})\,\vartheta(\boldsymbol{x}+\boldsymbol{r})\,\vartheta(\boldsymbol{x}+\boldsymbol{r}')} = \overline{u_j\vartheta'\vartheta''}$$

of the velocity and temperature fields, or its Fourier transform $F_{j\vartheta\vartheta}(\boldsymbol{k}, \boldsymbol{k}')$. The equation for $B_{j\vartheta\vartheta}(\boldsymbol{r}, \boldsymbol{r}')$ in the case of homogeneous turbulence is of the form

$$\begin{aligned}
\frac{\partial B_{j\vartheta\vartheta}(\boldsymbol{r}, \boldsymbol{r}')}{\partial t} &= \left(\frac{\partial}{\partial r_\alpha} + \frac{\partial}{\partial r'_\alpha}\right) B_{j\alpha,\vartheta,\vartheta}(\boldsymbol{r}, \boldsymbol{r}') - \\
&- \frac{\partial}{\partial r_\alpha} B_{\alpha\vartheta,j,\vartheta}(-\boldsymbol{r}, \boldsymbol{r}'-\boldsymbol{r}) - \frac{\partial}{\partial r'_\alpha} B_{\alpha\vartheta,j,\vartheta}(-\boldsymbol{r}', \boldsymbol{r}-\boldsymbol{r}') + \\
&+ \frac{1}{\rho}\left(\frac{\partial}{\partial r_j} + \frac{\partial}{\partial r'_j}\right) B_{p\vartheta\vartheta}(\boldsymbol{r}, \boldsymbol{r}') + \\
&+ \left[(\nu+\chi)\left(\frac{\partial^2}{\partial r_\alpha \partial r_\alpha} + \frac{\partial^2}{\partial r'_\alpha \partial r'_\alpha}\right) + 2\nu \frac{\partial^2}{\partial r_\alpha \partial r'_\alpha}\right] B_{j\vartheta\vartheta}(\boldsymbol{r}, \boldsymbol{r}'). \qquad (19.14)
\end{aligned}$$

If we apply the six-dimensional Fourier transformation to all the terms of the last equation, we obtain

$$\begin{aligned}
\frac{\partial F_{j\vartheta\vartheta}(\boldsymbol{k}, \boldsymbol{k}')}{\partial t} &= i\,[(k_\alpha + k'_\alpha) F_{j\alpha,\vartheta,\vartheta}(\boldsymbol{k}, \boldsymbol{k}') - k_\alpha F_{\alpha\vartheta,j,\vartheta}(-\boldsymbol{k}-\boldsymbol{k}', \boldsymbol{k}') - \\
&- k'_\alpha F_{\alpha\vartheta,j,\vartheta}(-\boldsymbol{k}-\boldsymbol{k}', \boldsymbol{k})] + \frac{i}{\rho}(k_j + k'_j) F_{p\vartheta\vartheta}(\boldsymbol{k}, \boldsymbol{k}') - \\
&- [(\nu+\chi)(k^2 + k'^2) + 2\nu \boldsymbol{k}\boldsymbol{k}']\, F_{j\vartheta\vartheta}(\boldsymbol{k}, \boldsymbol{k}'), \qquad (19.15)
\end{aligned}$$

where, as usual, $F_{j\alpha,\vartheta,\vartheta}$ and $F_{\alpha\vartheta,j,\vartheta}$ are the Fourier transforms of the functions

$$B^{(0)}_{j\alpha,\vartheta,\vartheta}(\boldsymbol{r}, \boldsymbol{r}') = B_{j\alpha,\vartheta,\vartheta}(\boldsymbol{r}, \boldsymbol{r}') - B_{j\alpha}(0)\, B_{\vartheta\vartheta}(\boldsymbol{r}'-\boldsymbol{r})$$

and

$$B^{(0)}_{\alpha\vartheta,j,\vartheta}(\boldsymbol{r}, \boldsymbol{r}') = B_{\alpha\vartheta,j,\vartheta}(\boldsymbol{r}, \boldsymbol{r}') - B_{\alpha\vartheta}(0)\, B_{j\vartheta}(\boldsymbol{r}'-\boldsymbol{r})$$

In the isotropic case the latter function is the same as $B_{\alpha\vartheta,j,\vartheta}(\boldsymbol{r}, \boldsymbol{r}')$. If the sum of the first three terms on the right-hand side of Eq. (19.15) is denoted by $\Gamma_{j\vartheta\vartheta}(\boldsymbol{k}, \boldsymbol{k}')$, then the vector field $\Gamma_{j\vartheta\vartheta}(\boldsymbol{k}, \boldsymbol{k}')$ will satisfy the relation

$$\int\int \Gamma_{j\vartheta\vartheta}(\boldsymbol{k}, \boldsymbol{k}')\, d\boldsymbol{k}\, d\boldsymbol{k}' = 0, \qquad (19.16)$$

which has the same significance as Eq. (19.5), and follows from the fact that the left-hand side of Eq. (19.16) is equal to $-\frac{\partial}{\partial x_\alpha}\overline{u_j(x)\,u_\alpha(x)\,\vartheta^2(x)}$.

The term in Eq. (19.15) which contains $F_{p\vartheta\vartheta}(k, k')$ can be expressed in terms of $F_{j\alpha,\vartheta,\vartheta}(k, k')$, since it is clear that

$$\frac{1}{\rho}F_{p\vartheta\vartheta}(k, k') = -\frac{(k_j + k'_j)(k_\alpha + k'_\alpha)\,F_{j\alpha,\vartheta,\vartheta}(k, k')}{(k + k')^2}. \tag{19.17}$$

In the case of isotropic turbulence, Eqs. (19.15) and (19.17) can be rewritten in the form of a single scalar equation for the function

$$\Gamma_\vartheta(k, k', \mu) = ik_jF_{j\vartheta\vartheta}(k, k'), \tag{19.18}$$

which is antisymmetric in $k$ and $k'$, and by Eqs. (12.155) and (12.151), determines the vector $F_{j\vartheta\vartheta}(k, k')$:

$$\begin{aligned}\left[\frac{\partial}{\partial t} + (\nu + \chi)(k^2 + k'^2) + 2\nu kk'\right]\Gamma_\vartheta(k, k', \mu) = \\ = k_jk''_\alpha\Delta_{j\beta}(k'')\,F_{\alpha\beta,\vartheta,\vartheta}(k, k') + k_jk_\alpha F_{\alpha\vartheta,j,\vartheta}(k'', k') + \\ + k_jk'_\alpha F_{\alpha\vartheta,j,\vartheta}(k'', k)\end{aligned} \tag{19.19}$$

where $k''$ and $\Delta_{j\beta}(k)$ have the usual significance defined by Eq. (19.9). We recall that, in view of Eq. (12.157), the function $\Gamma_{\vartheta\vartheta}(k)$ in Eq. (14.62) is related to $\Gamma_\vartheta(k, k', \mu)$ by the simple relation

$$\Gamma_{\vartheta\vartheta}(k) = 4\pi\int_0^\infty\int_{-1}^1 \Gamma_\vartheta(k, k', \mu)\,k'^2\,dk'\,d\mu. \tag{19.20}$$

When $\vartheta$ is the concentration of a decaying admixture, or of a chemically active admixture participating in a first-order chemical reaction with the reaction rate $\Phi[\vartheta] = \mu\vartheta$, where $\mu$ is a constant, then the right-hand side of Eq. (19.14) need only be augmented by the term $-2\mu B_{j\vartheta\vartheta}(r, r'$, and the right-hand side of Eqs. (19.15) and (19.19) by $-2\mu F_{j\vartheta\vartheta}(k, k')$ and $-2\mu\Gamma_\vartheta(k, k', \mu)$, respectively.

Similar equations can be obtained also for velocity and temperature moments of still higher order (see e.g., Orszag and Kruskal, 1968); however these equations will hardly be used below and we shall not consider them here.

## 19.2 Closure of the Moment Equations by the Moment Discard Assumption

Suppose that we know the equations for all the moments of the velocity field right up to the $n$-point moments of order $n$, inclusively. This set of equations is not closed, since the last of them contains terms which can be expressed in terms of the $n$-point and $(n + 1)$-point moments of order $n + 1$. The simplest assumption which enables us to close this system is that these terms are negligible in comparison with all other terms.

The ratio of the terms containing the $(n+1)$-th moments to the viscous terms in the equations for the $n$th moments is of the same order as the ratio of the nonlinear inertial terms to the viscous terms in the Navier-Stokes equation, which is characterized by the number Re. Therefore, the $(n+1)$-th moments in the equation for the $n$th moments can be neglected only in the case of weak turbulence with small Re. In particular, when $n = 2$, we must assume that the third-order moments are negligible in comparison with the second-order moments, i.e., we must adopt the assumption which has been discussed in detail in Sect. 15.3. Let us now consider what will happen if we apply the above hypothesis to the equations for the moments of order $n > 2$.

By neglecting moments of order $(n+1)$ in the equation for the $n$th moments, we are introducing an error of the same order as the error introduced into the von Kármán-Howarth equation by the rejection of the terms containing $B_{LL,L}(r, t)$. However, by setting the function $B_{LL,L}(r, t)$ equal to zero, we are prevented from saying anything about velocity moments of order higher than 2, while if we consider only the $(n+1)$-th moments as equal to zero, we can still determine (if only approximately) all the moments up to order $n$ inclusively. The least accurate values obtained in this way will obviously be the values of the $n$th moments (whose equations contained the rejected terms). The errors in the values of the $(n-1)$-th moments found from the resulting set of equations will probably be smaller (because the equations for these moments do in a way take into account the effects of the inertial terms of the dynamic equations). Errors in the values of moments of order $n-2$ will be smaller still, and so on. In other words, it may be supposed that weak turbulence will be described more accurately by the solution of the set of equations with $n > 2$ than by the solution of the single Eq. (15.35), and the accuracy will increase with increasing $n$.

Following Kraichnan (1962), we can make the above discussion more precise as follows. We shall measure all the lengths in terms of the typical length-scale $L$, and all the velocities in terms of the typical velocity scale $U$. As the unit of time we shall take the time scale $T = L^2/\nu$ of viscous damping and, finally, we shall express pressure in terms of velocity through the equation $\Delta p = -\rho \frac{\partial^2 u_i u_j}{\partial x_i \partial x_j}$. If we now transform to the dimensionless variables $y = x/L$, $v = u/U$ and $\tau = \nu t/L^2$, we can write down the equation for the $m$-point $m$th moments of the velocity field in the form

$$\left(\frac{\partial}{\partial\tau}+\Delta^{(3m)}\right)B_{i_1\ldots i_m}(y_1, \ldots, y_m, \tau)=$$
$$=\mathrm{Re}\ \mathscr{L}_{i_1\ldots i_m,\, j_1\ldots j_{m+1}}\left[B_{j_1\ldots j_{m+1}}(y_1, \ldots, y_m, \tau)\right], \quad (19.21)$$

where $\Delta^{(3m)}$ is the $3m$-dimensional Laplace operator, and $\mathscr{L}_{i_1\ldots i_m,\, j_1\ldots j_{m+1}}$ is a linear operator of the order of unity in magnitude, which transforms the tensor field $B_{j_1\ldots j_{m+1}}$ into a field of the same order of magnitude and acts on the set of $(m+1)$th-order velocity moments. The solution of Eq. (19.21) can formally be written as the power series

$$B_{i_1\ldots i_m}(y_1, \ldots, y_m, \tau)=\sum_{k=0}^{\infty}(\mathrm{Re})^k B^{(k)}_{i_1\ldots i_m}(y_1, \ldots, y_m, \tau), \quad (19.22)$$

which perhaps converges if Re is small enough [but very likely is in fact divergent for all Re; see Kraichnan (1967c)]. By substituting Eq. (19.22) into Eq. (19.21) we can readily see that if we use the exact equation (19.21) with $m = 2, \ldots\ldots, n-1$, and neglect the right-hand side in Eq. (19.21) only for $m = n$, we shall determine only the zero-order term in Eq. (19.22) for the $n$th moment, while in the expression for the $(n-1)$th moments we shall determine the zero- and first-order terms, etc., up to the second moments for which Eq. (19.22) cuts off at the terms of the order of $(\mathrm{Re})^{n-2}$.

In general, the solutions of Eq. (19.21) with $m = 2, \ldots\ldots, n$, and the right-hand side neglected in the last equation of the set, are inconsistent with the general properties of the moments which follow from the nonnegativeness of the probabilities. In particular, the moduli of the correlation coefficients corresponding to these solutions frequently exceed unity, and the spectral functions assume both positive and negative values [see Kraichnan (1962) and T. Lee (1966) for examples of the application of these methods to the study of a scalar field $\vartheta(x, t)$]. The reason for this is that the Taylor series for the characteristic functional cannot be arbitrarily cut off at some given term (see Vol. 1, the end of the Sect. 4.4). If, however, the values of all the moments up to order $n$ inclusive are chosen correctly at $t=t_0$, then the contradictions which were mentioned above may appear only for sufficiently large $t-t_0$, when departures of the moments from the initial values become dominant. Moreover, for very small $t - t_0$, the solutions of the set of equations for the moments, which has been cut off at the $n$th term, will be quite satisfactory even for $\mathrm{Re} \gg 1$. This is explained by the fact that, from the fluid mechanical equations, the quantity $\partial u_i / \partial t$ is represented by a sum of terms of zero and first order in Re, and therefore $\partial^n u_i / \partial t^n$, i.e., also the sum of the first $n+1$ terms of the Taylor series in powers of $t - t_0$ for any function of the velocity, will contain terms of the order of $(\mathrm{Re})^n$ or less. However, if both Re and $t - t_0$ are not small, then there is no reason to suppose that the solution of the truncated set of equations will be close to the true moments [see Kraichnan (1966c)].

Let us now consider some specific examples of the application of the assumption that moments of order $n + 1$ are zero. When $n = 3$, the general spectral equation (14.14) is augmented by the equation

$$\frac{\partial}{\partial t} F_{ijl}(\boldsymbol{k}, \boldsymbol{k}', t) = -2\nu (k^2 + k'^2 + \boldsymbol{k}\boldsymbol{k}') F_{ijl}(\boldsymbol{k}, \boldsymbol{k}', t). \tag{19.23}$$

This equation is obtained from Eq. (19.8) by neglecting terms containing the fourth-order spectral functions. It is readily integrated, and the only difficulty is to find the initial values of $F_{ijl}(\boldsymbol{k}, \boldsymbol{k}', t_0)$ for which there are almost no data at present. It is, of course, possible to base the choice of the initial value of the third moments on the requirement that the values of $E(k,t)$ for $t > t_0$ should be in agreement with existing data, but the scientific value of such a correspondence would not be great.[33]

Deissler (1958) took the function

$$\Gamma(k, k', \mu, t_0) = -\beta_0 (k^4 k'^6 - k^6 k'^4) \tag{19.24}$$

which is independent of $\mu$, as the initial value of the function $\Gamma(k, k', \mu) = ik_l F_{ljj}(\boldsymbol{k}, \boldsymbol{k}', t)$ which is antisymmetric in $k$ and $k'$.[34] Moreover, he assumed that $F(k, t_0) = \frac{\Lambda}{12\pi^2} k^2$ [see Eq. (15.64)]. Under these conditions, $\Gamma(k, k', \mu, t)$ can

[33] We recall however that, according to Sect. 5.3, the asymptotic behavior of the solution $B_{LL}(r,t)$ of Eq. (15.35) as $t \to \infty$ is determined only by the behavior of $F(k, t_0)$ for $k \to 0$, but is independent of the detailed form of the initial values $B_{LL}(r, t_0)$ or $F(k, t_0)$. There is a similar situation in the case of Eqs. (14.14), (19.23), (19.10), and (19.13). Therefore, if we are interested only in the behavior of the moments for $t \to \infty$, and restrict our attention to "ideal" isotropic turbulence in infinite space, then all that is necessary is to specify correctly the form of $F_{ijl}(\boldsymbol{k}, \boldsymbol{k}')$ near $\boldsymbol{k} = 0, \boldsymbol{k}' = 0$.

[34] In fact, Deissler used the function $\Gamma^{(1)}(k, k', \mu) = ik_l F_{jjl}(\boldsymbol{k}, \boldsymbol{k}')$ which was not antisymmetric in $k$ and $k'$, and assumed that its initial value was equal to the right-hand side of Eq. (19.24). It is readily seen, however, that the equations for the unknowns $F(k,t)$ and $\Gamma(k, k', \mu, t)$, and for $F(k,t)$ and $\Gamma^{(1)}(k, k', \mu, t)$, are the same when the fourth-order moments are neglected. Since the right-hand side of Eq. (19.24) is antisymmetric in $k$ and $k'$, it is more natural to assume that it is equal to the initial value of $\Gamma(k, k', \mu, t)$ and not $\Gamma^{(1)}(k, k', \mu, t)$.

readily be found from Eq. (19.23), and $\Gamma(k,t)$ and $F(k,t)$ from Eqs. (19.13) and (14.14). This gives a two-term expression for $\overline{u^2(t)} = \frac{8\pi}{3}\int_0^\infty k^2F(k,t)\,dk$, which contains both the "main term" (15.54) and an additional term proportional to $(t-t_0)^{-7}$ (with a coefficient depending on $\beta_0$ and $\nu$). By suitably choosing the numerical values of $\beta_0$ and $\Lambda$, Deissler succeeded in obtaining good agreement with the experimental curves deduced by Batchelor and Townsend (1948b) for the turbulence behind a grid at times approaching the final period of decay. Since, however, the "additional term" in this expression for $\overline{u^2(t)}$ decreases for $t-t_0 \to \infty$ much more rapidly than the term of the asymptotic expansion of $B_{LL}(r,t)$ at $r = 0$ which is the next after Eq. (15.54) [and is given by Eq. (15.56) with $s = 1$, i.e., it is proportional to $(t-t_0)^{-7/2}$], the agreement obtained by Deissler is apparently due to arbitrary adjustment of the two parameters.

The situation is even more arbitrary in the case of the set of equations for the second-, third-, and fourth-order moments which is obtained on the assumption that the fifth-order moments are zero. The solution of this set of equations is determined by the initial values of the second-order, three-point third-order, and four-point fourth-order moments of the velocity field. According to Deissler (1960) [his paper contains a number of easily corrected errors connected with the fact that the four-point moments were used instead of the four-order cumulants; see Ohji (1962) and Deissler (1965)], it is necessary to specify the initial values of $F(k,t)$, $\Gamma(k,k',\mu,t)$, and also of three scalar functions of six variables. Nothing is known at present about these initial values [possibly with the exception of $F(k,t_0)$], so that no precise experimental test can be made of resulting values of $E(k,t)$ and $\overline{u^2(t)}$. The apparent agreement between theory and existing very poor data reported by Deissler (1960, 1965) is subject to considerable doubt.

The assumption that high-order moments are zero can also be used to investigate temperature fluctuations in turbulent flows [see for example Leffler and Deissler (1961) and J. Lee (1966)]. Here again there are serious difficulties associated with the choice of the initial values of unknown functions. Some of the results are clearly unphysical (for example, the spectrum $E_{\vartheta\vartheta}(k,t)$ is not always positive), and there is no empirical confirmation of the remaining results. We shall not, therefore, stop to consider the corresponding equations.

Some applications of the same method to the multipoint multitime velocity correlations were considered by Deissler (1961) and Kumar and Dash (1961); we shall not analyze these either.

## 19.3 Closure of the Second- and Third-Order Moment Equations using the Millionshchikov Zero-Fourth-Cumulant Hypothesis

Closure of the moment or spectral function equations through the discard of moments of order higher than a given value is justified in the case of weak turbulence with small Reynolds number at times approaching the final period of decay. However, according to Sect. 15, this period of decay is very difficult to realize in the laboratory, and the corresponding motion of the fluid cannot, strictly speaking, be regarded as turbulent in the usual sense of this word. On the other hand, the opposite case of highly developed turbulence with large Reynolds number, in which turbulent mixing connected with inertial motion of the fluid particles plays a much greater role than viscous damping, is of principal interest in the theory of turbulence. Simple neglect of higher-order moments leads in this case to completely incorrect or even meaningless results, and different methods of closure must be employed for the above set of equations. A number of such methods have been developed (some of these will be discussed later in Sects. 19.6, 22.2 and Sect. 29), but so far none of them has been entirely satisfactory [cf., the discussion of the closure problem in the works of Kraichnan (1966c, 1969) and Orszag (1969, 1970)]. Nevertheless, to illustrate the fundamental features of the theories based on the different methods of closure of the moment equations, and to discuss the resulting conclusions, we must first consider in some detail the oldest [Millionshchikov (1941a, b)] and probably the simplest among the closure methods which do not involve complete neglect of higher order moments. Namely, we shall try to use the Millionshchikov hypothesis discussed in the last section, according to which the fourth-order velocity cumulants are zero, so that the fourth-order velocity moments can be expressed in terms of the second-order moments. This will enable us to close the equations for the second- and third-order moments.

Let us first say a few words about the general hypothesis that the velocity cumulants of a fixed order $n+1 \geqslant 4$ are zero, which enables us to develop a sequence of cumulant neglect (or "cumulant-discard") approximations of increasing complexity in which the use of Millionshchikov hypothesis is the first step.[35]

Consider a turbulence whose velocity field at the initial time $t_0$ is a Gaussian random field. It can be shown that the formal expansion of the cumulants of order $n+1$ in a series in powers of Re, begins with a term of the order of $(\text{Re})^{n-1}$ [see Kraichnan (1962)]. Therefore, if we neglect them in Eq. (19.21) with $m=n$, we obtain a series of the form of Eq. (19.22) for the moments of order $n$, and the terms of this series are exact up to the term of order $(\text{Re})^{n-2}$. For the $(n-1)$th-order moments the series is correct up to the term of order $(\text{Re})^{n-1}$, and so on. For the second moments the resulting expansion is exact up to the term of order $(\text{Re})^{2n-4}$. Therefore, when the rejection of terms of order $(\text{Re})^{2n-3}$ or higher results in a good approximation (i.e., for Re $\ll 1$ or very small $t-t_0$), a satisfactory result is obtained if the cumulants of order $n+1$ are also neglected. For large Re and $t-t_0$, the neglect of the higher-order cumulants is

[35] There are analogs of these procedures in other branches of theoretical physics [for example, Kirkwood's superposition approximation in statistical mechanics, and the Tamm-Dancoff method in quantum field theory; see, for example, Green (1952) and Schweber, Bethe and Hoffmann (1955)].

increasingly more justified than the neglect of the higher-order moments, although, in general, the results obtained in this case may also turn out to be in conflict with the general requirements of the theory of probability (see again the end of Sect. 4.4 in Vol. 1).

Let us now consider the particular consequences of the assumption that the cumulants of order $n+1$ are zero. When $n=2$ this is equivalent to neglecting the third-order moments in the von Kármán-Howarth and Corrsin equations of Sect. 15. Therefore, the first nontrivial application of the above assumptions occurs for $n=3$. In this case we use equations for the second- and third-order moments, where in the latter case the three-point fourth-order moments are replaced by certain combinations of the second moments [see below Eq. (19.25)]. In the next approximation, we take the equations for the second- and third-order moments without any simplifications, and must augment them by the equations for the fourth-order moments in which the fifth-order moments are replaced by certain combinations of the second- and third-order moments. This approximation is, however, so unwieldy that it has not been extensively investigated [see, however, the paper by Khazen (1963) on the evolution of eddies in shear flows with uniform shear and the paper by Tanaka (1969) on a simplified one-dimensional model of isotropic turbulence].

We shall therefore consider only the simplest *zero-fourth-cumulant* (or as it is often called, *quasi-normal*) *approximation*, i.e., we shall suppose that all the three-point fourth-order velocity cumulants are identically zero. In the homogeneous case, this means that in view of Eq. (4.21),

$$B_{lm,j,l}(\boldsymbol{r},\boldsymbol{r}')=B_{lm}(0)B_{jl}(\boldsymbol{r}'-\boldsymbol{r})+ \\ +B_{lj}(\boldsymbol{r})B_{ml}(\boldsymbol{r}')+B_{mj}(\boldsymbol{r})B_{ll}(\boldsymbol{r}') \qquad (19.25)$$

[(18.1′)]. Direct substitution of this equation into Eq. (19.3) leads to unwieldy relations which can be considerably simplified with the aid of the Fourier transformation. If we take the first term on the right-hand side of Eq. (19.25) on to the left-hand side, and then take the six-dimensional Fourier transform of both sides, we obtain

$$F_{lm,j,l}(\boldsymbol{k},\boldsymbol{k}')=F_{lj}(\boldsymbol{k})F_{ml}(\boldsymbol{k}')+F_{mj}(\boldsymbol{k})F_{ll}(\boldsymbol{k}'). \qquad (19.26)$$

Consequently, Eq. (19.8) will now read

$$\frac{\partial}{\partial t} F_{ijl}(\boldsymbol{k}, \boldsymbol{k}') = -i \sum_{\text{cyc}} k''_m \Delta_{ln}(\boldsymbol{k}'') \left[F_{jn}(\boldsymbol{k}) F_{lm}(\boldsymbol{k}') + \right.$$
$$\left. + F_{jm}(\boldsymbol{k}) F_{ln}(\boldsymbol{k}')\right] - \nu\left(k^2 + k'^2 + k''^2\right) F_{ijl}(\boldsymbol{k}, \boldsymbol{k}'). \quad (19.27)$$

This equation, together with the usual spectral Eq. (14.14), and the Eqs. (19.10) and (19.13), form a closed set of equations for the second- and third-order moments.

Since in accordance with Eq. (12.166), the isotropic tensor $F_{ijl}(\boldsymbol{k}, \boldsymbol{k}')$ can be expressed in terms of the two scalar functions $\Phi(k, k', k'')$ and $\Psi(k, k', k'')$, it follows that, in the case of isotropic turbulence, the tensor Eq. (19.27) can be reduced to two scalar equations. To derive them, we must replace $F_{ijl}(\boldsymbol{k}, \boldsymbol{k}')$ by Eq. (12.166) and $F_{jn}(\boldsymbol{k}), F_{lm}(\boldsymbol{k}')$, etc., by Eq. (12.73), where $E(k) = 4\pi k^2 F(k)$. Since $k_m \Delta_{jm}(\boldsymbol{k}) = 0$, we have

$$k''_m \Delta_{ln}(\boldsymbol{k}'') \left[F_{jn}(\boldsymbol{k}) F_{lm}(\boldsymbol{k}') + F_{jm}(\boldsymbol{k}) F_{ln}(\boldsymbol{k}')\right] =$$
$$= -\Delta_{lm}(\boldsymbol{k}'') \Delta_{jn}(\boldsymbol{k}) \Delta_{lp}(\boldsymbol{k}') \left(k_p \delta_{mn} + k'_n \delta_{mp}\right) F(k) F(k') \quad (19.28)$$

and

$$-i \sum_{\text{cyc}} k''_m \Delta_{ln}(\boldsymbol{k}'') \left[F_{jn}(\boldsymbol{k}) F_{lm}(\boldsymbol{k}') + F_{jm}(\boldsymbol{k}) F_{ln}(\boldsymbol{k}')\right] =$$
$$= i\Delta_{lm}(\boldsymbol{k}'') \Delta_{jn}(\boldsymbol{k}) \Delta_{lp}(\boldsymbol{k}') \left\{F(k'') \left[F(k) - F(k')\right] k_m \delta_{np} + \right.$$
$$\left. + F(k) \left[F(k') - F(k'')\right] k'_n \delta_{mp} + F(k') \left[F(k) - F(k'')\right] k_p \delta_{mn}\right\}. \quad (19.29)$$

Consequently, the scalar equations for $\Phi$ and $\Psi$ which follow from Eq. (19.27) are of the form

$$\frac{\partial}{\partial t} \Phi(k, k', k'', t) = F(k'', t) \left[F(k', t) - F(k, t)\right] -$$
$$- \nu\left(k^2 + k'^2 + k''^2\right) \Phi(k, k', k'', t), \quad (19.30)$$
$$\frac{\partial}{\partial t} \Psi(k, k', k'', t) = -\nu\left(k^2 + k'^2 + k''^2\right) \Psi(k, k', k'', t).$$

These two relatively simple equations were first derived by Proudman and Reid (1954). To obtain a closed system, they must be complemented by Eq. (14.14) and the relation expressing the function $\Gamma(k)$ in terms of $\Phi(k, k', k'')$ and $\Psi(k, k', k'')$. The last relation (which was also given by Proudman and Reid) is very complicated. It is therefore more convenient to use other unknown functions in place of $\Phi$ and $\Psi$.

Since the function $\Gamma(k)$ is most naturally expressed in terms of $\Gamma(k, k', \mu) = ik_l F_{ljj}(\boldsymbol{k}, \boldsymbol{k}')$, it is convenient to take $\Gamma(k, k', \mu)$ as one of the unknowns.[36] The equation for the function $\Gamma(k, k', \mu)$ can be obtained by expressing the fourth-order spectral tensors in Eq. (19.11) in terms of $F_{ij}(\boldsymbol{k})$ as given by Eq. (19.26). Having done this, and having substituted for the spectral tensors in accordance with Eq. (12.73), we can reduce Eq. (19.11) with the aid of some tedious algebraic transformations to the form

$$\frac{\partial}{\partial t}\Gamma(k, k', \mu, t) = (1-\mu^2)\Big\{F(k, t)F(k', t)\left(k'^2 - k^2\right)\frac{\mu k k'}{k''^2} -$$
$$- F(k, t)F(k'', t)\left[\mu k k' + \left(k'^2 - \mu k k'\right)\frac{k^2}{k''^2}\right] +$$
$$+ F(k', t)F(k'', t)\left[\mu k k' + \left(k^2 - \mu k k'\right)\frac{k'^2}{k''^2}\right]\Big\} -$$
$$- 2\nu\left(k^2 + k'^2 + \mu k k'\right)\Gamma(k, k', \mu, t). \quad (19.31)$$

The closed system consisting of this equation, the Eq. (14.14), and Eq. (19.13), is much more convenient than the system involving the unknowns $\Phi$, $\Psi$ and $F$.

If we are interested only in finding the spectral function $F(k, t)$, then Eq. (19.31) can be simplified still further. In fact, the calculation of $F(k, t)$ requires a knowledge not of the values of $\Gamma(k, k', \mu, t)$ itself, but only of the values of the integral of this function with respect to $\boldsymbol{k}'$. However, if we add an arbitrary function of $k, k', \mu, t$ to the right-hand side of Eq. (19.31) (or a function of $k, k', k'', t$), the integral of which with respect to $\boldsymbol{k}'$ vanishes for all $k$ and $t$, and if we retain the initial value $\Gamma(k, k', \mu, t_0)$ as before, then the integral with respect to $\boldsymbol{k}'$ of the solution $\Gamma(k, k', \mu, t)$ of Eq. (19.31) will remain the same. Consequently, the values of $F(k, t)$ will also remain the same. Therefore, when we investigate the spectrum $F(k, t)$ we can, for example, add to the right-hand side of Eq. (19.31) any function of $k, k', k'', t$ which is antisymemtric in $k'$ and $k''$, i.e., changes sign when $k'$ is replaced by $k''$, $k''$ by $k'$, and $\mu k k' = \boldsymbol{k}\boldsymbol{k}'$ by $\boldsymbol{k}\boldsymbol{k}'' = -k^2 - \mu k k'$, $1-\mu^2$ by $(1-\mu^2)k'^2 k''^{-2}$. In particular, the following function is antisymmetric in $k'$ and $k''$:

[36] We note that instead of the function $\Gamma(k, k', \mu)$ it is also possible to follow Tatsumi (1957a) and use the function $\Gamma^{(1)}(k, k', \mu) = ik_l F_{jjl}(\boldsymbol{k}, \boldsymbol{k}')$ for which the integral with respect to $\boldsymbol{k}'$ is equal to $\Gamma(k)$. However, the function $\Gamma^{(1)}(k, k', \mu)$ is not antisymmetric in $k, k'$, and it is therefore less convenient than $\Gamma(k, k', \mu)$.

$$(1-\mu^2)\left\{-F(k,t)F(k',t)\left[2k^2+\frac{(k'^2-k^2)\mu kk'}{k''^2}\right]+\right.$$
$$+F(k,t)F(k'',t)\left[-\mu kk'+(k'^2-\mu kk')\frac{k^2}{k''^2}\right]+$$
$$\left.+F(k',t)F(k'',t)\left[\mu kk'+2k^2-(k^2-\mu kk')\frac{k'^2}{k''^2}\right]\right\}.$$

Consequently, Eqs. (14.14) and (19.31) can be rewritten in the form

$$\frac{\partial F(k,t)}{\partial t}=2\pi\int_0^\infty\int_{-1}^1\Gamma_1(k,k',\mu,t)k'^2\,dk'\,d\mu-2\nu k^2F(k,t), \quad (19.32)$$

$$\frac{\partial\Gamma_1(k,k',\mu,t)}{\partial t}=-2\{F(k,t)[F(k',t)k^2+F(k'',t)\mu kk']-$$
$$-F(k',t)F(k'',t)(k^2+\mu kk')\}(1-\mu^2)-$$
$$-2\nu(k^2+k'^2+\mu kk')\Gamma_1(k,k',\mu,t) \quad (19.33)$$

where we have replaced $\Gamma(k, k', \mu, t)$ by $\Gamma_1(k, k', \mu, t)$, since the solution of Eq. (19.33) does not now coincide with the function $ik_lF_{ljj}(\mathbf{k},\mathbf{k}')$. The set of equations (19.32) and (19.33) was derived by Tatsumi (1957a) who also suggested the use of the equations

$$\frac{\partial F(k,t)}{\partial t}=2\pi\int_0^\infty\int_{-1}^1\Gamma_2(k,k',\mu,t)k'^2\,dk'\,d\mu-2\nu k^2F(k,t), \quad (19.34)$$

$$\frac{\partial\Gamma_2(k,k',\mu,t)}{\partial t}=-2[F(k,t)-F(k',t)]F(k'',t)\times$$
$$\times\left[\frac{k^2k'^2}{k''^2}+\mu kk'\right](1-\mu^2)-2\nu(k^2+k'^2+\mu kk')\Gamma_2(k,k',\mu,t), \quad (19.35)$$

which are obtained by adding to the right-hand side of Eq. (19.33) a further function of $k$, $k'$, $k''$, and $t$, which is antisymmetric in $k'$ and $k''$. Equations (19.32)–(19.33) and (19.34)–(19.35) are much simpler than Eqs. (14.14) and (19.31). Henceforth, we shall confine our attention to these particular equations.

The first-order equations (19.33) and (19.35) can readily be integrated:

$$\Gamma_1(k, k', \mu, t) = \Gamma(k, k', \mu, t_0)\, e^{-K(t-t_0)} - \\ -2(1-\mu^2)\int_{t_0}^{t} e^{-K(t-t')}\{F(k, t')[F(k', t')k^2 + \\ + F(k'', t')\mu k k'] - F(k', t')F(k'', t')(k^2+\mu k k')\}\, dt', \quad (19.36)$$

$$\Gamma_2(k, k', \mu, t) = \Gamma(k, k', \mu, t_0)e^{-K(t-t_0)} - 2(1-\mu^2)\int_{t_0}^{t} e^{-K(t-t')}[F(k,t') - \\ - F(k', t')]\, F(k'', t')\left(\frac{k^2k'^2}{k''^2} + \mu k k'\right) dt', \quad (19.37)$$

where $K = 2\nu(k^2 + k'^2 + \mu k k') = \nu(k^2 + k'^2 + k''^2)$ and it is assumed that $\Gamma_1(k, k', \mu, t_0) = \Gamma_2(k, k', \mu, t_0) = \Gamma(k, k', \mu, t_0)$. Substitution of Eqs. (19.36) and (19.37) in Eq. (19.32) and, correspondingly, in Eq. (19.34), leads to an integro-differential equation for the function $F(k,t)$. The structure of this equation is quite similar to the equations for $F(k, t)$ or $E(k, t)$ discussed in Sect. 17. However, it is much more complicated and its physical significance is less clear.

The spectral energy-transfer function $W(k_1, t)$ which plays an important role in the hypotheses of Sect. 17 can now, in view of Eqs. (19.13) and (19.37), be written in the form

$$W(k_1, t) = \int_{k_1}^{\infty}\int_{0}^{\infty} 8\pi^2\left\{\int_{-1}^{1} e(t-t_0)\Gamma(k, k', \mu, t_0)\, d\mu\right\} k^2k'^2\, dk\, dk' \\ - \int_{k_1}^{\infty}\int_{0}^{\infty} 16\pi^2\left\{\int_{t_0}^{t}\int_{-1}^{1} e(t-t')[F(k, t') - F(k', t')] \times \right. \\ \left. \times F(k'', t')\left(\frac{k^2k'^2}{k''^2} + \mu k k'\right)(1-\mu^2)\, dt'\, d\mu\right\} k^2k'^2\, dk\, dk', \quad (19.38)$$

where $e(t) = \exp\{-Kt\} = \exp\{-2\nu(k^2 + k'^2 + \mu k k')t\}$. Since the integrands in both the first and the second terms of Eq. (19.38) are antisymmetric in $k$ and $k'$, the condition $W(0, t) = 0$, which is equivalent to Eq. (14.18), is satisfied identically. Again, since the integrands are antisymmetric, the integrals with respect to $k'$ in Eq. (19.38) can be regarded as being evaluated between zero and $k_1$. Consequently, we have a representation of the form of Eq. (17.13):

$$W(k_1, t) = \int_{k_1}^{\infty}\int_{0}^{k_1} P(k, k', t)\, dk\, dk', \tag{19.39}$$

where the function

$$P(k, k', t) = 8\pi^2 \int_{-1}^{1} e(t-t_0)\Gamma(k, k', \mu, t_0) k^2 k'^2 d\mu -$$
$$-\int_{t_0}^{t} dt' \int_{-1}^{1} d\mu\, e(t-t')\left[E(k, t') k'^2 - E(k', t') k^2\right] E(k'', t') \times$$
$$\times \left(\frac{k^2 k'^2}{k''^2} + \mu k k'\right)\frac{1-\mu^2}{k''^2} \tag{19.40}$$

can be interpreted as the energy transfer density from the vicinity of the point $k'$ of the wave-number axis into the vicinity of the point $k$ at time $t$. This function depends not only on $E(k)$ and $E(k')$ but also on all the values of $E(k'') = E(|\boldsymbol{k}+\boldsymbol{k}'|)$, and passes continuously through zero at $k = k'$, in accordance with the properties of the "true" density function$P(k, k')$.

Consider now a case of turbulence with high enough Reynolds number, where viscous forces can be neglected in the first approximation. Substituting $\nu = 0$ in Eqs. (19.32) and (19.33), we obtain the following integro-differential equation:

$$\frac{\partial^2}{\partial t^2} F(k, t) = -4\pi \int_{0}^{\infty}\int_{-1}^{1} \{F(k, t)[F(k', t) k^2 + F(k'', t)\mu k k'] -$$
$$- F(k', t) F(k'', t)(k^2 + \mu k k')\}(1-\mu^2) k'^2 dk'\, d\mu. \tag{19.41}$$

Let us multiply both sides of this equation by $k^2$ and then integrate over all the space of the vectors $\boldsymbol{k}$. The final result (obtained after replacement of integration with respect to $\boldsymbol{k}$ and $\boldsymbol{k}'$ by integration with respect to $\boldsymbol{k}$ and $\boldsymbol{k}''$, or $\boldsymbol{k}'$ and $\boldsymbol{k}''$, in some cases) is

$$\frac{d^2}{dt^2}\int_{0}^{\infty} F(k, t) k^4 dk = \frac{8\pi}{3}\left[\int_{0}^{\infty} F(k, t) k^4 dk\right]^2. \tag{19.42}$$

It is clear that this equation can be rewritten in the form

$$\frac{d^2}{dt^2}\int_0^\infty E(k,\ t)\,k^2\,dk = \frac{2}{3}\left[\int_0^\infty E(k,\ t)\,k^2\,dk\right]^2, \tag{19.43}$$

or, using Eq. (11.83), in the form

$$\frac{d^2\overline{\omega^2}}{dt^2} = \frac{1}{3}\,(\overline{\omega^2})^2, \tag{19.44}$$

as suggested by Proudman and Reid (1954). It describes the generation of vorticity by the deformation flow field.

The first integral of Eq. (19.44) gives a first-order nonlinear differential equation whose solution is the elliptic Weierstrass function $\wp(x;\ 0,\ 1)$:

$$\overline{\omega^2} = 2^{2/3}\overline{\omega_1^2}\,\wp(x;\ 0,\ 1), \quad x = \frac{2^{1/6}}{3}\left(\overline{\omega_1^2}\right)^{1/2}(t - t_1), \tag{19.45}$$

where $\overline{\omega_1^2}$ and $t_1$ are integration constants. A graph of one period of the periodic function $\wp(x;\ 0,\ 1)$ is shown in Fig. 41. In this figure $x_0$ is one-half of the period, i.e., $x_0 \approx 1.53$ (see Jahnke, Emde, and Lösch, 1960). The initial values

$$\overline{\omega^2}\,|_{t_0} = 2\int_0^\infty E(k,\ t_0)\,k^2\,dk$$

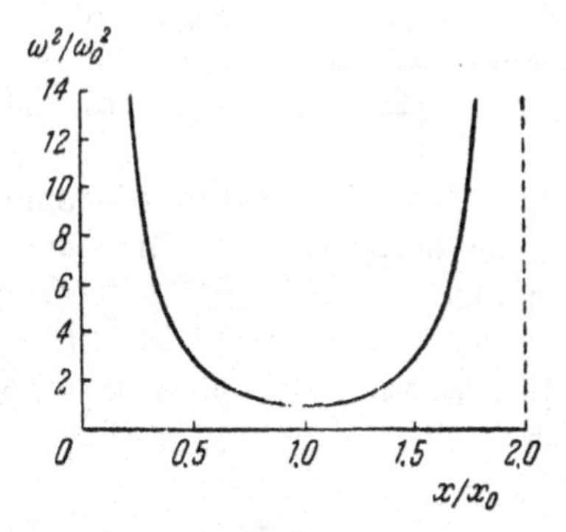


**Fig. 41** The ratio $\omega^2/\omega_0^2$ as a function of $x/x_0$ [Proudman and Reid (1954)].

and

$$\left.\frac{d\overline{\omega^2}}{dt}\right|_{t_0} = 2\int_0^\infty T(k, t_0)\, k^2\, dk$$

are determined by the values of $F(k, t_0)$ and $\Gamma(k, k', \mu, t_0)$, and can be quite arbitrary; hence the initial time $t = t_0$ can correspond to any point on the curve of Fig. 41. The subsequent variation of the vorticity is uniquely determined by the initial values, and under certain conditions (corresponding to points on the left half of the curve), the inertial motion of the fluid particles will at first give rise to a short-period decrease of vorticity, while under other conditions (corresponding to the right half of the curve), the vorticity will increase right from the beginning. If it were not for the viscosity, the vorticity would become infinitely great during a finite time interval (which also depends on the initial conditions) for any choice of the initial conditions. In reality, however, the infinite increase of vorticity is prevented by viscous energy dissipation. The final increase in the vorticity is in agreement with the physical idea of the stretching of vortex tubes due to inertial motion (see Sect. 15.1). Since

$$\frac{d\overline{\omega^2}}{dt} = 2\int_0^\infty T(k, t)\, k^2\, dk$$

this increase also shows that, in the case of isotropic turbulence with zero fourth-order cumulants, the energy transfer along the spectrum for any initial conditions will always eventually occur (on the average) from large-scale to small-scale eddies, and not the other way around.

It is readily seen that the zero-fourth-cumulant hypothesis is necessary only for the determination of the numerical coefficient on the right-hand side of Eq. (19.44), and its form depends only on the condition of isotropy. In fact, if we omit in Eq. (15.8) the term proportional to the viscosity, we obtain an equation which can be written in the form

$$\frac{d\overline{\omega^2}}{dt} = -\frac{7s}{3\sqrt{15}}\left(\overline{\omega^2}\right)^{3/2}, \tag{19.46}$$

where $s$ is the skewness of $\frac{\partial u_1}{\partial x_1}$. Differentiating Eq. (19.46) with respect to $t$, we obtain the relation

$$\frac{d^2\overline{\omega^2}}{dt^2} = \frac{49s^2}{90}(\overline{\omega^2})^2. \tag{19.47}$$

Therefore, it follows from the Millionshchikov hypothesis that

$$|s| = \frac{\sqrt{30}}{7} \approx 0{,}78. \tag{19.48}$$

This value of $|s|$ is of the same order of magnitude as the empirical results (see Sect. 18.1), but it is not in agreement with the inequality $|s| \leqslant 2/\sqrt{7} \approx 0.76$ which, in view of Betchov's general inequality (12.109), must be satisfied by $|s|$ for any solenoidal isotropic vector field with zero fourth-order cumulants. This discrepancy follows from the fact that, for given second- and third-order moments, the probability distribution with zero fourth-order cumulants may not actually exist. This is closely related to the inadmissibility of the arbitrary truncation of the Taylor series for the logarithm of the characteristic functional (see again the end of Sect. 4.4 in Vol. 1).

Another interesting consequence of Eq. (19.41) concerns the asymptotic behavior of the function $F(k, t)$ for $k \to 0$, which is naturally unaffected by viscosity. For very small $k$, the main contribution to the right-hand side of Eq. (19.41) is provided by the term in the integrand which does not contain $F(k, t)$. Since $k'' \approx k' + \mu k$ as $k \to 0$, it follows from Eq. (19.41) that

$$\frac{\partial^2 F(k, t)}{\partial t^2} = \frac{56\pi}{15}\int_0^\infty [F(k', t)]^2 k'^2\, dk' \cdot k^2 + o(k^2) \tag{19.49}$$

(see Proudman and Reid, 1954). Hence it is clear that even if $F(k, t_0) = o(k^2)$ for small $k$ (i.e., if the initial Loitsyanskii integral is zero), terms of the order of $k^2$ will still appear in $F(k, t)$ during the subsequent time, and if $F(k, t) = C(t)k^2 + o(k^2)$ then

$$\frac{d^2C(t)}{dt^2} = \frac{56\pi}{15}\int_0^\infty [F(k, t)]^2 k^2\, dk > 0,$$

so that the coefficient $C(t)$ is a function of time. According to Sect. 15.2, it follows that in isotropic turbulence with zero fourth-order cumulants, the function $B_{LL,L}(r)$ falls off as $r^{-4}$ when $r \to \infty$. If this were not so, the Loitsyanskii integral could not be convergent, though not constant.

Comparison of Eq. (19.33) with Eq. (19.41) shows that if $\Gamma(k, k', \mu, t_0) = 0$, then the derivative $\frac{d\Gamma(k, t)}{dt}$ at $t = t_0$ will be equal to the right-hand side of Eq. (19.41) at this time. Let us now consider the artificial example in which the velocity field has a normal probability distribution at the initial time $t = t_0$, so that the third-order moments of the field $\boldsymbol{u}(\boldsymbol{x})$ (through which the functions $\Gamma(k, k', \mu)$ and $\Gamma(k)$ are expressed) and the fourth-order cumulants are zero at $t = t_0$. For any initial value of the spectrum $E(k, t_0) = 4\pi k^2 F(k, t_0)$ we can then use Eq. (19.41) to determine the form of the energy transfer functions $\Gamma(k, t) \approx \frac{d\Gamma(k, t_0)}{dt_0}(t - t_0)$ and $T(k, t) = 4\pi k^2 \Gamma(k, t)$ for small positive values of $t - t_0$. Such calculations are clearly much easier than the complete solution of the system of zero-fourth-cumulant equations for $E(k, t)$ and $\Gamma(k, t)$. They were performed for a number of initial spectra among others by Proudman and Reid (1954) and Tatsumi (1957a). Thus, Tatsumi showed that, in the case of the idealized "white noise" initial spectrum $E(k, t_0) = E_0 k^2$, $F(k, t_0) = E_0/4\pi = \text{const}$ (which corresponds to an infinite energy), we have $\frac{d\Gamma(k, t)}{dt} = 0$, from which it is clear that, when $\nu = 0$ and the initial velocity distribution is normal, the "white noise" spectrum is time-independent. This result is in good agreement with the general conclusion of Hopf and Titt (1953) which we shall discuss in Chap. 10 (see the end of Sect. 29.1). Figure 42 shows the results for the case where the initial spectrum is of the form $E(k, t_0) = E_0 \left(\frac{k}{k_0}\right)^4 e^{-(k/k_0)^2}$ which is characteristic of the final period of decay, and for the idealized initial "linear" spectrum $E(k) = E_0\, \delta(k - k_0)$ where $\delta(k)$ is the Dirac function. We see that, in the first of these cases, the energy redistribution over the spectrum is quite reasonable for small $t - t_0$, and approaches the observed situation in real turbulent flows. In the second case, turbulent mixing tends to redistribute the energy contained in the "spectral line" $k = k_0$ over a continuous spectral interval. The energy transfer then occurs largely in the "correct" direction, i.e., toward shorter waves,

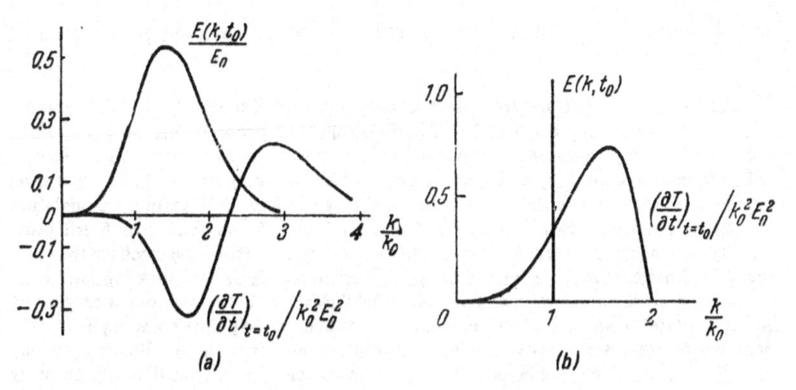


**Fig. 42** Spectral energy transfer at the initial time for two models of the spectrum $E(k, t_0)$, according to Proudman and Reid (1954) and Tatsumi (1957a).

but a proportion of the energy is transferred to the larger-scale disturbances with $k < k_0$.

The results shown in Fig. 42 refer only to the initial times, just after $t = t_0$, but they are based on the single assumption that the initial distribution is normal, and no additional hypotheses are required. The subsequent shape of the spectrum $E(k, t)$ can be interpreted in terms of the nonrigorous assumption that all the fourth-order velocity cumulants are identially zero. Tatsumi (1957a, b) used the Eqs. (19.32) and (19.33) following from this assumption to calculate a few terms in the expansion of the solution $E(k, t)$ [corresponding to the initial conditions $E(k, t_0) = E_0\delta(k - k_0)$, $\Gamma(k, k', \mu, t_0) = 0$] in a series in powers of $t - t_0$ and of the Reynolds number $\mathrm{Re} = (E_0)^{1/2}/\nu k_0$. A more complete solution was reported by Ogura (1962b, 1963) who carried out a numerical integration of Eqs. (19.32) and (19.33) (in the first paper for $\nu = 0$ and in the second for a number of finite values of $\nu$), subject to the initial conditions $E(k, t_0) = E_0(k/k_0)^4 e^{-(k/k_0)^2}$, $\Gamma(k, k', \mu, t_0) = 0$. He found that the calculated spectra $E(k, t)$ initially had a reasonable overall character, but for all $\nu$ that were not too large (i.e., Reynolds numbers that were not too small initially), the spectrum $E(k, t)$ eventually assumed small negative values over a range of $k$ (see, for example, Fig. 43). The Ogura data have thus shown once again that the use of the Millionshchikov hypothesis in calculations on the evolution of turbulence over a long period of time may lead to nonphysical results, although in calculations of the parameters of energy range disturbances at a given time, or over a short period of time, the errors involved are apparently not serious.

Negative parts of the spectrum were obtained also by Mirabel' (1969), who repeated Ogura's calculations for an artificial model of isotropic turbulence, having the same form of the initial energy spectrum, as in Ogura's example, but a nonzero initial energy-redistribution function $T(k)$ of a special form. In this case the negative part of the spectrum appears after even shorter time than in the case shown in Fig. 43; this is quite natural since the initial conditions used (with given functions $E(k)$ and $T(k)$ and zero fourth-order velocity cumulants) are very likely contradictory. Negative parts of the spectrum were also found by Khazen (1963) for homogeneous turbulence in a shear flow with constant shear.

Unphysical consequences of the Millionshchikov hypothesis were obtained also by Kraichnan (1957) who considered an example in which this hypothesis was applied to the two-time fourth-order velocity cumulants. Another similar example was discussed by Ohji (1961). A detailed discussion of the zero-fourth-cumulant approximation was given by Orszag (1969) [see also Orszag (1970) where some generalizations of the theory are also considered]. The appearance of negative values in the zero fourth cumulant approximation to the spectrum $E(k,t)$ for large $t$ was first discovered by Ogura (1962a) in the case of a two-dimensional isotropic turbulence model. It is interesting that the energy redistribution $T(k,t)$ in the case of two-dimensional turbulence was found to be quite different from the three-dimensional case. Instead of negative values for small $k$, and positive values for large $k$ (in accordance with the usual energy transfer from large- to small-scale disturbances), for the two-dimensional turbulence the function $T(k)$ was found to be positive, and its modulus a maximum, in the region of the smallest values of $k$, and negative for intermediate $k$; for very large $k$, it was again positive. This behavior of $T(k)$ can be explained by the fact that for two-dimensional turbulence with $\nu = 0$ (Ogura discussed this particular case) we have not only $\int_0^\infty T(k)\,dk = 0$ but also $\int_0^\infty k^2T(k)\,dk = 0$ because of conservation of vorticity. This shows that there is a basic difference between turbulent mixing in the two- and three-dimensional cases (for more details about this see Kraichnan, 1967).

We note further that Meecham (1965) has shown that the contradiction found by Kraichnan (1957) can readily be removed by modifying slightly the method of using the zero-fourth-cumulant hypothesis employed in Kraichnan's paper. As regards the appearance of negative spectra, this can apparently be avoided by generalization of the Eqs. (19.33) or (19.35) by analogy with the procedure adopted by J. Lee (1966) for the analogous moment equations of a scalar field mixed by turbulence. This is, however, equivalent to the introduction of a new semiempirical energy-transfer hypothesis, which is more complicated but also more accurate than the hypotheses given in Sect. 17. The accuracy of this new hypothesis can be deduced only by comparison with sufficiently extensive and detailed data on the decay of isotropic turbulence, but such data are not, as yet, available.

Finally let us briefly mention a comparatively new method of closing the moment equations which has a similarity to the cumulant discard method in its basic principles, but which has an advantage in that it cannot lead to negative spectra and other physically unrealizable predictions. The method is based on the so-called Wiener-Hermite (or Cameron-Martin-Wiener) expansion of the random fields, which was introduced by Wiener (1958) and some of his collaborators into random function theory as a quite natural generalization of the classical Hermite polynomial expansion. Let us recall that two Gaussian random functions are identical if they have the same first and second moments; hence it is easy to show that any Gaussian random function $u(x)$ can be represented in the form

$$u(x) = \overline{u(x)} + \int H(x, x_1)\, a(x_1)\, dx_1$$

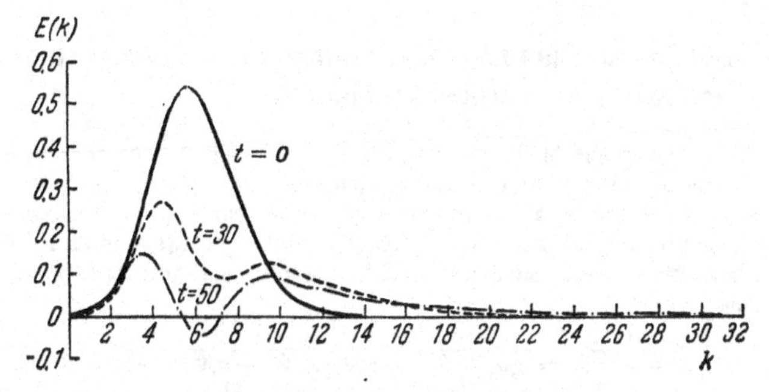


**Fig. 43** Example of the time evolution of the energy spectrum $E(k)$ calculated by Ogura (1963), using the Millionshchikov zero-fourth-cumulant hypothesis.

where $H(x,x_1)$ is an ordinary nonrandom function and $a(x)$ is a "white noise" random function, i.e., a Gaussian function with zero mean value and correlation function of the form $\overline{a(x_1)\,a(x_2)} = \delta(x_1 - x_2)$ (we shall not stop to discuss the strict mathematical meaning of the ideal white noise function and the integral over it which is not difficult to explain).

It seems reasonable to think that the random fields of turbulence theory have quite large Gaussian parts and can therefore be fruitfully expanded into a series whose first term is purely Gaussian while succeeding terms represent small corrections to the Gaussian field. Such an expansion can be obtained with the aid of a statistically orthogonal basis in the random function space constructed from polynomial functions of the white noise $a(x)$. It can be shown that the first few elements of such a basis have the form

$$a_0(x) = 1, a_1(x) = a(x), a_2(x_1, x_2) = a(x_1)a(x_2) - \delta(x_1 - x_2), a_3(x_1, x_2, x_3)$$
$$= a(x_1)a(x_2)a(x_3) - a(x_1)\delta(x_2 - x_3) - a(x_2)\delta(x_3 - x_1) - a(x_3)\delta(x_1 - x_2), \text{etc.}$$

(cf. Wiener, 1958). The expansion into an infinite series over the basis has the form

$$u(x) = H_0(x) + \int H_1(x, x_1)a(x_1)dx_1 + \int H_2(x, x_1, x_2)a(x_1, x_2)dx_1\,dx_2 + \ldots.$$

It contains an infinite number of the nonrandom coefficient functions $H_0(x)$, $H_1(x,x_1)$, $H_2(x,x_1,x_2)$, etc. If we expand all the components $u_\alpha$, $\alpha = 1, 2, 3$, of the random velocity field, compute the moments of the expanded fields and substitute these expressions into the dynamical moment equations, we shall obtain an infinite set of coupled differential equations for the unknowns $H_{\alpha i}$ where $\alpha$ is a vector index numbering the velocity components. The truncation of this set by neglecting all the functions $H_{\alpha i}$ with $i \geqslant n + 1$ results in a closed set of equations with the unknowns $H_{\alpha 0}, H_{\alpha 1}, \ldots, H_{\alpha n}$. However, now the truncated series for the random fields under investigation are clearly realizable; therefore we cannot obtain from them unphysical results.

This closing method was first applied to a model problem related to turbulence theory by Meecham and Siegel (1964); later the method and some of its modifications were discussed and developed further in a series of papers [see for example the works of Meecham and Jeng (1968), Meecham (1970), Canavan and Leith (1968), Crow and Canavan (1970), Bodner (1969), Saffman (1968, 1969), Siegel and Khang (1969), Khang and Siegel (1970), Clever and Meecham (1970), and Meecham and Clever (1971) where many additional references can also be found]. However the results are as yet of a preliminary character only; therefore we shall not discuss this method in more detail.

## 19.4 Zero-Fourth-Cumulant Approximation for Temperature Fluctuations in Isotropic Turbulence

Let us now apply the Millionshchikov zero-fourth-cumulant hypothesis to the joint three-point fourth-order cumulants of two components of the velocity $u$ and two values of the temperature $\vartheta$. The equations for the second-order temperature moments and the third-order joint temperature and velocity moments (quadratic in the temperature) will then form a closed system. The assumption that the fourth-order cumulants are zero is equivalent to equations

$$\begin{aligned} \overline{u_i u_j \vartheta' \vartheta''} &= \overline{u_i u_j} \cdot \overline{\vartheta' \vartheta''} + \overline{u_i \vartheta'} \cdot \overline{u_j \vartheta''} + \overline{u_i \vartheta''} \cdot \overline{u_j \vartheta'}, \\ \overline{u_i \vartheta u'_j \vartheta''} &= \overline{u_i \vartheta} \cdot \overline{u'_j \vartheta''} + \overline{u_i u'_j} \cdot \overline{\vartheta \vartheta''} + \overline{u_i \vartheta''} \cdot \overline{u'_j \vartheta}. \end{aligned} \tag{19.50}$$

For isotropic turbulence in which $\overline{u_i \vartheta'} = 0$ this yields

$$\begin{aligned} B_{ij,\vartheta,\vartheta}(\boldsymbol{r}, \boldsymbol{r}') &= B_{ij}(0)\, B_{\vartheta\vartheta}(\boldsymbol{r} - \boldsymbol{r}') = \delta_{ij} \overline{u^2}\, B_{\vartheta\vartheta}(\boldsymbol{r} - \boldsymbol{r}'), \\ B_{i\vartheta, j, \vartheta}(\boldsymbol{r}, \boldsymbol{r}') &= B_{ij}(\boldsymbol{r})\, B_{\vartheta\vartheta}(\boldsymbol{r}'). \end{aligned} \tag{19.50'}$$

In view of the relation between $\Delta p$ and the velocity derivatives, it follows from the first equation in Eq. (19.50') that

$$\sum_{m=1}^{3} \left( \frac{\partial}{\partial r_m} + \frac{\partial}{\partial r'_m} \right)^2 B_{p\vartheta\vartheta}(\boldsymbol{r}, \boldsymbol{r}') = 0. \tag{19.51}$$

Consequently, $B_{p\vartheta\vartheta}(\boldsymbol{r}, \boldsymbol{r}') = 0$ [this is the only bounded solution of the Laplace equation (19.51) which tends to zero as $\boldsymbol{r} \to \infty$ or $\boldsymbol{r}' \to \infty$]. Substituting Eqs. (19.50') and (19.51) in Eq. (19.14), we obtain

$$\begin{aligned} \frac{\partial}{\partial t} B_{j\vartheta\vartheta}(\boldsymbol{r}, \boldsymbol{r}', t) = &- \left[ B_{jm}(\boldsymbol{r}, t) \frac{\partial}{\partial r_m} + B_{jm}(\boldsymbol{r}', t) \frac{\partial}{\partial r'_m} \right] B_{\vartheta\vartheta}(\boldsymbol{r}' - \boldsymbol{r}, t) + \\ &+ \left[ (\nu + \chi) \left( \frac{\partial^2}{\partial r_m \partial r_m} + \frac{\partial^2}{\partial r'_m \partial r'_m} \right) + 2\nu \frac{\partial^2}{\partial r_m \partial r'_m} \right] B_{j\vartheta\vartheta}(\boldsymbol{r}, \boldsymbol{r}', t). \end{aligned} \tag{19.52}$$

If the functions $B_{jm}(\boldsymbol{r}, t)$ are known, Eq. (19.52) and the Corrsin equation (14.59) form a closed system. When $\nu = \chi = 0$, it follows from this system that

$$\frac{\partial^2 B_{\vartheta\vartheta}(r, t)}{\partial t^2} = 2 \left( \frac{\partial}{\partial r} + \frac{2}{r} \right) \left\{ \frac{\partial B_{\vartheta\vartheta}(r, t)}{\partial r} [B_{LL}(0, t) - B_{LL}(r, t)] \right\}. \tag{19.53}$$

If we transform from correlation functions to spectral functions, the equations in Eq. (19.50') assume the form

$$F_{ij,\vartheta,\vartheta}(\boldsymbol{k}, \boldsymbol{k}') = 0, \quad F_{i\vartheta, j, \vartheta}(\boldsymbol{k}, \boldsymbol{k}') = F_{ij}(\boldsymbol{k})\, F_{\vartheta\vartheta}(\boldsymbol{k}'). \tag{19.54}$$

Consequently, if we use the Millionshchikov hypothesis, the spectral equation (19.15) becomes

$$\frac{\partial}{\partial t} F_{j\vartheta\vartheta}(\boldsymbol{k}, \boldsymbol{k}') = -ik_m F_{jm}(-\boldsymbol{k}-\boldsymbol{k}') F_{\vartheta\vartheta}(\boldsymbol{k}') -$$
$$-ik'_m F_{jm}(-\boldsymbol{k}-\boldsymbol{k}') F_{\vartheta\vartheta}(\boldsymbol{k}) - [(\nu+\chi)(k^2+k'^2)+2\nu\boldsymbol{k}\boldsymbol{k}'] F_{j\vartheta\vartheta}(\boldsymbol{k}, \boldsymbol{k}'). \tag{19.55}$$

If we now transform from vector equation (19.55) to the scalar equation for the function

$$\Gamma_\vartheta(k, k', \mu) = ik_j F_{j\vartheta\vartheta}(\boldsymbol{k}, \boldsymbol{k}') = -ik'_j F_{j\vartheta\vartheta}(\boldsymbol{k}, \boldsymbol{k}'),$$

and replace the variables $k, k', \mu$ by $k, k', k'' = |\boldsymbol{k}+\boldsymbol{k}'|$ we obtain after some rearrangement

$$\frac{\partial}{\partial t}\Gamma_\vartheta(k, k', k'', t) = \frac{Q(k, k', k'')}{4k''^2}[F_{\vartheta\vartheta}(k', t) - F_{\vartheta\vartheta}(k, t)] F(k'', t) -$$
$$-[\nu(k^2+k'^2)+\chi k''^2]\Gamma_\vartheta(k, k', k'', t), \tag{19.56}$$

where

$$Q(k, k', k'') = 2k^2k'^2 + 2k^2k''^2 + 2k'^2k''^2 - k^4 - k'^4 - k''^4. \tag{19.57}$$

Equation (19.56) together with the spectral equation for the temperature field Eq. (14.62) or (14.63), and the relation given by Eq. (19.20) which expresses $\Gamma_{\vartheta\vartheta}(k)$ in terms of $\Gamma_\vartheta(k, k', k'')$, form a closed set. This set can readily be rewritten in the form of a single integro-differential equation for the temperature spectrum $E_{\vartheta\vartheta}(k, t)$. The resulting equation has the same form as the general spectral equation (14.63), but the redistribution of the temperature inhomogeneities over the wave-number spectrum is now given by

$$T_{\vartheta\vartheta}(k, t) = \int_0^\infty P_\vartheta(k, k', t)\, dk', \tag{19.58}$$

where

$$P_\vartheta(k, k', t) = 16\pi^2 \int_{-1}^{1} e_1(t-t_0)\Gamma_\vartheta(k, k', \mu, t_0)\, k^2k'^2\, d\mu -$$
$$-\int_{t_0}^{t} dt' \int_{-1}^{1} d\mu\, e_1(t-t')[E_{\vartheta\vartheta}(k, t')k'^2 -$$
$$-E_{\vartheta\vartheta}(k', t')k^2]\, E(k'', t)\frac{Q(k, k', k'')}{4k''^4}, \tag{19.59}$$

and $e_1(t) = \exp\{-[\nu(k^2+k'^2)+\chi k''^2]\, t\}$. The function (19.59) is antisymmetric in $k$ and $k'$, and can naturally be interpreted as the density of the transfer of the temperature fluctuation intensity between the vicinities of the points $k'$ and $k$ on the wave-number axis.

In the special case when we can neglect the effects of viscosity and thermal conductivity, we obtain, after differentiating both sides of Eq. (19.53) twice with respect to $r$ and then substituting $r = 0$,

$$-\frac{d^2}{dt^2} B''_{\vartheta\vartheta}(0) = 10 B''_{\vartheta\vartheta}(0) B''_{LL}(0). \tag{19.60}$$

This equation can also be rewritten in the spectral form

$$\frac{d^2}{dt^2} \int_0^\infty E_{\vartheta\vartheta}(k, t) k^2 \, dk = \frac{4}{3} \int_0^\infty E_{\vartheta\vartheta}(k, t) k^2 \, dk \int_0^\infty E(k, t) k^2 \, dk \tag{19.61}$$

or, finally, in the form

$$\frac{d^2 \overline{(\text{grad } \vartheta)^2}}{dt^2} = \frac{2}{3} \overline{\omega^2} \cdot \overline{(\text{grad } \vartheta)^2}. \tag{19.62}$$

which was put forward by Reid (1955) and can be used, for a given dependence of $\overline{\omega^2}$ on $t$, to investigate the increase of $\overline{(\text{grad } \vartheta)^2}$ in isotropic turbulence with very high Reynolds and Peclet numbers and zero fourth-order cumulants. In particular, if we consider the mixing of temperature inhomogeneities in stationary turbulence, which were generated at time $t = t_0$, we may suppose that $\overline{\omega^2} = \text{const}$, in which case the quantity $\overline{(\text{grad } \vartheta)^2}$ will increase exponentially for large values of $t - t_0$:

$$\overline{(\text{grad } \vartheta)^2} \sim \exp\left\{ \sqrt{\frac{2}{3} \overline{\omega^2}} \, t \right\} \tag{19.63}$$

(in reality, this exponential rise will continue only until thermal conduction becomes important). Still more rapid increase of the mean temperature gradient is obtained if we take into account the increase in the vorticity, produced by turbulent mixing. For example, if we assume that $\overline{\omega^2}$ increases in accordance with Eq. (19.44), then the solution of Eq. (19.62) will be

$$\overline{(\text{grad } \vartheta)^2} = \overline{\theta_0^2} \, \wp'(x), \qquad x = \frac{2^{1/6}}{3} \left(\overline{\omega_0^2}\right)^{1/2} (t - t_0), \tag{19.64}$$

where $\wp(x)$ is the elliptic Weierstrass function, and $\overline{\theta_0^2}$, $\overline{\omega_0^2}$, and $t_0$ are integration constants. This function will increase to infinity in a finite time.

If we omit the thermal diffusivity term in Eq. (15.13), which describes the increase of $\overline{(\text{grad } \vartheta)^2}$ in arbitrary isotropic turbulence, we obtain

$$\frac{d \overline{(\text{grad } \vartheta)^2}}{dt} = -10 B'''_{L\vartheta, \vartheta}(0) = -15 \overline{\left(\frac{\partial u_1}{\partial x_1}\right) \left(\frac{\partial \vartheta}{\partial x_1}\right)^2}. \tag{19.65}$$

Transforming the right-hand side of this equation with the aid of Eq. (12.110), differentiating with respect to $t$ (assuming that $s_\vartheta = \text{const.}$), and taking into account Eq. (19.47), we have

$$\frac{d^2}{dt^2} \overline{(\text{grad } \vartheta)^2} = \left(\frac{5}{3} s_\vartheta + \frac{7}{18} s\right) s_\vartheta \overline{(\text{grad } \vartheta)^2} \cdot \overline{\omega^2}, \tag{19.66}$$

where

$$s_\vartheta = \frac{\overline{\frac{\partial u_1}{\partial x_1}\left(\frac{\partial \vartheta}{\partial x_1}\right)^2}}{\left[\overline{\left(\frac{\partial u_1}{\partial x_1}\right)^2}\right]^{1/2}\overline{\left(\frac{\partial \vartheta}{\partial x_1}\right)^2}} = -\frac{2}{3}\frac{B'''_{L\vartheta,\vartheta}(0)}{\left[-B''_{LL}(0)\right]^{1/2} B''_{\vartheta\vartheta}(0)}. \tag{19.67}$$

Thus, the requirement that the joint fourth-order cumulants of velocity and temperature are zero, leads to the relation

$$\frac{5}{3}s_\vartheta^2 + \frac{7}{18}ss_\vartheta = \frac{2}{3}. \tag{19.68}$$

If we now substitute $s \approx -0.4$ into this equation (in accordance with the data of Townsend; see Sect. 18.1), we obtain $s_\vartheta \approx -0.55$, while for $s \approx -0.78$ (this value follows from the velocity zero-fourth-cumulant assumption) we have $s_\vartheta \approx -0.6$. Neither of these values contradicts the general estimate of $s_\vartheta$ given by Eq. (12.118). We note also that equations (19.62) and (19.56) do not contradict the invariance of the Corrsin integral (19.20). These equations show that, when $k$ is very small, the main term of $\frac{\partial^2 F_{\vartheta\vartheta}(k,t)}{\partial t^2}$ is of the order of $k^4$, which is consistent with the invariance of the coefficient of $k^2$ in the Taylor expansion of the function $F_{\vartheta\vartheta}(k,t)$ which is proportional to the Corrsin integral.

When the probability distributions for the initial fields $\boldsymbol{u}(\boldsymbol{x}, t_0)$ and $\vartheta(\boldsymbol{x}, t_0)$ (with $t_0$ fixed) are normal, the values of $\Gamma_\vartheta(k, k', \mu)$ and of all the fourth-order cumulants of the velocity and temperature fields at $t = t_0$ are zero. In view of Eqs. (19.20) and (19.56), it follows that

$$\left.\frac{\partial T_{\vartheta\vartheta}(k,t)}{\partial t}\right|_{t=t_0} = \int_0^\infty \int_{-1}^1 \frac{Q(k,k',k'')}{4k''^4}\left[k^2 E_{\vartheta\vartheta}(k',t_0) - k'^2 E_{\vartheta\vartheta}(k,t_0)\right] E(k'',t_0)\,dk'\,d\mu.$$

Having specified in some way the values of $E(k, t_0)$ and $E_{\vartheta\vartheta}(k, t_0)$, we can use this formula to calculate the spectral transfer of the temperature fluctuation intensity

$$T_{\vartheta\vartheta}(k,t) \approx \frac{\partial T_{\vartheta\vartheta}(k,t_0)}{\partial t_0}(t - t_0)$$

for small positive values of $t - t_0$. A number of examples of such calculations have been collected by O'Brien (1963).

The complete set of equations (14.62) and (19.56) was integrated numerically by O'Brien and Francis (1962) on the assumption that the energy spectrum $E(k)$ is time-independent (stationary turbulence) and is given (in the proper units) by the formula $E(k) = \frac{4}{(3\pi)^{1/2}}k^4 e^{-k^2}$ while $E_{\vartheta\vartheta}(k, t_0) = B_1 k^2 e^{-k^2}$, $\Gamma_\vartheta(k, k', \mu, t_0) = 0$. When $\nu = \chi = 0$ (infinitely large Reynolds and Peclet numbers), the integration of the above set of equations led to results which were qualitatively similar to those obtained by Ogura (1962b, 1963) for the velocity field (see Sect. 19.3). In particular, the form of the temperature spectral redistribution density $T_{\vartheta\vartheta}(k, t)$ again shows that the transfer of the mean square of the temperature fluctuations takes place from large- to small-scale disturbances, as expected. However, the spectrum $E_{\vartheta\vartheta}(k, t)$ is found to assume negative values after a certain interval of time has elapsed. O'Brien and Francis have also considered the case where $\nu$ and $\chi$ were nonzero and, moreover, assumed that $\vartheta$ is the concentration of an admixture which participates in first-order chemical reactions (see the final part of the Sect. 19.2). The results were again found to be qualitatively similar to those shown in Fig. 43. Similar results were obtained by J. Lee (1966) who considered a further example of the same type (with nonzero $\nu$ and $\chi$ but without chemical reactions), and compared his results with those deduced on the basis of other closure approximations [namely, the Heisenberg hypothesis (17.8), the "fifth-order moment discard" approximation, the modified Millionshchikov hypothesis (which produces no negative regions in the spectrum), and the Kraichnan "direct-interaction approximation" which will be considered in Sect. 19.6]. Finally, O'Brien and Pergament (1964) showed that results of the same kind are obtained when the maxima of $E(k)$ and $E_{\vartheta\vartheta}(k, t_0)$ are appreciably shifted relative to each other.

The fact that the application of the Millionshchikov zero-fourth-cumulant hypothesis to the time evolution of the field $\vartheta(x, t)$ must finally lead to the appearance of negative parts of the spectrum $E_{\vartheta\vartheta}(k, t)$, was also predicted by Kraichnan (1962). Following Roberts (1957), Kraichnan applied this hypothesis to the admixture diffusion problem from an instantaneous point source in a field of stationary turbulence, and showed that, when the diffusion time is large, the calculated concentration could assume negative values (see Sect. 24.1 below) Kraichnan also considered the zero fifth- and even sixth-order cumulant approximations. Here again unphysical results were obtained; such results can apparently not be avoided by truncating the cumulant sequence at any order.

## 19.5 Space-Time Correlation Functions. The Case of Stationary Isotropic Turbulence

So far we have discussed only the spatial correlation functions at particular instants. However, the space-time correlation functions are the more complete statistical characteristics of turbulence. They describe the correlation between the various flow variables at different points and different times. The dynamic equations for the space-time correlation functions are even simpler to derive than the corresponding equations for the purely spatial statistical characteristics. However, these more general correlation functions involve a greater number of variables and are, therefore, more difficult to investigate.

The space-time correlation tensor of the velocity field is defined by

$$B_{ij}(x, t; x', t') = \overline{u_i(x, t)\, u_j(x', t')}. \tag{19.69}$$

If we write down the Navier-Stokes equation for the component $u_i(x, t)$ and multiply all its terms by $u_j(x', t')$, we obtain, after taking the average, the following dynamic equation:

$$\frac{\partial B_{ij}(x, t; x', t')}{\partial t} + \frac{\partial B_{ik,j}(x, t; x', t')}{\partial x_k} = -\frac{1}{\rho}\frac{\partial B_{pj}(x, t; x', t')}{\partial x_i} + \nu \frac{\partial^2 B_{ij}(x, t; x', t')}{\partial x_k \partial x_k}, \tag{19.70}$$

where $B_{ik,j}(x, t; x', t') = \overline{u_i(x, t)\, u_k(x, t)\, u_j(x', t')}$ and $B_{pj}(x, t; x', t') = \overline{p(x, t)\, u_j(x', t')}$. In the case of isotropic turbulence in an incompressible fluid we have $B_{pj} = 0$, and the tensors $B_{ij}$ and $B_{ik,j}$ are expressed in the usual way in terms of the scalar functions $B_{LL}(r,t,t')$ and $B_{LL,L}(r,t,t')$. Equation (19.70) then becomes

$$\frac{\partial B_{LL}(r, t, t')}{\partial t} = \frac{1}{2}\left[\frac{\partial B_{LL,L}(r, t, t')}{\partial r} + \frac{4}{r} B_{LL,L}(r, t, t')\right] + \\ + \nu\left[\frac{\partial^2 B_{LL}(r, t, t')}{\partial r^2} + \frac{4}{r}\frac{\partial B_{LL}(r, t, t')}{\partial r}\right], \quad (19.71)$$

which is very similar to the von Kármán-Howarth equation (14.9). The equation

$$\frac{\partial B_{LL}(r, t, t')}{\partial t'} = \frac{1}{2}\left[\frac{\partial B_{LL,L}(r, t', t)}{\partial r} + \frac{4}{r} B_{LL,L}(r, t', t)\right] + \\ + \nu\left[\frac{\partial^2 B_{LL}(r, t, t')}{\partial r^2} + \frac{4}{r}\frac{\partial B_{LL}(r, t, t')}{\partial r}\right]. \quad (19.72)$$

can be derived in a similar way.

Equations (19.71) and (19.72) each contain two unknown functions and are not, therefore, closed. Although the function $B_{LL}(r,t,t')$ can be readily eliminated from these two equations (see Repnikov, 1967), this results in an equation for $B_{LL,L}(r,t,t')$ which contains $B_{LL,L}(r,t't)$, i.e., it merely relates the values of $B_{LL,L}(r,t,t')$ for $t' < t$ with those for $t' > t$. As in the case of the von Kármán-Howarth equation, the addition of equations for the higher-order correlation functions again does not result in a closed system, and closure requires additional hypotheses. The simplest method to adopt is to assume that the third-order moments are small in comparison with the second-order moments. From Eqs. (19.71) and (19.72) we then have

$$\frac{\partial B_{LL}(r, t, t')}{\partial t} = \frac{\partial B_{LL}(r, t, t')}{\partial t'} = \nu\left[\frac{\partial^2 B_{LL}(r, t, t')}{\partial r^2} + \frac{4}{r}\frac{\partial B_{LL}(r, t, t')}{\partial r}\right]. \quad (19.73)$$

Transforming from the variables $t$ and $t'$ to $t_1 = (t + t')/2$ and $\tau = t' - t$) we find that

$$\frac{\partial B_{LL}(r, t_1, \tau)}{\partial \tau} = 0, \quad \frac{\partial B_{LL}(r, t_1, \tau)}{\partial t_1} = 2\nu\left[\frac{\partial^2 B_{LL}(r, t_1, \tau)}{\partial r^2} + \frac{4}{r}\frac{\partial B_{LL}(r, t_1, \tau)}{\partial r}\right]. \quad (19.74)$$

Hence, it is clear that, in the linear theory, where the third-order velocity moments are neglected, the space-time correlation function depends only on the "mean time" $t_1 = (t + t')/2$ (but not on $\tau = t' - t$). Accordingly,

$$\overline{u_L(x, t)\, u_L(x', t')} = B_{LL}\left(r, \frac{t + t'}{2}\right) = B_{LL}(r, t_1). \quad (19.75)$$

The space-time correlation coefficient

$$R_{LL}(r, t, \tau) = \frac{B_{LL}(r, t_1)}{\left[B_{LL}\left(0, t_1 - \frac{\tau}{2}\right) B_{LL}\left(0, t_1 + \frac{\tau}{2}\right)\right]^{1/2}}$$

will then be dependent on $\tau$ also and, for example, when the turbulent energy decays in accordance with a power law, it will decrease with increasing $\tau$ for a fixed $t_1$. The dependence of the function $B_{LL}(r, t_1)$ on $t_1$ should be the same as the time dependence of the space correlation function $B_{LL}(r, t) = B_{LL}(r, t, t)$, so that in view of Eq. (15.44),

$$B_{LL}(r, t, t') = \frac{e^{-\frac{r^2}{4\nu(t+t'-2t_0)}}}{2\nu(t+t'-2t_0)\, r^{3/2}} \times$$
$$\times \int_0^\infty B_{LL}(\xi, t_0, t_0)\, \xi^{5/2} I_{3/2}\left[\frac{r\xi}{2\nu(t+t'-2t_0)}\right] e^{-\frac{\xi^2}{4\nu(t+t'-2t_0)}}\, d\xi \qquad (19.76)$$

This formula was put forward by Millionshchikov (1939a). It follows that the values of $B_{LL}(r, t, t')$ can be determined for all $t$ and $t'$ if we know the initial value of only the space correlation function $B_{LL}(r, t_0)$. In the case of an initial value $B_{LL}(r, t_0)$ with finite and nonzero Loitsyanskii integral $\Lambda$, we have for large $t + t'$,

$$B_{LL}(r, t, t') \approx \frac{\Lambda}{24\sqrt{2\pi}\,[\nu(t+t'-2t_0)]} e^{-\frac{r^2}{4\nu(t+t'-2t_0)}}. \qquad (19.77)$$

The data of Batchelor and Townsend (1948b), which were discussed in Sect. 15.4, suggest that Eq. (19.77) should be adequately satisfied in the turbulence behind a fine grid at a sufficient distance from it. This, however, has not as yet been verified.

The results given by Eqs. (19.76) and (19.77) can be applied only to the final period of decay (for small values of Re). Somewhat more accurate results can be obtained by assuming that moments of a particular higher order are identically zero [see, for example, Deissler (1961) where fourth-order space-time moments are neglected]. In this case the correlation function $B_{LL}(r, t, t') = B_{LL}(r, t_1, \tau)$ is also found to depend on the variable $\tau$, and the spectral energy redistribution density $T(k, t, t') = T(k, t_1, \tau)$ satisfies the condition $\int_0^\infty T\, dk = 0$ only for $\tau = 0$.

The study of the functions $B_{LL}(r, t, t')$, $B_{LL,L}(r, t, t')$, etc., depending on the three variables is in general a very complicated problem. Some quite preliminary results were published by Rosen (1967) and Rosen, Okolowski, and Eckstut (1969) concerning the space-time zero-fourth-cumulant approximation for the dynamic equation for the function $B_{LL}(r,t,t')$ [or, more precisely, for the function

$$\xi(r) = -\frac{1}{2} \int_r^\infty B_{LL}(y, t, t')\, y\, dy$$

which determines $B_{LL}(r,t,t')$ uniquely). The derivation of the equation was obtained by these authors with the aid of the functional formulation of turbulence theory which will be considered in Chapt. 10, but this formulation is in fact not essential for their analysis. The results obtained are rather cumbersome and we shall not discuss them here. Instead, we shall restrict our attention to the simplest case where the turbulence is not only isotropic but also stationary. The space-time correlation functions will then depend only on the two variables $r$ and $\tau = t' - t$. In viscous fluid, stationary turbulence is clearly possible only in the presence of external forces producing an influx of energy compensating the viscous energy dissipation. To ensure that the stationary turbulence is also isotropic, the external forces must themselves be isotropic. Hence, it is clear that this type of turbulence is a highly artificial mathematical idealization (see the description of the isotropic external forces in Sect. 14.1). Nevertheless, the model of stationary isotropic turbulence may be useful for the description of natural high Reynolds number turbulent flows having the properties of local isotropy and quasi-stationarity, which we shall discuss in detail in the next chapter.

Following Chandrasekhar (1955) let us consider stationary isotropic turbulence in which the fourth-order space-time cumulants are identically zero. Since the function $B_{LL}(r, t, t') = B_{LL}(r, t', t) = B_{LL}(r, \tau)$ is an even function of $\tau$, we may suppose that $t' > t$, i.e., $\tau > 0$. Equation (19.72) can then be used only to express the function $B_{LL, L}(r, -\tau)$ in terms of $B_{LL,L}(r, \tau)$ and $B_{LL}(r,\tau)$.[37] If we are not interested in the values of $B_{LL,L}(r,\tau)$ for $\tau < 0$ then it is sufficient to restrict our attention to Eq. (19.71). In the stationary case this equation assumes the form

$$-\left[\frac{\partial}{\partial \tau} + \nu\left(\frac{\partial^2}{\partial r^2} + \frac{4}{r}\frac{\partial}{\partial r}\right)\right] B_{LL}(r, \tau) = \frac{1}{2}\left(\frac{\partial}{\partial r} + \frac{4}{r}\right) B_{LL, L}(r, \tau). \quad (19.78)$$

Multiplying the Navier-Stokes equation for the component $u_l(\boldsymbol{x}', t')$ by the product $u_i(\boldsymbol{x}, t)\, u_j(\boldsymbol{x}, t)$, and taking the average, we obtain

$$\frac{\partial B_{ij, l}(\boldsymbol{r}, \tau)}{\partial \tau} + \frac{\partial B_{ij, kl}(\boldsymbol{r}, \tau)}{\partial r_k} = -\frac{1}{\rho}\frac{\partial B_{ij, p}(\boldsymbol{r}, \tau)}{\partial r_l} + \nu \frac{\partial^2 B_{ij, l}(\boldsymbol{r}, \tau)}{\partial r_k \partial r_k}, \quad (19.79)$$

This equation relates the third- and fourth-order space-time moments. Since the fourth-order space-time cumulants (on which there are no data at present) have been assumed to be zero, we have

$$B_{ij, kl}(\boldsymbol{r}, \tau) = B_{ik}(\boldsymbol{r}, \tau)\, B_{jl}(\boldsymbol{r}, \tau) + B_{il}(\boldsymbol{r}, \tau)\, B_{jk}(\boldsymbol{r}, \tau) + B_{ij}(0, 0)\, B_{kl}(0, 0). \quad (19.80)$$

Substituting this into Eq. (19.79), and expressing the tensors $B_{ij}$, $B_{ij, l}$ and $B_{ij, p}(\boldsymbol{r}, \tau) = B_{p, ij}(-\boldsymbol{r}, -\tau)$ in the usual way in terms of the scalar functions $B_{LL}(r, \tau)$, $B_{LL, L}(r, \tau)$, $B_{p', LL}(r, \tau)$ and $B_{p', NN}(r, \tau)$, we obtain

$$\left[\frac{\partial}{\partial \tau} - \nu\left(\frac{\partial^2}{\partial r^2} + \frac{4}{r}\frac{\partial}{\partial r} - \frac{4}{r^2}\right)\right] B_{LL, L}(r, \tau) =$$
$$= -\left[2B_{LL}(r, \tau) + r\frac{\partial B_{LL}(r, \tau)}{\partial r}\right]\frac{\partial B_{LL}(r, \tau)}{\partial r} +$$
$$+ \frac{2}{\rho}\frac{\partial B_{p', NN}(r, -\tau)}{\partial r} = X(r, \tau). \quad (19.81)$$

Using Eq. (18.8), which is a consequence of the Millionshchikov hypothesis, we can readily verify that the right-hand side $X(r, \tau)$ of Eq. (19.81) satisfies the relation

[37] We note that it is erroneounly assumed by Chandrasekhar (1955) that

$$B_{LL, L}(r, -\tau) = B_{LL, L}(r, \tau),$$

but this assumption is not used by him.

$$\frac{\partial}{\partial r}\left(\frac{\partial}{\partial r}+\frac{4}{r}\right)X(r,\tau)=-2B_{LL}(r,\tau)\frac{\partial}{\partial r}\left(\frac{\partial^2}{\partial r^2}+\frac{4}{r}\frac{\partial}{\partial r}\right)B_{LL}(r,\tau). \qquad (19.82)$$

We have thus obtained a closed set of equations, namely, Eqs. (19.78), (19.81), and (19.82), for the three unknown functions $B_{LL}$, $B_{LL,L}$, and $X$ of the two variables $r$ and $\tau$.

If we apply the operation $-\frac{\partial}{\partial \tau}+\nu\left(\frac{\partial^2}{\partial r^2}+\frac{4}{r}\frac{\partial}{\partial r}\right)$ to both sides of Eq. (19.78), and remember that

$$\left(\frac{\partial^2}{\partial r^2}+\frac{4}{r}\frac{\partial}{\partial r}\right)\left(\frac{\partial}{\partial r}+\frac{4}{r}\right)=\left(\frac{\partial}{\partial r}+\frac{4}{r}\right)\left(\frac{\partial^2}{\partial r^2}+\frac{4}{r}\frac{\partial}{\partial r}-\frac{4}{r^2}\right)$$

we can use Eq. (19.81) to eliminate the unknown function $B_{LL,L}$ and obtain

$$\left[\frac{\partial^2}{\partial \tau^2}-\nu^2\left(\frac{\partial^2}{\partial r^2}+\frac{4}{r}\frac{\partial}{\partial r}\right)^2\right]B_{LL}(r,\tau)=-\frac{1}{2}\left(\frac{\partial}{\partial r}+\frac{4}{r}\right)X(r,\tau). \qquad (19.83)$$

Finally, if we differentiate both sides of Eq. (19.83) with respect to $r$, and use Eq. (19.82), we can eliminate $X(r, \tau)$ and obtain the following single equation for the function $B_{LL}(r, \tau)$:

$$\frac{\partial}{\partial r}\left[\frac{\partial^2}{\partial \tau^2}-\nu^2\left(\frac{\partial^2}{\partial r^2}+\frac{4}{r}\frac{\partial}{\partial r}\right)^2\right]B_{LL}(r,\tau)=$$
$$=B_{LL}(r,\tau)\frac{\partial}{\partial r}\left(\frac{\partial^2}{\partial r^2}+\frac{4}{r}\frac{\partial}{\partial r}\right)B_{LL}(r,\tau). \qquad (19.84)$$

This equation was found by Chandrasekhar (1955). Its accuracy and range of validity are still not clear. All that is known is that it must be augmented by terms describing the effect of external forces. From this point of view, a preferable derivation is that due to Wyld (1961) who used the general methods of perturbation theory (see Sect. 19.6). In any case, Eq. (19.84) is quite difficult to apply. It is of a relatively high order, and its solution demands the use of a large number of boundary and initial conditions which are very difficult to choose. Equation (19.84) and its "spectral form" has been investigated by Chamberlain and Roberts (1955), Chandrasekhar (1956), Backus (1957), Wentzel (1958), and others, but no concrete results with clear physical interpretations have yet been obtained from it.

The above analysis can be extended to the case of the temperature space-time correlation function $B_{\vartheta\vartheta}(r,t,t')$. In particular, for stationary isotropic turbulence with zero space-time fourth-order cumulants, the Corrsin equation (14.59) becomes

$$-\left[\frac{\partial}{\partial \tau}+\chi\left(\frac{\partial^2}{\partial r^2}+\frac{2}{r}\frac{\partial}{\partial r}\right)\right]B_{\vartheta\vartheta}(r,\tau)=\left(\frac{\partial}{\partial r}+\frac{2}{r}\right)B_{L\vartheta,\vartheta}(r,\tau), \qquad (19.85)$$

and the equation for the third-order moments [obtained by multiplying the thermal conduction equation for $\vartheta(x', t')$ by $u_l(x, t)\,\vartheta(x, t)$, taking the average of the result, and replacing the term $B_{l\vartheta,j\vartheta}$ by a formula analogous to Eq. (19.80)] is of the form

$$\left[\frac{\partial}{\partial \tau}-\chi\frac{\partial}{\partial r}\left(\frac{\partial}{\partial r}+\frac{2}{r}\right)\right]B_{L\vartheta,\vartheta}(r,\tau)=-B_{LL}(r,\tau)\frac{\partial B_{\vartheta\vartheta}(r,\tau)}{\partial r}. \qquad (19.86)$$

If we apply the operation $-\frac{\partial}{\partial\tau}+\chi\left(\frac{\partial^2}{\partial r^2}+\frac{2}{r}\frac{\partial}{\partial r}\right)$ to both sides of Eq. (19.85), and take Eq. (19.86) into account, we obtain

$$\left[\frac{\partial^2}{\partial\tau^2}-\chi^2\left(\frac{\partial^2}{\partial r^2}+\frac{4}{r}\frac{\partial}{\partial r}\right)^2\right]B_{\vartheta\vartheta}(r,\tau)=\frac{1}{r^2}\frac{\partial}{\partial r}\left(r^2B_{LL}(r,\tau)\frac{\partial B_{\vartheta\vartheta}(r,\tau)}{\partial r}\right). \quad (19.87)$$

Derivations of this equation have been discussed by Panchev (1960) and Jain (1962). In both cases, however, the final result was incorrect because it was erroneously assumed that $B_{L\vartheta,\vartheta}(r,\tau)=B_{L\vartheta,\vartheta}(r,-\tau)$. Equation (19.87) is similar to Eq. (19.84), but it requires a preliminary determination of $B_{LL}(r,\tau)$, and is therefore even more difficult to use than Eq. (19.84).

## 19.6 Application of Perturbation Theory and the Diagram Technique

Closure of the velocity moment equations by neglecting moments of order $n+1$ is equivalent to the expansion of the moments in powers of Re, and the retention of only the zero-order terms for the $n$th moments, the zero-order and first-order terms for the $(n-1)$th moments, and so on down to the second moments, inclusive, for which terms of order up to $n-2$ in Re are retained. A similar situation was noted in Sect. 19.3 for the various cumulant discard approximations.

Series in powers of Re are very useful in establishing the nature of the approximation involved in the closure of the moment equations with the aid of any particular hypothesis (although for large Re, which are typical for developed turbulence, the series are found to diverge and are therefore of purely formal character). Moreover the series representation is a starting point for many approximate methods which use either a summation over some parts of the expansion or rearrangements and some additional transformation of the primary series (cf., for example, Kraichnan, 1970a). We shall therefore devote the present section to an analysis of series in powers of Re, and will begin by using these series to solve the fluid dynamic equations, prior to the introduction of any statistical procedures.

We recall that the nonlinear terms in the Navier-Stokes equations (including the pressure gradient which is a quadratic function of the velocity field) describe the forces of inertial interaction between the spatial inhomogeneities of the velocity field. If we transform these equations to the dimensionless variables $y=x/L$, $v=u/U$ and $\tau=\nu t/L^2$, where $L$ and $U$ are typical length and velocity scales in the particular flow under consideration, then the nonlinear terms will acquire the factor $\text{Re}=UL/\nu$, which will thus play the role of the inertial interaction constant. When Re is small, the inertial interaction forces will produce only a small perturbation of the "main flow" described by the linear equations (obtained from the Navier-Stokes equations by rejecting the nonlinear terms). The solution of the complete Navier-Stokes equations by a power series will then simply be an application of the usual perturbation-theory method, and we shall be able to use all the general results, including the graphical "diagram representation" of the series terms in powers of the interaction constant, developed in quantum field theory [see, for example, Schweber, Bethe and Hoffmann (1955)]. If, on the other hand, Re is large so that the inertial interactions are very strong, the direct use of expansions in powers of the interaction constant will be ineffective (as is always the case in theories with strong interaction), but the formal series in powers of Re will still be useful for the purposes noted above.

Since we shall be concerned in this section primarily with stationary isotropic turbulence, which is possible only in the presence of external forces (which themselves must be statistically stationary and isotropic, and can conveniently be assumed to be solenoidal as well), we shall write the Navier-Stokes equation in the form

$$\left(\frac{\partial}{\partial t}-\nu\Delta\right)u_j=X_j+\left(\frac{\partial}{\partial x_j}\Delta^{-1}\frac{\partial^2}{\partial x_\alpha\,\partial x_\beta}-\delta_{j\alpha}\frac{\partial}{\partial x_\beta}\right)u_\alpha u_\beta, \qquad (19.88)$$

where $X_j$ are the components of the external force vector. We shall use the fact that, since the fluid is incompressible, the pressure can be expressed in terms of the velocity field by the formula $p/\rho=-\Delta^{-1}\dfrac{\partial^2 u_\alpha u_\beta}{\partial x_\alpha\,\partial x_\beta}$ where $\Delta^{-1}$ is the integral operator reciprocal to the Laplace operator $\Delta$. Henceforth, we shall denote the four-dimensional space-time points $(\boldsymbol{x}, t), (\boldsymbol{x}_1, t_1)$, and so on, by the letters $M$, $M_1$, etc., and will use the abbreviated notation $dx_1\,dx_2\,dx_3\,dt=dM$.

It will be convenient to write down the second term on the right-hand side of Eq. (19.88) in the form

$$P_{j\alpha\beta}(M, M_1)\,u_\alpha(M_1)\,u_\beta(M_1)=\int P_{j\alpha\beta}(M-M_1)\,u_\alpha(M_1)\,u_\beta(M_1)\,dM_1. \qquad (19.89)$$

Writing $u_\alpha(M)\,u_\beta(M)$ in the form of an integral of $\delta(M-M_1)u_\alpha(M_1)u_\beta(M_1)$ with respect to $M_1$, we can verify that the kernal $P_{j\alpha\beta}(M)$ is the result of the application of the operator in round brackets on the right-hand side of Eq. (19.88) to the singular function $\delta(M)$. If we represent this function by the Fourier integral $(2\pi)^{-4}\int e^{i(\boldsymbol{kx}-\omega t)}\,dp$ [we shall be denoting the four-dimensional "wave vector" $(\boldsymbol{k}, \omega)$ by the symbol $\boldsymbol{p}$, and use the abbreviated notation $dk_1\,dk_2\,dk_3\,d\omega=dp$), we obtain

$$P_{j\alpha\beta}(M)=-\frac{i}{2}(2\pi)^{-4}\int e^{i(\boldsymbol{kx}-\omega t)}\Delta_{j\alpha\beta}(\boldsymbol{k})\,dp, \qquad (19.90)$$
$$\Delta_{j\alpha\beta}(\boldsymbol{k})=k_\alpha\Delta_{j\beta}(\boldsymbol{k})+k_\beta\Delta_{j\alpha}(\boldsymbol{k}),$$

where, as above, $\Delta_{jl}(\boldsymbol{k})=\delta_{jl}-k_jk_l/k^2$, and we have written down only that part of the result which is symmetric in $\alpha$ and $\beta$ [only this part enters into the definition of the operator $P_{j\alpha\beta}$ by Eq. (19.89)].

Let us also introduce the Green operator (or propagator) $G(M, M_1)$, which is the reciprocal of $\left(\dfrac{\partial}{\partial t}-\nu\Delta\right)$, on the left-hand side of the Navier-Stokes Eq. (19.88). In infinite space, and for zero initial conditions at $t\to-\infty$, the operator $G$ is given by

$$G(M, M_1)\,f(M_1)=\int G(M-M_1)\,f(M_1)\,dM_1,$$

where

$$G(M)=(2\sqrt{\pi\nu t})^{-3}\exp\left\{-\frac{\boldsymbol{x}^2}{4\nu t}\right\}E(t), \qquad (19.91)$$

The function $E(t)$ is equal to unity for $t > 0$ and zero for $t < 0$. The result of the application of this operator to the external force field $X_j(M)$, i.e., the solution of Eq. (19.88) linearized by rejection of the nonlinear terms, will be written in the form

$$G(M, M_1) X_j(M_1) = u_j^{(0)}(M). \tag{19.92}$$

It will be convenient later to use the tensor operator $G_{jl} = \delta_{jl} G$. The result of the application of the operator $G$ to Eq. (19.88) can then be written in the form

$$u_j(M) = u_j^{(0)}(M) + G_{ja}(M, M_1) P_{\alpha\beta\gamma}(M_1, M_2) u_\beta(M_2) u_\gamma(M_2). \tag{19.93}$$

This equation is merely the symbolic integral form of the Navier-Stokes equations. We shall seek its solution $u_j(M)$ in the form of a series in powers of Re. We recall that, when the Navier-Stokes equations are written in terms of dimensionless variables, the nonlinear terms are preceded by Re as a factor, so that the operator $P_{\alpha\beta\gamma}$ in Eq. (19.93) can be regarded as being of the order of Re. Consequently, assuming that Re = 0 in Eq. (19.93), we can verify that the series for $u_j(M)$ in powers of Re begins with the term $u_j^{(0)}(M)$:

$$u_j(M) = u_j^{(0)}(M) + u_j^{(1)}(M) + u_j^{(2)}(M) + \ldots, \tag{19.94}$$

where $u_j^{(n)}(M)$ is the term of order $(\text{Re})^n$. Substituting this series into Eq. (19.93), and collecting terms of the same order in Re, we have for the first powers of Re

$$u_j^{(1)}(M) = G_{ja}(M, M_1) P_{\alpha\beta\gamma}(M_1, M_2) u_\beta^{(0)}(M_2) u_\gamma^{(0)}(M_2). \tag{19.95}$$

Since the operator $P_{\alpha\beta\gamma}$ is symmetric in $\beta$ and $\gamma$, we have for the second power of Re

$$u_j^{(2)}(M) = 2G_{ja}(M, M_1) P_{\alpha\beta\gamma}(M_1, M_2) u_\beta^{(0)}(M_2) u_\gamma^{(1)}(M_2) \tag{19.96}$$

or, if we express $\boldsymbol{u}^{(1)}$ in terms of $\boldsymbol{u}^{(0)}$ by means of Eq. (19.95), we obtain

$$u_j^{(2)}(M) = 2G_{ja}(M, M_1) P_{\alpha\beta\gamma}(M_1, M_2) u_\beta^{(0)}(M_2) G_{\gamma\delta}(M_2, M_3) \times \\ \times P_{\delta\mu\nu}(M_3, M_4) u_\mu^{(0)}(M_4) u_\nu^{(0)}(M_4). \tag{19.97}$$

The expressions for the higher-order terms become increasingly more complicated, but all of them can be described graphically in the form of very simple "diagrams." Following Wyld (1961) (this technique is expounded also in the book by Beran, 1968) we can represent the operator $G_{jl}(M, M_1)$ by a line, one end of which is given by the subscript $j$ and the coordinate $M$, and the other end the subscript $l$ and the coordinate $M_1$. The operator $P_{\alpha\beta\gamma}(M, M_1)$ is represented by a point (called the *vertex*) which is assigned all the subscripts (i.e., $\alpha, \beta$, and $\gamma$) and both coordinates $M$ and $M_1$. Finally, the function $u_j^{(0)}(M)$ is represented by a broken line drawn from the vertex and assigned the subscript $j$ and the coordinate $M$. Equations (19.95), (19.97), and the formula for $u_j^{(3)}$ which we have not reproduced here, are represented by the diagrams shown in Fig. 44. These diagrams have the property that three lines are joined together at each vertex: either one solid line and two broken lines, or two solid lines and one broken line or, again, three solid lines (since a triplet of subscripts corresponds to each vertex). This property allows us to draw only the solid lines on the diagrams, and to "keep in mind" all the broken lines; we shall use such a procedure in what follows.

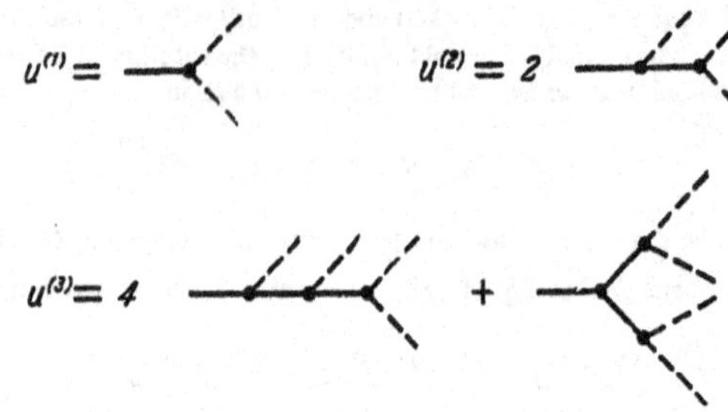

$u^{(1)} =$ $\qquad$ $u^{(2)} = 2$

$u^{(3)} = 4$ $\qquad + \qquad$

Fig. 44

It is now clear that the term $u^{(n)}$ in Eq. (19.94) can be represented by the sum of all the possible diagrams with $n$ vertices, having the structure of a "tree" lying on its side (this means that as we move along the diagram from left to right, a double branch is possible at each vertex). It is readily verified that each diagram enters this sum with the factor $2^{m_1+m_2}$ where $m_1$ is the number of vertices on which two solid lines converge, and $m_2$ is the number of vertices on which three solid lines converge but with an asymmetric branching of the diagram. As an illustration, Fig. 45 gives the diagrams for $u^{(4)}$ and $u^{(5)}$.

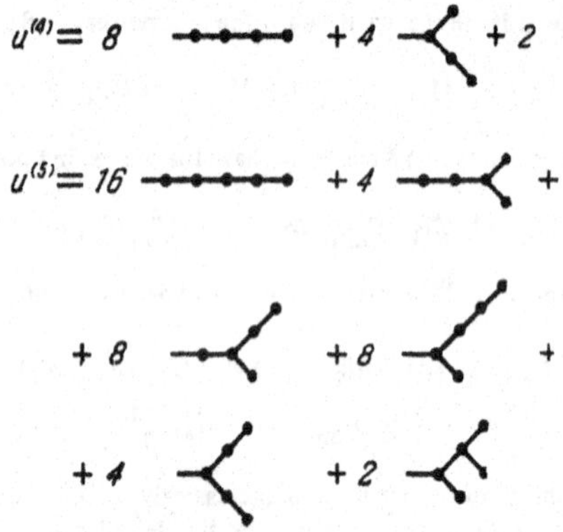

$u^{(4)} = 8 \qquad +4 \qquad +2$

$u^{(5)} = 16 \qquad +4 \qquad +$

$+8 \qquad +8 \qquad +$

$+4 \qquad +2$

Fig. 45

Let us now use the series in powers of Re to calculate the statistical characteristics of the velocity field. The most complete statistical characteristic of a random field is its characteristic functional (see Sect. 3.4, Vol. 1). In Chapt. 10 we shall derive a number of dynamic equations for the characteristic functional of the velocity field, while in Sect. 29.3 we shall investigate the solutions of these equations with the aid of series in powers of Re. Here, we shall follow Wyld (1961) and demonstrate how the series (19.94) can be used to calculate the second order velocity moments in stationary isotropic turbulence.

In this case, the mean velocity is zero, and the second moments of the velocity have the form

$$\overline{u_j(M)\,u_l(M_1)} = B_{jl}(M - M_1). \tag{19.98}$$

The second moments of the external forces $X_j(M)$ should have an analogous form in the present case, and therefore, in accordance with Eqs. (19.92)–(19.93) and the zero-order approximation for the velocity field $u_j^{(0)}(M)$, we have:

$$\overline{X_j(M)\,X_l(M_1)} = C_{jl}(M - M_1), \tag{19.99}$$

$$\overline{u_j^{(0)}(M)\,u_l^{(0)}(M_1)} = B_{jl}^{(0)}(M - M_1). \tag{19.100}$$

We shall evaluate the correlation function (19.98) using the series given by Eq. (19.94), and formulas (19.95), (19.97), and so on. If we multiply together, term by term, two series of the form of Eq. (19.94), take the average, and group together terms of the same order in Re, we can verify that terms of order $(\mathrm{Re})^n$ contain $(n + 2)$-order moments of the function $u^{(0)}$. We shall suppose that the external force field, and by Eq. (19.92) the field $u^{(0)}$ also, are Gaussian with zero means [this assumption is equivalent to using the maximal randomness principle put forward by Kraichnan (1959)]. All the odd-order moments of $u^{(0)}$ will then be zero, while even-order moments can be expressed in terms of the second moments by the formulas for the multidimensional normal probability distributions [see Eqs. (4.27), Vol. 1]: The mean value of any product of $2n$ factors of the form $u_j^{(0)}(M)$ will consist of $1\cdot 3\cdot 5 \ldots (2n - 1)$ terms which are all the possible different products of $n$ second moments made up of the above $2n$ factors.

Thus, the expansion of the correlation function given by Eq. (19.98) in powers of Re will be of the form

$$\begin{aligned} B_{jl}(M - M_1) = B_{jl}^{(0)}(M - M_1) &+ \left\{\overline{u_j^{(0)}(M)\,u_l^{(2)}(M_1)} + \overline{u_j^{(1)}(M)\,u_l^{(1)}(M_1)} + \right.\\ &\left. + \overline{u_j^{(2)}(M)\,u_l^{(0)}(M_1)}\right\} + \left\{\overline{u_j^{(0)}(M)\,u_l^{(4)}(M)} + \overline{u_j^{(1)}(M)\,u_l^{(3)}(M_1)} + \right.\\ + \overline{u_j^{(2)}(M)\,u_l^{(2)}(M_1)} + \overline{u_j^{(3)}(M)\,u_l^{(1)}(M_1)} &\left. + \overline{u_j^{(4)}(M)\,u_l^{(0)}(M_1)}\right\} + \ldots . \end{aligned} \tag{19.101}$$

To evaluate each of the terms $\overline{u_j^{(m)}(M)\,u_l^{(n)}(M_1)}$ we must "multiply" all the diagrams forming both factors term by term. Each such "product" of two diagrams will contain the $(m + n + 2)$-order moment of the function $u^{(0)}$ which, according to the foregoing analysis, is either zero (for odd $m + n$) or splits into the sum of $1\cdot 3\cdot 5 \ldots (m + n + 1)$ terms containing all the possible products of the second moments (for even $m + n$). Each of these terms can be represented graphically by drawing two tree-like diagrams to be multiplied end to end (with the branches facing each other) and joining in pairs all the broken lines (which were "kept in mind"). The broken lines joining the pairs of vertices will now denote "unperturbed" correlation functions of the form $\overline{u_j^{(0)}(M)\,u_l^{(0)}(M_1)}$; their ends will correspond to subscripts $j$ and $l$, and coordinates $M$ and $M_1$. Thus, the terms of zero order in Re in Eq. (19.101) will be represented simply by a broken line. The three second-order terms will be represented by the diagrams shown in Fig. 46 (the numerical factors arise when we take the factors of Fig. 44 into account, and also add up all the topologically equivalent diagrams representing identical terms).

Since the mean velocity in our model is zero, we have $\overline{u^{(n)}} = 0$ for any $n$. In particular, the quantity $\overline{u_j^{(1)}} =$ —◌ is zero, so that all the diagrams containing the element —◌ are zero. Consequently, the second terms in the three sums of Fig. 46 are all zero. An analogous result is valid for the higher-order terms in Eq. (19.101): All the diagrams containing elements such as $\overline{u_j^{(n)}}$ are zero and we shall, therefore, ignore them henceforth.

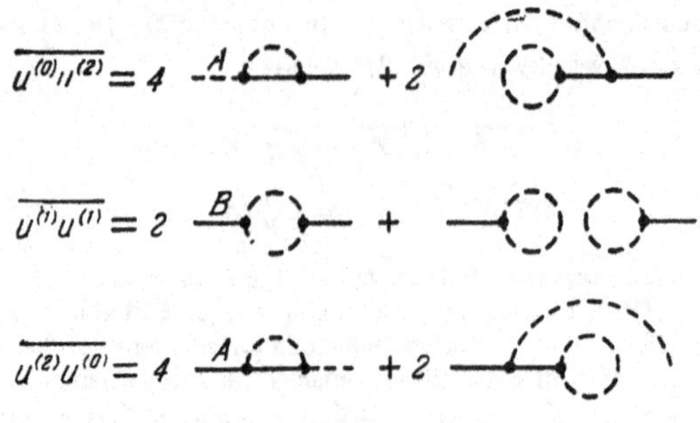

$\overline{u^{(0)}u^{(2)}} = 4$ A $+ 2$

$\overline{u^{(1)}u^{(1)}} = 2$ B $+$

$\overline{u^{(2)}u^{(0)}} = 4$ A $+ 2$

Fig. 46

The diagrams for the fourth-order terms in Eq. (19.101) based on this rule are shown in Fig. 47 (diagrams for $\overline{u_j^{(1)}u_l^{(3)}}$ and $\overline{u_j^{(0)}u_l^{(4)}}$ are, of course, obtained from the diagrams for $\overline{u_j^{(3)}u_l^{(1)}}$ and $\overline{u_j^{(4)}u_l^{(0)}}$ by reflection in the vertical line).

Instead of the correlation functions $B_{jl}(M - M_1)$ we can use their Fourier transforms, i.e., the spectral functions

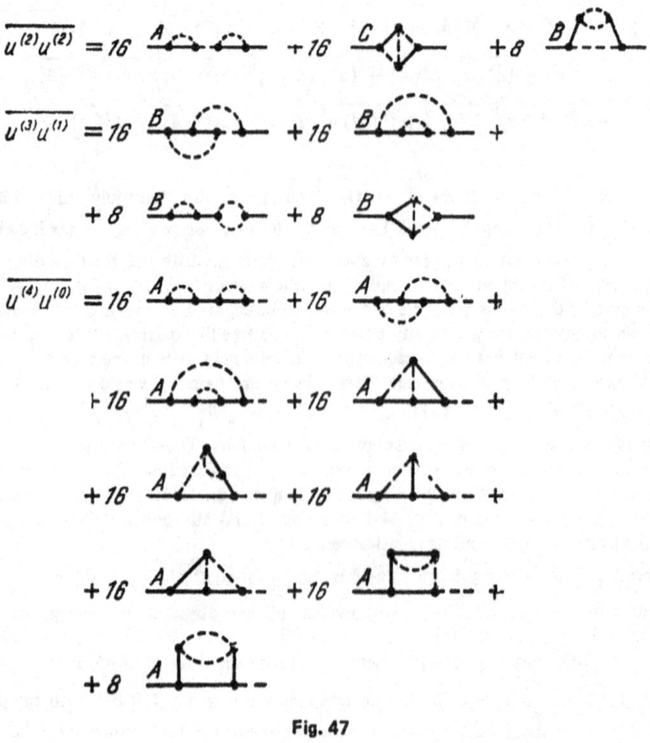

$\overline{u^{(2)}u^{(2)}} = 16$ A $+ 16$ C $+ 8$ B

$\overline{u^{(3)}u^{(1)}} = 16$ B $+ 16$ B $+$

$+ 8$ B $+ 8$ B

$\overline{u^{(4)}u^{(0)}} = 16$ A $+ 16$ A $+$

$+ 16$ A $+ 16$ A $+$

$+ 16$ A $+ 16$ A $+$

$+ 16$ A $+ 16$ A $+$

$+ 8$ A

Fig. 47

$$F_{jl}(p)=(2\pi)^{-4}\int e^{-i(kr-\omega\tau)}B_{jl}(r,\tau)\,dr\,d\tau. \qquad (19.102)$$

It is readily verified that the diagram for these functions will have exactly the same form as for $B_{jl}(M-M_1)$ provided we interpret a solid line with ends corresponding to the subscripts $j$ and $l$ as the operator representing multiplication by a function

$$\Gamma_{jl}(p)=\delta_{jl}(-i\omega+\nu k^2)^{-1}, \qquad (19.103)$$

a broken line with ends corresponding to $j$ and $l$ as the spectral function of the "unperturbed" velocity field $F^{(0)}_{jl}(p)$ and, finally, if we interpret a vertex which earlier corresponded to the operator $P_{j\alpha\beta}(M, M_1)$ as the operator $\Pi_{j\alpha\beta}(p, p_1)$ which associates the quantity

$$\Pi_{j\alpha\beta}(p,p_1)\varphi_\alpha(p_1)\psi_\beta(p_1)=-\frac{i}{2}(2\pi)^{-4}\Delta_{j\alpha\beta}(k)\int\varphi_\alpha(p_1)\psi_\beta(p-p_1)\,dp_1. \qquad (19.104)$$

with each pair of functions $\varphi_j(p)$ and $\psi_j(p)$. If we draw the vector $p$ from the end $j$ to the end $l$ of the lines corresponding to $\Gamma_{jl}(p)$ and $F^{(0)}_{jl}(p)$ [so that $\Gamma_{jl}(p)=\Gamma_{lj}(-p)$ and $F^{(0)}_{jl}(p)=F^{(0)}_{lj}(-p)$], while for $\Pi_{j\alpha\beta}(p,p_1)$ the vector $p$ is drawn along the line entering the vertex through the end $j$, then the integration in Eq. (19.104) can be interpreted so that the sum of the wave vectors of the three lines entering each vertex is zero, and the integration is carried out over all wave vectors which are not fixed by this rule.

For large Re the series (19.101) will apparently diverge, so that the sum of its first few terms could hardly be used as an approximation to the velocity correlation (or spectral) function. We can however, re-group the terms of this series in the hope that the sum of a finite number of terms in the rearranged series will be an approximation to the true correlation (or spectral) function. With this in view, let us first introduce the "generalized Green operator" $G'_{jl}(M, M_1)$ and the "generalized vertex" $P'_{j\alpha\beta}(M, M_1)$ (or in the spectral form $\Gamma'_{jl}(p)$ and $\Pi'_{j\alpha\beta}(p, p_1)$]. Let us determine the first of these operators as the sum of diagrams terminating in solid lines at both ends, and containing a sequence of solid lines running all through the entire diagram. The corresponding zero-, second-, and fourth-order diagrams are shown in Fig. 48. The second operator will be determined as the sum of "triangular" diagrams containing three "terminal" vertices (to which a solid or broken line can be attached), joined by a sequence of solid lines. The corresponding first-, third- and fifth-order diagrams are shown in Fig. 49 (in which the fifth-order diagrams must be augmented by diagrams obtained from them as a result of rotation through 120° and 240°.

The operators $G'_{jl}$ and $P'_{j\alpha\beta}$ are generalizations of the operators $G_{jl}$ and $P_{j\alpha\beta}$ in the same sense as the "true" correlation function $B_{jl}$ is a generalization of the "unperturbed" correlation function $B^{(0)}_{jl}$: The lower-order terms in the expansion in powers of Re for each of the three "generalized" quantities are the corresponding "ungeneralized" quantities. We shall represent the quantities $B_{jl}$, $G'_{jl}$ and $P'_{j\alpha\beta}$ by, respectively, a thick broken line, a thick solid line, and a full circle.

If we take one of the diagrams for the spectral function $F_{jl}(p)$ (or any other of our three generalized quantities), and replace in it one or a number of elements (i.e., solid or

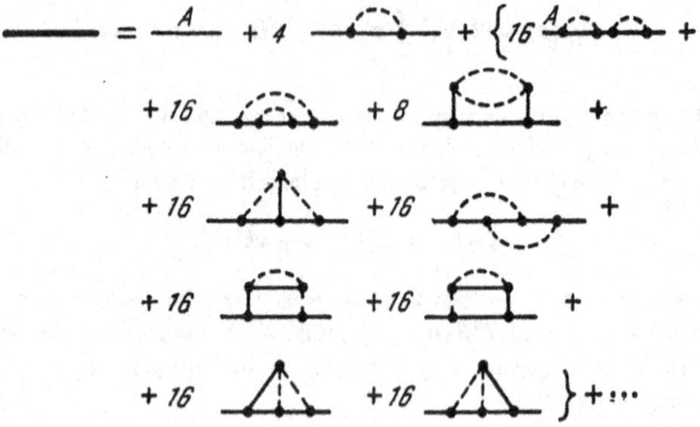

Fig. 48

broken lines, or points) by one of the diagrams for the corresponding generalized quantities, we obtain a higher-order diagram for $F_{jl}(p)$ (or another of the quantities considered). Diagrams which can be obtained by this method from lower-order diagrams will be called reducible, while the remaining diagrams will be called irreducible. It is clear that all reducible diagrams can be obtained by this method from irreducible diagrams, so that knowledge of the latter is sufficient to enable us to obtain all the diagrams. Next, a diagram in which one or several of the elements are replaced by the corresponding generalized

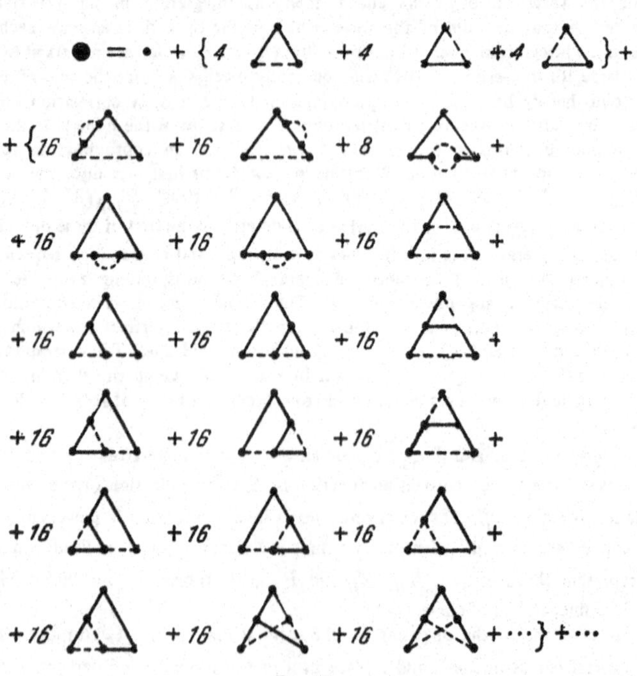

Fig. 49

elements will represent the sum of an infinite subsequence of diagrams in the expansion in powers of Re for $F_{jl}(p)$. One would, therefore, expect that this expression be represented by the sum of all the irreducible diagrams (with the appropriate numerical factors) in which all the elements are replaced by the corresponding generalized elements. The justification of this expectation was sketched by Wyld (1961); we shall not dwell on it here.

In the case of $F_{jl}(p)$ it is necessary to use the right-hand side of the equation

$$- - - = ——\square——, \qquad (19.105)$$

as the irreducible zero-order diagram, where the symbol $\square$ represents the correlation function of the external forces (19.99), or its Fourier transform which we shall denote by $\Phi_{jl}(p)$. This equation can also be written in the form

$$B_{jl}^{(0)}(M - M_1) = G_{j\alpha}(M, M')\, G_{l\beta}(M_1, M_1')\, C_{\alpha\beta}(M' - M_1'), \qquad (19.106)$$

which is readily verified with the aid of Eqs. (19.92), (19.99), and (19.100); or in the spectral form

$$F_{jl}^{(0)}(p) = \Gamma_{j\alpha}(p)\, \Phi_{\alpha\beta}(p)\, \Gamma_{\beta l}(p). \qquad (19.107)$$

It can be verified that the reducible diagrams for $F_{jl}(p)$, obtained from the irreducible diagram on the right-hand side of Eq. (19.105) by replacing the solid lines with the diagrams of Fig. 48, and then the rectangle together with two adjacent solid lines with a broken line are distinguished by the fact that they can be split into two pieces by severing a single broken line. In Figs. 46 and 47, all such diagrams are indicated by the letter $A$. The letter $B$ in Fig. 46 denotes the only irreducible second-order diagram, while in Fig. 47, this letter represents the fourth-order reducible diagrams which are obtained from it. Finally, letter $C$ in Fig. 47 represents the only irreducible fourth-order diagram.

Using the above results, we can now represent the spectral function as the sum of the "generalized" irreducible diagrams shown in Fig. 50. The right-hand side of the equation shown in Fig. 50 is, in fact, the rearranged series for $F_{jl}(p)$, which we mentioned above. The equation itself relates the three quantities $F_{jl}(p)$, $\Gamma'_{jl}(p)$ and $\Pi'_{j\alpha\beta}(p, p_1)$. The second equation relating these quantities is obtained by analogy with Fig. 50 by establishing the irreducible diagrams for $\Pi'_{j\alpha\beta}(p, p_1)$. The second equation is shown in Fig. 51 (where the fifth-order diagrams must be augmented by diagrams obtained from them as a result of rotation through 120° or 240°).

Unfortunately the third equation relating the "generalized" quantities cannot be obtained simply by finding all the irreducible diagrams for $\Gamma'_{jl}(p)$. The sum of the "generalized" irreducible diagrams in fact leads here to a series where some of the diagrams of Fig. 48 are counted twice; in other words we obtain the series with some of the numerical coefficients differing from those shown in Fig. 48. For example it is easy to see that all the fourth order diagrams in Fig. 48 appear to be reducible to the second order diagrams;

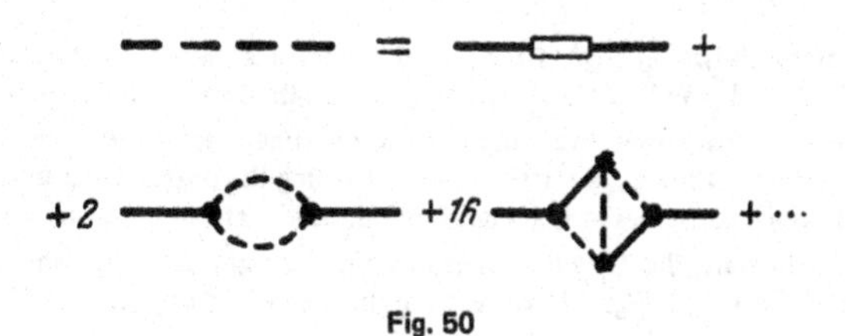


Fig. 50

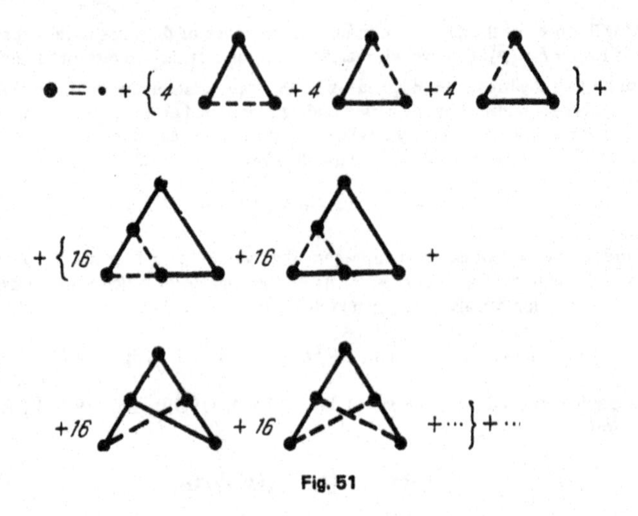


Fig. 51

however if we replace the second order term in Fig. 48 by a generalized diagram we find that the first fourth order diagram of Fig. 48 is counted twice. In order to avoid this difficulty, Wyld defined a new function and used a process leading to equations corresponding to the so-called Ward's Identities of quantum field theory. However, the possibility of using such a procedure for the three-dimensional fluid dynamics equations was questioned by L. Lee (1965) (see also Beran, 1968). L. Lee showed also that double counting does not occur if we replace the second order diagram in Fig. 48 by the nonsymmetric diagram containing only one generalized vertex and two generalized Green operators. Not all the fourth order diagrams in Fig. 48 are reducible to this new second order form; and those that cannot be so reduced must be now considered as the irreducible diagrams of fourth order. According to L. Lee the true equation for the "generalized" variable $\Gamma'_{jl}(p)$ has the form shown in Fig. 52. The general rule for constructing the terms of any order in this equation is not quite clear (it was not formulated precisely in L. Lee's paper); however the higher order terms will be unnecessary for us.

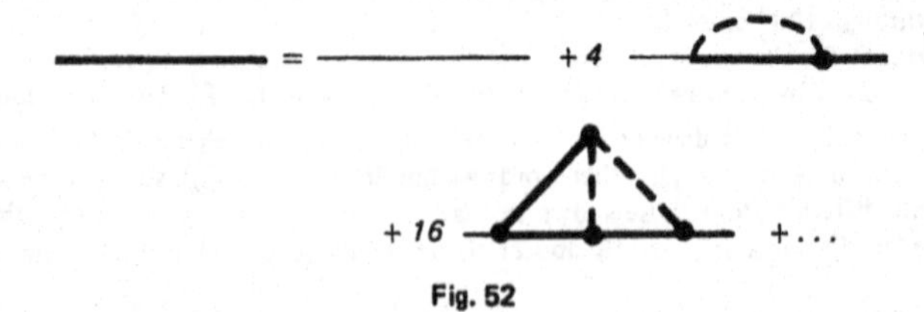


Fig. 52

The equations shown in Figs. 50, 51, and 52 form a closed set of equations for the quantities $F_{jl}(p)$, $\Gamma'_{jl}(p)$ and $\Pi'_{j\alpha\beta}(p, p_1)$. If we retain only a finite number of terms in the series in these equations, we obtain an approximate set of equations for the three unknown functions (although we have no way of telling the degree of accuracy). One of the simplest approximate schemes of this kind is obtained if, first, in the equation for $F_{jl}(p)$ in Fig. 50 we retain only the first two diagrams on the right-hand side and, second, in the equations of Fig. 51 and Fig. 52 we retain only the zero-order diagrams (i.e., we replace

$\Pi'_{j\alpha\beta}(p, p_1)$ and $\Gamma'_{jl}(p)$ by $\Pi_{j\alpha\beta}(p, p_1)$ and $\Gamma_{jl}(p)$ throughout). The equation for $F_{jl}(p)$ now assumes the form

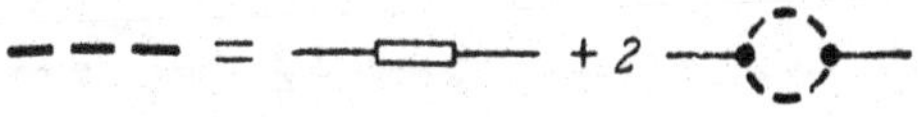

$$= + 2 \qquad (19.108)$$

or, in analytic form,

$$F_{jl}(p) = \Gamma_{j\alpha}(p)\,[\Phi_{\alpha\beta}(p) + 2\Pi_{\alpha mn}(p, p_1)\,F_{m\mu}(p_1)\times \\ \times F_{n\nu}(p_1)\,\Pi_{\beta\mu\nu}(-p, -p_1)]\,\Gamma_{\beta l}(-p). \quad (19.109)$$

If the fluid is incompressible, the turbulence isotropic and the force field solenoidal, we can substitute

$$F_{jl}(p) = F(k, \omega)\,\Delta_{jl}(k), \quad \Phi_{jl}(p) = \Phi(k, \omega)\,\Delta_{jl}(k) \qquad (19.110)$$

and, if we recall Eqs. (19.103)–(19.104), we can reduce Eq. (19.109) to the form

$$(\omega^2 + \nu^2 k^4)\,F(k, \omega) = \\ = \Phi(k, \omega) + \frac{k^2}{(2\pi)^4}\int a(k, k')\,F(k', \omega')\,F(|k - k'|, \omega - \omega')\,dp', \qquad (19.111)$$

where

$$a(k, k') = \frac{1}{2}\left[1 - 2\,\frac{(k\cdot k')^2(k\cdot k'')^2}{k^4 k'^2 k''^2} + \frac{(k\cdot k')(k\cdot k'')(k'\cdot k'')}{k^2 k'^2 k''^2}\right], \qquad (19.112)$$
$$k'' = k - k'.$$

Using Eqs. (19.102) and (19.110) and the formula

$$B_{\alpha\alpha}(r) = \left(3 + r\frac{\partial}{\partial r}\right) B_{LL}(r),$$

which is valid for an incompressible fluid, we can express $F(k, \omega)$ in terms of $B_{LL}(r, \tau)$:

$$F(k, \omega) = \frac{1}{2(2\pi)^4}\int e^{-i(k\cdot r - \omega\tau)}\left(3 + r\frac{\partial}{\partial r}\right) B_{LL}(r, \tau)\,dr\,d\tau. \qquad (19.113)$$

If we use this formula, and the analogous formula for the external force spectral function $\Phi(k, \omega)$, we can verify, after some quite cumbersome transformations, that Eq. (19.111) is equivalent to the Chandrasekhar equation (19.84) with an additional external force term. Therefore, the approximate scheme given by Eq. (19.108) corresponds to the Chandrasekhar equation.

Let us now consider a more complicated approximate scheme. As before, let us retain in the equation of Fig. 50 the first two diagrams, and in the equation of Fig. 51 only the zero-order diagram (i.e., we shall always replace $\Pi'_{j\alpha\beta}(p, p_1)$ by $\Pi_{j\alpha\beta}(p, p_1)$. However, in contrast to the foregoing scheme, we shall not now replace the generalized Green function $\Gamma'_{jl}(p)$ by $\Gamma_{jl}(p)$, but will evaluate it with the aid of the equation of Fig. 52 in which two terms on the right are retained. In other words, in our new scheme, the equations of Figs. 50 and 52 will be of the form

(19.114)

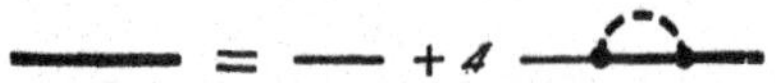

(19.115)

(the difference between the two vertices of the second diagram on the right of Fig. 52 is now lost since we have no generalized vertices in this scheme). If we replace the thick solid line on the left of the two right-hand side diagrams of Eq. (19.114) by the right-hand side of Eq. (19.115), and note that the replaced thick line appears again as the right element in the second diagram on the right-hand side of Eq. (19.115), we can reduce Eq. (19.114) to the form

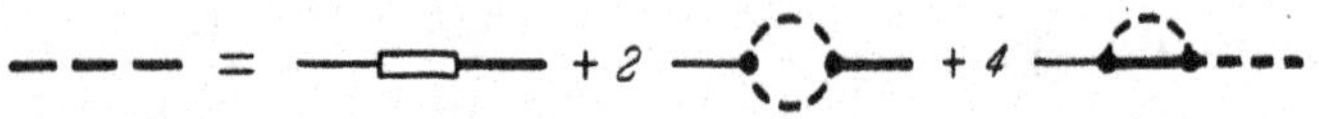

(19.116)

which is linear in the generalized Green's function.

From Eq. (19.115) it is clear that the tensor $\Gamma'_{jl}(p)$ is solenoidal in both indices, so that we can set

$$\Gamma'_{jl}(p) = \Gamma'(k, \omega)\,\Delta_{jl}(k). \tag{19.117}$$

Using Eqs. (19.110) and (19.117), and proceeding by analogy with the above discussion, we can reduce Eqs. (19.115) and (19.116) to the following analytic form:

$$\begin{aligned}(-i\omega + \nu k^2)\,\Gamma'(k, \omega) = \\ = 1 - \Gamma'(k, \omega)\frac{k^2}{(2\pi)^4}\int b(k, k')\,F(k', \omega')\,\Gamma'(|k - k'|, \omega - \omega')\,dp',\end{aligned} \tag{19.118}$$

$$\begin{aligned}(-i\omega + \nu k^2)\,F(k, \omega) = \Phi(k, \omega)\,\Gamma'(k, -\omega) + \frac{k^2}{(2\pi)^4}\Gamma'(k, -\omega)\times \\ \times\int a(k, k')\,F(k', \omega')\,F(|k - k'|, \omega - \omega')\,dp' - \\ -F(k, \omega)\frac{k^2}{(2\pi)^4}\int b(k, k')\,F(k', \omega')\,\Gamma'(|k - k'|, \omega - \omega')\,dp',\end{aligned} \tag{19.119}$$

where $a(k, k')$ is determined by Eq. (19.112), and $b(k, k')$ is of the form

$$b(k, k') = \frac{(k\cdot k'')^3}{k^4 k''^2} - \frac{(k\cdot k')(k'\cdot k'')}{k^2\cdot k'^2}. \tag{19.120}$$

These equations were first obtained by Kraichnan (1959) in a completely different way [see also Kraichnan (1961, 1962), Kraichnan (1970a) and Herring and Kraichnan (1971)]. In particular, Kraichnan took into account the character of the nonlinearity in the fluid dynamic equations due to which each three Fourier components of the velocity field (corresponding to co-planar wave vectors $k + k' + k'' = 0$) interact both directly and indirectly through other wave vectors. He assumed, moreover, that the indirect interactions are of minor importance as compared with direct interactions (this is the reason why Kraichnan's scheme is often called the *direct-interaction approximation*). Accordingly, Kraichnan put forward a definite perturbation-theory procedure which leads to Eqs. (19.118) and (19.119). A by-product of this approach is the elucidation of the physical significance of the generalized Green function: The tensor $\Gamma'_{jl}(p)$ describes the reaction of the velocity field to an infinitesimal external perturbation.

Kraichnan's approximate scheme was derived from various sets of assumptions, reconsidered, examined and criticized in an extensive series of works; see, for example, Proudman (1962), Betchov (1966, 1967), O'Brien (1968), Orszag (1970) and Phythian (1969). Here we shall only consider a useful physical interpretation of Kraichnan's scheme given in a work by Kadomtsev (1965) where it is referred to as the "weak coupling approximation." To develop this interpretation consider, for simplicity, the model equation

$$[\omega-\omega(\boldsymbol{k})+\gamma(p)]\,C(p)=\gamma(p)\,C(p)+\int V(p,\,p_1)\,C(p_1)\,C(p-p_1)\,dp_1, \tag{19.121}$$

which retains the main features of the Navier-Stokes equations subjected to Fourier transformation with respect to $\boldsymbol{x}$ and $t$. In this equation $[\omega-\omega(\boldsymbol{k})]$ is a linear operator (for the Navier-Stokes equation $\omega(\boldsymbol{k})=-i\nu k^2$), and the term $\gamma(p)\,C(p)$ describes the effect of direct interactions on the evolution of a wave with wave-number vector $\boldsymbol{k}$, frequency $\omega$, and amplitude $C(p)$, which reduces mainly to the damping of the wave. The right-hand side of Eq. (19.121) describes the effect of indirect interactions, which are assumed to be weak. We can therefore substitute $C(p)=C_0(p)+C_1(p)$ where $C_0(p)$ are the amplitudes of the main oscillations, $C_1(p)$ are the amplitudes of the forced oscillations produced by the indirect interactions between the main waves, and $|C_1(p)|\ll|C_0(p)|$. Hence,

$$C_1(p)=[\omega-\omega(\boldsymbol{k})+\gamma(p)]^{-1}\int V(p,\,p_1)\,C_0(p_1)\,C_0(p-p_1)\,dp_1. \tag{19.122}$$

The quantity $[\omega-\omega(\boldsymbol{k})+\gamma(p)]^{-1}$ is the analog of the generalized Green function. Let us multiply Eq. (19.121) by the complex conjugate amplitude $C^*(p)$, and take the average of the result, assuming that the amplitudes $C(p)$, with different $p$ are uncorrelated. In the expression for $\overline{C^*(p)\,C(p_1)\,C(p-p_1)}$ we must retain only terms of the order of $C_1(p)$. There are three such terms. Two of them are porportional to $\overline{|C(p)|^2}$, and can be removed by a suitable choice of $\gamma(p)$. The resulting equation for $\gamma(p)$, and the remaining equation for $\overline{C(p)\,C^*(p)}$, will be the analogs of Eqs. (19.118) and (19.119).

Let us now return to the true equations (19.118) and (19.119), and establish the asymptotic behavior of the first of them for $k\to\infty$, $|\omega|\to\infty$. The main contribution to the integral with respect to $p'$ on the right-hand side of this equation for large $k$, $|\omega|$ will be due to the region $k'\ll k$, $|\omega'|\ll\omega$ in which $\Gamma'(|\boldsymbol{k}-\boldsymbol{k}'|,\,\omega-\omega')\approx\Gamma'(k,\,\omega)$ and $b(\boldsymbol{k},\,\boldsymbol{k}')\approx\sin^2\alpha$, where $\alpha$ is the angle between $\boldsymbol{k}$ and $\boldsymbol{k}'$. Consequently, this integral can be written in the form $\Gamma'(k,\,\omega)\int F(k',\,\omega')\sin^2\alpha\,dp'$ for $k\to\infty$, $|\omega|\to\infty$. Using Eq. (19.102), we can verify that this integral is equal to $\frac{2}{3}\int F\,dp'=v_0^2=\frac{2}{3}\mathscr{E}$ where $\mathscr{E}$ is the turbulent kinetic energy per unit mass, determined mainly by the large-scale velocity fluctuations from the energy range of the spectrum. Consequently, Eq. (19.118) has the following asymptotic form:

$$(-i\omega + \nu k^2)\,\Gamma'(k, \omega) = 1 - [\Gamma'(k, \omega)]^2 \frac{k^2}{(2\pi)^4} v_0^2. \qquad (19.123)$$

Therefore, for large $k$, $|\omega|$, i.e., in the small-scale region, the function $\Gamma'(k, \omega)$ is found to depend on the parameter $v_0$ which characterizes large-scale motions. The same conclusion is reached for the spectral function $F(k, \omega)$. In particular, in the case of the spatial energy spectrum $E(k)$ and large $k$ (but not too large, so that the spectrum is still independent of molecular viscosity), Kraichnan showed from Eqs. (19.118) and (19.119) that the asymptotic expression is of the form $E(k) \sim (\bar{\varepsilon} v_0)^{1/2} k^{-3/2}$, which is in conflict with the Kolmogorov self-preservation in the small-scale region (see Sect. 16.5). It will be shown in detail in Sect. 23 that the Kolmogorov self-preservation is confirmed by extensive experimental data. This means that Kraichnan's "weak coupling" (or "direct-interaction") approximation is unsuitable for the description of the small-scale components of developed turbulence. It seems to be more natural to use this approximation to describe the evolution of the large-scale components (corresponding to the "energy range" of the spectrum) of isotropic turbulence with relatively small Re [Kraichnan (1964a), J. Lee (1965)]. However, even in this case, it is difficult to estimate the accuracy of the approximation.

The shortcomings of the Kraichnan scheme were analyzed by Proudman (1962) who argued that the assumption of relatively minor importance of indirect interactions is equivalent to the assumption that the $n$th order cumulants of a random spectral measure of the velocity field in a small volume (of the order of $V$) of wave-number space are small quantities (of the order of $V^{n-1}$), so that the successive approximations of Kraichnan's perturbation theory are similar to the approximations based on neglecting the cumulants of the corresponding orders (in fact, for example, when the fourth-order cumulants are neglected, the third moments may be nonzero only as a result of direct interaction between the corresponding wave-number triplets). Moreover the indirect interactions must play a very important part in the small-scale dynamics. It may be that this is precisely the reason why the energy spectrum $E(k)$ in the first Kraichnan approximation for large $k$ was found to depend explicitly on the mean square velocity $v_0$ of large-scale turbulent motions.

Kadomtsev (1964) has explained why the weak-coupling approximation is unsuitable for the description of the small-scale components of developed turbulence. The point is, that Kraichnan's scheme overestimates the effect of large-scale fluctuations (waves with small $k'$, $|\omega'|$) on the evolution of small-scale inhomogeneities (waves with large $k$, $|\omega|$). In fact, this effect can be reduced to a simple inertial transfer of small-scale inhomogeneities accompanied with small deformation. This type of interaction of a wave packet, having a mean wave number $k$ and mean frequency $\omega$, with a large scale wave $(k', \omega')$ was called *adiabatic* by Kadomtsev. It cannot be looked upon as a resonant excitation of a wave $(k, \omega)$ by a wave $(k - k', \omega - \omega')$ close to it, since both waves belong to the same wave packet and, therefore, their amplitudes [in the terminology of the model Eq. (19.121, $C(p)$ and $C(p - p_1)$] cannot be regarded as uncorrelated, as was the case in the weak-coupling approximation.

To see how this defect of the weak-coupling approximation could be corrected, let us rewrite the model Eq. (19.121), without separating the term $\gamma(p)\,C(p)$. We shall divide the region of integration with respect to $p_1$ into three parts: The main region $\Omega_1$ in which $k_1 \sim k$, $\omega_1 \sim \omega$, the long-wave region $\Omega_2$, and the short-wave region $\Omega_3$. In the region $\Omega_2$ we can expand the unknown functions in powers of $k_1/k$ and $\omega_1/\omega$, while in region $\Omega_3$ we can expand them in powers of $k/k_1$ and $\omega/\omega_1$. If we retain only zero-order terms in the integral on the right-hand side of Eq. (19.121) in these expansions, evaluated over the region $\Omega_2$, then we must substitute $C(p - p_1) \approx C(p)$, so that the integral assumes the form $\gamma C(p)$, i.e., leads only to a shift $\gamma$ of the frequency $\omega(k)$. In the integral over the region $\Omega_3$, we must set $C(p_1)\,C(p - p_1) \approx |C(p_1)|^2$ in the zero-order approximation. It is readily verified that when Eq. (19.121) is divided by the equations for the mean value $\overline{C}(p)$

and for $C'(p)$, this integral will not contribute to the equation for $C'(p)$ and can, therefore, be omitted. Therefore, in the above approximation, we can use Eq. (19.121) but the integral in this equation must be confined to the region $\Omega_1$. This is equivalent to taking into account only the adiabatic interaction between the small-scale wave packet and the large-scale wave.

To obtain the improved weak-coupling approximation, Kadomtsev recommends that both zero-order and first-order terms of the expansion for the unknown functions in powers of the small quantities $k_1/k$ and $\omega_1/\omega$ be taken into account in the integral over the long-wave region $\Omega_2$. We shall not consider the corrections to Eqs. (19.118) and (19.119) which result from this [in Kadomtsev's paper they are written down only for the model Eq. (19.121)]. Instead, let us briefly discuss the improved weak-coupling approximation put forward by Shut'ko (1964) who took into account certain additional diagrams in the equations of Figs. 50, 51, and 52, over and above those included in Eqs. (19.114) and (19.115) [this approximation, in fact, turns out to be in some respects less satisfactory than the original one; see Kraichnan (1966)]. In this procedure, the generalized vertices $\Pi'_{j\alpha\beta}(p, p_1)$ are not replaced by $\Pi_{j\alpha\beta}(p, p_1)$ but are represented by the sum of the first two terms on the right of the equation of Fig. 51. The equations used by Shut'ko are shown in Fig. 54.

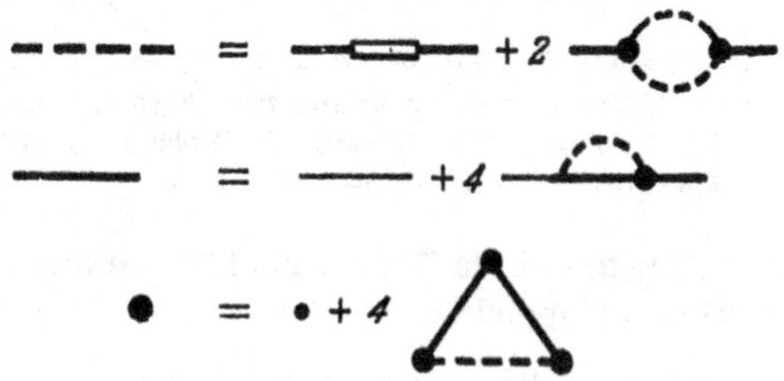

= +2

= +4

= + 4

Fig. 54

Since the fluid is incompressible and the turbulence is isotropic, the operator $\Pi'_{j\alpha\beta}(p, p_1)$ will differ from the operator $\Pi_{j\alpha\beta}(p, p_1)$, whose effect is indicated by Eq. (19.104), only by the additional factor $\Pi(p, p_1)$ under the integral with respect to $p_1$. The appearance of this factor is the only difference between the equations of Fig. 54 and those given by Eqs. (19.114) and (19.115). We shall show that the factor $\Pi(p, p_1)$ removes the above difficulty of the Kraichnan scheme, namely, the overestimation of the effect of large-scale fluctuations (with small $k_1$, $|\omega_1|$) on the evolution of small-scale inhomogeneities (with large $k$, $|\omega|$), since the factor tends to zero for both $k$, $|\omega| \to \infty$ and small $k_1$, $|\omega_1|$. In fact the equation for $\Pi(p, p_1)$, which is obtained from the third equation of Fig. 54, is of the form

$$\Pi(p, p_1) = 1 - \frac{k^2}{(2\pi)^4} \int c(k, k_1, k') \Pi(p, p-p') \Gamma'(p-p') \times$$
$$\times \Pi(p-p', p_1) \Gamma'(p-p_1-p') \Pi(p-p_1-p', p-p_1) F(p') \, dp', \quad (19.124)$$

where

$$c(k, k_1, k') = \frac{k_m}{2k''} \Delta_{\alpha\beta\gamma}(k) \Delta_{\beta mn}(k'') \Delta_{n\alpha p}(k''-k_1) \Delta_{\gamma p}(k'), \qquad k'' = k - k'. \quad (19.125)$$

For $k$, $|\omega| \to \infty$ and $k_1$, $|\omega_1|$ we have $c(k, k_1, k') \approx \sin^2 \alpha$ where $\alpha$ is the angle between $k$ and $k'$. Moreover, in Eq. (19.124) we can then substitute $\Pi(p, p-p') \approx \Pi(p-p_1-p', p-p_1) \approx \Pi(p, p)$, $\Pi(p-p', p_1) \approx \Pi(p, p_1)$ and $\Gamma'(p-p') \approx \Gamma'(p-p_1-p') \approx \Gamma'(p)$, so that the equation assumes the asymptotic form

$$\Pi(p, p_1) = 1 - \Pi(p, p_1)[\Pi(p, p)]^2 [\Gamma'(p)]^2 \frac{k^2}{(2\pi)^4} \int \sin^2 \alpha F(p')\, dp',$$

and, hence,

$$\Pi(p, p_1) \sim \left\{ 1 + [\Pi(p, p)]^2 [\Gamma'(p)]^2 \frac{k^2 v_0^2}{(2\pi)^4} \right\}^{-1}. \quad (19.126)$$

Again, from Eq. (19.126) with $p_1 = p$ we can verify that $\Pi(p, p) \to 1$ for $k$, $|\omega| \to \infty$ [since, $c(k, k, k') \to 0$]. If, finally, we recall that $\Gamma'(p) \sim \frac{1}{\omega} = \frac{1}{kv}$ for $|\omega| \to 0$, where $v$ is the mean square velocity of small-scale fluctuations, we obtain $\Pi(p, p_1) \sim (2\pi)^4 \frac{v^2}{v_0^2} \to 0$, so that, in fact, the function $\Pi(p, p_1)$ is a cut-off factor for large-scale fluctuations.

Further discussion of the methods for constructing closed systems of equations for isotropic turbulence, which are consistent with the Kolmogorov self-preservation of small-scale turbulent perturbations, is given in Sect. 22.2.

## 19.7 Equations for the Finite-Dimensional Probability Distributions of Velocities

Up until now consideration has been given to the velocity moment equations of turbulent flow, i.e., Friedmann-Keller equations which for $m$-point moments have generally the form (19.21) [see also Sect. 6.2)]. However, the preassignment of all the finite-dimensional probability distributions of the velocity fluctuations at every possible finite set of space-time points gives us a more satisfactory statistical description of the random velocity field than the determination of all its moments. Under rather general conditions, the probability distributions are fully determined by all their moments [cf. Sect. (4.10)], and since there are dynamical moment equations, one may expect that equations can as well be derived which describe the time evolution of the finite-dimensional probability distributions.

Though Friedmann-Keller equations were obtained as far back as 1924, strange as it may seem, the dynamical equations describing the time evolution of the finite-dimensional probability distributions were not deduced before 1967. They were published for the first time in Monin's papers (1967a, 1967b). Simultaneously (and independently of Monin), the dynamical equations for one-point and two-point probability distributions were published by T. S. Lundgren (1967). In the same year, though somewhat earlier, similar equations for the finite-dimensional distributions of the vorticity fluctuations were published by Novikov (1967).

We shall follow the above-mentioned papers by Monin (1967a, b) in deriving the dynamical equations for $n$-point probability distributions of the turbulent velocity field of incompressible fluid. Assume that $u_1 = u(M_1), \ldots, u_n = u(M_n)$ are the random velocities at $n$ space-time points $M_1 = (x_1, t_1), \ldots, M_n = (x_n, t_n)$ whose spatial coordinates $x_1, \ldots, x_n$ and times $t_1, \ldots, t_n$ are all different. Let us consider the characteristic function of the random variables $u_1, \ldots, u_n$:

$$\varphi_{M_1 \ldots M_n}(\vartheta_1, \ldots, \vartheta_n) = \overline{\exp\left\{i \sum_{m=1}^{n} \vartheta_m \cdot u(M_m)\right\}} \tag{19.127}$$

Let us differentiate it with respect to one of the times on which it depends, say, with respect to $t_k$:

$$\frac{\partial}{\partial t_k} \varphi_{M_1 \ldots M_n}(\vartheta_1, \ldots, \vartheta_n) = i\vartheta_{k\alpha} \overline{\frac{\partial u_\alpha(M_k)}{\partial t_k} \exp\left\{i \sum_{m=1}^{n} \vartheta_m \cdot u(M_m)\right\}} \tag{19.128}$$

We use here the designations $(\vartheta_{k1}, \vartheta_{k2}, \vartheta_{k3})$ for the Cartesian components of the vector $\vartheta_k$; similar designations $x_{k\alpha}$ and $u_{k\alpha}$ will be used, where required, for the components of the vectors $x_k$ and $u_k$. The usual repeated indices summation convention will be used in this subsection only in the case of Greek indices. Let us express the derivative $\partial u_\alpha(M_k)/\partial t_k$ in Eq. (19.28) as the right-hand side of the Navier-Stokes equation which we shall rewrite in the form

$$\frac{\partial u_\alpha(M_k)}{\partial t_k} = -\frac{\partial u_\alpha(M_k) u_\beta(M_k)}{\partial x_{k\beta}} + \Delta^{-1}(x_k, x) \frac{\partial^3 u_\beta(M) u_\gamma(M)}{\partial x_\alpha \partial x_\beta \partial x_\gamma} + \nu E(x_k, x) \frac{\partial^2 u_\alpha(M)}{\partial x_\beta \partial x_\beta} \tag{19.129}$$

Here $M = (x, t_k)$; $\Delta^{-1}(x_k, x)$ is the integral operator inverse to the Laplacian $\partial^2/\partial x_{k\beta} \partial x_{k\beta}$ and containing the integration with respect to the vectors $x$ (in the case of unbounded flow its form is given by Eqs.

(1.9)–(1.9′) of Vol. 1; if rigid walls are present in viscous incompressible fluid the operator $\Delta^{-1}$ will contain an additional term which we will not explicitly indicate here); $E(x_k, x)$ is the operator for replacing $x$ by $x_k$ introduced for later convenience. By substituting Eq. (19.129) in the right-hand side of Eq. (19.128) it is easy to see that the equation reduces to the form

$$\left(\frac{\partial}{\partial t_k} - i\,\frac{\partial^2}{\partial x_{k\alpha}\partial\vartheta_{k\alpha}}\right)\varphi_{M_1\ldots M_n}(\vartheta_1,\ldots,\vartheta_n) =$$

$$-\,i\vartheta_{k\alpha}\Delta^{-1}(x_k,x)\,\frac{\partial^3}{\partial x_\alpha\,\partial x_\beta\,\partial x_\gamma}\left[\frac{\partial^2}{\partial\vartheta_\beta\partial\vartheta_\gamma}\,\varphi_{M_1\ldots M_n M}(\vartheta_1,\ldots\vartheta_n,\vartheta)\right]_{\vartheta=0}$$

$$+\,\nu\vartheta_{k\alpha}E(x_k,x)\,\frac{\partial^2}{\partial x_\beta\partial x_\beta}\left[\frac{\partial}{\partial\vartheta_\alpha}\,\varphi_{M_1\ldots M_n M}(\vartheta_1,\ldots,\vartheta_n,\vartheta)\right]_{\vartheta=0} \qquad (19.130)$$

Let us consider the Fourier transform of the last equation with respect to the variables $\vartheta_1,\ldots,\vartheta_n$ and take into account that such a transform of the characteristic function $\varphi_{M_1\ldots M_n}(\vartheta_1,\ldots,\vartheta_n)$ is the probability density $P_{M_1\ldots M_n}(u_1,\ldots,u_n)$ of random variables $u_1,\ldots,u_n$. Then we shall obtain the following equation for this probability density

$$\left(\frac{\partial}{\partial t_k} + u_{k\alpha}\,\frac{\partial}{\partial x_{k\alpha}}\right)P_{M_1\ldots M_n}(u_1,\ldots,u_n) =$$

$$-\,\frac{\partial}{\partial u_{k\alpha}}\Bigg[\Delta^{-1}(x_k,x)\,\frac{\partial^3}{\partial x_\alpha\,\partial x_\beta\,\partial x_\gamma}\int u_\beta u_\gamma P_{M_1\ldots M_n M}(u_1,\ldots,u_n,u)\,du$$

$$+\,\nu E(x_k,x)\,\frac{\partial^2}{\partial x_\beta\partial x_\beta}\int u_\alpha P_{M_1\ldots M_n M}(u_1,\ldots,u_n,u)\,du\Bigg]. \qquad (19.131)$$

Equations (19.130) or (19.131) are linear, though they are not closed; the time derivative of the $n$-point probability density is determined not only by the latter function but also by the $(n+1)$-point probability density. Thus we obtain once more an infinite set of coupled differential equations (to a large extent similar to Bogolyubov's chain of equations for many-particle distribution functions in the kinetic theory of gases).

The solutions of Eqs. (19.130) and (19.131) must possess all the necessary properties of the finite-dimensional probability distributions. Namely they must satisfy the normalization condition, conditions on symmetry in arguments and compatibility conditions which connect distributions of a set and of a subset of random variables. Besides, since the points $M_1, M_2, \ldots$ are the parameters of random variables $u_1, u_2, \ldots$, specific compatibility conditions should be fulfilled with respect to these parameters. In the first place the following continuity conditions as $M_2 \to M_1$ must be valid

$$\varphi_{M_1 M_2 M_3 \cdots}(\vartheta_1, \vartheta_2, \vartheta_3, \ldots) \to \varphi_{M_1 M_3 \cdots}(\vartheta_1 + \vartheta_2, \vartheta_3, \ldots);$$

$$P_{M_1 M_2 M_3 \cdots}(u_1, u_2, u_3, \ldots) \to \delta(u_1 - u_2) P_{M_1 M_3 \cdots}(u_1, u_3, \ldots). \tag{19.132}$$

Second, the following differential conditions must be satisfied as $M_2 \to M_1$:

$$\frac{\partial}{\partial x_{k\alpha}} \varphi_{M_1 \ldots M_n}(\vartheta_1, \ldots, \vartheta_n) = E(x_k, x) \vartheta_{k\beta} \frac{\partial}{\partial x_\alpha} \left[ \frac{\partial}{\partial \vartheta_\beta} \varphi_{M_1 \ldots M_n M}(\vartheta_1, \ldots, \vartheta_n, \vartheta) \right]_{\vartheta = 0};$$

$$\frac{\partial}{\partial x_{k\alpha}} P_{M_1 \ldots M_n}(u_1, \ldots, u_n) = -E(x_k, x) \frac{\partial^2}{\partial u_{k\beta} \partial x_\alpha} \int u_\beta P_{M_1 \ldots M_n M}(u, \ldots, u_n, u) du \tag{19.133}$$

the first of which is easily verified from the definition of characteristic functions (19.127), and the second one can be derived by the Fourier transformation of the first. Condition (19.133) for many-point probability densities was established by Ulinich and Lyubimov (1968).

Finally, as a consequence of the continuity equation for incompressible fluid, the following conditions should be fulfilled

$$\frac{\partial}{\partial x_{k\alpha}} \left[ \frac{\partial}{\partial \vartheta_{k\alpha}} \varphi_{M_1 \ldots M_n}(\vartheta_1 \ldots, \vartheta_n) \right]_{\vartheta_k = 0} = 0;$$

$$\frac{\partial}{\partial x_{k\alpha}} \int u_{k\alpha} P_{M_1 \ldots M_n}(u_1, \ldots, u_n) du_k = 0 \tag{19.134}$$

Quite often the single-time statistical turbulence quantities alone are of interest, though they give a less complete description of the turbulence. The equations for the time evolution of the single time characteristic functions or probability densities (at $t_1 = \cdots = t_n = t$) are derived from Eqs. (19.130) or (19.131) by summing with respect to all $k$ if the relationship $\sum_k (\partial/\partial t_k) = (\partial/\partial t)$ is taken into account. Such equations for the single time probability density $f_n = P_{x_1 \ldots x_n}(u_1, \ldots, u_n; t)$ can be written in the form

$$\frac{\partial f_n}{\partial t} = -\sum_{k=1}^{n} \left( \frac{\partial u_{k\alpha} f_n}{\partial x_{k\alpha}} + \frac{\partial C_{k\alpha} f_n}{\partial u_{k\alpha}} \right), \tag{19.135}$$

where the $C_{k\alpha}$ are the "mean conditional accelerations"

$$C_{k\alpha} = \Delta^{-1}(x_k, x) \frac{\partial^3 \langle u_\beta u_\gamma \rangle_n}{\partial x_\alpha \partial x_\beta \partial x_\gamma} + \nu E(x_k, x) \frac{\partial^2 \langle u_\alpha \rangle_n}{\partial x_\beta \partial x_\beta} \tag{19.136}$$

and the symbol $\langle \ \rangle_n$ denotes a conditional mathematical expectation of a function of the random velocity at a point $(x, t)$ provided that values of the variables $u_1 = u(x_1, t), \ldots, u_n = u(x_n, t)$ are fixed. If both sides of Eq. (19.135) are multiplied by any function $F(u_1, \ldots, u_n)$ and integrated with respect to all values of its arguments, we shall obtain a generalized Friedmann-Keller equation

$$\frac{\partial \overline{F}}{\partial t} = \sum_{k=1}^{n} \left( -\frac{\partial \overline{u_{k\alpha} F}}{\partial x_{k\alpha}} + \overline{C_{k\alpha} \frac{\partial F}{\partial u_{k\alpha}}} \right) \tag{19.137}$$

This form has been given by Ievlev (1970). He has also given the following identities:

$$\int C_{k\alpha} f_n \, du_l = C^l_{k\alpha} f^l_{n-1}; \tag{19.138}$$

$$\int u_{l\alpha} \frac{\partial C_{k\beta} f_n}{\partial x_{l\alpha}} \, du_l = 0 \quad \text{if} \quad l \neq k \tag{19.139}$$

In the first of which the upper index $l$ means that the variable

$u_l = u(x_l, t)$ drops out in the corresponding quantity (i.e., it depends upon all the points except the $l$th). Ievlev suggested closing Eqs. (19.135) by replacing the unknown coefficients $C_{k\alpha}$ whose precise values are determined by the $(n + l)$-point distribution functions, by their approximate values of the form

$$C'_{k\alpha} = \sum_{l=1}^{n} \psi_{k\alpha}^{(l)} + \sum_{l \neq k} u_{l\beta} \frac{\partial \lambda_{k\alpha}^{(l)}}{\partial x_{l\beta}} \tag{19.140}$$

(where the superscript $(l)$ means that the dependence upon the variable $u_l$ drops out but the dependence upon $x_l$ remains), while the functions $\psi_{k\alpha}^{(l)}$ and $\lambda_{k\alpha}^{(l)}$ are determined from the equations obtained by substituting $C'_{k\alpha}$ for $C_{k\alpha}$ in the left-hand sides of the identities (19.138)–(19.139). Ievlev has proved that the corresponding approximate Eq. (19.135) is transformed into a precise equation for $f_{n-1}$ after integrating with respect to any $u_k$, which gives precise Friedmann-Keller equations for moments of all orders $m \leqslant n$, and that its solution $f_n$ satisfies the normalization and continuity conditions at any time and is always real and nonnegative.

Another closing scheme was suggested in the above-mentioned work by Ulinich and Lyubimov (1968). These authors have introduced relative coordinates $\xi_k = x_k - x_1$ and relative velocities $u_k = u_k - u_1$ into Eqs. (19.131), and assumed that at large Re probability distributions for relative velocities at small separations $\xi_k$ rapidly come to be in quasi-equilibrium, i.e., they depend on $x_1$ and $t$ only due to their functional dependence on two slowly changing variables $f_1(u_1; x_1, t) = P_{x_1}(u_1, t)$ and the rate of energy dissipation $\epsilon(x_1, t)$. The derivation of equations for slowly changing variables is similar to the derivation of equations of fluid dynamics and to finding kinetic transport coefficients from the sequence of coupled equations for many-particle distribution functions based on the assumption of their near equilibrium form (with the use of a small parameter $\mathrm{Re}^{-1/4}$). This procedure has allowed Ulinich and Lyubimov to obtain a Gaussian solution for $f_1(u_1, t)$ in the isotropic case with a power law in time for energy decay.

A similar problem for a nonhomogeneous flow was dealt with in Lundgren's paper (1969).

Two other closure hypotheses for the probability distribution equations of the single-time values of the turbulent velocities were investigated recently by Fox (1971) and Lundgren (1971). Fox tried to apply a closure hypothesis similar to the linearized form of the so-called superposition approximation of Kirkwood which is widely used for distribution functions in statistical mechanics [see, e.g., Green (1952)]. The hypothesis can be written as follows:

$$P_{x_1,x_2,x_3}(u_1,u_2,u_3) = P_{x_1}(u_1)P_{x_2,x_3}(u_2,u_3) + P_{x_2}(u_2)P_{x_1,x_3}(u_1,u_3) + P_{x_3}(u_3)P_{x_1,x_2}(u_1,u_2) - 2P_{x_1}(u_1)P_{x_2}(u_2)P_{x_3}(u_3). \quad (19.141)$$

It permits one to express three-point distributions in terms of one- and two-point distributions and thus to close the equations for one-point and two-point distribution functions.

In a short note by Fox (1971) results are outlined which were obtained from a numerical computation of the time evolution of the normalized correlation function $B_{LL}(r,t)/B_{LL}(O,t)$ with the aid of the closed system of equations for the functions $P_x(u)$ and $P_{x_1,x_2}(u_1,u_2)$ (both depending on time $t$) and a particular model of initial conditions. The results seem to be reasonable but they are clearly insufficient to justify the applicability of the hypothesis to real turbulent flows. On the other hand Lundgren (1971) called into question the usefulness of the closure hypothesis (19.141) since it violates many of the necessary properties of the probability density functions (e.g., the consistency property which requires that integration of a many-point density with respect to one velocity variable yields a density of order reduced by one). For this reason he used the following apparently more complicated closure in his investigation:

$$\hat{p}_{x_1,x_2,x_3}(u_1,u_2,u_3) = \hat{p}^{(n)}_{x_1,x_2,x_3}(u_1,u_2,u_3) \quad (19.142)$$

where $\hat{p}$ is the difference of the left-hand and right-hand sides of Eq. (19.141) and $\hat{p}^{(n)}$ is the similar difference of the normal density functions constructed with the correct (but unknown to us) velocity correlation tensor $B_{ij}(x_1,x_2)$. This new closure hypothesis satisfies most (although not all) the properties of the probability density functions; it provides apparently a satisfactory approximation for turbulence with nearly normal probability distributions but is probably not so good for turbulence which is far from normal.

Equations for the one- and two-point density functions become closed if the assumption (19.142) is used. Equation for the two-point density can be solved with appropriate initial and boundary conditions as if the one-point density and tensor $B_{ij}$ were known functions. The substitution of the solution obtained into the equation for the one-point density and the expression of $B_{ij}$ in terms of the two-point density gives then two coupled equations for $P_x(u)$ and $B_{ij}$. The resulting equations are however quite complicated even in the case of an isotropic turbulence. The only concrete result obtained by Lundgren from these equations relates to high Reynolds number locally isotropic turbulence; it will be formulated later in this book (in Sect. 22.2).

Some general approaches to the closure problem for the multi-dimensional probability density equations were also sketched by Lyubimov and Ulinich (1970b). However this paper contains no examples of application of the methods discussed to any specific problem. Finally, let us point out the papers by Lyubimov (1969) and Lyubimov and Ulinich (1970a), containing the derivation of equations describing the time evolution of the probability distributions of coordinates and velocities of a finite number of "fluid particles" in a turbulent flow. In the first of these papers the author managed to express the time derivative of the $n$-particle probability distribution in terms of the $n, n + 1, \ldots, n + 4$-particle distributions, whereas in the second one, only in terms of $n$ and $(n + 1)$-particle distributions.

## 20. TURBULENCE IN COMPRESSIBLE FLUIDS

### 20.1 Invariants of Isotropic Compressible Turbulence

So far, we have been concerned only with isotropic turbulence in incompressible fluids. Since in ordinary flows, with characteristic velocities much smaller than the sound velocity, the compressibility of the medium is of minor importance, we were quite justified in neglecting compressibility during the initial development of the theory of turbulence. This does not mean, however, that the effect of compressibility on turbulence can be ignored altogether, because compressibility does in fact lead to the appearance of certain new physical phenomena which are of considerable theoretical and applied interest (see Sect. 20.3 and also Sect. 26). Here, we shall briefly review some of the properties which distinguish turbulence in compressible fluids from ordinary incompressible turbulence.

The complete description of compressible fluids flow requires the specification of six variables related by three balance equations, namely, the momentum balance equation (1.3) [or Eq. (1.4)], the continuity equation (mass balance) (1.1) [or Eq. (1.2)], the heat equation (energy balance) (1.60) [or Eq. (1.65) or Eq. (1.65′)], and by the equation of state (1.63) (as in Sect. 1 of Vol. 1, we shall assume that the fluid is a perfect gas with constant specific heat). The six unknown functions in these equations can clearly be chosen in different ways, hence the correlation and spectral equations of "compressible turbulence" can be written in many different ways. Moreover, in view of the complexity of compressible turbulent flows, it is usual to introduce certain additional assumptions to describe such flows (for example, assumptions about the dependence of the coefficients $\mu$, $\zeta$, $\varkappa$, or

$\nu = \mu/\rho$, $\nu_1 = \zeta/\rho$, $\chi = \varkappa/c_p\rho$, on temperature and pressure, and the magnitude of the ratios of these coefficients), which increases the number of possibilities still further.

Let us begin with the exact equations obtained when the following are taken as the main variables: Three components of momentum density $\mathcal{U}_i = \rho u_i$, $i = 1, 2, 3$, mass density $\rho$, and energy density $\mathcal{E} = \rho\left(\frac{u^2}{2} + e\right)$ (in view of the algebraic nature of the equation of state, one of the unknowns in the dynamic equations can be eliminated, so that only five independent variables will remain). In terms of these variables the Eqs. (1.1), (1.3), and (1.60) in the absence of external forces $\boldsymbol{X}$ can be rewritten in the form

$$\frac{\partial \rho}{\partial t} = -\frac{\partial \mathcal{U}_j}{\partial x_j}; \quad \frac{\partial \mathcal{U}_i}{\partial t} = -\frac{\partial \Pi_{ij}}{\partial x_j}, \quad i = 1, 2, 3; \quad \frac{\partial \mathcal{E}}{\partial t} = -\frac{\partial E_j}{\partial x_j}, \tag{20.1}$$

where $\Pi_{ij}(\boldsymbol{x}, t) = \Pi_{ji}(\boldsymbol{x}, t)$ and $E_j(\boldsymbol{x}, t)$ are new random fields which are related to $\rho(\boldsymbol{x}, t)$, $\mathcal{U}_i(\boldsymbol{x}, t)$ and $\mathcal{E}(\boldsymbol{x}, t)$ and the coefficients $\mu$, $\zeta$, and $\varkappa$ by relatively complicated formulas which for the moment we shall not require. In the case of homogeneous turbulence, in which all the statistical characteristics are independent of the coordinates, we have from Eq. (20.1)

$$\frac{\partial \bar{\rho}}{\partial t} = \frac{\partial \overline{\mathcal{U}_i}}{\partial t} = \frac{\partial \overline{\mathcal{E}}}{\partial t} = 0, \quad \text{i.e.,} \quad \bar{\rho} = \text{const}, \quad \overline{\mathcal{U}} = \text{const}, \quad i = 1, 2, 3, \quad \overline{\mathcal{E}} = \text{const}. \tag{20.2}$$

In Eqs. (20.1) we can therefore replace the quantities $\rho$, $\mathcal{U}_i$ and $\mathcal{E}$ by the fluctuations $\rho'$, $\mathcal{U}_i'$ and $\mathcal{E}'$ of the corresponding fields, i.e., by the departures from the mean values. If the turbulence is not only homogeneous but also isotropic, so that there are no preferred directions, we can set $\overline{\mathcal{U}}_i = 0$, i.e.,$\mathcal{U}_i' = \mathcal{U}_i$ for all $i$.

The relations (20.2) are the consequences of the physical laws of mass, momentum, and energy conservation. Certain other less obvious consequences are known to follow from these laws. In fact, let $a(x, t)$ and $\beta(x, t)$ be two homogeneous and homogeneously correlated random fields with zero means, such that

$$\frac{\partial a}{\partial t} = -\frac{\partial C_j}{\partial x_j}, \quad \frac{\partial \beta}{\partial t} = -\frac{\partial D_j}{\partial x_j}, \tag{20.3}$$

where $C_j$ and $D_j$ ($j = 1, 2, 3$) are certain homogeneous random fields. Multiplying the first equation in Eq. (20.3) taken at the point $\boldsymbol{x}$ by $\beta(\boldsymbol{x}+\boldsymbol{r}, t) = \beta(\boldsymbol{x}', t)$, and the second equation taken at the point $\boldsymbol{x}' = \boldsymbol{x}+\boldsymbol{r}$ by $a(x,t)$, and adding the resulting equations we have, after taking the average,

$$\frac{\partial B_{a\beta}(\boldsymbol{r}, t)}{\partial t} = \frac{\partial}{\partial r_j}\left[B_{C_j\beta}(\boldsymbol{r}, t) - B_{aD_j}(\boldsymbol{r}, t)\right], \tag{20.4}$$

where $B_{a\beta}(\boldsymbol{r}, t) = \overline{a(\boldsymbol{x}, t)\beta(\boldsymbol{x}+\boldsymbol{r}, t)}$ and

$$B_{C_j\beta}(\boldsymbol{r}, t) = \overline{C_j(\boldsymbol{x}, t)\beta(\boldsymbol{x}+\boldsymbol{r}, t)}, \quad B_{aD_j}(\boldsymbol{r}, t) = \overline{a(\boldsymbol{x}, t) D_j(\boldsymbol{x}+\boldsymbol{r}, t)}.$$

Hence, it follows that

$$\frac{\partial}{\partial t}\int\limits_{|\boldsymbol{r}|<R} B_{a\beta}(\boldsymbol{r}, t)\, d\boldsymbol{r} = \int\limits_{|\boldsymbol{r}|=R} \left[B_{C_j\beta}(\boldsymbol{r}, t) - B_{aD_j}(\boldsymbol{r}, t)\right]\frac{r_j}{r}\, d\sigma(\boldsymbol{r}), \tag{20.5}$$

where $d\sigma(r)$ is an area element on the surface of the sphere $|r| = R$. If the correlation functions $B_{\alpha\beta}(r,t)$, $B_{C_j\beta}(r)$, and $B_{\alpha D_j}(r)$ fall off at infinity more rapidly than $r^{-3}$, then it follows from Eq. (20.5) that

$$\lim_{R\to\infty} \int\limits_{|r|<R} B_{\alpha\beta}(r, t)\, dr = \int B_{\alpha\beta}(r, t)\, dr = \Lambda_{\alpha\beta} = \text{const.} \tag{20.6}$$

In view of Eq. (20.1), the variables $\alpha$ and $\beta$ can be identified with any of the five fields $\rho'$, $\mathcal{U}_l'$ and $\mathscr{E}'$. Therefore, if the correlation functions of compressible homogeneous turbulence fall off at infinity more rapidly than $r^{-3}$, then 25 integrals of the form of Eq. (20.6), in which $\alpha$ and $\beta$ assume the values $\rho'$, $\mathcal{U}_1'$, $\mathcal{U}_2'$, $\mathcal{U}_3'$ and $\mathscr{E}'$ will be constants.[38] The invariance of these integrals also expresses the conservation of mass, momentum, and energy [see Eqs. (20.7').]

Since $B_{\alpha\beta}(r) = B_{\beta\alpha}(-r)$, 10 of the 25 invariants (20.6) will be encountered twice. Consequently in the case of an arbitrary homogeneous turbulence in compressible fluids we have 15 different invariants. If, on the other hand, the turbulence is also isotropic, the number of invariants will be considerably reduced. Since

$$B_{\mathcal{U}_l\rho'}(r) = B_{L\rho'}(r)\frac{r_l}{r}, \quad B_{\mathcal{U}_l\mathscr{E}'}(r) = B_{L\mathscr{E}'}(r)\frac{r_l}{r},$$

$$B_{\mathcal{U}_l\mathcal{U}_j}(r) = [\mathscr{B}_{LL}(r) - \mathscr{B}_{NN}(r)]\frac{r_l r_j}{r} + \mathscr{B}_{NN}(r)\,\delta_{lj}$$

many of them are zero while others are identical with each other. As a result, only four independent invariants remain:

$$\begin{aligned}
&\int_0^\infty B_{\rho'\rho'}(r, t)\, r^2\, dr = \Lambda_1,\\
&\int_0^\infty [\mathscr{B}_{LL}(r, t) + 2\mathscr{B}_{NN}(r, t)]\, r^2\, dr = \Lambda_2,\\
&\int_0^\infty B_{\mathscr{E}'\mathscr{E}'}(r, t)\, r^2\, dr = \Lambda_3,\\
&\int_0^\infty B_{\rho'\mathscr{E}'}(r, t)\, r^2\, dr = \Lambda_4.
\end{aligned} \tag{20.7}$$

In incompressible fluids, only the invariant $\Lambda_3$ can be nonzero [in this case it is expressed in terms of the fourth moments of the velocity $u(x)$, the joint moments of $u^2(x)$ and $T(x)$,

[38] In the case of homogeneous incompressible turbulence, it is natural to assume (see Sect. 15.5) that the correlation functions fall off at infinity in accordance with a power law (but at any rate more rapidly than $r^{-3}$). This relatively slow decrease is explained by the effectively instantaneous propagation of interactions in incompressible fluids. In compressible fluids, which are characterized by a finite sound velocity, the equations for the correlation functions are always differential equations (and not integro-differential), and therefore the decrease of the correlation functions can even be assumed to be exponential.

and the correlation function for the temperature field]. The invariant $\Lambda_1$ was first established by Chandrasekhar (1951) and the invariants $\Lambda_2$, $\Lambda_3$, $\Lambda_4$ by Sitnikov (1958) (the derivation of these invariants given here is in fact due to Sitnikov). Sitnikov also showed that, in the case of compressible fluids, the initial values of the fields $\rho'(x)$, $\mathcal{U}_i(x)$ and $\mathscr{E}'(x)$ can be chosen so that the invariants $\Lambda_1$, $\Lambda_2$ and $\Lambda_3$ have arbitrary nonnegative values while the invariant $\Lambda_4$ has an arbitrary value whose modulus does not exceed $(\Lambda_1\Lambda_3)^{1/2}$.

The invariants (20.7) are all very similar to the Corrsin invariant (15.20) in the theory of temperature-inhomogeneous incompressible turbulence. The similarity is in fact quite fundamental: both Eq. (15.20) and all the equations in Eq. (20.7) express the conservation of the total amounts of certain flow parameters. In point of fact, and in complete analogy with Eq. (15.30), the equations (20.7) can also be written in the form

$$\begin{aligned}
\lim_{V\to\infty}\frac{1}{V}\overline{\left[\int_V \rho'(x)\,dx\right]^2} &= 4\pi\Lambda_1 = \text{const},\\
\lim_{V\to\infty}\frac{1}{V}\overline{\left[\int_V \mathcal{U}(x)\,dx\right]^2} &= 4\pi\Lambda_2 = \text{const},\\
\lim_{V\to\infty}\frac{1}{V}\overline{\left[\int_V \mathscr{E}'(x)\,dx\right]^2} &= 4\pi\Lambda_3 = \text{const},\\
\lim_{V\to\infty}\frac{1}{V}\overline{\int_V \rho'(x)\,dx\int_V \mathscr{E}'(x)\,dx} &= 4\pi\Lambda_4 = \text{const},
\end{aligned}\tag{20.7'}$$

from which the relationship between the invariants $\Lambda_i$ and the corresponding conservation laws is quite obvious. The spatial means in Eqs. (20.7') can be replaced by weighted means with the weight function tending to unity at any fixed space point. For example, we can set

$$4\pi\Lambda_1 = \lim_{\alpha\to 0}\frac{\alpha^{3/2}}{\pi^{3/2}}\overline{\left[\int \rho'(x)\,e^{-\alpha x^2}\,dx\right]^2}\tag{20.7''}$$

and similarly for the other invariants $\Lambda_i$ [cf. Eqs. (15.27'), (15.28) and (15.32)].

We saw in Sect. 15.2 that, in incompressible fluids with sufficiently rapidly decreasing velocity correlations there is one further invariant, namely, the Loitsyanskii invariant (15.16), which can be derived from the physical law of conservation of angular momentum [i.e., the tensor $m_{ij}(x)=x_i\mathcal{U}_j(x)-x_j\mathcal{U}_i(x)$]. In compressible liquids, this conservation law will of course be valid again [since, from Eq. (20.1) and the fact that $\Pi_{ij}=\Pi_{ji}$, it follows that $\frac{\partial}{\partial t}m_{ij}(x,t)$ is equal to the divergence of the tensor field $-x_i\Pi_{jk}+x_j\Pi_{ik}$]. Here, however, this conservation law no longer leads to a new invariant for isotropic turbulence. In fact, if we repeat the calculation used for isotropic incompressible fluids, which leads to Eq. (15.28), we readily find that

$$\frac{\alpha^{3/2}}{\pi^{3/2}}\overline{[M(\alpha)]^2} = \frac{2\sqrt{2}\,\pi\Lambda_2}{\alpha} + O(1),$$

where as before

$$\frac{\alpha^{3/2}}{\pi^{3/2}}\overline{[M(\alpha)]^2}=\frac{\alpha^{3/2}}{\pi^{3/2}}\sum_{i<j}\overline{\left[\int m_{ij}(x)\,e^{-\alpha x^2}\,dx\right]^2}\sim\frac{1}{V}\sum_{i<j}\overline{\left[\int_V m_{ij}(x)\,dx\right]^2}.$$

Hence, it is clear that, roughly speaking, the mean square of the integral of the angular momentum, evaluated over a large volume $V$ in the case of a compressible fluid, increases as $V^{5/3}$ and not as $V$ when $V \to \infty$, and the corresponding proportionality factor, which, in accordance with Sect. 15.2, is conveniently taken to be $\frac{\alpha^{5/2}}{\pi^{5/2}}\overline{[M(\alpha)]^2}$, differs by only a numerical factor from the invariant $\Lambda_2$. Thus, in the case of isotropic turbulence in compressible fluids, the laws of conservation of momentum and of angular momentum lead to the same invariant $\Lambda_2$.

## 20.2 Linear Theory; Final Period of Decay of Compressible Turbulence

In the above derivation of the invariants for isotropic compressible turbulence, we used only the fact that the dynamic equations written in terms of the variables $\rho$, $\mathcal{U}_i$, and $\mathcal{E}$ have the special form indicated by Eq. (20.1). The explicit form of the right-hand sides of these equations was of no significance. However, if we are interested in correlation equations which generalize the von Kármán-Howarth equation (14.9) to the case of a compressible fluid, then we must use the precise form of the dynamic equations. However, we then come up against the difficulty that the equations contain viscosity and thermal conductivity, which are functions of the dynamic variables (above all the temperature). Moreover, the derivatives of the main fields (for example, $\partial u_i/\partial x_k$) must be regarded as new unknowns. There are also a number of moments of nonlinear functions of the initial fields, which must be introduced because of the nonlinearity of the equation of state (1.63). All this results in a complicated system containing more unknowns than equations. The system itself is readily written down [see, for example, Krzywoblocki (1952)], but is very difficult to use. Let us therefore consider, following Yaglom (1948), the simple case of "very weak turbulence" (which can be described by linearized dynamic equations), and then try to take approximately into account the main nonlinear effects.

Consider isotropic turbulent fluctuations in an infinite gaseous medium at rest, and suppose that the mean density $\bar{\rho}$ and mean temperature $\bar{T}$ are constants. We shall assume that the fluctuations are so weak that the third-order moments of all the variables are negligible in comparison with the corresponding second-order moments. In other words, we assume that the turbulence has reached the final period of decay (see Sect. 15.3). We note that the final period of decay of compressible turbulence with relatively low (in comparison with the sound velocity) characteristic velocity is more interesting than this period in incompressible turbulence. The point is that compressibility gives rise to only small corrections of the ordinary incompressible motions, and these corrections can frequently be described by linearized equations.

Since, in the linear approximation, fluctuations of an arbitrary function of the flow variables are expressed in the form of linear combinations (with constant coefficients) of fluctuations in the original fields, it follows that, in this approximation, any change of variables in the dynamic equations reduces to a trivial linear transformation of the set of correlation equations. As the main variables we shall use the quantities $u_i$, $i = 1, 2, 3$, $P = p/\gamma\bar{p}$, where $\gamma = c_p/c_v$ and $S = \ln(p^{1/\gamma}/\rho)$. These variables were introduced by Kovasznay (1953) and were used in Sect. 1.7 of Vol. 1 in the linearized dynamic equations for compressible fluids. Other sets of variables were used by Yaglom (1948) and Moyal (1952).

Linear equations for the fields $u_l(\boldsymbol{x}, t)$, $P(\boldsymbol{x}, t)$ and $S(\boldsymbol{x}, t)$ were given in Vol. 1 by Eq. (1.90) (where $D = \partial u_l/\partial x_l$), Eq. (1.86) and Eq. (1.87). These equations include the constant coefficient $a_0 = \sqrt{\gamma \bar{p}/\bar{\rho}}$ (equal to the undisturbed sound velocity). The coefficients $\nu = \mu/\bar{\rho}$, $\eta = \zeta/\bar{\rho}$ and $\chi = \varkappa/c_p\bar{\rho}$ must now be regarded as constants, since their fluctuations will generate only negligibly small nonlinear corrections. By taking the average of these equations we find that, in low-intensity homogeneous turbulence, the mean values of $\bar{P}$ and $\bar{S}$ are time-independent (the increase in the mean entropy $\bar{S}$ due to heating of the medium by kinetic-energy dissipation and conductivity effects will be a second-order phenomenon). We can, therefore, replace the quantities $P$ and $S$ in Eqs. (1.86), (1.87), and (1.90) by corresponding fluctuations $P' = P - \bar{P}$ and $S' = S - \bar{S}$.

To obtain the equations for the correlation functions for the above fields, we must multiply each of the equations containing the derivative $\partial\alpha(\boldsymbol{x}, t)/\partial t$ on the left (where $\alpha = u_l$, $P'$ or $S'$) by $\beta(\boldsymbol{x}', t) = \beta(\boldsymbol{x} + \boldsymbol{r}, t)$ (where $\beta$ is one of the same five quantities), add the resulting equation to the equation for $\partial\beta(\boldsymbol{x}', t)/\partial t$ multiplied by $\alpha(\boldsymbol{x}, t)$, and then take the average of the result. In the case of homogeneous turbulence, in which all the correlation functions $\overline{\alpha(\boldsymbol{x}, t)\beta(\boldsymbol{x}', t)} = B_{\alpha\beta}(\boldsymbol{r}, t)$ depend only on $\boldsymbol{r} = \boldsymbol{x}' - \boldsymbol{x}$ and $t$, we must replace $\partial/\partial x_l$ in the resulting equations throughout by $-\partial/\partial r_l$, and $\partial/\partial x'_j$ by $\partial/\partial r_j$. The final set of equations is as follows:

$$\frac{\partial B_{lj}}{\partial t} = -a_0^2\left(\frac{\partial B_{lP'}}{\partial r_j} - \frac{\partial B_{P'j}}{\partial r_l}\right) + 2\nu\,\Delta B_{lj} + \left(\frac{\nu}{3} + \eta\right)\left(\frac{\partial^2 B_{ll}}{\partial r_j \partial r_l} + \frac{\partial^2 B_{lj}}{\partial r_l \partial r_l}\right),$$

$$\frac{\partial B_{lP'}}{\partial t} = a_0^2 \frac{\partial B_{P'P'}}{\partial r_l} + [\nu + (\gamma - 1)\chi]\,\Delta B_{lP'} + \left(\frac{\nu}{3} + \eta\right)\frac{\partial^2 B_{lP'}}{\partial r_l \partial r_l} + \chi\,\Delta B_{lS'} - \frac{\partial B_{ll}}{\partial r_l},$$

$$\frac{\partial B_{lS'}}{\partial t} = a_0^2 \frac{\partial B_{P'S'}}{\partial r_l} + (\nu + \chi)\,\Delta B_{lS'} + \left(\frac{\nu}{3} + \eta\right)\frac{\partial^2 B_{lS'}}{\partial r_l \partial r_l} + (\gamma - 1)\chi\,\Delta B_{lP'}, \tag{20.8}$$

$$\frac{\partial B_{P'P'}}{\partial t} = 2(\gamma - 1)\chi\,\Delta B_{P'P'} + \chi[\Delta B_{P'S'} + \Delta B_{S'P'}] + \frac{\partial B_{lP'}}{\partial r_l} - \frac{\partial B_{P'l}}{\partial r_l},$$

$$\frac{\partial B_{P'S'}}{\partial t} = \gamma\chi\,\Delta B_{P'S'} + \chi\,\Delta B_{S'S'} + (\gamma - 1)\chi\,\Delta B_{P'P'} + \frac{\partial B_{lS'}}{\partial r_l},$$

$$\frac{\partial B_{S'S'}}{\partial t} = 2\chi\,\Delta B_{S'S'} + (\gamma - 1)\chi(\Delta B_{P'S'} + \Delta B_{S'P'}).$$

In these equations, $i$ and $j$ assume the values 1, 2, 3, $\Delta = \frac{\partial^2}{\partial r_1^2} + \frac{\partial^2}{\partial r_2^2} + \frac{\partial}{\partial r_3^2}$, and, as usual, the subscript $u_l$ in the correlation functions is replaced by $i$ throughout. It is readily seen that this system is closed in the sense that the number of unknowns is equal to the number of equations.

In the case of isotropic turbulence, the set of equations (20.8) can be reduced to seven equations for the seven scalar functions $B_{LL}(r, t)$, $B_{NN}(r, t)$, $B_{LP'}(r, t)$, $B_{LS'}(r, t)$, $B_{P'P'}(r, t)$, $B_{P'S'}(r, t)$ and $B_{S'S'}(r, t)$. The resulting system is, however, quite complicated and difficult to interpret physically. It will, therefore, be more convenient to start by transforming to the spectral equations, and then use the isotropy condition. If we apply Fourier transformation to all the Eqs. (10.8), we obtain

$$\begin{aligned}
\frac{\partial F_{ij}}{\partial t} &= -ia_0^2(k_jF_{iP} - k_iF_{Pj}) - 2\nu k^2F_{ij} - \\
&\qquad - \left(\frac{\nu}{3} + \eta\right) k_l(k_jF_{il} + k_iF_{lj}), \\
\frac{\partial F_{iP}}{\partial t} &= ia_0^2k_iF_{PP} - [\nu + (\gamma - 1)\chi]k^2F_{iP} - \\
&\qquad - \left(\frac{\nu}{3} + \eta\right) k_ik_lF_{lP} - \chi k^2F_{iS} - ik_lF_{il}, \\
\frac{\partial F_{iS}}{\partial t} &= ia_0^2k_iF_{PS} - (\nu + \chi)k^2F_{iS} - \left(\frac{\nu}{3} + \eta\right) k_ik_lF_{lS} - (\gamma - 1)\chi k^2F_{iP}, \\
\frac{\partial F_{PP}}{\partial t} &= -2(\gamma - 1)\chi k^2F_{PP} - \chi k^2(F_{PS} + F_{SP}) + ik_l(F_{lP} - F_{Pl}), \\
\frac{\partial F_{PS}}{\partial t} &= -\gamma\chi k^2F_{PS} - \chi k^2F_{SS} - (\gamma - 1)\chi k^2F_{PP} + ik_lF_{lS}, \\
\frac{\partial F_{SS}}{\partial t} &= -2\chi k^2F_{SS} - (\gamma - 1)\chi k^2(F_{PS} + F_{SP}).
\end{aligned} \tag{20.9}$$

We can now express the tensor $F_{ij}(k, t)$ and the vectors $F_{iP}(k, t)$ and $F_{iS}(k, t)$ in terms of the scalar functions $F_{LL}(k, t)$, $F_{NN}(k, t)$, $F_{LP}(k, t)$ and $F_{LS}(k, t)$ through the relations given by Eqs. (12.31) and (12.55). Since $F_{PL}(k, t) = -F_{LP}(k, t)$, $F_{SL}(k, t) = -F_{LS}(k, t)$ and $F_{SP}(k, t) = F_{PS}(k, t)$ [see Eqs. (12.53) and (12.54)], the result of all this is the following set of linear ordinary equations describing the evolution of weak isotropic turbulence in a compressible gas:

$$\begin{aligned}
\frac{\partial F_{NN}}{\partial t} &= -2\nu k^2F_{NN}, \\
\frac{\partial F_{LL}}{\partial t} &= -2\left(\frac{4\nu}{3} + \eta\right) k^2F_{LL} + 2a_0^2kF_{LP}, \\
\frac{\partial F_{LP}}{\partial t} &= -kF_{LL} - \left(\frac{4\nu}{3} + \eta + (\gamma - 1)\chi\right) k^2F_{LP} - \chi k^2F_{LS} + a_0^2kF_{PP}, \\
\frac{\partial F_{LS}}{\partial t} &= -(\gamma - 1)\chi k^2F_{LP} - \left(\frac{4\nu}{3} + \eta + \chi\right) k^2F_{LS} + a_0^2kF_{PS}, \\
\frac{\partial F_{PP}}{\partial t} &= -2kF_{LP} - 2(\gamma - 1)\chi k^2F_{PP} - 2\chi k^2F_{PS}, \\
\frac{\partial F_{PS}}{\partial t} &= -kF_{LS} - (\gamma - 1)\chi k^2F_{PP} - \gamma\chi k^2F_{PS} - \chi k^2F_{SS}, \\
\frac{\partial F_{SS}}{\partial t} &= -2(\gamma - 1)\chi k^2F_{PS} - 2\chi k^2F_{SS}.
\end{aligned} \tag{20.10}$$

The function $F_{NN}$ is present only in the first equation of the system. The system, therefore, splits into six equations in the six unknowns $F_{LL}$, $F_{LP}$, $F_{LS}$, $F_{PP}$, $F_{PS}$ and $F_{SS}$, and a separate equation for the function $F_{NN}$ [which is the same as Eq. (15.37) in the case of incompressible turbulence]. We know from Sect. 12.3 that the function $F_{NN}(k, t)$ describes the rotational component of the velocity field $u(x, t)$, while the function $F_{LL}(k, t)$ describes the potential component. Consequently, in the linear approximation the eddy component of the velocity does not interact either with the potential component or with the pressure and temperature (entropy) fluctuations, and evolves quite independently in accordance with the same equations as for an incompressible fluid. In other words, low-intensity velocity fluctuations in compressible fluids differ from the corresponding fluctuations in incompressible fluids only by the fact that the set of random eddies, which are unaffected by compressibility, have random waves superimposed upon them. These waves are described by a potential velocity field and, at first sight, are intimately connected with pressure and temperature fluctuations. It was shown in Sect. 1.7 of Vol. 1 that this is not specific for isotropic turbulence, and has a general explanation which follows from the linearized system of dynamic equations.

Let us now suppose that all the turbulent correlation functions decrease sufficiently rapidly at infinity. All the spectral functions in Eqs. (20.10) can then be expanded near the origin into series in powers of $k$, and in the case of the functions $F_{LL}(k)$, $F_{NN}(k)$, $F_{PP}(k)$, $F_{PS}(k)$ and $F_{SS}(k)$ these series begin with zero-order terms [which are the same for $F_{LL}(k)$ and $F_{NN}(k)$], while in the case of the functions $F_{LP}(k)$ and $F_{LS}(k)$ they begin with first-order terms (see Sect. 12.2). Substituting the corresponding expansions into Eqs. (20.10), we find that all the coefficients of the zero-order terms are time independent:

$$f_0^{(NN)} = f_0^{(LL)} = \text{const}, \quad f_0^{(PP)} = \text{const}, \quad f_0^{(PS)} = \text{const}, \quad f_0^{(SS)} = \text{const} \tag{20.11}$$

where $f_0^{(\alpha\beta)} = F_{\alpha\beta}(0, t)$. It is readily seen, however, that to within terms that are quadratic in the turbulent fluctuations, $\mathcal{U}_l = \bar{\rho} u_l$, $\rho' = \bar{\rho}(P' - S')$ and $\mathcal{E}' = (\rho u^2)' + (\rho c_v T)' = \frac{\gamma}{\gamma - 1} \bar{p} P'$. Therefore, within the framework of the linear theory, the correlation functions $B_{LL}$, $B_{NN}$, $B_{P'P'}$, $B_{P'S'}$ and $B_{S'S'}$ are readily expressed in terms of the functions $\mathcal{B}_{LL}$, $\mathcal{B}_{NN}$, $B_{\rho'\rho'}$, $B_{\rho'\mathcal{E}'}$ and $B_{\mathcal{E}'\mathcal{E}'}$. Substituting these expressions into formulas of the form of Eqs. (12.44), and (12.64), which give the coefficients (20.11), we obtain

$$\begin{aligned}
f_0^{(NN)} &= f_0^{(LL)} = \frac{\Lambda_2}{6\pi^2 \bar{\rho}^2}, \quad f_0^{(PP)} = \left(\frac{\gamma - 1}{\gamma \bar{p}}\right)^2 \frac{\Lambda_3}{2\pi^2}, \\
f_0^{(PS)} &= \left(\frac{\gamma - 1}{\gamma \bar{p}}\right)^2 \frac{\Lambda_3}{2\pi^2} - \frac{\gamma - 1}{\gamma \bar{p} \bar{\rho}} \frac{\Lambda_4}{2\pi^2}, \\
f_0^{(SS)} &= \left(\frac{\gamma - 1}{\gamma \bar{p}}\right)^2 \frac{\Lambda_3}{2\pi^2} + \frac{1}{\bar{\rho}^2} \frac{\Lambda_1}{2\pi^2} - \frac{\gamma - 1}{\gamma \bar{p} \bar{\rho}} \frac{\Lambda_4}{\pi^2},
\end{aligned} \tag{20.12}$$

where the quantities $\Lambda_1, \ldots, \Lambda_4$ have the same significance as in Eq. (20.7). The spectral meaning of these invariants throws additional light on the unsuccessful attempts to generalize the Loitsyanskii invariant to the case of compressible fluids. As we have seen in Sect. 15.6, this invariant appears because the coefficient $f_0^{(LL)}$ vanishes, while in the case of a compressible medium the coefficients of $k^2$ in the Taylor expansions of the spectral functions are not constants, not even in the linear theory.

Let us now consider the solution of Eqs. (20.10). The solution of the first equation of this set was discussed in Sect. 15.5, so that we need only consider the remaining six equations in six unknowns. These equations form a set of ordinary linear differential equations with constant coefficients for each fixed $\boldsymbol{k}$. To solve them, we must first set up the corresponding characteristic (or secular) equation (giving the special solutions proportional to $e^{\Omega t}$). If we evaluate the sixth-order determinant which gives this characteristic equation, we find that it has the form

$$\begin{aligned}\{\Omega^3+2(\nu_1+\gamma\chi)k^2\Omega^2+4(\gamma\nu_1\chi+a_0^2/k^2)k^4\Omega+8\chi a_0^2k^4\}\times\\ \times\{\Omega^3+2(\nu_1+\gamma\chi)k^2\Omega^2+(\nu_1^2+3\gamma\nu_1\chi+\gamma^2\chi^2+a_0^2/k^2)k^4\Omega+\\ +(\gamma\nu_1^2\chi+\gamma^2\nu_1\chi^2+\nu_1a_0^2/k^2+(\gamma-1)\chi a_0^2/k^2)k^6\}=0,\end{aligned} \quad (20.13)$$

where $\nu_1=\frac{4}{3}\nu+\eta$ is the total viscosity which governs internal friction in potential flows.

Equation (20.13) is a reducible sixth-degree equation: its left-hand side is equal to the product of two third-degree polynomials. Let us denote the three roots of the first of these polynomials by $2\Omega_1$, $2\Omega_2$, and $2\Omega_3$. It can then be readily seen that the three roots of the second equation are respectively $\Omega_1+\Omega_2$, $\Omega_1+\Omega_3$, and $\Omega_2+\Omega_3$. In other words, the six roots of Eq. (20.13) are equal to all the possible sums of two roots (which may coincide) of the cubic

$$\Omega^3+(\nu_1+\gamma\chi)k^2\Omega^2+(\gamma\nu_1\chi+a_0^2/k^2)k^4\Omega+\chi a_0^2k^4=0. \quad (20.14)$$

If we replace the time $t$ by the dimensionless time $\tau=a_0kt$, then $\Omega$ will be replaced by the dimensionless quantity $\lambda=\Omega/a_0k$, and Eq. (20.14) will be replaced by

$$\lambda^3+\frac{(\nu_1+\gamma\chi)k}{a_0}\lambda^2+\left(1+\frac{\gamma\nu_1\chi k^2}{a_0^2}\right)\lambda+\frac{\chi k}{a_0}=0, \quad (20.15)$$

which is the same as Eq. (1.94) of Chapt. 1. We recall that, according to Sect. 1.7, the spatially periodic solutions of the linearized dynamic equations (with given wave vector $\boldsymbol{k}$), which correspond to "particularly compressible" irrotational motions, are of the form

$$(A_1e^{\lambda_1a_0kt}+A_2e^{\lambda_2a_0kt}+A_3e^{\lambda_3a_0kt})e^{i\boldsymbol{kx}}=(A_1e^{\Omega_1t}+A_2e^{\Omega_2t}+A_3e^{\Omega_3t})e^{i\boldsymbol{kx}}, \quad (20.16)$$

where $A_1$, $A_2$, and $A_3$ are certain constants (different for different fields), $\lambda_1$, $\lambda_2$ and $\lambda_3$ are the three roots of Eq. (20.15), and $\Omega_1$, $\Omega_2$ and $\Omega_3$ are the corresponding three roots of Eq. (20.14). In the case of turbulent flows, the complex-valued coefficients $A_1$, $A_2$, and $A_3$, which determine the initial amplitudes and phases of the corresponding plane waves, will be random variables. The contribution of the plane wave Eq. (20.16) to the correlation function

$$B_{\alpha\beta}(\boldsymbol{x},\boldsymbol{x}',t)=\overline{\alpha(\boldsymbol{x},t)\beta(\boldsymbol{x}',t)}=\overline{\alpha^*(\boldsymbol{x},t)\beta(\boldsymbol{x}',t)}$$

will then be given by

$$\left[\sum_{m,n=1}^{3}\overline{A_m^{(\alpha)*}A_n^{(\beta)}}e^{(\Omega_m^*+\Omega_n)t}\right]e^{i\boldsymbol{kr}}=\sum_{m,n=1}^{3}C_{mn}^{(\alpha\beta)}e^{(\Omega_m^*+\Omega_n)t+i\boldsymbol{kr}},$$

where $r = x' - x$ and $A_m^{(\alpha)}$ is the value of $A_m$ in the representation of the field $\alpha(x, t)$ given by Eq. (20.16) [we have used the fact that $\Omega_m^*$ is present together with $\Omega_m$ among the roots of Eq. (20.14)]. In the case of homogeneous turbulence, the generalized Fourier transforms $dZ_\alpha(k)$ of the turbulent fluctuations will have a form analogous to Eq. (20.15):

$$dZ_\alpha(k) = dA_1^{(\alpha)}(k)\, e^{\Omega_1 t} + dA_2^{(\alpha)}(k)\, e^{\Omega_2 t} + dA_3^{(\alpha)}(k)\, e^{\Omega_3 t}, \tag{20.17}$$

where $dA_m^{(\alpha)*}(k)$ and $dA_n^{(\beta)}(k')$ will be mutually uncorrelated for $k \neq k'$. Hence, the spectral density $F_{\alpha\beta}(k, t)$ at a point $k$ in wave number space is given by

$$F_{\alpha\beta}(k, t) = \sum C_{mn}^{(\alpha\beta)}(k)\, e^{(\Omega_m + \Omega_n)t}, \tag{20.18}$$

where $C_{mn}^{(\alpha\beta)}(k)\,\delta(k - k')\, dk\, dk' = \overline{dA_{m_1}^*(k)\, dA_n(k')}$, $\Omega_{m_1} = \Omega_m^*$. This corresponds to the fact that the roots of the characteristic Eq. (20.13) are equal to all the possible sums of pairs of roots of Eq. (20.14).

In the case of isotropic turbulence, the coefficients $C_{mn}^{(\alpha\beta)}(k)$ corresponding to scalar fields $\alpha$ and $\beta$ should depend only on $k = |k|$, and the corresponding coefficients for vector fields (for $\alpha = u_j$, or $\alpha = u_j$ and $\beta = u_l$) should be equal to functions of $k$ multiplied by $k_j$ or, correspondingly, by $k_j k_l$ [we recall that we are considering only the potential component of the field $u(x)$]. Roughly speaking, we can say that, according to the first equation in Eqs. (20.10) and Eq. (20.18), a weak isotropic compressible turbulence consists of an isotropic set of random eddies [described by the function $F_{NN}(k, t)$], and a set of plane waves (20.16) which is uncorrelated with them, or with each other, and has random amplitudes and phases.

The classification of small oscillations in a compressible fluid given in Sect. 1.7, and based on the fact that the parameter $\delta_1 = \nu_1 k / a_0$ is small, can be used to improve the general formula (20.18). In fact, it was shown in Sect. 1.7 that, to within terms of the order of $\delta_1 \ll 1$, the entropy fluctuations $S'(x, t)$ form a set of standing damped waves of the form $A_3(k)\, e^{\lambda_3 a_0 k t + ikx} = A_3(k)\, e^{\Omega_3 t + ikx}$, while fluctuations in the velocity divergence $D(x, t) = \partial u_j / \partial x_j$ [which completely determines the potential component of the field $u(x, t)$] and in the pressure $P(x, t)$ form a set of progressive sound waves described by

$$\begin{aligned} D(x, t) &= a_0 k\, [A_1(k)\, e^{\Omega_1 t} + A_2(k)\, e^{\Omega_2 t}]\, e^{ikx}, \\ P(x, t) &= i\, [A_1(k)\, e^{\Omega_1 t} - A_2(k)\, e^{\Omega_2 t}]\, e^{ikx}, \end{aligned} \tag{20.19}$$

where $\Omega_1 = a_0 k \lambda_1$, $\Omega_2 = a_0 k \lambda_2 = \Omega_1^*$ [see Eqs. (1.99) and (1.100)] and $\lambda_1, \lambda_2, \lambda_3$ have the same meaning as in Eq. (1.98).[39] Hence it follows that, to within terms of the order of $\delta_1$, the spectral densities (20.18) are given by the equations

[39] We see that the coupling between velocity and entropy (temperature) fluctuations arises only when terms of the order of $\delta_1$ are taken into account. This is what we had in mind when we noted above that the potential component of the velocity is only "at first sight" closely coupled to the pressure and entropy fluctuations.

$$
\begin{aligned}
&F_{NN}(k, t) = C_0(k)\, e^{-2\nu k^2 t},\\
&F_{LL}(k, t) = C_1(k)\, a_0^2 e^{2\Omega_1 t} + C_2(k)\, a_0^2 e^{2\Omega_2 t} + C_3(k)\, a_0^2 e^{(\Omega_1+\Omega_2)\, t},\\
&F_{LP}(k, t) = iC_1(k)\, a_0 e^{2\Omega_1 t} - iC_2(k)\, a_0 e^{2\Omega_2 t},\\
&F_{LS}(k, t) = C_4(k)\, a_0 e^{(\Omega_1+\Omega_3)\, t} + C_5(k)\, a_0 e^{(\Omega_2+\Omega_3)\, t},\\
&F_{PP}(k, t) = -C_1(k)\, e^{2\Omega_1 t} - C_2(k)\, e^{2\Omega_2 t} + C_3(k)\, e^{(\Omega_1+\Omega_2)\, t},\\
&F_{PS}(k, t) = iC_4(k)\, e^{(\Omega_1+\Omega_3)\, t} - iC_5(k)\, e^{(\Omega_2+\Omega_3)\, t},\\
&F_{SS}(k, t) = C_6(k)\, e^{2\Omega_3 t}
\end{aligned}
\tag{20.20}
$$

These formulas can also be obtained directly from Eq. (20.10) by reducing them to a dimensionless form and solving them to within terms of the order of $\delta_1$. In these expressions, $C_0(k), C_1(k), \ldots, C_6(k)$ are seven functions of $k$ such that $C_2(k) = [C_1(k)]^*$, $C_5(k) = [C_4(k)]^*$, while the remaining $C_j$ are real. The solution (20.20) of Eq. (20.10) thus depends on seven arbitrary real functions which are determined by the initial values of the seven functions $F_{NN}(k, t), F_{LL}(k, t), \ldots, F_{SS}(k, t)$.

If we do not neglect terms of the order of $\delta_1$ or higher, the first formula in Eqs. (20.20) will remain, but the other formulas will assume the more complicated form

$$
\begin{aligned}
F_{\alpha\beta}(k, t) = C_1(k)\, p_{\alpha\beta}^{(1)} e^{2\Omega_1 t} + C_2(k)\, p_{\alpha\beta}^{(2)} e^{2\Omega_2 t} &+ \\
+ C_3(k)\, p_{\alpha\beta}^{(3)} e^{(\Omega_1+\Omega_2)\, t} + C_4(k)\, p_{\alpha\beta}^{(4)} e^{(\Omega_1+\Omega_3)\, t} &+ \\
+ C_5(k)\, p_{\alpha\beta}^{(5)} e^{(\Omega_2+\Omega_3)\, t} + C_6(k)\, p_{\alpha\beta}^{(6)} e^{2\Omega_3 t}&,
\end{aligned}
\tag{20.21}
$$

where the functions $C_1(k), \ldots, C_6(k)$ satisfy the same conditions as above, while the coefficients $p_{\alpha\beta}^{(1)}, \ldots, p_{\alpha\beta}^{(6)}$ are proportional to the values of the corresponding cofactors of the characteristic determinant for Eq. (20.10) when $\Omega = 2\Omega_1, \ldots, 2\Omega_3$. Thus, here again, the values of the spectral functions for a given $t$ depend on seven arbitrary real functions determined by the initial values of the spectral functions.

In view of Eq. (1.98), all six exponents $2\Omega_1, \ldots, 2\Omega_3$ have negative real parts which increase with $k$ for any $k > 0$. Therefore, the corresponding integrals in Eqs. (12.34), (12.59), and (12.4), evaluated between $k = \epsilon$ and infinity for any $\epsilon > 0$, will add only an exponentially decreasing term to the expressions for the corresponding correlation function. It follows that the asymptotic behavior of all the correlation functions for $t \to \infty$ will depend only on the values of the integral on the right-hand sides of Eqs. (12.34), (12.59), and (12.4), evaluated between zero and $\epsilon$. Consequently, it will depend only on the behavior of the corresponding spectral density $F_{\alpha\beta}(k, t)$ near $k = 0$. We note that, if the fluid is compressible, there is no reason to doubt that the spectral densities are regular near the origin (see the footnote in Sect. 20.1). Therefore, the case of a compressible fluid is in fact more favorable than that of an incompressible fluid for the investigation of the asymptotic behavior of the correlation functions as $t \to \infty$. When we investigate the main term of the spectral functions near $k = 0$, we can use the simplified formulas of Eqs. (20.20), since these ignore only terms of the order of $\delta_1 = \nu_1 k/a_0$ which vanish as $k \to 0$. Moreover, for the same reason, we can use the simplified formulas (1.98) to determine the exponents $\Omega_i = a_0 k \lambda_i$, $i = 1, 2, 3$, while the coefficients $C_0(k), C_1(k), \ldots, C_6(k)$ can be assumed equal to their values $C_0(0), C_1(0), \ldots, C_6(0)$ at $k = 0$.

To obtain simple final expressions it is best to introduce the above simplifications into Eqs. (12.34), (12.59), and (12.4), and then again replace the integral between $k = 0$ and $k = \epsilon$ by an integral between 0 and $\infty$. The result of this will be that the correlation function will acquire an exponentially decreasing term which does not affect its asymptotic behavior. If we apply this discussion to the functions $B_{LL}(r, t)+2B_{NN}(r, t)$ [the results for the individual functions $B_{LL}(r, t)$ and $B_{NN}(r, t)$ are more complicated], $B_{p'p'}(r, t)$ and $B_{S'S'}(r, t)$, and if we retain on the right-hand side only the leading terms, we obtain

$$\begin{aligned} B_{LL}(r, t)+2B_{NN}(r, t) &\approx 2C_0(0)\frac{\pi^{3/2}}{(2\nu t)^{3/2}}e^{-\frac{r^2}{8\nu t}}+ \\ &+a_0^2 C_3(0)\frac{\pi^{3/2}}{\{[\nu_1+(\gamma-1)\chi]t\}^{3/2}}e^{-\frac{r^2}{4[\nu_1+(\gamma-1)\chi]t}}, \\ B_{p'p'}(r, t) &\approx C_3(0)\frac{\pi^{3/2}}{\{[\nu_1+(\gamma-1)\chi]t\}^{3/2}}e^{-\frac{r^2}{4[\nu_1+(\gamma-1)\chi]t}}, \\ B_{S'S'}(r, t) &\approx C_6(0)\frac{\pi^{3/2}}{(2\chi t)^{3/2}}e^{-\frac{r^2}{8\chi t}} \end{aligned} \tag{20.22}$$

and hence

$$\begin{aligned} \overline{u^2(t)} &= \frac{2C_0(0)\pi^{3/2}}{(2\nu t)^{3/2}}+\frac{a_0^2 C_3(0)\pi^{3/2}}{\{[\nu_1+(\gamma-1)\chi]t\}^{3/2}}+o(t^{-3/2}), \\ \overline{P'^2(t)} &= \frac{C_3(0)\pi^{3/2}}{\{[\nu_1+(\gamma-1)\chi]t\}^{3/2}}+o(t^{-3/2}), \quad \overline{S'^2(t)}=\frac{C_6(0)\pi^{3/2}}{(2\chi t)^{3/2}}+o(t^{-3/2}). \end{aligned} \tag{20.23}$$

In view of Eqs. (20.20), we have $C_0(0)=f_0^{(NN)}$, $a_0^2[C_1(0)+C_2(0)+C_3(0)]=f_0^{(LL)}$, $-C_1(0)-C_2(0)+C_3(0)=f_0^{(PP)}$ and $C_6(0)=f_0^{(SS)}$, so that from Eq. (20.12) we find that

$$\begin{aligned} C_0(0) &= \frac{\Lambda_2}{6\pi^2\bar{\rho}^2}, \quad C_3(0)=\left(\frac{\gamma-1}{\gamma\bar{p}}\right)^2\frac{\Lambda_3}{4\pi^2}+\frac{1}{\gamma\bar{p}\,\bar{\rho}}\,\frac{\Lambda_2}{12\pi^2}, \\ C_6(0) &= \left(\frac{\gamma-1}{\gamma\bar{p}}\right)^2\frac{\Lambda_3}{2\pi^2}+\frac{1}{\bar{\rho}^2}\,\frac{\Lambda_1}{2\pi^2}-\frac{\gamma-1}{\gamma\bar{p}\bar{\rho}}\,\frac{\Lambda_4}{\pi^2}. \end{aligned} \tag{20.24}$$

Hence, it follows that the coefficients in Eqs. (20.22) and (20.23) can also be expressed in terms of the invariants $\Lambda_1, \ldots, \Lambda_4$, and, moreover, we can transform from the variables $P$ and $S$ to $\rho$ and $p$, or $\rho$ and $T$ [in the latter form, the formulas of Eq. (20.23) were essentially established in the paper by Sitnikov (1958) where, however, some misprints are present in the final expressions].

## 20.3 Quadratic Effects; Generation of Sound by Turbulence

The linear theory discussed above is valid for any very weak turbulence during its final period of decay. Let us now consider the case where the turbulence is relatively weak, but not weak enough to allow us to neglect the nonlinear terms in the dynamic equations. In that case, we must use the next perturbation-theory approximation which takes into account both the leading linear terms and corrections of the order of Re. In this approximation, which has already been discussed in Sect. 1.7 of Vol. 1, the nonlinear terms are retained in the dynamic equations, but the turbulent fluctuations are taken to be given by the solutions of the set of linear equations. In specific calculations, the nonlinear terms are conveniently regarded as additional "influxes" of mass, momentum, and energy, generating definite "additions" to the solutions of the linearized equations. These "additions" govern certain new physical phenomena such as the generation of eddies, sound, and entropy waves as a result of bilinear and quadratic interactions.

All this results in eighteen new second-order effects, half of which arise only through terms containing the small parameter $\delta_1 \sim \frac{l_{\rm mol}}{\Lambda} \ll 1$ (where $l_{\rm mol}$ is the mean free path of the molecules and $\Lambda$ is the wavelength of the perturbation), and cannot therefore play an appreciable role. The remaining nine effects were classified in Vol. 1 and are not all equally important under real conditions. The point is that in the usual turbulent flows of a constant-temperature medium, whose characteristic velocity is much less than the sound velocity, the eddy (i.e., "incompressible") flows always have much higher intensity than sound and entropy waves. Therefore, the most important interaction is the interaction of the eddies with themselves. The next in order of importance is the interaction between the eddies and sound and entropy waves. The remaining interactions are of much less importance. In the case of turbulence in a slightly compressible, temperature-inhomogeneous medium, entropy waves (which in this case are largely equivalent to disturbances of a scalar "passive admixture field" $\vartheta$) may become highly developed. However, here again, the sound intensity is usually low. Therefore, scattering of sound by sound, which occurs as a result of the quadratic interaction between sound waves, can only occur under certain very special conditions.

We know from Vol. 1 that the interaction of an "eddy component" with itself, which under usual conditions is the most important, can generate only either eddies or sound, apart from effects of the order of $\delta_1$. The generation of the "eddy component" through this interaction is explained by the stretching of the eddy tubes during convective, i.e., inertial, displacement of the fluid particles. In the case of isotropic turbulence, this effect is described by terms of the von Kármán-Howarth equation (14.9) which contain the function $B_{LL,L}(r, t)$ [or the term $T(k, t)$ of the spectral equation (14.15)]. Most of the present chapter has been devoted to the study of this effect. The generation of sound by the interaction of eddies with themselves will, of course, appear only when compressibility is taken into account. This important effect will be considered in this section.

The self-interaction of "entropy waves" is, in general, an effect of the order of $\delta_1$. On the other hand, the interaction of these waves with eddies, which is very important in the case of a temperature-inhomogeneous medium, in fact generates only entropy waves. This last effect should of course, also appear in an incompressible fluid where it reduces to convective mixing of temperature inhomogeneities during the inertial motion of the fluid particles. This motion is described by those terms of the Corrsin equation which contain the function $B_{L\vartheta,\vartheta}(r, t)$ [or the corresponding term $T_{\vartheta\vartheta}(k, t)$ of the spectral equation (14.63)]. We have already frequently encountered this effect and need not consider it again. Among effects due to the interaction of sound waves with the eddy and entropy components of the motion, the most important is the sound-generation effect which is usually interpreted as scattering of sound by velocity and temperature fluctuations. The interaction of sound waves with eddies may also lead to the generation of eddies, while their interaction with the entropy component may lead to the generation of the entropy component. However, the corresponding "convection of eddies and temperature inhomogeneities by acoustic waves" is usually very small in comparison with the analogous convection produced by the eddy component of the velocity field. Finally, the last effect, which does not contain the factor $\delta_1$, describes the generation of vorticity through the interaction of entropy waves, producing an entropy (density) gradient, and of sound waves, producing a pressure gradient. This effect (described by the so-called Bjerknes term in the vorticity balance equation in a compressible fluid) is important only in connection with the origin of large-scale circulation processes in the earth's atmosphere, but can usually be neglected for small-scale turbulence.

Thus, the most important second-order effects in a compressible fluid are the generation of sound by turbulence, and the scattering of sound by velocity and temperature inhomogeneities. The scattering of sound can be naturally postponed to Chapt. 9, which is concerned with the wave propagation in a turbulent medium. Here, we shall restrict our attention to the generation of sound by turbulence.

The general Lighthill equation [Lighthill (1952, 1954)], which describes the generation of sound by the incompressible velocity fluctuations to within terms of the order of $\delta_1$, was given in Sect. 1.7 of Vol. 1. It can be written in the form

$$\frac{\partial^2 P}{\partial t^2} - a_0^2 \Delta P = -\frac{\partial^2 T_{ij}}{\partial x_i \partial x_j}, \tag{20.25}$$

where

$$T_{ij}(\boldsymbol{x}, t) = u_i^{(s)}(\boldsymbol{x}, t)\, u_j^{(s)}(\boldsymbol{x}, t), \tag{20.26}$$

[$u^{(s)}(\boldsymbol{x}, t)$ is the incompressible component of the velocity field], and, as above, $P = p/\gamma\bar{p}$, $a_0^2 = \gamma\bar{p}/\bar{\rho}$. We recall that when we derived this equation we neglected viscosity and thermal conductivity effects, while the density fluctuations $\rho' = \rho - \bar{\rho}$ were assumed to be small in comparison with $\bar{\rho}$ (which implied that the velocity divergence was also small). Moreover, typical values of the incompressible velocity fluctuations $u^{(s)}$ were assumed to be small in comparison with the mean sound velocity $a_0$. Equation (20.25) can therefore be regarded as describing the generation of sound by arbitrary turbulence with small Mach number $\mathrm{Ma} = U/a_0$, and not merely by a turbulence approaching the final period of decay.[40] In the limit as $a_0 \to \infty$, Eq. (20.25) becomes identical with Eq. (1.9), i.e., with the usual equation relating the velocity and pressure fields in an incompressible fluid.

Equation (20.25) leads to a number of important consequences. For example, the right-hand side of this equation is a combination of the second derivatives of the field $T_{ij}(\boldsymbol{x})$ (namely, the divergence of this field with respect to both its indices), which means that, in the absence of solid walls, the generation of sound waves by the turbulence is equivalent to radiation by a set of acoustic quadrupoles (and not the usual sound sources or dipole sources), This fact was pointed out by Lighthill (1952, 1954). Hence it follows that, if there are no walls, a low Mach number turbulence is not an effective source of sound. This is also supported by the character of the dependence of the total intensity of the radiated acoustic waves on the characteristic velocity scale $U$. The general solution of Eq. (20.25) can be written in terms of the retarded potentials:

$$P'(\boldsymbol{x}, t) = P(\boldsymbol{x}, t) - \bar{P} = \frac{1}{4\pi a_0^2} \int \left. \frac{\partial^2 T_{ij}(\boldsymbol{y}, \tau)}{\partial y_i \partial y_j} \right|_{\tau = t - \frac{|\boldsymbol{x}-\boldsymbol{y}|}{a_0}} \frac{d\boldsymbol{y}}{|\boldsymbol{x}-\boldsymbol{y}|}. \tag{20.27}$$

[40] For Mach numbers that are not too low, Eq. (20.25) can be replaced by [Phillips (1960)]

$$\frac{d^2 P}{dt^2} - a_0^2 \Delta P = \frac{\partial^2 u_i u_j}{\partial x_i \partial x_j}, \qquad \frac{d}{dt} = \frac{\partial}{\partial t} + u_i \frac{\partial}{\partial x_i}$$

We shall suppose that acoustic "noise" is generated by a bounded volume of the fluid in which there are velocity fluctuations, while the surrounding medium is at rest. In that case, the "turbulent region" will emit acoustic waves in all directions, giving rise to the pressure fluctuations described by Eq. (20.27) outside this region. If we apply the integral Gauss formula twice to the right-hand side of Eq. (20.27), and bear in mind that the integrals over the bounding surface, which lies outside the turbulence region, are identically equal to zero, we can replace differentiation with respect to the running coordinates $y_i$ in Eq. (20.27) by differentiation of the integrand with respect to the coordinates of the point of observation $\boldsymbol{x}$:

$$P'(\boldsymbol{x}, t) = \frac{1}{4\pi a_0^2} \frac{\partial^2}{\partial x_i\, \partial x_j} \int T_{ij}\left(\boldsymbol{y}, t - \frac{|\boldsymbol{x}-\boldsymbol{y}|}{a_0}\right) \frac{d\boldsymbol{y}}{|\boldsymbol{x}-\boldsymbol{y}|} \tag{20.28}$$

[We note that the appearance of the second derivative of the integral on the right-hand side of Eq. (20.28) corresponds to the fact that the radiation is a quadrupole radiation]. Let us consider the field $P(\boldsymbol{x}, t)$, well away from the "noisy" volume, i.e., at a distance much greater than the characteristic wavelength of the emitted sound. This will be the so-called wave-zone approximation. The leading contribution to the right-hand side of Eq. (20.28) will then be due to terms obtained by differentiating the tensor $T_{ij}\left(\boldsymbol{y}, t - \frac{|\boldsymbol{x}-\boldsymbol{y}|}{a_0}\right)$ with respect to $x_i$ and $x_j$, and consequently

$$P'(\boldsymbol{x}, t) \approx \frac{1}{4\pi a_0^4} \int \frac{r_i r_j}{r^3} \frac{\partial^2}{\partial \tau^2} T_{ij}(\boldsymbol{y}, \tau)\Big|_{\tau = t - \frac{r}{a_0}} d\boldsymbol{y}, \quad r = |\boldsymbol{x}-\boldsymbol{y}|. \tag{20.29}$$

Since, for Mach numbers much less than unity, the tensor $T_{ij}$ is proportional to the square of the turbulent velocity fluctuations, and the velocity $\boldsymbol{u}^{(s)}$ can be assumed to be equal to the total velocity fluctuation $\boldsymbol{u}$, we see that the amplitude of the pressure fluctuation in the wave zone is proportional to the square of the characteristic value $U$ of the velocity fluctuations, multiplied by the square of the characteristic frequency. However, the characteristic fluctuation frequency is proportional to $U/L$, where $L$ is the turbulence length scale. The amplitude of the pressure fluctuations is therefore proportional to $U^4$, and the total emitted energy (intensity of the sound waves) is proportional to $U^8$. This means that the emitted sound intensity will be very low when the velocity $U$ is low.

The above result can be made more precise as follows. It is readily shown that the energy flux density carried by the sound waves in the wave zone is given by

$$j_E = \frac{a_0^3 \overline{\rho'^2}}{\bar{\rho}} = a_0^3 \bar{\rho} \overline{P'^2} \tag{20.30}$$

[see, for example, Landau and Lifshits (1953), Sect. 64]. In the wave zone, i.e., at distances from the "turbulent volume" that are much greater than the linear dimensions $D$ of the volume, the factor $r_i r_j / r^3$ in Eq. (20.29) can be regarded as constant and taken outside the integral sign. Hence in this zone the energy carried by the sound waves is given by

$$j_E(\boldsymbol{x}) = \frac{\bar{\rho}}{16\pi^2 a_0^5} \frac{r_i r_j r_k r_l}{r^6} \times$$
$$\times \overline{\int\int \frac{\partial^2}{\partial \tau^2}[u_i(\boldsymbol{y}, \tau) u_j(\boldsymbol{y}, \tau)]_{\tau = t - \frac{|\boldsymbol{x}-\boldsymbol{y}|}{a_0}} \frac{\partial^2}{\partial \tau'^2}[u_k(\boldsymbol{y}', \tau') u_l(\boldsymbol{y}', \tau')]_{\tau' = t - \frac{|\boldsymbol{x}-\boldsymbol{y}'|}{a_0}}} \times$$
$$\times d\boldsymbol{y}\, d\boldsymbol{y}'. \tag{20.31}$$

It is readily seen that we can neglect in this formula the difference between $\tau$ and $\tau'$. In fact, since the correlation between the tensors $T_{ij}(y, \tau)$ and $T_{kl}(y', \tau')$ is appreciable only for $|y-y'|<L$, we have $|\tau-\tau'|\lesssim\frac{L}{a_0}=\frac{L}{U}\frac{U}{a_0}$ and, consequently, for $\mathrm{Ma}=\frac{U}{a_0}\ll 1$ the difference $\tau-\tau'$ will be much less than the time scale $L/U$ of the turbulence fluctuations. The mean value under the integral sign in Eq. (20.31) can therefore be evaluated on the assumption that $\tau=\tau'$. The integral with respect to $y'$ in Eq. (20.31) can then be roughly estimated by assuming that for any fixed $y$, it is equal to the quantity

$$\overline{\frac{\partial^2 u_i(y,\tau)\,u_j(y,\tau)}{\partial\tau^2}\,\frac{\partial^2 u_k(y,\tau)\,u_l(y,\tau)}{\partial\tau^2}},$$

which is of the order of

$$U^4\left(\frac{U}{L}\right)^4=\frac{U^8}{L^4}$$

multiplied by the characteristic volume $L^3$ of the fluid in which the velocity fluctuations are appreciably correlated with the velocity fluctuations at the given point $y$. This shows that the total energy $\mathscr{E}$ emitted per unit time by a unit mass of the turbulent medium is of the order of

$$\mathscr{E}\sim U^8/a_0^5 L \tag{20.32}$$

[since the quantity $\mathscr{E}$ is equal to the integral of the right-hand side of Eq. (20.31) evaluated over a sphere of radius $r$, divided by $\bar{\rho}$ and the total volume of the turbulent region]. If we recall that the rate of turbulent-energy dissipation is of the order of $U^3/L$ (see Sect. 16.5), we can rewrite Eq. (20.32) in the form

$$\mathscr{E}\sim\bar{\varepsilon}\frac{U^5}{a_0^5}=\bar{\varepsilon}\,(\mathrm{Ma})^5. \tag{20.33}$$

Consequently, the ratio of the energy emitted in the form of sound waves to the energy converted into heat as a result of viscous friction, is proportional to the fifth power of the Mach number. The Mach number may, of course, be preceded in this expression by a relatively large numerical factor [since the estimates leading to Eq. (20.33) are very approximate], but even then the fact that the Mach number is raised to the fifth power ensures that turbulence is not a very effective source of sound for small $U/a_0$.

The above qualitative analysis (due to Lighthill) is valid for arbitrary turbulence with low Mach number and zero mean velocity (so that the characteristic velocity scale $U$ is equal to the root mean square velocity fluctuation). In the special case when the turbulence can be regarded as homogeneous and isotropic within the limits of the radiating region (which implies that the diameter $D$ of this region is much greater than the length scale $L$ of the turbulence), these considerations enable us to estimate the numerical coefficient $\alpha=\mathscr{E}/\bar{\varepsilon}(\mathrm{Ma})^5$. In fact, let us suppose that the turbulence is isotropic, and try to calculate the energy flux $j_E(x)$ at points on the $x$ axis lying at large distances from the turbulent region (the origin will be taken inside this region). The factor $r_i r_j r_k r_l$ in Eq. (20.31) will then be zero, provided only that not all the indices $i, j, k, l$ correspond to the $x$ axis, so that

$$j_E(x)=\frac{\bar{\rho}}{16\pi^2 a_0^5 x^2}\int\int\overline{\frac{\partial^2}{\partial\tau^2}u_x^2(y,\tau)\frac{\partial^2}{\partial\tau^2}u_x^2(y',\tau)}\,dy\,dy' \tag{20.34}$$

where we have taken into account the fact that the times $\tau$ and $\tau'$ can be assumed to be equal in Eq. (20.31).[41]

The right-hand side of Eq. (20.34) was subsequently transformed by Proudman (1952) who assumed that the Millionshchikov zero-fourth-cumulant hypothesis can be applied not only to the vector field $u(x, t)$, but also to the nine-dimensional field

$$\left\{ u(x, t),\ \frac{\partial u(x, t)}{\partial t},\ \frac{\partial^2 u(x, t)}{\partial t^2} \right\}.$$

This is partly justified by the fact that, as will be seen below, the main contribution to the energy radiated by the turbulence in the form of sound waves is due to large-scale disturbances in the energy range of the wave-number spectrum for which the Millionshchikov hypothesis is in adequate agreement with experimental data. Using the above assumption, we obtain

$$\overline{\frac{\partial^2 u_x^2(y, t)}{\partial t^2}\, \frac{\partial^2 u_x^2(y', t)}{\partial t^2}} = 4\overline{u_x u_x'}\; \overline{\frac{\partial^2 u_x}{\partial t^2}\, \frac{\partial^2 u_x'}{\partial t^2}} + 12\left(\overline{\frac{\partial u_x}{\partial t}\, \frac{\partial u_x'}{\partial t}}\right)^2 + \ldots, \tag{20.35}$$

where $u_x = u_x(y)$, $u_x' = u_x(y')$, where the dots on the right of Eq. (20.35) represent terms containing time derivatives of the correlation functions, for example, $\frac{\partial \overline{u_x u_x'}}{\partial t}$, $\frac{\partial}{\partial t}\left(\overline{\frac{\partial u_x}{\partial t}\frac{\partial u_x'}{\partial t}}\right)$, $\frac{\partial^2 \overline{u_x u_x'}}{\partial t^2}$, etc. These terms are obviously zero in the case of stationary turbulence, while in decaying turbulence they can be approximately estimated by assuming that the correlation function $B_{LL}(r, t)$ undergoes a self-preserving variation with time [see Proudman (1952)]. It then turns out that the terms omitted from Eq. (20.35) are of minor importance, and can be neglected in the first approximation.

If we replace $\frac{\partial u_x}{\partial t}$ by the right-hand side of the corresponding Navier-Stokes equation, we can express the correlation function $\overline{\frac{\partial u_x}{\partial t}\frac{\partial u_x'}{\partial t}}$ in the limiting case $\nu = 0$ (i.e., for very large Reynolds numbers) in terms of the tensor $B_{ij}(r, t)$ by a repeated application of the

[41] We note that the formula (20.34) is, strictly speaking, valid only for stationary isotropic turbulence [this is also the case for Eq. (20.31)]. In fact, the formula (20.29) will determine the quantity $P'(x, t) = P(x, t) - \overline{P}$ only for $\overline{P} = \text{const}$. If, on the other hand, the turbulence is not stationary, then the quantity $\frac{\partial^2}{\partial \tau^2} T_{ij}(y, \tau)$ on the right-hand side of this formula must be replaced by $\frac{\partial^2}{\partial \tau^2}[T_{ij}(y, \tau) - \overline{T}_{ij}]$ (since $\overline{T_{ij}(y, \tau)}$ depends on $\tau$). The quantities $\partial^2 u_x^2 / \partial \tau^2$ in Eq. (20.34) will then transform to $\frac{\partial^2}{\partial \tau^2}\left(u_x^2 - \overline{u_x^2}\right)$. The next equation, namely Eq. (20.35), will then be valid only if its right-hand side is replaced by the corresponding fourth moment

$$\overline{\frac{\partial^2}{\partial t^2}\left[u_x^2(y) - \overline{u_x^2}\right] \frac{\partial^2}{\partial t^2}\left[u_x^2(y') - \overline{u_x^2}\right]},$$

so that the final results are unaltered.

Millionshchikov hypothesis, i.e., in the final analysis we can replace it by the longitudinal correlation function $B_{LL}(r, t)$. A similar procedure can be applied to the correlation function $\overline{\frac{\partial^2 u_x}{\partial t^2} \frac{\partial^2 u_x'}{\partial t^2}}$ (using the equations obtained as a result of differentiation with respect to time of the Navier-Stokes equations). As a result, we can also express it in terms of $B_{LL}(r, t)$. If, moreover, we assume that $B_{LL}(r, t) = \overline{u^2}\varphi(r/L)$, i.e., that $B_{LL}(r, t)$ undergoes a self-preserving variation, so that the energy $\overline{u^2(t)}$ falls off as given by Eq. (16.11), then the final expression for the total energy $\mathscr{E}$ emitted per unit time by a unit mass of the turbulent volume is given by

$$\mathscr{E} = \frac{\alpha}{a_0^5} \frac{(\overline{u^2})^4}{L} = \alpha \bar{\varepsilon} (\mathrm{Ma})^5 \tag{20.36}$$

[cf. also Eqs. (20.32) and (20.33)]. This, however, results in an explicit, though quite complicated, formula for $\alpha$ which contains only the normalized correlation functions $\varphi(x)$ [see Proudman (1952)]. Proudman has estimated the numerical value of the coefficient $\alpha$ for two special forms of the function $\varphi(x)$. These were: $\varphi(x)$ corresponding to the fully self-preserving spectrum of the Heisenberg theory with Re $= \infty$ (see Fig. 31), and $\varphi(x) = \exp(-\pi x^2/4)$. He found that $\alpha = 38$ for the self-preserving spectrum of the Heisenberg theory, and $\alpha = 13$ for $\varphi(x) = \exp(-\pi x^2/4)$. In view of the assumptions employed (self-preservation of decay, large Reynolds numbers), the first of these values of $\alpha$ must be regarded as more realistic then the second. Actually, the case $\varphi(x) = \exp(-\pi x^2/4)$ was analyzed only to establish the stability of the values of $\alpha$ against a considerable change in the form of the correlation function $B_{LL}(r, t)$. The approximate calculation given by Proudman enables us to conclude that the true value of $\alpha$ for isotropic turbulence should apparently lie between 10 and 100. We emphasize once again that, although this coefficient may turn out to be relatively large, the presence of $(\mathrm{Ma})^5$ in Eq. (20.36) still ensures that, for small Mach numbers, turbulence is a very inefficient source of sound energy. Hence it is not surprising that, for example, Golitsyn (1961) has found that the energy flux in the upper atmosphere due to the emission of sound by the turbulent lower atmosphere is much smaller than the flux due to thermal conductivity for a realistic variation of the mean temperature with height.

Similar results were subsequently obtained by Müller and Matschat (1958). These workers considered isotropic turbulence decaying in accordance with the Heisenberg equation (17.50). Instead of the quantity $\mathscr{E} = \mathscr{E}(t)$, they calculated (using the same assumptions as Proudman) the total energy $E = \int_0^\infty \mathscr{E}(t)\, dt$ emitted per unit volume by a unit mass throughout the entire evolution of the turbulence. They then compared the energy $E$ with the specific initial kinetic energy of the turbulence $\frac{\overline{\mathbf{u}^2}}{2} = \frac{3}{2}\overline{u^2}$. Since

$$\frac{3}{2}\overline{u^2} = \int_0^\infty [\bar{\varepsilon}(t) + \mathscr{E}(t)]\, dt$$

(in the final analysis, all the initial kinetic energy is either converted into heat as a result of viscosity or, to a lesser extent, is emitted in the form of sound waves), the coefficient

$$\beta = \frac{E}{\frac{3}{2}\overline{u^2}(\mathrm{Ma})^5}, \tag{20.37}$$

determined by Müller and Matschat is, in fact, a measure of the time average of the coefficient $\alpha$ considered by Proudman (reduced by the fact that the Mach number is taken only for the initial instant of time). Müller and Matschat again neglected the effect of viscosity (and assumed that $B_{LL}(r, t)$ was self-preserving). Nevertheless, the value of $\nu$ did enter into the calculations, since all the values of the correlation function $B_{LL}(r, t)$ for $t > 0$ depended on this value. Assuming that the initial turbulence spectrum was

$$E(k, 0) = \begin{cases} E_0 & \text{when } k_0 \leqslant k \leqslant ak_0, \\ 0 & \text{when } k < k_0 \text{ and } k > ak_0, \end{cases}$$

and that $\gamma_H = 0.45$ in the Heisenberg equation (17.50), Müller and Matschat used numerical integration to determine a series of values of $\beta$ corresponding to different $\mathrm{Re}_0 = \frac{L_0 v_0}{\nu}$ and different $a$; $L_0$ is the initial longitudinal turbulence length scale determined from $E(k, 0)$ with the aid of Eq. (12.91), and $v_0^2 = 2\int_0^\infty E(k, 0)\, dk$. It was found that the coefficient $\beta$ increased with increasing $\mathrm{Re}_0$ as expected. For $\mathrm{Re}_0 = 5{,}000$ and $a = 2$, the result was $\beta = 18$, while for $\mathrm{Re}_0 = 50{,}000$ and $a$ varying between 1.25 and 5, the values of $\beta$ were found to lie between 19 and 24. The numerical values of $\beta$ for relatively large $\mathrm{Re}_0$ are thus very stable, and although they are only very approximate, they are also in agreement with the fact that for large Reynolds numbers the coefficient $\alpha$ should be of the order of 100.

Since appreciable emission of sound by turblence is observed only for sufficiently high mean velocity $U$, this effect cannot be measured in grid turbulence which is nearly isotropic. Comparison between theory and experiment must, therefore, be carried out by using examples of essentially anisotropic turbulence for which the calculations are even more approximate. However, the theory developed above can be improved to some extent (for example, by the inclusion of the fact that the mean flow velocity is not zero). It then provides some information about the emission of sound by turbulent boundary layers [Laufer (1962)], fast turbulent jets of fluids or gases, and other types of anisotropic turbulence encountered in practice. Since aerodynamic noise is important in aviation (especially in connection with jet planes), the theory of turbulence-generated sound has recently attracted considerable attention. Both theoretical and experimental aspects of this problem have been reviewed by Lighthill (1962) who has also given an extensive list of references. A more recent review of the same problem was published by Ffowcs Williams (1969).

Sound generation can also be considered from a purely fluid dynamic standpoint, independently of any connection with turbulence [see Lighthill (1952)]. In particular, Klyatskin (1966a) has calculated the emission of sound by a pair of vortex rings and found that the intensity at large distances from the vortices was proportional to $(\mathrm{Ma})^5$. Since a pair of circular vortices is not very similar to the above system of random velocity disturbances in a turbulent flow, this result suggests the existence of a universal rule governing the emission of sound by different fluid dynamical systems. In this connection, Obukhov noted that the formula $\mathscr{E} \sim \bar{\varepsilon}\, (\mathrm{Ma})^5 \sim U^8$ can be regarded as an analog of the well known Stefan-Boltzmann law in the theory of blackbody radiation if we assume that the turbulent energy is $E \sim U^2$ and can serve as a measure of the "effective turbulence temperature" $T_e$ (so that $U^2 \sim T_e$ and, consequently, $\mathscr{E} \sim T_e^4$).

Klyatskin (1966b) has investigated turbulence in a slightly compressible medium, taking into account the nonlinear interaction between the random acoustic field and the vortex component of the turbulence. Assuming that the turbulence was homogeneous and isotropic, Klyatskin linearized the dynamic equations in the quantities describing the random acoustic field, and obtained a relation giving an expansion of the energy balance equation in terms of the small parameter $\beta = U^{(p)}/U^{(s)}$ where $U^{(p)}$ and $U^{(s)}$ are the characteristic velocities of the solenoidal and potential components of the velocity fluctuations. In the absence of extraneous sources of acoustic waves, the parameter $\beta$ was found to be proportional to $(\mathrm{Ma})^2 = \left(U^{(s)}/a_0\right)^2$.

# 8 LOCALLY ISOTROPIC TURBULENCE

## 21. GENERAL DESCRIPTION OF THE SMALL-SCALE STRUCTURE OF TURBULENCE AT LARGE REYNOLDS NUMBERS

### 21.1 A Qualitative Scheme for Developed Turbulence

It was noted in Chapt. 7 that the concept of isotropic turbulence is a mathematical idealization having very little connection with real turbulent flows encountered either in nature or in the laboratory. Nevertheless, the theory of isotropic turbulence does have some practical importance. It will be shown below that there is reason to suppose that a set of sufficiently small-scale disturbances in any developed turbulent flow (with Reynolds number much greater than $Re_{cr}$) in small space-time regions is always practically homogeneous and isotropic. Real turbulence can therefore frequently be regarded as having certain properties quite similar to isotropy. This means that some of the results and methods of Chapt. 7 will be applicable to real high Reynolds number turbulence. We shall discuss this in detail in the present chapter.

The assumption that small-scale disturbances in a turbulent flow with sufficiently large Reynolds number can be regarded as isotropic has its origin in the purely intuitive ideas put forward by L. F. Richardson in the early twenties. According to Richardson's scheme, developed turbulence consists of a set of disordered disturbances ("eddies") of various orders and differing in their characteristic length and velocity scales. If we gradually increase the Reynolds number and pass from the laminar state to developed turbulence, we find that disturbances of different orders do not appear simultaneously. As the Reynolds number $\mathrm{Re} = UL/\nu$ passes through the value $\mathrm{Re}_{cr}$, the large-scale fluctuations are the first to appear in the form of continuous waves similar to those discovered by Schubauer and Skramstad in the boundary layer (see Vol. 1, Sect. 2.8). As the Reynolds number increases further, these first-order disturbances generate smaller second-order disturbances which draw their energy from the kinetic energy of the large-scale disturbances. The second-order disturbances then generate third-order disturbances, and so on. A hierarchy of disturbances thus arises in which each disturbance draws its energy from larger-scale disturbances, and transfers its own energy to smaller-scale disturbances. The smallest scale disturbances are characterized by the largest values of the local velocity gradients. This is why the direct conversion of kinetic energy into heat under the influence of viscosity is largely localized in them.

Let us consider the above cascade transfer of energy down the disturbance hierarchy in somewhat greater detail. The largest disturbances in developed turbulent flow have a length scale $l_1$, which is of the same order of magnitude as the length scale $L$ of the flow as a whole (the scale $l_1$ can be regarded as equal to Prandtl's mixing length and also the integral length scale obtained by integrating the normalized velocity correlation function). Both scales, $L$ and $l_1$, can in general be defined in slightly different ways; nevertheless, $l_1$ is usually smaller than $L$. The largest-scale fluctuations have the largest amplitudes: Their velocity scale $v_1$ approaches $(\overline{u'^2})^{1/2}$ and is of the same order of magnitude as the changes $\Delta U$ in the mean flow velocity over distances of the order of $L$ (in fact, $v_1$ is somewhat smaller than $\Delta U$). The largest-scale fluctuations draw their energy directly from the mean motion, and this affects all its characteristics, i.e., the fluctuations are inhomogeneous and anisotropic almost to the same extent to which the mean velocity field is inhomogeneous and anisotropic.

The Reynolds number $\mathrm{Re}_1 = v_1 l_1/\nu$, which characterizes the larger-scale fluctuations, is usually several times smaller than $\mathrm{Re} = UL/\nu$, but in developed turbulent flows it is also very high. Therefore, first-order disturbances in a flow of this kind are unstable and break down, generating turbulent motions of smaller length scales $l_2$. The velocity scale $v_2$ of the secondary motions is smaller than $v_1$, but the corresponding Reynolds number $\mathrm{Re}_2 = v_2 l_2/\nu$ for sufficiently large Re is still quite high (much greater than $\mathrm{Re}_{cr}$). Therefore, the secondary motions are also unstable and should break down, producing further motions with smaller length scales $l_3$, and so on. This process terminates when such a length scale $l_N = \eta$ is reached that the corresponding Reynolds number $\mathrm{Re}_N$ is of the order of unity (more precisely, of the order of $\mathrm{Re}_{cr}$). The motions of such a scale are hydrodynamically stable and practically do not break down. Since their Reynolds number is small, viscosity becomes important for them. Their energy is expended mainly in overcoming frictional forces, and is thus converted directly into heat. At the same time, the fluid viscosity for motions with scales much greater than $\eta$ is unimportant (since the Reynolds numbers are high). Therefore, they are not accompanied by appreciable direct energy dissipation. So long as the instability of the mean motion leads to the appearance of new first-order disturbances, the successive break down of all the sufficiently large-scale disturbances does not terminate, and there is a continuous energy flux along the spectrum from large-scale motions to motions with a minimum scale of the order of $\eta$. Since developed turbulence is accompanied by the dissipation of kinetic energy, it is clear that it can only be maintained by a continuous supply of energy from some external source.

Since the energy transfer to smaller-scale motion is random, the effect of anisotropy, inhomogeneity, and nonstationarity of the mean motion on the statistical state of the turbulent fluctuations should decrease with decreasing scale. In fact, an important role in the energy-transfer mechanism from motions of length scale $l_n$, to motions of length scale $l_{n+1}$ is played by the pressure fluctuations produced as a result of the inhomogeneity of the velocity field of length scale $l_n$, leading to the appearance of fluctuations of scale $l_{n+1}$ not only along the velocity of the original motion but in all other directions. Pressure fluctuations thus affect the redistribution of the energy of the original motions along all the possible directions (this was already mentioned in Sect. 6.2 of Vol. 1). One effect of this is that the directional effect of the mean motion, i.e., of the geometry of the flow as a whole, should decrease during each transition to lower-scale fluctuations. It may, therefore, be assumed that the directional effect practically ceases to influence disturbances of relatively low order, i.e., in the case of developed turbulence, the set of all the disturbances with the exception of the largest-scale disturbances will be statistically isotropic. Similarly, the variation in space of the velocity $\boldsymbol{U} = \overline{\boldsymbol{u}}$ of the mean motion, which characterizes the inhomogeneity of this motion, should be important at distances of the order of $L \sim l_1$ where it is of the order of $\Delta U \sim v_1$, but at much smaller distances it should have no effect on the small-scale fluctuations produced as a result of the multiple breakdown of disturbances of length scale $l_1$. Therefore, in regions whose linear dimensions are much less than $L$, small-scale velocity fluctuations, and also fluctuations of other flow variables, should be statistically homogeneous. Finally, it is very probable that as the length scale $l_n$ of the turbulent motions decreases, not only the Reynolds number $\mathrm{Re}_n$ but also the corresponding characteristic time scale $T_n = l_n/v_n$[1] will decrease. Consequently, the time scale $T_0$ which characterizes the possible nonstationarity of the mean motion, i.e., the relative time rate of change of the mean velocity $\overline{\boldsymbol{u}}(t)$ is much greater than the time scale $T_n$ of the $n$-order disturbances for $n$ not too small. In other words, when $n$ is not too small, the time variation of the integral characteristics of the mean motion, which determine the statistical state of the fluctuations of order $n$, is very slow in comparison with the periods of these fluctuations, and the above statistical state can be regarded as quasi-stationary, i.e., not depending explicitly on time, and varying only as a result of its dependence on the slowly changing integral flow characteristics.

In the case of an almost isotropic turbulence behind a grid $T_0 \sim L/U$ where $U = (\overline{u'^2})^{1/2}$ and $L$ is the integral turbulence scale (see the beginning of Sect. 16.5). In the case of turbulence with a mean velocity which varies in space, we usually have $T_0 \sim L/U$ where $L$ and $U$ are typical length and velocity scales for the mean motion (otherwise, the term involving the time derivative in the Reynolds equations would differ by an order of magnitude from the terms

[1] If we use the fact that for isotropic disturbances the energy distribution over the various scales $l = 1/k$ can be specified by the spectral density $E(k)$, then dimensional considerations or simple physical ideas can be used to show that $v_n \sim [kE(k)]^{1/2}$, $T_n \sim [k^3E(k)]^{-1/2}$ where $k = 1/l_n$ [see, for example, Onsager (1949) and Corrsin (1958b)]. Hence it is clear that the time scale $T_n$ will decrease with decreasing $l$ provided $E(k)\,k^3$ increases with increasing $k$. This condition is always satisfied over a large range of values of $k$.

containing the space derivatives). We shall, therefore, always assume that $T_0 = L/U$. We note that, even in the case of a steady turbulent shear flow in which $\bar{u} = \bar{u}(\boldsymbol{x})$ is independent of $t$, the necessary condition for fluctuations of length scale $l_n$ to be statistically isotropic is $T_n \ll \frac{L}{U} \approx |\nabla \bar{u}|^{-1}$. Unless this is so, the fluctuations will be deformed nonisotropically by the mean velocity field [the deformation rate is characterized by a time scale of the order of $|\nabla \bar{u}|^{-1}$; see Uberoi (1957) and Corrsin (1958b), where the minimum Reynolds numbers for which isotropic fluctuations can exist are estimated].

Thus, the above description of the mechanism of high Reynolds number turbulence naturally leads to the conclusion that the statistical state of small scale fluctuations (with length scales much less than the external turbulence length scale $L \sim l_1$ and with time scales which are small compared with $T_0 = L/U$) in suffieiently small space-time regions (with space dimensions much less than $L$ and time intervals much less than $T_0$) will be homogeneous, isotropic, and practically steady. This conclusion was first clearly formulated by Kolmogorov (1941a). It is implicitly contained also in the results obtained independently by Obukhov (1941a, b), Onsager (1945, 1949), Weizsäcker (1948), and Heisenberg (1948a).

### 21.2 Definition of Locally Isotropic Turbulence

We shall now try to develop a mathematical description of velocity fields $\boldsymbol{u}(\boldsymbol{x}, t)$ whose small-scale fluctuations are statistically homogeneous, isotropic, and stationary. To do this, we must first determine the characteristics of these fields which are independent of the large-scale motion. The values of $\boldsymbol{u}(\boldsymbol{x}, t)$ itself cannot be taken as these characteristics because they are largely determined by the mean flow. The separation of the velocity $\boldsymbol{u}$ into the mean and fluctuating components $\bar{\boldsymbol{u}}$ and $\boldsymbol{u}' = \boldsymbol{u} - \bar{\boldsymbol{u}}$ enables us to isolate the velocity component $\boldsymbol{u}'(\boldsymbol{x}, t)$ which is independent of the mean flow. The values of $\boldsymbol{u}'(\boldsymbol{x}, t)$ are determined in the first instance by the large-scale energy containing disturbances (scale length $l_1 \sim L$) which have the largest amplitudes. It is natural to try to isolate the small-scale fluctuations in which we are interested through a Fourier expansion (this was the procedure which we adopted in Sect. 16.5). However, since the field $\boldsymbol{u}(\boldsymbol{x}, t)$ is not assumed to be homogeneous, it is not so easy to ascribe a precise meaning to this expansion. The simplest method to adopt, therefore, is to start with the fact that the small-scale properties of turbulence are reflected only in the relative motion of particles in small regions of space and over short intervals of time. They are unrelated to the absolute motion of the individual volumes of the fluid, which is mainly determined by the mean flow and the largest-scale disturbances. It is reasonable to follow Kolmogorov (1941a) in the mathematical study of small-scale components and consider only the relative motion of the fluid particles, i.e., motion relative to some fixed fluid particle which is contained in a given small region.

Consider a small space-time region and a given "central point" $(\boldsymbol{x}_0, t_0)$ in this region. We can then define a moving inertial set of Cartesian coordinates traveling with a constant velocity $\boldsymbol{u}(\boldsymbol{x}_0, t_0)$ relative to the fixed ("absolute") set such that at time $t = t_0$ its origin lies at the point $\boldsymbol{x}_0$. Transition to this set of coordinates means that the usual coordinates $\boldsymbol{x} = (x_1, x_2, x_3)$ and time $t$ are replaced by

$$\boldsymbol{r} = \boldsymbol{x} - \boldsymbol{x}_0 - \boldsymbol{u}(\boldsymbol{x}_0, t_0)(t - t_0), \qquad \tau = t - t_0 \tag{21.1}$$

The first of these quantities clearly depends on $\boldsymbol{u}(\boldsymbol{x}_0, t_0)$ and is, therefore, random. The velocity $\boldsymbol{u}(\boldsymbol{x}, t)$ is now replaced by the relative velocity

$$\boldsymbol{v}(\boldsymbol{r}, \tau) = \boldsymbol{u}(\boldsymbol{x}, t) - \boldsymbol{u}(\boldsymbol{x}_0, t_0). \tag{21.2}$$

We can now formulate the following basic definition:

*A given turbulence in a space-time region $G$ is called locally isotropic if, for any fixed value $\boldsymbol{u}(\boldsymbol{x}_0, t_0) = \boldsymbol{u}_0$, the multidimensional probability distribution for each finite set of relative velocities $\boldsymbol{v}(\boldsymbol{r}_k, \tau_k)$, $k = 1, \ldots, n$, which consists of the values of the velocity $\boldsymbol{u}(\boldsymbol{x}, t)$ at the $n+1$ points $(\boldsymbol{x}_0, t_0)$, $(\boldsymbol{x}_1, t_1)$, ..., $(\boldsymbol{x}_n, t_n)$ of $G$, is* (1) *independent of $\boldsymbol{u}_0$,* (2) *stationary (independent of the time $t_0$ in $G$),* (3) *homogeneous (independent of the choice of $\boldsymbol{x}_0$ in $G$), and* (4) *isotropic (i.e., invariant under rotations and reflections in the space of the vectors $\boldsymbol{r}$).*

The discussion given in Sect. 21.1 enables us to assume that turbulence with sufficiently large Reynolds number is always locally isotropic in any region whose linear dimensions are much less than $L$, and whose time intervals are less than $T_0 = L/U$. In other words, if we restrict our attention to vectors $\boldsymbol{r}_k$ and time intervals $\tau_k$ such that $|\boldsymbol{r}_k| = r_k \ll L$ and $|\tau_k| \ll T_0$, then for $\mathrm{Re} \gg \mathrm{Re}_{cr}$ the probability distribution for any finite set of $\boldsymbol{v}_k(\boldsymbol{r}_k, \tau_k)$ can be expected to be independent of $u_0$ as well as stationary, homogeneous, and isotropic.[2]

If we know the probability distribution for the relative velocities (21.2), we can also determine the probability distribution for the velocity differences at sufficiently close space-time points. This can then be used to determine the probability distributions for any derivatives of the field $u(x, t)$ with respect to the coordinates and the time. In the case of an incompressible fluid, knowledge of the space derivatives of the velocity enables us to determine the pressure to within a constant term. Consequently, the probability distribution for the velocity differences can be used to determine all the possible statistical characteristics of the pressure differences at neighboring points.[3] If, however, the incompressible fluid is temperature-inhomogeneous, the situation is more complicated. The main fluid dynamical fields should then include the temperature field which cannot be expressed in terms of the velocity. The definition of locally isotropic turbulence must then be extended by including not only the relative velocities, but also temperature differences between pairs of space-time points $(x_0, t_0)$ and $(x_k, t_k)$, $k=1, 2, \ldots, n$, and by demanding that the conditions for stationarity, homogeneity, and isotropy are satisfied for the joint probability distribution for the velocity and temperature differences. Finally, in the case of turbulence in a compressible fluid, the definition of locally isotropic turbulence must be extended by including, for example, the density and pressure differences. We shall not pause to consider this here, because we shall not discuss locally isotropic turbulence in a compressible fluid.

Since the definition of locally isotropic turbulence includes the probability distribution for only the differences between flow

[2] We note that the condition $|\tau_k| \ll T_0$ ensures that the inequality $|\tau_k| \ll T$ is also satisfied, where $T$ is the Lagrangian time scale, which is usually greater than the Eulerian time scale $T_0$ (see Sect. 9.5, Vol. 1). This is why the velocity difference $v(r_k, \tau_k)$ can be approximately interpreted as the relative velocity, i.e., the velocity of motion relative to a fixed fluid particle which lies at the point $x_0$ at the time $t_0$. In general, however, the condition $|\tau_k| \ll T$ is not less important than $|\tau_k| \ll T_0$.

[3] This statement is in fact not quite certain since the pressure is determined by the integral of the velocity derivative with a comparatively slowly decreasing weighting function. Therefore the application of the theory of locally isotropic turbulence to the study of pressure fluctuations is slightly more questionable than other applications of the theory; it is possible that comparatively far regions of the flow make nonnegligible contributions to the pressure fluctuations at a given point.

variables at sufficiently close points, we need not worry about the "evolution of the mean values of the meteorological fields," noted in Sect. 7.1 of Vol. 1, when we apply this idea to natural turbulence. In fact, this "evolution" ensures that only the mean values of the meteorological fields themselves are statistically unstable (i.e., depend on the choice of the period over which the average is taken), and these do not enter the mean values of functions of their differences between closely lying points. It is also clear that the choice of the time interval over which the average is taken, and which must be sufficient to enable us to deduce reliable estimates of the parameters of locally isotropic turbulence, is a very simple matter. Since its definition involves only the high-frequency fluctuations, with time scales much less than $L/U$, it is sufficient to take this time interval to be longer than $L/U$.

From the theoretical point of view, on the other hand, the restriction to a finite space-time region $G$ is very inconvenient, because in a rigorous mathematical description of locally isotropic fluctuations, we must therefore include the determination of the precise form of this region and must specify a complete set of corresponding boundary conditions. At the same time, physical considerations, which have led us to the above definition of locally isotropic turbulence, lead us to the conclusion that the statistical state of locally isotropic fluctuations should be independent of the form of $G$, and that the boundary conditions should have a significant effect only through the values of one of their parameters, i.e., the parameter which determines the supply of energy to the small-scale fluctuations (we shall discuss this in greater detail below). Therefore, in the theoretical analysis of locally isotropic turbulence we may extend the homogeneous, isotropic, and stationary probability distributions for fluid dynamical fields specified in $G$ to the entire infinite four-dimensional space $(\boldsymbol{x}, t)$, i.e., consider the idealized model of a locally isotropic turbulence with an infinitely large external scale $L$. In this model, the flow fields are specified in all space, and are locally isotropic random fields in the sense defined in Sect. 13. It follows that we can now use the mathematical formulas given in that section. In particular, we shall be able to use the spectral representation of fluid dynamical fields without any restrictions connected with the distorting effect of macroscopic-flow irregularities, and without assuming that the turbulence is also isotropic in the ordinary sense. The supply of energy from external sources which is necessary to maintain the stationary turbulence can be achieved in our model by continuous excitation of motions with infinite wavelengths, i.e., zero wave vectors. It is thus hoped that the theoretical results of the idealized locally isotropic turbulence in infinite space will be applicable to small-scale components of real locally isotropic turbulence in sufficiently small space-time regions.

The concept of locally isotropic turbulence is not, of course, simply an extension of the concept of isotropic turbulence. In fact, although the conditions for the homogeneity and isotropy of the turbulent fluctuations in the region $G$ are automatically satisfied in the case of isotropic turbulence for any choice of this region, the stationarity condition will not, by any means, be always satisfied. In particular, in the case of almost isotropic grid turbulence only the smallest-scale fluctuations will be locally isotropic (i.e., also stationary), and the condition that such locally isotropic fluctuations will be present is that the Reynolds number $\mathrm{Re}_M = UM/\nu$ should be sufficiently large (at any rate, of the order of at least $10^5$; see Sect. 16.6). The concept of locally isotropic turbulence is thus both broader and narrower than the concept of isotropic turbulence; locally isotropic turbulence may not be isotropic, but at the same time, isotropic turbulence may not, in general, be locally isotropic in any space-time region $G$.

## 21.3 The Kolmogorov Similarity Hypotheses

The mathematical description of locally isotropic random fields is relatively simple. Their main statistical characteristics depend on a relatively small number of variables and are therefore readily interpreted. Nevertheless, the set of all the possible locally isotropic random fields is quite extensive. It is therefore important to establish whether all such fields can appear as the field of small-scale fluctuations in real turbulent flows, or whether the probability distributions for the turbulent fluctuations always belong to some subset of the locally isotropic distributions, which is defined by a small number of parameters.

To settle this problem we must establish the parameters on which the statistical state of small-scale fluctuations can depend. It is natural to expect that as the fluctuation scale is reduced, there is a reduction in the orienting effect of the mean flow, and also in the effect of all its geometric and kinematic properties. It is reasonable to assume that the mean-motion parameters (for example, the characteristic length scale $L$ and the characteristic velocity scale $U$) will not directly determine the statistical state of small-scale fluctuations. In that case, however, the statistical state of these fluctuations will not depend on the specific form of the mean motion, and will be determined by its own internal properties. Such properties should, of course, be determined by the inertial energy transfer from large-scale motions to smaller-scale motions (i.e., by the work against the Reynolds stresses), by terminating the viscous dissipation of energy into heat. This can be translated into the language of ordinary mechanics by considering developed turbulent flow as a dynamic system with a very large number of degrees of freedom, and then isolating the degrees of freedom corresponding to small-scale (and high-frequency) components of the motion. We must then conclude that the inertial and frictional forces which correspond to the isolated degrees of freedom should be in statistical equilibrium independently of the properties of the large-scale components of the motion.

It does not follow, however, that the statistical state of small-scale fluctuations will not in general depend on the properties of the mean motion, i.e., it will always be the same in all flows. The mean motion *will* affect the state of small-scale fluctuations but only indirectly, i.e., through the energy flux associated with the mean motion, which is transferred from the mean flow through the entire hierarchy of disturbances of various orders, and is finally transformed into heat. We shall suppose that the Reynolds number is so high that the homogeneity, isotropy, and stationarity of the statistical state is achieved for relatively large-scale disturbances which are not directly affected by viscosity, and for which the Reynolds number is much greater than $\mathrm{Re}_{cr}$. The mean energy dissipation rate $\bar{\varepsilon}$, i.e., the mean amount of energy converted into heat per unit mass of the fluid per unit time, will then be equal to the mean amount of energy transferred to the largest-scale disturbances from the mean flow per unit time per unit mass of the fluid. Consequently, the energy $\bar{\varepsilon}$ will be the particular characteristic of large-scale motions which will affect the statistical state of small-scale fluctuations (in the special case of isotropic turbulence, this conclusion was already formulated in Sect. 16.5). The energy $\bar{\varepsilon}$ is given by[4]

[4]We note that, since the above discussion shows that the quantity $\bar{\varepsilon}$ can be defined in terms of only the mean-motion characteristics, its order of magnitude can be estimated by dimensional reasoning in terms of a typical length $L$ of this motion and its typical velocity scale $U$ [or, in the absence of solid walls on which the velocity is zero, typical mean-velocity differences $\Delta(U)$]

(Continued on page 347)

$$\bar{\varepsilon}=\frac{1}{2}\nu\sum_{i,\,j}\overline{\left(\frac{\partial u_i}{\partial x_j}+\frac{\partial u_j}{\partial x_i}\right)^2}. \tag{21.3}$$

[see Eqs. (1.69), Vol. 1].

Since we are concerned with the case of very large Reynolds number turbulence, in which direct dissipation of the energy of mean motion under the action of molecular viscosity is negligible, we may suppose that

$$\bar{\varepsilon}=\frac{1}{2}\nu\sum_{i,\,j}\overline{\left(\frac{\partial u'_i}{\partial x_j}+\frac{\partial u'_j}{\partial x_i}\right)^2}, \qquad u'_i=u_i-\bar{u}_i \tag{21.3'}$$

This definition of $\bar{\varepsilon}$ was used in Sects. 7.5 and 8.5 of Vol. 1.

Let us now return to the question of which parameters govern the probability distribution for the relative velocities $\boldsymbol{v}(\boldsymbol{r},\tau)$ with sufficiently small values of $|\boldsymbol{r}|$ and $|\tau|$. In accordance with the above discussion, the properties of the large-scale motion may affect this distribution only through the value of $\bar{\varepsilon}$. In addition, this distribution may depend on the parameters characterizing the properties of the medium. In the case of an incompressible fluid, the properties of the medium can be described by two parameters, namely, $\rho$ and $\nu$. Since, however, the values of the velocity are independent of the chosen unit of mass, the distribution which we are considering cannot depend on $\rho$. We are therefore forced to conclude that the distribution depends only on $\bar{\varepsilon}$ and $\nu$. This was formulated by Kolmogorov (1941a) in the form of the following basic hypothesis.

*Kolmogorov's first similarity hypothesis. In the case of turbulence with a sufficiently large Reynolds number, the multidimensional probability distributions for the relative velocities* $\boldsymbol{v}(\boldsymbol{r},\tau)=\boldsymbol{u}(\boldsymbol{x}_0+\boldsymbol{r},t_0+\tau)-\boldsymbol{u}(\boldsymbol{x}_0,t_0)$, *in a space-time region G in which the turbulence is locally isotropic, are unambiguously defined by the values of* $\bar{\varepsilon}$ *and* $\nu$.

---

$$\bar{\varepsilon}\sim U^3/L \quad \text{or} \quad \bar{\varepsilon}\sim(\Delta U)^3/L$$

[compare this with the case of isotropic turbulence for which we have the formula (16.34)]. By analogy with the Stokes formula (21.3), the last formula can be rewritten in the form $\bar{\varepsilon}=K\left(\frac{\Delta U}{L}\right)^2$ where $K\sim\Delta U\cdot L$ is the effective eddy viscosity. Consequently, in developed turbulence we have $K/\nu\sim\Delta U\cdot L/\nu=\text{Re}$ or, more precisely, $K/\nu\sim\text{Re}/\text{Re}_{cr}$, since the values of $K$ and $\nu$ should be equal for $\text{Re}=\text{Re}_{cr}$.

This hypothesis (provided it is valid) restricts quite considerably the set of probability distributions corresponding to locally isotropic random fields $\boldsymbol{v}(\boldsymbol{r}, \tau)$ in different turbulent flows with large Re. In fact, we cannot construct a dimensionless combination out of the quantities $\bar{\varepsilon}$ and $\nu$ alone. Apart from unimportant numerical factors, we can only obtain unique combinations of $\eta$, $v_\eta$, and $\tau_\eta = \eta/v_\eta$ which have the dimensions of length, velocity and time. Namely

$$\eta = (\nu^3/\bar{\varepsilon})^{1/4}, \quad v_\eta = (\nu\bar{\varepsilon})^{1/4}, \quad \tau_\eta = (\nu/\bar{\varepsilon})^{1/2} \tag{21.4}$$

(the first two of these formulas were given in Sect. 16.5). Dimensional arguments then show that if the probability distributions for the relative velocity field $\boldsymbol{v}(\boldsymbol{r}, \tau)$ depend only on $\bar{\varepsilon}$ and $\nu$, then if we transform to the scales $\eta$, $v_\eta$, and $\tau_\eta$, the probability distributions for the dimensionless random field $\boldsymbol{w}(\boldsymbol{\xi}, s)$ defined by

$$\boldsymbol{w}(\boldsymbol{\xi}, s) = \frac{\boldsymbol{v}(\boldsymbol{\xi}\eta, s\tau_\eta)}{v_\eta} = \frac{\boldsymbol{u}(\boldsymbol{x}_0 + \boldsymbol{\xi}\eta, t_0 + s\tau_\eta) - \boldsymbol{u}(\boldsymbol{x}_0, t_0)}{v_\eta}, \tag{21.5}$$

should be universal. The first Kolmogorov similarity hypothesis can therefore be reformulated as follows: *The finite-dimensional probability distributions for the random field* $\boldsymbol{w}(\boldsymbol{\xi}, s)$ *defined by Eq.* (21.5), *and corresponding to values of* $r = |\boldsymbol{\xi}|\eta$ *and* $\tau = |s|\tau_\eta$ *that are not too large, are the same for all turbulent flows with sufficiently large Reynolds number*. Therefore, by investigating the corresponding universal probability distributions, we can determine the statistical laws which are valid for any turbulence with sufficiently large Re.

Disturbances of length scale $\eta$ will clearly be characterized by a velocity scale of the order of $v_\eta$, i.e., a Reynolds number of the order of $\mathrm{Re}_\eta = \eta v_\eta/\nu = 1$. Therefore, the length scale $\eta$ is of the same order of magnitude as the length scale of the largest of the disturbances on which viscosity still has an appreciable effect (and also of the order of the length scale of disturbances corresponding to the maximum energy dissipation which is several times smaller). Therefore, this length scale is an important physical characteristic of developed turbulence. It is usually called the *internal or Kolmogorov length scale* (in contrast to the scale $L \sim l_1$ which is called *the external length scale*). We have already seen that in the special case of isotropic turbulence $\eta \sim \lambda^{3/2}L^{-1/2} \sim \lambda/\mathrm{Re}_\lambda^{1/2}$ (see Sect. 16.6). In general, it is natural to expect that in geometrically similar flows, the scale $\eta$ will decrease with increasing $\mathrm{Re}=UL/\nu$ (or, more precisely, increasing $\mathrm{Re}=\Delta U\cdot L/\nu$, since in view of the Gallilean invariance of the dynamic equations, the local characteristics of turbulence should be determined by a typical velocity difference and not the absolute flow velocity). Following Landau and Lifshitz (1953) we can estimate the dependence of $\eta$ on Re if we recall that $\bar{\varepsilon}\sim(\Delta U)^3/L$. According to Eq. (21.4), it follows that

$$\eta\sim L\cdot\mathrm{Re}^{-3/4} \tag{21.6}$$

Therefore, for a fixed $L$ and an increasing Reynolds number of the basic flow, the internal turbulence scale, i.e., the length scale of the largest perturbations still affected by viscosity, decreases as $\mathrm{Re}^{-3/4}$. Similarly, the scales $v_\eta$ and $\tau_\eta$ are of the same order of magnitude as the typical velocity and period of the motions with which most of the energy dissipation is associated. These scales also decrease with increasing Reynolds number, but more slowly than $\eta$. From Eq. (21.4) we have

$$v_\eta\sim\Delta U\cdot\mathrm{Re}^{-1/4},\qquad \tau_\eta\sim\frac{L}{\Delta U}\cdot\mathrm{Re}^{-1/2}. \qquad (21.6')^5$$

Let us now consider the case where the Reynolds number of a turbulent flow is so high that the internal length scale $\eta$ is very small in comparison with the external length scale $L$ (which is of the order of the length scale of the largest-scale disturbances). In this kind of turbulence, we can isolate an extensive subrange of motions of length scale $l$ much smaller than $L$ (and consequently, homogeneous, isotropic, and quasi-stationary), but much greater than $\eta$. The characteristic relative velocities $v_l$ of these motions will be much greater than the velocity scale $v_\eta$ of motions of length scale $\eta$. Therefore, the corresponding Reynolds number $\mathrm{Re}_l = lv_l/\nu$ will be very large in comparison with the number $\eta v_\eta/\nu=1$ (and, for

[5] We note that similar considerations have enabled Landau and Lifshitz to estimate the number of degrees of freedom in a developed turbulent flow. Since the number of degrees of freedom $n$ per unit volume of the fluid has the dimensions of unity divided by a volume, we find from dimensional considerations that $n\sim\eta^{-3}\sim L^{-3}\mathrm{Re}^{9/4}$. The total number of degrees of freedom $N$ is obtained by multiplying $n$ by the volume of the flow, which is of the order $L^3$. Therefore, $N\sim\mathrm{Re}^{9/4}$ or, more precisely, $N\sim(\mathrm{Re}/\mathrm{Re}_{cr})^{9/4}$ (the numerical factor $\mathrm{Re}_{cr}$ must be included, since $N$ is of the order of unity when $\mathrm{Re}\sim\mathrm{Re}_{cr}$ and not when $\mathrm{Re}\sim 1$). This result was already noted in Sect. 13.3 in Vol. 1.

sufficiently large $l/\eta$, even in comparison with $\mathrm{Re}_{cr}$). In other words, the dominating process in the subrange is the inertial transfer of energy to small-scale disturbances but without any appreciable conversion of energy directly into heat. The statistical properties corresponding to this scale interval should not, therefore, depend on the viscosity $\nu$. These considerations have led Kolmogorov (1941a) to his second basic hypothesis:

*The second Kolmogorov similarity hypothesis. In the case of turbulence with sufficiently large Reynolds number* Re, *the multidimensional probability distributions for the relative velocities* $\boldsymbol{v}(\boldsymbol{r}_k, \tau_k)$, $k=1, \ldots, n$ *in sufficiently small space and time intervals* $|\boldsymbol{r}_k| \ll L$ *and* $|\tau_k| \ll L/U$ *which satisfy the additional conditions*

$$\begin{aligned} &|\boldsymbol{r}_k| \gg \eta, \quad |\boldsymbol{r}_j - \boldsymbol{r}_k| \gg \eta \quad \text{when} \quad j \neq k; \\ &|\tau_k| \gg \tau_\eta, \quad |\tau_j - \tau_k| \gg \tau_\eta \text{when} \quad j \neq k, \end{aligned} \tag{21.7}$$

*are unambiguously determined by the value of* $\bar{\varepsilon}$ *and are independent of* $\nu$.

Since we cannot construct length, velocity, and, time scales from the single quantity $\bar{\varepsilon}$, the statistical state of flows with space scales $l$ much greater than $\eta$, and time scales $\tau$ much greater than $\tau_\eta$, should be self-preserving if the above hypothesis is correct.

Subsequent progress in the theory of turbulence and its applications has shown that the Kolmogorov hypotheses provide a satisfactory explanation of many properties of turbulent flows, and that the predictions based on it are in good agreement with experimental data. Nevertheless, it must be remembered that the Kolmogorov hypotheses have not (and cannot) be rigorously proved, i.e., derived purely analytically from the general laws of mechanics. Moreover, Landau has pointed out shortly after the formulation of the hypotheses that these hypotheses cannot be absolutely exact. Much later, Kolmogorov (1962a, b) and Obukhov (1962a, b) outlined an improved theory of developed turbulence which introduces small corrections to the older formulas. These corrections are, however, outside the scope of existing experimental techniques. The improved theory will be discussed in detail in Sect. 25. For the moment, let us emphasize once again that, according to all the existing data, the Kolmogorov hypotheses correctly describe many of the real features of the local structure of developed turbulence. The consequences of these hypotheses (especially those which can be verified directly by experiment) are therefore of undoubted interest. We shall now proceed to derive some of these properties.

## 21.4 Local Structure of the Velocity Fluctuations

The range of scales in which the first Kolmogorov hypothesis is valid will be called the *quasi-equilibrium range*. In the idealized model of locally isotropic turbulence considered at the end of Sect. 21.2 this interval is infinite, but in any real turbulence with finite $L$ and $U$ it has an upper bound. The idealized model of locally isotropic turbulence is, however, very convenient since it enables us to use the spectral representations which are most naturally determined for random fields specified in all space (or at all times). When the spectral theory is applied to real turbulent flows we must remember that all the resulting universal formulas will be valid only for wave numbers $k = |\boldsymbol{k}|$ that are not too small ($k \gg 1/L$), and for frequencies $\omega$ that are not too small ($\omega \gg U/L$). The corresponding ranges on the wave number or frequency spectra will also be referred to as the *quasi-equilibrium ranges*.

The second similarity hypothesis states that in the ranges $L \gg r \gg \eta$ and $L/U \gg \tau \gg \tau_\eta$ (or in the equivalent ranges $1/L \ll k \ll 1/\eta$ and $U/L \ll \omega \ll 1/\tau_\eta$), all statistical laws should be determined by the single parameter $\bar{\varepsilon}$ (i.e., they cannot depend on $\nu$). The corresponding subranges of $r$ and $\tau$, or of the wave numbers $k$ and frequencies $\nu$, are usually referred to as the *inertial subranges* (since inertial forces play the main role as far as the energy balance of the corresponding disturbances is concerned).

*Statistical Characteristics of Spatial Velocity Differences*

Let us begin by considering the statistical characteristics of the velocity differences between two points $\boldsymbol{x}+\boldsymbol{r}$ and $\boldsymbol{x}$ at a fixed time $t$:

$$\Delta_r \boldsymbol{u} = \boldsymbol{u}(\boldsymbol{x}+\boldsymbol{r},\ t) - \boldsymbol{u}(\boldsymbol{x},\ t). \tag{21.8}$$

For sufficiently small $r = |\boldsymbol{r}|$ there are homogeneous, isotropic, and stationary probability distributions for $\Delta_r \boldsymbol{u} = \boldsymbol{v}(\boldsymbol{r},\ 0)$ for which we can use the first and second Kolmogorov hypotheses. However, we shall not consider here these distributions themselves, and will restrict our attention to the first three moments of the vector $\Delta_r \boldsymbol{u}$.

According to Sect. 13.3, for a locally isotropic turbulence we have $\overline{\Delta_r \boldsymbol{u}} = 0$, and the tensor of the second moments of $\Delta_r \boldsymbol{u}$ in the entire quasi-equilibrium range (i.e., for $r \ll L$) can be expressed in terms of two scalar functions, namely, the longitudinal and lateral structure functions $D_{LL}(r)$ and $D_{NN}(r)$ of the velocity field:

$$\overline{\Delta_r u_i \Delta_r u_j} = D_{ij}(r) = \frac{D_{LL}(r) - D_{NN}(r)}{r^2} r_i r_j + D_{NN}(r)\delta_{ij} \quad (21.9)$$

The two structure functions are given by

$$D_{LL}(r) = \overline{(\Delta_r u_L)^2}, \quad D_{NN}(r) = \overline{(\Delta_r u_N)^2}, \quad (21.10)$$

where $u_L$ and $u_N$ are the components of $\boldsymbol{u}$ respectively parallel and perpendicular to $\boldsymbol{r}$. Moreover,

$$D_{NN}(r) = D_{LL}(r) + \frac{r}{2}\frac{dD_{LL}(r)}{dr}, \quad (21.11)$$

which is a consequence of the continuity equation. By the first Kolmogorov hypothesis

$$D_{LL}(r) = v_\eta^2 \beta_{LL}\left(\frac{r}{\eta}\right), \quad D_{NN}(r) = v_\eta^2 \beta_{NN}\left(\frac{r}{\eta}\right), \quad (21.12)$$

where $\beta_{LL}(x)$ and $\beta_{NN}(x)$ are certain universal functions related by

$$\beta_{NN}(x) = \beta_{LL}(x) + \frac{x}{2}\frac{d\beta_{LL}(x)}{dx}, \quad (21.13)$$

which follows from Eq. (21.11). Similarly, the tensor of the third moments $D_{ijk}(r) = \overline{\Delta_r u_i \Delta_r u_j \Delta_r u_k}$ can be expressed in terms of the single scalar function

$$D_{LLL}(r) = \overline{(\Delta_r u_L)^3}\,. \quad (21.14)$$

This possibility follows from symmetry considerations and the continuity equation. From the first Kolmogorov hypothesis we then have

$$D_{LLL}(r) = v_\eta^3 \beta_{LLL}\left(\frac{r}{\eta}\right), \quad (21.15)$$

where $\beta_{LLL}(x)$ is a further universal function. According to Eqs.

(21.12) and (21.15), the functions $D_{LL}$, $D_{NN}$ and $D_{LLL}$ do not explicitly depend on time (but depend implicitly through their dependence on the slowly-varying quantity $\bar{\varepsilon} = \bar{\varepsilon}(t)$.

In a region of diameter $r \ll \eta$, viscous friction has a demanding dominating effect on the relative motion, and velocity components are described by a smooth function of the space coordinates. For $r \ll \eta$ we can therefore use the Taylor expansion $u_L(\boldsymbol{x}+\boldsymbol{r}) = u_L(\boldsymbol{x}) + \nabla u_L \cdot \boldsymbol{r} + \ldots$(and similarly for $u_N(\boldsymbol{x}+\boldsymbol{r})$]. Consequently,

$$D_{LL}(r) \approx Ar^2, \quad D_{NN}(r) \approx A'r^2, \quad D_{LLL}(r) \approx Br^3 \quad \text{for} \quad r \ll \eta, \tag{21.16}$$

where $A = \overline{\left(\frac{\partial u_1}{\partial x_1}\right)^2}$, $A' = \overline{\left(\frac{\partial u_1}{\partial x_2}\right)^2}$, $B = \overline{\left(\frac{\partial u_1}{\partial x_1}\right)^3}$ and, correspondingly,

$$\beta_{LL}(x) \approx ax^2, \quad \beta_{NN}(x) \approx a'x^2, \quad \beta_{LLL}(x) \approx bx^3 \quad \text{for} \quad x \ll 1. \tag{21.16'}$$

In the inertial subrange, i.e., for $\eta \ll r \ll L$, the second Kolmogorov hypothesis shows that the parameter $\nu$ should cancel out from Eqs. (21.12)–(21.15). In the case of Eq. (21.4) this is possible only if

$$\beta_{LL}(x) \approx Cx^{2/3}, \quad \beta_{NN}(x) \approx C'x^{2/3} \quad \text{for} \quad x \gg 1, \tag{21.17}$$

$$\beta_{LLL}(x) \approx Dx, \quad \text{for} \quad x \gg 1, \tag{21.18}$$

which lead to

$$D_{LL}(r) \approx C\bar{\varepsilon}^{2/3} r^{2/3}, \quad D_{NN}(r) \approx C'\bar{\varepsilon}^{2/3} r^{2/3} \quad \text{for} \quad \eta \ll r \ll L, \tag{21.17'}$$

$$D_{LLL}(r) \approx D\bar{\varepsilon}r \quad \text{for} \quad \eta \ll r \ll L. \tag{21.18'}$$

In view of Eqs. (21.11) and (21.13), the constants $A$ and $A'$ (or $a$ and $a'$) and $C$ and $C'$ should be related by

$$A' = 2A, \quad a' = 2a, \quad C' = \frac{4}{3}C. \tag{21.19}$$

To find the values of $A$ and $a$, we must evaluate the expression in brackets in Eq. (21.3), and express the terms $\overline{\left(\frac{\partial u_l}{\partial x_j}\right)^2}$ and $\overline{\frac{\partial u_l}{\partial x_j}\frac{\partial u_j}{\partial x_l}}$ in terms of the values of the second derivatives of the structure tensor $D_{lj}(r)$ at the origin. The result, due to Kolmogorov (1941c), is

$$a=\frac{1}{15}, \quad a'=\frac{2}{15}, \quad A=\frac{\bar{\varepsilon}}{15\nu}, \quad A'=\frac{2\bar{\varepsilon}}{15\nu} \qquad (21.19')$$

Equations (21.17) and (21.17'), which follow from the first and second Kolmogorov hypotheses are particularly important. They show that, *in any turbulent flow with sufficiently large Reynolds number, the mean square of the velocity difference between two points separated by a distance r should be proportional to* $r^{2/3}$, *provided r is neither too small nor too large*. This result, due to Kolmogorov (1941a), expresses one of the more important laws governing small-scale turbulent motions, and is usually referred to as the "two-thirds law."

In a real turbulent flow one would expect the two-thirds law to be obeyed up to a distance $L_1$, comparable with a typical length scale of the corresponding mean flow. The subsequent rise of the structure functions should, of course, be slower, and as $r \to \infty$, both functions in Eq. (21.10) usually approach some finite (but not universal) values.

A law equivalent to the two-thirds law can also be formulated in spectral language. According to Sect. 13, the tensor of the second moments of $\Delta_r \boldsymbol{u}$ for a locally isotropic solenoidal vector field can be characterized by the longitudinal structure function $D_{LL}(r)$, the spectrum $E(k)$, the longitudinal one-dimensional spectrum $E_1(k)$, or the lateral one-dimensional spectrum $E_2(k)$. The formulas of Eq. (21.12), which follow from the first Kolmogorov hypothesis, turn out to be equivalent to the following relations (which are valid for $k \gg k_L = 1/L$):

$$E(k)=\eta v_\eta^2\,\varphi(\eta k), \quad E_1(k)=\eta v_\eta^2\,\varphi_1(\eta k), \quad E_2(k)=\eta v_\eta^2\,\varphi_2(\eta k), \qquad (21.20)$$

where $\varphi(\xi)$, $\varphi_1(\xi)$ and $\varphi_2(\xi)$ are new universal functions (the first two have already been encountered in Sect. 16.5 in connection with isotropic turbulence with large Re). In view of Eqs. (13.9) and (12.86)–(12.88), these functions are related to $\beta_{LL}(x)$ and $\beta_{NN}(x)$ by

$$\beta_{LL}(x)=4\int_0^\infty \left[\frac{1}{3}+\frac{\cos\xi x}{(\xi x)^2}-\frac{\sin\xi x}{(\xi x)^3}\right]\varphi(\xi)\,d\xi=$$

$$=2\int_0^\infty [1-\cos\xi x]\,\varphi_1(\xi)\,d\xi, \qquad (21.21)$$

$$\beta_{LL}(x) + 2\beta_{NN}(x) = 4\int_0^\infty \left[1 - \frac{\sin \xi x}{\xi x}\right] \varphi(\xi)\, d\xi. \tag{21.22}$$

$$2\varphi(\xi) = \xi^2\varphi_1''(\xi) - \xi\varphi_1'(\xi), \quad \varphi_1(\xi) = \int_\xi^\infty \left(1 - \frac{\xi^2}{\xi_1^2}\right) \frac{\varphi(\xi_1)}{\xi_1}\, d\xi_1,$$
$$\varphi_2(\xi) = \frac{1}{2}\int_\xi^\infty \left(1 + \frac{\xi^2}{\xi_1^2}\right) \frac{\varphi(\xi_1)}{\xi_1}\, d\xi_1 = \frac{1}{2}\varphi_1(\xi) - \frac{1}{2}\xi\varphi_1'(\xi). \tag{21.23}$$

In the inertial subrange of the spectrum, the parameter $\nu$ should cancel out of Eq. (21.20), so that for $1/L \ll k \ll 1/\eta$

$$\varphi(\xi) = C_1 \xi^{-5/3}, \quad \varphi_1(\xi) = C_2 \xi^{-5/3}, \quad \varphi_2(\xi) = C_2' \xi^{-5/3} \text{ for } \xi \ll 1, \tag{21.24}$$

$$E(k) = C_1 \bar{\varepsilon}^{2/3} k^{-5/3}, \quad E_1(k) = C_2 \bar{\varepsilon}^{2/3} k^{-5/3}, \quad E_2(k) = C_2' \bar{\varepsilon}^{2/3} k^{-5/3} \tag{21.24'}$$

We have already indicated in Chapt. 7 that the formula given by Eq. (21.24′) for the three-dimensional spectrum $E(k)$ was first obtained by Obukhov (1948a, b). It was subsequently derived independently by Onsager (1945), Weizsäcker (1948), and Heisenberg (1948). The relations given by Eqs. (21.24) and (21.24′) express the very important "five-thirds law" which is the spectral equivalent of the two-thirds law.

The equivalence of the two-thirds and five-thirds laws is quite obvious in the idealized case of locally isotropic turbulence with infinitely large external length scale $L$ and zero internal length scale $\eta$. This corresponds to a locally isotropic velocity field with structure functions of the form $D_{LL}(r) \sim r^{2/3}$ and $D_{NN}(r) \sim r^{2/3}$, and therefore a power spectrum of the form $E(k) \sim k^{-5/3}$. The idealized model of turbulence with $L = \infty$ and $\eta = 0$ is also very convenient for establishing the relation between the constants $C$, $C_1$, $C_2$, $C_2'$. In fact, in this case, the functions $\varphi(\xi), \varphi_1(\xi)$, and $\varphi_2(\xi)$ are given by Eq. (21.24) for all $\xi$, and from Eqs. (21.21)–(21.23) we have [see also Eq. (13.100)]

$$C_1 = \frac{55}{27\Gamma(1/3)} C \approx 0{,}76\, C, \quad C_2 = \frac{18}{55} C_1, \quad C_2' = \frac{24}{55} C_1 = \frac{4}{3} C_2 \tag{21.25}$$

so that

$$C_2 = \frac{2}{3\Gamma(1/3)} C \approx \frac{C}{4}, \quad C_2' \approx \frac{C'}{4}. \tag{21.25'}$$

In fact, the formulas (21.24) are not valid for all $\xi$ but only for $\xi_1 < \xi < \xi_2$ where $\xi_2 \ll 1$ and $\xi_1 \gg \eta/L$. If, however, $x\xi_1 \ll 1$ (so that $r = x\eta \ll L$) and $x\xi_2 \gg 1$ (so that evidently $x \gg 1$), then the contribution to the integrals on the right-hand side of Eqs. (21.21) and (21.22), evaluated between 0 and $\xi_1$, will be very small because the function in the square brackets is small in this region. The contribution of these integrals taken between the limits $\xi_2$ and $\infty$ for real spectra (which tend to zero as $\xi \to \infty$ more rapidly than $\xi^{-5/3}$) will be less than

$$\text{const} \cdot \int_{\xi_2}^{\infty} \xi^{-5/3} d\xi = \text{const} \cdot \xi_2^{-2/3} \ll x^{2/3}$$

[see Webb (1964), where this analysis is given for the one-dimensional case]. Therefore, for $r = x\eta$ from the inertial subrange, the main contribution to the integrals in Eqs. (21.21) and (21.22) corresponds to the inertial subrange of wave numbers $k = \xi/\eta$. Therefore, for a real spectrum $E(k)$, described by the first formula in Eq. (21.24') only when $L^{-1} \ll k \ll \eta^{-1}$, the structure functions will also satisfy the two-thirds law (21.17') [with $C$ related to $C_1$ by the first formula in Eq. (21.25)] but only when $L \gg r \gg \eta$. The validity of the remaining formulas in Eqs. (21.25) and (21.25') for real turbulent spectra can be established in a similar way.

According to the first formula in Eq. (21.23), if the one-dimensional spectrum $\varphi_1(\xi)$ in the neighborhood of a given $\xi = k\eta$ is exactly proportional to $\xi^{-5/3}$, the three-dimensional spectrum $\varphi(\xi)$ will also satisfy here the five-thirds law. Since, however, the three-dimensional spectral density is obtained from the one-dimensional density by double differentiation, a small departure of the function $\varphi_1(\xi)$ from $C_2\xi^{-5/3}$ may be sufficient to ensure that $\varphi(\xi)$ will differ sharply from the power function $C_1\xi^{-5/3}$. On the other hand, according to Eq. (21.23), the one-dimensional spectra $\varphi_1(\xi)$ and $\varphi_2(\xi)$ at the point $\xi = k\eta$ are obtained by integrating the three-dimensional spectrum $\varphi(\xi_1)$ throughout the range $\xi < \xi_1 < \infty$, where for $\varphi_1(\xi)$ the corresponding weighting function vanishes at both ends of this interval, and for $\varphi_2(\xi)$ it decreases monotonically with increasing $\xi_1$. Therefore, for some values of $\xi$ that are not too small but still lie on the wave-number axis to the left of the inertial subrange of the three-dimensional spectrum [in which $\varphi(\xi)$ is well approximated by the function $C_1\xi^{-5/3}$], the function $\varphi_2(\xi)$, will still be determined mainly by the values of $\varphi(\xi)$ with $\xi$ in the inertial subrange, i.e., it will approach $C_2\xi^{-5/3}$. For the function $\varphi_1(\xi)$, on the other hand, this should be valid even for still larger values of $\xi$. The validity of the above discussion is confirmed in particular by the numerical results of Gifford (1959a) and Alekseev and Yaglom (1967) who calculated the one-dimensional spectra $\varphi_1(\xi)$ and $\varphi_2(\xi)$ corresponding to the particular three-dimensional spectrum

$$\varphi(\xi)=\frac{\xi}{(1+\xi^2)^{4/3}}, \tag{21.26}$$

which satisfies the five-thirds law (with $C_1=1$) for $\xi>4$ (and tends to zero for $\xi\to 0$). In this case, $\varphi_2(\xi)$ is quite well represented by $\frac{24}{55}\xi^{-5/3}$ even for $\xi\gtrsim 3$, while $\varphi_1(\xi)$ approaches $\frac{18}{55}\xi^{-5/3}$ over a still greater range of values of $\xi$ extending up to $\xi\approx 1.5$ [see Fig. 55, taken from the paper by Alekseev and Yaglom (1967), in which the function $\varphi_3(\xi)$ corresponds to the case of a scalar field which we shall discuss in Sect. 21.6]. Hence, it may be concluded that the subrange of $\xi$ (or $k=\xi/\eta$) for which the lateral one-dimensional spectrum satisfies the five-thirds law will, as a rule, extend considerably further into the region of small wave numbers than the subrange of $\xi$ (or $k$) for which the five-thirds law is valid for the three-dimensional spectrum. Similarly for the one-dimensional longitudinal spectrum the corresponding subrange extends even further into the region of small wave numbers than for the one-dimensional lateral spectrum.

Let us now apply a similar analysis to the upper limit of the inertial subrange in the region of large wave numbers. We should then conclude that, for a gradual increase of the wave number $k=\xi/\eta$, the departure of the lateral one-dimensional spectrum $E_2(k)$ or $\varphi_2'(\xi)$, determined by the three-dimensional spectrum in the entire range $k_1\geqslant k$ (or $\xi_1\geqslant\xi$), from the five-thirds law should become appreciable before the departure of the corresponding three-dimensional spectrum from the same law. In the case of the longitudinal one-dimensional spectrum, on the other hand (which is mainly determined by the three-dimensional spectrum at still greater $k_1$ or $\xi_1$), these departures should appear still earlier. In fact, numerical calculations of one-dimensional spectra corresponding to $E(k)$ given by Eq. (22.73) (see the final part of Sect. 22.3), which have been

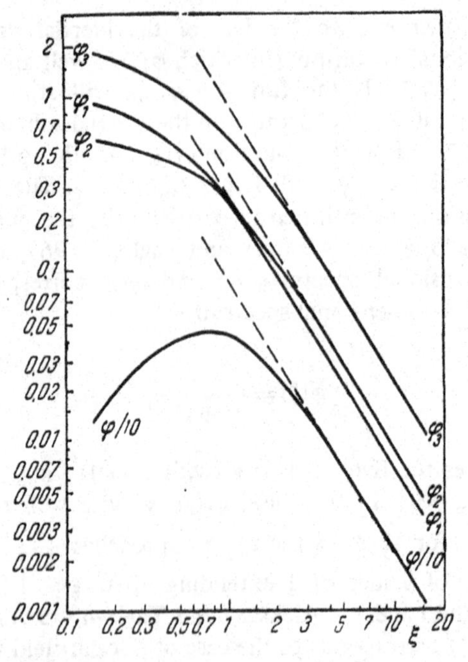


**Fig. 55** The functions $\varphi_1$, $\varphi_2$, and $\varphi_3$ defining the longitudinal, lateral, and one-dimensional scalar spectra corresponding to the three-dimensional spectrum $\varphi(\xi)$. Instead of $\varphi(\xi)$ the figure shows the function $\varphi(\xi)/10$ in order to reduce the number of intersecting curves.

carried out by Alekseev and Yaglom (1967), completely confirm this conclusion.

*The Lagrangian Statistical Characteristics of Velocity Differences*

Let us now consider the statistical characteristics of velocity differences for a given fluid particle at two successive times $t_0$ and $t=t_0+\tau$. Let

$$\Delta_\tau V = V(x_0, t) - u(x_0, t_0) = u[X(x_0, t), t] - u(x_0, t_0), \quad (21.27)$$

where $V(x_0, t)$ is the velocity and $X(x_0, t)$ is the coordinate of the fluid particle which at time $t=t_0$ occupies the point $x_0$ (see Vol. 1, Sect. 9). If $t-t_0=\tau$ is much less than the Lagrangian time scale $T$ [i.e., the time interval during which the change in the Lagrangian velocity $V(x_0, t)$ becomes comparable with $|V|]$, then $X(x_0, t) \approx x_0 + u(x_0, t_0)(t-t_0)$ so that $X - x_0 - u(x_0, t_0)(t-t_0) \approx 0$ [see Eqs. (21.1) and (21.2)]. Therefore, when $\tau \ll T$ and $\tau \ll T_0 = L/U$ it may be considered that $\Delta_\tau V$ has an isotropic probability distribution which is independent of $x_0$, $t_0$ and $u(x_0, t_0)$, and for which the first and second Kolmogorov hypotheses are valid. The condition $\tau \ll T_0$ is necessary because otherwise the macroscopic conditions for our fluid particle may change during the time $\tau$ (for example, this may happen as a result of a change in the mean velocity either in time or in space along the particle trajectory). We shall assume henceforth that $T_0 \lesssim T$, and will indicate only that $\tau \ll T_0$ without explicitly mentioning the condition $\tau \ll T$ which is also very important.

For the second moments of the random vector $\Delta_\tau V$ we have from the first Kolomogorov hypothesis

$$\overline{\Delta_\tau V_i \Delta_\tau V_j} = D^{(L)}(\tau)\delta_{ij}, \quad D^{(L)}(\tau) = v_\eta^2 \beta_0\left(\frac{\tau}{\tau_\eta}\right) \quad \text{for} \quad \tau \ll T_0, \tag{21.28}$$

where $\beta_0(x)$ is a new universal function. The function $D^{(L)}(\tau)$ is called the Lagrangian velocity structure function. When $\tau \ll \tau_\eta = (\nu/\bar{\varepsilon})^{1/2}$, the velocity difference $\Delta_\tau V$ for each particular realization of the turbulent motion is, of course, proportional to $\tau$ (for the same reason that $|\Delta_r u| \sim r$ when $r = |r| \ll \eta$). Therefore,

$$\beta_0(x) = a_0 x^2 \qquad \text{for} \quad x \ll 1, \tag{21.29}$$

$$D^{(L)}(\tau) = a_0 \bar{\varepsilon}^{3/2} \nu^{-1/2} \tau^2 \quad \text{for} \quad \tau \ll \tau_\eta. \tag{21.29'}$$

In the other limiting case $\tau_\eta \ll \tau \ll T_0$, which corresponds to the inertial subrange on the time axis, the parameter $\nu$ should drop out of the relation (21.28). Therefore,

$$\beta_0(x) = C_0 x \quad \text{for} \quad x \gg 1, \tag{21.30}$$

$$D^{(L)}(\tau) = C_0 \bar{\varepsilon} \tau \quad \text{for} \quad \tau_\eta \ll \tau \ll T_0, \tag{21.30'}$$

where $C_0$ is a universal constant. Thus, in the inertial subrange the Lagrangian velocity structure function is linear in $\tau$. The result given by Eqs. (21.30)–(21.30') was independently established by Obukhov and Landau just after the appearance of Kolmogorov's paper (9141a), and was first published in 1944 in the first edition of the book by Landau and Lifshitz (1963). It was subsequently rediscovered in various forms by a number of workers [see, in particular, Inoue (1950–1951, 1952a)]. Another derivation of this result was given by Lin (1960a). We shall discuss it again below (see Sect. 24.2). If we write this result in the form $\overline{(\Delta_\tau V_1)^2} \sim \bar{\varepsilon}\tau$, it becomes completely analogous to the well-known formula $\overline{(\Delta_\tau x_1)^2} \sim \chi\tau$ for the variance of the coordinate increment of a Brownian particle. Therefore, the motion of fluid particles in the field of locally isotropic turbulence can be looked upon as a form of Brownian motion (diffusion) in velocity space with the quantity $\bar{\varepsilon}$ replacing the molecular diffusivity $\chi$.

Since the probability distribution for $\Delta_\tau \boldsymbol{V}$ is isotropic, it follows that all the odd moments of this random vector are zero, and $\Delta_\tau \boldsymbol{V}$ and $-\Delta_\tau \boldsymbol{V} = \Delta_{-\tau} \boldsymbol{V}$ have the same distributions. For the higher-order even moments of $\Delta_\tau \boldsymbol{V}$ we can readily obtain formulas analogous to Eqs. (21.28)–(21.30′). Thus, for example,

$$\overline{\Delta_\tau V_i \, \Delta_\tau V_j \, \Delta_\tau V_k \, \Delta_\tau V_l} = (\delta_{ij}\,\delta_{kl} + \delta_{ik}\,\delta_{jl} + \delta_{il}\,\delta_{jk})\, D_4^{(L)}(\tau), \quad (21.31)$$

where

$$D_4^{(L)}(\tau) = v_\eta^4 \, \beta_4\left(\frac{\tau}{\tau_\eta}\right), \quad \beta_4(x) \sim \begin{cases} x^4 & \text{for} \quad x \ll 1, \\ x^2 & \text{for} \quad x \gg 1. \end{cases} \quad (21.32)$$

The function $D^{(L)}(\tau)$ is the structure function for the random process $V_1(\boldsymbol{x}_0, t)$ with stationary increments. Therefore, in view of the results of Sect. 13.1, it has the spectral representation (13.19), i.e.,

$$D^{(L)}(\tau) = 2 \int_0^\infty (1 - \cos \omega\tau)\, E^{(L)}(\omega)\, d\omega. \quad (21.33)$$

According to Eq. (21.28), this formula can also be rewritten in the form

$$\begin{gathered} \beta_0(x) = 2 \int_0^\infty (1 - \cos \xi x)\, \varphi_0(\xi)\, d\xi, \\ E^{(L)}(\omega) = \tau_\eta v_\eta^2 \, \varphi_0(\omega\tau_\eta) = \nu\, \varphi_0(\nu^{1/2}\bar{\varepsilon}^{-1/2}\omega). \end{gathered} \quad (21.34)$$

In the inertial frequency subrange $1/T_0 \ll \omega \ll 1/\tau_\eta$, the Lagrangian time spectrum $E^{(L)}(\omega)$ can depend only on $\omega$ and $\bar{\varepsilon}$. Consequently,

$$\varphi_0(\xi) = B_0 \xi^{-2} \quad \text{for} \quad \xi \ll 1, \tag{21.35}$$

$$E^{(L)}(\omega) = B_0 \bar{\varepsilon} \omega^{-2} \quad \text{for} \quad 1/T_0 \ll \omega \ll 1/\tau_\eta \tag{21.35'}$$

[Inoue (1951d)]. To establish the relation between the constants $C_0$ and $B_0$ it is sufficient to consider the idealized model in which $T_0 = T = \infty$ and $\tau_\eta = 0$, so that Eqs. (21.30′) and (21.35′) are valid for all $\tau$ and $\omega$. In view of Eq. (21.34) we then have

$$B_0 = \frac{C_0}{\pi} \approx 0{,}32\, C_0, \quad C_0 = \pi B_0 \approx 3{,}1 B_0 \tag{21.36}$$

[compare this with the general equation (13.24′)].

*Temporal Velocity Increments.*
*Taylor's Hypothesis*

The situation is more complicated in the case of the Eulerian velocity difference at a fixed point at two instants $t_0$ and $t = t_0 + \tau$:

$$\Delta_\tau \boldsymbol{u} = \boldsymbol{u}(\boldsymbol{x}_0, t) - \boldsymbol{u}(\boldsymbol{x}_0, t_0). \tag{21.37}$$

For sufficiently small $\tau = t - t_0$ this quantity can be written in the form

$$\Delta_\tau \boldsymbol{u} \approx \boldsymbol{v}(-\boldsymbol{u}(\boldsymbol{x}_0, t_0)\tau, \tau).$$

where we have used Eqs. (21.1) and (21.2). From this we can only conclude that there is a conditional probability distribution for $\Delta_\tau \boldsymbol{u}$ provided that the value $\boldsymbol{u}(\boldsymbol{x}_0, t_0)$ is fixed (equal to, say, $\boldsymbol{u}_0$) and for $\tau \ll T_0$ and $|\boldsymbol{u}_0|\tau \ll L$ it can depend only on the vector $\boldsymbol{r} = \boldsymbol{u}_0\tau$, the value of $\tau$, and the parameters $\bar{\varepsilon}$ and $\nu$. Because of the additional dependence on $\boldsymbol{u}_0$, exact results for $\Delta_\tau \boldsymbol{u}$ are much more complicated and much more difficult to verify than the results for $\Delta_\tau \boldsymbol{V}$. Thus, for example, the tensor $\overline{\Delta_\tau u_i \Delta_\tau u_j} = D^*_{ij}(\tau)$ for fixed $\boldsymbol{u}_0$ is again determined by the two scalar functions $D^*_{LL}(\tau)$ and $D^*_{NN}(\tau)$ through a formula of the form of Eq. (21.9). However, the corresponding dimensionless functions $\beta^*_{LL}$ and $\beta^*_{NN}$ will now depend on two dimensionless arguments, namely, $r/\eta = |\boldsymbol{u}_0|\tau|\eta$ and $\tau/\tau_\eta$. In the inertial subrange, i.e., for $\eta \ll |\boldsymbol{u}_0|\tau \ll L$ and $\tau_\eta \ll \tau \ll T_0 = L/U$, we now have

$$D^*_{LL}(\tau) = \bar{\varepsilon}\tau\,\beta^*_{LL}\left(\frac{|\boldsymbol{u}_0|^2}{\bar{\varepsilon}\tau}\right), \quad D^*_{NN}(\tau) = \bar{\varepsilon}\tau\,\beta^*_{NN}\left(\frac{|\boldsymbol{u}_0|^2}{\bar{\varepsilon}\tau}\right), \tag{21.38}$$

where $\beta^*_{LL}(x)$ and $\beta^*_{NN}(x)$ are new universal functions of a single variable for which we have no exact predictions. To verify these formulas, and to establish the form of the functions $\beta^*_{LL}(x)$ and $\beta^*_{NN}(x)$, we must have extensive observations referring to situations in which $\boldsymbol{u}(\boldsymbol{x}_0, t_0)$ assumes the same value $\boldsymbol{u}_0$, and this is quite difficult to achieve.

Nevertheless, we can obtain some approximate results for the differences $\Delta_\tau \boldsymbol{u}$ which are quite simply verified and are valid in a very broad range. Namely, we can use the fact that, for both artificially generated turbulent flows (grid-generated turbulence in wind tunnels, turbulent jets, flows in tubes, channels, boundary layers, etc.) and for atmospheric turbulence, the velocity fluctuations are, as a rule, much smaller than the typical mean velocity. It is therefore reasonable to suppose that in all such cases we can use the approximate result $\boldsymbol{u}(\boldsymbol{x}_0, t_0) \approx \bar{\boldsymbol{u}}(\boldsymbol{x}_0, t_0)$, i.e., we can replace $\boldsymbol{u}_0$ by the mean velocity $\bar{\boldsymbol{u}}$ at the point $(\boldsymbol{x}_0, t_0)$. Next, the turbulent fluctuations at a fixed point $\boldsymbol{x}_0$ and over a small time interval $(t_0, t_0+\tau)$ can be approximately represented as the transport of turbulent disturbances past this point with a constant velocity $\bar{\boldsymbol{u}}(\boldsymbol{x}_0, t_0) = \bar{\boldsymbol{u}}$ and without distortion, where at the initial time the disturbances lay on the ray $\mathscr{L}$ passing through $\boldsymbol{x}_0$ and directed in opposition to the vector $\bar{\boldsymbol{u}}$. As already noted in Sect. 11.1, this was first used by Taylor (1938b) for grid-generated turbulence in a wind tunnel. It has since been referred to as *Taylor's hypothesis* or the *frozen turbulence hypothesis* (since, according to this hypothesis, the turbulent formations are assumed to be "frozen," i.e., time-independent, in the system reference frame moving with velocity $\bar{\boldsymbol{u}}$). In fact, the turbulent disturbances are not, of course, transported by the mean flow without distortion, but gradually evolve during the transfer process and change their form. The point about Taylor's hypothesis is that, in many cases, this evolution is relatively slow and, therefore, the error introduced by using the idea of frozen turbulence is small (and decreases with decreasing time $\tau$, i.e., with decreasing scales of the turbulence components under consideration).

The "frozen turbulence" hypothesis enables us to express the statistical characteristics of the time differences $\Delta_\tau \boldsymbol{u}$ in terms of the space differences $\Delta_r \boldsymbol{u}$ corresponding to a fixed time $t = t_0$. In particular, it follows from it that the time structure functions $D^{\bullet}_{LL}(\tau) = \overline{(\Delta_\tau u_L)^2}$ and $D^{\bullet}_{NN}(\tau) = \overline{(\Delta_\tau u_N)^2}$ (where $u_L$ and $u_N$ are the components of the vector $\boldsymbol{u}$ in the direction of the mean velocity $\overline{\boldsymbol{u}}$ and in a perpendicular direction) can be obtained from the space structure functions (21.10) by replacing $r$ with $|\overline{\boldsymbol{u}}|\tau = \overline{u}\tau$:

$$D^{\bullet}_{LL}(\tau) = D_{LL}(\overline{u}\tau), \quad D^{\bullet}_{NN}(\tau) = D_{NN}(\overline{u}\tau). \tag{21.39}$$

When $\tau \to 0$ we have

$$\overline{\left(\frac{\partial u}{\partial t}\right)^2} = \overline{u}^2 \overline{\left(\frac{\partial u}{\partial x}\right)^2}, \quad \overline{\left(\frac{\partial v}{\partial t}\right)^2} = \overline{u}^2 \overline{\left(\frac{\partial v}{\partial x}\right)^2}, \tag{21.40}$$

where the $x$ axis lies along $\overline{\boldsymbol{u}}$, and $u = u_x$, $v = u_y$ (or $v = u_z$). We have already used these equations for grid turbulence (see, for example, Sect. 15.1). The time spectrum $F_{ij}(\omega) = \frac{1}{2} E_{ij}(\omega)$ of the fluctuations $\boldsymbol{u}(\boldsymbol{x}, t)$ for fixed $\boldsymbol{x}$ (regarded as a random vector function of time with stationary increments) can be expressed in a simple way with the aid of Taylor's hypothesis in terms of the one-dimensional longitudinal and lateral spectra

$$E_1(k_1) = \int_{k_1}^{\infty} \left(1 - \frac{k_1^2}{k^2}\right) \frac{E(k)}{k}\, dk \text{ and } E_2(k_1) = \frac{1}{2} \int_{k_1}^{\infty} \left(1 + \frac{k_1^2}{k^2}\right) \frac{E(k)}{k}\, dk$$

of a locally isotropic field $\boldsymbol{u}(\boldsymbol{x}, t)$ for fixed $t$. In fact, this hypothesis ensures that a sinusoidal wave with wave-length $l$ and wave number $k = 2\pi/l$ along the line $\mathscr{L}$ is equivalent to a harmonic oscillation at the point $\boldsymbol{x}_0$ with period $\tau = l/\overline{u}$ and angular frequency $\omega = k\overline{u}$. The time spectra of the fluctuations $u(\boldsymbol{x}, t)$ and $v(\boldsymbol{x}, t)$ are therefore given by

$$E_1(\omega) = \frac{1}{\overline{u}} E_1\left(\frac{\omega}{\overline{u}}\right) \text{ and } E_2(\omega) = \frac{1}{\overline{u}} E_2\left(\frac{\omega}{\overline{u}}\right). \tag{21.41}$$

respectively. In the inertial subrange along the time axis, i.e., for $\eta/\overline{u} \ll \tau \ll L/\overline{u}$), we have from Eqs. (21.39), (21.17′), and (21.19)

$$D^{*}_{LL}(\tau) \approx C\,(\overline{\varepsilon}\overline{u}\tau)^{2/3}, \quad D^{*}_{NN}(\tau) \approx \frac{4}{3}C\,(\overline{\varepsilon}\overline{u}\tau)^{2/3}, \qquad (21.42)$$

from which it is clear that in the approximation of frozen turbulence $\beta^{*}_{LL}(x) \approx x^{1/3}$ and $\beta^{*}_{NN}(x) \sim x^{1/3}$. Similarly, in the inertial frequency subrange $\overline{u}/L \ll \omega \ll \overline{u}/\eta$ we have from Eqs. (21.41), (21.24′), and (21.25)

$$E_1(\omega) \approx C_2(\overline{\varepsilon}\overline{u})^{2/3}\omega^{-5/3}, \quad E_2(\omega) \approx \frac{4}{3}\,C_2(\overline{\varepsilon}\overline{u})^{2/3}\omega^{-5/3} \qquad (21.43)$$

where $C_2 \approx C/4$. Comparison of these formulas with Eq. (21.35′) shows that the Lagrangian time spectrum $E^{(L)}(\omega)$ decreases with increasing frequency $\omega$ somewhat more rapidly than the velocity time spectrum at a fixed point, so that the relative importance of high-frequency components in the fluctuation of the fluid-particle velocity is smaller than in the case of velocity fluctuations at a fixed point.

It is quite difficult to establish the precise conditions under which the Taylor hypothesis is valid if this is to be based on an accurate estimate of the validity of Eqs. (21.39)–(21.43). Let us consider, to begin with, the simple case of homogeneous turbulence with constant (in time and space) mean velocity $\overline{\boldsymbol{u}}(\boldsymbol{x}, t) = \boldsymbol{U}$ directed along the $x_1 = x$ axis. In this case, the Navier-Stokes equations for the velocity fluctuations $u_l$, $l = 1, 2, 3$ are of the form

$$\frac{\partial u_l}{\partial t} + U\frac{\partial u_l}{\partial x_1} + u_j\frac{\partial u_l}{\partial x_j} = -\frac{1}{\rho}\frac{\partial p}{\partial x_l} + \nu\Delta u_l, \qquad U = |\boldsymbol{U}|. \quad (21.44)$$

The Taylor hypothesis, according to which $\frac{\partial u_l}{\partial t} = -U\frac{\partial u_l}{\partial x_1}$, reduces here to the proposition that the first two terms on the left-hand side of Eq. (21.44) are the leading terms. Let us now transform to a new set of coordinates moving with a mean velocity $U$. In this set $\overline{u} = 0$ and the quantity $\frac{\partial u_l}{\partial t} + U\frac{\partial u_l}{\partial x_1}$ transforms to the local (Eulerian) acceleration $a_l = \frac{\partial u_l}{\partial t}$. Therefore, in the case of homogeneous turbulence, estimates of the mean square of the relative error of the Taylor hypothesis, $\overline{\left(\frac{\partial u_l}{\partial t} + U\frac{\partial u_l}{\partial x_1}\right)^2} \Big/ U^2\overline{\left(\frac{\partial u_l}{\partial x_1}\right)^2}$, reduce to estimates of the quantity $\overline{a_l^2}$ for turbulence with zero mean velocity. We have already noted in Sect. 18.3 that, when the Reynolds number is large, the main term in the total local acceleration $a_i = -u_j \frac{\partial u_i}{\partial x_j} - \frac{1}{\rho}\frac{\partial p}{\partial x_i} + \nu \Delta u_i$ is the inertial acceleration $\alpha_i = -u_j \frac{\partial u_i}{\partial x_j}$, and that, in the case of isotropic turbulence, Uberoi and Corrsin (1953) have used the Schwarz inequality to obtain the exact inequality $\overline{\alpha_i^2} < 14.6\,(\overline{u_i^2})^2/\lambda^2$ for $\overline{\alpha_i^2}$. In view of Eq. (15.5), it follows from this that for isotropic turbulence with large Re, which is transported by the mean flow with constant velocity $U$, we have

$$\begin{aligned} \frac{\overline{\left(\frac{\partial u}{\partial t} + U\frac{\partial u}{\partial x}\right)^2}}{U^2\overline{\left(\frac{\partial u}{\partial x}\right)^2}} &< 14.6\,\frac{\overline{u^2}}{U^2} \approx 5\left(\frac{U'}{U}\right)^2, \\ \frac{\overline{\left(\frac{\partial v}{\partial t} + U\frac{\partial v}{\partial x}\right)^2}}{U^2\overline{\left(\frac{\partial v}{\partial x}\right)^2}} &< 7.3\,\frac{\overline{v^2}}{U} \approx 2.5\left(\frac{U'}{U}\right)^2, \end{aligned} \tag{21.45}$$

where $U' = (\overline{u_i^2})^{1/2} = \sqrt{3}\,(\overline{u^2})^{1/2}$. Consequently, the condition $U'/U \ll 1$ is now sufficient to ensure the validity of the Taylor hypothesis, as expected. The numerical coefficients in the inequalities given by Eq. (21.45) are, of course, very approximate, but their order of magnitude must be correct. In fact, if instead of the Schwarz inequality we use the approximate equation (18.29), obtained by Lin (1953a) with the aid of the Millionshchikov hypothesis, then instead of the first inequality in Eq. (21.45) we obtain

$$\overline{\left(\frac{\partial u}{\partial t} + U\frac{\partial u}{\partial x}\right)^2} \Big/ U^2\overline{\left(\frac{\partial u}{\partial x}\right)^2} \approx 5\,\frac{\overline{u'^2}}{U^2} = \frac{5}{3}\left(\frac{U'}{U}\right)^2.$$

See also the paper by Comte-Bellot and Corrsin (1971), appendix D, where the question of the validity of Taylor's hypothesis in homogeneous grid-generated turbulence is analyzed in more detail.

Another crude method of estimating theoretically the accuracy of Taylor's hypothesis for isotropic turbulence transported at constant velocity $U$ was put forward by Ogura (1953, 1955). This method is based on the use of a relatively artificial semi-empirical hypothesis [similar in some respects to the hypothesis of Ogura and Miyakoda Eq. (17.12)] on the evolution of the components of isotropic turbulence with different wave numbers during their transport by the mean flow. Numerical calculations of the structure function $D^*_{LL}(\tau)$ carried out by Gifford (1956) with the aid of this hypothesis for a special model of isotropic turbulence have shown that, if the hypothesis is valid, then for $U'/U \leqslant 0.3$ the function $D^*_{LL}(\tau)$ is practically the same as $D_{LL}(U\tau)$ for all $\tau < L/U$, and even when $U'/U = 3$ it differs from $D_{LL}(U\tau)$ but by not more than 10%. However, it must be remembered that the Ogura hypothesis used in this calculation has not been verified directly, and no information is, as yet, available about its accuracy.

There is no doubt that for homogeneous turbulence with $\bar{u} = U =$ const and $U'/U$ of the order that is typical for grid turbulence, the Taylor hypothesis is satisfied with high accuracy for all values of $\tau$ which correspond to scales of turbulent disturbances not exceeding the scales of the largest-scale disturbances containing an appreciable fraction of the turbulent energy. This conclusion can be justified theoretically with the aid of estimates such as Eq. (21.45), and is confirmed by experimental verifications of Taylor's hypothesis for grid turbulence [see, for example, Favre et al. (1952, 1954), Frenkiel and Klebanoff (1966), Comte-Bellot and Corrsin (1971)]. This experimental confirmation of the hypothesis has already been mentioned in Sect. 11.1. However, in the general case of inhomogeneous turbulence with a variable mean velocity $\bar{\boldsymbol{u}}(\boldsymbol{x}, t)$, the situation is more complicated. Here the single condition $U'/\bar{u} \ll 1$, which ensures a low level of turbulence, is insufficient. In fact, since in the Taylor hypothesis the transport velocity $\bar{\boldsymbol{u}} = \bar{\boldsymbol{u}}(\boldsymbol{x}_0, t_0)$ is assumed to be constant, it is clear that we should restrict our attention to time intervals $t_0 \leqslant t \leqslant t_0 + \tau$ for which we can neglect both the change in $\bar{\boldsymbol{u}}(\boldsymbol{x}_0, t) = \bar{\boldsymbol{u}}(x_0, y_0, z_0, t)$ with time, and the change in $\bar{\boldsymbol{u}}(x, y_0, z_0, t_0)$ in the range $x_0 - \bar{u}(\boldsymbol{x}_0, t_0)\tau \leqslant x \leqslant x_0$ on the $x$ axis [directed along $\boldsymbol{u}(\boldsymbol{x}_0, t_0)$]. Consequently, when $\bar{\boldsymbol{u}}(\boldsymbol{x}, t) \neq$ const, the Taylor hypothesis can be valid only for time intervals that are small in comparison with the period $T^* = |\bar{u}| / \left|\frac{\partial \bar{u}}{\partial t}\right|$, and for space scales that are small in comparison with the typical scale $|\bar{u}| / \left|\frac{\partial \bar{u}}{\partial x}\right|$ which is determined by the longitudinal mean-velocity gradient. Let us even suppose that $\frac{\partial \bar{u}}{\partial t} = 0$ and $\frac{\partial \bar{u}}{\partial x} = 0$, and following Lin (1953a), let us consider a steady shear flow in which the mean velocity $\bar{u} = \bar{u}(z)$ lies along the $x$ axis and is a function of $z$. In this case, the Taylor hypothesis will be appreciably less accurate than for flows with $\bar{u} = \text{const}$, since here a disturbance located at the point $z \neq z_0$ at time $t_0$ may pass through the point $x_0$ at time $t_0 + \tau$, and is, therefore transported with a mean velocity that is different from $\bar{u}(z_0)$. Mathematically, this is reflected in the fact that the Navier-Stokes equation for the component $u = u_x$ will now differ from the corresponding equation (21.44) by the additional term $w \frac{\partial \bar{u}}{\partial z}$ on the left-hand side, and if the Taylor hypothesis is to be valid, this term must be much less than the two first terms, i.e., we must have

$$\left| w \frac{\partial \bar{u}}{\partial z} \right| \ll \left| \bar{u} \frac{\partial u}{\partial x} \right|. \tag{21.46}$$

If we suppose that the velocity fluctuation components $u$ and $w$ are of the same order of magnitude and belong to a disturbance with a wave number $k$ (and frequency $\omega = k|\bar{u}|$), we find that

$$\left| \frac{\partial \bar{u}}{\partial z} \right| \ll k|\bar{u}| = \omega. \tag{21.46'}$$

Therefore, for shear flows we must expect that, even for a very low level of turbulence, the Taylor hypothesis will be satisfied only for frequencies appreciably exceeding the characteristic frequency which is determined by the shear.

Of course Lin's condition (21.46′) is only necessary but not sufficient for the validity of Taylor's hypothesis. A much more extensive theoretical analysis of the problem was presented by Lumley (1965), who listed several conditions for applicability of the hypothesis and also suggested corrections for more accurate determination of the spatial characteristics from the temporal data. Heskestad (1965) also investigated the same problem and suggested a rough correction to the usual Taylor's hypothesis when the latter is nearly applicable.

So far, the inertial frequency subrange has been established empirically only under the conditions under which the Taylor hypothesis is valid for all the frequencies in this range. It is then possible to use Eqs. (21.39) and (21.41), and the theoretical two-thirds and five-thirds laws can be conveniently verified for the velocity fluctuations at a fixed point [i.e., in the form of Eqs. (21.42) and (21.43)]. The results of such verifications will be reviewed in Sect. 23.

It does not, however, follow that the hypothesis of frozen turbulence can always be applied to any statistical characteristics of locally isotropic turbulence without restriction. Since the above theoretical estimates, which refer to the case of inhomogeneous turbulence, are very approximate and can serve only for order-of-magnitude calculations, there is considerable interest in the direct verification of the Taylor hypothesis for inhomogeneous flows through simultaneous measurements of different space and time statistical characteristics. Such verifications have recently been carried out both for laboratory flows and for atmospheric turbulence [see, for example, Gifford (1955), Favre et al. (1958, 1962), Panofsky, Cramer and Rao (1958), Gossard (1960a), Tsvang (1963), Lappe and Davidson (1963), Fisher and Davis (1964), Wills (1964), Koprov and Tsvang (1965) and Champagne, Harris and Corrsin (1970)]. It was found that for locally isotropic turbulent disturbances the frozen turbulent hypothesis is usually satisfied with high accuracy, although for larger disturbances the Taylor hypothesis is certainly not always valid (see, in particular, Sect. 23.1 below).

## 21.5 Statistical Characteristics of Acceleration, Vorticity, and Pressure Fields

*Acceleration-field Characteristics*

The fluid particle acceleration $A(x_0, t_0)$ in a turbulent flow can be defined by

$$A(x_0, t_0) = \lim_{t-t_0 \to 0} \frac{V(x_0, t) - u(x_0, t_0)}{t - t_0} = \lim_{\tau \to 0} \frac{\Delta_\tau V}{\tau}, \qquad (21.47)$$

where $V = V(x_0, t)$ is the Lagrangian fluid particle velocity. Since for locally isotropic turbulence, the probability distribution for $\Delta_\tau V$ is universal and isotropic in the case of sufficiently small $\tau$, the probability distribution for $A(x_0, t_0)$ will also have these properties. Therefore, in sufficiently small space-time regions the field $A(x, t)$ will be isotropic and its probability distribution will be stationary and dependent only on $\bar{\varepsilon}$ and $\nu$.

Consider the Lagrangian time correlation tensor for the acceleration field

$$B_{0,ij}(\tau) = \overline{A_i(t \mid x_0) A_j(t + \tau \mid x_0)} \qquad (21.48)$$

where $A(t \mid x_0) = A[X(x_0, t), t]$ is the acceleration of the fluid particle which occupies the point $x_0$ at time $t_0$. In view of isotropy and similarity in the quasi-equilibrium range of $\tau$, this tensor should be of the form

$$B_{0,\,ij}(\tau)=B_0(\tau)\,\delta_{ij},\quad B_0(\tau)=\frac{v_\eta^2}{\tau_\eta^2}\,a\,(\tau/\tau_\eta)=\bar{\varepsilon}^{3/2}\nu^{-1/2}a\,(\bar{\varepsilon}^{1/2}\nu^{-1/2}\tau), \tag{21.49}$$

where $a(x)$ is a universal function. The function $B_0(\tau)$ can be called the *Lagrangian acceleration correlation function*. If we then use Eq. (21.47), we can readily express this function in terms of the Lagrangian velocity structure function $D^{(L)}(\tau)$:

$$B_0(\tau)=\frac{1}{2}\frac{d^2}{d\tau^2}\,D^{(L)}(\tau), \tag{21.50}$$

so that

$$a\,(x)=\frac{1}{2}\,\beta_0''(x). \tag{21.50'}$$

According to Eq. (21.29), we have $\beta_0''(0)=2a_0$. The mean square of the turbulent acceleration is then given by

$$A'^2=\overline{A^2}=K\bar{\varepsilon}^{3/2}\nu^{-1/2}, \tag{21.51}$$

where $K=3a_0$ is a universal numerical constant [Yaglom (1949a)]. Since $\bar{\varepsilon}\sim U^3/L$ (see Sect. 21.3), we have

$$A'=(\overline{A^2})^{1/2}\sim v^{9/4}/L^{3/4}\nu^{1/4}. \tag{21.52}$$

Therefore, the turbulent acceleration increases rapidly with increasing characteristic velocity scale $U$ (it is in fact proportional to $U^{9/4}$) and, therefore, it should be very large for high mean velocities. In the nearly neutral atmospheric surface layer, the quantity $\bar{\varepsilon}$ can be expressed in terms of the wind velocity, the height $z$, and the roughness factor $z_0$ by using the formulas of the theory of the logarithmic layer given in Sect. 5 (Vol. 1). If we then assume that $K$ is of the order of unity, the turbulent acceleration $A'$ for a strong wind turns out to be of the order of the gravitational acceleration $g$ [Yaglom (1949a), Obukhov and Yaglom (1951, 1958); see also Sect. 22.2].

The formula (21.51) shows that the turbulent acceleration depends on the fluid viscosity $\nu$. For large Re, the viscosity affects only the very small-scale turbulent motions which correspond to the high-frequency end of the quasi-equilibrium range. Hence, it may be concluded that the turbulent acceleration is determined largely by the very small-scale motions with $l \lesssim \eta$. This conclusion is confirmed by the fact that, in the inertial subrange, the acceleration scale of the disturbances of length scale $l$ is proportional to $w_l = \bar{\varepsilon}^{2/3} l^{-1/3}$, i.e., it increases with decreasing $l$ (whereas the corresponding velocity scale is proportional to $v_l = (\bar{\varepsilon} l)^{1/3}$, i.e., it decreases with decreasing $l$). Therefore, when we consider the acceleration field of a locally isotropic turbulence, we can neglect the effect of nonisotropic large-scale motions, which play the dominant role in the formation of the velocity field.

When $\tau$ is small in comparison with $\tau_\eta$ we have $x = \tau/\tau_\eta \ll 1$, and the function $a(x)$ is not very different from $a(0) = a_0$. Using the first two terms in the expansion for $\beta_0(x)$ in powers of $x^2$, we find that $a(x) \approx a_0(1 - a_1 x^2)$ for $x \ll 1$, and

$$B_0(\tau) \approx a_0 \bar{\varepsilon}^{3/2} \nu^{-1/2} (1 - a_1 \bar{\varepsilon} \nu^{-1} \tau^2) \quad \text{for} \quad \tau \ll \tau_\eta = (\nu/\bar{\varepsilon})^{1/2}, \tag{21.53}$$

where $a_1$ is a numerical constant. In the second limiting case when $\tau \gg \tau_\eta$, i.e., $x \gg 1$, we can use the similarity hypothesis and demand that the function $B_0(\tau)$ should be independent of $\nu$. We then obtain

$$B_0(\tau) \approx D_0 \bar{\varepsilon}/\tau \text{ when } \tau \gg \tau_\eta, \text{ i.e., } \quad a(x) \approx D_0 x^{-1} \text{ when } x \gg 1, \tag{21.54}$$

where $D_0$ is a numerical constant [this result corresponds to a more accurate formula for $\beta_0(x)$ than the single-term expression given by Eq. (21.30)]. Consequently, $a(x) \ll 1$ for $x \gg 1$, i.e., the correlation coefficient between $A_i(t|x_0)$ and $A_i(t+\tau|x_0)$ (which is equal to $B_0(\tau)/B_0(0) = a(\tau/\tau_\eta)/a_0$) is very small in the inertial subrange. Therefore, the accelerations of a fixed fluid particle at times $t$ and $t+\tau$ for $\tau \gg \tau_\eta$ are practically uncorrelated, although there is an appreciable correlation for the Lagrangian velocities $V(x_0, t)$ up to values of $\tau$ which are of the order of the Lagrangian time scale $T$ (so that for $\tau \ll T$, i.e., throughout the entire quasi-equilibrium range, the velocity correlation coefficient is very close to unity). This is explained by the fact that the Lagrangian acceleration is largely determined by the smallest-scale motions, while the Lagrangian velocity is governed by the largest-scale motions.

The Lagrangian acceleration spectral density is defined as the spectrum of the stationary random process $A_1(t|x_0) = \frac{\partial V_1(x_0, t)}{\partial t}$. It is given by

$$\Phi(\omega) = \omega^2 E^{(L)}(\omega) = \bar{\varepsilon}\varphi^{(1)}(\omega\tau_\eta), \qquad \varphi^{(1)}(\xi) = \xi^2\varphi_0(\xi), \tag{21.55}$$

where the functions $E^{(L)}(\omega)$ and $\varphi_0(\xi)$ are the same as in Eq. (21.34). According to Eq. (21.35), in the inertial frequency subrange we have

$$\Phi(\omega) = B_0\bar{\varepsilon}, \qquad \varphi^{(1)}(\xi) = B_0 = \text{const}, \tag{21.56}$$

as expected, since the dimensions of $\Phi(\omega)$ are the same as those of $\bar{\varepsilon}$. A constant spectral density corresponds to a $\delta$-correlated "white noise," and therefore the formula (21.56) is in good agreement with the fact that the accelerations are practically uncorrelated in the inertial subrange.

Let us now consider the space correlation tensor for the acceleration field:

$$B_{ij}^{(A)}(r) = \overline{A_i(x, t)\, A_j(x+r, t)} = \\ = \left[B_{LL}^{(A)}(r) - B_{NN}^{(A)}(r)\right]\frac{r_i r_j}{r^2} + B_{NN}^{(A)}(r)\,\delta_{ij}, \tag{21.57}$$

which was investigated by Obukhov and Yaglom (1951). Dimensional analysis shows that for $r \ll L$ we have

$$B_{LL}^{(A)}(r) = \bar{\varepsilon}^{3/2}\nu^{-1/2}a_{LL}(r/\eta), \quad B_{NN}^{(A)}(r) = \bar{\varepsilon}^{3/2}\nu^{-1/2}a_{NN}(r/\eta), \tag{21.58}$$

where $a_{LL}(x)$ and $a_{NN}(x)$ are certain universal functions. It is clear that $a_{LL}(0) = a_{NN}(0) = a_0$. Next, $a_{LL}(x) \approx a_0(1 - b_1x^2)$ and $a_{NN}(x) \approx a_0(1 - b_2x^2)$ for $x \ll 1$, so that

$$B_{LL}^{(A)}(r) \approx B^{(A)}(0) - B_1r^2, \quad B_{NN}^{(A)}(r) \approx B^{(A)}(0) - B_2r^2 \quad \text{for} \quad r \ll \eta, \tag{21.59}$$

where $B^{(A)}(0) = A'^2/3$, and $B_1$ and $B_2$ are dimensional constants. In the inertial subrange, the correlation functions $B_{LL}^{(A)}(r)$ and $B_{NN}^{(A)}(r)$ should be independent of $\nu$. Therefore, $a_{LL}(x) \approx K_1x^{-2/3}$, $a_{NN}(x) \approx K_2x^{-2/3}$ for $x \gg 1$, and

$$B_{LL}^{(A)}(r) = K_1\,\bar{\varepsilon}^{4/3}r^{-2/3}, \quad B_{NN}^{(A)}(r) = K_2\,\bar{\varepsilon}^{4/3}r^{-2/3} \quad \text{for} \quad \eta \ll r \ll L, \tag{21.60}$$

where $K_1$ and $K_2$ are numerical constants. In view of Eq. (21.60), the correlation coefficients between acceleration components at two points separated by a distance $r \gg \eta$ will be of the order of $(r/\eta)^{-2/3}$ i.e., they will be very small.

Since the field $A(x)$ is not solenoidal or potential, its spectral tensor is determined by the two scalar functions $E_{LL}^{(A)}(k)$ and $E_{NN}^{(A)}(k)$, where $E_{LL}^{(A)}(k)$ is the longitudinal spectrum corresponding to the potential acceleration component $A^{(1)}(x) = -\frac{1}{\rho}\nabla p(x)$ produced by pressure fluctuations, and $E_{NN}^{(A)}(k) = \nu^2 k^4 E(k)$ is the lateral spectrum corresponding to the solenoidal component $A^{(2)}(x) = \nu \Delta u(x)$ produced by viscous forces [$E(k)$ is the velocity spectrum]. In the inertial subrange $k \ll 1/\eta$ and we can neglect the lateral spectrum $E_{NN}^{(A)}(k)$ generated by viscous friction (after transformation to dimensionless variables this spectrum turns out to be proportional to $(k\eta)^{7/3} \ll 1$). As regards the spectrum $E_{LL}^{(A)}(k)$, it is found that it is independent of $\nu$ for $k \ll 1/\eta$, i.e., it is of the form

$$E_{LL}^{(A)}(k) = G\bar{\varepsilon}^{4/3} k^{-1/3}, \tag{21.61}$$

where $G$ is a numerical constant. If we use the identity $\int_0^\infty k^{-1/3} \cos kr\, dk = \frac{1}{2}\Gamma\left(\frac{2}{3}\right) r^{-2/3}$ and Eqs. (12.39), (12.36), and (12.70), we can readily show that the dimensionless coefficients $K_1$, $K_2$, and $G$ are related by

$$K_1 = \frac{3}{7}\Gamma\left(\frac{2}{3}\right) G \approx 0{,}58G, \quad K_2 = 3K_1. \tag{21.62}$$

The slow decrease of the spectrum (21.61) with increasing $k$ is consistent with the fact that the principal contribution to the acceleration $A^{(1)}(x)$ produced by the pressure fluctuations is due to the smallest-scale disturbances. It is very likely that this acceleration is always considerably greater than the viscous acceleration $A^{(2)}(x)$ in locally isotropic turbulence (see Sect. 22.2 below). If this is so, then in the first approximation we can, in general, neglect the spectral density $E_{NN}^{(A)}(k)$, i.e., assume that the field $A(x) \approx -\frac{1}{\rho}\nabla p$ is potential (hence, it follows, in particular, that $b_2 \approx b_1/3$ and $B_2 \approx B_1/3$).

Let us now consider acceleration fluctuations at a fixed point $x$ (or at a point moving with a constant velocity, i.e., fixed in some inertial set of coordinates). The study of these fluctuations reduces to an analysis of the mixed Euler-Lagrange turbulence characteristics (since we are interpreting acceleration in the Lagrangian sense) which depend on the mean velocity $\bar{u} = \bar{u}(x)$. When the Taylor hypothesis is valid, the time correlation tensor $B_{ij}^{(A)}(\tau) = \overline{A_i(x, t) A_j(x, t+\tau)}$ can be expressed in terms of two scalar functions, namely,

$$B_{LL}^{(A)}(\tau) = \overline{A_L(x, t) A_L(x, t+\tau)}, \qquad B_{NN}^{(A)}(\tau) = \overline{A_N(x, t) A_N(x, t+\tau)} \tag{21.63}$$

where the subscripts $L$ and $N$ refer to components respectively parallel and perpendicular to $\bar{u}$, which are obtained from $B_{LL}^{(A)}(r)$ and $B_{NN}^{(A)}(r)$ by replacing $r$ by $\bar{u}\tau = |\bar{u}|\tau$. In particular, in the inertial subrange

$$B_{LL}^{(A)}(\tau) = K_1 \bar{\varepsilon}^{4/3} \bar{u}^{-2/3} \tau^{-2/3}, \quad B_{NN}^{(A)}(\tau) = 3K_1 \bar{\varepsilon}^{4/3} \bar{u}^{-2/3} \tau^{-2/3}. \tag{21.64}$$

The time spectral densities $E_1^{(A)}(\omega)$ and $E_2^{(A)}(\omega)$ of the fluctuations $A_L(x, t)$ and $A_N(x, t)$ for fixed $x$ are equal to $\frac{1}{\bar{u}} E_1\left(\frac{\omega}{\bar{u}}\right)$ and $\frac{1}{\bar{u}} E_2\left(\frac{\omega}{\bar{u}}\right)$ respectively when the Taylor hypothesis is valid, where $E_1(k)$ and $E_2(k)$ are the longitudinal and lateral one-dimensional space spectra of the field $A(x)$. When applied to the inertial subrange this shows that

$$E_1^{(A)}(\omega) = \frac{6}{7} G \bar{\varepsilon}^{4/3} \bar{u}^{-2/3} \omega^{-1/3}, \quad E_2^{(A)}(\omega) = \frac{18}{7} G \bar{\varepsilon}^{4/3} \bar{u}^{-2/3} \omega^{-1/3}. \tag{21.65}$$

These formulas may be of interest, for example, in the study of the effect of turbulent accelerations on the motion of an aircraft.

Finally, in the case of purely Eulerian statistical characteristics of the local acceleration $a(x, t) = \frac{\partial u(x, t)}{\partial t}$ at a fixed point $x$, the Kolmogorov similarity hypotheses can be successfully used only in conjunction with the frozen turbulence hypothesis. In fact, it is readily shown that

$$B_{ij}^{(a)}(\tau) = \overline{a_i(x, t) a_j(x, t+\tau)} = \frac{1}{2} \frac{d^2}{d\tau^2} [D_{ij}^{*}(\tau)],$$

where $D_{ij}^{*}(\tau) = \overline{\Delta_\tau u_i \Delta_\tau u_j}$. However, the tensor $D_{ij}^{*}(\tau)$ has a simple form only in the frozen-turbulence approximation in which it can be replaced by $D_{ij}(\bar{u}\tau)$. Consequently, in this approximation, the tensor

$B_{ij}^{(a)}(\tau)$ can be expressed in terms of the following scalar functions

$$B_{LL}^{(a)}(\tau)=\frac{1}{2}\bar{u}^2 D_{LL}''(\bar{u}\tau),\quad B_{NN}^{(a)}(\tau)=\frac{1}{2}\bar{u}^2 D_{NN}''(\bar{u}\tau), \qquad (21.66)$$

where $D_{LL}$ and $D_{NN}$ are the structure functions given by Eq. (21.10). In precisely the same way, the time spectra $E_1^{(a)}(\omega)$ and $E_2^{(a)}(\omega)$ of the longitudinal [relative to the direction of the vector $\bar{u}=\bar{u}(x)$] and lateral components of $a(x, t)$ are respectively equal to $\frac{\omega^2}{\bar{u}}E_1(\omega/\bar{u})$ and $\frac{\omega^2}{\bar{u}}E_2(\omega/u)$ when the Taylor hypothesis is valid. In particular, in the inertial subrange

$$B_{LL}^{(a)}(\tau)=-\frac{1}{9}C\bar{\varepsilon}^{2/3}\bar{u}^{2/3}\tau^{-4/3} \quad \text{and} \quad E_1^{(a)}(\omega)=C_2\bar{\varepsilon}^{2/3}\bar{u}^{2/3}\omega^{1/3}$$

and the functions $B_{NN}^{(a)}(\tau)$ and $E_2^{(a)}(\omega)$ are obtained from $B_{LL}^{(a)}(\tau)$ and $E_1^{(a)}(\omega)$ by replacing $C$ by $C'$, and $C_2$ by $C_2'$.

*Characteristics of the Vorticity Field*

The vorticity field of locally isotropic turbulence $\omega(x, t)=\operatorname{rot} u(x, t)=\nabla\times u(x, t)$ is evidently isotropic, and stationary in a sufficiently small space-time region. The longitudinal and lateral correlation functions of the vorticity field, $B_{LL}^{(\omega)}(r)$ and $B_{NN}^{(\omega)}(r)$, can be expressed in terms of the structure functions $D_{LL}(r)$ and $D_{NN}(r)$ through the formulas (11.81) and (13.71). In view of these formulas, and also the formulas (21.16), (21.17'), (21.19), and (21.19'), we have

$$B_{LL}^{(\omega)}(0)=B_{NN}^{(\omega)}(0)=5A=\frac{\bar{\varepsilon}}{3\nu},\quad \overline{\omega^2}=15A=\frac{\bar{\varepsilon}}{\nu}, \qquad (21.67)$$

$$B_{LL}^{(\omega)}(r)=\frac{11}{9}C\bar{\varepsilon}^{2/3}r^{-4/3},\quad B_{NN}^{(\omega)}(r)=\frac{11}{27}C\bar{\varepsilon}^{2/3}r^{-4/3} \quad \text{when } L\gg r\gg\eta. \qquad (21.68)$$

The longitudinal spectral function of the vorticity field is of course equal to zero, while the lateral spectral function $E^{(\omega)}(k)$ can be expressed in terms of the velocity spectrum $E(k)$ by the formula $E^{(\omega)}(k)=k^2E(k)$. In particular, in the inertial subrange,

$$E^{(\omega)}(k)=C_1\bar{\varepsilon}^{2/3}k^{1/3}. \qquad (21.69)$$

*Characteristics of the Pressure Field*

It is likely that the pressure field $p(\boldsymbol{x}, t)$ of a locally isotropic turbulence is locally isotropic and stationary in a sufficiently small region. The probability distribution for the pressure differences between neighboring points is then determined by the same parameters as the distribution of the velocity differences (apart from the addition of the parameter $\rho$). It follows that the Kolmogorov similarity hypotheses can be used to investigate the local statistical characteristics of the pressure field.

We shall restrict our attention to purely spatial pressure differences

$$\Delta_r p = p(\boldsymbol{x}+\boldsymbol{r}, t) - p(\boldsymbol{x}, t).$$

By the first Kolmogorov hypothesis, the probability distribution for $\Delta_r p$ in the quasi-equilibrium range of distances $r = |\boldsymbol{r}| \ll L$ can depend only on $r, \bar{\varepsilon}, \nu$, and $\rho$. Hence, the pressure structure function is given by

$$D_{pp}(r) = \overline{(\Delta_r p)^2} = \rho^2 \bar{\varepsilon} \nu \pi\left(\frac{r}{\eta}\right), \tag{21.70}$$

where $\pi(x)$ is a universal function. When $r \ll \eta$, the difference $\Delta_r p$ can be regarded as a linear function of $r$. Therefore

$$\begin{aligned} D_{pp}(r) &= c_p \rho^2 \bar{\varepsilon}^{3/2} \nu^{-1/2} r^2 \quad \text{for} \quad r \ll \eta, \\ \pi(x) &= c_p x^2 \quad \text{for} \quad x \ll 1. \end{aligned} \tag{21.71}$$

It is clear that $3c_p = \overline{(\nabla p)^2}/\rho^2 \bar{\varepsilon}^{3/2} \nu^{-1/2}$. Therefore, the constant $c_p$ determines the mean pressure gradient, and the important acceleration component $\overline{[A^{(1)}]^2} = \frac{1}{\rho^2} \overline{(\nabla p)^2}$ associated with it. This coefficient has been estimated approximately by Heisenberg (1948a), Yaglom (1949a), Batchelor (1951), and Golitsyn (1963) (see Sect. 22.2 below).

In the inertial subrange, the function $D_{pp}(r)$ can depend only on $r, \bar{\varepsilon}$, and $\rho$. Hence, it follows that

$$\begin{aligned} D_{pp}(r) &= C_p \rho^2 \bar{\varepsilon}^{4/3} r^{4/3} \quad \text{for} \quad \eta \ll r \ll L; \\ \pi(x) &= C_p x^{4/3} \quad \text{when } x \gg 1. \end{aligned} \tag{21.72}$$

If we recall the formula for $D_{LL}(r)$ given by Eq. (21.17′), we can rewrite Eq. (21.72) in the form

$$D_{pp}(r) = C'_p \rho^2 [D_{LL}(r)]^2. \tag{21.73}$$

The formulas (21.72) and (21.73) were published by Obukhov (1949b) who also suggested an approximate method for estimating the coefficient $C'_p$.

Instead of the pressure structure function $D_{pp}(r)$ we can consider the corresponding three-dimensional pressure spectrum $E_{pp}(k) = E^{(p)}(k) = 2\rho^2 k^{-2} E^{(A)}_{LL}(k)$ [where $E^{(A)}_{LL}(k)$ is the longitudinal acceleration spectrum], or the one-dimensional pressure spectrum $E^{(p)}_1(k)$. These spectra are given by

$$E^{(p)}(k) = \rho^2 \bar{\varepsilon}^{3/4} \nu^{7/4} \varphi^{(p)}(k\eta), \quad E^{(p)}_1(k) = \rho^2 \bar{\varepsilon}^{3/4} \nu^{7/4} \varphi^{(p)}_1(k\eta), \tag{21.74}$$

where

$$\pi(x) = 2 \int_0^\infty \left(1 - \frac{\sin x\xi}{x\xi}\right) \varphi^{(p)}(\xi)\, d\xi = 2 \int_0^\infty (1 - \cos x\xi)\, \varphi^{(p)}_1(\xi)\, d\xi,$$
$$\varphi^{(p)}(\xi) = -\xi \frac{d}{d\xi} \varphi^{(p)}_1(\xi), \quad \varphi^{(p)}_1(\xi) = \int_\xi^\infty \varphi^{(p)}(\xi_1)\, \xi_1^{-1}\, d\xi_1. \tag{21.75}$$

In the inertial subrange,

$$\varphi^{(p)}(\xi) = B^{(p)} \xi^{-7/3}, \quad \varphi^{(p)}_1(\xi) = B^{(p)}_1 \xi^{-7/3} \quad \text{for} \quad \xi \ll 1, \tag{21.76}$$

i.e.,

$$E^{(p)}(k) = B^{(p)} \rho^2 \bar{\varepsilon}^{4/3} k^{-7/3}, \quad E^{(p)}_1(k) = B^{(p)}_1 \rho^2 \bar{\varepsilon}^{4/3} k^{-7/3} \tag{21.76′}$$
$$\text{for} \quad 1/L \ll k \ll 1/\eta,$$

where

$$B^{(p)} = 2G = \frac{7C_p}{3\Gamma(-4/3)} \approx 0{,}77C_p, \quad B^{(p)}_1 = \frac{3}{7} B^{(p)} \approx 0{,}33C_p. \tag{21.77}$$

We see that the pressure spectrum decreases with increasing wave number more rapidly than the velocity spectrum, so that the small-scale motions are of lesser importance in the pressure fluctuations than in the velocity fluctuations.

According to the results of Sect. 13.3, the pressure differences $\Delta_r p$ in an incompressible fluid are not correlated with the velocity differences $\Delta_r \boldsymbol{u}$. This, however, does not mean that the pressure and velocity fluctuations are statistically independent, because the higher-order mixed moments of the fields $\Delta_r p$ and $\Delta_r \boldsymbol{u}$ are, in general, nonzero. In particular, even the third-order moments

$$D_{ppl}(\boldsymbol{r}) = \overline{(\Delta_r p)^2 \Delta_r u_l}. \tag{21.78}$$

are nonzero, although the other third-order moments $D_{plj}(\boldsymbol{r}) = \overline{\Delta_r p \, \Delta_r u_l \, \Delta_r u_j}$ are equal to zero by virtue of Eq. (13.76′). Owing to local isotropy, the vector $D_{ppl}(\boldsymbol{r})$ is fully defined by the single scalar function $D_{ppL}(r) = \overline{(\Delta_r p)^2 \Delta_r u_L}$. The general form of this function is readily established from dimensional considerations. For example, in the inertial subrange,

$$D_{ppL}(r) \sim \rho^2 \bar{\varepsilon}^{5/3} r^{2/3}$$

i.e.,

$$D_{ppL}(r) = A_{p} \rho^2 [D_{LL}(r)]^{5/2}. \tag{21.79}$$

## 21.6 Local Structure of the Temperature Field for High Reynolds and Peclet Numbers

Let us now consider the structure of the concentration field $\vartheta(\boldsymbol{x}, t)$ for a dynamically passive admixture, mixed by locally isotropic turbulence. To be specific, we shall suppose that $\vartheta(\boldsymbol{x}, t)$ is the temperature transported by the wandering fluid particles and that it has no appreciable effect on the turbulence. In other words, we shall consider forced convection in temperature-inhomogeneous fluids in the presence of developed turbulence of dynamic origin.

The similarity hypotheses are based on physical ideas indicating that, for sufficiently high Reynolds numbers, the statistical state of velocity fluctuations in each sufficiently small space-time region is isotropic and quasi-stationary, and is completely defined by the parameters $\bar{\varepsilon}$ and $\nu$. It is natural to expect that the temperature fluctuations due to the mixing of portions of the fluid with different initial temperatures will then be isotropic and stationary in small space-time regions. Consequently, the scalar field $\vartheta(\boldsymbol{x}, t)$ can be regarded as locally isotropic. However, there is no reason to suppose that its statistical parameters will depend only on $\bar{\varepsilon}$ and $\nu$. In fact, the evolution of the temperature field is described by the heat transfer equation containing the molecular temperature diffusivity $\chi = \varkappa/c_p\rho$. Therefore, it is clear that the value of $\chi$ may affect the local structure of the field $\vartheta(\boldsymbol{x}, t)$. This effect cannot be neglected: in the case of intensive turbulent mixing, molecular thermal conductivity plays an important role, since the turbulent motion may lead to the approach of fluid volumes with very different temperatures, i.e., to a rapid increase of temperature gradients (see Vol. 1, Sect. 10.2). However, even if we add the quantity $\chi$ to the two parameters $\bar{\varepsilon}$ and $\nu$, we still will not obtain the complete set of parameters defining the statistical state of small-scale temperature fluctuations. To achieve this aim, we must augment $\bar{\varepsilon}$, $\chi$, and $\nu$ by one further quantity, which we shall now define.

When we were investigating the local structure of the velocity field $\boldsymbol{u}(\boldsymbol{x}, t)$ we assumed that $\mathrm{Re} = L\Delta U/\nu$ (where $\Delta U$ is a typical velocity difference over a distance $L$) was sufficiently large, and considered the "cascade" process of fragmentation of macrostructural inhomogeneities. It was noted that, of all the quantities characterizing large-scale turbulent motions, the only one which affects sufficiently small disturbances is the rate of energy transfer from large-scale motions to small-scale motions. This finally results in conversion into heat through molecular viscosity. The temperature field can be considered in the same way. We must, however, assume that both the Reynolds number Re and the Peclet number $\mathrm{Pe} = L_\vartheta \Delta_\vartheta U/\chi$ are large, where $L_\vartheta$ is the length over which there is an appreciable change in the mean temperature $\bar{\vartheta}(\boldsymbol{x})$, and $\Delta_\vartheta U$ is a typical change in the mean velocity over the distance $L_\vartheta$ (moreover, for $L_\vartheta > L$, we can replace $L_\vartheta$ and $\Delta_\vartheta U$ by the usual parameters $L$ and $\Delta U$). The cascade fragmentation of velocity-field disturbances will also lead to the fragmentation of macrostructural temperature inhomogeneities of scale $L_\vartheta$ into smaller-scale disturbances of the field $\vartheta(\boldsymbol{x}, t)$. In regions with linear dimensions much smaller than $L_\vartheta$, the mean temperature $\bar{\vartheta}$ can be regarded as practically constant. The typical temperature fluctuation $\vartheta' = \vartheta - \bar{\vartheta}$ will therefore be a measure of the degree of inhomogeneity of the temperature field in such regions. Following Obukhov (1949a), it is convenient to take the quantity

$$H=\frac{1}{2}\int_V \rho\overline{(\vartheta(x,t)-\overline{\vartheta})^2}\,dx=\frac{1}{2}\rho\int_V \overline{\vartheta'^2}\,dx, \tag{21.80}$$

as a measure of the temperature inhomogeneity in a volume $V$, i.e., to characterize the degree of temperature inhomogeneity of a unit mass by the quantity $\overline{\vartheta'^2}/2$.[6] In that case, one half of the mean square of the temperature deviation at a point from the average temperature in a sphere of diameter $l$ (or, what is practically the same thing, the quantity $\overline{(\Delta_l\vartheta)^2}$, where $\Delta_l\vartheta$ is the temperature difference between two points separated by the distance $l$) will characterize the contribution due to disturbances of given length scale $l$ to the specific inhomogeneity of the temperature field. Fragmentation of temperature inhomogeneities will thus result in the fact that the total measure of the temperature inhomogeneity will increasingly concentrate in small-scale disturbances. However, the quantity $\overline{\vartheta'^2}/2$ will, of course, remain the same for all types of displacement of the fluid particles which are not accompanied by a change in the temperature. In other words, the quantity $\overline{\vartheta'^2}/2$ satisfies a definite "conservation law" which is a consequence of the fact that the temperature, like kinetic energy, does not change during inertial motion of the fluid particles. A change in the degree of temperature inhomogeneity can be produced only by molecular thermal conduction, leading to an equalization of temperatures at neighboring points, i.e., to a reduction in $\overline{\vartheta'^2}/2$.

Let $\overline{N}$ be the mean "dissipation rate of temperature inhomogeneities," i.e., the rate of reduction in the measure of temperature inhomogeneity $\overline{\vartheta'^2}/2$ due to molecular thermal conduction. For large values of Re and Pe, the "temperature dissipation rate" will be almost entirely concentrated in the smallest-scale disturbances, and will be equal to the "transport of the temperature-inhomogeneity measure over the length scale spectrum," i.e., it will be equal to the increase per unit time in the contribution to $\overline{\vartheta'^2}/2$ associated with small-scale disturbances of scale $l \ll \min(L_0, L)$ and due to the

[6] The numerical factor 1/2 is introduced here by analogy with the definition of the kinetic energy $\overline{u'^2}/2$. It is also possible to eliminate this factor, i.e., to use $\overline{\vartheta'^2}$ as a measure of the temperature inhomogeneity of a unit mass; this last convention is adopted by many authors writing in English. In such a case the mean temperature dissipation rate (see below) increases by a factor of two and therefore the numerical coefficients in the universal equations of this subsection will also differ from the values given here.

fragmentation of large-scale temperature-field inhomogeneities by the turbulent motions. It is clear that the quantity $\bar{N}$ may remain constant in time if the large-scale inhomogeneities of scale $L_\vartheta$ are maintained by external heat sources which produce a fixed distribution of mean temperature. Unless this is so, $\bar{N}$ will depend on $t$. However, the time variation of $\bar{N}$ will be very slow in comparison with the characteristic time scales of sufficiently small-scale turbulent motions. Therefore, when we consider the statistical properties of small-scale temperature disturbances, the quantity $\bar{N}$ can be regarded as constant. It will, in fact, characterize the macrostructural inhomogeneities, and will have an important effect on local isotropic temperature fluctuations.

The quantity $\bar{N}$ is proportional to the coefficient $\chi$ and to the mean square of the temperature gradient. In fact, we can use the heat transfer equation

$$\frac{\partial \vartheta}{\partial t}+u_j\frac{\partial \vartheta}{\partial x_j}=\chi\,\Delta\vartheta \tag{21.81}$$

to show that for a sufficiently large volume $V$, such that we can neglect convective heat transfer through its boundary, the reduction in the temperature-inhomogeneity measure will be described by the equation

$$\frac{dH}{dt}=\frac{d}{dt}\left[\frac{1}{2}\int\limits_V \rho\overline{(\vartheta'(\boldsymbol{x},t))^2}\,d\boldsymbol{x}\right]=-\chi\int\limits_V \rho\,\overline{(\nabla\vartheta'(\boldsymbol{x},t))^2}\,d\boldsymbol{x}. \tag{21.82}$$

Consequently, the specific (per unit mass) rate of reduction is the temperature-inhomogeneity measure is given by

$$\bar{N}=\chi\overline{(\nabla\vartheta')^2}=\chi\sum_{i=1}^{3}\overline{\left(\frac{\partial\vartheta'}{\partial x_i}\right)^2}. \tag{21.83}$$

For homogeneous turbulence, the mean convective transport of the fluctuations $\vartheta'$ is zero not only after averaging over a large volume $V$, but also at each point. In the case of locally isotropic turbulence, therefore, the contribution of convective transport to the rate of change of $\overline{\vartheta'^2}/2$ will be determined only by the space derivatives of very smooth large-scale components of the flow variables, i.e., it will be negligible throughout. This is associated with the fact that for large Re and Pe we can neglect the effect of molecular thermal conduction on the mean flow, so that the formula given by Eq. (21.83) can also be written in the form

$$\overline{N}=\chi\overline{(\nabla\vartheta)^2}=\chi\sum_{i=1}^{3}\overline{\left(\frac{\partial\vartheta}{\partial x_i}\right)^2}. \tag{21.83$'$}$$

This definition of $\overline{N}$ will be used below.

The form of Eq. (21.83′) is very close to the expression for the energy dissipation rate $\overline{\varepsilon}=\frac{\nu}{2}\sum\overline{\left(\frac{\partial u_i}{\partial x_j}+\frac{\partial u_j}{\partial x_i}\right)^2}$. The quantities $\overline{\varepsilon}$ and $\overline{N}$ are also very close to each other in their physical significance. We recall the entropy balance equation in a temperature inhomogeneous fluid [see Eq. (1.62) of Vol. 1]. If we use the identity

$$\frac{1}{\vartheta}\frac{\partial^2\vartheta}{\partial x_i^2}=\frac{\partial}{\partial x_i}\left(\frac{1}{\vartheta}\frac{\partial\vartheta}{\partial x_i}\right)+\frac{1}{\vartheta^2}\left(\frac{\partial\vartheta}{\partial x_i}\right)^2$$

and the fact that the mean absolute temperature $\overline{\vartheta}=T_0$ can be practically always regarded as constant, we may conclude that $\overline{\varepsilon}$ is equal to $T_0$ multiplied by the mean rate of increase in entropy per unit mass due to internal friction (i.e., molecular viscosity), and $\overline{N}$ is equal to $T_0^2/c_p$ multiplied by the mean rate of increase in the entropy due to molecular thermal conduction. The quantity $H$ in Eq. (21.80) is also found to have a simple physical meaning. Obukhov (1949a) has shown that it is equal to $T_0/c_p$ multiplied by the maximum work which can be extracted from an inhomogeneously heated volume $V$ through a reversible transition of this volume into a state of thermodynamic equilibrium (i.e., constant temperature). This provides an additional justification for comparing the temperature-inhomogeneity measure $H$ with the kinetic energy of turbulence, and the temperature dissipation rate $\overline{N}$ with the energy dissipation rate $\overline{\varepsilon}$.

The quantity $\overline{N}$ can also be determined from the quantities characteristic of the large-scale mean motion which is independent of the molecular transport coefficients (this was noted in Sect. 7.5 of Vol. 1). Since $\overline{N}$ has the dimensions of a square of temperature divided by time, the order of magnitude of $\overline{N}$ can be estimated from the relation

$$\bar{N} \sim \frac{\Delta_\theta U \cdot (\Delta\bar{\vartheta})^2}{L_\theta},$$

where $\Delta_\theta U$ and $\Delta\bar{\vartheta}$ are typical changes in the mean velocity and mean temperature over the distance $L_\theta$ (see the analogous relations for $\bar{\varepsilon}$ in Sect. 21.3). If we suppose that $L$ and $L_\theta$ are of the same order of magnitude (this is usually the case), and set $\bar{N} \approx K_\theta \left(\frac{\Delta\bar{\vartheta}}{L}\right)^2$ where $K_\theta$ is interpreted as the effective eddy temperature diffusivity, then it will follow that $K_\theta \sim L\,\Delta U \sim K$, as expected.

We shall use the length scale $L_0 = \min(L, L_\theta)$ which can be identified with $L$ and with $L_\theta$ if $L$ and $L_\theta$ are of the same order of magnitude. For the statistical turbulence characteristics containing temperature, the quasi-equilibrium range of $l$ will then be defined by the inequality $l \ll L_0$. The first similarity hypothesis will now assume the following form (we are restricting our attention to purely spatial probability distributions).

*First similarity hypothesis. In developed turbulence with sufficiently high values of* Re *and* Pe, *the multidimensional probability distributions for velocity and temperature differences at an arbitrary set of points in a spatial region $V$ of diameter $l \ll L_0$ are invariant under all rotations and translations of this set of points, provided these transformations do not take it beyond the limits of the region $V$, and are uniquely determined by the parameters $\bar{\varepsilon}$, $\nu$, $\bar{N}$, and $\chi$.*

Let us now generalize the second similarity hypothesis to the case of probability distributions containing temperature differences. We note that, for sufficiently large Pe, the molecular thermal conductivity, which is characterized by $\chi$, may play an appreciable role only for very small-scale disturbances. In fact, the ratio of typical values of the terms in the heat-transfer equation which describe the convection of heat and molecular thermal conduction is equal to the Peclet number. Therefore, the molecular thermal conduction is important only for disturbances with $\text{Pe} \leqslant 1$. It is natural to suppose that the Peclet number decreases monotonically with decreasing length scale of the disturbances. Therefore, for a sufficiently large Peclet number of the mean flow, there should exist a subrange of length scales which are small in comparison with $L_0$ and for which the Peclet number is much greater than unity. In this subrange, all the statistical characteristics should be independent of $\chi$. It can be referred to as the *convective subrange*.

If we attempt to determine the order of magnitude for the lower limit of the convective subrange, we encounter a specific difficulty connected with the presence of the two quantities $\nu$ and $\chi$ which have the same dimensions. It follows that the dimensionless parameters of small-scale turbulence which contain temperature will, in general, be functions of the dimensionless parameter $\mathrm{Pr} = \nu/\chi$. In particular, the ratio of the length scales of the smallest-scale disturbances which are still appreciably effected by molecular thermal conductivity to the Kolmogorov internal turbulence length scale $\eta = (\nu^3/\bar{\varepsilon})^{1/4}$ will also be the function of the Prandtl number. Therefore, the convective subrange of length scales is determined by inequalities of the form $L_0 \gg l \gg \lambda(\mathrm{Pr})\cdot\eta$ where $\lambda(z)$ is a universal function. Instead of the length scale $\eta$ we can use the so-called internal temperature length scale

$$\eta_\theta = (\chi^3/\bar{\varepsilon})^{1/4} = \eta(\mathrm{Pr})^{-3/4}. \tag{21.84}$$

However, our conclusion that for $l \gg \eta_\theta$ we are necessarily within the limits of the convective subrange cannot be regarded as justified, since the possibility of neglecting molecular conductivity in comparison with convection will also depend on the velocity field which is affected by the viscosity $\nu$.

However, these considerations are important only in the limiting cases $\nu \ll \chi$ and $\nu \gg \chi$ (which appear to be relatively exotic but are in fact, of definite practical importance). We shall discuss these cases in detail in Sect. 22.4. It will be shown there that when $\nu \ll \chi$, i.e., $\eta_\theta \gg \eta$, the convective subrange extends only to scales $l \gg \eta_\theta$, and for $\nu \gg \chi$ it extends to scales $l \gg (\nu\chi^2/\bar{\varepsilon})^{1/4} = (\eta\eta_\theta^2)^{1/3}$. In the more usual situation, $\nu/\chi = \mathrm{Pr}$ is of the order of unity, the scales $\eta$, $\eta_\theta$, and $(\eta\eta_\theta^2)^{1/3}$ are of the same order of magnitude, and $\lambda(\mathrm{Pr})$ becomes a numerical coefficient of the order of unity. Therefore, for Pr of the order of unity, the upper limits of the subranges in which molecular friction or molecular thermal conductivity are still important may be regarded as coincident.

Let us now consider the *inertial-convective subrange* of length scales $L_0 \gg l \gg \eta_0 = \max(\eta, \eta_\theta)$ which is the intersection of the convective and inertial subranges. For disturbances with length scales lying in this subrange we can neglect both internal friction and molecular thermal conductivity. In other words, we have the following hypothesis.

*The second similarity hypothesis. In developed turbulence with sufficiently high values of* Re *and* Pe, *the multidimensional* *probability distributions for the velocity and temperature differences at a set of points, such that all distances $r_k$ between the points satisfy the inequalities $L_0 \gg r_k \gg \eta_0 = \max(\eta, \eta_\vartheta)$, are uniquely determined by the values of the parameters $\bar{\varepsilon}$ and $\bar{N}$.*

The similarity hypotheses lead to certain simple consequences with regard to the statistical characteristics of the spatial temperature differences

$$\Delta_r\vartheta = \vartheta(x+r) - \vartheta(x)$$

in turbulent flows with sufficiently large Re and Pe. In particular, it follows from the first similarity hypothesis that, in the quasi-equilibrium range $r \ll L_0$, the spatial temperature structure function $D_{\vartheta\vartheta}(r) = \overline{(\Delta_r\vartheta)^2}$ depends only on $r = |r|$, and should be of the form

$$D_{\vartheta\vartheta}(r) = \bar{N}\bar{\varepsilon}^{-1/2}\chi^{1/2}\, h\left(\frac{r}{\eta_\vartheta};\ \frac{\nu}{\chi}\right), \tag{21.85}$$

where $h(x;\ z)$ is a universal function of the two variables. For sufficiently small $r$, the difference $\Delta_r\vartheta$ can be regarded as an approximately linear function of $r$, so that $D_{\vartheta\vartheta}(r) \sim r^2$ for small $r$ and $h(x;\ z) \sim x^2$ for small $x$. We shall assume for simplicity that the number Pr is of the order of unity and, consequently, the scales $\eta$ and $\eta_\vartheta$ are of the same order of magnitude. We then have $h(x;\ \mathrm{Pr}) \approx h_0 x^2$ for $x \ll 1$, and $D_{\vartheta\vartheta}(r) \approx h_0 \frac{\bar{N}}{\chi} r^2$ for $r \ll \eta_\vartheta$. It is readily shown that the numerical coefficient $h_0 = \frac{1}{2} \left.\frac{d^2h(x;\ \mathrm{Pr})}{dx^2}\right|_{x=0}$ is independent of Pr. In fact, since according to Eq. (21.83′) $\overline{(\nabla\vartheta)^2} = \frac{3}{2} D''_{\vartheta\vartheta}(0) = \frac{\bar{N}}{\chi}$ we have

$$D_{\vartheta\vartheta}(r) \approx \frac{\bar{N}}{3\chi} r^2 \quad \text{for} \quad r \ll \eta_\vartheta \tag{21.86}$$

i.e., $h_0 = 1/3$ and $h(x;\ \mathrm{Pr}) \approx x^2/3$ for $x \ll 1$. In the other limiting case $r \ll \eta_\vartheta$ (but $r \ll L_0$) we can use the second similarity hypothesis, which leads to the result that, for such $r$,

$$D_{\vartheta\vartheta}(r) = C_\vartheta \bar{N}\bar{\varepsilon}^{-1/3} r^{2/3} \tag{21.87}$$

i.e., $h(x;\ \mathrm{Pr}) \approx C_\vartheta x^{2/3}$ for $x \gg 1$ where $C_\vartheta$ is a universal constant.

In addition to the structure function $D_{\vartheta\vartheta}(r)$ we can also consider the spectrum of the locally isotropic temperature field, $E_{\vartheta\vartheta}(k) = E^{(\vartheta)}(k)$, or the corresponding one-dimensional spectrum $E_1^{(\vartheta)}(k)$. Similar to Eq. (21.85), spectral equations have the form

$$\begin{aligned} E^{(\vartheta)}(k) &= \overline{N}\,\bar{\varepsilon}^{-3/4}\chi^{5/4}\varphi^{(\vartheta)}(k\eta_\vartheta;\ \mathrm{Pr}), \\ E_1^{(\vartheta)}(k) &= \overline{N}\,\bar{\varepsilon}^{-3/4}\chi^{5/4}\varphi_1^{(\vartheta)}(k\eta_\vartheta;\ \mathrm{Pr}), \end{aligned} \tag{21.88}$$

where the function $\varphi^{(\vartheta)}(\xi;\ \mathrm{Pr})$ and $\varphi_1^{(\vartheta)}(\xi;\ \mathrm{Pr})$ are related to $h(x,\ \mathrm{Pr})$ through the same relations through which $\varphi^{(p)}(\xi)$ and $\varphi_1^{(p)}(\xi)$ are related to $\pi(x)$ [see Eq. (21.75)]. The equation (21.87) is equivalent to either of the two following equations:

$$E^{(\vartheta)}(k) = B^{(\vartheta)}\overline{N}\bar{\varepsilon}^{-1/3}k^{-5/3}, \quad E_1^{(\vartheta)}(k) = B_1^{(\vartheta)}\overline{N}\,\bar{\varepsilon}^{-1/3}k^{-5/3} \tag{21.89}$$

which are valid for $1/L \ll k \ll 1/\eta_\vartheta$). In view of Eqs. (13.59) and (12.13), we then have

$$B^{(\vartheta)} = \frac{10C_\vartheta}{9\Gamma(1/3)} \approx 0.4C_\vartheta, \qquad B_1^{(\vartheta)} = \frac{3}{5}B^{(\vartheta)} \approx 0.25C_\vartheta. \tag{21.90}$$

Since the function $\varphi_1^{(\vartheta)}(\xi)$ is obtained from $\varphi^{(\vartheta)}(\xi_1)$ by integration with respect to all values of $\xi_1$ between $\xi$ and infinity (with the weighting function $\xi_1^{-1}$) it follows that, as in the case of the velocity spectra, the one-dimensional spectrum $E_1^{(\vartheta)}(k)$ will, in general, be given by Eq. (21.89) up to appreciably smaller values of $k$ than the three-dimensional spectrum $E^{(\vartheta)}(k)$ [see, in particular, Fig. 55 in Sect. 21.4 where the function $\varphi(\xi)$ represents the model form of the three-dimensional (dimensionless) spectrum $\varphi^{(\vartheta)}$, and the function $\varphi_3(\xi)$ represents the corresponding one-dimensional spectrum $\varphi_1^{(\vartheta)}$]. For larger $k$, the three-dimensional spectrum $E^{(\vartheta)}(k)$ begins to depart from that predicted by Eq. (21.89) later than the one-dimensional spectrum$E_1^{(\vartheta)}(k)$.

The formulas (21.86) and (21.87) [the first of these represents the *temperature two-thirds law*] is due to Obukhov (1949a). The *temperature five-thirds law* (21.89), which is equivalent to Eq. (21.87), was given by Corrsin (1951b).[7]

[7] **Almost simultaneously with the publication of the papers containing Eqs. (21.87) and (21.89), Inoue (1950a, 1951a, 1952b) put forward a theory (based on certain disputable assumptions) according to which, in the inertial-convective subrange $D_{\vartheta\vartheta}(r) \sim r^{4/3}$ and**

**(Continued on page 386)**

The similarity hypotheses can also be used to obtain various expressions for the higher-order moments of the differences $\Delta_r\vartheta$, and also the higher-order joint moments of $\Delta_r\vartheta$ and $\Delta_r\boldsymbol{u}$ (the second-order moment $\overline{\Delta_r\vartheta\,\Delta_r\boldsymbol{u}}$ is, of course, zero). Thus, for example, the third-order moment $D_{l\vartheta\vartheta}(r)=\overline{\Delta_r u_l(\Delta_r\vartheta)^2}$ [Yaglom (1949b)] is determined by the scalar function

$$D_{L\vartheta\vartheta}(r)=\overline{\Delta_r u_L(\Delta_r\vartheta)^2}=\overline{N}\,\bar{\varepsilon}^{-1/4}\chi^{3/4}\,l\,(r/\eta_\vartheta;\ \mathrm{Pr}),$$

where $l\,(x;\ \mathrm{Pr})$ is a universal function, and

$$D_{L\vartheta\vartheta}(r)\approx d_1\overline{N}\,\bar{\varepsilon}^{1/2}\chi^{-3/2}r^3 \qquad r\ll\min(\eta,\ \eta_\vartheta), \tag{21.91}$$

$$D_{L\vartheta\vartheta}(r)\approx D_1\overline{N}\,r \qquad L_0\gg r\gg\max(\eta,\ \eta_\vartheta). \tag{21.92}$$

However the two scalar functions $D_{LL\vartheta}(r)=\overline{(\Delta_r u_L)^2\Delta_r\vartheta}$ and $D_{NN\vartheta}(r)=\overline{(\Delta_r u_N)^2\Delta_r\vartheta}$, which define the third-order tensor $D_{ij\vartheta}(r)=\overline{\Delta_r u_i\Delta_r u_j\Delta_r\vartheta}$, are both equal to zero by virtue of the general equation (13.76′).

The situation is more complicated in the case of the time differences

$$\Delta_\tau\vartheta=\vartheta(\boldsymbol{x},\ t+\tau)-\vartheta(\boldsymbol{x},\ t).$$

If, however, we are concerned with time intervals $\tau$ (or frequencies $\omega$) for which the Taylor frozen turbulence hypothesis is valid, then the corresponding characteristics can be readily reduced to those of the space differences $\Delta_r\vartheta$, which were discussed above. In particular, for $\eta_\vartheta/\bar{u}\ll\tau\ll L_0/\bar{u}$ or $\bar{u}/L_0\ll\omega\ll\bar{u}/\eta_\vartheta$, we have from Eqs. (21.87), (21.89), and (21.92)

---

$E^{(\vartheta)}(k)=E_{\vartheta\vartheta}(k)\sim k^{-7/3}$. Subsequently, Villars and Weisskopf (1955) considered the statistical characteristics of the electron-density in the ionosphere, and put forward the hypothesis [developed later by Wheelon (1957, 1958)] according to which small-scale density fluctuations depend not on the parameter $\overline{N}$ (which they do not consider), but on the mean gradient $\nabla\bar{\vartheta}$ [which governs smooth changes in the field $\bar{\vartheta}(x)$ and is very small in comparison with the typical value of the gradient $\vartheta'$]. According to this hypothesis, in the inertial-convective subrange $D_{\vartheta\vartheta}(r)\sim(\nabla\bar{\vartheta})^2r^2$ and $E_{\vartheta\vartheta}(k)\sim(\nabla\bar{\vartheta})^2k^{-3}$ [this last formula was critically discussed by Bolgiano (1957, 1958a, b)]. At present, however, there is extensive experimental evidence indicating that the formulas (21.87) and (21.89) are valid (see Sect. 23.5 below).

$$D_{\vartheta\vartheta}(\tau) = C_\vartheta \overline{N}\,\overline{\varepsilon}^{-1/3}\,\overline{u}^{2/3}\,\tau^{2/3}, \quad E^{(\vartheta)}(\omega) = B_1^{(\vartheta)}\,\overline{N}\,\overline{\varepsilon}^{-1/3}\,\overline{u}^{2/3}\,\omega^{-5/3}, \tag{21.93}$$

$$D_{L\vartheta\vartheta}(\tau) = \overline{\Delta_\tau u_L (\Delta_\tau \vartheta)^2} = D_1 \overline{N}\,\overline{u}\,\tau. \tag{21.94}$$

In conclusion, we emphasize once again that all the above formulas are valid not only for the temperature, but also for the concentration of an arbitrary passive admixture. Therefore, the results given in this subsection will be valid, for example, for the humidity or concentration of carbon dioxide in the atmosphere. They will also be valid for the salinity of seawater or the density of electrons in the ionosphere (if the effect of the geomagnetic field is small). The parameters $\chi$ and $\overline{N}$ will, of course, have different values in all these cases.

The results of the present subsection can also be used to draw some conclusions with regard to the statistical characteristics of the refractive-index fields which determine the velocity of propagation of light, sound, and radiowaves in a turbulent atmosphere. In fact, fluctuations in the refractive index for ordinary light are very largely due to temperature fluctuations. For acoustic waves, there are also appreciable effects due to the wind velocity fluctuations, whereas in the case of radiowaves, humidity fluctuations are important (or the electron density fluctuations if we consider the propagation of radiowaves in the ionosphere). Since all these fluctuations are relatively small, it can be considered that the refractive index fluctuations are linear functions of fluctuations in the temperature, wind velocity, humidity, and electron density. Hence, it follows, in particular, that in the inertial-convective subrange, the two-thirds law should also be valid for the refractive index.

## 21.7 Local Characteristics of Turbulence in the Presence of Buoyancy Forces and Chemical Reactions. Effect of Thermal Stratification

### *Local Statistical Characteristics of Turbulence in a Thermally Stratified Fluid*

It was assumed above that temperature behaves as if it were a passive admixture, i.e., it has no appreciable effect on the dynamics of turbulence. However, in the important case of a temperature-inhomogeneous fluid in a gravitational field, the temperature cannot be considered as a passive substance. In fact, in this case, temperature fluctuations give rise to density fluctuations which are in turn affected by buoyancy. Therefore, the temperature distribution generates a field of buoyant accelerations, i.e., it affects the flow dynamics. Consequently, in the case of a thermally stratified fluid, the theory of similarity for small-scale properties of turbulence must be generalized in some way.

As in Chapt. 4, Vol. 1, we shall assume that the temperature inhomogeneities are small in comparison with the mean temperature of the medium $\overline{T} = T_0$,[8] and that the motion of the medium is described by the Boussinesq equations of free convection (Sect. 1.5, Vol. 1). The equations of the Bousinesq approximation differ from the usual fluid dynamic equations for a temperature-homogeneous medium only by the presence (on the right-hand side of the equation for the vertical velocity) of an additional term describing the buoyant acceleration, i.e., $-g\beta T'$, where $T' = T - T_0$ is the temperature fluctuation, $g$ is the gravitational acceleration, and $\beta$ is the thermal expansion coefficient (to be specific, we shall assume that the thermal expansion coefficient is equal to $1/T_0$, which corresponds to the case of an ideal gas). The presence of this additional term leads to two important consequences. First, the vertical direction is a special one, and since the buoyant accelerations appear in motions of all scales, we may suspect that motions of all scales will be anisotropic. Second, to the dimensional parameters characterizing the motion of the fluid, we must now add the buoyancy parameter $g\beta = g/T_0$ (whose dimension is $LT^{-2}\Theta^{-1}$, where $L$, $T$, and $\Theta$ are the dimensions of length, time, and temperature).

The effect of anisotropy, i.e., the dependence of the statistical quantities containing the argument $\boldsymbol{r}$ or $\boldsymbol{k}$ on the angle between $\boldsymbol{r}$ or $\boldsymbol{k}$ and the vertical, can be eliminated by integrating the corresponding statistical quantities with respect to all the possible directions of $\boldsymbol{r}$ or $\boldsymbol{k}$ (i.e., over the sphere $|\boldsymbol{r}| = r$ or $|\boldsymbol{k}| = k$). This will, of course, result in only the mean data, which will not enable us to establish unambiguously the corresponding three-dimensional parameters. Instead of integration over a sphere, we can consider the characteristics of two-dimensional fluid dynamical fields in the fixed horizontal plane $z =$ const, which are definitely isotropic. The conclusions drawn from dimensional considerations, which are applied below to the three-dimensional structure or spectral functions averaged over a sphere, will also be valid for the corresponding characteristics of two-dimensional fields in the $z =$ const plane. In a stratified fluid with large values of Re and Pe, all the turbulence fields themselves can be regarded as locally axially symmetric (i.e., locally homogeneous in all directions, and locally isotropic along horizontal directions). When we investigate their three-dimensional local structure, we can use the results of the general theory of axially symmetric random fields due to Batchelor (1949b, 1953) and Chandrasekhar (1950) [see Bolgiano (1962)]. We shall not, however, consider this approach in our book.

Let us now list the dimensional parameters which affect the small-scale structure of the velocity field $\boldsymbol{u}(\boldsymbol{x}, t)$ and the temperature field $T(\boldsymbol{x}, t)$ in a stratified fluid. The quasi-stationary temperature fluctuations $T'(\boldsymbol{x}\ t)$ in the quasi-equilibrium range are determined, as in a homogeneous fluid, by a constant influx $\overline{N}$ of the mean "fluctuation intensity" $\overline{T'^2}/2$ from the large-scale disturbance region, which is balanced by an equal outflow of mean intensity $\overline{T'^2}/2$ as a result of the "smoothing out" of the field $T(\boldsymbol{x}, t)$ by molecular thermal conduction. This process is characterized by the parameters $\overline{N}$ and $\chi$ (together with the parameters which determine the velocity field producing convective mixing). The time evolution of velocity inhomogeneities is described by the Boussinesq equations, which contain the dimensional parameters $\nu$ and $g/T_0$. However, the "energy flux" transferred from disturbances of given length scale to disturbances of smaller length scale will now no longer be constant along this length scale spectrum. In fact, for a stable temperature stratification we have, in addition to energy transfer from one type of disturbance to another, a loss of kinetic energy through work done against buoyancy forces in a broad range of length scales (which leads to conversion of a proportion of the kinetic

[8] We recall that we have agreed to use the letter $\Theta$ to represent the temperature only when it is regarded as a passive admixture. We shall therefore use the symbol $T$ to denote the temperature in the present subsection.

energy into potential energy). In the case of unstable stratification, turbulent motion of different scales will draw additional kinetic energy from the potential energy of the medium (the buoyancy forces will, on the average, accelerate the fluid particles). It is important to note however that the mutual transformation of potential and kinetic energies in a given range of length scales $l$ is determined by the same spectral components of the fluctuations $u'(x, t)$ and $T'(x, t)$ that are responsible for the transport of the temperature-inhomogeneity measure along the spectrum in this range. We therefore expect that, if the values of Re and Pe are sufficiently high, then for $l \ll L_0$ the statistical source of the above energy conversions will not depend on the quantitative characteristics of the mean fields $\bar{u}(x, t)$ and $\bar{T}(x, t)$ having the scale $L_0$, and will be homogeneous in space and quasi-stationary in time.

However, one would still expect that the presence of stratification may have an effect for scales much less than $L_0$ by producing an anisotropy in the probability distributions, owing to the special role of the direction of the force of gravity. The sign of the vertical mean-temperature gradient may also be important since it influences the character of the mean mutual conversions of kinetic and potential energies. All this does not prevent, however, the range of scales $l \ll L_0$ from becoming a quasi-equilibrium range in the sense in which we have defined this term for turbulence in a nonstratified fluid. The kinetic-energy distribution over the length scale spectrum in this range will now be determined from the balance of inertial transport, the transformation into potential energy (positive or negative), and viscous dissipation. The total energy dissipation $\bar{\varepsilon}$ will then no longer be equal to the energy reaching the upper end of the length scale range under consideration. Nevertheless, it is reasonable to expect that the quantity $\bar{\varepsilon}$ will still influence the energy distribution for scales $l \ll L_0$, i.e., it will be an important energy parameter. In other words, it is probable that *for the turbulence in a stratified fluid with large* Re *and* Pe *there is a quasi-equilibrium range of length scales* $l \ll L_0$ *in which the multidimensional probability distributions for velocity and temperature differences can be regarded as stationary and homogeneous* (but not isotropic, and axially symmetric only relative to the vertical), *and are uniquely determined by the parameters* $\bar{\varepsilon}$, $\bar{N}$, $g/T_0$, $\nu$ *and* $\chi$. This is in fact a generalization of the first Kolmogorov similarity hypothesis to the case of turbulence in a stratified fluid. It was derived independently by Bolgiano (1959) and Obukhov (1959a).

The presence of the additional parameter $g/T_0$ complicates the use of dimensional reasoning in the study of local statistical properties of turbulence in a stratified medium. We shall use the fact that there is only one combination of $\bar{\varepsilon}$, $\bar{N}$, and $g/T_0$ which has the dimensions of length, namely,

$$L_* = \frac{\bar{\varepsilon}^{5/4}}{\bar{N}^{3/4}(g/T_0)^{3/2}} \tag{21.95}$$

(the length scale $L_*$ was also independently introduced by Obukhov and Bolgiano). Hence, it follows that, in a stratified medium, the general form of, for example, the longitudinal velocity structure function $D_{LL}(r)$ averaged over all the directions of $r$ for $|r| = r \ll L_0$ should be of the form

$$D_{LL}(r) = \frac{1}{4\pi r^2} \int_{|r|=r} D_{LL}(r)\, dr = \nu^{1/2}\bar{\varepsilon}^{1/2}\beta_{LL}(r/\eta, \eta/L_*, \nu/\chi), \tag{21.96}$$

where $\eta = \nu^{3/4}\bar{\varepsilon}^{-1/4}$ and $\beta_{LL}(x, y, z)$ is a universal function of the three variables. Analogous formulas are obtained for other structure and spectral functions in the quasi-equilibrium range [some of these are given by Bolgiano (1959)].

Further specification of these formulas can be achieved by using the second Kolmogorov similarity hypothesis according to which *the multidimensional distributions for the velocity and temperature differences at arbitrary pairs of points cannot depend on the molecular constants* $\nu$ *and* $\chi$ *provided only that the distances between the points are much greater than a certain fixed length* $\eta_0$. This hypothesis will also be valid in the case of thermal stratification, and for the same reasons as in the previous cases. In general, the length $\eta_0$ is given by relations of the form $\eta_0 = \eta \cdot \lambda\,(\eta/L_*,\ \nu/\chi)$ where $\lambda(y, z)$ is a function of the two variables. We shall show below, however, that $\eta_0$ will nearly always be independent of $g/T_0$. Consequently, $\lambda = \lambda(\mathrm{Pr})$ is independent of $\eta/L_*$, and the length $\eta_0$ in a stratified medium can be chosen in the same way as in the previous subsection. For the sake of simplicity, we shall suppose henceforth that $\eta/\chi = \mathrm{Pr}$ is of the order of unity (for air $\mathrm{Pr} \approx 0.7$). In that case, $\eta_0$ can be simply identified with $\eta$ (or with the length $\eta_T = \chi^{3/4}\,\bar{\varepsilon}^{-1/4}$ which is of the same order of magnitude).

Consider the longitudinal and lateral velocity structure functions $D_{LL}(r)$ and $D_{NN}(r)$, the temperature structure function $D_{TT}(r)$, and the joint structure function for the temperature and vertical velocity $D_{Tw}(r)$ for $L_0 \gg r \gg \eta_0$. All these functions are assumed to have been averaged with respect to the directions of the vector $r$. In the above range, all these quantities will depend only on $r$, $\bar{\varepsilon}$, $\bar{N}$, and $g/T_0$, and therefore dimensional considerations indicate that

$$D_{LL}(r) = C\,\bar{\varepsilon}^{2/3} r^{2/3} f_{LL}(r/L_*), \qquad D_{NN}(r) = C'\,\bar{\varepsilon}^{2/3} r^{2/3} f_{NN}(r/L_*),$$
$$D_{TT}(r) = C_\vartheta \bar{N}\,\bar{\varepsilon}^{-1/3} r^{2/3} f_{TT}(r/L_*), \quad D_{Tw}(r) = C''\,\bar{N}^{1/2}\,\bar{\varepsilon}^{1/6} r^{2/3} f_{Tw}(r/L_*), \tag{21.97}$$

where $C$, $C'$, $C_\vartheta$, and $C''$ are dimensionless constants, and $f_{LL}(x)$, $f_{NN}(x)$, $f_{TT}(x)$, and $f_{Tw}(x)$ are universal functions [the formula (21.97) for $D_{TT}(r)$ was first established by Obukhov (1959a)]. The numerical coefficients $C$, $C'$, $C_\vartheta$, and $C''$ can be chosen arbitrarily. It will be convenient to assume that the first three of these coincide with the corresponding coefficients in the two-thirds law of Eqs. (21.17) and (21.87). We note that for stable and unstable stratification, i.e., for different signs of $d\bar{T}/dz$, the functions $f_{LL}(x), \ldots, f_{Tw}(x)$ may turn out to be different.

Since the random fields $u(x, t)$ and $T(x, t)$ are locally homogeneous, we can define the spectral densities $F(k) = \frac{1}{2} F_{ii}(k)$, $F_{TT}(k)$ and $F_{Tw}(k)$ for them when $1/\eta_0 \gg k = |k| \gg 1/L_0$. If we write

$$E(k) = \int_{|k|=k} F(k)\,dk$$

and, similarly, for the functions $F_{TT}(k)$ and $F_{Tw}(k)$, then dimensional considerations show that

$$E(k) = C_1 \bar{\varepsilon}^{2/3} k^{-5/3} \psi(kL_*), \qquad E_{TT}(k) = B^{(\vartheta)} \bar{N}\,\bar{\varepsilon}^{-1/3} k^{-5/3} \psi_{TT}(kL_*),$$
$$E_{Tw}(k) = B'\,\bar{N}^{1/2}\,\bar{\varepsilon}^{1/6}\,k^{-5/3} \psi_{Tw}(kL_*), \tag{21.98}$$

where $\psi(\xi)$, $\psi_{TT}(\xi)$ and $\psi_{Tw}(\xi)$ are universal functions (which also can be different for stable and unstable stratification), $B'$ is an arbitrary constant, and the constants $C_1$ and $B^{(\vartheta)}$

can be conveniently regarded as equal to the coefficients in the five-thirds laws given by Eqs. (21.24) and (21.89).

When $g/T_0 = 0$, i.e., in the absence of gravitational forces which give rise to stratification, the formulas (21.97) and (21.98) should become identical with the usual formulas for the structure and spectral functions of locally isotropic turbulence in a nonstratified fluid. However, because of Eq. (21.95), $L_* \to \infty$ when $g/T \to 0$. Consequently, with the above choice of numerical coefficients,

$$f_{LL}(0) = f_{NN}(0) = \psi(\infty) = \psi_{TT}(\infty) = 1; \; f_{Tw}(0) = \psi_{Tw}(\infty) = 0$$

where the last two equations follow from the fact that in a locally isotropic turbulence $\overline{\Delta_r u \, \Delta_r T} = 0$. If, on the other hand, $g/T_0 \neq 0$ but $r/L_* \ll 1$, i.e., $r \ll L_*$ (but $r \gg \eta_0$), then the values of the correction functions in Eq. (21.97) can be replaced with good accuracy by their values at zero, i.e., we can use the ordinary two-thirds laws which are valid for nonstratified media. Similarly, when $k \gg 1/L_*$ (but $k \ll 1/\eta_0$), the values of the correction functions in Eq. (21.98) can be approximately replaced by their values at infinity, i.e., we can use the usual five-thirds laws. In other words, *the length scale $L_*$ characterizes the minimum length scale of inhomogeneities beyond which the effect of the stratification is appreciable.*[9] When $L_* \gg \eta$ and $L_* \gg \eta_T$ (and this is nearly always the case), we can ignore stratification for $l \ll L_*$ and use the usual form of the first and second Kolmogorov similarity hypotheses (and, in particular, we can determine the length scale $\eta_0$ in the usual way). The above analysis provides additional support for the use of the concept of locally isotropic turbulence and the associated similarity hypotheses for stratified media, but shows that the upper limit of the inertial and inertial-convective length scale subranges must satisfy both the inequalities $l \ll L$ and $l \ll L_T$ as well as $l \ll L_*$. If $L_* \ll L_0 = \min(L, L_T)$ then the inertial-convective subrange will be followed by the buoyancy subrange $L_0 \gg l \gtrsim L_*$ in which the probability distributions for $\Delta_r u$ and $\Delta_r T$ can still be regarded as quasi-stationary and homogeneous but no longer isotropic.

The values of the dimensionless correction functions in Eqs. (21.97) and (21.98) can, in principle, be determined empirically, but the necessary experimental data are not available at present. In the case of stable stratification Bolgiano (1959, 1962) has put forward certain hypotheses with regard to the asymptotic form of these functions for wave numbers much less than $1/L_*$, i.e., length scales much greater than $L_*$. For a stable stratification the energy transferred from disturbances of length scale $l \gg L_*$ to smaller-scale disturbances should be much greater than $\bar{\varepsilon}$, since most of this energy is spent in work against the buoyancy forces, and only a very small fraction of it reaches the small-scale disturbances in which viscous dissipation is concentrated. On this basis, one would expect that even a considerable change in $\bar{\varepsilon}$ will have very little effect on the shape of turbulence spectra in the region $k \ll 1/L_*$. This has lead Bolgiano to propose that the asymptotic form of the spectra $E(k)$, $E_{TT}(k)$ and $E_{Tw}(k)$ for $k \ll 1/L_*$ in the case of stable stratification should

[9] The minimum length scale for which thermal stratification must still be taken into account is, of course, equal to $L_*$, but only to within a constant factor. This factor may even turn out to be appreciably less than unity [compare this with the analogous situation encountered in Chapt. 4 of Vol. 1 where we introduced the length scale $L$ which characterized the thickness of the layers in which we could neglect the influence of thermal stratification on the mean profiles $\bar{u}(z)$ and $\bar{T}(z)$].

depend only on the values of the parameters $\overline{N}$ and $g/T_0$. From this, dimensional considerations yield

$$E(k) = c_1 \overline{N}^{2/5}\left(\frac{g}{T_0}\right)^{4/5} k^{-11/5}, \qquad E_{TT}(k) = b^{(\vartheta)} \overline{N}^{4/5}\left(\frac{g}{T_0}\right)^{-2/5} k^{-7/5},$$
$$E_{Tw}(k) = b' \overline{N}^{3/5}\left(\frac{g}{T_0}\right)^{1/5} k^{-9/5}, \tag{21.99}$$

for $c_1$, $b^{(\vartheta)}$ and $b'$ are universal functions. Consequently, one would expect that in the case of stable stratification $\psi(\xi) \sim \xi^{-8/15}$, $\psi_{TT}(\xi) \sim \xi^{4/15}$ and $\psi_{Tw}(\xi) \sim \xi^{-2/15}$ for $\xi \ll 1$. Similar considerations lead to the following hypotheses [which are equivalent to Eq. (21.99)] about the asymptotic form of the structure functions averaged over all directions for turbulence in a stably stratified medium:

$$D_{LL}(r) = c\,\overline{N}^{2/5}\left(\frac{g}{T_0}\right)^{4/5} r^{6/5}, \qquad D_{NN}(r) = c'\,\overline{N}^{2/5}\left(\frac{g}{T_0}\right)^{4/5} r^{6/5},$$
$$D_{TT}(r) = c_\theta\,\overline{N}^{4/5}\left(\frac{g}{T_0}\right)^{-2/5} r^{2/5}, \qquad D_{Tw}(r) = c''\,\overline{N}^{3/5}\left(\frac{g}{T_0}\right)^{1/5} r^{4/5} \tag{21.100}$$

when $r \gg L_*$ [so that $f_{LL}(x) \sim x^{8/15}$, $f_{NN}(x) \sim x^{8/15}$, $f_{TT}(x) \sim x^{-4/15}$, $f_{Tw}(x) \sim x^{2/15}$ for $x \gg 1$; see Teptin (1965)].

The above formulas are meaningful only if the length scale $L_*$ is much less than the external turbulence length scale $L_0$. If, on the other hand, $L_*$ approaches or even exceeds $L_0$, then not only the Bolgiano hypothesis about the asymptotic form of the correction functions in Eqs. (21.97) and (21.98), but the entire similarity theory developed in this subsection will be invalid. In this last case, the inertial-convective subrange of the spectrum will be followed by a subrange of scales in which quantities characterizing the mean fields $\overline{u}(x, t)$ and $\overline{T}(x, t)$ will play an important role. This refers in particular to the hypothesis put forward by Shur (1962) and Lumley (1964) about the turbulence spectrum in a free atmosphere according to which the region of validity of the five-thirds laws is followed in a stably stratified medium by a region of smaller wave numbers where the velocity spectrum is proportional to $k^{-3}$ and is determined by the parameter $g/T_0$ and $d\overline{T}(z)/dz$, in accordance with the formula $E(k) \sim \frac{g}{T_0}\frac{d\overline{T}}{dz}k^{-3}$.[10]

The length scale $L_*$ cannot be precisely defined for most real turbulent flows since the values of $\overline{\varepsilon}$ and $\overline{N}$ are unknown. A very rough estimate of the order of magnitude of the length scale for turbulence in a stratified medium occupying the half space $z > 0$ can be obtained from the relations $\overline{\varepsilon} \sim (\Delta\overline{u})^3/L$ and $\overline{N} \sim \frac{\Delta\overline{u}\,(\Delta\overline{T})^2}{L_T}$ (see Sects. 21.3 and 21.6). Typical length scales $L$ and $L_T$ of the fields $\overline{u} = \overline{u}(z)$ and $\overline{T} = \overline{T}(z)$ in the present case are of the same order of magnitude as the distance $z$ to the wall. Therefore,

$$L_* \sim \frac{(\Delta\overline{u})^3}{(\Delta\overline{T})^{3/2}(g/T_0)^{3/2} z^{1/2}}.$$

[10] We note that $T$ represents not the usual but the potential temperature (see the end of the Sect. 2.4 of Vol. 1). Since in the free atmosphere (at the height of 1-2 km and right up to about 10 km) the potential temperature increases with height, it follows that $d\overline{T}(z)/dz > 0$.

Near the wall, the most rapidly varying factor on the right-hand side is $z^{-1/2}$. Consequently, in the first approximation, we can assume that $L_* \sim z^{-1/2}$. We thus see that the length scale $L_*$ decreases with height quite rapidly, whereas the external length scale $L \sim z$ increases rapidly with height. Therefore, one would expect that at sufficiently great heights, the length scale $L_*$ will be appreciably less than the external length scale of the turbulence, and hence the effect of the thermal stratification will begin to appear earlier than the effect of the mean-flow parameters. In the case of the atmospheric layer near the earth's surface, very approximate numerical estimates due to Obukhov (1959a) show that the ratio $z/L_*$ becomes equal to unity even at heights of the order of 10 m. It is therefore probable that the formulas given above will be valid beginning with heights of the order of 100 m. The verification of this conclusion will require special observations which one hopes will be carried out in the future.

## *Statistical Characteristics of the Concentration Field for a Decaying or a Chemically Active Admixture*

Considerations similar to those given above are also valid for a number of other cases in which in addition to inertial transport of energy and the inhomogeneity measure of the field $\vartheta$ or $T$ along the length scale spectrum, there are certain other effects. A useful illustration is the case of turbulent mixing of an admixture which decays (i.e., is radioactive) or participates in some chemical reaction (which does not affect the flow dynamics). We shall restrict our attention to the case of decay or first-order reactions for which the reaction rate is proportional to the first power of the concentration. According to Sect. 14.5 the diffusion equation for the admixture concentration $\vartheta(x, t)$ is then replaced by the equation

$$\frac{\partial \vartheta}{\partial t} + u_j \frac{\partial \vartheta}{\partial x_j} = \chi \Delta \vartheta - \mu \vartheta, \tag{21.101}$$

where $\mu$ is a constant whose dimensions are reciprocal to the dimension of time. Therefore, the number of dimensional parameters which determine the statistical properties of the field $\vartheta(x, t)$ is greater than for nonreacting mixtures, owing to the addition of the parameter $\mu$.

Since it is assumed that the chemical reaction has no effect on flow dynamics, the statistical properties of the velocity and pressure fields will be the same as in the absence of the admixture. Therefore, all the results of Sects. 21.4 and 21.5 will remain valid in the present case without any modification. On the other hand, the admixture concentration $\vartheta(x, t)$ will decrease exponentially in the course of the chemical reaction, and will fall by a factor of $e$ in a time $T_r = 1/\mu$. It is clear that the reduction in the concentration will be reflected in the fluctuations $\vartheta'$ of all scales. As a result, the transport of the "inhomogeneity measure" $\overline{\vartheta'^2}/2$ along the length scale spectrum of fluctuations of the field $\vartheta$ will no longer be constant even when molecular diffusion is negligible.

Inhomogeneities in the field $\vartheta(x, t)$ of a given length scale $l \ll L_0$ are produced by eddies with similar scales, and are therefore characterized by the same typical time scale $T_l$ as the corresponding inhomogeneities in the field $u(x, t)$. Since $T_l$ decreases monotonically with increasing $l$, it is possible for sufficiently large Re and Pe to isolate a region of the smallest-scale disturbances of the field $\vartheta(x, t)$ for which $L \ll L_0$ and $T_l \ll T_r$. For such disturbances, the reduction in the concentration as a result of chemical reactions will be very slow in comparison with convective mixing and molecular diffusion. We shall suppose that the length scale $l$ for which the typical time scale $T_l$ is equal to $T_r$ falls in the inertial subrange (the necessary condition for this is that Re must be sufficiently large). This length scale must then be uniquely determined by the parameters $\mu = 1/T_r$ and $\bar{\varepsilon}$, i.e., it can differ only by a constant factor from the length scale

$$L_r = \bar{\varepsilon}^{1/2}\mu^{-3/2}, \tag{21.102}$$

which was introduced by Corrsin (1961).[11] Hence, it is clear that the convective subrange of length scales for the field $\vartheta(x, t)$ in the presence of radioactive decay or first-order chemical reactions must satisfy the condition $l \ll L_0$ and also the equally important condition $l \ll L_r$. All the discussions and conclusions of Sect. 21.6 will then be valid for the concentration fluctuations. The presence of decay or chemical reactions will then result only in an additional and relatively slow change in the parameter $\bar{N}$.

We thus see that *the length scale $L_r$ determines the minimum scale of inhomogeneities beyond which decay or chemical reactions begin to have an appreciable effect on the probability distributions for the concentration differences.* We shall suppose that there are admixture sources which produce a stationary distribution of the mean concentration $\bar{\vartheta}(x)$ (or, at any rate, ensure that the field $\bar{\vartheta}$ varies appreciably only in time intervals that are very large in comparison with $T_r$). Let $G$ be a space-time region with length scales that are small in comparison with $L_0$ and time scales that are small in comparison with $T_0 = \min(L_0/U,\ T_u,\ T_\vartheta)$ where $T_u$ is the time for an appreciable change in the field $\bar{u}(x, t)$ and $T_\vartheta$ is the time for an appreciable change in the field $\bar{\vartheta}(x, t)$. It is natural to expect that in this region *the field $\vartheta(x, t)$ will be locally isotropic and will have stationary increments, and that the probability distributions for the differences of this field will be uniquely determined by the parameters $\bar{\varepsilon}$, $\bar{N}$, $\mu$, $\nu$, and $\chi$.* Moreover, *when $L_0 \gg \eta_0 = \max(\eta, \eta_\vartheta)$ the probability distributions for space differences in the field $\vartheta(x, t)$ at pairs of points in $G$ will be determined only by the parameters $\bar{\varepsilon}$, $\bar{N}$, and $\mu$, provided only that the distances between these points are much greater than $\eta_0$.* These statements are a generalization of the Kolmogorov similarity hypotheses to the case of turbulent mixing of a decaying admixture, or a chemically active admixture participating in first-order reactions.

According to the above hypotheses, in the case of a decaying or chemically active admixture

$$D_{\vartheta\vartheta}(r) = C_\vartheta \bar{N}\bar{\varepsilon}^{-1/3} r^{2/3} f\left(\frac{r}{L_r}\right) \quad \text{when} \quad L_0 \gg r \gg \eta_0 \tag{21.103}$$

and

$$E_{\vartheta\vartheta}(k) = B^{(\vartheta)}\bar{N}\bar{\varepsilon}^{-1/3} k^{-5/3}\psi(kL_r) \quad \text{when} \quad 1/\eta_0 \gg k \gg 1/L_0, \tag{21.104}$$

where $C_\vartheta$ and $B^{(\vartheta)}$ are constants, and $f(x)$ and $\psi(\xi)$ are universal functions. If we choose the constants $C_\vartheta$ and $B^{(\vartheta)}$ to be the same as in Eqs. (21.87) and (21.89), then $f(0) = \psi(\infty) = 1$.

The formulas (21.103) and (21.104) are similar to those which were obtained earlier for turbulence in a stratified medium. A similar discussion can be carried out for other cases, for example, the case of several admixtures connected by one or more first-order chemical reactions [Corrsin (1962b) and Pao (1964)]. We shall not pause, however, to consider these aspects in detail.

[11] This formula is consistent with the fact that, in the inertial subrange, $T_l \sim \bar{\varepsilon}^{-1/3} l^{2/3}$ (see Sect. 21.2).

# 22. DYNAMIC THEORY OF THE LOCAL STRUCTURE OF DEVELOPED TURBULENCE

## 22.1 Equations for the Structure and Spectral Functions of Velocity and Temperature

In the previous section, the statistical properties of locally isotropic turbulence were investigated by similarity and dimensional methods which do not require the introduction of the explicit form of the fluid dynamic equations. To some extent, this was justified by the fact that the fluid dynamic equations cannot be used to obtain a closed set of equations for any finite set of statistical turbulence characteristics, and therefore these equations must be augmented by some special assumptions which we have tried to avoid. It does not follow, however, that the dynamic equations are, in general, useless for studying local structures. These equations do not uniquely define the local statistical properties, but they do lead to a number of important relations among them, which provide an important extension of the predictions based on similarity and dimensionality, and are of considerable interest.

When we develop the dynamic theory of locally isotropic turbulence we must first transform the dynamic equations for the moments of the main flow variables to a form containing only the local characteristics. This is not at all an easy task because the equations for the moments are rather complicated. It is therefore best to start with the following heuristic approach. We shall use the fact that the statistical state of small-scale turbulence components for large Re is independent of the macro-structural properties of the flow which affect only the parameter $\bar{\varepsilon}$. Hence it follows that the dynamic equations for the parameters of locally isotropic turbulence cannot depend on the nature of the large-scale turbulent motions. It is therefore sufficient to derive these equations for only one turbulent flow with sufficiently large Re, so that we can then confine our attention to the simplest case of isotropic turbulence in infinite space. Having found for this case the relation between the local characteristics, and having taken into account the fact that by the Kolmogorov similarity hypotheses the above characteristics must be the same for all turbulent flows with sufficiently large Re and the same values of $\bar{\varepsilon}$ and $\nu$, we can regard the resulting relations as universal, i.e., equally valid for any locally isotropic turbulence. It will then, of course, be interesting to try to derive these relations for the general case (i.e., without assuming that the turbulence is isotropic). This more general derivation will be considered at the end of the present subsection.

Let us begin with the simple equation for velocity structure functions. We shall use the fact that for isotropic turbulence the corresponding second- and third-order correlation functions $B_{LL}(r, t)$ and $B_{LL,L}(r, t)$ must satisfy the von Kármán-Howarth equation (14.9). However, an isotropic random field $\boldsymbol{u}(\boldsymbol{x}, t)$ is also locally isotropic and its structure functions, in this case, are determined by the formulas

$$D_{LL}(r, t) = 2[B_{LL}(0, t) - B_{LL}(r, t)] \text{ and } D_{LLL}(r, t) = 6B_{LL,L}(r, t)$$

(see Sect. 13.3). Let us assume that the Reynolds number for a particular isotropic turbulence under investigation is so large that its small-scale components with length scales $r \ll L$, where $L = \int_0^\infty B_{LL}(r)\,dr/B_{LL}(0)$, form locally isotropic turbulence in the sense of the definition given in Sect. 21.2. The functions $D_{LL} = D_{LL}(r)$ and $D_{LLL} = D_{LLL}(r)$ for $r \ll L$ can then be regarded as not being explicit functions of the time $t$ (and uniquely determined by the parameters $\bar{\varepsilon}$ and $\nu$). Let us express the correlation functions in the von Kármán-Howarth equation in terms of the structure functions and the constant $B_{LL}(0, t)$, and use the fact that $\frac{\partial D_{LL}}{\partial t} = 0$ for $r \ll L$, and that for isotropic turbulence $\frac{\partial B_{LL}(0, t)}{\partial t} = \frac{2}{3}\frac{d}{dt}\frac{\overline{u_i u_i}}{2} = -\frac{2}{3}\bar{\varepsilon}$. In that case, we have

$$-\frac{2}{3}\bar{\varepsilon} = \frac{1}{6}\left[\frac{dD_{LLL}(r)}{dr} + \frac{4D_{LLL}(r)}{r}\right] - \nu\left[\frac{d^2D_{LL}(r)}{dr^2} + \frac{4}{r}\frac{dD_{LL}(r)}{dr}\right] \quad (22.1)$$

which is valid for $r \ll L$. If we multiply this equation by $r^4$ and integrate with resepct to $r$, we obtain the velocity structure equation

$$D_{LLL}(r) - 6\nu\frac{dD_{LL}(r)}{dr} = -\frac{4}{5}\bar{\varepsilon}r. \quad (22.2)$$

This equation was first found by Kolmogorov (1941d); it relates the second- and third-order longitudinal structure functions of locally isotropic turbulence, corresponding to isotropic turbulence with sufficiently large Reynolds number. If the Kolmogorov similarity hypotheses are valid, then the structure equation (22.2) must also be satisfied for any locally isotropic turbulence, independently of whether or not the turbulence as a whole is isotropic.

In view of the continuity equation, the functions $D_{LL}(r)$ and $D_{LLL}(r)$ uniquely determine the structure tensors $D_{ij}(r)$ and $D_{ijl}(r)$. In particular, the functions $D_{NN}(r)$ and $D_{LNN}(r)$ can be expressed in terms of them in a simple way [see Eqs. (13.87), (13.90), and (13.91)]. The functions $D_{LL}(r)$ and $D_{LLL}(r)$ themselves cannot, of course, be unambiguously defined by means of the dynamic equation (22.2) alone. Nevertheless, this equation leads to a number of important consequences regarding these functions. Thus, for example, since the function $D_{LLL}(r)$ is of the third order in $r$ in the neighborhood of the point $r=0$, for very small $r$ (and, in particular, for $r \ll \eta$) it will be negligible in comparison with the remaining terms of Eq. (22.2). Consequently, $D_{LL}(r)=\frac{1}{15}\frac{\bar{\varepsilon}}{\nu}r^2$ for $r \ll \eta$. In view of the von Kármán relation (13.87) it also follows that $D_{NN}(r)=\frac{2}{15}\frac{\bar{\varepsilon}}{\nu}r^2$ for $r \ll \eta$. These relations have already been given above [see Eq. (21.19′)]. In the other limiting case, i.e., when $r \gg \eta$, which corresponds to the inertial subrange, viscous friction should not play an appreciable role, i.e., the term $6\nu\frac{dD_{LL}}{dr}$ should be negligible (this is also clear from the fact that $D_{LLL}(r) \sim r$, $D_{LL}(r) \sim r^{2/3}$ when $r \gg \eta$). Consequently,

$$D_{LLL}(r)=-\frac{4}{5}\bar{\varepsilon}r \text{ when } r \gg \eta. \tag{22.3}$$

This result is an important generalization of Eq. (21.18′) which follows from dimensional considerations. It shows that $D=-4/5$. At intermediate values of $r$, i.e., $r \sim \eta$, the universal functions $\beta_{LL}(x)$ and $\beta_{LLL}(x)$ in Eqs. (21.17) and (21.18) can be related by

$$\beta_{LLL}(x)-6\frac{d\beta_{LL}(x)}{dx}=-\frac{4}{5}x, \tag{22.4}$$

which enables us to determine one of these functions if the other is known.

According to Eq. (22.3), the probability distribution for $\Delta_r u_L = u_L(x+r) - u_L(x)$ should have a negative skewness in the inertial subrange:

$$S = \frac{D_{LLL}(r)}{[D_{LL}(r)]^{3/2}} < 0. \tag{22.5}$$

It can be shown with the aid of the formula relating $D_{LLL}(r)$ with the energy transport $T(k)$ along the spectrum that this property is closely related to the fact that fluctuations of a given scale should, on the average, transfer energy to fluctuations of smaller scales, and take energy from fluctuations of larger scales. Since $\overline{\Delta_r u_L} = 0$, the negative skewness indicates that the negative values of $\Delta_r u_L$ should, on the average, be encountered less frequently than positive values, but should exceed them in absolute magnitude. If we convert from variations in $\boldsymbol{u}(\boldsymbol{x})$ along a straight line $\mathscr{L}$ parallel to $\bar{\boldsymbol{u}}$ to variations in the values of $\boldsymbol{u}(t) = \boldsymbol{u}(\boldsymbol{x}_0, t)$ by using the frozen turbulence hypothesis, we may conclude that short invervals of rapid increase in the flow velocity at a fixed point should alternate with longer intervals of gradual decrease of the velocity.

The skewness $S = D_{LLL}(r)\,[D_{LL}(r)]^{-3/2}$ is constant for $r \ll \eta$ and $r \gg \eta$, while for $r \sim \eta$ it is a universal function of $x = r/\eta$. We shall denote this function by $S(x)$. The constant value of this skewness in the inertial subrange $r \gg \eta$ will also be denoted by the symbol $S$ for simplicity. If we then use Eqs. (22.3), (22.5), and the von Kármán relation which connects $D_{NN}(r)$ with $D_{LL}(r)$ we obtain for $r \gg \eta$

$$D_{LL}(r) = \left(-\frac{4}{5S}\right)^{2/3} (\bar{\varepsilon} r)^{2/3}, \quad D_{NN}(r) = \frac{4}{3}\left(-\frac{4}{5S}\right)^{2/3} (\bar{\varepsilon} r)^{2/3}$$

i.e.,

$$C = \left(-\frac{4}{5S}\right)^{2/3}, \quad C' = \frac{4}{3}\left(-\frac{4}{5S}\right)^{2/3}. \tag{22.6}$$

These formulas, which are also due to Kolmogorov (1941d), show the statistical significance of the coefficients $C$ and $C'$ in Eq. (21.17′).

If we use the formulas relating $D_{LL}(r)$ with $E(k)$ and $D_{LLL}(r)$ with $T(k)$ we can transform from the structure equation (22.2) to the spectral equation containing the unknown functions $E(k)$ and $T(k)$. It is simpler, however, to assume in this case also that the turbulence is completely isotropic and use the spectral form of the von Kármán-Howarth equation given in Sect. 14.3. The consequences of this equation with regard to the spectral characteristics in the range $k \gg 1/L$ should be valid for any locally isotropic turbulence because of the similarity hypotheses. The most important of these consequences has already been discussed in Sect. 16.5. It is of the form

$$W(k) = \int_k^\infty T(k')\,dk' = 2\nu \int_k^\infty k'^2 E(k')\,dk' \qquad (22.7)$$

or, in other words,

$$-\frac{dW(k)}{dk} = 2\nu k^2 E(k)$$

[see Eq. (16.39)]. The equation (22.7), where $k \gg 1/L$ and $\int_k^\infty k'^2 E(k')\,dk' = \bar{\varepsilon}/2\nu$ for $k \ll 1/\eta$, is in fact the spectral form of the Kolmogorov equation (22.2). In the inertial subrange $1/L \ll k \ll 1/\eta$ it assumes a particularly simple form, namely,

$$W(k) = \int_k^\infty T(k)\,dk = \bar{\varepsilon} = \text{const}; \qquad T(k) = 0. \qquad (22.7')$$

Similar equations can be obtained for the local properties of the temperature field $\vartheta(\boldsymbol{x}, t)$ (or the concentration of a passive admixture) for arbitrary turbulence with sufficiently large Reynolds and Peclet numbers. All that is necessary is to assume at the outset that the turbulence under consideration is isotropic, and then use the Corrsin equation (14.59). Having expressed in this equation the correlation functions $B_{\vartheta\vartheta}$ and $B_{L\vartheta,\vartheta}$ in terms of the structure functions $D_{\vartheta\vartheta}$ and $D_{L\vartheta\vartheta}$ and the constant $B_{\vartheta\vartheta}(0)$ with the aid of the relation $D_{\vartheta\vartheta}(r) = 2[B_{\vartheta\vartheta}(0) - B_{\vartheta\vartheta}(r)]$, $D_{L\vartheta\vartheta}(r) = 4B_{L\vartheta\vartheta}(r)$, and bearing in mind that $\frac{\partial}{\partial t} D_{\vartheta\vartheta}(r) = 0$ for $r \ll L$ and sufficiently large Re and Pe, and that $\frac{\partial}{\partial t} B_{\vartheta\vartheta}(0) = -2\bar{N}$, we obtain

$$-2\overline{N}=\frac{1}{2}\left[\frac{dD_{L\vartheta\vartheta}(r)}{dr}+\frac{2D_{L\vartheta\vartheta}(r)}{r}\right]-\chi\left[\frac{d^2D_{\vartheta\vartheta}(r)}{dr^2}+\frac{2}{r}\frac{dD_{\vartheta\vartheta}(r)}{dr}\right]. \tag{22.8}$$

If we multiply this equation by $r^3$, and integrate it once with respect to $r$, we obtain the temperature structure equation

$$D_{L\vartheta\vartheta}(r)-2\chi\frac{dD_{\vartheta\vartheta}(r)}{dr}=-\frac{4}{3}\overline{N}r, \tag{22.9}$$

which is analogous to the Kolmogorov equation (22.2). By the similarity hypotheses of Sect. 21.6, this equation should be satisfied for any turbulence with sufficiently large Re and Pe for $r\ll L$. Equation (22.9) together with its consequences given by Eqs. (22.10) and (22.12) below was first established by Yaglom (1949b).

When $r\ll\eta$ we can neglect the term $D_{L\vartheta\vartheta}(r)$ in Eq. (22.9), so that it leads to $D_{\vartheta\vartheta}(r)\approx\frac{\overline{N}}{3\chi}r^2$ which has already been noted in Sect. 21.6. In the inertial range $r\gg\eta$, the term $2\chi\frac{dD_{\vartheta\vartheta}(r)}{dr}$ turns out to be negligible in Eq. (22.9), since it describes the effect of molecular thermal conduction. Consequently,

$$D_{L\vartheta\vartheta}(r)=-\frac{4}{3}\overline{N}r \quad \text{when } r\gg\eta. \tag{22.10}$$

Therefore, we have $D_1=-4/3$ in Eq. (21.92). Let us now introduce the dimensionless quantity

$$F=\frac{D_{L\vartheta\vartheta}(r)}{D_{\vartheta\vartheta}(r)\,[D_{LL}(r)]^{1/2}}, \tag{22.11}$$

which has a clear statistical interpretation, and assumes a constant negative value in the inertial subrange $r\gg\eta$. In view of Eqs. (22.10) and (22.6), we have for $r\gg\eta$

$$D_{\vartheta\vartheta}(r)=-\frac{4}{3F}\left(-\frac{5S}{4}\right)^{1/3}\overline{N}\,\overline{\varepsilon}^{-1/3}r^{2/3}, \quad C_\vartheta=-\frac{4}{3F}\left(-\frac{5S}{4}\right)^{1/3}. \tag{22.12}$$

The spectral equation equivalent to Eq. (22.9) can be written in the form

$$W_\vartheta(k) = \int_k^\infty T_{\vartheta\vartheta}(k')\,dk' = 2\chi \int_k^\infty k'^2 E_{\vartheta\vartheta}(k')\,dk' \quad \text{when} \quad k \gg 1/L,$$
$$W_\vartheta(k) = 2\overline{N} = \text{const}, \quad T_{\vartheta\vartheta}(k) = 0 \qquad \text{when} \quad k \ll 1/\eta. \tag{22.13}$$

## *Derivation of the Structure Equations without the Isotropic Assumption*

To derive the Kolmogorov structure equation (22.2) without any special assumption about the isotropy of the large-scale structure of the flow, we must transform the Navier-Stokes equations so that they contain only the velocity differences and their derivatives. With this aim in view, let us consider both a fixed set of coordinates $\mathscr{S}$ and a set $\mathscr{S}^*$ whose origin moves together with a given fluid particle. Let $x_i$ and $u_i = u_i(x, t)$ represent the coordinates and the velocity components in the fixed system $\mathscr{S}$, and let $x_{0i} = x_{0i}(t)$ and $u_{0i} = u_i[\boldsymbol{x}_0(t), t]$ represent the corresponding quantities of the origin of the moving frame $\mathscr{S}^*$. We then have

$$r_i = x_i - x_{0i}(t)$$

$$v_i = v_i(\boldsymbol{r}, t) = u_i[\boldsymbol{x}_0(t) + \boldsymbol{r}, t] - u_i[\boldsymbol{x}_0(t), t].$$

where $r_i$ and $v_i$ are the coordinates and velocity components in the frame $\mathscr{S}^*$. The components of the relative acceleration of the fluid particle, i.e., the acceleration in the frame $\mathscr{S}^*$, can now be written in the form

$$w_i^* = \frac{d}{dt} v_i(\boldsymbol{r}, t) = \frac{\partial v_i}{\partial t} + v_k \frac{\partial v_i}{\partial r_k}.$$

Therefore, the difference between the Navier-Stokes equations at the points $\boldsymbol{x}$ and $\boldsymbol{x}_0$ assumes the form

$$\frac{\partial v_i}{\partial t} + v_k \frac{\partial v_i}{\partial r_k} = -\frac{1}{\rho}\frac{\partial p}{\partial x_i} + \frac{1}{\rho}\frac{\partial p_0}{\partial x_{0i}} + \nu(\Delta + \Delta_0)(u_i - u_{0i}), \tag{22.14}$$

where the left-hand side is referred to $\mathscr{S}^*$ and the right-hand side to $\mathscr{S}$. In this equation $p_0$ represents $p(\boldsymbol{x}_0, t)$, and $\Delta$ and $\Delta_0$ are the Laplace operators in the variables $x_i$ and $x_{0i}$ (we have taken into account the fact that $u_i$ depends only on $\boldsymbol{x}$, and $u_{0i}$ only on $\boldsymbol{x}_0$).

Let us multiply both sides of Eq. (22.14) by $v_j = u_j - u_{0j}$, and both sides of the analogous equation for the component $v_j$ by $v_i = u_i - u_{0i}$. Let us then add the two equations, and take the average of the sum over the ensemble of flows corresponding to the fixed value $\boldsymbol{u}_0 = \boldsymbol{u}(\boldsymbol{x}_0, t)$ (where $\boldsymbol{x}_0$ and $t$ are a given point and a given time respectively). Finally, let us assume that $r \ll L$, and that the Reynolds number for the flow is so large that the statistical properties of the random vector $\boldsymbol{v}(\boldsymbol{r}, t)$ satisfy the Kolmogorov similarity hypotheses (i.e., they do not depend on the value of $\boldsymbol{u}_0$). The terms containing the time derivatives in this sum will then yield the expression $\frac{\partial \overline{v_i v_j}}{\partial t} = \frac{\partial D_{ij}}{\partial t}$ which vanishes because, according to the similarity hypotheses, the statistical characteristics of the field $\boldsymbol{v}$ are independent of time $t$. The nonlinear terms on the left-hand side of Eq. (22.14) and the equation for $v_j$ yield

$$\overline{v_j v_k \frac{\partial v_l}{\partial r_k}} + \overline{v_l v_k \frac{\partial v_j}{\partial r_k}} = \frac{\partial \overline{v_l v_j v_k}}{\partial r_k} = \frac{\partial D_{ljk}}{\partial r_k}.$$

in view of the continuity equation $\frac{\partial v_k}{\partial r_k} = 0$. The term $\frac{1}{\rho}\overline{(u_j - u_{j0})\frac{\partial p}{\partial x_l}}$ and all the other terms containing pressure will vanish because, as was shown in Sect. 13.3, the velocity differences are uncorrelated with differences of any scalar in locally isotropic turbulence. Finally, terms containing the viscosity $\nu$ will provide a nonzero contribution to the above relation. We shall discuss this contribution in greater detail below.

Let us divide the terms containing $\nu$ into two groups, the first of which is of the form

$$\nu\,\overline{(u_j - u_{0j})\,\Delta\,(u_l - u_{0l})} + \nu\,\overline{(u_l - u_{0l})\,\Delta\,(u_j - u_{0j})},$$

and the second differs from this expression only in that the operator $\Delta$ must be replaced by $\Delta_0$. If we use the identity $a\,\Delta b + b\,\Delta a = \Delta\,ab - 2\frac{\partial a}{\partial x_k}\frac{\partial b}{\partial x_k}$, we can transform the first group of terms so that it reads

$$\nu\,\Delta\,\overline{(u_l - u_{0l})(u_j - u_{0j})} - 2\nu\,\overline{\frac{\partial\,(u_l - u_{0l})}{\partial x_k}\frac{\partial\,(u_j - u_{0j})}{\partial x_k}} = \nu\,\Delta D_{lj} - 2\nu\,\overline{\frac{\partial u_l}{\partial x_k}\frac{\partial u_j}{\partial x_k}}.$$

The quantity $a_{lj} = 2\nu\,\overline{\frac{\partial u_l}{\partial x_k}\frac{\partial u_j}{\partial x_k}}$ must be equal to $a\delta_{lj}$, because of local homogeneity and isotropy, where $a = \frac{1}{3}a_{ll} = \frac{2}{3}\nu\sum_{l,\,k}\overline{\left(\frac{\partial u_l}{\partial x_k}\right)^2}$. Since

$$\bar{\varepsilon} = \frac{\nu}{2}\sum_{l,\,k}\overline{\left(\frac{\partial u_l}{\partial x_k} + \frac{\partial u_k}{\partial x_l}\right)^2} = \nu\sum_{l,\,k}\overline{\left(\frac{\partial u_l}{\partial x_k}\right)^2}\left(\text{since } 2\,\overline{\frac{\partial u_l}{\partial x_k}\frac{\partial u_k}{\partial x_l}} = 2\,\frac{\partial^2 \overline{u_l u_k}}{\partial x_l\,\partial x_k} = 0\right),$$

in view of the equation, it follows that $a = \frac{2}{3}\bar{\varepsilon}$. The second group of terms which depend on $\nu$ can be transformed in precisely the same way and, of course, turns out to be equal to the first group. Therefore, our final conclusion is the following equation which relates the tensors $D_{lj}$ and $D_{ljk}$:

$$\frac{\partial D_{ljk}}{\partial r_k} = 2\nu\,\Delta D_{lj} - \frac{4}{3}\bar{\varepsilon}\delta_{lj}. \tag{22.15}$$

We note that, in the isotropic case, the equation (22.15) can be obtained from Eq. (14.7) in the same way as Eq. (22.2) is obtained from Eq. (14.9). Hence it is clear that the Eqs. (22.15) and (22.2) must be equivalent to each other. To prove this rigorously we note that the tensor on the right-hand side of Eq. (22.15) and, consequently, on the left-hand side also, is symmetric and solenoidal in both its indices. Therefore, both sides of Eq. (22.15) are determined by a single scalar function, and we can therefore transform to a scalar equation by taking the sum of all the terms in Eq. (22.15) with respect to the indices $i = j$. If we then substitute Eqs. (13.69), (13.87), and (13.91) into the resulting scalar equation, we obtain

$$\left(\frac{d^2}{dr^2} + \frac{7}{r}\frac{d}{dr} + \frac{8}{r^2}\right)\left(D_{LLL}(r) - 6\nu\frac{dD_{LL}(r)}{dr}\right) = -\frac{12\bar{\varepsilon}}{r}.$$

Hence, it follows that,

$$D_{LLL}(r) - 6\nu \frac{dD_{LL}(r)}{dr} = -\frac{4}{5}\bar{\varepsilon}r + \frac{C_1}{r^2} + \frac{C_2}{r^4},$$

and, since the left-hand side of this equation must be regular at the origin, we obtain Eq. (22.2).

The above derivation of Eq. (22.2) is largely due to Monin (1959a) and can be readily extended to the case of Eq. (22.9) for the temperature structure function. The derivation can also be used to deduce certain further dynamic equations for the structure functions. Thus, for example, we can readily verify that the tensor $D_{lj,k}(\boldsymbol{r}, \boldsymbol{r}') = \overline{v_l(\boldsymbol{r}) v_j(\boldsymbol{r}) v_k(\boldsymbol{r}')}$ must satisfy the equation

$$\frac{\partial D_{lj,k}(\boldsymbol{r}, \boldsymbol{r}')}{\partial r_j} + \frac{\partial D_{kj,l}(\boldsymbol{r}', \boldsymbol{r})}{\partial r'_j} =$$
$$= \nu \Delta D_{lk}(\boldsymbol{r}) + \nu \Delta D_{lk}(\boldsymbol{r}') - \nu \Delta D_{lk}(\boldsymbol{r} - \boldsymbol{r}') - \frac{2}{3}\bar{\varepsilon}\delta_{lk}. \quad (22.16)$$

where we have used Eq. (22.14). This shows that the combination of structure functions of two variables on the left-hand side of this equation is in fact the sum of functions of a single variable.

## 22.2 Closure of the Dynamic Equations

The structure and spectral dynamic equations for locally isotropic turbulence are not closed, i.e., each of them contains two unknown functions. To close these equations we must introduce some special hypotheses which enable us to express one of the unknown functions in terms of the other. Despite the lack of rigor and of physical basis of this approach, it is attractive in its simplicity and has therefore been comparatively widely used.

### *Hypothesis of Constant Skewness of Velocity Differences*

A very simple closing hypothesis was put forward by Obukhov (1949c) [see also Obukhov and Yaglom, 1951] with regard to the Kolmogorov structure equation (22.2) containing the functions $D_{LL}(r)$ and $D_{LLL}(r)$. We recall that the skewness of the longitudinal velocity difference $S(r) = D_{LLL}(r)[D_{LL}(r)]^{-3/2}$ should be constant (in fact, a negative constant) both for $r \ll \eta$ and $r \gg \eta$. The existing data do not contradict the suggestion that the values of $S(r)$ for $r \ll \eta$ and $r \gg \eta$ are equal. Obukhov introduced therefore as a working hypothesis the assumption that $S(r) =$ const $< 0$ for all $r$ in the quasi-equilibrium range $r \ll L$. In that case, Eq. (22.2) assumes the form

$$6\nu \frac{dD_{LL}(r)}{dr} + |S|[D_{LL}(r)]^{3/2} = \frac{4}{5}\bar{\varepsilon}r, \quad (22.17)$$

i.e., it becomes a first-order nonlinear differential equation for the unknown function $D_{LL}(r)$. The constant $|S|$ can then be eliminated simultaneously with the elimination of the dimensional parameters $\nu$ and $\bar{\epsilon}$ if the universal dimensionless function $\beta_{LL}(r)$ is normalized so that the normalization coefficient depends on $|S|$. In particular, if we substitute

$$D_{LL}(r) = \gamma_2 (\nu\bar{\varepsilon})^{1/2} \tilde{\beta}_{LL}(r/\gamma_1\eta), \quad \gamma_1 = \left(\frac{5}{2}\right)^{1/4} \frac{4}{|S|^{1/2}}, \quad \gamma_2 = \left(\frac{2}{45}\right)^{1/2} \frac{16}{|S|}, \tag{21.18}$$

the result is the universal differential equation

$$\frac{d\tilde{\beta}_{LL}}{dx} + \left[\frac{4}{3}\tilde{\beta}_{LL}(x)\right]^{3/2} = x. \tag{22.19}$$

It follows from this equation that

$$\begin{aligned} &\tilde{\beta}_{LL}(x) \approx \frac{1}{2}x^2, \quad \tilde{\beta}_{NN}(x) \approx x^2 \quad \text{when} \quad x \ll 1, \\ &\tilde{\beta}_{LL}(x) \approx \frac{3}{4}x^{2/3}, \quad \tilde{\beta}_{NN}(x) \approx x^{2/3} \quad \text{when} \quad x \gg 1 \end{aligned} \tag{22.20}$$

where $\tilde{\beta}_{NN}(x)$ represents a function analogous to $\tilde{\beta}_{LL}(x)$ and determined by the lateral structure function $D_{NN}(r)$. To find the values of $\tilde{\beta}_{LL}(x)$ at intermediate values of the argument, Obukhov integrated Eq. (22.19) numerically. The function $\tilde{\beta}_{LL}(x)$ obtained by him is plotted in Fig. 56 together with the corresponding function $\tilde{\beta}_{NN}(x)$ found from the von Kármán relation [the asymptotes of Eq. (22.20) are shown by the broken curves].

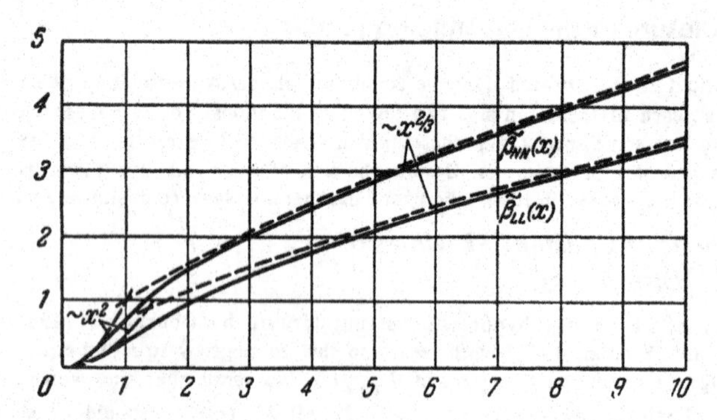


**Fig. 56** Longitudinal and lateral velocity structure functions corresponding to the constant skewness hypothesis.

The assumption that $S(r)$ is a constant is simpler than most of the energy transfer hypotheses considered in Sect. 17. Like other hypotheses formulated in terms of structure or correlation functions, however, this assumption suffers from the fact that it does not guarantee that the spectrum will be nonnegative. In fact, the numerical evaluation of the energy spectrum $E(k)$ carried out by Golitsyn (1960) based on the structure function $D_{LL}(r)$ satisfying Eq. (22.17) has shown that $E(k)$ is slightly negative in the neighborhood of the point $k = 8/\gamma_1\eta$. Hence it is clear that the Obukhov hypothesis cannot be exact. This is not of major importance because the form of the structure function $D_{LL}(r)$ is rather insensitive even to a considerable change in the form of the spectrum $E(k)$ for $k \gtrsim 1/\eta$. We therefore expect that the replacement of the Golitsyn spectrum by a similar but nonnegative function $E(k)$ would lead to such a small change in $D_{LL}(r)$ that it would not be noticeable in Fig. 56.

A similar hypothesis was introduced by Yaglom (1949b) in connection with the temperature structure equation (22.9) which relates to structure functions $D_{\vartheta\vartheta}(r)$ and $D_{L\vartheta\vartheta}(r)$. He assumed that $F(r) = D_{L\vartheta\vartheta}(r)/D_{\vartheta\vartheta}(r)\,[D_{LL}(r)]^{1/2} = \text{const} < 0$. In this case, for $D_{\vartheta\vartheta}(r)$ we have the linear equation

$$2\chi \frac{dD_{\vartheta\vartheta}(r)}{dr} + |F| \sqrt{D_{LL}(r)}\, D_{\vartheta\vartheta}(r) = \frac{4}{3} \overline{N} r, \tag{22.21}$$

whose solution satisfying the condition $D_{\vartheta\vartheta}(0) = 0$ is of the form

$$D_{\vartheta\vartheta}(r) = \frac{2}{3}\frac{\overline{N}}{\chi} \int_0^r \exp\left\{ -\frac{1}{2\chi} \int_y^r |F| \,[D_{LL}(x)]^{1/2}\, dx \right\} y\, dy. \tag{22.22}$$

This equation enables us to find $D_{\vartheta\vartheta}(r)$ in terms of the known values of $D_{LL}(r)$. In particular if we assume that $S(r) = \text{const}$, then $D_{LL}(r)$ is determined by the universal curve of Fig. 56 to within a scale factor, while $D_{\vartheta\vartheta}(r)$ can be represented by a one-parameter family of curves corresponding to different values of $F\nu/S\chi = \mathrm{Pr}\cdot F/S$. The spectrum $E_\vartheta(k)$ corresponding to these curves usually has slightly negative portions [Golitsyn, 1960].

## *Hypotheses on the Spectral Energy Transfer*

To obtain a strictly nonnegative spectrum we must use hypotheses formulated in terms of the spectral functions $E(k)$ and $T(k)$ [or $E_{\vartheta\vartheta}(k)$ and $T_{\vartheta\vartheta}(k)$] alone. The most widely used of these hypotheses is concerned with the spectral energy transfer, and was discussed in detail in Sect. 17. Each of these hypotheses can be used to calculate the statistical properties of locally isotropic turbulence in the quasi-equilibrium range (see Sect. 17.2). Moreover, the number of them can be appreciably increased within the framework of the theory of locally isotropic turbulence because we can suppose that, in principle, the energy transfer function $W(k)$ for $k \gg 1/L$ is explicitly dependent on the dimensional parameters $\overline{\epsilon}$ and $\nu$ which determine the statistical state of small-scale velocity fluctuations (for $k \ll 1/\eta$ the dependence on $\nu$ should, of course, disappear).

Thus, for example, Pao (1965, 1968) assumed that $W(k)$ was the product of $E(k)$ and a certain "rate of transfer of energy spectral elements through the point $k$ of the wave number spectrum," and that this rate of transfer depends only on $\overline{\epsilon}$ and $k$ for all $k \gg 1/L$. On this basis, he found that

$$W(k) = 2\gamma_p \overline{\varepsilon}^{1/3} k^{5/3} E(k), \qquad k \gg 1/L, \tag{22.23}$$

which is similar to the Kovasznay hypothesis (17.3). When this relation is substituted into the equation $\frac{d}{dk} W(k) + 2\nu k^2 E(k) = 0$ we obtain an ordinary differential equation for $E(k)$, and the solution of this which satisfies the normalization condition

$$\int_0^\infty k^2 E(k)\, dk = \overline{\varepsilon}/2\nu,$$

$$E(k) = Ak^{-5/3} \exp\left(-\frac{3}{4\gamma_P}\bar{\varepsilon}^{-1/3}\nu k^{4/3}\right), \quad \text{where} \quad A = \frac{1}{2\gamma_P}\bar{\varepsilon}^{2/3}. \tag{22.24}$$

To determine $E_{\vartheta\vartheta}(k)$ Pao used the analogous hypothetical formula

$$W_{\vartheta}(k) = 2\gamma_P' \bar{\varepsilon}^{1/3} k^{5/3} E_{\vartheta\vartheta}(k), \qquad k \gg 1/L_{\vartheta}, \tag{22.25}$$

which yielded

$$E_{\vartheta\vartheta}(k) = A' k^{-5/3} \exp\left(-\frac{3}{4\gamma_P'}\bar{\varepsilon}^{-1/3}\chi k^{4/3}\right), \quad \text{where} \quad A' = \frac{1}{2\gamma_P'}\bar{N}\bar{\varepsilon}^{-1/3}. \tag{22.26}$$

Similar rough reasoning was also used by Corrsin (1964) who proposed an equation of the form (22.26) earlier than Pao.

A natural generalization of the Pao energy transfer hypothesis (22.23) was proposed independently by Tennekes (1968a) and by Yaglom (1969). These authors suggested that

$$W(k) = 2\gamma\bar{\varepsilon}^{1-2n/3} k^{5n/3} [E(k)]^n, \; n > 0. \tag{22.23'}$$

At $n = 1$ this hypothesis coincides evidently with Pao's approximation to turbulent energy transfer, and at $n = 3/2$ it coincides with Kovasznay's approximation (17.3). The corresponding energy spectrum $E(k)$ turns out to vanish at some finite wave number $k_0$ and is identically zero for $k > k^0$ if $n > 1$; if $n < 1$ the spectrum $E(k)$ has power-law decay [with exponent $-5/3-4/3(1-n)$] as $k \to \infty$. Pao's case $n = 1$ is therefore the only one with exponential decay of the spectrum as $k \to \infty$.

Yaglom (1969) has investigated also the even more general family of spectral energy transfer hypotheses of the form

$$W(k) = 2\gamma\bar{\varepsilon}^{1-2n/3} k^{5n/3} [E(k)]^n \varphi(k\eta) \tag{22.23''}$$

where $\varphi(0) = 1$, and $\varphi(\xi) \sim \xi^{\alpha}$ for $\xi \gg 1$. With $n = 1$ and $\alpha < 4/3$ the spectrum $E(k)$ has exponential decay as $k \to \infty$; in all other cases it turns out to be either cut off at a finite wave number $k_0$ or decays no faster than some power of $k$. One particular case of the hypothesis (22.23'') was also considered by Panchev (1969) who assumed that $W(k) = 2\gamma[\nu + cE^{1/2}k^{-1/2}]Ek^3$ (this is asymptotically equivalent to Eq. (22.23'') with $n = 1$, $\varphi(\xi) = \xi^{4/3}$). Another model of the same kind was investigated by Panchev and Kesich (1969) and by J.-T. Lin (1972) who independently studied the same case $n = 1$, $\varphi(\xi) = (1 + c\xi^{2/3})^{-1}$.

A large number of new hypotheses on spectral energy transfer can be obtained from the von Kármán assumption that $W(k)$ can be expressed by the double integral (17.17) of the "spectral energy transfer density" $P(k', k'')$ which depends on $k'$, $k''$, $E(k')$, and $E(k'')$. In the framework of the theory of locally isotropic turbulence, it is permissible to suppose that the formula for $P(k', k'')$ can explicitly contain the parameters $\bar{\varepsilon}$ and $\nu$. If, for example, we restrict our attention to cases where the function $P(k', k'')$ is a simple power function of $E(k')$ and $E(k'')$, then we can write

$$P(k', k'') = \bar{\varepsilon}^{l} [E(k')]^n [E(k'')]^{n_1} (k')^m (k'')^{m_1} \varphi(k'\eta, k''\eta), \tag{22.27}$$

where $n$ and $n_1$ are arbitrary constants such that $n + n_1 = 3/2 - 3l/2 = 3(m + m_1)/5 + 6/5$, and $\varphi(\xi, \zeta)$ is an arbitrary function of two variables such that

$\varphi(\xi, \eta) \approx \varphi_1(\xi/\eta)$ for $\xi \ll 1$, $\eta \ll 1$. The above formulas agree well with dimensional considerations and the requirement that $W(k)$ must be independent of $\nu$ in the inertial subrange. In the special case when $\varphi(\xi, \zeta) = \varphi(\xi/\eta)$, the asymptotic behavior of the spectrum $E(k)$ for $k \to \infty$ will be the same as indicated in Sect. 17.2 for an analogous model (which differed only by the absence of the unimportant factor $\bar{e}^{l}$). It is also possible to consider the case when $\varphi(\xi, \zeta) = \varphi(\xi - \zeta)$, $0 < \varphi(0) < \infty$ (Yaglom, 1969). A special example of this kind, in which it was assumed that $\varphi(\xi - \zeta) = e^{-a(\xi-\zeta)}$ and $n + n_1 = 3/2$, was analyzed earlier by Dugstad (1962). In the general case, when $\varphi(\xi, \eta) = \varphi(\xi - \eta)$, $\varphi(x) \to 0$ as $x \to \infty$, and $n_1 > 1 - n > 0$, the spectrum $E(k)$ falls to zero quite rapidly at infinity: For example, we have $E(k) \sim \exp(-Bk^s)$ if $\varphi(x) \sim e^{-ax^s}$, and $E(k) \sim \exp(-e^{Bk})$ if $\varphi(x) = 1$ for $0 < x < a$ and $\varphi(x) = 0$ for $x > a$.

## *Applications of the Millionshchikov Zero-Fourth-Cumulant Hypothesis*

The Millionshchikov zero-fourth-cumulant hypothesis has already been discussed in Sects. 18 and 19. We saw there that the application of this hypothesis to the general space-time moments, designed to result in a closed set of moment equations can lead to nonphysical solutions which are in conflict with the requirement that the spectrum must be nonnegative. On the other hand, the use of this hypothesis for purely spatial properties does not appear to lead to appreciable errors in the energy range of the spectrum of isotropic turbulence. It is not excluded that it may also be approximately valid in the inertial subrange. Since no more accurate methods of calculation are available at present, let us briefly consider the simplest applications of the Millionshchikov hypothesis, which are concerned with the calculation of the pressure structure function $D_{pp}(r)$.

Equation (14.29), which relates the correlation function of the pressure Laplacian with the fourth-order moment of the velocity derivatives, can be rewritten in a form containing only the structure functions (of the second and fourth orders) of the fields $p(x, t)$ and $u(x, t)$. If we apply the zero-fourth-cumulant hypothesis to the right-hand side of this equation, and assume local isotropy, we obtain

$$\frac{d^4 D_{pp}(r)}{dr^4} + \frac{4}{r}\frac{d^3 D_{pp}(r)}{dr^3} = -\rho^2 \Phi(r), \tag{22.28}$$

where

$$\Phi(r) = \frac{\partial^2 D_{ij}(r)}{\partial r_m \partial r_n}\frac{\partial^2 D_{mn}(r)}{\partial r_i \partial r_j} = \frac{6}{r^2}\left[\frac{dD_{LL}(r)}{dr}\right]^2 + \frac{20}{r}\frac{dD_{LL}(r)}{dr}\frac{d^2 D_{LL}(r)}{dr^2} + 4\left[\frac{d^2 D_{LL}(r)}{dr^2}\right]^2 + 4\frac{dD_{LL}(r)}{dr}\frac{d^3 D_{LL}(r)}{dr^3} \tag{22.29}$$

[Obukhov and Yaglom, (1951); see also Eqs. (14.34) and (18.9)].

In the inertial subrange, the function $D_{LL}(r)$ satisfies the two-thirds law (21.17′), while the four-thirds law (21.72) is valid for $D_{pp}(r)$. Substituting Eqs. (21.17′) and (21.72) in Eqs. (22.28) and (22.29), and equating the dimensionless coefficients on either side, we find that

$$C_p = C^2, \qquad D_{pp}(r) = \rho^2 [D_{LL}(r)]^2. \tag{22.30}$$

This result is due to Obukhov (1949b).

The function $D_{pp}(r)$ can be expressed for all $r \ll L$ in terms of $D_{LL}(r)$ if the Green function for Eq. (22.28) is taken to be analogous to Eq. (14.36), where the boundary conditions are

$$D_{pp}(0) = 0, \quad \left.\frac{dD_{pp}(r)}{dr}\right|_{r=0} = 0 \text{ and } \frac{D_{pp}(r)}{r^2} \to 0 \quad \text{for} \quad r \to \infty$$

The last of these conditions is in fact equivalent to the requirement that the correlation coefficient between pressure fluctuations at two points must tend to zero as the distance between these points tends to infinity [see Eq. (13.22) in Sect. 13.1]. The explicit form of the corresponding solution of Eq. (22.28) was indicated by Obukhov and Yaglom (1951) who gave a numerical calculation of the structure function $D_{pp}(r)$ corresponding to the $D_{LL}(r)$ model of Fig. 56 for which $S(r) = \text{const}$ for all $r \ll L$. Subsequently, Golitsyn (1963) calculated $D_{pp}(r)$ by a similar method, using the interpolation formula for the spectrum $E(k)$ put forward by Novikov and discussed below [see Eq. (22.73)]. In both cases, the leading term of the asymptotic expansion obtained for $D_{pp}(r)$ for large $r$ was found to be $\rho^2 [D_{LL}(r)]^2 \sim r^{4/3}$, as expected in view of Eq. (22.30). The second term, on the other hand, of the asymptotic expansion was found to be a linear function of $r$ and, finally, the third term was found to be a constant.

Near the origin, the function $D_{pp}(r)$ is quadratic in $r$. Considerable errors may be introduced by the use of the Millionshchikov hypothesis to calculate the dimensionless coefficient

$$c_p = D''_{pp}(0)/2\rho^2\bar{\varepsilon}^{3/2}\nu^{-1/2}$$

(which is determined largely by the small-scale disturbances in the dissipation range of the spectrum). Since, however, there is no other method at present for estimating this coefficient, we shall summarize the results of such calculations. The first approximate calculation of the coefficient $c_p$ using the zero-fourth-cumulant hypothesis was carried out by Heisenberg (1948a) who also assumed that the spectral energy transfer $W(k)$ could be taken in the form (17.9). Heisenberg's result was subsequently improved somewhat by Chandrasekhar (1949) who made more precise calculations. It states that $c_p \approx 0.45/\gamma_H$, where $\gamma_H$ is the constant in Eq. (17.9). When applied to isotropic turbulence, the last formula can also be rewritten in the form $\lambda_p^2/\lambda^2 \approx \frac{\gamma_H}{26}\,\text{Re}_\lambda$, where $\text{Re}_\lambda = (\overline{u^2})^{1/2}\lambda/\nu$ and the relation $\bar{\varepsilon} = 15\nu\overline{u^2}/\lambda^2$ has been used.

Yaglom (1949a) used Eq. (22.28) with $S(r) = S = \text{const}$ (i.e., he used the model of Fig. 59). He found that $c_p \approx 0.4/|S|$ [see also Obukhov and Yaglom (1951)]. Batchelor (1951) used a similar method to calculate the value of $c_p$ corresponding to the simple interpolation formula for $\beta_{LL}(x) = (\nu\bar{\varepsilon})^{-1/2}D_{LL}(x\eta)$ of the form

$$\beta_{LL}(x) = (15)^{-1}x^2[1 + (15C)^{-3/2}x^2]^{-2/3}$$

which is valid for both $x \ll 1$ and $x \gg 1$. According to his results, $c_p \approx 0.46C^{3/2}$. Finally, Golitsyn (1963) repeated all these calculations using the spectrum $E(k)$ given by Eq. (22.73), and he found that $c_p \approx 0.3\alpha$. The discrepancy between these results is connected with the use of different models for $D_{LL}(r)$ or $E(k)$, and is probably appreciably less than the error caused by the assumption that the fourth-order velocity cumulants are zero.

Having determined the pressure structure function $D_{pp}(r)$ [or the spectrum $E^{(p)}(k)$] in some way, we can determine the correlation functions

$$B_{LL}^{(A_1)}(r) = \frac{1}{2\rho^2}D''_{pp}(r) \text{ and } B_{NN}^{(A_1)} = \frac{1}{2\rho^2 r}D'_{pp}(r)$$

and the spectrum $E_{LL}^{(A_1)}(k)$ of the field $A^{(1)}(x) = -\frac{1}{\rho}\nabla p(x)$. The vector $A^{(1)}$ is a most important component of the turbulent acceleration

$$A(x) = -\frac{1}{\rho}\nabla p + \nu\,\Delta u.$$

The viscous acceleration $A^{(2)} = \nu\,\Delta u$ is uncorrelated with $A^{(1)}$, therefore we can calculate the correlation functions and the spectrum of the acceleration field, if we have data on both $D_{pp}(r)$ [or $E^{(p)}(k)$] and on $D_{LL}(r)$ [or $D_{NN}(r)$ or $E(k)$]. Calculations of the acceleration correlation functions, which correspond to the above calculations of $D_{pp}(r)$, were carried out by Yaglom (1949a), Batchelor (1951) [in both these papers only the values of $\overline{A^2} = B_{LL}^{(A)}(0) + 2B_{NN}^{(A)}(0)$ were evaluated], Obukhov and Yaglom (1951), and Golitsyn (1963). The results obtained in all these papers are in agreement in the sense that they all indicate that the mean square viscous acceleration in locally isotropic turbulence is only a small fraction of the mean square of the total acceleration.

## *The Closure Problem in the Theory of Locally Isotropic Turbulence. Kraichnan's Lagrangian-History Direct-Interaction Approximation*

Various methods of closure of the dynamic equations for isotropic turbulence were discussed in Sects. 17 and 19. Any such method which enables us to consider separately the evolution of large-scale and small-scale components (i.e., to eliminate the nonlocal direct interaction between large-scale and small-scale turbulence components) can be also regarded as a method of closure for locally isotropic turbulence. However, the first nontrivial approximation of perturbation theory, i.e., Kraichnan's direct-interaction approximation (see Sect. 19.6) does not satisfy this condition. We have already seen that, in this approximation, the large-scale properties of turbulent flows directly affect the small-scale disturbances. The result of this is that the turbulence spectrum in the small-scale region does not satisfy the five-thirds law, and depends on the mean square velocity fluctuation—a typically large-scale characteristic.

Kadomtsev (1965) (published in Russian in 1964) and Kraichnan (1964c) explained the discrepancy in the direct interaction approximation by pointing out that the transport of small-scale inhomogeneities by large-scale velocity components can not be correctly taken into account within the framework of this approximation, nor can this transport be separated from deformation of small-scale inhomogeneities which governs their evolution (see also Orszag 1969, 1970). The second approximation of the method of stochastic models (in which the direct interaction approximation is the first approximation), put forward by Kraichnan (1961), suffered from the same defect. This higher approximation improves the original direct interaction approximation to some extent, but it also leads to an incorrect form for the turbulence spectrum in the inertial subrange [see Kraichnan (1964c)]. The simplest method of eliminating the nonlocal interaction between large-scale turbulence components and small-scale inhomogeneities, which is associated with the incorrectness of the direct-interaction description of the large-scale convective transport, is to introduce an artificial cutoff into the Navier-Stokes equations themselves. This cutoff factor eliminates the interaction between Fourier components with wave numbers $\boldsymbol{k}$ and $\boldsymbol{k}'$ such that $|\boldsymbol{k}-\boldsymbol{k}'|$ is very small. A cutoff factor of this type was introduced by Kraichnan (1964c) who rejected nonlinear terms with $|\boldsymbol{k}-\boldsymbol{k}'| < k/\alpha$ or $|\boldsymbol{k}-\boldsymbol{k}'| < k'/\alpha$, where $\alpha > 1$ is a fixed cutoff parameter, in the spectral form of the Navier-Stokes equations. It was found that, for sufficiently large Re, the application of the direct interaction approximation to the resulting "model fluid dynamic equations" leads to the five-thirds law (21.24') for the spectrum in the inertial subrange (with the coefficient $C_1$ depending on the chosen model, i.e., on the cutoff parameter $\alpha$).

A less artificial method of improving the direct interaction approximation, which also eliminates the parasitic nonlocal interactions, was put forward by Kadomtsev (1965) who applied it to a special form of the model equation (19.121). Kadomtsev's approach can apparently be used to close the dynamic equations for locally isotropic turbulence. The Shut'ko approximation (1964), which was briefly reviewed in Sect. 19.6, belong to the same category. The Shut'ko equations can be considered as an approximate set of closed equations for locally isotropic turbulence. Shut'ko suggests that his approximate equations should lead to a spectrum proportional to $k^{-5/3}$ in the inertial subrange, and should thus provide a theoretical estimate for the coefficient $C_1$. This was questioned by Kraichnan (1966c) who pointed out that any justified approximation must exhibit many of the properties of the exact fluid dynamic equations. Thus, for example, the approximate expressions for the nonlinear interactions should not lead to the violation of the conservation of energy and momentum, and should have the stationary solution found by Hopf and Titt (1953) for the exact fluid dynamic equations (see Sect. 29.1). They should not lead to the appearance of nonphysical results such as negative spectra and correlation coefficients greater than unity. The approximate equations should be invariant under arbitrary Galilean transformations, and so on. According to Kraichnan, the Shut'ko approximation does not satisfy many of these requirements and should not, therefore, lead to the five-thirds law.

The whole hierarchy of the coupled dynamical equations for various order spectra of the velocity field (which are Fourier transforms of the velocity cumulants of different orders) was studied by Orszag and Kruskal (1968) (see also Orszag, 1970) from the point of view of the theory of locally isotropic turbulence. They found that the infinite set of equations has a special self-preserving solution in the inertial subrange of wave numbers; the solution depends on the parameter $\bar{\epsilon}$ only and the corresponding energy spectrum $E(k)$ has a "correct" five-thirds form. Orszag and Kruskal showed also the consistency of this inertial subrange solution of the whole set of dynamical equations and investigated the connection of the solution with some general properties of the equations, first of all with their Galilean invariance. They also suggested that only Galilean-invariant closure approximations can lead to results consistent with Kolmogorov similarity hypotheses; this deduction coincides entirely with Kraichnan's conclusions. Based on these ideas Orszag and Kruskal (1966) put forward a series of comparatively complicated closure hypotheses which satisfied all the necessary properties and must lead therefore to the five-thirds law in the inertial subrange. The first of the Orszag-Kruskal hypotheses can be formulated as follows

$$\overline{dZ_l(\boldsymbol{k})\,dZ_j(\boldsymbol{l})\,dZ_m(\boldsymbol{p})\,dZ_n(\boldsymbol{q})} = \begin{cases} \dfrac{\overline{dZ_l(\boldsymbol{k})\,dZ_j(\boldsymbol{l})\,dZ_r(-\boldsymbol{k}-\boldsymbol{l})}\cdot\overline{dZ_m(\boldsymbol{p})\,dZ_n(\boldsymbol{q})\,dZ_r(-\boldsymbol{p}-\boldsymbol{q})}}{\overline{dZ_s(-\boldsymbol{k}-\boldsymbol{l})\,dZ_s(-\boldsymbol{p}-\boldsymbol{q})}} + \dots, & \text{if } \boldsymbol{k}+\boldsymbol{l}+\boldsymbol{p}+\boldsymbol{q}=0, \\ 0, & \text{if } \boldsymbol{k}+\boldsymbol{l}+\boldsymbol{p}+\boldsymbol{q}\neq 0, \end{cases}$$

where the dots on the right-hand side represent two further terms of the same kind, which are obtained for the two other subdivisions of the vector quartet ($\boldsymbol{k}$, $\boldsymbol{l}$, $\boldsymbol{t}$, $\boldsymbol{q}$) and the corresponding subscripts ($i$, $j$, $m$, $n$) into two pairs. This hypothesis corresponds to a special statistical model which appears to have a number of properties exhibited by real turbulence. According to Orszag and Kruskal, it is the first step in a sequence of increasingly complicated models which, one hopes, will provide increasingly more accurate descriptions of turbulent motion.

Unfortunately even the simplest among the Orszag-Kruskal closure approximations leads to a very complicated set of equations which is very difficult to use. To make calculations easy Orszag (1967) considered also a much simpler closure hypothesis which expresses the third moment $\overline{dZ_i(k)dZ_j(l)dZ_m(-k-l)}$ in terms of the energy spectrum values $E(k)$, $E(l)$, $E(|k+l|)$, vectors $k$, $l$ and $-k-l$, and an artificial "typical relaxation time for triple moments" $\tau(k, l, -k-l)$. Using a special equation for $\tau(k, l, -k-l)$ containing the dimensional parameter $\bar{\epsilon}$ and two dimensionless constants $n$ and $A$, Orszag showed that the final equations for this approximation are satisfied by the five-thirds spectrum and that the corresponding coefficient $C_1$ is a slowly varying function of $n$ and $A$ taking values comparatively close to 1.5 for all reasonable values of these two numerical parameters.

Another approach was used by Kraichnan (1965a, b; 1966a, b) who obtained a Galilean-invariant scheme for the approximate closure of dynamic equations based on the application of the direct-interaction approximation to the equations of motion in the Lagrangian form. This approximation was called by Kraichnan the *Lagrangian-history direct-interaction* approximation (in short LHDI approximation). The full equations of the LHDI approximation are quite complicated (cf. the application of the approximation to a simplified one-dimensional model equation at infinite Reynolds number in Kraichnan, 1968); hence Kraichnan was forced to supplement his general scheme by a nonrigorous abridgment procedure which simplified the equations substantially.[12] As a result he obtained a closed set of equations; however the set is based on several relatively crude assumptions, so that its accuracy is not at all clear. Nor is it clear either whether the resulting equations correspond to any realizable physical model. However, all the main properties of real fluid dynamic equations are exhibited by the Kraichnan equations. Therefore, the properties of the local structure of turbulence, which are a consequence of the Kolmogorov similarity hypotheses, are also exhibited by the "motions" described by these equations. It even turns out that some of the numerical coefficients and functions of the Kraichnan theory (where they are unambiguously determined by the main equations of this theory) assume values quite close to those for real turbulence (where they can be determined experimentally). Kraichnan was principally concerned with the Lagrangian velocity $\boldsymbol{v}(\boldsymbol{x}, t \mid \tau)$, of a fluid particle measured at time $\tau$, where $\boldsymbol{x}$ is the position of the particle at time $t$. When $\tau = t$, this velocity is equal to the Eulerian velocity $\boldsymbol{u}(\boldsymbol{x}, t)$. Moreover, it is clear that

$$\left[\frac{\partial}{\partial t} + \boldsymbol{u}(\boldsymbol{x}, t) \cdot \nabla\right] \boldsymbol{v}(\boldsymbol{x}, t \mid \tau) = 0. \tag{22.31}$$

Let the mean value of the velocity field be zero. The simplest statistical characteristic of the random field $\boldsymbol{v}(\boldsymbol{x}, t \mid \tau)$ will then be the Lagrangian velocity correlation function

$$B_{ij}(\boldsymbol{x}, t \mid \tau; \boldsymbol{x}', t' \mid \tau') = \overline{v_i(\boldsymbol{x}, t \mid \tau)\, v_j(\boldsymbol{x}', t' \mid \tau')}. \tag{22.32}$$

The Kraichnan theory also involves the Green tensor of the velocity field, which for $\tau \geqslant \tau'$ is given by

$$G_{ij}(\boldsymbol{x}, t \mid \tau; \boldsymbol{x}', t' \mid \tau') = \delta v_i(\boldsymbol{x}, t \mid \tau)/\delta f_j(\boldsymbol{x}', t' \mid \tau') \tag{22.33}$$

and is zero for $\tau < \tau'$. This tensor describes the reaction of the velocity field $\boldsymbol{v}(\boldsymbol{x}, t \mid \tau)$ to the infinitesimal force disturbance $\delta \boldsymbol{f}(\boldsymbol{x}, t \mid \tau)$ introduced into the right-hand side of Eq.

[12] Kraichnan considered in fact two different sets of approximate abridged equations, but we shall review here only the simpler of the two.

(22.31) for $\tau \neq t$, and into the right-hand side of the Navier-Stokes equation for $u(x, t)$ for $\tau = t$.[13]

If we take the variational derivative of the fluid dynamic equations with the above right-hand side with respect to the components of the external forces, we can verify that the Green tensor satisfies the following dynamic equations:

$$\left[\frac{\partial}{\partial t}+u^s_\alpha(x, t)\frac{\partial}{\partial x_\alpha}\right]G_{ij}(x, t\,|\,\tau;\ x', t'\,|\,\tau') =$$
$$= -G^s_{\alpha j}(x, t\,|\,t;\ x', t'\,|\,\tau')\frac{\partial v_i(x, t\,|\,\tau)}{\partial x_\alpha}, \tag{22.34}$$

$$\left(\frac{\partial}{\partial t}-\nu\Delta_x\right)G_{ij}(x, t\,|\,t;\ x', t'\,|\,\tau') =$$
$$= -\left[P_{i\beta}(\nabla_x)\frac{\partial}{\partial x_\alpha}+P_{i\alpha}(\nabla_x)\frac{\partial}{\partial x_\beta}\right]\left[u^s_\alpha(x, t)\,G^s_{\beta j}(x, t\,|\,t;\ x', t'\,|\,\tau')\right], \tag{22.35}$$

where the superscript $s$ over a particular subscript indicates that only the solenoidal component of the field is taken in this subscript, i.e., that the subscript is contracted with the tensor operator $P_{ij}(\nabla)$ which is defined as the operator whose Fourier transform in infinite space is the tensor $\Delta_{ij}(k) = \delta_{ij} - k_i k_j/k^2$). The Green operator $G_{ij}$ itself is random. The mean Green tensor will be denoted by the symbol $\overline{G}_{ij}$.

In the case of homogeneous turbulence, to which we shall restrict our attention henceforth, the functions $B_{ij}$ and $\overline{G}_{ij}$ will depend only on $x' - x$. In the case of isotropic turbulence, their Fourier transforms with respect to $x' - x$ can be written in the form

$$B_{ij}(k;\ t\,|\,\tau;\ t'\,|\,\tau') = \frac{1}{2}\Delta_{ij}(k)\,B^s(k;\ t\,|\,\tau;\ t'\,|\,\tau') + \frac{k_i k_j}{k^2}B^c(k;\ t\,|\,\tau;\ t'\,|\,\tau'), \tag{22.36}$$

$$\overline{G}_{ij}(k;\ t\,|\,\tau;\ t'\,|\,\tau') = \Delta_{ij}(k)\,G^s(k;\ t\,|\,\tau;\ t'\,|\,\tau') + \frac{k_i k_j}{k^2}G^c(k;\ t\,|\,\tau;\ t'\,|\,\tau'), \tag{22.37}$$

so that each of these tensors is determined by two scalar functions.

The dynamic equations for $B_{ij}$ can be obtained from Eq. (22.31) by the usual Friedmann-Keller method, while the equations for $\overline{G}_{ij}$ can be obtained by taking the average of Eqs. (22.34) and (22.35). Kraichnan first considers the direct-interaction approximation of these dynamic equations. This he achieves by (1) introducing the zero-order approximations $B^{(0)}_{ij}$, $G^{(0)}_{ij}$, $v^{(0)}_i$, i.e., the solutions of the dynamic equations, linearized by rejecting nonlinear terms, with the initial velocity field having a Gaussian probability distribution, (2) expanding all the functions in the dynamic equations into functional power series in terms of the zero-order approximations, (3) retaining in the resulting equations only the leading terms having the lowest order in the zero-order approximations, and (4) replacing the zero-order approximations in these leading terms by the complete functions.

The dynamic equations for $B_{ij}$ and $\overline{G}_{ij}$ in the direct-interaction approximation contain nonlinear terms (due to nonlinear terms in the fluid dynamic equations) in the form of

[13] We note that the incompressibility of the fluid imposes certain restrictions on the admissible disturbances $f(x, t\,|\,\tau)$, but these limitations can be avoided if, in addition to the solenoidal part of the Eulerian velocity field $u^s(x, t)$, which enters only Eq. (22.31) and the Navier-Stokes equations, we introduce a fictitious potential part $u^c(x, t)$, which satisfies the "equation of motion" $\left(\frac{\partial}{\partial t}-\nu\Delta\right)u^c(x, t) = 0$.

integrals with respect to time $\tau$ (and the space coordinates) of double and triple products of unknown functions. The time $\tau$ appears both after the vertical line (in which case it is the time at which the velocity of the fluid particle is measured and corresponds to integration along the trajectory) and in front of the vertical line (where it is the "labeling time" for the fluid particle, and the integration with respect to $\tau$ takes into account the time correlation between the Eulerian velocity fields). However, the presence of the Eulerian correlation times in the approximate dynamic equations, where the former depend on the velocity of transport of inhomogeneities past fixed space points, violates the invariance under random Galilean transformations of space and time, which the exact dynamic equations are known to satisfy.

To restore the Galilean invariance of the approximate dynamic equations Kraichnan replaced, somewhat arbitrarily, the integration variable $\tau$ in front of the vertical line by the fixed argument $t$ (or $t'$, depending on which particular argument occurred in the corresponding place before the average of the dynamic equations was taken). The equations for $B_{ij}$ and $\overline{G}_{ij}$ obtained in this way are, in fact, the final result of the Kraichnan LHDI closure method. These equations are, however, very complicated, and Kraichnan introduced the following—also somewhat arbitrary—simplification. He considered only the equations for the Fourier transforms

$$B_{ij}(\boldsymbol{k};\ t\,|\,t;\ t\,|\,\tau)\ \text{and}\ \overline{G}_{ij}(\boldsymbol{k};\ t\,|\,t;\ t\,|\,\tau),$$

and replaced $B_{kl}(\boldsymbol{a};\ t\,|\,\tau_1;\ t\,|\,\tau_2)$ on the right-hand sides by $B_{kl}(\boldsymbol{a};\ \tau_1\,|\,\tau_1;\ \tau_1\,|\,\tau_2)$ for $\tau_1 \geqslant \tau_2$, and by $B_{lk}(\boldsymbol{a};\ \tau_2\,|\,\tau_2;\ \tau_2\,|\,\tau_1)$ for $\tau_1 < \tau_2$; the function $\overline{G}_{kl}(\boldsymbol{a};\ t\,|\,\tau;\ t\,|\,\tau_2)$ was replaced by $\overline{G}_{kl}(\boldsymbol{a};\ \tau_1\,|\,\tau_1;\ \tau_1\,|\,\tau_2)$ where $\boldsymbol{a}$ is the space wave-number vector. If we now consider the isotropic case and use the notation

$$B(k;\ t\,|\,\tau) = B^s(k;\ t\,|\,t;\ t\,|\,\tau), \qquad G(k;\ t\,|\,\tau) = G^s(k;\ t\,|\,t;\ t\,|\,\tau), \tag{22.38}$$

where the functions $B^s$ and $G^s$ are defined by Eqs. (22.36)–(22.37), then the first of the simplified Kraichnan equations assumes the form

$$\left(\frac{\partial}{\partial t} + 2\nu k^2\right) B(k;\ t\,|\,t) = 2 \iint\limits_{\Delta(k)} B_{kpq}\,dp\,dq \int\limits_{t_0}^{t} B(q;\ t\,|\,\tau)\,[G(k;\ t\,|\,\tau) \times B(p;\ t\,|\,\tau) - G(p;\ t\,|\,\tau)\,B(k;\ t\,|\,\tau)]\,d\tau. \tag{22.39}$$

In this equation the range of integration $\Delta(k)$ represents the set of values $(p, q)$ which can be the lengths of the sides of a triangle, the third side being $k$. Moreover, $B_{kpq} = \pi p^2 q\,(xy + z^3)$ where $x, y, z$ are the cosines of the angles of this triangle lying opposite the sides $k, p, q$, respectively. The simplified equation for $B(k;\ t\,|\,\tau)$ for $t \geqslant \tau$ is even more complicated. It is of the form

$$\begin{aligned}\left(\frac{\partial}{\partial t} + \nu k^2\right) B(k;\ t\,|\,\tau) = &-\frac{4\pi k^2}{3} B(k;\ t\,|\,\tau) \int\limits_0^{\infty} q^2\,dq \int\limits_{\tau}^{t} B(q;\ t\,|\,s)\,ds + \\ &+ \iint\limits_{\Delta(k)} D_{kpq} B(p;\ t\,|\,\tau)\,dp\,dq \int\limits_{\tau}^{t} B(q;\ t\,|\,s)\,ds + \\ &+ \iint\limits_{\Delta(k)} dp\,dq \int\limits_{t_0}^{\tau} B(q;\ t\,|\,s)\,[B_{kpq} G(k;\ \tau\,|\,s)\,B(p;\ t\,|\,s) -\end{aligned} \tag{22.40}$$

$$-D'_{kpq}G(p;\,\tau\,|\,s)\,B(k;\,t\,|\,s)]\,ds-$$

$$-\iint\limits_{\Delta(k)} dp\,dq \int\limits_{t_0}^{t} B(q;\,t\,|\,s)\,[B_{kpq}G(p;\,t\,|\,s)\,B(k;\,\tau\,|\,s)-$$

$$-D_{kpq}G(k;\,t\,|\,s)\,B(p;\,\tau\,|\,s)]\,ds-$$

$$-\iint\limits_{\Delta(k)} dp\,dq \int\limits_{t_0}^{t} \left(B_{kpq}-D'_{kpq}\right)G(p;\,t\,|\,s)\,B(k;\,t\,|\,s)\,B(q;\,\tau\,|\,s)\,ds, \qquad \text{(20.40) (Cont.)}$$

where

$$D_{kpq}=\frac{B_{kpq}+B_{kqp}}{2}+\frac{1}{2}\pi kpq\,(z^2-y^2) \text{ and } D'_{kpq}=\frac{p^2}{k^2}D_{pkq}.$$

Finally, the equation for $G(k;\,t\,|\,\tau)$ has the analogous form

$$\left(\frac{\partial}{\partial t}+\nu k^2\right)G(k,\,t\,|\,\tau)=-\frac{4\pi k^2}{3}G(k;\,t\,|\,\tau)\int\limits_{0}^{\infty} q^2\,dq\int\limits_{\tau}^{t} B(q;\,t\,|\,s)\,ds+$$

$$+\iint\limits_{\Delta(k)} D_{kpq}G(p;\,t\,|\,\tau)\,dp\,dq\int\limits_{\tau}^{t} B(q;\,t\,|\,s)\,ds+$$

$$+\iint\limits_{\Delta(k)} (D_{kpq}-B_{kpq})\,G(p;\,t\,|\,\tau)\,B(q;\,t\,|\,\tau)\,dp\,dq\int\limits_{\tau}^{t} G(k,\,s\,|\,\tau)\,ds-$$

$$-\iint\limits_{\Delta(k)} dp\,dq\int\limits_{\tau}^{t} B(q;\,t\,|\,s)\,\Big[B_{kpq}G(p,\,t\,|\,s)\,G(k;\,s\,|\,\tau)-$$

$$-D'_{kpq}G(k;\,t\,|\,s)\,G(p;\,s\,|\,\tau)\Big]\,ds. \qquad \text{(22.41)}$$

We note further that Eq. (22.39) can also be written as the equation for the energy spectrum $E(k,\,t)=2\pi k^2 B(k;\,t\,|\,t)$ in the form of Eq. (14.15):

$$(\partial/\partial t+2\nu k^2)\,E(k,\,t)=T(k,\,t), \qquad \text{(22.39')}$$

where $T(k,\,t)/2\pi k^2$ is the same as the right-hand side of Eq. (22.39). Since $k^2B_{kpq}=p^2B_{pkq}$, it follows that $\int\limits_0^\infty T(k,\,t)\,dk=0$, which, as we know, represents the "energy conservation during inertial mixing" (described by the nonlinear terms of the fluid dynamic equations). It is also clear that $T(k,\,t)$ can, in this case, be written in the form

$$T(k,\,t)=\iint\limits_{\Delta(k)} T(k,\,p,\,q,\,t)\,dp\,dq, \qquad \text{(22.42)}$$

where $T(k,\,p,\,q,\,t)/2$ is equal to the symmetric part of the integrand (in the integral with respect to $p$ and $q$) on the right-hand side of Eq. (22.39); the part which is antisymmetric in $p$ and $q$ can be ignored, since it does not contribute to the integral. The function $T(k,\,p,\,q,\,t)$ satisfies Eq. (17.25) and can therefore be regarded as the density of the rate of energy transfer into the element $dk$ in wave-number space, which is due to the interaction of the spectral components $dZ(p)$ and $dZ(q)$ of the velocity field. For the function $W(k_1)$, which determines the rate of energy transfer through the point $k_1$ in the wave number spectrum, this yields

$$W(k_1, t) = \int_{k_1}^{\infty} \iint_{\Delta(k)} T(k, p, q, t)\, dp\, dq\, dk. \tag{22.42'}$$

To solve Eqs. (22.39), (22.40), and (22.41) we must either specify the initial value (for $t = t_0$) of the spectral density $E(k, t)$, or consider a steady state in which the functions $B(k; t \mid \tau)$ and $G(k; t \mid \tau)$ depend only on $k$ and the difference $t - \tau = s$. Kraichnan (1965b, 1966a) paid particular attention to the second of these problems, and was especially interested in the results referring to the inertial range, where we could assume that $\nu = 0$. In complete correspondence with the conclusions drawn from the similarity hypotheses, Eqs. (22.39), (22.40), and (22.41) (in which we can now set $t_0 = -\infty$) will then have solutions of the form

$$E(k) = C_1 \bar{\varepsilon}^{2/3} k^{-5/3}, \quad G(k; t \mid t-s) = G(\bar{\varepsilon}^{1/3} k^{2/3} s),$$
$$B(k; t \mid t-s)/[B(k; t \mid t)\, B(k; t-s \mid t-s)]^{1/2} = R(\bar{\varepsilon}^{1/3} k^{2/3} s),$$

where $\bar{\varepsilon}$ is equal to the right-hand side of Eq. (22.42') for arbitrary $k_1$ from the inertial subrange [defined by the condition that in this subrange $T(k_1) = 0$ and $W(k_1) = \text{const}$]. If we now substitute

$$G\left(\bar{\varepsilon}^{1/3} k^{2/3} s / C_1^{1/2}\right) = G_1(s) \text{ и } R\left(\bar{\varepsilon}^{1/3} k^{2/3} s / C_1^{1/2}\right) = R_1(s),$$

then Eqs. (22.40) and (22.41) will assume a universal form (independent of $C_1$), and their numerical integration will yield the unknown functions $G_1(s)$ and $R_1(s)$. If we then substitute the resulting solutions into the formula (22.42') for $W(k_1) = \bar{\varepsilon}$, we can also determine the constant $C_1$. Kraichnan (1966a) found that $C_1 \approx 1.77$, which was quite close to the value $C_1 \approx 1.5$ obtained from an analysis of measurements of real turbulent flows (see Sect. 23.3 below).

In another paper, Kraichnan (1966b) applied an analogous method to calculations of the spectrum $E^{(\vartheta)}(k)$ of the concentration $\vartheta(\boldsymbol{x}, t)$ of a passive admixture mixed by turbulence. Here again, the simplified LHDI equations lead to the five-thirds law for the spectrum $E^{(\vartheta)}(k)$ in the inertial-convective subrange of wave numbers $k$, while the dimensionless coefficient $B^{(\vartheta)}$ in this law is found to be $B^{(\vartheta)} \approx 0.4$, which corresponds to $C^{(\vartheta)} \approx 1$ where $C^{(\vartheta)}$ is the dimensionless coefficient of the two-thirds law (21.87) for the structure function $D_{\vartheta\vartheta}(r)$. This last value is not in good agreement with existing data on the local structure of the temperature and concentration fields (see Sect. 23.5 below).

A simpler Galilean-invariant closure approximation was constructed recently by Kraichnan (1971a) with the aid of a suitable modification of the representation of the direct-interaction approximation considered, e.g., in the paper of Kraichnan (1970a). The representation consists in a linear model equation for the velocity amplitude in which the nonlinear terms of the real fluid dynamic equation are replaced by a dynamical damping term, with memory, and a random forcing term. The specific change in both these terms makes the approximation Galilean-invariant and consistent with all the predictions of the theory of locally isotropic turbulence. We shall not discuss here this recent closure hypothesis in any detail restricting ourselves to the references to papers by Kraichnan (1971a, b) and Herring and Kraichnan (1971). Let us only note that the corresponding model has an adjustable parameter $g$ and the Kolmogorov constant $C$ is given by the expression $C = 1.76g^{2/3}$ and hence may take any value. When $g$ is chosen to yield the same value of $C$ as the LHDI approximation, the two approximations give as a rule very similar results.

Finally let us mention the closure hypothesis for the hierarchy of equations for velocity probability density functions proposed by Lundgren (1971) and briefly outlined at the end of Sect. 19.7. It was noted in this section that the hypothesis leads to very complicated equations for the one-point density function and correlation tensor [i.e., the function $B_{LL}(r, t)$]. However the equations can be simplified for the case of inertial subrange statistics where time derivatives and viscous terms may be neglected. Lundgren obtained for this case the "two-thirds law" of Kolmogorov with a coefficient $C \approx 1.7$. The agreement of this value with the existing experimental data is even better than might have been anticipated.

## *Equations for the Concentration Spectrum of a Chemically-Active or Decaying Admixture*

We shall now consider some cases in which the statistical state of small-scale fluctuations is affected by certain additional factors. Let us begin by investigating the properties of the concentration field $\vartheta(\boldsymbol{x}, t)$ of a passive admixture which either decays or undergoes a first-order chemical reaction during the turbulent mixing process (see Sect. 21.7). We shall suppose that sources of the admixture are present in the flow and ensure that the mean field $\bar{\vartheta}(\boldsymbol{x}, t)$ undergoes very little change in the time $T_r = \mu^{-1}$ and in times $\tau_\eta = (\nu/\bar{\varepsilon})^{1/2}$ and $\tau_\vartheta = (\chi/\bar{\varepsilon})^{1/2}$. We shall also assume that the typical length scale $L_0$ of the fields $\bar{\vartheta}(\boldsymbol{x}, t)$ and $\boldsymbol{u}(\boldsymbol{x}, t)$ is much greater than the lengths $L_r = \bar{\varepsilon}^{1/2}\mu^{-3/2}$, $\eta = (\nu^3/\bar{\varepsilon})^{1/4}$ and $\eta_\vartheta = (\chi^3/\bar{\varepsilon})^{1/4}$. The statistical state of the fluctuations in the field $\vartheta(\boldsymbol{x}, t)$ with length scales $l \ll L_0$ (or wave numbers $k \gg 1/L_0$) can be regarded as locally isotropic and quasi-stationary. From the main dynamic equation (21.101) which the field $\vartheta(\boldsymbol{x}, t)$ must satisfy it follows that the equation for the spectrum $E_{\vartheta\vartheta}(k) = E^{(\vartheta)}(k)$ for $k \gg 1/L_0$ is of the form

$$W_\vartheta(k) = 2\mu \int_k^\infty E^{(\vartheta)}(k')\, dk' + 2\chi \int_k^\infty k'^2 E^{(\vartheta)}(k')\, dk' \tag{22.43}$$

[this equation differs from Eq. (22.13) only by the presence of the additional first term on the right-hand side]. If, moreover, $k \ll 1/\eta_0 = 1/\max(\eta, \eta_\vartheta)$ then the last term on the right-hand side of Eq. (22.43) can be regarded as constant and, consequently

$$\frac{dW_\vartheta(k)}{dk} = -2\mu E^{(\vartheta)}(k) \quad \text{when} \quad 1/L_0 \ll k \ll 1/\eta_0. \tag{22.44}$$

Therefore, before we can determine the form of the spectrum $E^{(\vartheta)}(k)$ we must again use *some* hypothesis about the function $W_\vartheta(k)$.

Let $1/L_0 \ll k \ll 1/\eta_0$, so that Eq. (22.44) is valid. Following Kovasznay (1948), Onsager (1949), and Corrsin (1964), let us first assume that $W_\vartheta(k)$ depends only on the value of $E^{(\vartheta)}(k)$ but not on the values of $E^{(\vartheta)}(k')$ with $k' \neq k$. In that case, the form of $W_\vartheta(k)$ in the range $k \ll 1/\eta_0$, in which the molecular coefficients $\nu$ and $\chi$ can play no significant role, is uniquely determined by dimensional considerations, and is found that $W_\vartheta(k) = 2\gamma_1\bar{\varepsilon}^{1/3}k^{5/3}E^{(\vartheta)}(k)$, where $\gamma_1$ is a dimensionless coefficient [we already know that, subsequently, Pao (1965) suggested the use of the same hypothetical formula for a much broader spectral region]. If we substitute the above equation for $W_\vartheta(k)$ into Eq. (22.44), we can readily show that

$$E^{(\vartheta)}(k) \sim k^{-5/3} \exp\left\{\frac{3}{2\gamma_1}(kL_r)^{-2/3}\right\}, \quad L_r = \mu^{-3/2}\bar{\varepsilon}^{1/2} \tag{22.45}$$

This result is due to Corrsin (1961, 1964).

We know that, in the absence of radioactive decay and chemical reactions, all the reasonable hypotheses with regard to the form of $W_\vartheta(k)$ in the inertial-convective subrange $k \ll 1/\eta_0$ always lead to the same universal five-thirds law. It is clear, however, that for a decaying or chemically active admixture, for which the form of the spectrum in the range $1/L_0 \ll k \ll 1/\eta_0$ cannot be established from dimensional considerations alone, the consequences of different hypotheses with regard to the form of $W_\vartheta(k)$ in the above spectral range may, in fact, be different. Let us, for example, replace the Kovasznay-Onsager-Corrsin hypothesis by the Heisenberg hypothesis, according to which $W_\vartheta(k) = 2K_\vartheta(k)\int_{1/L_0}^{k} k'^2 E^{(\vartheta)}(k')\,dk'$, where $K_\vartheta(k)$ is the eddy diffusivity. In the inertial subrange, dimensional considerations suggest that $K_\vartheta(k) = \gamma_2\bar{\varepsilon}^{1/3}k^{-4/3}$ where $\gamma_2$ is a new dimensionless coefficient. Substituting the expression for $W_\vartheta(k)$ which results from this into Eq. (22.44), we can readily show that

$$[\gamma_2\bar{\varepsilon}^{1/3}k^{-4/3} + \mu k^{-2}]\frac{dH_\vartheta(k)}{dk} - \frac{4}{3}\gamma_2\bar{\varepsilon}^{1/3}k^{-7/3}H_\vartheta(k) = 0,$$

$$H_\vartheta(k) = \int_{1/L_0}^{k} k'^2 E^{(\vartheta)}(k')\,dk'.$$

Hence, after some simple rearrangement, we find that

$$E^{(\vartheta)}(k) \sim k^{-5/3}[1 + \gamma_2^{-1}(kL_r)^{-2/3}], \quad L_r = \mu^{-3/2}\bar{\varepsilon}^{1/2}. \tag{22.46}$$

This formula is in agreement with Eq. (22.45) for $kL_r \gg 1$ but for $kL_r \lesssim 1$ the two formulas are different. Therefore, semiempirical hypotheses about the form of $W_\vartheta(k)$ do not unambiguously establish the form of the spectrum $E^{(\vartheta)}(k)$ of the nonconservative admixture concentration [in the case of a chemical reaction which is not of the first order the situation is, of course, more complicated; see, for example, Corrsin (1964)]. Predictions based on semiempirical hypotheses about the form of $W_\vartheta(k)$ for the smallest-scale fluctuations with $k \gtrsim 1/\eta_0$ are even less satisfactory. We shall not, therefore, pause to consider these results here (another approach to the problem of determination of $E^{(\vartheta)}(k)$ for $k > 1/\eta_0$ will be discussed in Sect. 22.3).

## *Turbulence in Thermally Stratified Media*

Let us now consider the equations for the spectral functions of turbulent velocity and temperature fluctuations in a thermally stratified fluid in a gravitational field. We shall suppose that the random velocity field $\boldsymbol{u}(\boldsymbol{x}, t)$ and the temperature field $T(\boldsymbol{x}, t)$ are locally homogeneous. For sufficiently large wave numbers $k = |\boldsymbol{k}| \gg 1/L_0$(where $L_0$ is the external length scale of turbulence determined by the smaller of the distances $L$ and $L_T$ over which the mean velocity and temperature fields are substantially changed) we shall be able to consider the energy and temperature spectral densities $F(k)$ and $F_{TT}(k)$, as well as the cross spectral density of the temperature and vertical-velocity fluctuations $F_{Tw}(\boldsymbol{k})$ [the last density is determined by the condition that $c_p \rho F_{Tw}(\boldsymbol{k})\, d\boldsymbol{k}$ is the contribution to the turbulent heat flux $q = c_p \rho \overline{T'w'}$ from the region $d\boldsymbol{k}$ in wave-number space]. The spectra integrated over all directions of the wave number vector $\boldsymbol{k}$, i.e., over a sphere of radius $k$ in wave-number space, will be denoted by $E(k)$, $E_{TT}(k)$ and $E_{Tw}(k)$.

In the equation (22.7) for the energy spectrum $E(k)$ we must now also take into account the work done by the buoyancy forces. It is readily seen that we must add the term $\frac{g}{T_0}\int_k^\infty E_{Tw}(k')\, dk'$ to the left-hand side of Eq. (22.7), so that this equation assumes the form

$$W(k) + \frac{g}{T_0}\int_k^\infty E_{Tw}(k')\, dk' = 2\nu \int_k^\infty k'^2 E(k')\, dk'. \tag{22.47}$$

Let us also write down Eq. (22.13) for the temperature spectrum:

$$W_T(k) = 2\chi \int_k^\infty k'^2 E_{TT}(k')\, dk'. \tag{22.48}$$

The two equations, (22.47) and (22.48), are not closed: In addition to the three spectra $E(k)$, $E_{TT}(k)$ and $E_{Tw}(k)$ in which we are directly interested, they contain also the functions $W(k)$ and $W_T(k)$ which describe the spectral transfer of kinetic energy and temperature fluctuation intensity. It is also possible to introduce in the usual way the additional equation for the cross spectrum $E_{Tw}(k)$, but this would then include an additional unknown function, namely, the third-order spectral transfer function $W_{Tw}(k)$, and the difficulties connected with the unclosed nature of the equations would remain. It is therefore better to try to close the equations given by Eqs. (22.47)–(22.48) with the aid of some semiempirical hypotheses, which would enable us to express any three of the five unknown functions in these equations in terms of the remaining two.

An attempt of this kind was undertaken by Monin (1962c) who used the semiempirical Heisenberg formula (17.7) for the energy transfer along the spectrum, i.e., he assumed that $W(k) = K(k)\,\Omega_k^2$ where $K(k)$ is the eddy viscosity produced by small-scale turbulence components (with wave numbers greater than $k$), and $\Omega_k^2$ is the mean square vorticity of large-scale motions (with wave numbers smaller than $k$). The analogous formula $W_T(k) = K_T(k)\,|\nabla_k T|^2$ was used for the spectral transfer of the intensity of the temperature fluctuations. In these expressions $K_T(k)$ is the eddy thermal diffusivity [which, for simplicity, was assumed equal to $\alpha K(k)$, where $\alpha$ is a constant], and $|\nabla_k T|^2$ is the contribution of large-scale inhomogeneities (with wave numbers less than $k$) to the mean square of the temperature gradient. Finally, by analogy with the representation of the moment $\overline{w'T'}$ by the semiempirical formula $\overline{w'T'} = -K_T \frac{\partial \overline{T}}{\partial z}$ [see, for example, the formula (7.3) in Vol. 1], it was assumed that the integral $\int_k^\infty E_{Tw}(k')\, dk'$ could be written in the form $\pm K_T(k)|\nabla_k T|$ where the upper sign refers to unstable $\left(\frac{\partial \overline{T}}{\partial z} < 0\right)$ and the lower to stable $\left(\frac{\partial \overline{T}}{\partial z} > 0\right)$ thermal stratification. For $k \ll 1/\eta$, the right-hand sides of Eqs. (22.47) and (22.48) can, of course, be replaced by $\overline{\varepsilon}$ and $\overline{N}$, respectively, so that these equations now become

$$K(k)\,\Omega_k^2 \pm \frac{g}{T_0}\,\alpha K(k)\,|\nabla_k T| = \overline{\varepsilon}, \qquad \alpha K(k)\,|\nabla_k T|^2 = \overline{N}. \tag{22.49}$$

The second of these two equations can be used to eliminate the quantity $|\nabla_k T|$ from the first, so that we then obtain one equation in the two unknown functions $K(k)$ and $\Omega_k^2$. However, the function $\Omega_k^2$ is related to the energy spectrum $E(k)$ by the obvious formula $\frac{d\Omega_k^2}{dk} = 2k^2E(k)$, and in the approximate theory which we are discussing the function $K(k)$ can be expressed in terms of $E(k)$ through some semiempirical formula which is consistent with dimensional reasoning and gives $K(k)$ in the form of an integral over the spectral region $k' \geqslant k$ [see, for example, Eqs. (17.8) and (17.10)–(17.12)]. Since none of these formulas can be regarded as better justified than the others, the expression for $K(k)$ is best chosen so that the resulting equation has the simplest solution. Having this end in view, Monin assumed for $K(k)$ the Stewart-Townsend formula (17.10) with $c = 1/2$, and then used Eq. (22.49) to show that

$$k^4 = \frac{\gamma_{ST}^2}{4}\left(\overline{\varepsilon} \mp \frac{1}{2}\frac{g}{T_0}\sqrt{\alpha\overline{N}K}\right)K^{-3}, \quad E = \frac{4\sqrt{2}}{3\gamma_{ST}^{3/2}}K^{5/4}\frac{\left(\overline{\varepsilon} \mp \frac{1}{2}\frac{g}{T_0}\sqrt{\alpha\overline{N}K}\right)^{5/4}}{\overline{\varepsilon} \mp \frac{5}{12}\frac{g}{T_0}\sqrt{\alpha\overline{N}K}}, \tag{22.50}$$

which gives the function $E(k)$ in parametric form. Since $\frac{d}{dk}|\nabla_k T|^2 = k^2E_{TT}(k)$, we have from Eqs. (22.49) and (22.50)

$$E_{TT} = \frac{8\sqrt{2}\,\overline{N}}{3\alpha\gamma_{ST}^{3/2}}K^{5/4}\frac{\left(\overline{\varepsilon} \mp \frac{1}{2}\frac{g}{T_0}\sqrt{\alpha\overline{N}K}\right)^{1/4}}{\overline{\varepsilon} \mp \frac{1}{12}\frac{g}{T_0}\sqrt{\alpha\overline{N}K}}. \tag{22.51}$$

Finally, $E_{Tw}$ turns out to be proportional to $(E \cdot E_{TT})^{1/2}$, so that the correlation coefficient between the spectral components of the vertical velocity and the temperature throughout the whole range $k \ll 1/L_0$ is independent of $k$. This is an obvious defect in the theory because in the limit as $g/T_0 \to 0$, i.e., for $k \gg 1/L_* = (g/T_0)^{3/2}\,\overline{N}^{3/4}\overline{\varepsilon}^{-5/4})$, we should in fact obtain all the properties of the usual locally isotropic turbulence for which $E_{Tw}(k) = 0$. This is explained by the fact that the formula for $\int_k^\infty E_{Tw}(k')\,dk'$ is inaccurate for $k \gtrsim 1/L_*$. The five-thirds law for $E(k)$ and $E_{TT}(k)$ (and for $E_{T\omega}(k)$ also) is obtained from Eqs. (22.50) and (22.51) both for $k \gg 1\ L_*$ and for $g/T_0 \to 0$. In the case of stable stratification (lower sign in front of $g/T_0$) and small $k$, we have the Bolgiano formulas (21.99), whereas in the case of an unstable stratification (upper sign in front of $g/T_0$) the spectra at first reach maxima as $k$ is reduced, and then fall to zero.

The above results remain valid even when the mean velocity gradient $\partial\bar{u}/\partial z$ and the mean temperature gradient $\partial\bar{T}/\partial z$ are taken into account in Eq. (22.49), i.e., if we assume that

$$\Omega_k^2 = \left(\frac{\partial\bar{u}}{\partial z}\right)^2 + \int_{k_0}^{k} 2k^2 E(k)\,dk, \tag{22.52}$$

$$|\nabla_k T|^2 = \left(\frac{\partial\bar{T}}{\partial z}\right)^2 + \int_{k_0}^{k} k^2 E_{TT}(k)\,dk, \tag{22.53}$$

where $k_0$ is the wave number corresponding to the maximum scale of the turbulent inhomogeneities allowed by the flow geometry (the presence of the earth's surface in the case of the atmospheric surface layer). The expression given by Eq. (22.52) was first employed by Obukhov (1941b). If we use the above formulas we do, in fact, take into account on the left-hand side of the first equation in Eq. (22.49) the work done by the Reynolds stresses, i.e., the turbulence energy production by the mean motion $K(k)\left(\frac{\partial\bar{u}}{\partial z}\right)^2$, while on the left-hand side of the second equation in Eq. (22.49) we take into account the production of temperature mean square fluctuations by the mean temperature gradient $\alpha K(k)\left(\frac{\partial\bar{T}}{\partial z}\right)^2$. If we do not use the above semiempirical formulas for these quantities, then Eqs. (22.47) and (22.48) can be written after differentiation with respect to $k$ in the form

$$E_{uw}(k)\frac{\partial\bar{u}}{\partial z} + \frac{\partial W(k)}{\partial k} - \frac{g}{T_0}E_{Tw}(k) = -2\nu k^2 E(k), \tag{22.54}$$

$$E_{Tw}(k)\frac{\partial\bar{T}}{\partial z} + \frac{\partial W_T(k)}{\partial k} = -2\chi k^2 E_{TT}(k), \tag{22.55}$$

if we take the mean velocity gradient and mean temperature gradient into account, where $E_{uw}(k)$ is the spectrum of the turbulent shear stress. Equations related to these were in fact used by Lumley (1964) [see also a subsequent expository paper of Phillips (1967)]. Lumley (and Phillips too) neglected the effects of molecular viscosity and thermal conductivity, i.e., assumed that $k \ll 1/\eta$ (they also neglected the term containing the mean velocity gradient, but did take into account the mean temperature gradient, which is not completely logical). Lumley's derivation of the results of Shur (1962) is based on the stated equations augmented with the Kovasznay hypothesis (17.3), according to which $E(k) \sim [W(k)]^{2/3} k^{-5/3}$, and with the assumption that the heat-flux spectrum $E_{Tw}(k)$ is proportional to the mean temperature gradient, while the ratio $E_{Tw}(k)\Big/\frac{\partial\bar{T}}{\partial z}$ is determined only by the quantities $k$ and $W(k)$, again in the spirit of the Kovasznay hypothesis. From these assumptions and dimensional considerations, Lumley showed that

$$E_{Tw}(k) = -c\,[W(k)]^{1/3} k^{-7/3}\frac{\partial\bar{T}}{\partial z}. \tag{22.56}$$

This formula, and also Eq. (22.47) written in the form $\frac{\partial W}{\partial k} = \frac{g}{T_0}E_{Tw}(k)$ and the formula for $E(k)$ given by Eq. (17.3), can be used to deduce an interpolation formula joining the five-thirds law for large $k$ with the Shur formula $E(k) \sim \frac{g}{T_0} \frac{\partial \overline{T}}{\partial z} k^{-3}$ for small $k$. It is also worth noting that $E_{T\omega}(k)$ decreases as $k^{-7/3}$ in the inertial subrange, according to formula (22.56), i.e., it decreases faster than the five-thirds velocity and temperature spectra; this deduction seems quite reasonable. However the derivation of Eq. (22.56) is obviously unrigorous and its accuracy is not at all clear.

## 22.3 Behavior of the Turbulent Energy Spectrum in the Far Dissipation Range

Semiempirical hypotheses, which enable us to close the dynamic structure or spectral equations, lead to definite predictions with regard to the asymptotic behavior of the turbulence spectrum for $k \to \infty$. However, the use of such hypotheses in the spectral range $k > 1/\eta$ is quite unjustified, and the resulting asymptotic formulas (which are different for different initial hypotheses) are therefore very unreliable.

We shall now consider another approach to the behavior of the turbulence spectrum at very high wave numbers which is based on certain approximate but easily interpreted assumptions about the nature of real physical processes governing the dynamics of turbulence in the far dissipation range. Unfortunately, certain nonrigorous hypotheses must still be introduced. The results which we shall now review are therefore not entirely rigorous. However, in contrast to the purely speculative hypotheses about the functional form of the spectral energy transfer $W(k)$ in the range $k \gtrsim 1/\eta$, the hypotheses which we shall discuss below have a definite physical meaning and, to some extent, are confirmed by existing data.

We shall consider disturbances whose scales are small in comparison with the Kolmogorov scale $\eta = (\nu^3/\overline{\varepsilon})^{1/4}$. We have already noted in the previous section that, as a result of viscosity, the velocity field varies smoothly over distances $l \ll \eta$. Therefore, within a space region of diameter $l$, the field $\boldsymbol{u}(\boldsymbol{x}, t)$ can be expanded in a Taylor series in the coordinates, and can be approximated by a linear vector function which is the sum of the zero and first-order terms in this expansion. The coefficients of this linear function—the derivatives $\partial u_l/\partial x_j$—will, of course, fluctuate in time and space [with $\overline{\left(\frac{\partial u_l}{\partial x_j}\right)^2} = \frac{1}{15}\frac{\overline{\varepsilon}}{\nu}$ for $l = j$ and $\overline{\left(\frac{\partial u_l}{\partial x_j}\right)^2} = \frac{2}{15}\frac{\overline{\varepsilon}}{\nu}$ for $l \neq j$; see Eq. (21.19′)]. However, for distances $r \ll \eta$ the instantaneous values of these derivatives will be practically constant.[14]

[14] We note that the mean square of the change in the derivatives $\partial u_l/\partial x_j$ over a distance $r$ can be estimated from the obvious relation

(Continued on page 422)

We shall start with the vorticity equation in an incompressible fluid

$$\frac{\partial \omega_j}{\partial t} + u_l \frac{\partial \omega_j}{\partial x_l} - \omega_l \frac{\partial u_j}{\partial x_l} = \nu \Delta \omega_j, \tag{22.57}$$

where $\omega_j = \varepsilon_{jlm} \frac{\partial u_m}{\partial x_l}$. Consider a moving fluid particle, i.e., a small volume $\delta V$ of the fluid with linear dimensions $l < \eta$. According to the above discussion, the velocity field can be regarded in the first approximation as a linear function of the coordinates within this volume, i.e., we may suppose that $u_l(\boldsymbol{x}_0 + \boldsymbol{r}, t) = u_{0l} + a_{lk} r_k$ where $u_{0l} = u_l(\boldsymbol{x}_0, t), a_{lk} = \partial u_l(\boldsymbol{x}_0, t)/\partial x_{0k}$, and $\boldsymbol{x}_0$ is a fixed point inside $\delta V$. The term $u_{0l}$ will then describe the transport of the chosen fluid particle as a whole, the antisymmetric part $\frac{1}{2}(a_{lk} - a_{kl})$ of the tensor $a_{lk}$ will describe its rotation as a solid body, and the symmetric part $\frac{1}{2}(a_{lk} + a_{kl})$ will describe its strain rate. This strain, in general, takes the form of extensions or compressions along three mutually

---

$$\sum_{l,j} \overline{\left( \frac{\partial u_l'}{\partial x_j'} - \frac{\partial u_l}{\partial x_j} \right)^2} = 2 \frac{\overline{\varepsilon}}{\nu} - \Delta D_{ll}(r).$$

If we now express $D_{ll}(r)$ in terms of the scalar function $D_{LL}(r) = \frac{1}{15} \frac{\overline{\varepsilon}}{\nu} r^2 + dr^4 + \ldots$, we obtain

$$\sum_{l,j} \overline{\left( \frac{\partial u_l'}{\partial x_j'} - \frac{\partial u_l}{\partial x_j} \right)^2} \approx 140 dr^2.$$

using the Kolmogorov structure equation (22.2), we can transform the last result so that it reads

$$\sum_{l,j} \overline{\left( \frac{\partial u_l'}{\partial x_j'} - \frac{\partial u_l}{\partial x_j} \right)^2} \Big/ \sum_{l,j} \overline{\left( \frac{\partial u_l}{\partial x_j} \right)^2} \approx \frac{7|s|}{18\sqrt{15}} \frac{r^2}{\eta^2} \approx \frac{|s|}{10} \frac{r^2}{\eta^2};$$

$$|s| = -\overline{\left( \frac{\partial u_1}{\partial x_1} \right)^3} \Big/ \left[ \overline{\left( \frac{\partial u_1}{\partial x_1} \right)^2} \right]^{3/2} > 0$$

This result is due to Batchelor (1959). Existing data on the quantity $|s|$ (which refer to grid turbulence at moderate Reynolds numbers) is of the order of 0.5–0.4 and, as Re increases, the value of $|s|$ even appears to decrease (see Sect. 18.1). Therefore, we may suppose that the change in the derivatives $\partial u_l/\partial x_j$, even over distances of the order of $\eta$ will still be relatively small.

perpendicular axes (principal axes of strain rate). Therefore, if we transform to a new set of coordinates which moves and rotates together with the fluid particle, we can transform the linear part of the velocity field to the form $u_i(\boldsymbol{x}_0+\boldsymbol{r})=a_{ik}r_k$, where $a_{ik}$ is a symmetric tensor (equal to the symmetric part of the original tensor $a_{ik}$). If we denote by $a_1$, $a_2$, and $a_3$ the principal values of the symmetric tensor then, since the fluid is incompressible, we obtain

$$a_1+a_2+a_3=0,$$

so that at least one of the values of $a_i$ is positive and one is negative. We shall always assume henceforth that $0<a_1\geqslant a_2\geqslant a_3<0$, i.e., we shall use $a_1$ to denote the maximum positive strain rate (maximum rate of extension), and use $a_3$ to denote the maximum negative strain rate (maximum rate of compression).

A constant vorticity field will, of course, correspond to a linear velocity field. Therefore, when we investigate the statistical structure of the vorticity field we must take into account the departure of the true velocity field $\boldsymbol{u}(\boldsymbol{x}, t)$ from the linear field. Since, however, these departures are very small inside the small volume $\delta V$, it is natural to expect that, when we investigate small-scale disturbances of the vorticity field (with scales that are small in comparison with $\eta$), the equation (22.57) can be linearized, i.e., we can neglect the quadratic combinations of the above small quantities in this equation.[15] The linearized equation (describing the interaction of a small variable component of the vorticity with the main linear velocity field) will be used below. In the noninertial set of coordinates introduced above, which moves with the fluid particle, the vorticity $\boldsymbol{\omega}$ is itself a small quantity (since the antisymmetric part of the tensor $a_{ik}$ is now zero). Accordingly, the linearized equations (22.57) in such a system assume the form

$$\frac{\partial\omega_j}{\partial t}+a_{kl}x_l\frac{\partial\omega_j}{\partial x_k}-a_{jl}\omega_l=\nu\Delta\omega_j. \tag{22.58}$$

The coefficients $a_{kl}$ within the volume $\delta V$ are independent of the coordinates but, in general, cannot be regarded as time-independent. General considerations of the local structure then merely show that

[15]We note that the validity of this linearization procedure has been disputed by Kraichnan [see Batchelor (1962)].

the time dependence of $a_{kl}$ can be neglected for time intervals that are small in comparison with $\tau_\eta = \nu^{1/2}\bar{\varepsilon}^{-1/2}$. However, there are some indirect (and not very convincing) reasons to suppose that the change in the coefficients $a_{kl}$ with time is much slower and therefore it can be neglected in the first approximation, i.e., we may consider that $a_{kl}$ is constant even for time intervals of the order of $\tau_\eta$. The point is that if we accept this assumption we can readily calculate the effect of small-scale convection (determined by the linear velocity field) on the diffusion of a passive admixture initially concentrated in a very small volume. This calculation was carried out by Townsend (1951b) who compared the results with his own observations on the evolution of "heat spots" produced in the turbulence behind a grid by pulsed current discharges. The calculations were found to agree quite satisfactorily with experimental data, indicating that the method employed in the calculation must not be too far from reality. Visual observations of the behavior of dyes introduced into a turbulent flow lead to the same conclusion [data of this kind were published among others by Welander (1955) for the two-dimensional case]. In particular, the colored spots usually become extended into long bands of irregular shape (and increasing length), showing continuing elongation of fluid volume elements in the same direction. The colored bands do not exhibit small curls and rapid changes in curvature, which would indicate short-period rotations of small regions of the fluid relative to the remainder of the fluid.

The fact that the coefficients $a_{kl}$ are constant means that, over times of the order of $\tau_\eta$, the principal strain axes do not appreciably rotate relative to the fluid particle under consideration, and the rate of strain along these axes is practically constant. This is the main hypothesis which enables us to determine the asymptotic shape of the turbulence spectrum for $k \to \infty$. We note, moreover, that the hypothesis which we have just formulated is too restricted and can be replaced by a less stringent requirement. Thus, for example, it is quite sufficient to follow Novikov (1961a, c) and assume that, in a time interval of the order of $\tau_\eta$, the principal strain axes do not rotate relative to the fluid particle (i.e., the axis of maximum extension remains the axis of maximum extension, while the axis of maximum compression remains the axis of minimum compression), but the strain rates along the principal axes may vary. In that case, the quantities $a_1$, $a_2$ and $a_3$ will be functions of time, and expressions of the form $a_j t$ will be replaced by $\int^t a_j(s)\,ds$. In other respects, however, the discussion below remains valid. It is also possible to follow Saffman (1963) and assume that the principal strain axes can rotate relative to the fluid particle but in such a way that the rate of this rotation in a time of the order of $\tau_\eta$ is practically constant. In that case, in the set of coordinates attached to the principal strain axes we again obtain an equation of the form of (22.58) (but now with an asymmetric tensor $a_{kl}$), which can then be used together with certain simple assumptions to obtain the results which we shall outline below. However, some assumption must be introduced about the constancy of some parameters of the smallest-scale motions of the fluid over time intervals of the order of $\tau_\eta$. We shall, however, not try to select the least restrictive assumptions, and consider the consequences of the simplest hypothesis, namely, that the characteristic time of variation of all the coefficients $a_{kl}$ is large in comparison with $\tau_\eta$.[16]

Since we are assuming that the principal strain axes rotate together with the fluid particle, we can take these axes as our coordinate axes to which the equation (22.58) is referred.[17] The tensor $a_{kl}$ will be diagonal in this set of coordinates, i.e., $a_{kl}=a_k\delta_{kl}$, so that the linearized equation (22.58) now assumes the form

$$\frac{\partial\omega_j}{\partial t}+\sum_{l=1}^{3}a_l x_l\frac{\partial\omega_j}{\partial x_l}-a_j\omega_j=\nu\Delta\omega_j,\qquad j=1,\,2,\,3 \qquad (22.59)$$

(no summation over $j$ in the third term on the left-hand side).

Equation (22.59) (with constant coefficients $a_l$) was first used to determine the asymptotic behavior of the turbulence spectrum by Townsend (1951a). Townsend showed that, for certain values of $a_1$, $a_2$, and $a_3$, these equations have a stationary solution describing a vortex sheet of finite thickness, while for other values of $a_l$ the

[16] Let us emphasize at this point that the hypothesis is, in fact, not very well grounded and is used here only since no more accurate methods of calculation are available at present. See in this respect the paper by Lumley (1972) where arguments are given indicating that the time scale of the variation of the tensor $a_{kl}$ may be close to $\tau_\eta$ (but not larger) and the consequences of these arguments are analyzed.

[17] The corresponding coordinate system is of course not inertial and hence the dynamical equations here must contain some additional terms. However it is easy to show that the main additional term will be the Coriolis term $(1/2)\epsilon_{jln}\omega_l^0\omega_n$ where $\epsilon_{jln}$ is a completely skew-symmetric tensor and $\omega_l^0=\omega_l(0)$ is the vorticity at the coordinate origin (which determines the angular velocity of the new coordinate system). This term vanishes at the origin and is negligibly small if $|x|\ll\eta$; we are considering only such values of $|x|$.

solution takes the form of a vortex line. Next, he assumed that this stationary solution described the asymptotic state to which arbitrary initial disturbances would tend under the simultaneous action of the quasi-stationary linear field of large-scale (of the order of $\eta$) velocity components and the viscous forces. If this is so then the form of the solutions appears to indicate that the small-scale disturbances should collect together into concentrated vortex sheets or lines (which exist beyond the time intervals of the order of $\tau_\eta$ or more). Using this, Townsend calculated the one-dimensional longitudinal velocity spectrum $E_1(k)$ corresponding to a set of random and randomly oriented independent vortex sheets (or lines) and assumed that the result would describe the asymptotic form of the real spectrum for $k \gg 1/\eta$. In the vortex sheet model the formula for $E_1(k)$ is found to be of the form

$$E_1(k) = Ak^{-2} \int_0^1 (1 - \xi^2) \exp\{-\nu k^2/a\xi^2\}\, d\xi, \qquad (22.60)$$

where $a$ is an empirical constant (having the dimensions of the reciprocal of time). In the case of the model consisting of individual vortex lines, the formula for $E_1(k)$ is of the same type but is more complicated [it also shows that $E_1(k)$ falls off approximately as $\exp\{-ck^2\}$ at infinity, where $c$ is a constant]. By suitably selecting the parameter $a$, it was found that the formula given by Eq. (22.60) was in adequate agreement with the experimental data of Stewart and Townsend on the function $k^6E_1(k)$ for $0.2 \leqslant k\eta \leqslant 1.6$ (see Sect. 16.6). This agreement, however, can hardly be regarded as a reliable indication that Townsend's analysis was correct. The data of Stewart and Townsend were obtained for relatively small values of the Reynolds number and values of $k$ that were not sufficiently high. Moreover, recent precise measurements by Kuo and Corrsin (1972) agree best with the line model.

Subsequently, Pearson (1959) carried out certain calculations which, in principle, could be used to justify Townsend's analysis, but unexpectedly led to results which threw doubt on the entire approach based on Eq. (22.59). Namely, Pearson considered the general solution of the initial-value problem for the equations (22.59), and investigated the asymptotic behavior of this solution for $t \to \infty$. He found that $\overline{\omega^2} = \overline{\omega_l \omega_l} \to \infty$ for $t \to \infty$, i.e., in this particular approximation the mean vorticity increased without limit with time, in spite of the viscosity effects [a simplified derivation of this can also be found in Saffman (1963)]. Hence it follows that, in a constant linear velocity field, weak disturbances will in general be unstable, i.e., in the linear approximation they will increase exponentially, and will not tend to any stationary state determined by the linearized equations.

Pearson's work has led to attempts to replace Eq. (22.59) by more general equations, taking into account the evolution of the large-scale linear velocity field (for example, using the assumption that the principal strain axes rotate relative to the fluid particle). However, the divergence found by Pearson is unrelated to the behavior of the smallest-scale disturbances in real turbulent flow. It follows from the calculations of Pearson and Saffman that an indefinite increase in the mean square vorticity in the case of a linear velocity field is produced by an increase of the intensity of the largest-scale spectral components with time (for $|\boldsymbol{k}| < k_0$, where $k_0$ is a fixed number which we can choose as small as desired), and these components are least affected by viscosity. In other words, the conclusion that the vorticity increases is essentially based on the assumption that the basic linear velocity field occupies infinite space. Moreover, in the case of developed turbulence the velocity field can be regarded as approximately linear only over distances of the order of $\eta$ but not otherwise.

Hence it follows that Pearson's result should not prevent us from using Eq. (22.59) to investigate the asymptotic behavior of the turbulence spectrum, and merely shows that we must first "filter off" the large-scale motions which distort the situation in which we are interested. Moreover, it turns out that the asymptotic form of the spectrum can then be estimated even without introducing any artificial assumptions about the decay of the entire turbulent flow into independent vortex sheets or lines, and that this more rigorous development leads to results which are somewhat different from those of Townsend (1951a).

In fact, let us try, following Novikov (1961a), to investigate the fate of a weak sinusoidal disturbance imposed on the linear velocity field, which does not vary in the system of coordinates rotating together with the fluid particle. To do this, we must consider the solution of Eq. (22.59) corresponding to the initial condition

$$\omega_j(\boldsymbol{x}, 0) = A_{0j} \exp\{i\boldsymbol{k}_0\boldsymbol{x}\} = A_{0j} \exp\{i(k_{01}x_1 + k_{02}x_2 + k_{03}x_3)\}. \quad (22.61)$$

We shall seek this solution in the form

$$\omega_j(\boldsymbol{x},\ t) = A_j(t)\exp\{i\boldsymbol{k}(t)\boldsymbol{x}\} = A_j(t)\exp\{i(k_1x_1 + k_2x_2 + k_3x_3)\}. \tag{22.62}$$

Substituting this solution into Eq. (22.59), we obtain ordinary differential equations for the functions $k_j(t)$ and $A_j(t)$, whose solutions satisfying the conditions $k_j(0) = k_{0j},\ A_j(0) = A_{0j}$ are given by

$$\begin{gathered} k_j(t) = k_{0j}\exp\{-a_jt\}; \\ A_j(t) = A_{0j}\exp\left\{\int_0^t [a_j - \nu k^2(t_1)]\,dt_1\right\},\quad k^2(t) = k_j(t)\,k_j(t). \end{gathered} \tag{22.63}$$

Since we are assuming that $0 < a_1 \geqslant a_2 \geqslant a_3 < 0$, it is clear from the first formula in Eq. (22.63) that $k_1(t)$ decreases and $k_3(t)$ increases with time (provided only that $k_{03} \neq 0$). The function $k_2(t)$ increases or decreases with time, depending on the sign of $a_2$. However, in any case, $|k_2(t)| \ll |k_3(t)|$ for sufficiently large $t$, while in the first approximation it may be considered that asymptotically $k^2(t) \approx k_3^2(t)$.[18]

Next, the second formula in Eq. (22.63) shows that if $A_{01} \neq 0$, then whatever the relation between the initial values $A_{01} \neq 0,\ A_{02}$ and $A_{03}$, the quantity $A_1(t)$ will be much greater than $A_2(t)$ and $A_3(t)$ for sufficiently large $t$, so that asymptotically $A^2(t) \approx A_1^2(t)$.[19] In other words, during the evolution of the initial disturbance, the wave vector $\boldsymbol{k}(t)$ rotates and becomes asymptotically parallel to the axis of maximum compression of the fluid particle (i.e., the disturbance retains only its dependence on the coordinate $x_3$ along the axis of maximum compression) and the vorticity vector $\boldsymbol{\omega}(\boldsymbol{x},\ t)$ is asymptotically parallel to the axis of maximum extension. The characteristic

[18] We note that if $a_2 < 0$ the component $k_2(t)$ will also increase, and a more accurate approach is to consider that $k^2(t) \approx k_3^2(t) + k_2^2(t)$ for large $t$. In that case, we must replace the range $\delta k$ on the $k = k_3$ axis by the area element $\delta k_3 \cdot \delta' k_2$, so that instead of Eq. (22.64) the expression in the square brackets on the right-hand side of this formula must be augmented by the additional term $-a_2/a_3$. However, this term is negligible in comparison with $2\nu k^2/a_3$ which gives rise to the exponential factor on the right-hand side of Eq. (22.64). Moreover, the case $a_2 < 0$ is rarely encountered for real flows since $\overline{a_1a_2a_3} < 0$ (see the end of the present subsection). A similar remark can be made with regard to the evolution of small-scale temperature disturbances, which we shall discuss in the next subsection.

[19] This analysis must be corrected only in the exceptional case when $a_2 = a_1$, for which $A^2(t) \approx A_1^2(t) + A_2^2(t)$. We then have, however, $dA_1^2/dt = dA_2^2/dt$ and therefore the final result remains the same.

time of this process, given by Eq. (22.63), is determined by the reciprocal of typical strain rates, i.e., it is of the order of $\tau_\eta$. We shall consider quasi-stationary turbulence whose mean characteristics can change appreciably only over time intervals much greater than $\tau_\eta$. Moreover, we shall adopt the above assumption that the linear velocity field in the neighborhood of a given fluid particle remains practically constant over time intervals of the order of $\tau_\eta$. In that case, the spectral components of the vorticity field can be regarded as "adjusted" to the strain field of the fluid particle in accordance with the above asymptotic relations, i.e., we may suppose that the wave vectors of these components are parallel to the axis of maximum compression of the fluid particle, while the vorticity vectors are parallel to the maximum extension axis. Allowance for the dependence of the disturbance on the coordinate $x_3$ only in the direction of which the perturbation gradients increase rapidly with time (owing to the compression of the particle) in effect acts as a "filter" for the large-scale motions in which we are not interested. The asymptotic form of the turbulent spectrum for $k \gg 1/\eta$ can then be determined from the condition that it should be unaffected by the simultaneous inflow of energy from the region of larger-scale disturbances (associated with the strain of the fluid particle) and attenuation due to viscosity.

In fact, let us suppose that the spectral components of the vorticity field have already become "adjusted" to the strain field, i.e., $\omega_1(\boldsymbol{x}) = Ae^{ikx_3}$, $\omega_2 = \omega_3 = 0$. Let $\delta A = \delta A(t)$ be the amplitude of the spectral component corresponding at a time $t$ to the range $\delta k$ along the wave-number axis and containing the point $k$. By Eq. (22.63), this component will be converted into the disturbance $\delta A(t + dt)$ $\exp\{ik(t + dt)x_3\}$ in a time $dt$, where

$$\delta A(t + dt) = \delta A + d(\delta A) = \delta A[1 + (a_1 - \nu k^2)dt],$$
$$k(t + dt) = k + dk = k(1 - a_3 dt).$$

However, the kinetic energy of the disturbances per unit length along the wave-number axis is $\mathscr{E}(k, t) = (\delta A)^2/2k^2\,\delta k$ (the factor $2k^2$ in the denominator appears when we transform from the square of the vorticity $\omega^2$ to the kinetic energy $u^2/2$). Hence, when the range $\delta k$ along the wave-number axis is transformed into the range $\delta k(t + dt) = \delta k(1 - a_3 dt)$ in a time $dt$, we have

$$d\mathscr{E}(k, t)=\mathscr{E}(k, t)\left[\frac{2d(\delta A)}{\delta A}-\frac{2\,dk}{k}-\frac{d(\delta k)}{\delta k}\right]=$$
$$=\mathscr{E}(k, t)(2a_1-2\nu k^2+3a_3)\,dt$$

or (since $dk=-a_3k\,dt$)

$$d\mathscr{E}(k, t)=\mathscr{E}(k, t)\left[-2a_1/a_3-3+2\nu k^2/a_3\right]k^{-1}\,dk. \quad (22.64)$$

After taking a statistical average of the quantity $\mathscr{E}(k, t)$ we find that it is identical with the energy spectrum $E(k)$ which we are assuming to be independent of time $t$. When we take the statistical average of Eq. (22.64) we must remember, of course, that $\mathscr{E}(k, t)$, $a_1/a_3$, and $1/a_3$ are all random variables. However, the variables $a_1/a_3$ and $1/a_3$ are determined by the large-scale motion of scale $\eta$, and they vary very slowly. The variation becomes appreciable only over time intervals that are large in comparison with $\tau_\eta$ (or at any rate not less than the order of $\tau_\eta$, even if we suppose that only the *orientation* of the principal strain axes relative to the fluid particle does not change in a time $\tau_\eta$ while the strain rate along these axes can vary). At the same time, $\mathscr{E}(k, t)$ characterizes the small-scale motions of scale $1/k$ which are subject to strong influence of viscosity and have therefore a typical time of the order of $(\nu k^2)^{-1}\ll\tau_\eta$ (since $k\gg 1/\eta$). It is natural to expect that we can neglect the correlation between the rapidly fluctuating random variable $\mathscr{E}(k, t)$ and the much more slowly changing variables $a_1/a_3$ and $a_3^{-1}$, i.e., that we can replace the mean values of the products of these variables by the product of the corresponding means. If we accept this assumption and take the average of both sides of Eq. (22.64), we obtain

$$\frac{d\overline{\mathscr{E}(k, t)}}{dk}=\frac{\overline{\mathscr{E}(k, t)}}{k}\left(-2\,\overline{a_1a_3^{-1}}-3+2\nu k^2\,\overline{a_3^{-1}}\right)$$

or, in other words,

$$\frac{d\ln E(k)}{dk}=\frac{2\sigma-3}{k}-2\alpha\eta^2k, \quad (22.65)$$

where

$$\sigma=-\overline{a_1/a_3}, \quad \alpha=-\overline{1/\tau_\eta a_3}=-(\bar{\varepsilon}/\nu)^{1/2}\overline{a_3^{-1}}. \quad (22.66)$$

Integrating Eq. (22.65), and taking into account the dimensions of

$E(k)$, we find that

$$E(k) = Ak^{2\sigma-3}\exp\{-\alpha(k\eta)^2\} = C\bar{\varepsilon}^{2/3}k^{-5/3}(k\eta)^{2\sigma-4/3}\exp\{-\alpha(k\eta)^2\}, \tag{22.67}$$

where $C$ is a numerical integration constant of the order of unity. The formula (22.67) is due to Novikov (1961a) and, subject to the above assumptions, covers the asymptotic behavior of the turbulence spectrum for $k\eta \gg 1$. By substituting Eq. (22.67) into the general formula (12.86), which relates the spectrum $E(k)$ to the longitudinal one-dimensional spectrum $E_1(k)$, we can readily show that

$$E_1(k) \sim k^{2\sigma-3}\int_0^1 (1-\xi^2)\,\xi^{2-2\sigma}e^{-\alpha(k\eta)^2/\xi^2}\,d\xi. \tag{22.68}$$

This last relation is of the same kind as (but is not identical with) the Townsend formula (22.60) put forward ten years earlier, but it contains the additional parameter $\sigma$.

Following Saffman (1963), we can justify Eq. (22.67) by the following general considerations. From the vorticity equation (22.57) it follows in the usual way that the scalar correlation function $B^{(\omega)}(r, t) = B_{\omega_l\omega_l}(r, t) = \overline{\omega_l\omega_l'}$ for the vorticity field $\omega(x, t)$ satisfies Eq. (22.69).

$$\frac{\partial B^{(\omega)}}{\partial t} - 2\nu\left(\frac{\partial^2 B^{(\omega)}}{\partial r^2} + \frac{2}{r}\frac{\partial B^{(\omega)}}{\partial r}\right) = 2\overline{\omega_l\omega_j'\frac{\partial u_l'}{\partial x_j'}} + \overline{(u_j - u_j')\frac{\partial(\omega_l\omega_l')}{\partial r_j}}. \tag{22.69}$$

Let us consider this equation for distances $r \ll \eta$. The first term on the right is linear in the product $\omega_l\omega_j'$ and in the components of the tensor $\partial u_l/\partial x_j$. If we again assume that disturbances with very different scales are practically uncorrelated, and take into account the requirement of local isotropy, then it would seem that the only possible form of this term for $r \ll \eta$ is

$$\overline{\omega_l\omega_j'\frac{\partial u_l'}{\partial x_j'}} = aB^{(\omega)}(r), \tag{22.70}$$

where $a$ is a constant (having the dimensions of the reciprocal of time) which characterizes the mean values of the components of the tensor $\partial u_l/\partial x_j$, i.e., it is of the order of $\bar{\varepsilon}^{1/2}\nu^{-1/2}$. Similarly, for the second term on the right-hand side of Eq. (22.69) the only expression which for $r \ll \eta$ is consistent with both the linearity in the derivatives of the correlation tensor of the vorticity and the requirement of isotropy is apparently

$$\overline{(u_j - u_j')\frac{\partial(\omega_l\omega_l')}{\partial r_j}} = cr\,\frac{\partial B^{(\omega)}(r)}{\partial r}, \tag{22.70'}$$

where $c$ is a further constant (having the same dimensions and the same order as $a$). The physical significance of Eqs. (22.70) and (22.70') is readily established by comparison with the general equations for vorticity balance: The first of them describes the enhancement of vorticity due to the extension of the vortex lines, and the second the redistribution over the spectrum due to convection. Hence it is readily concluded that both $a$ and $c$ must be positive. If we substitute Eqs. (22.70) and (22.70') into Eq. (22.69) we have for $r \ll \eta$

$$\frac{\partial B^{(\omega)}}{\partial t} = \left(2aB^{(\omega)} + cr\frac{\partial B^{(\omega)}}{\partial r}\right) + 2\nu\left(\frac{\partial^2 B^{(\omega)}}{\partial r^2} + \frac{2}{r}\frac{\partial B^{(\omega)}}{\partial r}\right). \tag{22.71}$$

It is readily verified that, in the stationary state, the last equation with $c = (\alpha\tau_\eta)^{-1}$ and $a/c = \sigma$ is exactly equivalent to Eq. (22.65). We note that if we start with Eq. (22.59) and assume that the vorticity field is "adjusted" to the strain of the fluid particle, i.e., it is directed along the $x_1$ axis and depends only on the coordinate $x_3$, then we readily obtain an equation of the form of Eq. (22.71), where $a \approx \overline{a_1}$, $c \approx -\overline{a_3}$ [see below the analogous derivation of Eq. (22.83)].

If we suppose that the parameters $\sigma$ and $\alpha$ are exactly determined by Eq. (22.66), we can obtain an estimate of the values of these parameters. To begin with, from the inequalities $0 < a_1 \geqslant a_2 \geqslant a_3 < 0$ and the condition $a_1 + a_2 + a_3 = 0$ (which follows from incompressibility) we have $1/2 \leqslant -a_1/a_3 \leqslant 2$ and, consequently, $1/2 \leqslant \sigma \leqslant 2$. Next, we can use the relations

$$\overline{a_1^2} + \overline{a_2^2} + \overline{a_3^2} = \bar{\varepsilon}/2\nu, \quad \overline{a_1 a_2 a_3} = -\frac{35}{8}\overline{\left(\frac{\partial u_1}{\partial x_1}\right)^3} \tag{22.72}$$

which we have proved in Sect. 12.4 [see Eq. (12.108) and the result $a_1^3 + a_2^3 + a_3^3 = 3a_1a_2a_3$ in Sect. 12.4]. The first of these relations can be rewritten in the form $\overline{a_3^2(1 + a_1/a_3 + a_1^2/a_3^2)} = \bar{\varepsilon}/4\nu$. Since $1/2 \leqslant -a_1/a_3 \leqslant 2$ this shows that $\overline{a_3^2} \leqslant \bar{\varepsilon}/3\nu$ but $\overline{a_3^2} > \bar{\epsilon}/12\nu$. Moreover the application of the Schwarz inequality $\overline{|xy|} < (\overline{x^2}\,\overline{y^2})^{1/2}$ at first to $x = |a|^{1/2}$, $y = |a|^{-1/2}$, and then to $x = a$, $y = 1$ implies that $\overline{a^{-1}} \geqslant 1/(\overline{a^2})^{1/2}$ $\overline{a^{-1}} > 1/(\overline{a^2})^{1/2}$; hence we have $\alpha = -\sqrt{\bar{\varepsilon}/\nu}\,\overline{a_3^{-1}} \geqslant \sqrt{3}$ [Novikov (1961a)]. On the other hand, since both theory and experiment shows that the quantity $\overline{(\partial u_1/\partial x_1)^3}$ is negative, we see from the second equation in Eq. (22.72) that cases with $a_2 > 0$ should predominate on the average. Since $a_1 + a_2 + a_3 = 0$, it follows that, as a rule, $a_1$ is smaller than $|a_3| = a_1 + a_2$ and, consequently, $-a_1/a_3$ assumes values less than unity. Therefore, it is natural to expect that $\sigma = \overline{-a_1/a_3} < 1$. Moreover $a_3^2 > (a_1^2 + a_2^2 + a_3^2)/2$ if $-a_1/a_3 < 1$; hence it is natural to expect that $\overline{a_3^2}$ is much greater than $\bar{\epsilon}/12\nu = (\overline{a_1^2} + \overline{a_2^2} + \overline{a_3^2})/6$. On the other hand the difference between $\overline{a_3^{-1}}$ and $(\overline{a_3^2})^{-1/2}$ hardly can be very great; therefore it is very likely that $\alpha < \sqrt{12} = 2\sqrt{3}$ (C. H. Gibson, 1968).

If we assume that $\sigma = 2/3$ (this value lies in the range $1/2 < \sigma < 1$, which is the most probable) then the formula (22.67) assumes the form

$$E(k) = C\bar{\varepsilon}^{2/3}k^{-5/3}\exp\{-\alpha(k\eta)^2\}. \tag{22.73}$$

In this case, for $k \ll 1/\eta$ this formula is obviously in agreement with the five-thirds law for the inertial subrange. This formula with $C$ determined by the normalization condition $\int_0^\infty E(k)\,k^2\,dk = \bar{\varepsilon}/2\nu$, i.e., with $C = \alpha^{2/3}/\Gamma(2/3)$, is at first sight an attractive interpolation formula having the correct asymptotic behavior both for $k\eta \ll 1$ and $k\eta \gg 1$. Unfortunately, existing data on the turbulence spectrum for large Re are not in good agreement with this formula (see Sect. 23.4 below). This fact is however not of great importance, since the data must be greatly influenced by the effect of fluctuations of the rate of energy dissipation (see Sect. 25 below) which masks the original form of the spectrum $E(k)$. Moreover, the simplicity of Eq. (22.73) justifies its use as a model of the true spectrum $E(k)$ which in many cases enables us to obtain a closed analytic solution for turbulent-flow problems (see, in particular, Chapt. 9 below).

## 22.4 Behavior of the Temperature Spectrum at Very Large Wave Numbers

The analysis given in the previous section in connection with the asymptotic behavior of the energy spectrum $E(k)$ can also be used to investigate the temperature spectrum (or spectrum of the concentration of a passive admixture) $E^{(\vartheta)}(k) = E_{\vartheta\vartheta}(k)$. In this case, however, the dynamic equations contain an additional dimensional parameter, namely, the molecular diffusivity of temperature (or of mass) $\chi$. Therefore, the form of the spectrum $E^{(\vartheta)}(k)$ will now also depend on the (thermal or diffusion) Prandtl number $\mathrm{Pr} = \nu/\chi$.

Following Batchelor (1959), let us begin by considering the case $\mathrm{Pr} \gg 1$ (i.e., $\nu \gg \chi$) which, in some respects, is the simplest of all. Here, the velocity field for $l < \eta = (\nu^3/\bar{\varepsilon})^{1/4}$ can be regarded as practically linear in the coordinates, whereas the temperature field $\vartheta(x, t)$ undergoes appreciable turbulent fluctuations over such distances. When we investigate the statistical properties of these small-scale fluctuations we can replace the velocity components $u_j$ in the thermal conduction equation

$$\frac{\partial\vartheta}{\partial t} + u_j \frac{\partial\vartheta}{\partial x_j} = \chi\Delta\vartheta \tag{22.74}$$

by linear functions of the coordinates. If we then again transform to the set of coordinates moving through space and rotating together with the given fluid particle, and assume that the directions of the coordinate axes lie along the principal strain axes of the particle, we obtain

$$\frac{\partial\vartheta}{\partial t} + a_1x_1\frac{\partial\vartheta}{\partial x_1} + a_2x_2\frac{\partial\vartheta}{\partial x_2} + a_3x_3\frac{\partial\vartheta}{\partial x_3} = \chi\Delta\vartheta, \tag{22.75}$$

which is analogous to Eq. (22.59). If we also assume that the principal strain axes undergo practically no rotation in a time of the order of $\tau_\eta$ (and that, for simplicity, the corresponding strain rates $a_1$, $a_2$, and $a_3$ remain practically constant during such time intervals; see Sect. 22.3), we can readily determine from the above equation the asymptotic form of the spectrum $E^{(\vartheta)}(k)$. In fact, consider a sinusoidal disturbance of the field $\vartheta$ corresponding to the initial condition

$$\vartheta(x, 0) = B_0 \exp\{i(k_{01}x_1 + k_{02}x_2 + k_{03}x_3)\}.$$

If we seek the corresponding solution of Eq. (22.75) in the form

$$\vartheta(x, t) = B(t) \exp\{i(k_1(t)x_1 + k_2(t)x_2 + k_3(t)x_3)\}, \quad (22.76)$$

we readily obtain

$$k_j = k_{0j} \exp\{-a_j t\}, \quad B(t) = B_0 \exp\left\{-\chi \int_0^t k^2(t_1)\, dt_1\right\}, \quad (22.77)$$

where $k^2 = k_j k_j$.

Since we have agreed that $0 < a_1 \geqslant a_2 \geqslant a_3 < 0$, it is clear from Eq. (22.77) that asymptotically $k_1(t) \to 0$, $k_3(t)$ increases, and $k^2(t) \approx k_3^2(t)$. Therefore, the isothermal surfaces $\vartheta =$ const tend to rotate and approach the direction perpendicular to the $x_3$ axis, i.e., the direction of maximum compression of the particle. At the same time, the increase in $k_3$ shows that the separation between any two such surfaces decreases, i.e., the temperature gradients increase under the influence of convection.

We shall suppose that the spectral components of the temperature field have already become "adjusted" to the straining of the fluid particle, i.e., they have assumed the asymptotic form for which the isothermal surfaces are perpendicular to the $x_3$ axis. The asymptotic form of $E^{(\vartheta)}(k)$ can then be determined from the condition that it must be invariant under the joint effect of convection and molecular diffusion by analogy with the spectrum $E(k)$ in Sect. 22.3. In fact, let $\delta B(t)$ be the amplitude of the spectral component which at time $t$ corresponds to an interval $\delta k$ along the wave-number axis containing the point $k$. During the time $dt$ this component will become

$$\delta B(t + dt) \exp\{ik(t + dt)x_3\} = [\delta B + d(\delta B)] \exp\{i(k + dk)x_3\},$$

where $d(\delta B) = -\chi k^2 \delta B\, dt, dk = -a_3 k\, dt$. Since, moreover, $d(\delta k) = -a_3 \delta k\, dt$, the contribution of unit length of the wave-number axis to the temperature variance which is equal to $\mathscr{E}^{(\vartheta)}(k, t) = (\delta B)^2/\delta k$, varies in time according to the law

$$d\mathscr{E}^{(\vartheta)}(k, t) = \mathscr{E}^{(\vartheta)}(k, t)\left(-2\chi k^2 + a_3\right) dt = \mathscr{E}^{(\vartheta)}(k, t)\left[\frac{2\chi k^2}{a_3} - 1\right]\frac{dk}{k}. \tag{22.78}$$

By analogy with Sect. 22.3 we shall suppose that the random variable $\mathscr{E}^{(\vartheta)}(k, t)$, which characterizes the temperature fluctuations with wave number $k$, is practically uncorrelated with $a_3^{-1}$. Then after averaging Eq. (22.78) we obtain

$$\frac{dE^{(\vartheta)}(k)}{dk} = \frac{E^{(\vartheta)}(k)}{k}\left(-2\frac{\chi}{\nu}\alpha\eta^2k^2 - 1\right), \tag{22.78'}$$

i.e.,

$$E^{(\vartheta)}(k) = Ak^{-1}\exp\left\{-\frac{\chi}{\nu}\alpha(k\eta)^2\right\} \quad \text{for} \quad k \gg 1/\eta, \tag{22.79}$$

where $E^{(\vartheta)}(k) = \overline{\mathscr{E}^{(\vartheta)}(k, t)}$ and $\alpha = -\overline{(a_3\tau_\eta)^{-1}}$, i.e., $\alpha$ is the same quantity as in Eqs. (22.65) and (22.66), so that, in particular, $\alpha > \sqrt{3}$ and apparently $\alpha < 2\sqrt{3}$ (see the end of the Sect. 22.3). The integration constant $A$ in Eq. (22.79) has the dimensions of the square of $\vartheta$, i.e., it can be written in the form $A = C\,\overline{N}\tau_\eta$, where $C$ is a dimensionless constant. The value of this constant can be determined by using the fact that, for Pr $\gg 1$, molecular thermal diffusivity begins to play an appreciable role only for $k \gg (\bar{\varepsilon}/\nu^3)^{1/4}$, i.e., for values of $k$ for which Eq. (22.79) is valid. We can therefore substitute Eq. (22.79) into the normalization condition

$$2\chi\int_0^\infty k^2 E^{(\vartheta)}(k)\, dk = 2N,$$

from which it follows that $A = 2\alpha\overline{N}\tau_\eta$, i.e., $C = 2\alpha$ and

$$E^{(\vartheta)}(k) = \frac{2\alpha\overline{N}\tau_\eta}{k}\exp\left\{-\frac{\alpha}{\mathrm{Pr}}(k\eta)^2\right\}, \qquad k \gg 1/\eta. \tag{22.80}$$

This formula is due to Batchelor (1959) and provides information about the general shape of the temperature spectrum $E^{(\vartheta)}(k)$ for $\Pr \gg 1$ throughout the whole range $k \gg 1/L_0$ in which the statistical state of the fluctuations is quasi-universal, i.e., the Kolmogorov similarity hypotheses can be used. In the range $1/L_0 \ll k \ll 1/\eta$ which, in this case, can be referred to as the inertial-convective subrange, molecular viscosity and thermal diffusivity have no effect on the statistical state of the fluctuations. Therefore, both spectra, $E(k)$ and $E^{(\vartheta)}(k)$, take the form of the five-thirds law. When $k \sim 1/\eta$, molecular viscosity begins to play an appreciable role, and the spectrum $E(k)$ falls off more rapidly than $k^{-5/3}$. After a short transition region, the reduction of $E(k)$ with increasing $k$ becomes exponential [in accordance with Eq. (22.67)]. Since, however, $\alpha$ is of the order of unity and $\Pr \gg 1$, the exponential factor on the right-hand side of Eq. (22.80) is practically equal to unity for $k \sim 1/\eta$. This factor begins to play an appreciable role only for much higher wave numbers [of the order of $k \sim (\Pr)^{1/2}/\eta \sim (\bar{\varepsilon}/\nu\chi^2)^{1/4}$]. The formula (22.80) thus shows that the inertial-convective subrange of the spectrum is followed by an appreciable subrange [extending right up to $k \sim (\Pr)^{1/2}/\eta \sim (\bar{\varepsilon}/\nu\chi^2)^{1/4}$] in which molecular viscosity plays an appreciable role but molecular thermal diffusivity still has no effect on the spectrum $E^{(\vartheta)}(k)$. In this range, after a short transition region, the spectrum $E^{(\vartheta)}(k)$ begins to be inversely proportional to $k$:

$$E^{(\vartheta)}(k) = \frac{2\alpha \bar{N} \tau_\eta}{k}, \tag{22.81}$$

i.e., it falls with increasing $k$ even more slowly than the five-thirds law (this is illustrated schematically in Fig. 57). The subrange

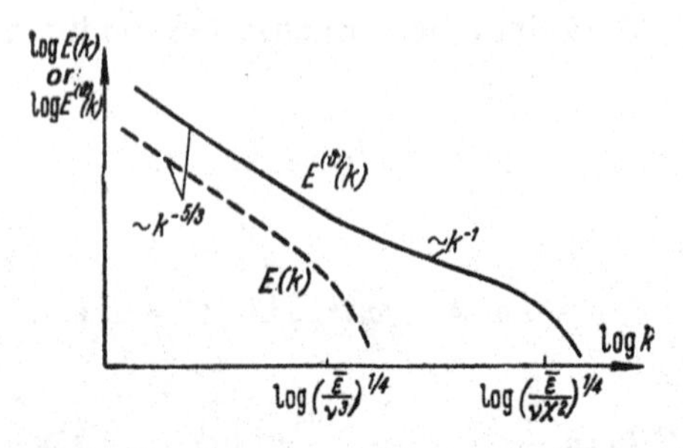


**Fig. 57** Schematic shape of the velocity and temperature spectra for Pr ≫ 1.

$1/\eta \ll k \ll (\mathrm{Pr})^{1/2}/\eta$ can be called *the viscous-convective subrange* [the fact that, for small $\chi$, the effect of molecular diffusion of heat begins to be appreciable only after the length scale $l \sim \eta/(\mathrm{Pr})^{1/2} \sim (\nu\chi^2/\bar{\varepsilon})^{1/4}$ has been reached, was implicit in Batchelor's work (1952b)]. Finally, molecular thermal conduction begins to have an effect for $k \sim (\bar{\varepsilon}/\nu\chi^2)^{1/4}$, and the form of $E^{(\vartheta)}(k)$ changes again. For $k \gg (\bar{\varepsilon}/\nu\chi^2)^{1/4}$ (this range can be called *the viscous-diffusive* subrange) the decrease of $E^{(\vartheta)}(k)$ is also exponential in view of Eq. (22.80) [although it is slower than the decrease of $E(k)$].

The formula (22.81) for the temperature spectrum in the viscous-convective subrange $(\bar{\varepsilon}/\nu^3)^{1/4} \ll k \ll (\bar{\varepsilon}/\nu\chi^2)^{1/4}$ can be established (to within a numerical factor of the order of unity) from dimensional considerations alone. To do this, we must take into account the fact that the quantity $\bar{\varepsilon}$ cannot have a direct effect on $E^{(\vartheta)}(k)$ in the above spectral range (since most of the energy dissipation is localized at smaller wave numbers). At the same time, convective mixing for $k \gg (\bar{\varepsilon}/\nu^3)^{1/4}$ can be reduced to the rotation and the mutual approach of isothermal surfaces under the action of the straining motion specified by the strain rate tensor which has a universal probability distribution when $\tau_\eta$ is taken as the unit of time. Hence, it follows that for $k \gg (\bar{\varepsilon}/\nu^3)^{1/4}$ and right up to values of $k$ for which molecular diffusion of heat begins to have an effect, the spectrum $E^{(\vartheta)}(k)$ can depend only on the two dimensional parameters $\tau_\eta$ and $\bar{N}$, i.e., it should be given by a formula of the form of Eq. (22.81) (with an undetermined numerical factor $a$). We note also that the fact that the spectrum (22.81) cannot be integrated for $k \to \infty$ means that the total contribution of the viscous-convective range to $\overline{\vartheta'^2}$ for large Pr can be very considerable [this contribution is, clearly, of the order of $\frac{a}{2}\bar{N}\tau_\eta \ln\frac{\nu}{\chi} \sim \bar{N}\tau_\eta \ln(\mathrm{Pr})$]. Therefore, it is possible in principle that, for $\mathrm{Pr} \gg 1$, most of the temperature fluctuation variance will be localized not in the large-scale fluctuations with scales $l \sim L_0$ but in the small-scale fluctuations with $l \ll \eta$. The establishment of the universal statistical state will then require a relatively long time [according to Batchelor (1959), this time is of the order of $\tau_\eta \ln(\mathrm{Pr})$], so that if the above theory is to be valid, the typical time for an appreciable change in the mean temperature field should be sufficiently long [much longer than $\tau_\eta \ln(\mathrm{Pr})$].

Since the study of the spectrum $E^{(\vartheta)}(k)$ is equivalent to the study of the structure function $D_{\vartheta\vartheta}(r)$, it is natural to expect that results of the kind discussed above can be obtained by starting with the equation for $D_{\vartheta\vartheta}(r)$. Thus, we can begin with Eq. (22.8) and use the fact that

$$D'_{L\vartheta\vartheta}(r)+\frac{2}{r}D_{L\vartheta\vartheta}(r)=\frac{\partial}{\partial r_j}(D_{j\vartheta\vartheta}(r))=\frac{\partial}{\partial r_j}\overline{\Delta_r u_j(\Delta_r\vartheta)^2}=\overline{\Delta_r u_j\frac{\partial}{\partial r_j}(\Delta_r\vartheta)^2},$$

where $\Delta_r u_j$ and $\Delta_r\vartheta$ are the velocity and temperature differences between points $x+r$ and $x$. For $|r|=r\ll\eta$ the field $u_j(x)$ can be regarded as linear in the coordinates, so that $\Delta_r u_j\approx\frac{\partial u_j}{\partial x_l}r_l$. On the other hand, If we suppose that the field $\vartheta(x)$ is throughout completely "adjusted" to the fluid-particle straining motion, i.e., if we take the asymptotic formula for which the isothermal surfaces are strictly perpendicular to the $x_3$ axis (maximum compression axis), then for each concrete realization of a turbulent flow the derivative $\frac{\partial}{\partial r_j}(\Delta_r\vartheta)^2$ will be nonzero only for $j=3$. Next, the derivative $\frac{\partial u_j}{\partial x_l}$ can be expanded into terms corresponding to the rotation of the fluid particle and those corresponding to its strain. The first of these terms will be uncorrelated with $\frac{\partial}{\partial r_j}(\Delta_r\vartheta)^2$ and, therefore, cancels out after taking the average. The second term in the system of coordinates defined by the principal strain axes will be a diagonal tensor. Therefore,

$$\overline{\Delta_r u_j\frac{\partial}{\partial r_j}(\Delta_r\vartheta)^2}\approx\overline{a_3 r\cos\theta\frac{\partial}{\partial r_3}(\Delta_r\vartheta)^2}=\overline{a_3 r\frac{\partial}{\partial r}(\Delta_r\vartheta)^2},$$

where $a_3$ is the maximum rate of compression of the fluid particle, and $\theta$ is the angle between the vector $r$ and the $x_3$ axis (so that $r_3=r\cos\theta$). If we assume that for $r\ll\eta$ the correlation between the random variables $a_3$ and $\frac{\partial}{\partial r}(\Delta_r\vartheta)^2$ can be neglected in the first approximation, we obtain

$$\overline{\Delta_r u_j\frac{\partial}{\partial r_j}(\Delta_r\vartheta)^2}\approx-\frac{r}{\alpha_1\tau_\eta}\frac{\partial}{\partial r}D_{\vartheta\vartheta}(r), \tag{22.82}$$

where $(\alpha_1\tau_\eta)^{-1}$ is an effective average strain rate $-a_3$, i.e., $\alpha_1$ will have roughly the same significance as the factor $\alpha$ in Eqs. (22.67) and (22.79). Consequently, for $r\ll\eta$ Eq. (22.80) assumes the form

$$2\bar{N}=\frac{1}{2\alpha_1\tau_\eta}r\frac{dD_{\vartheta\vartheta}}{dr}+\chi\left[\frac{d^2D_{\vartheta\vartheta}}{dr^2}+\frac{2}{r}\frac{dD_{\vartheta\vartheta}}{dr}\right]. \tag{22.83}$$

If we use Eq. (13.55) which relates $D_{\vartheta\vartheta}(r)$ and $E^{(\vartheta)}(k)$, or if we write $D''_{\vartheta\vartheta}(r)+2D'_{\vartheta\vartheta}(r)/r$ as $\Delta D_{\vartheta\vartheta}(r)$ and $rD'_{\vartheta\vartheta}(r)$ as $r_j\frac{\partial}{\partial r_j}D_{\vartheta\vartheta}(r)$, and use Eqs. (13.44) and (11.48) we can readily show that Eq. (22.83) is equivalent to

$$\frac{1}{2\alpha_1\tau_\eta}\left(E^{(\vartheta)}(k)+k\frac{d}{dk}E^{(\vartheta)}(k)\right)+\chi k^2E^{(\vartheta)}(k)=0, \tag{22.84}$$

so that $\alpha_1=\alpha$ [since Eq. (22.84) becomes identical with Eq. (22.78′) for $\alpha_1=\alpha$]. The solution of Eq. (22.83) with $\alpha_1=\alpha$ can be sought in the form

$$D_{\vartheta\vartheta}(r) = 2\bar{N}\alpha\tau_\eta h\left(\frac{\nu r^2}{\alpha\chi\eta^2}\right), \tag{22.85}$$

where $h(x)$ is the solution of the ordinary differential equation

$$4xh'' + (6 + x)h' = 1, \quad h(0) = 0, \quad |h'(0)| < \infty \tag{22.86}$$

This solution can be written explicitly in the form of a double integral of elementary functions [see Batchelor (1959)]. For $x \ll 1$, i.e., when $r \ll \eta(\mathrm{Pr})^{-1/2}$, the function $h(x)$ can be sought in the form of a power series the first term of which is of the form $h(x) \approx x/6$, i.e., $D_{\vartheta\vartheta}(r) \approx \frac{\bar{N}}{3\chi} r^2$, as expected. On the other hand, when $x \gg 1$, i.e., when $r \gg \eta(\mathrm{Pr})^{-1/2}$ but $r \ll \eta$, so that we can still use Eq. (22.83), the second term on the right-hand side of Eq. (22.83) can be neglected (it describes the effect of molecular thermal conduction) in comparison with the first term (describing convective mixing). In that case, the equation for $h(x)$ assumes the form $xh'(x) = 1$, so that $h(x) \approx \ln x +$ const and the leading term of the asymptotic formula for $D_{\vartheta\vartheta}(r)$ is

$$D_{\vartheta\vartheta}(r) \approx 2\bar{N}\alpha\tau_\eta \ln\frac{\nu r^2}{\alpha\chi\eta^2}. \tag{22.87}$$

This result is easily seen to correspond to the asymptotic relation (22.81) for the spectrum $E^{(\vartheta)}(k)$.

The above derivation of Eq. (22.83) is only partly based on the assumption that $\mathrm{Pr} \gg 1$. It can readily be demonstrated that an equation of this kind should be valid even when $r \ll \eta$, but for Prandtl numbers that are not too high (the coefficient $\alpha_1$ in this case can depend on the degree of orientation of the isothermal surfaces perpendicular to the local axis of maximum compression, i.e., it is a function of the Prandtl number). In fact, the equation (22.82) is fully analogous to Eq. (22.70′). It can therefore be regarded as a consequence of only the isotropy conditions and the fact that the left-hand side of Eq. (22.82) for $r \ll \eta$ is linear in the velocity derivatives (which are of the order of $\tau_\eta^{-1}$) and the derivatives $\frac{\partial}{\partial r_j}(\Delta_r\vartheta)^2$. However, if this is so, Eq. (22.83) must be valid even for finite values of Pr, and the coefficient $\alpha_1$ should be of the order of unity but may depend on Pr (it must be positive for all Pr since, otherwise, we would find that convective mixing would lead to the transport of the mean square of temperature fluctuations from small-scale to large-scale disturbances and not vice versa).

The validity of Eq. (22.82) for finite values of Pr should also affect the spectrum $E^{(\vartheta)}(k)$. To investigate this spectrum for Pr of the order of unity it is convenient to follow Novikov (1961c) and apply the Laplace operator to all the terms of the thermal conduction equation (22.74). Since in a compressible fluid $\Delta u_j = -\varepsilon_{jlm}\frac{\partial\omega_m}{\partial x_l}$, where $\omega_m$ are the components of the vorticity and $\varepsilon_{jlm}$ is a completely antisymmetric tensor, the resulting equation can be written in the form

$$\frac{\partial\Delta\vartheta}{\partial t} + u_j\frac{\partial\Delta\vartheta}{\partial x_j} + 2\frac{\partial u_j}{\partial x_l}\frac{\partial^2\vartheta}{\partial x_j\partial x_l} - \varepsilon_{jlm}\frac{\partial\omega_m}{\partial x_l}\frac{\partial\vartheta}{\partial x_j} = \chi\Delta^2\vartheta. \tag{22.88}$$

We shall use the fact that if Pr is of the order of unity, then for $l \ll \eta \sim (\chi^3/\bar{\varepsilon})^{1/4}$ the velocity field consists of the basic linear field $u_j = u_{0j} + \frac{\partial u_j}{\partial x_k} r_k$ (or $u_1 = a_1 x_1$, $u_2 = a_2 x_2$, $u_3 = a_3 x_3$, if we transform to the coordinate system $x_1 x_2 x_3$ defined by the principal strain axes of the fluid particle) and weak disturbances of this linear field, while the field $\vartheta(x)$ consists in the same way of the linear field $\vartheta = \vartheta_0 + \beta_k x_k$ and superimposed weak disturbances. If we linearize Eq. (22.88) with respect to the disturbances, we have in the $x_1 x_2 x_3$ system

$$\frac{\partial \Delta\vartheta}{\partial t} + \sum_{j=1}^{3} \left( a_j x_j \frac{\partial \Delta\vartheta}{\partial x_j} + 2a_j \frac{\partial^2 \vartheta}{\partial x_j^2} \right) - \varepsilon_{jlm} \beta_j \frac{\partial \omega_m}{\partial x_l} = \chi \Delta^2 \vartheta, \qquad (22.89)$$

where $\beta_j$ is the temperature gradient along the direction of the $j$th principal strain axis, and $\vartheta$ can now be interpreted as the disturbance of the linear temperature field. When applied to the temperature field, Eq. (22.89) plays the same role as did Eq. (22.59) for the velocity field.

Let us expand the initial disturbance into its Fourier components, and consider special solutions of Eqs. (22.59) and (22.89) corresponding to the initial conditions

$$\omega_j(x, 0) = A_{j0} e^{ik_0 x}, \qquad \vartheta(x, 0) = B_0 e^{ik_0 x}. \qquad (22.90)$$

As above, we shall suppose that the principal strain axes undergo practically no rotation relative to the fluid particle in a time of the order of $\tau_\eta$, and that the strain rates $a_1$, $a_2$, and $a_3$ can also be regarded as constants during this time interval (this latter assumption is not strictly necessary). If we seek our solution in the form

$$\omega_j(x, t) = A_j(t) e^{ik(t)x}, \qquad \vartheta(x, t) = B(t) e^{ik(t)x}, \qquad (22.91)$$

then from Eq. (22.59) alone we obtain the relations given by Eq. (22.63) for $k_j(t)$ and $A_j(t)$, which we already know, so that we can then use Eq. (22.89) to determine the function $B(t)$ as well.

The main asymptotic result for $B(t)$ can be established by considering the asymptotic state for which the vorticity field $\omega(x)$ is already completely "adjusted" to the fluid-particle strain. In other words, we can start by supposing that the vector $k(t)$ is parallel to the $x_3$ axis, where $k_3(t) = k(t) = k_0 \exp(-a_3 t)$ and the vector $\omega(x)$ is parallel to the $x_1$ axis. We then have

$$\omega_1(\boldsymbol{x}) = \omega_1(x_3) = A(t)\exp(k(t)x_3),$$
$$A(t) = A_0 \exp\left\{\int_0^t (a_1 - \nu k^2(t_1))\,dt_1\right\}.$$

In that case, Eq. (22.89) assumes the form

$$\frac{dB}{dt} + \chi k^2 B = -i\beta_2 \frac{A}{k}. \qquad (22.92)$$

Using the above formulas for $k(t)$ and $A(t)$, we can write the solution of this linear equation in the form

$$B(t) = \left\{B_0 - i\frac{A_0}{k_0}\int_0^t \beta_2 \cdot e^{\int_0^{t_1}(a_1 + a_3 + (\chi - \nu)k^2(t_2))\,dt_2}\,dt_1\right\} e^{-\chi\int_0^t k^2(t_1)\,dt_1}. \qquad (22.93)$$

Let us consider, to begin with, the case when $\nu > \chi$, i.e., $\mathrm{Pr} > 1$. For $k^2(t) \gg (\nu - \chi)^{-1}\tau_\eta^{-1}$, i.e., $k\eta \gg [\mathrm{Pr}/(\mathrm{Pr} - 1)]^{1/2} = \nu^{1/2}/(\nu - \chi)^{1/2}$, the second term in the braces on the right-hand side of Eq. (22.93) will then be damped out in a time of the order of $\tau_\eta$ and, consequently, we have the asymptotic result $B(t) \approx B_0 \exp\left\{-\chi\int_0^t k^2(t_1)\,dt_1\right\}$. Therefore, we have again obtained the relations (22.77) which, as we know, lead to Eq. (22.79) for $E^{(\vartheta)}(k)$. We see that a formula of this form for $E^{(\vartheta)}(k)$ can be justified for any $k \gg 1/\eta$ (subject to the usual assumptions that the strain component of the velocity field is quasi-stationary). Since, moreover, $a_1 + a_3 = -a_2$ is usually negative and of the order of $\tau_\eta^{-1}$, and $e^{-m}$ is quite small for $m$ of the order of unity, Novikov (1961c) has suggested that Eq. (22.79) should be satisfied with adequate accuracy even for $k \sim 1/\eta$, and not merely for $\mathrm{Pr} > 1$ but even for $\mathrm{Pr} = 1$. It is clear, however, that when Pr is of the order of unity, the viscous-convective subrange of the spectrum [in which $E^{(\vartheta)}(k) \sim k^{-1}$] will be absent, and we shall not be able to estimate the dimensionless coefficient $A/\bar{N}\tau_\eta = \gamma(\mathrm{Pr})$ which enabled us to rewrite Eq. (22.79) in the form of Eq. (22.80).

Let us now consider the case when $\nu < \chi$, i.e., $\mathrm{Pr} < 1$. Here, for very large wave numbers $k\eta \gg (\mathrm{Pr}/(1-\mathrm{Pr}))^{1/2} = \nu^{1/2}/(\chi-\nu)^{1/2}$, the second term in the braces on the right-hand side of Eq. (22.93) will become much greater than the constant first term after a time interval of the order of $\tau_\eta$ [i.e., for any initial value, the function $B(t)$ will be determined asymptotically only by the vorticity field $\boldsymbol{\omega}(\boldsymbol{x})$]. The asymptotic behavior of the second term will then be determined by the maximum value of the integrand, which is achieved at the upper limit of integration, i.e., at the point $t_1 = t$. The contribution of this maximum (which provides the asymptotic estimate for the value of the entire integral) can, as usual, be equated to the value of the integrand at the maximum, divided by the derivative of the integrand at this point.[20]

Hence, it follows that asymptotically

$$B(t) \approx -i\,\frac{A_0\beta_2 \exp\left\{\int\limits_0^t (a_1 + a_3 - \nu k^2(t_1))\,dt_1\right\}}{k_0\,[a_1 + a_3 + (\chi-\nu)\,k^2(t)]} \approx -i\,\frac{\beta_2}{(\chi-\nu)\,k^3}\,A(t)$$

$$\text{for } k\eta \gg \left(\frac{\mathrm{Pr}}{1-\mathrm{Pr}}\right)^{1/2}. \tag{22.94}$$

Since $E^{(\vartheta)}(k) = \dfrac{\overline{|\delta B|^2}}{\delta k}$ and $E(k) = \dfrac{\overline{|\delta A|^2}}{2k^2\,\delta k}$, the last equation shows that

$$E^{(\vartheta)}(k) = \frac{2\overline{\beta_2^2}}{(\chi-\nu)^2 k^4}\,E(k) \quad \text{for} \quad k\eta \gg \left(\frac{\mathrm{Pr}}{1-\mathrm{Pr}}\right)^{1/2}. \tag{22.95}$$

[20] This can be rigorously proved if we use the usual form of integration by parts

$$\int\limits_0^t \beta_2 \exp\{\varphi(t_1)\}\,dt_1 = \frac{\beta_2}{\varphi'(t_1)}\exp\{\varphi(t_1)\}\Big|_0^t - \int\limits_0^t \exp\{\varphi(t_1)\}\,d\left(\frac{\beta_2}{\varphi'(t_1)}\right),$$

$$\varphi(t) = \int\limits_0^t (a_1 + a_3 + (\chi-\nu)\,k^2(t_1))\,dt_1,$$

and by verifying that, for $t \sim \tau_\eta$ and $(\chi-\nu)\,k^2(t) \gg \tau_\eta^{-1}$, the main contribution on the right-hand side will be the term $\dfrac{\beta_2}{\varphi'(t)}\exp\{\varphi(t)\}$.

The quantity $\overline{\beta_2^2}$ is clearly of the order of the mean temperature gradient. If we suppose that the temperature fluctuations, which determine the value of $\overline{(\nabla\vartheta)^2}$, are isotropic not only in a given set of coordinates at rest [which is always valid for locally isotropic turbulence) but also in a set of coordinates defined by the moving principal strain axes, then

$$\overline{\beta_2^2} = \frac{1}{3}\overline{(\nabla\vartheta)^2} = \overline{N}/3\chi. \tag{22.96}$$

However, the last assumption is not necessary. If we do not admit it, then on the right-hand side of Eq. (22.96) we must add a numerical factor $c = c(\mathrm{Pr})$ of the order of unity. The formula given by Eq. (22.95) then assumes the form

$$E^{(\vartheta)}(k) = \frac{2c\overline{N}}{3\chi(\chi-\nu)^2 k^4} E(k) \tag{22.97}$$

[see Novikov (1961b, c)].

Comparison of Eqs. (22.80), (22.97), and (22.67) shows that, in the approximation which we are discussing here, the decrease of the temperature spectrum in the range of very high wave numbers is different for $\mathrm{Pr} > 1$ and $\mathrm{Pr} < 1$. Namely, in the second case, it is approximately proportional to the function $\exp\{-\alpha(k\eta)^2\}$, i.e., it has the same character as the decrease of the spectrum $E(k)$, while in the first case it is proportional to $\exp\left\{-\frac{\alpha}{\mathrm{Pr}}(k\eta)^2\right\}$, i.e., it is slower than the decrease of $E(k)$. It is important to note, however, that this decrease is very rapid in all cases, i.e., it is of the same type as the decrease at infinity of the Gaussian probability density.

In conclusion, let us briefly consider the case $\nu \ll \chi$, i.e., $\mathrm{Pr} \ll 1$. Here, in spatial regions of diameter $l \ll \eta_\vartheta = (\chi^3/\bar{\varepsilon})^{1/4}$ the temperature field will consist of a main linear field of the form $\vartheta_0 + \beta_j x_j$, $\beta_j = \frac{\partial\vartheta}{\partial x_j}$ and superimposed small disturbances of more complicated form. Therefore, for fluctuations with scales $l \ll \eta_\vartheta$, i.e., with wave numbers $k \gg 1/\eta_\vartheta$), the main contribution to the components $u_j \frac{\partial\vartheta}{\partial x_j}$ of the thermal conduction equation will be the term $u_j\beta_j$. Moreover, in the range of length scales such that $\eta_\vartheta \gg l \gg \eta = \eta_\vartheta(\mathrm{Pr})^{3/4}$, i.e., wave numbers $1/\eta_\vartheta \ll k \ll 1/\eta$), the velocity fluctuations $u_j$ will be considerable, and will obey the usual relations which are valid for the inertial subrange. These velocity fluctuations will give rise to fluctuations in convective heat transfer, which are the main source of small-scale temperature disturbances. Consequently, in the range $\eta_\vartheta \gg l \gg \eta$, i.e., $1/\eta_\vartheta \ll k \ll 1/\eta$), the time scale of the temperature and the velocity fluctuations will be roughly the same, i.e., it will be of the order of $(l^2/\bar{\varepsilon})^{1/3} \sim (k^2\bar{\varepsilon})^{-1/3}$ (see Sect. 21.2). Hence it follows that, when $\mathrm{Pr} \ll 1$, the term $\partial\vartheta/\partial t$ in the thermal-conduction equation for fluctuations with $\eta_\vartheta \gg l \gg \eta$ and $1/\eta_\vartheta \ll k \ll 1/\eta$ will be of the order of $l^{-2/3}\bar{\varepsilon}^{1/3}\,\vartheta$, i.e., $k^{2/3}\bar{\varepsilon}^{1/3}\vartheta$, and will be negligible in comparison with $\chi\Delta\vartheta$, which is of the order of $\chi l^{-2}\vartheta$ (or $\chi k^2\vartheta$). Therefore, for such fluctuations the thermal-conduction equation (22.74) will assume the very simple form

$$\beta_j u_j = \chi\Delta\vartheta, \tag{22.98}$$

where $\beta_j = \frac{\partial\vartheta}{\partial x_j}$ is the temperature gradient due in the first instance to fluctuations of length scale of the order of $\eta_\vartheta$.

Small-scale temperature fluctuations described by Eq. (22.98) are quite quickly smoothed out by molecular thermal conduction and have only a slight distorting effect on the main linear field $\vartheta_0(x) = \vartheta_0 + \beta_j x_j$. The temperature structure function $D_{\vartheta\vartheta}(r) = \overline{(\Delta_r\vartheta)^2}$ for $r \ll \eta_\vartheta$ will therefore be roughly proportional to $r^2$. However, these small-scale fluctuations provide the major contribution to the spectral density $E^{(\vartheta)}(k)$ for $k \gg 1/\eta_\vartheta$ for which Eq. (22.98) plays the major role. If we again assume that fluctuations of scale of the order of $\eta_\vartheta$ are uncorrelated with those of scale much less than $\eta_\vartheta$, we can regard the coefficients $\beta_j$ on the left-hand side of Eq. (22.98) as constant random variables (with equal variances $\overline{\beta_1^2} = \overline{\beta_2^2} = \overline{\beta_3^2} = \bar{N}/3\chi$). Let us now replace the locally isotropic random fields $u_j$ and $\vartheta$ in Eq. (22.98) by their spectral representation, and then take the square of both sides of this equation and average the result (at first over all the possible realizations of the small-scale fluctuations with $k \gg 1/\eta_\vartheta$, and then over all the possible values of the random vector $\beta_j$). The final result is

$$\frac{2\bar{N}}{3\chi}E(k) = \chi^2 k^4 E^{(\vartheta)}(k) \quad \text{for} \quad 1/\eta_\vartheta \ll k \ll 1/\eta. \tag{22.99}$$

Since the five-thirds law must be valid for the energy spectrum in the inertial subrange, this relation can be rewritten in the form

$$E^{(\vartheta)}(k) = \frac{2}{3} C_1 \overline{N} \bar{\varepsilon}^{2/3} \chi^{-3} k^{-17/3} \quad \text{for} \quad 1/\eta_\theta \ll k \ll 1/\eta. \quad (22.100)$$

This result is due to Batchelor, Howells, and Townsend (1959) who obtained it in a somewhat different way. It corresponds to the general shape of the spectra $E(k)$ and $E^{(\vartheta)}(k)$ for $\Pr \ll 1$, which is shown schematically in Fig. 58.

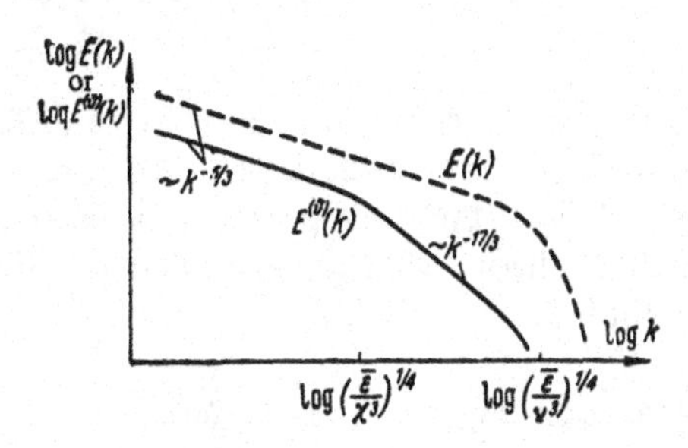


**Fig. 58** Schematic shape of the velocity and temperature spectra for $\Pr \ll 1$.

The relation (22.99) was derived on the assumption that $\chi \gg \nu$, and is obviously in agreement with the Novikov formula (22.97) which is valid for $\chi > \nu$ and $k\eta \gg \nu^{1/2}/(\chi - \nu)^{1/2}$, provided we assume that $c = 1$ (which seems very probable). We have therefore some reason to assume that, for $\Pr < 1$, the formula (22.97) is valid for all small-scale fluctuations with $k \gg 1/\eta_\theta$.

Let us finally note that the result of Batchelor, Howells, and Townsend was recently questioned by C. H. Gibson (1968) who emphasized that Batchelor et al. did not take into account zero-gradient points and minimal gradient surfaces of the temperature (or admixture concentration) field. In the neighborhood of these "singular" points and surfaces the linear approximation of the scalar field $\vartheta(\boldsymbol{x}, t)$ is inapplicable and hence the equation (22.98) is not valid. According to Gibson the regions of the flow where the scalar gradient is small or zero play the principal role in the process of the temperature or concentration mixing (by means of local stretching of the $\vartheta$ distribution where the gradient is small); however these regions are excluded from consideration in the Batchelor-Howells-Townsend theory. Gibson suggested (without any strict proof) that this new effect implies that the probability distributions of the differences of the field $\vartheta(x)$ over distances much smaller than $\eta_\vartheta = (\chi^3/\bar{\epsilon})^{1/4}$ are uniquely determined by quantities $\bar{N}, \chi$ and $\tau_\eta$ (but not $\bar{\epsilon}$) for $\mathrm{Pr} \ll 1$. On the other hand these distributions cannot depend on $\tau_\eta$ if all the distances considered are much larger than $(\chi\tau_\eta)^{1/2} = (\chi^2\nu/\bar{\epsilon})^{1/4}$. From these similarity arguments Gibson found the result

$$E^{(\vartheta)}(k) = \beta\bar{N}\chi^{-1}k^{-3} \quad \text{for } \mathrm{Pr} \ll 1 \text{ and } \frac{1}{\eta_\vartheta} \ll k \ll \frac{1}{\eta_\vartheta}(\mathrm{Pr})^{1/4} \tag{22.101}$$

where $\beta$ is a dimensionless constant; this is quite different from the Batchelor-Howells-Townsend formula (22.100). The complete resolution of the problem of the discrepancy between these two results clearly needs further theoretical and experimental work; we shall not linger on this point here.

The formula (22.99) can, clearly, be rewritten in the form

$$2\chi\int_k^\infty k'^2E^{(\vartheta)}(k')\,dk' = \frac{4\bar{N}}{3\chi^2}\int_k^\infty \frac{E(k')}{k'^2}\,dk' \approx 2\int_k^\infty \frac{2E(k')}{3\chi k'^2}\,dk' \int_0^k k''^2E^{(\vartheta)}(k'')\,dk''.$$

Hence, for the eddy thermal diffusivity $K_\vartheta(k)$ due to small-scale fluctuations with wave numbers $\gg k$, which can be determined following Heisenberg (1948a) from the condition that the rate of transfer of the temperature fluctuation intensity through the point $k$ on the wave-number spectrum $W_\vartheta(k)$ is equal to $2K_\vartheta(k)\int_0^k k''^2E_\vartheta(k'')\,dk''$) we have

$$K_\vartheta(k) = \frac{2}{3}\int_k^\infty \frac{E(k')}{\chi k'^2}\,dk' \quad \text{when } k \gg 1/\eta_\vartheta. \tag{22.102}$$

Howells (1960) suggested the use of the interpolation formula

$$K_\vartheta(k) = \left[\chi^2 + \frac{4}{3}\int_k^\infty \frac{E(k')}{k'^2}\,dk'\right]^{1/2} - \chi, \tag{22.103}$$

for the eddy diffusivity $K_\vartheta(k)$ for $\mathrm{Pr} \ll 1$ at any $k \gg 1/L_\vartheta$. This formula becomes identical with Eq. (22.102) for $\int_k^\infty E(k')k'^{-2}\,dk' \ll \chi^2$, and agrees with the semiempirical formula (17.10) with $c = 1/2$ in the convective range where $\int_k^\infty E(k')k'^{-2}\,dk' \gg \chi$. However, for $\mathrm{Pr} \gg 1$ this formula is in conflict with Batchelor's result (22.81). Howells therefore proposed the use of a much more complicated semiempirical formula for $W_\vartheta(k)$ at arbitrary Prandtl number. In addition to the term $2K_\vartheta(k)\int_0^k k'^2 E^{(\vartheta)}(k')\,dk'$ [where $K_\vartheta(k)$ is defined by Eq. (22.103)], this formula contains a term which cannot be written in the form $2K_\vartheta(k)\times\int_0^k k'^2 E^{(\vartheta)}(k')\,dk'$ for any $K_\vartheta(k)$.

## *Asymptotic Behavior of the Spectrum $E^{(\vartheta)}(k)$ in the Presence of a First Order Chemical Reaction or Radioactive Decay*

The above analysis of the asymptotic behavior of the spectrum $E^{(\vartheta)}(k)$ can be readily extended to the case of the concentration $\vartheta(\boldsymbol{x}, t)$ of a nonconservative admixture participating in a first-order chemical reaction or undergoing radioactive decay. Of course, if $\mu^{-3/2}\bar{\varepsilon}^{1/2} = L_r \gg \eta = (\nu^3/\bar{\varepsilon})^{1/4}$ and $L_r \gg \eta_\vartheta = (\chi^3/\bar{\varepsilon})^{1/4}$, i.e., if the typical reaction or decay time $\mu^{-1}$ is much greater than both time microscales $\tau_\eta = (\nu/\bar{\varepsilon})^{1/2}$ and $\tau_\vartheta = (\chi/\bar{\varepsilon})^{1/2}$), then the behavior of the spectrum $E^{(\vartheta)}(k)$ for $k \gg 1/\max(\eta, \eta_v)$ will be practically the same as in the case of the usual conservative admixture (as we know, the presence of chemical reactions or decays affects only the statistical properties of concentration fluctuations with length scales greater than or of the order of $L_r$; see Sect. 21.7). If, however, the length scale $L_r$ does not exceed in order of magnitude both microscales $\eta$ and $\eta_\vartheta$, the effect of reactions or decays on the asymptotic behavior of $E^{(\vartheta)}(k)$ may be quite appreciable.

In the presence of first-order chemical reactions the usual diffusion equation (22.74) is replaced by Eq. (21.101) which contains on the right the additional term $-\mu\vartheta$. If we assume from the start that $\nu \gg \chi$ and $k \gg 1/\eta$, we can use Eq. (22.75), which must also be augmented by the term $-\mu\vartheta$ on the right-hand side. Hence it follows that the second relation in Eq. (22.77) will now be of the form

$$B(t) = B_0 \exp\left\{-\int_0^t [\chi k^2(t_1) + \mu]\,dt_1\right\}.$$

Bearing this in mind in all our subsequent calculations we have, instead of Eq. (22.79), for $k \gg 1/\eta$

$$E^{(\vartheta)}(k) = Ak^{-(1+\beta)}\exp\left\{-\frac{\chi}{\nu}\alpha(k\eta)^2\right\}, \quad \beta = 2\alpha\mu\tau_\eta = 2\alpha(\eta/L_r)^{2/3}, \qquad (22.104)$$

where, as in Eq. (22.79), $\alpha = -\overline{(a_3\tau_\eta)}^{-1}$. In particular, for the viscous-convective subrange $1/\eta \ll k \ll \nu^{1/2}/\chi^{1/2}\eta$, we again obtain the power spectrum

$$E^{(\vartheta)}(k) \sim k^{-(1+\beta)}, \qquad (22.105)$$

but the exponent is now less than $-1$. Formulas (22.104) and (22.105) are due to Corrsin (1961). For $\mu\tau_\eta \ll 1$, i.e., $L_r \gg \eta$, they are practically the same as Eqs. (22.79) and (22.81), as expected.

When $\nu/\chi$ is of the order of unity, it is convenient to use Eq. (22.89) augmented by the additional term $-\mu\,\Delta\vartheta$ on the right-hand side. The relations (22.92) and (22.93) will then be of the form

$$\frac{dB}{dt} + (\chi k^2 + \mu) B = -i\beta_2 \frac{A}{k},$$

$$B(t) = \left\{ B_0 - i\frac{A_0}{k_0} \int_0^t \beta_2 \exp\left[ \int_0^{t_1} [a_1 + a_3 + \mu + \right.\right.$$

$$\left.\left. + (\chi - \nu) k^2(t_2)]\, dt_2 \right] dt_1 \right\} \cdot \exp\left[ -\int_0^t (\mu + \chi k^2(t_1))\, dt_1 \right].$$

If $\nu > \chi$ then for sufficiently large wave numbers [such that $(\nu - \chi) k^2 \gg \mu + \tau_\eta^{-1}$] we can neglect the second term in the braces on the right-hand side of the last equation. Consequently, for such wave numbers we can assume that

$$B(t) = B_0 \exp\left\{ -\int_0^t (\chi k^2 + \mu)\, dt_1 \right\}.$$

Hence, as above, we have the asymptotic formula (22.104) for $E^{(\vartheta)}(k)$ [but the formula (22.105) will no longer be valid for any $k$]. If, on the other hand, $\nu < \chi$ (or at least $\nu = \chi$ but $\mu$ is greater than $\tau_\eta^{-1}$ by a substantial factor) then for sufficient large $k$ we can neglect the term $B_0$ in the expression for $B(t)$. Proceeding as in the derivation of Eq. (22.94), we obtain the relation

$$B(t) = -i \frac{\beta_2}{[(\chi - \nu) k^2 + \mu] k} A(t).$$

Hence it follows that for $\nu < \chi$ and $(\chi - \nu) k^2 + \mu \gg \tau_\eta^{-1}$ (or $\chi = \nu$, $\mu \gg \tau_\eta^{-1}$ and $k \gg 1/\eta$) we have

$$E^{(\vartheta)}(k) = \frac{2c\overline{N}}{3\chi[(\chi - \nu) k^2 + \mu]^2} E(k), \tag{22.106}$$

where $E(k)$ is the energy spectrum and $c$ is a constant close to unity (in general, depending on $\chi/\nu$ and equal to unity for $\chi \gg \nu$). When $k \ll 1/\eta$ and $\chi \gg \nu$ we can use Eq. (22.106), or the considerations which have led us to Eq. (22.100), to show that

$$E^{(\vartheta)}(k) = \frac{2C_1 \overline{N}\, \bar{\varepsilon}^{2/3}}{3\chi(\chi k^2 + \mu)^2 k^{5/3}}, \qquad 1/\eta_\vartheta \ll k \ll 1/\eta \tag{22.107}$$

This result is also due to Corrsin (1961). We note that for $\mu\tau_\vartheta \ll 1$, i.e., $L_r \gg \eta_\vartheta$, the formulas (22.106) and (22.107) again become identical with Eqs. (22.97) and (22.100), as expected.

# 23. EXPERIMENTAL DATA ON THE FINE SCALE STRUCTURE OF DEVELOPED TURBULENCE

## 23.1 Methods of Measurement; Application of Taylor's Frozen-Turbulence Hypothesis

Measuring instruments are usually capable of recording time variations of flow variables such as the components of velocity, temperature, pressure, admixture concentration, etc., at a fixed space point. The flow sweeps past such points with a mean velocity $\bar{u}$. It is also not difficult to record the variables at a point moving relative to the flow which some fixed velocity $U$ (for example, an aircraft in the atmosphere, or a ship on the sea). There are, of course, some more sophisticated methods based, for example, on the practically instantaneous sounding of spatial flow inhomogeneities with radio waves or sound waves, but we shall not consider these here (possible measurements of the spatial turbulence spectra through sound, light or radiowave scattering are discussed in the next chapter).

Experimental studies of small-scale turbulence components are restricted both by the inertia of the measuring device (characterized by its time constant $\tau_0$) and the linear dimensions $l_0$ of the sensor. It is clear that an instrument with parameters $\tau_0$, $l_0$ takes a time average over the time interval $\tau_0$ and space interval $l_0$. This can usually be regarded as equivalent to taking an average over a cylindrical region of space with axial length $\bar{u}\tau_0$ in the direction of the mean flow and lateral diameter $l_0$. To anticipate, we note that, for example, in the atmospheric surface layer, the Kolmogorov turbulence microscale $\eta = (\nu^3/\bar{\varepsilon})^{1/4}$ is of the order of a few millimeters, so that to investigate the turbulence components responsible for energy dissipation, the time constant of the measuring device must be such that $\tau_0 < \eta/\bar{u}$. This means that for a mean wind velocity $\bar{u}$ of the order of 1 m/sec, $\tau_0$ should not exceed a figure of the order of $10^{-3}$ sec.

When we determine the statistical characteristics of turbulence from the experimental data, the average over the statistical ensemble must usually be replaced by a time average (assuming that the fluctuations recorded are the realizations of ergodic random processes; see Sects. 3.3 and 4.7 of Vol. 1). Moreover, in practice one always has to take the average of the measurement data over a finite interval of time $T$. It is important to note that if this period $T$ is not long enough, the mean values will fluctuate, i.e., they will vary from one measurement to the next. These variations are caused by turbulence components whose characteristic times are not small in comparison with $T$, and they will lead to an evolution of the level of the flow variables which we have already discussed in Sect. 7.1 of Vol. 1. Moreover, if $T$ lies in a range in which the spectral density of the time fluctuations is not small, the mean value over $T$ will be appreciably dependent on this time interval.

However, for $T \gg L/\bar{u}$ where $L$ is the integral length scale of the turbulence in the mean flow direction, the time average taken over a time interval of length $T$ will, as a rule be statistically stable and practically independent of the chosen value of $T$. More precisely this will be the situation for turbulence in wind tunnels and other laboratory devices, since turbulence components with scales well in excess of $L$ will then always have small amplitudes, and will not therefore have an appreciable effect on the average values. The situation is more complicated in the case of atmospheric turbulence whose spectrum extends well into the region of large scales and contains many local maxima (associated with processes affecting the weather, the diurnal and annual variation of meteorological fields, and so on; see Sect. 23.6 below). It is important to note, however, that according to the data which will be reviewed in Sect. 23.6, the time spectra of meteorological fields usually have a deep minimum ("gap") in the range of periods between a few minutes and a few hours. The mean values of meteorologic variables, obtained by taking an average over any particular period $T$ in this range, will not therefore be too sensitive to the precise value chosen for $T$. It is true that these mean values will not, strictly speaking, be absolutely statistically stable, since oscillations with much longer periods are always present in the atmosphere, but the effect of these oscillations can be reduced to a minimum by fixing the time of the year, the time of the day, and the general meterologic conditions, i.e., the weather. In empirical determinations of the statistical characteristics of atmospheric turbulence it is therefore usual to take an average over time intervals of the order of 10 to 20 minutes, and statistical stability is then interpreted in a somewhat conventional fashion, i.e., it refers to measurements performed at the same time of the year, at the same time of the day, and under the same weather conditions.

If we record the oscillations of some given flow variable at a fixed space point, we can determine the time structure functions for this quantity by taking a time average. The time spectrum can then be found by calculating the Fourier transform of the structure function, or by transmitting the recorded oscillations through the filters of a spectral analyzer. The spatial structure of turbulent fields is much more difficult to determine. One way of doing this is to record oscillations of a variable simultaneously at a number of fixed space points (this is a more difficult practical problem than the recording of fluctuations at a single point but the difficulties can be overcome). Then we can use the data to determine the corresponding spatial structure function $D(\boldsymbol{r})$ for a number of values of $\boldsymbol{r}$ corresponding to the differences between the position vectors of the observation points. However such values are, as a rule, insufficient to determine the general form of the function $D(\boldsymbol{r})$ for a large range of values of $\boldsymbol{r}$, and to calculate the spatial spectrum of the field by taking the Fourier transform of the function $D(\boldsymbol{r})$. Taylor's frozen-turbulence hypothesis (see Sect. 21.4) thus becomes of considerable importance for empirical estimates of the spatial characteristics of turbulence. The Taylor hypothesis enables us to convert the time structure functions and spectra into spatial one-dimensional structure functions and spectra in the mean flow direction through formulas such as (21.39) and (21.41).

We have already listed a number of works devoted to the empirical verification of the Taylor hypothesis for turbulent flows, both under laboratory and atmospheric conditions (Sect. 21.4). The atmospheric case will be of particular interest since the atmosphere is the most convenient laboratory for verifying the properties of the local structure of developed turbulence. Let us therefore consider this in somewhat greater detail. The most satisfactory data for the verification of the Taylor hypothesis in the case of atmospheric turbulence were obtained by comparing tethered-balloon or tower data on time variations with the approximate values of the corresponding spatial statistical quantities obtained from aircraft measurements at the same altitudes. In fact, since the aircraft velocity $U$ is very high in comparison with typical fluctuation velocities, the recorded values of any field $a(x, t) = a(x_0 + r, t_0 + r/U)$ for distances $r$ that are not too large can be approximately replaced by $a(x_0 + r, t_0)$, i.e., we obtain a practically instantaneous "photograph" of the field along the flight trajectory, which enables us to determine directly its spatial statistical characteristics (in the direction of flight). Comparison of the spatial parameters of atmospheric turbulence obtained in this way with the time dependence obtained at a given point (at the flight altitude) have been carried out by Gifford (1955), Gossard (1960a), Lappe and Davidson (1963), Tsvang (1963), and Koprov and Tsvang (1965). As an illustration, Fig. 59, which is taken from

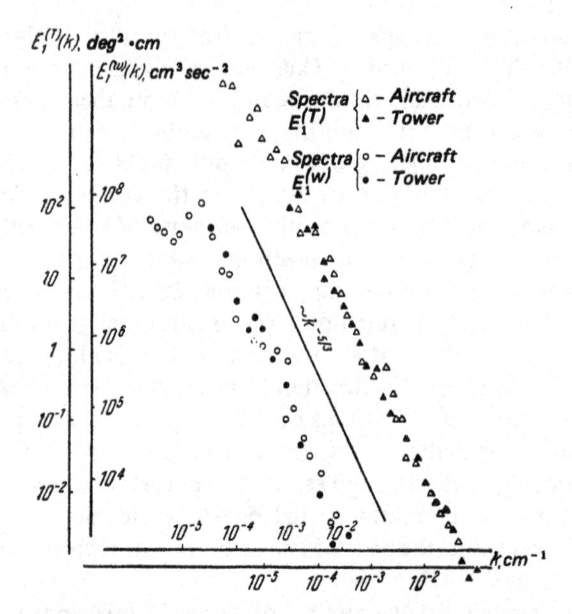


Fig. 59 Comparison of the spatial spectra of vertical velocity and temperature, measured from an aircraft and calculated from tower measurements using the Taylor frozen turbulence hypothesis.

the review paper by Gurvich, Koprov et al. (1967), shows two examples of a comparison of the one-dimensional spatial spectra $E_1^{(w)}(k)$ and $E_1^{(T)}(k)$ of the vertical wind velocity and temperature, measured from an aircraft flying at a height of 70 m around a 70 m meteorological tower (the aircraft velocity was 50 m/sec), with the spectra calculated from Eq. (21.41) using the time spectra measured from the tower. These examples show that, at heights of the order of 100 m, the Taylor hypothesis (21.41) is verified with high accuracy, at least up to wave numbers of the order of $k = 10^{-5}$ cm$^{-1}$. It does not, of course, follow from this that the individual turbulent disturbances in the atmosphere can be regarded as frozen with the same accuracy, so that we can use equations such as $a(x, t+\tau) = a(x - \bar{u}\tau, t)$, where $a$ represents the fluctuations of one of the wind-velocity components or temperature. However, henceforth we shall apply the Taylor hypothesis only to mean quantities such as the spectra and the structure functions. Therefore we must not be put off by the fact that it is not clear at present whether this hypothesis is valid for the instantaneous fluctuations.

## 23.2 Verification of the Local Isotropy Assumption

The most readily verifiable consequences of the local isotropy of developed turbulence are the conclusion that the velocity structure tensor $D_{ij}(r)$ for $r \ll L$ has the isotropic form of Eq. (13.69). It is also easy to verify the implied relations between the squares of space derivatives of the various velocity components, for example,

$$\overline{\left(\frac{\partial u_1}{\partial x_1}\right)^2} = \frac{1}{2}\overline{\left(\frac{\partial u_2}{\partial x_1}\right)^2} = \frac{1}{2}\overline{\left(\frac{\partial u_3}{\partial x_1}\right)^2}. \tag{23.1}$$

If instruments enabling us to determine the spectra are available, then we can readily verify the conclusion that the spectral tensor $F_{ij}(\mathbf{k})$ for $k \gg 1/L$ takes the isotropic form of Eq. (12.31), which is equivalent to Eq. (13.69) [more precisely, this would verify the resulting relations between the measured one-dimensional spectra such as $E_{12}(k) = 0$ or $E_2(k) = \frac{1}{2}(E_1(k) - kE_1'(k)$; see Eq. (12.87)].

Verifications of predictions of this kind were obtained by many experimenters for the grid-generated turbulence in wind tunnels [see, for example, the review paper by Batchelor (1947)], although the accuracy of these predictions for grid turbulence is still a matter of some dispute. We shall not, however, be especially interested in grid turbulence, since local isotropy in such cases can be explained simply by the fact that the turbulence iteself is approximately isotropic.[21] We shall therefore confine ourselves tò the verification of the above predictions for turbulence which is definitely known to be anisotropic.

[21] We recall, moreover, that, as noted in Sect. 14.1, the measurements of Uberoi (1963), Uberoi and Wallis (1966, 1967, 1969), Comte-Bellot and Corrsin (1967), and Van Atta and Chen (1968, 1969a) indicate the presence of appreciable anisotropy in grid-generated turbulence (but this conclusion was questioned by Portfors and Keffer, 1969a). At the same time, Uberoi and Wallis (1969) and Van Atta and Chen (1969a) have measured both the one-dimensional spectra $E_1(k)$ and $E_2(k)$, and showed that, in spite of the observed anisotropy, the relation $E_2(k) = \frac{1}{2}\left(E_1(k) - kE_1'(k)\right)$ is satisfied with high accuracy in the range of large wave numbers, indicating that the grid-generated turbulence is, in any case, locally isotropic. The results of Kistler and Vrebalovich (1966), who were concerned with grid-generated turbulence for large values of $\mathrm{Re}_M$ (see Sect. 16.6) have shown that the relation $E_2(k) = \frac{1}{2}\left(E_1(k) - kE_1'(k)\right)$ is violated for values of $k$ that are not too large (less than $0.03/\eta$ where $\eta$ is the Kolmogorov length), but are in good agreement with this relation for small-scale disturbances with $k\eta \geqslant 0.03$ (however these results are also controversial; see Sect. 16.6).

One of the first verifications of local isotropy in the case of anisotropic turbulence was performed by Townsend (1948a) who showed that the relations (23.1) were satisfied with good accuracy in the turbulent wake behind a circular cylinder in a wind tunnel. The agreement of these relations with the data of Townsend (1951c) for the turbulent boundary layer on a flat plate was also satisfactory, but the degree of agreement was somewhat lower. Both relations (23.1) were also confirmed quite accurately by the measurements of Champagne, Harris and Corrsin (1970) in the artificial nearly homogeneous shear flow produced in a wind tunnel.

The relation $E_{12}(k) = 0$, which is also a consequence of local isotropy, was verified by Corrsin (1949) and Tani and Kobayashi (1952) for an axially symmetric turbulent jet, by Laufer (1951) for turbulent channel flow, by Klebanoff (1955) for the turbulent boundary layer on a flat plate and by Champagne, Harris and Corrsin (1970) for nearly homogeneous wind tunnel turbulent flow which had practically constant shear. In all these cases it was found that, for small wave numbers, the spectral density $E_{12}(k)$ was quite large (from which it follows that the large-scale turbulence components are definitely anisotropic), but as $k$ increases, the spectrum $E_{12}(k)$ falls much more rapidly than $E_1(k)$ and $E_2(k)$, so that one can expect the isotropic state will be established in the small-scale region of the spectrum. As an illustration of this result, Fig. 60, taken from the rather old review paper by Corrsin (1957), shows the dimensionless *correlation coefficient spectrum* (i.e., the *normalized cospectrum*) $R_{12}(k) = |E_{12}(k)|[E_1(k)E_2(k)]^{-1/2}$ as a function of the wave number $k$ (or, more

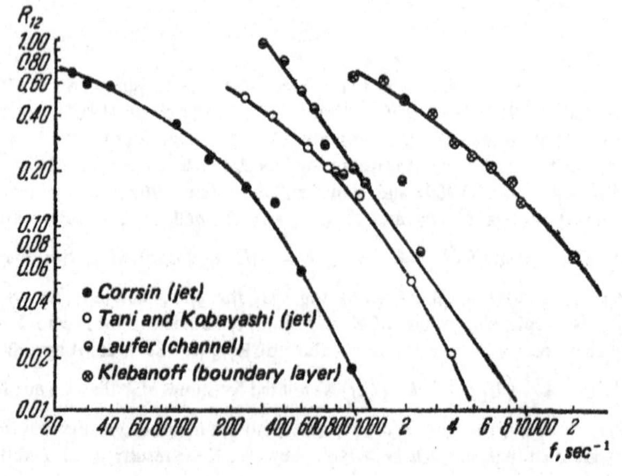


**Fig. 60** Correlation coefficient spectrum $R_{12}$ as a function of frequency $f = Uk/2\pi$.

precisely, the frequency $f = Uk/2\pi$) for flows in a channel, a boundary layer, and a jet.[22] These data show that, in all these flows, $R_{12}(f)$ falls rapidly with increasing frequency, so that the velocity fluctuations $u$ and $v$ are practically uncorrelated in the high-frequency range. More recently many measurements of the turbulent momentum flux and heat flux spectra were performed in the atmosphere, i.e., the cospectra of the horizontal and vertical wind component $E_{uw}(k)$ and of the vertical wind component and the temperature $E_{wT}(k)$ were measured (e.g., Panofsky and Mares, 1968, and Zubkovskii and Koprov, 1969). We shall discuss these measurements in more detail in Sect. 23.6 where additional references will also be given; here we wish only to emphasize that almost all the data show that $E_{uw}(k)$ and $E_{wT}(k)$ decrease much more rapidly than the velocity and temperature spectra. This is in full agreement with the prediction of the theory of locally isotropic turbulence.

Let us also recall the experiments of Uberoi (1957) who investigated the effect on isotropic turbulence of an axially symmetric lateral compression (in the ratio 4 to 1) obtained by joining a wide part of a wind tunnel to a narrower but coaxial tunnel, the two being connected by a smooth transition section. The turbulence obtained turns out to be anisotropic (the mean squares of the lateral velocity fluctuations $\overline{v'^2} = \overline{w'^2}$ are appreciably greater than $\overline{u'^2}$. However, as the turbulence decays (the distance $x$ from the entrance to the narrow tunnel increases), the ratio $\overline{v'^2}/\overline{u'^2}$ gradually approaches unity. Local isotropy is then indicated by the fact that the isotropy of small-scale turbulence components (the violation of which is characterized by the departure from unity of the ratio $\overline{\omega_1^2}/\overline{\omega_2^2}$ of the mean squares of the longitudinal and lateral vorticity components) is, in this case, reestablished appreciably more rapidly than the isotropy of the turbulence as a whole. This is indicated by Uberoi's data.

We note, moreover, that Uberoi (1957) has reported the violation of local isotropy in the boundary layer on the wind tunnel walls, and in a subsequent paper [Uberoi (1963)] has suggested that there is some deviation from strict local isotropy behind the grid in the central part of the wind tunnel. However, the results of these measurements are slightly doubtful (see Portfors and Keffer, 1969a)

[22] The *correlation coefficient spectrum* is closely related to the so-called *coherence* of two variables which is obtained if we replace the cospectrum $E_{12}(k)$ by the modulus of the complex cross-spectrum of two variables (quite often the coherence is also defined as the ratio of the squared modulus of the cross-spectrum to the product of both spectra). The main difference is that the quadrature spectrum contributes nothing to the *correlation coefficient spectrum*, but increases coherence.

and the corresponding Reynolds numbers are relatively low. In the other experiments mentioned above, the Reynolds numbers were also not sufficiently high. However, recent improvements in experimental techniques have ensured that measured turbulence spectra have become available for large Re for which one definitely expects local isotropy to appear. Many of the results obtained under these conditions have confirmed excellently the theoretical prediction of local isotropy for developed turbulence; however in some cases the results were contradictory. Thus, for example, simultaneous measurements of the spectra of vertical and horizontal wind-velocity components carried out in Tsimlyansk, USSR, from a 70 m meteorological tower have resulted in a mean value of the ratio of the one-dimensional spectra $E_2(k)$ and $E_1(k)$ in the inertial subrange which is in satisfactory agreement with the deductions from the local isotropy assumption (see Sect 23.3 below). The ratio of the structure functions $D_{NN}(r)$ and $D_{LL}(r)$ derived from measurements of velocity fluctuations in seawater (Dunn, 1965) is also in good agreement with these deductions. The quite scattered data on the time spectra of the three wind components at Round Hill, Mass., were analyzed by Busch and Panofsky (1968) who found that the data do not, at any rate, contradict isotropic relations for spectral ratios if we limit ourselves to high enough frequency. The data of airborne measurements of the one-dimensional spectra and cross-spectra for the components of wind fluctuation at a height of about 100 m published by Sheih, Tennekes, and Lumley (1971) are rather scattered but on the whole do not contradict the assumption of isotropy of the small scale turbulence. Finally the thorough measurements by Wyngaard and Coté (1971) from a tower in southwest Kansas lead on the average to a very accurate confirmation of the value 4/3 of the ratio of inertial subrange levels of vertical and longitudinal wind velocity spectra at heights of 11.3 and 22.6 meters; the value of the same ratio at the lowest height of 5.66 m was found to be slightly lower, but the authors suspect that this was due to the use of an insufficiently high frequency range. These results were also confirmed by more complete analysis of all the Kansas data of Kaimal, Wyngaard et al. (1972): The existence of a high frequency range with 4/3 spectral ratio was here observed for all values of $\zeta = z/L$ (where $L$ is the stratification length scale introduced in Chapt. 4 of Vol. 1) with the exception of the greatest values of $\zeta > +0.1$. However extensive data of sonic, thrust, and especially hot-wire anemometer measurements of the wind-velocity fluctuations in the surface layer over the sea by the Vancouver group lead to a confirmation of the 5/3 law of the locally isotropic turbulence but do not confirm the isotropic relation between one-dimensional spectra of different wind components in the five-thirds region (Smith, 1967; Weiler and Burling, 1967; Stewart, 1969; Miyake, Stewart and Burling, 1970). Some of these data also do not confirm the rapid decrease of the correlation coefficient spectrum with increase of the frequency (Stewart, 1969). Therefore the question of the value of the low-frequency limit of the isotropic region in the wind-velocity spectra is still not clear and it seems quite probable that it is considerably higher than the corresponding limit for the five-thirds region (for the longitudinal one-dimensional spectrum at any rate; this deduction agrees well with the results of Kistler and Vrebalovich, 1966). Summing up all the available information Panofsky (1969a) came to the conclusion that the isotropic limit is situated in the atmospheric surface layer near a wavelength approximately equal to 1/7 of the height above ground (i.e., a dimensionless frequency $fz/\bar{u} \approx 7$). Kaimal, Borkowski et al. (1969) recommended an even more conservative estimate $fz/\bar{u} \approx 10$ and also cast doubt on the existence of a true inertial subrange in the first few meters above ground. However later Kaimal, Wyngaard et al. (1972) proposed the following approximate estimate of the low-frequency limit of the locally isotropic range: $fz/\bar{u} \approx 1 \div 1.5$ under unstable and neutral conditions and $fL/\bar{u} \approx 10$ under stable conditions (i.e., the limiting wavelength has the order of $z$ under unstable conditions and of $L/10$ under stable conditions). All the abovementioned estimates of the isotropic limit are higher than a theoretical estimate for the neutrally stratified boundary layer by Pond, Stewart and Burling (1963): $fz/\bar{u} > 4.5/2\pi \approx 0.7$. Some preliminary estimates of the isotropy limit of length scales for turbulence in the free atmosphere were also indicated by Kaimal et al.; they are lower than has usually been assumed.

It is also stated in the report by Kaimal et al. (1969) that the data from three- or two-dimensional arrays of refractometers as well as radiowave scattering experiments indicate the isotropy of the refractive index fine structure in the atmosphere at scales of the order of 10 meter and less for not too small heights above the ground and at dimensionless frequencies $\geqslant 10$ in the surface layer. Definite data on the isotropy of the small-scale temperature structure in the atmospheric surface layer are provided by light scattering experiments (Gurvich, Pakhomov and Cheremukhin, 1971; see also Sect. 9.5 below). At the same time it was found both by the Vancouver group (Stewart, 1969; Boston, 1970) and the group at San Diego, Calif. (Gibson, Stegen and Williams, 1970; Gibson, Friehe and McConnell, 1971) that the skewness for the temperature time derivative in the atmospheric surface layer is strictly positive (of the order of 0.5); the same skewness for oceanic turbulence was found to be close to zero (Nasmyth, 1970). The nonzero value of the skewness of the temperature derivative is evidently incompatible with the isotropy of temperature fluctuations at scales small enough to be dominated by molecular conductivity. However Wyngaard (1971) made an attempt to explain the observed positive values of the skewness for the temperature derivative by the existence of a slight velocity sensitivity of fine wire resistance thermometers. According to Boston's estimate (1970) this effect is nonnegligible but it hardly can be responsible for the whole observed skewness. Moreover Gibson, Friehe and McConnell (1971) found that the skewness of the temperature derivative in a heated laboratory jet is strictly negative and such a skewness cannot be produced by contamination of the temperature signal by velocity sensitivity. Therefore we must conclude that the question of the true nature of the observed nonzero values of the skewness of the temperature derivative is not settled until now. Nevertheless we must remember that there is an enormous amount of data confirming the predictions of the theory of locally isotropic turbulence (many of them will be referred to below in this section); this fact forces us to be quite suspicious of all results which seem to refute the existence of local isotropy. At present we can only say that the exact limit of the isotropic range of length scales (or wave numbers) of atmospheric turbulence is very poorly known in all cases and additional experimental work in this direction is strongly needed.

Let us now return to the laboratory measurements. Particular attention to the verification of local isotropy was paid by M. Gibson (1962, 1963) who used a precise hot-wire anemometer to determine the one-dimensional spectra $E_1(k)$ and $E_2(k)$ of the longitudinal and lateral velocity components in an axially symmetric turbulent air jet with Reynolds number $Ud/\nu = 5 \cdot 10^5$, where $d$ is the initial jet diameter and $U$ is the velocity on the jet axis at a distance $50d$ from its beginning (see Fig. 61a). The spectrum $E_1(k)$ and the dissipation spectrum $k^2E_1(k)$ measured by Gibson show practically no overlap, so that one would expect not only local isotropy but the existence of an inertial subrange in this case. The existence of local isotropy in these experiments is demonstrated by Fig. 61b where the circles show direct measurements of $E_1(k)$ and the solid line shows the values of $E_1(k)$ calculated from measurements of $E_2(k)$ with the aid of the formula

$$\frac{dE_2(k)}{dk} = -\frac{k}{2}\frac{d^2E_1(k)}{dk^2}, \tag{23.2}$$

This formula is a consequence of Eq. (12.87), which in turn is based on the condition for local isotropy. It is clear from Fig. 61b that the requirement of local isotropy was satisfied satisfactorily for $k > 0.1$ cm$^{-1}$ under the conditions of Gibson's experiments.

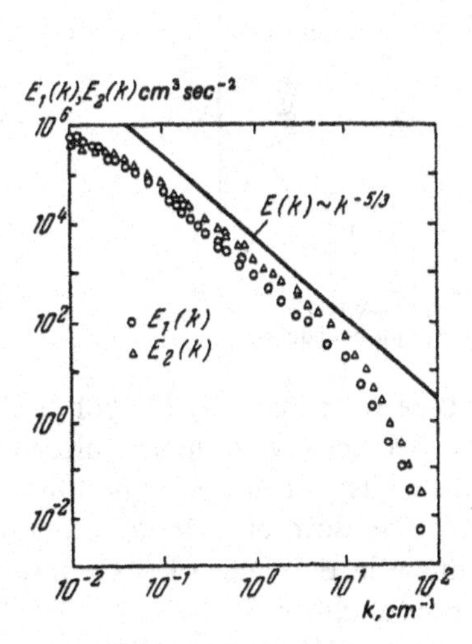


Fig. 61a Longitudinal and lateral one-dimensional velocity spectra in a turbulent jet.

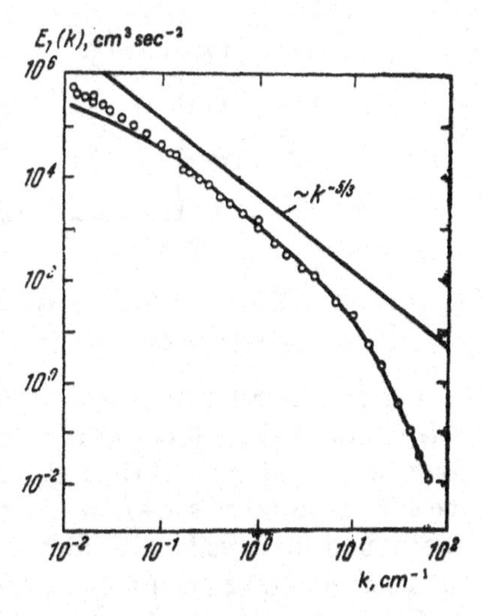


Fig. 61b Comparison of measured and calculated [from Eq. (23.2)] longitudinal velocity spectra $E_1(k)$.

Similar measurements were performed by Uberoi and Freymuth (1969) in a wake behind a circular cylinder of diameter $d$ placed perpendicular to a uniform flow of velocity $U$. The measured spectra, $E_u$, $E_v$, and $E_w$, of the three velocity components in the center of the wake are shown in Fig. 61c; the distance of the point of measurement from the cylinder was $x = 200\,d$, and the Reynolds number was $\mathrm{Re} = Ud/\nu = 2160$. The spectra and the wave numbers $k$ were made dimensionless with the aid of the velocity scale $[\overline{u^2}]^{1/2}$ and some suitably chosen length scale $l$. The solid line in the figure represents the longitudinal spectrum $E_u$ and the dotted line gives the values of the lateral spectrum computed from the measured longitudinal spectrum with the aid of the isotropic relation (12.87). We see that the isotropic relation is valid with high accuracy in the high wave number range (but not at low wave numbers).

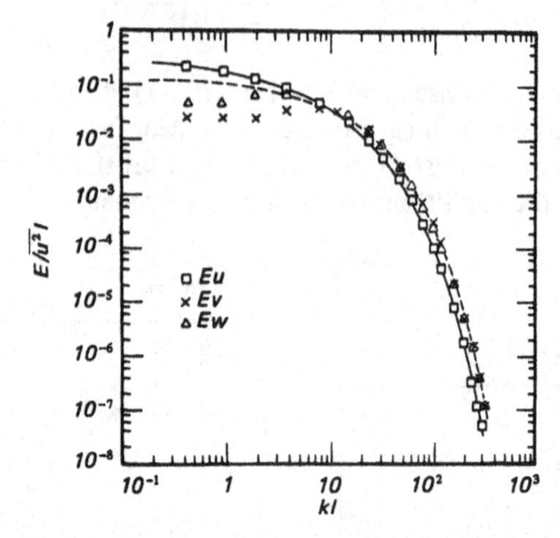


**Fig. 61c** One-dimensional spectra of the velocity fluctuations $u, v$, and $w$ in the center of the cylinder wake.

Later measurements of the same type were made by Uberoi and Freymuth (1970) also for the wake behind a solid sphere placed outside a wind tunnel. Here also the results turned out to be quite analogous to those shown in Fig. 61c: The isotropic relation (23.2) was found to be valid with high accuracy in the range of large wave numbers but to be incorrect at small wave numbers.

Finally Champagne, Harris, and Corrsin (1970) measured the time spectra (i.e., spectra in the mean flow direction) of the three velocity components in the nearly homogeneous turbulent shear flow in a wind tunnel at $\mathrm{Re} = \frac{\lambda}{\nu}\left(\frac{\overline{u_i'u_i'}}{3}\right)^{1/2} \approx 130$. They also found that the isotropic relations $E_{u_2} = E_{u_3}$ and Eq. (12.87) are valid with good accuracy for all $k\eta \geqslant 0.05$ (some small discrepancies at the largest $k\eta \geqslant 0.5$ are apparently due to instrument error). However, the simultaneous measurements of $E_{12}(k)$ and $R_{12}(k)$ showed that the value of $R_{12}(k)$ is appreciable until $k\eta = 0.15$, i.e., over a considerable part of the region where the isotropic relations for the component spectra are valid. Therefore Champagne et al. concluded that the validity of the isotropic relations for velocity component spectra is not enough for reliable indication of isotropy. It is natural to assume that this unexpected conclusion has some relation to the discrepancies between the results on the verification of local isotropy obtained by different authors. However this conclusion must be considered as quite preliminary at the present time. Hence we may say that the existence of isotropy of the smallest scale turbulent motions seems quite certain now, but the question of the limit of the isotropic region clearly needs additional investigation for laboratory fluid flows as well.

## 23.3 Verification of the Second Kolmogorov Similarity Hypothesis for the Velocity Fluctuations

Let us first list the consequences of the second Kolmogorov hypothesis which are most readily verified experimentally. These include the relations (21.17′), which expresses the two-thirds law for the velocity structure functions in the inertial subrange of distances $r$. These relations can be rewritten in the form

$$D_{LL}(r) = C(\bar{\varepsilon}r)^{2/3}, \quad D_{NN}(r) = \frac{4}{3}C(\bar{\varepsilon}r)^{2/3}. \tag{23.3}$$

when the third equation (21.19) is taken into account. More difficult to verify is the relation $D_{LLL}(r) = -\frac{4}{5}\bar{\varepsilon}r$ (this has, however, the advantage that it does not contain undetermined numerical factors), and the conclusion that the skewness

$$S = D_{LLL}/D_{LL}^{3/2} = -\frac{4}{5}C^{-3/2} \tag{23.4}$$

and the other similar dimensionless statistical properties of the velocity differences at distances $r$ are constant in the inertial subrange.

Second, we can readily verify formulas (21.43) which express the five-thirds law for the time spectra in the inertial subrange of wave numbers $k = \omega/\bar{u}$, which can be replaced with high accuracy by the formulas

$$E_1(\omega) \approx \frac{1}{4}C(\bar{\varepsilon}\bar{u})^{2/3}\omega^{-5/3}, \qquad E_2(\omega) \approx \frac{1}{3}C(\bar{\varepsilon}\bar{u})^{2/3}\omega^{-5/3}. \tag{23.5}$$

if we use Eq. (21.25). We note also that verification of the fact that the ratio of the numerical factors in Eq. (23.3) [or Eq. (23.5)] is equal to 4/3 is an additional piece of evidence for the existence of local isotropy.

*Measurements of the Velocity Structure Functions*

The first attempt to compare the two-thirds law with data, and to estimate the value of the coefficient $C$, is due to Kolmogorov (1941) who used the measurements of the longitudinal and lateral correlation functions of the velocity $B_{LL}(r)$ and $B_{NN}(r)$, and the energy dissipation rate $\bar{\varepsilon} = \frac{3}{2}\frac{d\overline{u'^2}}{dt} = \frac{3U}{2}\frac{d\overline{u'^2}}{dx}$ in the grid-generated turbulence in a wind tunnel for relatively low values of Re, which were reported by Dryden et al. (1937). Kolmogorov found that $C = 1.5$. The same data were later analyzed by Batchelor (1947) together with similar measurements obtained by British workers, and led to very similar conclusions. The validity of the consequences of the Kolmogorov theory for grid turbulence $\left(\mathrm{Re}_M = \frac{UM}{\nu}\right.$ of the order of $\left.10^4 - 10^5\right)$ were studied by Townsend (1948b) who measured the lateral correlation function $B_{NN}(r)$ and the skewness $S(r)$. Townsend at first succeeded in obtaining good agreement between his results and the theoretical predictions for locally isotropic turbulence (on the assumption that $C$ was approximately 1.6). Subsequently, however, Stewart (1951) found that some of the Townsend results were incorrect, and that generally the values of $\mathrm{Re}_M$ used by Townsend and other investigators were too low to ensure the existence of an appreciable inertial subrange. According to the estimates of Proudman (1951), Stewart and Townsend (1951), and Gibson and Schwarz (1963b), which were mentioned in Sect. 16.6, an appreciable inertial subrange will only appear provided the Reynolds number $\mathrm{Re}_M$ is of the order of $10^6$ or more. These estimates also throw doubt on the von Kármán (1948b) interpretation of the experimental data of Liepmann et al. (1951) on grid-generated turbulence for $\mathrm{Re}_M = 10^5$ and $3 \cdot 10^5$; this interpretation led to the conclusion that the corresponding one-dimensional spectra contained a five-thirds region (see Sect. 16.6).

The estimates of Proudman and of Stewart and Townsend have shown that a satisfactory verification of the consequences of the second similarity hypothesis, as far as the inertial subrange is concerned, is most simply carried out for the natural turbulent flows in the atmosphere and in the oceans. These are characterized by much larger Reynolds numbers than flows produced under laboratory conditions. The first attempt to determine the wind velocity structure functions by means of a hot-wire anemometer was made under the direction of Obukhov soon after the publication of the theory [see Obukhov (1942)]. World War II prevented the completion of these experiments, and only an estimate of the order of magnitude of the dimensional coefficient $b = C\bar{\varepsilon}^{2/3}$ was obtained. Obukhov (1949c) later returned to the verification of the two-thirds law using old [Gödecke (1935)] but very accurate hot-wire anemometer data. These data concern the mean absolute values of the wind velocity differences $\overline{|\Delta_r u|}$ at right angles to the wind direction (which corresponded to the lateral structure function $D_{NN}$) for $r = |\mathbf{r}|$ between 0.1 and 80 cm at a height of 1 m above the surface. As a result of an analysis of these data Obukhov obtained the graph in Fig. 62. This shows that the empirical lateral structure function $D_{NN}(r)$ is satisfactorily described by the two-thirds law for $r > 1$ cm.

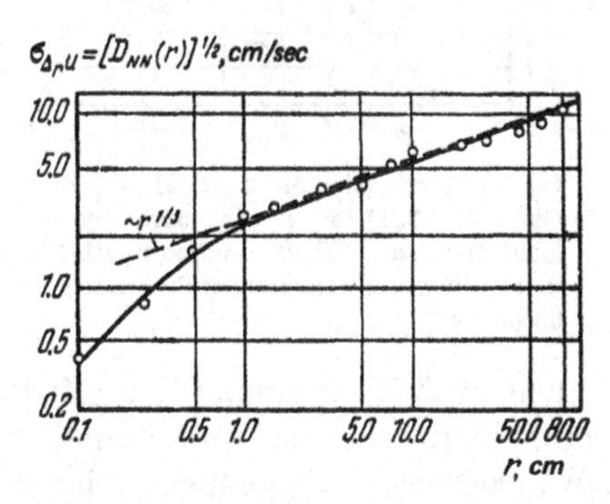


**Fig. 62** Measured lateral velocity structure function [Gödecke (1935)].

In a subsequent paper [see also the review papers by Monin (1958) and Obukhov and Yaglom (1958, 1959)], Obukhov (1951) used his own measurements of the quantity $\overline{(\Delta_r u)^2} = D_{NN}(r)$ for $r$ between 2 and 60 cm. These measurements were carried out with a differential hot-wire anemometer at heights $z = 1.5$, 3, and 15 m in the surface layer of the atmosphere with a near-neutral stratification. The energy dissipation rate under such stratification could be estimated from the formula $\bar{\varepsilon} = u_*^3/\varkappa z$ where $u_*$ is the friction velocity determined from the mean wind-velocity profile, and $\varkappa \approx 0.4$ is the von Kármán constant [see Eq. (7.93) of Vol. 1]. According to this estimate, the two-thirds law can be transformed to read

$$\frac{\overline{[(\Delta_r u)^2]}^{1/2}}{u_*} = a\left(\frac{r}{\varkappa z}\right)^{1/3}, \tag{23.6}$$

where $a = (4C/3)^{1/2}$. Figure 63 shows Obukhov's results on the verification of Eq. (23.6). It is clear that the data are consistent with the theoretical predictions. These data can also be used to estimate very approximately the value of the coefficient $a$ (since the estimate of $\bar{\varepsilon}$ from existing data is quite inexact). The result was $a \approx 1.1$ which is equivalent to $C \approx 0.9$.

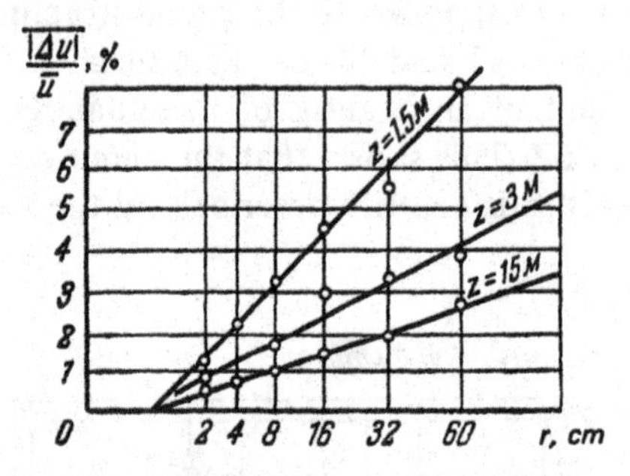


**Fig. 63** Verification of Eq. (23.6) using the Obukhov (1951) data. $\bar{u}$ is the mean wind velocity at height $z$, $\overline{|\Delta u|}/\bar{u}$ is proportional to $[\overline{(\Delta u)^2}]^{1/2}/u_*$, and the scale of the $r$ axis is linear in $r^{1/3}$.

Subsequent extensive data confirming the validity of the second Kolmogorov hypothesis for the wind velocity field were reported by many workers. We can note for example the hot-wire anemometer measurements of the time structure functions due to MacCready (1953) in the USA, Shiotani (1955, 1957, 1963) in Japan, and R. Taylor (1955, 1961) in Australia, and also the bivane and bead-thermistor anemomenter measurements by H. E. Cramer (1959) and Record and Cramer (1966) in the USA.[23]

[23] Record and Cramer also measured the quantities $-\overline{u'w'} = u_*^2$ and $\partial\bar{u}/\partial z$, simultaneously with the structure functions. They used these data to estimate the universal coefficient $C$ in the two-thirds law. Having this in view, they employed the relation $\bar{\varepsilon} \approx -\overline{u'w'} \cdot \partial\bar{u}/\partial z$ which is exact in the case of neutral stratification (under the conditions explained in Chapt. 4, of Vol. 1; see especially Sects. 7.1 and 7.5); according to the Record and Cramer data, it is approximately correct also in the case of moderate deviation from neutrality. As a result they found that $C \approx 2$, which is in good agreement with the results of other workers which we shall discuss below.

The unanimous conclusion is that the empirical time structure functions are in satisfactory agreement with the two-thirds law (21.42) over a considerable range of values of $\tau$. Therefore, such data can also be regarded as confirming both the Taylor frozen-turbulence hypothesis, and the second Kolmogorov similarity hypothesis. We note also that the validity of the two-thirds law for $D_{LL}(r)$ and $D_{NN}(r)$ [and of the ratio $D_{LL}(r)/D_{NN}(r) = 3/4$] was established by Dunn (1965) for turbulence in seawater at depths between 3 and 10 m. Finally, let us note precise measurements of the time structure functions of longitudinal velocity fluctuations of the second, third, and fourth orders, carried out by Van Atta and Chen (1970) in the atmospheric surface layer over the ocean at heights of 3, 23, and 31 m. These measurements were done with a fine hot-wire anemometer having a wire 5$\mu$ in diameter and 1 mm. long. The second order structure functions (transformed to space coordinates with Taylor's frozen-turbulence hypothesis) are shown in Fig. 63a; the solid lines in this figure have a slope of 2/3; the discussion of the dashed lines will be postponed until Sect. 25. It is seen from an examination of the solid lines in the figure that all the data are fairly well fitted by the two-thirds law over a considerable range of separations $r$ between 5 cm and about twice the height above sea level. The time constant of the probe and its size were small enough to permit reliable determination of the velocity derivative $\partial u/\partial t = -\bar{u}\partial u/\partial x$ and

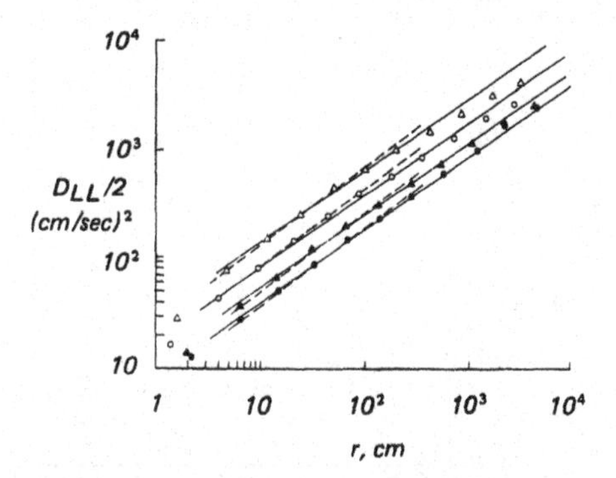


**Fig. 63a** Second-order structure functions (after Van Atta and Chen (1970). Here, and in Figs. 63b and 63c, the meaning of the symbols is as follows: ○, $z$ = 3 m, $\bar{u}$ = 7.2 m/s; △, $z$ = 3 m, $\bar{u}$ = 8.4 m/s; ▲, $z$ = 23 m, $\bar{u}$ = 11 m/s; ●, $z$ = 31 m, $\bar{u}$ = 11.3 m/s. The solid lines have a slope of 2/3. The dashed lines have a slope of 0.722

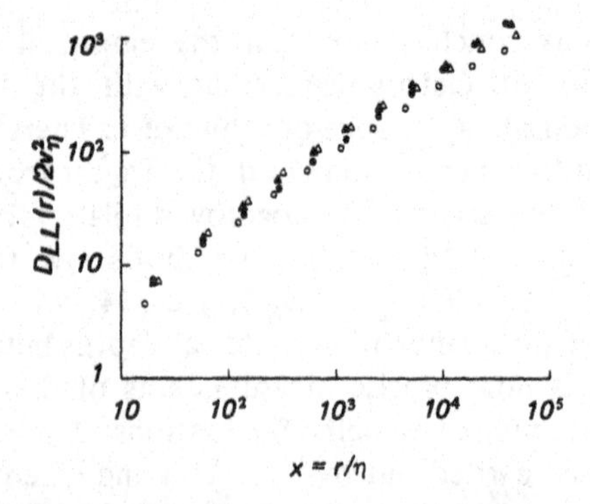


**Fig. 63b** Second-order structure functions normalized with Kolmogorov scales (after Van Atta and Chen, 1970).

subsequent evaluation of the mean rate of energy dissipation $\bar{\epsilon} = 15\nu\overline{(\partial u/\partial x)^2}$. Hence Kolmogorov's length scale $\eta$ could be calculated; after the change to dimensionless variables $x = r/\eta$ and $\beta_{LL} = D_{LL}/v_\eta^2$ all data in Fig. 63a collapse satisfactorily onto a single curve (Fig. 63b). The resulting curve agrees with the two-thirds law for $x = r/\eta \geqslant 50$; the corresponding value of $C$ is determined by the data with moderate precision and its average value is about 2.3. The measured dimensionless third-order structure functions $\beta_{LLL}(r/\eta) = D_{LLL}(r/\eta)\overline{v}_\eta^3$ cluster around the theoretical equation $\beta_{LLL}(x) = -0.8x$ of the Kolmogorov theory for all $x = r/\eta < 10^3$ (see Fig. 63c where

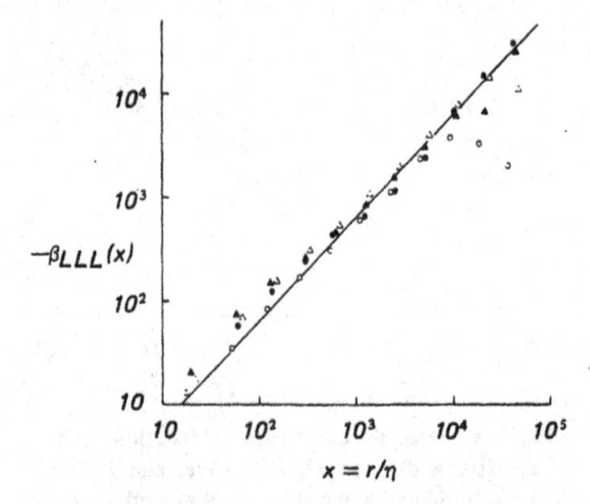


**Fig. 63c** Third-order structure functions normalized with Kolmogorov scales (after Van Atta and Chen, 1970). The solid line is the theoretical equation $\beta_{LLL}(x) = -0.8x$.

the solid line corresponds to the theoretical equation). The data of Van Atta and Chen on fourth order structure functions will be considered in Sect. 25 below. Less accurate data on the second and third order structure functions of the wind velocity fluctuation in the marine surface layer of the atmosphere were obtained by Paquin and Pond (1971); They are also in a satisfactory agreement with the predictions of Kolmogorov's theory.

The data collected in Figs. 63a–63c lend substantial support to the theory of locally isotropic turbulence considered in the preceeding two sections. However, the most satisfactory confirmations of the validity of the similarity hypothesis in the case of atmospheric and oceanic turbulence were obtained from measurements of the velocity spectra, which have led to extensive data of high accuracy. We shall now describe these measurements in detail.

*Measurements of Wind-Velocity Spectra*

The time spectra of the velocity field can be determined numerically (from records of velocity fluctuations, or by taking the Fourier transform of the measured time structure functions), but they can also be found directly by feeding electrical signals proportional to the velocity fluctuations into the filters of a spectral analyzer. An extensive program of such measurements of wind velocity spectra was undertaken in particular at the Institute of Atmospheric Physics, USSR Academy of Sciences. The measuring instruments were the sonic anemometers described briefly in Sect. 8.3 of Vol. 1. The anemometer signals were fed into the 30 filters (with widths of half an octave) of the spectral analyzer described by Bovsheverov et al. (1959). The program included surface layer measurements (at a height of 1 and 4 m above the steppe) of the vertical wind fluctuations $w'$ [Gurvich (1960a, b, 1962)] and the horizontal wind fluctuations $u'$ (in the direction of the mean wind) [Zubkovskii (1962)]. Extensive measurements were also carried out of the $w'$ spectra at different heights (up to 3–4 km) from an aircraft [Zubkovskii (1963), Koprov (1965)], and also individual measurements of the $w'$ and $u'$ spectra from the 300 m meteorological tower at Obnisk [Tsvang, Zubkovskii et al. (1963)] as well as the 70 m meteorological tower near Tsimlyansk. Certain additional spectral measurements, which will be discussed below, were carried out in parallel.

According to the results obtained from all this work, the spectra of $u'$ and $w'$ always contain an appreciable region in which the five-thirds law is valid and, when a sufficiently long average is taken, this law is satisfied to high accuracy. In Fig. 59 above we have already given an example of this at a height of 70 m, showing clearly the validity of the five-thirds law. One further example is provided by Fig. 64, which is taken from the paper by Tsvang et al. (1963). These spectra were obtained at a fixed point at a height of 300 m, and were converted to space spectra with the aid of the Taylor frozen-turbulence hypothesis (21.41). They were averaged over a number of groups (two in the case of $u'$, and three in the case of $w'$), characterized by roughly the same stratification conditions. Analogous results have also been reported by Gurvich, by Zubkovskii and Koprov (see Figs. 68 and 72 below).

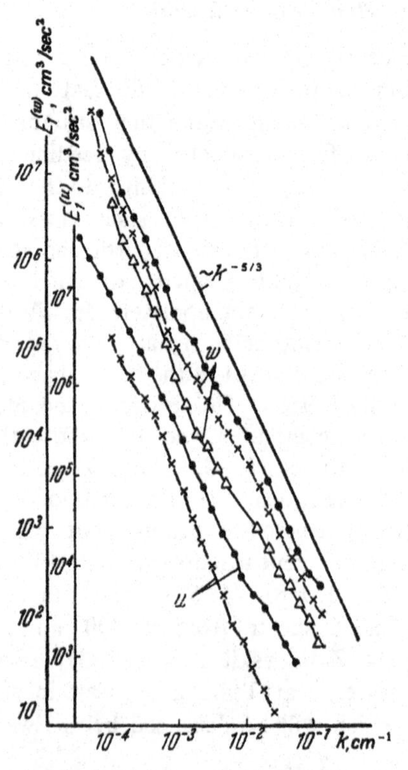


**Fig. 64** Measured one-dimensional spectra of the horizontal ($u$) and vertical ($w$) wind-velocity components at a height of 300 m.

Data confirming the validity of the five-thirds law for the time spectra of wind-velocity fluctuations have been reported by many other workers as well. For example, Panofsky and van der Hoven (1956) have shown that the spectra of the longitudinal component $u'$ carried out from a 100 m meteorological tower at Brookhaven satisfied the five-thirds law over a considerable range of frequencies $f$ (this range is even longer than expected from the condition $k \gg 1/L$, which is involved in the definition of the inertial subrange of the spectrum). Good agreement with the 5/3 law was also found for the time spectra of the fluctuations $w'$ at heights between 2 and 12 m above the earth's surface, which were measured by Businger and Suomi (1958), using a spectral analyzer and a sonic anemometer with a comparatively large base. The data reported by Panofsky and McCormick (1960) on the time spectra of $w'$ at different heights between 20 cm and 600 m show a considerable spread, but as a whole they are in adequate agreement with the five-thirds law for frequencies $f > 0.7\bar{u}/z$ in the case of stable stratification, and $f > 0.4$ $\bar{u}/z$ in the case of unstable stratification. The spectra of the velocity fluctuations in the surface layer above the sea reported by Weiler and Burling (1967), Volkov (1969), Gibson and Williams (1969) and Stewart (1969) all contain appreciable five-thirds regions, although some of these data do not confirm local isotropy (see the preceding subsection). Also the spectra of wind-velocity components at heights of between a few hundred meters and 3.5 km obtained by MacCready (1962a) with hot-wire anemometers carried by a sail-plane, as well as the extensive aircraft measurements reported by Shur (1962, 1964), Burns (1964), Pinus (1966), Payne and Lumley (1966), Vinnichenko; Pinus and Shur (1967), Reiter and Burns (1967), Vinnichenko, Pinus, Shmeter and Shur (1968), Sheih, Tennekes and Lumley (1971) and other workers, are also in satisfactory agreement with the existence of a spectral range (which is different at different heights and under different conditions) in which the five-thirds law is valid [see for example Fig. 65 which is taken from the paper by Payne and Lumley (1966)]. There is thus little doubt at present that the five-thirds law is valid for the velocity spectra of atmospheric turbulence over a considerable range of wave numbers.

Simultaneous determinations of the spectra of different wind-velocity components have usually been too approximate to verify satisfactorily the relations $E_1^{(u)}(k)/E_1^{(w)}(k) = E_1^{(u)}(k)/E_1^{(v)}(k) = 3/4$ between the longitudinal and lateral one-dimensional spectra of locally

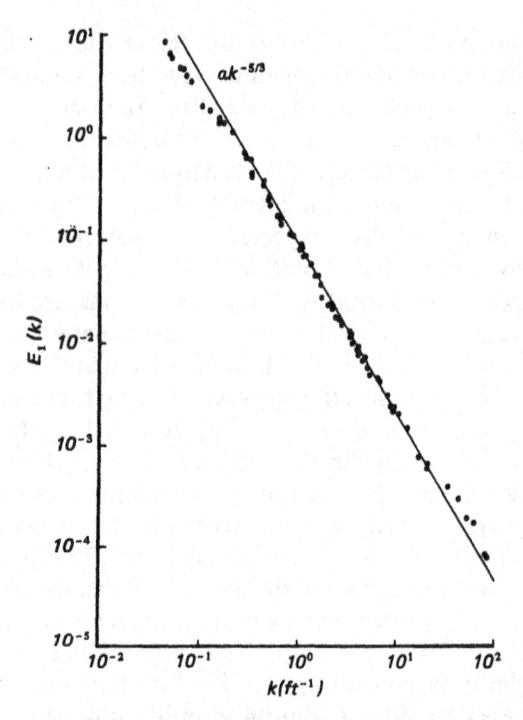


**Fig. 65** One-dimensional longitudinal velocity spectra derived from aircraft measurements of Payne and Lumley (1966).

isotropic turbulence in the inertial range. Usually, such data merely show that the ratio of the spectral densities of the various components is of the order of unity [see, for example, MacCready (1962a), Shiotani (1963), Berman (1965), Reiter and Burns (1967)]. However, the measurements of Burns (1964) show that $E_1^{(w)}(k) > E_1^{(u)}(k)$ in the inertial subrange. Analysis of the data of simultaneous determinations of the spectra of $u'$ and $w'$ fluctuations from the 70 m meteorological tower near Tsimlyansk, lead to the conclusion that the ratio $E_1^{(u)}(k)/E_1^{(w)}(k)$ is rather scattered but its mean value is $0.77 \pm 0.08$ in the inertial range, which is very close to the theoretical result of 0.75 [see Gurvich, Koprov et al. (1967)]. We have also mentioned in the previous subsection that the data published by Busch and Panofsky (1968), Wyngaard and Coté (1971) and Sheih, Tennekes and Lumley (1971) satisfactorily confirm this theoretical result. However recent data of hot-wire anemometer measurements near Vancouver (Stewart, 1969) definitely disagree with this result; this fact, which has already been mentioned in the preceding subsection, stands in need of explanation.

Gurvich (1960a, b; 1962) and Zubkovskii (1962) used the estimated mean energy dissipation rate $\bar{\varepsilon}$, deduced from mean wind-velocity and temperature profiles near the ground, and also their own data on wind-velocity spectra to determine the universal dimensionless coefficient $C$ in Eqs. (23.3) and (23.5) (see the subsequent passage in small print for further details). Their results, corrected for some small errors of the authors, are: $C \approx 1.1 - 1.3$ (Gurvich) and $C \approx 1.6 - 1.9$ (Zubkovskii).[24] The last result will be seen later to approach the estimate of $C$ that is regarded as the best value at present.

Another method of estimating the coefficient $C$ is based on the exact formula (23.4) and was employed by Gurvich (1960c), Stewart, Wilson and Burling (1970), Van Atta and Chen (1970) and Paquin and Pond (1971). Gurvich used two sonic anemometers whose outputs were automatically subtracted and transformed into an electrical signal proportional to the velocity difference at two points near the ground (at the same height $z = 1.8$ m). The signal was then squared and cubed by analog circuits. Gurvich deduced the spectral functions $D_{LL}(r)$ and $D_{LLL}(r)$ for two values of $r$ in the inertial subrange, and then used this to calculate the skewness factor $S = D_{LLL}(r)/[D_{LL}(r)]^{3/2}$. He found that $S = -0.45 \pm 0.05$ for $r = 25$ cm, and $S = -0.40 \pm 0.06$ for $r = 50$ cm. Subsequently Stewart (1963) pointed out, however, that instrumental averaging over the sonic anemometer base has a different effect on $D_{LL}(r)$ and $D_{LLL}(r)$, so that Gurvich's results will be modified when this is taken into account. Subject to certain natural assumptions about the effect of instrumental averaging on the third moments, Stewart succeeded in estimating the corresponding correction, and then found that Gurvich's corrected results were as follows: $S = -0.39 \pm 0.04$ for $r = 25$ cm and $S = -0.36 \pm 0.05$ for $r = 50$ cm. In view of Eq. (33.4), the former value of $S$ corresponds to $C \approx 1.6$, and the latter

[24] The main table in Zubkovskii's paper (on page 1431) is in fact based on measured values of the spectrum $E_1(f)$, where $f = \omega/2\pi$, normalized by the condition $\int_0^\infty E_1(f)\, df = \overline{u'^2}$ and differing by a factor of $4\pi$ from the spectrum $F(\omega)$ in the formulas quoted in this paper. Some of the scales in Gurvich's paper are shown incorrectly; the estimate of $C$ which we have just quoted is based on the numerical data in Table 4 of the most complete paper by Gurvich (1962).

corresponds to $C \approx 1.7$. As noted above, these values of $C$ are in satisfactory agreement with subsequent determinations which will be discussed at the end of this subsection.

The data on $S(r) = D_{LLL}(r)/[D_{LL}(r)]^{3/2}$ obtained by Paquin and Pond (1971) from wind fluctuation measurements in the marine atmosphere are roughly constant from two to five meter lags for most runs but are characterized by considerable scatter from run to run. The average value of skewness $S$ (corrected according to Stewart's suggestion) was $-0.25$ which corresponds to a value 2.2 for $C$. The value of $S$ reported by Stewart et al. (1970) is $S \approx -0.26$ (i.e., $C \approx 2.1$). The value of $-S$ found by Van Atta and Chen (1970) is smaller (i.e., the value of $C$ is larger): $-S \approx 0.2$. Keeping in mind however the low accuracy of the experimental determination of $S$ we can conclude that the existing data on $S$ are not in serious disagreement with each other. They also agree satisfactorily with various determinations of $C$ which will be discussed later in the present subsection.

The general similarity theory of turbulence in a thermally stratified surface layer given in Chapt. 4, Vol. 1, has been found useful in the analysis of wind-velocity spectra in the atmospheric surface layer above plane and homogeneous underlying surfaces.

As far as the structure functions and spectra are concerned, the predictions of this theory for values of $r$ that are not too small and wave numbers $k$ that are not too large (so that viscosity can be neglected) were summarized by Monin (1958, 1959b; 1962a, b). The formula for the spectral density of the vertical velocity fluctuations $E_w(\omega) = E_2(\omega)$, which follows from this theory, is given by

$$E_w(\omega) = \frac{u_*^2 z}{\bar{u}} \psi_w\left(\frac{\omega z}{\bar{u}}, \mathrm{Ri}\right), \tag{23.7}$$

where $\psi_w$ is a universal function of the dimensionless circular frequency $\omega z/\bar{u}$ and the Richardson number Ri [which can be replaced by the dimensionless height $\zeta = z/L$ of Eq. (7.11)]. Similar formulas are valid for the spectra of the horizontal velocity fluctuations $u'$ and $v'$. It follows from these formulas that the frequencies corresponding to any characteristic points in the spectra (maxima, low-frequency limits of the inertial subrange, etc.) will be of the form $\omega = \frac{\bar{u}}{z}\,\Omega(\mathrm{Ri}) = \frac{\bar{u}}{z}\,\Omega_1(\zeta)$ where $\Omega$ and $\Omega_1$ are certain universal functions.

For strong instability ($\zeta \ll -1$), the similarity formulas should not include the parameter $u_*$, i.e., the dimensionless functions $\psi\left(\frac{\omega z}{\bar{u}}, \zeta\right)$ should be asymptotically proportional to $|\zeta|^{2/3}$, and the functions $\Omega_1(\zeta)$ should approach constant values. In the case of strong stability ($\zeta \gg 1$) these formulas should apparently not contain the height $z$, so that in this case $\psi\left(\frac{\omega z}{\bar{u}}, \zeta\right) \sim \frac{1}{\zeta}\psi_1\left(\frac{\omega L}{\bar{u}}\right)$ and $\Omega_1(\zeta) \sim \zeta$, i.e., $\omega = \bar{u}\Omega_1/z \sim \bar{u}/L$. As the stability increases, the total intensity of turbulent fluctuations falls, in the first instance, as a result of the decay of large-scale inhomogeneities, i.e., the maximum of the spectrum shifts toward small-scale disturbances. The dependence of the shape of the spectra on the type of thermal stratification in the atmospheric surface layer, corresponding to these predictions, is illustrated by Fig. 66 which is taken from the paper by Monin (1959b). Moreover, comparison of the similarity-theory formulas with the five-thirds law (23.5) will show that, in the inertial subrange, we have

$$\psi_w\left(\frac{\omega z}{\bar{u}},\ \zeta\right)=\frac{1}{3}C\left(\frac{\bar{\varepsilon} z}{u_*^3}\right)^{2/3}\left(\frac{\omega z}{\bar{u}}\right)^{-5/3}, \tag{23.8}$$

and in the analogous formula for $\psi_u$ the numerical coefficient 1/3 is replaced by 1/4.

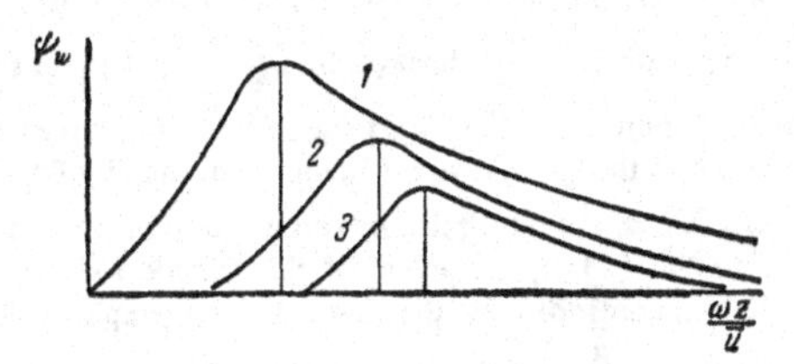


**Fig. 66** Typical shape of the velocity spectrum for different stratification. 1–Unstable; 2–neutral; 3–stable

Similarity theory was used in particular by Gurvich (1968b, 1962) to analyze his vertical velocity spectra at heights of 1 and 4 m. The measured spectra were split into groups in accordance with the values of the Richardson number Ri. An example of one such group ($z$ = 4 m and Ri = –0.09) is shown in Fig. 67a. Figure 67b shows a plot of $\tilde{\psi}_w = \frac{\bar{u}}{\tilde{u}_*^2 z} E_w$ as a

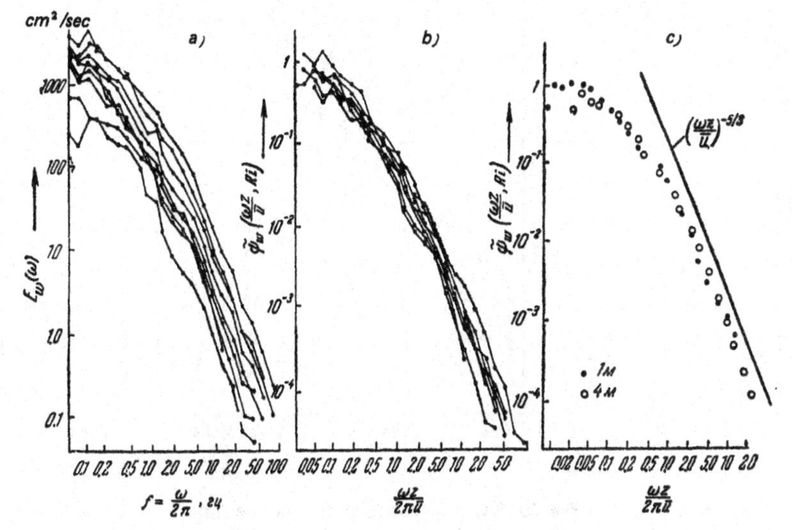


**Fig. 67** Vertical velocity spectra for Ri = –0.09 [Gurvich (1960a, b)].

function of the dimensionless frequency $fz/\bar{u}$ corresponding to the spectra of Fig. 67a, where $\tilde{u}_* = \varkappa z(\partial\bar{u}/\partial z)$ approaches $u_*$ for neutral stratification, and is equal to $u_*\varphi(\zeta)$ in the general case [see formula (7.15) of Vol. 1]. It is readily seen that the spread of individual spectra is considerably reduced by transformation to the dimensionless coordinates of Fig. 67b. Figure 67c shows the mean spectra $\tilde{\psi}_w$ for Ri = −0.09, and two heights $z$ = 1 m and 4 m. It is clear that these spectra satisfy the five-thirds law to high accuracy, and are practically the same at different heights as the similarity theory predicts. The predictions of the similarity theory were also verified by Weiler and Burling (1967) from data on wind-velocity spectra in the surface layer over the sea under neutral thermal stratification; the results were similar to those of Gurvich. More extensive similarity analysis of wind velocity spectra was performed by Busch and Panofsky (1968); some of their results will be considered below [see, in particular, Eq. (23.20) in Sect. 23.6].

Figure 68 shows the mean dimensionless spectra $\tilde{\psi}_w\left(\frac{\omega z}{\bar{u}}, \text{Ri}\right)$ for seven different values of Ri. They all have an appreciable five-thirds region, and show a regular dependence on Ri which is in agreement with the qualitative predictions indicated by Fig. 66. Figure 69 shows a plot of $(\omega z/\bar{u})^{5/3}\,\tilde{\psi}_w(\omega z/\bar{u};\,\text{Ri})$ multiplied by certain constants which we shall not specify here. According to to Eq. (23.8), the values of this function should be constant in the inertial subrange. This is confirmed by the shape of these graphs which also enable us to estimate the lower limit $\omega = \frac{\bar{u}}{z}\,\Omega\,(\text{Ri})$ of the inertial frequency subrange. The approximate

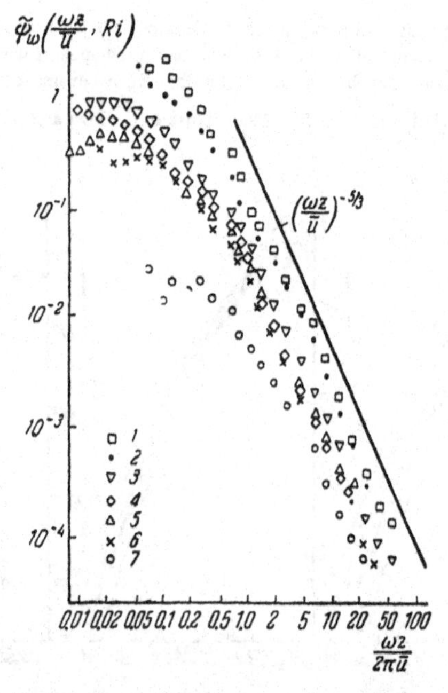


**Fig. 68** Vertical velocity spectra for seven different Richardson numbers. 1–Ri = −0.76; 2–Ri = −0.36; 3–Ri = −0.09; 4–Ri = −0.02; 5–Ri = 0; 6– Ri = 0.03; 7–Ri = 0.28.

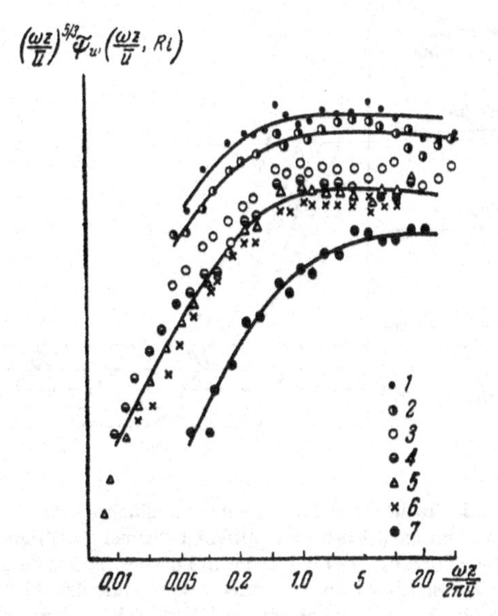


**Fig. 69** Values of the function $(\omega z/\overline{u})^{5/3}\,\tilde{\psi}_w(\omega z/\overline{u}, \mathrm{Ri})$ for different Ri. 1–7 correspond to the same values of Ri as in Fig. 68.

values of $\Omega$ were found to be as follows: $\Omega \approx 2.5$ for considerable instability (Ri = −0.76), $\Omega \approx 4.5$ for neutral stratification (Ri = 0), and $\Omega \approx 12$ for considerable stability (Ri = + 0.28). The order of magnitude of all these values is in agreement with the summarizing graph reported by Panofsky and McCormick (1960). Priestley (1959a) has found that approximately $\Omega \approx 0.6 \cdot 2\pi \approx 4$. All these estimates are considerably lower than the estimates of the low-frequency limit of the isotropic region given in Sect. 23.2. It must be remembered, however, that the very definition of the low-frequency limit of the inertial subrange is not very clear-cut. In fact the limits of validity of the five-thirds law for the one-dimensional spectra of $u'$ and $w'$ should not be the same, and may considerably exceed the limit of validity of this law for the three-dimensional spectrum $E(k)$. The values of $\omega$ below which practically the entire spectrum of the momentum flux $\tau = -\rho\overline{u'w'}$ or the heat flux $q = c_p\rho\overline{T'w'}$ is concentrated are not equal to any of the above limits, and so on [this problem is discussed by MacCready (1962b)]. A qualitative summary of existing data on the possible "low-frequency limits" of the inertial subrange in the atmospheric surface layer is shown in Fig. 70 (the weak dependence of the scale used for $f_z/\overline{u}$ on height is not inconsistent with the above discussion, since typical unstable and stable situations at different heights are characterized by different values of Ri and $\zeta = z/L$.

In conclusion of our review of Gurvich's results let us consider Fig. 71 which shows a plot of $\frac{f}{\sigma_w^2} E_w\left(\frac{fz}{\overline{u}}, \mathrm{Ri}\right)$ as a function of $f = \omega/2\pi$. This figure gives a clear picture of the distribution of the fluctuation intensity $\sigma_w^2$ over the frequency range for different Ri, in accordance with the formula

$$\sigma_w^2 = \int_0^\infty E_w(\omega)\,d\omega = \int_0^\infty \omega E_w(\omega)\,d\ln\omega, \tag{23.9}$$

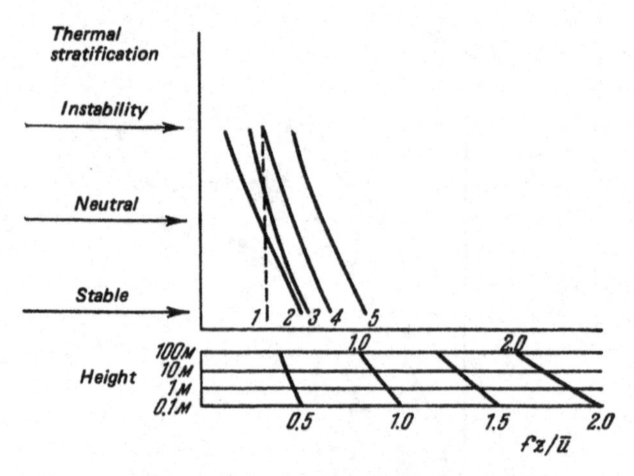


**Fig. 70** Low-frequency limits of the inertial subrange of the frequency spectrum at different heights and different thermal stratification [MacCready (1962b)]. 1–Limit of the 5/3 law for the longitudinal wind velocity; 2–frequency below which 80% of the turbulent heat flux is concentrated; 3–frequency below which 80% of the momentum flux is concentrated; 4–limit of the 5/3 law for the vertical and lateral velocity components; 5–limit of the 5/3 law for the three-dimensional velocity spectrum.

showing that $\ln 2 \cdot \omega E_w(\omega)$ is the contribution to the total intensity $\sigma_w^2$ of the fluctuations $w'$ generated in one octave. It is clear from these graphs that the greatest contribution to $\sigma_w^2$ in the case of unstable stratification is due to frequencies $\omega \approx 1{,}5\bar{u}/z$, while in the case of neutral and stable stratification the corresponding frequencies are $\omega \approx 2\bar{u}/z$ and $\omega \approx 6\bar{u}/z$ respectively. The relative contribution of the inertial subrange to $\sigma_w^2$ is about 30%.

Data on the spectra of longitudinal wind fluctuations $u'$ (at a height of 4 m above the steppe) interpreted in terms of Eq. (23.7) were published by Zubkovskii (1962). Figure 72 illustrates his results. It shows the mean dimensionless spectra $\tilde{\psi}_u(fz/\bar{u}, \mathrm{Ri}) = \bar{u}E_u(2\pi f)/u_*^2 z$ for six values of Ri. It is clear that these spectra are also in good agreement with the five-thirds law over a considerable range of frequencies. At the same time, they are in satisfactory agreement with the similarity equation (23.7), since the use of the variables $\tilde{\psi}_u$ and $fz/\bar{u}$ again enables us to *reduce* the spread of the individual spectra corresponding to a fixed value of Ri. According to Zubkovskii's data shown in Fig. 72, the lower limit of the inertial subrange in the $u'$ spectrum varies between $\omega \approx 1{,}5\bar{u}/z$ and $\omega \approx 45\bar{u}/z$ for the indicated values of Ri.

The data of numerous measurements of the vertical-velocity and longitudinal-velocity spectra were analyzed by Busch and Panofsky (1968); still later new extensive measurements of spectra of all three wind-velocity components were performed by Kaimal, Wyngaard et al. (1972). The generalized results of Kaimal, Wyngaard et al. measurements for vertical and longitudinal spectra are shown in Figs. 72a and 72b. The spectra $fE_w(f)$ and $fE_u(f)$ are normalized here by $u_*^2[\varphi_\epsilon(\zeta)]^{2/3}$ where $f = \omega/2\pi$, $\zeta = z/L$ and $\varphi_\epsilon(\zeta) = \bar{\epsilon}\kappa z/u_*^3$. It is easy to see that the normalization must bring all spectra, regardless of $z/L$, into coincidence in the inertial subrange. The data agree very well with this prediction. All the spectra converge to the same $-2/3$ line at the high frequency end, but at lower frequencies

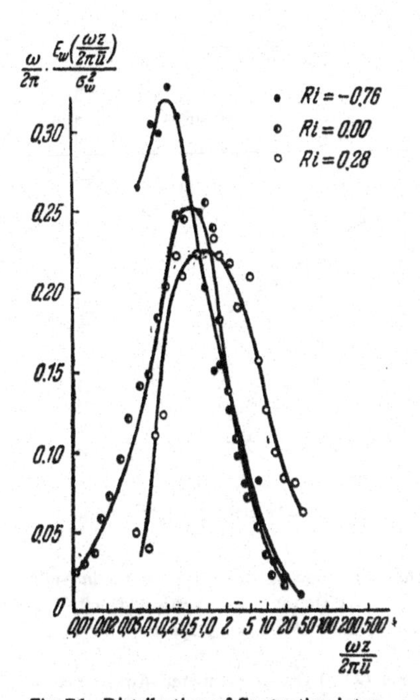


Fig. 71 Distribution of fluctuation intensity in the vertical velocity over the frequency spectrum for different stratifications.

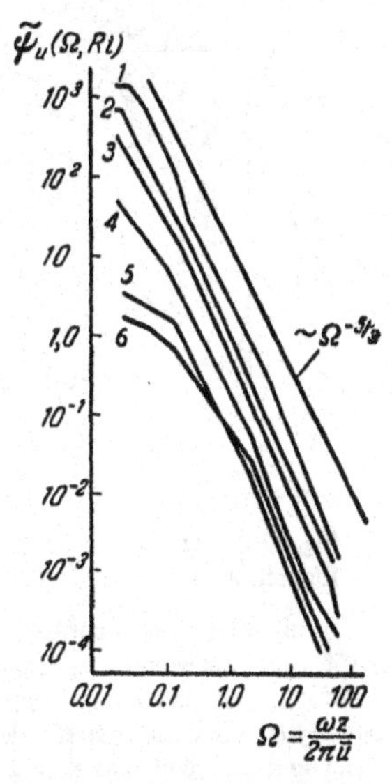


Fig. 72 Averaged dimensionless spectra of the horizontal wind velocity in the surface layer for different values of Ri. 1–Ri = −1.07; 2–Ri = −0.38; 3–Ri = −0.12; 4–Ri = 0.05; 5–Ri = 0.12; 6–Ri = 0.33.

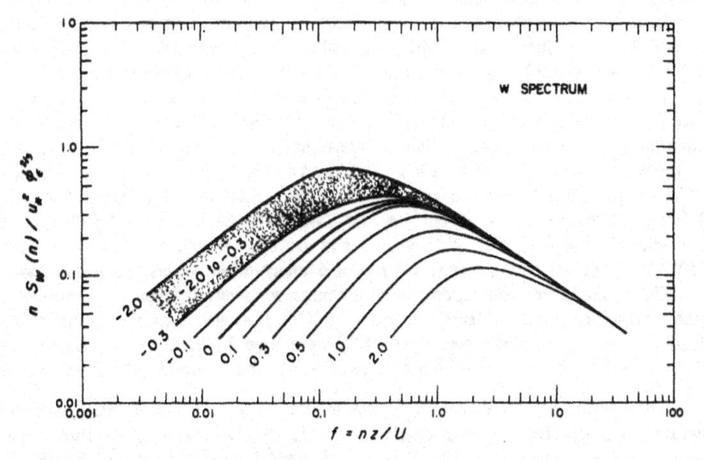


Fig. 72a Generalized dimensionless vertical velocity spectra according to Kaimal, Wyngaard et al. (1972).

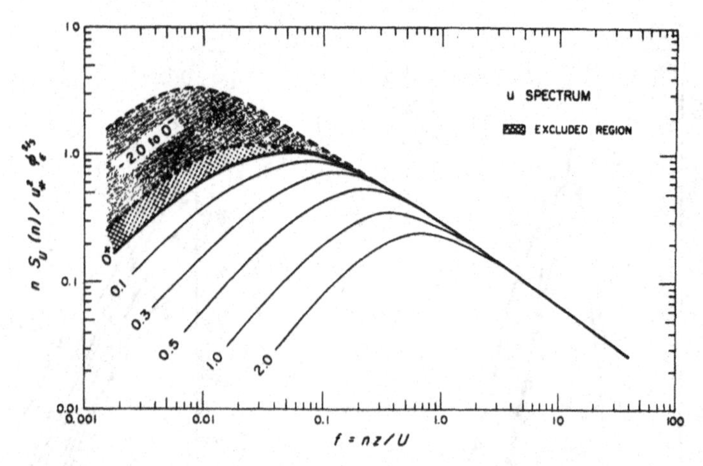


**Fig. 72b** Generalized dimensionless longitudinal velocity spectrum according to Kaimal, Wyngaard et al. (1972).

there is a separation according to $z/L$. The separation is quite systematic in stable stratification; however, in the range $-0.3 > z/L > -2.0$ (in the case of vertical velocity) or $0 > z/L > 2.0$ (in the case of longitudinal velocity) the low-frequency spectra form a quite disorderly cluster within the stippled area.

The universal constant $C$ in Eqs. (23.3) and (23.4) can be estimated from measured values of the dimensionless spectrum $\widetilde{\psi}_w(fz/\overline{u}, \text{ Ri})$ or $\widetilde{\psi}_u(fz/\overline{u}, \text{ Ri})$ in the inertial subrange and wind and temperature profiles. This can be done by using the formula $\overline{\varepsilon} \approx \frac{u_*^3}{\varkappa z}\varphi_\varepsilon(\zeta)$, where, in the first approximation $\varphi_\varepsilon(\zeta) = \zeta f' - \zeta$, and $f(\zeta)$ is a universal function of the dimensionless mean wind-velocity profile in stratified atmosphere [see Eq. (7.93) and (7.103) in Sect. 7.5 of Vol. 1]. Estimates corresponding to Ri = 0 must be regarded as particularly reliable since they involve only the formula $\overline{\varepsilon} = u_*^3/\varkappa z$ whose derivation does not involve the use of any additional hypothesis. Such a determination of $C$ was carried out by Gurvich and Zubkovskii and led to the values $C \approx 1.1 - 1.3$ (Gurvich) and $C \approx 1.6 - 1.9$ (Zubkovskii), which we have already noted above in this subsection.[25]

If the constant $C$ in Eqs. (23.3) and (23.5) is known, we can use data on structure functions and spectra in the inertial subrange to determine the mean rate of energy dissipation $\overline{\epsilon}$ in the atmosphere. Such a determination of $\overline{\epsilon}$ can be found in particular in MacCready (1953, 1962b), Ball (1961), Zubkovskii (1962), Ivanov (1962, 1964), Koprov (1965), Frenzen (1965), Record and Cramer (1966), Koprov and Tsvang (1966), Kaimal and Haugen (1967), Volkov, Kucharets, and Tsvang (1968), Gibson and Williams (1969), Trout and Panofsky (1969), Vinnichenko and Dutton (1969), Readings and Raymont (1969), Volkovitskaya and Ivanov (1970), and many other papers. Such data, and also values of $\overline{\epsilon}$ obtained from measurements of the mean wind-velocity and temperature profiles by the method described in Sects. 7.5 and 8.5 of Vol. 1, as well as measurements of relative diffusion in the atmosphere (see Sect 24.3) were used by Ball (1961), Ivanov (1962), Wilkins (1960, 1963), Record and Cramer (1966), and Zimmerman (1965) to obtain

[25] A similar method of estimating $\overline{\epsilon}$ was used by Takeuchi (1962) using data on the structure function $D_{uu}(\tau)$ calculated by R. Taylor (1961) who in turn used the measurements of Swinbank (1955). Takeuchi obtained the value $C \approx 1.25$ which is close to Gurvich's estimate but is too low according to all the other data.

approximate summarizing graphs of the energy dissipation rate $\bar{\epsilon}$ as a function of $z$, showing that, on the average, $\bar{\epsilon}$ decreases relatively rapidly with increasing $z$. Later Ellsaesser (1969b) found that climatological data on wind variability are generally consistent with the two-thirds law for time structure functions (21.42) for lag periods $\tau$ up to six hours (cf., the end of Sect. 23.6 of this book). This fact was used by Ellsaesser (1969a) to prepare experimental climatological maps of $\bar{\epsilon}$ for seven heights for the whole Northern Hemisphere and also maps of total dissipation $E_f$ in the free atmosphere (i.e., above 1 km height), total dissipation in the boundary layer (below 1 km) $E_b$, and total atmospheric dissipation $E_f + E_b$.

Unfortunately the methods available for determining $\bar{\epsilon}$ frequently involve the use of rather crude approximations; hence it is not surprising that they have in many cases low reliability, and result in a considerable spread of the final results. In the atmospheric surface layer, over a plane and homogeneous underlying surface, and neutral stratification, we can use the theoretical formula $\bar{\varepsilon} = u_*^3/\varkappa z$ to estimate $\bar{\epsilon}$, while in the case of free convection we can use the relation $\bar{\epsilon} \propto (g/T_0)(q/c_p\rho_0) = \text{const}$ where $q$ is the turbulent heat flux. Under other conditions, the theory does not allow us to obtain a formula for $\bar{\epsilon}$, but it does indicate that in the surface layer the values of $z$, $u_*$, and $q$ are always sufficient to determine $\bar{\epsilon}$ unambiguously (see Chapt. 4, Vol. 1). At large heights in the atmosphere, the situation is more complicated, but even here the values of $\bar{\epsilon}$ should, in principle, be determined by the height $z$ and the general meteorological conditions (above all, the type of thermal stratification [see, for example, Ivanov (1962, 1964), Kaimal and Haugen (1967), and Readings and Rayment (1969)] and also the nature of the underlying surface). According to aircraft measurements of Zubkovskii (1963), Koprov (1965), Koprov and Tsvang (1966), Volkov, Kucharets, and Tsvang (1968), the values of $\bar{\epsilon}$ under typical summer conditions of well-established convection over the steppe are constant in the 50–200 m layer, while above this layer they begin to decrease slightly up to heights of the order of 1000–2000 m (see Fig. 73). At greater heights the rate of decrease of $\bar{\epsilon}$ with increasing $z$ usually increases quite rapidly. The approximate constancy of $\bar{\epsilon}$ in the 50–300 m layer in well-established convection is also confirmed by Kaimal and Haugen (1967) tower observations, but the absolute values of $\bar{\epsilon}$ found in this work are considerably higher. In the 50–100 m layer, the values of $\bar{\epsilon}$ can be roughly estimated from the formula $\bar{\varepsilon} \approx \frac{g}{T_0} \frac{q}{c_p\rho}$ according to Russian data. At greater heights, however, this type of estimate cannot be justified. Under other meteorological conditions, the $\bar{\epsilon}$ profile above the steppe may be quite different, but in all cases turns out to be approximately the same under similar conditions [see Fig. 73 and also the papers of Kaimal and Haugen (1967), Volkov, Kucharets, and Tsvang (1968), and Readings and Rayment (1969), where additional data of this type can be found].

### *Field and Laboratory Measurements in a Broader Spectral Range and the Values of the Universal Constants*

The instruments used in almost all the above measurements of atmospheric turbulence spectra have such inertia and size, as to prevent their use for recording the smallest-scale components of turbulence beyond the high frequency (or wave number) limit of the inertial subrange. This difficulty has been overcome by the development of low-inertia, small-size velocity sensors which can be used to determine the turbulent spectra under natural and laboratory conditions not only in the inertial subrange but also in the

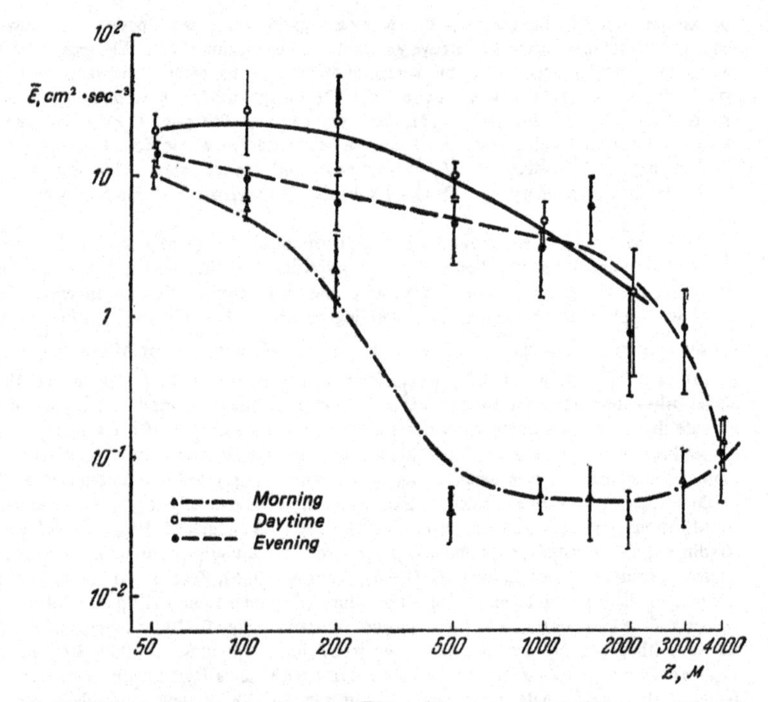


**Fig. 73** Morning, daytime, and evening $\bar{\epsilon}(z)$ profiles, according to Koprov and Tsvang.

dissipation range. In this connection we note above all the relatively early paper by Betchov (1957) who used a hot-wire anemometer with a platinum wire, 1.25 $\mu$m in diameter and 1 mm long, to determine the spectrum of turbulence generated under quite special conditions, namely, inside a tube into which air was drawn through 80 apertures in the front and lateral walls, producing intensive mixing of the resulting air jets. The Reynolds number corresponding to the mean velocity and diameter of one such jet was $3.5 \cdot 10^4$, but the turbulence was of much higher intensity than in a wind tunnel behind a grid for Re of the same order, and was characterized by much higher values of the "fluctuation Reynolds number" $\mathrm{Re}_\lambda = [\overline{u'^2}]^{1/2}\lambda/\nu$. According to Betchov's results, the one-dimensional longitudinal spectrum $E_1(k)$ is proportional to $k^{-5/3}$ over a relatively broad range of values of $k$, whereas for large wave numbers this spectrum falls off much more rapidly (the precise form of the spectrum in this region, obtained by Betchov, is not in very good agreement with subsequent measurements which we shall review below).

Betchov's work was followed by the hot-wire anemometer measurements of M. M. Gibson (1962, 1963), which were already mentioned in Sect. 23.2. This author examined the velocity spectra in round turbulent air jets for Re $\approx 5 \cdot 10^5$. Similar measurements were also performed by Brådbury (1965) in a turbulent plane jet, by Gibson, Friehe and McConnell (1971) in an axisymmetric heated jet, by Gibson, Chen and Lin (1968) and Uberoi and Freymuth (1970) in wakes behind solid spheres, and by Uberoi and Freymuth (1969) in the wake behind a circular cylinder. We also recall the work of Grant, Stewart, and Moilliet (1962), Stewart and Grant (1962), and Grant and Moilliet (1962), who determined the turbulent spectra in a tidal channel near Vancouver at a Reynolds number corresponding to channel depth and mean velocity of $3 \cdot 10^8$, using a precise hot-film anemometer (platinum film $10^{-6}$ cm thick, linear size less than 0.05 cm, resistance $5\Omega$) mounted on the bow of a special research vessel. The same instrument was used by Grant, Hughes, Vogel and Moilliet (1968) and Grant, Moilliet and Vogel (1968), who repeated the measurements by Grant, Stuart and Moilliet in the tidal channel from the research vessel and supplemented them by measurements in the offshore ocean waters (at depths between 15 and 90 m) from a submarine. Finally, Nasmyth (1970) made some modifications in the towed body carrying the same equipment, and this permitted him to make observations at depths exceeding 300 meters in deep water off the west coast of Vancouver Island. A slightly less refined velocity probe of the same type was used by Gibson and Schwarz (1963b) in a flow of saline water behind a grid in a circular water tunnel [Re $= (1-7) \cdot 10^4$]. Pond, Stewart, and Burling (1963) and Pond, Smith, Hamblin, and Burling (1966) have carried out measurements on the atmospheric boundary layer over sea waves (at height of 1–2 m with wave amplitudes of the order of 0.3 m and wind velocity of about 3 m/sec), using a hot-wire anemometer incorporating a platinum wire 8 $\mu$m thick and 0.5–1 mm long [see also Pond's thesis (1965)]. Similar hot-wire anemometer measurements of wind-velocity spectra over the sea or over a tidal mud flat were also performed by Gibson and Williams (1969), Gibson, Stegen and Williams (1970), Van Atta and Chen (1970) and Boston (1970, 1971), while Sheih, Tennekes and Lumley (1971) published the results of analogous airborne measurements at a height of about 100 m. Wind-velocity spectral measurements in the atmospheric surface layer above a flat earth surface have been reported by Wyngaard and Coté (1971) and Kaimal, Wyngaard et al. (1972). These measurements with a sonic anemometer do not include the dissipation range, but they were accompanied by simultaneous hot-wire measurements of a dissipation value $\bar{\epsilon}$ and can therefore be included in our list. The mentioned hot-wire measurements were also treated by Wyngaard and Pao (1971) independently from the sonic anemometer data. Finally, Sandborn and Marshall (1965) [see also Cermak and Chuang, 1967 and Plate and Arya, 1969] measured the spectra of the longitudinal velocity component in the boundary layer at three heights near the wall of a huge wind tunnel (having a square cross section of $4m^2$ in the working region, and a length of about 30 m) and Comte-Bellot (1965) measured the spectra of three velocity components at different points of a plane channel (having a rectangular cross section of 0.18 m · 2.4 m and a length of 12 m). Both these series of measurements were carried out also with a hot-wire anemometer.

Detailed analysis of the results reported in most of these papers will be given in the next section. Here, we shall merely note that it was reliably established that the inertial subrange of wave numbers in which the five-thirds law is satisfied does, in fact, exist. In the experiments of Gibson and Schwarz and of Comte-Bellot, in which the Reynolds number was relatively low, the inertial subrange was very short, while in the measurements of Grant, Stewart, and Moilliet, in which the Reynolds number was of the order of $3 \cdot 10^8$, the wave number $k$ varied by three orders of magnitude within the inertial subrange. This is clear from Fig. 74, which shows on a bilogorithmic scale the dimensionless one-dimensional longitudinal

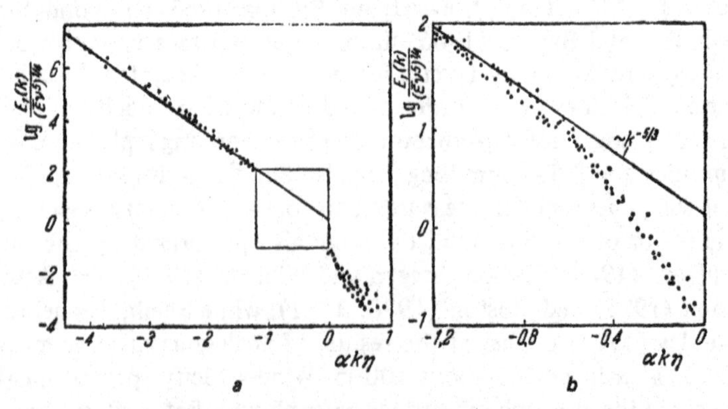


**Fig. 74** a–Dimensionless longitudinal velocity spectrum [Grant, Stewart, and Moilliet (1962)]; b–data inside the square of Fig. 74a shown on a larger scale.

spectrum $(\bar{\varepsilon}\nu^5)^{-1/4} E_1(k)$ plotted against the dimensionless wave number $x = \alpha k \eta$ where $\alpha$ is a numerical constant which was included in the expressions for $x$ for the sake of subsequent convenience of comparison with the semiempirical Heisenberg formula for the turbulence spectrum.

We note that measurements of spectra in the inertial subrange enable us to estimate the universal constant $C$ in Eqs. (23.5) and (23.3), provided only we can determine the corresponding energy dissipation rate $\bar{\varepsilon}$ in some way. Recent measurements are of particular interest from this point of view, since many of these contain data on the spectrum $E_1(k)$ covering most of the dissipation range. This enables us to determine $\bar{\varepsilon}$ from the exact relation [cf., Eq. (12.94)]

$$\bar{\varepsilon} = 15\nu \int_0^{\infty} k^2 E_1(k)\, dk$$

Grant, Stewart and Moilliet found by this method that $C = 1.90 \pm 0.08$. A very similar result ($1.97 \pm 0.16$) was obtained by Pond, Smith, Hamblin, and Burling (1966). Almost all the data of Grant, Hughes, Vogel, and Moilliet (1968) are in excellent agreement with those of Grant, Stewart, and Moilliet (1962) and hence imply the same value of $C$; the few exceptions can be easily explained by instrument errors or by the absence of high-Reynolds-number turbulence at the measurement point. However the very clean oceanic turbulence data of Nasmyth (1970) do not agree very well with the previous data of Grant et al. obtained with the same equipment and his analysis yields the value 2.3 as the best estimate of $C$. At the same time the excellent atmospheric turbulence data by Boston (1970, 1971) were found to agree very well with the Grant et al. and Pond et al. results and according to this $C \approx 2$. The extensive data of the Kansas measurements of longitudinal spectra by sonic anemometers in the atmospheric surface layer reported by Kaimal, Wyngaard et al. (1972) lead to an estimate $C = 2.0 \pm 0.1$ while the treatment of a portion of this data by Wyngaard and Coté (1971) gave a similar result $C = 2.1 \pm 0.2$. The estimates of $C$ from the data of Kansas hot-wire anemometer measurements of longitudinal spectra is $C = 2.1 \pm 0.1$ [Wyngaard and Pao (1971)]. The sonic anemometer measurements of $v$ and $w$ wind spectra in Kansas lead to very close estimates of $C$ since the isotropic relations for spectra of velocity components are valid with high accuracy in these measurements. M. M. Gibson (1962, 1963) also found a mean value of $C$ close to 2.1, while Sandborn and Marshall (1965) and Gibson, Chen and Lin (1968) did not report the results of direct determinations of $C$, but since the universal function describing the one-dimensional spectrum was found in both these works to be in excellent agreement with the curves reported by Grant, Stewart, and Moilliet, and Pond, Stewart, and Burling (see, for example, Fig. 76), there is no doubt that these data also led to a value for $C$ very close to that obtained by the other workers.

We have already mentioned that the measurements by Comte-Bellot (1965) in a plane channel were at insufficiently high values of Re and therefore the inertial subrange was quite narrow in her data. However the estimate of $C$ obtained from these data is in satisfactory agreement with the results of other authors: $C \approx 2$. The measurements of Bradbury (1965) in a plane jet, of Gibson, Friehe and McConnell (1971) in an axisymmetric jet and of Uberoi and Freymuth (1969, 1970) in two types of wakes lead to the following estimates: $C \approx 2.0$, $C \approx 2.3$, $C \approx 1.8$, and again $C \approx 2.0$ respectively. The spectrum $E_1(k)$ measured in the water tunnel by Gibson and Schwarz (1963b) did not lead to the establishment of an appreciable inertial subrange or to the possibility of obtaining a reliable spectral estimate for $\bar{\varepsilon}$, which was, in fact, determined (apparently with moderate accuracy only) from the rate of decay of the mean square velocity fluctuation with increasing distance from the grid. Using the values of $\bar{\varepsilon}$ found in this way and the rather restricted data on the inertial subrange in the measured spectra, Gibson and Schwarz estimated $C$ to be $C \approx 1.7 \pm 0.08$. Finally the spectral measurements by Gibson, Stegen and Williams (1970) in the wind over the ocean unexpectedly led to a considerably greater estimate $C \approx 2.8$. Estimates of the order 2.6–2.8 were also obtained by Van Atta and Chen (1970) from their spectral measurements (although the analysis of structure function values computed from the same data yields the lower value $C \approx 2.3$ as we have already mentioned) and by Sheih, Tennekes and Lumley (1971) from airborne spectral measurements (although the data of simultaneous measurements on the 100 m tower near which the aircraft was flying gave a value $C \approx 2.0$). The reason for the discrepancy between these last estimates and all mentioned above is still unclear. Nevertheless if we keep in mind the difficulty of precise turbulence measurements the spread of the existing data on $C$ will not seem too great.

The satisfactory internal agreement between the overwhelming majority of the above results which were obtained for very different types of turbulent flows, and the good coincidence between the universal curves in the dissipation range, which we shall discuss in the next subsection, are undoubtedly a major achievement in the experimental study of turbulence. It confirms, with high accuracy, the validity of the predictions of the Kolmogorov theory about the quasi-universal statistical state of the small-scale components of any turbulence with sufficiently high Re. The above data suggest that it is reasonable to assume at present that the universal coefficient $C$ is approximately equal to 2. By Eqs. (21.25), (21.25′), and (23.4) this corresponds to the coefficient $C_1 \approx 1.5$ in the five-thirds law (21.24′) for the three-dimensional spectrum, and $C_2 \approx 0.5$ in the five-thirds law for the one-dimensional longitudinal spectrum. The skewness factor for the longitudinal velocity difference turns out to be $S \approx 0.28$, and the coefficients in the two-thirds and five-thirds laws for the lateral structure function and lateral one-dimensional spectrum are $C' \approx 2.7$ and $C_2' \approx 0.7$. The accuracy of these estimates cannot be reliably established at present, but it is unlikely that the uncertainty exceeds 10–20%. It is interesting that these estimates were not very different from the earliest estimate of Kolmogorov, namely, $C \approx 1.5$. They are also in relatively good agreement with the estimate of Zubkovskii (1963) and the corrected estimate of Gurvich (1960b) (see above).

Once the universal coefficients in the two-thirds and five-thirds laws have been established with adequate accuracy, they can be used for reliable determinations of the mean energy dissipation rate $\bar{\varepsilon}$, using measured spectra or structure functions for the inertial range. In particular, the above estimates of the dimensionless coefficients were used by Stewart and Grant (1962) and Grant, Moilliet and Vogel (1968) to determine the rate of energy dissipation $\bar{\varepsilon}$ for the ocean and by many others to determine $\bar{\varepsilon}$ for the atmosphere [see, for example, Panofsky and Pasquill (1963), Koprov (1965), Koprov and Tsvang (1966), Record and Cramer (1966), Kaimal and Haugen (1967), Byzova, Ivanov, and Morozov (1967), Volkov, Kucharets, and Tsvang (1968), Trout and Panofsky (1969), Readings and Rayment (1969), and Volkovitskaya and Ivanov (1970)].

## 23.4 Verification of the First Kolmogorov Similarity Hypothesis for the Velocity Field

The most readily verifiable consequences of the first Kolmogorov similarity hypothesis are those given by Eq. (21.20) and relating to one-dimensional spectra. They can be written in the form

$$E_1(k) = (\bar{\varepsilon}\nu^5)^{1/4}\varphi_1(k\eta), \quad E_2(k) = (\bar{\varepsilon}\nu^5)^{1/4}\varphi_2(k\eta), \tag{23.10}$$

where $\eta = (\nu^3/\bar{\varepsilon})^{1/4}$ is the Kolmogorov length scale of turbulence and $\varphi_1(\xi)$ and $\varphi_2(\xi)$ are universal functions. When $k\eta \ll 1$ these formulas become identical with the five-thirds law which, as we have seen, is in satisfactory agreement with the experimental data. We shall now consider the verification of these formulas for $k\eta \gtrsim 1$, i.e., for wave numbers for which molecular viscosity has an appreciable effect.

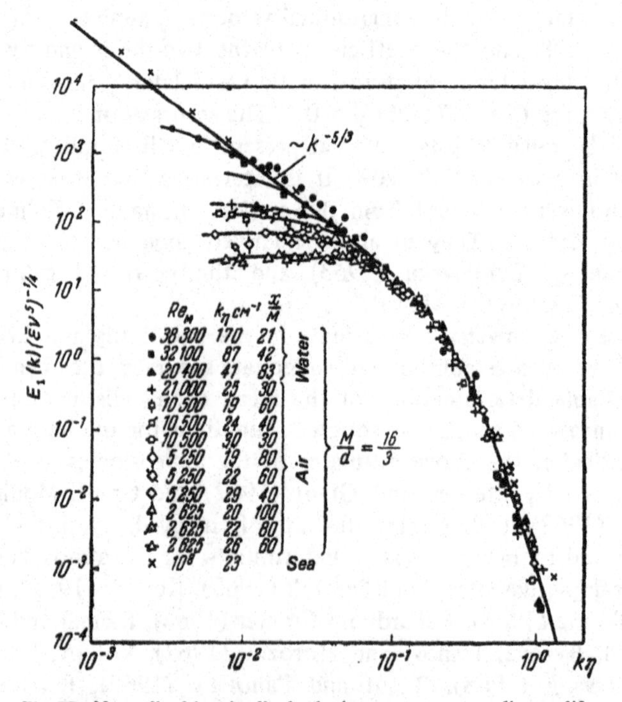


**Fig. 75** Normalized longitudinal velocity spectrum, according to different authors. The measurements in water are due to Gibson and Schwarz, in air–Stewart and Townsend, and in the sea–Grant, Stewart, and Moilliet.

In the last part of the previous subsection we listed various papers in which direct measurements of the velocity spectra were reported not only for the inertial subrange of wave numbers $k$, but also for a substantial part of the dissipation range. Examples of such spectra were given in Fig. 61 [Gibson (1962, 1963)] and 74 (Grant, Stewart, and Moilliet, 1962). Figure 75, which is taken from Gibson and Schwarz (1963b), shows the longitudinal normalized spectra $\varphi_1(k\eta)$ obtained for water behind a grid for different values of Re, together with the spectrum of Grant, Stewart, and Moilliet and the spectra calculated from the correlation measurements of Stewart and Townsend (1951) in practically isotropic grid turbulence. It is clear from this figure that the extent of the inertial subrange is different for different Re (in fact only the data of Grant, Stewart and Moilliet, and Gibson and Schwarz, which correspond to the highest Re, indicate the existence of this range). Below this subrange the spectra are quite different from each other and substantially dependent on Re. However, within the inertial subrange, and beyond its short-wave end, all the normalized spectra lie on a universal curve with a relatively small spread and in complete agreement with Eq. (23.10). It is interesting to note that this conclusion is valid even for the data of Stewart and Townsend which correspond to relatively low values of Re for which the energy range and the dissipation range apparently overlap substantially, in contrast to the requirement of the Kolmogorov theory of locally isotropic turbulence. Gibson and Schwarz have noted in this connection that the quasi-universal statistical state of the extreme small-scale turbulence components (lying well outside the energy range of the spectrum) is, in general, quite natural even for turbulence with not very high Re, except that the true mean energy-dissipation rate $\bar{\varepsilon}$ must now be replaced by a corrected value $\bar{\varepsilon}_1 \neq \bar{\varepsilon}$. Namely, instead of $\bar{\varepsilon}$ we must now use the dissipation rate $\bar{\varepsilon}_1$ for flows with very high Re, in which the high-frequency part of the spectrum (beyond the value of $k$ at which turbulent disturbances of our moderate Re flow have already become isotropic, quasi-stationary, and independent of large-scale properties) coincides with the high-frequency part of the spectrum which we are considering. This corrected value can, in principle, be established if we know the universal function $\varphi_1(\xi)$. Since, however, the universal normalization of $E_1$ and of the wave number $k$ involves only $\bar{\varepsilon}^{1/4}$, it can be shown, using the data of Stewart and Townsend (1951), that for the grid flows investigated by Gibson and Schwarz, and for all turbulent flows for which the energy range of the spectrum does not contain wave numbers $k > 0.2/\eta$, the effect of the necessary $\bar{\varepsilon}$ correction on the universal curve $\varphi_1(\xi)$ should not be appreciable in practice. According to rough estimates of Pao (1965) the corrected value $\bar{\varepsilon}_1$ could be larger than the true $\bar{\varepsilon}$ by about 10% for moderate Reynolds number flows; this gives a negligible difference between $\bar{\varepsilon}^{1/4}$ and $\bar{\varepsilon}_1^{1/4}$ [but is more important for the dimensionless forms of $k^2E_1(k), k^4E_1(k)$ and especially $k^6E_1(k)$ which are in proportion to $\bar{\epsilon}^{-3/4}$, $\bar{\epsilon}^{-5/4}$ and $\bar{\epsilon}^{-7/4}$ respectively]. This enables us to explain the agreement between the Stewart-Townsend data, which refer to small Re but relatively high $k\eta$, and the universal function $\varphi_1(\xi)$ for larger Re, without coming into conflict with the Kolmogorov theory. This reasoning explains also the agreement between the universal curve of Fig. 75 and the form of the dimensionless longitudinal spectra at $k\eta \geqslant 0.05$ of the moderate Re grid produced turbulence investigated by Uberoi (1963) and Van Atta and Chen (1968, 1969a) [see the papers of the last two authors]. Finally the dependence of $E_1(k)(\bar{\epsilon}\nu^5)^{-1/4}$ on $k\eta$ and Re in the cylinder and spheres wakes found by Uberoi and Freymuth (1969, 1970) are quite similar to those shown in Fig. 75 and also indicate that for $k\eta > 0.1$ all spectra tend to a universal curve (see, for example, Fig. 76b, taken from the second of cited papers). Data from various sources relating to different turbulent flows are collected in Fig. 76c [also from Uberoi and Freymuth (1970)]; they also show that the universal curve exists and agree well with the plots in Figs. 75 and 76b.

Consider now Fig. 76 taken from Sandborn and Marshall (1965), in which data on the normalized one-dimensional longitudinal velocity spectrum $\varphi_1(k\eta)$ obtained by these workers in the boundary layer on the wind-tunnel wall are compared with the data of Grant, Stewart, and Moilliet (1962) on tidal flows in the sea, and the data of Pond, Stewart, and Burling (1963) on turbulence in the atmosphere over seawater. It is clear that there is excellent agreement between data referring to the three quite different types of flow, and this supports the existence of a universal relationship. Similarly in Fig. 76a, data of the normalized longitudinal spectrum $\varphi_1(k\eta)$ in the turbulent wake of a sphere coincide with Grant, Stewart, and Moilliet's tidal channel spectrum within experimental error over two decades of wavenumber. The measured velocity spectra of Grant, Hughes, Vogel and Moilliet (1968) also agree quite well with the data of Grant, Stewart and Moilliet (see, for example, Fig. 89a below). Satisfactory agreement with the previous results on the universal function $\varphi_1(k\eta)$ was also found by Wyngaard and Pao (1971). Less

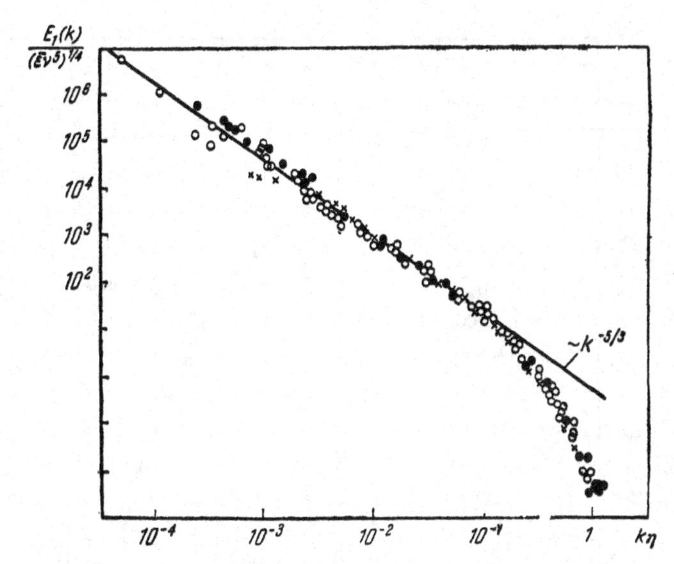


**Fig. 76** Normalized longitudinal velocity spectrum. ×–Sandborn and Marshall •–Grant, Stewart, and Moilliet; ○–Pond, Stewart, and Burling.

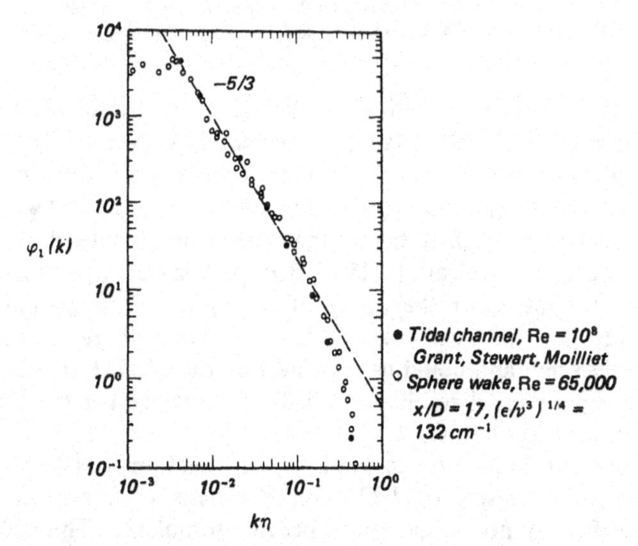


**Fig. 76a** Velocity spectra (after Gibson, Chen and Lin, 1968) normalized with Kolmogorov scales.

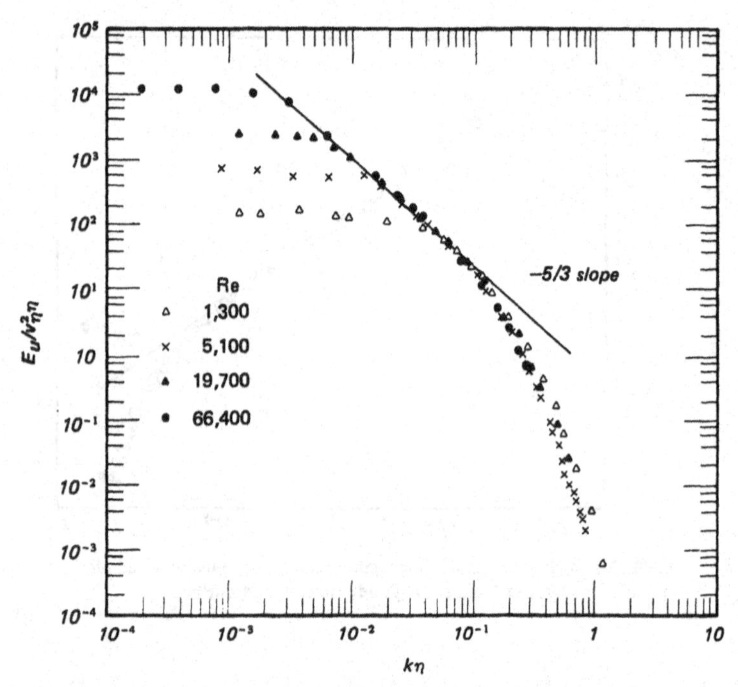


**Fig. 76b** Dependence of normalized spectra $E_u/v_\eta^2\eta$ on Reynolds number Re in the center of the sphere wake. Re is based on the $x$-dependent characteristic velocity $U$ and scale $L$ in terms of which the wake is self-preserving (after Uberoi and Freymuth, 1970).

satisfactory agreement with these data is shown by the results of M. M. Gibson (1962, 1963) (some of which are shown in Fig. 61). We have already seen that the value of the coefficient $C$ which determines the function $\varphi_1(k\eta)$ in the inertial range $k\eta \ll 1$ was found by Gibson to be slightly higher than the value obtained by Grant et al. (1962) and Pond et al. (1963, 1966). A similar discrepancy was found by Gibson for the values of $\varphi_1(k\eta)$ in the dissipation range. However, in this case again, the observed discrepancies were relatively small and could be explained by the usual statistical spread of the experimental results, especially if we remember the complexity of these measurements.

A new approximation to the universal function $\varphi_1(\xi)$, $\xi = k\eta$ was produced by Nasmyth (1970) on the basis of very clean (i.e., undisturbed by noise) spectra of oceanic turbulence. The new curve lies rather near the old curve of Grant et al., but it has a slightly sharper "knee" and falls slightly faster in the range of large wave numbers. Both curves were compared by Boston (1970, 1971) with his atmospheric spectral measurements. It was found that the best run of the Boston data (i.e., that with the greatest signal to noise ratio) agrees very well with the curve of Nasmith, while the spectra which were not as clean agree better with the old curve of Grant et al. However the difference between the two curves is rather insignificant. The difference between both these curves and the curve by Gibson, Stegen, and Williams (1970) is much more noticable, but we have already mentioned that the data of Gibson et al. disagree with the majority of the other data.

The use of the logarithmic scale along both axes in Fig. 76 may, to some extent, mask the spread of the experimental points relative to the universal curve. It is therefore desirable to consider the appearance of this plot with linear scales. To show this, Fig. 77 gives

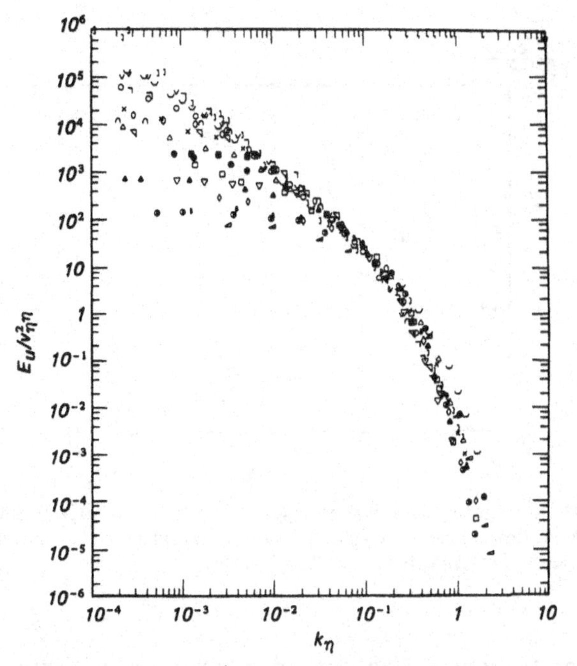


**Fig. 76c** Comparison of normalized spectra $E_u/v_\eta^2\eta$ measured in various flows with different values of $Re_\lambda = \lambda(\overline{u'^2})^{1/2}/\nu$. Laufer O $Re_\lambda = 170$; Grant, Stewart, Moilliet ]; Gibson ⎵ $Re_\lambda = 780$; Kistler, Vrebalovich V $Re_\lambda = 670$; ⅂ $Re_\lambda = 520$; ◁ $Re = 133$; Uberoi, Freymuth (cylinder wake) ○ $Re_\lambda = 308$; X $Re_\lambda = 186$; △ $Re_\lambda = 132$; ● $Re_\lambda = 93$; □ $Re_\lambda = 66$; ▲ $Re_\lambda = 46$; ◇ $Re_\lambda = 33$; D $Re_\lambda = 23$; Δ $Re_\lambda = 18$; Uberoi, Freymuth (sphere wake) ∩ $Re_\lambda = 260$; ■ $Re_\lambda = 140$; ▽ $Re_\lambda = 72$; ● $Re_\lambda = 36$.

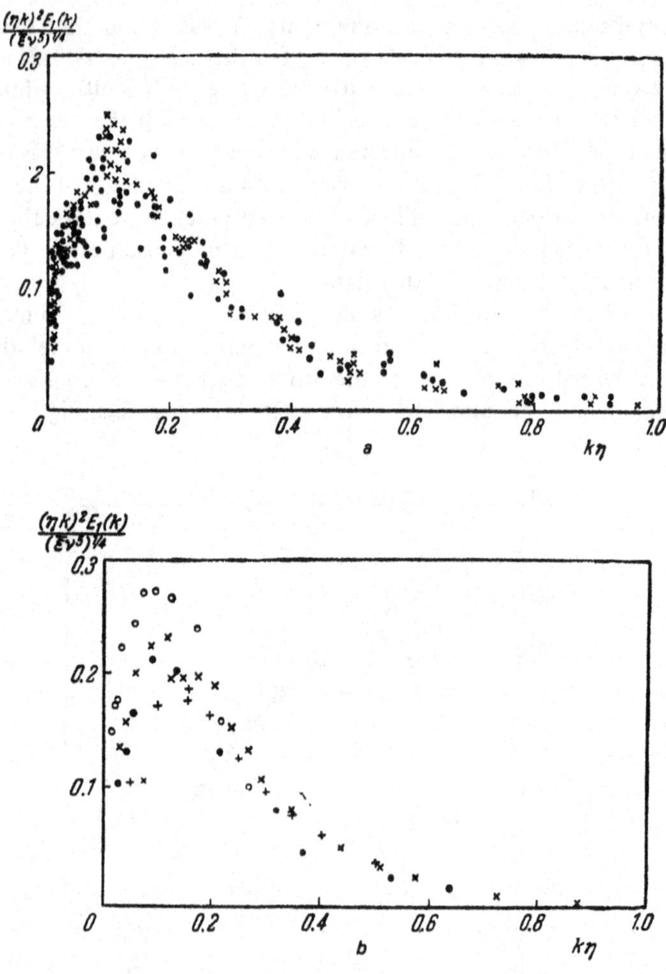


**Fig. 77** Normalized energy dissipation spectra. a–Grant, Stewart, and Moilliet (•) and Pond, Stewart, and Burling (x); b–Laufer for a channel (x), Laufer for a tube (○), Klebanoff (•), and Stewart and Townsend (+).

graphs of the normalized one-dimensional longitudinal energy dissipation spectra $(k\eta)^2\varphi_1(k\eta)$. Figure 77a refers to the measurements of Grant, Stewart, and Moilliet and those of Pond, Stewart, and Burling, while the data of Fig. 77b, which are taken from a paper by Grant, Stewart, and Moilliet, refer to earlier measurements on turbulent flows in a channel [Laufer (1951)], a tube [Laufer (1954)], in a boundary layer on a flat plate [Klebanoff (1955)], and in a wind tunnel behind a grid [Stewart and Townsend (1951)]. It is clear from this graph that, in spite of the somewhat greater spread of the individual measurements as compared with Fig. 76, these very different experimental data obtained by different authors are in satisfactory agreement with each other. In particular, they all show that the maximum of the energy dissipation spectrum is reached at $k \approx 1/8\eta$. The data of Boston (1970, 1971) also agree very well with all the data in Fig. 76. Similar plots of the longitudinal energy dissipation spectrum were given by Sheih, Tennekes and Lumley (1971) and Wyngaard and Pao (1971).

The above data enable us to estimate the small-scale (i.e., high wave number) limit of the inertial subrange (or, what amounts to the same thing, the large-scale or low wave number limit of the viscous subrange of the spectrum), i.e., the scale $l_0 = a_0\eta$ or the wave number $k_0 = a_1/\eta$ beyond which the statistical characteristics of turbulence become appreciably dependent on molecular viscosity. Using the crude Heisenberg model of the turbulent spectrum given by Eq. (17.40), MacCready (1953, 1962a) estimated that 90% of the energy dissipation was concentrated in the wavelength region below $l_0 \approx 15\eta$, so that it is reasonable to substitute $a_0 = a_1^{-1} = 15$ [see also Priestley (1959b) in this connection]. MacCready has also given certain other definitions of the small-scale limit of the inertial subrange, yielding other values for $a_0$. It is best, however, to base our considerations on the data of Fig. 77, according to which appreciable departures of one-dimensional longitudinal spectra from the five-thirds law begin with approximately the same wave number $k_0 \approx 1/8\eta$ at which the one-dimensional longitudinal dissipation spectrum reaches its maximum. Of course these numerical results are related only to the one-dimensional longitudinal spectrum. The corresponding limits for the one-dimensional lateral velocity spectrum or, even more, for the three-dimensional spectrum will evidently be higher. On the other hand the small-scale limit of the inertial subrange for structure functions will apparently be situated in the region of greater lengths: According to the Van Atta and Chen data in Fig. 63b the deviation of the longitudinal structure function $D_{LL}(r)$ from the two-thirds law is noticeable already at $r \approx 50\eta$. Nevertheless we must remember that the most usual measurement is of a spectrum of longitudinal velocity, and using this we shall have at least two different reasons for assuming that $a_0 = 1/a_1 = 8$. Using this definition and remembering that at heights of a few meters above ground level in the atmosphere, the quantity $\bar{\varepsilon}$ is often of the order of 100–1000 cm²/sec³ (such values were obtained in the summer over southern-Russian steppes), we find that $\eta$ = 0.4–0.7 mm and $l_0$ = 3–6 mm. As the height increases, the parameter $\bar{\varepsilon}$ decreases, the viscosity $\nu$ increases in inverse proportion to the air density and, consequently, $l_0$ increases. At a height of 1 km, $l_0$ reaches a few centimeters, while at 100 km it is of the order of a few tens of meters.

The above data on the spectrum $E_1(k)$ refer to relatively small values of $k$ which do not exceed $1/\eta$. They cannot, therefore, be used to determine the function $\varphi(\xi)$ for $\xi > 1$, nor do they enable us moreover to verify any conclusions about the asymptotic behavior of this function for $\xi \to \infty$. They do, however, enable us to draw certain conclusions about the source of the proposed formulas for the turbulence spectrum, which are in poor agreement even with the restricted experimental material available at present. For example, it can be shown that the model formulas resulting from the semiempirical hypothesis of Heisenberg (17.9) and Kovasznay (17.3) are in very poor agreement with the data of Fig. 76 and must, therefore, be regarded as unsatisfactory (although they can, of course, be used for order-of-magnitude calculations). The same is true in the case of the simple interpolation formula (22.73) [independently of whether we shall follow Neumann (1969) and try to apply it to the three-dimensional spectrum $E(k)$ or to the one-dimensional longitudinal spectrum $E_1(k)$]. Better agreement with data is achieved if we take the three-dimensional spectrum $E(k)$ in the form given by Eq. (22.24), which is due to Pao (1965). Still better agreement is obtained if we follow Tatarskii (1967) and substitute $\varphi_1(\xi) = C_2\xi^{-5/3}\exp(-a\xi^{1/2})$. On the other hand, Uberoi and Freymuth proposed to approximate their universal dimensionless longitudinal wake spectrum by an expression of the form $\varphi_1(\xi) = C_2\xi^{-5/3}\exp(-a\xi^{3/2})$. However, the last expressions for $\varphi_1(\xi)$ have the drawback that they do not agree with the asymptotic behavior of the spectrum at infinity, which is expected on the basis of the discussion given in Sect. 22.3. A simple empirical formula for $\varphi_1(\xi)$ which contains three undetermined parameters which must be adjusted in accordance with experimental data was proposed by Gorshkov (1966) [see also Gurvich, Koprov, Tsvang, and Yaglom (1967)]; it fits all the observations very well and decreases at infinity as $\exp(-a\xi^2)$, in accordance with the conclusions of Sect. 22.3. However we must note that the necessity of the data to agree with the asymptotic equations of Sect. 22.3 seems questionable at present. In fact the behavior of spectra at very large wave numbers must be influenced very much by the effect of fluctuations in energy dissipation rate (see Sect. 25 below); therefore the conclusions of Sect. 22.3 could disagree with the true data even if the theory of this subsection is quite accurate.

## 23.5 Data on the Local Structure of the Temperature and other Scalar Fields Mixed by Turbulence

*Basic Data on the Structure of the Temperature Field*

According to similarity theory, the structure function $D_{\vartheta\vartheta}(r)$ of the temperature field, or the concentration field of some passive substance, is described by the two-thirds law (21.87), and the one-dimensional spectrum $E_1^{(\vartheta)}(k)$ is given by the five-thirds law (21.89) if we are dealing with the inertial-convective subrange of the length scale or wave number spectrum in which we can neglect both molecular viscosity and molecular thermal conductivity. Using Eq. (21.90), we can rewrite these formulas in the form

$$D_{\vartheta\vartheta}(r) = C_\vartheta \overline{N}\,\overline{\varepsilon}^{-1/3} r^{2/3},\ E_1^{(\vartheta)}(k) \approx \frac{1}{4} C_\vartheta \overline{N}\,\overline{\varepsilon}^{-1/3} k^{-5/3}. \quad (23.11)$$

Moreover, according to Eq. (22.10) the third-order structure function $D_{L\vartheta\vartheta}(r)$ in the inertial-convective subrange of $r$ is given by $D_{L\vartheta\vartheta}(r) = -\frac{4}{3}\overline{N}r$. Therefore,

$$F = D_{L\vartheta\vartheta} D_{LL}^{-1/2} D_{\vartheta\vartheta}^{-1} = -\frac{4}{3C_\vartheta}\left(-\frac{5}{4}S\right)^{1/3}, \quad (23.12)$$

and the dimensionless quantity $F = F(r)$ is independent of $r$ [the quantity $S$ has here the same meaning as in Eq. (23.4)].

The first data confirming the two-thirds law (23.11) were obtained by Krechmer (1952), who used two resistance thermometers to measure the spatial structure function $D_{\vartheta\vartheta}(r)$ for a number of values of $r$ in the surface layer of the atmosphere (see Fig. 78, which is taken from Krechmer's paper). Similar results were obtained later by Shiotani (1955, 1963), R. Taylor (1961) (who used the data of Swinbank), Record and Cramer (1966), Martin (1966) and Paquin and Pond (1971), who measured the time structure functions of the temperature field. The time structure function is evidently obtained from $D_{\vartheta\vartheta}(r)$ with the replacement of $r$ by $\bar{u}\tau$ if we assume that the

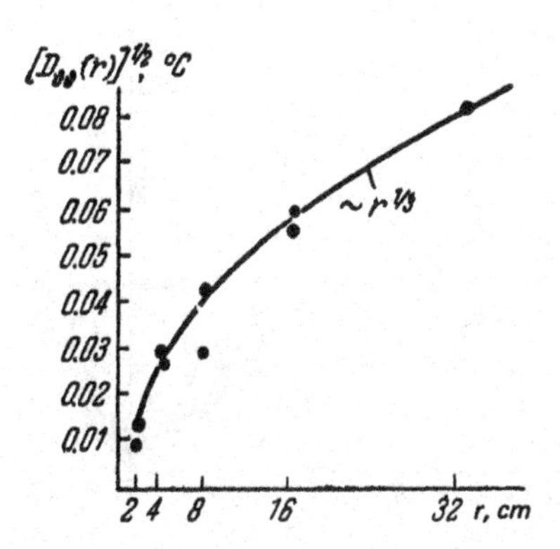


**Fig. 78** Temperature structure function [Krechmer (1952)].

hypothesis of frozen turbulence is valid. All the above mentioned authors found that the measured time structure functions were proportional to $\tau^{2/3}$ in a broad range of values of $\tau$. The most detailed measurements of the spatial structure function $D_{\vartheta\vartheta}(r)$ in the surface layer of the atmosphere (at heights of 1.5, 16, and 22 m, and $r$ varying between 3 and 100 cm) were carried out by Tatarskii (1956a) using two resistance thermometers and a special squaring device which automatically gives the mean square of the temperature difference at two points. Using the approximation $D_{\vartheta\vartheta}(r) \sim r^{\alpha}$, Tatarskii found a mean value of 0.81 for $\alpha$ (with a relatively large spread of the individual values) which is close to the theoretical result $\alpha = 2/3 \approx 0.67$.

Tatarskii paid particular attention to the dimensional coefficient $B^2 = C_\vartheta \overline{N} \bar{\varepsilon}^{-1/3}$ in the two-thirds law (23.11) for $D_{\vartheta\vartheta}(r)$. Using the rough relations $\overline{N} \approx \varkappa u_* T_*^2/z$ and $\bar{\varepsilon} = u_*^3/\varkappa z$ which are valid for nearly neutral thermal stratification and assuming that the ratio of the eddy diffusivities $K$ and $K_\vartheta$ is equal to unity [see Eqs. (7.93) and (7.109) in Vol. 1), Tatarskii found that

$$B \approx C_\vartheta^{1/2} \varkappa^{2/3} z^{-1/3} |T_*|. \tag{23.13}$$

He therefore plotted a graph of the measured values of $B$ as a function of $\varkappa^{2/3} z^{-1/3} |T_*|$, where $T_*$ was estimated from the measured temperature profiles using the very approximate formula $\overline{T}(z_2) - \overline{T}(z_1) = T_* \ln(z_2/z_1)$. This graph is shown in Fig. 79 in which the right-hand half corresponds to unstable and the left-hand part to

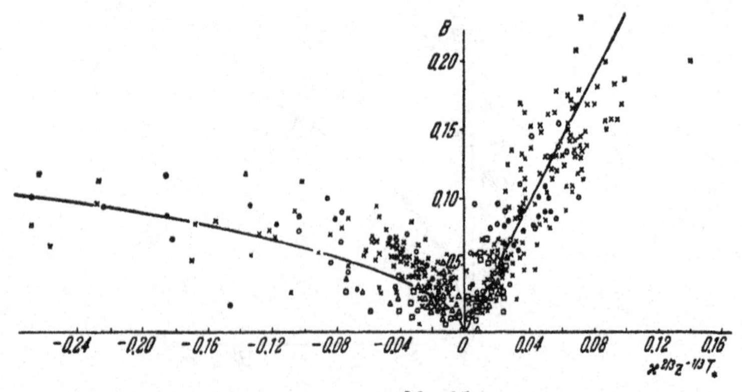


Fig. 79 Dependence of the coefficient $B$ on $\varkappa^{2/3} z^{-1/3} T_*$ [Tatarskii (1956a)]. Different symbols indicate measurements at different times and heights.

stable stratification. The fact that the mean curve in Fig. 79 passes through the origin is justified by the presence of a data point almost at the origin. This point corresponds to some tens of individual measurements which show the almost complete absence of temperature fluctuations in the case of isothermal stratification (in complete accordance with the expectation that, in the case of a temperature-homogeneous medium corresponding to strictly isothermal stratification, turbulent mixing cannot destroy this homogeneity and produce temperature fluctuations). By approximating the right-hand half of the graph in Fig. 79 by Eq. (23.13), Tatarskii found that $C_\theta^{1/2} \approx 2.4$, i.e., $C_\theta \approx 5.8$ [the violation of Eq. (23.13) in the case of stable stratification was explained by Tatarskii by the strong effect of this stratification on the turbulence, which does not allow one to regard the temperature as "dynamically passive"]. It is clear, however, that, even in the case of unstable stratification, the relation (23.13) (and derived on the basis of hypotheses that are valid only for nearly neutral stratification) cannot be very accurate. This is, in fact, confirmed by the considerable spread of the experimental points in Fig. 79. The value obtained by Tatarskii for $C_\theta$ must, therefore, be considered as quite unreliable. The satisfactory agreement of this oldest estimate of $C_\theta$ with recent estimates based on precise high frequency spectra measurements and considered below is evidently quite accidental.

A similar method of very roughly estimating $C_\theta$ (which was also based on approximate estimates of $\overline{N}$ and $\overline{\varepsilon}$ from measurements of the mean wind and temperature profiles or vertical turbulent heat and momentum fluxes using approximate formulas valid only for nearly neutral stratification) was used by R. Taylor (1961), Takeuchi (1962), and Martin (1966). The first two authors used the data on the time structure function for the temperature field $D_{\vartheta\vartheta}(\tau)$ (which can be transformed to the spatial structure function with the aid of the Taylor frozen turbulence hypothesis) obtained from the measurements of Swinbank (1955), while the third author used his own measurements of $D_{\vartheta\vartheta}(\tau)$. Taylor and Takeuchi found that $C_\theta \approx$ 1.1–1.2, while Martin found that $C_\theta \approx 1.4$. As can be seen, these values are appreciably lower than the Tatarskii estimate. The accuracy of these estimates, however, is no greater than in the case of the Tatarskii calculations, and can hardly enable us to conclude anything more than that $C_\theta$ is of the order of unity.

More reliable results can be obtained from the data of Tsvang (1960a, b, 1962, 1963), who verified (apparently for the first time) the five-thirds law (23.11) for the temperature spectrum by direct measurements of the spectra $E_\vartheta(\omega)$ [or $E_1^{(\vartheta)}(k)$] in the atmospheric surface layer at heights of between 50 and 3000 m from an aircraft [see also the paper by Tsvang, Zubkovskii, Ivanov, Klinov, and Kravchenko (1963) who reported measurements of $E_1^{(\vartheta)}(k)$ from a 300 m meteorological tower]. All the measurements were carried out with a low-inertia resistance thermometer and a spectral analyzer. One of the results obtained by Tsvang which illustrates the validity of the temperature five-thirds law has already been given in Fig. 59 in connection with the discussion of the validity of the Taylor frozen-turbulence hypothesis. The mean aircraft data corresponding to different heights at mid-day in the summer over the Russian Steppe, reported by Tsvang (1963), are also in good agreement with the five-thirds law: The exponent $\alpha$ in the formula $E_1^{(\vartheta)}(k) \sim k^{-\alpha}$ differs from the five-thirds law by not more than 10% for all these spectra. Tsvang used the similarity formula put forward by Monin (1958)

$$E_\vartheta(\omega) = \frac{T_*^2 z}{\bar{u}} \psi_\vartheta\left(\frac{\omega z}{\bar{u}},\ \mathrm{Ri}\right) \tag{23.14}$$

(where $T_*$ has the same significance as in Chapt. 4, Vol. 1), which is the analog of Eq. (23.7), to analyze his time spectra for the atmospheric surface layer. More precisely, Tsvang determined the dimensionless spectra $\tilde{\psi}_\vartheta = \bar{u} E_\vartheta / \tilde{T}_*^2 z$ where $\tilde{T}_* = z \frac{\partial \bar{T}}{\partial z}$ is a simply measured quantity the ratio of which, to $T_*$ tends to unity as we approach neutral stratification (while under other conditions it is a universal function of Ri). According to Eq. (23.14), the dimensionless spectrum $\tilde{\psi}_\vartheta(\omega z/\bar{u})$ should be independent of the height $z$ for given Ri, and in view of Eq. (23.11) the function $\tilde{\psi}_\vartheta$ should be proportional to $(\omega z/\bar{u})^{-5/3}$ in the inertial-convective subrange of $\omega$. Both these conclusions are satisfactorily confirmed by Fig. 80 [taken from Tsvang (1960a)], which gives the mean values of $\tilde{\psi}_\vartheta$ as a function of $fz/\bar{u} = \omega z/2\pi\bar{u}$ for $z = 1$ and 4 m, which correspond to measurements under almost identical temperature stratification (Ri $\approx -0.12$). Figure 81 shows values of $\tilde{\psi}_\vartheta(fz/\bar{u})$ obtained by Tsvang for six different values of Ri (between $-0.54$ and $+0.42$). Here again, we have a satisfactory confirmation of the five-thirds law for the one-dimensional temperature spectra.

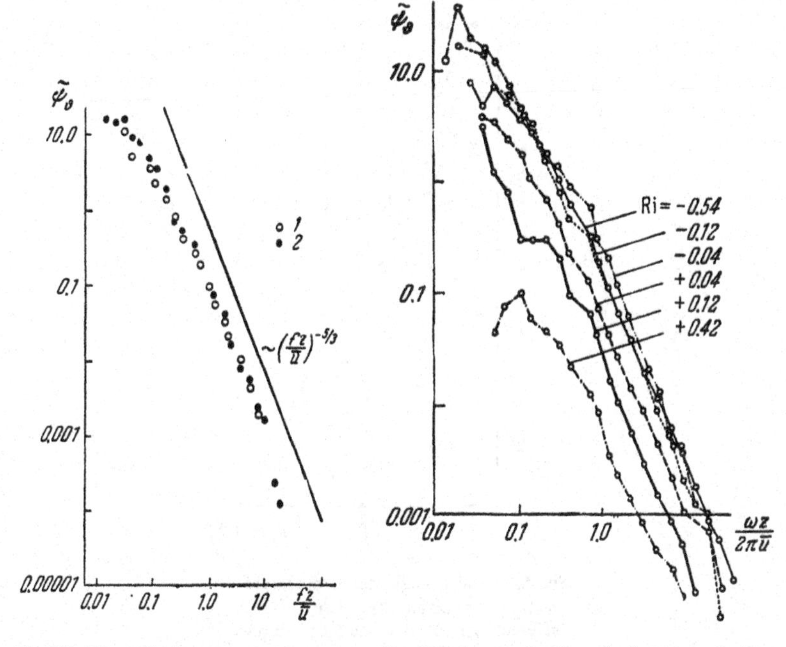


**Fig. 80** Dimensionless temperature spectra at heights of 1 m (1) and 4 m (2) for Ri ≈ −0.12.

**Fig. 81** Averaged dimensionless temperature spectra in the atmospheric surface layer for six different values of Ri.

Let us also recall a paper of Gurvich and Kravchenko (1962), who measured temperature spectra in the atmospheric surface layer using a special low-inertia resistance thermometer (with a tungsten wire 5 $\mu$ thick and 12 mm long bent so that the linear dimensions of the probe were only 3 mm; the time constant of this device was of the order of 0.001 sec). The results of these measurements are shown in Fig. 82 and indicate that the five-thirds law for the spectrum $E_1^{(\vartheta)}(k)$ is satisfied with high accuracy in the atmospheric surface layer at least up to length scales of the order of 3 mm. Excellent confirmation of the five-thirds law for the temperature spectra in the case of air over seawater were found by Pond (1965) who used an extremely low-inertia resistance thermometer (see also Pond, Smith, Hamblin, and Burling, 1966). These data are shown in Fig. 83. The spectra of temperature fluctuations in the atmosphere at a height 2 m above the sea surface were analyzed by Volkov (1969); they follow a 5/3 law with good accuracy. Numerous air temperature spectra were measured by H. E. Cramer et al. from two towers at Round Hill, Mass.,

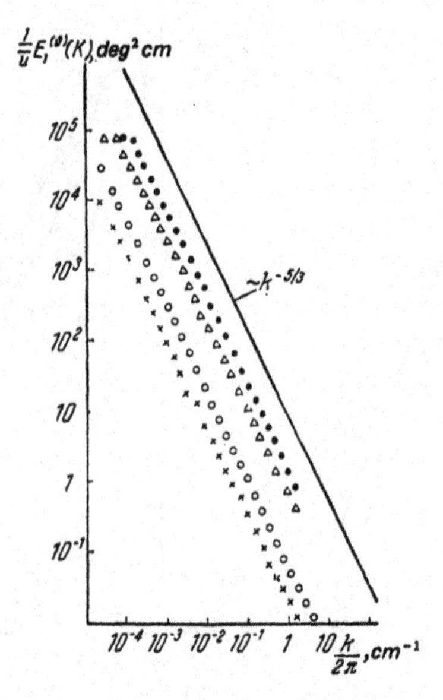


**Fig. 82** One-dimensional temperature spectra [Gurvich and Kravchenko (1962)]. Different symbols indicate different measurements.

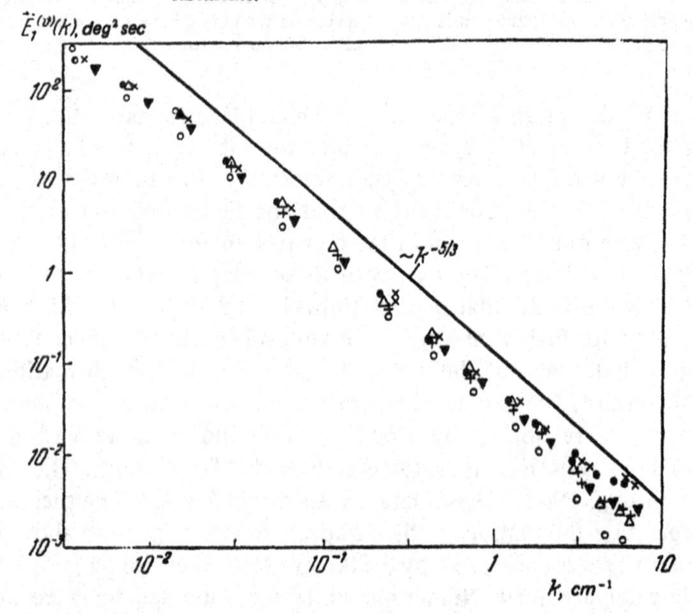


**Fig. 83** Temperature spectra according to Pond (1965). Different symbols show different measurements.

USA, for 5 heights between 15 and 91 m; these data were treated by Panofsky (1969b) who found that they also fit the 5/3 law well. Data on the temperature spectra in the stratosphere (heights 15–19 km) were collected by Vinnichenko and Dutton (1969); they are rather scattered but on the average are well represented by the five-thirds law. Laboratory temperature spectra in the boundary layer along the floor of the 27-meter long meteorological wind tunnel of Colorado State University also contain a five-thirds range (Plate and Arya, 1969). Quite reliable temperature spectra containing an extended range proportional to $k^{-5/3}$ were obtained by Grant, Hughes, Vogel, and Moilliet (1968) in the sea by resistance thermometer measurements from a research vessel and a submarine; these data will be discussed in more detail below (see in particular Fig. 89a). Extensive measurements of temperature spectra in the first 32 m of the atmospheric surface layer were performed during the 1968 Kansas expedition (see Wyngaard and Coté, 1971, and Kaimal, Wyngaard et al., 1972). All the spectra have a noticeable –5/3 range (generally in the band 0.5–5 Hz). Finally some very precise temperature fluctuation measurements were done in the atmospheric surface layer above a sea or a land surface by Gibson, Stegen and Williams (1970), by Boston (1970, 1971; see also Stewart, 1969) and in a heated laboratory jet by Gibson, Friehe and McConnell (1971). These authors used extremely fine resistance thermometers with a wire only $0.6\mu$ or even $0.25\mu$ in diameter and succeeded not only in confirming the 5/3 law but also in measuring the temperature dissipation (i.e., molecular diffusive) range of the spectrum. The data of Boston and Gibson et al. gave for the first time the possibility of determining the universal dimensionless temperature spectrum in air (i.e., for $\Pr = 0.7$) over a wide range of wave numbers. This universal spectrum will be considered later in our book.

We thus see that there is already a substantial volume of data confirming the validity of the five-thirds law for the temperature spectrum in the atmosphere and in the ocean. However there are considerable difficulties in obtaining reliable estimates of $\overline{N}$ and therefore the numerical value of the temperature structure and spectral constant $C_\vartheta$ is presently known less precisely than the value of the corresponding velocity constant $C$. Let us now consider various recent attempts to determine this constant.

Gurvich and Zubkovskii (1966) used the extensive material on the spectra $E_1^{(\vartheta)}(k)$ in the atmospheric surface layer for different stratifications collected by Tsvang (1960a) for approximate estimation of $C_\vartheta$. Tsvang's data give us the value of the function

$a^2(\text{Ri}) = A\tilde{\psi}_\vartheta(\omega z/2\pi\bar{u})^{5/3}$ where $\tilde{\psi}_\vartheta$ is the dimensionless temperature spectrum defined below Eq. (23.14) at a frequency $\omega$ from the inertial-convective subrange and $A \approx 10.5$ is a numerical constant (the solid line in Fig. 83b). If we use the balance equations for the energy and intensity of temperature fluctuations with the diffusion terms omitted (see the end of Sect. 7.5 in Vol. 1) then it is easy to show that $C_\vartheta = a^2(\text{Ri})\varphi_1(\zeta)[\varphi(\zeta) - \zeta]^{1/3}$, where $\zeta = z/L$, and $\varphi(\zeta)$ and $\varphi_1(\zeta)$ have the same meaning as in Chapt. 4 of Vol. 1. With the aid of the data collected in Sect. 8 of Vol. 1 Gurvich and Zubkovskii transformed the values of $a^2(\text{Ri})$ into values of $C_\vartheta$ and found that Tsvang's data led to roughly the same result for all Ri, namely $C_\vartheta \approx 2.7$ (the dotted line in Fig. 83b). Let us recall that this result is intermediate between the older very approximate values proposed by Tatarskii (1956a) and those of R. Taylor, Takeuchi, and Martin (see above).

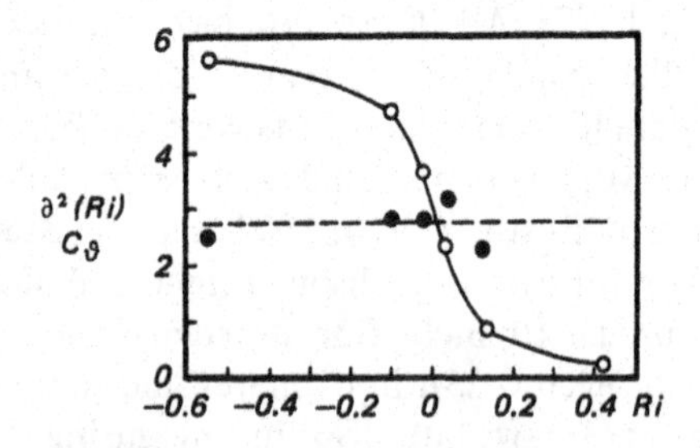


**Fig. 83b** The empirical function $a^2$ (Ri) of Tsvang (1962) (solid line) and the values of $C_\vartheta$ calculated from this function by Gurvich and Zubkovskii (1966) (dashed line).

A similar treatment of the Round Hill temperature spectra was carried out by Panofsky (1969b). He actually used two different, though not completely independent, procedures for estimating $C_\vartheta$; in one case, the energy dissipation rate was inferred from a simplified energy balance equation (as was done by Gurvich and Zubkovskii also), and, in the other case, it was determined from data on simultaneous values of the inertial subrange velocity spectrum. In addition, Panofsky made computations both with, and without the diffusion terms in the balance equations. The results show moderate scatter; the mean value of $C_\vartheta$ according to these estimates was close to 2.7, i.e., it coincides with the value derived from Tsvang's measurements near Tsimlyansk, USSR. Wyngaard and Coté (1971) performed the direct measurements of all the terms in Eq. (6.55) in Vol. 1, which is the balance equation for the mean square temperature (i.e., intensity of temperature fluctuations), with the exception of the dissipative term $\overline{N}$. These measurements were made in the atmospheric surface layer above the Kansas plains and were accompanied by simultaneous measurements of the wind velocity and temperature spectra in the $-5/3$ range. The data permit one to determine the values of $\overline{\varepsilon}$ and $\overline{N}$ and hence to compute the value of the constant $C_\vartheta$. With this method the preliminary estimate $C_\vartheta \approx 3.2$ was obtained with a standard deviation of 0.4. This estimate was based on a considerable portion of the Kansas data. Later Kaimal, Wyngaard et al. (1972) treated by the same method all the data of the 1968 Kansas expedition and obtained almost the same results: $C_\vartheta = 3.3 \pm 0.3$.

Moreover, Gurvich and Zubkovskii (1966) have also analyzed a relatively short synchronous record of the time variation in the temperature and longitudinal (in the mean wind direction) velocity fluctuations at two (practically coinciding) neighboring points in the atmosphere. Having Eq. (23.12) in view, they used the frozen turbulence hypothesis and calculated the parameter $F$; their result was $F \approx -0.28$. By using Eq. (23.12) with $|S|^{1/3} \approx 0.7$ they then found that $C_\vartheta \approx 3.5$. This result is rather close to that given above if we recall the relatively low accuracy which is involved in the latter (based on time averaging only over about 3.5 minutes). Much more extensive data of the same type were collected in a marine surface layer from FLIP (floating instrument platform) during February and May, 1969. These data were treated by Paquin and Pond (1971) who computed the second and third order structure functions and the ratio $F$. The spread of the values of $F$ for different runs is rather great, but the values of $F(r)$ are fairly constant in the range of $r = \overline{u}\tau$ from two to five meters for most runs. The average value $F = -0.27$ is quite close to that of Gurvich and Zubkovskii and leads to a similar estimate $C_\vartheta \approx 3.3$. On the other hand, Gibson and Schwarz (1963b) measured the velocity, temperature, and salinity fluctuation spectra in a laboratory water tunnel behind a grid, and determined the values of $\overline{N}$ (in fact, they used the quantity $\frac{d}{dt}\overline{\vartheta'^2}$, which is higher by a factor of two) and $\overline{\varepsilon}$ from data on the decay of the mean squares of the corresponding fluctuations with distance from the grid. They found that $C_\vartheta \approx 2.8$, which is practically identical with the results derived from Tsimlyansk and Round Hill spectra of the air temperature. Later Gurvich and Meleshkin (1966) used the theory of the subsequent Sect. 26 to estimate the constant $C_\theta$ from laboratory data on fluctuations in light beam intensity due to temperature fluctuations along its path. Their values were found to lie between 2 and 3. An estimate of $C_\theta$ was also obtained by Grant, Hughes et al. (1968), who determined $\bar{\varepsilon}$ in seawater from data on velocity spectra, but used an indirect (not very reliable) method to determine the values of $\bar{N}$. As a result they concluded that $C_\theta = 2.5 \pm 0.5$ (but the accuracy of the method is apparently overstated in this estimate).

All the above mentioned results obtained by very different methods and referring to different sets of data are in more or less satisfactory agreement with each other. They give some reason, for the moment, to assume that the value of the universal temperature structure function constant $C_\theta$ for the inertial-convective range is approximately 3. In accordance with Eq. (21.90), this corresponds to $B^{(\theta)} \approx 1.2$ in the five-thirds law for the three-dimensional temperature spectrum, and $B_1^{(\theta)} \approx 0.75$ in the similar law for the one-dimensional spectrum. It must be remembered, however, that all the above mentioned estimates are indirect and do not have very high accuracy. Unfortunately the results of a few recent more direct determinations of $C_\vartheta$ do not agree satisfactorily with each other and strongly contradict the group of data which was discussed above. The lowest estimate of $C_\vartheta$ obtained from measurements of the temperature spectrum in a wave number range including most of the dissipation range and the highest part of the $-5/3$ range (and simultaneous measurement of the velocity spectrum) is that by Gibson, Friehe and McConnell (1971) for a heated laboratory jet of a moderately high Reynolds number. This estimate is $B_1^{(\theta)} \approx 1.4$, i.e., $C_\theta \approx 5.6$. It is rather high, but the authors suggested that it may be slightly decreased by corrections for the effect of the finite wire length of the resistance thermometer. Nevertheless it can hardly be made consistent with the above estimates. However the other two estimates obtained with the same method are even higher. Gibson, Stegen and Williams (1970) measured the fine scale spectra of temperature and wind-velocity fluctuations in the marine surface layer of the atmosphere and came to the conclusion that $B_1^{(\theta)} \approx 2.3$, i.e., $C_\theta \approx 9$. Extensive measurements of the same type were also made by Boston (1970, 1971) at a height of 4 m in the atmosphere over a tidal mud flat. The results for 16 runs do not vary too strongly from run to run and lead to an average estimate $B_1^{(\theta)} \approx 1.62$ with a standard deviation of 0.04 (i.e., $C_\theta \approx 6.5$ with a standard deviation of 0.16). We have no explanation of the evident discrepancy between the estimates of $C_\vartheta$ from the high frequency spectrum measurements and all the other estimates. Of course the spectrum estimates are based on a considerably higher frequency portion of the $-5/3$ range than all the previous estimates, but it is quite unlikely to have two different $-5/3$ regions in the temperature spectrum with two constants differing by a factor of at least two. The assumption that the "5/3 law" is only approximately valid seems to be more natural. In this case the approximation of various spectral regions by power function with the exponent $-5/3$ may yield various factors of proportionality depending upon the frequency range. However, the observed differences in $C_\vartheta$ values are clearly too great to be easily explained in this manner. In any case these spectral results warn us to be quite cautious with the above estimated values of the temperature constants. Further investigation of this question is evidently strongly needed.

In order to give an idea of the distribution of the total temperature fluctuation intensity over the frequency spectrum, or the wave number spectrum, we reproduce in Figs. 84 and 85 the data of Tsvang (1962) on the function $kE_1^{(\vartheta)}(k)$ for the atmospheric

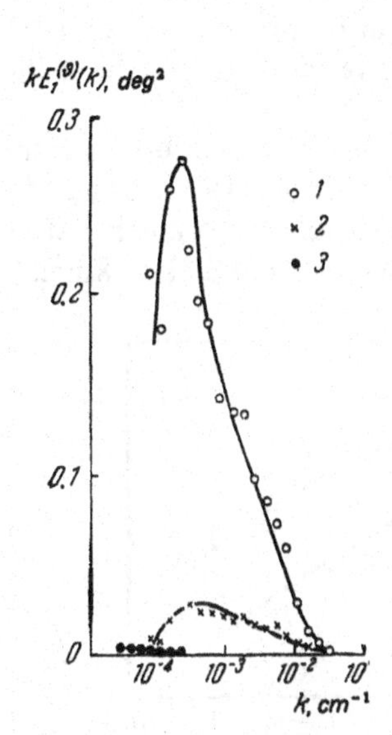


Fig. 84 The function $kE_1^{(\vartheta)}(k)$, according to Tsvang (1962). 1–$z = 1$ m, unstable stratification; 2–$z = 1$ m, stable stratification; 3–$z = 300$ m.

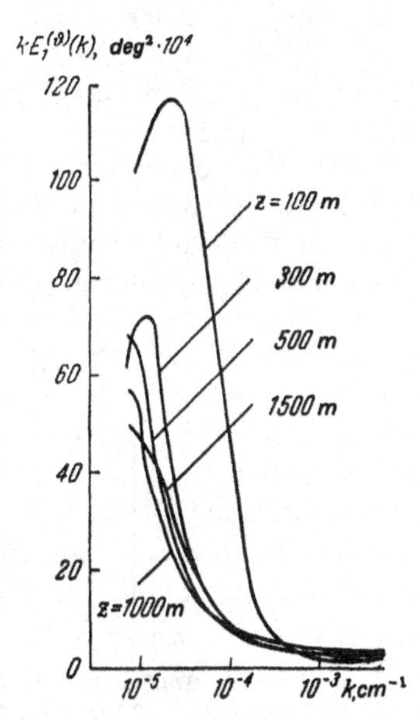


Fig. 85 Values of $kE_1^{(\vartheta)}(k)$ at different heights.

surface layer and the 300-meter meteorological tower (Fig. 84), and at heights between 100 and 1500 m (aircraft measurements; Fig. 85). It is clear from these data that the temperature spectra shift toward smaller wave numbers with increasing height, while the total temperature fluctuation intensity $\sigma_\vartheta^2$, which is equal to the area under the curve $k\,E_1^{(\vartheta)}(k)$, falls rapidly.

The dimensionless temperature spectrum $\psi_\vartheta(fz/\bar{u}, \mathrm{Ri})$, $f = \omega/2\pi$, of Eq. (23.14) was computed by Panofsky (1969b) for all the Round Hill data. Since $E_1^{(\vartheta)}$ and $T_*$ both become very small under near-neutral conditions, the near-neutral spectra were excluded from the computations. All other spectra were divided into two groups: Stable (with $0.42 \geqslant z/L \geqslant 0.05$) and unstable (with $-0.05 \geqslant z/L \geqslant -1.12$). The two groups of functions both collapse with satisfactory accuracy into a single curve; both these curves are shown in Fig. 85a where both scales on the coordinate axes are logarithmic and $fE_1^{(\vartheta)}(f)/T_*^2$ is taken as the ordinate. The two curves in Fig. 85a can be described by the following empirical equations: $fE_1^{(\vartheta)}(f)/T_*^2 = 4.6Y_m\hat{f}/\hat{f}_m \times (1 + 1.5\hat{f}/\hat{f}_m)^{-5/3}$ in unstable air, and $fE_1^{(\vartheta)}(f) = 2.5Y_m\hat{f}/\hat{f}_m \times 1 + 1.5(\hat{f}/\hat{f}_m)^{5/3-1}$ in stable air. Here $\hat{f} = fz/\bar{u}$ is the dimensionless frequency, $\hat{f}_m$ is the abscissa of the maximum of the corresponding curve in Fig. 85a, and $Y_m$ is the ordinate of this maximum. It is clear that both the equations satisfy the five-thirds law in the high-frequency range.

Extensive computations of dimensionless temperature spectra were also performed by Kaimal, Wyngaard et al. (1972) using the data of the 1968 Kansas expedition, which related to much flatter and more homogeneous terrain than that at Round Hill. Kaimal,

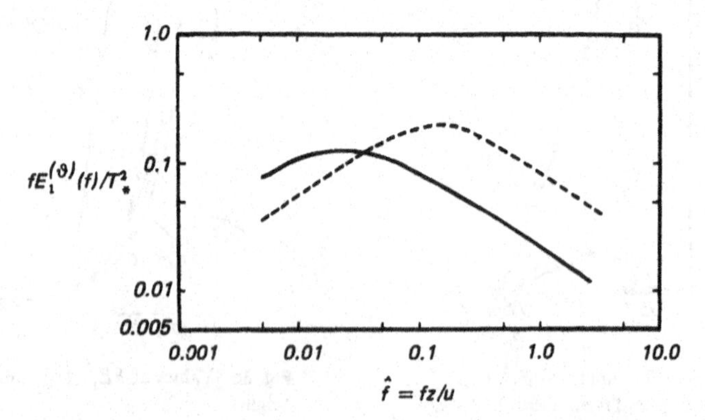


**Fig. 85a** Averaged, normalized logarithmic spectra of temperature at Round Hill (from Panofsky, 1969b). a–Stable, heights 15–91 meters; b–unstable, heights 15–46 meters.

Wyngaard et al. normalized their spectra $fE_1^{(\vartheta)}(f)$ by a product of $T_*^2$ and a dimensionless function $g(\zeta) = g(z/L)$ which is proportional to $\overline{N}\bar{\varepsilon}^{-1/3}$. Hence the ratio obtained $\dfrac{fE_1^{(\vartheta)}(f)}{T_*^2 g} = \hat{\psi}_\vartheta\left(\dfrac{fz}{u}, \dfrac{z}{L}\right)$ proves to be independent of $z/L$ (or Ri) in the inertial subrange. However, at lower dimensionless frequencies $fz/\bar{u}$ the dependence on $z/L$ becomes quite appreciable (see Fig. 85b). The stable temperature spectra in Fig. 85b like those of velocity in Figs. 72a and 72b, separate into distinct curves according to $z/L$ while all the unstable spectra form a rather disorderly crowd in the relatively narrow band indicated in the figure by the stippled area.

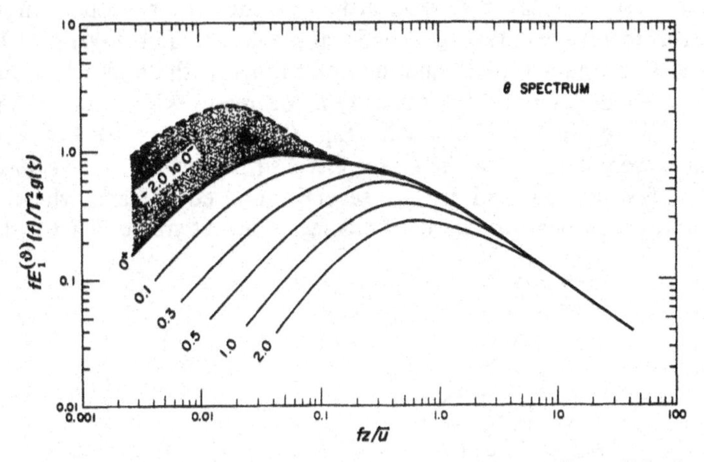


**Fig. 85b** Generalized dimensionless temperature spectrum according to Kaimal, Wyngaard et al. (1972).

If the numerical coefficients $C$ and $C_\vartheta$ are approximately known, then data of simultaneous measurement of the velocity and temperature spectra in the inertial (or inertial-convective) subrange can be used to estimate the mean rate of dissipation of temperature fluctuations $\overline{N}$. Such a procedure was used, in particular, by Vinnichenko and Dutton (1969) for air in the stratosphere at heights 15–19 km (the values of $\overline{N}$, changing with the change from light to severe turbulence, were in the range $4\cdot10^{-6}$–$5\cdot10^{-4}$ °C² sec⁻¹) and by Grant, Moilliet, and Vogel (1968) for turbulent regions in the open sea at depths 15–90 m (the values of $\overline{N}$ were in the range $4\cdot10^{-8}$–$3\cdot10^{-6}$ °C² sec⁻¹). We must note, however, that for many practical applications of the temperature two- and five-thirds laws we must have in fact information not about the variable $\overline{N}$ and/or the dimensionless coefficient $C_\vartheta$, but rather about the dimensional coefficient $B^2 = C_\vartheta \overline{N}\,\overline{\varepsilon}^{-1/3}$ itself, which unambiguosuly determines the spectrum $E_1^{(\vartheta)}(k)$ and the structure function $D_{\vartheta\vartheta}(r)$ in the inertial subrange. Approximate data of Tatarskii on the values of $B^2$ in the atmospheric surface layer have already been given in the beginning of this section (Fig. 79). Similar analysis was later performed also by Panofsky (1968) and by Wyngaard, Izumi and Collins (1971). The latter authors treated numerous data of the 1968 Kansas expedition and obtained experimental values of the dimensionless function $C(\mathrm{Ri}) = \frac{B^2}{z^{4/3}}\left(\frac{d\overline{T}}{dz}\right)^2$ which permit one to determine $B^2$ in the atmospheric surface layer from routine profile observations. Aircraft measurements by Tsvang (1963) and Koprov and Tsvang (1966) show that under typical summer conditions, with developed convection over the steppes, the quantity $B^2$ decreases with height roughly as $z^{-4/3}$ [i.e., in accordance with Eqs. (7.110) and (1.104) of Vol. 1 which were derived for the atmospheric surface layer!] above heights of a few meters and up to several hundred meters, where the potential temperature gradient changes sign. At greater heights there

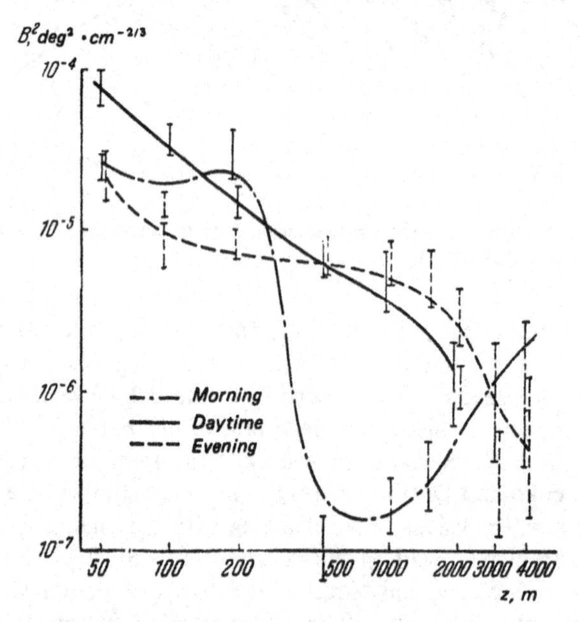


Fig. 86 The coefficient $B^2$ as a function of height at different times of day.

is an appreciably faster fall of $B^2$ (see Fig. 86). During early morning (under the conditions of developing convection, when very stable temperature stratification is still observed in the 50–250 m layer) and during the later afternoon hours (when the stratification is nearly neutral up to heights of about 1500 m), the $B^2$ profile is quite different (it is shown schematically in Fig. 86), but even in these cases it remains the same for different days, i.e., it is determined by the mean meteorologic conditions. Additional data on the profiles of $B^2(z)$ under the various weather conditions above the steppe (south-Russian prairie) and the sea are collected in Volkov, Kucharets, and Tsvang (1968) and also reproduced in Tsvang (1969).

*Data on the Structure of other Scalar Fields and on the Small-Scale Structure of the Temperature Field outside the Inertial-Convective Subrange*

The formulas (23.11) for the inertial-convective range can be applied not only to the temperature but also to the concentration $\vartheta(\boldsymbol{x}, t)$ of any passive admixture, provided only that we replace $\overline{N}$ by the mean rate $\overline{N_\vartheta}$ at which the concentration inhomogeneities are smoothed out by molecular diffusion. A typical example of such a concentration field is the absolute humidity field in the atmosphere, or the salinity field in the sea (or in any turbulent flow of saline water). Next, since fluctuations in the refractive index of the air for light waves can be regarded as proportional to the temperature fluctuations in view of Eq. (26.2′) (see below), these fluctuations should also obey the two-thirds and five-thirds laws of Eq. (23.11). Finally, small fluctuations in the refractive index of air for radio waves can be represented by a linear combination of temperature and humidity fluctuations in accordance with Eq. (26.2) (see below). These fluctuations must therefore again follow the two-thirds and five-thirds laws.

Existing data confirm the validity of the above laws in the case of humidity and the refractive index for radio waves in the atmosphere. We recall, for example, the data of Gossard (1960a), who calculated the humidity and refractive-index spectra using the readings from balloon- and aircraft-mounted "dry-bulb" and "wet-bulb" thermometers (or direct measurements of the refractive index by means of a refractometer). Most of Gossard's results are in adequate agreement with the five-thirds law [see also Gossard (1960b)]. Similarly, Bolgiano (1958a), Thompson et al. (1960), Edmonds (1960), Stilke

(1964), Bull (1966, 1967) and others have reported extensive measurements of the time spectra of the refractive index $E_\vartheta(\omega)$ (where $\vartheta$ now represents the refractive-index for radio waves), indicating that the exponent $\alpha$ in $E_\vartheta(\omega) \sim \omega^{-\alpha}$ is usually very close to 5/3 [see, for example, Fig. 87, taken from Bull (1967)]. The data on refractive index spectra of Bean, Emmanuel, and Krinks (1967) are characterized by considerable scatter of the exponent $\alpha$; however the average exponent is quite close to 5/3. Moderate departures from the five-thirds law were observed by Lane (1969), but it is difficult to decide at present whether they are statistically significant or not. Extensive data and references confirming the validity of the five-thirds law for refractive index fluctuations can be found in the book by Tatarskii (1967) and the collection edited by Yaglom and Tatarskii (1967).

Elagina (1963, 1970) has measured the humidity fluctuation spectra in the atmospheric surface layer using an infrared absorption hygrometer. These spectra were found to be approximately proportional to the −5/3 power of frequency in a broad range of

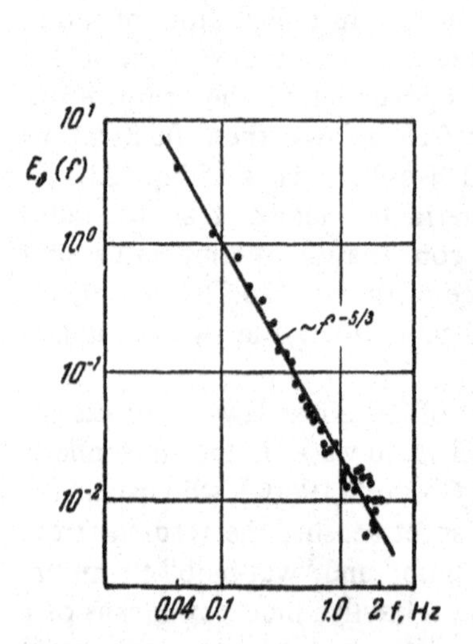


**Fig. 87** Time spectrum $E_\vartheta(f)$ (where $f = \omega/2\pi$ of the refractive index [Bull (1967)].

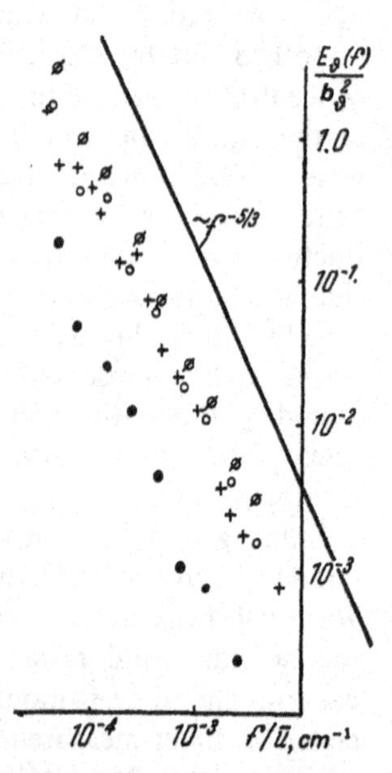


**Fig. 88** Spectrum of humidity fluctuations in the atmospheric surface layer. Different symbols represent different measurements.

frequencies (see Fig. 88). Humidity fluctuation spectra obeying a five-thirds law were also obtained by Sahashi (1967) with fine dry-bulb and wet-bulb thermocouples, by Chen and Mitsuta (1967) with infrared hygrometer and by Miyake and McBean (1970) and Miyake, Donelan and Mitsuta (1970) with ultraviolet (Lyman-α) absorption hygrometer. Martin (1966) has reported measurements in the atmospheric surface layer of the time structure function of humidity fluctuations $D_{\vartheta\vartheta}(r)$, using a combination of a refractometer and a resistance thermometer. His data indicated that this function was satisfactorily represented by the two-thirds law (23.11) for $r$ up to 2.5 m [see also Hay (1967)]. Extensive values of the humidity time structure functions $D_{\vartheta\vartheta}(\tau)$ and also of joint velocity-humidity functions $D_{L\vartheta\vartheta}(\tau)$ for the marine surface layer of the atmosphere were reported by Paquin and Pond (1971) who used the results of measurements with a sonic anemometer and ultraviolet (Lyman-Alpha) hygrometer in their computations. These two functions were satisfactorily represented (if $\tau$ is not too short) by laws of power two-third and of power one, respectively.

Becker, Hottel, and Williams (1967a, b) developed a special light-scatter technique for fast response measurements of the concentration fluctuations of colloidal particles in turbulent flows. The technique was used for obtaining the spectrum of oil smoke particle concentration in a turbulent air jet in the wave number range $0.5 \text{ cm}^{-1} \leqslant k \leqslant 40 \text{ cm}^{-1}$. It was found that the spectrum is described by a five-thirds law in the range $2 \text{ cm}^{-1} \leqslant k \leqslant 40 \text{ cm}^{-1}$ with very high precision (see Fig. 88a).

Gibson and Schwarz (1963a, b) have described measurements of $E_1^{(\vartheta)}(k)$ for temperature and salinity fluctuations behind a grid in a water tunnel up to values of $k$ appreciably exceeding the upper limit of the inertial-convective range in which the five-thirds law is satisfied. According to similarity theory, the spectrum $E_1^{(\vartheta)}(k)$ for locally isotropic turbulence and such large values of $k$ should be of the form (21.88). This formula can also be rewritten in the form

$$E_1^{(\vartheta)}(k) = \frac{\bar{N}}{\bar{\varepsilon}} (\bar{\varepsilon}\nu^5)^{1/4} \varphi_\vartheta(k\eta, \nu/\chi), \qquad (23.15)$$

where $\nu/\chi = \mathrm{Pr}$ was equal to 7 in the experiment of Gibson and Schwarz in the case of temperature fluctuations, and to 700 in the case of salinity fluctuations. According to Batchelor's theory (1959), discussed in Sect. 22.4, in the range $k > 1/\eta = (\bar{\varepsilon}/\nu^3)^{1/4}$ the spectrum $E_1^{(\vartheta)}(k)$ should depend for large $\nu/\chi$ not on $\bar{\varepsilon}$ and $\nu$ separately but

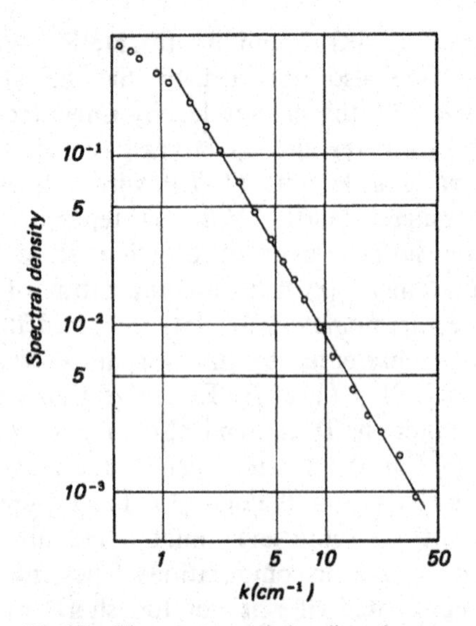


**Fig. 88a** The spectrum of the oil-smoke particle concentration in a turbulent airjet according to Becker, Hottel and Williams (1967a). The straight line represents a five-thirds law.

only on their combination $(\bar{\varepsilon}/\nu)^{1/2} = \tau_\eta^{-1}$ (or, more precisely, on the combination $(\alpha\tau_\eta)^{-1}$, where $\alpha$ is a universal constant representing a typical value of the maximum rate of compression of an infinitesimal fluid particle). It is readily seen that this means that

$$\varphi_\vartheta\left(k\eta,\ \frac{\nu}{\chi}\right) = \left(\frac{\nu}{\chi}\right)^{-1/2} \Phi\left[k\eta\left(\frac{\nu}{\chi}\right)^{-1/2}\right], \tag{23.16}$$

where the function $\Phi(x)$ can be found by applying Eq. (12.13) to the three-dimensional spectra (22.80) found by Batchelor for this case. Finally, for sufficiently large $\nu/\chi$, there should be a viscous-convective range of wave numbers $(\bar{\varepsilon}/\nu^3)^{1/4} < k < (\bar{\varepsilon}/\nu\chi^2)^{1/4}$ in which the spectrum (23.16) ceases to depend on $\chi$, i.e., $\Phi(x) = \beta/x$, where $\beta = 2\alpha$ is a constant which, according to a very approximate estimate of Batchelor, is not very different from 4. It is therefore clear that $\varphi_\vartheta(k\eta,\ \nu/\chi) = \frac{\beta}{k\eta} \approx \frac{4}{k\eta}$.

All the above theoretical predictions were confirmed by Gibson and Schwarz. Figure 89 shows six dimensionless salinity spectra $\varphi_\vartheta(k\eta,\ 700)$ corresponding to $\mathrm{Re}_M = UM/\nu$ between 11,700 and 65,550, together with the temperature spectrum $\varphi_\vartheta(k\eta,\ 7)$ at $\mathrm{Re}_M =$

35,300, the theoretical curves of Batchelor, and the empirical function $\varphi_1(k\eta)$ of Eq. (23.10) which describes the dimensionless one-dimensional longitudinal spectrum of the velocity field. The temperature spectrum is in good agreement with Batchelor's theoretical curve, while the salinity spectra lie below the theoretical curve (this can be explained by the low sensitivity of the salinity probe to very high-frequency salinity fluctuations) but above both the velocity and the temperature spectra. It is interesting that the Batchelor theory which refers to $\mathrm{Pr} \gg 1$ turns out to be quite accurate even for $\mathrm{Pr} \approx 7$ according to the Gibson and Schwarz data.

The last conclusion was confirmed also by results of the measurements of temperature fluctuation spectra in seawater carried out by Grant, Hughes, Vogel, and Moilliet (1967). All their temperature spectra have regions described fairly well by the five-thirds law and also higher wave number regions where the

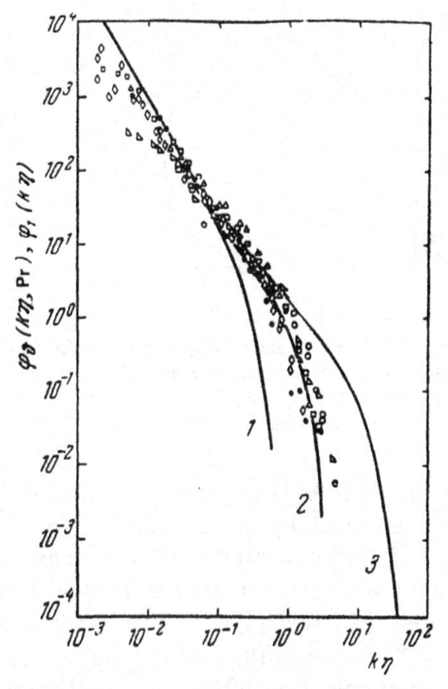


**Fig. 89** Dimensionless one-dimensional spectra of salinity (open circles) and temperature (points) [Gibson and Schwarz (1963b)]. 1–Experimental curve for $\varphi_1(k\eta)$; 2–Batchelor's curve for $\varphi_\vartheta(k\eta, 7)$; 3–Batchelor's curve for $\varphi_\vartheta(k\eta, 700)$.

decrease of the spectra is approximately proportional to $k^{-1}$ (one example of the one-dimensional temperature and velocity spectra measured by Grant et al. in a tidal channel is shown in Fig. 89a). The temperature spectra were fitted to Batchelor's theoretical curve by translations along both axes, and the condition of best fit was used together with the values of $\bar{\varepsilon}$ from simultaneous velocity spectral measurements to estimate the unknown parameters $C_\vartheta$ and $\alpha = \beta/2$. The result for $C_\vartheta$ was stated in the preceding subsection; the estimates of $\alpha$ appear to be quite scattered and suggest only that apparently $\alpha = 3.9 \pm 1.5$.

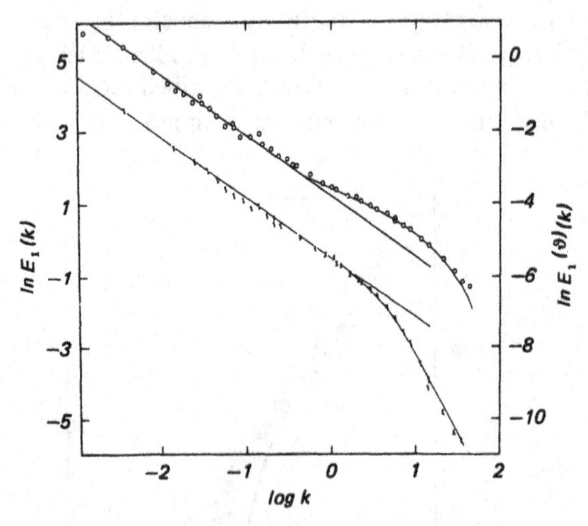


**Fig. 89a** Temperature and velocity spectra at a depth of 15 m in Discovery Passage near Vancouver, after Grant, Hughes, Vogel and Moilliet (1968). (○)–$E_1^{(\vartheta)}(k)$, °C²/unit wave number; (x) $E_1(k)$, (cm/sec)²/unit wave number.

Nye and Brodkey (1967) have also reported the existence of a spectral region described by Batchelor's "−1 power law" in their data on spectra of dye concentration fluctuations in water obtained with a refined optical measurement technique. Unfortunately the Reynolds number of their pipe flow was too low to allow the existence of a $k^{-5/3}$ region in the measured spectra and the data were insufficient for determination of the numerical constants entering in the theoretical predictions. Measurements of the high-frequency region of the temperature spectrum in air (i.e., at Pr = 0.7) were done by Plate and Arya (1969) in the huge meteorological wind tunnel of the Colorado State University. The values of $\bar{\varepsilon}$ and $\bar{N}$ were estimated very roughly in this work; the data suggest that the universal function $\varphi_\vartheta(k\eta, 0.7)$ of Eq. (23.15) is approximately proportional to the function $\varphi_1(k\eta)$ where $\varphi_1$ is the universal dimensionless one-dimensional longitudinal spectrum of Eq. (23.10), when the Prandtl number is close to unity.

There are very few measurements of temperature spectra in fully developed turbulence at low Prandtl number. However, the data of Granatstein, Buchsbaum and Bugnalo (1966) on the electron density spectrum in plasma pipe flow (Pr = 0.07) agree satisfactorily with the assumption that these spectra contain after a very short $-5/3$ range a considerable $-17/3$ range corresponding to Eq. (22.100). Also, the measurements of Clay (1973) of temperature spectra in turbulent mercury flow (Pr = 0.02) agree with the existence of a $-17/3$ range in the region of very high wave numbers.

Precise measurements of the dissipation range of one-dimensional temperature spectra in air (i.e., in fact of the function $\varphi_\vartheta(k\eta, 0.7)$) were made by Gibson, Stegen and Williams (1970) and Boston (1970, 1971) in an atmospheric surface layer and by Gibson, Friehe and McConnell (1971) in a heated laboratory jet. The results of these works do not agree very well with each other, but the observed disagreement can be explained quite naturally by the effect of instrumental filtering of very high frequencies which was significant in Gibson's experiments. The high-frequency portion of Boston's data is therefore more reliable than that of Gibson et al. According to these data the temperature spectrum has no region with a slope of minus one, when Pr = 0.7. The function $\varphi_\vartheta(k\eta, 0.7)$ obtained by Boston is shown in Fig. 89a; its shape is not very different from the shape of $\varphi_1(k\eta)$. The corresponding dimensionless temperature dissipation spectrum $(k\eta)^2\varphi_\vartheta(k\eta, 0.7)$ is shown in Fig. 89b; this figure is similar to Fig. 77 related to velocity spectra.

All the existing estimates of temperature constants in the inertial-convective range were reported above in this subsection. There are also two attempts to estimate the same constants for the humidity fluctuations. Miyake, Donelan and Mitsuta (1970) used the estimate of $\bar{\epsilon}$ from the velocity spectrum data and the estimate of $\bar{N}$ from a simplified balance equation for $\overline{\vartheta'^2}$ where $\vartheta'$ are humidity fluctuations at a height of about 20 m above the ocean measured from aircraft. They obtained from these data the approximate relation $B_1^{(\vartheta)} \approx 0.63$, i.e., $C_\vartheta \approx 2.5$. Paquin and Pond (1971) have computed the values of $F(r) = D_{L\vartheta\vartheta}(r)/[D_{LL}(r)D_{\vartheta\vartheta}^2(r)]^{1/2}$ from records of the wind-velocity and humidity fluctuations at a height of about 8 m over the ocean. They found that the values of $F$ vary considerably

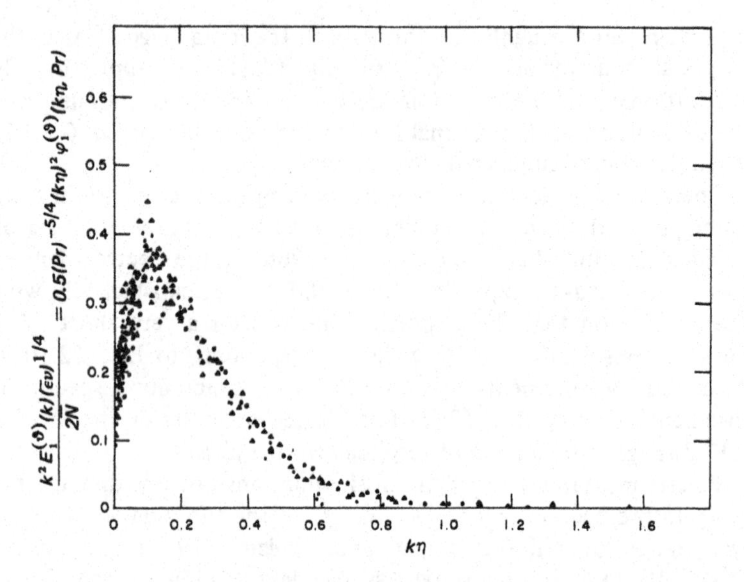


**Fig. 89b** Dimensionless temperature dissipation spectrum according to Boston (1970, 1971).

from run to run with the average $F = -0.29$ which leads to the conclusion that $C_\vartheta \approx 3.2$ for the humidity fluctuations. Both these estimates of $C_\vartheta$ agree satisfactorily with the estimates of $C_\vartheta$ for the temperature fluctuations obtained by similar methods (but not with the estimates of $C_\vartheta$ from fine-scale temperature spectrum measurements).

In conclusion, we must say a few words about the pressure fluctuation spectrum. According to the theoretical ideas of Sect. 21.5, we may expect that the time spectrum of these fluctuations in the inertial subrange should be proportional to $\omega^{-7/3}$, i.e., it should decrease more rapidly than the velocity or temperature spectra. However, order-of-magnitude calculations based on the formulas of Sect. 21.5 show that turbulent pressure fluctuations in the inertial subrange are very small and can be measured only with considerable difficulty. The situation is even more complicated by the fact that any pressure probe is also sensitive in some degree to the much greater velocity fluctuations whose influence must almost unavoidably mask the true pressure fluctuations. It is, therefore, not surprising that there is very little information about the statistical properties of small-scale pressure fluctuations in the inertial subrange. We note, however, that Gossard's measurements (1960b) have led to the conclusion that the pressure spectrum decreases with frequency more rapidly than $\omega^{-5/3}$, namely, on the average, it is proportional to $\omega^{-2}=\omega^{-6/3}$ in a considerable range of frequencies (apparently mostly to the left of the inertial subrange). We shall return to these measurements in the next subsection. The fact that the exponent $\alpha$ in the formula $E_p(\omega)\sim\omega^{-\alpha}$ was found to be somewhat less than 7/3 of course may be explained by the influence of velocity fluctuations on the pressure probe, the spectrum of which decreases more slowly than $E_p(\omega)$. Similar conclusions were reached by Gorshkov (1967), who measured the atmospheric pressure fluctuation spectrum in a considerably higher frequency range and found that, as a rule, this spectrum decreases with frequency more rapidly than $\omega^{-5/3}$ (which, in some cases, is exactly proportional to $\omega^{-7/3}$). All these results must, however, be regarded as preliminary for the time being.

## 23.6 Data on the Turbulence Spectra in the Atmosphere beyond the Low-Frequency Limit of the Inertial Subrange

Examples of the spectra of turbulent fluctuations in the atmospheric surface layer beyond the low-frequency limits of the inertial subrange were already given in Figs. 71, 84, 85, and 85a. Data on the statistical characteristics of fluctuations in this low-frequency part of the spectrum cannot, of course, be used to verify the predictions of the Kolmogorov similarity hypotheses. However, for many atmospheric turbulence problems, relatively low-frequency fluctuations which do not lie in the inertial subrange are of considerable interest. Moreover, even the accurate determinations of many of the quantities frequently encountered in this book (for example, the variances of the turbulent fluctuations or turbulent fluxes of heat, momentum, and water vapor), require data on the behavior of the spectra in the low-frequency region. At the same time the formulation of precise requirements which must be satisfied by instruments designed for measurements of practically any of the statistical characteristics of turbulence, is also impossible without such data. We shall, therefore, devote this subsection to a brief review of existing (and, unfortunately, rather scant) data on the low-frequency components of atmospheric turbulence [see also Lumley and Panofsky (1964), Chapt. 5 and Panofsky (1969a)].

### *Data on the Spectra of Turbulent Fluxes*

We shall begin by reviewing the spectra of vertical turbulent fluxes of heat $q=c_p\rho\overline{T'w'}$, momentum $\tau=-\rho\overline{u'w'}$ and water vapor $j=\overline{\rho\vartheta'w'}$ in the atmospheric surface layer. We recall that, in the region where the similarity hypotheses are valid, the fluctuations $T'$ and $w'$, $u'$ and $w'$, and $D'$ and $w'$ will be uncorrelated because of local isotropy. Therefore, nonzero values of $q$, $\tau$, and $j$ already indicate that we are beyond the low-frequency limit of the inertial subrange.

The first indirect estimation of the spectra of momentum and heat fluxes was due to Deacon (1959). Later the spectra of $q$ and $\tau$ were directly measured apparently for the first time by Gurvich and Tsvang (1960) and by Gurvich (1961), respectively using the following method. Let $u(\Delta\omega)$, $w(\Delta\omega)$ and $T(\Delta\omega)$ be the fluctuations $u'$, $w'$, and $T'$ transmitted through the narrow bandwidth $\Delta\omega$ of a frequency filter centered on the frequency $\omega$. The mean product $\overline{u(\Delta\omega)\,w(\Delta\omega)}$ will then be the contribution of the frequency interval $\Delta\omega$ to the vertical turbulent momentum flux per unit mass, i.e., to the quantity $u_*^2 = \tau/\rho$, while $\overline{w(\Delta\omega)\,T(\Delta\omega)}$ will be the corresponding contribution to the normalized vertical heat flux $q/c_p\rho$. Hence, the functions

$$E_{uw}(\omega) = -\frac{1}{\Delta\omega}\,\overline{u(\Delta\omega)\,w(\Delta\omega)}, \quad E_{wT}(\omega) = \frac{1}{\Delta\omega}\,\overline{w(\Delta\omega)\,T(\Delta\omega)} \tag{23.17}$$

will be the spectral densities of the turbulent momentum and heat fluxes, respectively (their integrals over all $\omega$ will be equal to $u_*^2$ and $q/c_p\rho$, respectively).[26]

According to similarity theory, in the case of turbulence in the atmospheric surface layer, the functions of Eq. (23.17) are given by the following formulas which are quite analogous to Eqs. (23.7) and (23.14):

$$E_{uw}(\omega) = \frac{u_*^2 z}{\bar{u}}\,\psi_{uw}\left(\frac{\omega z}{\bar{u}}, \mathrm{Ri}\right), \quad E_{wT}(\omega) = \frac{u_*|-T_*|\,z}{\bar{u}}\,\psi_{wT}\left(\frac{\omega z}{\bar{u}}, \mathrm{Ri}\right) \tag{23.18}$$

[Monin (1958, 1962a, b)]. The integral of the function $\psi_{uw}(\Omega)$ with respect to dimensionless circular frequency $\Omega = \Omega z/\bar{u}$ must, of course, be unity, while the integral of $\psi_{wT}(\Omega)$ must be equal to von Kármán's constant $\varkappa$. These formulas were compared by Gurvich and Tsvang (1960) and by Gurvich (1961, 1965) with the data of measurements of the functions (23.17) in the atmospheric surface layer near Tsimlyansk at a height of 1 m. The amount of data was clearly insufficient for determination of the dependence of the universal functions $\psi_{uw}$ and $\psi_{wT}$ on Ri, but it permits us to determine the general shape of these functions for moderately unstable conditions (all the measurements were carried out in the range $-0.5 < \mathrm{Ri} < -0.07$ of values of Ri). The main contribution to the turbulent fluxes is due to the dimensionless frequencies $\omega z/2\pi\bar{u} \approx 0.02 \div 1.0$ according to these data (the frequencies correspond of course to a horizontal wavelength range of the turbulent inhomogeneities considerably greater than the height of the observations).

More recent data on the cospectra of heat and momentum fluxes in the atmospheric surface layer above land and sea surfaces were collected by Weiler and Burling (1967), Smith (1967), Panofsky and Mares (1968), Zubkovskii and Koprov (1969), Stewart (1969), Miyake, Stewart, and Burling (1970), Miyake and McBean (1970), Miyake, Donelan and Mitsuta (1970), and Kaimal, Wyngaard et al. (1972). The data of Weiler and Burling, Smith, and Stewart refer to the momentum flux (i.e., stress) spectra over water, those of Miyake, Stewart and Burling, and of Miyake, Donelan and Mitsuta to the stress and heat flux spectra over water, and in the other three above mentioned papers the spectra of momentum and heat fluxes over land are analyzed. Data on the heat flux spectra in the free atmosphere (at heights 50–1000 m) were obtained by Kucharets and Tsvang (1969). Almost all these data indicate that the cospectra $E_{uw}(k)$ and $E_{wT}(k)$ decrease considerably faster than $k^{-5/3}$ at high enough wave numbers and hence the corresponding correlation coefficient spectra $E_{uw}/[E_{uu}E_{ww}]^{1/2}$ and $E_{wT}/[E_{ww}E_{TT}]^{1/2}$ come near to vanishing.[27] As an example we

[26] We recall that a filter with a narrow passband $\Delta\omega$ isolates the component $u(\Delta\omega) = 2\mathrm{Re}\left\{e^{i\omega t} Z(\Delta\omega)\right\}$ of the oscillation $u(t)$ transmitted by it [see Eq. (11.12)]. Hence, we can readily show that the functions $E_{uw}(\omega)$ and $E_{wT}(\omega)$ are equal to the real parts (cospectra) of the complex cross-spectra of the fluctuations $u'$ and $w'$, and, correspondingly, $w'$ and $T'$.

[27] We must recall however that the data of hot-wire anemometer measurements by the Vancouver group (Stewart, 1969) seem to give the impression that $E_{uw}$ also follows approximately a 5/3 law and the correlation coefficient spectrum for $u$ and $w$ fluctuations is approximately constant at high frequencies. This conclusion coincides with the results of Gurvich (1965) which were based on rather rough early measurements of cospectra.

show in Fig. 90 several collections of simultaneous data on the spectra $E_{uu}(k)$, $E_{ww}(k)$, and $E_{uw}(k)$ and of $E_{TT}(k)$, $E_{ww}(k)$, and $E_{wT}(k)$ presented on logarithmic scales; the solid straight lines in this figure correspond to the five-thirds law. Quite similar graphs were also obtained (for heat flux spectra only) by Kucharets and Tsvang (1969) in the free atmosphere. The dependence of the heat flux spectrum $E_{wT}(k)$ on the stability is illustrated by Fig. 91 (also taken from Zubkovskii and Koprov's paper) where the curves for $kE_{wT}(k)$ are shown for three different thermal stratifications (characterized by different values of Richardson number). A preliminary attempt was made by Panofsky and Mares (1968) to determine the general form of the universal functions $\psi_{uw}$ and $\psi_{wT}$ of Eq. (23.18). According to their data the function $\hat{f}\psi_{uw}(\hat{f})$ of the dimensionless frequency $\hat{f} = fz/\bar{u} = \omega z/2\pi\bar{u}$ can be roughly approximated by the empirical expression

$$\hat{f}\psi_{uw}(\hat{f}) = \frac{\hat{f}/\hat{f}_m}{(1 + 0.6\hat{f}/\hat{f}_m)^{8/3}} \tag{23.19}$$

for any stratification; here $\hat{f}_m$ is dimensionless frequency, which represent the value of $\hat{f}$ at the maximum of $\hat{f}\psi_{uw}(\hat{f})$ (only this frequency is directly dependent on stratification). The frequency $\hat{f}_m$ changes very little in all neutral and unstable conditions and may be taken under those conditions as 0.08, while under stable conditions it is greater (about 0.2 in moderate stabilities). However, the data of Zubkovskii and Koprov indicate a considerably stronger dependence of $\hat{f}_m$ on Ri (cf., Fig. 91). The function $\varkappa^{-1}\hat{f}\psi_{wT}(\hat{f}) = fE_{wT}(f)/\overline{w'T'}$ is fairly close to $\hat{f}\psi_{uw}(\hat{f})$ according to the Panofsky and Mares data and hence may be approximated by the same empirical expression. Let us note that the given expression implies that both cospectra $E_{uw}(k)$ and $E_{wT}(k)$ decrease as $k^{-8/3}$ at high wave numbers, i.e., much faster than $k^{-5/3}$.

Extensive measurement of the spectra of momentum and heat fluxes (and also of spectra of horizontal heat flux) under various stabilities were performed in the atmospheric surface layer during the 1968 Kansas expedition [see Kaimal, Wyngaard et al. (1972)]. The generalized flux spectra obtained have many features similar to those of the velocity and temperature spectra shown in Figs. 72a, 72b, and 85b; they also agree more or less satisfactorily with all previous data on turbulence cospectra. The $E_{uw}(k)$ and $E_{wT}(k)$ cospectra decrease as $k^{-7/3}$ in the high wave number range according to Kaimal, Wyngaard et al. data, while $E_{uT}(k)$ decreases approximately as $k^{-5/2}$. Both these decrease laws are considerably faster than the $k^{-5/3}$ law.

The free atmosphere data of Kucharets and Tsvang (1969) on the function $fE_{wT}(f)$ also show that this function can be approximated with moderate accuracy by an expression of the form $fE_{wT}(f) = Y_m\Phi(f/f_m)$ for all heights between 50 and 500 m above the steppe under typical mid-day summer conditions. Here $f_m$ is the frequency at the maximum of $fE_{wT}(f)$ and $Y_m$ is the maximum value of this function at $f = f_m$. The parameter $Y_m$ is rapidly decreasing with increase of height, while $f_m$ is also decreasing with height (the typical wavelength which contributes most effectively to turbulent heat flux under midday summer conditions increases from approximately 35 m at height $z$ = 2.5 m to approximately 1000 m at $z$ = 500 m). The exponent in the power law for the cospectra $E_{wT}(k)$ in the high wavenumber region is close to $-2.6$ according to Kucharets and Tsvang, i.e., it coincides with the value $-8/3$ which was recommended by Panofsky and Mares for the atmospheric surface layer.

There are no reasons to doubt that the cospectra of the absolute humidity and vertical velocity fluctuations $\vartheta'$ and $w'$ (i.e., the spectra of the turbulent flux of water vapor) must have the same general form as the cospectra $E_{uw}$ and $E_{wT}$. However the data on the spectra $E_{w\vartheta}$ are still extremely scarce; only a few preliminary results of this type were obtained by Chen and Mitsuta (1967), McBean (1968), Miyake and McBean (1970), Miyake, Donelan and Mitsuta (1970) and Elagina and Koprov (1971).

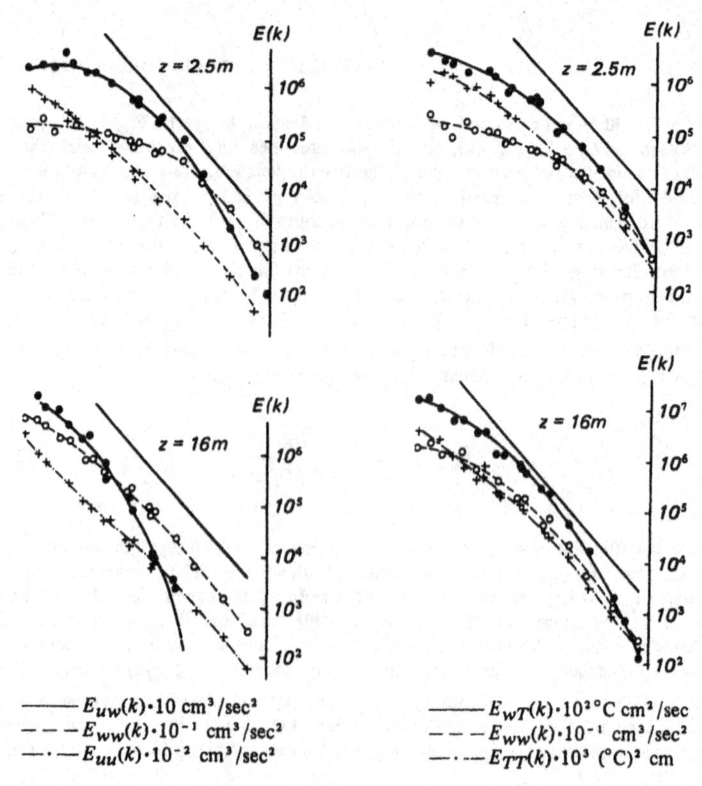


**Fig. 90** Examples of simultaneous data on spectra and cospectra of atmospheric turbulence (after Zubkovskii and Koprov, 1969).

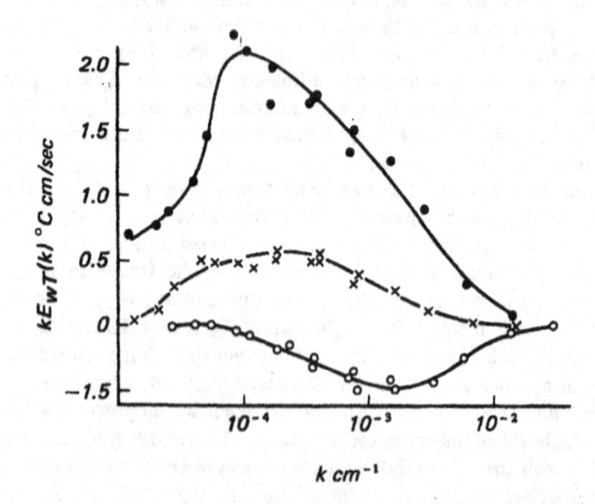


**Fig. 91** The forms of the function $kE_{wT}(k)$ at height 2.5 m for different thermal stratifications according to Zubkovskii and Koprov (1969). •–Ri = −0.08; x–Ri = −0.016; o–Ri = +0.1.

## *Data on the Low-Frequency Spectra of Meteorological Quantities*

The data of Figs. 71, 84, and 85a show the vertical velocity and temperature spectra $\omega E(\omega)$ for the atmospheric surface layer, which reflect the contributions of different frequency ranges to the total fluctuation variance. These spectra increase in proportion to $\omega^{-2/3}$ as the angular frequency $\omega$ decreases within the inertial subrange. After the lower limit of this subrange, which is of the form $\omega_l = \frac{\overline{u}}{z} \Omega_l(\mathrm{Ri})$ has been reached, the increase of the spectra slows down, and at certain frequencies $\omega_m = \frac{\overline{u}}{z} \Omega_m(\mathrm{Ri})$ (corresponding to periods $\tau_m = 2\pi/\omega_m$ of the order of 1 min) the spectra approach a maximum (occasionally referred to as the *micrometeorological* or *microscale maximum*). As the frequency decreases further, the spectra begin to fall again. According to the data of Fig. 91, and Eq. (23.19) the spectra of the turbulent heat and momentum fluxes behave in a similar way. Experimental data collected, for example, by Kolesnikova and Monin (1965) show that, for periods between a few minutes and several hours (referred to as the *mesometeorological* or *mesoscale range*), the spectra of most meteorological quantities have a broad and deep minimum, after which further decrease of frequency is accompanied by a rise of the spectra, reaching a maximum at about 4 days (this is the so-called *synoptic* or *macroscale maximum*), and then a further fall.

The main feature of the above behavior of the spectrum is the existence of the mesometeorological minimum, apparently first discovered by Panofsky and Van der Hoven (1955) and Van der Hoven (1957). Figure 92 shows the spectrum $fE_u(f)$ of the horizontal wind velocity at frequencies $f=\omega/2\pi$ between $7 \cdot 10^{-4}$ and 900 cycles/hour, deduced by Van der Hoven from experimental wind-velocity data obtained from the 125-meter meteorological tower at Brookhaven, USA. This spectrum shows a well-defined synoptic maximum at $\tau = 4$ days (the ordinate varies substantially between seasons), which is explained by the passage of large-scale atmospheric disturbances (responsible for weather changes), and the micrometeorological maximum at $\tau \approx 1$ min (the height of this maximum may vary during the day and from one day to another) due to small-scale turbulence of dynamic and convective origin (the small intermediate maximum at $\tau \approx 12$ h was

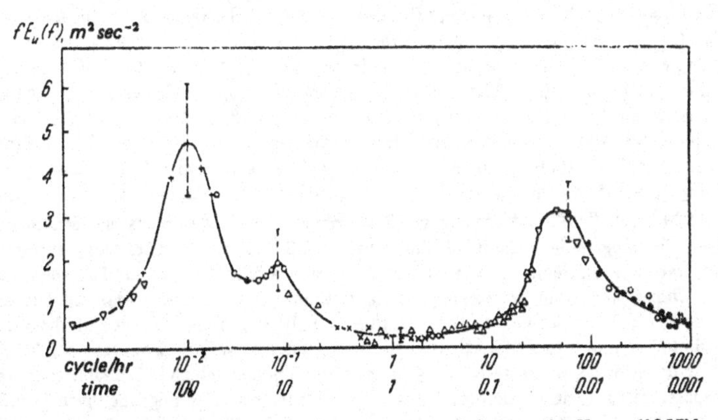


**Fig. 92** A plot of $fE_u(f)$ in a broad frequency range [after van der Hoven (1957)].

interpreted by Van der Hoven as insignificant). A mesometeorological minimum analogous to that in Fig. 92 was obtained by Panofsky and Van der Hoven as a result of measurements at heights between 30 and 125 m. The center of the minimum lay at periods $\tau = 7-60$ min (i.e., wavelengths $\lambda = \bar{u}\tau = 2.4-25$ km) and at its center $fE_u(f) = 0.1-0.2\ \text{m}^2\,\text{sec}^{-2}$. Other examples of velocity spectra in a wide frequency range were obtained by Byzova, Ivanov, and Morozov (1967) and by Oort and Taylor (1969). A discussion of these data can be found in the survey paper of Panofsky (1969a). In this paper a comparison of upper-air wind spectra (heights 3–20 km) with a surface-layer wind spectrum in the frequency range $10^{-3}-10^{5}$ cycles/day is also given. The main difference is that the ratios of the typical microscale energies to macroscale energies are much larger in the upper atmosphere. Additional data on the mesometeorological minimum of atmospheric spectra can be found in the paper by Fiedler and Panofsky (1970).

The low frequency spectra of the vertical wind velocities have apparently never been evaluated. Nevertheless it seems quite clear that such spectra would usually contain significant energy in the microscale range only and hence would have no definite mesoscale minimum and no peaks at larger scales.

A rather large number of works discuss the details of the microscale range of the velocity spectra beyond the low-frequency limit of the inertial subrange. In stable conditions it is quite often found that the spectral density begins to follow a power law with an exponent between $-2$ and $-3$ below the low-frequency limit of the inertial subrange (Shur, 1962; Vinnichenko, Pinus, and Shur, 1967; Vinnichenko, Pinus, Shmeter, and Shur, Busch, 1968; Myrup, 1968, 1969; Busch and Panofsky, 1968; Panofsky, 1969a). One possible explanation is the existence in such conditions of a so-called buoyant range of the spectrum which was briefly considered in Sects. 21.7 and 22.2 (with references to works of Bolgiano, Shur, Lumley and others). However we must bear in mind that the theoretical results concerning the existence of such a range and the slope of the spectra within it are far from being completely clear at present (cf. also the discussion of these data by Stewart, 1969).

The question of the form of the velocity spectra in the energy range has also been widely discussed in the literature. Here we shall only present the empirical expression of Busch and Panofsky (1968) for the vertical velocity spectra $E_{ww}(f)$ in the atmospheric surface layer. It has the form

$$\frac{fE_{ww}(f)}{u_*^2} = \hat{f}\psi_{ww}(\hat{f}) = \frac{1.075\,\hat{f}/\hat{f}_m}{1 + 1.5(\hat{f}/\hat{f}_m)^{5/3}} \qquad (23.20)$$

where $\hat{f} = fz/\bar{u}$ is the dimensionless frequency and $\hat{f}_m$ is the value of $\hat{f}$ at the maximum of $fE_{ww}(f)$. The parameter $\hat{f}_m$ is a function of Ri and can be approximately taken as 0.32 under neutral and moderately unstable conditions. The equation (23.20) is, of course, rather similar to the equation (23.19) for the spectra of turbulent fluxes and the empirical equations for $fE_{TT}(f)$ used in the construction of Fig. 85a. For more details about the energy range of the wind velocity spectrum we refer to papers by Busch and Panofsky (1968), Panofsky (1969a), and, especially, Kaimal, Wyngaard et al. (1972).

Figure 93 shows the spectrum $fE_T(f)$ of temperature fluctuations in the frequency range between 0.002 and 1000 cycles/hour obtained by Kolesnikova and Monin (1965) (from thermographic records in the range 0.002–0.1 cycles/hour, from approximate temperature measurements using a crude resistance thermometer in the range 0.1–15 cycles/hour, and from precise low-inertia resistance–thermometer measurements in the range 15–1000 cycles/hour). The spectrum clearly shows the presence of the synoptic maximum at $\tau \approx 4$ days (the height of this maximum may vary between seasons). There is a very clear strong spike corresponding to the diurnal period, which is natural because the air temperature has a clearly defined diurnal variation (strictly speaking, the diurnal period in the spectrum should correspond not to a narrow band but to an isolated discrete line). There is also a clear broad and deep mesometeorological minimum (centered on $\tau = 30$ min and corresponding to $fE_T(f) \approx 0.02°C^2$) and a relatively small micrometeorological maximum near $\tau = 1$ min with $fE_T(f) \approx 0.4°C^2$. The value of $fE_T(f)$ in the high-frequency part of the spectrum varies substantially from case to case. According to Palowchak and Panofsky (1968) (see also Panofsky, 1969), the synoptic maximum of $fE_T(f)$ occurs at periods varying from 4 to 20 days, depending on the climate of the observation point and the general weather conditions. Temperature spectra usually also have a definite gap between the synoptic range and the one-year peak.

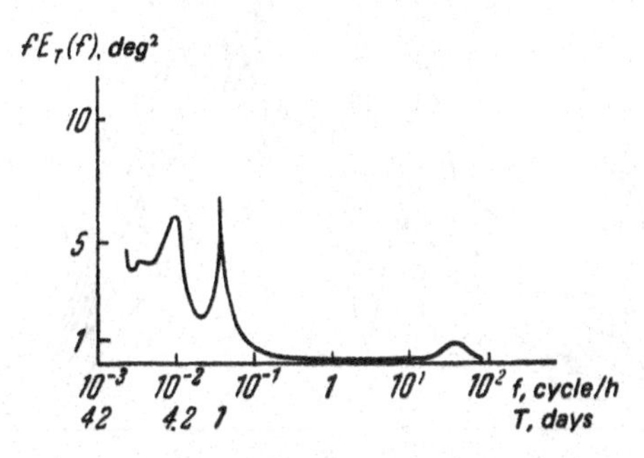


**Fig. 93** The function $fE_T(f)$, according to Kolesnikova and Monin (1965).

The spectrum of atmospheric pressure fluctuations looks quite different. One of the reasons for this difference may be that high-frequency pressure fluctuations may be generated both by turbulence and by the gravity and acoustic waves propagating through the atmosphere. According to the theoretical calculations of Monin and Obhukov (1958) and Dikii (1964, 1965), the spectra of gravity and acoustic waves practically do not overlap, since the former have periods largely in excess of 300 sec, while acoustic waves have periods below this value. This prediction has been confirmed by pressure spectra evaluated by Golitsyn (1964) from data of microbarographic recordings. The normalized mean pressure spectrum $\omega E_p(\omega)/\sigma_p^2$ in the range of periods between 500 and 10 sec, obtained by Golitsyn, is shown in Fig. 94 (the pressure fluctuation variance $\sigma_p^2$ is of the order of 0.01 mbar). This spectrum has a minimum at about $\tau = 300$ sec, i.e., precisely in the region where calculations predict the absence of both gravity and acoustic waves, and a maximum at $\tau \approx 50$ sec.

Figure 95 shows the atmospheric pressure spectrum in a much broader frequency range (between 0.001 and 1000 cycles/hour) calculated by Gossard (1960b) from microbarographic recordings augmented by recordings of fine-scale pressure fluctuations obtained with the aid of small microphones. The spectrum clearly shows the synoptic maximum at about $\tau = 4$ days, and strong spikes corresponding to the diurnal and semidiurnal periods (due to atmospheric tides), but the high-frequency fluctuations were found to be so small that the micrometeorologic maximum was very weak (its height is lower by four orders than the height of the synoptic maximum). At higher frequencies the spectrum falls by a further order of magnitude, so that throughout this frequency range the spectral density $E_p(f)$ changes by more than 5 orders of magnitude. The mesometeorological minimum is also only just seen in Fig. 95. More precisely, instead of the broad minimum for periods between a few minutes and a few hours, which is characteristic of Figs. 92 and 93, there is only a small minimum near 10 cycles/hour, which is the analog of the minimum in the Golitsyn spectrum and possibly corresponds to the interval between dominant frequencies of the gravity and acoustic waves. Between 1 and 10 cycles/hour the pressure spectrum sometimes shows the presence of peaks (two examples of this kind are shown in Fig. 95 by the broken

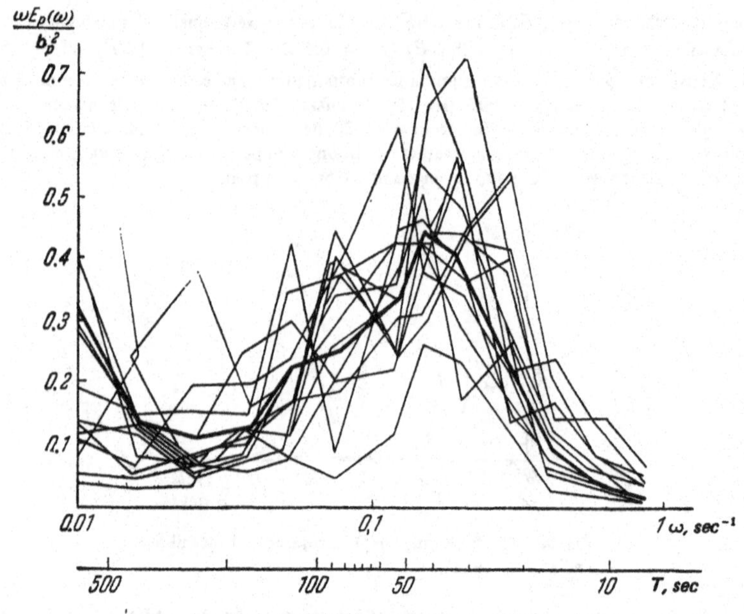


**Fig. 94** The function $\omega E_p(\omega)$ according to Golitsyn (1964).

lines). These are usually explained by the propagation of gravity waves with large amplitudes. This phenomenon is, however, relatively rare.

The mesometeorological minimum ("the gap") in the spectra of meteorological variables was also found by a number of other workers. We note, for example, the horizontal wind velocity spectra published by Davenport (1961), Walker (1964), Byzova, Ivanov, and Morozov (1967), Oort and Taylor (1969) and Panofsky (1969a). Kolesnikova and Monin (1965) have pointed out that the presence of the synoptic and micrometeorological maxima separated by the gap in the spectra of meteorological variables is apparently connected with the fact that the earth's atmosphere is relatively thin. Its effective thickness (e.g., the

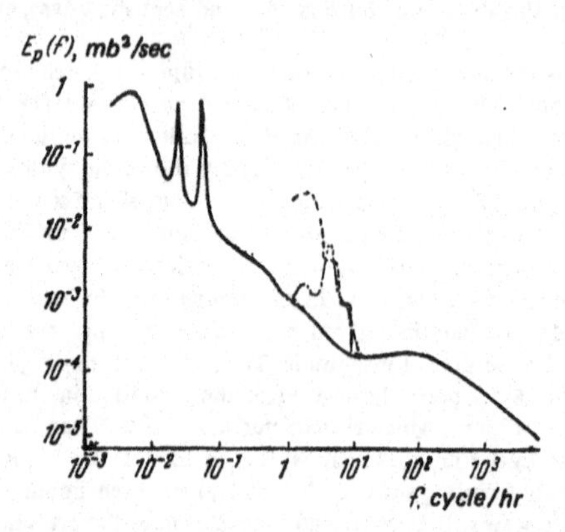


**Fig. 95** Pressure spectrum, according to Gossard (1960b).

thickness $H$ of the layer containing 80% of the mass, which is about 10 km) is small in comparison with its extent in the horizontal direction (which is determined by the diameter of the earth). Therefore, in the synoptic part of the spectrum, which involves space scales that are large in comparison with $H$, the inhomogeneities in meteorological fields are quasi-two-dimensional (horizontal), while in the micrometeorological region, involving scales that are small in comparison with $H$, they are essentially three-dimensional. The limiting scale $L \sim H$ corresponds to $\tau \approx 20$ min. according to the frozen-turbulence formula $\tau = L/\bar{u}$ for typical wind velocity $\bar{u} \approx 10$ m/sec. This is in moderate agreement with the position of the center of the mesometeorological minimum on the time spectra.

Data on the low-frequency region of the spectra of meteorological fields may be of interest in connection with the theory of locally isotropic turbulence because in the high-frequency region of the synoptic range one finds, unexpectedly, that relationships similar in form to those valid in the inertial subrange of the microturbulent spectrum are again valid. Thus, Richardson (1926) showed that the four-thirds law for the effective eddy diffusivity for an admixture cloud, which is now explained as a consequence of the second Kolmogorov similarity hypothesis for the inertial subrange (see Sect. 24.3) is, in fact, satisfied up to scales of the order of 100 or even 1000 km (we shall return to this in the next section). Syono and Gambo (1952) obtained the expression $E_1^{(p)}(k) \sim k^{-7/3}$ for the one-dimensional spatial spectrum of pressure fluctuations in a wavelength region smaller than 1/4 of the constant latitude circle, and this expression corresponds to the pressure spectrum in the inertial subrange. Similar results on the pressure spectrum were obtained by Inoue and Imai (1955) and some other workers [see, for example, references in Ogura (1958)]. The pressure spectrum found by Gossard (Fig. 95) is quite well represented by the formula $E_p(f) \sim f^{-7/3}$ in the synoptic and mesometeorological ranges. Ogura (1958) has analyzed the large-scale wind-velocity spectra found by Benton and Kahn (1958) and has obtained some evidence for the two-dimensional isotropy of large-scale macroturbulence. Hutchings (1955) found that the laws describing the behavior of time structure functions for the velocity and pressure fields in the inertial subrange of time lags are in some cases quite well satisfied for lag periods up to 1–2 days. Later Ellsaesser (1969b) examined the climatological data on wind variability and found that they are generally consistent with the predictions of the theory of locally isotropic turbulence (in the frozen turbulence approximation) for time lags up to 6 hr; these lags are shorter than that given by Hutchings, but they are also much larger than the lags for which three-dimensional isotropy can exist. Finally, there are numerous data showing that velocity and temperature five-thirds laws are often approximately valid in the atmosphere for wave number ranges which correspond to far too great length scales for isotropy to be possible [many examples of this kind may be found in the works of Pinus (1966, 1969), Vinnichenko, Pinus, Shmeter, and Shur (1969) and Vinnichenko and Dutton (1969)]; the situation with oceanic turbulence is apparently quite similar to this.

All these data can be explained by supposing that there is the usual cascade energy transfer along the length scale spectrum toward smaller scales in the synoptic range for scales that are small in comparison with the size of the continents, oceans, and large-scale cloud systems, i.e., in comparison with the length scales of the main inhomogeneities in the external energy influx (which are the same as the length scales in the region of which there is intensive direct conversion of solar energy into macroscale turbulence energy). Moreover it is natural to suppose that this energy transfer is not accompanied in this scale range by a substantial external energy influx or substantial dissipation. By energy dissipation we now of course understand not the direct conversion of kinetic energy into heat by molecular viscosity, which cannot be of any significance in this range, but the conversion of macroscale turbulence energy directly into microscale turbulence energy without involvement of the intermediate scales. Therefore, instead of energy losses associated with the molecular viscosity $\nu$ we must now consider energy losses associated with the eddy viscosity $\nu_{turb} = K$ (corresponding, say, to turbulence associated with the micrometeorological maximum). Let the specific rate of energy transfer along the macroturbulence spectrum be denoted by $\bar{\varepsilon}$ (this should not differ from the energy dissipation $\bar{\varepsilon}$ which plays the analogous role in the microturbulent range, since in the final analysis all the macroturbulent energy is converted into microturbulent energy). We can then construct "the internal length scale" $\eta_M = \left(\nu_{\text{turb}}^3/\bar{\varepsilon}\right)^{1/4}$, and in the range $L_M \gg l \gg \eta_M$ (where $L_M$ is the length scale of disturbances associated with most of the external energy influx, and is apparently of the order of 1000 km) we can apply the usual Kolmogorov similarity hypotheses, remembering only that the corresponding locally-isotropic turbulence will be two- rather than three-dimensional. Let us now note that the generation of microturbulence by synoptic formations occurs mainly as a result of instability of vertical inhomogeneities in the wind-velocity and temperature fields with scales $L_m$ that are small in comparison with the effective thickness $H$ of the atmosphere. The decay of synoptic formations will therefore lead mainly to the appearance of disturbances with scales comparable with $L_m$, so that the scale $L_m$ should correspond to the micrometeorologic maximum in the spectrum. The presence of the mesometeorologic minimum appears to signify that $\eta_M > L_m$. In fact, when this condition is satisfied, there would be a sharp decrease of the spectral density for $l \sim \eta_M$ (which is analogous to the situation in the microturbulent range for $l \sim \eta$), and this explains the gap between the synoptic range and the maximum at $l \sim L_m$. If we recall that the inequality $\eta_M > L_m$ can be rewritten in the form

$$\nu_{\text{turb}} > \bar{\varepsilon}^{1/3} L_m^{4/3}, \tag{23.21}$$

and if we substitute as typical values $\bar{\varepsilon} \approx 1\ \text{cm}^2 \cdot \text{sec}^{-3}$ and $L_m \approx 100$ m, we find that the right-hand side of this inequality is of the order of a few tens of m²/sec, while the value of $\nu_{\text{turb}}$ estimated from the variation of wind velocity with height in the planetary boundary layer usually yields 100 m²/sec. Therefore, the proposed explanation of the position of the mesometeorological minimum in the atmospheric turbulence spectra [established by Kolesnikova and Monin (1965)] is not inconsistent with existing data.

It is also possible, that turbulent flows exist in which there are more than two spectral ranges characterized by an appreciable influx of external energy, such that in the gaps between them both the energy influx and dissipation play a minor role (and, consequently, the Kolmogorov hypotheses are valid so that an inertial subrange becomes observable). In particular, Ozmidov (1965) has shown that in the spectra of oceanic turbulence one should expect the presence of three ranges of energy influx: At global scales $L_0 \approx 1000$ km, inertial and tidal scales $L_1 \approx 10$ km, and wind-wave scales $L_2 \approx 10$ m. The associated inertial subranges are characterized by different mean values $\bar{\varepsilon}$ since, in contrast to the above mentioned atmospheric turbulence situation, the small-scale maxima now have their own

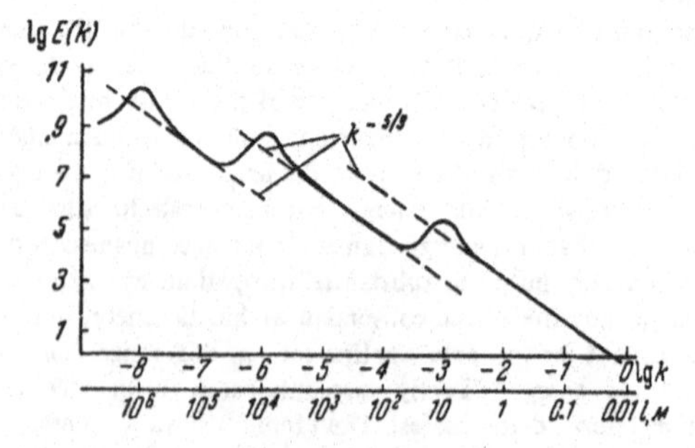


**Fig. 96** Schematic shape of the oceanic turbulence spectrum [Ozmidov (1965)].

energy influx sources which are added to the energy influxes from the large-scale regions of the wave-number spectrum. The corresponding oceanic turbulence spectra are shown schematically in Fig. 96 which is taken from Ozmidov's paper. We must also note that it is quite probable that similar effects can sometimes be exhibited in atmospheric turbulence spectra too. In fact there are data which show the existence of atmospheric wind velocity spectra having two five-thirds ranges with different proportionality constants divided by spectral regions where spectra have a more complicated shape (see, e.g., Pinus and Shcherbakova, 1966, or Vinnichenko, Pinus, Shmeter, and Shur, 1968). More detailed discussion of this question may be found in the paper of Pinus (1969).

## 24. DIFFUSION IN LOCALLY ISOTROPIC TURBULENCE

### 24.1 Diffusion in Isotropic Turbulence. Statistical Characteristics of the Motion of a Fluid Particle

The diffusion of admixtures in turbulent flows, and the closely related problem of the Lagrangian statistical characteristics of turbulence, were considered in Chapt. 5 of Vol. 1. We must now return to this problem again in order to consider new approaches which have become possible in view of the material which we have presented in this volume.

Let us begin with the simple case of diffusion in a stationary homogeneous turbulence field. We recall the classical Taylor formula for the covariance tensor

$$D_{ij}(\tau) = \overline{[Y_i(\tau) - \overline{Y_i(\tau)}]\,[Y_j(\tau) - \overline{Y_j(\tau)}]} = \overline{Y'_i(\tau)Y'_j(\tau)}$$

of the fluid particle displacement vector $\boldsymbol{Y}(\tau) = \{Y_1(\tau), Y_2(\tau), Y_3(\tau)\}$ in a time $\tau$:

$$D_{ij}(\tau) = \int_0^\tau (\tau - s)\left[B_{ij}^{(L)}(s) + B_{ji}^{(L)}(s)\right] ds \qquad (24.1)$$

[see Eq. (9.30′)]. In this expression $B_{ij}^{(L)}(s) = \overline{V'_i(\boldsymbol{x}, t+s)V'_j(\boldsymbol{x}, t)}$ is the Lagrangian correlation tensor for the velocity $\boldsymbol{V}(\boldsymbol{x}, t)$ of a fluid particle located at a fixed point $\boldsymbol{x}$ at the initial time $t = t_0$. In the stationary case the tensor $B_{ij}^{(L)}(s)$ has a spectral representation of the form (11.32). Let us substitute this into Eq. (24.1), denote by $F_{ij}^{(L)}(\omega)$ the spectral density tensor of the Lagrangian velocity, and remember that $F_{ij}^{(L)}(\omega) + F_{ji}^{(L)}(\omega) = E_{ij}^{(L)}(\omega)$ is an even function of $\omega$; then we readily obtain from Eq. (24.1) the following spectral form of the Taylor formula [first established by Kampé de Feriet (1939) and Batchelor

(1949b)]:

$$D_{ij}(\tau) = 4\int_0^\infty E_{ij}^{(L)}(\omega)\frac{\sin^2(\omega\tau/2)}{\omega^2}\,d\omega. \tag{24.2}$$

Hence it is clear that, for example, the variances of the displacement coordinates $D_{ll}(\tau)=\overline{[Y_l'(\tau)]^2}$ (no summation over $i$) for any $\tau$ can be obtained by integrating the same Lagrangian spectra $E_{ll}^{(L)}(\omega)$ multiplied by the weighting function $4\sin^2\left(\frac{\omega\tau}{2}\right)/\omega^2$ which is dependent on $\tau$. This dependent weighting function decreases as $\omega^{-2}$ when $\omega\to\infty$. When $\tau$ is sufficiently small, so that $\omega\ll\tau^{-1}$ for all frequencies $\omega$ contributing appreciably to $\overline{V_i'V_j'}$, we have from Eq. (24.2)

$$D_{ij}(\tau)\approx\tau^2\int_0^\infty E_{ij}^{(L)}(\omega)\,d\omega \tag{24.3}$$

which is a special case of Eq. (9.28). In this case, the entire spectrum of Lagrangian frequencies contributes to the variance of the displacement, and frequencies from the "energy range" play the leading role in this process. As $\tau$ increases there is evidently an increase in the relative contribution of lower frequencies. When $\tau$ is very large, so that $E_{ij}(\omega)$ is practically constant in an interval of length $\Delta\omega\sim\tau^{-1}$ near $\omega=0$, which makes the principal contribution to the integral $\int_0^\infty \frac{\sin^2\omega\tau/2}{\omega^2}\,d\omega=\pi\tau/4$, we have

$$D_{ij}(\tau)\approx\pi E_{ij}^{(L)}(0)\,\tau. \tag{24.4}$$

This is identical with Eq. (9.35). The quantity $D_{ij}(\tau)$ is now determined only by the lowest frequencies in the Lagrangian velocity spectrum. However, the quantities $D_{ij}(\tau)$ are never determined by high frequencies alone. The Kolmogorov theory which is valid only for sufficiently high frequencies cannot, therefore, be used to calculate the variances of the displacements of diffusing particles.

In the case of diffusion in isotropic turbulence, the values of components of the tensor $D_{ij}(\tau)$ and other statistical characteristics of a particle cloud emitted by an instantaneous source at the point $\boldsymbol{x}$ at time $t_0$ can be approximately calculated by using the closure methods for the moment equations discussed in Sect. 19. In fact, the concentration field $\vartheta(X, t)$ (as in Chapt. 5 we shall now denote running coordinates by capital letters and reserve lower-case letters for the initial coordinates of the particle) will satisfy the usual diffusion equation

$$\frac{\partial \vartheta}{\partial t} + u_j \frac{\partial \vartheta}{\partial X_j} = \chi \Delta \vartheta \tag{24.5}$$

and the initial condition

$$\vartheta(\boldsymbol{X}, t_0) = Q\delta(\boldsymbol{X} - \boldsymbol{x}), \tag{24.6}$$

where $Q$ is the total amount of admixture. Molecular diffusion (including the interaction between molecular and turbulent diffusion which we discussed in Sect. 10.2) can often be neglected in Eq. (24.5) in comparison with turbulent diffusion, i.e., we can set $\chi = 0$. However, independently of whether or not we assume that $\chi$ is zero, Eq. (24.5) together with the Navier-Stokes equations and the continuity equation for the velocity field can be used to set up an infinite chain of coupled equations for the moments and the joint moments of the fields $\vartheta(\boldsymbol{X}, t)$ and $\boldsymbol{u}(\boldsymbol{X}, t)$. This infinite set will, of course, be the same as the set of equations for the moments of the velocity and temperature fields in the case of isotropic turbulence, except that now the field $\vartheta(\boldsymbol{X}, t)$ must be regarded as nonisotropic and invariant only under rotations about a given point $\boldsymbol{x}$, and must satisfy the initial condition Eq. (24.6). If we truncate our infinite system after any finite number of equations and apply one of the closure methods discussed in Sect. 19, then having solved the remaining equations subject to the initial conditions ensuing from Eq. (24.6) (and the initial conditions for the moments of $\boldsymbol{u}(\boldsymbol{X}, t)$ which describe the statistical properties of the velocity field at time $t_0$), we obtain in particular an approximate value for the normalized first moment $\overline{\vartheta(\boldsymbol{X}, t)}/Q = p(\boldsymbol{X}, t)$ which can be interpreted as the probability density for the coordinates $\boldsymbol{X}$ of the diffusing particles. If we then use the probability $p(\boldsymbol{X}, t)$ (which depends on the parameters $\boldsymbol{x}$ and $t_0$) we can estimate the elements of the covariance tensor

$$D_{ij}(\tau) = \int X_i X_j \, p(\boldsymbol{X}, t_0 + \tau) \, d\boldsymbol{X} \tag{24.7}$$

(the mean values $\bar{X}_j$ are all zero in the case of isotropic turbulence) and other statistical parameters of the cloud of diffusing particles. This approach to the diffusion problem in the case of isotropic turbulence was used among others by Deissler (1961), Roberts (1957, 1961), Kraichnan (1962, 1965a, 1966b, 1970b), and Saffman (1969). The method put forward by Deissler for solving the diffusion problem during the final stage of decay of isotropic turbulence is based on the simplest method of closure of the moment equations, which involves neglecting all moments of order higher than two. Deissler also carried out more complicated calculations, taking into account some third-order moments but also introducing certain new quite crude approximate hypotheses. The resulting adequate agreement with the data of Uberoi and Corrsin is probably fortuitous. Roberts (1957) used the generalized Millionshchikov hypothesis to close the equations for the moments of $\vartheta(\boldsymbol{X}, t)$ and $\boldsymbol{u}(\boldsymbol{X}, t)$. By equating to zero all the fourth-order cumulants of these fields he obtained a linear integro-differential equation which, in principle, determined the probability density $p(\boldsymbol{X}, t)$. In a subsequent paper, Roberts (1961) applied Kraichnan's direct-interaction approximation (1959, 1961) to the closure of the moment equations. If we assume that $t_0 = 0$ and $\boldsymbol{x} = 0$, this method leads to the following linear integro-differential equation

$$\left(\frac{\partial}{\partial t} - \chi\Delta\right) p(\boldsymbol{X}, t) =$$

$$= \int_0^t dt' \int d\boldsymbol{X}' B_{lj}(\boldsymbol{X} - \boldsymbol{X}', t, t') \frac{\partial p(\boldsymbol{X} - \boldsymbol{X}', t - t')}{\partial X_l} \frac{\partial p(\boldsymbol{X}', t')}{\partial X'_j}, \qquad (24.8)$$

which contains the Eulerian space-time correlation function $B_{lj}(\boldsymbol{X} - \boldsymbol{X}', t, t') = \overline{u_l(\boldsymbol{X}, t) u_j(\boldsymbol{X}', t')}$ of the velocity field [Roberts confined his attention to the special case of Eq. (24.8) with $\chi = 0$]. Related nonlinear integro-differential equations containing the Lagrangian space-time correlation function are obtained by using the more complicated Lagrangian-history direct-interaction approximation of Kraichnan (1965a, 1966a, b). Kraichnan (1962) has reviewed various closure methods for the moment equations in the model problem involving the diffusion of an admixture by a random velocity field $\boldsymbol{u}(\boldsymbol{X})$ which was assumed isotropic, Gaussian, and time-independent (so that $B_{lj}(\boldsymbol{X} - \boldsymbol{X}', t, t')$ was independent of $t$ and $t'$). When all moments of order higher than two are neglected in this problem, the resulting function $p(X, t)$ is nonzero only in a small neighborhood of the point $x$ and, moreover, takes on unphysical negative values. Neglecting fourth-order cumulants turns out to be slightly more successful in some respects but again leads to a function $p(X, t)$ with negative parts. The direct-interaction approximation equation (24.8) leads to a strictly nonnegative probability but the shape of the admixture distribution turns out to be quite peculiar and corresponds apparently to a rather crude description of the real diffusion process. This last conclusion does not agree well, however, with the results of Kraichnan's more recent work (1970b). Here the diffusion of the fluid particles (having zero molecular diffusivity) by a random velocity field $u(X, t)$ was studied for the case when $u(X, t)$ is a solenoidal (i.e., incompressible), homogeneous, isotropic, statistically stationary (but time-dependent) Gaussian field in both two and three dimensions. Such a field is fully specified by its longitudinal space-time correlation function $B_{LL}(r, \tau)$ [or the function $B_0(r, \tau) = \overline{u(X, t)u(X + r, t + \tau)} = B_{LL}(r, \tau) + 2B_{NN}(r, \tau)$] and Kraichnan considered four particular examples of the time-space correlation function (two for a three-dimensional case and two for a two-dimensional one; the parameters of the correlation function varied in the course of the calculations). Two methods of calculation were used and compared with one another: Numerical solution of the approximate direct-interaction equation (24.8) and direct computer simulation of the diffusion process in which a random Gaussian velocity field $u(X, t)$ with given space-time correlation function was constructed numerically with the aid of computer produced pseudo-random uniformly distributed numbers. Two thousand realizations of the diffusion process by a constructed field $u$ were created and theh the ensemble averages over all 2000 realizations of the particle trajectory were compared with the values of the corresponding mean statistical diffusion quantities (Lagrangian velocity correlation function, particle displacement variance and effective eddy diffusivity) deduced from the approximate direct-interaction theory. The comparison showed that direct-interaction approximations for all the parameters agree well with the computer experiments except for some particular two-dimensional situations where special effects are responsible for the disagreement.

Finally let us note Saffman's attempt (1969) to apply the Wiener-Hermite expansion method which was sketched briefly at the end of Sect. 19.3 to the problem of diffusion from a point source in a field of isotropic turbulence. Saffman assumed that the isotropic velocity field $u(X, t)$ is Gaussian and therefore can be exactly represented by a two-term Wiener-Hermite expansion. Moreover he truncated the Wiener-Hermite expansion of the concentration field $\vartheta(X, t)$ on the first two terms also, i.e., he used an approximation which is based on retaining only the Gaussian part of the concentration field. Such an approximation can be easily formulated without any use of the Wiener-Hermite expansion: If we write $\vartheta(X, t) = \bar{\vartheta}(X, t) + \vartheta'(X, t)$ where $\bar{\vartheta}$ is a nonrandom mean concentration and $\vartheta'$ is a random concentration fluctuation and substitute this in the diffusion equation (24.5), then the Gaussian approximation of $\vartheta'(X, t)$ will correspond to the use of the exact equation for $\bar{\vartheta}(X, t)$, but the neglect of the nonlinear functions of the fluctuations $u_j$ and $\vartheta'$ of the form $u_j \partial\vartheta'/\partial X_j - \overline{u_j \partial\vartheta'/\partial X_j}$ in the equation for $\vartheta'(X, t)$. When the effect of molecular diffusivity is also neglected (i.e., it is assumed that $\chi = 0$) the approximation leads to a simple linear integro-differential equation for the probability $p(X, t)$:

$$\frac{\partial p(X, t)}{\partial t} = \int_0^t dt' B_{ij}(t - t') \frac{\partial^2 p(X, t)}{\partial X_i \partial X_j}$$

where $B_{ij}(t - t') = \overline{u_i(X, t) u_j(X, t')}$ is the Eulerian time correlation function of the velocity field (which is supposed here to be stationary for the sake of simplicity). The solution of this equation has a plausible form for large time values (much greater than the Eulerian integral time scale), but for small $t$ it is physically unreasonable (the mean concentration distribution is of a very strange shape and takes both positive and negative values).

We have seen that the covariance tensor $D_{ij}(\tau)$ cannot be determined from the theory of locally isotropic turbulence for any $\tau$. However, the Lagrangian velocity structure tensor $D_{ij}^{(L)}(\tau) = \overline{\Delta_\tau V_i \Delta_\tau V_j}$ for $\tau \ll T_0 = L/U$ can be described with the aid of the Kolmogorov similarity hypotheses (see Sect. 21.4). Therefore, the difference

$$B_{ij}(0)\tau^2 - D_{ij}(\tau) = \int_0^\tau (\tau - s) D_{ij}^{(L)}(s)\, ds \qquad (24.9)$$

for $\tau \ll T_0$ is a local statistical quantity which is of universal form for all turbulent flows with sufficiently high Reynolds number.

Another general method of obtaining the statistical characteristics of fluid particle motion, which involves the theory of locally isotropic turbulence, is to transform from a fixed set of coordinates $\mathscr{S}_0$ to a moving inertial system $\mathscr{S}$, the velocity of the latter being $\boldsymbol{u}(\boldsymbol{x}, t_0)$ (different for different realizations of the turbulence) whose origin at time $t=t_0$ is at the point $\boldsymbol{x}$. The coordinates $\boldsymbol{Y}^{(s)}$ and velocities $\boldsymbol{V}^{(s)}$ in the system $\mathscr{S}$ are related to the coordinates $\boldsymbol{x}$ and velocities $\boldsymbol{u}$ in the original system $\mathscr{S}_0$ by the simple relations $\boldsymbol{Y}^{(s)}=\boldsymbol{X}-\boldsymbol{x}-\boldsymbol{u}(\boldsymbol{x}, t_0)\tau$, $\boldsymbol{V}^{(s)}=\boldsymbol{u}-\boldsymbol{u}(\boldsymbol{x}, t_0)$, where $\tau=t-t_0$. A fluid particle which is located at the point $\boldsymbol{x}$ at time $t_0$ is found at the point $\boldsymbol{Y}^{(s)}(\tau)=\boldsymbol{X}(\boldsymbol{x}, t)-\boldsymbol{x}-\boldsymbol{u}(\boldsymbol{x}, t_0)\tau$ at time $t_0+\tau$ and has a velocity $\boldsymbol{V}^{(s)}(\tau)=\boldsymbol{V}(\boldsymbol{x}, t_0+\tau)-\boldsymbol{u}(\boldsymbol{x}, t_0)=\Delta_\tau \boldsymbol{V}$. Since the statistical characteristics of the field $\boldsymbol{V}^{(s)}(\tau)=\Delta_\tau \boldsymbol{V}$ must obey the Kolmogorov similarity hypotheses, and $\boldsymbol{Y}^{(s)}(\tau)=\int_0^\tau \boldsymbol{V}^{(s)}(\tau)\,d\tau$, all the statistical characteristics of the fluid particle motion in the system $\mathscr{S}$ have a universal form governed by $\bar{\varepsilon}$ and $\nu$ alone for $\tau \ll T_0$, provided Re is sufficiently large. Such parameters can be called the local parameters of the fluid particle motion. In particular,

$$\overline{V_i^{(s)}(\tau)\,V_j^{(s)}(\tau)}=D_{ij}^{(L)}(\tau)=D^{(L)}(\tau)\,\delta_{ij}, \tag{24.10}$$

$$\overline{V_i^{(s)}(\tau_1)\,V_j^{(s)}(\tau_2)}=\frac{1}{2}\left[D^{(L)}(\tau_1)+D^{(L)}(\tau_2)-D^{(L)}(|\tau_1-\tau_2|)\right]\delta_{ij} \tag{24.10'}$$

where the second formula is a consequence of the first and of the identity (13.33). Hence, it follows that

$$\begin{aligned}\overline{Y_i^{(s)}(\tau)\,V_j^{(s)}(\tau)}&=\int_0^\tau \overline{V_i^{(s)}(\tau_1)\,V_j^{(s)}(\tau)}\,d\tau_1=\frac{1}{2}\tau\,D^{(L)}(\tau)\,\delta_{ij},\\ \overline{Y_i^{(s)}(\tau)\,Y_j^{(s)}(\tau)}&=\int_0^\tau\int_0^\tau \overline{V_i^{(s)}(\tau_1)\,V_j^{(s)}(\tau_2)}\,d\tau_1\,d\tau_2=\delta_{ij}\int_0^\tau \tau' D^{(L)}(\tau')\,d\tau'.\end{aligned} \tag{24.11}$$

For very small $\tau \ll \tau_\eta$ we can set $D^{(L)}(\tau)=a_0\bar{\varepsilon}^{3/2}\nu^{-1/2}\tau^2=\frac{1}{3}\overline{A^2}\tau^2$, where $\overline{A^2}=3a_0\bar{\varepsilon}^{3/2}\nu^{-1/2}$ is the mean square acceleration of the fluid particle. Consequently,

$$\overline{V_i^{(s)}(\tau)\,V_j^{(s)}(\tau)} \approx \frac{1}{3}\overline{A^2}\tau^2\delta_{ij}, \quad \overline{Y_i^{(s)}(\tau)\,V_j^{(s)}(\tau)} \approx \frac{1}{6}\overline{A^2}\tau^3\delta_{ij},$$
$$\overline{Y_i^{(s)}(\tau)\,Y_j^{(s)}(\tau)} \approx \frac{1}{12}\overline{A^2}\tau^4\delta_{ij}. \tag{24.12}$$

for $\tau \ll \tau_\eta$. We see that, when $\tau \ll \tau_\eta$, the correlation coefficient between $Y_i^{(s)}(\tau)$ and $V_i^{(s)}(\tau)$ is unity, and the "relative displacement" variance $\overline{Y_i^{(s)}(\tau)\,Y_i^{(s)}(\tau)}$ is very much less than the variance of the total displacement $Y_i(\tau)$ [which for small $\tau$ is of the order of $\tau^2$ in view of Eq. (24.3)]. The last result is explained by the fact that the variance of the total displacement for small $\tau$ is determined largely by fluctuations in the initial particle velocity $\boldsymbol{u}(\boldsymbol{x}_0, t_0)$ which are eliminated by transformation to the moving system $\mathscr{S}$. Moreover for values of $\tau$ in the inertial subrange $\tau_\eta \ll \tau \ll T_0$ we can use Eq. (21.30′) for $D^{(L)}(\tau)$, from which it follows that, in this case,

$$\overline{V_i^{(s)}(\tau)\,V_j^{(s)}(\tau)} = C_0\bar{\varepsilon}\tau\delta_{ij}, \quad \overline{Y_i^{(s)}(\tau)\,V_j^{(s)}(\tau)} = \frac{1}{2}C_0\bar{\varepsilon}\tau^2\delta_{ij},$$
$$\overline{Y_i^{(s)}(\tau)\,Y_j^{(s)}(\tau)} = \frac{1}{3}C_0\bar{\varepsilon}\tau^3\delta_{ij}. \tag{24.13}$$

Therefore, for $\tau_\eta \ll \tau \ll T_0$ the correlation coefficient between $Y_i^{(s)}(\tau)$ and $V_i^{(s)}(\tau)$ is equal to $\sqrt{3}/2 = 0.866\ldots$. The above formulas are very similar to Eqs. (9.58) and (9.59) of Vol. 1, which describe the variances of displacement along different coordinate axes in the case of turbulent diffusion in a homogeneous shear flow with constant mean velocity gradient. An explanation of this similarity is given in Sect. 24.4.

The theory of locally isotropic turbulence enables us to establish the general form of the probability distributions for the parameters of a fluid particle motion in the moving system $\mathscr{S}$. Thus, for example, the probability density for the vectors $\boldsymbol{Y}^{(s)} = \boldsymbol{Y}^{(s)}(\tau)$ and $\boldsymbol{V}^{(s)} = \boldsymbol{V}^{(s)}(\tau)$ when $\tau \ll T_0$ can depend only on the arguments $Y^{(s)} = |\boldsymbol{Y}^{(s)}|$, $V^{(s)} = |\boldsymbol{V}^{(s)}|$ and $\boldsymbol{Y}^{(s)}\boldsymbol{V}^{(s)}$ (because it is invariant under rotations of $\mathscr{S}$) and on the parameters $\tau$, $\bar{\varepsilon}$, and $\nu$. Dimensional considerations then lead us to conclude that this density should have the form

$$p(\boldsymbol{Y}^{(s)}, \boldsymbol{V}^{(s)}) = (\eta v_\eta)^{-3}\,P^{(1)}\left(\frac{Y^{(s)}}{\eta}, \frac{\boldsymbol{Y}^{(s)}\boldsymbol{V}^{(s)}}{\eta v_\eta}, \frac{V^{(s)}}{v_\eta}; \frac{\tau}{\tau_\eta}\right), \tag{24.14}$$

where $P^{(1)}$ is a universal function of the four variables. For $\tau$ in the inertial subrange $\tau_\eta \ll \tau \ll T_0$ the dependence of the probability density (24.14) on $\nu$ is unimportant and, therefore, for such values of $\tau$

$$p\left(\mathbf{Y}^{(s)}, \mathbf{V}^{(s)}\right) = \left(\bar{\varepsilon}\tau^2\right)^{-3} P^{(2)}\left(\frac{\mathbf{Y}^{(s)}}{\left(\bar{\varepsilon}\tau^3\right)^{1/2}}, \frac{\mathbf{Y}^{(s)}\mathbf{V}^{(s)}}{\bar{\varepsilon}\tau^2}, \frac{\mathbf{V}^{(s)}}{\left(\bar{\varepsilon}\tau\right)^{1/2}}\right), \quad (24.15)$$

where $P^{(2)}$ is a universal function of the three variables. If we integrate this formula over all $\mathbf{V}^{(s)}$, we can verify that the probability density for $\mathbf{Y}^{(s)}$ is

$$p\left(\mathbf{Y}^{(s)}\right) = \left(\bar{\varepsilon}\tau^3\right)^{-3/2} P^{(3)}\left(\frac{\mathbf{Y}^{(s)}}{\bar{\varepsilon}^{1/2}\tau^{3/2}}\right), \quad (24.16)$$

where $P^{(3)}$ is a universal function of the single variable, and so on.

According to Eq. (24.2), the values of $D_{ij}(\tau) = \overline{Y'_i(\tau)\,Y'_j(\tau)}$ are affected by the turbulent disturbances of all frequencies which contribute to fluctuations in the Lagrangian velocity. Consequently, to obtain a reliable estimate of $\overline{Y'_i(\tau)\,Y'_j(\tau)}$ through time averaging of the data, the averaging period must substantially exceed the periods of all the oscillations contributing appreciably to $\overline{[V'(x, t)]^2}$. This last point becomes particularly restrictive in the case of atmospheric or oceanic turbulence whose spectra extend far into the long-period range and practically always include periods exceeding the time of observation. As a result, the variance of the particle displacements in the case of diffusion in the atmosphere (or in an ocean) will always depend on the period of observation $T$ and will increase with increasing $T$. If, for example, we perform observations on an admixture issuing from a stationary source, we see that the admixture particles emitted in a finite interval $T$ form a jet elongated in the direction of the mean flow velocity $\bar{u}$ but containing a number of irregular inflections, due to the long-period oscillations in the Lagrangian velocity (its limits are schematically indicated by the broken curves in Fig. 97). Particles emitted during a different time interval of the same duration $T$ form a different jet of roughly the same thickness but with different inflections. If we take the average of the data over a very long period, the supposition of a large number of such jets results in the broad cone shown by the solid lines in Fig. 97. The width of this cone at a distance $x$ from the source is directly related to the variance $D_{22}(x/\bar{u})$ given by Eq. (24.2). The mean width of the individual jets

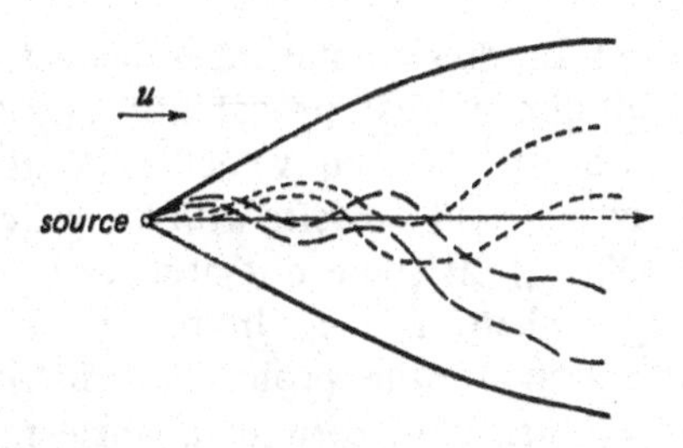


**Fig. 97** Schematic drawing of admixture jets in a turbulent flow.

at distance $x = \bar{u}\tau$ from the source cannot be, however, calculated from Eq. (24.2) and must be compared with the quantity

$$D_{22}^{(T)}(\tau) = \overline{\left[Y_2(t_0, \tau) - \frac{1}{T}\int_{-T/2}^{T/2} Y_2(t_0+s, \tau)\,ds\right]^2} = \overline{[Y_2^{(T)}(t_0, \tau)]^2}, \quad (24.17)$$

where $Y_l(t_0, \tau)$ is the displacement in a time $\tau$ of a fluid particle leaving the origin at time $t_0$. It is natural to expect that all the statistical characteristics of the vector

$$Y^{(T)}(t_0, \tau) = Y(t_0, \tau) - \frac{1}{T}\int_{-T/2}^{T/2} Y(t_0+s, \tau)\,ds$$

will depend only on disturbances with periods shorter than $T$ and, consequently, when $T$ is not too long, they will be determined only by turbulent disturbances satisfying the Kolmogorov hypotheses. Hence, it follows that when $\tau_\eta \ll T \ll T_0, \tau_\eta \ll \tau \ll T_0$ we have

$$D_{ij}^{(T)}(\tau) = \bar{\varepsilon}\,T^3 f(\tau/T)\,\delta_{ij}, \quad (24.18)$$

where $f$ is a universal function so that

$$D_{ij}^{(\tau)}(\tau) = C_0'\,\bar{\varepsilon}\,\tau^3\,\delta_{ij}, \quad C_0' = f(1). \quad (24.18')$$

The similarity between Eq. (24.18') and the last formula in Eq. (24.13) is explained by the fact that the quantities $D_{ij}^{(\tau)}(\tau)$ and $\overline{Y_i^{(s)}(\tau)\,Y_j^{(s)}(\tau)}$ have the same dimensions and depend on the same dimensional parameters.

A more detailed study of the statistical characteristics of the vector $Y^{(T)}(t_0, \tau)$ requires a knowledge of the statistical characteristics of the motion of a pair of fluid particles leaving a fixed point at different times. All the existing theoretical calculations of $D_{ij}^{(T)}(\tau)$ [see, for example, Ogura (1957, 1959) and Shimanuki (1961)] are based on one or another nonrigorous hypothesis, and lead to only approximate results which we shall not consider here.

## 24.2 Statistical Characteristics of the Motion of a Pair of Fluid Particles

The most general Lagrangian statistical characteristics of turbulence are the multidimensional probability distributions for the values of the coordinates $X$ and velocities $V$ at arbitrary times $t_1, t_2, \ldots, t_m$ of given $n$ fluid particles with known coordinates $x_1 = X_1(t_{01}), \ldots, x_n = X_n(t_{0n})$ at some not necessarily identical "initial times" $t_{01}, \ldots, t_{0n}$, where $t_{0l} < t_l$. In this book we have so far confined our attention to the probability distributions for the coordinates $X(t) = \{X(t), Y(t), Z(t)\}$ of a single fluid particle at a fixed time $t > t_0$. Occasionally we have also mentioned the joint probability distributions and joint moments for coordinates $\boldsymbol{X}(t)$ and velocities $\boldsymbol{V}(t)$ of a single fluid particle. In this subsection we shall consider certain statistical quantities specifying the motion of a *pair* of fluid particles.

To simplify the problem we shall confine our attention, to begin with, to the probability distributions for the coordinates $\boldsymbol{X}_1=\boldsymbol{X}_1(t)$ and $\boldsymbol{X}_2=\boldsymbol{X}_2(t)$ of a pair of fluid particles (leaving the given points $\boldsymbol{x}_1$ and $\boldsymbol{x}_2$ at time $t_0$) at a single time $t>t_0$. It will be convenient to take $t_0$ as the origin of time (i.e., replace $t$ by $\tau=t-t_0$), and instead of the vectors $\boldsymbol{X}_1(t)$ and $\boldsymbol{X}_2(t)$ use the displacement of the first particle $\boldsymbol{Y}(\tau)=\boldsymbol{X}_1(t_0+\tau)-\boldsymbol{x}_1$ and the vector $\boldsymbol{l}(\tau)=\boldsymbol{X}_2(t_0+\tau)-\boldsymbol{X}_1(t_0+\tau)$ which determines the relative position of the two particles. The probability density $p(\boldsymbol{Y}, \boldsymbol{l})$ of the random vectors $\boldsymbol{Y}(\tau)$ and $\boldsymbol{l}(\tau)$ depends on the paramters $\boldsymbol{x}_1$, $\boldsymbol{l}_0=\boldsymbol{l}(0)$, $\tau$ and $t_0$. We shall therefore frequently denote it by the symbol $p(\boldsymbol{Y}, \boldsymbol{l}\,|\,\boldsymbol{x}_1, \boldsymbol{l}_0, \tau, t_0)$. We note that in the case of homogeneous turbulence there will be no $\boldsymbol{x}_1$ dependence, while in the case of stationary turbulence the density $p(\boldsymbol{Y}, \boldsymbol{l})$ will not depend on $t_0$. Having integrated the probability density $p(\boldsymbol{Y}, \boldsymbol{l})$ over all $l$, we obtain the probability density for the displacement $\boldsymbol{Y}(\tau)$ of a single particle (independent of $\boldsymbol{l}_0$) which we were concerned with in Chapt. 5 and Sect. 24.1. Here, however, we shall be particularly interested in the probability density for the vector $\boldsymbol{l}(\tau)$, which is given by

$$p(\boldsymbol{l})=p(\boldsymbol{l}\,|\,\boldsymbol{x}_1, \boldsymbol{l}_0, \tau, t_0)=\int p(\boldsymbol{Y}, \boldsymbol{l}\,|\,\boldsymbol{x}_1, \boldsymbol{l}_0, \tau, t_0)\,d\boldsymbol{Y}. \tag{24.19}$$

This density was first introduced into the theory of turbulent diffusion (in the one-dimensional case) by Richardson (1926) who called it the distance-neighbor function. Its general properties were subsequently analyzed by Batchelor (1952a), Batchelor and Townsend (1956), and Monin (1960).

We shall distinguish cases corresponding to different diffusion-time intervals $\tau$. Let us suppose, to begin with, that $\tau$ is very large. We can assume that our two particles will be separated by a large distance at time $\tau$, so that their velocities $\boldsymbol{V}_1(t_0+\tau) = \boldsymbol{u}(\boldsymbol{X}_1(t_0+\tau), t_0+\tau)$ and $\boldsymbol{V}_2(t_0+\tau)=\boldsymbol{u}(\boldsymbol{X}_2(t_0+\tau), t_0+\tau)$ will be practically independent. Hence it follows that for very large $\tau$ the particles will wander independently, so that

$$\begin{aligned} p(\boldsymbol{Y}, \boldsymbol{l}\,|\,\boldsymbol{x}_1, \boldsymbol{l}_0, \tau, t_0)&= \\ &=p(\boldsymbol{Y}\,|\,\boldsymbol{x}_1, \tau, t_0)\,p(\boldsymbol{Y}+\boldsymbol{l}-\boldsymbol{l}_0\,|\,\boldsymbol{x}_1+\boldsymbol{l}_0, \tau, t_0) \end{aligned} \tag{24.20}$$

and

$$p(\boldsymbol{l} \mid \boldsymbol{x}_1, \boldsymbol{l}_0, \tau, t_0) =$$
$$= \int p(\boldsymbol{Y} \mid \boldsymbol{x}_1, \tau, t_0)\, p(\boldsymbol{Y} + \boldsymbol{l} - \boldsymbol{l}_0 \mid \boldsymbol{x}_1 + \boldsymbol{l}_0, \tau, t_0)\, d\boldsymbol{Y}, \quad (24.21)$$

where $p(\boldsymbol{Y} \mid \boldsymbol{x}, \tau, t_0)$ is the probability density for the displacement of one fluid particle. Therefore, in the case which we are considering, the probability distribution for $\boldsymbol{l}(\tau)$ is simply determined from the probability distribution $p(\boldsymbol{Y})$ for a single particle. In particular, if $p(\boldsymbol{Y})$ satisfies a parabolic diffusion equation with constant diffusivities $K_{ij}$, then $p(\boldsymbol{l})$ will also satisfy this equation but with doubled diffusivities $2K_{ij}$. If $|\boldsymbol{l}_0| \gg L$, where $L$ is the external turbulence length scale, then Eqs. (24.20) and (24.21) can be used for any (not necessarily large) values of $\tau$. If, on the other hand, $\boldsymbol{l}_0$ does not satisfy the above condition, these formulas will be valid if $\tau$ is sufficiently large to ensure that we can neglect the probability that the condition $|\boldsymbol{l}(\tau)| \gg L$ will be violated [in practice this is ensured if $[\overline{l^2(\tau)}]^{1/2} \gg L$]. Since the mean distance between diffusing particles in a turbulent flow evidently increases with time, and in the absence of boundaries can become as large as desired, we can always find a time $\tau_0$ (which depends on $\boldsymbol{l}_0$ and on the characteristics of large-scale turbulence components from the energy range) after which we can use Eqs. (24.20) and (24.21).

Let us assume that $|\boldsymbol{l}_0| \ll L$. Since $\boldsymbol{l}(\tau)$ varies continuously, we can then always find a value $\tau_1 < \tau_0$ such that when $\tau < \tau_1$ the inequality $|\boldsymbol{l}(\tau)| \ll L$ is almost surely satisfied. Disturbances with scales of the order of or greater than $L$ for $\tau < \tau_1$ will then transport our two fluid particles together without affecting their relative displacement. Therefore, the relative motion of the two particles is affected only by turbulent disturbances with scales much smaller than $L$, for which the Kolmogorov similarity hypotheses are valid, provided Re is sufficiently large. Hence, it follows that, for $\tau < \tau_1$ and sufficiently large Re, the probability density $p(\boldsymbol{l} \mid \boldsymbol{x}_1, \boldsymbol{l}_0, \tau, t_0)$ will not directly depend on $\boldsymbol{x}_1$ and $t_0$ (because the local statistical structure is homogeneous and stationary), but will be an isotropic function of the vectors $\boldsymbol{l}$ and $\boldsymbol{l}_0$, depending only on $\tau$ and on the parameters $\nu$ and $\bar{\varepsilon}$. Next, since the displacement $\boldsymbol{Y}(\tau)$ of the first particle is determined mainly by the large-scale turbulence disturbances from the energy range (with scales of the order of $L$), we would expect that for $\tau < \tau_1$ and large Re the random vectors $\boldsymbol{Y}(\tau)$ and $\boldsymbol{l}(\tau)$ will be statistically independent. Consequently,

$$p(Y, l \mid x_1, l_0, \tau, t_0) = p(Y \mid x_1, \tau, t_0)\, p(l \mid l_0, \tau), \qquad (24.22)$$

where the first factor on the right-hand side is the probability density for the displacement of a single particle, and the notation used for the second factor already indicates that it is independent of $x_1$ and $t_0$. It will be more convenient from now on to consider the vector $\Delta l = l(\tau) - l_0$ instead of $l = l(\tau)$. Let us denote its probability density by the symbol $p_1(\Delta l \mid l_0, \tau)$. It is then clear that $p_1(\Delta l \mid l_0, \tau) = p(\Delta l + l_0 \mid l_0, \tau)$. In view of the above discussion, for $\tau < \tau_1$ and sufficiently large Re we must have

$$p_1(\Delta l \mid l_0, \tau) = \eta^{-3} \Phi^{(1)}\left(\frac{|\Delta l|}{\eta}, \frac{\Delta l \cdot l_0}{\eta^2}, \frac{|l_0|}{\eta}, \frac{\tau}{\tau_\eta}\right), \qquad (24.23)$$

where $\eta = (\nu^3/\bar{\varepsilon})^{1/4}$, $\tau_\eta = (\nu/\bar{\varepsilon})^{1/2}$ are the length and time microscales, and $\Phi^{(1)}$ is a universal function of the four variables. In the limiting case of very small $\tau$, the expression given by Eq. (24.23) can be simplified. Namely, we can find a time $\tau_2 < \tau_1$ such that the particles continue to move practically rectilinearly during $\tau_2$ with initial velocities $u(x_1, t_0)$ and $u(x_1 + l_0, t_0)$. In that case, $\Delta l \approx \Delta_{l_0} u \cdot \tau$ for $\tau < \tau_2$ where $\Delta_{l_0} u$ is the space difference between the (Eulerian) velocities at a distance $l_0$. Consequently, for $\tau < \tau_2$

$$p_1(\Delta l \mid l_0, \tau) = \tau^{-3} P\left(\frac{\Delta l}{\tau} \middle| l_0\right) = \tau^{-3} v_\eta^{-3} \Phi^{(2)}\left(\frac{|\Delta l|}{\tau v_\eta}, \frac{\Delta l \cdot l_0}{\tau \nu}, \frac{|l_0|}{\eta}\right), \qquad (24.24)$$

where $v_\eta$ is the velocity microscale, and

$$P(U \mid l_0) = v_\eta^{-3} \Phi^{(2)}\left(\frac{|U|}{v_\eta}, \frac{U l_0}{v_\eta \eta}, \frac{|l_0|}{\eta}\right)$$

is the probability density for the space difference $\Delta_{l_0} u = U$ between the Eulerian velocities. Therefore, for $\tau < \tau_2$ the function $\Phi^{(1)}$ is expressed in terms of the function $P(U \mid l_0)$. The latter has a very simple significance and, in principle, can readily be determined from data. The formula (24.24) is similar to Eq. (9.25) of Vol. 1, which gives the probability density $p(Y \mid x, \tau, t_0)$ for small $\tau$. However, it must be remembered that the condition that $\tau$ must be small has a different significance in the two cases [since disturbances leading to changes in $Y(\tau)$ and $\Delta l(\tau)$ have different characteristic time scales.

If $\tau < \tau_2$ and, moreover, $l_0 = |\mathbf{l}_0| \gg \eta$, then in view of the second Kolmogorov hypothesis the density $P(U \mid \mathbf{l}_0)$ cannot depend on $\nu$. Hence, it should be of the form

$$(\bar{\varepsilon} l_0)^{-1} \Phi^{(3)}\left( |U|/(\bar{\varepsilon} l_0)^{1/3},\ U\mathbf{l}_0/(\bar{\varepsilon} l_0^4)^{1/3}\right).$$

Consequently, when $\tau < \tau_2$ and $l_0 \gg \eta$, formula (24.24) can be simplified further:

$$p_1(\Delta \mathbf{l} \mid \mathbf{l}_0, \tau) = (\bar{\varepsilon} l_0 \tau^3)^{-1} \Phi^{(3)}\left(\frac{|\Delta \mathbf{l}|}{\tau(\bar{\varepsilon} l_0)^{1/3}},\ \frac{\Delta \mathbf{l} \cdot \mathbf{l}_0}{\tau(\bar{\varepsilon} l_0^4)^{1/3}}\right). \quad (24.25)$$

Simplification analogous to transition from Eq. (24.24) to Eq. (24.25) is also possible in the general case $l_0 \ll L$, $\tau < \tau_1$. It is sufficient to assume that $l_0 \gg \eta$, i.e., that $\mathbf{l}_0$ lies in the inertial subrange, and then disturbances with scales of the order of $\eta$ or less will not affect the probability distribution for $\Delta \mathbf{l}$ for any $\tau$ (this is clear, for example, from the formula $\Delta \mathbf{l} = \int_0^\tau \Delta_{\mathbf{l}(s)} \mathbf{u}\, ds$, which shows that $\Delta \mathbf{l}$ is expressed in terms of the differences $\Delta_{\mathbf{l}} \mathbf{u}$, for $\mathbf{l}(s)$, $0 < s < \tau$ belonging to the inertial subrange). Consequently, when $\tau < \tau_1$ and $\eta \ll l_0 \ll L$, the molecular viscosity will be absent from Eq. (24.23), i.e., this formula should assume the form

$$p_1(\Delta \mathbf{l} \mid \mathbf{l}_0, \tau) = (\bar{\varepsilon} \tau^3)^{-3/2} \Phi^{(4)}\left(\frac{|\Delta \mathbf{l}|}{(\bar{\varepsilon}\tau^3)^{1/2}},\ \frac{\Delta \mathbf{l} \cdot \mathbf{l}_0}{\bar{\varepsilon} \tau^3},\ \frac{|\mathbf{l}_0|}{(\bar{\varepsilon}\tau^3)^{1/2}}\right), \quad (24.26)$$

where $\Phi^{(4)}$ is a universal function of the three variables. The formula (24.25) is a special case of this (for $\tau < \tau_2$). We note that, even when $l_0 \lesssim \eta$, the influence of the viscosity $\nu$ on the probability distribution for $\Delta \mathbf{l}$ will be important only during a limited time interval. The separation of the particles will become much greater than $\eta$ after some time $\tau$, and their subsequent relative motion will be influenced only by turbulent disturbances from the inertial range. Therefore, when $l_0 \lesssim \eta$, and beginning with some value $\tau(l_0)$ (which is of the order of $\tau_\eta$ for all not too small $l_0/\eta$ by virtue of dimensional arguments), we can use Eq. (24.26) which, as we shall see, can be simplified further in this case. We may therefore conclude that one of the two conditions, $l_0 \gg \eta$ or $\tau > \tau(l_0)$, must be satisfied if Eq. (24.26) is to be valid.

The analysis showing that the probability distribution for $\Delta l$ is unaffected by $l_0$ being small after some time has elapsed can be generalized further. It is clear intuitively that the dependence of the probability density for $\Delta l$ on $l_0$ can be important only so long as $|\Delta l|$ does not become very large in comparison with $l_0$ with a probability close to unity. For very large $\tau$ the dependence of the distribution for $\Delta l$ on $l_0$ disappears altogether. This is clear from Eq. (24.21), according to which the quantity $l_0$ for $\tau > \tau_0$ (when the particle motions can be regarded as independent) can enter the probability distribution for $l(\tau)$ only in the form of the difference $l - l_0 = \Delta l$ (under the natural assumption that the pair of points $x_1$ and $x_2 = x_1 + l_0$ lies in the region where the turbulence is practically homogeneous). However, the condition $\tau > \tau_0$ is not necessary for this purpose. It is sufficient to select a value $\tau_3$ such that the inequality $|l(\tau)| \gg |l_0|$ will be surely satisfied for $\tau > \tau_3$ (for example, $\tau_3$ may be determined from the condition $[\overline{l^2(\tau_3)}]^{1/2} \gg |l_0|$). We can then suppose that the particles have "forgotten" their initial relative separation for $\tau > \tau_3$, and move practically as if they have left the same point at time $t_0$. The time $\tau_3$ depends on $l_0$, and it is natural to expect that if $|l_0|$ is not too large, then $\tau_3 < \tau_0$, i.e., the dependence of the distribution for $\Delta l$ on $l_0$ will disappear before the motion of the particles becomes practically independent. The existence of a time range $\tau_3 < \tau < \tau_0$ for sufficiently small $l_0$ was put forward by Batchelor (1950, 1952a), who called the motion of a pair of particles in this range the *quasi-asymptotic motion* (in contrast to the asymptotic motion for $\tau > \tau_0$ when the particles move independently of each other).[28]

We shall be particularly interested in the case when $l_0$ is so small in comparison with $L$ that the quasi-asymptotic motion is reached before the probability of escape $l(\tau)$ beyond the inertial subrange of length scales becomes appreciable, i.e., when $\tau_3 < \tau_1$ [the diffusion times $\tau$ in the range $\tau_3 < \tau < \tau_1$ were called by Batchelor (1950) the intermediate times in contrast to the small ($\tau < \tau_2$) and large ($\tau > \tau_0$) times]. When $\tau_3 < \tau < \tau_1$ the dependence on $l_0$ should disappear from Eq. (24.23), so that

[28] Batchelor (1952a) has also noted that the value of $\tau_3$ can apparently be reduced further (replacing it by $\tau_3' < \tau_3$) if we assume that the dependence of the distribution of $\Delta l$ on $l_0$ during the quasi-asymptotic motion does not vanish completely, but can be replaced by a small increase in the diffusion time $\tau$ by an amount $t_1 = t_1(l_0)$ which is mainly a function of $|l_0|$. In that case, $p_1(\Delta l \mid l_0, \tau) = p_1(\Delta l \mid 0, \tau + t_1)$ for $\tau > \tau_3'$. In other words, when $\tau > \tau_3'$ the particles move as if they left the same point at some time $t_0 - t_1$, where $t_1$ is determined by $l_0$.

$$p_1(\Delta \boldsymbol{l} \mid \boldsymbol{l}_0, \tau) = p_1(\Delta \boldsymbol{l} \mid \tau) = \eta^{-3} \Phi^{(5)}\left(\frac{|\Delta \boldsymbol{l}|}{\eta}, \frac{\tau}{\tau_\eta}\right). \tag{24.27}$$

If in addition to the condition $\tau_3 < \tau < \tau_1$ we have either $l_0 \gg \eta$ or $\tau > \tau(l_0)$, the density $p_1(\Delta \boldsymbol{l} \mid \tau)$ will no longer depend on $\nu$. The above formula will then become

$$p_1(\Delta \boldsymbol{l} \mid \tau) = (\bar{\varepsilon}\tau^3)^{-3/2} \Phi^{(6)}\left(\frac{|\Delta \boldsymbol{l}|}{(\bar{\varepsilon}\tau^3)^{1/2}}\right). \tag{24.28}$$

The same result can be obtained by demanding that the right side of Eq. (24.26) be independent of $\boldsymbol{l}_0$. We note that $\tau(l_0)$ is of the order of $\tau_\eta$ and $\tau_3 \gg \tau_\eta$ for $l_0$ of the order of $\eta$. Therefore, the formula (24.27) is useful only for those cases where the initial distance between the particles is much less than $\eta$ and the diffusion time is very small, i.e., less than $\tau(l_0)$. Instead of the length $(\bar{\varepsilon}\tau^3)^{1/2}$ in Eq. (24.28) we can use any other length which is uniquely determined by $\bar{\varepsilon}$ and $\tau$, for example, the length $[\overline{\Delta l^2(\tau)}]^{1/2}$, which, in this case, is proportional to $(\bar{\varepsilon}\tau^3)^{1/2}$ [see Eq. (24.36) below]. Therefore, Eq. (24.28) can also be rewritten in the form

$$p_1(\Delta \boldsymbol{l} \mid \tau) = [\overline{\Delta l^2}]^{-3/2} F\left(\frac{\Delta l^2}{\overline{\Delta l^2}}\right); \tag{24.28'}$$

In this form it should evidently be already valid for $\tau_3' < \tau_3$ (see the footnote above).

Similar results can be obtained for the joint probability distribution for the vector $\Delta \boldsymbol{l}(\tau)$ or $\boldsymbol{l}(\tau)$, and the relative velocity of the particles $\Delta \boldsymbol{V}(\tau) = \boldsymbol{V}_2(t_0+\tau) - \boldsymbol{V}_1(t_0+\tau)$. In particular, for $\tau_3 < \tau < \tau_1$ and $l_0 \gg \eta$ [or $\tau > \tau(l_0)$] the probability density for the pair of vectors $\Delta \boldsymbol{l}(\tau)$ and $\Delta \boldsymbol{V}(\tau)$ will depend only on $\tau$, so that it can be denoted by $p_1(\Delta \boldsymbol{l}, \Delta \boldsymbol{V} \mid \tau)$. This density is given by formula

$$p_1(\Delta \boldsymbol{l}, \Delta \boldsymbol{V} \mid \tau) = (\bar{\varepsilon}\tau^2)^{-3} \Phi_1^{(6)}\left(\frac{|\Delta \boldsymbol{l}|}{(\bar{\varepsilon}\tau^3)^{1/2}}, \frac{\Delta \boldsymbol{l} \cdot \Delta \boldsymbol{V}}{\bar{\varepsilon}\tau^2}, \frac{|\Delta \boldsymbol{V}|}{(\bar{\varepsilon}\tau)^{1/2}}\right). \tag{24.28''}$$

When $\tau_3 < \tau_1$ the time $\tau_3$ should be fully specified by the parameters $l_0 = |\boldsymbol{l}_0|$, $\bar{\varepsilon}$, and $\nu$. Consequently, $\tau_3 = (l_0^2/\bar{\varepsilon})^{1/3} f(l_0/\eta)$ where $f(x)$ is a universal function such that $\lim_{x\to\infty} f(x) = f(\infty)$ will be a finite constant and therefore $\tau_3 \sim (l_0^2/\bar{\varepsilon})^{1/3}$ for $l_0 \gg \eta$. At the same time,

$\tau_3 \sim \tau_\eta$ for $l_0 \ll \eta$ (see Sect. 24.5 below). Similarly, the time $\tau_2$ for $l_0 \ll L$ is also determined by the parameters $l_0$, $\bar{\varepsilon}$ and $\nu$, i.e., it is given by a formula of the same form (but with a different function $f$) and again $\tau_2 \sim (l_0^2/\bar{\varepsilon})^{1/3}$ for $l_0 \gg \eta$ and $\tau_2 \sim \tau_\eta$ for $l_0 \ll \eta$. However, the times $\tau_0$ and $\tau_1$ cannot be determined in a similar way since they depend on the characteristics of the large-scale motions (in the first instance, on the length scale $L$).

Formulas (24.21) and (24.23)–(24.28) for the probability density of a random vector $\Delta \boldsymbol{l}$ can readily be used to obtain a number of results on the moments of the random vectors $\Delta \boldsymbol{l}$ and $\boldsymbol{l} = \boldsymbol{l}_0 + \Delta \boldsymbol{l}$. Here, we shall restrict our attention to the behavior of the two particle relative dispersion tensor $D_{ij}^{(r)}(\tau) = \overline{l_i(\tau) l_j(\tau)}$, or the closely connected tensor $\overline{\Delta l_i(\tau) \Delta l_j(\tau)}$, and the mean square distance between the particles $\overline{l^2}(\tau) = D_{ii}^{(r)}(\tau)$ [or the length $[\overline{\Delta l^2(\tau)}]^{1/2}$]. This behavior was investigated by Batchelor (1950, 1952a). When $\tau > \tau_0$, the motion of the two particles is independent. Therefore we find in this case from the equation $\boldsymbol{l}(\tau) = \boldsymbol{l}_0 + \boldsymbol{Y}_2(\tau) - \boldsymbol{Y}_1(\tau)$, where $\boldsymbol{Y}_i(\tau)$ is the displacement of the $i$th particle, that for homogeneous turbulence

$$\begin{gathered} D_{ij}^{(r)}(\tau) = l_{0i} l_{0j} + \overline{Y_{1i}(\tau) Y_{1j}(\tau)} + \overline{Y_{2i}(\tau) Y_{2j}(\tau)}, \\ \overline{l^2}(\tau) = l_0^2 + \overline{Y_1^2}(\tau) + \overline{Y_2^2}(\tau). \end{gathered} \tag{24.29}$$

Using the results of Sect. 9.3, we find that for very large $\tau$

$$D_{ij}^{(r)}(\tau) \approx 2\overline{u_i'^2}\, T_i \delta_{ij} \tau, \quad \overline{l^2}(\tau) \approx 2\left(\overline{u_1'^2} T_1 + \overline{u_2'^2} T_2 + \overline{u_3'^2} T_3\right)\tau, \tag{24.29'}$$

so that the distance between the particles increases approximately in proportion to $\tau^{1/2}$ for large times $\tau$. When $\tau < \tau_2$, $l_0 \ll L$ we have $\Delta \boldsymbol{l} \approx \Delta_{l_0} \boldsymbol{u} \cdot \tau$ and, consequently,

$$\overline{\Delta l_i \Delta l_j} \approx D_{ij}(\boldsymbol{l}_0)\tau^2, \quad \overline{|\Delta \boldsymbol{l}|^2} \approx [D_{LL}(l_0) + 2D_{NN}(l_0)]\tau^2, \tag{24.30}$$

where $D_{ij}$, $D_{LL}$, $D_{NN}$ are the structure tensor and the structure functions of the Eulerian velocity field introduced in Sect. 21.4. Since at the same time $\overline{\Delta \boldsymbol{l}} = \overline{\Delta_{l_0} \boldsymbol{u}} \cdot \tau = 0$, we find that for $\tau < \tau_2$

$$D_{ij}^{(r)}(\tau) = l_{0i} l_{0j} + D_{ij}(\boldsymbol{l}_0)\tau^2, \quad \overline{l^2} = l_0^2 + [D_{LL}(l_0) + 2D_{NN}(l_0)]\tau^2. \tag{24.30'}$$

When $l_0 \gg \eta$ we have $D_{LL}(l_0) \sim D_{NN}(l_0) \sim (\bar{\varepsilon} l_0)^{2/3}$ and, therefore, the condition $\tau < \tau_2 \sim (l_0^2/\bar{\varepsilon})^{1/3}$ is now surely satisfied for $(\overline{|\Delta \boldsymbol{l}|^2})^{1/2} \ll l_0$.

In the more general case $\tau < \tau_1$, $l_0 \ll L$, when the probability density for $\Delta l$ is of the form (24.23), symmetry and dimensional considerations lead to a formula of the form

$$D_{ij}^{(r)}(\tau) = D_0^{(r)}\left(\frac{l_0}{\bar{\varepsilon}^{1/2}\tau^{3/2}}, \frac{\tau}{\tau_\eta}\right) l_{0i} l_{0j} + D_1^{(r)}\left(\frac{l_0}{\bar{\varepsilon}^{1/2}\tau^{3/2}}, \frac{\tau}{\tau_\eta}\right) \bar{\varepsilon}\tau^3 \delta_{ij}. \quad (24.31)$$

Here $D_0^{(r)}$ and $D_1^{(r)}$ are universal functions of the two variables $\xi = l_0/\bar{\varepsilon}^{1/2}\tau^{3/2}$ and $\zeta = \tau/\tau_\eta = \tau\bar{\varepsilon}^{1/2}/\nu^{1/2}$ (or, what amounts to the same thing, the variables $l_0/\eta = \xi\zeta^{3/2}$ and $\tau/\tau_\eta$). A formula of the same form but with different functions $D_0^{(r)}$ and $D_1^{(r)}$ can be written for $\overline{\Delta l_i \cdot \Delta l_j}$ in this case. It follows from Eq. (24.31) that

$$\overline{l^2(\tau)} = \bar{\varepsilon}\tau^3 D_2^{(r)}\left(\frac{l_0}{\bar{\varepsilon}^{1/2}\tau^{3/2}}, \frac{\tau}{\tau_\eta}\right), \quad (24.32)$$

where $D_2^{(r)}(\xi, \zeta) = D_0^{(r)}(\xi, \zeta)\xi^2 + 3D_1^{(r)}(\xi, \zeta)$. It is also possible to follow Batchelor (1950) and Batchelor and Townsend (1956) and use instead of Eq. (24.32) the equivalent formula

$$\frac{d\overline{l^2}}{d\tau} = \bar{\varepsilon}\tau^2 D_3^{(r)}\left(\frac{l_0}{\bar{\varepsilon}^{1/2}\tau^{3/2}}, \frac{\tau}{\tau_\eta}\right), \qquad D_3^{(r)} = 3D_2^{(r)} + \zeta\frac{\partial}{\partial\zeta}D_2^{(r)} - \frac{3}{2}\xi\frac{\partial}{\partial\xi}D_2^{(r)}. \quad (24.32')$$

When $l_0 \gg \eta$ or $\tau > \tau(l_0)$, we can replace Eqs. (24.31)–(24.32′) by the much simpler formulas

$$D_{ij}^{(r)} = E_0^{(r)}\left(\frac{l_0}{\bar{\varepsilon}^{1/2}\tau^{3/2}}\right) l_{0i} l_{0j} + E_1^{(r)}\left(\frac{l_0}{\bar{\varepsilon}^{1/2}\tau^{3/2}}\right) \bar{\varepsilon}\tau^3 \delta_{ij}, \quad (24.33)$$

$$\overline{l^2} = \bar{\varepsilon}\tau^3 E_2^{(r)}\left(\frac{l_0}{\bar{\varepsilon}^{1/2}\tau^{3/2}}\right), \quad \frac{d\overline{l^2}}{d\tau} = \bar{\varepsilon}\tau^2 E_3^{(r)}\left(\frac{l_0}{\bar{\varepsilon}^{1/2}\tau^{3/2}}\right). \quad (24.34)$$

Finally, when $\tau_3 < \tau < \tau_1$, we have

$$D_{ij}^{(r)}(\tau) = \frac{1}{3}\bar{\varepsilon}\tau^3 G_0^{(r)}(\tau/\tau_\eta)\delta_{ij}, \quad \overline{l^2} = \bar{\varepsilon}\tau^3 G_0^{(r)}(\tau/\tau_\eta), \quad (24.35)$$

$$\frac{d\overline{l^2}}{d\tau} = \bar{\varepsilon}\tau^3 G_1^{(r)}(\tau/\tau_\eta), \quad G_1^{(r)}(\zeta) = 3G_0^{(r)}(\zeta) + \zeta\frac{\partial}{\partial\zeta}G_0^{(r)}(\zeta). \quad (24.35')$$

Moreover, if in addition $l_0 \gg \eta$ [or $l_0 \approx \eta$ but $\tau \gg \tau_\eta$, or $l_0 \ll \eta$ but

$\tau > \tau(l_0)]$, then

$$D_{ij}^{(r)}(\tau) = \frac{1}{3} g \bar{\varepsilon} \tau^3 \delta_{ij}, \quad \overline{l^2} = g \bar{\varepsilon} \tau^3, \tag{24.36}$$

$$\frac{d\overline{l^2}}{d\tau} = 3 g \bar{\varepsilon} \tau^2, \tag{24.36'}$$

where $g = G_0^{(r)}(\infty)$ is a universal constant. The very simple formulas (24.36) and (24.36') have a broad range of validity and form the most important result of the present subsection. They were known to Obukhov and Landau as far back as the beginning of the 1940's [see, in particular, the solution of the problem on p. 123 of the book by Landau and Lifshitz (1963) which was already given in the first Russian edition of 1944]. They were subsequently found independently by Batchelor (1950). Other derivations of the same results have also been reported by several Japanese workers [see, for example, Ogura, Sekiguchi, and Miyakoda (1953)]. In view of Eq. (24.36) we can rewrite Eq. (24.36') also in the form

$$\frac{d\overline{l^2}}{d\tau} = g_1 \bar{\varepsilon}^{-1/3} (\overline{l^2})^{2/3}, \tag{24.37}$$

where $g_1 = 3g^{1/3}$. This last formula was given in a slightly different (but equivalent) form by Obukhov (1941a, b). It provides a particularly clear illustration of the accelerating nature of the process of relative separation of two fluid particles which is connected with the fact that an increase in the distance $l$ between particles is produced by turbulent disturbances with length scales of the order of $l$ or less only (since larger disturbances transport both particles together).

Since $\frac{d\boldsymbol{l}(\tau)}{d\tau} = \frac{d\,\Delta\boldsymbol{l}(\tau)}{d\tau} = \Delta\boldsymbol{V}(\tau)$, for the joint second moments of the components of the relative velocity $\Delta\boldsymbol{V}(\tau)$, or the components of $\Delta\boldsymbol{V}(\tau)$ and $\boldsymbol{l}(\tau)$ for $\tau_3 < \tau < \tau_1$ and $l_0 \gg \eta$, we have from Eq. (24.36)

$$\overline{\Delta V_i \Delta V_j} = g \bar{\varepsilon} \tau \delta_{ij}, \qquad \overline{\Delta V_i(\tau) l_j(\tau)} = \frac{1}{2} g \bar{\varepsilon} \tau^2 \delta_{ij}. \tag{24.36''}$$

Equations (24.36) and (24.36'') are very similar to Eq. (24.13), while Eq. (24.28'') is very similar to Eq. (24.15). This is so because the statistical characteristics of the relative motion of a pair of fluid particles can be regarded as the characteristics of the motion of one of the particles relative to a noninertial system of coordinates $\mathscr{S}^*$, the origin of which moves together with the other particle. For small times $\tau$ (small in comparison with the time for an appreciable change in the Lagrangian velocity), such statistical quantities will not be very different from those for the motion of a particle relative to the inertial system $\mathscr{S}$ moving with a constant velocity equal to the initial velocity of the other particle at time $\tau = 0$. This is precisely why the statistical characteristics of relative motion corresponding to the limiting case $l_0 = 0$, i.e., the case $\tau > \tau_3$, are described by the same formulas as given in Sect. 24.1 for the local statistical characteristics of the motion of a single particle. Similarly, for the statistical characteristics of the motion of a particle located at the point $\boldsymbol{x}_0 + \boldsymbol{l}_0$ at time $t_0$ relative to the inertial system $\mathscr{S}$, moving with a constant velocity $\boldsymbol{u} = \boldsymbol{u}(\boldsymbol{x}_0, t_0)$, we have formulas that are similar in form to those given above for the relative motion of a pair of particles whose initial positions differ by the vector $\boldsymbol{l}_0$. We shall not, however, pause to consider this in detail.

It is also possible to establish a direct connection between the statistical characteristics of the motion of a particle relative to the inertial coordinate system $\mathscr{S}$ and the noninertial system $\mathscr{S}^*$. Suppose that $l_0 \gg \eta$, and hence $|\boldsymbol{l}(\tau)| \gg \eta$ for $\tau > 0$. Let us also suppose that $\tau_3 < \tau < \tau_1$, so that the two particles can be regarded as having left the same point at time $\tau = 0$. We shall use the fact that, as shown in Sect. 21.5, the accelerations of the fluid particles in a turbulent flow with large Re are practically uncorrelated for $l \gg \eta$. The accelerations $\boldsymbol{A}_1(\tau')$ and $\boldsymbol{A}_2(\tau'')$ of the two particles for all $0 \leqslant \tau' \leqslant \tau$ and $0 \leqslant \tau'' \leqslant \tau$ can therefore be regarded as uncorrelated. Since $\tau > \tau_3$, we may suppose that

$$\boldsymbol{V}_1^{(s)}(\tau) = \int_0^\tau \boldsymbol{A}_1(\tau')\,d\tau', \quad \boldsymbol{X}_1^{(s)}(\tau) = \int_0^\tau \boldsymbol{V}_1^{(s)}(\tau')\,d\tau' = \int_0^\tau (\tau - \tau')\boldsymbol{A}_1(\tau')\,d\tau',$$

$$\boldsymbol{V}_2^{(s)}(\tau) = \int_0^\tau \boldsymbol{A}_2(\tau')\,d\tau', \quad \boldsymbol{X}_2^{(s)}(\tau) = \int_0^\tau \boldsymbol{V}_2^{(s)}(\tau')\,d\tau' = \int_0^\tau (\tau - \tau')\boldsymbol{A}_2(\tau')\,d\tau'.$$

Therefore, in the case under consideration,

$$\Delta \boldsymbol{V}(\tau) = \boldsymbol{V}_2^{(s)}(\tau) - \boldsymbol{V}_1^{(s)}(\tau), \qquad \boldsymbol{l}(\tau) = \boldsymbol{X}_2^{(s)}(\tau) - \boldsymbol{X}_1^{(s)}(\tau), \quad (24.38)$$

where $(\boldsymbol{V}_1^{(s)}(\tau), \boldsymbol{V}_2^{(s)}(\tau))$ and $(\boldsymbol{X}_1^{(s)}(\tau), \boldsymbol{X}_2^{(s)}(\tau))$ are two pairs of independent and identically distributed random vectors. Hence the probability density $\Phi^{(6)}(\xi)$ in Eq. (24.28) can be expressed in terms of the density $P^{(3)}(\xi)$ given by Eq. (24.16), and the density $\Phi_1^{(6)}(\xi, \eta, \zeta)$ of Eq. (24.28″) can be expressed in terms of the density $P^{(2)}(\xi, \eta, \zeta)$ given by Eq. (24.15). Particularly simple relations are obtained between the moments of the vectors $\Delta V(\tau)$ and $V^{(s)}(\tau)$, or $l(\tau)$ and $X^{(s)}(\tau)$. Thus, for example, $\overline{|\Delta V(\tau)|^2} = 2\overline{|V^{(s)}(\tau)|^2}$ and $\overline{l^2(\tau)} = 2\overline{|X^{(s)}(\tau)|^2}$ from which it is clear that

$$g = 2C_0 \tag{24.39}$$

This last result was given by Novikov (1963a), who derived it by similar means.

Let us now consider the various methods of deriving Eq. (24.36) which are formally independent of similarity hypotheses. We shall begin with the hypothesis of Obukhov (1959b), according to which the variation of the coordinate $X$ and velocity $V$ of a given fluid particle can be regarded as a Markov process in the six-dimensional phase space $(X, V)$ (see Vol. 1 for the definition of a Markov process). This hypothesis will be discussed in detail in Sect. 24.4, but for the moment we note only that it automatically leads to relations

$$\overline{Y_i^{(s)}(\tau)\, Y_j^{(s)}(\tau)} = \frac{2}{3} D\tau^3\delta_{ij}, \quad \overline{Y_i^{(s)}(\tau)\, V_j^{(s)}(\tau)} = D\tau^2\delta_{ij}, \quad \overline{V_i^{(s)}(\tau)\, V_j^{(s)}(\tau)} = 2D\tau\delta_{ij}. \tag{24.40}$$

which is similar to Eq. (24.13). In this expression $Y^{(s)}$ and $V^{(s)}$ are the coordinates and velocities relative to the inertial system $\mathscr{S}$, and $D$ is a constant having the same dimensions as $\bar{\varepsilon}$ (it is the "diffusivity in velocity space"). The above result is consistent with derivations based on similarity and dimensional considerations, and this can be looked upon as an argument for the validity of the Obukhov hypothesis for large Re and not too small (but also not too large) $\tau$ as a reasonable first approximation (if it is assumed that $D = \frac{1}{2} C_0\bar{\varepsilon}$, where $C_0$ is a numerical constant). On the other hand, this result justifies the application of formulas such as Eq. (24.13) or Eqs. (24.36)–(24.36″) independently of the similarity hypotheses in all cases where we have some reason to assume that the variation in the state of the fluid particle is described by a Markov process.

This fact has led Lin (1960a, b) [see also Corrsin (1962a) and Lin and Reid (1963)] to a more detailed analysis of the assumptions necessary for the derivation of formulas such as Eqs. (24.36)–(24.36″) and Eq. (24.13). It was found that the Markov-process hypothesis for the six-dimensional random process $(X(\tau), V(\tau))$ or $(\Delta l(\tau), \Delta V(\tau))$ can be replaced by certain assumptions that, at first sight, appear to be simpler and broader although, in reality, they are apparently quite close.

Before we review Lin's work on the relative motion of a pair of fluid particles, let us consider his derivation of the expression for the Lagrangian correlation function for the velocity of a single particle, which provides a good illustration of his approach to the problem. For the sake of simplicity, let us consider the motion of a particle along a line (say, the $Ox_1$ axis). The increase in the corresponding component of the Lagrangian velocity $V_1(x, t) = V(t)$ in a time $\tau$ can be written in the form

$$V(t_0+\tau)-V(t_0)=\int\limits_{t_0}^{t_0+\tau} A(t')\,dt',$$

where $A(t)=A_1(x,t)$ is the acceleration of the particle which has left the point $x$ at time $t_0$ in the direction of the $Ox_1$ axis. Consequently,

$$\overline{[V(t_0+\tau)-V(t_0)]^2}=\int\limits_0^\tau\int\limits_0^\tau B_{AA}^{(L)}(\tau'-\tau'')\,d\tau'\,d\tau''=2\int\limits_0^\tau(\tau-s)\,B_{AA}^{(L)}(s)\,ds, \tag{24.41}$$

where $B_{AA}^{(L)}(s)=\overline{A(t+s)\,A(t)}$ is the Lagrangian acceleration correlation function [cf. the derivation of Eq. (9.30) in Vol. 1]. If the integral of $B_{AA}^{(L)}(s)$ evaluated between $s=0$ and $s=\tau$ converges for $\tau\to\infty$ to a finite limit $\mathcal{A}$, then for $\tau \gg T_A$, where $T_A$ is a typical correlation time for the Lagrangian accelerations, we have

$$D^{(L)}(\tau)=\overline{[V(t_0+\tau)-V(t_0)]^2}\approx 2\mathcal{A}\tau \tag{24.42}$$

[see the derivation of Eq. (9.35) in Vol. 1]. Thus, we seem to have shown that the Lagrangian velocity correlation function is a linear function of time [i.e., we have obtained a result analogous to Eq. (21.30′) which follows from the Kolmogorov similarity hypothesis] by assuming only that the correlation function for the accelerations decreases sufficiently rapidly at infinity, but without demanding that the Reynolds number be large and $\tau$ much smaller than $T_0$. In fact, however, the above simple analysis is erroneous and needs modification. It is clear from Eq. (24.41) that $B_{AA}^{(L)}(\tau)=\frac{d^2}{d\tau^2}D^{(L)}(\tau)$ and, consequently,

$$\int\limits_0^\infty B_{AA}^{(L)}(s)\,ds=\left.\frac{dD^{(L)}(\tau)}{d\tau}\right|_0^\infty=0. \tag{24.43}$$

Therefore, the constant $\mathcal{A}>0$ in Eq. (24.42) cannot be equated to the integral (24.43). Hence, instead of demanding the existence of the (nonzero) improper integral $\int\limits_0^\infty B_{AA}^{(L)}(s)\,ds$, it is necessary to demand that, after a certain $s=\tau_0$, the function $B_{AA}^{(L)}(s)$ must become so small that

$$\left|\int\limits_{\tau_0}^{\tau} B_{AA}^{(L)}(s)\,ds\right| \ll \left|\int\limits_0^{\tau_0} B_{AA}^{(L)}(s)\,ds\right| \tag{24.44}$$

in a broad range of values of $\tau>\tau_0$. In that case, $\int\limits_0^\tau B_{AA}^{(L)}(s)\,ds\approx\int\limits_0^{\tau_0} B_{AA}^{(L)}(s)\,ds=\mathcal{A}=$ const, and hence $\frac{dD^{(L)}(\tau)}{d\tau}\approx 2\mathcal{A}$, $D^{(L)}(\tau)\approx 2\mathcal{A}\tau$ for values of $\tau$ for which Eq. (24.44) is valid. We have thus again obtained an equation of the form of Eq. (21.30′), but we can now justify it only for a finite interval of not too small but also not too large $\tau$ [since for very large $\tau$ the condition (24.44) cannot be satisfied in view of Eq. (24.43)]. The condition

(24.44) does not seem to be so natural that it can be readily accepted in all cases. When Re is sufficiently large, we know from Sect. 21.5 that for $\tau_\eta \ll \tau \ll T_0$ the Lagrangian accelerations are practically uncorrelated and, consequently, for such $\tau$ the condition (24.44) can be regarded as satisfied. However, the conclusion that the correlation $B^{(L)}_{AA}(\tau)$ is small was deduced in Sect. 21.5 from the similarity hypotheses and is, in fact, a consequence of the linearity of the function $D^{(L)}(\tau)$ in $\tau$. Another proof that, for very large Re, the typical time scale of the acceleration correlation is quite small (of the order of $\tau_\eta$ or even less) was proposed by Lin and Reid (1963). However, their proof is also based on certain special hypotheses. Let us also emphasize that all such proofs do not enable us to conclude that the correlation between accelerations is *exactly zero* for $\tau \gg \tau_\eta$. Their significance is limited to the statement that this correlation falls rapidly to very small values, but there is a long tail of small but nonzero negative values in the region of relatively large $\tau$ (connected, as it is possible to show, with viscous accelerations of the fluid particles) which, in the end, ensures the validity of Eq. (24.43). Nevertheless, it does follow from this that, for very large Re, the condition (24.44) is satisfied (to the same degree of certainty to which the Kolmogorov similarity hypotheses are valid) in a broad range of values of $\tau$. However, if Re is relatively small, the condition (24.44) is unlikely to be satisfied, so that Lin's proof is no longer convincing.[29] We note also that if instead of demanding that the accelerations be uncorrelated we demand that the Lagrangian accelerations at different times be independent (for example, if we assume additionally that the acceleration field is Gaussian), it will follow that the random process $(X(t), V(t))$ is a Markov process, in complete accordance with the Obukhov hypothesis.

Assumptions of the form of Eq. (24.44) can also be applied to the relative motion of two fluid particles. Let us again restrict our attention to one-dimensional motion along the $Ox_1$ axis, and consider two fluid particles whose projections on this axis at time $t_0$ are separated by a distance $l_0$ and whose initial relative velocity is $\Delta V_0$. We then have

$$\Delta V(\tau) = \Delta V_0 + \int_0^\tau \Delta A(\tau')\,d\tau', \quad l(\tau) = l_0 + \Delta V_0 \tau + \int_0^\tau (\tau - \tau')\,\Delta A(\tau')\,d\tau',$$

where $\Delta V(\tau)$ is the relative velocity, and $\Delta A(\tau)$ is the relative acceleration at time $t_0 + \tau$. If $\overline{\Delta A(\tau)} = 0$ and $\overline{\Delta A(\tau')\Delta A(\tau'')} = B^{(r)}_{AA}(\tau' - \tau'')$, where $B^{(r)}_{AA}(s)$ is the correlation function for the relative accelerations, then if we substitute $\tau' - \tau'' = s$, $\tau' + \tau'' = 2s_1$ we find that

$$\overline{[\Delta V(\tau)]^2} = \Delta V_0^2 + 2\tau \int_0^\tau B^{(r)}_{AA}(s)\,ds - 2\int_0^\tau s B^{(r)}_{AA}(s)\,ds,$$

$$\overline{l(\tau)\,\Delta V(\tau)} = l_0 \Delta V_0 + \Delta V_0^2 \tau + \int_0^\tau \int_0^\tau (\tau - \tau')\,B^{(r)}_{AA}(\tau' - \tau'')\,d\tau'\,d\tau'' =$$

$$= l_0 \Delta V_0 + \Delta V_0^2 \tau + \tau^2 \int_0^\tau B^{(r)}_{AA}(s)\,ds - \tau \int_0^\tau s B^{(r)}_{AA}(s)\,ds, \tag{24.45}$$

[29] In this respect it is not clear which particular situations correspond to the analysis of Lin (1960b), who applied conditions such as Eq. (24.44) to all the components of the correlation tensor for the accelerations, and assumed simultaneously that the small-scale turbulence producing the relative motion of the particle pair was anisotropic.

$$\overline{l^2(\tau)} = l_0^2 + 2l_0\Delta V_0\tau + \Delta V_0^2\tau^2 + \\ + \int_0^\tau\int_0^\tau (\tau-\tau')(\tau-\tau'')B_{AA}^{(r)}(\tau'-\tau'')\,d\tau'\,d\tau'' = \\ = l_0^2 + 2l_0\Delta V_0\tau + \Delta V_0^2\tau^2 + \frac{2}{3}\tau^3\int_0^\tau B_{AA}^{(r)}(s)\,ds - \\ -\tau^2\int_0^\tau sB_{AA}^{(r)}(s)\,ds + \frac{1}{3}\int_0^\tau s^3B_{AA}^{(r)}(s)\,ds. \tag{24.45 (Cont.)}$$

Hence, it follows that if for a certain $\tau_0$, and a broad range of values of $\tau > \tau_0$,

$$\left[\int_{\tau_0}^{\tau} B_{AA}^{(r)}(s)\,ds\right] \ll \left|\int_0^{\tau_0} B_{AA}^{(r)}(s)\,ds\right| = \mathcal{A}^{(r)},$$

$$\tau \gg \int_0^\tau sB_{AA}^{(r)}(s)\,ds\left[\int_0^\tau B_{AA}^{(r)}(s)\,ds\right]^{-1}; \quad \tau^3 \gg \int_0^\tau s^3B_{AA}^{(r)}(s)\,ds\left[\int_0^\tau B_{AA}^{(r)}(s)\,ds\right]^{-1} \tag{24.46}$$

then

$$\overline{[\Delta V(\tau)]^2} \approx 2\mathcal{A}^{(r)}\tau, \quad \overline{\Delta V(\tau)\,l(\tau)} \approx \mathcal{A}^{(r)}\tau^2, \quad \overline{l^2(\tau)} \approx \frac{2}{3}\mathcal{A}^{(r)}\tau^3, \tag{24.47}$$

in complete accord with Eqs. (24.36)–(24.36″). The relations (24.47) clearly refer to values of $\tau$ for which the initial distance $l_0$ between the particles is no longer important, i.e., the relative motion is quasi-asymptotic. Moreover, these formulas do not include the initial velocity $\Delta V_0$, since the values of $\Delta V(\tau)$ and $l(\tau)$ for large $\tau$ are determined mainly by the cumulative effect of the relative accelerations. As regards the conditions (24.46), which play the main role in the derivation of Eq. (24.47), all the discussions referring to Eq. (24.44) will again be valid. Thus, for small and moderate values of Re, these conditions are unlikely to be valid, while for large Re they are valid to the extent to which the Kolmogorov similarity hypotheses are valid. In fact, it follows from these similarity hypotheses that the spectral density of a stationary process $\Delta A(\tau)$ in the inertial subrange should be proportional to $\bar{\epsilon}$, i.e., it should be equal to a constant. For $\tau_\eta \ll \tau_3 < \tau < \tau_1$ the process $\Delta A(\tau)$ is therefore a $\delta$-correlated white noise, and hence we have the conditions given by Eq. (22.46). If the similarity hypotheses are valid, then $\mathcal{A}^{(r)} = g\bar{\epsilon}/2$ where $g$ is the universal dimensionless constant which we have already encountered in Eqs. (24.36)–(24.36″).

By analogy with the derivation of Eq. (24.47), we can establish the expressions for the correlation functions of random processes $\Delta V(\tau)$ and $l(\tau)$ relating their values at different times. Technically, the simplest procedure to adopt is to assume right from the beginning that $B_{AA}^{(r)}(s) = 2\mathcal{A}^{(r)}\delta(s)$, from which it follows immediately that, when $\tau_\eta \ll \tau_3 < \tau' \leqslant \tau'' < \tau_1$,

$$\overline{\Delta V(\tau')\,\Delta V(\tau'')} \approx 2\mathcal{A}^{(r)} \int_0^{\tau'}\int_0^{\tau''} \delta(s'-s'')\,ds'\,ds'' = 2\mathcal{A}^{(r)}\tau',$$

$$\overline{l(\tau')\,l(\tau'')} \approx 2\mathcal{A}^{(r)} \int_0^{\tau'}\int_0^{\tau''} (\tau'-s')(\tau''-s'')\,\delta(s'-s'')\,ds'\,ds'' =$$

$$= \frac{1}{3}\mathcal{A}^{(r)}\tau'^2(3\tau''-\tau'), \tag{24.48}$$

$$\overline{l(\tau')\,\Delta V(\tau'')} \approx 2\mathcal{A}^{(r)} \int_0^{\tau'}\int_0^{\tau''} (\tau'-s')\,\delta(s'-s'')\,ds'\,ds'' = \mathcal{A}^{(r)}\tau'^2,$$

$$\overline{l(\tau'')\,\Delta V(\tau')} \approx 2\mathcal{A}^{(r)} \int_0^{\tau'}\int_0^{\tau''} (\tau''-s'')\,\delta(s''-s')\,ds'\,ds'' = \mathcal{A}^{(r)}\tau'(2\tau''-\tau')$$

The first two of these relations were given by Novikov (1963a). We note that the processes $\Delta V(\tau)$ and $l(\tau)$ are evidently not stationary [in contrast to $\Delta A(\tau)$]. As already noted in Sect. 21.5, it is natural to assume that, for large Re, the acceleration field of the fluid particles is uncorrelated not only in time but also in space (at any rate for distances much greater than $\eta$). We then have $B^{(r)}_{AA}(\tau-\tau') = \overline{\Delta A(\tau)\,\Delta A(\tau')} = 2B^{(L)}_{AA}(\tau-\tau')$ and, consequently, $\mathcal{A}^{(r)} = 2\mathcal{A}$ [for large Re the last equation is clearly identical with Eq. (24.39)].

## 24.3 Relative Diffusion and Richardson's Four-Thirds Law

The results obtained in the last section can be applied to the study of relative diffusion, i.e., spreading of an admixture cloud consisting of a large number of particles. If we could label one of the particles in the cloud, the concentration distribution relative to this particle at time $t=t_0+\tau$ would be the same as the distribution of the values of $l(\tau)$ for all the possible pairs consisting of this particular particle and any of the other particles. Consequently, in a noninertial set of coordinates $\mathcal{S}^*$, whose origin is labeled and moves together with the fluid particle, the distribution $\vartheta^*(l,\ \tau)$ of the admixture concentration at time $t=t_0+\tau$ will satisfy the relation

$$\overline{\vartheta^*(l,\ t)} = \int p(l\,|\,l_0,\ \tau)\,\vartheta_0(l_0)\,dl_0, \tag{24.49}$$

where $p(l\,|\,l_0,\ \tau)$ has the same significance as in the preceding section, and $\vartheta_0(X) = \vartheta(x+X,\ 0) = \vartheta^*(X,\ 0)$ is the initial concentration distribution [compare this with the analogous formula (10.5) of Vol. 1, which refers to a fixed set of coordinates]. Similarly, if we denote the admixture concentration at time $t_0+\tau$ relative to the inertial set of coordinates $\mathscr{S}$ by the symbol $\vartheta^{(s)}(Y, \tau)$, where the inertial set moves with constant velocity $u(x, t_0)$ and its origin at time $t_0$ is at the point $x$, then

$$\overline{\vartheta^{(s)}(Y, \tau)} = \int p(Y|Y_0, \tau)\,\vartheta_0(Y_0)\,dY_0, \qquad (24.50)$$

where $p(Y|Y_0, \tau)$ is the probability density function at time $t_0+\tau$ for the "local coordinate" $Y = Y^{(s)} = X - x - u(x, t_0)\tau$ of the fluid particle occupying the point $x + Y_0$ at time $t_0$. However, the formulas (24.49) and (24.50) are of no great practical value. In fact, it is not usually possible to label any of the fluid particles, and it is very inconvenient to use the set of coordinates $\mathscr{S}$ moving with the random velocity $u(x, t_0)$. Therefore, when we investigate relative diffusion it is best to specify the diffusion characteristics relative to the fixed set of coordinates $\mathscr{S}_0$.

One of these quantities is the function $q(l, \tau)$ introduced by Richardson (1926). To define it, consider a cloud of $N$ diffusing particles and, as usual, let us denote the coordinates of these particles at time $t = t_0 + \tau$ by $X^{(1)}(\tau) = (X_1^{(1)}(\tau), X_2^{(1)}(\tau), X_3^{(1)}(\tau)), \ldots, X^{(N)}(\tau) = (X_1^{(N)}(\tau), X_2^{(N)}(\tau), X_3^{(N)}(\tau))$. We shall consider at the first stage only dispersion along the $X_1$ axis. Let $\Delta l$ be a small segment; then $Nq(l, \tau)\Delta l$ is equal to the number of particles (i.e., of indexes $i$) for which the interval $X_1^{(i)}(\tau) + l \leqslant X_1 \leqslant X_1^{(i)}(\tau) + l + \Delta l$ along the $X_1$ axis contains at least one of the coordinates $X_1^{(j)}(\tau)$. The analogous three-dimensional quantity is the function $q(\boldsymbol{l}, \tau)$ such that $Nq(\boldsymbol{l}, \tau)\Delta\boldsymbol{l}$ is equal to the number of particles for which a small fixed volume $\Delta\boldsymbol{l}$ surrounding the point $X^{(i)}(\tau) + \boldsymbol{l}$ at time $t_0+\tau$ contains at least one of the cloud particles. Instead of the discrete cloud of $N$ particles it is often more convenient to follow Batchelor (1952a) and consider a continuous cloud of a passive admixture (differing from the ambient medium by, say, its color), filling a given volume $V$ at time $t_0$. We shall not take into account the effect of molecular diffusion. In that case, the shape of the set of colored particles will deform with time (see schematic Fig. 79 of Vol. 1), but the volume of the set will remain constant and equal to $V$. If we define the function $q(\boldsymbol{l}, \tau)$ through the condition that $V^2 q(\boldsymbol{l}, \tau)$ is equal to the volume of the set of colored points $X$ at time $t_0+\tau$ for which the point $X+\boldsymbol{l}$ is also colored, then $q(\boldsymbol{l}, \tau)$ will be a continuous function and will not depend on the choice of the elementary volume $\Delta\boldsymbol{l}$. Let $\vartheta(X, \tau)$ be the admixture concentration density at the point $X$ at time $t_0+\tau$, normalized so that $\int \vartheta(X, \tau)\,dX = 1$. We can then readily show that

$$\overline{q(l, \tau)} = \int \vartheta(X, \tau)\,\vartheta(X+l, \tau)\,dX. \tag{24.51}$$

This equation can also be taken as a general definition of the function $q(l, \tau)$ for an arbitrary admixture cloud. In the case of a discrete cloud of $N$ particles for which $\vartheta(X, \tau)$ is a linear combination of $\delta$-functions, the function $\vartheta(X, \tau)$ is conveniently first smoothed out in some way [see Richardson (1926, 1952)]. The function $q(l, \tau)$ will also satisfy the normalization condition

$$\int q(l, \tau)\,dl = 1. \tag{24.52}$$

It is clearly equal to the relative number of particle pairs in the cloud, whose coordinates differ by the vector $l$. Hence it follows that

$$\overline{q(l, \tau)} = \int p(l \mid l_0, \tau)\,q_0(l_0)\,dl_0, \tag{24.53}$$

where $p(l \mid l_0, \tau)$ has the same significance as in Sect. 24.2, and $q_0(l) = q(l, 0)$ is the initial value of $q(l)$ [for example, in the case of an admixture cloud which initially occupies a sphere of radius $R$, it is readily shown that $\frac{4}{3}\pi R^3 \cdot q_0(l) = 1 - \frac{3}{4}\frac{l}{R} - \frac{l^3}{16R^3}$, where $l = |l|$]. Therefore, data on the values of $q(l, \tau)$ can be used to estimate the probability density $p(l \mid l_0, \tau)$.

Since the determination of $q(l, \tau)$ from the results of diffusion experiments is nevertheless quite complicated, it is usual to employ some simpler numerical parameters instead of $\overline{q(l, \tau)}$. The most important of these are the components of the relative dispersion tensor $\Sigma_{ij}(\tau) = \overline{\langle l_i(\tau)\,l_j(\tau)\rangle}$ of the admixture cloud; the angular brackets here represent averaging over all the particle pairs of the cloud. This tensor is defined by

$$\Sigma_{ij}(\tau) = \int l_i l_j\,\overline{q(l, \tau)}\,dl. \tag{24.54}$$

In view of Eq. (24.53), the expressions for $\Sigma_{ij}(\tau)$ can also be rewritten in the form

$$\Sigma_{ij}(\tau) = \int D_{ij}^{(r)}(\tau \mid l_0)\, q_0(l_0)\, dl_0, \tag{24.54'}$$

so that the quantities $\Sigma_{ij}(\tau)$ are simply related to the components of the relative dispersion tensor $D_{ij}^{(r)}(\tau \mid l_0) = \overline{l_i l_j}$ of a pair of particles. In the case of a discrete cloud consisting of $N$ particles, the tensor $\Sigma_{ij}(\tau)$ will, of course, be given by

$$\Sigma_{ij}(\tau) = \frac{1}{N(N-1)} \sum_{m,\, n=1}^{N} \overline{\left(X_i^{(n)}(\tau) - X_i^{(m)}(\tau)\right)\left(X_j^{(n)}(\tau) - X_j^{(m)}(\tau)\right)} =$$

$$= \frac{1}{N(N-1)} \sum_{m,\, n=1}^{N} D_{ij}^{(r)}\left(\tau \mid X^{(n)}(0) - X^{(m)}(0)\right). \tag{24.55}$$

From the practical point of view, the simplest method of describing relative diffusion is to transform to a moving set of coordinates $\mathscr{S}_c$ whose origin lies at the center of gravity of the cloud for any $\tau$. If we again denote by $\vartheta(X, \tau)$ the concentration density of the cloud (which varies from one experiment to another), normalized by the condition $\int \vartheta(X, \tau)\, dX = 1$, the mean distribution of the cloud concentration relative to its center of gravity will be described by

$$\overline{\vartheta_c(x, \tau)} = \overline{\vartheta\left(x + \int X\vartheta(X, \tau)\, dX,\ \tau\right)}. \tag{24.56}$$

Theoretical investigation of this formula is very complicated because it depends on the Lagrangian characteristics of all the diffusing particles. However, the cloud dispersion tensor relative to its center of gravity

$$\sigma_{ij}(\tau) = \int x_i x_j \overline{\vartheta_c(x, \tau)}\, dx = \overline{\int (X_i - c_i)(X_j - c_j)\vartheta(X, \tau)\, dX},$$
$$c_i = \int X_i \vartheta(X, \tau)\, dX, \tag{24.57}$$

is very simply expressed in terms of $\Sigma_{ij}(\tau)$. In fact, by Eq. (24.51) we can rewrite Eq. (24.54) in the form

$$\Sigma_{ij}(\tau)=\int (X''_i-X'_i)(X''_j-X'_j)\overline{\vartheta(X',\tau)\vartheta(X'',\tau)}\,dX'\,dX''=$$
$$=2\int \overline{X_iX_j\vartheta(X,\tau)\,dX}-2\overline{c_ic_j}=2\sigma_{ij}(\tau), \quad (24.58)$$

i.e., *the dispersion tensor of an admixture cloud evaluated relative to its center of gravity is equal to one-half of the relative dispersion tensor of the cloud* [Brier (1950), Batchelor (1952a)].

The quantity

$$l_*(\tau)=[\Sigma_{ii}(\tau)]^{1/2}=[\overline{\langle l^2(\tau)\rangle}]^{1/2}=\sqrt{2}\,[\sigma_{ii}(\tau)]^{1/2} \quad (24.59)$$

can be regarded as the *effective diameter of the admixture cloud* at time $t_0+\tau$, and the quantity

$$K=\frac{1}{6}\frac{d}{d\tau}\,l_*^2(\tau) \quad (24.60)$$

as *the effective eddy diffusivity for the cloud* [similarly, $K_1=\frac{1}{2}\frac{d}{d\tau}\Sigma_{11}(\tau)=\frac{1}{2}\frac{d}{d\tau}\overline{\langle l_1^2(\tau)\rangle}$ can be taken as the effective eddy diffusivity in the direction of the $X_1$ axis]. If the initial separations between the cloud particles are all much less than the length scale $L$, then for $\tau\gg\tau_\eta$ and $\tau_3(l_{*0})<\tau<\tau_1(l_{*0})$, where $l_{*0}$ is the initial cloud diameter, the function $D_{ij}^{(r)}(\tau)$ will be given by Eq. (24.36). Therefore $l_*^2(\tau)\sim\tau^3$, and

$$K=\alpha\,l_*^{4/3}, \quad (24.61)$$

where $\alpha=(g\bar{\varepsilon})^{1/3}/2$ (and $K_1=K$). In other words, in the case which we are considering, *the effective eddy diffusivity for the admixture cloud is proportional to the effective cloud radius raised to the power 4/3*. This important law is often called the four-thirds law or Richardson's law, since it was established (purely empirically) by Richardson (1926). The four-thirds law is a direct consequence of the general physical description of the small-scale structure of turbulence. In fact, in all cases of turbulent mixing produced by eddies (i.e., turbulent inhomogeneities) with length scales limited by some typical scale $l_*$ from the inertial subrange $L\gg l_*\gg\eta$, the "exchange coefficient" $K$ which characterizes the rate of mixing will be determined only by $\bar{\varepsilon}$ and $l_*$. Since the exchange coefficient has the dimensions $L^2T^{-1}$, it follows that

$$K(l_*) = a\bar{\varepsilon}^{1/3} l_*^{4/3}, \tag{24.61'}$$

where $a$ is a dimensionless constant (whose value depends on the exact definition of $K$ and $l_*$ and can in principle be different for mixing substances of different natures). We have, in fact, already used Eq. (24.61') in Sect. 22.2. The derivation given here was first published by Obukhov (1941a, b) (this paper also contains references to the original experimental work of Richardson), and was then independently established by Weizsäcker (1948) and Heisenberg (1948a).

*Experimental Data on Relative Diffusion*

Let us now consider experimental data on the quantities $K(l_*)$, $\Sigma_{ij}(\tau)$ and $\sigma_{ij}(\tau)$, and compare these data with theoretical predictions. The first analysis of data of this kind was due to Richardson (1926). Richardson used atmospheric dispersion data for pilot balloons and volcanic ashes available at the time, and assumed also that the vertical eddy viscosity and thermal diffusivity in the atmosphere at a height $z$ were of the same order as the effective eddy diffusivity for an admixture cloud of diameter $z$ (since, in both cases, only eddies with scales of the order of $z$ or less can be effective in the diffusion process). This assumption enabled him to use data of W. Schmidt, F. Akerblom and G. I. Taylor on the values of the eddy viscosity and eddy diffusivity for heat at heights 15, 140, and 500 m. Additionally the data of A. Defant on the rate of horizontal macroturbulent mixing in the course of the general atmospheric circulation were also used. In this way, Richardson was able to estimate the values of $K = K(l_*)$ for six fixed distances $l_*$ in the range between 15 m and 1000 km. The resulting values of $K$ were found to increase by a factor of $10^7$ in this range, clearly showing the accelerating nature of the relative diffusion process, which we have already mentioned above. The function $K(l_*)$ is reasonably well approximated by the function (see Fig. 98)

$$K(l_*) = 0{,}2\, l_*^{4/3}$$

Only the last point in Fig. 98 departs appreciably from the straight line. It corresponds to macroturbulent mixing with $l_* \approx 1000$ km. The data shown in Fig. 98 are of course very approximate, and do not enable us to determine the exponent $m$ in the law $K(l_*) \sim l_*^m$ with an accuracy exceeding one or two tenths. The fact that Richardson chose the exponent $m = 4/3$ indicates his faith in the existence of a universal physical law of sufficiently simple form. Nevertheless, the agreement between the empirical law of Richardson and the theoretical formula (24.61) of Obukhov can be regarded as an argument in favor of the theory. We note that the inhomogeneity of the data employed should not be particularly important, since it could only affect the factors $a$ and $\bar{\varepsilon}^{1/3}$ in Eq. (24.61′), corresponding to different points in Fig. 98. This would not be noticeable against a background of the much more rapid variation of $l_*^{4/3}$ and $K(l_*)$. A more important argument against the identification of Richardson's empirical law with the theoretical formula (24.61) is that the range of values $l_*$ in Fig. 98 substantially exceeds the range $l_* < L$ for which the four-thirds law can be justified theoretically (we recall that an analogous situation was encountered at the end of Sect. 23.6).

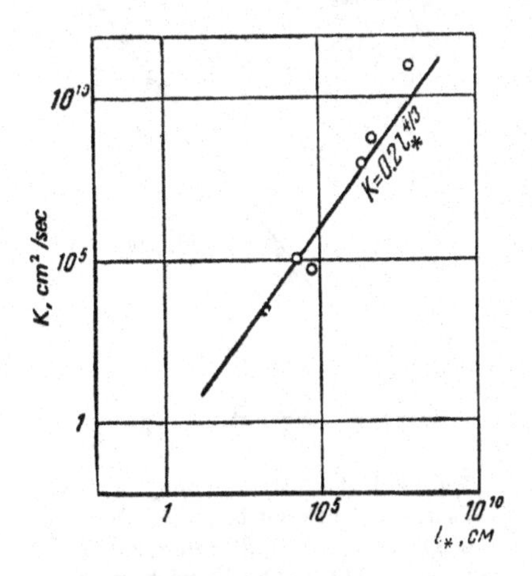


**Fig. 98** Eddy diffusivity $K$ as a function of the length scale $l_*$ [Richardson (1926)].

Additional data on the effective eddy diffusivity in the atmosphere for $l_*$ varying between 1 cm and a few tens of kilometers, which were later analyzed by Richardson (1929), were again in adequate agreement with the four-thirds law. These results were disregarded by almost all researchers for a long time but, since the late 1940's, a large number of papers have appeared on the verification of the four-thirds law (24.61) for two-dimensional horizontal diffusion of clusters of discrete particles of various kinds or of continuous clouds of marked fluid on the surface of the sea or a large lake [see, for example, Richardson and Stommel (1948), Stommel (1949), Inoue (1950b), Hanzawa (1953), Defant (1954), Ozmidov (1957, 1959a), Pearson (1958), Olson and Ichiye (1959), Gunnerson (1960), Csanady (1963), Kilezhenko (1964), Kullenberg (1972), and others; cf. also the surveys by Bowden (1962) and Ozmidov (1968)]. The data collected in these papers do, as a rule, agree with (24.61) in a broad range of scales $l_*$, which in some cases extends from some tens of centimeters to hundreds of kilometers [see, for example, Fig. 99, which is taken from the paper by Olson and Ichiye (1959)]. Similar summarizing graphs can be found in Orlob (1959) and Gunnerson (1960). However, the values of the dimensional constant $\alpha$ as reported by different workers are appreciably different, varying from 0.002 to 0.07 $\text{cm}^{2/3}\,\text{sec}^{-1}$.

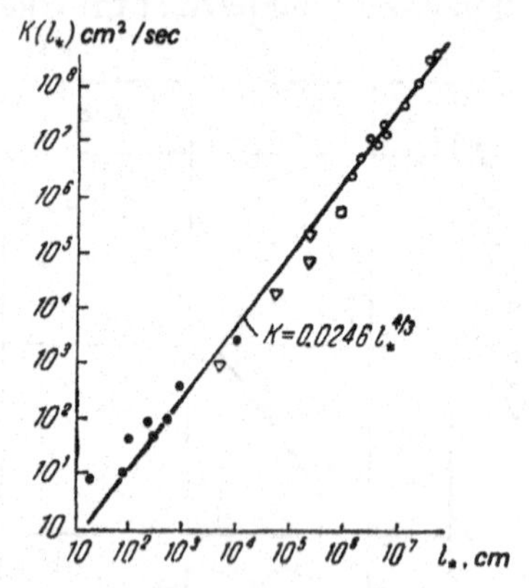


**Fig. 99** Dependence of $K$ on $l_*$ for two-dimensional diffusion in the sea [Olson and Ichiye (1959)]. Different symbols show data obtained by different workers.

The spread of the values of the coefficient $\alpha$ can be explained by differences in the energy dissipation $\bar{\varepsilon}$, and by differences in the exact definitions of the effective eddy diffusivity $K$ and the effective length scale $l_*$. Moreover, not all the determinations of $K=K(l_*)$ used in the above papers are theoretically justified. Many workers have uncritically followed Richardson and Stommel (1948) and assumed that relative diffusion can be described by a parabolic diffusion equation of the form $\frac{\partial p}{\partial \tau}=\frac{\partial}{\partial l}\left(K(l)\frac{\partial p}{\partial l}\right)$ for all $\tau$. Having solved this equation on the assumption that for small time intervals $\tau$ (when $\overline{[l(\tau)-l_0]^2}/l_0^2 \ll 1$) the diffusivity $K(l)=a\bar{\varepsilon}^{1/3}l^{4/3}$ can be regarded as practically constant, and equating the variance of the resulting Gaussian distribution to $\overline{[l(\tau)-l_0]^2}$ (where the average is taken over a large number of particle pairs with practically identical initial separation $l_0$), we find that

$$\overline{[l(\tau)-l_0]^2}/2\tau = K(l_*) = a\,\bar{\varepsilon}^{1/3} l_*^{4/3}, \quad l_* = \overline{[l_0 + l(\tau)]}/2, \quad (24.62)$$

where $\overline{[l(\tau)-l_0]^2}/l_0 \ll 1$. This formula has been widely used in the literature although its derivation does not follow from dimensional considerations and is based on special hypotheses which are, in fact, inconsistent with such considerations [see Eq. (24.63) below]. The point is that the four-thirds law is valid only for the "intermediate" values of $\tau$ in the range $\tau_3 < \tau < \tau_1$ and, therefore, if we apply Eq. (24.62) to relatively long time intervals $\tau$, such that $\overline{[l(\tau)-l_0]^2}/l_0^2 \gg 1$ [but $\left[\overline{l^2(\tau)}\right]^{1/2} \ll L$], we obtain $\overline{l^2(\tau)} \sim \tau\bar{\varepsilon}^{1/3}\left[\overline{l^2(\tau)}\right]^{2/3}$, which is equivalent to Eq. (24.36) and, consequently, follows from dimensional considerations [the last result enables us to justify the estimated values of $K(l_*)$ due to Ichiye and shown by the open circles in Fig. 99]. However, if $l_0 \gg \eta$ and $\overline{[l(\tau)-l_0]^2}/l_0^2 \ll 1$, so that $\tau < \tau_2$ (see Sect. 24.2), then the four-thirds law will no longer be valid. For such values of $\tau$ we have Eq. (24.30) and, consequently,

$$\overline{[l(\tau)-l_0]^2}/2\tau \sim \bar{\varepsilon}^{2/3}\tau\, l_0^{2/3}. \quad (24.63)$$

which is different from Eq. (24.62). Nevertheless, Ichiye and Olson (1960) have used the formula (24.63) to explain why many workers have, in fact, obtained positive results when using small $\tau$ data for the experimental verification of Eq. (24.62). Actually, the condition $\overline{[l(\tau)-l_0]^2} \ll l_0^2$ means that $\overline{[l(\tau)-l_0]^2}/l_0^2$ is chosen to be some small fraction $b$. Since, however, it is desired to measure the difference $l(\tau)-l_0$ with sufficient accuracy and reliability, any value of $b$ that is too small is also inconvenient. Therefore, considerations of convenience frequently led to the choice of values of $\tau$ for different $l_0$ such that $\overline{[l(\tau)-l_0]^2}/l_0^2 = b$ varies relatively little [i.e., $\tau$ is approximately proportional to $l_0^{2/3}$; see, for example, the tabulated values of $\tau$ which were recommended by Stommel (1949) for different $l_0$ and yield satisfactory agreement with the last condition]. When $\overline{[l(\tau)-l_0]^2}/l_0^2 = b \approx \text{const}$, the formula (24.63) is equivalent to Eq. (24.62). If in a broad range of values of $l_0$ we take the same value of $\tau$ for which $\overline{[l(\tau)-l_0]^2} \ll l_0^2$, then the formula given by Eq. (24.62) will no longer be valid. This last remark can possibly explain those of the results reported by Ozmidov (1957, 1959b, 1968) which contradict the four-thirds law. Moreover, in Ozmidov's experiments which were performed in an artificial reservoir, the value of Re was probably insufficient to ensure the presence of an appreciable inertial subrange in the turbulence spectrum. Still smaller values of Re were employed by Orlob (1959), who found that the eddy diffusivity for a single particle in a broad and shallow laboratory channel is proportional to the Lagrangian integral length scale raised to the power 4/3. This, however, can hardly be related to Richardson's law given by Eq. (24.61).

Let us also note that it is at present not completely clear whether it is permissible to use the above theoretical deduction to explain the data on horizontal turbulent diffusion on a sea or lake surface. It is usually assumed that the theory of locally isotropic turbulence can be applied to two-dimensional sea turbulence provided only that there are no substantial vertical motions which occasionally lead to a rapid distortion of the diffusion process [cf., e.g., Csanady (1963)]. However it is quite possible that the interaction between horizontal diffusion and vertical shear (studied in Sects. 10.4 and 10.5 of Vol. 1) and between horizontal and vertical turbulent diffusion can in many cases significantly affect the rate of horizontal admixtures dispersion in the surface layer of a sea [see e.g., Bowden (1965), Csanady (1966), Okubo (1967, 1968a), Monin (1969), Kullenberg (1969, 1972), Schuert (1970) where additional references can be found]. It is also clear that the greatest length scales in the diagram of Fig. 99 and others similar to it are substantially outside the range of scales of three-dimensional locally isotropic turbulence in the sea. We must recall in this respect the discussion at the end of Sect. 23.6 where it was explained that there are apparently several five-thirds ranges in the spectra of horizontal oceanic turbulence which are characterized by different values of the spectral energy transfer rate $\bar{\epsilon}$ (cf. Fig. 96 above). Having this remark in view Okubo and Ozmidov (1970) reconstructed the graph of the function $K(l_*)$ on the basis of numerous data on the diffusion of clouds of marked water in the ocean analyzed by Okubo (1962b, 1968b). The resulting diagram is shown in Fig. 99a; it shows two four-thirds ranges with different values of coefficient $\alpha$ and is in a very good qualitative agreement with the schematic graph in Fig. 96. The estimates of the value of $\bar{\epsilon}$ from the data in Fig. 99a for the range of length scales 10m–1km and 10 km–1000 km are in satisfactory agreement with the values recommended by Ozmidov (1968). However there are evidently some discrepancies between the graph in Fig. 99a, the previous diagram of Olson and Ichiye in Fig. 99, and the similar diagrams published by Kullenberg (1972); these discrepancies still need explanation.

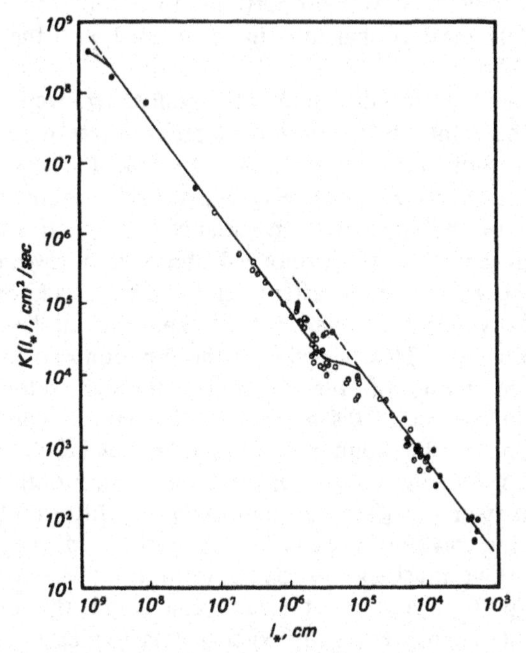


**Fig. 99a** The dependence of the horizontal eddy diffusivity $K$ on the length scale $l_*$ according to Okubo and Ozmidov (1970). The different symbols correspond to different data sets.

Experiments on the mean square separations between pairs of particles have also been carried out in the atmosphere. Thus, Durst (1948) has described the relative motion of twenty pairs of smoke-puffs ejected from an aircraft at heights between 1 km and 2 km, while Wilkins (1958) investigated the separation $l = l(\tau)$ between two balanced pilot balloons released simultaneously in the surface layer from a pair of points separated by a distance of 10 or 100 m (in the direction perpendicular to the mean wind). Durst's data were analyzed by Charnock (1951) who calculated the ratios $\overline{[l(\tau) - l_0]^2}/\tau$

for $\tau$ between 10 and 50 seconds. His results show a considerable spread but, on the whole, they are not inconsistent with Eq. (24.63), yielding reasonable values of $\bar{\varepsilon}$. Wilkins determined the quantity $\overline{[\Delta l(\tau)]^2} = \overline{[l(\tau) - l_0]^2}$ for $\tau$ in the range between fractions of a minute and ten minutes, and found that for $l_0 =$ 10 m the quantity $\overline{[\Delta l(\tau)]^2}$ remained approximately proportional to $\tau^2$ throughout this range, and for $l_0 =$ 100 m it was proportional to $\tau^2$ for small $\tau$, and then approximately proportional to $\tau^3$ in accordance with the theoretical formula (24.36).

More data on relative diffusion in the atmosphere were obtained as a result of observations on continuous puffs or jets of admixture. In the case of short time observations of smoke-puffs in the atmosphere, the observable radius $r = r(\tau)$ is usually proportional to the square root of the concentration variance $\sigma^2(\tau)$ relative to the puff center of gravity. However, prolonged observations revealed that the radius $r(\tau)$ began to decrease after a certain value $\tau$ had been reached, and eventually became equal to zero, i.e., the smoke-puff disappeared completely. This behavior of the function $r(\tau)$ can be easily explained in terms of the "visibility through smoke" theory developed in the early 20's by O. F. T. Roberts. According to this theory the observable boundary of a smoke cloud is determined by the condition that the integral of the smoke concentration along the line of sight assumes a given constant value [cf. Gifford (1957a)]. To determine the relation between the observable radius $r(\tau)$ and the variance $\sigma^2(\tau)$ we must specify the concentration distribution within the smoke-puff. The simplest assumption about this distribution, which is not inconsistent with existing data (which so far are very approximate; see the end of Sect. 24.4 below) is that the distribution $\overline{\vartheta_c(\boldsymbol{x}, \tau)}$ in Eq. (24.56) is Gaussian for all $\tau$, i.e.,

$$\overline{\vartheta_c(\boldsymbol{x}, \tau)} = Q\,[2\pi\sigma^2(\tau)]^{-3/2} \exp\{-\boldsymbol{x}^2/2\sigma^2(\tau)\}, \tag{24.64}$$

where $\sigma^2(\tau) = \sigma_{ll}(\tau)/3 = \Sigma_{ll}(\tau)/6$. In that case, the observable radius $r(\tau)$ for the case of observation from below is given by the condition

$$\int_{-\infty}^{\infty} \overline{\vartheta_c(x, y, z, \tau)}\, dz = Q\,[2\pi\sigma^2(\tau)]^{-1} \exp\{-r^2(\tau)/2\sigma^2(\tau)\} = \text{const.} \tag{24.65}$$

This equation permits us to find $\sigma^2(\tau)$ from given values of $r(\tau)$ and of the constant in the right-hand side of Eq. (24.66) and gives quite definite forms of dependence of $r^2(\tau)$ on $\tau$ for the various time ranges studied in Sect. 24.2 [see, for example, Zimmerman (1965)]. Having determined the maximum observable radius $r_{max}$ as the value of $r(\tau)$ for which $dr^2(\tau)/d\tau = 0$, we can also readily show that

$$\sigma^2(\tau) = r^2(\tau)\Big/2\left[\ln\frac{r_{max}^2 e}{2} - \ln\sigma^2(\tau)\right]. \qquad (24.66)$$

This equation does not contain any empirical constant and enables us to determine uniquely the value of $\sigma^2(\tau)$ and, consequently, $\Sigma_{ll}(\tau) = \overline{\langle l^2(\tau)\rangle}$ from the measured values of $r(\tau)$.

Gifford (1957a, b) has used Eq. (24.66) to analyze a number of experiments on atmospheric dispersion of smoke-puffs [Frenkiel and Katz (1956), Seneca (1955), and Kellogg (1956)], which were carried out at heights of the order of a few meters or tens of meters, some hundreds of meters, and between 7.5 km and 22 km, respectively. His main result is that the values of $\sigma^2(\tau) = \overline{l^2(\tau)}/2$ above some value of $\tau$ are, in fact, usually approximately proportional to $\tau^3$, whereas for smaller $\tau$ they grow more slowly (see Fig. 100). This result is in good agreement with the theoretical predictions given in Sect. 24.2. The same data were reexamined later by Zimmerman (1965) who also found that they agree with the prediction from the theory of locally isotropic turbulence. Gifford also analyzed the data of Tank (1957), who produced an admixture cloud near the ground, and measured the admixture concentration at a number of points arranged in the form of a regular grid. His analysis was based on the assumption that the time integral of the concentration on the cloud axis at a distance $X$ downwind from the source can, in the first approximation, be regarded as inversely proportional to the variance $\sigma^2(X/\bar{u})$, where $\bar{u}$ is the mean wind velocity. [This assumption was justified by Gifford by equations similar to Eqs. (10.6′) and (10.90) of Vol. 1, in the second of which he substituted $Y = Z = H = 0$, $K_{yy}X = K_{zz}X = \bar{u}\,\sigma^2(X/\bar{u})$]. Starting with this Gifford found values of $\sigma^2(\tau)$ for the experimental data of Tank, and discovered that the values of $\overline{l^2(\tau)}$ for sufficiently large $\tau$ were often approximately proportional to $\tau^3$. However later Zimmerman (1965) criticized Gifford's assumption and derived another equation for the time integral of the concentration according to which this integral must be proportional to $\tau^{-7/2}$ (and not to $\tau^{-3}$) for the "intermediate" values of $\tau$. He discovered also that this last dependence agrees well with

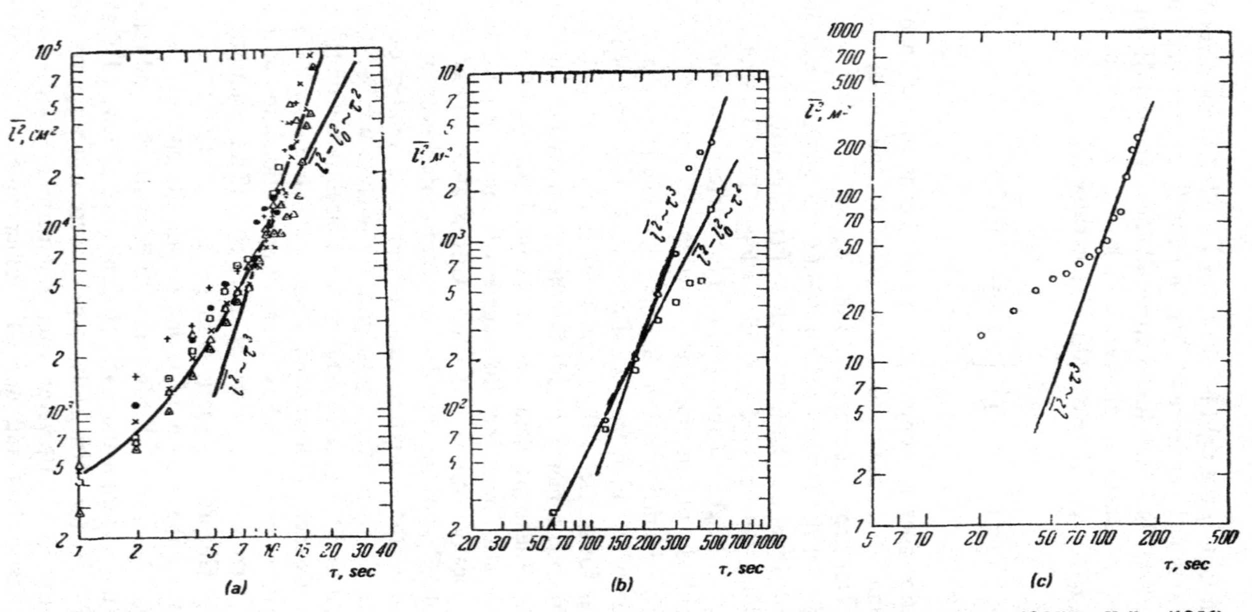


**Fig. 100** Dependence of $\overline{l^2}$ on the time $\tau$ according to Gifford (1957a, b). a—Frenkiel and Katz (1956); b—Seneca (1955); c—Kellog (1956).

Tank's data for the case of medium wind velocity (but disagrees strongly with data for stable conditions with very light wind). Experiments similar to those of Tank were also carried out by Smith and Hay (1961). However, here the size of the cloud was not small in comparison with the turbulence length scale $L$, and the results were less definite.

Data on admixture jets propagating in a turbulent medium can be analyzed in a similar way. Thus, Greenhow (1959), who followed the suggestion by Booker and Cohen (1956), used meteor-trail data to estimate $\bar{\varepsilon}$ in the lower ionosphere (at heights of the order of 80–100 km). According to his results, in the case of persistent trails (with lifetimes in excess of 100 sec) the trail radius for sufficiently large values of $\tau$ is approximately proportional to $\tau^3$, as expected from Eq. (24.36). A more accurate estimate of $\sigma^2(\tau)$ and $\overline{l^2(\tau)}=2\sigma^2(\tau)$ from data on the observable width $y=y(\tau)$ of a jet for different $\tau$ can be obtained from the formula $\sigma^2(\tau)=y^2(\tau)\,[\ln(ey^2_{\max}/\sigma^2(\tau))]^{-1}$, which is analogous to Eq. (24.66) [see Gifford (1959b)]. This formula, with $\tau$ replaced by $X/\bar{u}$), was used by Bowne (1961) to analyze data on a smoke plume issuing from a stationary source at a height of 50 m on a meteorological tower. The same analysis was later undertaken by Garger (1969) and Byzova and Garger (1970) who treated numerous photographs of smoke plumes from sources at several heights of about 100 m on a meteorological tower. In both analyses, the result was $\sigma^2(\tau)\sim\tau^3$ for sufficiently large $\tau$. Data confirming that the square of the width of a dye jet issuing into seawater or a lake is proportional to the cube of the distance $X$ from the source (for sufficiently large $X$) can be found in Okubo (1962a), Csanady (1963) and a number of other papers. We recall that in the Taylor theory of single particle turbulent diffusion in homogeneous turbulence, a jet of admixture emitted by a stationary source should at first expand in proportion to $\tau=X/\bar{u}$ (i.e., it should have a conical form), while for large $\tau$ it should be proportional to $\tau^{1/2}=(X/\bar{u})^{1/2}$ [i.e., it should have the form of a paraboloid of revolution; see formulas (9.28) and (9.35) in Vol. 1]. The theory of relative diffusion developed in this section shows that, for sufficiently large Re, the admixture jet emitted by a stationary source (with linear size exceeding the scale $\eta$) should at first expand in proportion to $\tau=X/\bar{u}$, then, in proportion to $\tau^{3/2}=(X/\bar{u})^{3/2}$, and only finally, in the homogeneous case, it should follow the law $\sigma\sim\tau^{1/2}=(X/\bar{u})^{1/2}$ [see Eqs. (24.30), (24.36), and (24.29′)]. It is important, however, that the width of the jet be measured not relative to a fixed axis, directed along the mean velocity $\bar{u}$ (which is also the symmetry axis of the field of the mathematical expectation of the concentration distribution; see Fig. 97), but relative to the center of gravity of the cross section of the jet oscillating under the influence of long-period turbulence components and producing at each time a large number of inflections in the observed jet axis. If, on the other hand, we use a fixed set of coordinates, then the analysis of the statistical characteristics of the concentration fluctuations at a fixed space point will have to allow both for relative diffusion, which produces the expansion of the jet, and slow oscillations of the jet axis. Since these two processes are produced by turbulent disturbances with very different scales, they are naturally regarded as statistically independent. Calculations of the statistical characteristics of the concentration distribution in a jet have been made on the basis of this assumption by Gifford (1959c, 1960); see also Garger (1969) and Byzova and Garger (1970).

Observations of the relative diffusion of discrete particle clusters or continuous admixture clouds enable us to estimate the mean energy dissipation $\varepsilon$, since this quantity is present in many of the formulas given above. The most convenient for this purpose are the formulas (24.30), (24.36), and (24.37), and finally the expressions given in Sect. 24.2, namely, $\tau_2 \sim (l_0^2/\bar{\varepsilon})^{1/3}$, $\tau_3 \sim (l_0^2/\bar{\varepsilon})^{1/3}$. These expressions show that transition from $\overline{l^2(\tau)} = l_0^2 + c(\bar{\varepsilon} l_0)^{2/3}\tau^2$ to $\overline{l^2(\tau)} = g\bar{\varepsilon}\tau^3$ occurs, roughly speaking, at the time $t_1 = \varkappa l_0^{2/3}\bar{\varepsilon}^{-1/3}$ where $\varkappa$ is a dimensionless constant of the order of unity [close in magnitude to $c/g$ since $t_1$ can be estimated from the condition $c(\bar{\varepsilon} l_0)^{2/3} t_1^2 \approx g\bar{\varepsilon} t_1^3$]. The one-dimensional and two-dimensional versions of all these formulas can also be used for determination of $\bar{\varepsilon}$. There are many examples in the literature of the application of these formulas to the estimation of $\bar{\varepsilon}$ [see, for example, Charnock (1961), Gifford (1957a, b, 1962), Wilkins (1958, 1960, 1963), Greenhow (1959), Ozmidov (1960b), Tatarskii (1960), Ball (1961), Ivanov (1962), and Zimmerman (1965)]. However, the results obtained in this way are not in very good agreement with each other and are not entirely reliable because of the presence of undetermined numerical factors in the $m$. The only relatively reliable estimate of these factors is that obtained for $c$, namely, $c = C + 2C' = 11C/3 \approx 7$, which follows from the data of Sect. 23.3. As regards the constants $g$, $a = g^{1/3}/2$ and $\varkappa$, these can best be estimated from simultaneous measurements of $\bar{\varepsilon}$ based on diffusion data, and with the aid of any method using only formulas with well-established numerical factors. Unfortunately, almost no such data are available at present.[30] The coefficients $g$, $a$, and $\varkappa$ are therefore usually determined by comparing diffusion data with data on $\bar{\varepsilon}$ referring to different points and different times. For example, Obukhov (1941b) and, subsequently, Batchelor (1950) compared the data of Richardson with the approximate estimate of the mean dissipation rate $\bar{\varepsilon}$ for the entire atmosphere due to D. Brunt, and have concluded that the coefficient $a$ is not very different from unity. Later, Ozmidov (1960b) used the same very approximate material to conclude that $a \approx 0.1$. Gifford (1957a, b, 1962) and Zimmerman (1965) assumed that $\varkappa=1$ (and Zimmerman even showed that this estimate is consistent with the existing data), whereas, according to Ball, $\varkappa \gtrsim 2$. Finally, Tatarskii (1960) has put forward some arguments in favor of estimates $g \approx 0.06$, $a \approx 0.2$. Determination of more precise values of all these constants is a matter for the future.

## 24.4 Hypotheses on the Probability Distributions of Local Diffusion Characteristics

Similarity hypotheses enable us to determine the probability distributions for the local statistical characteristics of the motion of a given fluid particle, or the characteristics describing the relative motion of two fluid particles, but only to within certain universal functions whose form is not indicated by the hypotheses. In particular, similarity hypotheses often enable us to determine to within numerical factors the moments of the relative distribution of admixture concentration, but are of no help if we are interested in the concentration distribution within the cloud. Finally, theoretical determination of the numerical constants in the formulas of similarity theory given in Sect. 24.2 again cannot be done on the basis of dimensional considerations alone. Therefore, there is considerable interest in supplementing the theory of relative diffusion by additional hypotheses which enable us to extend the conclusions that follow from dimensional considerations alone.

### *Application of Dynamic Equations and of the Closure Methods for the Moment Equations*

We have already noted in Sects. 9.1–9.2 of Vol. 1 and in Sect. 24.1 of this volume that transformation from Eulierian variables to the Lagrangian description of the motion of a fluid particle (located at the point $x$ at time $t_0$) can be achieved by considering the scalar field $\vartheta(X, t)$ which satisfies the "transport equation"

$$\frac{\partial\vartheta}{\partial t} + u_j \frac{\partial\vartheta}{\partial X_j} = 0 \qquad (24.67)$$

[30] The only such attempt known to us is that of Byzova and Garger (1970). These authors compared estimates of $\bar{\epsilon}$ from data on the cubic dependence of smoke plume widths on time with simultaneous estimates from data on the time structure function of the longitudinal wind component. Using the value $C = 1.95$ in his second method Garger obtained the approximate relation $\sigma^2(\tau) \approx 0.3\,\bar{\epsilon}\tau^3$, i.e., $g \approx 0.3 \cdot 6 \approx 1.8$.

subject to the initial condition

$$\vartheta(X, t_0) = \delta(X - x). \tag{24.68}$$

By assuming the velocity field $u(X, t)$ to satisfy the incompressible continuity equation and the Navier-Stokes equation, and solving simultaneously with these equations the problem defined by Eqs. (24.67)–(24.68), we can express all the statistical characteristics of diffusion from a fixed source in terms of the resulting solutions.

A similar approach is possible in the problem of the relative motion of two fluid particles. The first step is to consider two scalar fields $\vartheta_1(X, t)$ and $\vartheta_2(X, t)$ satisfying the same transport equation (24.67) but with different initial conditions $\vartheta_1(X, t_0) = \delta(X - x_1)$ and $\vartheta_2(X, t_0) = \delta(X - x_1 - l_0)$. Having solved the equation for $\vartheta_1(X, t)$ and $\vartheta_2(X, t)$ simultaneously with the Navier-Stokes and continuity equations, we can express all the statistical characteristics of the motion of a pair of fluid particles in terms of these solutions. In particular, $\overline{\vartheta_1(X_1, t_1)\,\vartheta_2(X_2, t_2)} = p(X_1, X_2 \mid x_1, x_1 + l_0, t_1, t_2, t_0)$ can be interpreted as the joint probability density for the coordinate $X_1$ of the first particle at time $t_1$, and the coordinate $X_2$ of the second particle at time $t_2$. Second, we can consider a set of statistically independent pairs for which the initial coordinate $x_1$ of the first particle is distributed uniformly in space and the initial coordinate of the second particle is $x_1 + l_0$ where $l_0$ is a fixed vector. Let us now associate with each pair the solution $\vartheta(X, t)$ of the transport equation which is zero everywhere at the initial time $t_0$, except for singularities at $x_1$ and $x_1 + l$ which have the same sign at both points but differ in sign for different pairs (so that the sign of $\vartheta$ for each pair is positive with probability 1/2 and negative with equal probability). Then we can construct the statistical ensemble of realizations of the random function $\vartheta(X, t)$, whose values at time $t_0$ are independent of the velocity field $u(X, t_0)$ and satisfy the transport equation (24.67) together with the conditions

$$\overline{\vartheta(X, t_0)} = 0, \quad \overline{\vartheta(X + l, t_0)\,\vartheta(X, t_0)} = \delta(l) + \frac{1}{2}\left[\delta(l - l_0) + \delta(l + l_0)\right]. \tag{24.69}$$

Having solved the transport equation subject to these conditions simultaneously with the Navier-Stokes and continuity equations, we can determine the probability density $p(l \mid l_0, \tau)$ with the aid of the equation

$$\overline{\vartheta(X + l, t_0 + \tau)\,\vartheta(X, t_0 + \tau)} = \delta(l) + p(l \mid l_0, \tau) \tag{24.70}$$

[see Kraichnan (1966b)].

The above two methods enable us to obtain (after eliminating the pressure field) an infinite set of moment equations for all the possible moments and joint moments of the fields $u(X, t)$, $\vartheta_1(X, t)$ and $\vartheta_2(X, t)$ [or $u(X, t)$ and $\vartheta(X, t)$]. This set contains the statistical characteristics of the relative motion of a pair of fluid particles as the unknowns in which we are interested. In other words, these methods result in an analytic formulation of the problem of relative diffusion which is similar to the formulation of the turbulence problem. Thereafter, the theoretical determination of the relative diffusion characteristics is found to encounter the usual difficulties associated with the closure of the moment equations, which we have already discussed.

The reduction of the determination of the relative diffusion parameters to the determination of the moments of certain random fields enables us to apply the approximate closure methods discussed in Sects. 19 and 22. Here we shall confine our attention to the closure methods for moment equations developed by Kraichnan. Application of the old Kraichnan direct-interaction approximation (1959, 1961) to relative diffusion (by considering the two fields $\vartheta_1(X, t)$ and $\vartheta_2(X, t)$ was considered by Roberts (1961). A complicated integro-differential equation was found as a result of this analysis for $\overline{\vartheta_1(X, t_1)\vartheta_2(X_2, t_2)} = p(X_1, X_2 | x_1, x_1 + l_0, t_1, t_2, t_0) = p(X_1, X_2, t_1, t_2)$, in which the integrand contains both the first and second space derivatives of $p(X_1, X_2, t_1, t_2)$, the Eulicrians space-time velocity correlation function $B_{ij}(X - X', t, t')$, and the single-particle probability density $p(X, t)$ given by Eq. (24.8). Subject to certain additional assumptions which are acceptable for large Re, it is possible to obtain the following equation for $p(l | l_0, \tau)$:

$$\frac{\partial p(l | l_0, \tau)}{\partial \tau} = \frac{\partial^2}{\partial l_i \partial l_j}[K_{ij}(l, \tau) p(l | l_0, \tau)], \tag{24.71}$$

where

$$K_{ij}(l, \tau) = 2\int_0^\tau d\tau' \int dl' [B_{ij}(l', \tau') - B_{ij}(l + l_0 - l', \tau')]\tilde{p}(l', \tau'), \tag{24.72}$$

and $B_{ij}(l, \tau) = \overline{u_i'(x, t_0) u_j'(x + l, t_0 + \tau)}$, $\tilde{p}(l, \tau) = p(x + l, t_0 + \tau | x, t_0)$. Equation (24.71) has the form of the usual parabolic diffusion equation in which the diffusivity $K_{ij}(l, \tau)$ is a function of $l$ and $\tau$ [cf. also Eq. (24.83) below put forward as far back as 1926 by Richardson]. It is readily shown that, when $\tau < \tau_2$, equations (24.71)–(24.72) lead to the correct result (24.30), while for $\tau > \tau_0$ these equations are in agreement with Eq. (24.29'). However, for intermediate values of $\tau$ we do not obtain Eq. (24.36), which follows from dimensional considerations [since for all $\tau$ the diffusivities $K_{ij}(l, \tau)$ depend on the turbulent energy which is determined by large-scale disturbances]. This is not surprising because we already know that the direct-interaction approximation cannot be made to agree with the Kolmogorov similarity hypotheses (see Sect. 19.6).

The more complicated Lagrangian-history direct-interaction approximation, which is consistent with the similarity hypotheses, was developed by Kraichnan (1965a) (see Sect. 22.2) and was applied to the relative diffusion problem by Kraichnan (1966b). In this approximation the probability density $p(l | l_0, \tau)$ again obeys a diffusion equation of the form (24.71) but now

$$K_{ij}(l, \tau) = 2\int_{t_0}^{t_0+\tau} \{\overline{u_i'(x, t_0 + \tau) V_j'(t' | x, t_0 + \tau)} - \overline{u_i'(x + l, t_0 + \tau) V_j'(t' | x, t_0 + \tau)}\} dt', \tag{24.73}$$

where $V'(t' | x, s)$ is the fluctuation in the fluid particle velocity at time $t'$ which occupies a given point $x$ at time $s > t'$. Hence for $\tau < \tau_2$ and $\tau > \tau_0$ we again obtain the correct formulas (24.30) and (24.29') but, in addition, when $\tau_3 < \tau < \tau_1$ (and $l_0 \gg \eta$) Eqs. (24.71) and (24.73) now lead to results which are in agreement with the conclusions drawn from the similarity hypotheses. Namely, since for $\tau_3 < \tau < \tau_1$ and $l_0 \gg \eta$ the probability density $p(l | l_0, \tau) = p(l, \tau)$ is independent of $l_0$ but is a function of $l = |l|$ and $\tau$, Eqs. (24.71) and (24.73) now transform into the isotropic diffusion equation

$$\frac{\partial p(l, \tau)}{\partial \tau} = \frac{1}{l^2}\frac{\partial}{\partial l}\left[l^2 K(l, \tau)\frac{\partial p(l, \tau)}{\partial l}\right], \tag{24.74}$$

where

$$K(l, \tau) = \overline{\varepsilon}^{1/3} l^{4/3} a(\overline{\varepsilon}^{1/3} l^{-2/3} \tau),$$

$$a(s) = 4C_1 \int_0^s ds' \int_0^\infty \left(\frac{1}{3} + \frac{\cos x}{x^2} - \frac{\sin x}{x^3}\right) R(s' x^{2/3}) x^{-5/3} \, dx, \tag{24.75}$$

$C_1$ is the coefficient in the five-thirds law (21.24) for the three-dimensional spectrum, $R(\overline{\varepsilon}^{1/3} k^{2/3} \tau) = F(k, \tau)/F(k, 0)$, and $F(k, \tau)$ and $F(k, 0)$ are the three-dimensional Fourier transforms of the correlation functions $u_i'(x+r, t) V_i'(t-\tau | x, t)$ and $u_i'(x+r, t) u_i'(x, t)$ for $k = |k|$ and $\tau$ from the inertial subrange. The formulas (24.75) show that Richardson's four-thirds law is still valid but with a factor which is a function of $\tau$ (or, more precisely, a function of the dimensionless combination $\overline{\varepsilon}^{1/3} l^{-2/3} \tau$) which is not inconsistent with dimensional considerations. It is readily seen that, by the first equation (24.75), the solution $p(l, \tau)$ can be written in the form

$$p(l, \tau) = (\overline{\varepsilon} \tau^3)^{-3/2} \Phi(l (\overline{\varepsilon} \tau^3)^{-1/2}) \tag{24.76}$$

as expected [see Eq. (24.28), where the function $\Phi$ was indicated by the symbol $\Phi^{(6)}$]. Using the numerical values for $C_1$ and $R(s)$ obtained as indicated in Sect. 22.2, and integrating Eqs. (24.74)–(24.75) numerically subject to the normalizing condition

$$4\pi \int_0^\infty p(l, \tau) l^2 \, dl = 1,$$

we can obtain the values of the function $\Phi(\xi)$. [The result of this calculation is shown in Fig. 101 together with graphs of the analogous functions corresponding to the probability

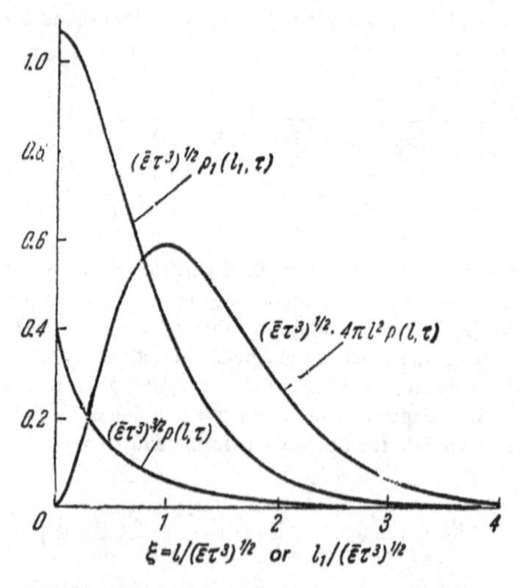


**Fig. 101** Graphs of universal functions describing the densities $p(l, \tau)$, $p_0(l, \tau) = 4\pi l^2 p(l, \tau)$ and $p_1(l_1, \tau)$ [Kraichnan (1966b)].

density $p_0(l, \tau) = 4\pi l^2 p(l, \tau)$ of the variable $l = |\boldsymbol{l}|$ and the probability density $p_1(l_1, \tau) = 4\pi \int_{l_1}^{\infty} p(l, \tau)\, l\, dl$ of the component $l_1$ of the random vector $\boldsymbol{l}$]. For very large $\xi$ the resulting function $\Phi(\xi)$ decreases as $\exp\{-b\xi^{4/3}\}$, $b = \text{const}$. For the first three moments of $l^2(\tau)$ the probability density $p(l, \tau)$ in Fig. 101 yields

$$\overline{l^2(\tau)} \approx 2{,}4\bar{\varepsilon}\tau^3,$$
$$\overline{l^4(\tau)} \approx 2{,}3\,[\overline{l^2(\tau)}]^2,$$
$$\overline{l^6(\tau)} \approx 8{,}9\,[\overline{l^2(\tau)}]^3,$$

so that, in particular, $g \approx 2.4$. Unfortunately, there are no reliable measurements of the corresponding quantities at the present time. Such data and comparisons with the above predictions should, in future, provide additional information about the validity of the hypotheses used by Kraichnan to close the equations for the moments.

## *The Obukhov Hypothesis of Markovian Diffusion in Phase Space*

There are many special semiempirical hypotheses in the literature which are much simpler than the one just discussed and are capable of giving concrete predictions which can be compared with existing (not very extensive) data on relative diffusion.

We shall begin by considering the local quantities characterizing the motion of a given fluid particle, namely, its coordinates $Y^{(S)}(\tau) = X(\tau) - x - u(x, t_0)\tau$ and velocities $V^{(S)}(\tau) = V(\tau) - u(x, t_0)$ relative to an inertial set of coordinates $\mathscr{S}$ moving with a constant velocity $u = u(x, t_0)$ whose origin at time $t_0$ lies at the point $x$. Let us denote by the symbol $p\left(Y^{(S)}, V^{(S)} \middle| Y_1^{(S)}, V_1^{(S)}, \tau\right)$ the probability density for the variables $Y^{(S)}(\tau)$ and $V^{(S)}(\tau)$ of a particle located at the point $X(0) = x + Y_1^{(S)}$ at time $t_0$ and having the velocity $V(0) = u(x, t_0) + V_1^{(S)}$. To obtain the explicit form of this function we must introduce certain additional hypotheses. One such hypothesis was introduced by Obukhov (1959b) who suggested that *the process of evolution of the coordinate $X(\tau)$ and velocity $V(\tau)$ of a given fluid particle is a random Markov process in a six-dimensional space which is invariant under the Galilean group of transformations* (including, in particular, all spatial translations and rotations). This assumption does not depend on the choice of the initial inertial set of coordinates and can therefore be used for the local quantities $Y^{(S)}(\tau)$, $V^{(S)}(\tau)$ which we shall discuss below.

The Obukhov hypothesis cannot be exact. In fact, one of its consequences is that the particle acceleration is infinite (compare this with the analogous discussion in Sect. 10.3 of Vol. 1 referring to semiempirical diffusion equations). Moreover, we shall show below that this hypothesis leads to relations such as Eq. (24.13) which are valid only for sufficiently large time intervals $\tau \gg \tau_\eta$ but are not valid for small $\tau$ [for which we have the formulas of Eq. (24.24)]. If, however, we consider a discrete sequence of times $t_0, t_0 + \tau_0, t_0 + 2\tau_0, \ldots$, where $\tau_0 \gg \tau_\eta$ (and belongs to the inertial subrange of time scales), then the sequence $\{Y^{(S)}(n\tau_0), V^{(S)}(n\tau_0)\}$, $n = 0, 1, 2, \ldots$, can apparently be regarded as a Markov sequence with satisfactory accuracy (since the "accelerations" will then be practically uncorrelated and, one hopes, practically independent; see Sect. 24.2 above). The Obukhov differential equation, whose derivation we shall now give, will therefore be the differential analog of some reasonable difference equation with a time interval $\tau_0$, and its solutions can be taken as the approximate formulas for values of $\tau$ from the inertial range (cf. the analogous explanation of the meaning of the semiempirical diffusion equation in Sect. 10.3 of Vol. 1).

We shall use the general Kolmogorov theory of Markov processes discussed in Sect. 10.3 of Vol. 1. Let us also note that for fixed values of the coordinate $Y^{(s)} = Y^{(s)}(\tau)$ and velocity $V^{(s)} = V^{(s)}(\tau)$ we have $\left[\frac{\partial}{\partial\theta}\overline{\Delta_\theta Y_j^{(s)}(\tau)}\right]_{\theta=0} = V_j^{(s)}$, where $\Delta_\theta Y_j^{(s)}(\tau) = Y_j^{(s)}(\tau+\theta) - Y_j^{(s)}(\tau)$, and that the "diffusivity for the coordinate" $K_{jk} = \frac{1}{2}\left[\frac{\partial}{\partial\theta}\overline{\Delta_\theta Y_j^{(s)}\Delta_\theta Y_k^{(s)}}\right]_{\theta=0}$ is now identically zero [cf. Vol. 1, Eq. (10.53)]. According to the Kolmogorov theory these facts together with the Obukhov hypothesis about the Markovian property of the six-dimensional process $\{Y^{(s)}(\tau), V^{(s)}(\tau)\}$ imply that under natural regularity assumptions the probability density $p(Y^{(s)}, V^{(s)}, \tau)$ should satisfy the differential equation

$$\frac{\partial p}{\partial\tau} + V_j^{(s)}\frac{\partial p}{\partial Y_j^{(s)}} + \frac{\partial(B_j p)}{\partial V_j^{(s)}} = \frac{\partial^2(D_{jk}p)}{\partial V_j^{(s)}\partial V_k^{(s)}}, \tag{24.77}$$

where

$$\begin{aligned} B_j &= B_j(Y^{(s)}, V^{(s)}, \tau) = \frac{\partial}{\partial\theta}\left[\overline{\Delta_\theta V_j^{(s)}(\tau)}\right]_{\theta=0}, \\ 2D_{jk} &= 2D_{jk}(Y^{(s)}, V^{(s)}, \tau) = \left[\frac{\partial}{\partial\theta}\overline{\Delta_\theta V_j^{(s)}(\tau)\,\Delta_\theta V_k^{(s)}(\tau)}\right]_{\theta=0}, \end{aligned} \tag{24.78}$$

and the mean values in Eq. (24.78) are evaluated for fixed $Y^{(s)}$ and $V^{(s)}$ [see, for example, Kolmogorov (1935)]. Since Eq. (24.77) should be invariant under translations of the origins in the space of the vectors $Y^{(s)}$ and on the time axis, it follows that the coefficients $B_j$ and $D_{jk}$ should be independent of $Y^{(s)}$ and $\tau$, while the requirement of Galilean invariance (i.e., invariance under transformations to a new inertial set of coordinates) demands that they should be independent of $V^{(s)}$ as well. Therefore these coefficients should be constants. If, in addition, we recall that Eq. (24.77) must be invariant under rotations of the set of coordinates in the space of the vectors $Y^{(s)}$, we find that the vector $B = (B_1, B_2, B_3)$ should be zero, and the tensor $D_{jk}$ should be of the form $D\delta_{jk}$, where $D$ = const. Therefore, Eq. (24.77) finally assumes the form

$$\frac{\partial p}{\partial\tau} + V_j^{(s)}\frac{\partial p}{\partial Y_j^{(s)}} = D\Delta_V p, \tag{24.79}$$

where $\Delta_V$ is the Laplace operator in the variables $V_1^{(s)}, V_2^{(s)}, V_3^{(s)}$, and $D$ is a dimensional characteristic of the process which has the same dimensions as the dissipation rate $\bar{\epsilon}$, i.e., $L^2T^{-3}$. If we suppose that the Obukhov hypothesis is valid in the inertial subrange, then it follows from the second Kolmogorov similarity hypothesis that

$$D = \frac{1}{2}C_0\bar{\varepsilon}, \tag{24.80}$$

where $C_0$ is a universal dimensionless constant. Conversely, if we suppose that Eq. (24.80) is valid, then the process $\{Y^{(s)}(\tau), V^{(s)}(\tau)\}$ will be uniquely determined by the parameter $\bar{\epsilon}$, and will satisfy the Kolmogorov similarity hypotheses for the inertial subrange. Since we have no quantities other than $D$ at our disposal, we may also conclude that the Obukhov hypothesis cannot be an acceptable approximation outside the inertial subrange.

Since Eq. (24.79) is invariant under Galilean transformations, it is sufficient to determine the function $p(Y^{(s)}, V^{(s)} \mid 0, 0, \tau)$, i.e., the solution of this equation subject to the initial condition $p(Y^{(s)}, V^{(s)}, 0) = \delta(Y^{(s)})\,\delta(V^{(s)})$ [the function $p\left(Y^{(s)}, V^{(s)} \mid Y_1^{(s)}, V_1^{(s)}, \tau\right)$ satisfying the initial condition $p(Y^{(s)}, V^{(s)}, 0) = \delta\left(Y^{(s)} - Y_1^{(s)}\right)\delta\left(V^{(s)} - V_1^{(s)}\right)$ is obtained from this solution by the simple replacement $Y^{(s)} \to Y^{(s)} - Y_1^{(s)} - V_1^{(s)}\tau$, $V^{(s)} \to V^{(s)} - V_1^{(s)}$]. This solution has long been known in the literature [see for example, Kolmogorov (1935)]. It is of the form

$$p(Y^{(s)}, V^{(s)}, \tau) = p(Y^{(s)}, V^{(s)} \mid 0, 0, \tau) = \\ = \left(\frac{\sqrt{3}}{2\pi D\tau^2}\right)^3 \exp\left\{-\left[\frac{3(Y^{(s)})^2}{D\tau^3} - \frac{3Y^{(s)}V^{(s)}}{D\tau^2} + \frac{(V^{(s)})^2}{D\tau}\right]\right\}. \tag{24.81}$$

The joint probability distribution for $(Y^{(s)}(\tau), V^{(s)}(\tau))$ in the Obukhov theory is therefore normal, and the coordinates $Y_i^{(s)}(\tau), i = 1, 2, 3$ are statistically independent and have equal variances $2D\tau^3/3$. The components of the relative velocity $V_i^{(s)}(\tau)$ are also statistically independent and have variances $2D\tau$. Finally, each $Y_i^{(s)}(\tau)$ is correlated with $V_i^{(s)}(\tau)$ [but independent of $V_j^{(s)}(\tau)$ with $j \neq i$], and the corresponding correlation coefficient is $\sqrt{3}/2$. Subject to Eq. (24.80), this result becomes identical with Eq. (24.13) from which it is clear that the coefficients $C_0$ in Eqs. (24.13) and (24.80) should be identical.

We are now in a position to explain the reason for the similarity between Eqs. (24.13) and (9.58), (9.59) of Vol. 1 which was noted in Sect. 24.1. We see that, having accepted a special additional hypothesis which is consistent with the similarity hypotheses, we can obtain formula (24.13) from the differential equation (24.79). The latter is the complete three-dimensional analog of Eq. (10.87) of Vol. 1, which leads to Eqs. (9.58) and (9.59). Hence it is clear that the formulas given by Eqs. (24.13) and (9.58), (9.59) cannot differ in form.

If we determine the "virtual eddy diffusivity" in the system $\mathscr{S}$ with the aid of the formula $K = \frac{1}{6}\frac{\partial\overline{(Y^{(s)})^2}}{\partial\tau}$, then if we start with Eq. (24.81) or Eq. (24.13), we find that $K = D\tau^2 = 2^{-2/3}D^{1/3}\left[\overline{(Y^{(s)})^2}\right]^{2/3}$. This relation is the analog of the four-thirds law for diffusion relative to the set of coordinates $\mathscr{S}$.

## *Hypotheses on the Probability Density* $p(l \mid l_0, \tau)$

The Obukhov hypothesis can also be applied to the evolution of the state of a fluid particle relative to the set of coordinates $\mathscr{S}^*$. In other words, we may suppose that the random process $\{X_2(\tau) - X_1(\tau), V_2(\tau) - V_1(\tau)\} = \{l(\tau), \Delta V(\tau)\}$, where $X_i(\tau), i = 1, 2$, and $V_i(\tau), i = 1, 2$ are the coordinates and velocities of two given fluid particles, is a Markov process. The corresponding probability density $p(l, \Delta V \mid l_0, \Delta V_0, \tau) = p(l, \Delta V, \tau)$, then will satisfy Eq. (24.79) with the coefficient $D$ greater by a factor of 2 than before [cf. Eq. (24.39) in Sect. 24.2]. This assumption was formulated in particular in the survey paper by Lin and Reid (1963). The resulting formula for $p(l, \Delta V \mid l_0, \Delta V_0, \tau)$ is analogous to Eq. (24.81) [with $Y^{(s)}$ replaced by $l - l_0 - \Delta V_0\tau$, and $V^{(s)}$ by $\Delta V - \Delta V_0$]. However, the hypothesis which we have adopted can be an acceptable approximation only in the case of the quasi-asymptotic state for $\tau_3 < \tau < \tau_1$, when the dependence of the probability distributions on $l_0$ (and $\Delta V_0$) ceases to be noticeable, but the process of relative motion of the two particles still remains isotropic and is determined only by the parameter

$\bar{\varepsilon}$. Hence the only meaningful solution is that for $l_0 = 0$ and $\Delta V_0 = 0$. This implies in particular, that

$$p(l, \tau) = p(l \mid l_0, \tau) = \left(\frac{3}{4\pi D\tau^3}\right)^{3/2} \exp\{-3l^2/4D\tau^3\}, \quad D = \frac{1}{2} g\bar{\varepsilon}, \tag{24.82}$$

and, therefore,

$$\Phi(\xi) \sim \exp(-b\xi^2), \qquad b = \text{const} \tag{24.82'}$$

where $\Phi(\xi)$ is defined in Eq. (24.76). The Obukhov hypothesis is therefore equivalent, in this case, to the assumption that the probability distribution for $l = l(\tau)$ is normal for $\tau_3 < \tau < \tau_1$ [with zero mean and the variance given by Eq. (24.36)]. This was also suggested by Batchelor (1952a) on the basis of other considerations (we shall return to this point below).

A different assumption about the function $p(l, \tau)$ was put forward by Richardson in 1926. Starting with the analogy with the usual (semiempirical) parabolic diffusion equation [see, for example, Eq. (10.76) of Vol. 1] and the four-thirds law established by him, Richardson suggested that the evolution of the function $q(l, \tau)$ could be described by the equation

$$\frac{\partial q}{\partial \tau} = \frac{\partial}{\partial l_j}\left(K(l)\frac{\partial q}{\partial l_j}\right), \quad K(l) = \alpha\, l^{4/3}, \quad l = |l| \tag{24.83}$$

[Richardson (1926) considered in fact only the one-dimensional analog of this equation]. Richardson's equation is reasonable only in connection with the mean function $\overline{q(l, \tau)}$, since for the individual realizations of the diffusion process we cannot expect simple regularities. However, by virtue of Eq. (24.53) the same equation will be satisfied by the probability density $p(l \mid l_0, \tau) = p(l, \tau)$, so that Eq. (24.83) can be rewritten in the form

$$\frac{\partial p}{\partial \tau} = \frac{\partial}{\partial l_j}\left(K(l)\frac{\partial p}{\partial l_j}\right), \quad K(l) = \alpha\, l^{4/3} = a\,\bar{\varepsilon}^{1/3} l^{4/3}. \tag{24.83'}$$

The solution of the last equation which satisfies the initial condition $p(l, 0) = \delta(l)$ depends only on $l = |l|$ and is of the form

$$p(l, \tau) = \frac{3}{35\pi^{3/2}}\left(\frac{3}{2}\right)^6 (\alpha\tau)^{-9/2} \exp\left\{-\frac{9l^{2/3}}{4\alpha\tau}\right\}, \tag{24.84}$$

so that, in this case,

$$\Phi(\xi) \sim \exp\{-b\xi^{2/3}\} \tag{24.84'}$$

[see Batchelor (1952a), Ozmidov (1960a)]. Thus, Richardson's semiempirical theory yields the probability distribution for the vector $l(\tau)$ when $\tau_3 < \tau < \tau_1$ which is non-Gaussian. The individual components $l_j(\tau)$ are uncorrelated in this theory, but not independent. However, here again we have $\overline{l^2} = \frac{143}{3}\left(\frac{2}{3}\alpha\tau\right)^3$, which is in agreement with Eq. (24.36).

Equation (24.83') was set up by analogy with equations describing quite different physical processes and cannot be justified in any rigorous fashion. The qualitative discussion given by Palm (1957), whose aim was to explain the significance of this equation, lead him to the conclusion that Eq. (24.83') might possibly be augmented by a further term so that, in the one-dimensional case,

$$\frac{\partial p}{\partial \tau} = \alpha \frac{\partial}{\partial l}\left(l^{4/3}\frac{\partial p}{\partial l}\right) + \beta \frac{\partial}{\partial l}(l^{1/3}p), \tag{24.85}$$

where $\alpha$ and $\beta$ are two constants proportional to $\bar{\varepsilon}^{1/3}$. On the other hand, Batchelor (1952a) indicated that a better description of relative diffusion is achieved by regarding the corresponding "diffusivity" as a function not of the particle pair separation but of the statistical quantities such as $\overline{l^2(\tau)} \sim \tau^3$. Hence he proposed to replace Eq. (24.83') by

$$\frac{\partial p}{\partial \tau} = K(\tau)\frac{\partial^2 p}{\partial l^2}, \qquad K(\tau) = D\tau^2. \tag{24.86}$$

The solution of this last equation is the spherically symmetric, normal, probability density with variances $\overline{l_1^2} = \overline{l_2^2} = \overline{l_3^2} = \frac{2}{3}D\tau^3$ given by Eq. (24.82).

One further hypothesis which enables us to determine the shape of the probability density $p(\boldsymbol{l} \mid \boldsymbol{l}_0, \tau) = p(\boldsymbol{l}, \tau)$ for $\tau_3 < \tau < \tau_1$ was investigated by Monin (1955, 1956). To explain this hypothesis it is convenient to take the Fourier transform and rewrite Eq. (24.49), which relates the concentration distribution in the coordinate frame $\mathscr{S}^*$ at times $t_0$ and $t_0 + \tau$, in the form

$$\tilde{\vartheta}(\boldsymbol{k}, \tau) = A_\tau \tilde{\vartheta}_0(\boldsymbol{k}). \tag{24.87}$$

In this expression the symbols $\tilde{\vartheta}_0(\boldsymbol{k})$ and $\tilde{\vartheta}(\boldsymbol{k}, \tau)$ represent the three-dimensional Fourier transforms of the functions $\overline{\vartheta^*(\boldsymbol{l}, t_0 + \tau)}$ and $\vartheta_0(\boldsymbol{l})$, and $A_\tau$ is a $\tau$-dependent linear operator in the space of the functions of $\boldsymbol{k}$. Since the turbulence is locally isotropic, the operation $A_\tau$ for $\tau < \tau_1$ should be invariant under rotations of the coordinate system. Therefore, by introducing the function $\tilde{\vartheta}_0(\boldsymbol{k}) = 1$, which corresponds to an initial "point" distribution $\vartheta_0(\boldsymbol{l}) = \delta(\boldsymbol{l})$, we obtain $A_\tau 1 = a(k, \tau)$ where $a$ is a numerical function of $k = |\boldsymbol{k}|$ and $\tau$. Since an arbitrary initial distribution can be written in the form of the convolution $\vartheta_0(\boldsymbol{l}) = \int \vartheta_0(\boldsymbol{l}_1)\,\delta(\boldsymbol{l}_1 - \boldsymbol{l})\,d\boldsymbol{l}_1$, it follows that for an arbitrary initial distribution $\vartheta_0(\boldsymbol{l})$ the application of the operation $A_\tau$ reduces to multiplication of $\tilde{\vartheta}_0(\boldsymbol{k})$ by the function $a(k, \tau)$:

$$\tilde{\vartheta}(\boldsymbol{k}, \tau) = a(k, \tau)\,\tilde{\vartheta}_0(\boldsymbol{k}). \tag{24.87'}$$

If we confine our attention to quasi-asymptotic diffusion times $\tau_3 < \tau < \tau_1$, we must allow the function $a(k, \tau)$ to depend on only one further dimensional quantity in addition to $k$ and $\tau$, namely $\bar{\varepsilon}$, so that

$$a(k, \tau) = a_0(\bar{\varepsilon}^{1/3}k^{2/3}\tau). \tag{24.88}$$

The choice of the function $a_0(x)$ of a single variable uniquely determines the probability density $p(\boldsymbol{l}, \tau)$ [the three-dimensional Fourier transform of this must be equal to $a_0(\bar{\varepsilon}^{1/3}k^{2/3}\tau)$]. In particular, the Gaussian probability density (24.82) corresponds to

$a_0(x) = \exp(-gx^3/6)$. Conservation of the total amount of admixture during the diffusion process evidently implies the condition $a_0(0) = 1$.

The Monin hypothesis can be stated as the assumption that the operators $A_\tau$ form a semigroup, i.e., they have the property $A_{\tau_1} A_{\tau_2} = A_{\tau_1+\tau_2}$. Hence, it follows that $a_0(x_1 + x_2) = a_0(x_1) \cdot a_0(x_2)$, and therefore $a(x) = e^{-cx}$, where $c$ is a numerical constant (obviously positive). If we take the Fourier transform of the function $\exp\{-c\bar{\varepsilon}^{1/3}|\boldsymbol{k}|^{2/3}\tau\}$ we can readily verify that the Monin hypothesis corresponds to the probability density

$$p(l, \tau) = \frac{A}{(\bar{\varepsilon}\tau^3)^{3/2}} \left(\frac{l^2}{\bar{\varepsilon}\tau^3}\right)^{-3/2} \exp\left(\frac{2c^3\bar{\varepsilon}\tau^3}{27l^2}\right) W_{-3/2, 1/6}\left(\frac{4c^3\bar{\varepsilon}\tau^3}{27l^2}\right), \tag{24.89}$$

where $W_{\lambda, \mu}(z)$ is the Whittaker confluent hypergeometric function, and $A$ is a normalizing constant. The probability density (24.89) decreases at infinity only as $|\boldsymbol{l}|^{-11/3}$ (so that here $\overline{l^2(\tau)} = \infty$; this shows that this theory cannot be valid for large $l = |\boldsymbol{l}|$). We note further that the function $\tilde{\vartheta}(k, \tau) = \exp(-c\bar{\varepsilon}^{1/3}k^{2/3}\tau)\,\tilde{\vartheta}_0(k)$ is the solution of the equation

$$\frac{\partial\tilde{\vartheta}}{\partial\tau} = -c\,\bar{\varepsilon}^{1/3}k^{2/3}\,\tilde{\vartheta}, \tag{24.90}$$

which, if we take the inverse Fourier transform, can be rewritten in the form $\dfrac{\partial\bar{\vartheta}^*}{\partial\tau} = B\,\bar{\vartheta}^*$ where $B$ is a linear operator proportional to the Laplace operator raised to the power of 1/3 (since, in the Fourier representation, the Laplace operator corresponds to multiplication by $-k^2$). In fact, $B$ is an integral operator with a complex kernel [in the two-dimensional case, this kernel can be explicitly expressed in terms of a Bessel function; see Monin (1955, 1956)]. If we differentiate Eq. (24.90) twice with respect to $\tau$ we obtain $\dfrac{\partial^3\tilde{\vartheta}}{\partial\tau^3} = -c^3\bar{\varepsilon}k^2\tilde{\vartheta}$, which shows that all the solutions of the integral equation $\dfrac{\partial\bar{\vartheta}^*}{\partial\tau} = B\bar{\vartheta}^*$ are included among the solutions of the differential equation

$$\frac{\partial^3\bar{\vartheta}^*}{\partial\tau^3} = c^3\,\bar{\varepsilon}\,\Delta\bar{\vartheta}^* \tag{24.91}$$

[in particular, it is readily verified that the probability density (24.89) satisfies Eq. (24.91)].

Results analogous to those found by Monin were subsequently obtained by Tchen (1959) who started with the spectral equation

$$\frac{\partial\tilde{\vartheta}}{\partial\tau} = -\nu(k)\,k^2\,\tilde{\vartheta}, \qquad \nu(k) = \gamma\int_k^\infty [E(k')]^{1/2}\,k'^{-3/2}\,dk' \tag{24.92}$$

[which, in turn, is based on the Heisenberg formula (17.9)]. By substituting $E(k) \sim k^\alpha$, where $\alpha = -5/3$, Tchen again obtained Eq. (24.90) (in addition, he also considered the cases $\alpha = -1$ and $\alpha = +1$).

The experimental data on the function $p(l, \tau)$ permitting the verification of the theoretical relations are very scanty at present. However Sullivan (1971) calculated approximate values of the function $p$ from observations of dye plumes on Lake Huron and compared them with the equations suggested by Batchelor-Obukhov and by Richardson. He concluded that Richardson's equation is very far from the real data but the Gaussian curve describes them with satisfactory accuracy.

### *Hypotheses on the Concentration Distribution Relative to the Center of Gravity of a Cloud*

The concentration distribution $\overline{\vartheta_c(\boldsymbol{x}, \tau)}$ in a noninertial frame $\mathscr{S}_c$, whose origin at all times lies at the center of gravity of the admixture cloud, is of the greatest practical interest since this distribution is readily determined from experimental data on diffusion. As already noted in Sect. 24.3, diffusion relative to the center of gravity of the cloud is more complicated from the standpoint of turbulence theory than diffusion relative to the coordinate systems $\mathscr{S}$ and $\mathscr{S}^*$ (since the concentration $\overline{\vartheta_c(\boldsymbol{x}, \tau)}$ depends on the coordinates of all the cloud particles). If, however, we do not use the exact equations of fluid dynamics, and confine our attention to approximate semiempirical hypotheses then, in this approach, the coordinate system $\mathscr{S}_c$ will not differ in any basic way from $\mathscr{S}$ and $\mathscr{S}^*$ Therefore, all semiempirical hypotheses applied above to the functions $p(\boldsymbol{Y}^{(s)}, \tau)$ [or $\overline{\vartheta^{(s)}(\boldsymbol{Y}, \tau)}$] and $p(\boldsymbol{l}, \tau)$[or $\overline{\vartheta^*(\boldsymbol{l}, \tau)}$] might be transferrable to the function $\overline{\vartheta_c(\boldsymbol{x}, \tau)}$ as well (in actual fact some of these hypotheses were formulated precisely for this last function). We note also that, since $2\sigma_{ij}(\tau) = \Sigma_{ij}(\tau)$, the fact that the variance $\overline{[Y^{(s)}(\tau)]^2}$ or $\overline{l^2(\tau)}$ is proportional to $\bar{\varepsilon}\tau^3$, which is a consequence of the above hypotheses, will be in agreement with the conclusions of the similarity theory (for $\tau_3 < \tau < \tau_1$) and after transformation from $\overline{[Y^{(s)}(\tau)]^2}$ or $\overline{l^2(\tau)}$ to the variance $\sigma^2(\tau)$ of the distribution $\overline{\vartheta_c(\boldsymbol{x}, \tau)}$.

We have already used (see Sect. 24.3 above) the Batchelor-Obukhov hypothesis about the normal distribution of the mean concentration in connection with the function $\overline{\vartheta_c(\boldsymbol{x}, \tau)}$ [see Eqs. (24.64) and (24.82)]. We have referred to Gifford's works; however the same hypothesis was also used by Inoue (1963) in his analysis of data on the concentration $\overline{\vartheta_c(\boldsymbol{x}, \tau)}$ in the case of two-dimensional diffusion on the sea surface. The Richardson hypothesis, according to which the concentration during relative diffusion satisfied Eq. (24.83′), was applied to the function $\overline{\vartheta_c(\boldsymbol{x}, \tau)}$ by Ozmidov (1958, 1960a, 1968) in the two- and three-dimensional cases. Application of the various semiempirical hypotheses to the description of relative diffusion of an admixture on the sea surface (relative to the center of the cloud) has been discussed in a number of papers. For example, MacEwen (1950) and Joseph and Sendner (1958) used the two-dimensional isotropic diffusion equation $\frac{\partial \bar{\vartheta}_c}{\partial \tau} = \frac{1}{r}\frac{\partial}{\partial r}\left(rK\frac{\partial \bar{\vartheta}_c}{\partial r}\right)$ with a diffusivity of the form $K(r) = \alpha r$ where $r = |\boldsymbol{x}|$ [subsequently, Joseph and Sendner (1962) used the same equation with the coefficient $K(r) = \alpha r^{1-\mu}$ where $\mu$ is an empirical constant]. Okubo and Pritchard [see Okubo (1962a)] discussed the same equation with a time-dependent diffusivity $K = K(\tau) = \alpha\tau$, and Okubo (1962a) has discussed this equation with $K = K(r, \tau) = \alpha r^{2/3}\tau$. Finally, Schönfeld (1962) put forward a more complicated integral diffusion equation [which takes the form of Eq. (24.87′) after the use of the Fourier transform]. Comparisons of the consequences of the various semiempirical equations with existing data can be found in the paper by Okubo (1962a) where the analytical solution of the diffusion equation with a diffusivity of the general form $K = K(r, \tau) = \alpha r^m \varphi(\tau)$ is also given [cf, Ozmidov (1968) and Ozmidov, Gezentsvei and Karabashev (1969)]. Ozmidov (1968), Ozmidov et al. (1969) and Schuert (1970) also presented some additional comparisons of recent observations with deductions from semiempirical diffusion equations. From the standpoint of the similarity theory of locally isotropic turbulence, the admissible semiempirical equations for the "intermediate" times $\tau_3 < \tau < \tau_1$ are the Batchelor-Obukhov equations with $K = \alpha\tau^2 = a\bar{\varepsilon}\tau^2$, the Richardson equation with $K = \alpha r^{4/3} = a\bar{\varepsilon}^{1/3}r^{4/3}$ and the Okubo equation with $K = \alpha r^{2/3}\tau = a\,\bar{\varepsilon}^{2/3}r^{2/3}\,\tau$. The fundamental solutions of the first two of these equations in the two-dimensional case are only slightly different from the corresponding three-dimensional solutions (24.82) and (24.84) [they also decrease at infinity as $\exp(-br^2)$ and $\exp(-br^{2/3})$ respectively]. The Okubo equation corresponds to a fundamental solution of the form $\overline{\vartheta_c(r,\tau)} \sim (\bar{\varepsilon}\tau^3)^{-2} \exp(-cr^{4/3}/\bar{\varepsilon}^{2/3}\tau^2)$ which decreases at infinity as $\exp(-br^{4/3})$, i.e., in the same way as the density $p(l, \tau)$ in the Lagrangian-history direct-interaction approximation of Kraichnan, which also involves the diffusion equation (24.74) with the diffusivity $K$ being a function of both time and space coordinates. In spite of the important difference in the behavior of the concentration field in the above three semiempirical theories, existing data are not sufficient to enable us to choose reliably between them. We can only note that the quite rough data analyzed by Okubo (1962a) and by Schuert (1970) give some slight indications that the third solution is in somewhat better agreement with observations than the first and particularly the second, while the observational data of Sullivan (1971) agree excellently with the Gaussian curve suggested by Batchelor and Obukhov.

The conclusions that can be drawn from dimensional considerations based on the assumptions of self-preservation of relative diffusion for $\tau_3 < \tau < \tau_1$, and on the dependence of this process on the single dimensional parameter $\bar{\varepsilon}$, turn out of course to be the same for all the semiempirical theories that are consistent with the Kolmogorov similarity hypothesis. These conclusions include the prediction that the maximum concentration at the center of the cloud, $\overline{\vartheta_c(0,\tau)}$, is proportional to $\tau^{-9/2}$ [more precisely, it is proportional to $(\bar{\varepsilon}\tau^3)^{-3/2}$] in the case of three-dimensional diffusion, and to $\tau^{-3}$ [or more precisely $(\bar{\varepsilon}\tau^3)^{-1}$] in the case of two-dimensional diffusion. Most (but not all) of the existing data on two-dimensional diffusion on the sea surface are in satisfactory agreement with the law $\overline{\vartheta_c(0,\tau)} \sim \tau^{-3}$ [see for example Ozmidov (1958, 1968), Hela and Voipio (1960), Okubo (1962a) and Kilezhenko (1964)]. This provides additional grounds for supposing that the use of the theory of locally isotropic turbulence for the description of relative diffusion yields correct results in a broad range of values of $\tau$.

## 24.5 Material Line and Surface Stretching in Turbulent Flows

The results of Sect. 24.2 show that the mean distance between two given fluid particles in a turbulent flow will increase with time in all time intervals which can be written explicitly in terms of this distance. Hence, it follows, that the mean length of any chord of a material line (i.e., a line consisting of given fluid particles) or material surface will usually increase. One would therefore expect that the mean length of a material line and the mean area of a material surface in a turbulent flow will be monotonically increasing functions of time. The physical reason for the stretching of material lines and surfaces is the complex twisting of these lines and surfaces produced by turbulent fluctuations (cf. Fig. 79 of Vol. 1 in which the boundary of an admixture cloud does in fact form a material surface). This stretching is interesting in itself, since it demonstrates clearly the diffusive property of any turbulent motion. It is also quite important for a number of applications since, for example, vortex lines or magnetic lines of force in a turbulent medium with low viscosity and very high electrical conductivity coincides with the material lines in the first approximation, whereas surfaces of constant temperature or constant concentration of a passive admixture coincide with the material surfaces when molecular diffusivity can be neglected.

Since each line of finite length (or surface of finite area) can be represented by the sum of a large number of small straight segments (or plane areas), it is possible to follow

Batchelor (1952b) and to reduce the general analysis of material line and surface stretching to the analysis of the variation in the distance between fluid particles for which the initial separation is very small (infinitesimal). The derivatives $\frac{\partial u_l}{\partial x_j}$ of the velocity components will then be practically constant over such an elementary segment (or area) and hence the time evolution of the material line element $l(\tau)$ will be described by the simple equation

$$\frac{dl_i(\tau)}{d\tau} = \frac{\partial u_i}{\partial x_j} l_j \tag{24.93}$$

where the velocity gradient tensor $\partial u_i/\partial x_j$ is evaluated at the moving point coinciding with any point of the line element.

It was noted by Batchelor (1952b) and especially emphasized by Saffman (1968) that even such a basic and intuitively clear fact as the fact of continuous increase of the length of a material line element in the course of its time evolution had no convincing mathematical proof at the time when their works were published. However recently Cocke (1969) gave a rigorous proof of some results which can naturally be considered as quite convincing support of the idea on the continuous stretching of any infinitesimal material line and surface element by isotropic incompressible turbulence (the isotropy requirement can in fact be replaced in Cocke's proof by the requirement of local isotropy if only the small and intermediate times $\tau \ll \tau_1$ are considered). Cocke's arguments were somewhat simplified later by Orszag (1970b) at the cost of a partial weakening of the result; we shall follow Orszag's paper below.

Equation (24.93) evidently implies that

$$l_i(\tau) = U_{ij} l_{oj}$$

where $l_{oj} = l_j(0)$ are the components of the initial separation vector and $U_{ij}$ is a random matrix which is determined by the values of the velocity gradient tensor along the trajectory of the infinitesimal line element and is a function of time and of the initial spacial position of the element but not of $l_{oj}$. Therefore,

$$l^2(\tau) = l_i(\tau) l_i(\tau) = W_{jk} l_{oj} l_{ok} \tag{24.94}$$

where $W_{jk} = U_{ij} U_{ik}$ is a real symmetric matrix. For fixed time $\tau$, denote the eigenvalues of this matrix by $w_1, w_2$, and $w_3$. Then it is clear that the turbulence will transform an infinitesimal fluid sphere of radius $\delta$ placed at the position of the vector $l_0$ at time $\tau = 0$ into a triaxial ellipsoid with principal axes of length $w_1\delta, w_2\delta$, and $w_3\delta$, respectively. Hence $w_1 > 0, w_2 > 0, w_3 > 0$, and (by incompressibility) $w_1 w_2 w_3 = 1$. Let us now take the eigenvector corresponding to the eigenvalue $w_3$ as the polar axis of a spherical coordinate system and let the latitude and longitude of the tip of the vector $l_0$ be $\theta$ and $\phi$, respectively. Then, it follows easily from Eq. (24.94) that

$$l^2(\tau) = [\sin^2\theta\,(w_1 \cos^2\phi + w_2 \sin^2\phi) + w_3 \cos^2\theta]\, l_0^2. \tag{24.95}$$

Both the angles $\theta, \phi$ and the eigenvalues $w_1, w_2, w_3$ in the right-hand side of Eq. (24.95) are random variables. Since $(W_{jk})$ is independent of $l_0$ the variables $w_1, w_2, w_3$ must be statistically independent of the angles $\theta, \phi$. Moreover if the turbulence is isotropic (or is locally isotropic and $\tau \ll \tau_1$), the eigenvectors of the matrix $(W_{jk})$ must have an isotropic space distribution. In other words, the tip of $l_0$ must in this case be homogeneously distributed over the unit sphere. Since the averages of $\sin^2\theta \cos^2\phi$, $\sin^2\theta \sin_2\phi$, and $\cos^2\theta$ over the unit sphere are each 1/3, it follows from Eq. (24.95) that in isotropic turbulence

$$\overline{l^2(\tau)} = \frac{1}{3}\,\overline{w_1 + w_2 + w_3\, l_0^2}. \tag{24.96}$$

But it is easy to show that $w_1 + w_2 + w_3 \geqslant 3$ under the constraints $w_1 w_2 w_3 = 1$ and $w_1 > 0, w_2 > 0, w_3 > 0$, with equality holding only if $w_1 = w_2 = w_3 = 1$. Therefore we have rigorously shown that

$$\overline{l^2(\tau)} \geqslant l^2(0) = l_0^2 \tag{24.97}$$

for all $\tau > 0$, with the equality holding only if $w_1 = w_2 = w_3 = 1$ for all realizations, i.e., only if there is no isotropic turbulence at all.[31]

Equation (24.97) gives the necessary support to the intuitive idea on continuous drifting apart of any pair of fluid particles in isotropic (or locally isotropic) turbulence. Cocke (1969) used similar reasoning to prove the even stronger statement that $\overline{\ln |l(\tau)|/l_0} \geqslant 0$ for all $\tau > 0$.

By virtue of the general inequality $\overline{\exp X} \geqslant \exp \bar{X}$, which holds for any random variable $X$, this last equation implies Eq. (24.97) and also a more general result of the form

$$\overline{|l(\tau)|^p} \geqslant l_0^p$$

for all $\tau > 0$ and $p > 0$. Let us note in particular that this inequality with $p = 1$ indicates that the mean distance between two fixed fluid particles increases monotonically with time, while Eq. (24.97) does not imply such a conclusion. Moreover, the monotonic growth of mean separation of two fluid particles and of mean length of fluid line was also demonstrated by Corrsin (1972) with the aid of a quite different method based on the simple assumption that two fluid points which are far enough apart move independently in any homogeneous stationary turbulence.

If now we denote the area of an infinitesimal material surface element at time $\tau$ by $s(\tau)$ then quite similar arguments can be used to prove that under the conditions guaranteeing Eq. (24.97) the following inequalities will also hold:

$$\overline{[s(\tau)]^2} \geqslant [s(0)]^2,\ \overline{[s(\tau)]^p} \geqslant [s(0)]^p,\ \overline{\ln \frac{s(\tau)}{s(0)}} \geqslant 0 \tag{24.98}$$

for all $\tau > 0$ and $p > 0$, with equality holding in all these relations only if there is no turbulence at all.

Later Cocke (1971) applied a version of a central limit theorem for dependent random variables to the line- and surface-stretching problem and derived with it some supplementary general results. For example, he found that under mild conditions the relations

$$\lim_{\tau\to\infty} \ln \frac{|l(\tau)|}{|l_0|} = 0,\ \lim_{\tau\to\infty} \ln \frac{|s(\tau)|}{|s(0)|} = 0$$

must be valid [in contradiction to an early suggestion of Batchelor (1952b)].

[31] The result (24.97) was given much earlier in the unpublished thesis of Lumley (1957). Here the following reasoning was used for its justification. It is clear that $\overline{l_i(\tau)} = l_{i0}$; hence $l_i(\tau) = (\delta_{ij} + V_{ij}) l_{j0}$ where $\overline{V_{ij}} = \overline{U_{ij}} - \delta_{ij} = 0$. It follows from the last relation that $l^2(\tau) = l_0^2 + \overline{V_{ij} V_{ik}\, l_{j0}\, l_{k0}}$. If we introduce the isotropic relation $\overline{V_{ij} V_{ik}} = \lambda \delta_{jk}$ where $\lambda = \frac{1}{3}\overline{V_{ij} V_{ij}} > 0$ for all $\tau > 0$, we have $\overline{l^2(\tau)} - l_0^2 = \lambda l_0^2 > 0$. Orszag's derivation differs from this by more detailed motivation of the result implied by isotropy.

The previous derivation uses only the condition of the isotropy (or local isotropy) of the turbulence. Let us now assume that the Reynolds number Re is sufficiently large to guarantee the validity of the Kolmogorov similarity hypotheses and the initial lengths of all the elementary material segments under consideration are very small in comparison with the Kolmogorov length scale $\eta$. Then the time evolution of a material line (or surface) can be investigated in more detail by using additional assumptions and arguments similar to those used in Sects. 22.3 and 22.4 for the study of the small-scale behavior of the turbulence spectra.

For distances small in comparison with $\eta$ the velocity field $u(X)$ can be assumed to be a linear function of $X$. As already noted in Sect. 22.3, motion corresponding to a linear velocity field consists of (1) translation of the small volume of the fluid (containing the elementary segment or area in which we are interested), (2) rotation of this volume relative to some instananeous axis, and (3) strain of the volume which can be reduced to extensions and compressions along three mutually perpendicular axes $X_1, X_2, X_3$. As in Sect 22.3, we shall suppose that, in the coordinate system moving and rotating together with the volume under consideration, the directions of the principal strain axes and the eigenvalues $a_1, a_2$, and $a_3$ of the strain rate tensor do not change significantly over time intervals of the order of $-\tau_\eta$ (possible relaxations of this assumption were discussed in Sect. 22.3). We shall take the coordinate system $X_1, X_2, X_3$ so that the $X_1$ axis lies along the axis of maximum expansion, while the $X_3$ axis lies along the axis of maximum contraction. In that case, $u_j(X) = a_j X_j$ (no summation over $j$) where $0 < a_1 > a_2 > a_3 < 0$, $a_1 + a_2 + a_3 = 0$.

The evolution of a material line element (or area) in the strain field $u_j(X) = a_j X_j$ at first depends on its form, size, and orientation relative to the coordinate axes. This evolution reduces to a uniform expansion along the $X_1$ axis at a rate $a_1$, expansion (or contraction when $a_2 < 0$) along the $X_2$ axis at the rate $a_2$, and rapid contraction along the $X_3$ axis at the rate $-a_3 = a_1 + a_2$. We shall now suppose that the initial size of the segment (or area) is so small that after a time of the order of $\tau_\eta$ the size is still small in comparison with $\eta$. For such intervals of time, therefore, its evolution will continue as at the beginning. Since typical values of $a_1$, $a_2$, and $a_3$ are of the order of $\tau_\eta^{-1}$, the size of our element along the $X_1$ axis will increase by a large factor in the time $\tau_\eta$, the size along the $X_2$ axis will be reduced by a large factor, while the size along the $X_3$ axis will change in an intermediate fashion (as a rule, this size will also appreciably increase but to a much smaller extent than along the $X_1$ axis since, usually, $a_1 > a_2 > 0$; see the final part of Sect. 22.3). For larger $\tau = t - t_0$, when the linear dimensions of the initial elementary segment or area are no longer small in comparison with $\eta$, the velocity field within the segment or area can no longer be regarded as linear, and the segment ceases to be straight and the area plane. Therefore, before this effect begins to play an appreciable role, we must subdivide the segment or area again into smaller parts, and apply the analysis to each of them.

It follows from the above discussion that the time $\tau_3$ after which $l(\tau) \gg l_0$ and further relative motion of a pair of fluid particles no longer depends on the vector $l_0$, is of the order of $\tau_\eta$ when $l_0 \ll \eta$. It may be considered in the first approximation that after the time $\tau_3$ has elapsed, the elementary material line element becomes practically parallel to the $OX_1$ axis (if we neglect the very rare cases when the initial vector $l_0$ is almost exactly perpendicular to $OX_1$), and the elementary material surface element becomes almost perpendicular to $OX_3$. Further evolution of the material line or surface will no longer depend on the initial size, shape, and orientation. In the case of the line element, it will take the form of a simple elongation with relative velocity $a_1$, while in the case of the surface element it will take the form of an elongation in one of the directions with relative velocity $a_1$, and in the perpendicular direction with relative velocity $a_2$ which is usually positive but very occasionally can be negative. In other words, when $\tau > \tau_3$ the length $l(\tau)$ will increase so that $\frac{1}{l}\frac{dl}{d\tau} = a_1$, and the area $s(\tau)$ will increase so that $\frac{1}{s}\frac{ds}{d\tau} = a_1 + a_2 = -a_3$. We thus see that, during most of the time, the length $l(\tau)$ will be proportional to $e^{a_1\tau}$, while the area $s(\tau)$ will be proportional to $e^{-a_3\tau}$ (where $a_3 < 0$). Hence it may be concluded that for $\tau = t - t_0 > \tau_\eta$ the mean length $\overline{L(t)}$ of the material line, and the mean area $\overline{S(t)}$ of the material surface will be respectively given by

$$\overline{L(t)} = L(t_0)\, e^{\xi(t-t_0)} = L(t_0)\, e^{\gamma(t-t_0)/\tau_\eta},$$
$$\overline{S(t)} = S(t_0)\, e^{\zeta(t-t_0)} = S(t_0)\, e^{\delta(t-t_0)/\tau_\eta}, \tag{24.99}$$

where $L(t_0)$ and $S(t_0)$ are the initial values, and $\xi = \gamma\tau_\eta^{-1} > 0$ and $\zeta = \delta\tau_\eta^{-1} > 0$ are parameters whose dimensions are reciprocal to that of time. These parameters are of the order of $\tau_\eta^{-1}$ and are the result of an averaging procedure applied to $a_1$ and $-a_3$.[32]

The formulas (24.99) were obtained by Batchelor (1952b) [his paper included an error which was subsequently rectified in the appendix to Reid (1955)]. These formulas can be used to obtain a number of additional results of the same kind. Consider, for example, an area in the form of a parallelogram formed by the vectors $\boldsymbol{l}_1$ and $\boldsymbol{l}_2$. In the course of time, the lengths $l_1 = l_1(\tau)$ and $l_2 = l_2(\tau)$ of the sides of the parallelogram, and its area $s = s(\tau) = l_1 l_2 \sin\varphi$, will increase exponentially in accordance with Eq. (24.99). Consequently, the mean angle between two intersecting material lines decreases exponentially in accordance with the formula $\overline{\sin\varphi} \approx \overline{\varphi} \sim e^{(\zeta - 2\xi)\tau}$ (we recall that $\zeta - 2\xi \approx -\overline{a}_3 - 2\overline{a}_1 = \overline{a}_2 - \overline{a}_1 < 0$ since $a_1 > a_2$). Similarly, since the volume of the material parallelopiped formed by the vectors $\boldsymbol{l}_1$, $\boldsymbol{l}_2$, and $\boldsymbol{l}_3$ must remain constant, it is readily verified that the mean sine of the angle between the material area and the intersecting material line decreases with time in proportion to $e^{-(\xi+\zeta)\tau}$. Finally, since the material volumes must be constant, we may conclude that the mean distance along the normal between two closely lying parallel material surfaces decreases with time in proportion to $e^{-\zeta\tau}$. In particular, if we consider the surfaces of constant concentration $\vartheta$ of some passive admixture, we can verify that until molecular diffusion becomes appreciable, the mean admixture gradient increases in proportion to $e^{\zeta\tau}$ (so that $\overline{|\nabla\vartheta|^2} = |\nabla\vartheta(t_0)|^2 e^{2\zeta(t-t_0)}$). In precisely the same way, we can consider the evolution of a vector field $\boldsymbol{F}(\boldsymbol{X}, t)$ such that the lines formed by it can be identified with material lines, and the flux of the vector over all cross-sections of the corresponding "tube" remains constant (for example, as for the vorticity field). So long as these conditions can be regarded as satisfied (i.e., until molecular effects set in), the quantity $\overline{|\boldsymbol{F}|}$ will evidently be inversely proportional to the cross-sectional area of the "vector tube," i.e., it will be directly proportional to the length of the material line acting as the "axis" of the tube (since the material volumes must be conserved). Therefore, for example

$$\overline{F^2(t)} = F^2(t_0)\, e^{2\xi(t-t_0)}.$$

Batchelor and Townsend (1956) have suggested the use of Eq. (22.72) for approximate estimates of the parameters $\xi$ and $\zeta$, i.e., of the numerical coefficients $\gamma$ and $\delta$. Moreover

[32] The parameters $\xi$ and $\zeta$ are not equal to the simple probabilistic means $\overline{a}_1$ and $-\overline{a}_3$. They are more correctly defined by the equations $\exp\xi = \overline{\exp a_1}$ and $\exp\zeta = \overline{\exp(-a_3)}$. However, this definition leads to values of $\xi$ and $\zeta$ that differ from $\overline{a}_1$ and $-\overline{a}_3$ only by numerical factors which are apparently rather close to unity. Since the values of $\overline{a}_1$ and $\overline{a}_3$ are not known accurately, and the entire theory is approximate, there is no point at present in distinguishing between $\xi$ and $\overline{a}_1$ and $\zeta$ and $-\overline{a}_3$.

they assumed that in the second of these equations we can to a first approximation neglect the terms associated with fluctuations in $a_l$ and replace $\overline{a_1 a_2 a_3}$ by $\bar{a}_1 \bar{a}_2 \bar{a}_3$. In that case, the two equations in Eq. (22.72) together with the equation $\bar{a}_1 + \bar{a}_2 + \bar{a}_3 = 0$ enable us to express $\xi \approx \bar{a}_1$ and $\zeta \approx -\bar{a}_3$ in terms of $\epsilon$, $\nu$, and the skewness $s = \overline{\left(\frac{\partial u_1}{\partial x_1}\right)^3} \Big/ \left[\overline{\left(\frac{\partial u_1}{\partial x_1}\right)^2}\right]^{3/2}$. Using the very approximate values of $s$ given by Batchelor (1953) (these in fact refer to flows with insufficiently high Re), it was found by Batchelor and Townsend that

$$\xi \approx 0{,}4\tau_\eta^{-1}, \ \zeta \approx 0{,}5\tau_\eta^{-1}, \ \text{i.e., } \gamma \approx 0{,}4, \quad \delta \approx 0{,}5. \tag{24.100}$$

This estimate is of course very crude and adds therefore very little to the general conclusion that the coefficients $\gamma$ and $\delta$ should be of the order of unity. However, the fact that the difference $\delta - \gamma \approx \bar{a}_2$ is positive and small in comparison with $\gamma$ agrees well with intuitive expectations. Another, also very approximate, method of extimating $\gamma$ and $\delta$ involves the use of the Millionshchikov zero-fourth-cumulant hypothesis and the comparison of the results obtained on the rate of increase of $\overline{|\nabla\vartheta|^2}$ and $\overline{F^2}$ (neglecting molecular effects) with the equations (24.99). Let us compare, for example, Reid's equations (1962) and (1963), which describe the rate of increase of $\overline{(\nabla\vartheta)^2}$ in the approximation in which $\chi = 0$ and all the fourth-order cumulants are neglected, with the relation $\overline{(\nabla\vartheta)^2} \sim e^{2\zeta(t-t_0)}$. If we assume that $\overline{\omega^2} = \bar{\varepsilon}/\nu = \tau_\eta^{-2} =$ const during time intervals for which molecular effects can be neglected, then in view of Eq. (19.63) we must assume that $\delta = 1/\sqrt{6} \approx 0.4$ [the good agreement between this estimate and Eq. (24.100) is apparently fortuitous]. A similar conclusion, derived from the Millionshchikov hypothesis applied to the equation for the vector field $F$, states that $\gamma = \delta = 1/\sqrt{6}$, i.e., $\xi = \zeta$ [see Reid (1955)]. The fact that the value of $\xi$ is in this approximation not smaller than $\zeta$ but is exactly equal to it is evidently due to the inaccuracy of the approximation.

In conclusion, let us briefly consider the relation between the results of this subsection with the general conclusions of Sect. 24.2. In view of the general equation (24.23), the quantity $\frac{1}{\overline{|l|}} \frac{d\overline{|l|}}{d\tau}$, where $l = l(\tau)$ is the distance between two given fluid particles, should be given by

$$\frac{1}{\overline{|l|}} \frac{d\overline{|l|}}{d\tau} = \frac{1}{\tau_\eta} \Psi\left(\frac{|l_0|}{\eta}, \frac{\tau}{\tau_\eta}\right), \tag{24.101}$$

where $\Psi$ is a universal function of two variables [see the analogous formula (24.32) which is also a consequence of Eq. (24.23)]. We shall assume that $l_0 = |l_0| \ll \eta$ and that the turbulence is homogeneous. It is then readily shown that $\Psi(l_0/\eta, 0) \approx \Psi(0, 0) = 0$. In fact, $\frac{d|l|}{d\tau}$ is equal to the projection of the velocity difference for the two fluid particles, $V_2(t) - V_1(t)$, $t = t_0 + \tau$, along the direction of the vector $l$. If $|l| \ll \eta$, we can use the equation

$$V_2(t) - V_1(t) = u[X_1(t) + l(\tau), t] - u[X_1(t), t] \approx l_i(\tau) \frac{\partial u[X_1(t), t]}{\partial X_{1i}},$$

and consequently

$$\frac{1}{|l(\tau)|}\frac{d|l(\tau)|}{d\tau} \approx \frac{\partial u_j[X_1, t]}{\partial X_{1l}} \frac{l_l(\tau)}{|l(\tau)|} \frac{l_j(\tau)}{|l(\tau)|}.$$

At the initial time $t_0$ the vector $l = l_0$ is fixed and, since $\overline{\frac{\partial u_j(X, t)}{\partial X_l}} = 0$ (the turbulence is homogeneous) we have $\frac{1}{\overline{|l|}}\frac{d\overline{|l|}}{d\tau} = 0$ when $\tau = 0$. As $\tau$ increases, the quantity (24.101) should increase at least for some time, since we know that the distance $\overline{|l|}$ should increase. After the quasi-asymptotic state has been reached (for $\tau \gtrsim \tau_s$), the dependence on $l_0$ in Eq. (24.101) will disappear, so that

$$\frac{1}{\overline{|l|}}\frac{d\overline{|l|}}{d\tau} = \frac{1}{\tau_\eta}\Psi_1\left(\frac{\tau}{\tau_\eta}\right). \tag{24.102}$$

For still larger $\tau$, when $\tau \gg \tau_\eta$, this formula will not contain the dependence on the molecular viscosity $\nu$ (which is normally present in the microscale $\tau_\eta$), so that

$$\frac{1}{\overline{|l|}}\frac{d\overline{|l|}}{d\tau} = \frac{A}{\tau}, \tag{24.103}$$

where $A$ is a numerical constant [equal to 3/2 in view of Eq. (24.36)]. Therefore, for $\tau = 0$ the quantity $\frac{1}{\overline{|l|}}\frac{d\overline{|l|}}{d\tau}$ is zero. It then increases for some time and, when $\tau \gg \tau_\eta$, it begins to fall quite rapidly. Consequently, it reaches a maximum for $\tau$ of the order of $\tau_\eta$. It is natural to expect that this quantity may be regarded as approximately constant in a certain neighborhood of this maximum. This conclusion is in fact confirmed by the first equation in Eq. (24.99), which refers to the special case when $|l_0|$ is so small that there is a range of values $\tau > \tau_s$ for which $|l(\tau)| \ll \eta$).

# 25. REFINED TREATMENT OF THE LOCAL STRUCTURE OF TURBULENCE, TAKING INTO ACCOUNT FLUCTUATIONS IN DISSIPATION RATE

## 25.1 General Considerations and Model Examples

Let us now return to the physical basis of the Kolmogorov similarity hypotheses, and discuss an important critical remark stated by Landau almost immediately after the formulation of Kolmogorov's theory (see, for example the footnote on p. 126 of the book by Landau and Lifshitz (1963) which was also given in the first Russian edition of this book in 1944). However it was only in recent years that this remark has attracted wide attention. It shows that the similarity hypotheses cannot be regarded as quite exact, so that the theory developed in the above section is in a sense only preliminary.

Landau's remark was concerned with the effect of fluctuations in energy dissipation rate on the small-scale properties of turbulence. We have always assumed in the above discussion that the energy transfer to small-scale disturbances was $\bar{\varepsilon}$, i.e., that it was the same throughout, and that the probability distributions for the increments in the turbulence fields in a sufficiently small space-time region depend only on this "mean flux" $\bar{\varepsilon}$. In fact, however, the energy dissipation rate per unit mass of a turbulent fluid is given by

$$\varepsilon = \frac{\nu}{2} \sum_{i,\, j} \left( \frac{\partial u_i}{\partial x_j} + \frac{\partial u_j}{\partial x_i} \right)^2 , \tag{25.1}$$

i.e., it is a random function of the coordinates $\boldsymbol{x}$ and time $t$, which fluctuates together with the field $\boldsymbol{u}(\boldsymbol{x}, t)$. These fluctuations may depend on the properties of the large-scale motion, above all on the Reynolds number Re. This follows from the fact that the number Re determines the ratio $L/\eta$, i.e., roughly speaking, it determines the number of "stages" in the hierarchy of eddies of different orders along which the kinetic energy is transferred before it is dissipated into heat. Since the statistical properties of the field $\varepsilon(\boldsymbol{x}, t)$ must evidently affect the probability distributions for the small-scale turbulence components, the distributions must apparently depend on Re and the other mean-flow parameters, i.e., they cannot be strictly universal. This was in fact the main reason why we have used the term "quasi-equilibrium range" in the above discussion of the Kolmogorov theory, and avoided speaking of "universal equilibrium," as most authors do.

In order to understand how the probability distribution for $\varepsilon$ can affect the local characteristics of the field $\boldsymbol{u}(\boldsymbol{x}, t)$, let us consider a simplified turbulence model in which the dissipation rate $\varepsilon(\boldsymbol{x}, t)$ does not vary within space-time regions $G$ with space dimensions much greater than $\eta$ and time dimensions much greater than $\tau_\eta$, but can take different values in different regions. For example, let us assume that $\varepsilon$ is equal to $\varepsilon_1 = (1-\gamma)\bar{\varepsilon}$ with probability 1/2, and to $\varepsilon_2 = (1+\gamma)\bar{\varepsilon}$ with an equal probability, where $\gamma$ is a numerical factor characterizing the intermittency of the turbulence. In each region $G$ the probability distribution for the increments of $\boldsymbol{u}(\boldsymbol{x}, t)$

must obviously depend only on the value of $\varepsilon$ in this region (which is equal to the energy transfer rate from large-scale turbulent disturbances to small-scale disturbances within $G$). Therefore, for example, the two-thirds and five-thirds laws for the velocity field in some regions $G$ will be of the form

$$D_{LL}(r) = C(1-\gamma)^{2/3}\varepsilon^{2/3}r^{2/3}, \quad E(k) = C_1(1-\gamma)^{2/3}\varepsilon^{2/3}k^{-5/3}, \tag{25.2}$$

whereas in other regions we shall have

$$D_{LL}(r) = C(1+\gamma)^{2/3}\varepsilon^{2/3}r^{2/3}, \quad E(k) = C_1(1+\gamma)^{2/3}\varepsilon^{2/3}k^{-5/3}. \tag{25.2'}$$

In these expressions $C$ and $C_1$ are the values of the numerical coefficients in the stated laws which correspond to the idealized turbulence model without fluctuations in the energy dissipation rate. The true values of $D_{LL}(r)$ and $E(k)$ in the turbulent flow will then be equal to the arithmetic mean of the right-hand sides of Eqs. (25.2) and (25.2′) (equal in this case to the probabilistic mean), so that

$$D_{LL}(r) = C(\gamma)\varepsilon^{2/3}r^{2/3}, \quad E(k) = C_1(\gamma)\varepsilon^{2/3}k^{-5/3}, \tag{25.3}$$

where

$$C(\gamma) = \frac{C}{2}\left[(1-\gamma)^{2/3} + (1+\gamma)^{2/3}\right],$$
$$C_1(\gamma) = \frac{C_1}{2}\left[(1-\gamma)^{2/3} + (1+\gamma)^{2/3}\right].$$

We thus see that the two-thirds and five-third laws retain their form but the numerical coefficients in them depend on $\gamma$, i.e., on the probability distribution for $\varepsilon$. It is true, however, that the dependence of these coefficients on $\gamma$ is quite weak in this case. Thus, the transition from $\gamma = 0$ (corresponding to constant $\varepsilon$) to $\gamma = 2/3$ (for which $\varepsilon_2/\varepsilon_1 = 5$) results in only a 6% reduction in the coefficients, and even for $\gamma = 0.9$, i.e., $\varepsilon_2/\varepsilon_1 = 19$, these coefficients are smaller by only 13% as compared with the case where there are no fluctuations in $\varepsilon$.

Similar results can be obtained for a broad class of local statistical characteristics of turbulence for more or less arbitrary probability distributions for the dissipation rate. As in the derivation of Eq. (25.3), we shall for the moment neglect possible fluctuations in the field $\varepsilon(\boldsymbol{x}, t)$ within the space-time region $G$ to which the particular statistical parameters refer, but we shall take into account the change in $\varepsilon$ between different regions. The analog of the first Kolmogorov similarity hypothesis will then be the assumption that, *for a given viscosity* $\nu$, *the conditional probability distributions for the relative-velocity field* $\boldsymbol{v}(\boldsymbol{r}, \tau)$ *in Eq.* (21.2) *are isotropic and depend only on* $\nu$ *and* $\varepsilon$ *provided that the energy dissipation rate in the corresponding region* $G$ *assumes a fixed value* $\varepsilon$. Hence it follows, for example, that the conditional value of the moment

$$B = B_{i_1 \dots i_N}^{k_1 \dots k_N} = \overline{[v_{i_1}(\boldsymbol{r}_1, \tau_1)]^{k_1} \dots [v_{i_N}(\boldsymbol{r}_N, \tau_N)]^{k_N}},$$

in a sufficiently small space-time region $G$ can be written in the form

$$B(\varepsilon) = (\nu\varepsilon)^m f(\boldsymbol{r}_1\varepsilon^{1/4}\nu^{-3/4}, \dots, \boldsymbol{r}_N\varepsilon^{1/4}\nu^{-3/4}, \tau_1\varepsilon^{1/2}\nu^{-1/2}, \dots, \tau_N\varepsilon^{1/2}\nu^{-1/2}), \tag{25.4}$$

where $\varepsilon$ is the energy dissipation in the region, $m = (k_1 + \dots + k_N)/4$, and $f$ is a universal function which does not change under arbitrary rotations and reflections of the coordinate system. The second similarity hypothesis can also be extended to the above conditional probability distributions, i.e., it may be assumed that $B(\varepsilon)$ should not depend on $\nu$, provided only that all the arguments of the function $f$ and all the differences of two vector or two scalar arguments are much greater than unity in absolute value. It is important, however, that to obtain the unconditional mean of the moment $B$, the function $B(\varepsilon)$ must be averaged over the possible values of $\varepsilon$, i.e., we must consider the integral

$$B = \int_0^\infty B(\varepsilon)\, p(\varepsilon)\, d\varepsilon, \tag{25.5}$$

where $p(\varepsilon)$ is the probability density for $\varepsilon$. This integral depends on $p(\varepsilon)$ which in turn is determined by the large-scale motion, i.e., it is not a universal quantity. Similar considerations are valid for the velocity spectra of various orders in the region of small scales, and for other local quantities which depend only on small-scale turbulence components.

Let us now suppose that the dependence of $B(\varepsilon)$ on $\varepsilon$ can be represented by the power law $B(\varepsilon) = C\varepsilon^n$ where $n \neq 0$ and $C$ is independent of $\varepsilon$ [this will occur in particular when purely spatial or purely temporal spectra or moments of the field $\boldsymbol{v}(\boldsymbol{r}, \tau)$ in the inertial subrange are considered]. In that case

$$B = K_n C (\bar{\varepsilon})^n, \quad \text{where} \quad K_n = \frac{\overline{\varepsilon^n}}{(\bar{\varepsilon})^n} = \frac{\int\limits_0^\infty \varepsilon^n p(\varepsilon)\, d\varepsilon}{\left[\int\limits_0^\infty \varepsilon p(\varepsilon)\, d\varepsilon\right]^n}. \tag{25.6}$$

Hence it is clear that the general form of the dependence of $B$ on $\bar{\varepsilon}$ and on all the remaining arguments is the same for all densities $p(\varepsilon)$. However, the value of the numerical coefficient in front of $(\bar{\varepsilon})^n$ is in general not universal but depends on $p(\varepsilon)$. The only exception is the case $n = 1$ (linear dependence) when $K_1 = 1$ and therefore the coefficient of $\bar{\varepsilon}$ is universal [this last conclusion refers in particular to the coefficients in Eq. (21.18′) for $D_{LLL}(r)$, and in formulas (21.30′) and (21.35′) for the Lagrangian structure function $D^{(L)}(\tau)$ and the Lagrangian spectrum $E^{(L)}(\omega)$]. It is important however that in the remaining cases also the dependence of the correction factor $K_n$ in Eq. (25.6) on $p(\varepsilon)$ is usually relatively weak. We have already verified this in the case of Eq. (25.3). The same conclusion follows from the calculations of Grant, Stewart and Moilliet (1962) who assumed that the density $p(\varepsilon)$ is constant in the range $(\varepsilon_1, \varepsilon_2) = [(1-\gamma)\varepsilon, (1+\gamma)\varepsilon]$ and zero outside this range, and who showed that, subject to this assumption, $K_{2/3} = 0.964$ and $K_2 = 1.23$ (i.e., $K_{2/3}$ and $K_2$ are both not very different from unity) even for $\gamma = 5/6$ (i.e., $\varepsilon_2/\varepsilon_1 = 11$). Similar results were obtained by Novikov (1963b) who considered an example of a continuous density function $p(\varepsilon)$ which was nonzero for $0 < \varepsilon < \infty$.

In the case of a function $B(\varepsilon)$ which cannot be written down as a power law, the general form of the dependence of $B$ on $\bar{\varepsilon}$ and other arguments may change when $p(\varepsilon)$ is changed, i.e., it will be essentially nonuniversal. A typical example of this kind was discussed by Novikov (1963b). He assumed that the conditional value of the spectrum $E(k)$ for fixed dissipation rate $\varepsilon$ has the form

$$E(k) = C\varepsilon^{2/3} k^{-5/3} \exp\{-\alpha \eta_\varepsilon^2 k^2\}, \quad \eta_\varepsilon = \varepsilon^{-1/4} \nu^{3/4} \tag{25.7}$$

throughout the quasi-equilibrium range $k \gg 1/L$ [see Eq. (22.73) at the end of Sect. 22.3]. Then he showed that the continuous density

$p(\varepsilon)$ (which depends on the positive numerical parameter $\gamma$) can be chosen in such a way that the mean value of the spectrum (25.7) is of the form

$$E(k) = C(\gamma)\bar{\varepsilon}^{2/3}k^{-5/3}\left[1 + d(\gamma)\eta^2 k^2\right]^{-\gamma}, \quad \eta = \bar{\varepsilon}^{-1/4}\nu^{3/4}, \quad (25.7')$$

where $C(\gamma)$ depends on $C$ and $\gamma$, and $d(\gamma)$ depends on $\alpha$ and $\gamma$. In the inertial subrange, the expressions (25.7) and (25.7′) both lead to the five-thirds law (but with different numerical coefficients), as expected. In the dissipation ranges $k \gg 1/\eta_\varepsilon$ and $k \gg 1/\eta$, the spectra given by Eqs. (25.7) and (25.7′) show a quite different behavior. The first of them falls exponentially, while the second falls as a finite negative power of $k$ (namely as $k^{-(2\gamma+5/3)}$).[33]

Another example of the same kind is provided by the results of Keller and Yaglom (1970) which will be formulated in Sect. 25.4 below.

All above considerations can be generalized without any difficulty to the study of the local structure of a scalar field $\vartheta(\boldsymbol{x}, t)$ (which we shall call "temperature" for the sake of simplicity). Here we must take into account fluctuations in both dissipations rates $\varepsilon(\boldsymbol{x}, t)$ and $N(\boldsymbol{x}, t) = \chi(\nabla\vartheta)^2$. These fluctuations will depend on the large-scale properties of the flow, above all on the Reynolds and Peclet numbers Re and Pe. If we suppose that $\varepsilon$ and $N$ both take constant values in sufficiently large regions $G$, but can assume different values in different regions, then the statistical averaging must include the averaging over the two-dimensional probability distribution for $\varepsilon$ and $N$ in addition to conditional averaging taken at fixed values of these two variables. If some statistical characteristic of the field $\vartheta$ (or fields $\vartheta$ and $\boldsymbol{u}$) is represented by a power function of $\varepsilon$ and $N$ for a definite range of the independent variables, then the functional form of the corresponding law will survive the averaging over any $(\varepsilon, N)$-distribution and only the numerical coefficient of the power law will change a little, depending on the distribution. This is true for example for the two-thirds and five-thirds laws of Obukhov and

[33] It may at first seem strange that when we take the average of a very smooth field with a spectrum, given by Eq. (25.7), which has derivatives of all orders, the final result is a field with the spectrum (25.7′) which can be differentiated only a finite number of times. This is however simply explained by the fact that the behavior of the spectrum at infinity determines only the existence of the mean square derivatives but not the instantaneous derivatives of the individual realizations of the random field. There is of course nothing strange in the fact that the additional ensemble averaging may lead to infinite mean square values of some of the derivatives of the field.

Corrsin [Eqs. (21.87) and (21.89) of Sect. 21.6] and for Batchelor's "minus one power law" (22.81). If the parameter is a linear function of a single fluctuating variable, then the averaging over its values will result in replacement of the variable by its mean value; hence for example, the numerical coefficient in the Eqs. (21.92) and (22.10) for $D_{L\vartheta\vartheta}(r)$ can be considered as universal even in the framework of the refined theory. However all the nonpower functional forms of the statistical laws probably do not survive averaging over the distribution for $\varepsilon$ and $N$. A typical example of such a kind was considered by Kraichnan (1968b) who studied the quite simple model of the influence of the fluctuations in the variable $\varepsilon$ on the form of Batchelor's temperature spectrum (22.80). Namely, it was supposed by Kraichnan that $N$ does not undergo fluctuations at all and $(\varepsilon/\nu)^{1/2}$ (i.e., $\varepsilon$) may assume two different values with some given (generally nonequal) probabilities. As a result he found after the averaging over the two values of $\varepsilon$, that the "minus one power law" (22.81) preserves its form (but the numerical coefficient $a$ changes its value), and that the exponential cutoff factor on the right-hand side of Eq. (22.80) is replaced by a more slowly decreasing function.

## 25.2 Refined Similarity Hypothesis

So far we have been concerned with artificial examples in which the density $p(\varepsilon)$ was chosen in an arbitrary fashion. However, to explain the effect of fluctuations in the dissipation rate on the local characteristics of real turbulent flows we must start with the true statistical properties of these fluctuations. We must then abandon the artificial assumption that the dissipation rate $\varepsilon(\boldsymbol{x}, t)$ is constant in space-time regions $G$ of diameter $L \gg \eta$, and varies only as a result of transition from one such region to another. However, if that is so then the assumption that the conditional probability distributions for the field $\boldsymbol{v}(\boldsymbol{r}, \tau)$ are universally given the value of the energy dissipation rate within the region $G$ must be replaced by some other assumption of a similar kind which generalizes the usual formulation of the Kolmogorov similarity hypotheses to the case of intermittent turbulence. This has been discussed by Obukhov (1962a, b) and Kolmogorov (1962a, b).

We shall begin by supposing that $\varepsilon(\boldsymbol{x}, t)$ is a random field in the usual probabilistic sense of this term. It is natural to suppose then that the local statistical characteristics of the velocity field will depend only on the local values of $\varepsilon(\boldsymbol{x}, t)$. More precisely, it is natural to assume that the statistical characteristics of small-scale motions deduced from the values of the velocity field at a given finite set of close space-time points will depend only on the values of $\varepsilon(\boldsymbol{x}, t)$ in some bounded four-dimensional region $G$ containing all these points. Let us take the simplest assumption, namely, that the only important characteristic of the field $\varepsilon(\boldsymbol{x}, t)$ is the mean value $\varepsilon_G = \int_G \varepsilon(\boldsymbol{x}, t)\, d\boldsymbol{x}\, dt \Big/ \int_G d\boldsymbol{x}\, dt$ of this field in the region $G$. This assumption is probably an oversimplification of the real physical world, but in any case it is already a substantial improvement on the initial form of the similarity hypothesis which involves only the probabilistic mean $\bar{\varepsilon}$. The formulation of the first similarity hypothesis given in the previous section can largely be used now without modification. All that is necessary is to replace the conditional probability distributions for a fixed constant dissipation rate $\varepsilon$ in $G$ by the *conditional distributions for a fixed mean value* $\varepsilon_G$ *of the energy dissipation rate in* $G$. Consequently, equations (25.4)–(25.6) will remain valid provided only that we replace $\varepsilon$ by $\varepsilon_G$, and interpret $p(\varepsilon_G)$ as the probability density function for the mean dissipation rate $\varepsilon_G$ (which may depend on the choice of $G$).

To obtain the final formulation, we must settle the question of the choice $G$. It is natural to suppose, however, that the final results will not be very dependent on the shape of this region. The only important point is that the linear dimensions of this region should be of the same order as the characteristic linear dimensions of the original set of points. Therefore, Obukhov considered the problem of the determination of the statistical characteristics of the velocity difference

$$\Delta_r \boldsymbol{u}(\boldsymbol{x}, t) = \boldsymbol{u}(\boldsymbol{x}+\boldsymbol{r}, t) - \boldsymbol{u}(\boldsymbol{x}, t) = \boldsymbol{u}(\boldsymbol{x}_0+\boldsymbol{r}/2, t) - \boldsymbol{u}(\boldsymbol{x}_0-\boldsymbol{r}/2, t)$$

(where $\boldsymbol{x}_0 = \boldsymbol{x}+\boldsymbol{r}/2$) and supposed that the most important factor for these characteristics is the dissipation rate averaged over a spherical volume of radius $|\boldsymbol{r}|/2 = r/2$ with $\boldsymbol{x}_0+\boldsymbol{r}/2$ and $\boldsymbol{x}_0-\boldsymbol{r}/2$ as the poles of this sphere, i.e., the quantity

$$\varepsilon_r(\boldsymbol{x}_0, t) = \frac{6}{\pi r^3} \int\limits_{|\boldsymbol{r}'| < r/2} \varepsilon(\boldsymbol{x}_0+\boldsymbol{r}', t)\, d\boldsymbol{r}'. \tag{25.8}$$

Developing this idea, Kolmogorov assumed that the quantity $\varepsilon_r = \varepsilon_r(\boldsymbol{x}_0, t)$ plays the leading role for the entire multidimensional probability distribution for the set of relative velocities $\boldsymbol{v}(\boldsymbol{r}_1, \tau_1), \ldots, \boldsymbol{v}(\boldsymbol{r}_n, \tau_n)$, constructed from the field $\boldsymbol{u}(\boldsymbol{x}, t)$ at the points $(\boldsymbol{x}_0, t_0), (\boldsymbol{x}_1, t_1), \ldots, (\boldsymbol{x}_n, t_n)$, provided only that (1) $r \ll L$, (2) all the values $|\boldsymbol{r}_k| = |\boldsymbol{x}_k - \boldsymbol{x}_0|$, $k = 1, \ldots, n$ are of the order of $r$, and (3) all the values $|\tau_k|$, $k = 1, \ldots, n$ are of the order of the time scale $T_r = r^{2/3}\varepsilon_r^{-1/3}$. The quantities $r$ and $\varepsilon_r$ can also be used to construct the velocity scale $U_r = (r\varepsilon_r)^{1/3}$, while $r$, $\varepsilon_r$, and $\nu$ can form the unique dimensionless combination

$$\mathrm{Re}_r = \frac{U_r r}{\nu} = \frac{r^{4/3}\varepsilon_r^{1/3}}{\nu} = \left(\frac{r}{\eta_r}\right)^{4/3}, \quad \eta_r = \nu^{3/4}\varepsilon_r^{-1/4}. \tag{25.9}$$

The value of this number for fixed $r$ and $\nu$ determines uniquely the value of the random variable $\varepsilon_r$, and for fixed $r$ and $\varepsilon_r$ it depends only on $\nu$. Kolmogorov used this to propose the following refined formulation of his two basic similarity hypotheses:

*The first refined similarity hypothesis. If* $r \ll L$ *then the conditional probability distribution for the dimensionless relative velocities*

$$\boldsymbol{w}(\boldsymbol{\xi}_k, \tau_k) = \frac{\boldsymbol{v}(\boldsymbol{\xi}_k r, \tau_k T_r)}{U_r}, \qquad k = 1, 2, \ldots, n, \tag{25.10}$$

*where* $|\boldsymbol{\xi}_k|$ *and* $|\tau_k|$ *are all of the order of unity and the value of the random variable* $\mathrm{Re}_r$ *is fixed, depends only* $\mathrm{Re}_r$ *and does not vary under arbitrary rotations and reflections of the set of vectors* $\boldsymbol{\xi}_k$, $k = 1, \ldots, n$.

*The second refined similarity hypothesis. If* $\mathrm{Re}_r \gg 1$ *then the conditional distribution indicated in the first hypothesis does not depend on* $\mathrm{Re}_r$, *i.e., it is universal.*

It follows from the first of the above two hypotheses that the values of the structure functions $D_{LL}(r)$ and $D_{LLL}(r)$ for $r \ll L$ and fixed $\varepsilon_r$ can be written in the form

$$D_{LL}(r) = C(\varepsilon_r r)^{2/3}\,\tilde{\beta}_{LL}\left(\frac{r^{4/3}\varepsilon_r^{1/3}}{\nu}\right), \quad D_{LLL}(r) = D\,\varepsilon_r r\,\tilde{\beta}_{LLL}\left(\frac{r^{4/3}\varepsilon_r^{1/3}}{\nu}\right), \tag{25.11}$$

where $C$ and $D$ are constants (which can be chosen arbitrarily), and

$\tilde{\beta}_{LL}(x)$ and $\tilde{\beta}_{LLL}(x)$ are universal functions of a single variable. These formulas can also be written in the form

$$D_{LL}(r) = (\varepsilon_r \nu)^{1/2} \beta_{LL}(r\varepsilon_r^{1/4}\nu^{-3/4}), \quad D_{LLL}(r) = (\varepsilon_r \nu)^{3/4} \beta_{LLL}(r\varepsilon_r^{1/4}\nu^{-3/4}), \tag{25.11'}$$

which is closer to Eqs. (21.12), (25.15), and (25.4). In these expressions $\beta_{LL}(x)$ and $\beta_{LLL}(x)$ are other universal functions which are simply related to $\tilde{\beta}_{LL}(x)$ and $\tilde{\beta}_{LLL}(x)$ [for example, $\beta_{LL}(x) = Cx^{2/3}\tilde{\beta}_{LL}(x^{4/3})$]. The form given by Eq. (25.11) is more convenient for large $r$, since it follows from the second refined similarity hypothesis that the functions $\tilde{\beta}_{LL}(x)$ and $\tilde{\beta}_{LLL}(x)$ tend to constant values as $x \to \infty$. By suitably choosing the constants $C$ and $D$ we can ensure that $\tilde{\beta}_{LL}(\infty) = 1$ and $\tilde{\beta}_{LLL}(\infty) = 1$. We shall assume that $C$ and $D$ are chosen in this way, in which case

$$D_{LL}(r) = C\varepsilon_r^{2/3} r^{2/3}, \quad D_{LLL}(r) = D\varepsilon_r r \quad \text{when } r \gg \eta_r = \nu^{3/4}\varepsilon_r^{-1/4}. \tag{25.12}$$

The first equation in Eq. (25.12) is very similar to the usual two-thirds law except that $D_{LL}(r)$ is now only the conditional value of the structure function subject to the condition that the random variable (25.8) has assumed a fixed value $\varepsilon_r$, and the range $r \gg \eta_r$ itself is now a function of $\varepsilon_r$.

Since the field $\boldsymbol{w}(\boldsymbol{\xi}) = \boldsymbol{w}(\boldsymbol{\xi}, 0)$ can be regarded as homogeneous and isotropic in accordance with the first similarity hypothesis, it is possible to define a spectrum for it. Consequently in our refined theory we can consider the conditional velocity spectrum $E(k)$ for fixed $\varepsilon_{1/k}$ and $k \gg 1/L$. It is clear that

$$E(k) = C_1 \varepsilon_{1/k}^{2/3} k^{-5/3} \varphi(k^{4/3} \nu \varepsilon_{1/k}^{-1/3}). \tag{25.13}$$

From the second similarity hypothesis we can show in the usual way that $\varphi(0) = 1$ for a suitably chosen constant $C_1$, and hence

$$E(k) = C_1 \varepsilon_{1/k}^{2/3} k^{-5/3} \quad \text{for} \quad 1/L \ll k \ll 1/\eta_{1/k}. \tag{25.14}$$

All the other results in Sects. 21.4–21.5 can be generalized in a similar way.

To obtain the unconditional statistical characteristics we must average the conditional values for fixed $\varepsilon_r$ over all the possible values of $\varepsilon_r$. Since the mathematical expectation $\bar{\varepsilon} = \bar{\varepsilon}(\boldsymbol{x}, t)$ is practically constant in regions whose linear dimensions are small in comparison with $L$, we have $\bar{\varepsilon}_r = \bar{\varepsilon}$ for $r \ll L$. Therefore, if, for example, $r$ is much greater than all the actually encountered values of $\eta_r$, i.e., all values of $\eta_r$ except, possibly, a small set of values having a negligible total probability, then the probabilistic averaging of the second formula in Eq. (25.12) results in the well known relation

$$D_{LLL}(r) = D\bar{\varepsilon}r. \tag{25.15}$$

However, in the case of statistical characteristics that are nonlinear functions of $\varepsilon_r$ the situation is considerably more complicated: Here, it is no longer sufficient to know only the mean value $\bar{\varepsilon}_r = \bar{\varepsilon}$, but we must also have information about the probability distribution for $\varepsilon_r$.

The situation regarding the theory of the local structure of the temperature (or any other scalar field mixed by turbulence) is quite analogous. Here we must only add the parameters $\chi$ and $N_r(\boldsymbol{x}_0, t)$ (the value of $N = \chi(\nabla\theta)^2$ averaged over a spherical volume of radius $r/2$) to $\nu$ and $\varepsilon_r$. Hence we shall have a natural temperature scale $\theta_r = r^{1/3}N_r^{1/2}\varepsilon_r^{-1/6}$ and two dimensionless combinations $\mathrm{Re}_r$ and $\mathrm{Pr} = \nu/\chi$. The precise formulation of the corresponding refined similarity hypotheses can now be established without any difficulty and we shall not linger on it.

## 25.3 Statistical Characteristics of the Dissipation

According to the definition (25.1) for the field $\varepsilon(\boldsymbol{x}, t)$, its statistical characteristics can be expressed in terms of the characteristics of the derivatives $\partial u_i/\partial x_j$. However, the probability distributions for the derivatives of the velocity which are determined largely by the small-scale components of the turbulence have not, as yet, been extensively investigated. Therefore, when we calculate the dissipation-field characteristics, we must start from some hypothetical statistical models, and verify their validity by comparing the results with data.

The simplest hypothesis on the multidimensional probability distributions for the field $\boldsymbol{u}(\boldsymbol{x}, t)$ is that the distributions are practically normal (i.e., Gaussian). Under this assumption the dissipation rate $\varepsilon(\boldsymbol{x}, t)$ will have a quite definite probability distribution, namely it will be a quadratic form of Gaussian variables. The correlation function for the dissipation rate fluctuations

$$B_{\varepsilon'\varepsilon'}(r) = b_{\varepsilon\varepsilon}(r) = \overline{[\varepsilon(\boldsymbol{x}+\boldsymbol{r}, t) - \bar{\varepsilon}][\varepsilon(\boldsymbol{x}, t) - \bar{\varepsilon}]}$$

and the corresponding spectrum $E_{\varepsilon\varepsilon}(k)$ can be found particularly simply in this case. To calculate $b_{\varepsilon\varepsilon}(r)$ we must use the formula relating the fourth-order moments of the random Gaussian field with its second moments, i.e., use the Millionshchikov zero-fourth-cumulant hypothesis. The corresponding second moments—the velocity structure functions—can be specified in the first approximation for $r \ll \eta = \bar{\varepsilon}^{-1/4}\nu^{3/4}$ and for $r \gg \eta$ by the formulas of Sect. 21.4. In the transition region $r \approx \eta$ we can use some interpolation formula. Having done this, we can try to estimate the effect of the dissipation fluctuations on the structure functions and the velocity spectrum, i.e., obtain the second-approximation formulas for $D_{LL}(r)$ and $E(k)$.

A substantial part of this program was completed by Golitsyn (1962) who showed, in particular, that in the approximation based on the zero-fourth-cumulant hypothesis $b_{\varepsilon\varepsilon}(0) = \overline{(\varepsilon - \bar{\varepsilon})^2} = 0.4\bar{\varepsilon}^2$, i.e., $\sigma_\varepsilon \approx 0.6\bar{\varepsilon}$. He also obtained formulas for the functions $b_{\varepsilon\varepsilon}(r)$ and $E_{\varepsilon\varepsilon}(k)$ in the inertial subrange, according to which $b_{\varepsilon\varepsilon}(r) \sim r^{-8/3}$ and $E_{\varepsilon\varepsilon}(k) \sim k^{5/3}$. Since the spectrum $E_{\varepsilon\varepsilon}(k)$ increases with increasing $k$, the one-dimensional spectrum $E_1^{(\varepsilon)}(k)$ which can be expressed in terms of $E_{\varepsilon\varepsilon}(k)$ by Eq. (12.13) will, in this approximation, be determined by the values of $E_{\varepsilon\varepsilon}(k)$ for $k \gtrsim 1/\eta$. In the inertial subrange it will not be proportional to $k^{5/3}$ but will be constant [Novikov (1965)]. Golitsyn (1962) also estimated the variance $\sigma^2_{\varepsilon_r}$ of the spherical average $\varepsilon_r$ of the dissipation rate. The quantity $\overline{(\varepsilon_r)^{2/3}}$ in the case of a Gaussian field $u(x, t)$ is more difficult to evaluate, but it can be found by numerical integration. This calculation has not been carried out by anyone, but it is clear that, in this case $\overline{(\varepsilon_r)^{2/3}} = K\bar{\varepsilon}^{2/3}$, i.e., the corresponding second-approximation formulas for $D_{LL}(r)$ and $E(k)$ in the inertial subrange (which take dissipation fluctuations into account) will differ from the original two-thirds and five-thirds laws only by constant factors independent of the properties of large-scale motions. This suggests that the statistical model we have adopted is apparently invalid since, as indicated in Sect. 25.1, it is natural to expect that fluctuations in dissipation should depend on the Reynolds number. This model is even less satisfactory in the light of the measurements of Townsend (1947), Batchelor and Townsend (1949) and some other authors discussed in Sect. 18.1. It was shown in these works that even the one-dimensional probability distribution for the velocity derivative in grid generated turbulence is substantially non-Gaussian, and it may be supposed that, as the Reynolds number increases, the departure from the Gaussian distribution should increase (see also the discussion of the existing data below in this section). Approximate measurements of the spectrum of dissipation-rate fluctuations performed by Gurvich and Zubkovskii (1963), Pond and Stewart (1965) and some other authors (we shall discuss all of these works again below) have also lead to results which are in sharp disagreement with the conclusions obtained by Golitsyn [see also Novikov (1965)]. All this forces us to conclude that a Gaussian model of the velocity fluctuation field is not suitable for calculations of statistical characteristics of dissipation, and to seek other models leading to better agreement with experiment.

We must now consider the problem of which particular properties of small-scale turbulence components ensure that their probability distributions are highly non-Gaussian. We recall the data of Batchelor and Townsend (1949) according to which the excess $\tilde{\delta}_n$ of the derivatives $\partial^n u_1/\partial x_1^n$ in grid-generated turbulence increases with increasing $n$, indicating an increasing departure of the corresponding probability distributions from normality. We also recall that Batchelor and Townsend's values of $\tilde{\delta}_n$ increase with increasing Reynolds number, although this increase is relatively small. Batchelor and Townsend (1949) also carried out similar measurements in the turbulent wake of a cylinder. Here again they obtained similar qualitative results and very close numerical values of $\tilde{\delta}_n$ [see Batchelor (1953), Chapt. 8, Sect. 1]. As the order $n$ increases, there is obviously an increase in the relative contribution of the smallest-scale turbulence components to the corresponding derivative. It may therefore be concluded from the results of Batchelor and Townsend that the excess for velocity components of a given scale increases with decreasing scale and increasing Reynolds number.

This conclusion is confirmed by the data of Sandborn (1959), Kennedy and Corrsin (1961) and Kuo and Corrsin (1971) who measured directly the excess of the different spectral components of the turbulence in boundary layers, turbulent jets and grid-generated turbulence. In all three papers a set of band-pass filters was used to isolate the contribution of different spectral regions to the fluctuations in the velocity $u_1$. In the first two papers only relatively low Reynolds number turbulence was considered while in the third the whole range of $\mathrm{Re}_\lambda$ values from 12 to 830 was studied. Sandborn determined the excess of fluctuations at the output of different filters for a number of points in a turbulent boundary layer, and found that for distances $z$ from the solid wall that were not too large, the excess of the total velocity $u_1$ was practically zero, whereas the excess of the fluctuation isolated by the filter was practically always positive and relatively high (reaching values of the order of 5 or more in individual cases). Kennedy and Corrsin performed analogous measurements in a turbulent jet. They also found that the excess of the total longitudinal velocity was nearly zero, while the excess of the individual spectral components was positive and monotonically increased with increasing frequency at the center of the isolated spectral band. Finally most complete results of the same type were obtained by Kuo and Corrsin. They found in particular that at a given midband frequency $f_m$ the excess of band-pass signal is a function of relative bandwidth $\Delta f/f_m$; it is small for very great $\Delta f/f_m$ (because the filtering is unimportant in such cases) and for very small $\Delta f/f_m$ (since narrow-band filtering introduces considerable smoothing and any smoothing of nonnormal stationary random process makes it tend toward normality) and takes a maximum value around $\Delta f/f_m = 0.3$. At a fixed $\Delta f/f_m$ the excess of a signal increases with increase of $f_m$ and at given $\Delta f/f_m$ and $f_m$ it increases with increase of $\mathrm{Re}_\lambda$ (but it is approximately constant at fixed values of $\Delta f/f_m$ and $f_m/f_\eta$ where $f_\eta = \bar{u}/2\pi\eta$). For $\Delta f/f_m = 0.52$ and $f_m/f_\eta = 1.5$ the excess of a filtered signal reaches a value close to 20. Conclusions of the same type were also obtained by Kuo and Corrsin from the measurement of the excess of derivatives $\partial u/\partial t$ and $\partial^2 u/\partial t^2$.

Similar measurements were performed by Pond and Stewart (1965) in air above a water surface at a height of 1 m, which is a typical example of a natural turbulent boundary layer. In this case, the excess of the horizontal velocity component was also close to zero, and the filtered signal corresponding to the wave number range between 0.03 and 10 $\mathrm{cm}^{-1}$ had an excess of 2.2.

Later Van Atta and Park (1971) measured the probability density functions of the time differences of longitudinal velocity at a fixed point of the atmospheric surface layer above the sea. The time intervals corresponded to spatial separations in the range from 1.4 cm to 28 m. The probability distribution was found to be practically Gaussian for the greatest separation of 28 m and therefore the excess of the corresponding velocity difference was close to zero; however, with the decrease of separation the distributions deviate more and more from a Gaussian and the excess of the velocity difference increases more and more.

It is natural to suppose that the excess of the velocity or temperature derivatives in air near a land or water surface should be even higher, since the spectrum of the derivatives has a maximum in the wavenumber region around 1 $\mathrm{cm}^{-1}$ (see the closing part of Sect.

23.4). In fact, Pond and Stewart (1965) measured the excess of the derivative $\partial u_1/\partial x_1$ in the atmospheric surface layer above a water surface, and found that its magnitude was 17. In their opinion, this result must be regarded as a lower bound for the true value of the excess, because the dynamic range of the instrument which they used was insufficient to reproduce without distortion the largest excursions of the signal. Similar values of the order of ten, or a few tens, were obtained also by Gurvich (1966, 1967), Kholmyanskii (1970), Stewart, Wilson and Burling (1970), Gibson, Stegen and Williams (1970), Sheih, Tennekes and Lumley (1971), Boston (1970), Gibson, Friehe and McConnell (1971), Kuo and Corrsin (1971) and several other experimenters, for excesses of the derivatives $\partial u_1/\partial x_1, \partial u_3/\partial x_1$, and $\partial T/\partial x_1$ in the atmospheric surface layer above a land or water surface, and in some laboratory flows. Here, $u_1, u_3$ and $T$ are velocity components in the streamwise (i.e., downwind) and vertical directions and temperature, or, in some cases, related variables obtained from these velocities and temperature by some averaging procedure over a region with a length scale of the order of 1 cm. All the values of the excess obtained are somewhat unreliable, since experimental determination of higher order moments involves considerable error. However, the values give us, in most cases, lower bounds of the true excess, and consequently suggest that for sufficiently large Re the true values may turn out to be very large.

The positive value of the excess shows that the corresponding distribution is less flat than the normal distribution, i.e., the graph of its density has a higher and sharper central part and more elongated tails than the graph of the normal density with the same variance. In other words, for a large positive excess, the values of a random variable tend to concentrate in separate regions: Very large or very small values, and also values close to the most probable value, are more probable for this case than for the normal distribution with the same variance, while the intermediate values are less probable. This fact is clearly illustrated by Fig. A taken from the paper by Van Atta and Chen (1970). In this figure measured values of the probability density for the normalized derivative $\sigma^{-1}\partial u_1/\partial t$ are compared with the normal density of variance unity. Similar graphs of probability densities for $\partial u_1/\partial t$ or $u_1(t) - u_1(t - \tau)$ where $\tau$ is sufficiently small can be also found in the papers by Sheih, Tennekes, and Lumley (1970), Wyngaard and Pao (1970), and Van Atta and Park (1971).

A simple model of a distribution with a positive excess is obtained if we assume that the given variable assumes zero values with finite

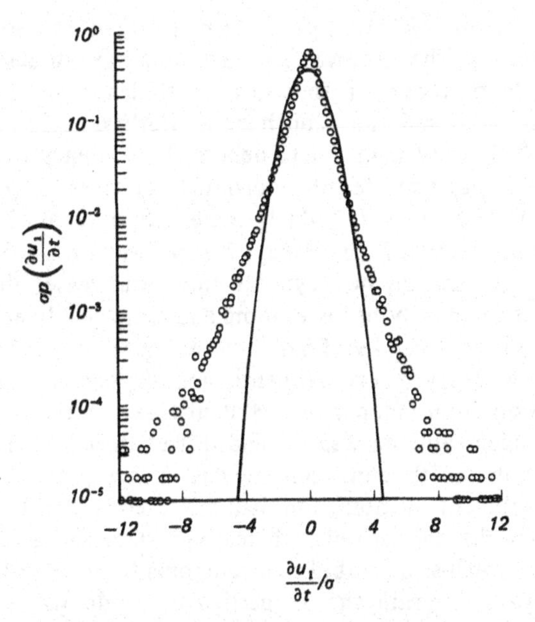


**Fig. A** Probability density of $\partial u_1/\partial t$ for $z = 3$, $u = 11$ m/s after Van Atta and Chen (1970). $\sigma$ is the standard deviation of $\partial u_1/\partial t$. Solid curve is the Gaussian distribution.

probability $p_0$ and nonzero values with probability $1 - p_0$, and that it has a normal probability distribution if it takes nonzero values. In this mixed discrete-continuous model, the excess is given by the formula $\tilde{\delta} = 3p_0/(1 - p_0)$, so that by choosing $p_0$ to be sufficiently close to unity, we can make $\tilde{\delta}$ as large as desired. The interpretation of positive values of excess in terms of a "mixed distribution" is in qualitative agreement with direct observations of Batchelor and Townsend (1949), Sandborn (1959) and Kuo and Corrsin (1971) on the fluctuations of various velocity derivatives and filtered band- and high-pass velocity signals on an oscillograph screen. These observations showed that high-frequency velocity oscillations are characterized by alternating relatively long periods of relative quiet (very small fluctuations) and periods of enhanced activity (large turbulent fluctuations). Such behavior is most clearly seen in band-passed high-frequency signals; the results for derivative signals are less definite [and do not agree between themselves in the works by Batchelor and Townsend (1948), Frenkiel and Klebanoff (1971) and

Kuo and Corrsin (1971)], but the general trend of the data is the same. Therefore, the observations which we have discussed suggest that the distributions of small-scale turbulence are highly non-uniform in space and time and have a clearly defined *intermittent character*. This word is meant to denote the tendency of small-scale turbulence to concentrate into individual "bunches" surrounded by extensive flow regions in which there are only much smoother large-scale disturbances (or perhaps no disturbances at all). As the scale decreases, and/or the Reynolds number increases, the intermittency effect apparently becomes more and more clearly defined.

At the end of Sect. 5.9 of Vol. 1 we emphasized that intermittent turbulence is observed very frequently and plays an important role in the transition from laminar to turbulent flows, in the outer regions of the turbulent boundary layer, and in free turbulent flows of any type. We now see that intermittency is in fact even a much more frequent turbulent phenomenon than indicated in Vol. 1, and is even more important than shown there. The data summarized above suggest that small-scale turbulence components are almost always, or simply always, intermittent. In particular, Sandborn's experiments show that the small-scale velocity components in the turbulent boundary layer are intermittent beginning at the wall itself, while the total velocity field has this character only at considerable distances from the wall. As the Reynolds number increases, the degree of intermittency increases too and simultaneously the scale range (or wave number range) for which there is appreciable intermittency continues to expand. This is clearly seen from the fact that in natural turbulent flows, characterized by especially high Re, and in particular in the free atmosphere and in the interior of the ocean, there was observed the existence of alternating regions of intense turbulence and nonturbulence, i.e., intermittency is observed here even for disturbances from the energy range [see, for example, Krechmer, Obukhov, and Pinus (1952), Grant, Stewart, and Moilliet (1962), Grant, Moilliet, and Vogel (1968), or Vinnichenko and Dutton (1969)]. It is also interesting to indicate that Grant, Stewart, and Moilliet's data (at Re of the order of $10^8$) are characterized by a variation of measured values of $\bar{\varepsilon}$ (which were obtained from runs of length between 4 min and 15 min) by a factor four or more even when there was no evidence of any change in the overall nature of turbulence. It is clear that the fluctuations in $\epsilon$ would be much larger if $\varepsilon$ were to be measured from considerably shorter records, or, the more so, to be defined as instantaneous value of the variable (25.1) and that these fluctuations are inevitable for any flow with high enough Reynolds number. On the other hand, Kraichnan (1967b) explained in a simple manner that the very small scales of turbulence (namely those in the far dissipation range with length scales small in comparison to internal scale) must be strongly intermittent at any Reynolds number. The explanation is based on the fact that small fluctuations of, say, a few percent in the scaling wave length $\eta$ will produce enormous fluctuations in the value of the spectral density at a fixed wave number in the range where the density decreases faster than any power of the wave number; this is closely related to reasons leading to the nonsurvival after averaging (over the dissipation rate fluctuations) of the spectral forms which decrease faster than any power of $k$.

The strong intermittency of the small-scale fluctuations ensures that the probability distributions for the small-scale turbulence components are highly non-Gaussian. This must undoubtedly be taken into account in any statistical model put forward for describing fluctuations in the energy dissipation rate $\varepsilon(\boldsymbol{x}, t)$. The first intermittent models of the probability distributions for small-scale velocity fluctuations were put forward by Townsend (1951a) [see the part of Sect. 22.3 adjacent to Eq. (22.60)]. In these models the small-scale turbulence is concentrated in thin vortex sheets or vortex tubes separated by much thicker regions of fluid from which the small-scale disturbances are absent. Subsequently, Corrsin (1962c) used the vortex sheets model to calculate dissipation fluctuations by assuming additionally that the layers in which there are no small-scale disturbances (and consequently there is no energy dissipation) have a thickness of the order of the external turbulence scale $L$ for large Re, while the vortex sheets have thickness of the order of the internal scale (i.e., Kolmogorov's microscale) $\eta = \nu^{3/4}\bar{\varepsilon}^{-1/4} \ll L$. Hence, it follows that the ratios of the volume in which practically the entire dissipation is concentrated for large Re to the total volume occupied by the fluid is of the order of $\eta/(L+\eta) \approx \eta/L$. Therefore, the mean dissipation rate $\tilde{\varepsilon}$ in the former volume should be of the order of $\bar{\varepsilon}(L/\eta)$. If we suppose that, within the vortex sheets, the velocity derivatives have a normal probability distribution, then the root mean square value of the dissipation fluctuations in the sheet will be approximately 0.6 $\tilde{\varepsilon}$ [see the discussion of Golitsyn's paper (1962) at the beginning of this section]. However, since the assumed normal distribution cannot be regarded as accurate, we may consider its root mean square value as equal to $A\tilde{\varepsilon}$ where $A$ is an unknown constant (probably having the order of unity). By taking the average of $\varepsilon^2(\boldsymbol{x}, t)$ over the entire volume of the fluid, we find that the dissipation variance can be estimated from the formula

$$\overline{(\varepsilon-\bar{\varepsilon})^2}=\overline{\varepsilon^2}-\bar{\varepsilon}^2\approx(1+A^2)\,\tilde{\varepsilon}^2(\eta/L)\approx(1+A^2)\,\bar{\varepsilon}^2(L/\eta),$$

so that

$$\sigma_\varepsilon=(\overline{\varepsilon^2}-\bar{\varepsilon}^2)^{1/2}\approx(1+A^2)^{1/2}\,\bar{\varepsilon}\,(L/\eta)^{1/2}$$

which is the main result given by Corrsin. Recalling that $L/\eta\sim\mathrm{Re}^{3/4}$ according to Eq. (21.6), we find that $\sigma_\varepsilon/\bar{\varepsilon}\sim\mathrm{Re}^{3/8}$, which illustrates the dependence of the dissipation fluctuation level on the Reynolds number in the Townsend-Corrsin model.

The fact that the statistical characteristics of the field $\varepsilon(x, t)$ depend on the parameters of the large-scale motion, namely, the length scale $L$ and the Reynolds number Re, is in agreement with general physical ideas and was to be expected. However, the model used here cannot be regarded as entirely satisfactory, since it is based on arbitrary assumptions which have not been verified experimentally. Moreover, some of the quantitative conclusions which follow from this model are in conflict with well established facts. For example, if we take a natural assumption that $\tilde{\varepsilon}$ is of the order of $\nu U'^2/\eta^2$ where $3U'^2=\overline{u_i'u_i'}$ then it is easy to calculate that $\bar{\varepsilon}$ would be of the order of $\frac{U'^3}{L}(\mathrm{Re}_L)^{-1/4}$, where $\mathrm{Re}_L=U'L/\nu$, although it is well known that $\bar{\varepsilon}$ must be of the order of $U'^3/L$. In order to remedy this discrepancy Tennekes (1968b) proposed a somewhat modified model of a Townsend-Corrsin type. This new model consists of a collection of vortex tubes of diameter $\eta$, which are stretched by eddies of length scale $\lambda=\overline{u_1'^2}\Big/\overline{\left(\frac{\partial u_1'}{\partial x_1}\right)^2}$. The whole dissipation is concentrated in the tubes which occupy a volume fraction of the order $\eta^2\lambda/\lambda^3=\eta^2/\lambda^2$. Hence if we assume that the order of the dissipation rate in the vortex tubes is $\nu U'^2/\eta^2$, we obtain a correct estimate of the volume-averaged dissipation rate. The dissipation variance $\sigma_\varepsilon^2$ in Tennekes' model will be evidently of the order of $\bar{\varepsilon}^2(\lambda/\eta)^2$, i.e., $\sigma_\varepsilon/\bar{\varepsilon}\sim(\mathrm{Re}_L)^{1/4}\sim(\mathrm{Re}_\lambda)^{1/2}$. Tennekes also used his model for estimating the skewness $s$ and the flatness factor $\delta$ of the velocity derivative $\partial u_1/\partial x_1$ (assuming that the contributions to the mean cube and fourth power of $\partial u_1/\partial x_1$ are only made in the vortex tubes). He found that according to the model $s$ is a constant (i.e., does not depend on Reynolds number) while $\delta \sim \lambda^2/\eta^2 \sim (\mathrm{Re}_L)^{1/2} \sim \mathrm{Re}_\lambda$. Tennekes concluded at first that these estimates were in rough agreement with the existing sparse data on the quantities $s$ and $\delta$. However, later Sheih, Tennekes and Lumley (1971) and Friehe, Van Atta and Gibson (1971) reconsidered all the data and came to a conclusion that the true dependence of $\delta$ on $\mathrm{Re}_\lambda$ is apparently much weaker than as predicted by this model.

The further development of the same line of thought is due to Saffman (1968). He began with the collection of energy-containing eddies which produce a local straining field and then assumed that the convergence associated with these motions tends to produce concentrated vortex sheets and tubes. It is possible to estimate that in such a case the thickness of the sheets or the radius of the tubes must be of order of $\lambda$. If we also assume that the typical velocity fluctuation within the sheets and tubes is of order $U'$, then it is clear that we will obtain neither the empirical result that $\bar{\varepsilon}$ is independent of Reynolds number nor the explanation of the important role played by the Kolmogorov length scale $\eta$ in small-scale turbulence mechanics. Therefore Saffman was forced to suggest that the curved vortex sheets of thickness $\lambda$ or tubes of radius $\lambda$ are themselves unstable (to a kind of so-called Taylor-Görtler instability) and they are continually splitting and generating a stable secondary motion, with a cellular structure whose length scale is of the order of the thickness or radius of the original vortex sheet or tube. He also estimated that the boundary layers between these secondary (Taylor-Görtler) cells have a thickness of order $(\nu\lambda/U')^{1/2} \sim \eta$. If we assume that these new boundary layers are the true vortex tubes within which all the energy dissipation is concentrated then we shall naturally obtain Tennekes' model discussed above. The model explains quite satisfactorily the independence of $\bar{\varepsilon}$ on Re and leads to a number of predictions that are in rough agreement (or at least are not in evident disagreement) with experiment. However, the model seems also to be artificial and oversimplified in some of its basic assumptions and there are several points in the above arguments which are quite vague and strongly need further clarification. Slightly more detailed discussion of the model can be found in the book by Orszag (1970); many additional remarks on it were made by Saffman himself (1968, 1970). Here we shall not continue to discuss the vortex sheet and tube models of small-scale turbulence structure, but shall pass on to consideration of the other existing data on the small-scale fluctuations.

The most interesting small-scale variable is the dissipation rate $\varepsilon$. To determine the instantaneous values of $\varepsilon$ experimentally, we must of course determine simultaneously the values of all the derivatives $\partial u_i/\partial x_j$ on the right-hand side of Eq. (25.1). This problem is technically very difficult. Only the derivatives in the direction of the mean flow can be measured relatively simply. We can then use the Taylor frozen-turbulence hypothesis and express them in terms of the time derivatives. We note that the velocity derivatives are very dependent on fluctuations having a length scale of the order of the Kolmogorov scale $\eta$, since the spectrum of these derivatives has a maximum near wave numbers of the order of $k_\eta = \eta^{-1} = (\bar{\varepsilon}/\nu^3)^{1/4}$. The determination of such small-scale motions requires the use of very small probes and highly sensitive and low-inertia measuring instruments.

Because simultaneous measurements of all the components of the tensor $\partial u_i/\partial x_j$ are very difficult, Gurvich and Zubkovskii (1963) have tried to determine the general shape of the spectrum of variable $\varepsilon(x, t)$ in the surface layer of the atmosphere (at a height of 4 m) from measurements of the single derivative $\partial u_3/\partial x_1 = \partial w/\partial x$, i.e., the derivative of the vertical velocity component in the direction of the mean wind. It is well known that, for locally isotropic turbulence, $\bar{\varepsilon} = \frac{15\nu}{2}\overline{(\partial w/\partial x)^2}$ [see Eqs. (21.16) and (21.19′)]. It is natural to suppose that all the statistical characteristics of the fluctuations in $\varepsilon$ and in the quantity $\varepsilon_w = \frac{\nu}{2}\left(\frac{\partial w}{\partial x}\right)^2 = \frac{\nu}{2}\left(\frac{\partial w}{\bar{u}\,\partial t}\right)^2$ (which is one of the components of $\varepsilon$) should be roughly the same, and should differ only by numerical factors. In particular, it is natural to suppose that the spectra of $\varepsilon$ and $\varepsilon_w$ should not be too different. This assumption was the basis of Gurvich and Zubkovskii's work.

Gurvich and Zubkovskii used the sonic anemometer, an electrical differentiating and squaring circuit (the current at the output was proportional to the square of the current at the input), and a spectral analyzer to determine the time spectrum of the fluctuations in $\varepsilon_w$:

$$E_{\varepsilon_w}(\omega) = \frac{2}{\pi}\int_0^\infty \overline{[\varepsilon_w(t) - \bar{\varepsilon}_w]\,[\varepsilon_w(t+\tau) - \bar{\varepsilon}_w]}\cos\omega\tau\, d\tau,$$

where the overbar represents a time average. Since the base of the sonic anemometer was finite, and so were the diameters of the transmitting and receiving microphones, and since the instrument had a finite inertia (time constants of the order of 0.01 sec), the resulting value of $w$ in fact corresponded to the mean vertical velocity within the limits of a parallelopiped with sides of 5, 0.5, and 4 cm along the $Ox$, $Oy$, and $Oz$ axes. Altogether eight spectra $E_{\varepsilon_w}(\omega)$ were obtained, using an averaging period of 10 minutes in each case. The spectra were then normalized to $\overline{\varepsilon_w^2}$, and a transformation was made from the frequency $\omega$ to the wave number $k=\omega/\bar{u}$. The resulting points clustered around a single curve corresponding to a power-law dependence on $k$ of the one-dimensional spectrum $E_1^{(\varepsilon_w)}(k)$ in a broad range of wave numbers (see Fig. 102 which shows the averaged data of all the eight measurements). The exponent (calculated by the method of least squares between $k=0.02$ and $k=100\ \text{m}^{-1}$) was found to be close to –0.6. Thus, according to these data

$$E_1^{(\varepsilon_w)}(k) \sim k^{-0,6}. \tag{25.16}$$

Practically the same results were obtained when the measurements were repeated in the summer of 1964 [Gurvich and Zubkovskii (1965)].

When these results are interpreted, the most unfavorable aspect is the instrumental averaging of the measured quantity $w$ over a volume whose linear dimensions exceed the internal turbulence length scale $\eta$

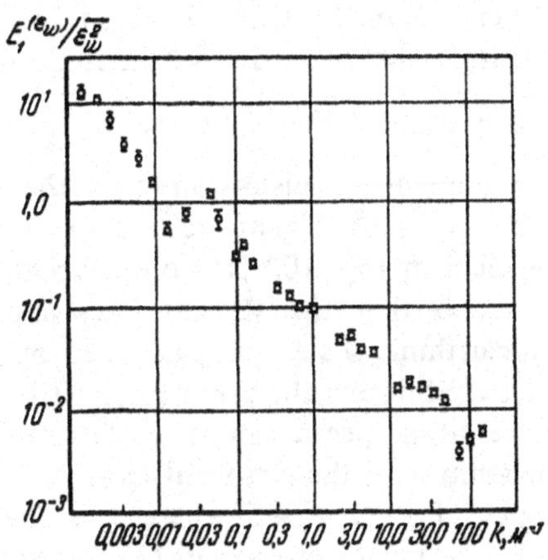


**Fig. 102** The spectrum $E_1^{(\varepsilon_w)}(k)$, according to Gurvich and Zubkovskii (1963).

in the surface layer. It is very difficult to take the effect of this averaging on the spectrum of $(\partial w/\partial x)^2$ into account, since this requires further information about the fourth-order moments of the velocity field which is not available at the present time. Nevertheless, it is not very likely that this effect would explain the substantial qualitative difference between the spectrum $E_1^{(\varepsilon_w)}(k)$ (Fig. 102), which is proportional to $k^{-0.6}$, and the spectrum of energy-dissipation rate fluctuations, which does not decrease with increasing $k$, and which is predicted by calculations based on the zero-fourth-cumulant assumption or on the Townsend-Corrsin model. This latter conclusion is also confirmed by results of more refined hot-wire measurements of the spectrum of $(\partial u_1/\partial x_1)^2$ performed by Pond and Stewart (1965) [see also Pond (1965)], Stewart, Wilson and Burling (1970), Gibson, Stegen and McConnell (1970), Van Atta and Chen (1970), Sheih, Tennekes and Lumley (1971), Wyngaard and Pao (1971), Gibson, Friehe and McConnell (1971), Friehe, Van Atta and Gibson (1971), and Kholmyanskii (1972).

All the above mentioned authors used a hot-wire anemometer with a miniature wire to determine the fluctuations in the longitudinal component of the wind velocity $u$ of the atmosphere at various heights in the range 1–100 m, or of the longitudinal velocity at the centerline of a laboratory turbulent jet. In these experiments the instrumental averaging, produced by the finite dimension of the probe and the inertia of the apparatus, was much smaller than in the experiments of Gurvich and Zubkovskii, and therefore the fluctuations of $\partial u/\partial t$ were recorded without significant distortion over practically the entire dissipation range of the frequency spectrum. It may therefore be concluded that the values of $\varepsilon_u = \frac{\nu}{2}\left(\frac{\partial u}{\partial x}\right)^2$ can be obtained without appreciable distortion from the above measurement data. The spectrum $E_1^{(\varepsilon_u)}(k)$ of the quantity $\varepsilon_u$ found by Pond and Stewart is shown in Fig. 103 [the ordinates are normalized to $E_1^{(\varepsilon_u)}(10^{-1})/10$, i.e., so that the straight lines approximating the spectra on the logarithmic scale, and calculated by the method of least squares, all pass through the point (0.1, 10)]. It is clear from Fig. 103 that the resulting spectra can be satisfactorily described by a power-law relationship with the exponent close to –0.6 in the wave number range such that $kz \gg 4.5$ where $z$ is the height of the observation point (this range corresponds to local isotropy according to most of the data discussed in Sect. 23.3). Therefore the data of

Pond and Stewart suggest that

$$E_1^{(\varepsilon_u)}(k) \sim k^{-0,6}. \tag{25.16'}$$

The precise agreement between the exponents in Eqs. (25.16) and (25.16′) is, of course, to some extent accidental because the measurements of Gurvich and Zubkovskii involved much greater instrumental averaging, and the wave number range for which Eq. (25.16) was obtained covered much greater scales (partly corresponding to nonisotropic disturbances, than in the case of Eq. (25.16′). To obtain better correspondence with the Gurvich and Zubkovskii data, Pond and Stewart fed the signal from the hot-wire anemometer, which is proportional to $\partial u/\partial x$, into a special low pass filter (producing an averaging effect similar to the instrumental averaging due to the sonic anemometer). They then squared the filtered signal and determined its spectrum. They found, however, that their new results were even more different from those described by Eq. (25.16) than the old ones (the power exponent was now about –0.4). It must be remembered, however, that Pond and Stewart did in fact measure a quantity which was different from that determined by Gurvich and Zubkovskii (namely $\partial u/\partial x$ and not $\partial w/\partial x$), and the analogy between their filter and the instrumental averaging of the sonic anemometer is not a quantitative one. In view of this, the difference between the two exponents is of course not surprising.

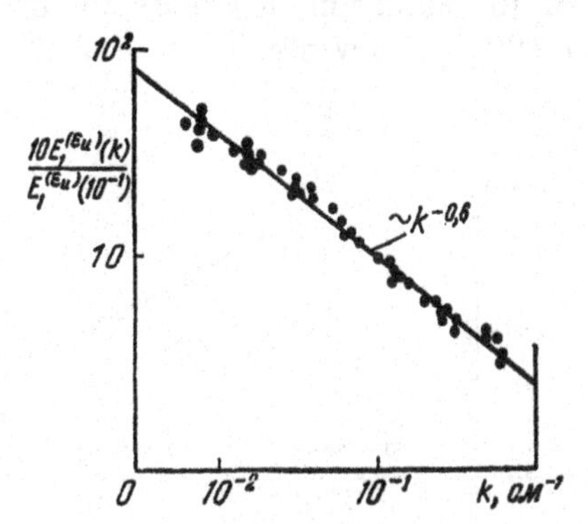


**Fig. 103** The spectrum $E_1^{(\varepsilon_u)}(k)$, according to Pond and Stewart (1965).

Similar results were also obtained by all other investigators, though their data on the values of the exponent in the power law for the spectrum $E_1^{(e_u)}(k)$ show a moderate scatter. According to the results of Stewart, Wilson, and Burling (1970) the spectrum $E_1^{(e_u)}$ can be approximated with satisfactory accuracy by a power law with exponent $-0.65$ over a considerable range of wave numbers (or frequencies). Gibson, Stegen, and McConnell (1970) measured the spectrum of the squared longitudinal velocity derivative at heights above the water of 12.25, 8.25, 4.25, and 2.25 meters and calculated the slope and standard deviation of the slope for each of the spectra (in logarithmic coordinates) for frequencies below the viscous cutoff. [It was found later by Gibson, Friehe and McConnell (1971) and Friehe, Van Atta and Gibson (1971) that practically the same slope can be obtained in laboratory measurements of velocity derivative fluctuations in a turbulent jet.] Very close results were also obtained by Wyngaard and Pao (1971), who treated the data of hot-wire anemometer measurements in the atmospheric surface layer over land. All the spectra turn out to obey a power law with moderate accuracy and the average value of the slope (i.e., of exponent) was $-0.493 \pm 0.022$. Van Atta and Chen (1970) measured the frequency spectrum $E_1^{(e_u)}$ by a numerical method using digitized velocity derivative data from a 26 min record obtained at a height of 31 m above the sea surface and at mean wind velocity $\bar{u} = 11.3$ m/sec. This spectrum is shown in Fig. B; the dashed line in it has a slope of $-0.5$. Sheih, Tennekes and Lumley (1971) published the data of airborne measurements of the spectrum of $(\partial u/\partial t)^2$ in the atmosphere at a height of about 100 m. They also found that the spectrum is well

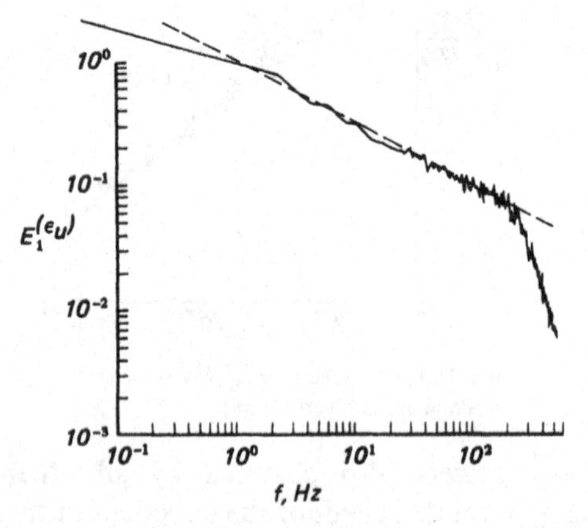


**Fig. B** Energy spectrum $(\partial u/\partial x)^2$ for $z = 31$ m, $\bar{u} = 11.3$ m/s after Van Atta and Chen (1970). The dashed line has a slope of $-0.5$.

represented by a power law over an extended range of wave numbers, but the exponent was close to –0.3 according to their data. It is clear that the exact value of the exponent depends upon the range selected for power-law approximation and the accuracy of its determination is not very high. Nevertheless all the above data with the exception of these by Sheih et al. agree quite satisfactorily with each other on the general shape of the spectrum.

The recent results of Kholmyanskii (1972) are slightly different from all the previous results. He evaluated two examples of $E_1^{(\varepsilon u)}(k)$ and found that for $\log_{10} k_\eta \leqslant -1.5$ (i.e., in the "inertial subrange" of the energy spectrum $E_1(k)$ measured by Kholmyanskii) the slope of the spectrum is close to –0.7 while for $-1.5 < \log_{10} k\eta < -0.6$ the slope is smaller (close to –0.5). Therefore he concluded that most of the data of other authors refer in fact to the high frequency range above the high frequency limit of the inertial subrange.

Thus all the existing data (with only one curious exception) do indicate that the one-dimensional spectrum $E_1^{(\varepsilon)}(k)$ can apparently be approximately represented by a power law in the inertial subrange, and that it decreases with increasing $k$, the corresponding exponent being somewhere between –0.7 and –0.5. Novikov and Stewart (1964) were the first who put forward a simple model of the spatial distribution of energy dissipation rate $\varepsilon(\boldsymbol{x})$ which takes into account the intermittency effect and can be used to explain this shape of the spectrum $E_1^{(\varepsilon)}(k)$. According to their scheme, if we take a cube of size $L_0$, which is of the order of the external length scale $L$, and split it into $n \gg 1$ equal but smaller cubes of length $L_1 = L_0 n^{-1/3}$, then practically the entire dissipation will be concentrated in $m \ll n$ of them, and will not be present in all the remaining ones. It was further assumed that the $m$ cubes in which the entire dissipation was concentrated were distributed in space in a random fashion, and that each contained the same fraction of total dissipation. Novikov and Stewart also assumed that the turbulence in each of these $m$ volumes was again distributed quite nonuniformly and in the same way as in the large cube of length $L_0$. Namely, within each of the $m$ first-order cubes, each of length $L_1$, the dissipation was assumed to be concentrated in randomly chosen $m$ second-order cubes with lengths $L_2 = L_1 n^{-1/3} = L_0 n^{-2/3}$, and in each of them it was again considered to be the same. This process was continued up to cubes of order $s$ for which molecular viscosity became important, i.e., cubes with lengths $L_s = L_0 n^{-s/3} = \eta_0$ for which the Reynolds number is of the order of unity. In these smallest cubes, the dissipation could be regarded as uniformly distributed.

The mean square dissipation rate in the Novikov-Stewart model satisfies the relation $\overline{\varepsilon^2} = \bar{\varepsilon}^2 \left(\frac{n}{m}\right)^s \sim \bar{\varepsilon}^2 (L_0/\eta_0)^{3q}$ where $q = 1 - \frac{\log m}{\log n}$, so that $0 < q < 1$. Similarly, $\overline{\varepsilon^{2/3}} = \bar{\varepsilon}^{2/3}\left(\frac{n}{m}\right)^{-s/3} \sim \bar{\varepsilon}^{2/3}(L_0/\eta_0)^{-q}$. However, Novikov and Stewart paid particular attention to the correlation function $b_{\varepsilon\varepsilon}(r) = B_{\varepsilon'\varepsilon'}(r)$ and the one-dimensional spectral density $E_1^{(\varepsilon)}(k)$ in the inertial subrange. With this in view, they considered a one-dimensional distribution of dissipation rate along the fixed straight line which can be taken as the $x$ axis. In the first approximation, this distribution was represented by a set of rectangular "pulses" of length $\lambda_1$ and amplitude $A_1$, the mean distance between the pulses being $l_1$. In the $j$th approximation, the dissipation was concentrated in a set of $j$th order intervals (and assumed a constant value $A_j$), each of which had a length $\lambda_j$, and were randomly distributed within each of the $(j-1)$th order intervals (of length $\lambda_{j-1}$), so that the mean distance between them was $l_j$. For all the approximations for which viscosity could be ignored, i.e., $\lambda_j > \eta_0$, it was assumed that the following similarity relations were satisfied:

$$\begin{gathered}\frac{\lambda_1}{l_1} = \ldots = \frac{\lambda_j}{l_j} = \ldots = \alpha \ll 1, \\ \frac{\lambda_2}{\lambda_1} = \ldots = \frac{\lambda_j}{\lambda_{j-1}} = \ldots = \beta \ll \alpha.\end{gathered} \tag{25.17}$$

In that case, $\bar{\varepsilon} = A_1\alpha = \ldots = A_j\alpha^j = \ldots$ и $\lambda_j = \lambda_1\beta^{j-1}$. The correlation function $b_{\varepsilon\varepsilon}(r)$ and the spectrum $E_1^{(\varepsilon)}(k)$ have the following form in the inertial subrange in this model:

$$b_{\varepsilon\varepsilon}(r) = C\bar{\varepsilon}^2\alpha^{-1}(r/\lambda_1)^{-\mu} \sim r^{-\mu}, \quad E_1^{(\varepsilon)}(k) = C_1\bar{\varepsilon}^2\alpha^{-1}\lambda_1(k\lambda_1)^{-1+\mu} \sim k^{-1+\mu}, \tag{25.18}$$

where $C$ and $C_1$ are certain constants, and

$$\mu = \frac{\log \alpha}{\log \beta}, \qquad 0 < \mu < 1. \tag{25.19}$$

The spectrum given by Eq. (25.18) is in agreement with the experimental results (25.16) and (25.16′) if we assume that $\mu \approx 0.4$.

The order of magnitude of the minimum length scale $\eta_0$ of dissipation inhomogeneities (limited by viscosity) when dissipation fluctuations are taken into account may be different from $\eta = \nu^{3/4}\bar{\varepsilon}^{-1/4}$. Novikov and Stewart determined $\eta_0 = \lambda_s$ from the requirement that the Reynolds number obtained from the length $\lambda_s$, the energy dissipation rate $A_s = \bar{\varepsilon}\alpha^{-s}$, and the viscosity $\nu$ was of the order of unity. Since $A_s \gg \bar{\varepsilon}$, the length $\eta_0 = \lambda_s = \nu^{3/4}A_s^{-1/4}$ was found to be appreciably smaller than $\eta$.

The Novikov-Stewart model is of course very approximate and cannot pretend to be in quantitative agreement with the real dissipation distribution in a turbulent flow. The various assumptions of the model were introduced for the sake of simplicity and are in fact not essential. It is possible to put forward a number of more general and more complicated models leading to quite similar results. Some of these models were considered by Novikov (1965, 1966). The random field $\varepsilon(x)$ on the straight line $-\infty < x < \infty$ was determined in these papers as the limit of a sequence of random functions $\varepsilon_j(x)$, $j = 1, 2, \ldots$, consisting of individual pulses. Under certain special assumptions about these pulses (including a self-similar assumption of some kind) it was found that formulas such as Eq. (25.18) were valid for this type of model. Another method for the mathematical description of intermittent self-similar random functions of the kind discussed by Novikov and Stewart (1964) and by Novikov (1965, 1966) was developed by Mandelbrot (1965, 1967a, b, 1972) in the form of the theory of "sporadically self-similar random functions" which was originally put forward in connection with other problems (cf. also Mandelbrot, 1967b). Similar formulas may, however, be obtained by another rather simple and natural reasoning. This follows from the model of "self-similar breakdown of turbulent eddies" described by Yaglom (1966) which is implicit in the hypotheses given by Obukhov (1962a, b) and Kolmogorov (1962a, b).

In the papers by Obukhov and Kolmogorov a suggestion was made which we have not mentioned before, that the dissipation rate $\varepsilon(\boldsymbol{x}, t)$ has a log-normal probability distribution, i.e., that $\ln \varepsilon$ is distributed in a normal fashion. Moreover, it was assumed by Kolmogorov that the variance of the quantity $\ln \varepsilon(\boldsymbol{x}, t)$ for large Reynolds numbers is given by

$$\sigma^2_{\ln \varepsilon} \approx A'(\boldsymbol{x}, t) + \mu' \ln(L/\eta), \tag{25.20}$$

where $L$ is the external turbulence length scale, $\eta$ is the appropriately defined internal length scale, $\mu$ is a universal constant, and the term

$A'(x, t)$ may depend on the parameters of the large-scale motions. Similarly, the logarithm of the dissipation rate $\varepsilon_r$ averaged over a sphere of radius $r/2$ was also assumed to be normally distributed with a variance of the form

$$\sigma^2_{\ln \varepsilon_r} \approx A(x, t) + \mu \ln(L/r), \qquad (25.20')$$

where $A(x, t)$ may depend on the macrostructure of the flow.

To verify the log-normal assumption, Obukhov noted that any probability distribution of a positive random variable can be approximated by a logarithmically normal distribution with correct values of the first two moments and also made passing reference to the paper of Kolmogorov (1941b) in which it was shown that the log-normal distribution corresponds asymptotically to the particle-size distribution obtained as a result of a sequence of successive and independent fragmentations. The reference is in fact quite appropriate to the problem since the process of successive fragmentation is a natural model of the cascade process of successive breakdown of turbulent eddies as described in Sect. 21.1. Let us now consider this question in more detail. One of the most important features of the Richardson-Kolmogorov scheme of developed turbulence is the existence of a one-directional energy flux described by a relation of the form $A \to B \to C$ (large eddies→small eddies→heat). However the word equilibrium is usually used in the physical sciences for a two-sided process of the form $A \leftrightarrow B$ or $A \leftrightarrow B \leftrightarrow C$ and so on where the principle of detailed balance is valid, i.e., there is mutual balance between direct and inverse transitions at every step. Therefore it is not reasonable to speak of equilibrium when Kolmogorov's theory is discussed; rather, it is preferable to use the term "quasi-equilibrium" (as is done in our book) or, for example, simply "steady state" (as was recommended by Pond, Stewart, and Burling, 1963). The consequences of this are quite significant. In the usual equilibrium process the fluctuations are smoothed out at every transition, whereas in the cascade quasi-equilibrium there is a successive amplification of any randomly appearing fluctuation. One may reasonably assume that each fixed band of scales (or wave numbers) within the inertial subrange will produce the same fractional increase in the fluctuations of the rate of energy transfer, quite similarly to Kolmogorov's assumption (1941b) that every fragmentation of particles produces the same fractional increase in the particle-size fluctuations. In other words one may expect that every band of scales or every individual fragmentation contributes the same amount to the logarithm of the corresponding fluctuating variable. Since the different contributions are not likely to be strongly correlated, this leads, in view of the central limit theorem of probability theory, to a limiting log-normal distribution for both types of problem.

These qualitative considerations were formalized in the model of eddy breakdown introduced by Yaglom (1966). Since the details of the cascade process of eddy breakdown are not known, he confined his attention to the simplest fragmentation scheme, bearing in mind the fact that his final results can also be obtained under much more general assumptions.

Let us follow the work of Yaglom and consider the following simple scheme. Let $\bar{\varepsilon}$ be the mean energy dissipation rate within a cube of length $L_0$ which is of the order of the external length scale $L$. The quantity $\bar{\varepsilon}$ is determined by the parameters of large-scale motions, and is of the order of $(\Delta U)^3/L$ (see Sect. 21.3). Let us subdivide the original cube into an arbitrary number $n$ of smaller cubes with lengths $L_1 = L_0 n^{-1/3}$. In contrast to the Novikov-Stewart treatment, we shall not assume that the dissipation is concentrated in a small number of these first-order cubes, but will suppose that the space-average value of the dissipation rate within each cube is a random variable with the same probability distribution (and mean value $\bar{\varepsilon}$). Each of these first-order cubes will be subdivided further into $n$ second-order cubes (with lengths $L_2 = L_0 n^{-2/3}$). A similar subdivision will be carried out in each second-order cube, and so on. This process of successive subdivision of cubes corresponds to a certain extent to the cascade process of breakdown of turbulent eddies. As the cascade process must be self-similar, we shall make the following assumption of self-similarity: *if $L_j = L_0 n^{-j/3} \ll L$, then the conditional probability distribution for the energy dissipation rate averaged over the volume of a jth order cube is the same for all cubes of order j and is independent of j until molecular viscosity becomes directly important, provided that the value of the dissipation averaged over the volume of the cube of order $j-1$, which includes the cube of order j, is fixed.* This assumption is similar to the Novikov and Stewart assumption (25.17), but is of course much more general.

Let $\varepsilon_j$ represent the dissipation rate averaged over the volume of the cube of order $j$, and let $e_j = e_j(\varepsilon_{j-1})$ denote the dimensionless random variable $\varepsilon_j/\varepsilon_{j-1}$ where $\varepsilon_{j-1}$ is fixed. According to the above assumption, the probability distribution for the variable $e_j$ for $n^{-j/3} \gg 1$ is independent of $j$, and is universal until viscosity becomes important, i.e., as long as $L_j = L_0 n^{-j/3} \gg \eta_0$ or, in other words, so long as $j \ll 3 \ln(L_0/\eta_0)/\ln n$ where $\eta_0$ is the scale of the largest-scale disturbances affected by viscosity. We then have

$$\varepsilon_j = \bar{\varepsilon} e_1 e_2 \ldots e_j, \tag{25.21}$$

where all the $e_1, \ldots, e_j$ except for the first few of them are independent, identically distributed, random variables when $j \ll 3 \ln(L_0/\eta_0)/\ln n$. We shall suppose that the random variable $\ln e_k$ has a finite mean value $m_e$ and a finite variance $\sigma_e^2$ (this assumption does not seem to be too restrictive although it is not satisfied for the Novikov-Stewart model and, in general, for any model for which $\varepsilon_k$ vanishes identically with finite probability). According to Eq. (25.21),

$$\ln \varepsilon_j = \ln \bar{\varepsilon} + \ln e_1 + \ldots + \ln e_j, \tag{25.22}$$

where all the variables $\ln e_1, \ldots, \ln e_j$, except for the first few of them, are independent random variables with a mean value $m_e$ and a variance $\sigma_e^2$. In view of the well-known central limit theorem of probability theory [see, e.g., Feller (1966)], it follows from this that, when $1 \ll j \ll 3 \ln(L_0/\eta_0)/\ln n$, the variable $\ln \varepsilon_j$ will have an approximately normal probability distribution with a mean $m_j$ and a variance $\sigma_j^2$, where

$$m_j = \ln \bar{\varepsilon} + A_1(\boldsymbol{x}, t) + j m_e, \quad \sigma_j^2 = A(\boldsymbol{x}, t) + j\sigma_e^2, \tag{25.23}$$

and the terms $A_1(\boldsymbol{x}, t)$ and $A(\boldsymbol{x}, t)$ appear because the first few terms have nonuniversal distributions. In other words, the variable $\varepsilon_j$ has the probability density

$$p_j(x) = \frac{1}{\sqrt{2\pi}\sigma_j x} \exp\left[-\frac{(\ln x - m_j)^2}{2\sigma_j^2}\right]. \tag{25.24}$$

The $q$-order moment of $\varepsilon_j$ (i.e., the mean value of the variable $\varepsilon_j^q$) is then given by

$$\overline{\varepsilon_j^q}=\int_0^\infty x^q p_j(x)\,dx=\exp\left(qm_j+\frac{q^2\sigma_j^2}{2}\right). \qquad (25.25)$$

However, when $q=1$ the variable $\bar{\varepsilon}_j=\bar{\varepsilon}=\exp(m_j+\sigma_j^2/2)$ should not depend on $j$, and hence it is clear that

$$m_e=-\sigma_{el}^2/2. \qquad (25.26)$$

The quantity $\varepsilon_j$ is the dissipation rate averaged over the cube of length $L_j=L_0 n^{-j/3}$, so that $j=3\,\frac{\ln L_0/L_j}{\ln n}$. It is natural to suppose, however, that similar considerations can be applied to volumes of other fixed form, and also not only to the energy dissipation rate $\varepsilon$ but to many other local turbulence characteristics as well. Let $\varepsilon^*=\varepsilon^*(x,t)$ be some local nonnegative variable determined only by small-scale turbulent fluctuations (for example, the square or absolute value of any space derivative of the velocity field or of the scalar temperature field, the energy or temperature variance dissipation rate, the square of the vorticity, or the square of the velocity or temperature difference at two sufficiently close space points). Let us consider two similar one-, two-, or three-dimensional space regions $V$ and $V'$ with length scales $l$ and $l'<1$ such that $V'$ is located within $V$. Then the intuitive notions on the purely accidental character and self-similarity of the cascade process of eddy breakdown make reasonable the following mathematical hypothesis: *If $l \ll L$ (i.e., if the region $V$ is small in comparison with a typical scale of the inhomogeneities of the mean flow) and $l' \gg \eta'$, where $\eta'=(\nu^3/\varepsilon')^{1/4}$ is Kolmogorov's microscale corresponding to the mean rate of energy dissipation $\varepsilon'$ within region $V'$ (i.e., the region $V'$ is large as compared to the distances at which the molecular viscosity begins to be significant) and the Reynolds number is large enough, then the probability distribution of the ratio of the values $\varepsilon^*$ averaged over $V'$ and over $V$ depends only on the ratio $l'/1$.* The length $\eta'$ is evidently itself a random variable, since it depends on the random variable $\varepsilon'$. It seems also clear that if we select a third region $V''$ within $V'$ with a length scale $l''$ such that $l'>l''\gg\eta''$ than *the probability distribution of the ratio of the values $\varepsilon^*$ averaged over $V''$ and over $V'$ must be independent of the ratio of the same values averaged over $V'$ and over $V$.* These general assumptions were stated by Gurvich and Yaglom (1967); later they were called by Novikov (1969, 1970) the *conditions of scale similarity of turbulent fields* (and the range of validity of the assumption was called the *scale similarity subrange* which is closely related to the notion of the inertial subrange of the original Kolmogorov theory).

Let us now consider a sequence of successively enclosed similar space regions $V_0, V_1, \ldots V_N$ with the same ratio of length scales $l_j/l_{j-1}$ for any two successive regions. Let us suppose that the scale $l_0$ of the largest region is of the order $L$ while within the smallest region fluctuations in the variable $\varepsilon^*(x,t)$ can be neglected. If $\varepsilon_j^*$ is the average value of $\varepsilon^*$ in the region $V_j$ and $e_j = \varepsilon_j^*/\varepsilon_{j-1}^*$, then $\varepsilon_0^*$ can be identified with the mathematical expectation $\overline{\varepsilon^*}$ and all the $e_j$ (where the variable $\varepsilon_{j-1}^*$ in $e_j$ is considered to be fixed, and only the variable $\varepsilon_j^*$ is random) will be identically distributed independent random variables for all indices $j$ such that $L \gg l_{j-1} > l_j \gg \eta_j = (\nu^3/\varepsilon_j)^{1/4}$. Moreover

$$\varepsilon_n^* = \overline{\varepsilon^*} e_1 e_2 \ldots e_n, \qquad \varepsilon^* = \overline{\varepsilon^*} e_1 e_2 \ldots e_N, \tag{25.21'}$$

$$\ln \varepsilon_n^* = \ln \overline{\varepsilon^*} + \sum_{j-1}^{n} \ln e_j \quad \ln \varepsilon^* = \ln \overline{\varepsilon^*} + \sum_{j=1}^{N} \ln e_j, \tag{25.22'}$$

where $\varepsilon^*$ is the value of the field $\varepsilon^*(x,t)$ at the point of the smallest region $V_N$.

In the case of fully developed turbulence the scale similarity range is wide enough and hence there are many identically distributed independent terms in the right-hand sides of relations (25.22′) which provide the main contribution to these sides. Consequently, the central limit theorem implies that both the variables $\varepsilon^*$ and $\varepsilon_n^*$ where $n$ is such that $l_n \ll L$ have approximately log-normal probability distributions with a probability density of the form (25.24). The mean values $m_n$ and $m$ and variances $\sigma_n^2$ and $\sigma^2$ of the variables $\ln \varepsilon_n^*$ and $\ln \varepsilon^*$ are evidently of the form

$$m_n = \ln \overline{\varepsilon^*} + A_1(x,t) + n m_1^*, \qquad m = \ln \overline{\varepsilon^*} + A_1'(x,t) + N m_1^*,$$
$$\sigma_n^2 = A(x,t) + n\mu', \qquad \sigma^2 = A'(x,t) + N\mu'.$$

Here $m_1^*$ and $\mu' > 0$ are the mean value and variance of the identically distributed variables $\ln e_j$, and all the terms $A$ with any subscript and superscript depend exclusively upon the parameters of the large-scale motion (and change notably only when the point $x$ is displaced a distance of the order $L$). It is also clear that if the ratio $l_j/l_{j-1} = \lambda < 1$ is fixed, then $n \sim \ln(L/l_n)$ and $N \sim \ln(L/\eta)$ where $\eta$ is a local value of the Kolmogorov microscale. Hence the equation for $m_n, m, \sigma_n^2$, and $\sigma^2$ can also be rewritten in the form

$$m_r = \ln \overline{\varepsilon^*} + A_1(x, t) + m^* \ln \frac{L}{r}, \qquad m = \ln \overline{\varepsilon^*} + A_1'(x, t) + m^* \ln \frac{L}{\eta},$$

$$\sigma_r^2 = A(x, t) + \mu \ln \frac{L}{r}, \qquad \sigma^2 = A'(x, t) + \mu \ln \frac{L}{\eta}, \tag{25.23'}$$

where $m_r$ and $\sigma_r^2$ are the mean value and the variance of the logarithm of the variable $\varepsilon_r^* = (1/V_r)\int_{V_r} \varepsilon^*(x, t)dx$, $V_r$ is a space region of fixed form with length scale $r$ and $m^*$ and $\mu$ are constants. For the case when $\varepsilon^*$ is the dissipation rate $\epsilon$ and $V_r$ is a spherical region of diameter $r$ the last two equations (25.23′) coincide with Kolmogorov's assumptions (25.20) and (25.20′). Using the fact that $\overline{\varepsilon_r^*} = \overline{\varepsilon^*}$ for all $r$ we easily obtain the following relations connecting $m^*$ with $\mu$, $A_1$ with $A$, and $A_1'$ with $A'$:

$$m^* = -\frac{1}{2}\mu, \quad A_1 = -\frac{1}{2}A, \quad A_1' = -\frac{1}{2}A'. \tag{25.26'}$$

It is also clear that $A'$ differs from $A$ only by a term which is due to the specific form of the last few summands on the right-hand side of the second equation (25.22′) and is independent of the parameters of the large-scale motion. On the other hand both the parameters $A$ and $A' - A$ may vary when the form of the space regions $V_l$ varies. However the value of the parameter $\mu$ does not depend on the form of $V_l$ as will be seen from the discussion below [but of course $\mu$ may take different values for different fields $\varepsilon^*(x, t)$]. Let us also note that in the models of dissipation rate fluctuations suggested by Novikov and Stewart (1964) and Novikov (1965, 1966) the variance of all the identically distributed variables $\ln \varepsilon_j$ on the right-hand side of Eq. (25.22′) proves to be infinite. It follows from this that the probability distribution for $\ln \varepsilon_r^*$ and $\ln \varepsilon^*$ coincides according to these models with one of the more special so-called "stable nonnormal distributions" (see, e.g., Feller, 1966). However the assumption about the infinite variance of $\ln \varepsilon_j$ does not seem to be particularly reasonable and we shall not try to analyze its consequences.

It follows from the expression (25.25) for the $q$-order moment of the log-normal distribution and equations (25.26′) that

$$\overline{\varepsilon^{*q}} = \overline{\varepsilon^{*}}^{q} \exp\left[\frac{1}{2}\, q(q-1)\sigma^2\right], \qquad \overline{\varepsilon_r^{*q}} = \overline{\varepsilon^{*}}^{q} \exp\left[\frac{1}{2}\, q(q-1)\sigma_r^2\right] \qquad (25.27)$$

[for the case when $\varepsilon^*$ is the energy dissipation rate these equations are due to Kolmogorov (1962a, b)]. By virtue of Eq. (25.23′) the equations (25.27) can also be rewritten in the form

$$\overline{\varepsilon^{*q}} = D_q' \overline{\varepsilon^{*}}^{q} \left(\frac{L}{\eta}\right)^{\mu q(q-1)/2}, \qquad \overline{\varepsilon_r^{*q}} = D_q \overline{\varepsilon^{*}}^{q} \left(\frac{L}{r}\right)^{\mu q(q-1)/2} \qquad (25.27')$$

where $D_q' = D_q'(x, t)$ and $D_q = D_q(x, t)$ are coefficients which may depend on the macrostructure of the flow. Moreover the above model of the cascade process of eddy breakdown permits one to evaluate the correlation functions $B_{\varepsilon^*\varepsilon^*}(r) = \overline{\varepsilon^*(x+r)\varepsilon^*(x)}$ and $b_{\varepsilon^*\varepsilon^*}(r) = \overline{[\varepsilon^*(x+r) - \overline{\varepsilon^*}]\,[\varepsilon^*(x) - \overline{\varepsilon^*}]} = B_{\varepsilon^*\varepsilon^*}(r) - \overline{\varepsilon^*}^2$ and the one-dimensional spectrum $E_1^{(\varepsilon^*)}(k)$ of the field $\varepsilon^*(x, t)$ in the scale-similarity subrange. The simplest method for deriving the equation for $B_{\varepsilon^*\varepsilon^*}(r)$ is that which is based on the special selection of the regions $V_i$ in the form of one-dimensional line segments. In this case we can use the well-known formula which relates the correlation function $B_{\varepsilon^*\varepsilon^*}(r)$ of the stationary process $\varepsilon^*(x)$ and the mean square of the variable $\varepsilon_r^* = \frac{1}{r}\int_0^r \varepsilon^*(x)dx$:

$$\overline{\varepsilon_r^{*2}} = \frac{2}{r^2}\int_0^r \int_0^{r_1} B_{\varepsilon^*\varepsilon^*}(r')dr'dr_1$$

[see Eq. (4.81) in Vol. 1 of this book]. Hence $B_{\varepsilon^*\varepsilon^*}(r) = \frac{1}{2}\frac{d^2}{dr^2}\left(r^2\overline{\varepsilon_r^{*2}}\right)$, i.e.,

$$B_{\varepsilon^*\varepsilon^*}(r) = \text{const}\; \overline{\varepsilon_r^{*2}} = D\overline{\varepsilon^*}^2 \left(\frac{L}{r}\right)^{\mu} \qquad (25.28)$$

where $D = D(x, t)$ and

$$b_{\varepsilon^*\varepsilon^*}(r) = B_{\varepsilon^*\varepsilon^*}(r) - \overline{\varepsilon^*}^2 \approx B_{\varepsilon^*\varepsilon^*}(r) = D\overline{\varepsilon^*}^2 \left(\frac{L}{r}\right)^{\mu} \qquad (25.28')$$

in the scale similarity subrange of the distances $r$. The same result can also be obtained based on the breakdown scheme with arbitrary space regions $V_i$. Let $x$ and $x' = x + r$ be two points at a distance $r$ apart where $L \gg r \gg (\nu^3/\bar{\varepsilon}_r)^{1/4}$. Then we can select a sequence of similar successively enclosed space regions $V_0, V_1, \ldots, V_n, V_{n+1}, \ldots$ with length scales $l_0 \approx L$, $l_1 = \lambda l_0, \ldots, l_n = \lambda^n l_0 = r$, $l_{n+1} = \lambda^{n+1} l_0, \ldots$, where $\lambda < l$, such that the point $x$ will lie inside all the regions $V_i$ but $x'$ will lie only inside the regions $V_0, \ldots, V_n$ and outside of the regions $V_{n+1}, V_{n+2}, \ldots$. We can also select a second sequence of similar regions $V_0, V_1, \ldots, V_n, V'_{n+1}, \ldots$ (differing from the first one only in terms beginning with $V'_{n+1}$) such that the point $x'$ will lie inside of all these regions. If we now use the representation of the variables $\varepsilon^* = \varepsilon^*(x, t)$ and $\varepsilon^{*\prime} = \varepsilon^*(x + r, t)$ in the form (25.21′), then we obtain

$$\varepsilon^* = \overline{\varepsilon^*} e_1 \ldots e_n e_{n+1} e_{n+2} \ldots, \varepsilon^{*\prime} = \overline{\varepsilon^*} e_1 \ldots e_n e'_{n+1} e'_{n+2} \ldots \tag{25.29}$$

where $e_i, i = 1, 2, \ldots, n, n+1, \ldots$, and $e'_j, j = n+1, n+2, \ldots$, are identically distributed different random variables, while at $j > n + 1$ the variables $e_j$ and $e'_j$ are even mutually independent (the variables $e_{n+1}$ and $e'_{n+1}$ are not independent since they determine the average values of the field $\varepsilon^*$ in two different parts $V_{n+1}$ and $V'_{n+1}$ of the same region $V_n$ ). Since $\overline{e_j} = 1$ for all $j$, it follows from Eq. (25.29) that the conditional mean value of the product $\varepsilon^*\varepsilon^{*\prime}$ given the value of $\varepsilon_r^* = \overline{\varepsilon^*} e_1 \ldots e_n$ differs from $\varepsilon_r^{*2}$ only by the constant factor $e_{n+1} e'_{n+1}$. After averaging this conditional mean value over all the possible values of $e_r^*$ we again obtain the equation

$$B_{\varepsilon^*\varepsilon^*}(r) = \overline{\varepsilon^*(x + r)\varepsilon^*(x)} = \text{const}\, \overline{\varepsilon_r^{*2}} = D\overline{\varepsilon^*}^2 \left(\frac{L}{r}\right)^\mu.$$

Since the exponent $\mu$ in this equation is a property of the field $\varepsilon^*$ it is clear that it cannot depend on the form of regions $V_i$.

It will be shown below that in all cases we can take for granted that $\mu < 1$ (and of course $\mu > 0$). Hence the one-dimensional spectrum of the field $\varepsilon^*(x, t)$

$$E_1^{(\varepsilon^*)}(k) = \frac{1}{\pi} \int_0^\infty b_{\varepsilon^*\varepsilon^*}(r) \cos kr\, dr$$

will have the form

$$E_1^{(\varepsilon^*)}(k) = D'\overline{\varepsilon^{*2}}L(kL)^{-1+\mu} \sim k^{-1+\mu} \tag{25.30}$$

in the scale similarity subrange of wave numbers, where $D' = (D/\pi)\int_0^\infty x^{-\mu}\cos x\,dx = D/2\cos(\pi\mu/2)\Gamma(\mu)$. [If $\mu > 1$ then the values of $E_1^{(\varepsilon^*)}(k)$ in the scale similarity subrange will be determined mainly by the values of $b_{\varepsilon^*\varepsilon^*}(r)$ beyond the small-scale limit of this subrange. Therefore in this case $E_1^{(\varepsilon^*)}(k) = \text{const}$ in the scale similarity subrange, i.e., Eq. (25.30) will not be valid.]

Using the representation (25.29) of the variables $\varepsilon^* = \varepsilon^*(x)$ and $\varepsilon^{*\prime} = \varepsilon^*(x + r)$ it is easy to show that the two-dimensional random vector $(\ln \varepsilon^*, \ln \varepsilon^{*\prime})$ will have a two-dimensional normal probability distribution with both variances equal to the $\sigma^2$ of equation (25.23′) and the correlation coefficient between two components of the vector very close to $\sigma_r^2/\sigma^2$. Hence the vector $(\varepsilon^*, \varepsilon^{*\prime})$ will have a two-dimensional log-normal distribution. One can easily deduce from these facts that

$$\overline{\varepsilon^{*p}(x)\varepsilon^{*q}(x + r)} = \overline{\varepsilon^{*p+q}} \exp\left\{\left[\frac{1}{2}p(p-1) + \frac{1}{2}q(q-1)\right]\sigma^2 + pq\sigma_r^2\right\} \sim \left(\frac{L}{r}\right)^{\mu pq}. \tag{25.31}$$

This equation generalizes Eq. (25.28).

The treatment given above was suggested by Yaglom (1966) and Gurvich and Yaglom (1967). Later Novikov (1969, 1970) generalized some of previous results and also offered another approach which does not require the introduction of the breakdown scheme consisting of a sequence of discrete fragmentations. Novikov considered the one-dimensional case only which is most directly related to experiment (since data are usually taken at a fixed point and can be transformed into data on a straight line in the mean flow direction with the aid of Taylor's "frozen turbulence hypothesis"). In other words he investigated the field $\varepsilon^*(x)$ on the line $-\infty < x < \infty$ where $\varepsilon^*$ is a nonnegative local variable of the same kind as in the above considerations. Following Novikov we shall now consider the random variable

$$e_{r,l}(x,x_1)=\frac{\varepsilon_r^*(x_1)}{\varepsilon_l^*(x)}=\frac{\frac{1}{r}\int_{x_1-r/2}^{x_1+r/2}\varepsilon^*(x')dx'}{\frac{1}{l}\int_{x-l/2}^{x+l/2}\varepsilon^*(x')\,dx'},\ r<l, \tag{25.32}$$

where both $l$ and $r$ belong to the scale similarity subrange and $|x_1-x|\leqslant(l-r)/2$ (i.e., the integration range in the numerator of the equation lies inside the integration range in the denominator) as a function of two continuously varying arguments $r$ and $l$.[34] According to hypotheses stated (and used extensively) above (which were called the conditions of scale similarity) the probability distribution of $e_{r,l}$ depends only on the ratio $l/r$ and moreover if $l>\rho>r$ are three lengths from the scale similarity subrange then the variables $e_{r,\rho}$ and $e_{\rho,l}$ are statistically independent. If we consider the moment

$$a_q\left(\frac{l}{r}\right)=\overline{e_{r,l}^q} \tag{25.33}$$

then the conditions of scale similarity and the obvious relation $e_{r,l}=e_{r,\rho}e_{\rho,l}$ imply that

$$a_q\left(\frac{l}{r}\right)=a_q\left(\frac{\rho}{r}\right)a_q\left(\frac{l}{\rho}\right)$$

Since $\rho$ is arbitrary all the continuous solutions of the above functional equation are of the form

$$a_q\left(\frac{l}{r}\right)=\left(\frac{l}{r}\right)^{\mu_q} \tag{25.34}$$

Clearly $\overline{e_{r,l}}=a_1(l/r)=1$ for any $l$ and $r$, so that $\mu_1=0$. Let us now divide the segment of a given length $l$ into $n$ equal parts and consider $n$ corresponding identically distributed variables

[34] Novikov considered in fact in his paper of 1970 a slightly more general scheme of the field $\varepsilon^*(x)$ satisfying a nonhomogeneous scale similarity condition. Namely he considered $e_{r,l}$ as a function of $r,l$, and $h=x-x_1$ and assumed that its probability distribution can depend on two ratios $r/l$ and $|h|/l$. However we shall not discuss this generalization in our book.

$e_{l/n,l}(x, x-(n-1-2i)l/2n) = e^{(i)}, i = 0, 1, \ldots, n-1$. Then $\sum_{i=0}^{n-1} e^{(i)} = n$ and after squaring and averaging we obtain

$$a_2(n) = n - \frac{1}{n} \sum_{i \neq j} \overline{e^{(i)} e^{(j)}}$$

Since all the variables $e^{(i)}$ are nonnegative and $l$ is arbitrary, $\sum_{i \neq j} \overline{e^{(i)} e^{(j)}} \geqslant 0$ and is equal to zero only when $\overline{\varepsilon^*(x)\varepsilon^*(x+r)} = 0$ for all $r$ from the scale similarity subrange. This case is of course uninteresting; hence we can always assume that

$$a_2(n) < n, \text{ i.e. } \mu_2 = \mu < 1.$$

Thus

$$\overline{(e_{r,l})^2} = \left(\frac{l}{r}\right)^{\mu}, 0 < \mu < 1. \tag{25.35}$$

Moreover $e_{r,l} \leqslant l/r$ since $\varepsilon^*(x)$ is nonnegative. Hence

$$\begin{gathered} \overline{e_{r,l}^q} = \overline{e_{r,l}^{q-2} e_{r,l}^2} \leqslant \left(\frac{l}{r}\right)^{q-2} \overline{e_{r,l}^2} = \left(\frac{l}{r}\right)^{\mu+q-2}, \text{ i.e.} \\ \mu_q \leqslant q + \mu - 2 \quad \text{for} \quad q > 2. \end{gathered} \tag{25.36}$$

Let us now investigate the characteristic function of the random variable $\ln e_{r,l}$

$$\varphi\left(\theta, \frac{l}{r}\right) = \overline{\exp(i\theta \ln e_{r,l})}. \tag{25.37}$$

Then it follows from the relation $e_{r,l} = e_{r,\rho} e_{\rho,l}$ and the independence of the two factors on the right-hand side that

$$\varphi\left(\theta, \frac{l}{r}\right) = \varphi\left(\theta, \frac{\rho}{r}\right) \varphi\left(\theta, \frac{l}{\rho}\right), l > \rho > r. \tag{25.38}$$

Since $\rho$ is arbitrary this equation yields the important result

$$\varphi\left(\theta, \frac{l}{r}\right) = \left(\frac{l}{r}\right)^{-\alpha(\theta)} = \exp\left\{-\alpha(\theta)\ln\frac{l}{r}\right\} \tag{25.39}$$

The function $\alpha(\theta)$ which satisfies the normalizing condition $\alpha(0) = 0$ characterizes the given field $\varepsilon^*(x, t)$. It determines all the parameters $\mu_q$ by the relation

$$\mu_q = -\alpha(-iq) \tag{25.40}$$

so that $\alpha(-i) = 0, \alpha(-2i) = \mu$. In view of the definition of the cumulants of a random variable (see Sect. 4.1 of Vol. 1) the equation (25.39) implies that all the cumulants of $\ln e_{r,l}$ are proportional to $\ln(l/r)$:

$$s_q\left(\frac{l}{r}\right) = (-i)^q \left.\frac{d^q \ln\varphi}{d\theta^q}\right|_{\theta=0} = -(i)^{-q}\alpha^{(q)}(0)\ln\frac{l}{r} \tag{25.41}$$

Consider instead of $\ln e_{r,l}$ the centered and normalized random variable

$$\zeta_{r,l} = \left[\ln e_{r,l} - s_1\left(\frac{l}{r}\right)\right] s_2^{-1/2}\left(\frac{l}{r}\right) \tag{25.42}$$

Then it is clear that the first cumulant of it will be zero, the second cumulant will be unity, and all the other cumulants will be proportional to some negative power of $\ln(l/r)$, i.e., they will tend to zero as $\ln(l/r) \to \infty$. Hence the probability distribution of $\zeta_{r,l}$ must tend to a normal (Gaussian) distribution (which is the only distribution with all cumulants of order higher than two equal to zero) as $\ln(l/r) \to \infty$ (i.e., that $l/r \to \infty$). This statement can be proved quite rigorously if we assume that $\alpha(\theta)$ is a twice differentiable function at $\theta = 0$ (i.e., that $\ln e_{r,l}$ has finite mean value and variance) and use the three-term Taylor expansion of $\alpha(\theta)$ in the form

$$\alpha(\theta) = \alpha(0) + \alpha'(0)\theta + \frac{1}{2}\alpha''(0)\theta^2 + o(\theta^2)$$

Then using Eq. (25.41) we obtain

$$\varphi_\zeta\left(t, \frac{l}{r}\right) = \exp\,(\text{it}\,\zeta_{r,l}) = \exp\left\{-\alpha\left(\frac{t}{s_2^{1/2}(l/r)}\right)\ln\frac{l}{r} - is_l\,(l/r)\,\frac{t}{s_2^{1/2}(l/r)}\right\}$$

$$= \exp\left\{-\frac{t^2}{2} + \ln\frac{l}{r}\,o\left(\frac{t^2}{\ln\,(l/r)}\right)\right\} \to \exp\left\{-\frac{t^2}{2}\right\} \text{ as } \ln\frac{l}{r} \to \infty$$

It follows from the last relation by the usual reasoning (cf. for example, Chapt. XV in Feller, 1966) that

$$P\{\zeta_{r,l} < z\} \to \frac{1}{\sqrt{2\pi}}\int_{-\infty}^{z} e^{-x^2/2}\,dx \qquad (25.43)$$

where $P$ denotes probability, i.e., that

$$P\{e_{r,l} < y\} - \frac{1}{\sqrt{2\pi s_2(l/r)}}\int_0^y \exp\left\{-\frac{[\ln x - s_1(l/r)]^2}{2s_2(l/r)}\right\}\frac{dx}{x} \to 0 \qquad (25.43')$$

as $\ln\,(l/r) \to \infty$. This means that $e_{r,l}$ has asymptotically a log-normal probability distribution as $l/r \to \infty$.

Novikov made the assumption that $e_{r,l} \to \varepsilon_r^*/\overline{\varepsilon^*}$ as $l \to \infty$ and used this assumption to derive the above results on $\varepsilon_r^*$ from Eqs. (25.35) and (25.43') by passing to the limit $l \to \infty$. Such derivation is of course slightly inaccurate since Eqs. (25.35), (25.43') and all the others in the last part of this subsection are valid only for values of $l$ from the scale similarity subrange. However we can introduce a new dimensionless variable $e_{r,L} = \varepsilon_r^*/\overline{\varepsilon^*}$ where the fixed subscript $L$ indicates that it is necessary to average over a segment of length of the order of the external scale $L$ to obtain a sufficiently accurate estimate of $\overline{\varepsilon^*}$. Then we have $e_{r,L} = e_{r,\rho}e_{\rho,L}$ if $\rho > r$ belongs to the scale similarity subrange. Moreover the two factors on the right-hand side of the last relation are statistically independent. It seems reasonable to assume that the moments of $e_{r,L}$ are of the form

$$\overline{e_{r,L}^q} = a_q\left(\frac{L}{r}, x, t\right)$$

where the inclusion of the variables $x$ and $t$ indicates the dependence of the functions $a_q$ on the macrostructure of the flow (hence these functions change notably only when space-time point $(x, t)$ is displaced a distance of the order of the external (integral) scale of the turbulence). Then it follows from the relations (25.33)–(25.35) that

$$\overline{e_{r,L}^q} = D_q(x,t)\left(\frac{L}{r}\right)^{\mu_q}, \qquad \overline{e_{r,L}^2} = D_2(x,t)\left(\frac{L}{r}\right)^{\mu}, \tag{25.44}$$

i.e.,

$$\overline{\varepsilon_r^{*q}} = D_q\overline{\varepsilon^{*q}}\left(\frac{L}{r}\right)^{\mu_q}, \qquad \overline{\varepsilon_r^{*2}} = D_2\overline{\varepsilon^{*2}}\left(\frac{L}{r}\right)^{\mu}. \tag{25.44'}$$

Similarly we can use Eq. (25.43) to justify that $\varepsilon_r^* = \overline{\varepsilon^*}e_{r,L}$ has asymptotically (at $\ln l/r \to \infty$) a log-normal probability distribution with the variance of $\ln \varepsilon_r^*$ approximately proportional to $\ln L/r$.

Let us now consider what is known about the experimental confirmation of the above results. It was already mentioned in this subsection that the data of Pond and Stewart (1965), Stewart, Wilson, and Burling (1970), Gibson, Stegen, and McConnell (1970), Van Atta and Chen (1970), Sheih, Tennekes and Lumley (1971), Wyngaard and Pao (1971), Gibson, Friehe and McConnell (1971), Friehe, Van Atta and Gibson (1971) and Kholmyanskii (1972) indicate that in the case when $\varepsilon^* = \varepsilon_u = (\nu/2)(\partial u/\partial x)^2$ the spectrum $E_1^{(\varepsilon_u)}(k)$ can be approximated by a power function of the form (25.30) over a considerable range of wave numbers $k$. The value of the parameter $\mu$ (for $\varepsilon^* = \varepsilon_u$) lies somewhere between 0.3 and 0.5 according to these data. Only the airborne measurements by Sheih et al. (1971) lead to a marked discrepancy with all the other data, since $\mu \approx 0.7$ according to these measurements. The reason for this discrepancy is not clear at the present time. Similar results (with $\mu \approx 0.4$) were obtained by Gurvich and Zubkovskii (1963, 1965) for $\varepsilon^* = \varepsilon_w = (\nu/2)(\partial w/\partial x)^2$, but with an instrument which introduced substantial instrumental averaging. Their result is nevertheless consistent with the intuitive expectation that the values of $\mu$ and $\mu_q$ for the fields $\varepsilon_u$ and $\varepsilon_w$ (as well as for the true dissipation rate $\varepsilon$) must be the same.

Finally, Gibson, Friehe and McConnell (1971) measured the spectra of flucutations in $(\partial T/\partial t)^2$ where $T$ is the temperature both in the atmosphere above the sea and in a heated laboratory turbulent jet. The spectra have a considerable range of power law behavior and the exponent of the law shows that $\mu \approx 0.5$ for $\varepsilon^* = (\partial T/\partial x)^2$ in both types of flow.

The values of the moments $\overline{\varepsilon_r^{*2}}$ were evaluated by Gurvich (see Gurvich and Yaglom, (1967) from data on $\varepsilon^* = (\partial T/\partial x)^2$ where $T$ is the temperature in the atmospheric surface layer and the $x$-axis coincides with the mean-wind direction, and by Kholmyanskii (1970) for $\varepsilon^* = (\partial u/\partial x)^2$. In both works only rather short samples were used (one 10-sec sample in Gurvich's measurements and three 15–20-sec samples in Kholmyanskii's) and these samples were chosen not quite at random, but from the condition that considerable fluctuations of the field are present in the sample. These circumstances cause the results to be somewhat unreliable. Nevertheless it is worth noting that in both cases the dependence of $\overline{\varepsilon_r^{*2}}$ on $r$ was found to be a power law, as it must be according to equations (25.27′) and (25.44′). The exponent $\mu$ was found by Gurvich to be close to 0.4 [which agrees satisfactorily with the estimate of the same parameter obtained by Gibson et al. (1971) with a different method] while Kholmyanskii found that $\mu \approx 0.35$ (this value also agrees with the estimates of the parameter $\mu$ for $\varepsilon^* = \varepsilon_u$ from the spectral data). Later Kholmyanskii (1972) repeated his evaluation for two considerably longer samples of duration of about 5 min. He found in this work that the dependence of $\overline{\varepsilon_r^{*2}}$ on $r$ was of a power-law form in the "inertial subrange" $2.3 < \log_{10}(r/\eta) < 4.1$ with an exponent $\mu$ close to the value of 0.3 obtained by him from the data on $E_1^{(\varepsilon_u)}(k)$.

The evaluation of the higher-order moments of $\varepsilon_r^*$ from a short sample is clearly even less reliable than the evaluation of the second moments. Therefore the data of Kholmyanskii (1970, 1972) on moments of orders 3–6 of the variables $\varepsilon_{u,r} = 1/r\int_0^r \varepsilon_u(x')dx'$ must be treated with great caution. This is especially true of the data from his first paper, where only very short samples were analyzed. According to these data the dependence of $\overline{\varepsilon_{u,r}^3}$ and $\overline{\varepsilon_{u,r}^4}$ on $r$ can be approximated by a power function over a considerable range of values of $r$ and the corresponding exponents are close to $-3\mu$ for the third moment and to $-6\mu$ for the fourth moment (where $\mu$ was determined for each of the three samples used from the dependence on $r$ of the mean square of $\varepsilon_{u,r}$). This result agrees well with the predictions (25.27′) from a log-normal approximation. In the second paper by Kholmyanskii (1972) the moments of $\varepsilon_{u,r}$ of order $q$ equal to 3, 4, 5, and 6 were evaluated from two different samples of duration about 5 min. It was found here that the dependence of the moments on $r$ is of a power form in the inertial subrange of $r$-values with exponents which do not differ significantly from the predictions (25.27′) with $\mu = 0.3$ for $q = 3$ and 4, but which are considerably smaller than the predicted values for $q = 5$ and, especially, 6. These last results agree well with the conclusions of Novikov (1970) which have already been considered above. In particular all the values of the exponents $\mu_q$ obtained by Kholmyanskii satisfy the inequality (25.36) which is not satisfied by the log-normal approximation to $\mu_q$ (this last fact will be discussed later in this section).

Another more crude approach to the verification of the theoretical predictions referring to fourth-order moments was used by Gibson, Stegen, and McConnell (1970). These authors started from the observation that the flatness factor $\delta_u = \overline{(\partial u/\partial x)^4}/[\overline{(\partial u/\partial x)^2}]^2$ of the velocity derivative must be proportional to $(L/\eta)^\mu \sim (L\bar{\varepsilon}^{1/4})^\mu$ where $\mu$ is referred to $\varepsilon^* = \varepsilon_u \sim (\partial u/\partial x)^2$. They examined therefore the data on the dependence of $\ln\delta_u$ at a height $z$ above the sea surface on $\ln(z\bar{\varepsilon}^{1/4})$ where $\bar{\varepsilon}$ is the dissipation rate averaged over a considerable time at a particular height $z$ in the marine surface layer. The dependence (deduced from data at four heights $z$, namely 2.25, 4.25, 8.25, and 12.25 meters) turns out to be quite scattered; however if we try to approximate it by a linear function, then the best value of the slope is 0.44 which agrees well with all other estimates of $\mu$ for $\varepsilon^* = \varepsilon_u$.

Let us also mention the work of Gibson and Masiello (1971) who investigated the probability distributions of the values of $\varepsilon_u = (\partial u/\partial x)^2$ averaged over some twenty different time intervals corresponding to longitudinal space intervals in the range from 0.05 cm to 700 cm. The data used for this investigation relate to the atmospheric surface layer at height 30 m above the sea. All the calculated probability distribution functions had the property that they could be approximated by log-normal distribution over a considerable range of intermediate probability values: this approximation permitted determination of the values of $\sigma^2_{\ln\varepsilon_r}$ where $\varepsilon = \varepsilon_u$ and $r$ ranged from 0.05 to 700 cm. It was found that in the range of $r$-values from about 0.1 cm to 20 cm the values of $\sigma^2_{\ln\varepsilon_r}$ satisfy Eq. (25.20) with good accuracy, if we suppose that $\mu = 0.5$ (and $A = -1.2$ while $L$ is defined as the height above water). This estimate of $\mu$ agrees with all the other estimates (the summary of various such estimates can be found also in the paper of Gibson and Masiello). Some other of the above mentioned theoretical predictions were also tested by Gibson and Masiello and the results were more or less satisfactory in all cases; however, we shall not linger on them here with the single exception discussed below.

All the above mentioned verifications of the consequences of the log-normal approximation of the probability distributions for small-scale turbulence are clearly indirect. However there are also many attempts to measure directly the probability distributions for nonnegative local fields $\varepsilon^*(x, t)$. The first data of this kind were obtained by Gurvich (1966, 1967) [see also Gurvich and Yaglom (1967)] who measured the probability distribution for the variable $(\Delta_l T)^2 = |T(x+l) - T(x)|^2$ (with $l \approx l \approx 2$ cm), $\left(\frac{\partial T}{\partial t}\right)^2 \approx u^2 \left(\frac{\partial T}{\partial x}\right)^2$, and $\left(\frac{\partial \tilde{w}}{\partial t}\right)^2$ (where $\tilde{w}$ is the slightly averaged value of the vertical velocity $w$, and the derivative $\frac{\partial \tilde{w}}{\partial t}$ was additionally averaged over a range of $x$-values, due to finite inertia of the instrument). The data in Gurvich (1967) refers to the difference between the readings of two miniature resistance thermometers $\xi(t)$ which is proportional to $\Delta_l T$. Two samples were analyzed, both obtained at a height of 4 m above the ground and having a duration of about 10 sec (corresponding to about $10^4$ independent measurements per sample). A special simple device was used to obtain graphically the probabilities $P(x) = P\{\xi^2(t) < x\}$ for different $x$. Gurvich found that the probability distribution for $\xi^2$ is satisfactorily approximated by a log-normal distribution (see Fig. 104 plotted on coordinates such that log-normal distribution fucntions follow straight lines). Gurvich felt that the range of probabilities $P$ obtained was insufficient for a direct and

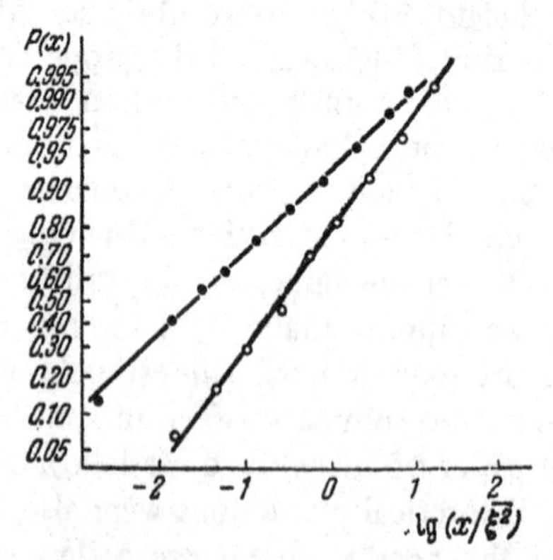


**Fig. 104** Probability distributions for the square of the temperature difference at two neighboring points [Gurvich (1967)].

reliable determination of the third and higher order moments of the variable $\xi$. By assuming, however, that the log-normal distribution was valid throughout the whole range of $\xi$, Gurvich obtained two estimates of the flatness factor $\delta$ for $\Delta_l T$ equal to 18 and 1400, respectively. However these estimates must be considered as very unreliable (see the discussion of this question below in this subsection).

Gurvich and Yaglom (1967) have reported the results of an analogous analysis of the values of $\frac{\partial T}{\partial t}=\xi_1$ obtained by a resistance thermometer in the atmosphere and a differentiating circuit. Here again, the distribution for $\xi_1$ was highly non-Gaussian, but the probability distribution for $\xi_1^2 \approx u^2\left(\frac{\partial T}{\partial x}\right)^2$ (which is similar to $N=\chi(\nabla T)^2$)] was also found to be approximately log-normal throughout the range investigated. Finally, Gurvich (1966) has analyzed data on the derivative $\frac{\partial \tilde{w}}{\partial t} \approx u\frac{\partial \tilde{w}}{\partial x}$ smoothed out by instrumental averaging, where $\tilde{w}$ is the signal from a sonic anemometer with a vertical base of about 5 cm. In this case, the empirical probability distribution for the resulting values of $\left(\frac{\partial \tilde{w}}{\partial t}\right)^2$ (which because of averaging were relatively close to $\Delta_l \tilde{w}$ where $l \approx 10$ cm) are not as well represented by the log-normal distribution, although the departures from this distribution were not very great.

More recently numerous measurements of the probability distributions for nonnegative small-scale turbulence parameters were performed by Gibson and Williams (1969), Kholmyanskii (1970, 1972), Stewart, Wilson, and Burling (1970), Gibson, Stegen, and Williams (1970), Van Atta and Chen (1970), Gibson, Friehe and McConnell (1971), Gibson and Masiello (1971), and Kuo and Corrsin (1971). Gibson and Williams (1969) [see also Gibson, Stegen, and Williams (1970)] calculated the probability distribution of the square of the longitudinal velocity difference $\Delta_l u = u(x+l) - u(x)$ where $l \approx 1$ cm (obtained from the hot-wire anemometer record by the "frozen turbulence" hypothesis) at three heights above the ocean surface, using approximately 40,000 sample values of $\Delta_l u$ at every height, recorded at the speed of 350 samples/sec. Their results are plotted in Fig. C on the same coordinates as were used in Fig. 104; the directly calculated values of skewness and excess of $\Delta_l u$ are also given in the figure. We see that the agreement of the data with straight lines is

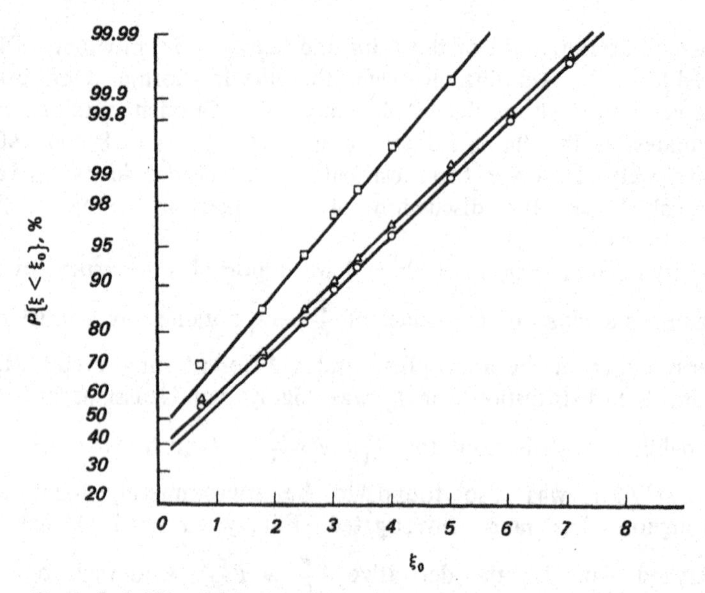


**Fig. C** Probability distribution of the square of the velocity difference according to Gibson and Williams (1969). $\xi = \ln(\Delta_l u)^2$ + const, $\Delta_l u = u(x + l) - u(x)$, $l \approx 1$ cm.

| | Height (m) | Skewness of $\Delta_l u$ | Excess of $\Delta_l u$ |
|---|---|---|---|
| □ | 7 | −0.52 | 16.6 |
| △ | 2 | −0.71 | 19.1 |
| ○ | 1 | −0.89 | 20.4 |

good enough at all not too small values of $(\Delta_l u)^2$. The curvature of the data on the left can be explained quite reasonably by the effects of count errors and instrument and tape recorder noise. The excess values were all about 20, but were different from the values inferred from the slopes of the straight lines in Fig. C on the assumption of unlimited validity of the log-normal assumption (i.e., by the method of Gurvich).

The probability distribution function of the variable $(\partial u/\partial x)^2 = \varepsilon_u$ in the surface layer of the atmosphere above land or water or in turbulent flows in the laboratory was calculated from the data of Kholmyanskii (1970, 1972), Stewart, Wilson and Burling (1970), Gibson, Stegen and Williams (1970), Van Atta and Chen (1970), Gibson, Friehe and McConnell (1971), Gibson and Masiello (1971) and Kuo and Corrsin (1971).[35] The results of all those authors agree

[35] In fact, Stewart et al. have studied the probability distribution of the absolute value of the longitudinal velocity derivative and not of the square of this derivative. However, $|\partial u/\partial x| = \varepsilon_u^*$ is also a local nonnegative variable which is closely related to $\varepsilon_u$. In particular,

well one with another and all show that the probability distribution for $\partial u/\partial x$ is highly non-Gaussian while the distribution of $(\partial u/\partial x)^2$ can be approximated by a log-normal distribution over a considerable range of probabilities. As typical examples the data of Kholmyanskii and of Stewart, Wilson, and Burling are shown in Figs. D and E. Similar results were also obtained by Gibson, Stegen and Williams (1970), Gibson, Friehe and McConnell (1971) and Gibson and Masiello (1971) for the probability distribution of $(\partial T/\partial x)^2$ (see, for example, Fig. F), and by Gibson and Masiello (1971) for the distributions of $(\partial w/\partial x)^2$ and of some squared derivatives averaged over a small interval.

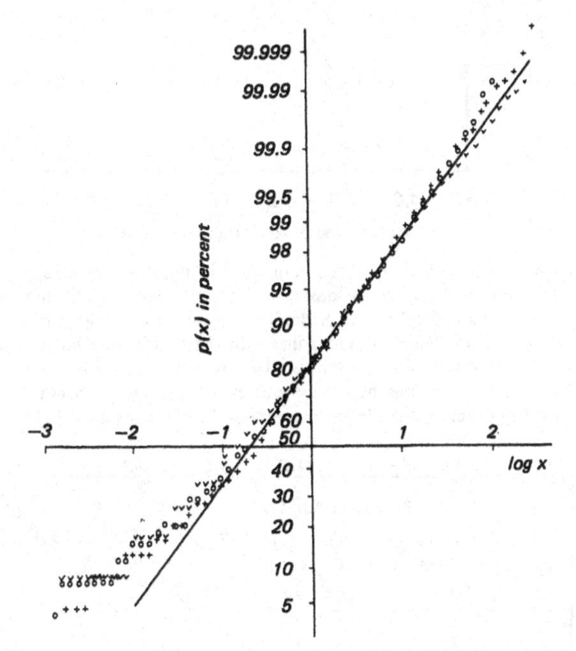


**Fig. D** The distribution function $P(x) = P\{\xi^2 < x\}$ of the variable $\xi^2 = \left(\frac{\partial u}{\partial x}\right)^2 \Big/ \overline{\left(\frac{\partial u}{\partial x}\right)^2}$ according to Kholmyanskii (1970) plotted in coordinates such that straight lines correspond to log-normal distributions. The different symbols refer to different data samples.

---

the probability distribution for either of the variables $\varepsilon_u$ and $\varepsilon_u^*$ can be easily reestablished from the probability distribution for the other and both these distributions will simultaneously be or not be of a log-normal type. Therefore, we shall not distinguish the variables $\varepsilon_u$ and $\varepsilon_u^*$ in the subsequent discussion.

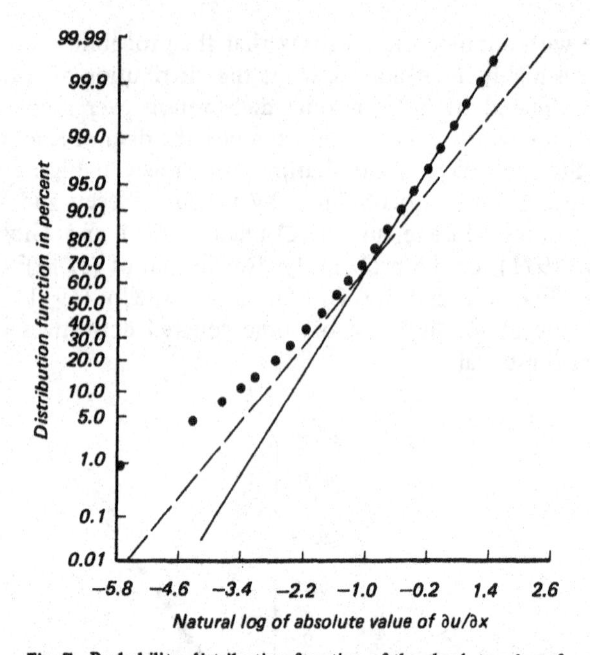


**Fig. E** Probability distribution function of the absolute value of the velocity derivative according to Stewart, Wilson and Burling (1970). The straight lines in the figure correspond to log-normal distributions. The continuous line is the best fit straight line for the probability distribution plotted by eye; the dotted line corresponds to the best least-squares fit for the probability density function (i.e., to the continuous smooth curve in Fig. H).

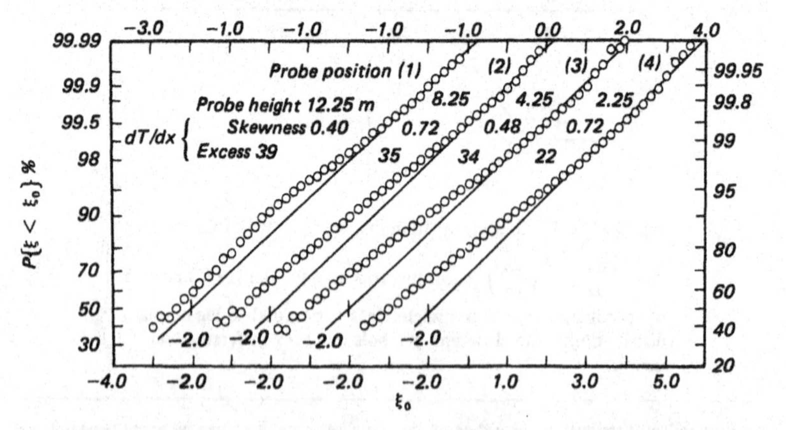


**Fig. F** Probability distributions of squared temperature derivative compared with log-normality, from Gibson, Stegen and Williams (1970). $10^5$ samples per plot. $\xi = \ln [(dT/dx)^2/\overline{(dT/dx)^2}]$.

All the data show noticeable departure of the measured distributions from log-normal ones at small values of the measured variables. These departures can be quite reasonably explained in all cases by the effect of instrumental noise. It was indicated by Stewart et al. in this respect that the probability density of a log-normal distribution (with parameters close to those typically observed in the atmosphere) has a very peculiar shape (see Fig. G). It shows small values of the probability in the neighborhood of zero and large values of the probability for very nearby values. This fact has a simple analytical explanation: it is very easy to show that the maximum value of a log-normal probability density of the form (25.23) with the parameters $\sigma^2$ and $m = -\sigma^2/2$ is equal to $p_{\max} = (2\pi\sigma^2)^{-1/2}e^{\sigma^2}$ and that this value is taken at $x = e^{-3\sigma^2}/2$. Hence $p_{\max} \to \infty$ and $x \to 0$ as $\sigma^2 \to \infty$ while at large finite values of $\sigma^2$ this leads to the shape of $p(x)$-curve which is shown schematically in Fig. G. Moreover the instrumental noise tends to fill in the gap about zero and produce an excess of small values; hence it leads to a curvature in the graphs of the measured distribution functions which is in agreement with the data in Figs. D, E, and F. Therefore the existing data do not

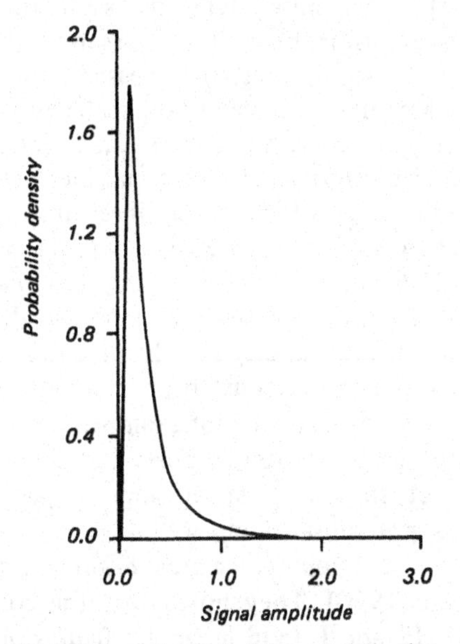


**Fig. G** Schematic graph of the probability density function for a log-normal distribution with parameters nearly equal to typical observed values (after Stewart, Wilson and Burling, 1970).

contradict the assumption that the true probability density of the nonnegative variables investigated coincides with a log-normal probability density of the type shown in Fig. G in the neighborhood of zero, although they do not, of course, confirm this assumption either. The data in Fig. D show also some small departures from the log-normal distribution at very high values of the probability (above $P = 0.999$). These departures can be explained by the deficiency of data at very large values of the variable which makes the measured values of the probability unreliable when $P$ is close to one. Let us also recall that according to the data of Grant, Stewart, and Burling (1962) (obtained at $\mathrm{Re} \approx 10^8$ which is considerably higher than Re in the atmospheric surface layer) the mean values of $\varepsilon$ over a period of about 10 min may change by a factor of four or more without any change in the overall conditions of the flow. This fact leads one to suspect that a sampling time of the order of a few tens or few hundreds of seconds (Kholmyanskii, Gibson, et al., Stewart et al.) or even of the order of several tens of minutes (Van Atta and Chen) is insufficient for the reliable determination of the probability distribution and a considerably longer sample (or a number of samples obtained at different points under the same flow conditions) is needed for this determination. It was assumed in this respect by Keller and Yaglom (1970) that an increase of the sample size may lead to an increase of the range of applicability of the log-normal approximation of the distribution of $\varepsilon^*$ and simultaneously to an increase of the variance of $\ln \varepsilon^*$, i.e., to an increase of the effective Reynolds number $\mathrm{Re}_{ef}$ which determines this variance $\sigma^2$ by a relation of the form $\sigma^2 \approx \mu \ln (L/\eta_{ef}) \approx 0.75\, \mu \ln (\mathrm{Re}_{ef})$ [cf. Eqs. (25.23′) and (21.6) above]. This assumption has not yet been verified and in general at present there are no data to determine satisfactorily the range of validity and the precision of the log-normal approximation. However we must bear in mind that the results of Van Atta and Chen for relatively long samples are similar to those of Kholmyanskii for much shorter samples and that there are general reasons to expect that at least in some respect the log-normal approximation is quite unsatisfactory.

To explain the last statement we must begin with the remark made by Stewart et al. (1970). They noted that the continuous straight lines in Figs. D, E, and F (and in similar figures obtained by other authors) give quite a good approximation for the integral distribution functions of the variables by a log-normal distribution function, but this can be very different from the best approximation of the probability density of the logarithm of these variables by a normal curve. When Stewart et al. tried to transform from the graph of the distribution function for $|\partial u/\partial x|$ to the graph of the probability density for $\ln|\partial u/\partial x|$ they found that the continuous line in Fig. E gives a very poor description of the measured probability density (the description is quite accurate at the highest absolute values of the velocity derivative, but gives very great errors at all values of the derivative which are close to, or smaller than, the most probable value of $|\partial u/\partial x|$). Therefore Stewart et al. plotted a graph of the measured probability density of $\ln|\partial u/\partial x|$ and then found the best fit of a curve of the normal type to points located in the center of each bar of the bar graph (Fig. H). At first glance the fit in Fig. H seems to be quite good. However, there are significant deviations from normality on the tails of the distribution. This is especially clearly seen if we transform again to the integral distribution function of $|\partial u/\partial x|$ (see the dotted line in Fig. E which corresponds to the best-fit curve in Fig. H). We have already noted that the excess of small values in the observed distribution can be reasonably explained by the noise effect and hence is expected. However, the observed noticeable deficiency of large values cannot be explained in the same way: it is easy to

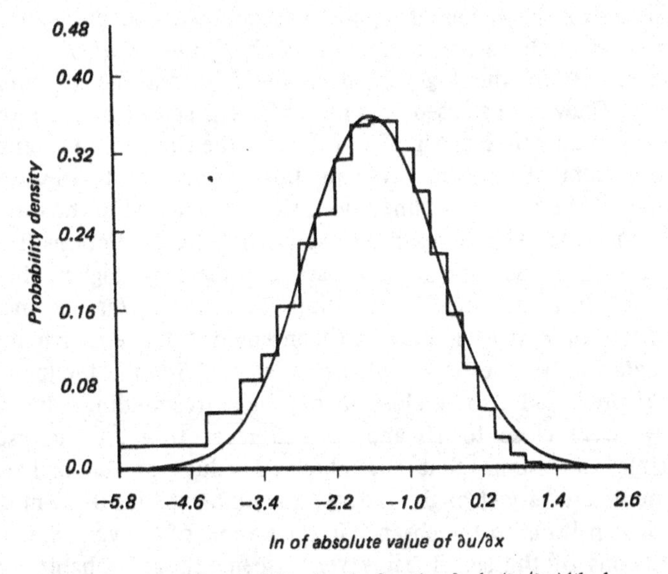


**Fig. H** The observed probability density function for $\ln|\partial u/\partial x|$ (the bar graph) and its approximation by the best (least squares) normal curve (the smooth continuous curve). After Stewart, Wilson and Burling (1970).

show that the effect of noise must be negligible on the right half of the continuous curve in Fig. H. The explanation of this deficiency suggested by Stewart et al. is the following. The arguments in favor of the log-normal distribution are based on the central limit theorem and the assumption that the number of identically distributed terms $\ln e_j$ on the right-hand side of equations (25.22′) is large enough. This number is proportional to ln Re and hence Re must be so high that even ln Re is quite a large number. The same is true with respect to Novikov's derivation of the log-normal distribution (25.43′) which is based on the assumption that $\ln l/r$ is very large. If the Reynolds number and therefore the number of the terms $\ln e_j$ is not high enough, then a deficiency of large values may be expected because of the limit on how large the sum (25.22′) can become. To illustrate this point more explicitly Stewart et al. consider the observed distribution of the product of ten random numbers generated on a computer and having a constant probability density inside the interval $x_a \leqslant x \leqslant x_b$ and zero probability density outside of this interval. The distribution of the logarithm of this product was approximated by a best least squares normal curve; the resulting graph turns out to be very similar to the graph in Fig. H.

The noticeable deficiency of large values implies an appreciable deviation of the values of higher-order moments of the distribution calculated in the usual manner from the observed data from those predicted from the log-normal model with the aid of equations (25.25). Stewart et al. used the values of the first and second moments of the normal curve in Fig. H to calculate the first, second, third, and fourth moments of the corresponding log-normal distribution and compared the values obtained with those calculated in the usual way from the data. The comparison shows that the predictions from the log-normal model are in all cases considerably higher than the estimate from the real observations and this difference increases rapidly with increasing order of the moment. For the fourth moment the data imply an estimate of the order of 0.5 while the log-normal model predicted a value close to 17; the corresponding values of the excess were close to 10 and to 186 respectively. Of course it is possible to assume that the observed values of the higher-order moments are very strongly underestimated by the insufficient sample size which leads to the absence in the sample of the very rare highest excursions of the signal. However it seems more probable that the observed considerable deficiency of large values in comparison to the prediction of the log-normal model is real and hence the model cannot be used for the calculation of high-order moments.

The last assumption was strictly confirmed later by Novikov (1970) who noticed that the accurate inequality (25.36) is inconsistent with equations (25.27′) which are implied by log-normality. In fact the inequality (25.36) requires that the exponents $\mu_q$ increase no faster than a linear function of the order of the moment while equations (25.27′) show that these exponents are represented by a quadratic function of $q$ in the log-normal approximation. Thus we see that the approach of the probability distribution for $\varepsilon^*$ (or for $\varepsilon_r^*$, or for $e_{r,l}$) to a log-normal probability distribution as Re increases is of such a kind that the integral distribution function tends to a log-normal one, but the moments do not tend to the moments of the limiting log-normal distribution.[36] Hence we cannot expect to be able to estimate the moments of local nonnegative small-scale turbulence characteristics from the log-normal approximation: the error of such an estimate may be enormous.

To illustrate these statements more clearly Novikov considered a very simple model which is indirectly related to the Stewart et al. model of the product of a number of random variables with constant probability density on a finite interval, but which permits an elegant analytical treatment. Namely he considered the random variable $e_{r,2r} = \varepsilon_r^*/\varepsilon_{2r}^*$ which can take only values in the interval $0 \leqslant y \leqslant 2$ and assumed that the variable $e_{r,2r}$ has a constant probability density $p(y) = 1/2$ in this interval. It is easy to see that for such a model

$$\mu_q = \log_2 \left[ \frac{1}{2} \int_0^2 y^q \, dy \right] = q - \log_2 (q+1) \tag{25.45}$$

In particular $\mu_1 = 0$ as expected, $\mu_2 = \mu \approx 0.41$ (which is close to the value of $\mu$ obtained in all the measurements of the fields $\varepsilon^*$),

[36] We must, of course, remember that the only completely rigorous result is the limiting relation of the form (25.43) for the centered and normalized random variables, while the original random variables do not tend to anything since all their moments tend to infinity as ln Re or ln $l/r$ tends to infinity. The exponents $\mu_q$ describe the rate of increase of the moments of the original variables and therefore the difference between their true values and the values implied by the limiting log-normal approximation is not completely extraordinary. The lack of justification for the calculation of the moments of $\varepsilon^*$ and related variables with the aid of log-normal approximation was noticed also by Mandelbrot (1972) who disputed in this respect the deductions of Orszag (1970c) which will be considered later in this section.

$\mu_{2/3} \approx -0.07, \mu_{4/3} \approx 0.11, \mu_3 = 1, \mu_4 = 1.68$. The characteristic function of the probability distribution is also evaluated quite easily and yields the following result:

$$\varphi(\theta, 2) = \frac{2^{i\theta}}{1 + i\theta}$$

so that

$$\alpha(\theta) = -\log_2 \varphi(\theta, 2) = -i\theta + \log_2(1 + i\theta). \tag{25.46}$$

By virtue of Eq. (25.40), Eq. (25.46) implies also Eq. (25.45). Moreover the log-normal approximation is obtained when the function $\alpha(\theta)$ is replaced by the two first terms of its Taylor expansion. Hence if we denote the quantities related to the log-normal approximation by primes then

$$\alpha'(\theta) = i\theta \, \frac{1 - \ln 2}{\ln 2} + \frac{1}{2 \ln 2} \, \theta^2$$

which implies that

$$\mu'_q = -\alpha'(-iq) = \frac{q^2 - 2(1 - \ln 2)q}{2 \ln 2} \tag{25.45'}$$

The estimates $\mu'_q$ of the exponents $\mu_q$ deviate significantly from their true values (25.45) and this deviation increases with growth of the order of the moment; in particular $\mu'_1 \approx 0.27$ (while $\mu_1 = 0$), $\mu'_2 = 1$ (while $\mu_2 \approx 0.4$), $\mu'_3 \approx 5$ (while $\mu_3 = 1$), and $\mu'_4 \approx 10$ (while $\mu_4 \approx 1.7$). It is also clear that in this model

$$\varphi\left(\theta, \frac{l}{r}\right) = \left(\frac{l}{r}\right)^{-\alpha(\theta)} = \frac{\exp\,[is \ln l/r]}{(1 + i\theta) \log_2 (l/r)} \tag{25.47}$$

The derivation of the limiting relations (25.43) and (25.43') can of course be applied to this particular example of the characteristic function $\varphi(\theta, l/r)$. Moreover it is also possible to obtain here explicit formulas for the probability density $p_1(z, l/r)$ of the variable $\ln e_{r,l}$ [this density is the Fourier transform of Eq. (25.47)] and for the probability density $p(y, (l/r))$ of the variable $e_{r,l}$ [this density satisfies the relation $yp\,[y, (l/r)] = p_1[\ln y, (l/r)]$. In particular

$$p\left(y, \frac{l}{r}\right) = \begin{cases} \dfrac{\left(\ln \dfrac{l}{ry}\right)^{\log_2 (l/r)-1}}{\dfrac{1}{r}\Gamma\left(\log_2 \dfrac{l}{r}\right)} & \text{for} \quad 0 \leqslant y \leqslant \dfrac{l}{r} \\ 0 & \text{for} \quad y < 0 \text{ or } y > \dfrac{l}{r} \end{cases} \tag{25.48}$$

This probability density is consistent with Eq. (25.43′) but it differs considerably from a density of log-normal type. In comparison with the log-normal density with the same values of the two first moments the density (25.48) is characterized by a noticeable deficiency of large values and an excess of small values and only in the range of moderate values it is well approximated by the log-normal equation; in other words it describes a deviation from log-normality of the same type which is observed in the experimental data. Let us also note that the density (25.48) is strictly equal to zero (i.e., the corresponding integral distribution function takes a constant value) in the region $y \geqslant l/r$ and, on the other hand, $p(0, l/r) = \infty$ if $l/r > 2$, although the log-normal probability density takes a zero value at the point $y = 0$. Hence not only the moments but the probability density function also do not tend in this model to a density of the log-normal distribution. The model evidently suggests that the gap near zero in Fig. G is absent in all existing data not only because of noise but also because of the fact that the gap itself is due to an inappropriate idealization.

The last remark is also related to another question of some importance. We have already seen that the value of the variance and of all the higher-order moments for small-scale nonnegative variables are very high for a high Reynolds number turbulence; therefore, the distribution of small-scale turbulence in space will be very nonuniform for high enough Re. This high degree of nonuniformity should give the impression that all the turbulent disturbances are concentrated in a few isolated space regions so that the turbulence is highly intermittent. However the model of log-normally distributed small-scale turbulence is in some respect even too intermittent. It was noted at the end of Sect. 4.1 of Vol. 1 that there are wide general conditions which guarantee that the probability distribution is uniquely determined by the values of all the moments. Now it is worthwhile stressing that these conditions require that the moments do not grow too rapidly with increasing order [the precise formulation can be found, e.g., in the books of Feller (1966, Chapt. VII, 3) or Orszag (1975, Chapt. 1)]. Although the conditions are relatively unrestrictive, the moments of the log-normal distribution grow so rapidly (as quadratic functions of the order) that they do not satisfy them. This fact was known to specialists in probability theory for some years and it was also recently established independently by Orszag (see, e.g., his paper (1970c) who noted its importance for turbulence theory.[37]

Orszag emphasized also that if the moment problem is indeterminate for a quantity satisfying some dynamical equations then the initial values of all the moments do not necessarily determine uniquely the values of all the moments at future times. Nonuniqueness of the future moment values can exist in such a case even when the corresponding dynamical problem has only one solution for any fixed initial values. Hence if the moment problem is indeterminate for the probability distributions of small-scale turbulence then the usual method of moments is unsuitable for describing such turbulence, i.e., the deductive theory of small-scale turbulence must inevitably use some more sophisticated statistical properties than the usual moments of various orders. This would make the future rigorous theory of turbulence even more complicated than at present seems natural. Therefore it is stimulating to know that Novikov's conditions (25.36) guarantee the uniqueness of the probability distribution with moments of the form (25.34) and hence show that the doubts regarding the sufficiency of the method of moments for describing small-scale turbulent fluctuations are due only to oversimplification in the mathematical model used.

## 25.4 Refined Expressions for the Statistical Characteristics of Small-scale Turbulence

Let us now return to the investigation of the effect of fluctuations in the energy dissipation rate on the statistical characteristics of

[37] Orszag gave also a simple example showing the nonuniqueness of the probability distribution with moments of the form (25.25). Namely he noted that the family of probability distributions with densities of the form

$$p_C(y) = (2\pi y\sigma^2)^{-1/2}\left[\exp\left\{-\frac{(\ln y - m)^2}{2\sigma^2}\right\} + C\exp(-ky^{1/3})\sin(k\sqrt{3}y^{1/3})\right]$$

has the same moments of all orders for any $k > 0$ and $C$ sufficiently small to imply that $p_C(y)$ is nonnegative and, therefore, is a probability density (the integral of $p_C(y)$ over the range from $y = 0$ to $y = \infty$ is equal to one for all values of $C$).

locally isotropic turbulence. To take this effect into account, we must use the generalized similarity hypotheses formulated in Sect. 25.2 and supplement them by some hypotheses of quite different kind concerning the statistical regime of dissipation fluctuations. It is only after this has been done that one can hope to obtain definite results which could possibly be compared with experimental data.

We shall use both the refined similarity hypotheses of Sect. 25.2 and the statistical hypotheses given in Sect. 25.3 which lead to log-normal or related probability distributions for the variables $\varepsilon$ (or $\varepsilon^*$) and $\varepsilon_r$ (or $\varepsilon_r^*$). Both these sets of hypotheses are not completely rigorous and correspond to an idealization and simplification of reality. The same is true however for any hypothesis used in the physical sciences since all of them are only approximately valid. Hence the only important thing is to understand the limitations and accuracy of the hypotheses when they are applied to some specific fluid flow. For example, we must remember that the log-normal distribution does not describe accurately the extreme tail of the true dissipation distribution and can be used to evaluate moments of relatively high order only with very great caution. However the central part of the distribution and its moments of moderately low order can be obtained from the log-normal approximation with satisfactory accuracy, if the Reynolds number of the flow is high enough and sufficiently extensive averaging is used.

After these introductory remarks let us pass over to the consideration of the consequences of the hypotheses formulated in Sects. 25.2 and 25.3. We shall begin with the corrections resulting from fluctuations in the energy dissipation rate which are necessary in the equations for the statistical characteristics in the inertial subrange of locally isotropic turbulence. If we take the average of the first formula of Eq. (25.12) over the log-normal probability distribution for the values of the random variable $\varepsilon_r$ (assuming that $r$ is considerably greater than all the values of $\eta_r = \nu^{3/4}\varepsilon_r^{-1/4}$ having nonnegligible probability), we find from Eq. (25.20′) together with Eqs. (25.25)–(25.26) and Eq. (25.27′) that

$$\overline{\varepsilon_r^{2/3}} \sim \bar{\varepsilon}^{2/3}\left(\frac{L}{r}\right)^{-\mu/9}, \quad D_{LL}(r) \approx C^{(1)}\bar{\varepsilon}^{2/3} r^{2/3}\left(\frac{r}{L}\right)^{\mu/9}, \qquad (25.49)$$

where the dimensionless coefficient $C^{(1)} = C^{(1)}(\boldsymbol{x}, t)$ and the dimensional combination $C^{(1)}L^{-\mu/9}$ both can be slightly dependent on the parameters of the large-scale motions. If we adopt the above experimental value $\mu \approx 0.4$–$0.5$, the exponent $\alpha$ in the equation $D_{LL}(r) \sim r^{\alpha}$ will be given by $\alpha \approx \frac{2}{3} + 0.05 \approx 0.72$, i.e., it will not be significantly different from two-thirds [this estimate of $\alpha$ was first indicated by Yaglom (1966)]. The experimental detection of variations of the order of a few hundredths in the exponent of a power law is, however, a very difficult task. This is clearly seen from Fig. 63a taken from the paper by Van Atta and Chen (1970) and reproduced in Sect. 23.3 of our book. The solid lines in this figure correspond to the usual two-thirds law, while the dashed lines represent a power law with the exponent $\alpha = (2/3) + (0.5/9) \approx 0.722$ which corresponds to Eq. (25.49) with $\mu = 0.5$. Van Atta and Chen themselves consider their data as producing a better fit to the refined law (25.49) than to the usual two-thirds law in the restricted range 5 cm $\leqslant r \leqslant$ 80 cm (which is considerably smaller than that for which the two-thirds law provides a fairly good fit to all the data). However this deduction can hardly be regarded as sufficiently reliable.

Similarly, the inertial subrange spectrum $E(k)$ is now given by

$$E(k) \approx C_1^{(1)} \bar{\varepsilon}^{2/3} k^{-5/3} (Lk)^{-\mu/9}. \tag{25.50}$$

and equations of the same form (with the exponent $\beta = -(5/3) - (\mu/9)$ are valid for the longitudinal and lateral one-dimensional spectra $E_1(k)$ and $E_2(k)$. Here again the difference between the exponent in the usual five-thirds law and in the refined law (25.50) is too small to be reliably detected experimentally. For the structure function of order $p$ we find from Eq. (25.27′) that in the inertial subrange

$$\overline{[\Delta_r u_L(x)]^p} \approx C^{(p)} (\bar{\varepsilon})^{\frac{p}{3}} r^{\frac{p}{3}} \left(\frac{L}{r}\right)^{\frac{p(p-3)\mu}{18}} \tag{25.51}$$

[this equation and also Eqs. (25.49) and (25.50) were first given by Kolmogorov (1962a, b)]. In the special case $p = 3$ we again obtain Eq. (25.15), while the dimensionless skewness and flatness factor of the longitudinal velocity difference are now given by the equations

$$S(r) = \frac{D_{LLL}(r)}{[D_{LL}(r)]^{3/2}} = S\left(\frac{L}{r}\right)^{\mu/6}, \delta(r) = \frac{D_{LLLL}(r)}{[D_{LL}(r)]^2} = Q\left(\frac{L}{r}\right)^{4\mu/9} \tag{25.52}$$

where $S = S(x, t)$ and $Q = Q(x, t)$ may depend on the macrostructure of the flow. We see that the corrections in the higher-order structure functions and in the corresponding dimensionless parameters [e.g., in the flatness factor $\delta(r)$] are considerably greater then the corrections for the usual structure functions of second order and for the turbulence spectra; therefore we must expect that the former corrections are considerably easier to detect experimentally.

Attempts to detect the deviations in shape of the higher-order structure functions in the inertial subrange from the predictions of the original Kolmogorov theory were made by Van Atta and Chen (1970) and Van Atta and Park (1971). In the first of these papers the fourth-order longitudinal structure function $D_{LLLL}(r)$ was measured at several heights in the atmospheric surface layer above the sea. Their data normalized with Kolmogorov length and velocity scales $\eta$ and $v_\eta$ are shown in Fig. I. We see that all the normalized data fall

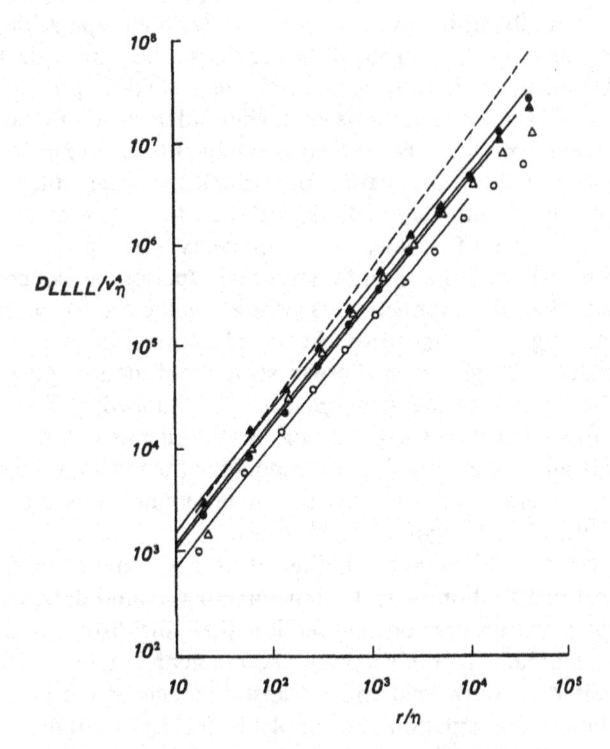


**Fig. I** Fourth-order structure functions normalized with Kolmogorov scales (after van Atta and Chen, 1970). The solid lines have a slope of 1.222, while the dashed line has a slope of 4/3. Different symbols correspond to different runs.

onto a single curve with sufficient accuracy with the single weak exception of the experimental points marked with light circles. The exceptional data refers to the lowest height of 3 m and is singular in some other respects too. The solid lines in the figure have a slope of $1.222 = 4/3 - 0.111 = 4/3 - 0.1/9$ which corresponds to the correct equation (25.51) when $p = 4$ and $\mu = 0.5$, while the dashed line has a slope of 4/3 which corresponds to the original Kolmogorov theory. We see that the dashed line is clearly steeper than all the data while the solid lines very closely follow the data. Van Atta and Park (1971) published the results of computations of the longitudinal structure functions of all orders up to ninth. A part of the data treated by Van Atta and Chen (1970) was used in these computations. The results obtained are not quite reliable since the accuracy of computation of moments higher than fourth order from experimental data is apparently rather low in many practical situations (this is especially so in the case of relatively small separations since then the corresponding probability densities have slowly decreasing tails). Therefore the investigation of the statistical precision and reliability of the computations of higher-order structure functions seems to be important. Nevertheless it is worth noting that all the graphs of Van Atta and Parks for various structure functions are relatively smooth and may be fairly well fitted by a power law over a considerable range of $r$-values. The exponents in the power laws for fourth-order through ninth-order structure functions are all considerably lower than the exponents $p/3$ (where $p$ is the order) corresponding to the original Kolmogorov theory. Moreover, the exponents for fourth-order through seventh-order structure function power laws agree fairly well with the values $p(15 - p)/36$ implied by Eq. (25.51) with $\mu = 0.5$. However, the eight- and ninth-order structure function exponents are higher than the corresponding theoretical values—they are in fact intermediate between the values predicted by the original and the refined (i.e., log-normal) theories.

It is reasonable to ask whether it is appropriate to use the log-normal distribution of $\varepsilon_r$ to explain experimental data, when we have shown in the previous subsection that this distribution yields incorrect estimates of the shape of the moment functions. However it is possible to show that more realistic models of the probability distribution for dissipation rate imply in fact quite often relatively close results for all moments of not too high order. Let us consider, for example, Novikov's model which is described by Eqs. (25.45)–(25.48), and which corresponds to small-scale turbulence statistics similar to those observed. We have already seen that $\mu \approx 0.41$ according to the model; this value is consistent with the majority of the existing small-scale observations. Moreover $\mu_{2/3} \approx -0.07$ in this model and hence it implies an increase in slope of second-order structure functions and a decrease of slope of velocity spectra in the inertial subrange by 0.07. These changes in the slopes are slightly greater than the change by $\mu/9 \approx 0.05$ predicted by the log-normal probability law, but nevertheless they are also too small to be reliably detectable experimentally.

Moreover $\mu_{4/3} \approx 0.11$ according to Novikov's model and this implies that the correction to the slope of the fourth-order structure function is the same according to this model and according to the log-normal assumption. The corrections to the slope (i.e., to the exponent) of the fifth-order structure function corresponding to these two theories differ only by about 0.02 and hence are also experimentally undetectable. The results for the sixth- through ninth-order structure functions implied by Novikov's and the log-normal theories differ slightly more than those for lower orders. However, first, the experimental results for structure functions of relatively high order are not very reliable; second, Novikov's model is a simple approximation only which does not pretend to be precise, and, third, the most important consequence of the general inequality (25.36) and of Novikov's model in particular is the fact that the exponents $\mu_p$ increase with the order $p$ slower than the log-normal assumption predicts, and this fact is confirmed by the results of Van Atta and Park relative to eighth- and ninth-order structure functions.

Similar considerations can also be applied to the analysis of the statistical characteristics of turbulence determined by the high frequency range above the high frequency limit of the inertial subrange. It is clear, for example, that the refined similarity hypotheses imply that

$$\overline{\left(\frac{\partial u_1}{\partial x_1}\right)^n} = a_n \nu^{-n/2} \overline{\varepsilon^{n/2}}$$

where $a_n$ is a universal constant. Hence if we accept the log-normality assumption and the relation (25.20), then

$$\overline{\left(\frac{\partial u_1}{\partial x_1}\right)^n} \sim \left(\frac{\bar{\varepsilon}}{\nu}\right)^{n/2} \left(\frac{L}{\eta}\right)^{\mu n(n-2)/8} \sim \left(\frac{\bar{\varepsilon}}{\nu}\right)^{n/2} \mathrm{Re}^{3\mu n(n-2)/32}$$

It is easy to deduce from these equations many relations for various moments of velocity derivatives. For example, if $s$ and $\delta$ are the skewness and flatness factor (i.e., excess increased by 3) of $\partial u_1/\partial x_1$, respectively, that it is easy to see that

$$s \sim \delta^{3/8}$$

[Wyngaard and Tennekes (1970)]. This last relation was tested by Wyngaard and Tennekes themselves and more completely by Wyngaard and Pao (1971) and Kholmyanskii (1972); the results of Wyngaard and Pao are shown in Fig. J. The use of Novikov's scale similarity assumption instead of the log-normality implies obviously the replacement of $\mu n(n-2)/8$ by $\mu_{n/2}$ in equations for $\overline{(\partial u_1/\partial x_1)^n}$; hence, for example, the exponent 3/8 in the last equation must be replaced by $\mu_{3/2}/\mu_2 = \mu_{3/2}/\mu$ (in the particular case of the simplified Novikov model this exponent will be about 0.44 instead of $3/8 = 0.375$).

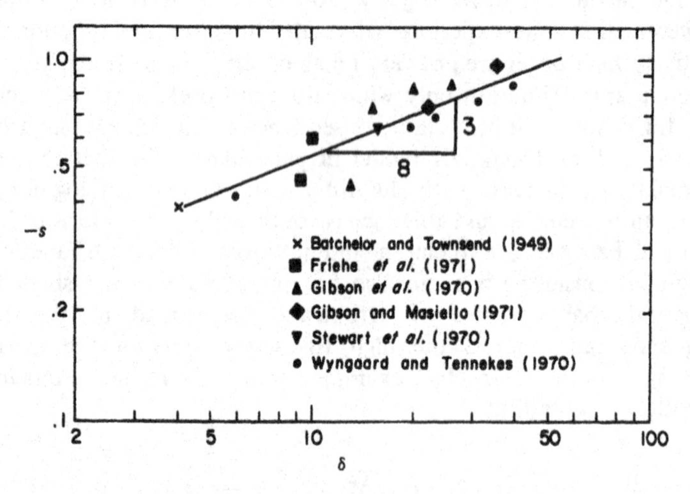


Fig. J The values of the skewness and flatness factor of $\partial u_1/\partial x_1$ from various sources according to Wyngaard and Pao (1971).

Let us now briefly consider the question of the influence of dissipation fluctuations on the shape of the spectra in the dissipation range. Since we have seen that the consideration of the deviations of the true small-scale probability distributions from a log-normal distribution does not lead to significant corrections for inertial subrange parameters, it is natural to use here also at the first stage the assumption of the log-normal distribution of $\varepsilon$. Attempts to analyze the consequences of this assumption for the shape of the spectra in the dissipation range were made by Ellison (1967) and by Keller and Yaglom (1970). Ellison considered the simplest model of the cutoff energy spectrum of the form

$$E(k) = \begin{cases} C_1 \varepsilon^{2/3} k^{-5/3} & \text{for } k < b^{-1}(\varepsilon/\nu^3)^{1/4} \\ 0 & \text{for } k > b^{-1}(\varepsilon/\nu^3)^{1/4} \end{cases} \tag{25.53}$$

where $b$ is a dimensionless constant. If we now suppose that $\varepsilon$ is in fact a log-normally distributed random variable and average the equation (25.53) over all the possible values of $\varepsilon$, then the averaged spectrum will be, of course, different from zero at any $k$ [since there is a positive probability that any given $k$ may be greater than $b(\varepsilon/\nu^3)^{1/4}$]. Ellison calculated the result of the average over the log-normal distribution of $\varepsilon$ on the spectrum (25.53) and found that the averaged spectrum will be described by an equation of the form

$$E(k) = 0.5 C_1 \bar{\varepsilon}^{2/3} k^{-5/3} e^{-\sigma^2/9} \left[1 - \Phi\left(\frac{2^{3/2}}{\sigma}\left\{\ln bk\eta - \frac{\sigma^2}{24}\right\}\right)\right] \tag{25.54}$$

where $\sigma^2$ is the variance of $\ln \varepsilon$, $\eta = (\nu^3/\bar{\varepsilon})^{1/4}$, and $\Phi(x) = \frac{2}{\sqrt{\pi}} \int_0^x e^{-t^2}\, dt$ is an error function. The spectrum (25.54) is close to $C_1 e^{-\sigma^2/9} \bar{\varepsilon}^{2/3} k^{-5/3}$ when $k_\eta$ is very small as must be expected; however, the cutoff jump discontinuity of the original spectrum (25.53) is smoothed out by the averaging over the probability distribution of $\varepsilon$. Ellison even found that a smooth curve plotted in accordance with Eq. (25.54) can be fitted more or less satisfactorily to the experimental data of Grant, Stewart, and Moilliet (1962) as transformed from one-dimensional to three-dimensional form [i.e., from $E_1(k)$ to $E(k)$]. It is interesting to note also that the spectrum (25.54) decreases relatively very slowly in the far dissipation range $k\eta \gg 1$, although the original spectrum (25.53) cuts off at a finite value $k$. In fact it is easy to show that according to Eq. (25.54)

$$E(k) \approx A \exp\left[-\frac{8}{\sigma^2}(\ln k\eta)^2\right] \quad \text{for very large } k\eta \tag{25.55}$$

where $A = A(k)$ is a slowly varying function of $k$ (which varies practically as a power of $k$). Equation (25.55) shows that the averaged spectrum $E(k)$ decreases at infinity more slowly than any exponential function of the form $\exp[-c(k\eta)^\alpha]$ however small $\alpha$ may be and even more slowly than the spectrum (17.45″) which has a rate of decrease that is intermediate between a power and an exponential law (but nevertheless it decreases more rapidly than any negative power of $k$).

This last result was generalized significantly by Keller and Yaglom (1970) who studied the asymptotic behavior of the averaged spectrum $E(k)$ in the far dissipation range $k\eta \gg 1$ for various shapes of the original ("nonaveraged") spectrum corresponding to a fixed value of the dissipation rate $\varepsilon$. They found that if $\varepsilon$ has a log-normal probability distribution then the asymptotic behavior of $E(k)$ can be obtained for most of the usual shapes of the original spectrum with the aid of the well-known Laplace's method of asymptotic analysis. The results of such calculations show that the asymptotic shape of the averaged spectrum is in fact practically independent of the shape of the original spectrum. Namely the asymptotic shape of the averaged spectrum at $k\eta \to \infty$ turns out to be described by Eq. (25.55) in all cases in which the original spectrum decreases at infinity more rapidly then any function of the form $\exp[-C(\ln k\eta)^2]$ however large $C$ may be. In other words, the smoothing of the original spectrum by averaging over the log-normal probability distribution of the dissipation rate $\varepsilon$ plays, as a rule, a much more important role than the original rate of decrease of the nonaveraged spectrum and it is precisely this smoothing that determines the ultimate rate of decrease of the averaged spectrum as $k$ tends to infinity. However the low-wavenumber limit of the applicability of the asymptotic relation (25.55) [i.e., the subrange of the smallest wavenumbers at the left end of the wavenumber range where Eq. (25.55) is valid] is strongly dependent on the original rate of decrease: The quicker this decrease is, the later Eq. (25.55) begins to be valid. Hence it is even possible in principle to determine the rate of decrease of the original spectrum from the low-wavenumber limit of applicability of Eq. (25.55) to the observed averaged spectrum.

If the original spectrum decreases at infinity more slowly than any function of the form $\exp[-C(\ln k\eta)^2]$ however small $C$ may be then the averaging over the log-normal distribution of the values of $\varepsilon$ does not change the ultimate rate of decrease of $E(k)$ at all. This is of course expected since the supplementary averaging cannot diminish the rate of decrease of $E(k)$ and here the original rate of decrease is even slower than that which is due to the averaging. Finally if the original spectrum decreases precisely as $\exp[-C(\ln k\eta)^2]$ then averaging over the dissipation distribution will not change the general law of decrease but will lead to a decrease of the effective value of the constant $C$ (namely $C$ will be replaced by a coefficient $C_1 = 8C/(C\sigma^2 + 8)$). However, both these last conclusions correspond to an unnaturally slow decrease of the original spectrum and seem therefore to be rather improbable.

The asymptotic law of decrease of $E(k)$ given by Eq. (25.55) is strongly dependent upon the Reynolds number of the flow and shows that the rate of spectral fall-off decreases with an increase of Re [since $\sigma^2 \approx 0.75\mu \ln \mathrm{Re}$ by virtue of Eqs. (25.23′) and (21.6)]. This is of course to be expected. We must also note that in applications to real spectral estimates, the parameter Re must apparently be replaced by its "effective value" $\mathrm{Re}_{ef}$ which depends also on the sample size and increases with an increase of this size (see the preceding subsection of the book).

The deviation of the true probability distribution of $\epsilon$ from a log-normal one (of the type described by the model equation (25.48) produces some important changes in the situation. We have already seen that the true probability distribution of $\varepsilon$ can be approximated with fair accuracy by the log-normal distribution in the range of moderate values, but the far tails of the distribution are in fact quite different from the log-normal law. Let us take into account that in the range of very large values of $\varepsilon$ the true probability density decreases much more rapidly then the log-normal law predicts. Let us also make the natural assumption that the nonaveraged spectrum corresponding to a strictly constant value of $\varepsilon$ decreases at infinity considerably more rapidly than the function $\exp[-(\ln k\nu^{3/4}\varepsilon^{-1/4})^2]$. If Re is large enough then the averaged spectrum $E(k)$ will follow the asymptotic law (25.55) over a rather extensive range of sufficiently high but not too high wavenumbers $k$. However as $k$ increases indefinitely, a critical value will ultimately be reached where the values of $\varepsilon$ beyond the log-normal range are most important in determining the rate of fall-off of $E(k)$. At this critical value the law (25.55) ceases to be valid and at higher values of $k$ the spectrum $E(k)$ begins to fall off much more rapidly than this law predicts. The critical value of $k$ strongly depends on the rate of decrease of the original spectrum; based on this it was even suggested by Keller and Yaglom that the high-wavenumber limit of applicability of the relation (25.55) can in principle, also be used to estimate the original rate of the spectral fall-off.

Similar analysis can be applied to the temperature-field characteristics (or the characteristics of the concentration of a passive admixture) which depend on the "temperature dissipation" $N = \chi(\nabla\vartheta)^2$. It is clear that the quantity $N$ will also be subject to considerable fluctuations, and that these fluctuations may affect all the statistical characteristics of the field $\vartheta(\boldsymbol{x}, t)$. The quantity $N(\boldsymbol{x}, t)$ is, of course, of the same type as the field $\varepsilon^*(x, t)$ studied in Sect. 25.3 and hence all the general results derived above for the field $\varepsilon^*(x, t)$ must be valid for $N(\boldsymbol{x}, t)$. The parameter $\mu$ for the field $N(\boldsymbol{x}, t)$ need not in general have the same value as for the field $\varepsilon(\boldsymbol{x}, t)$ or $\varepsilon_u(x, t)$. However the data of the very preliminary measurements of Gurvich (see Gurvich and Yaglom, 1967) seem to favor the impression that the values of $\mu$ for $N$ and for $\varepsilon$ are rather close one to another. When we investigate the structure function or spectral densities of the field $\vartheta(\boldsymbol{x}, t)$ we must keep in mind that all these functions may be affected both by fluctuations in $N$ and in $\varepsilon$, and there is no reason to suppose that they will be statistically independent. We can however use the fact that the analysis applied above to the field $\varepsilon^*(x, t)$ is apparently also valid for the field $N(\boldsymbol{x}, t)\varepsilon^{-1/3}(\boldsymbol{x}, t)$. Since, however there are no experimental data on this point at present, we shall not pursue this matter any further.

## 25.5 More General Form of the Refined Similarity Hypothesis

The new form of the similarity hypothesis used above is more accurate than the initial form but, in principle, it can be improved still further. Thus, the assumption that the probability distributions for the velocity field in a given small space-time region depend only on the integral of $\varepsilon(\boldsymbol{x}, t)$ taken over this region, is probably only approximate. The use of the dissipation $\varepsilon_r$ averaged over the volume of a sphere as the main statistical parameter, which was in fact chosen exclusively on the basis of convenience, may be objected to even more strongly. It is therefore, desirable to find a formulation of the similarity hypotheses which does not make use of special quantities such as $\varepsilon_r$.

The general approach which enables us to find such similarity hypotheses was given by Kolmogorov (1962a, b). He based his analysis on the replacement of the velocity differences $\Delta u_l = u_l(\boldsymbol{x}_k) - u_l(\boldsymbol{x})$, $k = 1, \ldots, n$ by the dimensionless ratios of such differences

$$\frac{u_l(\boldsymbol{x}_k) - u_l(\boldsymbol{x})}{u_l(\boldsymbol{x}_0) - u_l(\boldsymbol{x})}, \qquad l = 1, 2, 3; \qquad k = 1, \ldots, n \tag{25.56}$$

where, for simplicity, we are restricting our attention to purely spatial statistical characteristics at a given time $t$. For the arguments of the $3n$ independent random variables (25.56) we can take the dimensionless vectors $\boldsymbol{y}_k = \frac{\boldsymbol{x}_k - \boldsymbol{x}}{|\boldsymbol{x}_0 - \boldsymbol{x}|}$, $k = 0, 1, \ldots, n$ and the

Reynolds number $\mathrm{Re} = \frac{|u(x_0) - u(x)||x_0 - x|}{\nu}$ which depends on $u(x_0) - u(x)$ and is therefore a random variable. We can therefore take the following assumption as the analog of the first similarity hypothesis: *for sufficiently small (in comparison with the external turbulence length scale L) values of* $|x_k - x|$, $k = 0, 1, \ldots, n$, *such that all the* $|y_k|$, $k = 1, \ldots, n$ *do not exceed unity in order of magnitude, and a given* Re, *the conditional probability distribution for the random variables* (22.56) *will depend only on* $y_k$, $k = 0, 1, \ldots, n$ *and the Reynolds number* Re, *and will be invariant under arbitrary rotations and reflections in the space of the vectors* $y$. Next, it is natural to assume that *if all the* $|y_k|$ *are of the order of unity and* $\mathrm{Re} \gg 1$, *then the probability distribution for the variables* (25.56) *will be independent of* Re *as well*. This assumption is the analog of the second similarity hypothesis.

The two assumptions which we have just formulated will to some extent replace the two refined similarity hypotheses given in Sect. 25.2. However, it is clear that they do not in themselves lead to the log-normal distribution for the energy dissipation rate or to some other related statistical result indicating that the cascade breakdown of vortices is a purely random physical process which consists of a number of independent stages. Therefore, when we use the similarity hypotheses formulated in terms of the ratios of velocity differences, we must introduce at least one more (third) hypothesis ensuring that disturbances with very different scales are practically independent. This will naturally lead to the log-normal distribution (with variance proportional to the logarithm of the Reynolds number) for the quantity $\epsilon$ and other similar local nonnegative quantities. Kolmogorov suggested that this third hypothesis can for example be taken to be: *if* $r_1 \gg r_2$, *then two groups of variables defined by* (25.56) *which are such that in the first group* $|x_k - x| \gg r_1$ *for all* $k$, *and in the second* $|x_k - x| \ll r_2$ *for all* $k$, *are statistically independent*. We shall not, however, discuss in detail the various ways of using this third hypothesis, since the entire approach to the theory of turbulence based on analyzing the probability distributions of ratios of velocity differences is still in a preliminary stage, and a detailed development of this theory is a matter for the future. Let us only note that this approach has some points in common with the conditions of scale similarity discussed in Sect. 25.3. In fact, if we confine ourselves to the one-dimensional case [i.e., the field $u(x, t)$ on the line $-\infty < x < \infty$] and consider the ratio (25.32) for the variable $\epsilon^* = \partial u/\partial x$ (which is not of the nonnegative type) we obtain a ratio of velocity differences of the form (25.56). With this remark we finish the section on the refinements of the theory of locally isotropic turbulence.

# 9 WAVE PROPAGATION THROUGH TURBULENCE

## 26. PROPAGATION OF ELECTROMAGNETIC AND SOUND WAVES IN A TURBULENT MEDIUM

### 26.1 Foundations of the Theory of Electromagnetic Wave Propagation in a Turbulent Medium

The propagation of sound, light, and radio waves in a turbulent medium, for example, the earth's atmosphere or the sea, is accompanied by a number of fluctuational phenomena. These include the scattering of the waves by random (turbulent) inhomogeneities of the medium and wave amplitude and phase fluctuations. As a result the images of wave sources received by the detector undergo considerable scintillation and jitter, which are of major importance in many practical problems. Thus, stellar scintillation, and also scintillation of extra-terrestrial radio sources, tends to interfere with optical and radio astronomy and is quite harmful in this respect. Such interference may also occur in optical and radio communication with artificial satellites and space probes. Hydroacoustic communication in the sea is also affected by these phenomena. Conversely, scattering of short radio waves by irregular inhomogeneities in the troposphere provides us with a means for long-range television communication, and may therefore be useful. Moreover the characteristics of waves propagated in a turbulent medium and then received by a sensor (in particular, the characteristics of scattered waves) contain valuable information on medium inhomogeneities along the wave path and hence provide a quite useful possibility of remote probing of our environment (both atmosphere, including ionosphere, and ocean) at great distances.

Very extensive literature is available on the theoretical calculation of the above fluctuational phenomena and their experimental investigation. In particular, a detailed review of the problems considered in the present chapter can be found in two books by Tatarskii (1961a, 1971). A number of interesting problems concerning wave propagation through random media have been discussed in a book by Chernov (1960) who, however, mostly used a simplified statistical model of a random refractive-index field with a Gaussian correlation function of the form (12.5′) in his calculations. There are also several survey papers on the subject [see, for example, Hoffman (1964), Keller (1964), Frisch (1968), Barabanenkov, Kravtsov et al. (1970) and some others] where many additional references can be found (the survey by Barabanenkov et al. alone contains about 550 references). Finally, let us refer to the October 1955 special issue of the journal *Proceedings of the IRE* wholly devoted to the problem of long-range radio communication by tropospheric scattering and the April 1969 issue of the journal *Proceedings of the IEEE* (which is the new title of the former *Proc. of the IRE*) devoted to remote environment probing. In the present book we shall restrict our attention to a brief account of the physical aspects of the subject without going into technical details and shall consider mostly only the case of propagation in an atmosphere. We shall begin with a general description of the propagation of electromagnetic waves in a turbulent atmosphere.

We shall look upon the atmosphere as a nonconducting medium whose magnetic permeability is equal to unity. The propagation of electromagnetic waves in such a medium is described by the well-known Maxwell equations

$$\nabla \times \boldsymbol{E} = -\frac{1}{c}\frac{\partial \boldsymbol{H}}{\partial t}, \quad \nabla \times \boldsymbol{H} = \frac{1}{c}\frac{\partial \boldsymbol{D}}{\partial t}, \qquad (26.1)$$
$$\nabla \boldsymbol{D} = 0, \qquad \nabla \boldsymbol{H} = 0,$$

where the cross, as usual, represents the vector product, so that $\nabla \times \boldsymbol{E}$ is the curl of $\boldsymbol{E}$, $c$ is the velocity of light in empty space, $\boldsymbol{E}$ and $\boldsymbol{H}$ are the electric and magnetic fields, $\boldsymbol{D} = \varepsilon \boldsymbol{E}$ is the electric induction, and $\varepsilon$ is the permittivity. The quantity $n = \sqrt{\varepsilon}$ is called the *refractive index* of the medium. The refractive index of the atmosphere is in general a function of temperature, humidity, and pressure, but it also depends on wavelength. Thus, for example, for centimeter radio waves the refractive index of the atmosphere is given by

$$n = 1 + 10^{-6}\,\frac{79p}{T}\left(1 + \frac{7800\vartheta}{T}\right), \tag{26.2}$$

while for visible radiation (for $\lambda \approx 0.45\ \mu$)

$$n = 1 + 10^{-6}\,\frac{80p}{T}, \tag{26.2'}$$

where $p$ is the pressure in millibars, $T$ is the absolute temperature, and $\vartheta$ is the specific humidity (ratio of water-vapor density to the humid air density). In a turbulent atmosphere the quantities $p$, $T$ and $\vartheta$ will fluctuate, and this will result in a fluctuation in the refractive index, so that $n$ can be regarded as a random function of the space-time point. It is clear from Eq. (26.2) that the mathematical expectation $\bar{n}$ of this random function is close to unity. We shall assume henceforth that $\bar{n} = 1$ and $n = 1 + n'$ where $n'$ is the refractive index fluctuation. For example, for centimeter radio waves

$$n' \approx -10^{-6}\,\frac{79\bar{p}}{\bar{T}^2}\left(1 + \frac{15\,600\bar{\vartheta}}{\bar{T}}\right)\left(T' - \frac{7800\vartheta'}{1 + \frac{15\,600\bar{\vartheta}}{\bar{T}}}\right), \tag{26.3}$$

and for visible radiation

$$n' \approx -10^{-6}\,\frac{80\bar{p}}{\bar{T}^2}\,T' \tag{26.3'}$$

We have neglected pressure fluctuations because $p'/\bar{p}$ is small in comparison with $T'/\bar{T}$ and $\vartheta'/\bar{\vartheta}$. In the atmospheric surface layer $n'$ is of the order of $10^{-6}$ and therefore we can assume that $\varepsilon = n^2 \approx 1 + 2n'$.

Fluctuations in the permittivity ensure that the turbulent atmosphere behaves as an inhomogeneous medium with weak (random) inhomogeneities which give rise to scattering of electromagnetic waves, and to fluctuations in their amplitudes, phases, frequencies, and other parameters. We shall investigate the propagation of electromagnetic waves in a turbulent atmosphere at moderate distances $L$ such that

$$\frac{L}{c} \ll \frac{2\pi}{\omega_{\text{turb}}}, \tag{26.4}$$

where $\omega_{\text{turb}}$ are the frequencies of the turbulent fluctuations which contribute appreciably to fluctuations in the electromagnetic field. When this condition is satisfied, the fluctuation field $n'(\boldsymbol{x})$ of the refractive index is practically unaltered in the time $L/c$ taken by the waves to propagate through a distance $L$.[1]

Assuming that $n'$ is time-independent, we can look upon the solutions of the Maxwell equations as monochromatic waves with fixed frequency $\omega$. In other words, we can assume that the electric and magnetic fields and the electric induction are of the form $\operatorname{Re}(\boldsymbol{E}e^{-i\omega t})$, $\operatorname{Re}(\boldsymbol{H}e^{-i\omega t})$, and $\operatorname{Re}(\boldsymbol{D}e^{-i\omega t})$, respectively, where the complex amplitudes $\boldsymbol{E}, \boldsymbol{H}, \boldsymbol{D}$ are time-independent, and $\boldsymbol{D} = (1+2n')\boldsymbol{E}$. For the complex amplitudes we find that Eq. (26.1) assumes the form

$$\nabla \times \boldsymbol{E} = i\kappa \boldsymbol{H}, \quad \nabla \times \boldsymbol{H} = -i\kappa \boldsymbol{D}, \tag{26.5}$$

where $\kappa = 2\pi/\lambda = \omega/c$ is the wave number ($\lambda$ is the wavelength). The equations $\nabla \boldsymbol{D} = \nabla \boldsymbol{H} = 0$ follow from Eq. (26.5).

If we take the curl of the first equation in Eq. (26.5), and eliminate $\nabla \times \boldsymbol{H}$ with the aid of the second equation, we obtain

$$\kappa^2 \boldsymbol{D} = \nabla \times (\nabla \times \boldsymbol{E}), \quad \boldsymbol{D} = (1+2n')\boldsymbol{E}, \tag{26.6}$$

which does not contain $\boldsymbol{H}$. If necessary, $\boldsymbol{H}$ can be found from the formula $\boldsymbol{H} = -(i/\kappa)(\nabla \times \boldsymbol{E})$. We shall require the mean flux density of

[1] In fact the condition $U/c \ll n'$ where $U$ is a typical velocity of a medium (i.e., mean wind velocity) is also necessary for justification of the neglect of time variations of refractive index fluctuations [cf. Tatarskii (1971), Sect. 24]. However this condition is valid with good accuracy in the atmosphere, since here $n'$ is of the order of $10^{-6}$ and $U/c$ is of the order of $10^{-8}$.

electromagnetic energy (the Poynting vector) over one period $2\pi/\omega$, which is given by

$$s = \frac{c}{4\pi}\frac{\omega}{2\pi}\int_0^{\frac{2\pi}{\omega}} [\mathrm{Re}\,(\boldsymbol{E}e^{-i\omega t}) \times \mathrm{Re}\,(\boldsymbol{H}e^{-i\omega t})]\, dt.$$

Substituting $\mathrm{Re}\, F = \frac{1}{2}(F + F^*)$, where the asterisk represents the complex conjugate, we obtain after some simple rearrangement

$$s = \frac{c}{8\pi}\,\mathrm{Re}\,[\boldsymbol{E} \times \boldsymbol{H}^*]$$

or, after eliminating $\boldsymbol{H}$ with the aid of the first equation in (26.5), we have

$$s = \frac{c}{8\pi\kappa}\,\mathrm{Im}\,[\boldsymbol{E}^* \times (\nabla \times \boldsymbol{E})]. \tag{26.7}$$

We shall consider the propagation of electromagnetic waves in a turbulent atmosphere in the following simplified formulation. Let us suppose that the turbulence is concentrated within a particular volume $V$, so that $n'(\boldsymbol{x})$ is nonzero only for $\boldsymbol{x} \in V$. Let us suppose further that the plane wave $\boldsymbol{E}_0 = \boldsymbol{p}A_0 e^{i\boldsymbol{\kappa x}}$ is incident on this volume, where $\boldsymbol{p}$ is a unit polarization vector perpendicular to the direction of propagation (i.e., to the vector $\boldsymbol{\kappa}$), $A_0$ is the amplitude, and $s_0 = \boldsymbol{\kappa x}$ is the phase of the wave. The corresponding Poynting vector of the equation (26.7) then is given by

$$s_0 = \frac{cA_0^2}{8\pi}\frac{\boldsymbol{\kappa}}{\kappa}. \tag{26.8}$$

We shall use Eq. (26.6) in the first instance to describe the scattering of electromagnetic waves by turbulent inhomogeneities localized in $V$. The scattering of radio waves by a turbulent atmosphere is of practical importance, since it provides a means of using ultrashort waves for long-range communication purposes. In fact, observed propagation of ultrashort radio waves over long distances in the atmosphere, i.e., beyong the so-called radio horizon, can apparently in some cases be explained by the scattering of waves by turbulent inhomogeneities in the refractive index of the troposphere.

The explanation of long-range propagation of ultrashort waves by scattering on turbulent inhomogeneities of the tropospheric refractive index was first put forward by Booker and Gordon (1950). The theory developed by these workers was subsequently extensively investigated by Megaw (1950), Villars and Weisskopf (1954), Batchelor (1955), Staras (1955), Silverman (1956, 1957), Tatarskii and Goltisyn (1962) and many others. We have already noted that the October 1955 issue of the journal *Proceedings of the IRE* is specially devoted to the scattering of radio waves by tropospheric turbulence.

As a result of scattering of the incident wave $\boldsymbol{E}_0$ by inhomogeneities localized in $V$, the electric field will be distorted throughout space, and will be given by $\boldsymbol{E}=\boldsymbol{E}_0+\boldsymbol{E}'$, where $\boldsymbol{E}'$ corresponds to the scattered waves. Since $n(\boldsymbol{x})$ is a random function, the scattered wave field $\boldsymbol{E}'(\boldsymbol{x})$ will also be random. Our problem is to investigate the random function $\boldsymbol{E}'(\boldsymbol{x})$. Since $\boldsymbol{E}_0$ satsifies Eq. (26.6) for $n'=0$, we have

$$\kappa^2\boldsymbol{D}'=\nabla\times(\nabla\times\boldsymbol{E}'),\qquad \boldsymbol{D}'=\boldsymbol{E}'+2n'\boldsymbol{E}_0. \tag{26.9}$$

In the formula for $\boldsymbol{D}'$ we have here neglected the term $2n'\boldsymbol{E}'$, assuming that because $|n'(\boldsymbol{x})|\ll 1$, we must have

$$|\boldsymbol{E}'(\boldsymbol{x})|\ll|\boldsymbol{E}_0(\boldsymbol{x})|=A_0. \tag{26.10}$$

Eliminating $\boldsymbol{E}'$ from the right-hand side of the first equation in Eq. (26.9) with the aid of the second equation, and remembering that $\nabla\times(\nabla\times\boldsymbol{D}')=-\Delta\boldsymbol{D}'$ by virtue of $\nabla\boldsymbol{D}'=0$, we have

$$(\Delta+\kappa^2)\,\boldsymbol{D}'=-\nabla\times[\nabla\times(2n'\boldsymbol{E}_0)]. \tag{26.11}$$

We thus see that $\boldsymbol{D}'$ is determined by a wave equation with a random right-hand side. The solution of this equation which corresponds to scattered waves can be written in the form

$$\boldsymbol{D}'(\boldsymbol{x})=\frac{1}{4\pi}\nabla\times\left[\nabla\times\int_V 2n'(\boldsymbol{x}_1)\,\boldsymbol{E}_0(\boldsymbol{x}_1)\frac{e^{i\kappa|\boldsymbol{x}-\boldsymbol{x}_1|}}{|\boldsymbol{x}-\boldsymbol{x}_1|}\,d\boldsymbol{x}_1\right]=$$
$$=\frac{A_0}{2\pi}\nabla\times\left[\nabla\times\boldsymbol{p}\int_V n'(\boldsymbol{x}_1)\frac{e^{i\boldsymbol{\kappa}\boldsymbol{x}_1+i\kappa|\boldsymbol{x}-\boldsymbol{x}_1|}}{|\boldsymbol{x}-\boldsymbol{x}_1|}\,d\boldsymbol{x}_1\right]. \tag{26.12}$$

Let us take the origin of our coordinate system inside the scattering volume $V$, and try to determine the dependence of the scattered intensity on the direction $\boldsymbol{q} = \boldsymbol{x}/|\boldsymbol{x}|$ for $\boldsymbol{x}$ lying inside $V$ at distances $|\boldsymbol{x}|$ that are large in comparison with the linear dimensions $L$ of this volume and the wavelength $\lambda$ (more precisely, we shall consider the zone defined by the condition $\sqrt{\lambda|\boldsymbol{x}|} \gg L$; the field within small regions in this zone can be regarded as a plane wave). This formulation of the problem corresponds to the so-called Fraunhofer diffraction.

Since the refractive-index fluctuations are zero outside $V$, we can replace $\boldsymbol{D}'(\boldsymbol{x})$ by $\boldsymbol{E}'(\boldsymbol{x})$ on the left-hand side of Eq. (26.12). For values of $\boldsymbol{x}$ corresponding to the Fraunhofer diffraction zone, the integral on the right-hand side of this formula can readily be simplified. Since for $\boldsymbol{x}_1 \in V$ the values of $|\boldsymbol{x}_1|$ do not exceed the diameter $L$ of $V$, and $|\boldsymbol{x}| \gg L$ so that $|\boldsymbol{x}_1| \ll |\boldsymbol{x}|$, we can expand the quantity $|\boldsymbol{x} - \boldsymbol{x}_1|$ into a series in powers of $|\boldsymbol{x}_1|/|\boldsymbol{x}|$. The quadratic terms in this series will have the values of the order of $|\boldsymbol{x}|(|\boldsymbol{x}_1|/|\boldsymbol{x}|)^2 \sim L^2/|\boldsymbol{x}|$, and will contribute terms of the order of $\kappa L^2/|\boldsymbol{x}| \sim L^2/\lambda|\boldsymbol{x}|$ to the argument of the exponential function $\exp\{i\kappa|\boldsymbol{x} - \boldsymbol{x}_1|\}$. As noted above, we are assuming that

$$|\boldsymbol{x}| \gg \lambda, \quad \lambda|\boldsymbol{x}| \gg L^2, \tag{26.13}$$

where the second condition distinguishes Fraunhofer diffraction from the so-called Fresnel diffraction [in the calculation of the statistical characteristics of scattered waves, which we shall carry out below, this second condition will turn out to be too stringent and will be replaced by $\lambda|\boldsymbol{x}| \gg L_0^2$ where $L_0$ is the integral correlation length scale for the refractive index fluctuations, which is given by formulas such as Eq. (12.21)]. Subject to these conditions, it is sufficient to take into account only the linear terms in the series for $|\boldsymbol{x} - \boldsymbol{x}_1|$ in powers of $|\boldsymbol{x}_1|/|\boldsymbol{x}|$, which are of the form $|\boldsymbol{x}| - \boldsymbol{q}\boldsymbol{x}_1$, i.e., we can assume that

$$e^{i\boldsymbol{\kappa}\boldsymbol{x}_1 + i\kappa|\boldsymbol{x}-\boldsymbol{x}_1|} \approx e^{i\kappa|\boldsymbol{x}|} e^{i(\boldsymbol{\kappa} - \kappa\boldsymbol{q})\boldsymbol{x}_1}.$$

In the denominator of the integrand in Eq. (26.12) we can let $|\boldsymbol{x} - \boldsymbol{x}_1| \approx |\boldsymbol{x}|$, so that Eq. (26.12) will assume the form

$$\boldsymbol{E}'(\boldsymbol{x}) \approx \frac{A_0}{2\pi} \nabla \times \left[\nabla \times \frac{\boldsymbol{p} e^{i\kappa|\boldsymbol{x}|}}{|\boldsymbol{x}|} \int\limits_V n'(\boldsymbol{x}_1) e^{i(\boldsymbol{\kappa} - \kappa\boldsymbol{q})\boldsymbol{x}_1} d\boldsymbol{x}_1\right].$$

The integral in this formula is independent of $\boldsymbol{x}$, and can be taken outside the operator $\nabla \times (\nabla \times)$. Next, to within terms of the order of $\lambda/|\boldsymbol{x}|$, we may write

$$\nabla \times \left(\nabla \times \frac{\boldsymbol{p} e^{i\kappa|\boldsymbol{x}|}}{|\boldsymbol{x}|}\right) \approx \frac{\kappa^2 e^{i\kappa|\boldsymbol{x}|}}{|\boldsymbol{x}|} [(\boldsymbol{q} \times \boldsymbol{p}) \times \boldsymbol{q}].$$

The final expressions therefore become

$$\boldsymbol{E}'(\boldsymbol{x}) \approx \frac{\kappa^2 A_0 e^{i\kappa|\boldsymbol{x}|}}{2\pi|\boldsymbol{x}|} G[(\boldsymbol{q} \times \boldsymbol{p}) \times \boldsymbol{q}], \quad G = \int_V n'(\boldsymbol{x}_1) e^{i(\boldsymbol{\kappa}-\kappa\boldsymbol{q})\boldsymbol{x}_1} d\boldsymbol{x}_1. \tag{26.14}$$

The vector $[(\boldsymbol{q} \times \boldsymbol{p}) \times \boldsymbol{q}]$, whose length is $|\sin\alpha|$ where $\alpha$ is the angle between $\boldsymbol{p}$ and $\boldsymbol{q}$, characterizes the polarization of the scattered wave. It is coplanar with $\boldsymbol{p}$ and $\boldsymbol{q}$, and perpendicular to $\boldsymbol{q}$ (the wave is of course transverse). The Poynting vector for the scattered radiation which is given by Eqs. (26.7) and (26.14), again to within terms of the order of $\lambda/|\boldsymbol{x}|$, turns out to be

$$\boldsymbol{s}'(\boldsymbol{x}) = \frac{c\kappa^4 A_0^2 \sin^2\alpha}{32\pi^3|\boldsymbol{x}|^2} GG^*\boldsymbol{q}. \tag{26.15}$$

If we multiply $|\boldsymbol{s}'(\boldsymbol{x})|$ by the area $|\boldsymbol{x}|^2 d\Omega$ on the surface of a sphere of radius $|\boldsymbol{x}|$ corresponding to an infinitesimal solid angle $d\Omega$ (with the apex at the origin), which contains the direction of $\boldsymbol{q}$, we obtain the energy flux scattered by the volume $V$ into the solid angle element $d\Omega$. The ratio of this quantity to the energy flux density in the incident wave, i.e.,

$$d\sigma = \frac{|\boldsymbol{s}'(\boldsymbol{x})|\,|\boldsymbol{x}|^2 d\Omega}{|\boldsymbol{s}_0|} \tag{26.16}$$

is called the *effective scattering cross-section* of the volume $V$ in the direction of $\boldsymbol{q}$. Using Eqs. (26.15) and (26.8), we find that

$$d\sigma = \frac{\kappa^4 \sin^2\alpha}{4\pi^2} GG^* d\Omega. \tag{26.17}$$

We recall that the quantity $G$ in this expression is given by Eq.

(26.14) and is a random function of the scattering vector $\boldsymbol{\kappa} - \kappa\boldsymbol{q}$, i.e, the difference between the wave vectors of the incident and scattered waves, whose statistical characteristics are determined by the random field $n'(\boldsymbol{x})$.

We shall now use Eq. (26.6) to describe fluctuations in the amplitudes and phases of electromagnetic waves in a turbulent atmosphere. Because of these fluctuations, the electromagnetic waves emitted by any given bodies (including stars, cosmic radio sources, artificial earth satellites) or reflected by given objects (for example, in radar studies) are received by detectors in a distorted form. These distortions take the form of fluctuations in the spectral and integrated intensity of the received signals, and also of fluctuations in the angle of arrival. They are responsible, for example, for the scintillation, chromatic scintillation, and jitter of stellar images in telescopes.

Let us first simplify Eq. (26.6) somewhat. We shall use the fact that

$$\nabla \times (\nabla \times \boldsymbol{E}) = -\Delta \boldsymbol{E} + \nabla (\nabla \boldsymbol{E}) \approx -\Delta \boldsymbol{E} - 2\nabla (\boldsymbol{E} \nabla n').$$

The last equation is a consequence of the relation $\nabla \boldsymbol{D} = \nabla (\boldsymbol{E} + 2n'\boldsymbol{E}) = 0$, so that

$$\nabla \boldsymbol{E} = -2\nabla n'\boldsymbol{E} = -2\boldsymbol{E} \cdot \nabla n' - 2n'\nabla \boldsymbol{E} \approx -2\boldsymbol{E} \cdot \nabla n'$$

to within small terms of the order of $n'^2$. Bearing this result in mind, and eliminating the quantity $\boldsymbol{D}$ from the left-hand side of the first equation in Eq. (26.6) with the aid of the second equation, we obtain

$$(\Delta + \kappa^2) E_j = -2\kappa^2 n' E_j - 2 \frac{\partial E_\alpha}{\partial x_j} \frac{\partial n'}{\partial x_\alpha} - 2E_\alpha \frac{\partial^2 n'}{\partial x_j \partial x_\alpha}. \tag{26.18}$$

Let us now estimate the orders of magnitude of the terms on the right-hand side of this equation. Since the inhomogeneities in the field $n'(\boldsymbol{x})$ are due to the turbulence, we can assume that $\frac{\partial n'}{\partial x_j} \sim \frac{n'}{l_{\text{turb}}}$ where $l_{\text{turb}}$ represents the length scale of the turbulence components contributing appreciably to the electromagnetic field fluctuations [the quantity $l_{\text{turb}}$ will be defined more precisely below; its value is at any rate not less than the internal turbulence length scale $(\nu^3/\bar{\varepsilon})^{1/4}$)]. Next, we shall confine our attention to fluctuations in the characteristics of sufficiently short electromagnetic waves whose wave lengths are small in comparison with $l_{turb}$, i.e., we shall adopt the condition

$$\lambda \ll l_{turb}. \tag{26.19}$$

When this is satisfied, the values of $\partial E/\partial x_j$ are of an order not greater than the order of $|E|/\lambda$ so that the orders of magnitude of the three terms on the right-hand side of Eq. (26.18) are respectively $2n'|E|/\lambda^2$, $2n'|E|/\lambda l_{turb}$ and $2n'|E|/l^2_{turb}$. Therefore, the second and third terms, which are responsible for the change in the polarization of the electromagnetic wave, turn out to be smaller than the first term by factors of $\lambda/l_{turb}$ and $(\lambda/l_{turb})^2$ respectively, and can therefore be neglected in comparison with the first term.[2] Equation (26.18) then assumes the form

$$(\Delta + \kappa^2) E = -2\kappa^2 n' E, \tag{26.20}$$

where $E$ represents any of the Cartesian components of the vector $\boldsymbol{E}$.

As before, we shall consider the case of a volume $V$ with nonzero refractive-index fluctuations $n'$, which intercepts a plane wave $E_0 = A_0 e^{iS_0}$ where $S_0 = \boldsymbol{\kappa x}$. The incident electric field is distorted by the optical inhomogeneities in $V$, and assumes the form $E = Ae^{iS} = e^{\psi}$ where $A$ is the amplitude, $S$ is the phase, and $\psi = \ln A + iS$ is the so-called eikonal which is given by $\psi_0 = \ln A_0 + iS_0$ for the incident wave $E_0$ (occasionally the eikonal is also defined as $\psi/i$). We shall investigate fluctuations in the eikonal, i.e., $\psi' = \psi - \psi_0 = \ln(A/A_0) + i(S - S_0)$, and will use the notation

$$\psi' = \chi' + iS',$$
$$\chi' = \mathrm{Re}\,\psi' = \ln(A/A_0), \quad S' = \mathrm{Im}\,\psi' = S - S_0. \tag{26.21}$$

The quantities $\chi'$ and $S'$ describe the fluctuations in the amplitude and phase of the electromagnetic wave in which we are interested.

Substituting $E = e^{\psi_0 + \psi'} = A_0 e^{i\boldsymbol{\kappa x}} e^{\psi'}$ in Eq. (26.20), we obtain the following equation:

[2] According to the estimate of Tatarskii (1967a) the relative intensity of the depolarized light component [i.e., of the component produced by the neglected terms of Eq. (26.18)] is usually of the order of $10^{-18}$ at a distance of 1 km from a light source placed in an atmosphereic surface layer.

$$\Delta\psi' + (2i\kappa + \nabla\psi') \cdot \nabla\psi' = -2\kappa^2 n'.$$

This is a nonlinear equation which can be linearized by assuming that the absolute magnitude of the second term in parenthesis on the left-hand side is small in comparison with the absolute magnitude of the first term (which is equal to $2\kappa = 4\pi/\lambda$). This condition can be written in the form

$$\lambda|\nabla\psi'| \ll 4\pi. \tag{26.22}$$

It is equivalent to the requirement that the changes in $\psi'$ occurring over distances of the order of the wavelength $\lambda$ are small, i.e., that the field $\psi'(\boldsymbol{x})$ is sufficiently smooth. Let us note that it is not necessary to demand that the fluctuations $|\psi'|$ are small. If Eq. (26.22) is satisfied, we can then rewrite the equation for $\psi'$ in the form

$$\Delta\psi' + 2i\kappa\, \nabla\psi' = -2\kappa^2 n'. \tag{26.23}$$

The above linearization of the eikonal equation (sometimes referred to as the method of smooth perturbations) was put forward by Rytov (1937) in connection with the diffraction of light by ultrasonic waves. It was first used by Obukhov (1953) in connection with fluctuations in wave characteristics in a turbulent atmosphere. We note that more detailed analysis would show that when the statistical characteristics of fluctuations in the eikonal are calculated, it is not always possible to neglect the nonlinear term $|\nabla\psi'|^2$. Analysis of corrections due to this nonlinear term shows that the linearization procedure which we have employed is admissible only when the mean square of the quantity $\chi'$ is small, but according to observational data it is sufficient in practice to ensure that $\overline{\chi'^2} < 1/2$ (see below, Sect. 26.5).

Equation (26.23) can be rewritten in the form

$$(\Delta + \kappa^2)\, \psi' e^{i\kappa x} = -2\kappa^2 n' e^{i\kappa x}, \tag{26.24}$$

which is analogous to Eq. (26.11). Hence it is clear that the solution of Eq. (26.23) is

$$\psi'(\boldsymbol{x}) = \frac{\kappa^2}{2\pi} \int_V n'(\boldsymbol{x}_1) \frac{e^{-i\kappa(x - x_1) + i\kappa|\boldsymbol{x} - \boldsymbol{x}_1|}}{|\boldsymbol{x} - \boldsymbol{x}_1|}\, d\boldsymbol{x}_1. \tag{26.25}$$

If we demand that the fluctuations $E' = E - E_0$ should be small, we obtain a formula for $E'$ which is similar to Eq. (26.25). However, the condition that the fluctuations $E'$ should be small is a more stringent limitation than the requirement that $\psi'$ should be small, since the latter is the fluctuation in the logarithm of the field. We note also that in spite of the fact that the formulas for $\psi'$ and $E'$ are the same, the results given by these formulas are in fact different because the probability distributions for the amplitude fluctuations obtained from them are different (in particular, the so-called Rayleigh distribution is obtained if Eq. (26.25) is used for $E'$, and the log-normal distribution if Eq. (26.25) is used for $\psi'$). We shall see below that existing experimental data cannot be explained by the Rayleigh distribution, but are in good agreement with the log-normal distribution.

The formula (26.25) can be used to interpret fluctuations in the eikonal $\psi'$ at a given point $\boldsymbol{x}$ as the result of the superposition of scattered waves arriving at the point $\boldsymbol{x}$ from different parts of the volume $V$ For points $\boldsymbol{x}$ lying outside $V$ at large distances from it (in comparison with its linear dimensions), the integral in Eq. (26.25) can be simplified by analogy with the simplification introduced into the integral (26.12) subject to Eq. (26.13). We now shall consider in detail the amplitude and phase fluctuations at internal points $\boldsymbol{x}$ of $V$ at which this simplification is unjustified. However, at such points the integral (26.25) can still be simplified to some extent, since large-angle scattering can be neglected when Eq. (26.19) is satisfied, and when Eq. (26.25) is evaluated it is sufficient to include only contributions to $\psi'(\boldsymbol{x})$ due to waves scattered through angles not exceeding $\theta = \lambda/l_{\text{turb}} \ll 1$. In other words, the integration in Eq. (26.25) can be confined to that part of $V$ which lies within the cone $K(\boldsymbol{x})$ with the apex at $\boldsymbol{x}$ and angle $\theta$.

It will be convenient to take the direction of propagation of the incident wave as the $Ox$ axis (so that $\boldsymbol{\kappa x} = \kappa x$), and to assume that the volume $V$ lies within the region $x > 0$, the plane $x = 0$ being the boundary of this volume. The cone $K(\boldsymbol{x})$ will be the range of values $\boldsymbol{x}_1 = (x_1, y_1, z_1)$, defined by

$$x - x_1 \geqslant 0, \quad \rho/(x - x_1) \leqslant \theta \ll 1, \tag{26.26}$$

where $\rho = \sqrt{(y - y_1)^2 + (z - z_1)^2}$. Let us expand the quantity $|\boldsymbol{x} - \boldsymbol{x}_1|$ in a series in powers of $\rho^2/(x - x_1)^2$. We shall indicate the abscissa $x$ of the observation point $\boldsymbol{x}$ by the letter $L$, so that we then have the result that the quadratic terms in the above series will have values of the order of $(x-x_1)[\rho^2/(x-x_1)^2]^2 \leqslant L(\lambda/l_{turb})^4$ and will contribute to the argument of the exponential function $e^{i\kappa|\boldsymbol{x}-\boldsymbol{x}_1|}$ terms of the order of $\kappa L(\lambda/l_{turb})^4 \sim L\lambda^3/l_{turb}^4$. We shall use the condition

$$L \ll L_{crit}(l_{turb}/\lambda)^2, \quad L_{crit} = l_{turb}^2/\lambda \tag{26.27}$$

(the significance of the quantity $L_{crit}$ will be explained below). Subject to this condition, it will be sufficient to take into account only linear terms in the series for $|\boldsymbol{x}-\boldsymbol{x}_1|$ in powers of $\rho^2/(x-x_1)^2$, which are of the form $x-x_1+\frac{\rho^2}{2(x-x_1)}$, so that we can write

$$\exp\{-i\kappa(x-x_1)+i\kappa|\boldsymbol{x}-\boldsymbol{x}_1|\} \approx \exp\left\{\frac{i\kappa\rho^2}{2(x-x_1)}\right\}.$$

In the denominator of the integrand in Eq. (26.25) we can let $|\boldsymbol{x}-\boldsymbol{x}_1| \approx x-x_1$, so that if we denote the intersection of the volumes $V$ and $K(\boldsymbol{x})$ by the symbol $V \cdot K(\boldsymbol{x})$ then Eq. (26.25) will assume the form

$$\psi'(\boldsymbol{x}) = \frac{\kappa^2}{2\pi} \int_{V \cdot K(\boldsymbol{x})} n'(\boldsymbol{x}_1) \frac{\exp\left\{\frac{i\kappa\rho^2}{2(x-x_1)}\right\}}{x-x_1} d\boldsymbol{x}_1. \tag{26.28}$$

We note that the function $(x-x_1)^{-1}\exp\left\{\frac{i\kappa\rho^2}{2(x-x_1)}\right\}$ oscillates rapidly outside the cone $K(\boldsymbol{x})$, so that for a sufficiently smooth variation of $n'(\boldsymbol{x}_1)$, which is ensured by Eq. (26.19), integration over the region $x-x_1 \geqslant 0$, which lies outside the cone $K(\boldsymbol{x})$, provides a negligible contribution. The integral over the region $V \cdot K(\boldsymbol{x})$ can therefore be replaced by an integral over the part of $V$ which lies inside the plane layer $0 \leqslant x_1 \leqslant x$. The function $\psi'(\boldsymbol{x})$ will then be the exact solution of

$$\frac{\partial^2\psi'}{\partial y^2} + \frac{\partial^2\psi'}{\partial z^2} + 2i\kappa\frac{\partial\psi'}{\partial x} = -2\kappa^2 n', \tag{26.29}$$

which is obtained from Eq. (26.23) by neglecting the term $\frac{\partial^2\psi'}{\partial x^2}$ on the left-hand side.

The formula given by Eq. (26.25), and its simplified versions given by Eqs. (26.28) and by equation (26.32) below, show that the fluctuations $\psi'(\boldsymbol{x})$ are given by an integral of $n'$ multiplied by certain weighting functions and evaluated over the scattering volume $V$. If the linear dimensions $L$ of this volume are very large in comparison with the external turbulence length scale $L_0$, then the volume $V$ can be represented by the sum of a large number of parts with linear dimensions of the order of $L_0$. Under these conditions, the function $\psi'(\boldsymbol{x})$ can be written as the sum of a large number of practically uncorrelated terms, and by the central limit theorem of probability theory, it is reasonable to expect that the random variable $\psi'(\boldsymbol{x})$ (for fixed $\boldsymbol{x}$) will have a normal distribution. Consequently, in the case of fluctuations in the light intensity $I$, we may expect that the random variable $\ln(I/I_0) = 2\ln(A/A_0) = 2\chi'$ is normally distributed, and the intensity $I$ itself is log-normally distributed. This conclusion has been satisfactorily confirmed by measurements of the scintillations of an artificial light source in the surface layer of the atmosphere carried out by the 1956–57 expeditions of the USSR Institute of Atmospheric Physics, [Gurvich, Tatarskii and Tsvang (1958, 1959), Tatarskii (1971)]. The measurements were carried out at heights of between 1.5 and 5 m, and at distances of 250–2000 m from the light source (in conditions when $\chi'^2 < 0.5$ for all experiments). The integral distribution function for the light intensity at a given distance from the source, $F(I)$, was determined. The above discussion then suggests that $F(I) = \Phi(\ln(I/I_0)/\sigma)$ where $\Phi$ is the normalized normal integral probability distribution function, and $\sigma^2 = \overline{(\ln(I/I_0))^2} = 4\chi'^2$. In that case, $\Phi^{-1}[F(I)]$ should be linear in $\ln(I/I_0)$. The corresponding empirical graph, which is given in Fig. 105, does in fact confirm this conclusion.

If we replace the range of integration in Eq. (26.28) by that part of $V$ which lies inside the layer $0 < x < L$, we can rewrite this formula for the point $\boldsymbol{x} = (L, y, z)$ in the form

$$\psi'(L, y, z) = i\kappa \int_0^L dx_1 \iint_{-\infty}^{\infty} n'(x_1, y_1, z_1) \left[\frac{\exp(-\rho^2/2\sigma^2)}{2\pi\sigma^2}\right] dy_1\, dz_1, \qquad (26.30)$$

where $\sigma^2 = i(L - x_1)/\kappa$. Hence it is clear that the value of $\psi'$ at the point $x_1 = L$, $y_1 = y$, $z_1 = z$ depends on the refractive-index inhomogeneities not only on the segment $0 < x_1 < L$ of the straight line $y_1 = y$, $z_1 = z$, but also on a certain region around this segment. This is so because in the plane $x_1 = L$ we observe not simply the geometric shadows of the inhomogeneities lying in the layer $0 < x_1 < L$, but the shadows "spread" by diffraction. The factor in brackets in the integrand describes this diffraction spread. Its length scale is of the order of $|\sigma| \sim \sqrt{\lambda L}$. The length $\sqrt{\lambda L}$ is, in this case, the radius of the so-called Fresnel zone, i.e., a circular region on the surface of the incident wave front (in

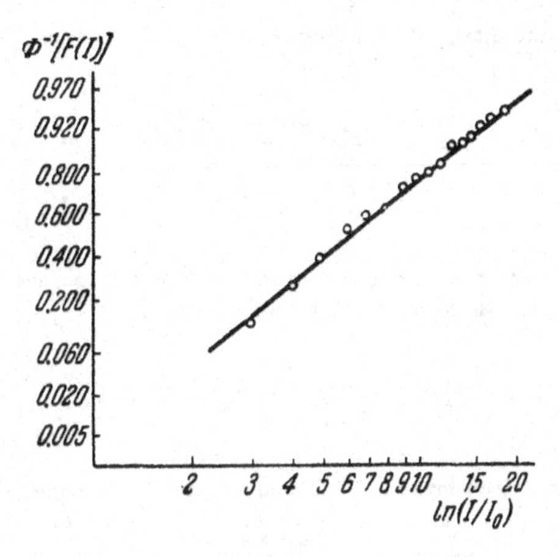


**Fig. 105** Probability distribution for fluctuations in the light intensity from a source located in the atmospheric surface layer.

our case, a plane wave), the center of which lies on the straight line joining the observation point and the light source (at a distance of $L$ from the observation point, the path difference between the edges and the center of the zone being $\lambda/2$. If the length scale of the diffraction "spread" $\sqrt{\lambda L}$ is small in comparison with the size of the geometric shadows, i.e., the scale $l_{turb}$ of the inhomogeneities in the refractive index, the diffraction distortion of the geometric shadows can be neglected. If we recall that $L_{crit}$ is given by Eq. (26.27), we can rewrite the condition $\sqrt{\lambda L} \ll l_{turb}$ in the form

$$L \ll L_{crit}. \tag{26.31}$$

Therefore, electromagnetic wave propagation in a turbulent medium through sufficiently small distances $L$ which satisfy Eq. (26.31) can be described by neglecting diffraction effects, i.e., using the geometric optics approximation. Fluctuations in the amplitude and phase of radio waves have been considered by Bergman (1946) in this approximation, and in more detail by Kasil'nikov (1947, 1949a).

The condition (26.31) was obtained above with the aid of Eq. (26.30), which enabled us to estimate the order of magnitude of the diffraction "spread" length scale $|\sigma|$. However, this condition can be derived in a more instructive way. Namely, for a diffraction angle $\theta \sim \lambda/l_{turb}$, the width of the diffraction "spread" of the boundaries of the geometric shadows at a distance $L$ from the objects producing these shadows is $\theta L \sim \lambda L/l_{turb}$. By requiring that this width should be much less than the size $l_{turb}$ of the shadows themselves, we again obtain Eq. (26.31).

If we neglect the diffraction effects, this is formally equivalent to going to the limit $\sigma \to 0$ in the inner integral in Eq. (26.30). However, the factor in the brackets in the integrand then takes the form of the delta function $\delta(y - y_1)\,\delta(z - z_1)$, and $\psi'$ turns out to be purely imaginary, i.e., we lose the amplitude fluctuations. To allow for these fluctuations, we must retain the next term in the expansion in powers of $\sigma$.

To do this, we must rewrite the inner integral in Eq. (26.30) in the form

$$I = \frac{1}{2\pi} \int\int n'(x_1, y - \sigma\eta, z - \sigma\zeta) \left\{ \exp\left( - \frac{\eta^2 + \zeta^2}{2} \right) \right\} d\eta\, d\zeta$$

and expand $n'$ in the integrand in a series in powers of $\sigma$, retaining only the linear and quadratic terms:

$$n'(x_1, y-\sigma\eta, z-\sigma\zeta) \approx \left[1-\sigma\left(\eta\frac{\partial}{\partial y}+\zeta\frac{\partial}{\partial z}\right)+\frac{\sigma^2}{2}\left(\eta^2\frac{\partial^2}{\partial y^2}+2\eta\zeta\frac{\partial^2}{\partial y\,\partial z}+\right.\right.$$
$$\left.\left.+\zeta^2\frac{\partial^2}{\partial z^2}\right)\right]n'(x_1, y, z).$$

The use of this expression enables us to integrate with respect to $\eta$ and $\zeta$ in the formula for $I$, so that the final expression can be written in the form

$$I \approx \left(1+\frac{\sigma^2}{2}\Delta_{y,z}\right)n'(x_1, y, z).$$

Substituting this last result into Eq. (26.30), and then substituting for $\sigma^2$, we finally obtain

$$\psi'(L, y, z) = i\kappa\int_0^L dx_1\left[n'(x_1, y, z)+\frac{i}{2\kappa}(L-x_1)\Delta_{y,z}n'(x_1, y, z)\right]. \tag{26.32}$$

This formula is valid when Eq. (26.31) is satisfied, and can be regarded as the definition of $\psi'$ in the geometric-optics approximation. Substituting $\psi' = \chi' + iS'$, we can readily verify that $\chi'$ and $S'$ satisfy the equations

$$\frac{\partial S'}{\partial x} = \kappa n', \quad \frac{\partial^2 S'}{\partial y^2}+\frac{\partial^2 S'}{\partial z^2}+2\kappa\frac{\partial \chi'}{\partial x} = 0, \tag{26.33}$$

which are linearized equations of geometric optics for our problem.

## 26.2 Sound Propagation in a Turbulent Atmosphere

The effect of turbulence on sound propagation in the atmosphere was considered by Obukhov (1941c), Blokhintsev (1945, 1946), Krasil'nikov (1945, 1947), Pekeris (1947), Ellison (1951), Kraichnan (1953), Mintzer (1953–1954), and others. Turbulence has a two-fold effect on propagation of sound waves in the atmosphere. First, since the velocity of sound $c$ depends on the air temperature $T$ (namely, $c = \sqrt{\gamma RT}$ where $R$ is the gas constant and $\gamma = c_p/c_v$ is the ratio of specific heats), the turbulent temperature fluctuations lead to fluctuations in the velocity of sound. Since the temperature fluctuations $T'$ are always small in comparison with the absolute temperature $\overline{T}$ (which is of the order of 300°K), we can assume that

$$c^2 \approx c_0^2(1+T'/\overline{T}), \tag{26.34}$$

where $c_0^2 = \gamma R\overline{T}$. Hence, the atmosphere behaves as if it were a medium with weak random inhomogeneities (which are, however, stronger than the optical inhomogeneities described by the refractive index fluctuations $n'$). Second, sound waves are transported by air motions, and therefore the presence of turbulent velocity fluctuations $u_i'$ produces additional random distortions of the sound wave front.

The propagation of sound waves in a turbulent atmosphere is described by the acoustic equations for a moving inhomogeneous medium. We shall derive these equations from the condition that the motion is adiabatic, and that it can be described by the nonviscous Euler equations.[3] The adiabatic equation will be written in the form

$$\frac{dp}{dt} = \frac{\gamma p}{\rho}\frac{d\rho}{dt} = -\gamma p \nabla \boldsymbol{u}.$$

where we have used the continuity equation $d\rho/dt = -\rho\nabla\boldsymbol{u}$. Next, it will be convenient to replace pressure $p$ by the function $\Pi = \ln p$, so that $dp = \gamma p\, d\Pi$ and $\nabla p/\rho = c^2\nabla\Pi$. By virtue of Eq. (26.34), we can rewrite the adiabatic and Euler equations in the form

$$\begin{aligned} &\frac{\partial \boldsymbol{u}}{\partial t} + c_0^2\nabla\Pi = -(\boldsymbol{u}\cdot\nabla)\,\boldsymbol{u} - c_0^2\frac{T'}{\overline{T}}\nabla\Pi, \\ &\frac{\partial\Pi}{\partial t} + \nabla\boldsymbol{u} = -(\boldsymbol{u}\cdot\nabla)\,\Pi. \end{aligned} \tag{26.35}$$

We shall use the notation $\boldsymbol{u} = \boldsymbol{u}' + \boldsymbol{v}$ and $\Pi = \Pi' + \pi$, where $\boldsymbol{u}'$ and $\Pi'$ refer to the turbulent motion, and $\boldsymbol{v}$ and $\pi$ to the sound vibrations. Since the amplitude of the sound wave is small, Eq. (26.35) can be linearized by neglecting quadratic combinations of the acoustic parameters. The linearized equations assume the form

$$\begin{aligned} &\frac{\partial \boldsymbol{v}}{\partial t} + c_0^2\nabla\pi = -(\boldsymbol{u}'\cdot\nabla)\,\boldsymbol{v} - (\boldsymbol{v}\cdot\nabla)\,\boldsymbol{u}' - c_0^2\frac{T'}{\overline{T}}\nabla\pi, \\ &\frac{\partial\pi}{\partial t} + \nabla\boldsymbol{v} = -(\boldsymbol{u}'\cdot\nabla)\,\pi - (\boldsymbol{v}\cdot\nabla)\,\Pi'. \end{aligned} \tag{26.36}$$

These are, in fact, the equations of acoustics for a moving

[3] We shall use the Euler equations only for the sound vibrations in a fluid (and will thus neglect the attenuation of sound due to molecular viscosity and thermal conductivity), but will not employ them to describe turbulent motions (which in general require the use of the Navier-Stokes equations).

inhomogeneous medium. The unknown quantities are the characteristics of the sound vibration $\boldsymbol{v}$ and $\pi$, while the coefficients $c_0^2$, $\bar{T}$, $\boldsymbol{u}'$, $\Pi'$, $T'$ are looked upon as given functions of the coordinates and time ($c_0^2$ and $\bar{T}$ are nonrandom, and $\boldsymbol{u}'$, $\Pi'$, $T'$ are random functions with known statistical properties.

We shall investigate the sound wave propagation in a turbulent atmosphere for distances $L$ that are not too large and satisfy the condition $L/c_0 \ll 2\pi/\omega_{turb}$ which is the analog of Eq. (26.4). The coefficients in Eq. (26.36) can then be regarded as time-dependent, and these equations will have special solutions in the form of monochromatic waves of fixed frequency $\omega$, which are described by the functions $\mathrm{Re}\,(\boldsymbol{v}e^{-i\omega t})$ and $\mathrm{Re}\,(\pi e^{-i\omega t})$ where $\boldsymbol{v}$ and $\pi$ are the time-independent complex amplitudes. We shall confine our attention to those cases in which the change in the mean temperature and mean wind velocity over distances $r \ll L$ can be regarded as negligible, and will hence assume that $\bar{T} = \text{const}$, $c_0^2 = \text{const}$, $\overline{\boldsymbol{u}'} = \text{const}$. in Eq. (26.36). It is also convenient to replace the frequencies $\omega$ by the wave numbers $\kappa = \omega/c_0$, so that from Eq. (26.36) we obtain the following equations for the complex amplitudes of the monochromatic sound waves

$$
\begin{aligned}
-i\kappa\boldsymbol{v} + c_0\nabla\pi &= -\frac{1}{c_0}(\boldsymbol{u}'\cdot\nabla)\boldsymbol{v} - \frac{1}{c_0}(\boldsymbol{v}\cdot\nabla)\boldsymbol{u}' - c_0\frac{T'}{\bar{T}}\nabla\pi, \\
-i\kappa\pi + \frac{1}{c_0}\nabla\boldsymbol{v} &= -\frac{1}{c_0}(\boldsymbol{u}'\cdot\nabla)\pi - \frac{1}{c_0}(\boldsymbol{v}\cdot\nabla)\Pi'.
\end{aligned}
\tag{26.37}
$$

Since the constant transport ($\boldsymbol{u}' = \text{const}$) leads merely to a Doppler frequency shift (replacement of $\omega = c_0\kappa$ by $\omega = c_0\kappa + \boldsymbol{u}'\cdot\boldsymbol{\kappa}$ where $\boldsymbol{\kappa}$ is the wave vector) we can restrict our analysis to the case where the mean wind velocity is zero, so that $\boldsymbol{u}'$, $T'$ and $\Pi'$ can be interpreted as the turbulent fluctuations with zero means). We note that $\Pi'$ is of the order of $p'/\bar{\rho}c_0^2 \sim (u'/c_0)^2$ (since the amplitude of the turbulent pressure fluctuations can be roughly estimated from the formula $p' \sim \bar{\rho}u'^2$). Hence it is clear that the second term on the right of the second equation in Eq. (26.37) is of a lower order of magnitude than the second term on the left by at least by a factor of $(u'/c_0)^2$, while the order of magnitude of the first term on the right is lower than the order of magnitude of the first term on the left by a factor of only $u'/c_0$. Therefore, the term which contains $\Pi'$ will be neglected below. Comparing the first two terms on the right of the first equation in Eq. (26.37) with the first term on the left, and the third term on the right with the second term on the left, will show that the terms on the right are smaller by a factor of at least $u'/c_0$ or $T'/\bar{T}$ (in the atmosphere, the quantities $u'/c_0$ and $T'/\bar{T}$ are of the same order of magnitude).

When there is no turbulence, the right-hand sides of Eq. (26.37) will vanish, and these equations will have the plane-wave solutions $\pi = Ae^{i\boldsymbol{\kappa x}}$, $\boldsymbol{v} = c_0\pi\boldsymbol{\kappa}/\kappa$ where $A$ is a constant (the quantity $\bar{\rho}c_0^2 A$ can be interpreted as the amplitude of pressure oscillations in the plane wave). In the ensuing analysis we shall need the energy flux density $\boldsymbol{s}$ averaged over one period of the oscillations $2\pi/\omega$. For a plane sound wave this is given by $\boldsymbol{s} = c_0\mathscr{E}\boldsymbol{\kappa}/\kappa$ where the mean energy density $\mathscr{E}$ is equal to twice the kinetic-energy density. Since the kinetic-energy density for a plane sound wave is

$$\frac{1}{2}\bar{\rho}\,|\,\mathrm{Re}\,(ve^{-i\omega t})\,|^2 = \frac{1}{2}\,\bar{\rho}c_0^2\,|\,\mathrm{Re}\,(\pi e^{-i\omega t})\,|^2,$$

we have

$$\boldsymbol{s} = \bar{\rho}c_0^3\,\frac{\omega}{2\pi}\int_0^{2\pi/\omega} |\,\mathrm{Re}\,(\pi e^{-i\omega t})\,|^2\,dt \cdot \frac{\boldsymbol{\kappa}}{\kappa},$$

and hence, after some simple rearrangement,

$$\boldsymbol{s} = \bar{\rho}c_0^3\pi\pi^*\boldsymbol{\kappa}/2\kappa. \tag{26.38}$$

By analogy with Sect. 26.1, we shall consider the propagation of sound waves in a turbulent atmosphere in the following simplified formulation. We shall assume that the turbulence is concentrated in a volume $V$, and that this volume intercepts the plane sound wave $\pi_0 = A_0e^{i\boldsymbol{\kappa x}}$, $\boldsymbol{v}_0 = c_0\pi_0\boldsymbol{\kappa}/\kappa$. The turbulence localized in $V$ will distort the sound field in all space, and this will take the form $\pi = \pi_0 + \pi'$, $\boldsymbol{v} = \boldsymbol{v}_0 + \boldsymbol{v}'$ where $\pi'$, $\boldsymbol{v}'$ represent the scattered waves. Since the fields of turbulent fluctuations $\boldsymbol{u}'(\boldsymbol{x})$, $T'(\boldsymbol{x})$, $\Pi'(\boldsymbol{x})$ are random, the field $\pi'(\boldsymbol{x})$, $\boldsymbol{v}'(\boldsymbol{x})$ which characterizes the scattered waves will also be random. Our problem will be to investigate the properties of the random functions $\pi'(\boldsymbol{x})$, $\boldsymbol{v}'(\boldsymbol{x})$.

Since the right-hand sides of Eq. (26.37) are much smaller (by a factor of at least $u'/c_0$ in order of magnitude) than the individual terms of the left-hand sides of these equations, we may suppose that

$$|\pi'(\boldsymbol{x})| \ll |\pi_0(\boldsymbol{x})| = A_0, \quad |\boldsymbol{v}'(\boldsymbol{x})| \ll |\boldsymbol{v}_0(\boldsymbol{x})| = c_0 A_0, \quad (26.39)$$

where the symbol "much smaller than" means "smaller by a factor of $u'/c_0$ in order of magnitude." Therefore, substituting $\pi = \pi_0 + \pi'$ and $\boldsymbol{v} = \boldsymbol{v}_0 + \boldsymbol{v}'$ in Eq. (26.37), we can neglect all terms containing $\pi'$, $\boldsymbol{v}'$ on the right of these equations. Moreover, since $\pi_0$, $\boldsymbol{v}_0$ satisfies the homogeneous equations given by Eq. (26.37), we obtain the following equations for $\pi'$, $\boldsymbol{v}'$:

$$\begin{aligned} -i\kappa\boldsymbol{v}' + c_0\nabla\pi' &= -\frac{1}{c_0}(\boldsymbol{u}' \cdot \nabla)\boldsymbol{v}_0 - \frac{1}{c_0}(\boldsymbol{v}_0 \cdot \nabla)\boldsymbol{u}' - c_0\frac{T'}{\overline{T}}\nabla\pi_0, \\ -i\kappa\pi' + \frac{1}{c_0}\nabla\boldsymbol{v}' &= -\frac{1}{c_0}(\boldsymbol{u}' \cdot \nabla)\pi_0, \end{aligned} \quad (26.40)$$

where the right-hand sides are known (random) functions. Substituting $\pi_0 = A_0 e^{i\boldsymbol{\kappa x}}$, $\boldsymbol{v}_0 = c_0\pi_0\frac{\boldsymbol{\kappa}}{\kappa}$ into the right-hand sides of these equations, we obtain

$$\begin{aligned} -i\kappa\boldsymbol{v}' + c_0\nabla\pi' &= -ic_0\kappa\left[\left(\frac{u_k'}{c_0} + \frac{T'}{\overline{T}}\right)\frac{\boldsymbol{\kappa}}{\kappa} - \frac{i}{\kappa}\frac{\partial}{\partial x_k}\frac{\boldsymbol{u}'}{c_0}\right]\pi_0, \\ -i\kappa\pi' + \frac{1}{c_0}\nabla\boldsymbol{v}' &= -i\kappa\frac{u_k'}{c_0}\pi_0, \end{aligned}$$

where $x_k$ is the coordinate in the direction of propagation of the incident wave (specified by the wave vector $\boldsymbol{\kappa}$), and $u_k'$ is the corresponding component of the velocity fluctuation $\boldsymbol{u}'$. Let us eliminate the variable $\boldsymbol{v}'$ from these equations. To do this, we must first calculate the divergence of the left- and right-hand sides of the first equation, using the formula $\nabla(\varphi\boldsymbol{a}) = \varphi\nabla\boldsymbol{a} + (\boldsymbol{a} \cdot \nabla)\varphi$. As a result of this calculation, one of the terms on the right will contain the quantity $\nabla\boldsymbol{u}'$. We shall neglect this term, assuming that $\nabla\boldsymbol{u}' = 0$, i.e., that the turbulent air motion is incompressible. Having eliminated $\boldsymbol{v}'$ we obtain the following equation for $\pi'$:

$$(\Delta + \kappa^2)\pi' = -2\kappa^2 n'\pi_0, \quad (26.41)$$

in which

$$n' = -\left(1 - \frac{i}{\kappa}\frac{\partial}{\partial x_k}\right)\left(\frac{u_k'}{c_0} + \frac{1}{2}\frac{T'}{\overline{T}}\right) \quad (26.42)$$

(no summation over $k$). Equation (26.41) is similar to Eq. (26.20), and it is clear that $n'$ plays the role of the fluctuation in the refractive index of the turbulent atmosphere for the sound waves. This equation was obtained by Monin (1961), using the above method of expansion in terms of the small parameter $u'/c_0$.

We shall now use Eq. (26.41) to describe the scattering of a plane sound wave by a turbulent volume $V$, and will confine our attention to the calculation of $\pi'(\boldsymbol{x})$ at large distances $x$ from the scattering volume, i.e., in the zone in which Eq. (26.13) is satisfied and the scattered wave can be regarded as plane in small regions of space. By analogy with the procedure adopted to obtain the solution (26.14) of Eq. (26.11), we have

$$\pi'(\boldsymbol{x}) = \frac{\kappa^2 A_0 e^{i\kappa|\boldsymbol{x}|}}{2\pi|\boldsymbol{x}|} G, \qquad G = \int_V n'(\boldsymbol{x}_1)\, e^{i(\boldsymbol{\kappa}-\kappa\boldsymbol{q})\boldsymbol{x}_1}\, d\boldsymbol{x}_1, \tag{26.43}$$

where $\boldsymbol{q} = \boldsymbol{x}/|\boldsymbol{x}|$ is a unit vector in the direction of scattering. In the above zone, the energy flux density in the scattered wave can be calculated as if it were plane, i.e., we can use Eq. (26.38). This yields

$$\boldsymbol{s}'(\boldsymbol{x}) = \frac{\bar{\rho} c_0^3 \kappa^4 A_0^2}{8\pi^2 |\boldsymbol{x}|^2} G G^* \boldsymbol{q}. \tag{26.44}$$

From Eqs. (26.16), (26.38), and (26.44) we obtain the following expression for the scattering cross-section of the volume $V$ in the direction of $\boldsymbol{q}$:

$$d\sigma = \frac{\kappa^4}{4\pi^2} G G^* \, d\Omega, \tag{26.45}$$

where $d\Omega$ is an infinitesimal solid angle with the apex in $V$ and containing the direction of $\boldsymbol{q}$.

Equation (26.41) can also be used to calculate fluctuations in the amplitude and phase of a sound wave in a turbulent atmosphere. Since $(\Delta + \kappa^2)\pi_0 = 0$, the left-hand side of (26.41) can be rewritten in the form $(\Delta + \kappa^2)\pi$, while the right-hand side can be replaced by $-2\kappa^2 n'\pi$, i.e., we can add the small term $-2\kappa^2 n'\pi'$ to it. This introduces a relative error of the order of $u'/c_0$ into the right-hand side, but we have allowed an error of this order before in the transition from Eq. (26.37) to Eq. (26.40). Therefore, equation

(26.41) can be rewritten in the form

$$(\Delta + \kappa^2)\,\pi = -2\kappa^2 n'\pi, \tag{26.46}$$

which is quite similar to Eq. (26.20). Therefore, fluctuations in the amplitude and phase of sound waves can be described in precisely the same way as in the case of electromagnetic waves (the only difference is in the form of the fluctuations $n'$ in the refractive index). Thus, substituting $\pi = \pi_0 e^{\psi'}$, and using Eq. (26.22), we can describe the fluctuations $\psi'(\boldsymbol{x})$ in the eikonal for the sound waves by the formula (26.25). When Eqs. (26.19) and (26.27) are satisfied, the quantity $\psi'(\boldsymbol{x})$ can be determined from Eq. (26.29).

## 26.3 Turbulent Scattering of Electromagnetic and Sound Waves

To describe the scattering of a plane electromagnetic wave or sound wave by the volume $V$ containing the turbulence, we must determine the mean scattering cross-section $\overline{d\sigma}$ of this volume (for any direction $\boldsymbol{q}$). The quantity $d\sigma$ is determined by Eq. (26.17) in the case of electromagnetic waves, and by Eq. (26.45) in the case of sound waves. According to these formulas, to determine $\overline{d\sigma}$ it is sufficient to calculate the quantity $\overline{GG^*}$ where

$$G = \int_V n'(\boldsymbol{x}) \exp\{i(\boldsymbol{\kappa} - \kappa\boldsymbol{q})\,\boldsymbol{x}\}\,d\boldsymbol{x} \tag{26.47}$$

is a random function of the wave vector $\boldsymbol{\kappa} - \kappa\boldsymbol{q}$. When we calculate $\overline{GG^*}$ we shall assume that the correlation function $\overline{n'(\boldsymbol{x}_1)\,n'^*(\boldsymbol{x}_2)}$ for the random field $n'(\boldsymbol{x})$, where $\boldsymbol{x}_1, \boldsymbol{x}_2 \in V$, can be approximately written in the form[4]

$$\overline{n'(\boldsymbol{x}_1)\,n'^*(\boldsymbol{x}_2)} = B_{nn}(\boldsymbol{x}_1 - \boldsymbol{x}_2) = \int \exp\{-i\boldsymbol{k}(\boldsymbol{x}_1 - \boldsymbol{x}_2)\}\,F_{nn}(\boldsymbol{k})\,d\boldsymbol{k}, \tag{26.48}$$

[4]We note that if the random field $n'(\boldsymbol{x})$ is homogeneous, the formulas (26.48) are valid for any $\boldsymbol{x}_1, \boldsymbol{x}_2$. However, a random field which is nonzero only within the given volume $V$, cannot be homogeneous. This is the reason why we demand only that (26.48) should be approximately satisfied for $\boldsymbol{x}_1, \boldsymbol{x}_2 \in V$. The results which will be obtained in this way for values of $|\boldsymbol{\kappa} - \kappa\boldsymbol{q}|$ that are not too small, will be valid also in the more general case of a locally homogeneous randim field $n'(\boldsymbol{x})$.

where the correlation function $B_{nn}(r)$ for the refractive index, and the corresponding spectral function $F_{nn}(k)$, are regarded as given. Using the first equation in (26.48), we obtain

$$\overline{GG^*} = \int_V \int_V B_{nn}(x_1 - x_2) \exp[i(k - kq)(x_1 - x_2)] dx_1 dx_2.$$

Let us now introduce a function $\Omega_V(x)$ which is equal to one if $x$ belongs to a volume $V$ and is equal to zero for points $x$ outside $V$. Then the last equation can be rewritten in a form

$$\overline{GG^*} = \int\int B_{nn}(x_1 - x_2)\Omega(x_1)\Omega(x_2) \exp[i(k - kq)(x_1 - x_2)] dx_1 dx_2 =$$

$$= \int F(r) B_{nn}(r) \exp[i(k - kq)r] dr$$

where integration is extended over all the space and

$$F(r) = \int \Omega(x)\Omega(x + r) dx, F(0) = V.$$

Using the spectral representation (26.48) of the correlation function $B_{nn}(r)$ we obtain

$$\overline{GG^*} = \int\int F(r) F_{nn}(k) \exp[i(k - kq - k)r] dr\, dk =$$

$$= 8\pi^3 V \int F_{nn}(k) f_V(k - kq - k) dk$$

or

$$\overline{GG^*} = 8\pi^3 V \tilde{F}_{nn}(\kappa - \kappa q), \tag{26.49}$$

where $\tilde{F}_{nn}(\kappa - \kappa q)$ is taken at the point $k = \kappa - \kappa q$, and the value of the spectral density for the refractive index, smoothed out with the weight

$$f_V(k) = \frac{1}{8\pi^3 V} \int\limits_V F(r) \exp(ikr) dr.$$

The weighting function $f_V(k)$ has the following properties: (1) the integral of $f_V(\boldsymbol{k})$ over all the wave-number space is equal to unity, (2) the function $f_V(\boldsymbol{k})$ has a maximum at $\boldsymbol{k}=0$, and the maximum value is $V/8\pi^3$, (3) the values of $f_V(\boldsymbol{k})$ are appreciably different from zero only for moderate values of $k$ (its order of magnitude must not exceed $2\pi/L$ where $L$ is a characteristic linear dimension of $V$).

Let $\theta$ be the angle of scattering, i.e., the angle between the direction of $\boldsymbol{\kappa}/\kappa$ in the incident wave and the direction $\boldsymbol{q}$ in the scattered wave, so that $\cos\theta=\boldsymbol{\kappa}\cdot\boldsymbol{q}/\kappa$. We then have $|\boldsymbol{\kappa}-\kappa\boldsymbol{q}|=2\kappa|\sin\theta/2|$. If we consider values of $2\kappa|\sin\theta/2|$ that are large in comparison with $2\pi/L$, i.e., if

$$2\kappa|\sin\theta/2|\gg 2\pi/L, \tag{26.50}$$

and if we suppose that the function $F_{nn}(\boldsymbol{k})$ is sufficiently smooth in the neighborhood of the point $\boldsymbol{k}=\boldsymbol{\kappa}-\kappa\boldsymbol{q}$, then smoothing in wave-number space can be neglected in Eq. (26.49). In that case,

$$\overline{GG^*}=8\pi^3 V F_{nn}(\boldsymbol{\kappa}-\kappa\boldsymbol{q}). \tag{26.51}$$

This formula shows that only one spectral component of the turbulence contributes to scattering through the angle $\theta$, namely, the component corresponding to the wave vector $\boldsymbol{k}=\boldsymbol{\kappa}-\kappa\boldsymbol{q}$. Turbulent inhomogeneities described by this spectral component form an infinite sinusoidal diffraction grating in space with grating spacing $l(\theta)$ which is given by

$$l(\theta)=\frac{2\pi}{|\boldsymbol{\kappa}-\kappa\boldsymbol{q}|}=\frac{\lambda}{2|\sin\theta/2|}, \tag{26.52}$$

where $\lambda=2\pi/\kappa$ is the wavelength of the incident wave. The expression given by Eq. (26.52) is the well-known Bragg law in crystal diffraction theory. It gives the angle $\theta$ of a principal maximum in the diffraction pattern of a three-dimensional crystal. We note that diffraction in the direction of the principal maximum can be described as "reflection" by crystal planes perpendicular to the vector $\boldsymbol{\kappa}-\kappa\boldsymbol{q}$, where the vectors $\boldsymbol{\kappa}$ and $\boldsymbol{q}$ are at equal angles of "incidence" and "reflection" to these planes. The interpretation of the scattering cross-section formula in terms of the Bragg equation was given in particular by Rytov (1937), Obukhov (1941c), and Tatarskii (1959a, b; 1971).

The turbulent inhomogeneities localized in $V$ do in fact form a set of sinusoidal diffraction gratings with finite linear dimensions of the order of $L$. Diffraction by each of these gratings results in the formation of a diverging bundle of rays with an effective angle of divergence $\delta\theta \sim \lambda/L$. The result of this is that the complete set of diffraction gratings (with finite dimensions and close grating spacings) participates in scattering through the angle $\theta$. However, when Eq. (26.50) is satisfied, and this can be rewritten in the form $\frac{\lambda}{L} \ll 2\sin(\theta/2)$, the angular broadening of the diffraction beams can be neglected. In other words, in this case, the gratings can be regarded as infinite, and this enables us to use the approximate formula (26.51) instead of the exact formula (26.49).

The formula (26.51) was derived above under the assumption that the random field $n'(\boldsymbol{x})$ is at least approximately homogeneous in $V$. However, we have verified that this requirement is in fact not necessary because scattering through angles $\theta$ that are not too small and satisfy Eq. (26.50) is determined by a single spectral component of the turbulence, and does not depend on the properties of the remaining spectral components. In particular, when

$$2\kappa\,|\sin(\theta/2)| \gg 2\pi/L_0, \qquad (26.53)$$

where $L_0$ is the external turbulence length scale, the formula (26.51) remains valid even in the case of a locally isotropic field $n'(\boldsymbol{x})$ which we shall now consider. This formula in particular, makes possible direct determination of the three dimensional refractive index spectrum from measurements of turbulent scattering.

Let us take as an example the scattering of an electromagnetic wave in the centimeter range by a volume $V$ containing turbulence. We shall assume that the refractive index $n'$ is given by Eq. (26.3), and that the turbulence is locally isotropic. We then have[5]

$$\begin{gathered} F_{nn}(\boldsymbol{\kappa}-\kappa\boldsymbol{q}) = A^2\left[F_{TT}\left(2\kappa\sin\frac{\theta}{2}\right) + B^2 F_{\vartheta\vartheta}\left(2\kappa\sin\frac{\theta}{2}\right)\right], \\ A = 10^{-6}\,\frac{79\bar{p}}{\bar{T}^2}\left(1+\frac{15\,600\,\bar{\vartheta}}{\bar{T}}\right), \quad B = \frac{7800}{1+\dfrac{15\,600\bar{\vartheta}}{\bar{T}}}, \end{gathered} \qquad (26.54)$$

[5] We note that we have neglected the contribution to the right-hand side of the possible correlation between temperature and humidity fluctuations. This may be avoided by considering fluctuations in the refractive index (measured by a refractometer), and using their spectral function.

where $F_{TT}(k)$ and $F_{\vartheta\vartheta}(k)$ are the three-dimensional spectral densities of the temperature and humidity fields which are respectively of the form $\frac{E_T(k)}{4\pi k^2}$ and $\frac{E_\vartheta(k)}{4\pi k^2}$ [where $E_T(k)$ and $E_\vartheta(k)$ are the spectra describing the distribution of the squares of fluctuation amplitudes over the wave numbers $k$].

In the inertial subrange of the turbulence spectrum, i.e., when.

$$2\pi/\eta \gg 2\kappa|\sin\theta/2| \gg 2\pi/L_0, \tag{26.55}$$

where $\eta$ is the internal turbulence length scale, we can assume $E_T(k)=B^{(\vartheta)}\overline{N}_T\overline{\varepsilon}^{-1/3}k^{-5/3}$ and $E_\vartheta(k)=B^{(\vartheta)}\overline{N}_\vartheta\overline{\varepsilon}^{-1/3}k^{-5/3}$ where $B^{(\vartheta)}$ is a numerical constant of the order of unity, and $\overline{N}_T=\chi_T\overline{(\nabla T)^2}$ and $\overline{N}_\vartheta=\chi_\vartheta\overline{(\nabla\vartheta)^2}$ are parameters characterizing the rate of dissipation of the temperature and humidity inhomogeneities due to molecular effects. Using this fact, together with Eqs. (26.17), (26.51), and (26.54), we obtain

$$\overline{d\sigma}=2^{-14/3}V\kappa^{1/3}\sin^2\alpha\left[A^2B^{(\vartheta)}(\overline{N}_T+B^2\overline{N}_\vartheta)\overline{\varepsilon}^{-1/3}\right]\cdot\left|\sin\frac{\theta}{2}\right|^{-11/3}d\Omega. \tag{26.56}$$

An equivalent formula has been given by Silverman (1956).

Let us investigate the following experiment on radio-wave scattering in the troposphere (Fig. 106). Suppose that the radio transmitter $A$ has a highly directional antenna with an angular beamwidth $\gamma$ (for example, we may suppose that the antenna radiates into a circular cone of angle $\gamma$), while the receiver $B$ is also highly directional and has the same parameters. Both antennas will be assumed to be pointing in the direction of the point $O$, which is at equal distances $D$ from $A$ and $B$. The angle between $AO$ and $OB$ will be denoted by $\theta$. The receiver will intercept radio waves scattered by $V$ (hatched in the

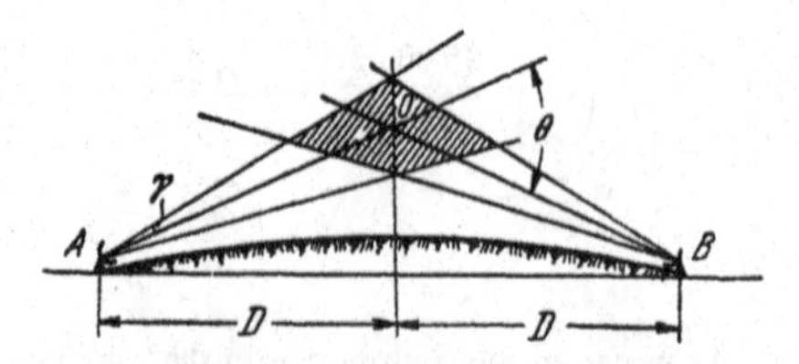


**Fig. 106** Schematic plot of the experiment on the scattering of radio waves in the troposphere.

figure) which is formed by the intersection of the two cones, and $\theta$ will be the mean angle of scattering. The energy flux density at the center of the scattering volume will be smaller by a factor of $D^2$ than in the neighborhood of the transmitter. Only the fraction $\overline{d\sigma}$ of this energy will be scattered into the solid angle $d\Omega$ in the direction of the receiver. In the neighborhood of the receiver, i.e., at the distance $D$ from the scattering volume, the scattered energy will be distributed over the area $D^2\,d\Omega$, so that the density $P'$ of the scattered energy flux near the receiver will be smaller by the factor $\frac{1}{D^4}\cdot\frac{\overline{d\sigma}}{d\Omega}$ than the energy flux density near the transmitter. At the same time, if there is no scattering the energy flux density $P_0$ at a distance $2D$ from the receiver (along the line $AO$) will be smaller by a factor of only $1/4D^2$ than the energy flux density near the transmitter. Therefore,

$$\frac{P'}{P_0}=\frac{4}{D^2}\frac{\overline{d\sigma}}{d\Omega}. \tag{26.57}$$

When $\gamma<\theta$ the scattering volume $V$ formed by the intersection of the two circular cones is given by

$$V=\frac{2}{3}D^3\sin\gamma\sin\frac{\gamma}{2}\left(1-\frac{\sin^2(\gamma/2)}{\sin^2(\theta/2)}\right)^{3/2}\frac{\beta}{1-\beta^2}\left(1+\frac{\arcsin\beta}{\beta\sqrt{1-\beta^2}}\right), \tag{26.58}$$
$$\beta=\frac{\sin\gamma}{\sin\theta}.$$

When $\gamma<\theta\ll 1$ we have the approximate result $V\approx 2D^3\gamma^3/3\theta$. Let us consider the case where the transmitting and receiving antennas point toward the horizon (so that $AO$ is perpendicular to the vertical at $A$). In this case, $D\approx R\theta/2$ and, consequently, $V\approx D^2R\gamma^3/3$ where $R$ is the earth's radius. Bearing this in mind, and substituting Eq. (26.56) in Eq. (26.57), we obtain

$$\frac{P'}{P_0}=\frac{2\sqrt[3]{2\pi}}{3}R\gamma^3\lambda^{-1/3}\left[A^2B^{(\vartheta)}\left(\overline{N}_T+B^2\overline{N}_\theta\right)\overline{\varepsilon}^{-1/3}\right]\theta^{-11/3}. \tag{26.59}$$

In this expression we have additionally taken into account the fact that, when $\theta\ll 1$, the angle $\alpha$ between the direction of polarization in the incident wave and the direction of the scattered wave is nearly 90°, so that $\sin\alpha\approx 1$.

Chisholm, Portman, deBettencourt, and Roche (1955) have published the results of systematic measurements of $P'/P_0$ which characterizes tropospheric scattering of radio waves, using the above scheme with $D \approx 150$, $\lambda = 8.17$, $\gamma = 0.012$, and $\theta = 0.048$. The height of the center $O$ of the scattering volume above the Earth's surface was 1.5 km. The linear dimensions of inhomogeneities producing scattering can be estimated from Eq. (26.52), i.e., $l \approx \lambda/\theta \approx 1.7$ m. Such length scales are from the inertial subrange of the atmospheric turbulence, so that the use of Eq. (26.56), and the resulting formula (26.59), for the calculation of the observed effects is fully justified. The measurements of Chisholm et al. show that tropospheric scattering has a clear annual variation with a maximum during the summer and a minimum during the winter, and the values of $P'/P_0$ vary from summer to winter by two or three orders of magnitude. The estimated values of the meteorologic parameter in Eqs. (26.56) and (26.59) (the factor in brackets) are in agreement to within an order of magnitude with other indirect estimates and with existing results of direct measurements of the parameters $\overline{N_T}$, $\overline{N_\vartheta}$, and $\overline{\varepsilon}$ of atmospheric turbulence. We note, however, that in addition to the scattering of radio waves by turbulent inhomogeneities, there are also other effects that are recorded during measurements of $P'/P_0$, for example, reflection of the radio waves from inversion layers, which in many cases may provide a dominating contribution to the ratio $P'/P_0$.

Let us now consider the scattering of sound waves by a turbulent volume $V$, and calculate to begin with the spectral function $F_{nn}(\boldsymbol{k})$ for the complex refractive index $n'$ given by Eq. (26.42), assuming that the turbulence is locally isotropic. This calculation can be carried out with the aid of the general spectral representations of the locally isotropic fields $\boldsymbol{u}'(\boldsymbol{x})$ and $T'(\boldsymbol{x})$. However, we shall use a less rigorous but simpler method, namely, we shall assume temporarily that the turbulence is not only locally isotropic but also isotropic, and begin by calculating the correlation function for the field $n'(\boldsymbol{x})$, and hence the spectral function. The final result will be valid for locally isotropic but not isotropic turbulence.

Using Eq. (26.42), we find that the correlation function for the field $n'(\boldsymbol{x})$ is given by

$$\overline{n'(\boldsymbol{x}_1)\, n'^{*}(\boldsymbol{x}_2)} = \left(1 - \frac{i}{\kappa}\frac{\partial}{\partial x_{1k}}\right)\left(1 + \frac{i}{\kappa}\frac{\partial}{\partial x_{2k}}\right)\left[\frac{\overline{u'_k(\boldsymbol{x}_1)\, u'_k(\boldsymbol{x}_2)}}{c_0^2} + \frac{\overline{T'(\boldsymbol{x}_1)\, T'(\boldsymbol{x}_2)}}{4\overline{T}^2}\right],$$

since the velocity and temperature fields can be regarded as uncorrelated. Substituting $x_1 - x_2 = r$, we can rewrite the last formula in the form

$$B_{nn}(r) = \left(1 - \frac{i}{\kappa}\frac{\partial}{\partial r_\kappa}\right)^2 \left[\frac{B_{kk}(r)}{c_0^2} + \frac{B_{TT}(r)}{4\bar{T}^2}\right]. \tag{26.60}$$

Using the spectral representations of correlation functions for the velocity and temperature fields

$$B_{jl}(r) = \int e^{-ikr}\left(\delta_{jl} - \frac{k_j k_l}{k^2}\right) F(k)\, dk,$$

$$B_{TT}(r) = \int e^{-ikr} F_{TT}(k)\, dk$$

and the analogous formula (26.48) for $B_{nn}(r)$, we obtain

$$F_{nn}(k) = \left(1 - \frac{k_\kappa}{k}\right)^2 \left[\left(1 - \frac{k_\kappa^2}{k^2}\right)\frac{F(k)}{c_0^2} + \frac{F_{TT}(k)}{4\bar{T}^2}\right], \tag{26.61}$$

where $k_\kappa = k \cdot \kappa/\kappa$ is the component of the wave vector $k$ in the direction of propagation of the incident wave, and $F(k) = E(k)/4\pi k^2$ and $F_{TT}(k) = E_T(k)/4\pi k^2$ are the spectral functions for the velocity and temperature fields. It is clear from Eqs. (26.60) and (26.61) that the complex refractive-index field $n'(x)$ is not locally isotropic for sound waves: its statistical characteristics will in general depend on the direction of propagation of the incident wave.

Substituting $k = \kappa - \kappa q$ in Eq. (26.61), and remembering that $k = |\kappa - \kappa q| = 2\kappa|\sin\theta/2|$ and

$$k_\kappa = \frac{(\kappa - \kappa q)\,\kappa}{\kappa} = \kappa(1 - \cos\theta),$$

we obtain

$$F_{nn}(\kappa - \kappa q) = \cos^2\theta\left[\cos^2\frac{\theta}{2}\frac{F(2\kappa\sin\theta/2)}{c_0^2} + \frac{F_{TT}(2\kappa\sin\theta/2)}{4\bar{T}^2}\right]. \tag{26.62}$$

When Eq. (26.53) is satisfied, this formula can be regarded as valid in the general case of locally isotropic turbulence. From Eqs. (26.45), (26.55), and (26.62) we then obtain the following expression for the mean effective scattering cross-section of $V$:

$$\overline{d\sigma} = 2\pi\kappa^4 V \cos^2\theta \left[ \cos^2\frac{\theta}{2} \frac{F\left(2\kappa \sin\frac{\theta}{2}\right)}{c_0^2} + \frac{F_{TT}\left(2\kappa \sin\frac{\theta}{2}\right)}{4\overline{T}^2} \right] d\Omega. \quad (26.63)$$

This formula shows, in particular, that scattering through $\theta = 90°$ does not occur, while backward scattering, i.e., scattering through $\theta = 180°$), is produced by temperature fluctuations but not by velocity fluctuations.[6]

In the inertial subrange of the wave numbers $k = 2\kappa\,|\sin(\theta/2)|$, i.e., when Eq. (26.55) is satisfied, we can assume that $E(k) = C_1\overline{\varepsilon}^{2/3}k^{-5/3}$ and $E_T(k) = B^{(T)}\overline{N}_T\overline{\varepsilon}^{-1/3}k^{-5/3}$, so that Eq. (26.63) takes the form [Kraichnan (1953), Monin (1961)]:

$$\overline{d\sigma} = 2^{-14/3} V \kappa^{1/3} \cos^2\theta \left[ \frac{C_1\,(\overline{\varepsilon}^{2/3}\cos^2(\theta/2))}{c_0^2} + \frac{B^{(T)}\overline{N}_T\overline{\varepsilon}^{-1/3}}{4\overline{T}^2} \right] \cdot \left| \sin\frac{\theta}{2} \right|^{-11/3} d\Omega$$
(26.64)

According to Eq. (7.93) of Vol. 1, in the atmospheric surface layer $\overline{\varepsilon} = \frac{u_*^3}{\varkappa z}\varphi_\varepsilon(\zeta)$ and $\overline{N}_T = \frac{\varkappa u_* T_*^2}{z}\varphi_N(\zeta)$ where $z$ is the height above ground, and $\zeta = \frac{z}{L}$ is the thermal stability parameter. When $\zeta \to 0$,

[6] More precisely, scattering through $\theta = 90°$ is very small and is determined by small terms of the order of $(u'/c_0)^2$ in the expression for $n'$, which we have neglected in the derivation of Eq. (26.42). In particular, if we retain in the second equation in (26.37) the last term which contains the turbulent pressure fluctuations (the quantity $\Pi' \approx p'/\overline{\rho}c_0^2$), then, in Eq. (26.42), we would have a term of the form $\frac{i}{2\kappa}\frac{\partial\Pi'}{\partial x_k}$, and in Eq. (26.63) we would have the term $2\pi\kappa^4 V \sin^4\frac{\theta}{2}\frac{F_{pp}\left(2\kappa\sin\frac{\theta}{2}\right)}{\overline{\rho}^2 c_0^4} d\Omega$, where $F_{pp}$ is the pressure spectral function. This term provides the largest contribution to backward scattering, but it is not zero even for $\theta = 90°$. In the inertial subrange of $2\kappa\left|\sin\frac{\theta}{2}\right|$ it is of the form $\frac{B^{(p)}\overline{\varepsilon}^{4/3} V\, d\Omega}{2^{16/3}c_0^4\,|\kappa\sin(\theta/2)|^{1/3}}$.

i.e., when neutral thermal stratification is approached, the values of $\varphi_\varepsilon(\zeta)$ and $\varphi_N(\zeta)$ approach unity. Accordingly, the formula (26.64) for the surface layer of the atmosphere reduces to

$$\frac{1}{V}\frac{\overline{d\sigma}}{d\Omega}=\frac{2^{-14/3}\sqrt[3]{2\pi}\,\lambda^{-1/3}}{(\varkappa z)^{2/3}}\left[C_1\frac{u_*^2}{c_0^2}\varphi_\varepsilon^{2/3}(\zeta)\cos^2\frac{\theta}{2}+\right.$$
$$\left.+\frac{B^{(T)}\varkappa^2}{4}\frac{T_*^2}{\bar{T}^2}\varphi_N(\zeta)\varphi_\varepsilon^{-1/3}(\zeta)\right]\cos^2\theta\left|\sin\frac{\theta}{2}\right|^{-11/3} \qquad (26.65)$$

The values of the meteorologic parameter in this formula (the factor in the brackets) for each fixed $\theta$ can be approximately determined from measurements of the wind-velocity and temperature profiles.

Experimental studies of the scattering of sound waves by turbulent inhomogeneities in the atmospheric surface layer were carried out by Kallistratova (1959, 1962). The experiments were as shown schematically in Fig. 106. For scattering angles $\theta=15^\circ\div 30^\circ$ the distance between the sound source and receiver was 140 m, for $\theta=20^\circ\div 50^\circ$ this distance was 80 m, for $\theta=30^\circ\div 70^\circ$ it was 40 m, and $\theta=70^\circ\div 130^\circ$ it was 20 m. In the case of backward scattering, the sound waves were emitted in a practically vertical direction, and waves scattered at the height of about 12 m were recorded. Specially designed transducers were used as sources and receivers. These were in the form of rectangular plates, having an area of about 1 $m^2$, and produced highly directional radiation with an angular beamwidth of a few degrees. Scattering of sound waves at 11 kHz ($\lambda=3$ cm) was investigated. The scales of turbulent inhomogeneities responsible for the scattering of such waves are shown by Eq. (26.52) to vary between 12 and 3 cm for $\theta=15^\circ\div 60^\circ$, and between 3 and 1.5 cm for $\theta=60^\circ\div 180^\circ$. Since the internal turbulence length scale $\eta$ in the surface layer of the atmosphere is of the order of a few millimeters, it may be assumed that scattering through angles $\theta<60^\circ$ was determined in the Kallistratova experiments by turbulent inhomogeneities in the inertial subrange of the spectrum, and could therefore be described by Eq. (26.65). When $\theta>60^\circ$, the relevant turbulence possibly had smaller scales, and hence the values of $\frac{1}{V}\frac{\overline{d\sigma}}{d\theta}$ calculated from Eq. (26.65) may be expected to be too high.

In the above experiments, the sound waves were emitted in the form of short pulses, 1.5 $\mu$sec long, at a repetition frequency of 30

Hz. At each instant only the area shown hatched in Fig. 106 contributed to this scattering, and the actual scattering volume was about 3 $m^3$. The receiving apparatus was capable of recording sound oscillations down to $10^{-14}$ of the radiated power. As an example, Fig. 107 shows photographs of the transmitted pulse and the received direct and scattered pulses on an oscillograph screen for $\theta = 20° \div 45$ and source-receiver separation of 40 m.

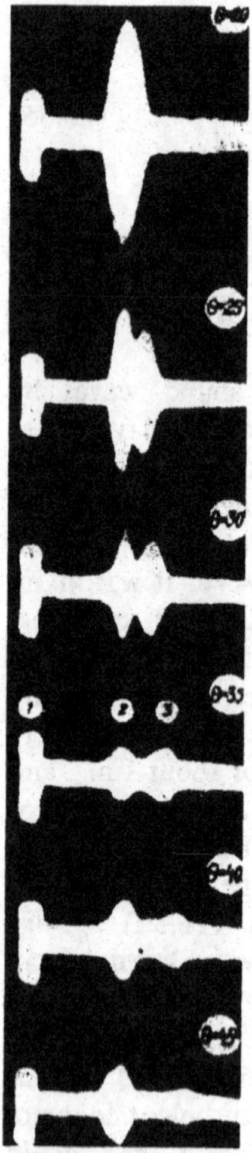

These experiments confirm the above theoretical predictions. The values of $\frac{1}{V}\frac{\overline{d\sigma}}{d\Omega}$ averaged over different meteorologic conditions were $0.6 \cdot 10^{-6}$ cm for $\theta = 25°$, which is in good agreement with the estimate based on Eq. (26.65), and $3.4 \cdot 10^{-12}$ $\text{cm}^{-1}$ for $\theta = 180°$, which is considerably smaller than predicted by Eq. (26.65).

Figure 108 shows the scattering indicatrix for the sound waves, i.e., the function $J = \frac{V(25°)}{V(\theta)} \frac{\overline{d\sigma}(\theta)}{\overline{d\sigma}(25°)}$ in decibels [1 db = 20 lg $(x/x_0)$] as a function of $\theta$. According to Eq. (26.55), the effect of temperature fluctuations on sound scattering can be neglected when $\theta$ is not too large, in which case

$$J = \frac{\cos^2 \frac{\theta}{2} \cos^2 \theta \left| \sin \frac{\theta}{2} \right|^{-11/3}}{\cos^2 \frac{25°}{2} \cos^2 25° \left| \sin \frac{25°}{2} \right|^{-11/3}}, \quad \text{дб}. \tag{26.66}$$

The corresponding curve is shown in Fig. 108 by the solid line. It is clear that there is good agreement with the experimental data. However, for $\theta > 60°$, and with the exception of the regions $\theta \approx 60°$ and $\theta \approx 180°$, the curve predicted by Eq. (26.66) lies somewhat higher than the experimental points, although it does not take into account the contribution of temperature fluctuations to sound scattering.

**Fig. 107** Transmitted (1), received direct (2) and received scattered (3) pulses shown on an oscillograph screen in the sound scattering experiments of Kallistratova.

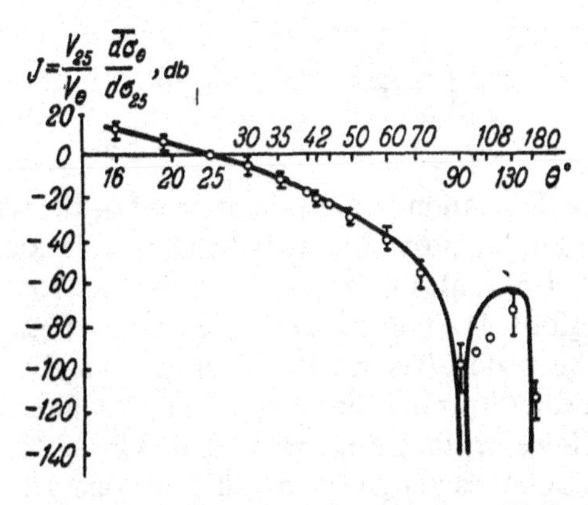


**Fig. 108** Sound-scattering indicatrix in the surface layer of the atmosphere [Kallistratova (1962)]. Solid curve represents the formula given by Eq. (26.66).

## 26.4 Fluctuations in the Amplitude and Phase of Electromagnetic and Sound Waves in A Turbulent Atmosphere

In this section we shall investigate the statistical characteristics of amplitude and phase fluctuations, $\chi'$ and $S'$, for plane electromagnetic or sound waves propagating in a turbulent atmosphere. We shall use Eq. (26.29) which is satisfied by the function $\psi' = \chi' + iS'$. Separating the real and imaginary parts, we obtain

$$\Delta\chi' - 2\kappa\frac{\partial S'}{\partial x} = -2\kappa^2 n_1, \quad \Delta S' + 2\kappa\frac{\partial \chi'}{\partial x} = -2\kappa^2 n_2, \tag{26.67}$$

and hence

$$\begin{aligned} \left(\Delta^2 + 4\kappa^2\frac{\partial^2}{\partial x^2}\right)\chi' &= -2\kappa^2\left(\Delta n_1 + 2\kappa\frac{\partial n_2}{\partial x}\right), \\ \left(\Delta^2 + 4\kappa^2\frac{\partial^2}{\partial x^2}\right)S' &= -2\kappa^2\left(\Delta n_2 - 2\kappa\frac{\partial n_1}{\partial x}\right), \end{aligned} \tag{26.68}$$

where the $Ox$ axis is parallel to the direction of propagation of the incident wave, $\Delta = \frac{\partial^2}{\partial y^2} + \frac{\partial^2}{\partial z^2}$, and $n_1 + in_2 = n'$ represents fluctuations in the refractive index. For electromagnetic waves $n_2 = 0$, and $n_1 = n'$ is given by Eq. (26.3) or (26.3'). For sound waves, we have from (26.42)

$$n_1 = -\left(\frac{u'}{c_0} + \frac{1}{2}\frac{T'}{\overline{T}}\right), \qquad n_2 = -\frac{1}{\kappa}\frac{\partial n_1}{\partial x}, \tag{26.69}$$

where $u'$ is the fluctuation in the $x$-component of the velocity.

We shall again suppose that the turbulence is localized in the region $x > 0$ (so that the plane $x = 0$ is the boundary of the turbulence region). The random field $n_1(\boldsymbol{x})$ will be regarded as locally isotropic in this region. The random fields $\chi'(\boldsymbol{x})$ and $S'(\boldsymbol{x})$ cannot be regarded as locally isotropic since, for them, the direction of $Ox$ is a special one. However, they can be regarded as locally isotropic in the planes $x = \text{const}$. Our main problem will be to calculate the structure functions for the fields $\chi'(\boldsymbol{x})$ and $S'(\boldsymbol{x})$ in a given plane $x = L$.

This calculation can be carried out with the aid of Eq. (26.68), using the spectral representations of two-dimensional locally isotropic fields $n_1(\boldsymbol{x})$, $\chi'(\boldsymbol{x})$ and $S'(\boldsymbol{x})$ in the $x = \text{const}$. planes, which are analogous to the spectral representations of locally isotropic fields in three-dimensional space described in Sect. 13. However, we shall use a less rigorous approach, and assume for the moment that these fields are homogeneous in the planes $x = \text{const}$ [and the field $n_1(\boldsymbol{x})$ is homogeneous in all space], and that there exists for them the correlation functions

$$\begin{aligned}
\overline{n_1(\boldsymbol{x}_1)\, n_1(\boldsymbol{x}_2)} &= B_{n_1 n_1}(\boldsymbol{\rho}_1 - \boldsymbol{\rho}_2;\ |x_1 - x_2|) = \\
&= \int e^{-i\boldsymbol{k}(\boldsymbol{\rho}_1 - \boldsymbol{\rho}_2)} F^{(2)}_{n_1 n_1}(\boldsymbol{k};\ |x_1 - x_2|)\, d\boldsymbol{k}, \\
\overline{\chi'(\boldsymbol{x}_1)\, \chi'(\boldsymbol{x}_2)} &= B_{\chi\chi}(\boldsymbol{\rho}_1 - \boldsymbol{\rho}_2;\ x_1,\ x_2) = \\
&= \int e^{-i\boldsymbol{k}(\boldsymbol{\rho}_1 - \boldsymbol{\rho}_2)} F^{(2)}_{\chi\chi}(\boldsymbol{k};\ x_1,\ x_2)\, d\boldsymbol{k}, \\
\overline{S'(\boldsymbol{x}_1)\, S'(\boldsymbol{x}_2)} &= B_{SS}(\boldsymbol{\rho}_1 - \boldsymbol{\rho}_2;\ x_1,\ x_2) = \\
&= \int e^{-i\boldsymbol{k}(\boldsymbol{\rho}_1 - \boldsymbol{\rho}_2)} F^{(2)}_{SS}(\boldsymbol{k};\ x_1,\ x_2)\, d\boldsymbol{k},
\end{aligned} \tag{26.70}$$

where $\boldsymbol{x} = (x, y, z)$, $\boldsymbol{\rho} = (y, z)$ is the radius vector in the plane $(y, z)$, $\boldsymbol{k}$ is the corresponding two-dimensional wave vector, and the superscript (2) carried by $F$ indicates that we are concerned with two-dimensional spectral densities. In the first formula in Eq. (26.7) we have additionally used the fact that the statistical characteristics of the field $n_1(\boldsymbol{x})$ should be invariant under a change in the direction of the $Ox$ axis.

Let us begin by considering the simple case of an electromagnetic wave for which $n_1 = n'$, $n_2 = 0$. Let us rewrite the first equation in

Eq. (26.68) for the points $x_1$ and $x_2$, then multiply together the left- and right-hand sides of the resulting two equations, and finally take an average. The final result is

$$\left(\Delta^2+4\kappa^2\frac{\partial^2}{\partial x_1^2}\right)\left(\Delta^2+4\kappa^2\frac{\partial^2}{\partial x_2^2}\right)B_{\chi\chi}(\rho_1-\rho_2;\ x_1,\ x_2)=$$
$$=4\kappa^4\Delta^2 B_{nn}(\rho_1-\rho_2;\ |x_1-x_2|).$$

Using the spectral representations of the correlation functions (26.70), we obtain

$$\left(k^4+4\kappa^2\frac{\partial^2}{\partial x_1^2}\right)\left(k^4+4\kappa^2\frac{\partial^2}{\partial x_2^2}\right)F_{\chi\chi}^{(2)}(\boldsymbol{k};\ x_1,\ x_2)=$$
$$=4\kappa^4 k^4 F_{nn}^2(\boldsymbol{k};\ |x_1-x_2|).$$

We are assuming that the $x=0$ plane is the boundary of the turbulent region, so that when $x=0$ the fluctuations $n'$, $\chi'$, and $S'$ are all zero and, therefore, in accordance with Eq. (26.67), the derivatives $\frac{\partial\chi'}{\partial x}$ and $\frac{\partial S'}{\partial x}$ are also zero. Hence, it follows that the function $F_{\chi\chi}^{(2)}(\boldsymbol{k};\ x_1,\ x_2)$ and its first derivatives with respect to $x_1$ and $x_2$ are identically zero for $x_1=0$ or $x_2=0$. With these boundary conditions, the above equation for $F_{\chi\chi}^{(2)}(\boldsymbol{k};\ x_1,\ x_2)$ can readily be solved, for example, by the Green function method. The corresponding solution written for the point $x_1=x_2=L$ is of the form

$$F_{\chi\chi}^{(2)}(\boldsymbol{k})=\kappa^2\int_0^L\int_0^L \sin\frac{k^2(L-\xi)}{2\kappa}\sin\frac{k^2(L-\zeta)}{2\kappa}F_{nn}^{(2)}(\boldsymbol{k};\ |\xi-\zeta|)\,d\xi\,d\zeta. \tag{26.71}$$

Henceforth, it will be convenient to use the symbol $F^{(2)}(\boldsymbol{k})$ instead of the complete designation $F^{(2)}(\boldsymbol{k};\ L,\ L)$. It is readily verified that an analogous formula for $F_{SS}^{(2)}(\boldsymbol{k})$ differs from that obtained for $F_{\chi\chi}^{(2)}(\boldsymbol{k})$ only by the fact that the sines are replaced by cosines under the integral sign. If we carry out the integration with respect to $\xi+\zeta$ in these formulas, we obtain

$$F_{\chi\chi}^{(2)}(\boldsymbol{k})=\kappa^2 L\int_0^L\left[\left(1-\frac{\xi}{L}\right)\cos\frac{k^2\xi}{2\kappa}+\frac{\kappa}{k^2L}\sin\frac{k^2\xi}{2\kappa}-\right.$$
$$\left.-\frac{\kappa}{k^2L}\sin\frac{k^2(2L-\xi)}{2\kappa}\right]F_{nn}^{(2)}(\boldsymbol{k};\ \xi)\,d\xi, \tag{26.72}$$

$$F_{SS}^{(2)}(\boldsymbol{k}) = \kappa^2 L \int_0^L \left[\left(1 - \frac{\xi}{L}\right)\cos\frac{k^2\xi}{2\kappa} - \frac{\kappa}{k^2 L}\sin\frac{k^2\xi}{2\kappa} + \right.$$

$$\left. + \frac{\kappa}{k^2 L}\sin\frac{k^2(2L-\xi)}{2\kappa}\right] F_{nn}^{(2)}(\boldsymbol{k};\ \xi)\, d\xi. \qquad (26.72)\ (\text{Cont.})$$

It can be shown that these formulas remain valid in the general case of locally isotropic turbulence. It follows from Eq. (26.72) that to ensure the isotropy of the amplitude fluctuation field in the $x = L$ plane, and the existence of a bounded correlation function $B_{\chi\chi}(\boldsymbol{r})$ for these fluctuations, it is sufficient to impose a very slight restriction on the spectral function of the refractive index. In fact, to ensure that $B_{\chi\chi}(\boldsymbol{r})$ exists, it is evidently sufficient for the spectral function $F_{\chi\chi}^{(2)}(\boldsymbol{k})$ to tend to a finite limit for $\boldsymbol{k} \to 0$. However, when $\boldsymbol{k} \to 0$, the factor in the brackets in the first formula in Eq. (26.72) will tend to zero as $k^4$, so that for $B_{\chi\chi}(\boldsymbol{r})$ to exist, it is sufficient that the spectral function of the refractive index should not increase more rapidly than $k^{-4}$ as $\boldsymbol{k} \to 0$. In the case of locally isotropic turbulence, this condition is satisfied because the spectral function for the refractive index does not increase more rapidly than $k^{-11/3}$ as $k$ decreases. At the same time, we are not justified in assuming that there exists a correlation function for the phase fluctuations in the case of locally isotropic turbulence, and we shall henceforth largely restrict our attention to the structure function for the phase fluctuations.

The formulas (26.72) can be substantially simplified by using the fact that for large values of $k\xi$ the quantity $F_{nn}^{(2)}(\boldsymbol{k};\ \xi)$ tends to zero very rapidly. This quantity is a measure of the correlation between refractive-index inhomogeneities with scales of the order of $1/k$ in two $x = \text{const}$ planes separated by a distance $\xi$. For sufficiently small scales, these inhomogeneities are approximately isotropic, so that the correlation remains appreciable only over distances $\xi$ that do not exceed the order of magnitude of the scale $1/k$ of the inhomogeneities. Therefore, for $k\xi \gg 1$ the quantity $F_{nn}^{(2)}(\boldsymbol{k};\ \xi)$ becomes very small in comparison with its values for $k\xi \leqslant 1$, so that the main contribution to the integrals in Eq. (26.72) is due to the region $\xi \leqslant 1/k$. In this region $\frac{k^2\xi}{2\kappa} \leqslant \frac{k}{2\kappa} \sim \frac{\lambda}{2l_{\text{turb}}} \ll 1$ in accordance with the condition (26.19) used in the above derivation of the equation for $\psi' = \chi' + iS'$. Consequently, we may suppose that $\sin\frac{k^2\xi}{2\kappa} \approx \frac{k^2\xi}{2\kappa}$ and $\cos\frac{k^2\xi}{2\kappa} \approx 1$. If $L < 1/k$, then this simplification is justified for any $\xi$ in the range $0 \leqslant \xi \leqslant L$. If, on the other hand, $1/k < L$ then the simplification is valid for $0 \leqslant \xi \leqslant 1/k$, but it can also be used in the range $1/k \leqslant \xi \leqslant L$, since this interval will not introduce any appreciable contribution to the integrals in Eq. (26.72) because $F_{nn}^{(2)}(\boldsymbol{k};\ \xi)$ is small for $\xi > 1/k$. When the above simplification is made, the formulas (26.72) assume the form

$$F_{XX}^{(2)}(\boldsymbol{k}) = \kappa^2 L \int_0^L \left[1 - \frac{\kappa}{k^2 L}\sin\frac{k^2 L}{\kappa} - \frac{\xi}{2L}\left(1 - \cos\frac{k^2 L}{\kappa}\right)\right] F_{nn}^{(2)}(\boldsymbol{k};\ \xi)\, d\xi,$$

$$F_{SS}^{(2)}(\boldsymbol{k}) = \kappa^2 L \int_0^L \left[1 + \frac{\kappa}{k^2 L}\sin\frac{k^2 L}{\kappa} - \frac{\xi}{2L}\left(3 + \cos\frac{k^2 L}{\kappa}\right)\right] F_{nn}^{(2)}(\boldsymbol{k};\ \xi)\, d\xi.$$

(26.73)

The last two formulas refer to the case of electromagnetic wave propagation. In the case of sound waves, for which $n_1$ and $n_2$ on the right-hand side of (26.63) are given by (26.69), a similar calculation shows that $F_{XX}^{(2)}(\boldsymbol{k})$ and $F_{SS}^{(2)}(\boldsymbol{k})$ are are given by formulas that differ from Eqs. (26.72) and (26.73) only in that $F_{n_1 n_1}^{(2)}(\boldsymbol{k};\ \xi$ is replaced by $\left(1 - \frac{k^2}{2\kappa^2}\right)^2 F_{n_1 n_1}^{(2)}(\boldsymbol{k};\ \xi)$. All our discussions are restricted by the condition (26.19), according to which $k/\kappa \ll 1$, so that $\left(1 - \frac{k^2}{2\kappa^2}\right)^2 \approx 1$ and, consequently, in the case of sound waves we need only replace $F_{nn}^{(2)}(\boldsymbol{k};\ \xi)$ in Eqs. (26.72) and (26.73) by $F_{n_1 n_1}^{(2)}(\boldsymbol{k};\ \xi)$.

If we consider the statistical structure of random fields $\chi'(\boldsymbol{x})$ and $S'(\boldsymbol{x})$ in the $x = L$ plane for scales that are small in comparison with $L$, i.e., if we suppose that

$$1/k \ll L, \tag{26.74}$$

then the formulas in Eq. (26.73) can be simplified further. The main contribution to the integrals in Eq. (26.73) is due to values $\xi \leqslant 1/k$ or, when Eq. (26.74) is satisfied, the values $\xi \ll L$. The expressions in brackets in Eq. (26.73) can then be respectively replaced by $1 - \frac{\kappa}{k^2 L}\sin\frac{k^2 L}{\kappa}$ and $1 + \frac{\kappa}{k^2 L}\sin\frac{k^2 L}{\kappa}$, and can be taken out from under the integral signs. The remaining integrals of the function $F_{nn}^{(2)}(\boldsymbol{k};\ \xi)$ must be evaluated over all values of $\xi$ for which this function is not too small. Without introducing any substantial error, we can evaluate these integrals over the range $0 \leqslant \xi < \infty$. If we use the spectral expansion

$$F_{nn}^{(2)}(\boldsymbol{k};\ \xi)=\int_{-\infty}^{\infty}\cos k_1\xi\, F_{nn}\left(\sqrt{k_1^2+k^2}\right)dk_1,$$

which is valid for the locally isotropic field $n'(\boldsymbol{x})$, where $F_{nn}(k)$ is the three-dimensional spectral function of the field $n'(\boldsymbol{x})$, we find that

$$F_{nn}\left(\sqrt{k_1^2+k^2}\right)=\frac{1}{\pi}\int_{0}^{\infty}\cos k_1\xi\, F_{nn}^{(2)}(\boldsymbol{k};\ \xi)\,d\xi.$$

Substituting $k_1=0$, we can readily verify that the integral of $F_{nn}^{(2)}(\boldsymbol{k};\ \xi)$ over the range $0\leqslant\xi<\infty$ is equal to $\pi F_{nn}(k)$. Therefore, in the case of locally isotropic turbulence we have

$$\begin{aligned} F_{\chi\chi}^{(2)}(\boldsymbol{k})&=\pi\kappa^2L\left(1-\frac{\kappa}{k^2L}\sin\frac{k^2L}{\kappa}\right)F_{nn}(k),\\ F_{SS}^{(2)}(\boldsymbol{k})&=\pi\kappa^2L\left(1+\frac{\kappa}{k^2L}\sin\frac{k^2L}{\kappa}\right)F_{nn}(k). \end{aligned}\qquad(26.75)$$

where we have assumed that Eq. (26.74) is valid. These formulas were first derived by Tatarskii (1961a). Two-dimensional spectra of the amplitude and phase fluctuations can be determined either by processing a number of photographs of the wave field image in a plane $x=L$ (cf. Sect. 26.5 below) or by measuring the statistical characteristics of the time variations of a wave field at a given point and applying Eq. (26.78) or some related equation. Hence Eq. (26.75) yields the possibility of experimentally determining the refractive index spectrum from wave propagation experiments. The paper of Gurvich (1968) is devoted to discussion of this possibility in more detail.

In the case of a sound wave, the field $n_1(\boldsymbol{x})$ is locally homogeneous, but is not locally isotropic because the $x$ direction of the incident wave is a special direction for it. The integral of $F_{nn}^{(2)}(\boldsymbol{k};\ \xi)$ evaluated for $0\leqslant\xi<\infty$ is then equal to the product of the three-dimensional spectral function of the field $n_1(\boldsymbol{x})$ for $k_1=0$ multiplied by $\pi$, where $k_1$ is the component of the wave vector in the $x$ direction. If we recall the derivation of Eq. (26.61), we can readily verify that, in the case of a sound wave, the function $F_{nn}(k)$ in Eq. (26.75) must be replaced by $\dfrac{F(k)}{c_0^2}+\dfrac{F_{TT}(k)}{4\overline{T}^2}$.

The form of the spectral functions (26.75) is very dependent on the value of the length scale $\sqrt{\lambda L}$ (the radius of the first Fresnel zone; see Sect. 26.1) Let this length scale be much smaller than the internal turbulence length scale $\eta$ [so that Eq. (26.31) is satisfied and we can use the geometric-optics approximation]. We shall restrict our attention to the structure of the fields $\chi'(\boldsymbol{x})$ and $S'(\boldsymbol{x})$ for scales that are large in comparison with $\sqrt{\lambda L}$ and therefore wave numbers $k \ll \frac{2\pi}{\sqrt{\lambda L}}$. The condition $\frac{k^2 L}{\kappa} \ll 1$ will then be satisfied, and hence the formulas (26.75) will be substantially simpler and will assume the form

$$
\begin{aligned}
F_{\chi\chi}^{(2)}(k) &= \frac{1}{6}\pi L^3 k^4 F_{nn}(k), \\
F_{SS}^{(2)}(k) &= 2\pi\kappa^2 L F_{nn}(k).
\end{aligned}
\tag{26.76}
$$

It is clear that when $\sqrt{\lambda L} \ll \eta$ the amplitude fluctuations are independent of frequency, and are proportional to the cube of the distance traversed by the wave in the turbulent medium. The phase fluctuations are proportional to the square of the frequency and the first power of the distance. The maximum of the function $F_{\chi\chi}^{(2)}(k) \sim k^4 F_{nn}(k)$ is reached near the point $k = 2\pi/\eta$, so that the characteristic length of inhomogeneities of the field $\chi'(\boldsymbol{x})$ is in this case close to the internal turbulence length scale $\eta$, while the characteristic length of inhomogeneities in the field $S'(\boldsymbol{x})$ is the same as for the field $n'(\boldsymbol{x})$, i.e., it is equal to the external turbulence length scale $L_0$.

Let us now consider the case where the length scale $\sqrt{\lambda L}$ belongs to the inertial subrange of the turbulence spectrum, i.e., $\eta \ll \sqrt{\lambda L} \ll L_0$. In this case, the factor $1 - \frac{\kappa}{k^2 L} \sin \frac{k^2 L}{\kappa}$ in the first equation (26.75), has a maximum near the point $k = 2\pi/\sqrt{\lambda L}$, i.e., it lies in the inertial subrange of the spectrum. The maximum of the function $F_{\chi\chi}^{(2)}(k)$ is also reached near this point, so that the length scale of inhomogeneities in the field $\chi'(\boldsymbol{x})$ in the $x = L$ plane is of the order $\sqrt{\lambda L}$ [and the length scale of inhomogeneities in the field $S'(\boldsymbol{x})$ is as before equal to $L_0$].

If, finally, $\sqrt{\lambda L} \gg L_0$ then the main contribution to the spectrum of fluctuations in the logarithm of the amplitude is due to large-scale inhomogeneities with scales $l$ in the range $L_0 \leqslant l \leqslant \sqrt{\lambda L}$. The statistical structure of the fields $\chi'(\boldsymbol{x})$ and $S'(\boldsymbol{x})$ for smaller scales, i.e., for $k > 2\pi/L_0$, can be judged from Eq. (26.75) if we remember that in the present case $\frac{k^2L}{\kappa} = 2\pi k^2\lambda L \gg 2\pi k^2 L_0^2 \gg 1$ so that $\left|\frac{\kappa}{k^2L}\sin\frac{k^2L}{\kappa}\right| \ll 1$. Therefore, when $k > 2\pi/L_0$ we have

$$F^{(2)}_{\chi\chi}(k) \approx F^{(2)}_{SS}(k) \approx \pi\kappa^2 L F_{nn}(k). \tag{26.77}$$

In other words, when $\sqrt{\lambda L} \gg L_0$ and $k > 2\pi/L_0$ the spectral functions for the logarithm of the amplitude and phase are equal to each other, and are proportional to the square of the frequency and to the first power of the distance traversed by the wave in the turbulent medium.

Formulas similar to Eq. (26.75) were also derived by Townsend (1965) on the basis of other considerations although he did not point out that the main length scale of the fluctuations is determined by the radius of the first Fresnel zone.

The properties of spectral functions of amplitude and phase fluctuations which were described above are not susceptible to direct experimental verification because the spatial spectral functions cannot be measured directly. However, the time spectral functions, which characterize the spectral composition of amplitude and phase fluctuations at a fixed point in the flow, can be measured directly. These fluctuatuions have the following origin. According to Eq. (26.4), which limits the distances $L$ considered, the turbulence fields are practically constant during the time $L/c$ spent by the wave in traversing the distance $L$, so that the fields $\chi'(\boldsymbol{x})$ and $S'(\boldsymbol{x})$ in the $x = L$ plane are determined by the instantaneous field of turbulent fluctuations in the layer $0 \leqslant x \leqslant L$. For time intervals much greater than $L/c$, the turbulence in the layer $0 \leqslant x \leqslant L$ will vary because of the transport of turbulent fluctuations as a whole by the mean wind, and also as a result of the evolution of individual turbulent elements. The values of $\chi'$ and $S'$ at points in the $x = L$ plane will therefore also change.

Let us consider the time spectral function of the logarithm of the wave amplitude at a fixed point in the $x = L$ plane, subject to the condition $\sqrt{\lambda L} \ll L_0$. When this condition is satisfied the characteristic length scale $l$ of inhomogeneities in the field $\chi'(\boldsymbol{x})$ will also be much less than $L_0$ (since $l \sim \eta$ when $\sqrt{\lambda L} \ll \eta$ and $l \sim \sqrt{\lambda L}$ for $\eta \ll \sqrt{\lambda L} \ll L_0$) and the velocity scale of the corresponding turbulent componets, $v_l \sim (\bar{\varepsilon} l)^{1/3}$, will be much less than the total flow velocity $v \sim (\bar{\varepsilon} L_0)^{1/3}$. Therefore, the time $\tau = l/v$ during which there is correlation between the values of $\chi'$ at a fixed point will be much shorter than the time scale $\tau_l = l/v_l$ for the evolution of individual inhomogeneities of the field $\chi'(\boldsymbol{x})$. Consequently, to determine the time spectral function of the logarithm of the amplitude for $\sqrt{\lambda L} \ll L_0$, it is sufficient to take into account only the transport of the turbulent field by the mean wind, and neglect the evolution of the individual turbulent elements (i.e., we can regard the turbulence as "frozen").

Let $\bar{u}$ be the component of the mean wind velocity in the direction $Ox$ of wave propagation, and let $\bar{v}$ be the component in the perpendicular direction. When $\bar{v}/\bar{u} \gg \sqrt{\lambda/L}$ we can neglect the transport of turbulence formations in the $x$ direction when we determine the time structure function of the variable $\chi'$, since in the time $\tau = l/\bar{v} \lesssim \sqrt{\lambda L}/\bar{v}$ the longitudinal displacement of turbulent inhomogeneities $\bar{u}\tau$ will be small in comparison with $L$, and will have little effect on the values of $\chi'(\boldsymbol{x})$ in the $x = L$ plane. Therefore, fluctuations in the variable $\chi'$ at a fixed point on the $x = L$ plane are determined by the inhomogeneities in the distribution of $\chi'$ on the straight line $\mathscr{L}$ in this plane, which corresponds to the mean wind component in the direction perpendicular to the $Ox$ axis. The one-dimensional spatial spectral function of $\chi'$ on the straight line $\mathscr{L}$ is given by

$$F_{\chi\chi}^{(\mathscr{L})}(k_2) = \int_{-\infty}^{\infty} F_{\chi\chi}^{(2)}\left(\sqrt{k_2^2 + k_3^2}\right) dk_3 = 2 \int_0^{\infty} F_{\chi\chi}^{(2)}\left(\sqrt{k_2^2 + k_3^2}\right) dk_3,$$

where $k_2$ is the component of the wave vector $\boldsymbol{k}$ in the direction of $\mathscr{L}$. Using the "frozen turbulence" hypothesis, we can determine the time spectral function of $\chi'$ from the formula

$$F_{\chi\chi}(\omega) = \frac{1}{\bar{v}} F_{\chi\chi}^{(\mathscr{L})}\left(\frac{\omega}{\bar{v}}\right) = \frac{2}{\bar{v}} \int_0^{\infty} F_{\chi\chi}^{(2)}\left(\sqrt{k^2 + \frac{\omega^2}{\bar{v}^2}}\right) dk, \qquad (26.78)$$

where $\omega$ is the frequency and $\bar{v}$ is the mean wind velocity in the direction $\mathscr{L}$. Using the first equation (26.75), we obtain

$$F_{\chi\chi}(\omega) = \frac{2\pi\kappa^3}{\bar{v}} \left(\frac{\lambda L}{2\pi}\right)^{1/2} \times$$
$$\times\, \Omega \int_0^{\infty} \left[1 - \frac{\sin \Omega^2 (1 + t^2)}{\Omega^2 (1 + t^2)}\right] F_{nn}\left[\Omega \sqrt{\frac{2\pi (1 + t^2)}{\lambda L}}\right] dt, \qquad (26.79)$$

where $\Omega = \frac{\omega}{\bar{v}} \sqrt{\frac{\lambda L}{2\pi}}$ is the dimensionless frequency. Since for small values of the argument $\Omega \sqrt{\frac{2\pi(1+t^2)}{\lambda L}}$ of the function $F_{nn}$ the first factor in the integrand is very small, the function $F_{\chi\chi}(\omega)$ is practically independent of the large-scale structure of the turbulence, and is determined only by the turbulent components from the quasi-equilibrium range of the spectrum. It is easy to show that the result obtained in this way for small values of the dimensionless frequency $\Omega$ is only slightly dependent both on the precise shape of the spectrum $F_{nn}$ in the "dissipation range" (i.e., beyond the small-scale boundary of the inertial subrange) and the precise value of a length scale describing the fall in the spectrum in this range (for example, on the value of the parameter $a$ in the specific model used below). In the high frequency range $\Omega \gg 1$ the spectrum $F_{\chi\chi}(\omega)$ and especially the spectrum $\omega^2 F_{\chi\chi}(\omega)$ of the time derivative $\partial\chi/\partial t$ are very strongly dependent on the shape and length scale of the spectrum $F_{nn}$ in the "dissipation range" of wave numbers (see Gurvich (1968) where this statement is illustrated by results of computations for several spectrum models). Since, however, the true form of the turbulence spectra beyond the small-scale boundary of the inertial subrange is not known at present, we shall consider here as an example only the simple model in which the fall in the spectrum of the refractive-index fluctuations for $k \gtrsim 1/\eta$ is specified by the factor $\exp\{-\alpha\eta^2k^2\}$ where $\alpha$ is a dimensionless constant of the order of unity [see the end of Sect. 22.3, and in particular the formula (22.73)]. In other words, we shall assume that $F_{nn}(k) = \frac{C_n^2}{4\pi} k^{-11/3} \times \exp\{-\alpha\eta^2k^2\}$ for $k \gg 1/L_0$ where for centimeter radio waves $C_n^2 = A^2B^{(\vartheta)}(\bar{N}_T + B^2\bar{N}_\vartheta)\bar{\varepsilon}^{-1/3}$ [see Eq. (26.56)] while for sound waves $C_n^2 = \frac{C_1\bar{\varepsilon}^{2/3}}{c_0^2} + \frac{B^{(T)}\bar{N}_T\bar{\varepsilon}^{-1/3}}{4\bar{T}^2}$. Substituting the above formula for $F_{nn}(k)$ into Eq. (26.79), we can rewrite the latter in the form

$$F_{\chi\chi}(\omega) = \frac{C_n^2 k^3}{4\bar{v}} \left(\frac{\lambda L}{2\pi}\right)^{7/3} \Omega^{-8/3} e^{-\delta^2\Omega^2} \times$$

$$\times \int_0^\infty \left[1 - \frac{\sin \Omega^2(1+t)}{\Omega^2(1+t)}\right] e^{-\delta^2\Omega^2 t} t^{-1/2} (1+t)^{-11/6}\, dt,$$

where $\delta = \eta\sqrt{2\pi\alpha/\lambda L}$. The integral in this formula can be expressed in terms of the confluent hypergeometric Whittaker functions, so that the spectrum is finally given by

$$F_{\chi\chi}(\omega) = \frac{C_n^2 \kappa^3 \sqrt{\pi}}{4\bar{v}} \left(\frac{\lambda L}{2\pi}\right)^{7/3} \Omega^{-8/3} e^{-\delta^2\Omega^2/2} \{\delta^{1/3}\Omega^{1/3} W_{-2/3,\,-2/3}(\delta^2\Omega^2) + \\ + \Omega^{-2/3} \operatorname{Im} (\delta^2 + i)^{2/3} e^{-i\Omega^2/2} W_{-7/6,\,-7/6} [\Omega^2(\delta^2 + i)]\}. \quad (26.80)$$

Using the properties of the Whittaker functions, we find that in the limit as $\omega \to 0$ we have

$$F_{\chi\chi}(0) = \frac{C_n^2 \kappa^3}{4\bar{v}} \left(\frac{\lambda L}{2\pi}\right)^{7/3} K(\delta), \\ K(\delta) = \Gamma\left(-\frac{4}{3}\right)\delta^{8/3} + \Gamma\left(-\frac{7}{3}\right) \operatorname{Im} (\delta^2 + i)^{7/3}. \quad (26.81)$$

Let us discuss in greater detail the properties of the spectral function (26.80) in the case where the length scale $\sqrt{\lambda L}$ belongs to the inertial subrange of the spectrum, i.e., for $L_0 \gg \sqrt{\lambda L} \gg \eta$ [the function $F_{\chi\chi}(\omega)$ for this case has been investigated by Tatarskii (1961a, 1971)]. In that case, $\delta \ll 1$. When $\Omega \ll 1$, the values of the function $F_{\chi\chi}(\omega)$ are close to the quantity (26.81) where $K(\delta) \approx K(0) = \frac{27}{56}\Gamma\left(\frac{2}{3}\right)$. If $1 \ll \Omega \ll 1/\delta$ and, consequently, $\omega \ll \bar{v}/\eta$), then it follows from Eq. (26.80) that

$$F_{\chi\chi}(\omega) \approx \frac{\sqrt{\pi}}{10} \frac{\Gamma(1/3)}{\Gamma(5/6)} \frac{C_n^2 \kappa^3}{\bar{v}} \left(\frac{\lambda L}{2\pi}\right)^{7/3} \Omega^{-8/3}. \quad (26.82)$$

Therefore, in the range of values of $\Omega$ which we are considering, the spectral function will decrease with increasing frequency. Since for $\Omega \ll 1$ it increases, it may be concluded that the maximum of the spectral function should be reached for values of $\Omega$ of the order of unity. Finally, if $\Omega \gg 1/\delta$ then it follows from Eq. (26.80) that

$$F_{\chi\chi}(\omega) \approx \frac{C_n^2 \kappa^3}{4\bar{v}\delta} \left(\frac{\lambda L}{2\pi}\right)^{7/3} \Omega^{-11/3} e^{-\delta^2\Omega^2} = \frac{\pi \kappa^2 L}{\bar{v}\eta\sqrt{a}} F_{nn}\left(\frac{\omega}{\bar{v}}\right). \quad (26.83)$$

Similar conclusions can be established when the length scale $\sqrt{\lambda L}$ belongs to the dissipation range, i.e., when $\sqrt{\lambda L} \ll \eta$. In that case, $\delta \gg 1$. If $\Omega \ll 1/\delta$ then the values of the function $F_{\chi\chi}(\omega)$ are close to

$F_{\chi\chi}(0)$ and we can assume in Eq. (26.81) that $K(\delta) \approx \frac{1}{6}\Gamma\left(\frac{2}{3}\right)\delta^{-4/3}$. A maximum of the function $F_{\chi\chi}(\omega)$ is reached for values of $\Omega$ of the order of $1/\delta$, i.e., for values of $\omega$ of the order of $\bar{v}/\eta$. If $\Omega \gg 1$ then Eq. (26.83) is valid.

Knowing the spectral function $F_{\chi\chi}(\omega)$, we can calculate the mean square of the amplitude fluctuations. Using Eq. (26.78), we obtain

$$\overline{\chi'^2} = \int_{-\infty}^{\infty} F_{\chi\chi}(\omega)\, d\omega = 2\pi \int_0^{\infty} F_{\chi\chi}^{(2)}(k)\, k\, dk. \tag{26.84}$$

If we recall the first formula in Eq. (26.75), and assume as above that $F_{nn}(k) = \frac{C_n^2}{4\pi} k^{-11/3} \exp\{-\alpha\eta^2 k^2\}$, we find that

$$\overline{\chi'^2} = \frac{\pi C_n^2 \kappa^3}{4}\left(\frac{\lambda L}{2\pi}\right)^{11/6} \mathscr{L}(\delta), \quad \mathscr{L}(\delta) = \int_0^{\infty}\left(1 - \frac{\sin x}{x}\right) x^{-11/6} e^{-\delta^2 x^2}\, dx. \tag{26.85}$$

The integral $\mathscr{L}(\delta)$ can be evaluated by expanding the factor $1 - \frac{\sin x}{x}$ in a power series, integrating term by term, and finally summing the resulting series. The final result is

$$\mathscr{L}(\delta) = \frac{36}{55}\Gamma\left(\frac{1}{6}\right) \operatorname{Im}(\delta^2 + i)^{11/6} - \frac{6}{5}\Gamma\left(\frac{1}{6}\right)\delta^{5/3}.$$

When $\delta \ll 1$, i.e., for values of $\sqrt{\lambda L}$ from the inertial subrange $L_0 \gg \sqrt{\lambda L} \gg \eta$, we obtain

$$\mathscr{L}(\delta) \approx \mathscr{L}(0) = \frac{36}{55}\Gamma\left(\frac{1}{6}\right) \times \sin\frac{11\pi}{12},$$

so that

$$\overline{\chi'^2} = \frac{9\pi}{55}\Gamma\left(\frac{1}{6}\right)\sin\frac{11\pi}{12} \cdot C_n^2 \kappa^3 \left(\frac{\lambda L}{2\pi}\right)^{11/6}. \tag{26.86}$$

In that case, the root mean square $\sigma_\chi$ of the fluctuations in the logarithm of the amplitude will increase with distance $L$ in proportion to $L^{11/12}$. This result was established by Tatarskii (1956b). Conversely, when $\delta \gg 1$, i.e., when $\sqrt{\lambda L} \ll \eta$, we have $\mathscr{L}(\delta) \approx \frac{1}{36}\Gamma\left(\frac{1}{6}\right)\delta^{-7/3}$, so that

$$\overline{\chi'^2} = \frac{\pi}{144}\Gamma\left(\frac{1}{6}\right)\alpha^{-7/6} \cdot C_n^2 \eta^{-7/3} L^3. \tag{26.87}$$

In this case $\sigma_\chi$ increases in proportion to $L^{3/2}$.

Let us now briefly consider the determination of the time spectral function of phase fluctuations. If we recall Eq. (26.75), and the derivation of Eq. (26.79), we readily see that for sufficiently high frequencies (for which the frozen-turbulence hypothesis is valid), the spectral function is determined by an expression which differs from Eq. (26.79) only by the replacement of the minus sign in the brackets in the integrand by the plus sign. At low frequencies $\omega$, the spectral function for the phase fluctuations is very dependent on the properties of the large-scale turbulence structure, but at high frequencies (for values of $\omega/\bar{v}$ from the quasi-equilibrium range of the wave number spectrum), the spectrum of the phase fluctuations is determined only by the turbulence compoennts from the quasi-equilibrium range, and can be determined from a formula which differs from Eq. (26.80) only by the replacement of the plus sign in the braces by the minus sign. In particular, for values of $\sqrt{\lambda L}$ belonging to the inertial subrange, and when $1 \ll \Omega \ll 1/\delta$, the spectral function for the phase fluctuations is given by Eq. (26.82), whereas for $\Omega \gg 1/\delta$ it is given by Eq. (26.83).

Let us now consider the calculation of the spatial structure functions for the random fields $\chi'(\boldsymbol{x})$ and $S'(\boldsymbol{x})$ on the $x = L$ plane. The structure functions can be measured directly. In the case which we are considering, i.e., that of locally isotropic turbulence for which, according to Eq. (26.75) the spatial spectra of the fields $\chi'(\boldsymbol{x})$ and $S'(\boldsymbol{x})$ depend only on the modulus of $\boldsymbol{k}$ (the two-dimensional wave vector), the spatial structure functions are expressed in terms of the spatial spectra through the formulas

$$\begin{aligned} D_{\chi\chi}(r) &= \overline{[\chi(\boldsymbol{x}_1) - \chi(\boldsymbol{x}_2)]^2} = 4\pi \int_0^\infty [1 - J_0(kr)]\, F_{\chi\chi}^{(2)}(k)\, k\, dk, \\ D_{SS}(r) &= \overline{[S(\boldsymbol{x}_1) - S(\boldsymbol{x}_2)]^2} = 4\pi \int_0^\infty [1 - J_0(kr)]\, F_{SS}^{(2)}(k)\, k\, dk, \end{aligned} \tag{26.88}$$

where $r$ is the distance between the points $x_1$ and $x_2$ in the $x=L$ plane, and $J_0(x)$ is the Bessel function of order zero.

Since for small $k$ the factor $1-J_0(kr)$ in the integrand of Eq. (26.88) is very small, so that the structure functions are practically independent of the behavior of the spatial spectra for small $k$, we need not consider the form of the spectra for $k \lesssim 1/L_0$. Let us now use again the model in which $F_{nn}(k)=\frac{C_n^2}{4\pi}k^{-11/3}\exp\{-\alpha\eta^2k^2\}$. The formulas for $F_{\chi\chi}^{(2)}(k)$ and $F_{SS}^{(2)}(k)$, which are given by Eq. (26.88) after the substitution of Eq. (26.75) into them, can be transformed to the form

$$D_{\chi\chi}(r)=C_n^2\pi\kappa^3\left(\frac{\lambda L}{2\pi}\right)^{11/6}\int_0^\infty[1-J_0(\rho t)]\left(1-\frac{\sin t^2}{t^2}\right)t^{-8/3}e^{-\delta^2t^2}\,dt,$$

$$D_{SS}(r)=C_n^2\pi\kappa^3\left(\frac{\lambda L}{2\pi}\right)^{11/6}\int_0^\infty[1-J_0(\rho t)]\left(1+\frac{\sin t^2}{t^2}\right)t^{-8/3}e^{-\delta^2t^2}\,dt. \tag{26.89}$$

where $\rho=r\sqrt{\frac{2\pi}{\lambda L}}$. If we expand the factor $1-J_0(\rho t)$ under the integral signs into power series, and integrate term by term, we obtain

$$D_{\chi\chi}(r)=\frac{3\pi}{5}\Gamma\left(\frac{1}{6}\right)C_n^2\kappa^3\left(\frac{\lambda L}{2\pi}\right)^{11/6}\left\{\delta^{5/3}\left[M\left(-\frac{5}{6},1,-\frac{\rho^2}{4\delta^2}\right)-1\right]-\right.$$
$$\left.-\frac{6}{11}\,\mathrm{Im}\,(\delta^2+i)^{11/6}\left[M\left(-\frac{11}{6},1,-\frac{1}{4}\frac{\rho^2}{\delta^2+i}\right)-1\right]\right\}, \tag{26.90}$$

where $M(\alpha,\gamma,x)=1+\frac{\alpha}{\gamma}\frac{x}{1!}+\frac{\alpha(\alpha+1)}{\gamma(\gamma+1)}\frac{x^2}{2!}+\cdots$ is the confluent hypergeometric function. The formula for $D_{SS}(r)$ will differ from Eq. (26.90) only by the opposite sign between the two terms in braces. Using the power expansions for the confluent hypergeometric functions for $\rho\ll\delta$ (i.e., for $r\ll\eta$), we find that to within terms of the order of $\rho^2/\delta^2$ we have

$$D_{\chi\chi}(r)=\frac{\pi}{8}\Gamma\left(\frac{1}{6}\right)\alpha^{-1/6}C_n^2\kappa^2L\eta^{-1/3}r^2\cdot\mathcal{M}(\delta),$$
$$\mathcal{M}(\delta)=1-\frac{6}{5}\delta^2\,\mathrm{Im}\left(1+\frac{i}{\delta^2}\right)^{5/6}. \tag{26.91}$$

A similar formula is valid for $D_{SS}(r)$ except that the sign in front of the second term in the function $\mathcal{M}(\delta)$ is now different.

Let us consider the properties of the structure functions in the case where the length scale $\sqrt{\lambda L}$ belongs to the inertial subrange (so that $\delta \ll 1$). When $r \ll \eta$ the functions $D_{\chi\chi}(r)$ and $D_{SS}(r)$ are given by Eq. (26.91) in which we can set $\mathcal{M}(\delta) = 1$. When $\eta \ll r \ll \sqrt{\lambda L}$ we find from Eq. (26.90) that

$$D_{\chi\chi}(r) = D_{SS}(r) = \frac{9\pi}{25\sqrt[3]{4}} \frac{\Gamma(1/6)}{\Gamma(5/6)} C_n^2 \kappa^2 L r^{5/3}. \tag{26.92}$$

Finally, when $r \gg \sqrt{\lambda L}$ and $D_{SS}(r)$ we obtain a formula which differs from Eq. (26.92) only by a factor of 2 [a formula of this kind was first obtained by Krasil'nikov (1945)], while $D_{\chi\chi}(r)$ reaches the limiting value $2\overline{\chi'^2}$.

When $\sqrt{\lambda L} \ll \eta$ (i.e., $\delta \gg 1$) and $r \ll \eta$ the formula for $D_{\chi\chi}(r)$ is given by Eq. (26.91) with $\mathcal{M}(\delta) = 7\delta^{-4}/216$, while $D_{SS}(r)$ is given by a similar formula with $\mathcal{M}(\delta) = 2$. When $r \gg \eta$ the function $D_{SS}(r)$ is equal to twice the value given by Eq. (26.92), and the function $D_{\chi\chi}(r)$ reaches the limiting value $2\overline{\chi'^2}$.

A number of the above theoretical results have been verified in experiments on the propagation of sound, light, and radio waves in a turbulent atmosphere. As an example, Fig. 109 shows the results of experiments by Krasil'nikov and Ivanov-Shits (1949) who measured the root mean square phase difference $\sigma_{\delta S}(r) = [D_{SS}(r)]^{1/2}$ and the root mean square fluctuations $\sigma_\chi = [\overline{\chi'^2}]^{1/2}$ in the logarithm of the amplitude of a sound wave at 3 kHz over distances $L = 22$, 45, and 67 m (the length scale $\sqrt{\lambda L}$ is then found to lie in the inertial subrange of the spectrum) and for a mean wind velocity $\bar{v} = 5$ m/sec. In fact, these workers measured not the spatial but the time structure function for the phase which is obtained from $D_{SS}(r)$ by replacing $r$ with $\bar{v}\tau$ where $\tau$ is the time interval (if we adopt the frozen-turbulence hypothesis). The measurements were carried out for $\tau = 0.04$, 0.08, and 0.2 sec which correspond to $r = 20$, 40, and 100 cm in the inertial subrange of the spectrum. Under these conditions the function $D_{SS}(r)$ should

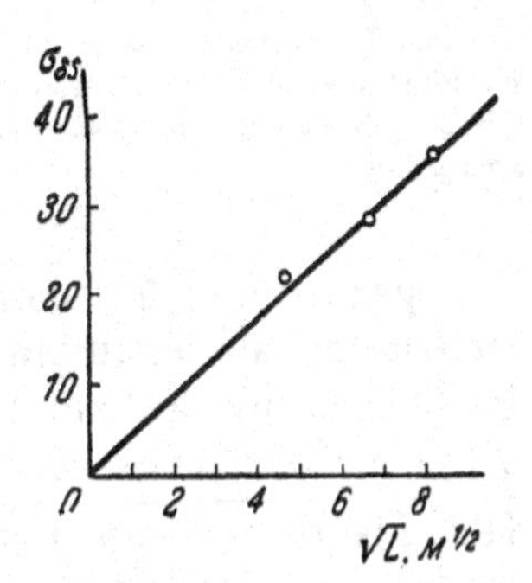


Fig. 109 Root mean square phase difference $\sigma_{\delta S}$ as a function of the square root of the distance $L$ traversed by a wave [Krasil'nikov and Ivanov-Shits (1949)]. Points correspond to values of $\sigma_{\delta S}$ averaged over $\tau = r/\bar{v}$.

be given by Eq. (26.92), so that $\sigma_{\delta S} \sim L^{1/2} r^{5/6}$. This prediction is in good agreement with the experimental data of Fig. 110 (and also with the measurements of Krasil'nikov (1953) for ultrasonic waves at frequencies up to 50 kHz). The quantity $\overline{\chi'^2}$ should then be given by Eq. (26.86) so that $\sigma_\chi \sim L^{11/12}$. This result is in satisfactory agreement with the data of Fig. 111. The same relationship was verified by Dunn (1965) who used measurements of fluctuations in the amplitude of a sound wave in sea water. Here again the agreement between theory and experiment was found to be completely satisfactory.

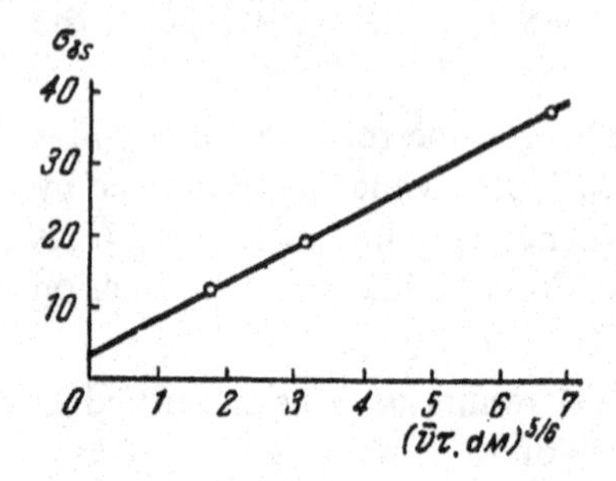


**Fig. 110** The dependence of $\sigma_{\delta s}$ on $r = \overline{v}\tau$, according to Krasil'nikov and Ivanov-Shits (1949). Points correspond to values of $\sigma_{\delta s}$ averaged over $L$.

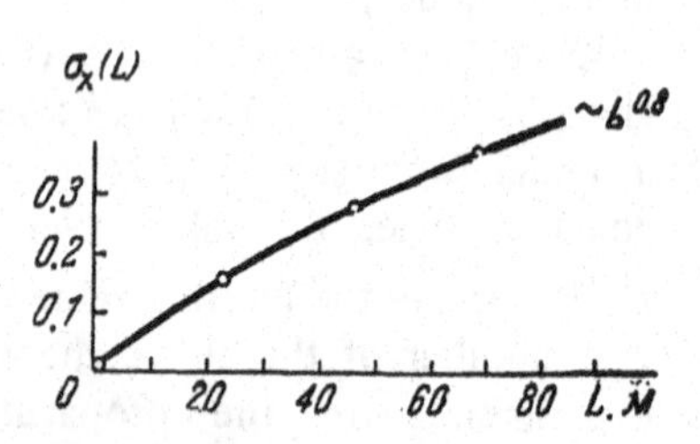


**Fig. 111** Dependence of $\sigma_\chi(L)$ on the distance $L$ traversed by a wave [Krasil'nikov and Ivanov-Shits (1949)].

Dependence of fluctuations in the amplitude of sound waves on meteorological conditions in the atmospheric surface layer was investigated by Suchkov (1958) who measured the quantity $\sigma_\chi$ for waves at frequencies between 3 and 76 kHz and small distances $L$ [the quantity $\overline{\chi'^2}$ should again be given by Eq. (26.86) in this case]. At the same time measurements were carried out on the temperature and wind velocity profiles, so that the meteorological parameter $C_n^2 = \frac{C_1 \overline{\varepsilon}^{2/3}}{c_0^2} + \frac{B^{(T)} \overline{N}_T \overline{\varepsilon}^{-1/3}}{4\overline{T}^2}$ could be approximately determined (from the formulas derived in Sect. 7.5 of Vol. 1 for the quantities $\overline{\varepsilon}$ and $\overline{N}_T$). The correlation coefficient between the measured and calculated values of $\lg \sigma_\chi$ using Eq. (26.86) was found to be 0.90 (see Fig. 112).

Measurements of the time spectra of fluctuations in the amplitude and phase difference of sound waves at 2.6 and 8.5 kHz in the surface layer of the atmosphere were carried out by Golitsyn,

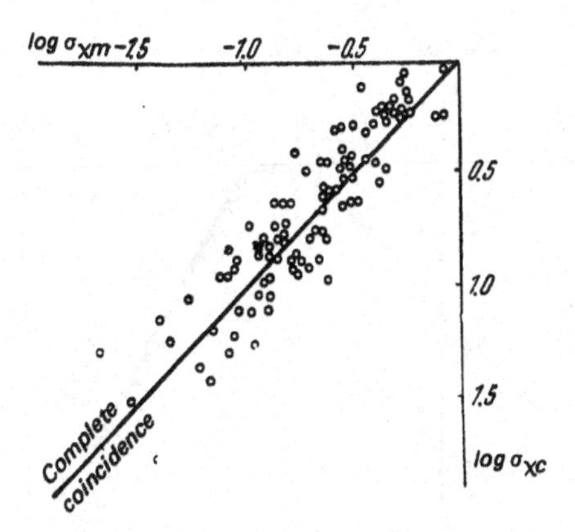


**Fig. 112** Comparison of the measured $\sigma_{\chi m}$ and calculated $\sigma_{\chi c}$ [from Eq. (26.86)] values of log $\sigma_\chi$ [Suchkov (1958)].

Gurvich, and Tatarskii (1960) for $L$ between 21 and 80 m. The spectra were determined in the frequency range between 0.05 and 1,160 Hz. To eliminate the effect of meteorological conditions they analyzed only the dimensionless spectral function

$$U_{\chi\chi}(\Omega) = \frac{\omega F_{\chi\chi}(\omega)}{\overline{\chi'^2}}$$

(and the analogous dimensionless spectral function for the phase differences). The length scale $\sqrt{\lambda L}$ in these experiments corresponded to the inertial subrange of the spectrum, so that formulas (26.81), (26.82), and (26.86) could be used for $F_{\chi\chi}(\omega)$ and $\overline{\chi'^2}$. It is then readily verified that, when $\Omega \ll 1$, the quantity $U_{\chi\chi}(\Omega)$ is proportional to $\Omega$, and when $1 \ll \Omega \ll 1/\delta$ it is proportional to $\Omega^{-5/3}$. These results are in good agreement with the experimental data in Fig. 113.

In the experiments by Tatarskii et al. mentioned at the end of Sect. 26.1, which were concerned with the scintillation of a terrestrial light source, the condition $\sqrt{\lambda L} \gg \eta$ was satisfied. These workers also verified the formula $\sigma_\chi \sim L^{11/12}$. Finally, they measured the correlation function for fluctuations in the logarithm of the amplitude

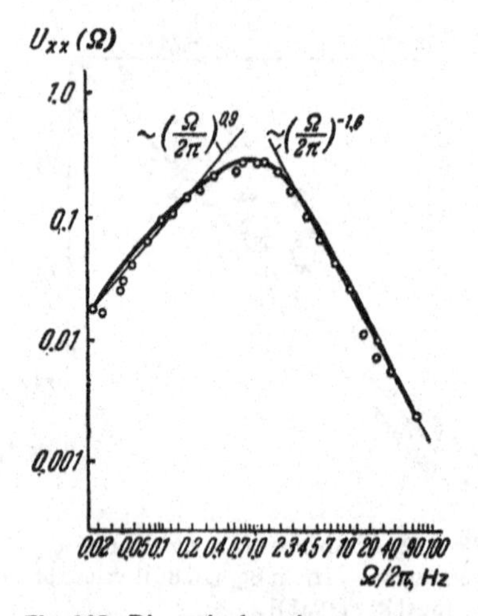


**Fig. 113** Dimensionless time spectrum of fluctuations in the logarithm of the wave amplitude $U_{\chi\chi}(\Omega)$ [Golitsyn, Gurvich, and Tatarskii (1960)].

$$R_{\chi\chi}(r) = \frac{B_{\chi\chi}(r)}{\overline{\chi'^2}} = 1 - \frac{D_{\chi\chi}(r)}{2\overline{\chi'^2}},$$

which for $\sqrt{\lambda L} \gg \eta$ and $r \gg \eta$ should depend only on $r/\sqrt{\lambda L}$ (when $\eta \ll r \ll L$ we should have $R_{\chi\chi}(r) = 1 - \text{const}\left(\frac{r}{\sqrt{\lambda L}}\right)^{5/3}$). This prediction was satisfactorily confirmed by data (see Fig. 114).

The statistical characteristics of fluctuations in the phase of decimeter and centimeter radio waves during their propagation

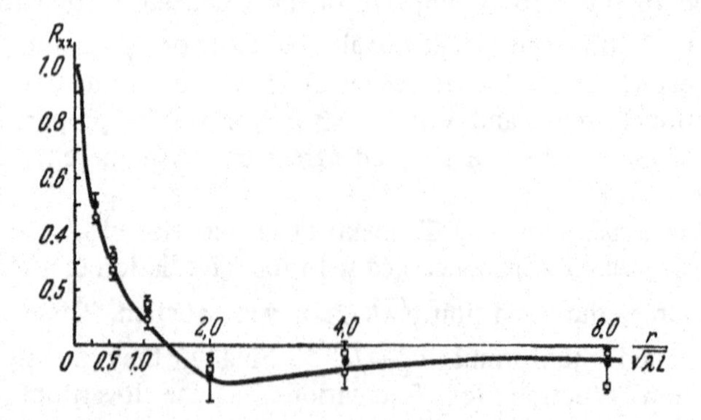


**Fig. 114** Correlation function for fluctuations in the light intensity in the surface layer $R_{\chi\chi}(r/\sqrt{\lambda L})$ [Tatarskii (1959)].

through the troposphere were investigated in a number of experimental studies. Thus, Herbstreit and Thompson (1955) verified the relation $D_{SS}(r) \sim \kappa^2$ which follows from Eq. (26.92). In these experiments the value of $D_{SS}(r)$ was measured with a base of $r = 150$ m, and the above relation was satisfactorily verified for frequencies between $10^8$ and $10^{10}$ Hz. Herbstreit and Thompson also measured the time spectrum of phase fluctuations for radio waves of 1046 Hz and $L = 18.5$ km between points at heights of 4300 and 1950 m above sea level. They found that in the frequency range between 1/3600 and 1/36 Hz the spectral density of the phase fluctuations was proportional to frequency raised to the power of $-8/3$ (Fig. 115). Similar spectra were obtained by Norton (1959) at a frequency of 9414 MHz and $L = 30$ km. We note that the relation $F(\omega) \sim \omega^{-8/3}$ for $\omega/\bar{v}$ in the inertial subrange of wave numbers is a consequence of theoretical considerations. However, in the above experiments this result was obtained not only for $\omega/\bar{v}$ from the inertial subrange, but also at much lower frequencies. Thus, in the experiments of Herbstreit and Thompson the mean velocity of transport of turbulent inhomogeneities in the refractive index by the wind was of the order of $\bar{v} \sim 3$ m/sec, and the frequency range for which the result $F(\omega) \sim \omega^{-8/3}$ was obtained corresponded to scales of turbulent components of between 0.1 and 10 km, which are well outside the limits of the intertial subrange.

The analysis of the experimental data on centimeter radio wave phase fluctuations makes it possible to estimate the atmospheric

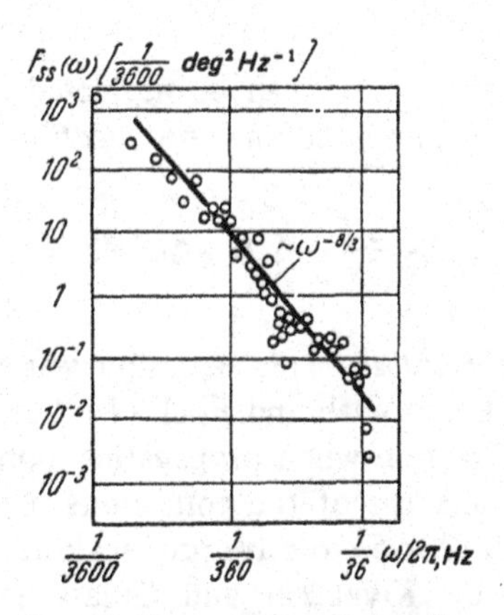


**Fig. 115** Time spectrum of phase fluctuations for decimeter waves propagating in the troposphere [Herbstreit and Thompson (1955)].

values of the dimensional coefficient $C_n$ entering the five-thirds law for refractive index spectra. According to Tatarskii (1960) in the low troposphere $C_n$ has usually the order of $10^{-8}$ cm$^{-3}$, though in some cases it can differ from this value by a factor of the order of 10 or 0.1.

### 26.5 Strong Fluctuations of Wave Amplitude

All the major results of the previous subsections of this section were obtained by Rytov's method of smooth perturbations. The basic equation (26.29) of this method can be derived from Eq. (26.20) in the following way. The solution of Eq. (26.20) is sought as a plane wave $E(x, \rho) = U(x, \rho) \exp(i\kappa x)$ with a slowly varying complex amplitude $U$. Substitution of this expression for $E$ into Eq. (26.20) yields the following equation for $U$

$$2i\kappa \frac{\partial U}{\partial x} + \frac{\partial^2 U}{\partial x^2} + \frac{\partial^2 U}{\partial y^2} + \frac{\partial^2 U}{\partial z^2} + 2\kappa^2 n' U = 0. \qquad (26.93)$$

The smallest characteristic length scale of the $U$ variation along the $x$-axis is the Kolmogorov scale $\eta = (\nu^3/\bar{\epsilon})^{1/4}$. Hence the first term in Eq. (26.93) is of order $\delta U/\lambda\eta$ where $\delta U$ is a typical change of the field $U$ over a distance $\eta$, and the second term is of order of $\delta U/\eta^2$. We see that if the condition

$$\lambda \ll \eta \qquad (26.94)$$

is fulfilled the term $\partial^2 U/\partial x^2$ can be neglected in Eq. (26.93). This leads to the so-called *parabolic wave propagation equation*

$$2i\kappa \frac{\partial U}{\partial x} + \Delta_\perp U + 2\kappa^2 n' U = 0 \qquad (26.95)$$

where $\Delta_\perp = \partial^2/\partial y^2 + \partial^2/\partial z^2$. This equation was used apparently for the first time by Leontovich and Fock (1948) in their investigation of radio wave diffraction when propagating along the earth surface. A more detailed analysis of the conditions of applicability of the parabolic equation (26.95) to wave propagation in a random medium was undertaken by Klyatskin and Tatarskii (1970a) [see also Tatarskii (1970, 1971)]; it shows that the condition of smallness of the total scattering at a wave-length distance must be valid in addition to the condition (26.94), and moreover the propagation distance $L$ must be bounded from above by a length of the order of $\lambda/\sigma_{n'}^2$.

If a solution of Eq. (29.95) is sought in the form $U(x, \rho) = U_0 \exp [\psi'(x, \rho)]$, a nonlinear equation

$$2i\kappa \frac{\partial \psi'}{\partial x} + \Delta_\perp \psi' + (\nabla_\perp \psi')^2 + 2\kappa^2 n' = 0 \qquad (26.96)$$

is obtained for $\psi'(x, \rho)$ (here $\nabla_\perp$ is a transverse two-dimensional gradient operator). The basic equation (26.29) of Rytov's method results from linearization of Eq. (26.96). The conditions justifying neglect of the $(\nabla_\perp \psi')^2$ term in Eq. (26.96) are discussed in the papers by Shirokova (1959), Pisareva (1959), and Tatarskii (1962a). The latter of these papers contains an evaluation of corrections to $\overline{\chi'^2}$ due to the term $(\nabla_\perp \psi')^2$ in equation (26.96). If $\sigma_1^2$ is a first-order approximation to $\overline{\chi'^2}$ evaluated by solving the linearized equation (26.96) for $\psi'$, the correction to $\overline{\chi'^2}$ due to the nonlinear term $(\nabla_\perp \psi')^2$ proves to be proportional to $(\sigma_1^2)^2$. Therefore the condition for applicability of the first-order Rytov approximation can be written as $\sigma_1^2 \ll 1$.

However the quantity $\sigma_1^2 \propto C_n^2 \kappa^{7/6} L^{11/6}$ [see Eq. (26.86)] often becomes large as compared to unity in real cases of light propagation in a turbulent atmosphere. Equation (26.86) is clearly incorrect in these cases. The experiments by Gracheva and Gurvich (1965), Gurvich, Kallistratova, and Time (1968), Gracheva, Gurvich, and Kallistratova (1970a, b), Livingston, Deitz, and Alcaraz (1970), and Mordukhovich (1970) show that the variable $\overline{\chi'^2}$ stops increasing with the growth of $L$ when $\sigma_1^2 \gtrsim 1$. Instead, a kind of rough saturation of the fluctuation intensity was observed by Gracheva and Gurvich (1965) while more recent experiments even show that there is a slow decrease of $\overline{\chi'^2}$ with distance when distance is large enough [see, for example, Fig. 116 taken from the paper of Gracheva, Gurvich, and Kallistratova (1970b)].

To explain such behavior of $\overline{\chi'^2}$ it is necessary to retain the nonlinear term $(\nabla_\perp \psi')^2$ in Eq. (26.96). However then we come to a very complicated nonlinear problem. The solution of the problem with the aid of the Feynman diagram method yields graphs coinciding with those considered by Wyld (1961) in his turbulence theory (see Sect. 19.6 above). Let us suppose that the condition $\eta \ll \sqrt{L\lambda} \ll L_0$ is valid and that the refractive index spectrum $F_{nn}(k)$ has an inertial subrange form, i.e., $F_{nn}(k) \propto k^{-11/3}$. Then the evaluation of any of the higher-order terms of the perturbation series yields a power function of the variable $\sigma_1^2$. Following Tatarskii (1967b) we can deduce from this that $\overline{\chi'^2} = f(\sigma_1^2)$ in this case (i.e., the true $\overline{\chi'^2}$ is a single-valued function of its first approximation $\sigma_1^2$). This deduction agrees well with all the experimental data from the above-mentioned sources (cf. Fig. 116).

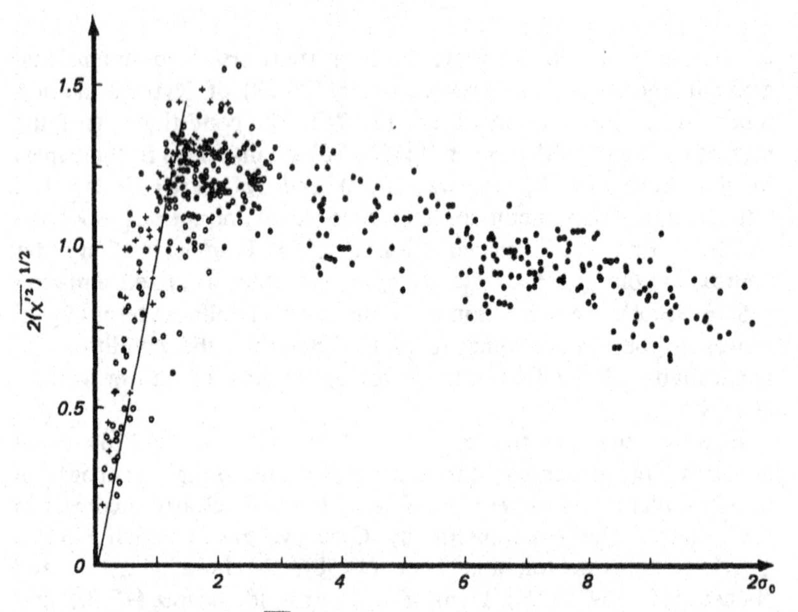


**Fig. 116** The dependence of $\overline{\chi'^2}$ on $\sigma_1^2$ according to experiments on light propagation in an atmospheric surface layer (after Gracheva, Gurvich and Kallistratova, 1970b). ○, measurements of 1967, $L = 250$ m; +, measurements of 1968, $L = 250$ m; •, measurements of 1968, $L = 1750$ m; —, $\overline{\chi'^2} = \sigma_1^2$.

Shishov (1968) established a simplified summation method for the perturbation theory series and by using it derived an approximate equation describing the wave intensity fluctuations in the strong fluctuation region $\sigma_1^2 \gg 1$. Similar results were also obtained at approximately the same time by a number of authors using various methods, in particular by Dolin (1968), Chernov (1969), Beran and Ho (1969), Tatarskii (1969, 1970), Klyatskin (1969), Klyatskin and Tatarskii (1970a, b). Below we shall briefly consider the variant of strong fluctuation theory developed in the works of Tatarskii and Klyatskin.

Let us first of all note that the transformation from the solution (26.71) of the equation for an amplitude fluctuation spectrum to the simplified formula (26.75) valid at $kL \gg 1$ is equivalent to the assumption that the spectral function $F_{nn}^{(2)}(k, |\xi - \zeta|)$ has the following form

$$F_{nn}^{(2)}(k, |\xi - \zeta|) \approx 2\pi F_{nn}(k)\delta(\xi - \zeta) \qquad (26.97)$$

In fact, substituting the right-hand side of Eq. (26.97) into Eq. (26.71) we obtain immediately Eq. (26.75) by simple integration. We also note that $\int_0^\infty F_{nn}^{(2)}(k, \xi)d\xi = \pi F_{nn}(k)$ and hence $F_{nn}(k) = F_{nn}(k)$ in Eq. (26.97) is a usual three-dimensional spectrum of refractive index fluctuations. The approximation (26.97) seems at first glance rather unnatural since it introduces an artificial anisotropy in the statistical description of the refractive index field [cf. Eq. (26.97′) below which is equivalent to Eq. (26.97)]. There is however a forcible physical argument in favor of the applicability of Eq. (26.97) in practical problems. It is the fact that there is a marked anisotropy in the formulation of the problem itself and the longitudinal inhomogeneities of the refractive index field play a quite special role. Moreover, the length scale of such inhomogeneities contributing mostly to the amplitude fluctuations is almost always much less than the wave path $L$ in a turbulent medium. This explains why the formal passage to the limit as the longitudinal correlation length of refractive index fluctuations tends to zero yields the quite useful approximation (26.75).

The zero-correlation-length assumption (26.97) plays an important part also in the Tatarskii-Klyatskin strong fluctuation theory. In the discussion below it will be of importance that this assumption can also be formulated in terms of the spatial refractive index correlation function:

$$\overline{n'(x_1, \rho_1)n'(x_2, \rho_2)} = \delta(x_1 - x_2)A(\rho_1 - \rho_2) \qquad (26.97')$$

The function $A(\rho_1 - \rho_1)$ on the right-hand side of Eq. (26.97′) can be determined from the condition requiring that the integral over the $x$-axis of the true correlation function coincides with the integral of the approximation (26.97′). This condition gives

$$A(\rho) = \int_{-\infty}^\infty B_{nn}(x, \rho)dx = 2\pi \int_{-\infty}^\infty\int F_{nn}(k)e^{ik\rho}dk$$

So that $A(\boldsymbol{\rho}) = A(|\boldsymbol{\rho}|) = A(\rho)$. In the following we shall assume the random field $n'(x, \boldsymbol{\rho})$ to be Gaussian, considering this assumption as the simplest nontrivial and relatively reasonable approximation (in fact this assumption can be considerably weakened; see below). Then the zero-correlation-length condition together with Eq. (4.27) of Vol. 1 imply the following equations for higher-order moments of the field $n'(x, \boldsymbol{\rho})$:

$$\overline{n'(x_1, \boldsymbol{\rho}_1)n'(x_2, \boldsymbol{\rho}_2)\dots n'(x_{2k}, \boldsymbol{\rho}_{2k})}$$
$$= \Sigma\delta(x_{\alpha_1} - x_{\alpha_2})\dots\delta(x_{\alpha_{2k-1}} - x_{\alpha_{2k}})A(\boldsymbol{\rho}_{\alpha_1} - \boldsymbol{\rho}_{\alpha_2})\dots A(\boldsymbol{\rho}_{\alpha_{2k-1}} - \boldsymbol{\rho}_{\alpha_{2k}}) \tag{26.98}$$

where the summation is taken over all the permutations of $2k$ integers $1, 2, \dots, 2k$ taken as indices $\alpha_1, \alpha_2, \dots, \alpha_{2k}$.

We shall proceed from the parabolic equation (26.95) which is equivalent to the *nonlinearized* equation (26.96). Let us show that the field $U(x, \boldsymbol{\rho})$ depends functionally only on the "preceding" values of the refractive index field, i.e., on values $n'(\xi, \boldsymbol{\rho})$ at $0 \leqslant \xi \leqslant x$. By using the identity

$$2i\kappa \frac{\partial U(x, \boldsymbol{\rho})}{\partial x} + 2\kappa^2 n'(x, \boldsymbol{\rho})U(x, \boldsymbol{\rho})$$
$$= 2i\kappa \left[\exp i\kappa \int_0^x n'(\xi, \boldsymbol{\rho})d\xi\right] \frac{\partial}{\partial x}\left\{U(x, \boldsymbol{\rho})\left[\exp - i\kappa \int_0^x n'(\xi, \boldsymbol{\rho})d\xi\right]\right\}$$

we can rewrite Eq. (26.95) in the form

$$2i\kappa \frac{\partial}{\partial x}\left\{U(x, \boldsymbol{\rho})\left[\exp - i\kappa \int_0^x n'(\xi, \boldsymbol{\rho})d\xi\right]\right\}$$
$$= - \left[\exp - i\kappa \int_0^x n'(\xi, \boldsymbol{\rho})d\xi\right]\Delta_\perp U(x, \boldsymbol{\rho})$$

Let us integrate this equation over $x$ in the range $(0, x)$ and designate by $U_0(\boldsymbol{\rho})$ the "initial condition" $U(0, \boldsymbol{\rho})$ of the solution $U(x, \boldsymbol{\rho})$. If we multiply the result of the integration by $\left[\exp i\kappa \int_0^x n'(\xi, \rho)d\xi\right]$, we obtain an integral equation

$$2i\kappa\, U(x,\rho) = 2i\kappa\, U_o(\rho)e^{i\kappa\int_0^x n'(\xi,\rho)d\xi} - \int_0^x dx' e^{i\kappa\int_{x'}^x n'(\xi,\rho)d\xi}\, \Delta_\perp U(x',\rho) \quad (26.99)$$

Now by iteration we can express the solution $U(x, \rho)$ in a series form

$$U(x,\rho) = U_o(\rho)e^{i\kappa\int_0^x n'(\xi,\rho)d\xi} - \frac{1}{2i\kappa}\int_0^x e^{i\kappa\int_{x'}^x n'(\xi,\rho)d\xi}\, \Delta_\perp\left[U_o(\rho)e^{i\kappa\int_0^{x'} n'(\xi_1,\rho)d\xi_1}\right]dx' + \cdots \quad (26.100)$$

This expression (and also the original equation (26.99)) show clearly that $U(x, \rho)$ depends functionally only on the values of $n'(\xi, \rho)$ at $0 \leqslant \xi \leqslant x$. Using the functional derivative formalism introduced in Sect. 4.4 of Vol. 1 we can rewrite this condition in the form

$$\frac{\delta U(x,\rho)}{\delta n'(x_1,\rho_1)dx_1} = 0 \quad \text{for} \quad x_1 > x \quad (26.101)$$

Of course, in the general case there is a *statistical correlation* between the variable $U(x, \rho)$ and the "subsequent" values of $n'(x_1, \rho_1)$ despite the validity of the nonstatistical independence condition (26.101), i.e., generally speaking, $\overline{U(x,\rho)n'(x_1,\rho_1)} \neq 0$ when $x_1 > x$. This can be easily confirmed by multiplication of Eq. (26.100) by $n'(x_1, \rho_1)$ and averaging the product. However, if we assume that the condition (26.98) is valid, then the variable $U(x, \rho)$ will also be *statistically independent* on the "subsequent" values of $n'(x_1, \rho_1)$, i.e.,

$$\overline{U(x_1,\rho_1)\ldots U(x_k,\rho_k)n'(\xi_1,\rho_1')\ldots n'(\xi_l,\rho_l')} = \overline{U(x_1,\rho_1)\ldots U(x_k,\rho_k)}\;\overline{n'(\xi_1,\rho_1')\ldots n'(\xi_l,\rho_l')} \quad (26.102)$$

when all $\xi_j, j = 1, \ldots, l$, are greater than all $x_i, i = 1, \ldots, k$. The validity of this relation becomes evident if we use the expression of $U(x_i, \rho_i)$ in the form of a power series of the "preceding" values of $n'(\xi, \rho')$ implied by Eq. (26.100) and then apply Eq. (26.98).

Let us now pass to the derivation of the moment equations for the complex random field $U(x, \rho)$. Let us select any $n + m$ points in the $x$ = const plane and write down Eq. (26.95) for each of $n$ points $(x, \rho_1) \ldots, (x, \rho_n)$ and complex conjugate equation for each of $m$ other points $(x, \rho_1'), \ldots, (x, \rho_m')$:

$$2i\kappa \frac{\partial U(x, \rho_1)}{\partial x} + \Delta_\perp(\rho_1) U(x, \rho_1) + 2\kappa^2 n'(x, \rho_1) U(x, \rho_1) = 0,$$

$$\cdots\cdots\cdots\cdots\cdots\cdots\cdots\cdots\cdots\cdots\cdots$$

$$2i\kappa \frac{\partial U(x, \rho_n)}{\partial x} + \Delta_\perp(\rho_n) U(x, \rho_n) + 2\kappa^2 n'(x, \rho_n) U(x, \rho_n) = 0,$$

$$2i\kappa \frac{\partial U^*(x, \rho_1')}{\partial x} - \Delta_\perp(\rho_1') U^*(x, \rho_1') - 2\kappa^2 n'(x, \rho_1') U^*(x, \rho_1') = 0,$$

$$\cdots\cdots\cdots\cdots\cdots\cdots\cdots\cdots\cdots\cdots\cdots$$

$$2i\kappa \frac{\partial U^*(x, \rho_m')}{\partial x} - \Delta_\perp(\rho_m') U^*(x, \rho_m') - 2\kappa^2 n'(x, \rho_m') U^*(x, \rho_m') = 0. \tag{26.103}$$

Multiply each of the equations (26.103) by a product of the amplitudes $U$ and $U^*$ entering all the other equations and sum the results. Then we shall obtain the equation

$$2i\kappa \frac{\partial \gamma_{n,m}}{\partial x} + [\Delta_\perp(\rho_1) + \cdots + \Delta_\perp(\rho_n) - \Delta_\perp(\rho_1') - \cdots - \Delta_\perp(\rho_m')]\gamma_{n,m}$$
$$+ 2\kappa^2 F(x; \rho_1, \ldots, \rho_n; \rho_1', \ldots, \rho_m')\gamma_{n,m} = 0 \tag{26.104}$$

where

$$\gamma_{n,m}(x; \rho_1, \ldots, \rho_n; \rho_1', \ldots, \rho_m') = U(x, \rho_1) \ldots U(x, \rho_n) U^*(x, \rho_1') \ldots U^*(x, \rho_m')$$

$$F(x; \rho_1, \ldots, \rho_n; \rho_1', \ldots, \rho_m') = n'(x, \rho_1) + \cdots + n'(x, \rho_n) - n'(x, \rho_1') - \cdots - n'(x, \rho_m')$$

Using now the identity

$$2i\kappa \frac{\partial \gamma_{n,m}}{\partial x} + 2\kappa^2 F\gamma_{n,m} = 2i\kappa\left\{\exp i\kappa \int_0^x F(\xi;\rho_1,\ldots,\rho_m')d\xi\right\}$$
$$\times \frac{\partial}{\partial x}\left[\gamma_{n,m}\left\{\exp - i\kappa \int_0^x F(\xi,\rho_1,\ldots,\rho_m')d\xi\right\}\right]$$

we can rewrite Eq. (26.104) in the form

$$2i\kappa \frac{\partial}{\partial x}\left[\gamma_{n,m}(x;\rho_1,\ldots,\rho_m')\left\{\exp - i\kappa \int_0^x F(\xi;\rho_1,\ldots,\rho_m')d\xi\right\}\right]$$
$$+\left\{\exp - i\kappa \int_0^x F(\xi;\rho_1,\ldots,\rho_m')d\xi\right\} [\Delta_\perp(\rho_1)$$
$$+ \cdots - \Delta_\perp(\rho_m')]\gamma_{n,m}(x;\rho_1,\ldots,\rho_m') = 0 \quad (26.105)$$

Let us integrate Eq. (26.105) with respect to $x$ in the range $(0, x)$ and then multiply the result by $\exp\left\{i\kappa\int_0^x F(\xi,\rho_1,\ldots,\rho_m')d\xi\right\}$. This procedure yields an equation

$$2i\kappa\gamma_{n,m}(x;\rho_1,\ldots,\rho_m') - 2i\gamma_{n,m}^{(0)}\left\{\exp i\kappa\int_0^x F(\xi;\rho_1,\ldots,\rho_m')d\xi\right\}$$
$$+\int_0^x dx'\left\{\exp i\kappa\int_{x'}^x F(\xi;\rho_1,\ldots,\rho_m')d\xi\right\}[\Delta_\perp(\rho_1)$$
$$+ \cdots - \Delta_\perp(\rho_m')]\gamma_{n,m}(x',\rho_1,\ldots,\rho_m') = 0 \qquad (26.106)$$

where

$$\gamma_{n,m}^{(0)} = U_o(\rho_1)\ldots U_o(\rho_n)U_o^*(\rho_1')\ldots U_o^*(\rho_m')$$

is the initial value of the function $\gamma_{n,m}$ in the $x = 0$ plane. Let us examine now the mean value of the equation (26.106). Let

$$\overline{\gamma_{n,m}(x;\rho_1,\ldots,\rho_m')} = \overline{U(x,\rho_1)\ldots U(x,\rho_n)U^*(x,\rho_1')\ldots U^*(x,\rho_m')}$$
$$= M_{n,m}(x;\rho_1,\ldots,\rho_m')$$

be the general higher-order moment of the complex field $U(x, \rho)$ and

$$\overline{\left[\exp i\kappa \int_{x'}^{x} F(\xi; \rho_1, \ldots, \rho'_m)\, d\xi\right]} = \Omega(x; x'; \rho_1, \ldots, \rho'_m) \quad (26.107)$$

When averaging the integrand in the third term on the left-hand side of Eq. (26.106) we shall keep in mind that the function $\exp i\kappa \int_{x'}^{x} F(\xi; \rho_1, \ldots, \rho'_m) d\xi$ contains only values of $n'(\xi, \rho)$ at $\xi \geqslant x'$, whereas the argument of the function $\gamma_{n,m}$ is equal to $x'$. Taking power series expansion of the exponent we may use the relation (26.102) for each of the expansion terms (the point $\xi = x'$ clearly yields zero contribution to the integral and hence can be disregarded). After summing again the exponent series we shall have the relation

$$\overline{\left\{\exp i\kappa \int_{x'}^{x} F(\xi; \rho_1, \ldots, \rho'_m) d\xi\right\} [\Delta_\perp(\rho_1) + \cdots - \Delta_\perp(\rho'_m)]\, \gamma_{n,m}(x'; \rho_1, \ldots, \rho'_m)}$$

$$= \overline{\left\{\exp i\kappa \int_{x'}^{x} F(\xi; \rho_1, \ldots, \rho'_m) d\xi\right\}} \cdot [\Delta_\perp(\rho_1) + \cdots - \Delta_\perp(\rho'_m)]\, \overline{\gamma_{n,m}(x'; \rho_1, \ldots, \rho'_m)}$$

Its substitution in the averaged equation (26.106) leads immediately to the following closed integral equation for the moment $M_{n,m}$:

$$2i\kappa M_{n,m}(x; \rho_1, \ldots, \rho'_m) - 2i\gamma^{(0)}_{n,m}\, \Omega(x; 0; \rho_1, \ldots, \rho'_m) + \int_0^x dx'\, \Omega(x; x'; \rho_1, \ldots, \rho'_m)[\Delta_\perp(\rho_1) + \cdots + \Delta_\perp(\rho_n) - \Delta_\perp(\rho'_1) - \cdots - \Delta_\perp(\rho'_m)] M_{n,m}(x'; \rho_1, \ldots, \rho'_m) = 0 \quad (26.108)$$

Let us now find an explicit expression for the function $\Omega$ in terms of the statistical characteristics of the field $n'(x, \rho)$. Consider the random variable

$$Z = \int_{x'}^{x} F(\xi; \rho_1, \ldots, \rho_n; \rho'_1, \ldots, \rho'_m) d\xi$$

This variable is an integral of a Gaussian random function and hence its probability distribution is also Gaussian.[7] On the other hand

$$\Omega = \overline{\exp i\kappa Z} = \varphi_Z(\kappa)$$

is a value at $\theta = \kappa$ of the characteristic function $\varphi_Z(\theta)$ of the random variable $Z$. But for a Gaussian random variable with mean value $m$ and variance $\sigma^2$, the characteristic function is equal to $\exp\{i\theta m - \theta^2\sigma^2/2\}$ by virtue of the general equation (4.30) of Vol. 1. Hence it is sufficient to determine $\overline{Z}$ and $\sigma_Z^2$. It is clear that $\overline{Z} = 0$ since $\overline{n'} = 0$. Moreover the variance of $Z$ is given by the equation

$$\sigma_Z^2 = \overline{\int_{x'}^{x} d\xi_1 \int_{x'}^{x} d\xi_2 \left[\sum_{i=1}^{n} n'(\xi_1, \rho_i) - \sum_{k=1}^{m} n'(\xi_1, \rho_k')\right]\left[\sum_{j=1}^{n} n'(\xi_2, \rho_j) - \sum_{l=1}^{m} n'(\xi_2, \rho_l')\right]}$$

Using Eq. (26.97′) we can easily transform this to the form $\sigma_Z^2 = (x - x')Q_{n,m}$, where

$$Q_{n,m} = \sum_{i=1}^{n}\sum_{j=1}^{n} A(\rho_i - \rho_j) - 2\sum_{i=1}^{n}\sum_{l=1}^{m} A(\rho_i - \rho_l') + \sum_{k=1}^{m}\sum_{l=1}^{m} A(\rho_k' - \rho_l') \tag{26.109}$$

Hence

$$\Omega(x; x'; \rho_1, \ldots, \rho_n; \rho_1', \ldots, \rho_m') = \exp\left\{-\frac{1}{2}\kappa^2(x - x')Q_{n,m}(\rho_1, \ldots, \rho_m')\right\} \tag{26.110}$$

Let us substitute Eq. (26.110) into Eq. (26.108) and multiply all the terms by $\exp(\kappa^2/2)xQ_{n,m}$. We obtain

[7] This is the main point where we use the assumption of the Gaussian nature of the refractive-index field $n'(x, \rho)$. We know, however, that the distribution of the integrals of random functions will be approximately Gaussian under rather wide conditions. Hence it is reasonable to hope therefore that the assumptions on $n'(x, \rho)$ can in fact be considerably weakened.

$$2i\kappa M_{n,m} e^{\frac{\kappa^2 x}{2} Q_{n,m}} - 2i\kappa \gamma_{n,m}^{(0)}$$

$$+ \int_0^x dx' e^{\frac{\kappa^2 x'}{2} Q_{n,m}} [\Delta_\perp(\rho_1) + \cdots - \Delta_\perp(\rho'_m)] M_{n,m}(x') = 0$$

Differentiating this result with respect to $x$ and then canceling the joint factor $\exp(\kappa^2 x/2) Q_{n,m}$ we obtain the following differential equation for the moment $M_{n,m}$:

$$2i\kappa \frac{\partial M_{n,m}}{\partial x} + i\kappa^3 Q_{n,m} M_{n,m}$$

$$+ [\Delta_\perp(\rho_1) + \cdots + \Delta_\perp(\rho_n) - \Delta_\perp(\rho'_1) - \cdots - \Delta_\perp(\rho'_m)] M_{n,m} = 0 \quad (26.111)$$

Note that these moment equations are not coupled, i.e, each of them is closed. The whole set of moment equations (26.111) turns out to be equivalent to a single functional derivative equation for a characteristic functional of the field $U$ [see Tatarskii (1969, 1970, 1971)]. All the equations (26.111) and also the corresponding functional derivative equation are second order partial (or functional) derivative equations of generalized parabolic type resembling the Fokker-Planck-Kolmogorov diffusion equation for Markov processes, and hence the field $U(x, \rho)$ itself resembles a diffusion Markov process with the space variable $x$ playing the role of time. For this reason Tatarskii and Klyatskin called the above approximation the *Markov (or diffusion) approximation of the wave propagation theory in a random medium.*

It should be emphasized that the small fluctuation condition was not used in the derivation of equations (26.111). However, a more detailed analysis shows that the conditions used (e.g., zero-correlation-length approximation and the validity of the parabolic wave propagation equation) place some rather weak restrictions on the path covered by a wave in a turbulent medium [see Klyatskin (1969), Klyatskin and Tatarskii (1970a), Tatarskii (1970, 1971)]. It is important that these restrictions do not hinder the application of equations (26.111) in the strong fluctuation region where $\sigma_1^2 \gtrsim 1$.

The above mentioned restrictions can be found most easily from an analysis of the successive approximations to the exact solution of the moment problem for a field $U(x, \rho)$. Such a sequence of approximations was developed by Klyatskin and Tatarskii (1971). The first approximation is given above. The second approximation is obtained when the zero-correlation-length approximation is applied to the equation for the variable:

$$\overline{F(x;\rho_1,\ldots,\rho_n;\rho_1',\ldots,\rho_m')\gamma_{n,m}} = L_{n,m}$$

which appears in the averaged equation (26.104), but it is not applied to the more elementary equation for the moment $M_{n,m} = \gamma_{n,m}$. It can be shown that such an application of a zero-correlation-length approximation yields a closed system of two coupled equations with unknowns $M_{n,m}$ and $L_{n,m}$ instead of a single closed equation (26.111) for $M_{n,m}$. The higher-order approximations are established quite similarly; the $s$th approximation leads to a closed system of $s$ equations with $s$ unknowns which permits evaluation of the moment $M_{n,m}$. Klyatskin and Tatarskii did not give a strict proof of the convergence of the sequence of approximations to an exact solution. Instead, they considered the application of the approximations to a model example of Eq. (26.95) with the term $\Delta_\perp U$ rejected, which has a simple analytical solution. In this special case the second approximation of the sequence gave an excellent description of the exact solution of all values of $x$. Hence Klyatskin and Tatarskii decided that the conditions for the closeness of the second approximation to the first one must also yield the restrictions necessary for the applicability of the first (i.e., Markov) approximation. As a result they obtained very simple conditions for the applicability of a Markov approximation to a mean field $\bar{U} = M_1$ and to the most important second moment $M_{1,1}$.

Let us now consider particular cases of the moment equation (26.111) for two simple low-order moments. If $m = n = 1$ then

$$M_{1,1}(x;\rho_1,\rho_1') = \overline{U(x,\rho_1)U^*(x,\rho_1')} = \Gamma_2(x,\rho_1,\rho_1'),$$
$$Q_{1,1}(\rho_1,\rho_1') = 2[A(0) - A(\rho_1 - \rho_1')] = 2\pi H(\rho_1 - \rho_1'),$$

where

$$H(\rho) = H(\rho) = \frac{1}{\pi}[A(0) - A(\rho)] = 2\int F_{nn}(k)[1 - \cos k\rho]\,dk \qquad (26.112)$$

(integration is taken over the whole two-dimensional $k$-space).

Equation (26.111) for the moment $\Gamma_2$ has the form

$$\frac{\partial \Gamma_2(x,\rho_1,\rho_1')}{\partial x} + \frac{1}{2i\kappa}[\Delta_\perp(\rho_1) - \Delta_\perp(\rho_1')]\Gamma_2 + \pi\kappa^2 H(\rho_1 - \rho_1')\Gamma_2 = 0. \tag{26.113}$$

It is worth noting that it contains a single statistical quantity $H(\rho)$ which depends only on small-scale turbulent fluctuations of the refractive-index field.

Equation (26.113) can be explicitly solved if we use the two-dimensional vectors $\boldsymbol{R} = (\rho_1 + \rho_1')/2$ and $\rho = \rho_1 - \rho_1'$ as new coordinates instead of $\rho_1$ and $\rho_1'$ and then apply Fourier transformation with respect to $\boldsymbol{R}$. The solution has the following form [see Dolin (1964)]

$$\Gamma_2\left(x, \boldsymbol{R} + \frac{\rho}{2}, \boldsymbol{R} - \frac{\rho}{2}\right) = \widetilde{\Gamma}_2(x,\boldsymbol{R},\rho)$$

$$= \frac{1}{4\pi^2}\int d\boldsymbol{R}'\int d\boldsymbol{p}\, U_0\left[\boldsymbol{R}' + \frac{1}{2}\left(\rho - \frac{\boldsymbol{p}x}{\kappa}\right)\right] U_0^*\left[\boldsymbol{R}' - \frac{1}{2}\left(\rho - \frac{\boldsymbol{p}x}{\kappa}\right)\right]$$

$$\times \exp\left\{i\boldsymbol{p}(\boldsymbol{R} - \boldsymbol{R}') - \frac{\pi\kappa^2}{4}\int_0^x H\left[\rho - \frac{\boldsymbol{p}}{\kappa}(x - \xi)\right] d\xi\right\}. \tag{26.114}$$

For a plane wave case when $U_o$ = const all integrals in Eq. (26.114) can be explicitly evaluated and we obtain

$$\widetilde{\Gamma}_2(x,\boldsymbol{R},\rho) = \widetilde{\Gamma}_2(x,\rho) = |U_o|^2 \exp\left\{-\frac{\pi\kappa^2 x}{4} H(\rho)\right\}. \tag{26.115}$$

If $\rho = |\rho|$ belongs to the inertial subrange of length scales we can assume that $F_{nn}(k) \sim C_n^2 k^{-11/3}$ in Eq. (26.112). This leads to a result of the form

$$\widetilde{\Gamma}_2(x,\boldsymbol{R},\rho) = |U_o|^2 \exp\left\{-\text{const}\ C_n^2 \kappa^2 x \rho^{5/3}\right\} \tag{26.116}$$

The power of the exponential function in this equation is

proportional to the sum of the structure functions $D_{XX}$ and $D_{SS}$ of Eqs. (26.88) and (26.92).

Equation (26.114) is examined in greater detail in the works of Klyatskin and Tatarskii (1970b) and Tatarskii (1970, 1971) where the relation of the function $\Gamma_2$ to a so-called radiation transfer equation is established (cf. also Dolin, 1964) and formula (26.114) is used for the analysis of the mean intensity distribution in beams of light.

Function $\Gamma_2(x, \rho_1, \rho_1')$ is called the *second-order coherence function*. It describes the mean structure of a diffraction image in the focal plane of a lens. If we have a lens of surface $\Sigma$ in the plane $x =$ const, and $F$ is its focal distance, then the field $U_F(\rho)$ in the focal plane will be expressed in terms of values of a random field $E(x, \rho) = U(x, \rho) \exp(i\kappa x)$ over the lens surface by the so-called Debye equation:

$$U_F(\rho) = \frac{\kappa \exp\left\{i\kappa(F + x) + \dfrac{i\kappa\rho^2}{2F}\right\}}{2\pi i F} \int_\Sigma U(x, \rho') \exp\left\{-\frac{i\kappa\rho\rho'}{F}\right\} d\rho'$$

Hence the field intensity in a focal plane is given by the equation

$$I_F(\rho) = |U_F(\rho)|^2$$

$$= \left(\frac{\kappa}{2\pi F}\right)^2 \int_\Sigma d\rho' \int_\Sigma d\rho'' U(x, \rho') U^*(x, \rho'') \exp\left\{-\frac{i\kappa\rho(\rho' - \rho'')}{F}\right\}$$

and for mean intensity we obtain the equation

$$\bar{I}_F(\rho) = \left(\frac{\kappa}{2\pi F}\right)^2 \int_\Sigma d\rho' \int_\Sigma d\rho'' \Gamma_2(x, \rho', \rho'') \exp\left\{-\frac{i\kappa\rho(\rho' - \rho'')}{F}\right\} \qquad (26.117)$$

If the incident wave is a plane wave then the second-order coherence function $\Gamma_2$ is given by Eq. (26.115). Let us consider the case when the lens surface $\Sigma$ is a circle of radius $a$ and the refractive-index fluctuations are isotropic in any $x =$ const plane, i.e., $H(\rho) = H(\rho)$. It is easy to show that in this case also $\bar{I}_F(\rho) = \bar{I}_F(\rho)$ is independent of the direction of $\rho$ and

$$\bar{I}_F(\rho) = \frac{\kappa^2 a^2}{\pi F^2} |U_o|^2 \int_0^{2a} \left[\cos^{-1}\frac{\rho'}{2a} - \frac{\rho'}{2a}\sqrt{1-\left(\frac{\rho'}{2a}\right)^2}\right]$$

$$\exp\left\{-\frac{\pi\kappa^2 x}{4} H(\rho')\right\} J_o\left(\frac{\kappa\rho\rho'}{F}\right)\rho' d\rho' \quad (26.118)$$

Conversely, the isotropic distribution of the mean intensity in the focal plane of a circular lens implies the isotropy of the refractive index fluctuations in the $x$ = const planes [i.e., the relation $H(\rho) = H(\rho)$] in the range of length scales $\rho$ mostly responsible for formation of the image.

Focal plane photographs of diffraction images of laser beams have been obtained by Gurvich, Pakhomov, and Cheremukhin (1971). In these experiments a beam propagated over a distance of 50 m at a height 1.3 m above a flat steppe before being photographed. The corresponding amplitude fluctuations were small (i.e.,$\chi'^2 \ll 1$) and the authors referred to Eq. (26.75) for $F^{(2)}_{XX}$, but their results can be as well interpreted by Eqs. (26.117) and (26.118) which are valid both for small and strong amplitude fluctuations. The mean images (i.e., mean intensity distribution determined by averaging over 80 individual photographs) were found to be completely isotropic (i.e., circularly symmetric) within the accuracy of the measurements. This fact permitted the assertion that possible anisotropy of the two-dimensional refractive index (i.e., temperature) spectrum does not exceed 10% in the range of length scales from 0.1 to 1 cm (and apparently to a few centimeters as well). In other words, these experiments give a direct confirmation of the isotropy of the statistical state of small scale temperature fluctuations in the atmospheric surface layer.

Formula (26.118) permits the determination of the distribution of the image mean intensity and also permits in principle the determination of the structure function $H(\rho)$ from known $\bar{I}_F(\rho)$ by its inversion. All the calculations become simpler if $\exp\left\{-(\pi\kappa^2 x/4)H(\rho')\right\}$ is small enough at $\rho' = 2a$ (i.e., $C_n^2\kappa^2 x a^{5/3} \gg 1$ in the case of $a$ belonging to the inertial subrange of length scales). Then the main contribution to the integral on the right of Eq. (26.118) is supplied by the range $\rho' \ll 2a$ and hence we can replace the expression in brackets by its value at $\rho'/2a = 0$ (i.e., $\pi/2$) and replace the integration limit $2a$ by $\infty$. In this case

$$\bar{I}_F(\rho) = \frac{\kappa^2 a^2 |U_0|^2}{2F^2} \int_0^\infty \exp\left\{-\frac{\pi\kappa^2 x}{4} H(\rho')\right\} J_0\left(\frac{\kappa\rho\rho'}{F}\right)\rho' d\rho' \qquad (26.119)$$

i.e., the function $\bar{I}_F(\rho)$ is here a zero order Hankel transform of $\exp\left\{-\frac{\pi\kappa^2 x}{4} H(\rho')\right\}$. It is also possible to use a narrow moving slit in a focal plane to isolate the field $U_F(y, z)$ along a line $y = \text{const}$ and to measure with such a device the integral $\bar{J}_F(y) = \int_{-\infty}^{\infty} \bar{I}_F(y, z)dz$; then it is easy to show that the function $\bar{J}_F(y)$ will be represented by the Fourier cosine transform of $\exp\left\{-\frac{\pi\kappa^2 x}{4} H(\rho')\right\}$ (cf. Artem'ev and Gurvich, 1971). Measurements of $\bar{I}_F(\rho)$ for a wide range of values of $\rho$ were performed by Vlasova, Markus, and Cheremukhin (1971) while the measurements of $\bar{J}_F(y)$ (in a more restricted range of $y$-values) were done by Artem'ev and Gurvich (1971). Both sets of data were compared with computations based on existing theoretical predictions about the refractive index spectrum $F_{nn}(k)$, but the results turn out to be somewhat different. The experimental points of Vlasova et al. agree more or less satisfactorily with the theoretical curve implied by Eq. (26.119) supplemented by the inertial range spectrum equation $F_{nn}(k) = C_n^2(4\pi)^{-1}k^{-11/3}$ (i.e., $H(\rho) \sim C_n^2\rho^{5/3}$). The data of Artem'ev and Gurvich show however significant deviations from the inertial range theoretical curve and the authors explained them by using a refined spectral equation of the form $F_{nn}(k) \sim C_n^2(\alpha k\eta)^{-11/6}K_{-11/6}(\alpha k\eta)$ where $K_m$ is the modified Bessel function of the third kind of order $m$ (the so-called Basset or Macdonald function); this spectrum decreases exponentially in the dissipation range $\alpha k\eta > 1$ and it implies that

$$H(\rho) \sim C_n^2(\alpha\eta)^{5/3}\left[(1 + \rho^2/(\alpha^2\eta^2)^{5/6} - 1\right]$$

According to Artem'ev and Gurvich such a form for $H(\rho)$ yields a theoretical $\bar{J}_F$-curve which satisfactorily agrees with the data if the length $\alpha\eta$ is taken of the order of 0.1–1 cm.

Let us now consider the fourth-order moment

$$\begin{aligned} M_{2,2}(x; \rho_1, \rho_2; \rho_1', \rho_2') &= \overline{U(x, \rho_1)U(x, \rho_2)U^*(x, \rho_1')U^*(x, \rho_2')} \\ &= \Gamma_4(x, \rho_1, \rho_1', \rho_2, \rho_2') \end{aligned}$$

This function is called the *fourth-order coherence function* and it describes intensity fluctuations of the radiation field. In fact, since $I(x,\rho) = |U(x,\rho)|^2 = U(x,\rho)U^*(x,\rho)$, the intensity correlation function is equal to

$$B_{II}(x,\rho_1,\rho_2) = \overline{I(x,\rho_1)I(x,\rho_2)} - \overline{I(x,\rho_1)}\,\overline{I(x,\rho_2)}$$
$$= \Gamma_4(x,\rho_1,\rho_2,\rho_1,\rho_2) - \Gamma_2(x,\rho_1,\rho_2)\Gamma_2(x,\rho_1,\rho_2),$$

i.e., it can be expressed in terms of the functions $\Gamma_4$ and $\Gamma_2$. If we substitute $n = m = 2$ in the general equations (26.111) and (26.109) we obtain

$$Q_{2,2}(\rho_1,\rho_2;\rho_1',\rho_2') = 2\pi F(\rho_1,\rho_2;\rho_1',\rho_2')$$

where

$$F(\rho_1,\rho_2;\rho_1',\rho_2') = H(\rho_1 - \rho_1') + H(\rho_1 - \rho_2') + H(\rho_2 - \rho_1') + H(\rho_2 - \rho_2') - H(\rho_1 - \rho_2) - H(\rho_1' - \rho_2') \quad (26.120)$$

and

$$\frac{\partial\Gamma_4(x,\rho_1,\rho_2,\rho_1',\rho_2')}{\partial x} + \frac{1}{2i\kappa}\left[\Delta_\perp(\rho_1) + \Delta_\perp(\rho_2) - \Delta_\perp(\rho_1') - \Delta_\perp(\rho_2')\right]\Gamma_4 + \pi\kappa^2 F(\rho_1,\rho_2,\rho_1',\rho_2')\Gamma_4 = 0 \quad (26.121)$$

The appropriate initial condition for Eq. (26.121) is

$$\Gamma_4(0,\rho_1,\rho_2,\rho_1',\rho_2') = U_o(\rho_1)U_o(\rho_2)U_o^*(\rho_1')U_o^*(\rho_2')$$

Unlike Eq. (26.113) for the second-order coherence function $\Gamma_2$, no explicit analytical solution could be found for Eq. (26.121). Tatarskii (1970, 1971) obtained an approximate (and apparently rather crude) solution of this equation for the case of a plane incident wave (i.e., $U_o$ = const) and a pure inertial subrange spectrum of refractive index [i.e., $F_{nn}(k) = C_n^2(4\pi)^{-1}k^{-11/3}$, $H(\rho) \sim C_n^2\rho^{5/3}$]. Let us introduce the notation $\beta^2 = (\overline{I^2} - \overline{I}^2)/\overline{I}^2$ for the relative intensity fluctuation and denote by $\beta_1^2$ the approximate value of the same quantity evaluated with the aid of Rytov's method of smooth perturbations. It is easy to show that in the small fluctuation case $\sigma_1^2 \ll 1$ we shall have $\beta_1^2 = 4\sigma_1^2 \sim C_n^2 \kappa^{7/6} L^{11/6}$. Tatarskii's results suggest that $\beta^2 = f(\beta_1^2)$ where $f(z) = z$, i.e., $\beta^2 = \beta_1^2$ for $\beta_1^2 \ll 1$ and $f = \beta^2$ tends to a constant $\beta_\infty^2 \approx 1.4$ as $\beta_1^2 \to \infty$ (subsequent more accurate asymptotic calculations of Klyatskin (1971a) confirms the existence of a limit $\beta_\infty^2$ but show that $\beta_\infty^2 = 1$). Such behavior of the function $f$ agrees qualitatively with the data presented in Fig. 116. In the strong fluctuation region $\beta_1^2 \gg 1$ the theory involves not a single length scale $\sqrt{\lambda L}$ characteristic of the small fluctuation region, but two scales $l_1 = (C_n^2 \kappa^3)^{-3/11} \sim \beta_1^{-6/11}\sqrt{\lambda L}$ and $l_2 = \beta_1^{6/5}\sqrt{\lambda L}$. Hence the condition $\eta \ll \sqrt{\lambda L} \ll L_0$ is now insufficient to justify the use of an inertial subrange spectrum of refractive index; rather, two conditions $\eta \ll l_1, l_2 \ll L_0$ must be satisfied.

In the paper by Dagkesamanskaya and Shishov (1970) the results of numerical solution of Eq. (26.121) are presented under the assumption that the refractive index correlation function has a "Gaussian" form (12.5′). The resulting dependence of $\beta^2$ on the propagation distance $L$ is characterized by a maximum at a finite value of $L$ followed by a slow decrease of $\beta^2$ in the region of greater $L$. This also agrees qualitatively with the data in Fig. 116. A quantitative comparison of the theory with data is however impossible at the present time for two reasons: first, the theory of Tatarskii is very approximate while the computations of Dagkesamanskaya and Shishov are based on a quite artificial model of the refractive index correlation function; second, the existing data relate to statistics of the logarithm of the amplitude (in accordance with the most usual form of the small fluctuation theory), while the strong fluctuation theory deals with the fluctuations of intensity, which is a squared amplitude. Therefore we must wait for a solution of Eq. (26.121) corresponding to a more realistic model of the statistical properties of refractive index fluctuations and also for future data related directly to intensity (and not logarithm of intensity) fluctuations.

## 27. STELLAR SCINTILLATION

### 27.1 Fluctuations in the Amplitude and Phase of Star Light Observed on the Earth's Surface

The results of the preceding section can be applied to many problems connected with the investigation of fluctuations in the characteristics of waves propagating through the earth's atmosphere or ocean. We have already mentioned in this section two important problems of this type: The problem of extra long-range propagation of centimeter radio waves due to atmospheric scattering and the remote atmospheric sensing problem. Now as a third typical example, let us consider the statistical description of the behavior of stellar images in telescopes. The images exhibit random displacements in the field of view (which we shall call *jitter*). These are associated with fluctuations in the angle of arrival of the light waves. Second, there are random variations in the intensity of the image (which we shall call *scintillations*). These are associated with fluctuations in the amplitudes of the waves. Third, for sufficiently large zenith distances there are also random changes in the color of the objects (we shall call these *chromatic scintillations*) which are associated with differences in the fluctuations of wave amplitude at different wavelengths.

To describe all these phenomena we can use the results of the preceding section subject, however, to two modifications. First, since starlight passes through the entire earth's atmosphere, we must take into account the variation with height in the statistical characteristics of the refractive index, which are produced by variations in the turbulent state of the atmosphere. Second, when we describe the amplitude fluctuations we must take into account the fact that because of its finite dimensions, the objective of the telescope tends to smooth out these fluctuations.

Let us first consider the dependence of the statistical characteristics of the refractive index on height in the atmosphere. We must replace the first of the formulas in Eq. (26.70) by the more general equation

$$\overline{n'(\boldsymbol{x}_1)\,n'(\boldsymbol{x}_2)} = B_{nn}\left(\boldsymbol{\rho}_1-\boldsymbol{\rho}_2;\ |x_1-x_2|;\ \frac{x_1+x_2}{2}\right) = \int e^{-i\boldsymbol{k}(\boldsymbol{\rho}_1-\boldsymbol{\rho}_2)} F_{nn}^{(2)}\left(\boldsymbol{k};\ |x_1-x_2|;\ \frac{x_1+x_2}{2}\right)d\boldsymbol{k}. \quad (27.1)$$

where $x$ is the vertical coordinate. Strictly speaking, the functions $B_{nn}$ and $F_{nn}^{(2)}$ [and also the other correlation and spectral functions in Eq. (26.70)] depend also on $(\boldsymbol{\rho}_1+\boldsymbol{\rho}_2)/2$. However, the correlation functions have a much weaker dependence on $(\boldsymbol{\rho}_1+\boldsymbol{\rho}_2)/2$ than on $\boldsymbol{\rho}_1-\boldsymbol{\rho}_2$, and when the derivatives of these functions with respect to the components of the vectors $\boldsymbol{\rho}_1$ and $\boldsymbol{\rho}_2$ are calculated it is sufficient to differentiate only with respect to the argument $\boldsymbol{\rho}_1-\boldsymbol{\rho}_2$, while derivatives with respect to the argument $(\boldsymbol{\rho}_1+\boldsymbol{\rho}_2)/2$ can be neglected. Our subsequent calculations will not be affected by the dependence on $(\boldsymbol{\rho}_1+\boldsymbol{\rho}_2)/2$, and we shall not consider it here.

When Eq. (27.1) is employed the formula (26.71) and the analogous formula for $F_{SS}^{(2)}(\boldsymbol{k})$ assume the form

$$\begin{aligned} F_{XX}^{(2)}(\boldsymbol{k}) &= \kappa^2 \int_0^L\!\!\int_0^L \sin\frac{k^2(L-\xi)}{2\kappa}\sin\frac{k^2(L-\zeta)}{2\kappa} F_{nn}^{(2)}\left(\boldsymbol{k};\ |\xi-\zeta|;\ \frac{\xi+\zeta}{2}\right)d\xi\,d\zeta, \\ F_{SS}^{(2)}(\boldsymbol{k}) &= \kappa^2 \int_0^L\!\!\int_0^L \cos\frac{k^2(L-\xi)}{2\kappa}\cos\frac{k^2(L-\zeta)}{2\kappa} F_{nn}^{(2)}\left(\boldsymbol{k};\ |\xi-\zeta|;\ \frac{\xi+\zeta}{2}\right)d\xi\,d\zeta. \end{aligned} \quad (27.2)$$

To the same accuracy as in the derivation of Eq. (26.73), these integrals can be reduced to the form

$$\kappa^2 \int_0^{L/2}\left[1 \mp \cos\frac{k^2(L-\zeta)}{\kappa}\right]d\zeta \int_0^{2\zeta} F_{nn}^{(2)}(\boldsymbol{k};\ \xi;\ \zeta)\,d\xi +$$

$$+\kappa^2 \int_{L/2}^{L} \left[1 \mp \cos \frac{k^2(L-\zeta)}{\kappa}\right] d\zeta \int_{0}^{2(L-\zeta)} F_{nn}^{(2)}(\boldsymbol{k};\, \xi;\, \zeta)\, d\xi,$$

where the upper sign corresponds to the function $F_{\chi\chi}^{(2)}(\boldsymbol{k})$ and the lower to the function $F_{SS}^{(2)}(\boldsymbol{k})$. Henceforth, we shall consider very high values $L \gg 1/k$, so that over most of the region of integration with respect to $\zeta$ the upper limits of the inner integrals will exceed $1/k$. Therefore, as in the transition from Eqs. (26.73) to (26.75), the inner integrals in the case of locally isotropic turbulence can be approximately replaced by $\pi F_{nn}(k;\, \zeta)$ where $F_{nn}(k;\, \zeta)$ is the three-dimensional spectral density function of the field $n'(x)$. The final result therefore is

$$F_{\chi\chi}^{(2)}(k) \approx 2\pi\kappa^2 \int_0^L \sin^2 \frac{k^2(L-\zeta)}{2\kappa} F_{nn}(k;\, \zeta)\, d\zeta,$$

$$F_{SS}^{(2)}(k) \approx 2\pi\kappa^2 \int_0^L \cos^2 \frac{k^2(L-\zeta)}{2k} F_{nn}(k;\, \zeta)\, d\zeta. \tag{27.3}$$

If $F_{nn}$ is independent of $\zeta$, then Eq. (27.3) leads directly to the corresponding formulas in Eq. (26.75). It is clear from Eq. (27.3) that inhomogeneities in the refractive index which lie close to the observation point $\zeta = L$ have very little effect on $F_{\chi\chi}^{(2)}(k)$, since the factor $\sin^2 \frac{k^2(L-\zeta)}{2\kappa}$ becomes very small near the observation point.

We shall be interested not in the spectral functions (27.3) themselves, but in the mean square of fluctuations in the logarithm of the amplitude

$$\overline{\chi'^2} = 2\pi \int_0^\infty F_{\chi\chi}^{(2)}(k)\, k\, dk$$

and in the structure functions $D_{\chi\chi}(r)$ and $D_{SS}(r)$ which can be defined in terms of the spectral density functions through Eq. (26.88).

The quantity $\overline{\chi'^2}$ is only slightly dependent on the properties of the large-scale structure of the refractive-index field, since for small $k$, corresponding to scales much greater than $\sqrt{\lambda L}$, the values of the function $F_{\chi\chi}^{(2)}(k)$ are small because the factor $\sin^2 \frac{k^2(L-\zeta)}{2\kappa}$ is small in Eq. (27.3) (we note that $L$ corresponds to the path of the wave in the essentially turbulent layer of the atmosphere, and its values are at most of the order of some tens of kilometers; because the wavelengths of visible light are small, the length scale $\sqrt{\lambda L}$ is small, i.e., of the order of 10 cm; see below). The quantities $D_{\chi\chi}(r)$ and $D_{SS}(r)$ for $r$ from the quasi-equilibrium range (we shall restrict our attention to these values of $r$ only) also depend only weakly on the large-scale structure of the refractive-index field because the factor $1 - J_0(kr)$ is small for $k \ll 1/r$. Therefore, when we calculate $\overline{\chi'^2}$, $D_{\chi\chi}(r)$ and $D_{SS}(r)$ we can use the formulas (27.3), which are valid for $k \gg 1/L$. It seems reasonable to take $F_{nn}(k;\, \zeta)$ in the approximate form

$$F_{nn}(k;\, \zeta) = \frac{C_n^2}{4\pi} k^{-11/3} \exp\{-\alpha\eta^2 k^2\},$$

which is a simple and convenient spectral model for $k \gg 1/L_0$. If we recall Eqs. (26.56) and (26.31) we find that $C_n^2 = A^2 B^{(\vartheta)}(\bar{N}_T + B^2 \bar{N}_\vartheta)\bar{\varepsilon}^{-1/3}$ for the case of centimeter-range radio waves while $C_n^2 = A_1^2 B^{(\vartheta)} \bar{N}_T \bar{\varepsilon}^{-1/3}$, $A_1 \approx 10^{-6}(80\bar{p}/\bar{T}^2)$ for visible light. The quantity $C_n^2$, and also the Kolmogorov length scale $\eta$, are functions of the height $h$ above the earth's surface. We recall that the integration variable $\zeta$ in Eq. (27.3) is the distance measured along the direction of propagation of the incident wave, measured from the beginning of the turbulent region (the upper boundary of the layer in which $C_n^2$ is not negligibly small), and $\zeta = L$ corresponds to the observation point which is on the earth's surface. The relation between $\zeta$ and $h$ is $h = (L - \zeta)\cos\theta$ where $\theta$ is the zenith angle of the light source (star). Substituting for $F_{nn}$ into the equation valid in the quasi-equilibrium range, transforming from $\zeta$ to $h$, and substituting the resulting spectra $F_{\chi\chi}^{(2)}$ and $F_{SS}^{(2)}$ into the formulas for $\overline{\chi'^2}$, $D_{\chi\chi}$ and $D_{SS}$, we find that

$$\overline{\chi'^2} = \pi\kappa^2 \sec\theta \int_0^\infty C_n^2(h)\, dh \int_0^\infty \sin^2\left(\frac{k^2 h \sec\theta}{2\kappa}\right) k^{-8/3} e^{-\alpha\eta^2 k^2}\, dk,$$

$$D_{\chi\chi}(r) = 2\pi\kappa^2 \sec\theta \int_0^\infty C_n^2(h)\, dh \int_0^\infty [1 - J_0(kr)] \times$$
$$\times \sin^2\left(\frac{k^2 h \sec\theta}{2\kappa}\right) k^{-8/3} e^{-\alpha\eta^2 k^2}\, dk, \quad (27.4)$$

$$D_{SS}(r) = 2\pi\kappa^2 \sec\theta \int_0^\infty C_n^2(h)\, dh \int_0^\infty [1 - J_0(kr)] \times$$
$$\times \cos^2\left(\frac{k^2 h \sec\theta}{2\kappa}\right) k^{-8/3} e^{-\alpha\eta^2 k^2}\, dk.$$

In these expressions we have replaced the upper limits $L \cos\theta$ in the outer integrals by infinity, since integration should in fact be carried out over the entire thickness of the atmosphere.

The integral for $\overline{\chi'^2}$ can be reduced to

$$\overline{\chi'^2} = \frac{3\pi}{10}\Gamma\left(\frac{1}{6}\right)\kappa^{7/6}(\sec\theta)^{11/6} \int_0^\infty C_n^2(h)\, h^{5/6}\left[\mathrm{Re}\,(\delta^2 + i)^{5/6} - \delta^{5/3}\right] dh, \quad (27.5)$$

where $\delta = \eta\sqrt{\dfrac{2\pi\alpha}{\lambda h \sec\theta}}$. To establish the variation of $\delta$ with the height $h$ we recall that $\eta = \nu^{3/4}\bar{\varepsilon}^{-1/4}$. The factor $\bar{\varepsilon}^{-1/4}$ varies relatively slowly with height, and $\nu$ is approximately inversely proportional to the density $\rho$ of the atmosphere, which in very approximate calculations can be taken to be $\rho(h) = \rho(0)\exp\{-h/h_0\}$ where $h_0$ is of the order of 7 km. Therefore, $\delta \sim h^{-1/2}\exp\{3h/4h_0\}$. In the case of light waves this quantity is greater than unity firstly in a thin layer near the earth's surface which has a thickness $h_1 = 2\pi\alpha\eta^2\cos\theta/\lambda$ (for $\eta \sim 10^{-1}$ cm and $\lambda \sim 1\mu$ we have $h_1 \sim 10$ m), and second, at large heights in the atmosphere where the level $h_2$ which can be determined from the condition $\dfrac{h_2}{h_1} = \exp\left\{\dfrac{3}{2}\,\dfrac{h_2 - h_1}{h_0}\right\}$ and which considerably exceeds $h_0$. When $\delta > 1$ we have

$$\text{Re}\,(\delta^2+i)^{5/6} - \delta^{5/3} \approx \frac{5}{72}\delta^{-7/3} \sim h^{7/6}\exp\left\{-\frac{7}{4}\frac{h}{h_0}\right\}.$$

This quantity is small for both $h < h_1$ and $h > h_2$. Therefore, the main contribution to the integral in Eq. (27.5) is due to the atmospheric layer for which $\delta < 1$. In this layer $\text{Re}\,(\delta^2+i)^{5/6} - \delta^{5/3} \approx \cos\frac{5\pi}{12}$. Since at large heights ($h > h_2$) the quantity $C_n^2(h)$ decreases quite rapidly, we can replace the factor $\text{Re}\,(\delta^2+i)^{5/6} - \delta^{5/3}$ by the constant $\cos(5\pi/12)$ for all $h$ as a first approximation, and this yields

$$\overline{\chi'^2} \approx \frac{3\pi}{10}\Gamma\left(\frac{1}{6}\right)\cos\frac{5\pi}{12}\cdot\kappa^{7/6}(\sec\theta)^{11/6}\int_0^\infty C_n^2(h)\,h^{5/6}\,dh.$$

This formula was given by Tatarskii (1958).

We shall now assume that $C_n^2(h) = C_n^2(0)\cdot f_n(h/H)$ where $H$ is the characteristic length scale of vertical variations in the refractive index (i.e., the typical thickness of a turbulent atmospheric layer), and $f_n(\zeta)$ is a decreasing function. We then have

$$\overline{\chi'^2} = a_0 C_n^2(0)\,\kappa^3(\lambda H\sec\theta/2\pi)^{11/6},$$
$$a_0 = \frac{3\pi}{10}\Gamma\left(\frac{1}{6}\right)\cos\frac{5\pi}{12}\int_0^\infty f_n(\zeta)\,\zeta^{5/6}\,d\zeta. \tag{27.6}$$

This formula differs from Eq. (26.86) only by a numerical factor and by the replacement of $L$ by $H\sec\theta$.

Let us now consider the structure functions $D_{\chi\chi}(r)$ and $D_{SS}(r)$. The inner integrals in Eq. (27.4) for these functions, which we shall denote by $I_1$ and $I_2$, are evaluated by analogy with Eq. (26.89) and lead to

$$I_{1,2} = \frac{3}{10}\Gamma\left(\frac{1}{6}\right)\left(\frac{\lambda h\sec\theta}{2\pi}\right)^{5/6}\left\{\delta^{5/3}\left[M\left(-\frac{5}{6},1,-\frac{\rho^2}{4\delta^2}\right)-1\right]\mp\right.$$
$$\left.\mp\,\text{Re}\,(\delta^2+i)^{5/6}\left[M\left(-\frac{11}{6},1,-\frac{1}{4}\frac{\rho^2}{\delta^2+i}\right)-1\right]\right\},$$

where the plus sign corresponds to $I_1$ and the minus sign to $I_2$. The quantity $\delta$ has the same significance as in Eq. (27.5), and $\rho = r\sqrt{2\pi/\lambda h\sec\theta}$. Using the properties of the confluent hypergeometric functions $M$ we can verify that when $\rho \ll \delta \ll 1$

$$I_1 \approx I_2 \approx \frac{1}{16}\Gamma\left(\frac{1}{6}\right)\alpha^{-1/6}\eta^{-1/3}r^2;$$

and when $\rho \ll \delta$, $\delta \gg 1$

$$I_1 \approx \frac{7}{1152}\Gamma(1/6)\,\alpha^{-13/6}\eta^{-13/3}\left(\frac{\lambda h\sec\theta}{2\pi}\right)^2 r^2,\qquad I_2 \approx \frac{1}{8}\Gamma(1/6)\,\alpha^{-1/6}\eta^{-1/3}r^2;$$

Moreover, when $\rho \gg \delta \gg 1$ and $\rho \gg 1 \gg \delta$

$$I_1 \approx \frac{1}{36\sqrt[3]{4}} \frac{\Gamma(1/6)}{\Gamma(5/6)} \left(\frac{\lambda h \sec\theta}{2\pi}\right)^2 r^{-7/3}; \qquad I_2 \approx \frac{9}{25\sqrt[3]{4}} \frac{\Gamma(1/6)}{\Gamma(5/6)} r^{5/3};$$

and when $1 \gg \rho \gg \delta$.

It is also easy to find the general asymptotic equations for $\delta \ll 1$ (i.e., for a pure inertial subrange turbulence spectrum) and arbitrary $\rho \gg \delta$. They will evidently include a single confluent hypergeometric function

$$M\left(-\frac{5}{6}, 1, i\rho^2/4\right)$$

[Tatarskii (1971), cf. also Fried (1969)].

When $r \ll \eta(0)$, where $\eta(0)$ is the Kolmogorov length scale at ground level, and since $\eta(h)$ increases monotonically with height (because of the increase in the kinematic viscosity $\nu$ in inverse proportion to the air density), we have $r \ll \eta(h)$ for any $h$, i.e., $\rho \ll \delta$. Since the layers $h < h_1$ and $h > h_2$ in which $\delta > 1$ do not contribute appreciably to the integrals in Eq. (27.4) [the first of these layers is very thin and in the second $C_n^2(h)$, $I_1$, and $I_2$ are all small and decrease rapidly with height], we can evaluate these integrals for all $h$ by using the formulas $I_{1,2}$ which are valid for $\delta < 1$. Since, moreover, the length $\eta$ varies relatively slowly with height for $h_1 < h < h_2$, we have

$$D_{\chi\chi}(r) \approx D_{SS}(r) \approx a_1 k^3 C_n^2(0)\, \tilde{\eta}^{-1/3} \left(\frac{\lambda H \sec\theta}{2\pi}\right) r^2, \qquad a_1 = \frac{\pi}{8}\Gamma\left(\frac{1}{6}\right)\alpha^{-1/6} \int_0^\infty f_n(\zeta)\, d\zeta. \tag{27.7}$$

In these expressions $\tilde{\eta}$ must be interpreted as the mean Kolmogorov length scale for the troposphere and the lower stratosphere.

When $r \gg \tilde{\eta}$ then in the layer $\delta < 1$, which provides the main contribution to the integrals in Eq. (27.4), we have $\rho \gg \delta$. The integral $I_2$ is then approximately proportional to $r^{5/3}$ (the proportionality factor is a slowly varying function of $\rho$), and for the function $D_{SS}(r)$ we have the formula of Tatarskii (1958)

$$D_{SS}(r) \approx a_2 k^3 C_n^2(0) \left(\frac{\lambda H \sec\theta}{2\pi}\right) r^{5/3}, \tag{27.8}$$

where $a_2$ is a dimensionless coefficient which can be approximately calculated from the equation

$$a_2 \approx \frac{9\pi}{25\sqrt[3]{4}} \frac{\Gamma(1/6)}{\Gamma(5/6)} \int_0^\infty f_n(\zeta)\, d\zeta$$

and which increases slowly with increasing $\rho_H = r\sqrt{2\pi/\lambda H \sec\theta}$. For $D_{\chi\chi}(r)$ similar results are obtained only if the relation $1 \gg \rho \gg \delta$ is satisfied in the layer $\delta < 1$ (or at least in most of it). In general, it may be shown that, when $r \gg \tilde{\eta}$,

$$R_{\chi\chi}(r) = 1 - D_{\chi\chi}(r)/2\overline{\chi'^2} = F(\rho_H). \tag{27.9}$$

Formula (27.8) can be used to explain the phenomenon of jitter of stellar images in the view field of a telescope. This phenomenon can be approximately evaluated as follows. Let us first assume that the light wave emitted by a star and propagated through the turbulent atmosphere is received not by the finite surface of a telescope objective but by an interferometer with a base of length $r$. Let us also assume that the mean position of the wave front is parallel to the interferometer base. The phase difference $\delta S$ along the interferometer base corresponds to rotation of the wave front by the angle $\delta\alpha = \delta S/\kappa r$. Fluctuations in the angle $\delta\alpha$ will evidently be perceived as image jitter. The mean square of these fluctuations is given by equation

$$\overline{(\delta\alpha)^2} = \frac{\overline{(\delta S)^2}}{\kappa^2 r^2} = \frac{D_{SS}(r)}{\kappa^2 r^2}$$

The phase difference $\delta S$ can be considered as a Gaussian random variable and hence $\delta\alpha$ is also a Gaussian variable. Therefore the mean square $\overline{(\delta\alpha)^2}$ gives a complete statistical description of the image jitter.

The situation is only slightly more complicated in the real case when the receiver of the wave is a circular telescope objective of diameter $2R$. It is clear that turbulent inhomogeneities of the field $S$ with length scales smaller than $R$ will not cause image displacement, but will be perceived as image diffusion (i.e., image deterioration). At the same time large-scale inhomogeneities of $S$ will produce image displacement, i.e., jitter. We can therefore use the above equation for $\overline{(\delta\alpha)^2}$ as an approximate result for telescope image jitter, if we substitute $r = 2R$ in it. Using Eq. (27.8) for $D_{SS}(r)$ we obtain

$$\overline{(\delta\alpha)^2} = a_2 C_n^2(0) H \sec\theta (2R)^{-1/3} \tag{27.10}$$

(the use of Eq. (27.8) is justified by the fact that telescope diameters are usually much greater than typical values of the turbulence microscale $\eta$ in the atmosphere).

The approximate qualitative theory of image jitter given above is due to Krasil'nikov (1949b). It shows that the mean square of the stellar-image jitter is proportional to the secant of the zenith angle, and the proportionality factor depends on meteorological conditions, and falls slowly with increasing telescope diameter. The dependence on the meteorological conditions is described by the factor $\int_0^\infty C_n^2(h)\,dh$ included in the product $a_2 C_n^2(0)$ on the right-hand side of Eq. (27.10). Since the maximum values of $C_n^2(h)$ are usually reached near the earth's surface, the main contribution to the above integral is due to the lower layers of the atmosphere, which are therefore mostly responsible for the stellar-image jitter. An experimental verification of the law (27.10) was carried out by Kolchinskii (1957), one of whose graphs is reproduced in Fig. 117.

A more accurate analysis of the stellar-image jitter and quality problems was carried out by Kon (1969a) and Tatarskii (1971). In these works a solution was given of the problem of fluctuations in the "image center of gravity" in the focal plane of a telescope. Namely, it was shown that the mean square of the corresponding angle $\delta\alpha$ can be obtained from the following equation:

$$\overline{(\delta\alpha)^2} = 0.97\, a_2 C_n^2(0) H \sec\theta (2R)^{-1/3} \left\{ 1 + \right.$$
$$\left. + \frac{\sqrt{\pi}\,\Gamma(11/6)}{\Gamma(4/3)} \alpha_R^{-5/3} \operatorname{Im} i^{-11/6} \left[ 1 - F\left(\frac{1}{2}, -\frac{11}{6}; 2; -i\alpha_R^2\right)\right]\right\} \tag{27.11}$$

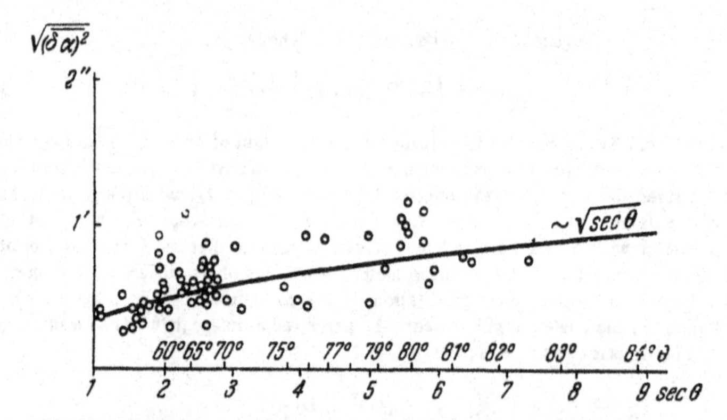


**Fig. 117** Dependence of the root mean square of fluctuations in the angle of arrival of stellar light on the zenith distance [Kolchinskii (1957)].

where the parameter $\alpha_R = (\kappa^{1/2}R/(H\sec\theta)^{1/2}) = R\sqrt{2\pi/\lambda H\sec\theta})^{1/2}$ characterizes the objective size, and $F(a, b; c; z)$ is the hypergeometric function of Gauss. The dependence of the right-hand side of Eq. (27.11) on $\alpha_R$ is relatively weak, since the sum in braces varies only in the range from 1 (for very small telescopes with $\alpha_R \ll 1$) to 2 (for $\alpha_R \gg 1$). We see that the accurate equation (27.11) differs only slightly from the approximate equation (27.10) (in the case of $\alpha_R \ll 1$ the difference is only in the numerical factor 0.97 instead of 1).

Let us also discuss briefly the question of stellar-image quality. "Pure image diffusion" can be described by the statistics of the stellar-image fluctuations in the coordinate system moving jointly with the "image center of gravity." These statistics are clearly determined only by the small-scale phase fluctuations on the surface of telescope lens, while the statistical structure of the image at a fixed point is dependent on both small-scale and large-scale (causing image jitter) fluctuations. This is clearly illustrated in Fig. 117a (due to Kon) where the dependence of mean image intensity on $D_{SS}(2R)$ is shown for the cases of intensity at the fixed "telescope axis" (curve 2) and at the "center of gravity of the image" (curve 1). Curve 2 gives the measure of "image quality" in the case of long-exposure photography when the exposure time is much greater than the typical correlation time

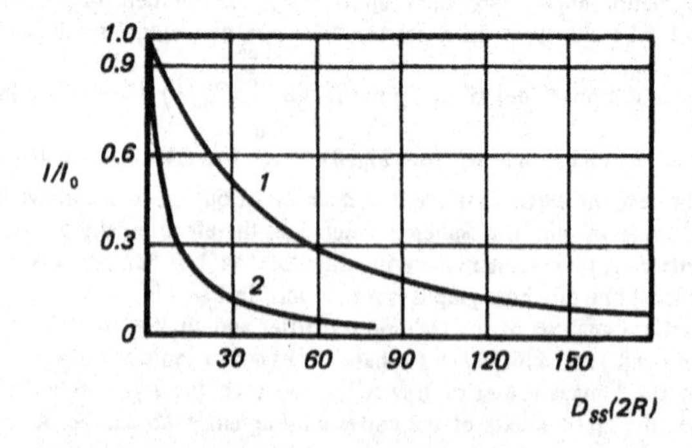


**Fig. 117a** The dependence of normalized mean intensity at the "center of gravity of the image" (curve 1) and at the telescope axis (curve 2) on mean square phase difference at the distance $2R$ (after Kon, 1969b).

$2R/U$ of the image jitter (here $U$ is a typical wind velocity in the direction perpendicular to the wave propagation direction). Curve 1 gives the measure of "image quality" for visual observations or short-exposure photography (with exposure time much smaller than the correlation time $2R/U$).

## 27.2 The Effect of Telescope Averaging and Scintillation of Stellar and Planetary Images

The interpretation of the data on wave propagation in the turbulent atmosphere strongly requires calculation of the dependence of the statistical characteristics of such waves on the sizes of the wave emitter and receiver. This is a consequence of the fact that real observations are carried out with instruments of very different sizes and often relate to wave sources of very different angular sizes.

In the present section we are dealing with astronomical scintillation and hence our main interest is here in telescope observations of various space objects. It is well known to astronomers that an increase of light source and telescope objective sizes causes additional smoothing (i.e., averaging) of the light fluctuations. This is illustrated by the reduction of fluctuation intensity when the observations are made with a larger telescope and by the known fact that visible planetary scintillations are much weaker than stellar scintillations. Both these effects are dut to mutual cancellation of the fluctuations of opposite sign as a result of averaging.

The problem of the averaging effect of a finite telescope objective was first considered by Krasil'nikov and Tatarskii (1953) and was later studied in more detail by Tatarskii (1959), Kon and Tatarskii (1964) and Kon (1969b). We shall follow here the work of Kon where the problems of the averaging effects of the finite sizes of the light source and the receiver are solved simultaneously. This is important since the effects of emitter and receiver averaging are, in general, not multiplicative and cannot be taken into account successively and independently.

Let us consider the light radiation from a circular source of angular diameter $2\gamma$ situated far outside the turbulent atmosphere. This radiation is received by a receiver of finite aperture, let us say, by a telescope with a circular objective of radius $R$. Each point of the objective will receive a beam of incoherent plane waves emitted by different points of the source. The intensities of all these waves will evidently be added together. If we denote by $i(y, z, \theta, \varphi)$ the intensity of light received at the point $(y, z)$ of the objective from the direction with angles $(\theta, \varphi)$ in spherical coordinates with a polar axis directed toward the source center, then the total intensity received at a point $(y, z)$ will be equal to

$$I(y, z) = \int_{\Omega} i(y, z, \theta, \varphi) d\omega$$

where $\Omega$ is the solid angle of a source observed at the receiver position and $d\omega = \sin\theta d\theta d\varphi$. The total light flux through the objective is given by the equation

$$P = \int_{\Sigma} I(y, z) d\rho = \int_{\Sigma} d\rho \int_{\Omega} d\omega i(y, z, \theta, \varphi)$$

where $\rho = (y, z)$ and $\Sigma$ is the objective surface. Hence $\bar{P} = 4\pi^2 R^2 \sin^2(\gamma/2)\bar{i}$,

$$P' = P - \bar{P} = \int_{\Sigma} d\rho \int_{\Omega} d\omega i'(y, z, \theta, \varphi)$$

and

$$\overline{(P')^2} = \int_\Sigma \int_\Sigma d\rho_1 d\rho_2 \int_\Omega \int_\Omega d\omega_1 d\omega_2 \overline{i'(y_1, z_1, \theta_1, \varphi_1) i'(y_2, z_2, \theta_2, \varphi_2)}$$

This yields the following equation for the relative variance of fluctuations of the total flux of stellar or planetary radiation $\sigma_P^2 = \overline{(P')^2}/\bar{P}^2$:

$$\sigma_P^2(R, \gamma) = \frac{1}{(4\pi^2 R^2 \sin^2 \gamma/2)^2} \int_\Sigma \int_\Sigma d\rho_1 d\rho_2 \int_\Omega \int_\Omega d\omega_1 d\omega_2 \widetilde{B}_{ii}(\rho_1, \rho_2, \omega_1, \omega_2)$$

where $\rho_1$ and $\rho_2$ are points of the objective, $\omega_1$ and $\omega_2$ are the directions in the solid angle $\Omega$ and $\widetilde{B}_{ii} = \overline{i'(\rho_1, \omega_1) i'(\rho_2, \omega_2)}/\bar{i}^2$ is the normalized correlation function of the field $i(y, z, \theta, \varphi)$.

The joint averaging effect of the objective and the source can be characterized by a dimensionless function

$$G(R, \gamma) = \frac{\sigma_P^2(R, \gamma)}{\sigma_P^2(0, 0)}$$

which determines the reduction in the relative fluctuations of the light flux from an extended source received by an objective of finite size in comparison with the fluctuations for a single plane wave received by a point receiver.

If we use an obvious relation between intensity fluctuations and fluctuations in the logarithm of the amplitude $\chi'$ and assume that the joint probability distribution of the random variables $\chi'(r_1)$ and $\chi'(r_2)$ is a Gaussian distribution and that $\overline{(\chi')^2} \ll 1$ (this last assumption is always valid for astronomical observations), we shall obtain the following equation for $G(R, \gamma)$:

$$G(R, \gamma) = \frac{1}{(4\pi^2 R^2 \sin^2 \gamma/2)^2 \overline{(\chi')^2}} \int_\Sigma \int_\Sigma d\rho_1 d\rho_2 \int_\Omega \int_\Omega d\omega_1 d\omega_2 B_{\chi\chi}(\rho_1, \rho_2, \omega_1, \omega_2) \tag{27.12}$$

where $B_{\chi\chi}(\rho_1, \rho_2, \omega_1, \omega_2) = \overline{\chi'(\rho_1, \omega_1)\chi'(\rho_2, \omega_2)}$ is the correlation function of the field $\chi'(\rho, \omega)$, so that $B_{\chi\chi}(\rho, \rho, \omega, \omega) = \overline{(\chi')^2}$. This correlation function was evaluated by Kon with the aid of Rytov's method of "smooth perturbations." For the case of waves with a given wave number $\kappa = 2\pi/\lambda$ his result is the following

$$B_{\chi\chi}(\rho_1, \rho_2, \omega_1, \omega_2) = 2\pi^2 \kappa^2 \int_0^L d\xi \int_0^\infty \left(1 - \cos \frac{k^2 \xi}{\kappa}\right) J_0(k|\rho - \xi p|) F_{nn}(k) k dk \tag{27.13}$$

where it is supposed that waves cover the path $L$ in a turbulent atmosphere with a refractive index spectrum $F_{nn}(k)$, $\rho = \rho_1 - \rho_2$, $p = p_1 - p_2$, and $p_1$ and $p_2$ are two-dimensional vectors which are projections of the unit vectors in the directions $\omega_1$ and $\omega_2$ on the $(y, z)$-plane. After substitution of Eq. (27.13) into Eq. (27.12), the substitution of the pure inertial-subrange power spectrum for $F_{nn}(k)$ (this is justified when $\eta \ll \sqrt{\lambda L} \ll L_0$) and transformation to new dimensionless variables $\zeta = \xi/L$ and $\kappa = k^2 L/\kappa$, we can explicitly evaluate some of the multiple integrals and finally obtain

$$G(R, \gamma) = \frac{16}{c_1 \alpha_R^2 \alpha_S^2} \int_0^1 d\zeta \int_0^\infty d\kappa \frac{1 - \cos \zeta\kappa}{\zeta^2} \kappa^{-23/6} J_1^2(\alpha_R \sqrt{\kappa}) J_1^2(\alpha_S \zeta \sqrt{\kappa}) \tag{27.14}$$

The parameter $\alpha_R = \kappa^{1/2}R/L^{1/2} = R(2\pi/\lambda L)^{1/2}$ in this formula is proportional to the ratio of the objective radius $R$ to the radius of the first Fresnel zone, $\alpha_S = \gamma\kappa^{1/2}L^{1/2}$ is the ratio of the angular size of the source to the "correlation angle" $\psi_0 = 1/\kappa^{1/2}L^{1/2}$ (which is of the order of $\sqrt{\lambda/L}$) and $c_1 = (3\pi/11\cos(\pi/12)\Gamma(11/6)) \approx 0.94$.

The function $G(R, \gamma)$ is equal to one at $\alpha_R = \alpha_S = 0$ and decreases when its arguments increase. The results of its numerical computation by Kon (1969b) are shown in Fig. 118 in the form of a family of curves corresponding to different values of the parameter $\alpha_S$.

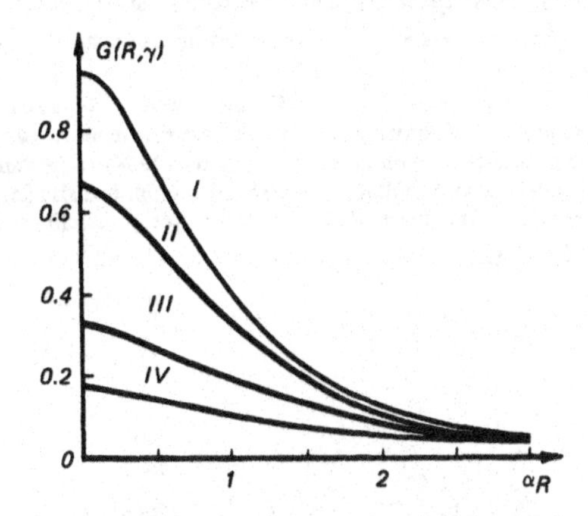


**Fig. 118** Dependence of the averaging effect of the receiver on its size for various sizes of source (after Kon, (1969b). I, $\alpha_s = 0.2$; II, $\alpha_s = 1$; III, $\alpha_s = 2$; IV, $\alpha_s = 3$.

Let us now consider two important particular cases of Eq. (27.14).

i) The case $\alpha_S = 0$ corresponds to the observation of a point source emitting plane waves with the aid of an extended objective. If the source is a star with a zenith angle $\theta$, then $\alpha_R = (\kappa^{1/2}R/(H\sec\theta)^{1/2})$ and the corresponding reduction coefficient is equal to

$$G(R,0) = G(\alpha_R)$$
$$= \frac{24\,\mathrm{ctn}\,\frac{\pi}{12}}{17\alpha_R^2}\left[1 + \mathrm{Im}\,\frac{i^{-5/6}}{\cos\frac{\pi}{12}}F\left(\frac{1}{2}, -\frac{17}{6}; 2; -i\alpha_R^2\right)\right] - \frac{44\,\Gamma\left(\frac{7}{3}\right)\Gamma\left(\frac{11}{6}\right)}{17\sin\frac{\pi}{12}\,\Gamma^2\left(\frac{17}{6}\right)\sqrt{\pi}}\alpha_R^{5/3}. \tag{27.15}$$

The function $G(\alpha_R)$ was evaluated numerically by Tatarskii (1959); in our Fig. 118 it is represented by curve I. Let us also note that Eq. (27.15) implies the following asymptotic equation for $\alpha_R \gg 1$:

$$G(\alpha_R) \sim \alpha_R^{-7/3} \tag{27.16}$$

We are now in a position to give a quantitative explanation of the phenomenon of stellar scintillation in telescopes. If we use the relative variance of the total flux of stellar radiation $\sigma_P^2$ as the main characteristic of the phenomenon, we have from Eq. (27.6) and the definition of $G(\alpha_R)$

$$\sigma_P^2 \approx 5.3C_n^2(0)\kappa^{7/6}(H\sec\theta)^{11/6}G(\alpha_R)\int_0^\infty f_n(\zeta)\zeta^{5/6}d\zeta \tag{27.17}$$

(since $\sigma_p^2(0,0) = \overline{(e^{2\chi'} - 1)^2} \approx 4\overline{\chi'^2}$ for $\overline{\chi'^2} \ll 1$ by virtue of the definitions of $p$ and $\chi'$). This equation can easily be used for practical computations.

Let us first explain the dependence of $\sigma_p^2$ on the zenith angle $\theta$ of the star. For small zenith angles, $\sec\theta \approx 1$ and the argument $\alpha_R$ of the function $G$ is large for telescopes that are not too small (it will be shown below that the parameter $\sqrt{\lambda H}$ is of the order of 8–10 cm). We can then use Eq. (27.16) and show that $\sigma_p \sim (\sec\theta)^{3/2}$. For large zenith angles $\sec\theta$ is large, the argument $\alpha_R$ of the function $G$ is small, and the function $G$ itself is close to unity. We then have $\sigma_p \sim (\sec\theta)^{11/12}$. These results are in satisfactory agreement with observational data.

Thus, Fig. 119 shows a plot of $\sigma_p$ as a function of $\sec\theta$ obtained by Butler with a 15-inch telescope. For $\theta < 60°$ Butler's data are satisfactorily represented by the formula $\sigma_p \sim (\sec\theta)^{3/2}$. Similar results were obtained by Protheroe (1954) at the Perkins Observatory, using the 12.5-inch telescope. Simultaneously Protheroe published also data obtained with a small 3-inch telescope and these data are satisfactorily described by the formula $\sigma_p \sim (\sec\theta)^{11/12}$ (Fig. 120). Figure 121 shows the data obtained by Zhukova (1958) at the

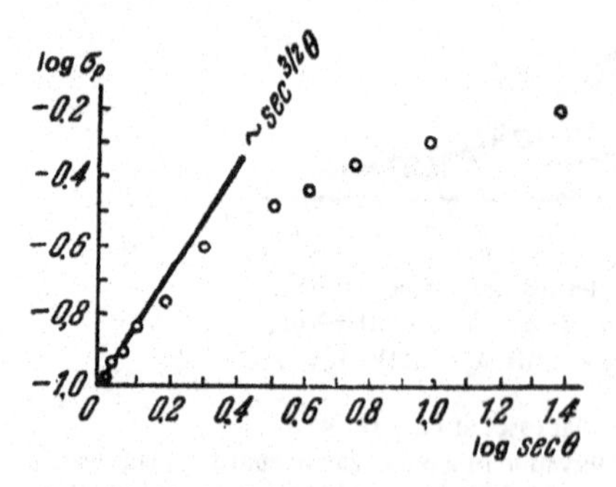


**Fig. 119** Dependence of stellar scintillation on the zenith distance of the stars for a 15" telescope [Butler (1954)].

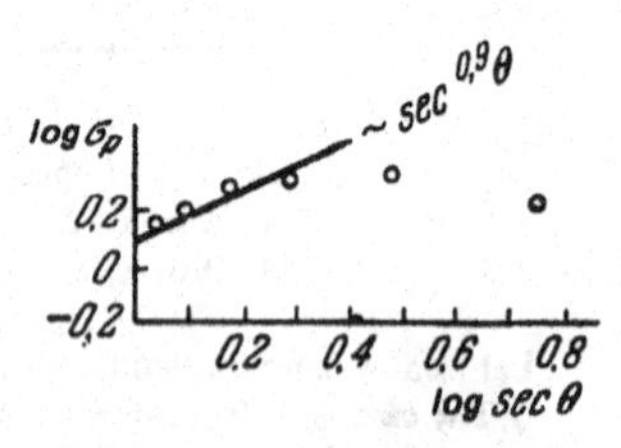


**Fig. 120** Dependence of stellar scintillation on the zenith distance of the stars for a 3" telescope [Protheroe (1954)].

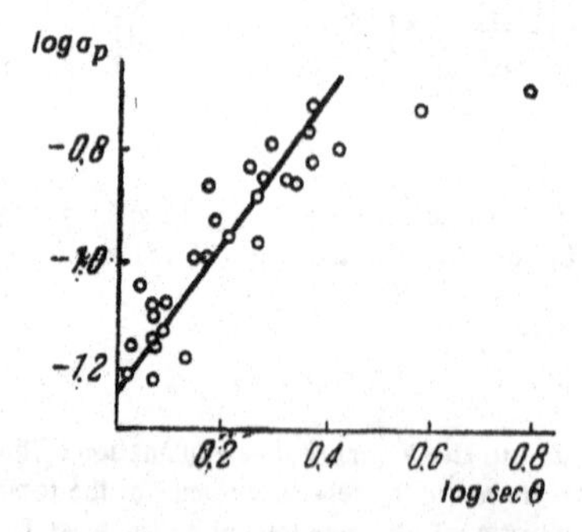


**Fig. 121** Comparison of theoretical predictions (27.17) for stellar scintillation with the experimental data of Zhukova (1958).

Pulkovo Observatory, using a 25-cm telescope. The solid line in this figure represents the theoretical curve calculated from Eq. (27.17) with $\sqrt{\lambda H} = 9$ cm.

ii) It is well known that planets exhibit smaller scintillations than stars. The various points on the planetary disk can be regarded as incoherent light sources, and thus the entire planetary disk produces a beam of incoherent plane waves within a finite solid angle $\gamma$. The case of observation of a planet with a very small telescope corresponds obviously to substitution of $\alpha_R = 0$ in Eq. (27.14). This case was analyzed in detail by Kon and Tatarskii (1964). Their calculations were then verified experimentally by Gurvich and Kon (1964). Figure 122 shows the results (the solid line is calculated and the points represent experimental results).

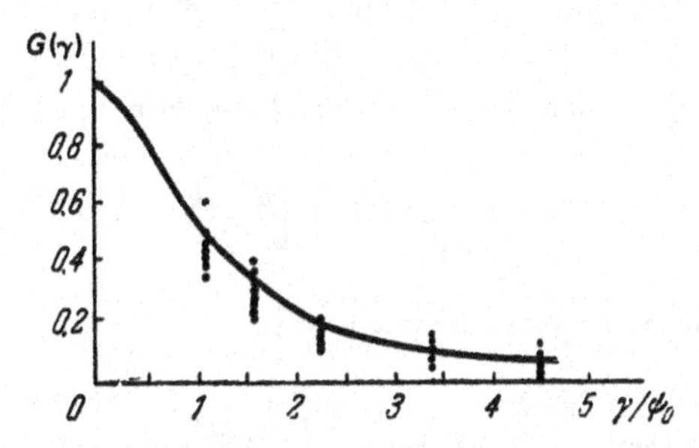


**Fig. 122** Reduction in the brightness fluctuations for a light source of angular size $\gamma$ (relative to fluctuations for a source with $\gamma = 0$) according to Kon and Tatarskii (1964) and Gurvich and Kon (1964).

Observations indicate that the reduction in the scintillation of extraterrestrial sources as their angular size increases is appreciable for $\gamma \sim 1'' = 0.5 \cdot 10^{-5}$ radian, and hence for such values of $\gamma$ the ratio $\gamma/\phi_0$ is apparently of the order of unity. Substituting $\sec\theta \sim 1$, $\lambda = 0.5 \cdot 10^{-4}$ cm, and $\gamma \sqrt{\frac{H}{\lambda}} \sim 1$, we find that $\sqrt{\lambda H} \sim 10$ cm. This estimate is confirmed by direct measurements of the correlation functions for fluctuations in the intensity of stellar light [see Eq. (27.9)] with the aid of two mobile telescopes as reported by Keller (1955). The result obtained for $\sqrt{\lambda H}$ was about 3.5 inches. Additional data confirming this estimate will be given below.

## 27.3 Time Spectra of Fluctuations in the Intensity of Stellar Images in Telescopes

Let us now consider the time spectra of brightness fluctuations for stellar images in telescopes, which are analogous to the spectra of Eq. (26.78). We shall restrict our attention to small fluctuations, so that the total light flux $p$ can be written in the form

$$P(t) = I_0 \int_{\Sigma} [1 + 2\chi'(\boldsymbol{x}, t)]\, d\boldsymbol{x}.$$

This yields the following expression for the normalized time correlation function for the light flux through the telescope objective:

$$R_{pp}(\tau) = \frac{\overline{[P(t) - \overline{P}][P(t+\tau) - \overline{P}]}}{\overline{P}^2} = \frac{4}{\Sigma^2} \int_{\Sigma} \int_{\Sigma} \overline{\chi'(\boldsymbol{x}_1, t)\chi'(\boldsymbol{x}_2, t+\tau)}\, d\boldsymbol{x}_1\, d\boldsymbol{x}_2.$$

If we now introduce the spectral representation of the space-time correlation function

$$\overline{\chi'(\boldsymbol{x}_1, t)\chi'(\boldsymbol{x}_2, t+\tau)} = \int e^{-i\boldsymbol{k}(\boldsymbol{x}_1-\boldsymbol{x}_2)} F^{(2)}_{\chi\chi}(\boldsymbol{k}, \tau)\, d\boldsymbol{k}$$

and use the formula

$$\int\limits_{\Sigma}\int\limits_{\Sigma} e^{-i\boldsymbol{k}(\boldsymbol{x}_1-\boldsymbol{x}_2)}\, d\boldsymbol{x}_1\, d\boldsymbol{x}_2 = \left[\frac{2J_1(kR)}{kR}\right]^2 \Sigma^2$$

which is valid for $\Sigma$ in the form of a circle of radius $R$, we obtain

$$R_{pp}(\tau) = 4\int F^{(2)}_{\chi\chi}(\boldsymbol{k}, \tau)\left[\frac{2J_1(kR)}{kR}\right]^2 d\boldsymbol{k}. \tag{27.18}$$

The final result for the time spectral function is thus

$$F_{pp}(\omega) = \frac{1}{2\pi}\int\limits_{-\infty}^{\infty} e^{i\omega\tau} R_{pp}(\tau)\, d\tau = \frac{2}{\pi}\int\left[\frac{2J_1(kR)}{kR}\right]^2 d\boldsymbol{k}\int\limits_{-\infty}^{\infty} e^{i\omega\tau} F^{(2)}_{\chi\chi}(\boldsymbol{k}, \tau)\, d\tau.$$

To conclude the calculation we must determine the space-time spectral density function $F^{(2)}_{\chi\chi}(\boldsymbol{k}, \tau)$. It can be approximately obtained from Eq. (27.3) by replacing the space spectral function $F_{nn}(\boldsymbol{k}; \zeta)$ with the function $F_{nn}(\boldsymbol{k}; \zeta)\exp\{-i\boldsymbol{k}\boldsymbol{v}\tau\}$ which corresponds to the frozen turbulence approximation for the space-time correlation function

$$\overline{n'(\boldsymbol{x}_1, t)\, n'(\boldsymbol{x}_2, t+\tau)} = \overline{n'(\boldsymbol{x}_1, t)\, n'(\boldsymbol{x}_2 - \boldsymbol{v}\tau, t)} = B_{nn}(\boldsymbol{x}_1 - \boldsymbol{x}_2 + \overline{\boldsymbol{v}}\tau),$$

where $\overline{\boldsymbol{v}}$ is the component of the wind velocity in the plane containing the vectors $\boldsymbol{x}_1$ and $\boldsymbol{x}_2$ and perpendicular to the direction of the light rays. In obtaining the above expressions we have taken into account the fact that the shift by $\overline{\boldsymbol{v}}\tau$ in the argument of the correlation function corresponds to the multiplication of the spectrum by $e^{(-i\boldsymbol{k}\overline{\boldsymbol{v}}\tau)}$. Substituting $h = (L-\zeta)\cos\theta$ in Eq. (27.3), and replacing $L$ by infinity, we obtain

$$F^{(2)}_{\chi\chi}(\boldsymbol{k}, \tau) = 2\pi\kappa^2\sec\theta\int\limits_0^{\infty}\sin^2\frac{k^2 h\sec\theta}{2\kappa}\, F_{nn}(\boldsymbol{k}, h)\, e^{-i\boldsymbol{k}\overline{\boldsymbol{v}}\tau}\, dh. \tag{27.19}$$

The wind velocity $\overline{\boldsymbol{v}}$ must be interpreted in this equation as a given function of height $h$. Substituting this result into the formula for $F_{pp}(\omega)$, we can first integrate with respect to $\tau$, using the formula

$$\int\limits_{-\infty}^{\infty}\exp\{i(\omega - \boldsymbol{k}\overline{\boldsymbol{v}})\tau\}\, d\tau = 2\pi\delta(\omega - \boldsymbol{k}\overline{\boldsymbol{v}}),$$

and then integrate with respect to $\boldsymbol{k}$ by substituting $d\boldsymbol{k} = k\,dk\,d\varphi$ where $\phi$ is the angle between $\boldsymbol{k}$ and $\overline{\boldsymbol{v}}$ and using the equation

$$\int_0^{2\pi} \delta(\omega - kv\cos\varphi)\,d\varphi = \begin{cases} \dfrac{2}{\sqrt{k^2v^2 - \omega^2}} & \text{for} \quad \omega^2 < k^2v^2, \\ 0 & \text{for} \quad \omega^2 > k^2v^2. \end{cases}$$

The final result is

$$F_{pp}(\omega) = 16\pi\kappa^2 \sec\theta \int_0^\infty dh \int_{|\omega|/\overline{v}}^\infty \left[\frac{2J_1(kR)}{kR}\right]^2 \sin^2\frac{k^2 h \sec\theta}{2\kappa} F_{nn}(k, h) \frac{k\,dk}{\sqrt{k^2v^2 - \omega^2}}. \tag{27.20}$$

If we take for $F_{nn}$ the same expression that was used in Eq. (27.4), and substitute $h = H\zeta$ and $\overline{v} = Vw$, where $H$ and $V$ are, respectively, the characteristic height and velocity scales, we can show with the aid of Eq. (27.6) that

$$\frac{\omega F_{pp}(\omega)}{4\overline{\chi'^2}} = U_{pp}(\Omega, \rho, \delta) = \frac{\Omega^{-5/3}}{a_0} \int_0^\infty \frac{f_n\,d\zeta}{w} \int_0^\infty \left[\frac{2J_1\left(\rho\Omega\sqrt{t^2 + \frac{1}{w^2}}\right)}{\rho\Omega\sqrt{t^2 + \frac{1}{w^2}}}\right]^2 \times$$

$$\times \sin^2\left[\frac{\Omega^2\left(t^2 + \frac{1}{w^2}\right)\zeta}{2}\right]\left(t^2 + \frac{1}{w^2}\right)^{-11/6} e^{-\delta^2\Omega^2(t^2 + 1/w^2)}\,dt, \tag{27.21}$$

where

$$\Omega = \frac{\omega}{V}\sqrt{\frac{\lambda H \sec\theta}{2\pi}}, \quad \rho = R\sqrt{\frac{2\pi}{\lambda H \sec\theta}}, \quad \delta = \eta\sqrt{\frac{2\pi a}{\lambda H \sec\theta}}. \tag{27.22}$$

(the parameter $\rho$ is identical to $\alpha_R$ of the previous subsection). For small $\Omega$ the quantity $U_{pp}$ increases in proportion to $\Omega$. It then reaches a maximum, and thereafter decreases (at first in accordance with a power law and then exponentially when $\Omega$ becomes very large). The quantity $\delta$ is a function of height and, in the troposphere and the lower stratosphere which provide the main contribution to stellar scintillation, it is found that $\delta \ll 1$. As a first approximation, therefore, we can use the function $U_{pp}(\Omega, \rho, 0)$ in comparisons with experimental data. Tatarskii (1961a) has plotted the function $U_{pp}(\Omega, \rho, 0)/\Omega$ with $\rho$ as a parameter, assuming for simplicity that $w(\zeta) = 1$, i.e., $\overline{v} = V =$ const and $f_n(\zeta) =$ const. These graphs are shown in Fig. 123. It is clear from them that as $\rho$ increases (i.e., as the diameter $D = 2R$ of the telescope objective increases), the spectrum of fluctuations in the brightness of stellar images is smoothed out and loses the high frequency components (which correspond to spatial-fluctuation scales smaller than the telescope diameter).

Figure 124 shows the values of $\sqrt{4\pi F_{pp}(\omega)}$ based on the data of Protheroe (1954) of the Perkins Observatory for 1-, 3-, 6-, and 12.5-inch telescopes. Figure 124a shows the winter means, and Fig. 124b the summer means. The spread in the values of the function $\overline{v}(h)$ and $C_n^2(h)$ in the individual spectra has, of course, led to a certain broadening and

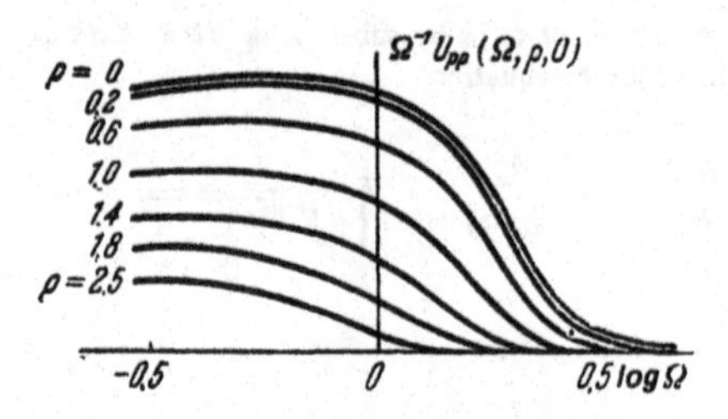


**Fig. 123** Time spectra of stellar-image intensity fluctuations $\Omega^{-1}U_{pp}(\Omega, \rho, 0)$ for different values of ρ [Tatarskii (1961a)].

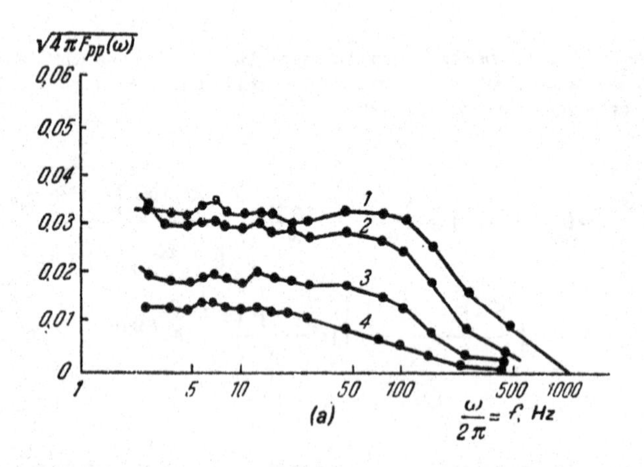


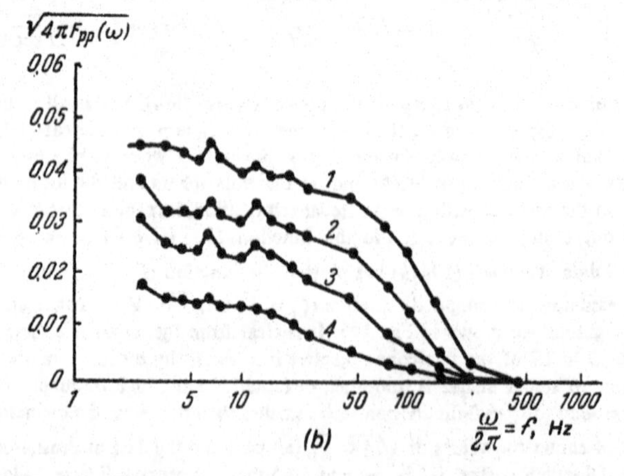


**Fig. 124** Time spectra of stellar-image intensity fluctuations in 1, 3, 6, and 12.5-inch telescopes [Protheroe (1954)] in winter (a) and summer (b).

smoothing of the average spectra. Moreover, bearing in mind the fact that significant simplifications were introduced in the calculations relating to the graphs of Fig. 124, one would not expect a quantitative agreement between the calculated and empirical spectra. However, there is an undoubted qualitative similarity between Figs. 123 and 124.

The values of $\sigma_P^2$ corresponding to the spectra of Fig. 124, were calculated by Tatarskii, and compared by him with the values of the function $G(\alpha_R) \equiv G(\rho) = \sigma_P^2(\rho)/\sigma_P^2(0)$ calculated from Eq. (27.15) for different values of $D = 2R$ and $\sqrt{\lambda L} = \sqrt{\lambda H \sec\theta}$. By choosing $\sqrt{\lambda L}$ so that the values of $\sigma_P^2(\rho)$ and $G$ varied in proportion to each other as $D$ was varied, Tatarskii found that $\sqrt{\lambda L} = 9.2$ cm (i.e., 3.6 inch) for winter spectra and $= 8.1$ cm (i.e., 3.2 inch) for summer spectra. The accuracy of these estimates is relatively high because a variation of $\sqrt{\lambda L}$ by only 1 cm gives rise to a violation of the linear relation between $\sigma_P^2$ and $G$ (see Fig. 125). These estimates are in good agreement with those given earlier.

Instead of studying the time spectra of brightness fluctuations it is possible to consider a more approximate estimate of the characteristic frequencies of the stellar-image scintillations. Thus, by recording the curve $P(t)$ with a low frequency oscillograph, it is possible to determine the mean frequency $\nu$ of the peaks on this curve, which are determined by the size and the speed of the "traveling shadows." Since the size of these shadows is of the order of the correlation radius for the amplitude fluctuations $\sqrt{\lambda H \sec\theta}$, we should have $\nu \sim \overline{v}/\sqrt{\lambda H \sec\theta}$ where $\overline{v}$ is the mean value of the wind velocity component at right angles to the ray. The formula $\nu \sim (\sec\theta)^{-1/2}$ has been verified by Zhukova (1958) whose data are shown in Fig. 126.

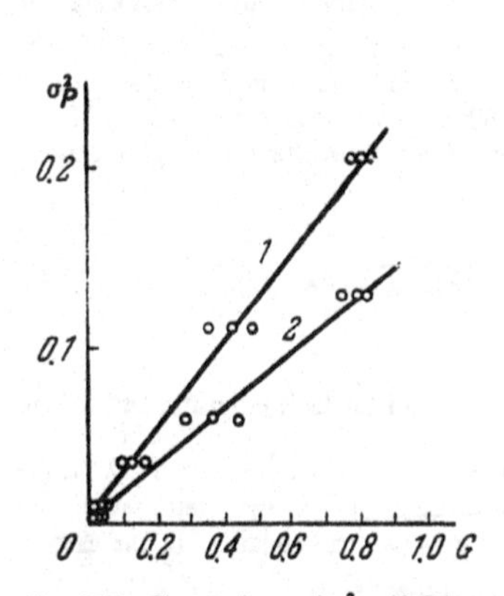


**Fig. 125** Comparison of $\sigma_P^2$ and $G$ for different values of $\sqrt{\lambda L}$ [Tatarskii (1971)]. Points corresponding to values of $\sqrt{\lambda L}$ changed by ±1 cm lie to the right and left of the two lines 1 (winter) and 2 (summer). These points do not lie on the straight lines.

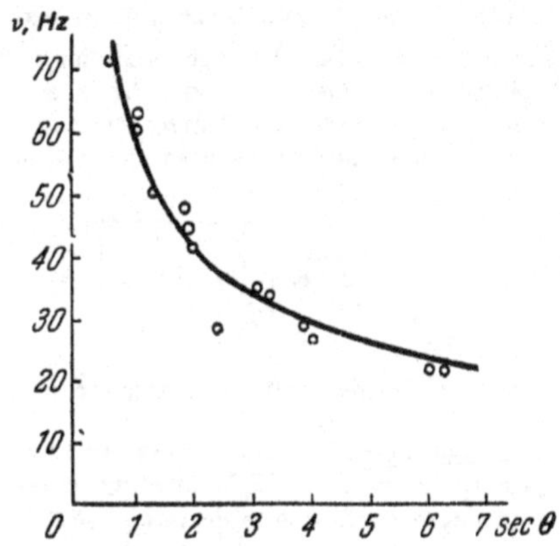


**Fig. 126** A typical frequency $\nu$ of stellar scintillation as a function of the angular distance [Zhukova (1958)]. The solid line shows the function $\nu \sim (\sec\theta)^{-1/2}$.

## 27.4 Chromatic Stellar Scintillation

So far, all our theoretical discussions were concerned with monochromatic light (of given wavelength $\lambda = 2\pi/\kappa$), while the observational data, in particular those of Figs. 119–121, referred to integrated light. At low zenith distances, differences in atmospheric refraction for different wavelengths has very little effect on the paths traversed by the rays in the atmosphere, and scintillations in monochromatic and integrated light are roughly the same. For large zenith distances, however, the differences in atmospheric refraction give rise to considerable path differences at different wavelengths. Thus, for example, according to Perntner and Exner (1910), the distance $r$ between rays with wavelengths $\lambda_1$ and $\lambda_2$ which converge on a point on the earth's surface at a height $h$ is given by

$$r = \psi(\theta)\,[n(\lambda_1) - n(\lambda_2)]\,(1 - e^{-h/H_0}), \tag{27.23}$$

where $H_0$ is the height of the homogeneous atmosphere, $\psi(\theta)$ is a function which vanishes for $\theta = 0$ and increases with $\theta$. When $h \gg H_0$ and $n(\lambda_1) - n(\lambda_2) = 5.9 \cdot 10^{-6}$, which corresponds to the extreme rays in the visible range, the values of $r$ are given by the following table:

| $\theta$ | $\leqslant 10°$ | 20° | 30° | 40° | 50° | 60° | 70° | 80° | 88° |
|---|---|---|---|---|---|---|---|---|---|
| $r$, cm | 0 | 0.2 | 0.8 | 2.1 | 4.9 | 11.6 | 31.2 | 128 | 1283 |

The main contribution to amplitude fluctuations is due to the upper troposphere and lower stratosphere in which the two rays of wavelength $\lambda_1$ and $\lambda_2$ are approximately parallel. If the distance between them exceeds the correlation radius for amplitude fluctuations $\sqrt{\lambda H \sec\theta}$ (which, according to the above table, occurs for $\theta > 60°$), the scintillations at these two wavelengths are uncorrelated, and hence scintillations in integrated light are reduced. This reduction for $\theta > 60°$ is clearly seen in Figs. 119–121.

Let us now calculate the mean square of intensity fluctuations for the integrated light:

$$\overline{J'^2} = \overline{\left[\int_{\lambda_1}^{\lambda_2} I'(\lambda)\,d\lambda\right]^2} = \int_{\lambda_1}^{\lambda_2}\int_{\lambda_1}^{\lambda_2} \overline{I'(\lambda)\,I'(\lambda')}\,d\lambda\,d\lambda',$$

where $I'(\lambda)$ represents fluctuations in the spectral density of the light intensity, and $\lambda_1$ and $\lambda_2$ are the limits of the visible range. Recalling that $I(\lambda) = I_0(\lambda)\,e^{2\chi'_\lambda}$ where $I_0(\lambda)$ is the spectral density of the light intensity at the upper boundary of the atmosphere, and restricting our attention to the case of small fluctuations (which is justified in the case of stellar scintillations) we have $I'(\lambda) \approx 2I_0(\lambda)\,\chi'_\lambda$ and, consequently,

$$\overline{J'^2} = 4\int_{\lambda_1}^{\lambda_2}\int_{\lambda_1}^{\lambda_2} I_0(\lambda)\,I_0(\lambda')\,\overline{\chi'_\lambda \chi'_{\lambda'}}\,d\lambda\,d\lambda'. \tag{27.24}$$

It follows from the above analysis that the quantity $\overline{\chi'_\lambda \chi'_{\lambda'}}$ can be approximately replaced by the mean product of $\chi'_\lambda$ and $\chi'_{\lambda'}$ at two points on the earth's surface, where the distance $r$ between them is given by Eq. (27.73). This mean product will be denoted by $B_{\chi\chi}(r;\,\lambda,\,\lambda')$. Let us express this quantity in terms of the correlation function $B_{\chi\chi}(r,\,\lambda)$ of the field $\chi'_\lambda(\boldsymbol{x})$, which we have calculated from the spectral function $F^{(2)}_{\chi\chi}(\boldsymbol{k},\,\lambda)$ given by the first formula in Eq. (27.2). It is readily shown that to calculate $B_{\chi\chi}(r;\lambda,\lambda')$ we must replace the spectral function $F^{(2)}_{\chi\chi}(k,\lambda)$ by the function

$$F^{(2)}_{\chi\chi}(k;\lambda,\lambda')=$$
$$=\kappa\kappa'\int_0^L\int_0^L \sin\frac{k^2(L-\xi)}{2\kappa}\sin\frac{k^2(L-\zeta)}{2\kappa'}F^{(2)}_{nn}\left(k;|\xi-\zeta|;\frac{\xi+\zeta}{2}\right)d\xi\,d\zeta. \quad (27.25)$$

Here and henceforth we are assuming that the fluctuations $n'$ in the refractive index are not very dependent on wavelength. As in the derivation of Eqs. (26.73) and (27.3), when we integrate with respect to $\xi-\zeta$ in Eq. (27.25) we need only consider the region $|\xi-\zeta|\lesssim 1/k$ in which it can be assumed that $\frac{k^2|\xi-\zeta|}{2\kappa}\ll 1$ and $\frac{k^2|\xi-\zeta|}{2\kappa'}\ll 1$. Under these conditions

$$\sin\frac{k^2(L-\xi)}{2\kappa}\sin\frac{k^2(L-\zeta)}{2\kappa'}\approx\sin\frac{k^2\left(L-\frac{\xi+\zeta}{2}\right)}{2\kappa}\sin\frac{k^2\left(L-\frac{\xi+\zeta}{2}\right)}{2\kappa'},$$

and the formula (27.25) becomes

$$F^{(2)}_{\chi\chi}(k;\lambda,\lambda')=2\pi\kappa\kappa'\int_0^L\sin\frac{k^2(L-\zeta)}{2\kappa}\sin\frac{k^2(L-\zeta)}{2\kappa'}F_{nn}(k;\zeta)\,d\zeta. \quad (27.26)$$

Substituting

$$\tilde{\lambda}_1=\frac{\lambda+\lambda'}{2},\qquad \tilde{\lambda}_2=\frac{\lambda-\lambda'}{2}, \quad (27.27)$$

using the identity

$$\sin\frac{k^2(L-\zeta)}{2\kappa}\sin\frac{k^2(L-\zeta)}{2\kappa'}=\sin^2\frac{k^2(L-\zeta)}{2\tilde{\kappa}_1}-\sin^2\frac{k^2(L-\zeta)}{2\tilde{\kappa}_2}$$

where $\tilde{\kappa}_i=2\pi/\tilde{\lambda}_i$), and comparing the formulas (27.26) and (27.3), we find that

$$F^{(2)}_{\chi\chi}(k;\lambda,\lambda')=\frac{\tilde{\lambda}_1^2}{\lambda\lambda'}F^{(2)}_{\chi\chi}(k;\tilde{\lambda}_1)-\frac{\tilde{\lambda}_2^2}{\lambda\lambda'}F^{(2)}_{\chi\chi}(k;\tilde{\lambda}_2).$$

It is clear that an analogous relation will be valid for the corresponding correlation functions, and we find that

$$B_{\chi\chi}(r;\lambda,\lambda')=\frac{(\lambda+\lambda')^2}{4\lambda\lambda'}B_{\chi\chi}\left(r,\frac{\lambda+\lambda'}{2}\right)-\frac{(\lambda-\lambda')^2}{4\lambda\lambda'}B_{\chi\chi}\left(r,\frac{\lambda-\lambda'}{2}\right). \quad (27.28)$$

Using the formulas (27.28), (27.6), and (27.9), and substituting $\varepsilon=(\lambda-\lambda')/(\lambda+\lambda')$, we find that the correlation coefficient between $\chi'_\lambda$ and $\chi'_{\lambda'}$ is

$$R_{\chi\chi}(\lambda, \lambda') = \frac{B_{\chi\chi}(r; \lambda, \lambda')}{\left(\overline{\chi_{\lambda}'^{2}} \cdot \overline{\chi_{\lambda'}'^{2}}\right)^{1/2}} = (1-\varepsilon^2)^{-5/12}\left[R_{\chi\chi}\left(r\sqrt{\frac{2\pi}{\frac{\lambda+\lambda'}{2}H\sec\theta}}\right) - \right.$$

$$\left. -\varepsilon^{5/6}R_{\chi\chi}\left(r\sqrt{\frac{2\pi}{\frac{\lambda+\lambda'}{2}\varepsilon H\sec\theta}}\right)\right]. \qquad (27.29)$$

We recall that, by Eq. (27.23), $r$ is a function of $\theta$, $\lambda$, and $\lambda'$. This formula was obtained by Tatarskii and Zhukova (1959). We note that the correlation coefficient $R_{\chi\chi}(\lambda, \lambda')$ for $\theta = 0$ when $r = 0$ is equal to $(1-\varepsilon^{5/6})(1-\varepsilon^2)^{-5/12}$, i.e., it is not equal to unity. This is explained by the fact that the "regions of influence" surrounding the ray, which have diameters of the order of $\sqrt{\lambda H}$ and $\sqrt{\lambda' H}$, are different when $\lambda \neq \lambda'$.

Using Eqs. (27.29) and (27.23) for $h = 10$ km, and adopting for $R_{\chi\chi}(\rho_H)$ the values given in Fig. 114, Tatarskii and Zhukova constructed graphs of the correlation coefficient $R_{\chi\chi}'(\lambda, \lambda')$ as a function of $\delta\lambda = \lambda - \lambda'$ for $\sqrt{\frac{\lambda+\lambda'}{2}H} \approx 10$ cm and different $\theta$ (Fig. 127). These graphs are in satisfactory agreement with the observational data obtained at the Pulkovo Observatory, which are also shown in the same figure.

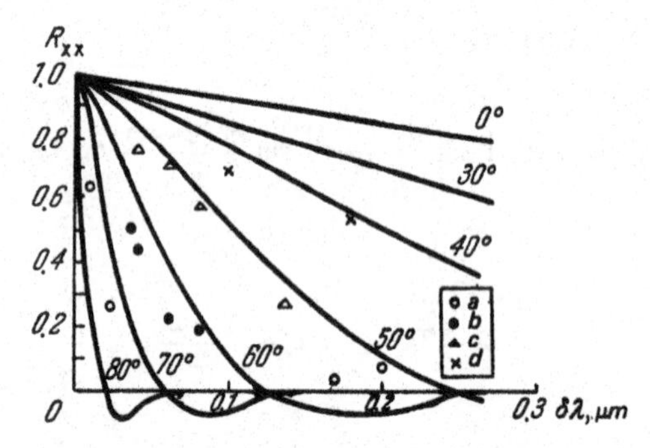


**Fig. 127** Correlation coefficient between fluctuations in the logarithm of the wave amplitude for different wavelengths [Tatarskii and Zhukova (1959)] at zenith angles $\theta$ equal to a–76°; b–65°; c–50°; d–40°.

Returning now to Eq. (27.24), we find from Eqs. (27.29) and (27.6) that

$$\frac{\overline{J'^2}}{J'^2\left(\frac{\lambda_1+\lambda_2}{2}\right)} = \int_{\lambda_1}^{\lambda_2}\int_{\lambda_1}^{\lambda_2} \frac{K(\lambda)K(\lambda')}{K^2\left(\frac{\lambda_1+\lambda_2}{2}\right)} R_{\chi\chi}(\lambda, \lambda')\,d\lambda\,d\lambda', \qquad (27.30)$$

where $K(\lambda) = \lambda^{-7/12}I_0(\lambda)$. The quantity given by Eq. (27.30) shows the factor by which scintillations in integrated light are weaker than in monochromatic light at the wavelength $(\lambda_1+\lambda_2)/2$, and depends on the zenith distance $\theta$ of the star (and to some extent on the meteorological conditions which determine the value of $H$). Let us denote this by $F(\theta)$. Figure 128 shows a graph of the function $F(\theta)/F(0)$ calculated by Tatarskii and Zhukova

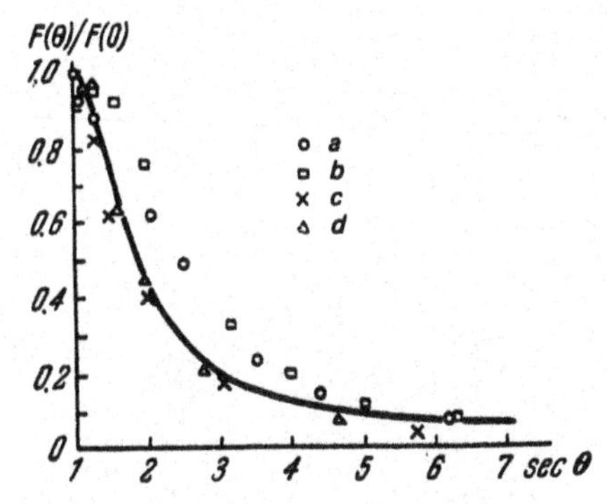


**Fig. 128** The function $F(\theta)/F(0)$ describing the reduction in the scintillation in integrated light in comparison with the scintillation in monochromatic light as a function of the zenith distance [Tatarskii and Zhukova (1959)] for telescopes of different diameter $D$. a–$D = 6$;b–$D = 10$; c–$D = 12$;d–$D = 15$ inches respectively.

with the aid of Eq. (27.30), together with the empirical points obtained by Zhukova (telescope diameter 10 inch), Protheroe (telescope diameters 6 and 12 inches) and Butler (telescope diameter 15 inches). The agreement between the theoretical curve and the various experimental data is satisfactory. This confirms the above theory that the reduction in stellar scintillation in integrated light with increasing zenith distance is a chromatic effect.

# 10 FUNCTIONAL FORMULATION OF THE TURBULENCE PROBLEM

## 28. EQUATIONS FOR THE CHARACTERISTIC FUNCTIONAL

### 28.1 Equations for the Spatial Characteristic Functional of the Velocity Field

It was shown in Sect. 3.4 of Vol. 1 that the complete statistical description of the velocity field $\boldsymbol{u}(\boldsymbol{x}, t)$ in a turbulent flow could be achieved by specifying its *characteristic functional*

$$\Phi[\theta(\boldsymbol{x}, t)] = \Phi[\theta_1(\boldsymbol{x}, t), \theta_2(\boldsymbol{x}, t), \theta_3(\boldsymbol{x}, t)] = \overline{\exp\left\{i \int\int_{-\infty}^{\infty}\int\int \sum_{k=1}^{3} \theta_k(\boldsymbol{x}, t) u_k(\boldsymbol{x}, t)\, d\boldsymbol{x}\, dt\right\}} \quad (28.1)$$

[see Eq. (3.27) of Vol. 1]. It was also explained that the values of the functional $\Phi[\theta(\boldsymbol{x}, t)]$ at the special "ideal points"

$$\theta(\boldsymbol{x}, t) = \sum_{k=1}^{N} \theta_k \delta(\boldsymbol{x} - \boldsymbol{x}_k) \delta(t - t_k)$$

of the space of the functions $\theta(\boldsymbol{x}, t)$ coincide with the characteristic functions of the probability distributions for the values $\boldsymbol{u}(\boldsymbol{x}_k, t_k)$ of the field $\boldsymbol{u}(\boldsymbol{x}, t)$ on finite sets of space-time points

$$(\boldsymbol{x}_1, t_1), (\boldsymbol{x}_2, t_2), \ldots, (\boldsymbol{x}_N, t_N);$$

and these "ideal points" are the limits of the sequences of usual points. Therefore all the finite-dimensional probability distributions for the field $\boldsymbol{u}(\boldsymbol{x}, t)$ can be unambiguously established from $\Phi[\theta(\boldsymbol{x}, t)]$. The *functional derivatives* of the functional $\Phi[\theta(\boldsymbol{x}, t)]$ defined in Sect. 4.4 can also be used to calculate directly any moments and cumulants of the velocity field $\boldsymbol{u}(\boldsymbol{x}, t)$ [see Eqs. (4.51) and (4.52)].

The characteristic functional $\Phi[\theta(\boldsymbol{x}, t)]$ of the velocity field $\boldsymbol{u}(\boldsymbol{x}, t)$ will be referred to as the *space-time functional.* A less complete statistical description of the velocity field can be obtained by specifying its *spatial characteristic functional*

$$\Phi[\theta(\boldsymbol{x}), t] = \overline{\exp\left\{i \int\int\int_{-\infty}^{\infty} \sum_{k=1}^{3} \theta_k(\boldsymbol{x}) u_k(\boldsymbol{x}, t) \, d\boldsymbol{x}\right\}} \qquad (28.2)$$

[see Eq. (3.28) of Vol. 1]. This functional includes the complete statistical description of the velocity field at a given time $t$, but it cannot be used to calculate the joint statistical characteristics at different times. Nevertheless, the spatial functional $\Phi[\theta(\boldsymbol{x}), t]$ contains very extensive information about the statistical properties of the velocity field, which is quite sufficient for a broad range of applications.

In this chapter we shall consider the possibility of explicit determination of the characteristic functional for the velocity field in turbulent flow. Particular attention will be paid to the time evolution of $\Phi[\theta(\boldsymbol{x}), t]$. We shall restrict our attention to turbulence in an incompressible homogeneous fluid, so that the density $\rho$ will be regarded as constant and the pressure field $p(\boldsymbol{x}, t)$ will satisfy the Poisson equation $\Delta p = -\rho\partial^2(u_\alpha u_\beta)/\partial x_\alpha \partial x_\beta$ and will be expressed in terms of the velocity field. Consequently, the velocity field will, in this case, provide a complete description of the fluid flow, and the characteristic functional of the velocity field will contain information about all the statistical properties, not only of the velocity field but generally of turbulence in incompressible fluids.

We must note the following points. The characteristic functional of the velocity field is a compact form of specifying information which, in general, is contained in the infinite set of all the possible moments of this field (assuming that these moments exist). The moments satisfy an infinite set of moment equations (expressing the limitations imposed by the conservation laws for mass and momentum, i.e., the fact that the velocity field satisfies the continuity equation and the Navier-Stokes equations). These moment equations were discussed in Sect. 19. The problem thus arises as to whether the characteristic functional of the velocity field satisfies some equations which provide a compact way of writing down the dynamic restrictions which follow from the continuity and the Navier-Stokes equations, and which are equivalent to the infinite set of dynamic equations for the moments.

The answer to this question is affirmative: The spatial characteristic functional $\Phi[\theta(\boldsymbol{x}), t]$ does satisfy certain dynamic equations (which are a consequence of the continuity and the Navier-Stokes equations) and can, in principle, be determined from these equations in terms of its initial value $\Phi[\theta(\boldsymbol{x}), t_0] = \Phi_0[\theta(\boldsymbol{x})]$ (a similar result is valid also for the space-time characteristic functional; see Sect. 28.3). The dynamic equations for $\Phi[\theta(\boldsymbol{x}), t]$ were first obtained by Hopf (1952) [see also Hopf (1957, 1962)]. They are derived below.

In addition to the notation introduced in Sects. 3 and 4 of Vol. 1, it will be convenient to employ the following abbreviated notation. We shall denote the "scalar product" in the functional space of the functions of the point $M$ [space point with coordinates $\boldsymbol{x}$, or space-time point with coordinates $(\boldsymbol{x}, t)$] by the symbol

$$(\theta \cdot \boldsymbol{u}) = \int \theta(M)\, \boldsymbol{u}(M)\, dM \tag{28.3}$$

where $dM = d\boldsymbol{x}$ for $M = \boldsymbol{x}$ and $dM = d\boldsymbol{x}\, dt$ for $M = (\boldsymbol{x}, t)$, and the functions $\theta(M)$ and $\boldsymbol{u}(M)$ can assume both vector and scalar values. The definition (28.2) can then be rewritten in the form

$$\Phi[\theta(\boldsymbol{x}), t] = \overline{\exp\{i(\theta \cdot \boldsymbol{u})\}}. \tag{28.2'}$$

Moreover, we shall use the following abbreviated notation for the functional derivative operator:

$$\mathscr{D}_j(M) = \frac{\delta}{\delta\theta_j(M)\,dM}. \tag{28.4}$$

In view of the general formula (4.45) of Vol. 1 we have

$$\mathscr{D}_j(M)\,\Phi = \overline{iu_j(M)\exp\{i(\theta\cdot u)\}}, \tag{28.5}$$

$$\mathscr{D}_j(M_1)\,\mathscr{D}_k(M_2)\,\Phi = -\overline{u_j(M_1)\,u_k(M_2)\exp\{i(\theta\cdot u)\}}. \tag{28.6}$$

Let us now consider the restrictions imposed on the characteristic functional by the continuity equation $\partial u_\alpha/\partial x_\alpha = 0$ in an incompressible fluid, i.e., by the fact that the velocity field $u(x)$ is solenoidal. Since the difference between the spatial and space-time characteristic functionals is unimportant in this problem, we shall denote the characteristic functional by the shorter symbol $\Phi[\theta(x)]$ without indicating the dependence on $t$. Consider, to begin with, the case where the fluid occupies a finite volume $V$ bounded by a smooth rigid surface $\Gamma$. On $\Gamma$ we have the obvious boundary condition $u_n|_\Gamma = 0$ where $u_n$ are the velocity components in the direction of the inward normal to the boundary. If now $\varphi(x)$ is an arbitrary scalar function, then from the continuity equation $\partial u_\alpha/\partial x_\alpha = 0$ we obtain

$$\int_V u_\alpha \frac{\partial\varphi}{\partial x_\alpha}\,dV = \int_V \frac{\partial(u_\alpha\varphi)}{\partial x_\alpha}\,dV = -\oint_\Gamma u_n\varphi\,d\Gamma = 0,$$

i.e.,

$$(\nabla\varphi\cdot u) = 0. \tag{28.7}$$

However, we then have

$$(\{\theta + \nabla\varphi\}\cdot u) = (\theta\cdot u) \tag{28.7'}$$

for any functions $\theta(x)$ and $\varphi(x)$. Consequently, by Eq. (28.2′) we have

$$\Phi[\theta(x) + \nabla\varphi(x)] = \Phi[\theta(x)]. \tag{28.8}$$

This is the desired restriction imposed on the characteristic functional by the incompressible continuity equation.

If the boundary condition $u_n = 0$ is not valid on some (perhaps disconnected) part $\Gamma_1$ of the boundary $\Gamma$, then Eq. (28.8) will nevertheless be valid, but now not for any scalar function $\varphi(x)$ but only for functions $\varphi(x)$ satisfying a special boundary condition on $\Gamma_1$. In particular, it is sufficient to impose on $\varphi$ the following restriction: $\varphi|_{\Gamma_1} = \text{const}$; since the incompressibility of the fluid implies that $\int\limits_{\Gamma_1} u_n \, d\Gamma = \int\limits_{\Gamma} u_n \, d\Gamma = 0$. The situation for the case of an unbounded region $V$ is quite similar: Here the condition (28.8) will also be valid for any scalar function $\varphi(x)$ satisfying some special boundary conditions at infinity (for example, for any function $\varphi(x)$ which vanishes outside of a bounded region).

Let us now consider an interesting transformation of the condition (28.8). Namely, we shall use the fact that any vector field $\theta(x)$ on a closed domain $V$ can be uniquely represented as a sum of a solenoidal vector field $\tilde{\theta}(x)$ with a vanishing normal component $\tilde{\theta}_n$ on the boundary $\Gamma$ of a domain $V$ and a potential vector field $\nabla\varphi(x)$:

$$\theta(x) = \tilde{\theta}(x) + \nabla\varphi(x), \quad \frac{\partial\tilde{\theta}_\alpha(x)}{\partial x_\alpha} = 0, \quad \tilde{\theta}_n|_\Gamma = 0. \tag{28.9}$$

In other words, there is a linear operator $\mathscr{L}$ which associates a definite solenoidal vector field $\mathscr{L}\theta(x) = \tilde{\theta}(x)$ having zero normal component on $\Gamma$ with each vector field $\theta(x)$ in $V$. From Eq. (28.8) it then follows that

$$\Phi[\theta(x)] = \Phi[\tilde{\theta}(x)], \tag{28.10}$$

i.e., the characteristic functional $\Phi$ is invariant under the transformation which is achieved with the aid of the operator $\mathscr{L}$ in the space of the vector functions $\theta(x)$. The relation (28.10) is another form of expressing the restrictions imposed on the characteristic functional by the fact that the velocity field is solenoidal. It is readily verified that it is not only a consequence of this condition, but, in a sense, it is equivalent to the condition that the velocity field is solenoidal. Namely, if the characteristic functional of a random but not necessarily solenoidal vector field $u(x)$ satisfies Eq. (28.10), then it must coincide with the characteristic functional of the solenoidal component $\tilde{u}(x)$ of the field $u(x)$ (i.e., a solenoidal field $\tilde{u}(x)$ such that $u - \tilde{u}$ is a potential field). This is clear from the relation

$$(\tilde{\theta} \cdot u) = (\tilde{\theta} \cdot \tilde{u}),$$

which can be proved in a way quite similar to the derivation of Eqs. (28.7) and (28.7').

Finally, there is also a differential analog of Eq. (28.8) or Eq. (28.10). If we differentiate both sides of Eq. (28.5) with respect to $x_j$, take into account the fact that $(\theta \cdot u)$ and therefore $\exp\{i(\theta \cdot u)\}$ are independent of $x$, and then sum over $j$, we obtain on the right-hand side the expression

$$i \overline{\frac{\partial u_j(x, t)}{\partial x_j} \exp\{i(\theta \cdot u)\}},$$

which must be zero because of the continuity equation $\partial u_j/\partial x_j = 0$. Consequently,

$$\frac{\partial}{\partial x_\alpha} \{\mathscr{D}_\alpha(x)\Phi\} = 0. \tag{28.11}$$

This is a differential form of the restrictions imposed on the characteristic functional by the condition that the velocity field is solenoidal. It can be derived directly from Eq. (28.8) without using its definition (28.2) and the continuity equation. Conversely, we can derive Eq. (28.8) from Eq. (28.11). Both these derivations can be carried out, for example, by using the Taylor expansion of $\Phi[\theta(x) + h\nabla\varphi(x)]$ in powers of $h$, or the equivalent expansion of $\Phi[\theta(x) + \nabla\varphi(x)]$ in a functional Taylor series such as Eq. (4.53) about the "point" $\theta(x)$. However, we shall not consider this here. Let us now go on to derive the dynamic equation for the characteristic functional which is a consequence of the fact that the velocity field satisfies the Navier-Stokes equations. The difference between spatial and space-time functionals will now be important. In this section we shall restrict our attention to the spatial form of the functional $\Phi[\theta(x), t]$. Differentiating (28.2') with respect to $t$, we obtain

$$\frac{\partial \Phi}{\partial t} = i \overline{\left(\theta \cdot \frac{\partial u}{\partial t}\right) \exp\{i(\theta \cdot u)\}}. \tag{28.12}$$

If we now substitute

$$\frac{\partial u}{\partial t} = -\frac{\partial u u_\alpha}{\partial x_\alpha} - \frac{1}{\rho}\nabla p + \nu\Delta u$$

into this equation (i.e., if we use the Navier-Stokes equations), we can rewrite the resulting expression in the form

$$\frac{\partial \Phi}{\partial t} = i\left(\theta \cdot \left\{-\frac{\partial}{\partial x_\alpha}\overline{u u_\alpha e^{i(\theta \cdot u)}} - \nabla \overline{\frac{p}{\rho} e^{i(\theta \cdot u)}} + \nu\,\Delta \overline{u e^{i(\theta \cdot u)}}\right\}\right), \quad (28.13)$$

where we have included the factor $\exp\{i(\theta \cdot u)\}$ inside the space derivatives because it is independent of $x$. Finally, using Eqs. (28.5) and (28.6), and also using the notation

$$\Pi = \frac{1}{\rho}\overline{p \exp\{i(\theta \cdot u)\}} = \Pi[\theta(x);\ x,\ t], \quad (28.14)$$

we can rewrite Eq. (28.13) in the form

$$\frac{\partial \Phi}{\partial t} = \left(\theta \cdot \left\{i\frac{\partial \mathscr{D}\mathscr{D}_\alpha \Phi}{\partial x_\alpha} + \nu\,\Delta \mathscr{D}\Phi - i\nabla\Pi\right\}\right), \quad (28.15)$$

where $\mathscr{D} = (\mathscr{D}_1, \mathscr{D}_2, \mathscr{D}_3)$ is the vector operator representing functional differentiation. Equations (28.15) and (28.11) [or Eqs. (28.8) and (28.10)] form a set of two equations for the two unknown functionals $\Phi[\theta(x), t]$ and $\Pi[\theta(x);\ x,\ t]$. The functional $\Pi$ can be eliminated from this system by noting that the expression in the braces in Eq. (28.15) is the vector field $i\overline{\frac{\partial u}{\partial t}\exp\{i(\theta \cdot u)\}}$ with vanishing divergence; since $\frac{\partial^2 u_\alpha}{\partial t\,\partial x_\alpha} = 0$ and $\exp\{i(\theta \cdot u)\}$ is independent of $x$. The fact that the divergence of this vector field is zero can be expressed in the form

$$\Delta\Pi = \frac{\partial^2 \mathscr{D}_\alpha \mathscr{D}_\beta \Phi}{\partial x_\alpha\,\partial x_\beta} \quad (28.16)$$

where the term $+i\nu\Delta\frac{\partial \mathscr{D}_\alpha \Phi}{\partial x_\alpha}$ is omitted on the right-hand side because it is zero by Eq. (28.11). If we write Eq. (28.16) in the symbolic form

$$\Pi = \Delta^{-1}\frac{\partial^2 \mathscr{D}_\alpha \mathscr{D}_\beta \Phi}{\partial x_\alpha\,\partial x_\beta}, \quad (28.17)$$

where $\Delta^{-1}$ is the reciprocal of the Laplace operator $\Delta$, and substitute this into Eq. (28.15), we obtain

$$\frac{\partial \Phi}{\partial t}=\left(\theta \cdot \left\{ i \frac{\partial \mathscr{D} \mathscr{D}_\alpha \Phi}{\partial x_\alpha} - i \nabla \Delta^{-1} \frac{\partial^2 \mathscr{D}_\alpha \mathscr{D}_\beta \Phi}{\partial x_\alpha \partial x_\beta} + \nu \Delta \mathscr{D} \Phi \right\}\right), \quad (28.18)$$

which contains only the functional $\Phi$.

The functional $\Pi$ can also be eliminated from Eq. (28.15) with the aid of Eq. (28.10). According to this equation, we can replace $\theta(x)$ in Eq. (28.15) by $\tilde{\theta}(x)=\mathscr{P}\theta(x)$ which satisfies Eq. (28.9). Using $(\tilde{\theta} \cdot \nabla \Pi)=0$, which can be proved similarly to the derivation of Eq. (28.7), we obtain

$$\frac{\partial \Phi}{\partial t}=i\left(\tilde{\theta} \cdot \frac{\partial \mathscr{D} \mathscr{D}_\alpha \Phi}{\partial x_\alpha}\right)+\nu(\tilde{\theta} \cdot \Delta \mathscr{D} \Phi). \quad (28.19)$$

This equation, or the equivalent equation (28.18), is the *Hopf equation* for the characteristic functional $\Phi[\theta(x), t]$. Since it is of the first order in time, it can, in principle, be used to find the functional $\Phi[\theta(x), t]$ from its initial value $\Phi[\theta(x), t_0]=\Phi_0[\theta(x)]$.[1]

The initial functional $\Phi_0[\theta(x)]$ should, of course, satisfy Eq. (28.10) [or Eq. (28.8), or Eq. (28.11)] which is a consequence of the continuity condition. If this requirement is fulfilled, the corresponding solution $\Phi[\theta(x), t]$ of the Hopf equation will automatically have the property indicated by Eq. (28.10) for any $t > t_0$. In fact, if $\Phi[\theta(x), t_0]=\Phi_0[\theta(x)]$ satisfies Eq. (28.10), then, in accordance with the foregoing discussion, the initial velocity field can be regarded as solenoidal, i.e., the divergence $\frac{\partial u_\alpha}{\partial x_\alpha}=D$ can be regarded as equal to zero at the initial time $t_0$. However, it is known that the solution of the Navier-Stokes equations for a solenoidal initial value is solenoidal for any $t > t_0$; this follows, for example, from the fact that the divergence of the vector Navier-Stokes equation has the form of the scalar equation $\frac{\partial D}{\partial t}=\nu \Delta D$. If, however,

[1] We must stress however that the mathematical integration theory for functional differential equations in infinite-dimensional spaces has not been established. Even questions on the existence and uniqueness of the solutions of such equations have at present no satisfactory answers [only very preliminary results of this type were recently given by Foias (1970, 1971)]. This is of course not surprising at all since the proof of strict existence and uniqueness theorems for solutions of the (nonstatistical) Navier-Stokes differential equations in three-dimensional space has also encountered serious difficulties.

for any $t > t_0$ the velocity field $\boldsymbol{u}(\boldsymbol{x}, t) = \boldsymbol{u}(t)$ is solenoidal, then for any $t > t_0$ the functional $\Phi[\theta(\boldsymbol{x}), t] = \overline{\exp\{i(\theta, \boldsymbol{u}(t))\}}$ which coincides with the solution of the Hopf equation will have the property (28.10). Therefore, when we consider the Hopf equation, it is sufficient to demand that Eq. (28.10), or the equivalent conditions (28.8) and (28.11), are satisfied only at the initial time, in which case they will automatically be satisfied at all subsequent times.

The problem of the determination of the characteristic functional $\Phi[\theta(\boldsymbol{x}), t]$ from the Hopf equation (28.18) [or Eq. (28.19)] for a given initial value $\Phi_0[\theta(\boldsymbol{x})]$ is the most compact formulation of the general *turbulence problem*, i.e., of the problem of determination of the statistical characteristics of turbulence from given statistical characteristics of the initial velocity field $\boldsymbol{u}(\boldsymbol{x}, t_0) = \boldsymbol{u}_0(\boldsymbol{x})$.

A striking feature of the Hopf equation (28.18) [or Eq. (28.19)] is that it is *linear*. Thus, although fluid dynamics is nonlinear, i.e., the evolution of an individual velocity field is described by nonlinear equations, the main problem of the *statistical dynamics* of a turbulent flow (the turbulence problem) is a linear problem. This means that we have a *superposition principle* for the characteristic functional $\Phi[\theta(\boldsymbol{x}), t]$: If the initial functional $\Phi_0[\theta(\boldsymbol{x})]$ is a linear combination of given functionals $\Phi_0^{(\gamma)}$, the functional $\Phi[\theta(\boldsymbol{x}), t]$ can be written for $t > t_0$ as a linear combination of the functionals $\Phi^{(\gamma)}[\theta(\boldsymbol{x}), t]$ which are the solutions of the Hopf equation for the initial data $\Phi_0^{(\gamma)}$.

We note, however, that while the transition from the discussion of the evolution of individual realizations of the turbulent flow to the statistical discussion, achieved with the aid of the characteristic functional, enables us to avoid difficulties connected with the nonlinearity of fluid dynamics, we now encounter other serious difficulties (of a purely mathematical rather than physical nature) which are connected with the necessity for developing mathematical tools for solving the *linear* (in the unknown functionals) *functional differential equations*. So far, only the first few steps have been taken in this direction.

## 28.2 Spectral Form of the Equations for the Spatial Characteristic Functional

In the preceding chapters we frequently verified that there are considerable advantages in transforming from the spatial description of the velocity field (and its moments, i.e., the various correlation functions) to the spectral (i.e., wave) representation of this field.

Transition to the spectral representation in the Hopf equation for the characteristic functional has comparable advantages.

The spectral description of an idealized, statistically homogeneous, or locally homogeneous random field $\boldsymbol{u}(\boldsymbol{x})$ in infinite space can be achieved with the aid of the random set function $\boldsymbol{Z}(\Delta \boldsymbol{k})$ in wave-number space which is determined from $\boldsymbol{u}(\boldsymbol{x})$ with the aid of the Fourier-Stieltjes inversion formula [see, for example, Eq. (11.45), which refers to a scalar homogeneous field]. Transition to the spectral representation is simpler when not the velocity field itself but the functional $\Phi[\theta(\boldsymbol{x}), t]$ of the velocity field is employed. Since the argument $\theta(\boldsymbol{x})$ of this functional is a nonrandom function which can be chosen practically arbitrarily, it can, in general, be written not in the form of the Fourier-Stieltjes integral but simply in the form of the usual Fourier integral

$$\theta(\boldsymbol{x}) = (2\pi)^{-3} \int e^{i\boldsymbol{k}\boldsymbol{x}} \boldsymbol{z}(\boldsymbol{k})\, d\boldsymbol{k} \tag{28.20}$$

which is evaluated over infinite space. By considering the functional $\Phi[\theta(\boldsymbol{x}), t]$ only on the set of functions $\theta(\boldsymbol{x})$ represented in the form of the Fourier integrals (28.20), we achieve the transition to the spectral representation if we transform in this functional from $\theta(\boldsymbol{x})$ to a new "independent variable" $\boldsymbol{z}(\boldsymbol{k})$ defined by[2]

$$\boldsymbol{z}(\boldsymbol{k}) = \int e^{-i\boldsymbol{k}\boldsymbol{x}}\, \theta(\boldsymbol{x})\, d\boldsymbol{x}. \tag{28.20'}$$

Since the function $\theta(\boldsymbol{x})$ is real, we have

$$\boldsymbol{z}(-\boldsymbol{k}) = \boldsymbol{z}^*(\boldsymbol{k}), \tag{28.21}$$

where, as usual, the asterisk represents the complex conjugate. The spatial characteristic functional of the velocity field in the spectral representation is given by

$$\Psi[\boldsymbol{z}(\boldsymbol{k}), t] = \Phi\left[(2\pi)^{-3} \int e^{i\boldsymbol{k}\boldsymbol{x}}\, \boldsymbol{z}(\boldsymbol{k})\, d\boldsymbol{k},\ t\right]. \tag{28.22}$$

[2] We note that, in practice, we can restrict our consideration to an even narrower set of functions $\theta(\boldsymbol{x})$ which vanish identically outside a finite volume [see, for example, Gel'fand and Vilenkin (1964)]. If the volume $V$ occupied by the fluid is finite, then instead of the Fourier integral (28.20) we can use the expansion of $\theta(\boldsymbol{x})$ in terms of any complete set of vector functions $\boldsymbol{f}_{\boldsymbol{k}}(\boldsymbol{x})$ which are orthogonal in $V$ and correspond to a discrete set of values of $\boldsymbol{k}$. The coefficients $\boldsymbol{z}_{\boldsymbol{k}}$ of this expansion then form the new independent variable.

We emphasize that this representation is unrelated to any statistical properties of the random velocity field itself and, in particular, it does not require that the field should be statistically homogeneous or locally homogeneous. When the field $\boldsymbol{u}(\boldsymbol{x})$ is, say, homogeneous, i.e., its spectral representation (11.52) is possible, then

$$(\theta \cdot \boldsymbol{u}) = \int \theta(\boldsymbol{x}) \boldsymbol{u}(\boldsymbol{x})\, d\boldsymbol{x} = \int \boldsymbol{z}^*(\boldsymbol{k})\, d\boldsymbol{Z}(\boldsymbol{k}, t)$$

and, consequently,

$$\Psi[\boldsymbol{z}(\boldsymbol{k}), t] = \overline{\exp\left\{i \int \boldsymbol{z}^*(\boldsymbol{k})\, d\boldsymbol{Z}(\boldsymbol{k}, t)\right\}}. \tag{28.23}$$

The last formula is also valid for a locally homogeneous field $\boldsymbol{u}(\boldsymbol{x})$ but, in this case, the characteristic functional $\Phi[\theta(\boldsymbol{x}), t]$ is defined only for $\theta(\boldsymbol{x})$ for which $\int \theta(\boldsymbol{x})\, d\boldsymbol{x} = 0$. Hence, the functional $\Psi[\boldsymbol{z}(\boldsymbol{k}), t]$ is defined in this case only for those $\boldsymbol{z}(\boldsymbol{k})$ for which $\boldsymbol{z}(0) = 0$. The functional $\Psi[\boldsymbol{z}(\boldsymbol{k}), t] = \Phi[\theta(\boldsymbol{x}), t]$ has the following obvious properties:

$$\Psi[0, t] = 1, \quad |\Psi[\boldsymbol{z}(\boldsymbol{k}), t]| \leqslant 1, \qquad \Psi^*[\boldsymbol{z}(\boldsymbol{k}), t] = \Psi[-\boldsymbol{z}(\boldsymbol{k}), t]. \tag{28.24}$$

It will be convenient to be able to express the functional derivatives

$$\mathscr{D}_j(\boldsymbol{k}) \Psi[\boldsymbol{z}(\boldsymbol{k})] = \frac{\delta \Psi[\boldsymbol{z}(\boldsymbol{k})]}{\delta z_j(\boldsymbol{k})\, d\boldsymbol{k}} \tag{28.25}$$

of the functional $\Psi$ in terms of the functional derivatives of $\Phi$. To find this expression, let us take the first variation of the right- and left-hand sides of the equation $\Phi[\theta(\boldsymbol{x})] = \Psi[\boldsymbol{z}(\boldsymbol{k})]$ in which $\boldsymbol{z}(\boldsymbol{k})$ is understood to be given by the Fourier integral (28.20′). Using the definition of functional derivatives [Eqs. (4.47) and (4.48) in Vol. 1], and taking into account only the principal linear parts of the variations, we obtain

$$\delta\Phi[\theta(\boldsymbol{x})] = \int \mathscr{D}_\alpha(\boldsymbol{x}) \Phi[\theta(\boldsymbol{x})]\, \delta\theta_\alpha(\boldsymbol{x})\, d\boldsymbol{x},$$

$$\delta\Psi[\boldsymbol{z}(\boldsymbol{k})] = \int \mathscr{D}_\alpha(\boldsymbol{k}) \Psi[\boldsymbol{z}(\boldsymbol{k})]\, \delta z_\alpha(\boldsymbol{k})\, d\boldsymbol{k}.$$

If we equate these expressions, subject to the condition

$$\delta z_\alpha(\boldsymbol{k}) = \int e^{-i\boldsymbol{k}\boldsymbol{x}} \delta\theta_\alpha(\boldsymbol{x})\, d\boldsymbol{x},$$

and use the fact that $\delta\theta_\alpha(\boldsymbol{x})$ is arbitrary, we find that

$$\mathscr{D}_j(\boldsymbol{x})\Phi = \int e^{-i\boldsymbol{k}\boldsymbol{x}} \mathscr{D}_j(\boldsymbol{k})\Psi\, d\boldsymbol{k}. \tag{28.26}$$

This formula enables us to derive quite easily the differential equation for the characteristic functional $\Psi[\boldsymbol{z}(\boldsymbol{k})]$, which follows from the fact that the velocity field is solenoidal. All that is required is to differentiate both sides of Eq. (28.26) with respect to $x_j$, sum over $j$, and use Eq. (28.11). The left-hand side is then zero and, consequently,

$$0 = -i \int e^{-i\boldsymbol{k}\boldsymbol{x}} \boldsymbol{k}\mathscr{D}(\boldsymbol{k})\Psi\, d\boldsymbol{k}.$$

However, since a function of which the Fourier transform is identically zero must itself be zero, we finally obtain

$$\boldsymbol{k}\mathscr{D}(\boldsymbol{k})\Psi = 0. \tag{28.27}$$

This equation is the analog of Eq. (28.11) in the spectral representation, and gives the differential form of the restrictions imposed on the characteristic functional due to the fact that the velocity field is solenoidal. It is also a relatively simple matter to obtain the spectral representation of Eqs. (28.8) and (28.10) which are equivalent to Eq. (28.11). Thus, replacing the functional arguments $\theta(\boldsymbol{x})$ and $\theta(\boldsymbol{x}) + \nabla\varphi(\boldsymbol{x})$ in Eq. (28.8) by their Fourier transforms $\boldsymbol{z}(\boldsymbol{k})$ and $\boldsymbol{z}(\boldsymbol{k}) + \boldsymbol{k}\psi(\boldsymbol{k})$, where $-i\psi(\boldsymbol{k})$ is the Fourier transform of the scalar function $\varphi(\boldsymbol{x})$, we can verify that for any scalar function $\psi(\boldsymbol{k})$ we have

$$\Psi[\boldsymbol{z}(\boldsymbol{k}) + \boldsymbol{k}\psi(\boldsymbol{k})] = \Psi[\boldsymbol{z}(\boldsymbol{k})]. \tag{28.28}$$

In particular, if we set

$$\psi(\boldsymbol{k}) = -\frac{\boldsymbol{k}\boldsymbol{z}(\boldsymbol{k})}{k^2},$$

i.e., introduce the function

$$\tilde{z}(\boldsymbol{k}) = z(\boldsymbol{k}) - \frac{\boldsymbol{k}z(\boldsymbol{k})}{k^2}\boldsymbol{k}, \tag{28.29}$$

which is uniquely and linearly determined by $z(\boldsymbol{k})$, and has the property $\boldsymbol{k}\tilde{z}(\boldsymbol{k}) = 0$, we have

$$\Psi[z(\boldsymbol{k})] = \Psi[\tilde{z}(\boldsymbol{k})]. \tag{28.30}$$

The formulas (28.28) and (28.30) are the spectral analogs of Eqs. (28.8) and (28.10). Both of them are obviously equivalent to Eq. (28.27).

We must go on to derive the spectral form of the Hopf equation (28.18). We shall use the equation $\Phi[\theta(\boldsymbol{x}), t] = \Psi[z(\boldsymbol{k}), t]$ and the well-known Parseval relation of Fourier integral theory, according to which the Fourier transform does not affect the functional "scalar product." This formula can be written in the form

$$(\theta \cdot \varphi) = (z \cdot A[\varphi]), \tag{28.31}$$

where

$$A[\varphi] = (2\pi)^{-3} \int e^{i\boldsymbol{k}\boldsymbol{x}} \varphi(\boldsymbol{x})\, d\boldsymbol{x}, \tag{28.32}$$

and the parentheses on the right indicate integration of the scalar product $z \cdot A[\varphi]$ over all $\boldsymbol{k}$.

In view of the above two properties, the spectral form of Eq. (28.18) is

$$\frac{\partial \Psi}{\partial t} = \left(z \cdot A\left\{i \frac{\partial \mathscr{D} \mathscr{D}_\alpha \Phi}{\partial x_\alpha} - i \nabla \Delta^{-1} \frac{\partial^2 \mathscr{D}_\alpha \mathscr{D}_\beta \Phi}{\partial x_\alpha \partial x_\beta} + \nu \Delta \mathscr{D} \Phi\right\}\right), \tag{28.33}$$

and all that remains is to find the Fourier transform of the expression in the braces. This is most simply established for the last term in the braces, which by Eq. (28.26) can be written in the form

$$\nu \Delta \mathscr{D} \Phi = \nu \Delta \int e^{-i\boldsymbol{k}\boldsymbol{x}} \mathscr{D}(\boldsymbol{k}) \Psi\, d\boldsymbol{k} = -\nu \int e^{-i\boldsymbol{k}\boldsymbol{x}} k^2 \mathscr{D}(\boldsymbol{k}) \Psi\, d\boldsymbol{k}.$$

Hence, by Eq. (28.32) we obtain

$$A[\nu \Delta \mathscr{D} \Phi] = -\nu k^2 \mathscr{D}(\boldsymbol{k}) \Psi. \tag{28.34}$$

If we now use Eq. (28.26) twice, we find that

$$\mathscr{D}_j(\boldsymbol{x})\mathscr{D}_l(\boldsymbol{x})\Phi = \iint e^{-i(\boldsymbol{k}'+\boldsymbol{k}'')\boldsymbol{x}}\mathscr{D}_j(\boldsymbol{k}')\mathscr{D}_l(\boldsymbol{k}'')\Psi\,d\boldsymbol{k}'\,d\boldsymbol{k}'', \quad (28.35)$$

so that, for example,

$$i\frac{\partial\mathscr{D}\mathscr{D}_\alpha\Phi}{\partial x_\alpha} = \iint e^{-i(\boldsymbol{k}'+\boldsymbol{k}'')\boldsymbol{x}}(k'_\alpha + k''_\alpha)\mathscr{D}_\alpha(\boldsymbol{k}')\mathscr{D}(\boldsymbol{k}'')\Psi\,d\boldsymbol{k}'\,d\boldsymbol{k}''.$$

When we apply the Fourier transformation $\boldsymbol{A}$ to this function of $\boldsymbol{x}$, the factor $e^{-i(\boldsymbol{k}'+\boldsymbol{k}'')\boldsymbol{x}}$ under the integral sign will transform to the delta-function $\delta(\boldsymbol{k}-\boldsymbol{k}'-\boldsymbol{k}'')$ and, consequently,

$$\boldsymbol{A}\left[i\frac{\partial\mathscr{D}\mathscr{D}_\alpha\Phi}{\partial x_\alpha}\right] = k_\alpha\int\mathscr{D}_\alpha(\boldsymbol{k}')\mathscr{D}(\boldsymbol{k}-\boldsymbol{k}')\Psi\,d\boldsymbol{k}'. \quad (28.36)$$

Similarly, we find that

$$\boldsymbol{A}\left[-i\nabla\Delta^{-1}\frac{\partial^2\mathscr{D}_\alpha\mathscr{D}_\beta\Phi}{\partial x_\alpha\,\partial x_\beta}\right] = -\boldsymbol{k}\frac{k_\alpha k_\beta}{k^2}\int\mathscr{D}_\alpha(\boldsymbol{k}')\mathscr{D}_\beta(\boldsymbol{k}-\boldsymbol{k}')\Psi\,d\boldsymbol{k}'. \quad (28.37)$$

Substituting Eqs. (28.34), (28.36), and (28.37) in Eq. (28.33), and evaluating the functional "scalar product," we have, using Eq. (28.29),

$$\frac{\partial\Psi}{\partial t} = \iint\tilde{z}_\alpha(\boldsymbol{k}'+\boldsymbol{k}'')k''_\beta\mathscr{D}_\beta(\boldsymbol{k}')\mathscr{D}_\alpha(\boldsymbol{k}'')\Psi\,d\boldsymbol{k}'\,d\boldsymbol{k}'' - \\ -\nu\int k^2 z_\alpha(\boldsymbol{k})\mathscr{D}_\alpha(\boldsymbol{k})\Psi\,d\boldsymbol{k}. \quad (28.38)$$

This is the spectral form of the Hopf equation. In the second term on the right of Eq. (28.38) we can replace $z_\alpha(\boldsymbol{k})$ by $\tilde{z}_\alpha(\boldsymbol{k})$, since $(z_\alpha - \tilde{z}_\alpha)\mathscr{D}_\alpha\Psi = \frac{\boldsymbol{k}z}{k^2}k_\alpha\mathscr{D}_\alpha\Psi = 0$ in view of Eq. (28.27). In this form, Eq. (28.38) is the spectral version of Eq. (28.19).

Hopf (1952) has derived an equation equivalent to Eq. (28.38) in another way. He has considered a velocity field $\boldsymbol{u}(\boldsymbol{x}, t)$ which is periodic in all space coordinates (with the same period $L$), and which can therefore be written in the form of the Fourier series $\sum_{\boldsymbol{k}}\boldsymbol{v}(\boldsymbol{k}, t)e^{i\boldsymbol{k}\boldsymbol{x}}$, where the wave vector is given by $\boldsymbol{k} = \frac{2\pi}{L}(n_1, n_2, n_3)$ and $n_1, n_2, n_3$ are integers. Using the dynamic equations for the Fourier coefficients $\boldsymbol{v}(\boldsymbol{k}, t)$ obtained from the Navier-Stokes equations after replacing $u(x, t)$ by the corresponding Fourier series, Hopf obtained the equation for the characteristic functional $\Psi$ of the sequence $\{v(k, t)\}$ by a procedure similar to that used in Sect. 28.1 to derive Eq. (28.18). Then he proceeded to the limit as $L \to \infty$. His final result was

$$\frac{\partial\Psi}{\partial t} = \int dk' \mathcal{D}_\beta(k') \int \tilde{z}_\alpha(k' + k'') k''_\beta \mathcal{D}_\alpha(k'')\Psi \, dk'' - \\ - \nu \int k^2 z_\alpha(k) \mathcal{D}_\alpha(k) \Psi \, dk, \qquad (28.38')$$

which differs from Eq. (28.38) by the inversion of the order of the function $\tilde{z}_\alpha(k' + k'')$ and the operator $\mathcal{D}_\beta(k')$ under the integral sign of the first term on the right. To verify that this interchange is justified in the present case, let us use the identity

$$z_\alpha(k) = \delta_{\alpha\beta} \int z_\beta(k') \delta(k - k') \, dk'.$$

Hence we obtain from the definition of the functional derivative

$$\mathcal{D}_\beta(k') z_\alpha(k) = \delta_{\alpha\beta} \delta(k - k'). \qquad (28.39)$$

Consequently, we have the commutation relation

$$\mathcal{D}_\beta(k') z_\alpha(k) - z_\alpha(k) \mathcal{D}_\beta(k') = \delta_{\alpha\beta} \delta(k - k'). \qquad (28.40)$$

Since $\tilde{z}_\alpha(k) = \left(\delta_{\alpha\beta} - \frac{k_\alpha k_\beta}{k^2}\right) z_\beta(k)$ and the factors of the form $\delta_{\alpha\beta} - k_\alpha k_\beta / k^2$ commute with any functional differentiation operators $\mathcal{D}_\gamma(k')$, we have from Eq. (28.40)

$$\mathcal{D}_\gamma(k')\tilde{z}_\alpha(k' + k'') = \tilde{z}_\alpha(k' + k'') \mathcal{D}_\gamma(k') + \left[\delta_{\alpha\gamma} - \frac{(k'_\alpha + k''_\alpha)(k'_\gamma + k''_\gamma)}{|k' + k''|^2}\right] \delta(k'').$$

Substituting this expression into the first term on the right of Eq. (28.38'), we obtain the first term on the right of Eq. (28.38) plus the term

$$\int dk' \int \left[\delta_{\alpha\beta} - \frac{(k'_\alpha + k''_\alpha)(k'_\beta + k''_\beta)}{|k' + k''|^2}\right] \delta(k'') k''_\beta \mathcal{D}_\alpha(k'') \Psi \, dk'',$$

This term is however zero because of $k''_\beta \delta(k'') = 0$. This proves the equivalence of (28.38) and (28.38').

Bass (1953) has also considered the derivation of Eq. (28.38) starting with Eq. (28.23), which is valid for a homogeneous velocity field. Later Tatarskii (1962b) used a heuristic device to the same end, assuming that $u(x, t)$ can be represented by an ordinary Fourier integral. The above general method of deriving Eq. (28.38) is due to Lewis and Kraichnan (1962), who used it to derive the equation for the space-time characteristic functional of the velocity field.

The Hopf equation (28.38) is, in many respects, analogous to certain equations used in quantum field theory [see, for example, Schweber, Bethe, and Hoffmann (1955)]. To exhibit this analogy let us rewrite Eq. (28.38) in the Schroedinger quantum mechanical form

$$i\frac{\partial\Psi}{\partial t}=(\mathscr{H}_0+\mathscr{H}_1)\Psi, \tag{28.41}$$

where

$$\mathscr{H}_0=-i\nu\int dk\, k^2 z_\alpha(k)\,\mathscr{D}_\alpha(k), \tag{28.42}$$

$$\mathscr{H}_1=\int\int dk'\,dk'' B_{\alpha,\beta\gamma}(k'+k'')\,z_\beta(k'+k'')\,\mathscr{D}_\alpha(k')\,\mathscr{D}_\gamma(k''), \tag{28.43}$$

and

$$B_{\alpha,\beta\gamma}(k)=ik_\alpha\left(\delta_{\beta\gamma}-\frac{k_\beta k_\gamma}{k^2}\right). \tag{28.44}$$

In Eq. (28.41) the functional $\Psi$ is analogous to the "state vector" in the Schroedinger representation of the quantized field. In particular, if $\Psi$ is a power functional of degree $n$

$$\Psi[z(k)]=\int\cdots\int z_{\alpha_1}(k_1)\cdots z_{\alpha_n}(k_n)\,\Psi_{\alpha_1\ldots\alpha_n}(dk_1\ldots dk_n), \tag{28.45}$$

then the corresponding state vector describes the $n$-particle state of the field, i.e., the state with $n$ quanta. Since multiplication by $z_j$ increases by unity the order of this functional and the application of the operator $\mathscr{D}_j$ reduces it by unity, the operator representing multiplication by $z_j(k)$ and the functional differentiation operator $\mathscr{D}_j(k)$ can be regarded as the analogs of the creation and annihilation operators for quanta with momentum $k$, respectively. The commutation relation (28.40) is then identical with the commutation relation for the creation and annihilation operators corresponding to the so-called Bose field, i.e., a quantized field whose quanta obey the Bose-Einstein statistics.

Next, the operator $\mathscr{H}_0$ is bilinear in the creation and annihilation operators, and is analogous to the Hamiltonian for a free field in quantum field theory, whereas $\mathscr{H}_1$, which is cubic in the creation and annihilation operators, is analogous to the interaction Hamiltonian in the Schroedinger representation. The interactions with which we have been concerned, are the inertial interactions between spatial velocity-field inhomogeneities, which are described in the Navier-Stokes equations by terms that are nonlinear in the velocity field $u$. The ratio of the typical values of such terms to the linear terms describing the effect of viscosity is equal to the Reynolds number Re, which is an "inertial interaction constant" (see Sect. 19.2). If we transform in Eq. (28.41) to dimensionless quantities, so that the Hamiltonian $\mathscr{H}_0$ for the free field is of the order of unity, then the interaction constant Re will be a factor in front of the interaction Hamiltonian $\mathscr{H}_1$. Since in the case of developed turbulence Re is large, the interaction described by the Hamiltonian $\mathscr{H}_1$ is a *strong* interaction.

Therefore, there is a close analogy between the Hopf equation and the Schroedinger equation for a quantized Bose field with strong interaction. This analogy is, of course, incomplete. In particular, the interaction Hamiltonian (28.43) has a very special form which is hardly typical for quantum field theory (it describes only the coalescence of two quanta into one). Nevertheless, the above analogy is useful since it enables us to use the methods of quantum field theory to help us in the solution of the Hopf equation [Tatarskii (1962b)].

Thus, for example, the transformation which in quantum field theory describes the transition from the Schroedinger representation to the interaction representation, may be useful in the solution of Hopf's equation. This transformation enables us to eliminate the second term on the right-hand side of Eq. (28.38), which corresponds to the free-field Hamiltonian $\mathscr{H}_0$ in Eq. (28.41). In this approach the functional $\Psi[z(\boldsymbol{k}), t]$ is sought in the form

$$\Psi[z(\boldsymbol{k}), t] = \tilde{\Psi}[e^{-\nu k^2 t} z(\boldsymbol{k}), t] \tag{28.46}$$

where, for the sake of simplicity, we shall assume that $t_0 = 0$. To within terms of the order of $dt$ we then have

$$\Psi[z(\boldsymbol{k}), t+dt] - \Psi[z(\boldsymbol{k}), t] = \tilde{\Psi}[e^{-\nu k^2 t} z(\boldsymbol{k}), t+dt] + \\ + \tilde{\Psi}[e^{-\nu k^2 t}(1 - \nu k^2\, dt)\, z(\boldsymbol{k}), t] - \Psi[z(\boldsymbol{k}), t].$$

From Eq. (28.46) we find that the second term on the right-hand side of the last equation can be written in the form $\Psi[z(\boldsymbol{k}) - \nu k^2 z(\boldsymbol{k})\, dt, t]$. If we regard $-\nu k^2 z(\boldsymbol{k})\, dt$ as a variation of $z(\boldsymbol{k})$, we can write the last two terms in the form

$$\int \{-\nu k^2 z_\alpha(\boldsymbol{k})\, dt\}\, \mathscr{D}_\alpha(\boldsymbol{k})\, \Psi\, d\boldsymbol{k}.$$

Hence, we can readily show that

$$\frac{\partial \Psi}{\partial t} = \frac{\partial \tilde{\Psi}}{\partial t} - \nu \int k^2 z_\alpha(\boldsymbol{k})\, \mathscr{D}_\alpha(\boldsymbol{k})\, \Psi\, d\boldsymbol{k},$$

so that Eq. (28.38) reduces to

$$\frac{\partial}{\partial t} \tilde{\Psi}[e^{-\nu k^2 t} z(\boldsymbol{k}), t] = \int\int d\boldsymbol{k}'\, d\boldsymbol{k}''\, \tilde{z}_\alpha(\boldsymbol{k}' + \boldsymbol{k}'')\, k''_\beta \mathscr{D}_\beta(\boldsymbol{k}') \mathscr{D}_\alpha(\boldsymbol{k}'') \times \\ \times \tilde{\Psi}[e^{-\nu k^2 t} z(\boldsymbol{k}), t].$$

Let us now replace $e^{-\nu k^2 t} z(\boldsymbol{k})$ by $y(\boldsymbol{k})$. We then have

$$\mathscr{D}_\alpha(\boldsymbol{k})\, \tilde{\Psi}[y(\boldsymbol{k})] = e^{-\nu k^2 t} \frac{\delta \tilde{\Psi}[y(\boldsymbol{k})]}{\delta y_\alpha(\boldsymbol{k})\, d\boldsymbol{k}},$$

and if the operator $\delta/\delta y_\alpha(\boldsymbol{k})\, d\boldsymbol{k}$ is again represented by the symbol $\mathscr{D}_\alpha(\boldsymbol{k})$, the equation for $\tilde{\Psi}[y(\boldsymbol{k}), t]$ will assume the form

$$\frac{\partial \tilde{\Psi}}{\partial t} = \int\int \tilde{y}_\alpha(\boldsymbol{k}' + \boldsymbol{k}'')\, k''_\beta\, \mathscr{D}_\beta(\boldsymbol{k}')\, \mathscr{D}_\alpha(\boldsymbol{k}'')\, \tilde{\Psi}\, d\boldsymbol{k}'\, d\boldsymbol{k}''. \tag{28.47}$$

This differs from Eq. (28.38) by the absence of the second term from the right-hand side. If we now substitute

$$\mathscr{D}_\alpha(\boldsymbol{k}, t) = e^{-\nu k^2 t} \mathscr{D}_\alpha(\boldsymbol{k}), \quad y_\alpha(\boldsymbol{k}, t) = e^{\nu k^2 t} y_\alpha(\boldsymbol{k}), \tag{28.48}$$

$$\mathscr{H}(t) = i \int \int dk'\, dk'' B_{\alpha,\beta\gamma}(k' + k'')\, y_\beta(k' + k'', t)\, \mathscr{D}_\alpha(k', t)\, \mathscr{D}_\gamma(k'', t), \tag{28.49}$$

we can rewrite Eq. (28.47) in the form

$$i \frac{\partial \widetilde{\Psi}}{\partial t} = \mathscr{H}(t)\, \widetilde{\Psi}. \tag{28.50}$$

This equation is the analog of the Schroedinger equation in the interaction representation (the functional $\widetilde{\Psi}$ and operator $\mathscr{H}(t)$ are the analogs of the state vector and the interaction Hamiltonian in the interaction representation, respectively).

## 28.3 Equations for the Space-Time Characteristic Functional

Let us now briefly consider the equations which are satisfied by the space-time characteristic functional of the velocity field. It will be useful to introduce the following three spectral representations

$$\Psi[z(k, t)] = \Phi\left[(2\pi)^{-3} \int e^{ikx} z(k, t)\, dk\right], \tag{28.51}$$

$$\Phi_1[\eta(x, \omega)] = \Phi\left[(2\pi)^{-1} \int e^{i\omega t} \eta(x, \omega)\, d\omega\right], \tag{28.52}$$

$$\Psi_1[\zeta(k, \omega)] = \Phi\left[(2\pi)^{-4} \int e^{i(kx+\omega t)} \zeta(k, \omega)\, dk\, d\omega\right]. \tag{28.53}$$

The representation (28.51) can be called the wave or the three-dimensional spectral representation. The representation (28.52) can be called the frequency representation, while that given by Eq. (28.53) can be called the frequency-wave or the four-dimensional spectral representation.

The equation for the characteristic functional, which is implied by the fact that the velocity field is solenoidal, is independent of the presence of the time (or frequency) as an argument. Therefore, for the functionals $\Phi[\theta(x, t)]$ and $\Phi_1[\eta(x, \omega)]$ this equation can be written in any of the three forms given by Eqs. (28.8), (28.10), or (28.11), while for the functionals $\Psi[z(k, t)]$ and $\Psi_1[\zeta(k, \omega)]$ it can be written in any of the three forms (28.27), (28.28), or (28.30) (with an obvious modification of the notation which is due to the presence of the additional argument $t$ or $\omega$). Thus, for example, the equations which follow from the solenoidal condition can be written in the differential form as follows:

$$\frac{\partial}{\partial x_\alpha}\{\mathcal{D}_\alpha(\boldsymbol{x}, t)\,\Phi[\boldsymbol{\theta}(\boldsymbol{x}, t)]\} = 0, \tag{28.54}$$

$$\boldsymbol{k}\cdot\mathcal{D}(\boldsymbol{k}, t)\,\Psi[\boldsymbol{z}(\boldsymbol{k}, t)] = 0, \tag{28.54'}$$

$$\frac{\partial}{\partial x_\alpha}\{\mathcal{D}_\alpha(\boldsymbol{x}, \omega)\,\Phi_1[\boldsymbol{\eta}(\boldsymbol{x}, \omega)]\} = 0, \tag{28.54''}$$

$$\boldsymbol{k}\cdot\mathcal{D}(\boldsymbol{k}, \omega)\,\Psi_1[\boldsymbol{\zeta}(\boldsymbol{k}, \omega)] = 0, \tag{28.54'''}$$

where the symbol $\mathcal{D}$ represents the functional derivative with respect to the corresponding functional argument.

Let us now consider the derivation of the main dynamic equation which is a consequence of the Navier-Stokes equations. We shall begin by considering Eq. (28.5) written for the functional $\Phi[\boldsymbol{\theta}(\boldsymbol{x}, t)]$. Differentiating with respect to $t$, and bearing in mind the fact that $(\boldsymbol{\theta}\cdot\boldsymbol{u})$ is now independent of $t$, we have

$$\frac{\partial}{\partial t}\{\mathcal{D}(\boldsymbol{x}, t)\,\Phi\} = i\,\overline{\frac{\partial \boldsymbol{u}}{\partial t}\exp\{i(\boldsymbol{\theta}\cdot\boldsymbol{u})\}}.$$

If we express $\partial\boldsymbol{u}/\partial t$ with the aid of the Navier-Stokes equations we have, after some simple transformations similar to those corresponding to the transition from Eq. (28.13) to Eq. (28.15),

$$\frac{\partial\mathcal{D}\Phi}{\partial t} = i\,\frac{\partial\mathcal{D}\mathcal{D}_\alpha\Phi}{\partial x_\alpha} + \nu\Delta\mathcal{D}\Phi - i\nabla\Pi, \tag{28.55}$$

where $\Pi$ is given by Eq. (28.14) [but the functional scalar product $(\boldsymbol{\theta}\cdot\boldsymbol{u})$ is now determined in a different way than in Sect. 28.1]. The functional $\Pi$ can be eliminated from Eq. (28.55) by taking, for example, the divergence of both sides (regarded as vector functions of $\boldsymbol{x}$) and using Eq. (28.54). This gives Eq. (28.17) for $\Pi$ which, in turn, transforms Eq. (28.55) to the form

$$\frac{\partial\mathcal{D}\Phi}{\partial t} = i\,\frac{\partial\mathcal{D}\mathcal{D}_\alpha\Phi}{\partial x_\alpha} - i\nabla\Delta^{-1}\,\frac{\partial^2\mathcal{D}_\alpha\mathcal{D}_\beta\Phi}{\partial x_\alpha\,\partial x_\beta} + \nu\Delta\mathcal{D}\Phi, \tag{28.56}$$

which contains only one unknown functional, namely, $\Phi$. It is obvious that we can reduce the order of functional differentiation in this equation by unity by taking as the new unknown the vector functional $\boldsymbol{F} = (F_1, F_2, F_3)$ of $\boldsymbol{\theta}(\boldsymbol{x}, t)$, which depends on the parameters $\boldsymbol{x}$ and $t$ and is defined by

$$\boldsymbol{F}[\boldsymbol{\theta}(\boldsymbol{x}, t);\ \boldsymbol{x}, t] = \mathcal{D}(\boldsymbol{x}, t)\,\Phi[\boldsymbol{\theta}(\boldsymbol{x}, t)]. \tag{28.57}$$

Equation (28.56) can then be rewritten in the form

$$\left(\frac{\partial}{\partial t}-\nu\Delta\right)F_j=i\left(\delta_{j\beta}-\frac{\partial}{\partial x_j}\Delta^{-1}\frac{\partial}{\partial x_\beta}\right)\frac{\partial\mathscr{D}_\alpha F_\beta}{\partial x_\alpha}, \qquad (28.58)$$

which contains only the first functional derivatives of $F$.

Another method of eliminating $\Pi$ from Eq. (28.55) involves scalar multiplication [in the sense of Eq. (28.3)] of both sides of this equation by an arbitrary vector function $\boldsymbol{y}(\boldsymbol{x}, t)$ which decreases sufficiently rapidly for $|\boldsymbol{x}|\to\infty$ and satisfies the solenoidal condition $\partial y_\alpha/\partial x_\alpha=0$ [in particular, $\boldsymbol{y}(\boldsymbol{x}, t)$ can be taken to be the solenoidal component $\tilde{\theta}(\boldsymbol{x}, t)$ of the functional argument $\theta(\boldsymbol{x}, t)$ of the functional $\Phi$]. Since $(\boldsymbol{y}\cdot\nabla\Pi)=0$, we have

$$\left(\boldsymbol{y}\cdot\frac{\partial\mathscr{D}\Phi}{\partial t}\right)=i\left(\boldsymbol{y}\cdot\frac{\partial\mathscr{D}\mathscr{D}_\alpha\Phi}{\partial x_\alpha}\right)+\nu(\boldsymbol{y}\cdot\Delta\mathscr{D}\Phi). \qquad (28.59)$$

The dynamic equation for $\Phi[\theta(\boldsymbol{x}, t)]$ is given in this form in the paper by Lewis and Kraichnan (1962). Equation (28.56) [or Eq. (28.59)] must be solved subject to the initial condition

$$\Phi[\theta(\boldsymbol{x})\delta(t-t_0)]=\Phi_0[\theta(\boldsymbol{x})], \qquad (28.60)$$

where $\Phi_0[\theta(\boldsymbol{x})]$ is the characteristic functional of the initial velocity field $\boldsymbol{u}(\boldsymbol{x}, t_0)$ [the possibility of defining $\Phi[\theta(\boldsymbol{x}, t)]$ for generalized functions of the form $\theta(\boldsymbol{x}, t)=\theta(\boldsymbol{x})\delta(t-t_0)$ was discussed in Sect. 3.4 of Vol. 1].

Transition in Eq. (28.56) to any of the three spectral representations given by Eqs. (28.51)–(28.53) can be achieved by analogy with the derivation of Eq. (28.38) given in the last section. We shall reproduce only the final results. The analogs of Eq. (28.56) for the functionals of Eqs. (28.51)–(28.53) are the equations

$$\left(\frac{\partial}{\partial t}+\nu k^2\right)\mathscr{D}_j(\boldsymbol{k}, t)\Psi=k_\alpha\Delta_{j\beta}(\boldsymbol{k})\int\mathscr{D}_\alpha(\boldsymbol{k}', t)\mathscr{D}_\beta(\boldsymbol{k}-\boldsymbol{k}', t)\Psi\,d\boldsymbol{k}', \qquad (28.61)$$

$$(i\omega+\nu\Delta)\mathscr{D}_j(\boldsymbol{x}, \omega)\Phi_1=$$
$$=-i\left(\delta_{j\alpha}-\frac{\partial}{\partial x_j}\Delta^{-1}\frac{\partial}{\partial x_\alpha}\right)\frac{\partial}{\partial x_\beta}\int\mathscr{D}_\alpha(\boldsymbol{x}, \omega')\mathscr{D}_\beta(\boldsymbol{x}, \omega-\omega')\Phi_1\,d\omega', \qquad (28.62)$$

$$(i\omega-\nu k^2)\mathscr{D}_j(\boldsymbol{k}, \omega)\Psi_1=$$
$$=-k_\alpha\Delta_{j\beta}(\boldsymbol{k})\int\mathscr{D}_\alpha(\boldsymbol{k}', \omega')\mathscr{D}_\beta(\boldsymbol{k}-\boldsymbol{k}', \omega-\omega')\Psi'\,d\boldsymbol{k}'\,d\omega'. \qquad (28.63)$$

where, as usual, $\Delta_{j\beta}(\boldsymbol{k}) = \delta_{j\beta} - \frac{k_j k_\beta}{k^2}$ and the functional derivatives are taken in all cases with respect to the functional arguments of the functional in the particular equation. The order of the functional differentiation in Eqs. (28.61)–(28.63) can also be reduced by unity by the method analogous to Eqs. (28.57)–(28.58).

Moreover, we can readily derive the dynamic equations for the representations (28.51)–(28.53) of the space-time characteristic functional in the form analogous to Eq. (28.59). Thus, for example, if we multiply both sides of Eq. (28.61) by the component $a_j(\boldsymbol{k}, t)$ of an arbitrary vector function $\boldsymbol{a}(\boldsymbol{k}, t)$ which satisfies the condition $\boldsymbol{k} \cdot \boldsymbol{a}(\boldsymbol{k}, t) = 0$, and if we then sum over $j$ and integrate with respect to $\boldsymbol{k}$ and $t$, we obtain

$$\left(\boldsymbol{a} \cdot \left\{\frac{\partial}{\partial t} + \nu k^2\right\} \mathcal{D}\Psi\right) = \left(\boldsymbol{a} \cdot k_\alpha \int \mathcal{D}_\alpha(\boldsymbol{k}', t)\, \mathcal{D}(\boldsymbol{k} - \boldsymbol{k}', t)\, \Psi \, d\boldsymbol{k}'\right). \tag{28.64}$$

This form of the dynamic equation for $\Psi[\boldsymbol{z}(\boldsymbol{k}, t)]$ was given by Lewis and Kraichnan (1962). Some forms of the dynamic equation for the space-time characteristic functional were given earlier by Bass (1953). We note finally that the dynamic equation for the characteristic space-time functional of a random function $\boldsymbol{\xi}(\boldsymbol{x}, t)$, which describes the displacements of fluid particles in turbulent flow and which follows from the Lagrangian equations of motion of an incompressible fluid (see Sect. 9.1 of Vol. 1), was derived by Monin (1962d).

## 28.4 Equations for the Characteristic Functional in the Presence of External Forces

Let us now consider the case when the fluid is located in the field of external forces described by the term $\boldsymbol{X}(\boldsymbol{x}, t)$ on the right-hand side of the vector Navier-Stokes equation. These forces are of interest in the first instance for the description of the influx of energy from external sources: The rate of influx of energy per unit mass of fluid at the point $(\boldsymbol{x}, t)$, which is due to the work done by the external forces, is equal to $\boldsymbol{u}(\boldsymbol{x}, t) \cdot \boldsymbol{X}(\boldsymbol{x}, t)$. When this rate is zero, i.e., there are no external forces, or they are everywhere perpendicular to the flow direction, the fluid receives no energy from the external sources. Since the dissipation of energy into heat under the action of molecular viscosity must always take place, the total kinetic energy of the fluid must decrease with time, i.e., the motion of the fluid will be damped (at any rate, in an incompressible fluid; in a compressible fluid, which we are not considering here, the influx of energy can be produced not only by external forces, but also by external sources of heat). We have encountered a situation of this kind in the case of isotropic turbulence (Chapt. 7).

Under real conditions the most usual external forces are non-random, for example, gravitational forces or surface forces produced by moving bodies. However, in some theoretical models of turbulent flows it is convenient to introduce random forces $\boldsymbol{X}(\boldsymbol{x}, t)$. Thus, turbulence in a temperature stratified medium (see Chapt. 4) can be described by the equations for an incompressible fluid in the field of random buoyancy forces proportional to the turbulent temperature fluctuations. Another interesting situation is that in the idealized model of stationary isotropic turbulence in which stationarity and isotropy is ensured by introducing an artificial stationary and isotropic field of random forces $\boldsymbol{X}(\boldsymbol{x}, t)$ [this model was used, for example, by Wyld (1961); see Sect. 19.6]. The model is, of course, fictitious because the forces $\boldsymbol{X}(\boldsymbol{x}, t)$ have no real analogs. However, if the forces $\boldsymbol{X}$ are introduced so that they ensure an appreciable mean influx of energy only for the large-scale turbulence components (in which case the small-scale components can receive energy practically exclusively from the large-scale components and not from the work done by the forces $\boldsymbol{X}$), then the theory of locally isotropic turbulence in which the statistical state of small-scale components is independent of the large-scale properties of the motion leads us to expect that the fictitious nature of the force field $\boldsymbol{X}(\boldsymbol{x}, t)$ will have no effect on the statistical properties of small-scale turbulence components. Therefore, the small-scale properties of turbulence can also be correctly described by the above fictitious model.

If the forces $\boldsymbol{X}(\boldsymbol{x}, t)$ are not random, there is no difficulty in taking them into account in the equations for the characteristic functional of the velocity field. Thus, on the right-hand side of Eq. (28.15) we must simply add the term $i(\theta \cdot \boldsymbol{X})\Phi$, while to the right-hand side of Eq. (28.18) we must add the term $i(\theta \cdot \tilde{\boldsymbol{X}})\Phi$, where $\tilde{\boldsymbol{X}}$ is the solenoidal component of the vector field $\boldsymbol{X}$. Moreover, in Eq. (28.19) we must add the term $i(\tilde{\theta} \cdot \boldsymbol{X})\Phi = i(\tilde{\theta} \cdot \tilde{\boldsymbol{X}})\Phi$, while in Eq. (28.38) we must add $i(\boldsymbol{z} \cdot \boldsymbol{Y})\Psi$, where $\boldsymbol{Y} = \boldsymbol{Y}(\boldsymbol{k}, t)$ is the Fourier transform of the function $\boldsymbol{X}(\boldsymbol{x}, t)$ of the form given by Eq. (28.32). Similarly, on the right-hand side of Eq. (28.55) we must add the term $i\boldsymbol{X}\Phi$, in Eq. (28.56) the term $i\tilde{\boldsymbol{X}}\Phi$, in

Eq. (28.59) the term $i(\boldsymbol{y}\cdot\boldsymbol{X})\Phi$, and in Eqs. (28.61)–(28.63) the terms $i\tilde{Y}\Psi$, $-i\tilde{X}_1\Phi_1$ and $-i\tilde{Y}_1\Psi_1$, where $\tilde{Y}$, $\tilde{X}_1$, and $\tilde{Y}_1$ are the Fourier transforms of the function $\tilde{X}(\boldsymbol{x}, t)$ with respect to $\boldsymbol{x}$, with respect to $t$, and with respect to $\boldsymbol{x}$ and $t$, respectively. Finally, we must add the term $i(\boldsymbol{a}\cdot Y)\Psi = i(\boldsymbol{a}\cdot\tilde{Y})\Psi$ to the right-hand side of Eq. (28.64).

If there are external forces then the velocity field at any time $t > 0$ will depend not only on the initial field $\boldsymbol{u}(\boldsymbol{x}, 0) = \boldsymbol{u}_0(\boldsymbol{x})$, but also on the values of the vector function $\boldsymbol{X}(\boldsymbol{x}, \tau)$, $0 \leqslant \tau < t$. In the statistical description of the velocity field in a turbulent flow the mean field characteristics do not depend on the initial velocity field (because the means are evaluated over the initial velocity fields, i.e., over a probability measure given on a set of possible initial velocity fields), but their dependence on the nonrandom external forces, i.e., on the function$\boldsymbol{X}(\boldsymbol{x}, \tau)$, $0 \leqslant \tau < t$, will remain. Therefore, it may be considered that the characteristic functional of the velocity field, $\Phi[\boldsymbol{\theta}(\boldsymbol{x}, t)]$, will depend on this function. The simplest characteristics of this type of dependence will be the "response tensors" (or "Green tensors") which are defined for $t > t', t'', \ldots$ by the equations

$$G^{\alpha}_{\beta}(\boldsymbol{x}, t \mid \boldsymbol{x}', t') = \mathscr{D}_{X_\beta}(\boldsymbol{x}', t')\overline{u_\alpha(\boldsymbol{x}, t)} = \\ = \frac{1}{i}\mathscr{D}_{X_\beta}(\boldsymbol{x}', t')\{\mathscr{D}_\alpha(\boldsymbol{x}, t)\Phi|_{\theta=0}\}, \quad (28.65)$$

$$G^{\alpha}_{\beta\gamma}(\boldsymbol{x}, t \mid \boldsymbol{x}', t'; \boldsymbol{x}'', t'') = \mathscr{D}_{X_\gamma}(\boldsymbol{x}'', t'')\mathscr{D}_{X_\beta}(\boldsymbol{x}', t')\overline{u_\alpha(\boldsymbol{x}, t)} = \\ = \frac{1}{i}\mathscr{D}_{X_\gamma}(\boldsymbol{x}'', t'')\mathscr{D}_{X_\beta}(\boldsymbol{x}', t')\{\mathscr{D}_\alpha(\boldsymbol{x}, t)\Phi|_{\theta=0}\} \quad (28.65')$$

and so on (where $\mathscr{D}_{X_j}(\boldsymbol{x}, t) = \delta/\delta X_j(\boldsymbol{x}, t)\,d\boldsymbol{x}\,dt$), which describe the mean response of the velocity fields to infinitesimal disturbances of the external force field.[3] These tensors have been used by Kraichnan (1959, 1964a–c and elsewhere) and Wyld (1961) (see Sect. 19.6 above), as well as by Lewis and Kraichnan (1962).

[3] If we define the response tensors by the formulas

$$G^{\alpha}_{\beta}(\boldsymbol{x}, t \mid \boldsymbol{x}', t') = \overline{\mathscr{D}_{X_\beta}(\boldsymbol{x}', t')\,u_\alpha(\boldsymbol{x}, t)},$$
$$G^{\alpha}_{\beta\gamma}(\boldsymbol{x}, t \mid \boldsymbol{x}', t'; \boldsymbol{x}'', t'') = \overline{\mathscr{D}_{X_\gamma}(\boldsymbol{x}'', t'')\,\mathscr{D}_{X_\beta}(\boldsymbol{x}', t')\,u_\alpha(\boldsymbol{x}, t)}$$

and so on (but do not express them in terms of $\Phi$), they will remain meaningful even for a random external force field $\boldsymbol{X}(\boldsymbol{x}, t)$.

Let us now suppose that the external force field $X(x, t)$ is random. We then have two random fields $u(x, t)$ and $X(x, t)$ which will, in general, be statistically related [because the velocity field $u(x, t)$ for $t > t_0$ in each individual realization of the fluid flow depends on the corresponding realization of the function $X(x, \tau)$ for $t_0 \leqslant \tau < t$]. The complete statistical description of the set of these two random fields is their joint space-time characteristic functional

$$\Omega[\theta(x, t), f(x, t)] = \overline{\exp\{i(\theta \cdot u) + i(f \cdot X)\}}, \quad (28.66)$$

where $(\theta \cdot u)$ and $(f \cdot X)$ represent the integrals of the corresponding scalar products with respect to all $x$ and $t$. A less complete but still frequently sufficient description is obtained with the aid of the characteristic velocity space and force space-time functional defined by

$$\Omega[\theta(x), f(x, t), t] = \overline{\exp\left\{i(\theta \cdot u) + i\int_{t_0}^{\infty}(f \cdot X)\,dt\right\}}, \quad (28.67)$$

[where $(\theta \cdot u)$ and $(f \cdot X)$ now represent integrals of the corresponding scalar products with respect to $x$ only]. Knowledge of this functional will provide us with information about the spatial (but not space-time) statistical state of the velocity field.

If the functional (28.66) is known, the mean rate of energy influx per unit mass of the fluid at the point $(x, t)$, which is due to the work done by the external forces, can be calculated from the formula

$$\overline{u(x, t)\,X(x, t)} = -\mathscr{D}_\alpha(x, t)\,\mathscr{D}_{f_\alpha}(x, t)\,\Omega\big|_{\theta=f=0}. \quad (28.68)$$

Similar expressions [with $\mathscr{D}_\alpha(x, t)$ replaced by $\mathscr{D}_\alpha(x)$] are obtained by using the functional $\Omega[\theta(x), f(x, t), t]$.

The condition of solenoidality on the velocity field imposes a restriction on the dependence of $\Omega$ on the functional argument $\theta$, which is the same as in the case of the characteristic functional of the velocity field, $\Phi$. Thus, the functional $\Omega[\theta(x, t), f(x, t)]$ should satisfy the equation

$$\frac{\partial}{\partial x_\alpha}[\mathscr{D}_\alpha(x, t)\,\Omega] = 0 \quad (28.69)$$

and the functional $\Omega[\theta(x), f(x, t), t]$ should satisfy a similar equation with $\mathscr{D}_\alpha(x, t)$ replaced by $\mathscr{D}_\alpha(x)$.

We must now derive the dynamic equation for $\Omega$, which follows from the fact that the components of the velocity field $\boldsymbol{u}(\boldsymbol{x}, t)$ satisfy the Navier-Stokes equations which include the external force field $\boldsymbol{X}(\boldsymbol{x}, t)$. To begin with, consider the functional $\Omega[\theta(\boldsymbol{x}), \boldsymbol{f}(\boldsymbol{x}, t), t]$. It is clear that

$$\frac{\partial \Omega}{\partial t} = \overline{i\left(\theta \cdot \frac{\partial \boldsymbol{u}}{\partial t}\right) \exp\left\{i(\theta \cdot \boldsymbol{u}) + i \int_{t_0}^{\infty} (\boldsymbol{f} \cdot \boldsymbol{X})\, dt\right\}}.$$

If we now express $\partial \boldsymbol{u}/\partial t$ with the aid of the Navier-Stokes equations, and carry out transformations analogous to the transition from Eq. (28.12) to Eq. (28.15), we obtain

$$\frac{\partial \Omega}{\partial t} = \left(\theta \cdot \left\{i \frac{\partial \mathscr{D} \mathscr{D}_\alpha \Omega}{\partial x_\alpha} + \nu \Delta \mathscr{D} \Omega - i \nabla \Pi + \mathscr{D}_f \Omega\right\}\right). \tag{28.70}$$

where $\mathscr{D}_f$ is the vector functional differentiation operator with components $\mathscr{D}_{f_\alpha}(\boldsymbol{x}, t)$, and the functional $\Pi$ is defined by

$$\Pi = \overline{\frac{1}{\rho} p \exp\left\{i(\theta \cdot \boldsymbol{u}) + i \int_{t_0}^{\infty} (\boldsymbol{f} \cdot \boldsymbol{X})\, dt\right\}} = \Pi[\theta(\boldsymbol{x}), \boldsymbol{f}(\boldsymbol{x}, t); \boldsymbol{x}, t].$$

Since the expression in the braces in Eq. (28.70) is the solenoidal vector field $\overline{\frac{\partial \boldsymbol{u}}{\partial t} \xi}$, where $\xi$ is a random quantity independent of $\boldsymbol{x}$, we can eliminate the functional $\Pi$ from Eq. (28.70) by using the solenoidal property, and proceeding by analogy with the transition from Eq. (28.15) to Eq. (28.18). This yields the equation

$$\begin{aligned} \frac{\partial \Omega}{\partial t} = \Bigg(\theta \cdot \Bigg\{ i \frac{\partial \mathscr{D} \mathscr{D}_\alpha \Omega}{\partial x_\alpha} - i \nabla \Delta^{-1} \frac{\partial^2 \mathscr{D}_\alpha \mathscr{D}_\beta \Omega}{\partial x_\alpha \partial x_\beta} + \\ + \nu \Delta \mathscr{D} \Omega + \mathscr{D}_f \Omega - \nabla \Delta^{-1} \frac{\partial \mathscr{D}_{f_\alpha} \Omega}{\partial x_\alpha} \Bigg\}\Bigg), \end{aligned} \tag{28.71}$$

which contains only the single functional $\Omega$. Since this equation is of the first order in time, we can use it, at least in principle, to find the functional $\Omega[\theta(\boldsymbol{x}), \boldsymbol{f}(\boldsymbol{x}, t), t]$ from a given initial value

$$\Omega[\theta(\boldsymbol{x}), \boldsymbol{f}(\boldsymbol{x}, t), t_0] = \Omega_0[\theta(\boldsymbol{x}), \boldsymbol{f}(\boldsymbol{x}, t)], \tag{28.72}$$

which provides a complete statistical description of the two random fields $\boldsymbol{u}(\boldsymbol{x}, t_0) = \boldsymbol{u}_0(\boldsymbol{x})$ and $\boldsymbol{X}(\boldsymbol{x}, t), t \geqslant t_0$. In particular,

$$\Omega_0[\boldsymbol{\theta}(\boldsymbol{x}), 0] = \Phi_0[\boldsymbol{\theta}(\boldsymbol{x})], \tag{28.73}$$

$$\Omega_0[0, \boldsymbol{f}(\boldsymbol{x}, t)] = F[\boldsymbol{f}(\boldsymbol{x}, t)], \tag{28.74}$$

where $\Phi_0$ is a given characteristic functional of the initial velocity field $\boldsymbol{u}_0(\boldsymbol{x})$, and $F$ is a given characteristic space-time functional of the external force field $\boldsymbol{X}(\boldsymbol{x}, \tau), \tau \geqslant t_0$. If we suppose that these two fields are statistically independent, then the functional (28.72) will, of course, be simply the product of the functionals (28.73) and (28.74).

We have noted that the introduction of a suitably chosen random force $\boldsymbol{X}(\boldsymbol{x}, t)$ may be of interest as a method of constructing an idealized model of stationary turbulence. In the definition (28.67) of the functional $\Omega$ it is then convenient to let $t_0 = -\infty$. Since the problem is stationary, this functional will not depend on $t$ and, therefore, it will satisfy Eq. (28.71) with the left-hand side set to zero. We may expect that this equation will have a single-valued solution for a given "boundary condition" (28.74).

It is occasionally convenient to use the joint characteristic functional of the velocity and external force fields in the spectral rather than the spatial representiation. This is given by the functional

$$\Lambda[\boldsymbol{z}(\boldsymbol{k}), \boldsymbol{g}(\boldsymbol{k}, t), t] = \\ = \Omega\left[(2\pi)^{-3}\int e^{i\boldsymbol{k}\boldsymbol{x}}\boldsymbol{z}(\boldsymbol{k})\,d\boldsymbol{k}, (2\pi)^{-3}\int e^{i\boldsymbol{k}\boldsymbol{x}}\boldsymbol{g}(\boldsymbol{k}, t)\,d\boldsymbol{k}, t\right]. \tag{28.75}$$

Proceeding as in Sect. 28.2, we can readily verify that this functional will satisfy the equations

$$\boldsymbol{k}\cdot\mathcal{D}(\boldsymbol{k})\Lambda = 0, \tag{28.76}$$

$$\frac{\partial\Lambda}{\partial t} = \int\int \tilde{z}_\alpha(\boldsymbol{k}'+\boldsymbol{k}'')\,k''_\beta\mathcal{D}_\beta(\boldsymbol{k}')\mathcal{D}_\alpha(\boldsymbol{k}'')\Lambda\,d\boldsymbol{k}'\,d\boldsymbol{k}'' - \\ -\nu\int k^2 z_\alpha(\boldsymbol{k})\mathcal{D}_\alpha(\boldsymbol{k})\Lambda\,d\boldsymbol{k} + \int \tilde{z}_\alpha(\boldsymbol{k})\mathcal{D}_{g_\alpha}(\boldsymbol{k}, t)\Lambda\,d\boldsymbol{k} \tag{28.77}$$

and the initial condition

$$\Lambda[\boldsymbol{z}(\boldsymbol{k}), \boldsymbol{g}(\boldsymbol{k}, t), t_0] = \Lambda_0[\boldsymbol{z}(\boldsymbol{k}), \boldsymbol{g}(\boldsymbol{k}, t)]. \tag{28.78}$$

In this last equation

$$\Lambda_0[z(k),\ 0] = \Psi_0[z(k)], \tag{28.79}$$

$$\Lambda_0[0,\ g[k,\ t] = G[g(k,\cdot t)], \tag{28.80}$$

where the right-hand sides contain the spectral representations of the functional $\Phi_0$ and $F$ of Eqs. (28.73) and (28.74).

Let us now reproduce the dynamic equations for the space-time functional $\Omega[\theta(x,\ t),\ f(x,\ t)]$ and its three-dimensional and four-dimensional spectral representations $\Lambda[z(k,\ t),\ g(k,\ t)]$ and

$$M[z(k,\omega), g(k,\omega)]$$

$$= \Omega[(2\pi)^{-4}\ e^{i(kx+\omega t)}z(k,\omega)dk d\omega, (2\pi)^{-4}\int e^{i(kx+\omega t)}g(k,\omega)dk d\omega]. \tag{28.75'}$$

The derivation of these equations is analogous to the derivation of Eqs. (28.56) and (28.61) and yields the results

$$\frac{\partial \mathscr{D}\Omega}{\partial t} = i\frac{\partial \mathscr{D}\mathscr{D}_\alpha\Omega}{\partial x_\alpha} - i\nabla\Delta^{-1}\frac{\partial^2 \mathscr{D}_\alpha\mathscr{D}_\beta\Omega}{\partial x_\alpha\,\partial x_\beta} + \nu\Delta\mathscr{D}\Omega + \mathscr{D}_f\Omega - \nabla\Delta^{-1}\frac{\partial \mathscr{D}_{f_\alpha}\Omega}{\partial x_\alpha}, \tag{28.81}$$

$$\left(\frac{\partial}{\partial t} + \nu k^2\right)\mathscr{D}_j(k,\ t)\Lambda = k_\alpha\Delta_{j\beta}(k)\int \mathscr{D}_\alpha(k',\ t)\,\mathscr{D}_\beta(k-k',\ t)\,\Lambda\,dk' + + \Delta_{j\alpha}(k)\,\mathscr{D}_{g_\alpha}(k,\ t)\,\Lambda. \tag{28.82}$$

$$(i\omega + \nu k^2)\mathscr{D}_j(k,\omega)M$$

$$= k_\alpha\Delta_{j\beta}(k)\int \mathscr{D}_\alpha(k',\omega)\,\mathscr{D}_\beta(k-k',\omega-\omega')M dk' d\omega' + \Delta_{j\alpha}(k)\mathscr{D}_{g_\alpha}(k,\omega)M. \tag{28.82'}$$

An equation equivalent to Eq. (28.81) [in the form analogous to Eq. (28.59)] is given by Lewis and Kraichnan (1962). This equation must be solved subject to the initial condition

$$\Omega[\theta(x)\,\delta(t-t_0),\ f(x,\ t)\,e(t-t_0)] = \Omega_0[\theta(x),\ f(x,\ t)], \tag{28.83}$$

where $e(t-t_0)$ is a function which is equal to zero for $t < t_0$ and to unity for $t \geqslant t_0$. The initial condition for Eq. (28.82) has an analogous form. Finally, Eq. (28.82') is unsuitable for application to initial condition problems; it is useful primarily when a steady statistical state is studied.

So far, we have used the joint characteristic functional of the velocity and external force fields for the description of the random fluid motion in a field of random external forces. The question is: To what extent is this joint characteristic functional necessary if we are interested in the statistical properties of the velocity field alone? Is it possible to derive an equation which would be satisfied by the characteristic functional of the velocity field in the presence of random external forces? Consider, for example, the spatial characteristic functional $\Psi[\boldsymbol{z}(\boldsymbol{k}), t]$. It can be obtained from the functional $\Lambda[\boldsymbol{z}(\boldsymbol{k}), \boldsymbol{g}(\boldsymbol{k}, t), t]$ defined by Eqs. (28.75) and (28.67) if we set $\boldsymbol{g}(\boldsymbol{k}, t) = 0$. Bearing this in mind, let us substitute $\boldsymbol{g}(\boldsymbol{k}, t) = 0$ into Eq. (28.77), so that

$$\frac{\partial \Psi}{\partial t} = \int\int \tilde{z}_\alpha(\boldsymbol{k}' + \boldsymbol{k}'') k''_\beta \mathscr{D}_\beta(\boldsymbol{k}') \mathscr{D}_\alpha(\boldsymbol{k}'') \Psi \, d\boldsymbol{k}' \, d\boldsymbol{k}'' - $$
$$- \nu \int k^2 z_\alpha(\boldsymbol{k}) \mathscr{D}_\alpha(\boldsymbol{k}) \Psi \, d\boldsymbol{k} + \int \tilde{z}_\alpha(\boldsymbol{k}) \left[\mathscr{D}_{g_\alpha}(\boldsymbol{k}, t) \Lambda\right]\Big|_{\boldsymbol{g}(\boldsymbol{k}, t)=0} d\boldsymbol{k} \equiv 0. \tag{28.84}$$

This differs from Eq. (28.38), which is valid in the absence of external forces, only by the presence of the third term on the right-hand side. This term contains however not the functional $\Psi$ but the general functional $\Lambda$, so that Eq. (28.84) is not as yet closed with respect to $\Psi$. Let us consider this term in greater detail. Using a relation of the form (28.26) and the definition (28.67) of $\Lambda$ we have

$$\left[\mathscr{D}_{g_\alpha}(\boldsymbol{k}, t) \Lambda\right]\Big|_{\boldsymbol{g}(\boldsymbol{k}, t)=0} = (2\pi)^{-3} \int d\boldsymbol{x} \, e^{i\boldsymbol{k}\boldsymbol{x}} \left[\mathscr{D}_{f_\alpha}(\boldsymbol{x}, t) \Omega\right]\Big|_{\boldsymbol{f}(\boldsymbol{x}, t)=0} =$$
$$= (2\pi)^{-3} \int e^{i\boldsymbol{k}\boldsymbol{x}} \, \overline{\left[i X_\alpha(\boldsymbol{x}, t) e^{i(\boldsymbol{\theta}\cdot\boldsymbol{u})}\right]} \, d\boldsymbol{x}. \tag{28.85}$$

We recall that $(\boldsymbol{\theta} \cdot \boldsymbol{u})$ represents the integral of the scalar product $\boldsymbol{\theta}(\boldsymbol{x}) \cdot \boldsymbol{u}(\boldsymbol{x}, t)$ with respect to $\boldsymbol{x}$ but not with respect to $t$. We thus see that we can obtain a closed equation for $\Psi$ from Eq. (28.84) only if we succeed in expressing the mathematical expectation of the product $X_\alpha(\boldsymbol{x}, t) e^{i(\boldsymbol{\theta}\cdot\boldsymbol{u})}$ in terms of $\Psi$. Novikov (1964) has shown that this can be done when the external random field $\boldsymbol{X}(\boldsymbol{x}, t)$ is Gaussian with zero mean, statistically stationary, and uncorrelated at different times, so that

$$\overline{X_j(\boldsymbol{x}_1, t_1) X_l(\boldsymbol{x}_2, t_2)} = \delta(t_1 - t_2) \hat{B}_{jl}(\boldsymbol{x}_1, \boldsymbol{x}_2). \tag{28.86}$$

Henceforth we shall confine our attention mainly to this simplest case.

The general equation for the mean value of the product of a Gaussian random vector field $\boldsymbol{X}(M)$ (where $M$ is a point of an arbitrary space) and a functional $F = F[\boldsymbol{X}(M)]$ of the same field plays a central part in Novikov's derivation. This equation was independently established by the physicists Furutsu (1963) and Novikov (1964) and by the mathematician Donsker (1964); it has the form

$$\overline{X_j(M) F} = \int \overline{X_j(M) X_\alpha(M_1)} \cdot \overline{\mathscr{D}_{X_\alpha}(M_1) F} \, dM_1 \tag{28.87}$$

This result is most easily proved if we assume that $F[\boldsymbol{X}(M)]$ can be expanded in a functional power series of the form (4.53) of Vol. 1 and then multiply both sides of this expansion by $X_j(M)$ and take the mean value using Eq. (4.27) for the evaluation of even order moments of $\boldsymbol{X}(M)$ (all odd order moments of $\boldsymbol{X}(M)$ are evidently zero). [Such a proof of Eq. (28.87) can also be found in Tatarskii (1971), (1965)].

Let us apply Eq. (28.87) to the case when $F[X(x,t)] = \exp\{i(\theta \cdot u)\}$. It is easily seen that then $\mathscr{D}_{X_\alpha}(x,t)F = iF\int \theta_\beta(x')\mathscr{D}_{X_\alpha}(x,t)u_\beta(x',t)dx'$. Moreover the Navier-Stokes equations imply that $\mathscr{D}_{X_\alpha}(x,t)u_\beta(x',t) = \delta_{\alpha\beta}\delta(x'-x)\int_{t_0}^{t}\delta(t-s)ds = \frac{1}{2}\delta_{\alpha\beta}\delta(x'-x)$. Hence Eqs. (28.87) and (28.86) yield the result

$$\overline{X_\alpha(t)\exp\{i(\theta\cdot u)\}} = \frac{i}{2}\int \theta_\beta(x')\,\hat{B}_{\alpha\beta}(x,x')\,dx'\cdot\Phi[\theta(x),t] \quad (28.88)$$

We shall consider below the case where the external force field is not only statistically stationary and uncorrelated in time, but is also statistically homogeneous in space, so that

$$\hat{B}_{\alpha\beta}(x,x') = \hat{B}_{\alpha\beta}(x'-x) = \int e^{-ik(x'-x)}\hat{F}_{\alpha\beta}(k)\,dk. \quad (28.89)$$

Equation (28.88) can then be reduced to

$$\overline{X_\alpha(x,t)\,e^{i(\theta\cdot u)}} = \frac{i}{2}\int e^{ikx}z_\beta(k)\,\hat{F}_{\alpha\beta}(k)\,dk\cdot\Psi[z(k),t]. \quad (28.90)$$

Substituting this result into Eq. (28.85), we obtain

$$\left[\dot{\mathscr{D}}_{g_\alpha}(k,t)\Lambda\right]_{g(k,t)=0} = -\frac{1}{2}z_\beta(-k)\,\hat{F}_{\alpha\beta}(-k). \quad (28.91)$$

Bearing all this in mind, we can finally rewrite Eq. (28.84) in the form

$$\frac{\partial\Psi}{\partial t} = \int\int \tilde{z}_\alpha(k'+k'')\,k''_\beta\,\mathscr{D}_\beta(k')\,\mathscr{D}_\alpha(k'')\,\Psi\,dk'\,dk'' - $$
$$-\nu\int k^2 z_\alpha(k)\,\mathscr{D}_\alpha(k)\,\Psi\,dk - \frac{1}{2}\int \tilde{z}_\alpha(-k)\,z_\beta(k)\,\hat{F}_{\alpha\beta}(k)\,dk\cdot\Psi, \quad (28.92)$$

which is closed with respect to the functional $\Psi$. This is the equation found by Novikov. We recall, however, that this equation is valid only for Gaussian stationary external forces with zero mean and delta-function correlation in time.

The procedure for taking into account random forces in the equation for the characteristic functional $\Psi[z(k),t]$, i.e., the transformation from the Hopf equation (28.38) to Eq. (28.92) is in a sense analogous to the procedure for taking into account the inertia of Browian particles in the diffusion equation, i.e., the Fokker-Planck-Kolmogorov equation for Brownian motion. The net effect is that the equation acquires an important additional term describing "diffusion in velocity space." The last term in Eq. (28.92) has a similar significance and shows that the spectral tensor of the external forces $\hat{F}_{\alpha\beta}(k)$ is in a sense the analog of the "diffusivity tensor in velocity space," which varies with the wave number of the spectral velocity component.

Let us now consider the stationary case, when $\frac{\partial\Psi}{\partial t} = 0$, and transform in Eq. (28.92) from the characteristic functional $\Psi[z(k)]$ in the spectral representation to the functional $\Phi[\theta(x)]$ in the spatial representation. Moreover, we apply the operator $\mathscr{D}_j(x)$ to the equation obtained, and then set $\theta(x) = 0$. The result is an equation relating the third and second velocity moments which can be reduced to the form

$$D_{LLL}(r) - 6\nu \frac{dD_{LL}(r)}{dr} = -\frac{2}{r^4} \int_0^r r'^4 \hat{B}_{\alpha\alpha}(r')\, dr', \tag{28.93}$$

where $\hat{B}_{jl}(r)$ is the correlation function for the external force field given by Eqs. (28.86) and (28.89). When $r = 0$ we have $\hat{B}_{\alpha\alpha}(0) = 2\bar{\varepsilon}$, where $\bar{\varepsilon}$ is the mean rate of turbulent energy dissipation. Let $\frac{\hat{B}_{\alpha\alpha}(0)}{\hat{B}''_{\alpha\alpha}(0)} = -L_0^2$ (so that $L_0$ is the differential or Taylor length scale for the external force field). We can then write

$$\hat{B}_{\alpha\alpha}(r) = 2\bar{\varepsilon}\, \hat{B}\left(\frac{r}{L_0}\right) = 2\bar{\varepsilon}\left[1 - \frac{1}{2}\frac{r^2}{L_0^2} + O\left(\frac{r^4}{L_0^4}\right)\right], \tag{28.94}$$

and Eq. (28.93) assumes the form

$$D_{LLL}(r) - 6\nu \frac{\partial D_{LL}(r)}{\partial r} = -\frac{4}{5}\bar{\varepsilon} r\left[1 - \frac{5}{14}\frac{r^2}{L_0^2} + O\left(\frac{r^4}{L_0^4}\right)\right]. \tag{28.95}$$

Equations (28.93) and (28.95) are also due to Novikov (1964). They are a generalization of the Kolmogorov structure equation (22.2) corresponding to our turbulence model excited by random external forces. It is clear that when $r \ll L_0$ both these equations become identical with the Kolmogorov equation. In particular, this will be so for any $r$ provided $L_0 \to \infty$, i.e., if the spectrum of the external forces is of the form $\hat{F}_{jl}(k) = \bar{\varepsilon}\,\delta(k)\,\Delta_{jl}(k)$. In this special case, the last term in Eq. (28.92) assumes the form $-\frac{1}{3}\bar{\varepsilon}\,|z(0)|^2\,\Psi$. Hence in the stationary case the corresponding equation (28.92) will contain only the two parameters $\bar{\varepsilon}$ and $\nu$ which, according to Kolmogorov, govern the statistical state of small-scale turbulence.

We note finally that one of the consequences of Eq. (28.93) is that if the function $\hat{B}(r/L_0)$ falls sufficiently rapidly at infinity, then for large $r$, for which the second term on the left of Eq. (28.93) can be neglected, we have

$$D_{LLL}(r) \sim -A\bar{\varepsilon}L_0\left(\frac{L_0}{r}\right)^4, \quad A = 4\int_0^\infty x^4 \hat{B}(x)\, dx, \tag{28.96}$$

i.e., $D_{LLL}(r)$ falls off as $r^{-4}$ as $r \to \infty$.

A closed equation for the characteristic functional of the velocity field of turbulence driven by a random field of external forces was also obtained by Hosokawa (1968) for the more general case of an arbitrary (non-Gaussian) force field which is $\delta$-correlated-in-time and has all higher-order cumulants containing only $\delta$-functions of all differences of its time arguments. It was found that in this case the equation for $\Psi$ contains on the left also summands expressed in terms of the Fourier transforms of all higher space cumulants of the force field. However, we shall not linger on this derivation in this book.

**A generalization of another type was considered by Klyatskin (1971b) who assumed that the external force field is Gaussian but that it is not $\delta$-correlated-in-time. He obtained a system of functional derivative equations for this case and the solution yields the expression for the characteristic functional of the turbulent velocity field.**

# 29. METHODS OF SOLVING THE EQUATIONS FOR THE CHARACTERISTIC FUNCTIONAL

## 29.1 Use of a Functional Power Series

When we tried to develop a complete statistical description of turbulence with the aid of the Hopf equation for the characteristic functional of the velocity field we found that no general mathematical formalism was available for solving linear equations in functional derivatives. There are also no rigorous theorems on the existence and uniqueness of the solutions of such equations [with the exception of some preliminary results of Foias (1970, 1971)]. Some special methods for solving some special types of such equations have been developed in particular by Tatarskii (1961b) and Novikov (1961d), but they are insufficient for solving Hopf's equation. The only general approach to the theory of integration of functional differential equations, which is based on the so-called functional integrals, will be discussed later (see Sect. 29.5). For the moment, let us consider some simpler approximate methods similar to those used to solve ordinary differential equations with the aid of power series.

Let us begin with the method which is analogous to the expansion of an unknown function in a series in powers of the independent variables. We shall restrict our attention to the spatial characteristic functional of the velocity field $\Phi[\theta(\boldsymbol{x}, t)]$ (or its spectral form $\Psi[\boldsymbol{z}(\boldsymbol{k}), t]$). The analog of the usual power series expansion is now the representation of the functional $\Phi$ by a "functional power series"

$$\Phi = 1 + \sum_{n=1}^{\infty} \Phi_n, \tag{29.1}$$

where $\Phi_n$ is a homogeneous power functional of degree $n$ given by

$$\Phi_n = \Phi_n[\theta(\boldsymbol{x}), t] = \int \ldots \int \theta_{\alpha_1}(\boldsymbol{x}_1) \ldots \theta_{\alpha_n}(\boldsymbol{x}_n) \times \\ \times \varphi_{\alpha_1 \ldots \alpha_n}(\boldsymbol{x}_1, \ldots, \boldsymbol{x}_n; t)\, d\boldsymbol{x}_1 \ldots d\boldsymbol{x}_n \tag{29.2}$$

(see pp. 241–242, Vol. 1). Since

$$\Phi = \overline{\exp\{i(\theta \cdot u)\}} = 1 + \sum_{n=1}^{\infty} \frac{i^n}{n!} \overline{(\theta \cdot u)^n},$$

so that $\Phi_n = \frac{i^n}{n!} \overline{(\theta \cdot u)^n}$, we have

$$\varphi_{\alpha_1 \dots \alpha_n}(x_1, \dots, x_n; t) = \frac{i^n}{n!} \overline{u_{\alpha_1}(x_1, t) \dots u_{\alpha_n}(x_n, t)}. \quad (29.3)$$

It follows that the determination of the functional $\Phi_n$ is equivalent to the determination of all the spatial moment functions of order $n$ of the velocity field. However, no finite segment of the functional power series (29.1) has the properties of the characteristic functional (see pp. 242–243, Vol. 1). Therefore, even in the approximate determination of the characteristic functional $\Phi$ we must, strictly speaking, specify all the terms in Eq. (29.1) (if only approximately).

Let us substitute Eq. (29.1) into the Hopf equation for $\Phi$ written in the form, for example, of Eq. (28.19). We note that functional differentiation reduces by unity the degree of the power functional of $\theta(x)$ given by Eq. (29.2), while scalar multiplication by $\theta$ or by $\tilde{\theta}$ increases it by unity. If we equate functionals of the same degree on the left and right of Eq. (28.19) we obtain

$$\frac{\partial \Phi_n}{\partial t} = i\left(\tilde{\theta} \cdot \frac{\partial \mathscr{D} \mathscr{D}_\alpha \Phi_{n+1}}{\partial x_\alpha}\right) + \nu(\tilde{\theta} \cdot \Delta \mathscr{D} \Phi_n), \qquad n = 1, 2, \dots \quad (29.4)$$

Hence, it is clear that the expression for $\partial \Phi_n / \partial t$ will always include not only $\Phi_n$ but also $\Phi_{n+1}$ for any $n$, so that a closed set of equations cannot be obtained for any finite number of terms in Eq. (29.1). This was, of course, to be expected because by Eqs. (29.2) and (29.3) the equation (29.4) is equivalent to the differential equations for the $n$th order moments of the velocity field, which as we know from Sect. 19, always contain the $(n+1)$-th order moments. The use of the functional power series (19.1) is thus seen to bring us back to the infinite set of coupled equations for the moments of the velocity field which we already know. The only advantage of the functional approach in this case is the compactness of the notation. For example, the Millionshchikov zero-fourth-cumulant hypothesis for the case of zero mean velocity $\overline{u(x, t)}$ (which is equivalent to the condition $\Phi_1 = 0$) can now be simply written in the form $\Phi_4 = \frac{1}{2}\Phi_2^2$.

Instead of Eq. (29.1) we can use the functional power series for the logarithm of the characteristic functional:

$$\ln \Phi = \sum_{n=1}^{\infty} \Phi^{(n)} \tag{29.5}$$

(see p. 242, Vol. 1), where $\Phi^{(n)}$ is again a homogeneous power functional of the form (29.2) except that it now includes the weight functions $\varphi_{\alpha_1 \ldots \alpha_n}(x_1, \ldots, x_n; t)$ which are proportional not to the moments but to the cumulants of the random variables $u_{\alpha_1}(x_1, t), \ldots, u_{\alpha_n}(x_n, t)$. As already noted on pp. 242–243 of Vol. 1, a finite segment of the series (29.5) will not, in general, be the logarithm of a characteristic functional [with the exception of the case where the field $u(x, t)$ is Gaussian and $\Phi^{(n)} = 0$ for $n \geqslant 3$]. Therefore, strictly speaking, the determination of $\Phi$ requires the specification of all the terms in the series. Substitution of Eq. (29.5) into the Hopf equation (29.19) leads to

$$\frac{\partial \Phi^{(n)}}{\partial t} = i\left(\tilde{\theta} \cdot \frac{\partial \mathscr{D} \mathscr{D}_\alpha \Phi^{(n+1)}}{\partial x_\alpha}\right) + i\left(\tilde{\theta} \frac{\partial}{\partial x_\alpha} \sum_{k=1}^{n} \mathscr{D} \Phi^{(k)} \cdot \mathscr{D}_\alpha \Phi^{(n+1-k)}\right) + \\ + \nu(\tilde{\theta} \cdot \Delta \mathscr{D} \Phi^{(n)}), \qquad n = 1, 2, \ldots, \tag{29.6}$$

These equations are also seen to be coupled together and form an infinite system. It is clear that the equation for $\Phi^{(n)}$ is equivalent to a certain combination of the equations for the velocity-field moments of the first $n$ orders. The Millionshchikov hypothesis can now be written simply in the form $\Phi^{(4)} = 0$.

Finally, we can use the power series

$$\Phi = e^{\tilde{\Phi}_1 + \tilde{\Phi}_2}\left(1 + \sum_{n=3}^{\infty} \tilde{\Phi}_n\right), \tag{29.7}$$

recommended by Hopf (1952), where $\tilde{\Phi}_k$ is a homogeneous power functional of degree $k$. This series is analogous to the Gram-Charlier series for the probability density [see, for example, Sect. 17.6 in Cramér (1946). The exponential factor in Eq. (29.7) is the characteristic functional for the Gaussian random field [compare this with Eq. (4.37)] which has the same first and second moments as the random velocity field which we are considering. Hence the series in parentheses describes the deviation of the real velocity field from the Gaussian field. Since each functional $\tilde{\Phi}_n$, $n \geqslant 3$ enters the equation for $\Phi$ in the form of the term $\tilde{\Phi}_n \exp\{\tilde{\Phi}_1 + \tilde{\Phi}_2\}$, the determination of $\tilde{\Phi}_n$ is equivalent to the determination of a combination of an infinite number of moments of the velocity field.

By substituting Eq. (29.7) into the Hopf equation (28.19) we again obtain an infinite set of coupled equations for the functionals $\tilde{\Phi}_k$. The first three of these are

$$\frac{\partial \tilde{\Phi}_1}{\partial t} = i\left(\tilde{\theta} \cdot \frac{\partial}{\partial x_\alpha}\{\mathscr{D}\mathscr{D}_\alpha \tilde{\Phi}_2 + \mathscr{D}\tilde{\Phi}_1 \mathscr{D}_\alpha \tilde{\Phi}_1\}\right) + \nu(\tilde{\theta} \cdot \Delta \mathscr{D}\tilde{\Phi}_1),$$

$$\frac{\partial \tilde{\Phi}_2}{\partial t} = i\left(\tilde{\theta} \cdot \frac{\partial}{\partial x_\alpha}\{\mathscr{D}\mathscr{D}_\alpha \tilde{\Phi}_3 + \mathscr{D}\tilde{\Phi}_1 \mathscr{D}_\alpha \tilde{\Phi}_2 + \mathscr{D}\tilde{\Phi}_2 \mathscr{D}_\alpha \tilde{\Phi}_1\}\right) + \nu(\tilde{\theta} \cdot \Delta \mathscr{D}\tilde{\Phi}_2),$$

$$\frac{\partial \tilde{\Phi}_3}{\partial t} = i\left(\tilde{\theta} \cdot \frac{\partial}{\partial x_\alpha}\{\mathscr{D}\mathscr{D}_\alpha \tilde{\Phi}_4 + \mathscr{D}\tilde{\Phi}_1 \mathscr{D}_\alpha \tilde{\Phi}_3 + \mathscr{D}\tilde{\Phi}_3 \mathscr{D}_\alpha \tilde{\Phi}_1 + \right.$$
$$\left. + \mathscr{D}\tilde{\Phi}_2 \mathscr{D}_\alpha \tilde{\Phi}_2\}\right) + \nu(\tilde{\theta} \cdot \Delta \mathscr{D}\tilde{\Phi}_3). \tag{29.8}$$

In particular, if $\overline{u(x, t)} = 0$, then $\tilde{\Phi}_1 = 0$ and, consequently, $\left(\tilde{\theta} \cdot \frac{\partial \mathscr{D}\mathscr{D}_\alpha \tilde{\Phi}_2}{\partial x_\alpha}\right) = 0$ in view of the first equation in Eq. (29.8). The second and third equations in Eq. (29.8) are then much simpler and assume the form

$$\frac{\partial \tilde{\Phi}_2}{\partial t} = i\left(\tilde{\theta} \cdot \frac{\partial \mathscr{D}\mathscr{D}_\alpha \tilde{\Phi}_3}{\partial x_\alpha}\right) + \nu(\tilde{\theta} \cdot \Delta \mathscr{D}\tilde{\Phi}_2),$$
$$\frac{\partial \tilde{\Phi}_3}{\partial t} = i\left(\tilde{\theta} \cdot \frac{\partial}{\partial x_\alpha}\{\mathscr{D}\mathscr{D}_\alpha \tilde{\Phi}_4 + \mathscr{D}\tilde{\Phi}_2 \mathscr{D}_\alpha \tilde{\Phi}_2\}\right) + \nu(\tilde{\theta} \cdot \Delta \mathscr{D}\tilde{\Phi}_3). \tag{29.9}$$

The Millionshchikov zero-fourth-cumulant hypothesis can now be written in the form $\tilde{\Phi}_4 = 0$.

Let us now consider the functional power series approach for the spectral form of the velocity characteristic functional $\Psi[z(k), t]$. In this case, the analog of Eq. (19.1) is

$$\Psi = 1 + \sum_{n=1}^{\infty} \Psi_n, \tag{29.10}$$

where $\Psi_n$ is a homogeneous power functional of degree $n$ in the function $z(\boldsymbol{k})$. However, we cannot now expect that the functionals $\Psi_n$ will have a simple integral form such as that given by Eq. (29.2). In general, they can be represented only in the more complicated form of the Stieltjes integrals

$$\Psi_n[z(\boldsymbol{k}), t] = \int \ldots \int z_{\alpha_1}(\boldsymbol{k}_1) \ldots z_{\alpha_n}(\boldsymbol{k}_n) \Psi_{\alpha_1 \ldots \alpha_n}(d\boldsymbol{k}_1, \ldots, d\boldsymbol{k}_n; t), \tag{29.11}$$

Where $\Psi_{\alpha_1 \ldots \alpha_n}(S_1, \ldots, S_n; t)$ are functions of the sets $S$ in the space of the wave vectors $\boldsymbol{k}$. In fact, using formulas such as those given by Eqs. (28.26) and (28.35), we can readily show that if we could substitute

$$\Psi_{\alpha_1 \ldots \alpha_n}(d\boldsymbol{k}_1, \ldots, d\boldsymbol{k}_n; t) = \psi_{\alpha_1 \ldots \alpha_n}(\boldsymbol{k}_1, \ldots, \boldsymbol{k}_n; t)\, d\boldsymbol{k}_1 \ldots d\boldsymbol{k}_n,$$

then the functions $\psi_{\alpha_1 \ldots \alpha_n}(\boldsymbol{k}_1, \ldots, \boldsymbol{k}_n; t)$ would be proportional to the Fourier transforms of the $n$th order moments of the velocity field. However, we know from Sect. 4.2 of Vol. 1 that the moments do not, in general, tend to zero as one or more of the arguments $\boldsymbol{x}_1, \ldots, \boldsymbol{x}_n$ tends to infinity (only the cumulants have this property) and, consequently, we cannot take their Fourier transforms.

Substituting Eq. (29.10) into the Hopf equation (28.38), we obtain the following infinite set of coupled equations for the functionals $\Psi_n$, $n = 1, 2, \ldots$,

$$\frac{\partial \Psi_n}{\partial t} = \int\int \tilde{z}_\alpha(\boldsymbol{k}' + \boldsymbol{k}'') k''_\beta \mathscr{D}_\beta(\boldsymbol{k}') \mathscr{D}_\alpha(\boldsymbol{k}'') \Psi_{n+1}\, d\boldsymbol{k}'\, d\boldsymbol{k}'' - \\ - \nu \int k^2 z_\alpha(\boldsymbol{k}) \mathscr{D}_\alpha(\boldsymbol{k}) \Psi_n\, d\boldsymbol{k}. \tag{29.12}$$

Let us consider in greater detail the case of homogeneous turbulence.[4] In this case, the characteristic functional of the velocity field should be invariant under translations of the system of space coordinates, i.e., for any $\theta(\boldsymbol{x})$ and any $\boldsymbol{a}$

$$\Phi[\theta(\boldsymbol{x}), t] = \Phi[\theta(\boldsymbol{x} + \boldsymbol{a}), t]. \tag{29.13}$$

[4] In the case of locally homogeneous turbulence we should only demand that, in addition, $\int \theta(\boldsymbol{x})\, d\boldsymbol{x} = 0$, and $z(0) = 0$ and, apart from this, all the subsequent discussions and conclusions remain valid. This is why we need not consider separately the case of locally homogeneous (or locally isotropic) turbulence.

This identity must, of course, be satisfied for all the power functionals in the series (29.1), (29.5), and (29.7). If now $z(k)$ is the Fourier transform of the function $\theta(x)$, the Fourier transform of $\theta(x+a)$ will be $e^{ika}z(k)$. Equation (29.13) can therefore also be written in the form

$$\Psi[z(k), t] = \Psi[e^{ika}z(k), t]. \qquad (29.14)$$

The same equation should be satisfied by all the power functionals $\Psi_n$ in Eq. (29.10). From Eq. (29.11) it follows that the condition (29.14) is equivalent to the demand that $\Psi_{\alpha_1 \ldots \alpha_n}(dk_1, \ldots, dk_n; t)$ be nonzero only for $k_1 + \ldots + k_n = 0$.

Let us establish the form of the first three terms $1 + \Psi_1 + \Psi_2$ in Eq. (29.10) in the case of homogeneous turbulence. Here, the function $\Psi_{\alpha_1}(dk_1, t)$ in terms of which $\Psi_1$ is expressed must be zero for $k_1 \neq 0$ and, therefore,

$$\Psi_1[z(k), t] = \int z_\alpha(k)\Psi_\alpha(dk, t) = z_\alpha(0)C_\alpha,$$

where $C_\alpha = \int \Psi_\alpha(dk, t)$. In view of Eq. (28.39) it follows that $\mathscr{D}_j(k)\Psi_1 = C_j\delta(k)$. According to Eq. (28.26) this function is the Fourier transform of $\mathscr{D}_j(x)\Phi|_{\theta(x)=0} = i\bar{u}_j$ which is independent of $x$ because the turbulence is homogeneous. Consequently, $C_j = i\bar{u}_j$, and hence

$$\Psi_1[z(k), t] = i\bar{u} \cdot z(0). \qquad (29.15)$$

However, in the case of homogeneous turbulence we can set $\bar{u} = 0$ without loss of generality, and we shall do this to determine the form of $\Psi_2$. We then have $\Psi_1 = 0$, and Eq. (29.12) with $n = 1$ assumes the form of the following restriction on $\Psi_2$:

$$\int\int \tilde{z}_\alpha(k' + k'')k''_\beta \mathscr{D}_\beta(k')\mathscr{D}_\alpha(k'')\Psi_2 dk' dk'' = 0. \qquad (29.16)$$

We note now that when $\bar{u} = 0$, the second moments of the velocity field are at the same time second cumulants and, consequently, allow Fourier transformation. Moreover, if we use Eq. (29.14) we can rewrite Eq. (29.11) with $n = 2$ in the form

$$\Psi_2[z(\boldsymbol{k}), t] = \int\int z_\alpha(\boldsymbol{k}_1) z_\beta(\boldsymbol{k}_2) \psi_{\alpha\beta}(\boldsymbol{k}_1, \boldsymbol{k}_2, t) \delta(\boldsymbol{k}_1 + \boldsymbol{k}_2) d\boldsymbol{k}_1 d\boldsymbol{k}_2. \tag{29.17}$$

Substituting $\boldsymbol{k}_1 = \boldsymbol{k}$, and integrating with respect to $\boldsymbol{k}_2$, we have

$$\Psi_2[z(\boldsymbol{k}), t] = -\frac{1}{2}\int z_\alpha(\boldsymbol{k}) z_\beta(-\boldsymbol{k}) F_{\alpha\beta}(\boldsymbol{k}, t) d\boldsymbol{k}. \tag{29.18}$$

where $F_{\alpha\beta}(\boldsymbol{k}, t) = -2\psi_{\alpha\beta}(\boldsymbol{k}, -\boldsymbol{k}, t)$. If, on the other hand, $\boldsymbol{k}_2 = \boldsymbol{k}$ in Eq. (29.17), then by integrating with respect to $\boldsymbol{k}_1$ and interchanging the summation indices $\alpha$ and $\beta$, we obtain the same formula (29.18), except that $F_{\alpha\beta}(\boldsymbol{k}, t)$ is replaced by $F_{\beta\alpha}(-\boldsymbol{k}, t)$. Therefore, $F_{\alpha\beta}(\boldsymbol{k}, t) = F_{\beta\alpha}(-\boldsymbol{k}, t)$.

To establish the physical significance of $F_{\alpha\beta}$ let us evaluate the functional derivatives of the functional (29.18). Using Eq. (28.39), we can readily show that

$$\begin{aligned} \mathscr{D}_\alpha(\boldsymbol{k}'')\Psi_2 &= -F_{\alpha\beta}(\boldsymbol{k}'', t) z_\beta(-\boldsymbol{k}''), \\ \mathscr{D}_\beta(\boldsymbol{k}')\mathscr{D}_\alpha(\boldsymbol{k}'')\Psi_2 &= -F_{\alpha\beta}(\boldsymbol{k}'', t)\delta(\boldsymbol{k}' + \boldsymbol{k}''). \end{aligned} \tag{29.19}$$

Substituting the second of these formulas into

$$\mathscr{D}_\beta(\boldsymbol{x}')\mathscr{D}_\alpha(\boldsymbol{x}'')\Phi_2 = \int\int \mathscr{D}_\beta(\boldsymbol{k}')\mathscr{D}_\alpha(\boldsymbol{k}'')\Psi_2 e^{-i(\boldsymbol{k}'\boldsymbol{x}' + \boldsymbol{k}''\boldsymbol{x}'')} d\boldsymbol{k}' d\boldsymbol{k}'',$$

which is the analog of Eq. (28.35), and is obtained by a double application of Eq. (28.26), and expressing the left-hand side of this relation in terms of the second moment of the velocity field, we obtain

$$\overline{u_\alpha(\boldsymbol{x}'', t) u_\beta(\boldsymbol{x}', t)} = \int F_{\alpha\beta}(\boldsymbol{k}, t) e^{-i\boldsymbol{k}(\boldsymbol{x}'' - \boldsymbol{x}')} d\boldsymbol{k}. \tag{29.20}$$

It follows that $F_{\alpha\beta}(\boldsymbol{k}, t)$ is the Fourier transform with respect to $\boldsymbol{r} = \boldsymbol{x}'' - \boldsymbol{x}'$ of the correlation tensor $B_{\alpha\beta}(\boldsymbol{r}, t)$, i.e., the spectral tensor of the velocity field. In the case of an incompressible fluid which we are considering here, this tensor satisfies the conditions $k_\alpha F_{\alpha\beta}(\boldsymbol{k}, t) = k_\beta F_{\alpha\beta}(\boldsymbol{k}, t) = 0$ and, consequently,

$$F_{\alpha\beta}(\boldsymbol{k}, t) = \Delta_{\alpha m}(\boldsymbol{k})\,\Delta_{\beta n}(\boldsymbol{k})\,f_{mn}(\boldsymbol{k}, t), \tag{29.21}$$

where $\Delta_{ij}(\boldsymbol{k}) = \delta_{ij} - k_i k_j/k^2$ [see Eq. (11.89)]. Substituting this formula into Eq. (29.18), and using the definition of $\tilde{z}(\boldsymbol{k})$ given by Eq. (28.29), we can rewrite Eq. (29.18) in the form

$$\Psi_2[\boldsymbol{z}(\boldsymbol{k}), t] = -\frac{1}{2}\int \tilde{z}_\alpha(\boldsymbol{k})\,\tilde{z}_\beta(-\boldsymbol{k})\,f_{\alpha\beta}(\boldsymbol{k}, t)\,d\boldsymbol{k}, \tag{29.22}$$

where the function $f_{\alpha\beta}$ satisfies the condition $f_{\alpha\beta}(\boldsymbol{k}, t) = f_{\beta\alpha}(-\boldsymbol{k}, t)$ [and also the Hermitian symmetry condition $f^*_{\alpha\beta}(\boldsymbol{k}, t) = f_{\beta\alpha}(\boldsymbol{k}, t)$, which is necessary if the tensor of the second moments (29.20) is to be real].

Let us now consider the restrictions, if any, which are imposed on the characteristic functional when the turbulence is not only homogeneous but also isotropic. In this case, the functional $\Phi[\theta(\boldsymbol{x}), t]$ should be invariant not only with respect to any translations [i.e., it must satisfy Eq. (29.13)] but also with respect to any rotations and mirror reflections in the space of the vectors $\boldsymbol{x}$. Let $\mathscr{L}$ be an arbitrary rotation or reflection transforming $\theta(\boldsymbol{x})$ into $\mathscr{L}\theta(\boldsymbol{x})$ (we emphasize that the transformation $\mathscr{L}$ applies not only to the argument $\boldsymbol{x}$ but also to the components of the vector field $\theta$). For any $\theta(\boldsymbol{x})$ and any $\mathscr{L}$ we must then have

$$\Phi[\theta(\boldsymbol{x}), t] = \Phi[\mathscr{L}\theta(\boldsymbol{x}), t]. \tag{29.23}$$

When we transform to the spectral representation, the analogs of $\mathscr{L}$ in the space of the vectors $\boldsymbol{x}$ are transformations of the same type (rotations and mirror reflections) in the space of the wave vectors $\boldsymbol{k}$. Therefore, in the case of homogeneous and isotropic turbulence the characteristic functional $\Psi[\boldsymbol{z}(\boldsymbol{k}), t]$ must satisfy both Eq. (29.14) and the additional condition

$$\Psi[\boldsymbol{z}(\boldsymbol{k}), t] = \Psi[\mathscr{L}\boldsymbol{z}(\boldsymbol{k}), t] \tag{29.24}$$

for all $\mathscr{L}$. All the power functionals $\Psi_n$ in Eq. (29.10) must also satisfy this condition. The functional $\Psi_1$ is then zero [in view of homogeneity it is of the form (29.15), where $\bar{\boldsymbol{u}} = 0$ because the turbulence is isotropic]. Moreover in the formulas (29.18) and (29.21) [or Eq. (29.22)] in the case of isotropic turbulence we can use the fact that the function $f_{mn}(\boldsymbol{k}, t)$ in Eq. (29.21) can then be specified by $f_{mn}(\boldsymbol{k}, t) = F(k, t)\delta_{mn}$, where $F(k, t) = E(k, t)/4\pi k^2$ is the kinetic-energy density in the space of the wave vectors $\boldsymbol{k}$. In this case, the velocity spectral tensor assumes the usual form

$$F_{\alpha\beta}(\boldsymbol{k}, t) = F(k, t)\Delta_{\alpha\beta}(\boldsymbol{k}), \tag{29.25}$$

and Eq. (29.22) becomes

$$\Psi_2[\boldsymbol{z}(\boldsymbol{k}), t] = -\frac{1}{2}\int |\tilde{\boldsymbol{z}}(\boldsymbol{k})|^2 F(k, t)\, d\boldsymbol{k}. \tag{29.26}$$

Let us finally note that the question of convergence of all the above considered series is quite difficult and presently unclear. It is even possible that all these series are in fact divergent at any value of the Reynolds number. Recently some new approaches to the construction of convergent series expansions for turbulent functions were proposed by Kraichnan (1970a) and some other authors; however, these approaches are not formulated in functional terms and we shall not consider them in our book.

Substituting the second formula in Eq. (29.19) into Eq. (29.16), and using Eq. (29.21) we can readily show that Eq. (29.16) is then satisfied identically. In other words, the functional $\Psi = \Psi_2[\boldsymbol{z}(\boldsymbol{k})]$ defined by Eq. (29.22) is the exact solution of

$$\int\int \tilde{z}_\alpha(\boldsymbol{k}' + \boldsymbol{k}'') k''_\beta \mathscr{D}_\beta(\boldsymbol{k}') \mathscr{D}_\alpha(\boldsymbol{k}'') \Psi \, d\boldsymbol{k}'\, d\boldsymbol{k}'' = 0, \tag{29.27}$$

which, by Eq. (28.38), is identical with the Hopf equation for the statistically stationary motion of an ideal incompressible fluid. However, Eq. (29.27) can be given a more extended interpretation. Namely, we know from Chapt. 8 that statistical stationarity and the possibility of neglecting viscous effects is a distinguishing feature of turbulent motions with wave numbers $k$ lying in the *inertial subrange* of the spectrum. One would therefore hope that Eq. (29.27) would be satisfied not only for stationary random motions of an ideal fluid but would also, to some extent, be applicable to developed turbulence in viscous fluids if we consider this equation on a set of functions $\boldsymbol{z}(\boldsymbol{k})$ differing from zero only for $k = |\boldsymbol{k}|$ in the inertial subrange.

It is possible, moreover, that even in the absence of this restriction on $\boldsymbol{z}(\boldsymbol{k})$ Eq. (29.27) will still retain its signfiicance as a model equation describing properties of turbulent motions corresponding to the inertial subrange of the spectrum. In fact, the lower limit of the inertial subrange is determined by the geometric size of the flow and by the length scale of external force-field inhomogeneities acting on the fluid. In an infinite space and in the absence of external forces we can, in principle, imagine stationary turbulence in a viscous fluid (with infinite mean kinetic-energy density) in which the inertial subrange extends down to wave numbers $k$ that are as small as desired. Such turbulence is described by the Hopf equation (28.38) with zero left-hand side. It may be expected that the solution $\Psi$ of this equation will be unique at any $\nu$. Let us suppose that this solution has the limit $\Psi_0$ as

$\nu \to 0$. In this limit the upper boundary of the inertial wave-number subrange will increase without limit and, therefore, the functional $\Psi_0$ should apparently satisfy the model equation (29.27).

The above solution $\Psi = \Psi_2$ of Eq. (29.27), which is given by Eq. (29.22), is not the characteristic functional of any random vector field but it can be used to construct a class of solutions which will include characteristic functionals. Namely we shall seek solutions of Eq. (29.27) in the form

$$\Psi = W(\Psi_2), \tag{29.28}$$

where $W(\xi)$, $\xi \leqslant 0$ is a twice continuously differentiable function and $\Psi_2$ is the above solution of Eq. (29.27). Let us note that in particular, when $W(\xi) = e^{\xi}$, the formula (29.28) will describe the characteristic functional of the solenoidal Gaussian velocity field with the spectral tensor $F_{\alpha\beta}(\boldsymbol{k})$. Since the turbulence is isotropic in the inertial subrange, we shall restrict our attention to isotropic solutions of the form (29.28), i.e., we shall suppose that $\Psi_2$ is given by Eq. (29.26).

We shall now prove the following theorem: *The functional* (29.28), *where* $W'' \neq 0$ *and* $\Psi$ *is given by Eq.* (29.26), *is the solution of Eq.* (29.27) *only if the function* $F(k)$ *in Eq.* (29.26) *is equal to a constant* [this theorem was formulated by Hopf (1952) and was proved by Hopf and Titt (1953)]. To prove it let us substitute Eq. (29.28) into Eq. (29.27). Using the formula

$$\mathscr{D}_\beta(\boldsymbol{k}')\mathscr{D}_\alpha(\boldsymbol{k}'')W(\Psi_2) = W''(\Psi_2)\mathscr{D}_\beta(\boldsymbol{k}')\Psi_2\mathscr{D}_\alpha(\boldsymbol{k}'')\Psi_2 + \\ + W'(\Psi_2)\mathscr{D}_\beta(\boldsymbol{k}')\mathscr{D}_\alpha(\boldsymbol{k}'')\Psi_2$$

and remembering that $\Psi_2$ satisfies Eq. (29.27) and $W''(\Psi_2) \neq 0$, we obtain the following equation:

$$\int\int \tilde{z}_\alpha(\boldsymbol{k}' + \boldsymbol{k}'')\, k''_\beta \mathscr{D}_\beta(\boldsymbol{k}')\Psi_2\mathscr{D}_\alpha(\boldsymbol{k}'')\Psi_2\, d\boldsymbol{k}'\, d\boldsymbol{k}'' = 0.$$

However, in view of the first formula in Eq. (29.19) and the formula (29.25), we have

$$\mathscr{D}_\alpha(\boldsymbol{k})\Psi_2 = -F(k)\Delta_{\alpha\beta}(\boldsymbol{k})z_\beta(-\boldsymbol{k}) = -F(k)\tilde{z}_\alpha(-\boldsymbol{k}),$$

so that after replacement of $\boldsymbol{k}'$ by $-\boldsymbol{k}'$ and $\boldsymbol{k}''$ by $-\boldsymbol{k}''$ the last equation assumes the form

$$\int\int \tilde{z}_\alpha(-\boldsymbol{k}' - \boldsymbol{k}'')\, k''_\beta \tilde{z}_\alpha(\boldsymbol{k}'')\,\tilde{z}_\beta(\boldsymbol{k}')\,F(k')\,F(k'')\, d\boldsymbol{k}'\, d\boldsymbol{k}'' = 0.$$

If we apply functional differentiation with respect to $\delta\tilde{z}_\gamma(\boldsymbol{k}_1)\, d\boldsymbol{k}_1\, \delta\tilde{z}_\gamma(\boldsymbol{k}_2)\, d\boldsymbol{k}_2$ to both sides of this equation we have, using Eq. (28.39),

$$k_{2\beta}\tilde{z}_\beta(\boldsymbol{k})F(k)[F(k_2) - F(k_1)] = 0,$$

where $\boldsymbol{k} = -\boldsymbol{k}_1 - \boldsymbol{k}_2$. If the three vectors $\boldsymbol{k}$, $\boldsymbol{k}_1$, and $\boldsymbol{k}_2$ form a nondegenerate triangle, i.e., if

$$|k_1 - k_2| < k < k_1 + k_2, \tag{29.29}$$

then $k_{2\beta}\tilde{z}_\beta(\boldsymbol{k}) \neq 0$ and, consequently,

$$F(k)\,[F(k_2)-F(k_1)]=0. \qquad (29.30)$$

It now remains to show that it follows from Eqs. (29.29) and (29.30) that $F(k)$ = const. The solution $F(k) = 0$ is of no interest to us. Let us suppose that $F(k_0) \neq 0$ for any $k_0$. We then have $|k_1 - k_2| < k_0 < k_1 + k_2$ for any $k_1$ and $k_2$ in the range $\left(\frac{4}{3}\right)^{-1} k_0 \leqslant k \leqslant \frac{4}{3} k_0$ and, therefore, by Eq. (29.30) we have $F(k_1) = F(k_2) = F(k_0) = \text{const} \neq 0$ throughout this range. Let us now replace $k_0$ by 3/4 $k_0$ or 4/3 $k_0$. The result $F(k) = F(k_0) = \text{const}$ can then be extended to the ranges $\left(\frac{4}{3}\right)^{-2} k_0 \leqslant k \leqslant k_0$ and $k_0 \leqslant k \leqslant \left(\frac{4}{3}\right)^{2} k_0$, i.e., we can prove that it is valid for $\left(\frac{4}{3}\right)^{-2} k_0 \leqslant k \leqslant \left(\frac{4}{3}\right)^{2} k_0$. If we continue in this way we can extend the result $F(k) = F(k_0)$ to all $k > 0$. This proves the theorem.

It follows from the above theorem that the Gaussian characteristic functional $\Psi = \exp \Psi_2$ is a solution of Eq. (29.27) only if it corresponds to a random velocity field with a uniform energy distribution in wave number space, i.e., with $F(k)$ = const. The statistical properties of such a velocity field are, in a certain sense, analogous to the statistical properties of a gas described by the classical Maxwell-Boltzmann canonical distribution. However, this velocity field is not self-preserving in the Kolomogorov sense [which would demand that $E(k) \sim k^{-5/3}$, i.e., $F(k) \sim k^{-11/3}$]. Consequently, the characteristic functional of the turbulent velocity field with the spectrum $F(k) \sim k^{-11/3}$ which satisfies Eq. (29.27) cannot be Gaussian and, in general, cannot be of the form given by Eq. (29.28).

It is readily verified that the Gaussian characteristic functional $e^{\Psi_2}$ with a constant spectral density $F(k)$ = const satisfies not only the model equation (28.27) but also the equation (28.92) derived by Novikov for the case of turbulence in a viscous fluid driven by statistically stationary and homogeneous $\delta$-correlated-in-time external forces, provided only that the spectral tensor $\hat{F}_{\alpha\beta}(\boldsymbol{k})$ of these forces is of the form $Ck^2\Delta_{\alpha\beta}(\boldsymbol{k})$ where $C$ is a positive constant. In other words, to ensure that the spectral kinetic-energy density is constant, the spectral density of the external forces must increase as $k^2$ with increasing $k$.

## 29.2 Zero-Order Approximation in the Reynolds Number

If we measure all distances in units of the length scale $L$, which is typical for the given flow, all velocities in units of the typical velocity scale $U$, and all time intervals in units of the time scale $L^2/\nu$ (which is the typical time for viscous damping of velocity-field inhomogeneities), then after transformation to the corresponding dimensionless variables in the Hopf equation, we find that the terms containing the second functional derivatives of the characteristic functional will be preceded by the factor Re (we have already mentioned this in Sect. 28.2). If we consider the solution of the Hopf equation (written in dimensionless variables) as a function of the parameter Re, we may try to represent it by a series in powers of Re in the neighborhood of the point Re = 0. Substituting this formal series into the Hopf equation, and equating terms of the same degree in Re, we obtain a set of equations for the coefficients of our series.

The equation for the coefficient of $(\mathrm{Re})^n$ will then include coefficients of the lower but not of the higher powers of Re. Consequently, these equations can be solved successively, beginning with the lower orders, so that the expansion in powers of the Reynolds number may in principle be an effective method of finding the formal solution of the Hopf equation. If the series found in this way converges, then its sum will be the exact solution. This type of solution can be useful at least for small Re, for which the series will converge sufficiently rapidly (and for which the expansion in powers of Re can be regarded as an application of the usual methods of perturbation theory; see Sect. 19).

However, it is unclear at present whether the series converges or not and moreover for small Re one would expect to find only very weak turbulence (for example, turbulence which appears just after the loss of stability of laminar flow, or very highly degenerate turbulence). On the other hand, the most interesting flows in the theory of turbulence are those with very high Re, for which one would expect the presence of developed turbulence. It is clear that expansion in powers of Re is not a convenient method of investigating solutions corresponding to large Re. Even if the resulting series does converge for large Re, (and this is quite doubtful) the convergence is probably extremely slow and the series is useful only if infinitely many terms of it can be summed (an analogous difficulty arises in all attempts to use perturbation theory methods to describe the dynamics of strongly interacting systems). Nevertheless, even for large Re the functional solution of the Hopf equation in the form of an expansion in powers of Re may be useful for some special purposes (for example, as a standard with which solutions obtained by approximate methods can be compared).

To construct the formal solution of Hopf's equation with the aid of an expansion in powers of Re we need not transform to dimensionless variables but we may seek a solution in the form

$$\Phi = \sum_{n=0}^{\infty} \Phi_n, \tag{29.31}$$

where $\Phi_n$ is a term of the order $(\mathrm{Re})^n$ (or in the form of an analogous power series for the characteristic functional in the spectral representation). In doing this we must remember that the terms in the Hopf equation containing the second functional derivatives are of an order in Re greater by unity than the order of the remaining terms in this equation. In particular, the zeroth term $\Phi_0$ will satisfy an equation obtained from the Hopf equation for which terms containing the second functional derivatives have been omitted. It is not difficult to solve this equation. Its solution will, of course, correspond to the case of "degenerate" turbulence at very small Re, which is described by Eq. (15.33).

The zeroth term in the expansion in powers of Re of the spatial characteristic functional of the velocity field is best found by using the Hopf equation in the form (28.47). If we omit from this equation terms containing the second functional derivatives (i.e., the right-hand side), we obtain $\frac{\partial \tilde{\Psi}}{\partial t} = 0$ and hence $\tilde{\Psi}[\boldsymbol{y}(\boldsymbol{k}), t] = \tilde{\Psi}[\boldsymbol{y}(\boldsymbol{k}), 0]$. Transforming from $\boldsymbol{y}(\boldsymbol{k})$ to $\boldsymbol{z}(\boldsymbol{k})$ with the aid of Eq. (28.46), we obtain

$$\Psi[\boldsymbol{z}(\boldsymbol{k}), t] = \Psi_0[e^{-\nu k^2 t} \boldsymbol{z}(\boldsymbol{k})], \tag{29.32}$$

where $\Psi_0$ is the characteristic functional of the initial velocity field in the spectral representation. Finally, transforming to the spatial representation in accordance with Eq. (28.22), we obtain

$$\Phi[\boldsymbol{\theta}(\boldsymbol{x}), t] = \Phi_0\left[(4\pi\nu t)^{-3/2} \int e^{-\frac{|\boldsymbol{x}-\boldsymbol{x}'|^2}{4\nu t}} \boldsymbol{\theta}(\boldsymbol{x}')\, d\boldsymbol{x}'\right], \tag{29.33}$$

where $\Phi_0[\boldsymbol{\theta}(\boldsymbol{x})]$ is the spatial characteristic functional of the initial velocity field. We can readily verify directly that Eq. (29.33) does, in fact, satisfy the Hopf equation (28.18) or Eq. (28.19) without the terms containing the second functional derivatives. The formula (29.33) shows that, in the zeroth approximation in Re, the time evolution of the spacial characteristic functional of the velocity field reduces to the smoothing of its functional argument $\boldsymbol{\theta}(\boldsymbol{x})$ by a Gaussian weight function with a variance $2\nu t$. It is readily seen that this is equivalent to the results obtained in Sect. 15.3.

Let us now consider the space-time characteristic functional. To determine the zeroth term of its expansion in powers of Re, we shall use the Hopf equation in the form (28.61) without terms containing the second functional derivatives. To increase the generality somewhat, we shall add to the right-hand side a term describing the effect of nonrandom external forces acting on the fluid (the form of this term is indicated in the beginning of Sect. 28.4). In other words, we shall solve the equations

$$\left(\frac{\partial}{\partial t}+\nu k^2\right)\mathscr{D}_j(\boldsymbol{k},\, t)\,\Psi\,[\boldsymbol{z}(\boldsymbol{k},\, t)] = i\widetilde{Y}_j(\boldsymbol{k},\, t)\,\Psi\,[\boldsymbol{z}(\boldsymbol{k},\, t)] \quad (29.34)$$

where $\widetilde{Y}_j = \Delta_{j\alpha}(\boldsymbol{k})\, Y_\alpha(\boldsymbol{k},\, t)$ and $\boldsymbol{Y}$ is the Fourier transform (28.32) of the external force field $\boldsymbol{X}$. This equation is to be solved subject to the initial condition $\Psi\,[\boldsymbol{z}(\boldsymbol{k})\,\delta(t)] = \Psi_0\,[\boldsymbol{z}(\boldsymbol{k})]$, where $\Psi_0$ is the characteristic functional of the initial velocity field. The required solution found by Lewis and Kraichnan (1962) is of the form

$$\Psi\,[\boldsymbol{z}(\boldsymbol{k},\, t)] = e^{i(\widetilde{\boldsymbol{h}}\cdot\boldsymbol{z})}\Psi_0\left[\int_0^\infty e^{-\nu k^2\tau}\,\widetilde{\boldsymbol{z}}(\boldsymbol{k},\, \tau)\, d\tau\right], \quad (29.35)$$

where

$$\boldsymbol{h}(\boldsymbol{k},\, t) = \int_0^t e^{-\nu k^2(t-\tau)}\boldsymbol{Y}(\boldsymbol{k},\, \tau)\, d\tau, \quad (29.36)$$

and $(\widetilde{\boldsymbol{h}}\cdot\boldsymbol{z})$ denotes, as usual, the integral of the scalar product $\widetilde{\boldsymbol{h}}\cdot\boldsymbol{z}$ with respect to $t$ evaluated between 0 and $\infty$ and with respect to all $\boldsymbol{k}$. The validity of this solution is readily verified by direct substitution of Eq. (29.35) into Eq. (29.34) if we use the following rule for functional differentiation of a functional (which is readily proved): If $\Psi$ is a functional of the function $\boldsymbol{\zeta}(\boldsymbol{k},\, t)$, which in turn is a functional of the function $\boldsymbol{z}(\boldsymbol{k},\, t)$, then

$$\mathscr{D}_{z_j}(\boldsymbol{k},\, t)\,\Psi = \int \mathscr{D}_{\zeta_\alpha}(\boldsymbol{k}',\, t')\,\Psi\cdot\mathscr{D}_{z_j}(\boldsymbol{k},\, t)\,\zeta_\alpha(\boldsymbol{k}',\, t')\, d\boldsymbol{k}'\, dt'. \quad (29.37)$$

The fact that the initial condition is satisfied by Eq. (29.35) can also be readily verified in a direct fashion.

Equations (29.35) and (29.36) can be used to calculate the "response tensors" (28.65), (28.65'), and so on for the zero-order approximation in the Reynolds number. Tensors of rank higher than two are zero in this approximation, since they describe nonlinear effects which are not taken into account in the zero-order approximation. As regards the tensor $G^\alpha_\beta(\boldsymbol{x}, t\,|\,\boldsymbol{x}', t')$, we can readily show, using Eq. (28.26), that

$$G^{\alpha}_{\beta}(\boldsymbol{x}, t \mid \boldsymbol{x}', t') = \frac{1}{i}\mathscr{D}_{X_\beta}(\boldsymbol{x}', t') \int [\mathscr{D}_\alpha(\boldsymbol{k}, t)\Psi]\,|_{\boldsymbol{z}=0}\, e^{-i\boldsymbol{k}\boldsymbol{x}}\, d\boldsymbol{k}.$$

According to Eq. (29.35), the value of the functional derivative $\mathscr{D}_\alpha(\boldsymbol{k}, t)\Psi$ at $\boldsymbol{z}=0$ differs from $i\tilde{h}_\alpha(\boldsymbol{k}, t)$ only by a term which is independent of $\tilde{\boldsymbol{h}}(\boldsymbol{k}, t)$ and hence of $\boldsymbol{X}(\boldsymbol{x}, t)$. Consequently,

$$G^{\alpha}_{\beta}(\boldsymbol{x}, t \mid \boldsymbol{x}', t') = \int \mathscr{D}_{X_\beta}(\boldsymbol{x}', t')\, \tilde{h}_\alpha(\boldsymbol{k}, t)\, e^{-i\boldsymbol{k}\boldsymbol{x}}\, d\boldsymbol{k}.$$

The functional derivative with respect to $X_\beta(\boldsymbol{x}', t')$ in this expression can readily be evaluated in view of the fact that the function $\boldsymbol{Y}(\boldsymbol{k}, t)$ in Eq. (29.36) is the Fourier transform of $\boldsymbol{X}(\boldsymbol{x}, t)$. The final result is the Lewis-Kraichnan formula

$$G^{\alpha}_{\beta}(\boldsymbol{x}, t \mid \boldsymbol{x}', t') = (2\pi)^{-3} \int \Delta_{\alpha\beta}(\boldsymbol{k})\, e^{-\nu k^2 (t-t') - i\boldsymbol{k}(\boldsymbol{x}-\boldsymbol{x}')}\, d\boldsymbol{k} \qquad (29.38)$$

where $t > t'$ by definition of $G^{\alpha}_{\beta}$. This formula shows that $G^{\alpha}_{\beta} = G^{\beta}_{\alpha}$, and that this tensor depends only on $t - t'$ and $\boldsymbol{x} - \boldsymbol{x}'$.

Finally, let us consider the joint space-time characteristic functional of the velocity field, and the field of random external forces. This functional satisfies Eq. (28.82) in its spectral representation. In the zero-order approximation in Re this equation assumes the form

$$\left(\frac{\partial}{\partial t} + \nu k^2\right) \mathscr{D}_j(\boldsymbol{k}, t)\, \Lambda[\boldsymbol{z}(\boldsymbol{k}, t), \boldsymbol{g}(\boldsymbol{k}, t)] = \\ = \Delta_{j\alpha}(\boldsymbol{k})\, \mathscr{D}_{g_\alpha}(\boldsymbol{k}, t)\, \Lambda[\boldsymbol{z}(\boldsymbol{k}, t), \boldsymbol{g}(\boldsymbol{k}, t)]. \qquad (29.39)$$

We shall solve it subject to the "initial condition"

$$\Lambda[\boldsymbol{z}(\boldsymbol{k})\delta(t - t_0),\ \boldsymbol{g}(\boldsymbol{k}, t)\, e(t - t_0)] = \Psi_0[\boldsymbol{z}(\boldsymbol{k})]\, G[\boldsymbol{g}(\boldsymbol{k}, t)] \qquad (29.40)$$

[compare this with Eq. (28.83)], where $\Psi_0$ and $G$ are the spectral forms of the characteristic functionals of the initial velocity field $\boldsymbol{u}(\boldsymbol{x}, t_0)$ and the field of random external forces $\boldsymbol{X}(\boldsymbol{x}, t)$, $t \geqslant t_0$ (assumed to be statistically independent of each other). Direct verification using Eq. (29.37) will readily show that the solution of Eq. (29.39) subject to the initial condition (29.40) is

$$\Lambda[\boldsymbol{z}(\boldsymbol{k}, t), \boldsymbol{g}(\boldsymbol{k}, t)] = \Psi_0[\tilde{\boldsymbol{\xi}}(\boldsymbol{k}, t_0)]\, G[\boldsymbol{g}(\boldsymbol{k}, t) + \tilde{\boldsymbol{\xi}}(\boldsymbol{k}, t)], \qquad (29.41)$$

where

$$\zeta(\boldsymbol{k}, t) = \int_t^\infty e^{-\nu k^2(\tau - t)} z(\boldsymbol{k}, \tau)\, d\tau. \tag{29.42}$$

The spatial representation of the functional (29.41) is of the form

$$\Omega[\theta(\boldsymbol{x}, t), \boldsymbol{f}(\boldsymbol{x}, t)] = \Phi_0[\boldsymbol{r}(\boldsymbol{x}, t_0)]\, F[\boldsymbol{f}(\boldsymbol{x}, t) + \boldsymbol{r}(\boldsymbol{x}, t)], \tag{29.43}$$

where

$$r_j(\boldsymbol{x}, t) = \int_t^\infty d\tau \int G_j^\alpha(\boldsymbol{x}', \tau | \boldsymbol{x}, t)\, \theta_\alpha(\boldsymbol{x}', \tau)\, d\boldsymbol{x}', \tag{29.44}$$

and $G_j^\alpha$ is given by Eq. (29.38). In these expressions $\Phi_0[\theta(\boldsymbol{x})]$ and $F[\boldsymbol{f}(\boldsymbol{x}, t)]$ are the characteristic functionals of the initial velocity field and the random external force field. The functional (29.43) is the solution (in the zero-order approximation in Re) of Eq. (28.81) (in which terms containing the second functional derivatives have been neglected), subject to the initial condition (28.83). In this representation the solution was first given by Lewis and Kraichnan (1962).

The joint space-time characteristic functional of the velocity field and the external force field gives the most complete statistical description of these fields. In particular, it is readily verified that Eqs. (29.41) and (29.42) [or Eqs. (29.43) and (29.44)] contain the previous results given by Eqs. (29.35), (29.36), and (29.32) [or Eq. (29.33)]. In fact, when $\boldsymbol{g}(\boldsymbol{k}, t) = 0$, the formula (29.41) takes the form

$$\Psi[\boldsymbol{z}(\boldsymbol{k}, t)] = \Psi_0[\tilde{\zeta}(\boldsymbol{k}, t_0)]\, G[\tilde{\zeta}(\boldsymbol{k}, t)]. \tag{29.45}$$

If the external force field is nonrandom, then $G[\tilde{\zeta}(\boldsymbol{k}, t)] = e^{i(\tilde{\zeta} \cdot Y)}$. If we then integrate by parts with respect to time, we can readily verify that $(\tilde{\zeta} \cdot \boldsymbol{Y}) = (\tilde{\boldsymbol{h}} \cdot \boldsymbol{z})$ and, consequently, when $t_0 = 0$ the formula (29.45) becomes identical with Eq. (29.35). When $\boldsymbol{Y} = 0$ and, therefore, $\tilde{\boldsymbol{h}} = 0$ and $\boldsymbol{z}(\boldsymbol{k}, \tau) = \boldsymbol{z}(\boldsymbol{k})\,\delta(\tau - t)$, Eq. (29.35) leads to Eq. (29.32).

It is natural to suppose that whatever the initial velocity field, the statistical state of the motion of a fluid in a given field of stationary random external forces will eventually approach a stationary state which is independent of the initial velocity field. Therefore, it we allow $t_0 \to -\infty$ in Eqs. (29.41) and (29.42) for a fixed $G[\boldsymbol{g}(\boldsymbol{k}, t)]$ and assume, for example, that $\lim_{t \to -\infty} \Psi_0[\tilde{\xi}(\boldsymbol{k}, t_0)] = 1$ (which corresponds to an initial "state of rest"), we should obtain the stationary solution of Eq. (29.39) in the limit as $t_0 \to -\infty$. This solution will obviously be of the form

$$\Lambda[\boldsymbol{z}(\boldsymbol{k}, t), \boldsymbol{g}(\boldsymbol{k}, t)] = G[\boldsymbol{g}(\boldsymbol{k}, t) + \tilde{\xi}(\boldsymbol{k}, t)]; \qquad (29.46)$$

and, in particular, the space-time characteristic functional of the velocity field will now be of the form $\Psi[\boldsymbol{z}(\boldsymbol{k}, t)] = G[\tilde{\xi}(\boldsymbol{k}, t)]$. The corresponding spatial functional will be

$$\Psi[\boldsymbol{z}(\boldsymbol{k})] = G[\exp(-\nu k^2 t)\, e(t)\, \tilde{\boldsymbol{z}}(\boldsymbol{k})]. \qquad (29.47)$$

As an example, consider the case where the external force field $\boldsymbol{X}(\boldsymbol{x}, t)$ is a stationary homogeneous and isotropic solenoidal Gaussian random field, described by the space-time correlation tensor

$$\overline{X_\alpha(\boldsymbol{x}', t') X_\beta(\boldsymbol{x}'', t'')} = \delta(t' - t'') \int e^{-i\boldsymbol{k}(\boldsymbol{x}'' - \boldsymbol{x}')} \Delta_{\alpha\beta}(\boldsymbol{k}) \hat{F}(k)\, d\boldsymbol{k}. \qquad (29.48)$$

This type of force field with delta-function time correlations is the idealized limiting case of a field in the form of a sequence of pulses with a very short aftereffect. The fact that the mean square $\overline{|\boldsymbol{X}(\boldsymbol{x}, t)|^2}$ does not exist in this case need not worry us because it will not lead to difficulty in the calculation of the statistical characteristics of the velocity field in which we are interested. The characteristic functional of the force field in the present case is given by

$$G[\boldsymbol{g}(\boldsymbol{k}, t)] = \exp\left\{-\frac{1}{2} \int |\tilde{\boldsymbol{g}}(\boldsymbol{k}, t)|^2 \hat{F}(k)\, d\boldsymbol{k}\, dt\right\}, \qquad (29.49)$$

so that by Eq. (29.46) the stationary solution of Eq. (29.39) is given by

$$\Lambda[\boldsymbol{z}(\boldsymbol{k}, t), \boldsymbol{g}(\boldsymbol{k}, t)] = \exp\left\{-\frac{1}{2} |\tilde{\boldsymbol{g}}(\boldsymbol{k}, t) + \tilde{\xi}(\boldsymbol{k}, t)|^2 \hat{F}(k)\, d\boldsymbol{k}\, dt\right\}. \qquad (29.50)$$

We can use this equation to verify that the mean rate of energy influx per unit mass of the fluid due to the work done by the external forces, which is given by Eq. (28.68), is

$$\overline{\boldsymbol{u}(\boldsymbol{x},t)\cdot\boldsymbol{X}(\boldsymbol{x},t)}=\int \hat{F}(k)\,d\boldsymbol{k}=4\pi\int_0^\infty k^2\hat{F}(k)\,dk. \quad (29.51)$$

When $\boldsymbol{g}(\boldsymbol{k},t)=0$ the formula (29.50) transforms to

$$\Psi[\boldsymbol{z}(\boldsymbol{k},t)]=\exp\left\{-\frac{1}{2}\int|\tilde{\boldsymbol{\xi}}(\boldsymbol{k},t)|^2\hat{F}(k)\,d\boldsymbol{k}\,dt\right\}. \quad (29.52)$$

This leads directly to the following expression for the space-time correlation function of the velocity field:

$$\overline{u_\alpha(\boldsymbol{x}',t')\,u_\beta(\boldsymbol{x}'',t'')}=\int e^{-i\boldsymbol{k}(\boldsymbol{x}''-\boldsymbol{x}')-\nu k^2|t''-t'|}\Delta_{\alpha\beta}(\boldsymbol{k})\frac{\hat{F}(k)}{2\nu k^2}\,d\boldsymbol{k}. \quad (29.53)$$

Finally, from Eqs. (29.47) and (29.49) we obtain the spatial characteristic functional of the velocity field in the form

$$\Psi[\boldsymbol{z}(\boldsymbol{k})]=\exp\left\{-\frac{1}{2}\int|\tilde{\boldsymbol{z}}(\boldsymbol{k})|^2\frac{\hat{F}(k)}{2\nu k^2}\,d\boldsymbol{k}\right\}. \quad (29.54)$$

Hence [or from Eq. (29.53) for $t'=t''$] it follows that the spectral density function of the kinetic-energy distribution is $E(k)=\frac{2\pi}{\nu}\hat{F}(k)$, i.e., it is proportional to the three-dimensional spectrum of the external force. The spectral density of energy dissipation is then equal to $2\nu k^2E(k)=4\pi k^2\hat{F}(k)$, i.e., by Eq. (29.51) it is equal to the spectral density of the energy influx due to the work done by external forces. This is not surprising because in the zero-order approximation in Re there is no interaction between the velocity-field components with different wave numbers, so that each component evolves independently of all the other components and, consequently, the steady-state influx of energy due to work done by the external forces should balance the energy dissipation in any infinitesimal region of wave-number space.

## 29.3 Expansion in Powers of the Reynolds Number

In the previous subsection we considered the zero-order term in the formal expansion of the solution of the Hopf equation in powers of Re. We shall now consider the next few terms of this expansion. Namely we shall try, following the paper by Monin (1964) to deduce some information about the general structure of this expansion as a whole.

Consider the most general characteristic functional $\Lambda[\boldsymbol{z}(\boldsymbol{k}, t), \boldsymbol{g}(\boldsymbol{k}, t)]$. We shall write it in the form of a series:

$$\Lambda = \sum_{n=0}^{\infty} \Lambda_n, \tag{29.55}$$

where $\Lambda_n$ is of the order of $(\mathrm{Re})^n$. If we recall that in Eq. (28.82) the order in Re of the terms containing the second functional derivative is higher by unity than all the remaining terms, we find that for $n \geqslant 1$

$$\left[\left(\frac{\partial}{\partial t} + \nu k^2\right) \mathscr{D}_j(\boldsymbol{k}, t) - \Delta_{j\beta}(\boldsymbol{k}) \mathscr{D}_{g_\beta}(\boldsymbol{k}, t)\right] \Lambda_n = \\ = \Delta_{j\beta}(\boldsymbol{k}) k_\alpha \int \mathscr{D}_\alpha(\boldsymbol{k}', t) \mathscr{D}_\beta(\boldsymbol{k} - \boldsymbol{k}', t) \Lambda_{n-1} d\boldsymbol{k}' \tag{29.56}$$

where the zero-order term $\Lambda_0$ has already been determined in the previous subsection. By integrating successively these first-order functional differential equations with a known right-hand side, we can, at least in principle, determine all the terms in the expansion (29.55).

To be specific, consider the stationary solution of Eq. (28.82) for a given stationary random external force field described by the characteristic functional (29.49). The zero-order term of Eq. (29.55) is now given by Eq. (29.50), and the subsequent terms can be sought in the form of functionals of the functions $\tilde{\boldsymbol{\xi}}(\boldsymbol{k}, t)$ and $\tilde{\boldsymbol{g}}(\boldsymbol{k}, t)$ by substituting $\Lambda_n = \Lambda_n[\tilde{\boldsymbol{\xi}}(\boldsymbol{k}, t), \tilde{\boldsymbol{g}}(\boldsymbol{k}, t)]$. Transforming in Eq. (29.56) from functional differentiation with respect to $z_j(\boldsymbol{k}, t)$ and $g_j(\boldsymbol{k}, t)$ to functional differentiation with respect to $\tilde{\xi}_j(\boldsymbol{k}, t)$ and $\tilde{g}_j(\boldsymbol{k}, t)$ with the aid of Eq. (29.37), we can reduce these equations to the form

$$\mathscr{L}_j(\boldsymbol{k}, t) \Lambda_n = \mathscr{C}_j(\boldsymbol{k}, t) \Lambda_{n-1}, \tag{29.57}$$

where

$$\mathscr{L}_j(\boldsymbol{k}, t) = \Delta_{j\alpha}(\boldsymbol{k})\left[\mathscr{D}_{\tilde{\zeta}_\alpha}(\boldsymbol{k}, t) - \mathscr{D}_{\tilde{g}_\alpha}(\boldsymbol{k}, t)\right], \qquad (29.58)$$

$$\mathcal{C}_j(\boldsymbol{k}, t) = \int d\boldsymbol{k}_1\, d\boldsymbol{k}_2\, dt_1\, dt_2 C_{jpq}(\boldsymbol{k}, t \,|\, \boldsymbol{k}_1, t_1 \,|\, \boldsymbol{k}_2, t_2) \times \\ \times \mathscr{D}_{\zeta_p}(\boldsymbol{k}_1, t_1)\mathscr{D}_{\zeta_q}(\boldsymbol{k}_2, t_2), \qquad (29.59)$$

$$C_{jpq}(\boldsymbol{k}, t \,|\, \boldsymbol{k}_1, t_1 \,|\, \boldsymbol{k}_2, t_2) = \Delta_{j\alpha}(\boldsymbol{k})\, k_\beta \Delta_{\beta p}(\boldsymbol{k}_1)\, \Delta_{\alpha q}(\boldsymbol{k}_2) \times \\ \times \delta(\boldsymbol{k}_1 + \boldsymbol{k}_2 - \boldsymbol{k})\, e^{\nu k_1^2(t_1 - t) + \nu k_2^2(t_2 - t)}\, e(t - t_1)\, e(t - t_2).$$

The solutions $\Lambda_n$ of Eq. (29.57) are conveniently sought in the form $\Lambda_n = \Lambda_0 M_n$, so that Eq. (29.55) becomes

$$\Lambda = \Lambda_0 \sum_{n=0}^{\infty} M_n, \qquad (29.60)$$

where $M_n$ is of the order $(\mathrm{Re})^n$ and $M_0 = 1$. Since $\Lambda_0$ is the characteristic functional of certain Gaussian random fields, this series is superficially similar to the Gram-Charlier series (29.7). There is an important difference, however, in that the second moments of the fields described by $\Lambda_0$ and $\Lambda$ do not coincide anywhere, so that the terms $M_1 + M_2 + \ldots$ describe not only the deviations of the field $\{\boldsymbol{u}(\boldsymbol{x}, t), \boldsymbol{X}(\boldsymbol{x}, t)\}$ from the Gaussian field, which are due to the nonlinear interactions between the Fourier components of the velocity field, but also corrections to the second moments due to these nonlinear interactions (in particular, corrections to the velocity spectrum which are of special interest in the theory of turbulence).

Substituting $\Lambda_n = \Lambda_0 M_n$ into Eq. (29.57), and using the definition (29.50) of the functional $\Lambda_0$, we can readily reduce Eq. (29.57) to the form

$$\mathscr{L}_j(\boldsymbol{k}, t)\, M_n = \{\mathscr{A}_j[\boldsymbol{\xi}(\boldsymbol{k}, t); \boldsymbol{k}, t] + \\ + \mathscr{B}_j[\boldsymbol{\xi}(\boldsymbol{k}, t); \boldsymbol{k}, t] + \mathcal{C}_j(\boldsymbol{k}, t)\}\, M_{n-1}, \qquad (29.61)$$

where $\boldsymbol{\xi}(\boldsymbol{k}, t) = \tilde{\boldsymbol{\zeta}}(\boldsymbol{k}, t) + \tilde{\boldsymbol{g}}(\boldsymbol{k}, t)$; $\mathscr{A}_j$ is the quadratic functional defined by Eq. (29.59) in which the operators $\mathscr{D}_{\zeta_j}(\boldsymbol{k}, t)$ are now replaced by the functions $\hat{F}(\boldsymbol{k})\xi_j(-\boldsymbol{k}, t)$, and $\mathscr{B}_j$ is an operator which depends on the function $\boldsymbol{\xi}(\boldsymbol{k}, t)$ and contains first-order functional differentiation. The expression for this last operator can be obtained from Eq. (29.59) by replacing $\mathscr{D}_{\zeta_p}(\boldsymbol{k}_1, t_1)\mathscr{D}_{\zeta_q}(\boldsymbol{k}_2, t_2)$

with

$$-\hat{F}(\mathbf{k}_2)\,\xi_q(-\mathbf{k}_2, t_2)\,\mathscr{D}_{\tilde{\xi}_p}(\mathbf{k}_1, t_1) - \hat{F}(\mathbf{k}_1)\,\xi_p(-\mathbf{k}_1, t_1)\,\mathscr{D}_{\tilde{\xi}_q}(\mathbf{k}_2, t_2).$$

The general solution $M_n[\tilde{\xi}(\mathbf{k}, t), \tilde{\mathbf{g}}(\mathbf{k}, t)]$ of the inhomogeneous equation (29.61) consists of the sum of the partial solution $M_n^*[\tilde{\xi}(\mathbf{k}, t), \tilde{\mathbf{g}}(\mathbf{k}, t)]$, corresponding to the right-hand side, and the general solution of the homogeneous equation $\mathscr{L}_j(\mathbf{k}, t)\,M_n = 0$, which has the form of an arbitrary (differentiable) functional $F[\xi(\mathbf{k}, t)]$. In fact, however, this functional must be chosen in a unique fashion. When $\xi = 0$ the functional $\Lambda$ must, of course, become identical with the characteristic functional of the external forces. Since this is also the case for the functional $\Lambda_0$, it follows that for any $n \geqslant 1$ we must have

$$M_n[0, \tilde{\mathbf{g}}(\mathbf{k}, t)] = 0. \tag{29.62}$$

To ensure that the general solution $M_n^*[\tilde{\xi}(\mathbf{k}, t), \tilde{\mathbf{g}}(\mathbf{k}, t)] + F[\xi(\mathbf{k}, t)]$ of Eq. (29.61) satisfies Eq. (29.62) we must set $F[\xi(\mathbf{k}, t)] = -M_n^*[0, \xi(\mathbf{k}, t)]$. If we denote the solution $M$ of $\mathscr{L}_j(\mathbf{k}, t)\,M = A_j$ which satisfies Eq. (29.62) by the symbol $M = \mathscr{L}_j^{-1}(\mathbf{k}, t)\,A_j$, the required solution $M_n$ of Eq. (29.61) can be written in the form

$$M_n = (\mathscr{S}_1 + \mathscr{S}_2 + \mathscr{S}_3)\,M_{n-1}, \tag{29.63}$$

where

$$\begin{aligned}
\mathscr{S}_1 &= \mathscr{L}_j^{-1}(\mathbf{k}, t)\,\mathcal{A}_j[\xi(\mathbf{k}, t); \mathbf{k}, t],\\
\mathscr{S}_2 &= \mathscr{L}_j^{-1}(\mathbf{k}, t)\,\mathscr{B}_j[\xi(\mathbf{k}, t); \mathbf{k}, t],\\
\mathscr{S}_3 &= \mathscr{L}_j^{-1}(\mathbf{k}, t)\,\mathcal{C}_j(\mathbf{k}, t).
\end{aligned} \tag{29.64}$$

In particular, it is readily verified that

$$M_1 = \mathscr{S}_1 1 = \int \tilde{\xi}_j(\mathbf{k}', t')\,\mathcal{A}_j[\xi(\mathbf{k}, t); \mathbf{k}', t']\,d\mathbf{k}'\,dt'. \tag{29.65}$$

We note that if $M[\tilde{\xi}(\mathbf{k}, t), \tilde{\mathbf{g}}(\mathbf{k}, t)]$ is a homogeneous functional of degree $N$, then $\mathcal{A}_jM$, $\mathscr{B}_jM$, $\mathcal{C}_jM$, $\mathscr{L}_j^{-1}M$ are, respectively, homogeneous functionals of degree $N+2$, $N$, $N-2$, and $N+1$. There are also the following exceptions in this rule: The operator $\mathscr{B}_j$

removes functionals which are independent of $\tilde{\xi}(\boldsymbol{k}, t)$, while the operator $\mathcal{C}_j$ has the same effect on functionals containing only the zero and first power of $\tilde{\xi}(\boldsymbol{k}, t)$. Similarly, if $M$ is a homogeneous functional of degree $N$, then $\mathcal{S}_1 M$, $\mathcal{S}_2 M$ and $\mathcal{S}_3 M$ are, respectively, homogeneous functionals of degree $N+3, N+1$, and $N-1$ with the following exceptions: The operator $\mathcal{S}_2$ removes functionals which are independent of $\tilde{\xi}(\boldsymbol{k}, t)$, while the operator $\mathcal{S}_3$ has the same effect on functionals containing the zero and first power of $\tilde{\xi}(\boldsymbol{k}, t)$. Since $M_0 = 1$, we find from Eq. (29.63) and the above rule that $M_1$ is a homogeneous functional of degree 3 [see Eq. (29.65)], $M_2$ is the sum of homogeneous functionals of degree 6, 4, and 2, and, in general,

$$M_{2n} = \sum_{m=1}^{3n} M_{2n}^{(2m)}; \quad M_{2n+1} = \sum_{m=0}^{3n+1} M_{2n+1}^{(2m+1)}, \tag{29.66}$$

where $M_n^{(m)}$ are homogeneous functionals of degree $m$.

The process of formation of these functionals is illustrated by Fig. 129, where $m$ and $n$ have the same meaning as in $M_n^{(m)}$ and as we move along the graph from left to right, lines with the greater upward slope represent the operator $\mathcal{S}_1$, while those with a smaller upward slope represent the operator $\mathcal{S}_2$. Those with a downward slope represent $\mathcal{S}_3$. The graph provides information not only about the set of functionals $M_n^{(m)}$ among those in $M_n$, but also of the method of evaluating any functional $M_n^{(m)}$. All that is required is to sum the contributions corresponding to all the possible broken lines, for

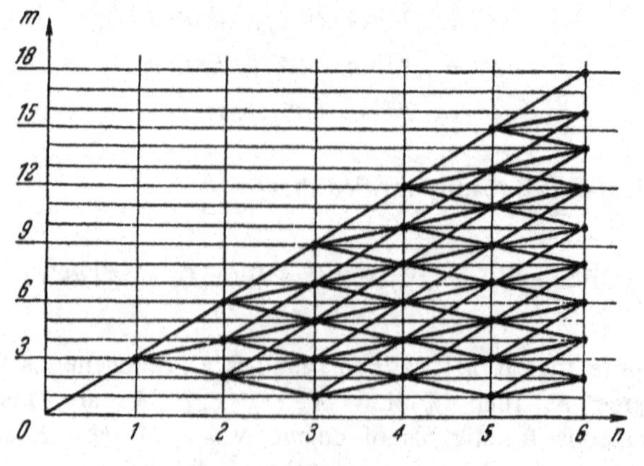


Fig. 129

which the horizontal coordinate of corner sites does not exceed $n$ and which end at the point $(n, m)$. Such broken lines will be referred to as the *diagrams* corresponding to the functional $M_n^{(m)}$. Thus, for example, the functional $M_2^{(2)}$ has the single diagram of Fig. 130 which represents the operator $\mathscr{S}_3\mathscr{S}_1$ so that $M_2^{(2)} = \mathscr{S}_3\mathscr{S}_1 1$. It is readily verified that this functional can be written in the form

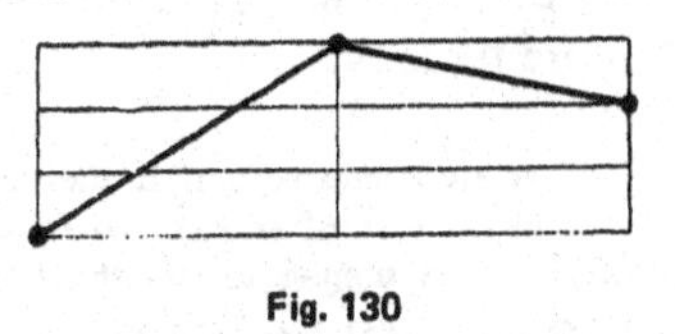

Fig. 130

$$M_2^{(2)} = \int dk\, dt\, dk'\, dt' \tilde{\zeta}_j(k, t) \Big\{ \mathcal{C}_j(k, t) \tilde{\zeta}_\gamma(k', t') - \\ - \frac{1}{2} \tilde{\zeta}_\gamma(k', t') \mathcal{C}_j(k, t) \Big\} \mathcal{A}_\gamma [\xi(k, t); k', t']. \quad (29.67)$$

The functional $M_4^{(2)}$ correspond to the three diagrams (Fig. 131) representing the operators $\mathscr{S}_3^2\mathscr{S}_2\mathscr{S}_1$, $\mathscr{S}_3\mathscr{S}_2\mathscr{S}_3\mathscr{S}_1$, and $\mathscr{S}_2\mathscr{S}_3^2\mathscr{S}_1$, so that

$$M_4^{(2)} = (\mathscr{S}_3^2\mathscr{S}_2\mathscr{S}_1 + \mathscr{S}_3\mathscr{S}_2\mathscr{S}_3\mathscr{S}_1 + \mathscr{S}_2\mathscr{S}_3^2\mathscr{S}_1) \cdot 1. \quad (29.68)$$

The entire expression given by Eq. (29.68) can be represented by the single resultant diagram of Fig. 132. The functional $M_6^{(2)}$ corresponds

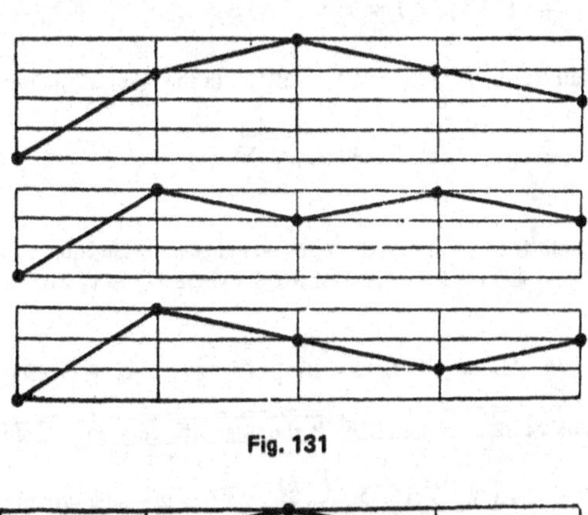

Fig. 131

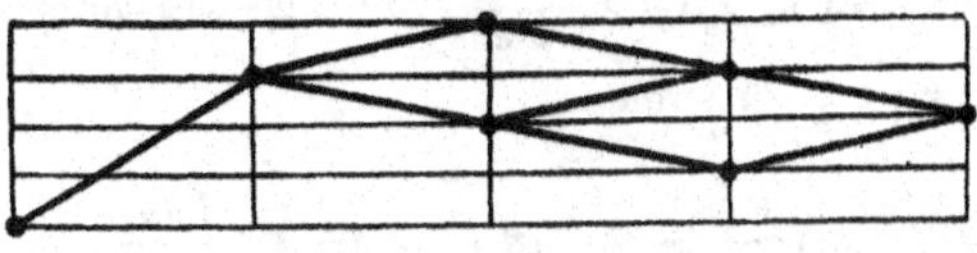

Fig. 132

to as many as 12 diagrams, and so on. Figure 129 gives a representation of the structure of the series (29.60) as a whole, and contains the specific prescriptions for evaluating any term of this series (although as $n$ increases such calculations become more and more tedious).

If we are interested only in the statistical properties of the velocity field at a given time then in the case of an external force field with the characteristic functions given by Eq. (29.49) it is simpler to use the Hopf equation not in the form of Eq. (28.82) but in Novikov's (1964) form (28.92). We can then set $\hat{F}_{\alpha\beta}(\boldsymbol{k}) = \Delta_{\alpha\beta}(\boldsymbol{k})\hat{F}(\boldsymbol{k})$, so that for the statistically stationary state equation (28.95) assumes the form

$$\nu \int k^2 \tilde{z}_\alpha(\boldsymbol{k}) \mathscr{D}_{\tilde{z}_\alpha}(\boldsymbol{k}) \Psi \, d\boldsymbol{k} + \frac{1}{2} \int |\tilde{z}(\boldsymbol{k})|^2 \hat{F}(\boldsymbol{k}) \, d\boldsymbol{k} \cdot \Psi =$$

$$= \int \int d\boldsymbol{k}' \, d\boldsymbol{k}'' \tilde{z}_\alpha(\boldsymbol{k}' + \boldsymbol{k}'') k''_\beta \Delta_{\beta p}(\boldsymbol{k}') \Delta_{\alpha q}(\boldsymbol{k}'') \mathscr{D}_{\tilde{z}_p}(\boldsymbol{k}') \mathscr{D}_{\tilde{z}_q}(\boldsymbol{k}'') \Psi. \quad (29.69)$$

This equation will definitely be satisfied provided we can satisfy the equation

$$\left[\nu k^2 \mathscr{D}_{\tilde{z}_j}(\boldsymbol{k}) + \frac{1}{2} \hat{F}(\boldsymbol{k}) \tilde{z}_j(-\boldsymbol{k})\right] \Psi = \mathscr{C}_j(\boldsymbol{k}) \Psi, \quad (29.70)$$

where

$$\mathscr{C}_j(\boldsymbol{k}) = \int \int d\boldsymbol{k}_1 \, d\boldsymbol{k}_2 C_{jpq}(\boldsymbol{k} \mid \boldsymbol{k}_1, \boldsymbol{k}_2) \mathscr{D}_{\tilde{z}_p}(\boldsymbol{k}_1) \mathscr{D}_{\tilde{z}_q}(\boldsymbol{k}_2),$$

$$C_{jpq}(\boldsymbol{k} \mid \boldsymbol{k}_1, \boldsymbol{k}_2) = k_\beta \Delta_{\beta p}(\boldsymbol{k}_1) \Delta_{jq}(\boldsymbol{k}_2) \delta(\boldsymbol{k}_1 + \boldsymbol{k}_2 - \boldsymbol{k}). \quad (29.71)$$

Similarly, we shall seek the solution of Eq. (29.70) in the form of the series

$$\Psi = \Psi_0 \sum_{n=0}^{\infty} M_n, \quad (29.72)$$

where $\Psi_0$ is the solution of Eq. (29.70) with the right-hand side equal to zero, which is given by Eq. (29.54), and $M_n$ is of the order of $(\mathrm{Re})^n$ (where $M_0 = 1$). For the functional $M_n$ we then have

$$\mathscr{L}_j(\boldsymbol{k}) M_n = \{\mathscr{A}_j[\tilde{z}(\boldsymbol{k}), \boldsymbol{k}] + \mathscr{B}_j[\tilde{z}(\boldsymbol{k}), \boldsymbol{k}] + \mathscr{C}_j(\boldsymbol{k})\} M_{n-1}, \quad (29.73)$$

where $\mathscr{L}_j(\boldsymbol{k}) = \nu k^2 \mathscr{D}_{\tilde{z}_j}(\boldsymbol{k})$ and $\mathscr{A}_j$ is the quadratic functional defined by Eq. (29.71) with $\mathscr{D}_{\tilde{z}_j}(\boldsymbol{k})$ replaced by $\tilde{z}_j(-\boldsymbol{k}) \dfrac{\hat{F}(\boldsymbol{k})}{2\nu k^2}$. $\mathscr{B}_j$ is the operator obtained from $\mathscr{C}_j$ by replacing $\mathscr{D}_{\tilde{z}_p}(\boldsymbol{k}_1) \mathscr{D}_{\tilde{z}_q}(\boldsymbol{k}_2)$ with

$$-\tilde{z}_q(-\boldsymbol{k}_2) \frac{\hat{F}(\boldsymbol{k}_2)}{2\nu k_2^2} \mathscr{D}_{\tilde{z}_p}(\boldsymbol{k}_1) - \tilde{z}_p(-\boldsymbol{k}_1) \frac{\hat{F}(\boldsymbol{k}_1)}{2\nu k_1^2} \mathscr{D}_{\tilde{z}_q}(\boldsymbol{k}_2).$$

The solution of Eq. (29.73) for $M_n$ must be sought subject to the condition $M_n(0) = 0$, $n \geqslant 1$. By analogy with the foregoing discussion, we can now introduce operators $\mathscr{S}_1$, $\mathscr{S}_2$ and $\mathscr{S}_3$, and use the same diagrams (Fig. 129). We shall employ this formalism to establish the form of the expansion for the kinetic energy spectral density $E(k)$ into a series in powers of the Reynolds number. Using the obvious formula

$$\overline{u_\alpha(\boldsymbol{x}_1, t)\, u_\beta(\boldsymbol{x}_2, t)} = -\int\int e^{-i(\boldsymbol{k}_1\boldsymbol{x}_1 + \boldsymbol{k}_2\boldsymbol{x}_2)}\, d\boldsymbol{k}_1\, d\boldsymbol{k}_2\, [\mathscr{D}_\alpha(\boldsymbol{k}_1)\, \mathscr{D}_\beta(\boldsymbol{k}_2)\, \Psi]\,|_{\boldsymbol{z}=0},$$

we can readily show that

$$E(k) = -\frac{k^2}{2}\int\int d\Omega(\boldsymbol{k})\, d\boldsymbol{k}_1\, [\mathscr{D}_\alpha(\boldsymbol{k})\, \mathscr{D}_\alpha(\boldsymbol{k}_1)\, \Psi]\,|_{\boldsymbol{z}=0}, \tag{29.74}$$

where $d\Omega(\boldsymbol{k})$ is an area element on a unit sphere in the space of the wave vectors $\boldsymbol{k}$. Substituting Eq. (29.72) for $\Psi$ in Eq. (29.74), we obtain

$$E(k) = \frac{2\pi}{\nu}\hat{F}(k) + \mathscr{S}(k)\sum_{n=1}^{\infty} M_{2n}^{(2)}, \tag{29.75}$$

where $M_{2n}^{(2)}$ is a homogeneous functional of order 2, which is part of $M_{2n}$, and $\mathscr{S}(k)$ is the operator defined by

$$\mathscr{S}(k) = -\frac{k^2}{2}\int\int d\Omega(\boldsymbol{k})\, d\boldsymbol{k}_1 \Delta_{\alpha p}(\boldsymbol{k})\, \Delta_{\alpha q}(\boldsymbol{k}_1)\, \mathscr{D}_{\tilde{z}_p}(\boldsymbol{k})\, \mathscr{D}_{\tilde{z}_q}(\boldsymbol{k}_1). \tag{29.76}$$

We note, first, that the above expansion of the function $E(k)$ into a series in powers of the Reynolds number contains only even powers of Re. Second, this expansion is also the expansion of the function $E(k)$ into a functional power series in $\hat{F}(k)$ in which the term of the order of $(\mathrm{Re})^{2n}$ [i.e., $\mathscr{S}(k)\, M_{2n}^{(2)}$] is a homogeneous functional of order $n+1$ in $\hat{F}(k)$. These two conclusions can readily be established with the aid of the diagrams of Fig. 129. The first conclusion follows directly from these diagrams, whereas the second can be proved by remembering that the operator $\mathscr{S}_1$ exceeds by 2, $\mathscr{S}_2$ by 1, and $\mathscr{S}_3$ by 0, the order of the functionals with respect to $\hat{F}(k)$.

The functional $\mathscr{S}(k)\, M_{2n}^{(2)}$ is, in a sense, analogous to the "convolution" of the $n+1$ functions $\hat{F}(k)$, i.e., the probability density for the sum of $n+1$ independent random variables, each of which has a probability density proportional to $\hat{F}(k)$. Such "convolutions" have the following property. If $\hat{F}(k)$ differs from 0 only for small wave numbers, say, those in the range $0 \leqslant k \leqslant k_0$, then for very large numbers $k \gg k_0$ the only "convolutions" which will differ from zero will be those corresponding to a very large number of functions $\hat{F}(k)$, and such convolutions will not be very dependent on the specific form of these functions (and will approach the universal Gaussian function). Similarly, we expect that if the spatial distribution of external forces has only large-scale inhomogeneities, i.e., if $\hat{F}(k)$ differs from zero only for $0 \leqslant k \leqslant k_0$, then for $k \gg k_0$ the only functionals which will differ from zero will be $\mathscr{S}(k)\, M_{2n}^{(2)}$ with very large $n$, and these functionals will not be very dependent on the specific form of the function $\hat{F}(k)$. Hence if Eq. (29.75) converges (which is quite questionable), its sum $E(k)$ for $k \gg k_0$ will apparently approach a universal function of $k$. A rigorous proof of the correctness or incorrectness of these statements and the determination of the corresponding universal function, if it exists, is of considerable interest for the theory of turbulence.

## 29.4 Other Expansion Schemes

We have already noted that solution of the equation for the characteristic functional of the velocity field with the aid of a series in powers of the Reynolds number is quite ineffective in the usual case of turbulence with large Re. In particular, in the case of the solution $\Psi$ of Eq. (29.69) in the form of a series (29.72), the zero-order approximation $\Psi_0$ is the characteristic functional of the Gaussian random velocity field with the spectral function $F_0(k) = \frac{E_0(k)}{4\pi k^2} = \frac{\hat{F}(k)}{2\nu k^2}$, which is very different from the true characteristic functional for the velocity field of developed turbulence. Edwards (1964) therefore proposed the replacement of Eq. (29.72) by Eq. (29.7), which is analogous to the Gram-Charlier series. In this series the zero-order approximation is also a characteristic functional of the Gaussian random field (which is a good approximation only for single-point characteristics of large-scale components of the random velocity field) but this time with the *true* spectral function $F(k)$.

The Edwards formalism differs from that given above mainly in the fact that, instead of the characteristic functionals, Edwards used the probability densities in the function space of the individual realizations of the random velocity field. These densities are not, in fact, strictly rigorous, since in an infinite-dimensional function space there are no volume elements. The Edwards formalism can therefore have in its original form only a heuristic significance. However, the principle ideas introduced by Edwards can also be expressed in terms of characteristic functionals and we shall do this now.

Consider, to begin with Eq. (29.70), and note that the first term on its left-hand side describes the energy loss by the spectral velocity component with wave-number vector $\boldsymbol{k}$, which is due to molecular viscosity, the second term represents the influx of energy into this component due to the work done by external forces, and the right-hand side represents energy transfer and "adiabatic interactions" between this particular component and all the remaining velocity components. The above energy transfer can be described (in the spirit of the semiempirical theories) as the sum of the energy losses involved in overcoming the "eddy viscosity" or "dynamic friction" $\left(\text{the kinematic eddy viscosity will be denoted by } K(k) = \frac{\Omega(k)}{k^2}\right)$ and the energy influx due to diffusion in the wave-number space [with diffusion coefficient $H(k)$]. Therefore, to take into account the energy transfer between a component with wave vector $\boldsymbol{k}$ and all the other components of the velocity field in the framework of the semiempirical approach we must replace the quantities $\nu k^2$ and $\frac{1}{2}\hat{F}(k)$ with $\omega(k)$ and $\eta(k)$ on the left-hand side of Eq. (29.70), where

$$\omega(k) = \nu k^2 + \Omega(k), \qquad \eta(k) = \frac{1}{2}\hat{F}(k) + H(k), \tag{29.77}$$

To preserve the equality we must now add the term

$$\left[\Omega(k)\,\mathscr{D}_{\tilde{z}_j}(\boldsymbol{k}) + H(k)\,\tilde{z}_j(-\boldsymbol{k})\right]\Psi;$$

to the right-hand side of Eq. (29.70). The new right-hand side will then describe only the "adiabatic interactions" with all the remaining spectral velocity components. Returning now to Eq. (29.69), we shall rewrite it in the form

$$\mathscr{H}_0\Psi = \mathscr{H}_1\Psi + \mathscr{H}_2\Psi, \tag{29.78}$$

where

$$\begin{aligned}\mathscr{H}_0 &= \int d\boldsymbol{k}\left\{\omega(k)\,\tilde{z}_\alpha(\boldsymbol{k})\,\mathscr{D}_{\tilde{z}_\alpha}(\boldsymbol{k}) + \eta(k)\,|\,\tilde{z}(\boldsymbol{k})\,|^2\right\},\\ \mathscr{H}_2 &= \int d\boldsymbol{k}\left\{\Omega(k)\,\tilde{z}_\alpha(\boldsymbol{k})\,\mathscr{D}_{\tilde{z}_\alpha}(\boldsymbol{k}) + H(k)\,|\,\tilde{z}(\boldsymbol{k})\,|^2\right\},\end{aligned} \tag{29.79}$$

and $\mathscr{H}_1$ is the operator on the right-hand side of Eq. (29.69). The procedure put forward by Edwards for the solution of Eq. (29.78) involves, first, the representation of $\Psi$ by a series $\Psi_0 + \Psi_1 + \Psi_2 + \ldots$ in powers of some parameter, where the operators $\mathscr{H}_0$, $\mathscr{H}_1$ and $\mathscr{H}_2$ are regarded as zero, first, and second-order quantities, respectively. Second, the procedure imposes the condition that the second moments of the velocity field should be completely determined by the zero-order approximation $\Psi_0$, so that the functional $\Psi_1 + \Psi_2 + \ldots$ should not contribute to the second moments. The terms $\Psi_n$ are thus determined by the equations

$$\mathscr{H}_0\Psi_0 = 0, \tag{29.80}$$

$$\mathscr{H}_0\Psi_1 = \mathscr{H}_1\Psi_0, \tag{29.81}$$

$$\mathscr{H}_0\Psi_n = \mathscr{H}_1\Psi_{n-1} + \mathscr{H}_2\Psi_{n-2}, \qquad n \geqslant 2, \tag{29.82}$$

and in the $n$th approximation the following supplementary condition must be valid:

$$[\mathscr{D}_j(\boldsymbol{k}_1)\,\mathscr{D}_l(\boldsymbol{k}_2)\,(\Psi_1 + \Psi_2 + \ldots + \Psi_n)]\,|_{z(\boldsymbol{k})=0} = 0 \tag{29.83}$$

We shall show below that Eqs. (29.80)–(29.83) do not define $\Psi$ unambiguously because the single equation given by Eq. (29.83) will then have to be used to determine the two functions $\Omega(k)$ and $H(k)$. Edwards therefore introduces one further condition which is, in fact, arbitrary.

The solution $\Psi_0$ will be the characteristic functional of the Gaussian random field, i.e., it will be given by Eq. (29.54) but with the spectral function $\dfrac{\hat{F}(k)}{2\nu k^2}$ replaced with

$$F(k) = \eta(k)/\omega(k). \tag{29.84}$$

The right-hand sides of Eqs. (29.81) and (29.82) will be the sums of terms of the form

$$\Psi_0 \int \cdots \int \prod_{j=1}^{N} [\tilde{z}_{\alpha_j}(\boldsymbol{k}_j)]^{n_j}\, \psi_{\alpha_1 \cdots \alpha_N}(\boldsymbol{k}_1, \ldots, \boldsymbol{k}_N)\, d\boldsymbol{k}_1 \ldots d\boldsymbol{k}_N. \tag{29.85}$$

It is readily verified that the application of the operator $\mathscr{H}_0$ to this functional results in the appearance of the factor $\sum_{j=1}^{N} n_j\,\omega(k_j)$ in the integrand. Consequently, the solution of the equation $\mathscr{H}_0\Psi = F$ with the right-hand side $F$ equal to the expression given by Eq. (29.85) can be obtained from Eq. (29.85) by introducing the factor $\left[\sum_{j=1}^{N} n_j\omega(\boldsymbol{k}_j)\right]^{-1}$ in the integrand. All this eventually leads to the following solution of Eq. (29.81):

$$\Psi_1 = -\Psi_0 \int\int\int \frac{\tilde{z}_\alpha(\boldsymbol{k}_1)\,\tilde{z}_\beta(\boldsymbol{k}_2)\,\tilde{z}_\gamma(\boldsymbol{k}_3)}{\omega(\boldsymbol{k}_1)+\omega(\boldsymbol{k}_2)+\omega(\boldsymbol{k}_3)}\,\delta(\boldsymbol{k}_1+\boldsymbol{k}_2+\boldsymbol{k}_3)\times$$
$$\times A_{\alpha\beta,\gamma}(\boldsymbol{k}_1,\boldsymbol{k}_2)\,F(\boldsymbol{k}_1)\,F(\boldsymbol{k}_2)\,d\boldsymbol{k}_1\,d\boldsymbol{k}_2\,d\boldsymbol{k}_3, \tag{29.86}$$

where

$$A_{\alpha\beta,\gamma}(\boldsymbol{k}_1,\boldsymbol{k}_2) = \frac{1}{2}(k_{1\lambda}+k_{2\lambda})\,[\Delta_{\alpha\lambda}(\boldsymbol{k}_1)\,\Delta_{\beta\gamma}(\boldsymbol{k}_2)+\Delta_{\alpha\gamma}(\boldsymbol{k}_1)\,\Delta_{\beta\lambda}(\boldsymbol{k}_2)]. \tag{29.87}$$

Using Eqs. (29.86) and (29.82) with $n = 2$, we obtain

$$\Psi_2 = \Psi_0\Big\{\int \frac{|\tilde{z}(\boldsymbol{k})|^2}{2\omega(\boldsymbol{k})}[H(\boldsymbol{k})-\Omega(\boldsymbol{k})F(\boldsymbol{k})]\,d\boldsymbol{k} - 2\int \frac{\tilde{z}_\lambda(-\boldsymbol{k})\,\tilde{z}_\mu(\boldsymbol{k})}{2\omega(\boldsymbol{k})}\times$$
$$\times\frac{\delta(\boldsymbol{k}+\boldsymbol{k}_1+\boldsymbol{k}_2)}{\omega(\boldsymbol{k})+\omega(\boldsymbol{k}_1)+\omega(\boldsymbol{k}_2)}\,A_{\alpha\beta,\lambda}(\boldsymbol{k}_1,\boldsymbol{k}_2)\,[F(\boldsymbol{k}_1)\,F(\boldsymbol{k}_2)\,A_{\alpha\beta,\mu}(\boldsymbol{k}_1,\boldsymbol{k}_2)+$$
$$+2F(\boldsymbol{k})\,F(\boldsymbol{k}_1)\,A_{\mu\alpha,\beta}(\boldsymbol{k},\boldsymbol{k}_1)]\,d\boldsymbol{k}\,d\boldsymbol{k}_1\,d\boldsymbol{k}_2+\ldots\Big\}, \tag{29.88}$$

where the running points represent higher-order terms in $z(\boldsymbol{k})$, which do not contribute to the second moments of the velocity field. Consider now the condition (29.83), and let us restrict our attention to the second approximation ($n = 2$). Since the functional $\Psi_1$ does not contribute to the second moments of the velocity field, we shall satisfy Eq. (29.83) by demanding that the second moments, calculated with the functional in the braces in Eq. (29.88), should vanish. This leads to the following equation:

$$H(\boldsymbol{k})-\Omega(\boldsymbol{k})F(\boldsymbol{k}) = \int \frac{\delta(\boldsymbol{k}+\boldsymbol{k}_1+\boldsymbol{k}_2)\,d\boldsymbol{k}_1\,d\boldsymbol{k}_2}{\omega(\boldsymbol{k})+\omega(\boldsymbol{k}_1)+\omega(\boldsymbol{k}_2)}\,[a(\boldsymbol{k},\boldsymbol{k}_1,\boldsymbol{k}_2)\,F(\boldsymbol{k}_1)\times$$
$$\times F(\boldsymbol{k}_2)-b(\boldsymbol{k},\boldsymbol{k}_1,\boldsymbol{k}_2)\,F(\boldsymbol{k})\,F(\boldsymbol{k}_1)], \tag{29.89}$$

where

$$\begin{aligned} a(\boldsymbol{k},\boldsymbol{k}_1\,\boldsymbol{k}_2) &= \Delta_{\lambda\mu}(\boldsymbol{k})\,A_{\alpha\beta,\lambda}(\boldsymbol{k}_1,\boldsymbol{k}_2)\,A_{\alpha\beta,\mu}(\boldsymbol{k}_1,\boldsymbol{k}_2),\\ b(\boldsymbol{k},\boldsymbol{k}_1,\boldsymbol{k}_2) &= -2\Delta_{\lambda\mu}(\boldsymbol{k})\,A_{\alpha\beta,\lambda}(\boldsymbol{k}_1,\boldsymbol{k}_2)\,A_{\mu\alpha,\beta}(\boldsymbol{k},\boldsymbol{k}_1). \end{aligned} \tag{29.90}$$

We thus find that we have to use Eq. (29.89) to determine the two functions $H(\boldsymbol{k})$ and $\Omega(\boldsymbol{k})$. Edwards assumes that

$$\begin{aligned} H(\boldsymbol{k}) &= \int \frac{\delta(\boldsymbol{k}+\boldsymbol{k}_1+\boldsymbol{k}_2)\,F(\boldsymbol{k}_1)\,F(\boldsymbol{k}_2)}{\omega(\boldsymbol{k})+\omega(\boldsymbol{k}_1)+\omega(\boldsymbol{k}_2)}\,a(\boldsymbol{k},\boldsymbol{k}_1,\boldsymbol{k}_2)\,d\boldsymbol{k}_1\,d\boldsymbol{k}_2,\\ \Omega(\boldsymbol{k}) &= \int \frac{\delta(\boldsymbol{k}+\boldsymbol{k}_1+\boldsymbol{k}_2)\,F(\boldsymbol{k}_1)}{\omega(\boldsymbol{k})+\omega(\boldsymbol{k}_1)+\omega(\boldsymbol{k}_2)}\,b(\boldsymbol{k},\boldsymbol{k}_1,\boldsymbol{k}_2)\,d\boldsymbol{k}_1\,d\boldsymbol{k}_2. \end{aligned} \tag{29.91}$$

These formulas, together with Eqs. (29.77) and (29.84), enable us, in principle, to determine the three functions $F(\boldsymbol{k})$, $H(\boldsymbol{k})$, and $\Omega(\boldsymbol{k})$ for a given external force spectrum

$\hat{F}(k)$. Equation (29.84) then reduces to the form

$$\nu k^2 F(k) + \int \frac{\delta(k + k_1 + k_2)\, dk_1\, dk_2}{\hat{\omega}(k) + \omega(k_1) + \omega(k_2)} [b(k, k_1, k_2) F(k) F(k_1) - \\ - a(k, k_1, k_2) F(k_1) F(k_2)] = \frac{1}{2} \hat{F}(k), \qquad (29.92)$$

which is the balance equation for the turbulent energy, and is the analog of the Boltzmann equation, with the basic difference that two-particle collisions $(v, v_1) \to (v', v_1')$ involving the conservation of total momentum [so that $v + v_1 = v' + v_1'$] and of total energy [which is represented by the factor $\delta(E + E_1 - E' - E_1')$ in the expression for the interaction cross section in the Boltzmann collision integral] are replaced in the fluid dynamic case by triple interactions between the Fourier components, $k \to (k_1, k_2)$, in which the wave number is "conserved" [this is expressed by the factor $\delta(k + k_1 + k_2)$ in the "cross section" in the integral in Eq. (19.92)].

The second Edwards approximation thus leads to a Boltzmann equation for the waves. We note that an analogous equation for phonons, i.e., the sound quanta in solids, was obtained earlier by Peierls (1955). It has been also used to describe weak nonlinear interactions between plasma waves by Camac et al. (1962), Galeev and Karpman (1963), Kadomtsev (1965), and others, to describe water wave interactions by Hasselmann (1967), and for the description of weak turbulence by Zakharov (1965).

A somewhat similar approach to the theory of turbulence was put forward by Herring (1965, 1966), who again took the equations for the probability density at a particular time in the space of the realizations of the velocity field as his starting point. However he used a different procedure for constructing the perturbation series which is also based on expansion about an assumed approximate statistical state but in some ways resembles the self-consistent field approach of the theory of many-body systems. This method is based on the replacement of the effect of all the particles on any one of them by an artificial external force field which satisfies some natural self-consistency conditions. After this replacement the equations of motion for each of the particles can be solved in isolation, and the solutions can be used to set up the next approximation for the self-consistent field. According to Herring, the application of this approach to the set of all possible spectral components of a turbulent velocity field leads to results which differ from those obtained by Edwards only by the somewhat different forms of $H(k)$ and $\Omega(k)$. However, the approximation obtained by Herring has not as yet been investigated to any great extent. Another self-consistent perturbation procedure was proposed later by Phythian (1969) [see also J. Lee (1971) and Phythian (1972)]. This procedure differs from all others by the fact that it yields Kraichnan's direct-interaction approximation as the simplest nontrivial approximation of zero-order. The procedure is formulated in terms of solutions of the Navier-Stokes equations and its reformulation in terms of characteristic functionals is of interest for turbulence theory. However we shall consider here neither Herring's nor Phythian's methods of self-consistent approximations.

Let us now return to the Edwards theory. The solution of Eqs. (29.77), (29.84), and (29.91) can readily be found if we neglect the effect of molecular viscosity and assume additionally that $\hat{F}(k) \sim k^{-\alpha}$. In that case,

$$F(k) \sim k^{-\frac{5+2\alpha}{3}}, \quad H(k) \sim k^{-\alpha}, \quad \Omega(k) \sim k^{\frac{5-\alpha}{3}}, \qquad (29.93)$$

and the next approximtions change only the numerical factors in these expressions. The solution is found to be satisfactory only for $2 > \alpha > -1$. Namely, for $\alpha < -1$ we must take into account molecular viscosity, and its effect is significant for all $k$; for $\alpha > 2$ the form of the function $\hat{F}(k)$ for small $k$ is quite important and this form must be changed so that the total energy influx due to the work done by the external forces remains finite. Therefore, the law $F(k) \sim k^{-11/3}$, which corresponds to the Kolmogorov self-preservation cannot be obtained from Eq. (29.83), nor can it be found when $\hat{F}(k)$ is specified as nonzero only for small wave numbers $k \leqslant k_0$, i.e., when the external forces form the so-called "red noise". For this case Edwards showed that $F(k)$ was intermediate between the Kolmogorov spectrum $(F(k) \sim k^{-11/3})$ and the Kraichnan spectrum $(F(k) \sim k^{-7/2})$ according to all the approximations. [Later Kraichnan (1971a) gave another derivation of a generalized version of Edwards' theory based on a model equation for the turbulent velocity field. An attempt to modify this equation in a manner permitting one to obtain the Kolmogorov spectrum led him to the formulation of a new closure hypothesis which we have already mentioned in Sect. 22.2. See also the paper by Herring and Kraichnan (1971) where both these theories are compared with several other approximations and with data of laboratory and computer experiments.]

The reason for the failure of these earlier attempts to obtain the Kolmogorov turbulence spectrum by the above method in the presence of external forces of the kind referred to as "red noise" was thought by Edwards (1965) to be the inadequacy of a purely spatial approach to the description of interactions between the Fourier components of the velocity field in developed turbulence. Therefore he introduced in this work a space-time expansion procedure and obtained in its framework the Kolmogorov spectrum, but at the cost of using some artificial and quite arbitrary assumptions. This treatment was later declared by Edwards and McComb (1969) to be unconvincing. Instead, Edwards and McComb introduced a measure of the "degree of randomness of the state" resembling the thermodynamic entropy of the system and also the probabilistic entropy entering in Shannon's information theory. Then they postulated the principle of maximum randomness (i.e., maximum entropy) as a basic principle for the selection of a steady state. Unfortunately it was found that the calculations involved in the use of this principle are tremendous and to make it easier, Edwards and McComb dropped somewhat arbitrarily the most complicated term. As a result they eventually succeeded in obtaining a Kolmogorov spectrum of the form

$$E(k) = 4\pi k^2 F(k) = C_1 \bar{\epsilon}^{2/3} k^{-5/3} \tag{29.94}$$

in all the approximations but the cost this time was also not low. Moreover the numerical value of the coefficient $C_1$ turns out to be about three times larger in the Edwards and McComb theory than the experimental value given in Sect. 23.3 of this book. Nevertheless a quite rigorous reformulation of all the assumptions of the Edwards and McComb theory in strict terms of characteristic functionals would apparently be (if only it is possible) of interest for the future development of turbulent theory.

## 29.5 Use of Functional Integrals

In the solution of linear differential equations both ordinary and partial, we often use the Fourier transform, i.e., the solution is sought in the form of Fourier integrals of some new function which is more simply determined. An analogous approach to the solution of linear equations in functional derivatives should lead to the representation of the required functional $\Phi[\theta(M)]$ by the *functional Fourier transform*:

$$\Phi[\theta(M)] = \int e^{i(\theta \cdot F)} \Psi[F(M)]\, d\mu[F(M)], \tag{29.95}$$

Here $\Psi$ is a new functional [of the function $F(M)$], and $(\theta \cdot F)$ is the functional scalar product, i.e., the integral of the product $\theta(M) \cdot F(M)$ over all the space of the points $M$. Finally, $\mu$ is a measure in the infinite-dimensional function space (whose "points" are the functions $F$), i.e., a function of the sets $S$ in this space, which has the additive property [so that $\mu\left(\sum_l S_l\right) = \sum_l \mu(S_l)$ for any nonintersecting sets $S_l$ for which the measure $\mu$ is defined] and certain further properties which we shall not discuss here.

In the special case when all the values of $\mu(S)$ are nonnegative and the measure of the whole function space is equal to unity, the quantity $\mu(S)$ can be interpreted as the *probability* that the value of $F$ will belong to the given set $S$, i.e., we can regard $F$ as a *random function*. In that case, the integral given by Eq. (29.95) will be identical with the probability mean value of $\overline{\exp\{i(\theta \cdot F)\}\Psi[F]}$, and a given measure $\mu$ in the functional space will be a further means of specifying the random function $F$, which is equivalent to specifying a set of finite-dimensional probability densities $p_{M_1 \ldots M_n}(F_{\alpha_1}, \ldots, F_{\alpha_n})$. In fact, these densities after multiplication by $dF_{\alpha_1} \ldots dF_{\alpha_n}$, determine the measure $\mu$ of the set of all functions $F(M)$ satisfying the inequalities

$$F_{\alpha_1} < F(M_1) < F_{\alpha_1} + dF_{\alpha_1}, \ldots, F_{\alpha_n} < F(M_n) < F_{\alpha_n} + dF_{\alpha_n},$$

i.e., they are equivalent to the values of $\mu$ for certain special sets $S$. We recall also that, in this special case, the integral (29.95) with $\Psi[F] = 1$ is equal to $\overline{\exp\{i(\theta \cdot F)\}}$, i.e., it is identical with the characteristic functional of the random function $F$ whose specification is a third method of defining this function.

In the general case of an arbitrary measure $\mu$, the significance of Eq. (29.95) is much more complicated. It is then reasonable to use the fact that on each finite-dimensional subspace of the infinite-dimensional space of all the functions $F$, i.e., on the set of functions $F$ depending only on a finite number of parameters (for example, specified by a finite segment of a Fourier series), the measure $\mu$ will be the ordinary measure in finite-dimensional space, and the integral (29.25) will be the usual finite-dimensional integral. Hence, it follows that the integral on the right-hand side of Eq. (29.95) can be interpreted as a symbolic representation of the limit of a particular sequence of finite-dimensional integrals. This sequence is obtained when the function $F$ is approximated with increasing accuracy by a sequence of functions each of which is determined by a finite number of parameters, and each time the integration with respect to $\mu[F]$ is carried out over a finite-dimensional subspace of the functions $F$. There are also certain other approaches to the evaluation of Eq. (29.95). We note, however, that the precise definition of measure in function space, and of the integrals of functionals with respect to this measure, as well as the properties of the above integrals, their transformation rules, and the connections between their different definitions, form a very subtle mathematical problem which, despite the existence of extensive literature on the subject, is still rather obscure [see, for example, the seminar notes of Friedrichs, Shapiro et al. (1957), the book of Gel'fand and Vilenkin (1964), and the survey papers by Daletskii (1962), Shilov (1963), McShane (1963), Sigal (1965), Streit (1965), and Hamilton and Schulman (1971), which contain reviews of the various approaches to the integration of functionals and an extensive list of references].

Functional (or "path") integrals, i.e., integrals of functionals with respect to the measure in function space, were apparently first used in connection with physical problems by Feynman (1948). This author obtained with their aid a new derivation of the Schroedinger equation of quantum mechanics, and put forward a symbolic path integral notation for the general solution of this equation [a detailed description of these results is given in a book of Feynman and Hibbs (1965)]. In another paper, Feynman (1951) made use of the functional Fourier transform (29.95) to solve the equations of quantum field theory. At approximately the same time it was shown by J. Schwinger that the basic equations of quantum field theory can be written as linear differential equations with functional derivatives. Later,

Edwards and Peierls (1954) and Gel'fand and Minlos (1954) independently found the general solution of Schwinger's equations in the form of a functional integral [see also the survey papers by Gel'fand and Yaglom (1960) and Gel'fand, Minlos, and Yaglom (1958), containing many additional references]. The functional integral representation of the solutions of a wide class of partial differential equations (and its relation to functional differential equations) was considered by Kac (1951, 1959), Donsker and Lions (1962), Donsker (1964) and some other scientists. The applications of functional integrals to the investigation of time evolution of solutions of some one-dimensional nonlinear differential equations resembling the equations of turbulence theory were considered, in particular, by Varadhan (1966), Hosokawa (1968) and Hosokawa and Yamomoto (1970). It was however important to all these works that a simple transformation exists which makes the equation linear and usually gives an explicit formula for the general solution of the equation. Therefore the generalization of these results to the three-dimensional case is very far from being obvious. Finally, in the theory of true three-dimensional turbulence the functional integral formalism was used, in particular, by Rosen [(1960) and some subsequent papers] and by Tatarskii (1962b). The present subsection is based mainly on these works. It is important to note, however, that a rigorous mathematical interpretation of the functional integrals, which appear both in quantum mechanics and in turbulence theory, has not so far been available. The formulas established in this context must therefore be regarded as heuristic, and a rigorous derivation and extensive application of these results is a matter for the future. Nevertheless, the basic idea that the application of functional integrals should yield a general approach for obtaining solutions of linear equations in functional derivatives, and could simplify considerably the study of these equations, is a very natural one, and will probably be successful in spite of the fact that, so far, the development of this program has only just begun, especially as far as real physical problems are concerned.

The complexity of the mathematical interpretation of functional integrals, which arise in the theory of turbulence and in quantum physics, is connected with the fact that the measures $\mu$ are not measures of bounded variation. This means that there is no constant $C$ such that the inequality $\sum_l |\mu(S_l)| \leqslant C$ is valid for any set of nonintersecting sets $S_l$ for which the measure $\mu$ is determined. Moreover the measure $\mu$ does not have some of the other general properties demanded of measures in mathematical analysis. Nevertheless, even for these measures $\mu$ it is possible to define the quantity $\int \Psi[F]\, d\mu[F] = \Lambda[\Psi]$ for a sufficiently broad class of functionals $\Psi = \Psi[F(M)]$, where $\Lambda[\Psi]$ has the following basic property of ordinary integrals: If for any two functionals $\Psi_1$ and $\Psi_2$ there exist the integrals $\Lambda[\Psi_1]$ and $\Lambda[\Psi_2]$, then for any numerical coefficients $c_1$ and $c_2$ the integral $\Lambda[c_1\Psi_1 + c_2\Psi_2]$ will exist and will be equal to $c_1\Lambda[\Psi_1] + c_2\Lambda[\Psi_2]$. In other words $\Lambda[\Psi]$ turns out to be a *linear functional* of $\Psi$.

Since the integral $\Lambda[\Psi]$ is frequently determined in applications more simply than the measure $\mu$, it may even be better to regard the concept of the integral as the primary concept, i.e., define the integral in function space of the functions $F$ as any linear functional $\Lambda[\Psi]$ defined on a sufficiently broad set of functionals $\Psi$ of $F$ (and having a further property of continuity in $\Psi$ which we shall not consider here for the sake of simplicity). By analogy with the well-known theory of generalized functions [see, e.g., Gel'fand and Shilov (1964)] we may then suppose that each of these functionals corresponds to its own *generalized measure* $\mu$ such that

$$\Lambda[\Psi] = \int \Psi[F]\, d\mu[F]. \tag{29.96}$$

It is clear that the generalized measures will then include all the existing true measures over which the functionals $\Psi$ can be integrated independently of both the properties of these measures and the rigorous definition of integrals. Let us now consider some examples of generalized measures in the space of the functions $\theta(M)$. The simplest of these measures is the $\delta$-measure $\mu(S) = \delta[S;\ \theta_0(M)]$ which corresponds to the functional $\Lambda[\Phi(\theta)] = \Phi[\theta_0]$, where $\theta_0(M)$ is a given function. This measure is completely analogous to the Dirac $\delta$-function (which is the simplest generalized function), and is defined by

$$\int \Phi[\theta(M)]\, d\delta[\theta(M);\ \theta_0(M)] = \Phi[\theta_0(M)] \tag{29.97}$$

It is the ordinary ("true") measure localized at the "point" $\theta_0(M)$ of the function space, so that $\delta[S;\ \theta_0(M)] = 1$ if $\theta_0(M)$ belongs to the set $S$, and $\delta[S;\ \theta_0(M)] = 0$ if it does not. Hence, it follows that the $\delta$-measure of the set $S_F(x)$ of the functions $\theta(M)$ satisfying the inequality $(\theta \cdot F) \leqslant x$ [the "half-space" of the functional space of the functions $\theta(M)$] is given by

$$\delta[S_F(x);\ \theta_0(F)] = \delta_F(x) = \begin{cases} 1 & \text{when } x \geqslant x_0 = (\theta_0 \cdot F), \\ 0 & \text{when } x < x_0 = (\theta_0 \cdot F), \end{cases} \tag{29.98}$$

Using the Fourier representation of the Dirac $\delta$-function, we can write this result in a form

$$d\delta_F(x) = \delta(x - x_0)\, dx = \frac{dx}{\sqrt{2\pi}} \int_{-\infty}^{\infty} e^{i(x_0 - x)y} \frac{dy}{\sqrt{2\pi}}. \tag{29.98'}$$

This formula can be used for the representation of the infinite-dimensional integral over the $\delta$-measure in the form of the limit of a sequence of finite-dimensional integrals. In fact, let us divide the domain of the argument $M$ into $k$ nonintersecting sets $\Pi_1, \ldots, \Pi_k$, and define the projection $P_k\theta$ of the function $\theta(M)$ corresponding to this subdivision as the new function which assumes the constant value $\frac{1}{V_j} \int_{\Pi_j} \theta(M)\varphi(M)\, dM$ on each of the $\Pi_j$ where $V_j = \int_{\Pi_j} \varphi(M)\, dM$ and $\varphi(M)$ is an arbitrary nonnegative weight function.

The operator $P_k$ will then transform the infinite-dimensional space of the functions $\theta(M)$ into the finite-dimensional space $R_n$ of the points $(x_1, \ldots, x_n)$, in which $P_k\theta(M)$ has the coordinates $\frac{1}{V_j} \int_{\Pi_j} \theta_l(M)\varphi(M)\, dM$, $j = 1, \ldots, k$, $l = 1, \ldots, s$ and $n = ks$ where $s$ is the number of components of the vector $\theta = (\theta_1, \ldots, \theta_s)$. It is clear that the $\delta$-measure $\delta[S;\ \theta_0(M)]$ induces in $R_n$ the measure $\delta[S;\ P_k\theta_0(M)]$, whose differential can be written in the form

$$d\delta[P_k\theta(M);\ P_k\theta_0(M)] = \prod_{j=1}^{n} \delta(x_k - x_k^0)\, dx_k =$$
$$= dm[P_k\theta(M)] \int e^{i((\theta - \theta_0) \cdot \varphi P_k\lambda)}\, dm[P_k\lambda(M)], \tag{29.99}$$

where

$$dm\,[P_k\theta(M)] = \prod_{l=1}^{n} \sqrt{\frac{V_l}{2\pi}}\, dx_l \tag{29.100}$$

[and $V_l$ is equal to the integral of $\varphi(M)$ over $\Pi_j$ for all the $s$ coordinates $x_l$ corresponding to a set $\Pi_j$]. Consider now a sequence of projections $P_k$ such that $\lim_{k\to\infty} P_k\theta(M) = \theta(M)$ for all functions $\theta(M)$ which are involved in the closure of the set of step functions. If we then pass to the limit in Eq. (29.99) as $k \to \infty$, we obtain the expression for the integral over $d\delta[\theta(M);\, \theta_0(M)]$ in the form of the limit of a sequence of finite-dimensional integrals. Symbolically, this limiting transition can be denoted in the form

$$d\delta\,[\theta(M);\, \theta_0(M)] = dm\,[\theta(M)] \int e^{i((\theta-\theta_0)\cdot\varphi\lambda)}\, dm\,[\lambda(M)] \tag{29.101}$$

[see Novikov (1961d)].

Let us now consider a special class of generalized measures which is of particular interest for our purposes. Consider the equation describing the time variation of the functional $\Phi[\theta(x), t]$, i.e., the *evolution equation* for $\Phi$, which is of the form

$$\frac{\partial\Phi\,[\theta(x), t]}{\partial t} = \mathcal{L}\Phi\,[\theta(x), t], \tag{29.102}$$

where $\mathcal{L}$ is a linear operator in the space of the functionals $\Phi$. We shall suppose that Eq. (29.102) has a unique solution corresponding to a given initial condition $\Phi[\theta_0(x), t_0] = \Phi_0[\theta_0(x)]$. In that case, since the equation is linear, the solution $\Phi[\theta(x), t]$ will depend linearly on the initial functional $\Phi_0[\theta_0(x)]$, so that the quantity $\Phi[\theta(x), t]$ will be a linear functional of $\Phi_0[\theta_0]$ for any given $\theta(x)$ and $t > t_0$. Consequently, we can associate a generalized measure $\mu[S, t_0;\, \theta(x), t]$ defined by

$$\Phi\,[\theta(x), t] = \int \Phi\,[\theta_0(x), t_0]\, d\mu\,[\theta_0(x), t_0;\, \theta(x), t]. \tag{29.103}$$

with the solution $\Phi[\theta, t]$ of Eq. (29.102) for given $\theta = \theta(x)$ and $t > t_0$. This generalized measure is naturally referred to as *Green's measure* of Eq. (29.102). It must have the following obvious properties:

$$\mu\,[S, t_0;\, \theta(x), t_0] = \delta\,[S;\, \theta(x)], \tag{29.104}$$

$$\int \mu\,[S, t_0;\, \theta_1(x), t_1]\, d\mu\,[\theta_1(x), t_1;\, \theta_2(x), t_2] = \mu\,[S, t_0;\, \theta_2(x), t_2], \quad t_2 > t_1 > t_0, \tag{29.105}$$

and for given $S$ it will be a solution of Eq. (29.102) satisfying the initial condition (29.104). If the operator $\mathcal{L} = \mathcal{L}_\theta$ in Eq. (29.102) is independent of $t$, then Green's measure will be of the form $\mu[S, t_0;\, \theta(x), t] = \mu[S;\, \theta(x), t - t_0]$. In this case, if we repeatedly apply Eq. (29.105) we obtain

$$d\mu\,[\theta_0(x);\, \theta(x), t] = \int_{\theta_1} \cdots \int_{\theta_{n-1}} \prod_{k=1}^{n} d\mu\,[\theta_{k-1}(x);\, \theta_k(x), \Delta t], \tag{29.106}$$

where $\Delta t = t/n$ and $\theta_n(x) = \theta(x)$. However, by Eqs. (29.102) and (29.104) the value of Green's measure for an infinitesimal time interval $\Delta t$ can be written in the form

$$\mu[S; \theta(x), \Delta t] = [1 + \Delta t \cdot \mathscr{L}_\theta]\, \delta[S; \theta(x)] + o(\Delta t); \tag{29.107}$$

Therefore, Eq. (29.106) can be rewritten in the form of the following symbolic expression for Green's measure (corresponding to the operator $\mathscr{L} = \mathscr{L}_\theta$ which is independent of $t$) in terms of the delta-measure:

$$d\mu[\theta_0(x); \theta(x), t] = \lim_{n\to\infty} \int\limits_{\theta_1} \cdots \int\limits_{\theta_{n-1}} \prod_{k=1}^{n} \left(1 + \frac{t}{n}\mathscr{L}_{\theta_k}\right) d\delta[\theta_{k-1}(x); \theta(x)]. \tag{29.108}$$

The above formulas can now be applied directly to the Hopf equation (28.19), which is a special case of the evolution equation. In this special case,

$$\mathscr{L}_\theta = i(\theta \cdot A[-i\mathscr{D}_\theta]), \tag{29.109}$$

where $A[u]$ is the integro-differential operator on the right-hand side of the Navier-Stokes equation $\frac{\partial u}{\partial t} = A[u]$ (after the pressure has been eliminated with the aid of the continuity equation). If we now use the symbolic expression (29.101) for the delta-measure, and assume that the operator $A[-i\mathscr{D}_\theta]$ can be taken under the integral sign over $dm[\lambda(x)]$, we obtain

$$\begin{aligned}\left(1 + \frac{t}{n}\mathscr{L}_{\theta_k}\right) d\delta[\theta_{k-1}(x); \theta_k(x)] = \\ = dm[\theta_{k-1}(x)] \int \left\{1 + \frac{it}{n}(\theta_k \cdot A[\varphi\lambda_k])\right\} e^{i((\theta_k - \theta_{k-1})\cdot\varphi\lambda_k)}\, dm[\lambda_k(x)] = \\ = dm[\theta_{k-1}(x)] \int e^{i((\theta_k-\theta_{k-1})\cdot\varphi\lambda_k) + \frac{it}{n}(\theta_k \cdot A[\varphi\lambda_k])}\, dm[\lambda_k(x)] + o\left(\frac{1}{n}\right).\end{aligned} \tag{29.110}$$

It will be convenient to introduce now the functions $\eta_n(x, \tau)$ and $\zeta_n(x, \tau)$ which have the values $\eta_n(x, \tau_k) = \theta_k(x)$ and $\zeta_n(x, \tau_k) = \lambda_k(x)$ at the points $\tau_k = kt/n$ for fixed $n$. If we then substitute Eq. (29.110) into Eq. (29.108), we can reduce Eq. (29.108) to the form

$$d\mu[\theta_0(x); \theta(x), t] = K[\theta_0(x); \theta(x), t]\, dm[\theta_0(x)], \tag{29.111}$$

where

$$\begin{aligned}K[\theta_0(x); \theta(x), t] = \lim_{n\to\infty} \int \cdots \int \prod_{k=1}^{n-1} dm[\eta_n(x, \tau_k)] \times \\ \times \int \cdots \int e^{i\Delta\tau \sum_{k=1}^{n} \left\{\left(\frac{\eta_n(x,\tau_k) - \eta_n(x,\tau_{k-1})}{\Delta\tau}\varphi\zeta_n(x,\tau_k)\right) + (\eta_n(x,\tau_k) A[\varphi\zeta_n(x,\tau_k)])\right\}} \times \\ \times \prod_{k=1}^{n} dm[\zeta_n(x, \tau_k)]\end{aligned} \tag{29.111'}$$

The above limit of the sequence of finite-dimensional integrals can be written symbolically in the form of the double functional integral

$$K[\theta_0(\boldsymbol{x}); \theta(\boldsymbol{x}), t] = \int\limits_{\eta(\boldsymbol{x}, 0)=\theta_0(\boldsymbol{x})}^{\eta(\boldsymbol{x}, t)=\theta(\boldsymbol{x})} dm[\eta(\boldsymbol{x}, \tau)] \int e^{i\int_0^t d\tau\left\{\left(\frac{\partial\eta}{\partial\tau}\cdot\varphi\zeta\right)+(\eta\cdot A[\varphi\zeta])\right\}} dm[\zeta(\boldsymbol{x}, \tau)]. \quad (29.112)$$

We can now use Eq. (29.103) to obtain the corresponding symbolic formula for the solution $\Phi[\theta(\boldsymbol{x}), t]$ of the Hopf equation subject to the initial condition $\Phi[\theta(\boldsymbol{x}), 0] = \Phi_0[\theta(\boldsymbol{x})]$:

$$\Phi[\theta(\boldsymbol{x}), t] = \int^{\eta(\boldsymbol{x}, t)=\theta(\boldsymbol{x})} \Phi_0[\eta(\boldsymbol{x}, 0)]\, dm[\eta(\boldsymbol{x}, \tau)] \int e^{i\int_0^t d\tau\left\{\left(\frac{\partial\eta}{\partial\tau}\cdot\varphi\zeta\right)+(\eta\cdot A[\varphi\zeta])\right\}} \times$$
$$\times\, dm[\zeta(\boldsymbol{x}, \tau)], \quad (29.113)$$

where the outer integral is evaluated over the functions $\eta(\boldsymbol{x}, \tau)$ satisfying the single boundary condition $\eta(\boldsymbol{x}, t) = \theta(\boldsymbol{x})$. This formula was obtained by Rosen (1960). It is valid for the spectral representation of the characteristic functional provided we replace $\boldsymbol{x}$ by $\boldsymbol{k}$, $\theta(\boldsymbol{x})$ by $z(\boldsymbol{k})$, $\Phi[\theta(\boldsymbol{x}), t]$ by $\Psi[z(\boldsymbol{k}), t]$, and define $A$ so that $\frac{\partial\Psi}{\partial t} = i(z\cdot A[-iD_z])\Psi$ becomes identical with the Hopf equation (28.38). If we then set

$$\eta(\boldsymbol{k}, \tau) = z(\boldsymbol{k})e^{-\nu k^2(t-\tau)} + \int_\tau^t e^{-\nu k^2(\tau'-\tau)}\xi(\boldsymbol{k}, \tau')\,d\tau' \quad (29.114)$$

and transform from integration with respect to $\eta$ to integration with respect to $\xi = \nu k^2\eta - \frac{\partial\eta}{\partial\tau}$, we shall reduce Eq. (29.113) to the form obtained by Tatarskii (1962b) by a completely different method [a method analogous to the "ordering" of the $S$ matrix in quantum field theory and applied to the Hopf equation in the "interaction representation" (28.50)].

We shall now derive formulas for the moments of the velocity field which are analogous to the symbolic formula (29.113). If we express these moments in terms of the functional derivatives of the characteristic functional $\Phi[\theta(\boldsymbol{x}), t]$ in accordance with Eq. (4.51) of Vol. 1, and use Eq. (29.113) for $\Phi[\theta(\boldsymbol{x}), t]$ with the differential of the Green measure given by Eq. (29.111), we obtain

$$\overline{u_{j_1}(\boldsymbol{x}_1, t)\ldots u_{j_n}(\boldsymbol{x}_n, t)} = (-i)^n\left\{\mathcal{D}_{j_1}(\boldsymbol{x}_1)\ldots\mathcal{D}_{j_n}(\boldsymbol{x}_n)\,\Phi[\theta(\boldsymbol{x}), t]\right\}\Big|_{\theta(\boldsymbol{x})=0} =$$
$$= (-i)^n\int\Phi_0[\theta_0(\boldsymbol{x})]\,dm[\theta_0(\boldsymbol{x})]\left\{\mathcal{D}_{j_1}(\boldsymbol{x}_1)\ldots\mathcal{D}_{j_n}(\boldsymbol{x}_n)K[\theta_0(\boldsymbol{x}); \theta(\boldsymbol{x}), t]\right\}\Big|_{\theta(\boldsymbol{x})=0}.$$
$$(29.115)$$

Let us first evaluate the first-order functional derivative $\mathcal{D}_{j_1}(\boldsymbol{x}_1)K$. To do this, we shall use Eq. (29.112) to calculate $\delta K = K[\theta_0(\boldsymbol{x}); \theta(\boldsymbol{x}) + \delta\theta(\boldsymbol{x}), t] - K[\theta_0(\boldsymbol{x}); \theta(\boldsymbol{x}), t]$. If we transform in this formula from integration with respect to $\eta(x, \tau)$ to integration with respect to $\eta(x, \tau)+\omega(x, \tau)$, where $\omega(x, \tau)$ is a given infinitesimally small vector function such that $\omega(x, 0)=0$, $\omega(x, t)=\delta\theta(x)$, we obtain

$$K[\theta_0(x); \theta(x)+\delta\theta(x), t]=$$

$$=\int\limits_{\eta(x,0)+\omega(x,0)=\theta_0(x)}^{\eta(x,t)+\omega(x,t)=\theta(x)+\delta\theta(x)} dm[\eta(x, \tau)+\omega(x, \tau)]\times$$

$$\times\int \exp\left\{i\int\limits_0^t d\tau\left[\left(\frac{\partial(\eta+\omega)}{\partial\tau}\cdot\varphi\zeta\right)+((\eta+\omega)\cdot A[\varphi\zeta])\right]\right\} dm[\zeta(x, t)]=$$

$$=\int\limits_{\eta(x,0)=\theta_0(x)}^{\eta(x,t)=\theta(x)} dm[\eta(x, \tau)]\int\exp\left\{i\int\limits_0^t d\tau\left\{\left(\frac{\partial(\eta+\omega)}{\partial\tau}\cdot\varphi\zeta\right)+((\eta+\omega)\cdot A[\varphi\zeta])\right\}\right\}\times$$

$$\times\, dm[\zeta(x, \tau)], \qquad (29.116)$$

where we have used the fact that $dm[\eta(x, \tau)+\omega(x, \tau)]=dm[\eta(x, \tau)]$ (invariance of the "measure" $dm$ under "translations" in functional space). The quantity $\delta K$ differs from that given by Eq. (29.116) by replacement of the exponential in the integrand by the difference

$$\exp\left\{i\int\limits_0^t d\tau\left\{\left(\frac{\partial(\eta+\omega)}{\partial\tau}\cdot\varphi\zeta\right)+((\eta+\omega)A[\varphi\zeta])\right\}\right\}-$$

$$-\exp\left\{i\int\limits_0^t d\tau\left\{\left(\frac{\partial\eta}{\partial\tau}\cdot\varphi\zeta\right)+(\eta\cdot A[\varphi\zeta])\right\}\right\}\approx i\exp\left\{i\int\limits_0^t d\tau\left[\left(\frac{\partial\eta}{\partial\tau}\cdot\varphi\zeta\right)+\right.\right.$$

$$\left.\left.+(\eta\cdot A[\varphi\zeta])\right]\right\}\left\{(\delta\theta\cdot\varphi\zeta(t))-\int\limits_0^t d\tau\left(\omega\cdot\left(\frac{\partial\varphi\zeta}{\partial\tau}-A[\varphi\zeta]\right)\right)\right\},$$

where we have retained only those terms which are of the first order in small quantities in $\omega(x, \tau)$. It is readily verified that the last term in the braces which contains the difference $\frac{\partial\varphi\zeta}{\partial\tau}-A[\varphi\zeta]$ does not contribute to $\delta K$ (it is sufficient to calculate the quantity $(\mathcal{D}_{\omega_j}K)\big|_{\omega=0}$ with $K$ given by Eq. (29.116), where $\omega(x, 0)=\omega(x, t)=0$, and take into account the fact that this quantity must be zero). We recall that, by definition of the functional derivative $\delta K=(\delta\theta\cdot\mathcal{D}_\theta K)$ and we have

$$\mathcal{D}_{j_1}(x_1)K=i\int\limits_{\eta(x,0)=\theta_0(x)}^{\eta(x,t)=\theta(x)} dm[\eta(x, \tau)]\int\varphi(x_1)\zeta_{j_1}(x_1, t)\times$$

$$\times\exp\left\{i\int\limits_0^t d\tau\left\{\frac{\partial\eta}{\partial\tau}\cdot\varphi\zeta\right)+(\eta\cdot A[\varphi\zeta])\right\}\right\} dm[\zeta(x, \tau)]. \qquad (29.117)$$

Therefore, functional differentiation of $K$ with respect to $\theta_{j_1}(x_1)$ leads to the appearance of the factor $i\varphi(x_1)\zeta_{j_1}(x_1, t)$ in the integrand. Analogous factors will appear after each successive differentiation so that, eventually, Eq. (29.115) will assume the form

$$\overline{\prod_{k=1}^{n} u_{j_k}(x_k, t)} = \int^{\eta(x, t)=0} \Phi_0[\eta(x, 0)]\, dm[\eta(x, \tau)] \int \prod_{k=1}^{n} \varphi(x_k)\zeta_{j_k}(x_k, t) \times$$
$$\times \exp\left\{i\int_0^t d\tau \left\{\left(\frac{\partial\eta}{\partial\tau}\cdot\varphi\zeta\right) + (\eta\cdot A[\varphi\zeta])\right\}\right\} dm[\zeta(x, \tau)]. \qquad (29.118)$$

The outermost integral in this expression is evaluated over the functions $\eta(x, \tau)$ satisfying the boundary condition $\eta(x, t) = 0$. This formula was also established by Rosen.

The inner integrals [with respect to $dm[\zeta(x, \tau)]$] in Eqs. (29.113) and (29.118) can be substantially simplified with the aid of a method used, for example, by Feynman (1951). Let us use the fact that the operator $A[u]$ on the right-hand side of the Navier-Stokes equation $\frac{\partial u}{\partial t} = A[u]$ can be written in the form

$$A[u(x)] = \nu\Delta u(x) - \int \mathscr{D}_{\alpha\beta}(x - x')\, u_\alpha(x')\, u_\beta(x')\, dx', \qquad (29.119)$$

where $\mathscr{D}_{\alpha\beta}(x) = \mathscr{D}_{\beta\alpha}(x)$. Substituting

$$a(x, \tau) = \frac{1}{2}\left[\frac{\partial\eta(x, \tau)}{\partial\tau} + \nu\Delta\eta(x, \tau)\right],$$
$$B_{\alpha\beta}(x', \tau) = \int \eta(x, \tau)\cdot\mathscr{D}_{\alpha\beta}(x - x')\, dx \qquad (29.120)$$

and using the relation $(\eta\cdot\Delta\varphi\zeta) = (\Delta\eta\cdot\varphi\zeta)$, we can write the inner integral in Eq. (29.113) in the form

$$I = \int \exp\left\{i\int_0^t d\tau\left\{2(a\cdot\varphi\zeta) - \int B_{\alpha\beta}\varphi\zeta_\alpha\varphi\zeta_\beta\, dx\right\}\right\} dm[\zeta(x, \tau)]. \qquad (29.121)$$

It is now clear that the inner integral in Eq. (29.118) can be obtained by $n$-fold functional differentiation of $I$ with respect to the components of the function $a(x, \tau)$. The method mentioned above, whereby these integrals can be simplified, involves the replacement of $\zeta$ in the integrand by $\zeta + \zeta_0$ (which is justified because the "measure" $dm$ is invariant under translations), and with $\zeta_0$ chosen so that the argument of the exponential does not contain the first power of $\zeta$. It is readily verified that this requires that $\varphi\zeta_{0j} = B^{-1}_{j\alpha}a_\alpha$, and that the formula (29.121) then takes the form

$$I = I_0 \exp\left\{i\int_0^t d\tau \int B^{-1}_{\alpha\beta}a_\alpha a_\beta\, dx\right\}; \qquad (29.122)$$

$$I_0 = \int \exp\left\{ -i \int_0^t d\tau \int B_{\alpha\beta}\varphi\zeta_\alpha\varphi\zeta_\beta \, dx \right\} dm\, [\zeta(x, \tau)]. \qquad \text{(Cont.)} \quad (29.122)$$

It is clear from Eq. (29.120) that $I_0$ is a functional of the function $\eta(x, \tau)$ and is not constant as assumed by Rosen. The integral $I_0$ can also be evaluated (see the above cited papers by Feynman and Tatarskii), but the final result is quite complicated, and we shall not reproduce it here.

The representation of the characteristic space-time functional $\Phi[\theta(x, t)]$ [i.e., of the general solution of the space-time functional derivative equation (28.57)–(28.58)] in the form of a double functional integral was given by Rosen (1970, 1971). The representation contains an arbitrary real functional and, an auxiliary characteristic functional which can be selected in a form making possible the explicit evaluation of one of the two functional integrals entering the expression for $\Phi[\theta(x, t)]$. The final result is nevertheless rather cumbersome and will be omitted in this book.

In conclusion, we shall show that, following Novikov's idea, the representation of the spatial characteristic function of the velocity field in the form of a functional integral can be obtained with the aid of the symbolic formula (29.101) without using the Hopf equation. For the sake of generality, we shall suppose that the field of external forces is present, i.e., we shall consider the Navier-Stokes equation in the form $\frac{\partial u}{\partial t} = A[u] + X$. For $t > t_0$ we then have

$$u(x, t) = u_0(x) + \int_{t_0}^{t-0} A[u(x, \tau)]\, d\tau + \int_{t_0}^{t} X(x, \tau)\, d\tau, \qquad (29.123)$$

where $u_0(x) = u(x, t_0)$ is the initial velocity field.[5]

Consider the following functional of $u(x, \tau)$;

$$\exp\left\{ i\left(\theta \cdot \left(u_0 + \int_{t_0}^{t-0} A[u]\, d\tau + \int_{t_0}^{t} X\, d\tau\right)\right)\right\} =$$

$$= \int \exp\left\{ i\left(\theta \cdot \left(u_0 + \int_{t_0}^{t-0} A[v]\, d\tau + \int_{t_0}^{t} X\, d\tau\right)\right)\right\} dm\,[v(x, \tau)] \times$$

$$\times \int \exp\left\{ i \int_{t_0}^{t} d\tau\, ((u - v) \cdot \varphi w) \right\} dm\,[w(x, \tau)]. \qquad (29.124)$$

We have used the formulas (29.97) and (29.101) in which the role of $M$ is played by the space-time point $(x, t)$ and the role of $\theta_0$, $\theta$, and $\lambda$ by the functions $u$, $v$, and $w$. From Eqs. (29.99) and (29.101) it is clear that the integral with respect to $m[v(x, \tau)]$ includes contributions due only to the function $v = u$. Using this equality on the right-hand side of Eq. (29.123), and substituting the resulting value

[5] The symbol $t - 0$ in Eq. (29.123) represents the fact that, when we transform to step functions, the last step in time is not taken into account in the corresponding integral.

$$u(x, \tau) = u_0(x) + \int_{t_0}^{\tau-0} A[v(x, \tau')]\, d\tau' + \int_{t_0}^{\tau} X(x, \tau')\, d\tau' \qquad (29.125)$$

into the inner integral in Eq. (29.124) we obtain, after integration by parts

$$e^{i(\theta \cdot u)} = \int\int \exp\left\{i\left(u_0 \cdot \left(\theta + \varphi \int_{t_0}^{t} w\, d\tau\right)\right)\right\} \exp\left\{i \int_{t_0}^{t} d\tau \left(X \cdot \left(\theta + \varphi \int_{\tau}^{t} w\, d\tau'\right)\right)\right\} \times$$

$$\times \exp\left\{i \int_{t_0}^{t} d\tau \left\{\left(A[v] \cdot \left(\theta + \varphi \int_{\tau+0}^{t} w\, d\tau'\right)\right) - (v \cdot \varphi w)\right\}\right\} dm[v]\, dm[w].$$

(29.126)

If we take the average of both sides of the equation (under the double integral sign on the right-hand side) and assume that the initial velocity field $u_0(x)$ and the external force field $X(x, \tau)$, $\tau > t_0$ are statistically independent, we obtain

$$\Phi[\theta(x), t] = \int\int \Phi_0\left[\theta + \varphi \int_{t_0}^{t} w\, d\tau\right] F\left[\theta + \varphi \int_{\tau}^{t} w\, d\tau'\right] \times$$

$$\times \exp\left\{i \int_{t_0}^{t} d\tau \left[\left(A[v] \cdot \left(\theta + \varphi \int_{\tau+0}^{t} w\, d\tau'\right)\right) - (v \cdot \varphi w)\right]\right\} dm[v]\, dm[w].$$

(29.127)

where $F[\eta(x, \tau)]$ is the space-time characteristic functional of the external force field. In the absence of external forces, the functional $F$ must be replaced by unity. In that case, after replacement of $\theta + \varphi \int_{\tau}^{t} w\, d\tau'$ by $\eta(x, \tau)$ and $v$ by $\varphi\xi$, we again obtain Eq. (29.113). Conversely, in the presence of statistically stationary external forces we obtain the characteristic functional for the associated statistically stationary velocity field by substituting $\Phi_0 = 1$, $t_0 \to -\infty$ and $t \to \infty$ in Eq. (29.127).

In conclusion, we emphasize once again that a precise justification of the formulas given in this subsection for the characteristic functional of the velocity field, and the precise significance of these formulas, is a matter for the future. For the moment the formulas merely indicate one of the possible directions of further investigation.

# BIBLIOGRAPHY


Alekseev, V. G., and A. M. Yaglom, (1967) Examples of a comparison between one-dimensional and three-dimensional velocity and temperature spectra, *Izv. Akad. Nauk SSSR*, Fiz. Atmosf. i Okeana, **3**, No. 8, 902–907.

Artem'ev, A. V., and A. S. Gurvich, (1971) Experimental investigation of coherence function spectra, *Izv. Vyssh. Uchebn. Zaved., Radiofizika*, **14**, No. 5, 734–738.

Backus, G., (1957) The existence and uniqueness of the velocity correlation derivative in Chandrasekhar's theory of turbulence, *J. Math. Mech.*, **6**, No. 2, 215–233.

Ball, F. K., (1961) Viscous dissipation in the atmosphere, *J. Meteor.*, **18**, No. 4, 553–557.

Barabanenkov, Yu. N., Yu. A. Kravtsov, S. M. Rytov, and V. I. Tatarskii, (1970) Theory of wave propagation in a randomly inhomogeneous medium, *Uspechi Fiz. Nauk (Adv. Phys. Sci.)*, **102**, No. 1, 3–42.

Bass, J., (1949) Sur les bases mathématiques de la théorie de la turbulence d'Heisenberg, *Compt. Rend.*, **228**, No. 3, 228–229; (1953) Sur les équations fonctionelles des fluides turbulents, *Compt. Rend.*, **237**, No. 13, 645–647.

Batchelor, G. K., (1946) The theory of axisymmetric turbulence, *Proc. Roy. Soc. (London)*, **A186**, No. 1007, 480–502; (1947) Kolmogoroff's theory of locally isotropic turbulence, *Proc. Cambr. Phil. Soc.*, **43**, No. 4, 533–559; (1948) Energy decay and self-preserving correlation functions in isotropic turbulence, *Quart. Appl. Math.*, **6**, No. 2, 97–116; (1949a) The role of big eddies in homogeneous turbulence, *Proc. Roy. Soc. (London)*, **A195**, No. 1043, 513–532; (1949b) Diffusion in a field of homogeneous turbulence, *Austr. J. Sci. Res.*, **A2**, No. 4, 437–450; (1950) The application of the similarity theory of turbulence to atmospheric diffusion, *Quart. J. Roy. Meteor. Soc.*, **76**, No. 328, 133–146; (1951) Pressure fluctuations in isotropic turbulence, *Proc. Cambr. Phil. Soc.*, **47**, No. 2, 359–374; (1952a) Diffusion in a field of homogeneous turbulence. II The relative motion of particles, *Proc. Cambr. Phil. Soc.*, **48**, No. 2, 345–362; (1952b) The effect of homogeneous turbulence on material lines and surfaces, *Proc. Roy. Soc. (London)*, **A213**, No. 1114, 349–366; (1953) The theory of homogeneous turbulence, The University Press, Cambridge; (1955) The scattering of radio waves in the atmosphere by turbulent fluctuations in refractive index, *Res.*

*Rep. No. 262,* School of Electrical Eng., Cornell Univ., USA; (1959) Small-scale variation of convected quantities like temperature in turbulent fluid. Part 1. General discussion and the case of small conductivity, *J. Fluid Mech.,* **5**, No. 1, 113–133; (1962) Discussion de la section: Transfert d'énergie en turbulence homogène, *Mécanique de la turbulence* (Coll. Intern. du CNRS à Marseille), Paris, Ed. CNRS, 123–126.

Batchelor, G. K., I. D. Howells, and A. A. Townsend, (1959) Small-scale variation of convected quantities like temperature in turbulent fluid. Part 2. The case of large conductivity, *J. Fluid Mech.*, **5**, No. 1, 134–139.

Batchelor, G. K., and I. Proudman (1956) The large-scale structure of homogeneous turbulence, *Phil. Trans. Roy. Soc.*, **A248**, No. 949, 369–405.

Batchelor, G. K., and R. W. Stewart (1950) Anisotropy of the spectrum of turbulence at small wave numbers, *Quart. J. Mech. Appl. Math.*, **3**, No. 1, 1–8.

Batchelor, G. K., and A. A. Townsend, (1947) Decay of vorticity in isotropic turbulence, *Proc. Roy. Soc. (London),* **A191**, No. 1021, 534–550; (1948a) Decay of isotropic turbulence in the initial period, *Proc. Roy. Soc. (London),* **A193**, No. 1035, 539–558; (1948b) Decay of turbulence in the final period, *Proc. Roy. Soc. (London),* **A194**, No. 1039, 527–543; (1949) The nature of turbulent motion at large wave-numbers, *Proc. Roy. Soc. (London),* **A199**, No. 1057, 238–255; (1956) Turbulent diffusion, *Surveys in Mechanics* (G. K. Batchelor and R. M. Davies, eds.), The University Press, Cambridge, 352–399.

Bean, B. R., C. B. Emmanuel, and R. W. Krinks, (1967) Some spectral characteristics of the radio refractivity in the surface layer of the atmosphere, *Radio Science,* **2**, No. 5, 503–510.

Becker, H. A., H. C. Hottel, and G. C. Williams, (1967a) On the light-scatter technique for the study of turbulence and mixing, *J. Fluid Mech.*, **30**, No. 2, 259–284; (1967b) The nozzle-fluid concentration field of the round, turbulent, free jet, *J. Fluid Mech.*, **30**,, No. 2, 285–303.

Benton, G. S., and A. B. Kahn, (1958) Spectra of large scale atmospheric turbulence at 300 mb, *J. Meteor.*, **15**, No. 4, 404–410.

Beran, M. J., (1968) *Statistical Continuum Theories.* Interscience Publ., New York-London-Sydney.

Beran, M. J., and T. L. Ho., (1969) Propagation of the fourth-order coherence function in a random medium (a nonperturbative formulation), *J. Opt. Soc. Amer.*, **59**, No. 9, 1134–1138.

Bergmann, P. G., (1946) Propagation of radiation in a medium with random inhomogeneities, *Phys. Rev.*, **70**, No. 7–8, 486–492.

Berman, S., (1965) Estimating the longitudinal wind spectrum near the ground, *Quart. J. Roy. Meteor. Soc.*, **91**, No. 389, 302–317.

Betchov, R., (1956) An inequality concerning the production of vorticity in isotropic *turbulence, J. Fluid Mech.*, **1**, No. 5, 497–504; (1957) On the fine structure of turbulent flows, *J. Fluid Mech.*, **3**, No. 2, 205–216; (1966) Introduction to the Kraichnan theory of turbulence. *Dynamics of Fluids* *and Plasmas*, (S. I. Pai, ed.), Academic Press, New York, 215–237; (1967) Review of Kraichnan's theory of turbulence, *Phys. Fluids*, **10** (Supplement), S17–S24.

Blanch, G., and H. Ferguson, (1959) Remarks on Chandrasekhar's results relating to Heisenberg's theory of turbulence, *Phys. Fluids*, **2**, No. 1, 79–84.

Blokhintsev, D. I., (1945) Sound propagation in a turbulent flow, *Dokl. Akad. Nauk SSSR*, **46**, No. 4, 150–153; (1952) The acoustics of an inhomogeneous moving medium. Trans. by R. T. Beyer and D. Mintzer, Research Analysis Group, Physics Dept., Brown Univ., Providence, R. I. Contract N7-ONR-35808.

Bodner, S. E., (1969) Turbulence theory with a time-varying Wiener-Hermite basis, *Phys. Fluids*, **12**, No. 1, 33-38.

Bolgiano, R., (1957) Spectrum of turbulent mixing, *Phys. Rev.*, **108**, No. 5, 1348; (1958a) The role of turbulent mixing in scatter propagation, *Trans. Inst. Radio Engrs. on Antennas and Propag.*, **AP-6**, No. 2, 161–168; (1958b) On the role of convective transfer in turbulent mixing, *J. Geophys. Res.*, **63**, No. 4, 851–853; (1959) Turbulent spectra in a stably stratified atmosphere. *J. Geophys. Res.*, **64**, No. 12, 2226–2229; (1962) Structure of turbulence in stratified media, *J. Geophys. Res.*, **67**, No. 8, 3015–3023.

Booker, H. G., and R. Cohen, (1956) A theory of long-duration meteor echoes based on atmospheric turbulence with experimental confirmation, *J. Geophys. Res.*, **61**, No. 4, 707–733.

Booker, H. G., and W. E. Gordon, (1950) A theory of radio scattering in the troposphere, *Proc. Inst. Radio Engrs.*, **38**, No. 4, 401–412.

Boston, N. E. J., (1970) *An investigation of high wave number temperature and velocity spectra in air*, Ph.D. Dissertation, Univ. of British Columbia, Vancouver, Canada.

Boston, N. E. J., and R. W. Burling, (1972) An investigtion of high wave number temperature and velocity spectra in air, *J. Fluid Mech.*, **55**, No. 3, 473–492.

Bovsheverov, V. M., A. S. Gurvich, V. I. Tatarskii, and L. R. Tsvang, (1959) Instruments for the statistical analysis of turbulence. Proc. Second All-Union Conf. on Stellar Scintillation, *An SSSR Press*, Moscow, 26–32

Bowden, K. F., (1962) Turbulence. *The Sea*, (M. N. Hill, ed.), Vol. 1, Interscience Publ., New York-London, 802–826; (1965) Horizontal mixing in the sea due to a shearing current, *J. Fluid Mech.*, **21**, No. 2, 83–95.

Bowne, N. E., (1961) Some measurements of diffusion parameters from smoke plumes, *Bull. Amer. Meteor. Soc.*, **42**, No. 2, 101–105.

Bradbury, L. J. S., (1965) The structure of a self-preserving turbulent plane jet, *J. Fluid Mech.*, **23**, No. 1, 31–64.

Bradshaw, P., (1969) Conditions for the existence of an inertial subrange in turbulent flow, *Aeronaut. Res. Council*, Repts. and Mem. No. 3603, 5.

Brier, G. W., (1950) The statistical theory of turbulence and the problem of diffusion in the atmosphere, *J. Meteor.*, **7**, No. 4, 283–290.

Bull, G., (1966) Power spectra of atmospheric refractive index from microwave refractometer measurements, *J. Atmosph. Terrest. Phys.*, **28**, No. 5,

513–519; (1967) Spectra of radio refractive index. *Atmospheric Turbulence and Radio Wave Propagation* (A. M. Yaglom and V. I. Tatarskii, eds) Nauka Press, Moscow; 206–214

Burns, A. B., (1964) Power spectra of low level atmospheric turbulence measured from aircraft, *Aeronaut. Res. Council Current Papers*, No. 733.

Busch, N. E., and H. A. Panofsky, (1968) Recent spectra of atmospheric turbulence, *Quart. J. Roy. Meteor. Soc.*, **94**, No. 400, 132–148.

Businger, J. A., and V. E. Suomi, (1958) Variance spectra of the vertical wind component derived from observations with the sonic anemometer at O'Neill, Nebraska in 1953, *Arch. Meteor. Geophys. Bioklimatol.*, **A10**, No. 4, 415–425.

Butler, M. E., (1954) Observations of stellar scintillations, *Quart. J. Roy. Meteor. Soc.*, **80**, No. 344, 241–245.

Byzova, N. L. and E. K. Garger, (1970) Experimental study of diffusion parameters with the aid of smoke plumes, *Izv. Akad. Nauk SSSR*, Fiz. Atmosf. i Okeana, **6**, No. 10, 996–1006.

Byzova, N. L., V. N. Ivanov, and S. A. Morozov, (1967) Turbulent characteristics of wind velocity and temperature in the atmospheric boundary layer. *Atmospheric Turbulence and Radio Wave Propagation* (A. M. Yaglom and V. I. Tatarskii, eds.), Nauka Press, Moscow, 76–92.

Camac, M., A. R. Kantrowitz, M. M. Litvak, R. M. Patrick, H. E. Petschek, (1962) Shock-waves in collision-free plasmas, *Nucl. Fusion*, Suppl., Pt. 2.

Canavan, G. H., and C. E. Leith, (1968) Lagrangian Wiener-Hermite expansion for turbulence, *Phys. Fluids*, **11**, No. 12, 2759–2761.

Cermak, J. E., and H. Chuang, (1967) Vertical velocity fluctuations in thermally stratified shear flows. *Atmospheric Turbulence and Radio Wave Propagation* (A. M. Yaglom and V. I. Tatarskii, eds) Nauka Press, Moscow, 93–104

Chamberlain, J. W., and P. H. Roberts, (1955) Turbulence spectrum in Chandrasekhar's theory, *Phys. Rev.*, **99**, No. 6, 1674–1677.

Champagne, F. H., V. G. Harris, and S. Corrsin, (1970) Experiments on nearly homogeneous turbulent shear flow, *J. Fluid Mech.*, **41**, No. 1, 81–139.

Champagne, F. H., and C. A. Sleicher, (1967) Turbulence measurements with inclined hot-wires. Part 2. Hot-wire response equations, *J. Fluid Mech.*, **28**, No. 1, 177–182; (1968) (Corrigenda), **31**, No. 2, 416.

Chandrasekhar, S., (1949) On Heisenberg's elementary theory of turbulence, *Proc. Roy. Soc.*, **A200**, No. 1060, 20–33; (1950) The theory of axisymmetric turbulence, *Phil. Trans. Roy. Soc.*, **A242**, No. 855, 557–577; (1951) Density fluctuations in isotropic turbulence, *Proc. Roy. Soc.*, **A210**, No. 1100, 18–24; (1955) A theory of turbulence, *Proc. Roy. Soc.*, **A229**, No. 1175, 1–19; (1956) Theory of turbulence, *Phys. Rev.*, **102**, No. 4, 941–952.

Charnock, H., (1951) Note on eddy diffusion in the atmosphere between one and two kilometers, *Quart. J. Roy. Meteor. Soc.*, **77**, No. 334, 654–658.

Chen, H. -S., and Y. Mitsuta, (1967) An infrared absorption hygrometer and its applications to the study of the water vapor flux near the ground. Special Contributions, *Geophys. Inst. Kyoto Univ.*, No. 7, 83–94.

Chernov, L. A., (1960) *Wave Propagation in a Random Medium,* McGraw-Hill, New York; (1969) Equations for statistical moments of a field in a randomly inhomogeneous medium, *Akust. Zh. (Acoustical Journal)*, **15**, No. 4, 594–603.

Chou, Pei-Yuan, (1940) On an extension of Reynolds method of finding apparent stress and the nature of turbulence, *Chin. J. Phys.*, **4**, No. 1, 1–33.

Chou, Pei-Yuan, and Tsai Shu-Tang, (1957) Vortex motion and the vorticity structure of homogeneous isotropic turbulence in its final period of decay, *Actes IX Congr. Internat. Mécan. Appl.*, **3**, Bruxelles, Univ. Bruxelles, 257–268.

Chisholm, J. H., P. A. Portmann, J. T. de Bettencourt, and J. E. Roche, (1955) Investigations of angular scattering and multipath properties of tropospheric propagation of short radio waves beyond the horizon, *Proc. Inst. Radio Engrs,* **43**, No. 10, 1317–1335.

Clay, J. P. (1973) *Turbulent mixing of temperature in water, air and mercury,* Ph. D. Dissertation, Engineering Physics, University of California, San Diego.

Clever, W. C., and W. C. Meecham, (1970) Comments on "Symmetry properties of Cameron-Martin-Wiener kernels," *Phys. Fluids,* **13**, No. 12, 3066–3067.

Cocke, W. Y., (1969) Turbulent hydrodynamic line stretching: Consequences of isotropy, *Phys. Fluids,* **12**, No. 12, Part 1, 2488–2492; (1971) Turbulent hydrodynamic line-stretching: The random walk limit, *Phys. Fluids,* **14**, No. 8, 1624–1628.

Collis, D. C., (1948) The diffusion process in turbulent flow, *Austr. Council Sci. Ind. Res.*, Div. Aero., Rep. A55.

Comte-Bellot, G., (1965) Ecoulement turbulent entre deux parois parallèles, *Publ. Sci. Techn. Min. Air,* No. 419, Paris.

Comte-Bellot, G., and S. Corrsin, (1966) The use of a contraction to improve the isotropy of grid-generated turbulence, *J. Fluid Mech.*, **25**, No. 4, 657–682; (1971) Simple Eulerian time correlation of full- and narrow-band velocity signals in grid-generated "isotropic" turbulence, *J. Fluid Mech.*, **48**, No. 2, 273–337.

Corrsin, S., (1949) An experimental verification of local isotropy, *J. Aeronaut. Sci.*, **16**, No. 12, 757–758; (1951a) The decay of isotropic temperature fluctuations in an isotropic turbulence, *J. Aeronaut. Sci.*, **18**, No. 6, 417–423; (1951b) On the spectrum of isotropic temperature fluctuations in an isotropic turbulence, *J. Appl. Phys.*, **22**, No. 4, 469–473; (1957) Some current problems in turbulent shear flows, Chapt. 15, Proc. 1st Symp. on Naval Hydro., *Nat. Acad. Sci., Nat. Res. Council,* 373–407; (1958a) Statistical behavior of a reacting mixture in isotropic turbulence, *Phys. Fluids,* **1**, No. 1, 42–47; (1958b) Local isotropy in turbulent shear flow, Nat. Adv. Com. Aeronaut., *Res. Memo.* RM 58B11; (1959) Outline of some topics in homogeneous turbulent flow, *J. Geophys. Res.*, **64**, No. 12, 2134–2150; (1961) Reactant concentration spectrum in turbulent mixing with a first-order reaction, *J. Fluid Mech.*, **11**, No. 3, 407–416; (1962a) Theories of turbulent dispersion, *Mécanique de la turbulence* (Coll. Intern.

du CNRS à Marseille), Paris, Ed. CNRS, 27–52; (1962b) Some statistical properties of the product of a turbulent, first-order reaction, *Proc. 1961 Symp. on Fluid Dyn. and Appl. Math.*, N. Y., Gordon and Breach, 105–124; (1962c) Turbulent dissipation fluctuations, *Phys. Fluids*, **5**, No. 10, 1301–1302; (1964) Further generalization of Onsager's cascade model for turbulent spectra, *Phys. Fluids*, **7**, No. 8, 1156–1159; (1972) Simple proof of fluid line growth in stationary homogeneous turbulence, *Phys. Fluids*, **15**, No. 8, 1370–1372.

Cramér, H., (1940) On the theory of stationary random processes, *Ann. Math.*, **41**, No. 1, 215–230; (1942) On harmonic analysis in certain functional spaces, *Ark. Mat. Astr. Fys.*, **28B**, No. 12, 1–17; (1946) Mathematical methods of statistics, University Press, Princeton, N. J.

Cramer, H. E., (1959) Measurements of turbulence structure near the ground within the frequency range from 0.5 to 0.01 cycles $sec^{-1}$, *Adv. Geophys.*, **6**, (Atmospheric diffusion and air pollution), 75–96.

Crow, S. C., and G. H. Canavan, (1970) Relationship between a Wiener–Hermite expansion and an energy cascade, *J. Fluid Mech.*, **41**, No. 2, 387–403.

Csanady, G. T., (1963) Turbulent diffusion in Lake Huron, *J. Fluid Mech.*, **17**, No. 3, 360–38; (1966) Accelerated diffusion in the skewed shear flow of lake currents, *J. Geophys. Res.*, **71**, No. 2, 411–420.

Curle, N., (1955) The influence of solid boundaries upon aerodynamic sound, *Proc. Roy. Soc., (London)*, **A231**, No. 1187, 505–514.

Dagkesamanskaya, I. M., and V. I. Shishov, (1970) Strong intensity fluctuations in wave propagation in a statistically homogeneous and isotropic medium, *Izv. Vyssh. Uchebn. Zaved., Radiofizika*, **13**, No. 1, 16–20.

Daletskii, Yu. L., (1962) Functional integrals related to evolutionary operator equations, *Ups. Mat. Nauk*, **17**, No. 5(107), 3–115.

Davenport, A. G., (1961) The spectrum of horizontal gustiness near the ground in high winds, *Quart. J. Roy. Meteor. Soc.*, **87**, No. 372, 194–211.

Deacon, E. L., (1959) The measurements of turbulent transfer in the lower atmosphere, *Adv. Geophys.*, **6** (Atmospheric diffusion and air pollution), 211–228.

Defant, A., (1954) Turbulenz und Vermischung im Meer, *Dtsch. hydrogr. Zs.*, **7**, No. 1–2, 2–14.

Deissler, R. G., (1958) On the decay of homogeneous turbulence, *Phys. Fluids*, **1**, No. 2, 111–121; (1960) A theory of decaying homogeneous turbulence, *Phys. Fluids*, **3**, No. 2, 176–187; (1961) Analysis of multipoint-multitime correlations and diffusion in decaying homogeneous turbulence, *Nat. Aeronaut. Space Adm.*, Tech. Rep. R-96; (1965) Some remarks on the approximations for moderately weak turbulence, *Phys. Fluids*, **8**, No. 11, 2106–2107.

Dikii, L. A., (1964) Free oscillation frequencies in the atmosphere, *Dokl. Akad. Nauk SSSR*, **157**, No. 3, 580–582; (1965) Earth's atmosphere as an oscillatory system, *Izv. Akad. Nauk SSSR*, Fiz. Atmosf. i Okeana, **1**, No. 5, 469–489.

Dolin, L. S., (1964) On the scattering of a light beam in a layer of turbid medium, *Izv. Vyssh. Uchebn. Zaved., Radiofizika,* 7, No. 2, 380–383; (1968) Equations for correlation functions of a wave beam in a chaotically inhomogeneous medium, *Izv. Vyssh. Uchebn. Zaved. Radiofizika,* **11**, No. 6, 840–849.

Donsker, M. D., (1964) On function space integrals. *Analysis in Function Space* (W. T. Martin and I. Segal, eds.), M.I.T. Press, Cambridge, Mass., 17–30.

Donsker, M. D., and J. L. Lions, (1962) Volterra variational equations, boundary value problems and function space integrals, *Acta Math.*, **108**, 147–228.

Doob, J. L., (1953) *Stochastic Processes.* Wiley, New York.

Dryden, H. L., (1943) A review of the statistical theory of turbulence, *Quart. Appl. Math.*, **1**, No. 1, 7–42.

Dryden, H. L., G. B. Schubauer, W. C. Mock, and H. K. Scramstad, (1937) Measurements of intensity and scale of wind-tunnel turbulence and their relation to the critical Reynolds number of spheres, *Nat. Adv. Com. Aeronaut.*, Rep. No. 581.

Dugstad, I., (1962) A hypothesis about the energy transfer in isotropic turbulence, *Meteorol. Ann.*, **4**, No. 17, 441–462.

Dunn, D. J., (1965) Turbulence and its effect upon the transmission of sound in water, *J. Sound Vib.*, **2**, No. 3, 307–327.

Durst, C. S., (1948) The fine structure of the wind in the free air, *Quart. J. Roy. Meteor. Soc.*, **74**, No. 321–322, 349–360.

Edmonds, F. N., Jr., (1960) An analysis of airborne measurements of tropospheric index of refraction fluctuations, *Proc. Symp. on Statist. Methods in Radio Wave Propogat.*, (W. C. Hoffman, ed.), London-Oxford-New York-Paris, Pergamon Press, 197–211.

Edwards, S. F., (1964) The statistical dynamics of homogeneous turbulence, *J. Fluid Mech.*, **18**, No. 2, 239–273; (1965) Turbulence in hydrodynamics and plasma physics, *Trieste 1964 Int. Conf. on Plasma Physics,* Vienna, International Atomic Energy Agency.

Edwards, S. F., and W. D. McComb, (1969) Statistical mechanics far from equilibrium, *J. Phys. A (Gen. Phys.)*, Ser. 2, **2**, No. 2, 157–171.

Edwards, S. F., and R. E. Peierls (1954) Field equations in functional form, *Proc. Roy. Soc.*, **A224**, No. 1156, 24–33.

Elagina, L. G., (1963) Measurement of frequency spectra of absolute humidity fluctuations in the atmospheric surface layer, *Izv. Akad. Nauk SSSR,* Ser. Geofiz., No. 12, 1859–1865; (1970) On the connection between frequency spectra of the humidity fluctuations and humidity fluxes, *Izv. Akad. Nauk SSSR,* Fiz. Atmosf. Okeana, **5**, No. 10, 1079–1081.

Elagina, L. G., and B. M. Koprov, (1971) Measurements of the turbulent humidity fluxes and their frequency spectra, *Izv. Akad. Nauk SSSR,* Fiz. Atmosf. i Okeana, **7**, No. 2, 115–120.

Ellison, T. H., (1951) The propagation of sound waves through a medium with very small random variations in refractive index, *J. Atmosph. Terr. Phys.*, **2**,

No. 1, 14–21; (1962) The universal small-scale spectrum of turbulence at high Reynolds number, *Mécanique de la turbulence* (Coll. Intern. du CNRS à Marseille), Paris, Ed. CNRS, 113–121; (1967) Kolmogoroff's similarity theory of small-scale motion and the effect of fluctuations in dissipation. Unpublished manuscript, 17.

Ellsaesser, H. W., (1969a) A climatology of epsilon (Atmospheric dissipation), *Monthly Weather Rev.*, **97**, No. 6, 415–423; (1969b) Wind variability as a function of time, *Monthly Weather Rev.*, **97**, No. 6, 424–428.

Eschenroeder, A. Q., (1965) Solution for the inertial energy spectrum of isotropic turbulence, *Phys. Fluids*, **8**, No. 4, 598–602.

Favre, A., J. Gaviglio, and R. Dumas (1952) Quelques mesures de corrélation dans le temps et l'espace en soufflerie, *La Rech. Aéro.*, No. 32, 21–28; (1954) Quelques fonctions d'autocorrélation et de répartitions spectrales d'énergie, pour la turbulence en aval des diverses grilles, *C. r. Acad. Sci.*, **283**, No. 15, 1561–1563; (1958) Further space-time correlations of velocity in a turbulent boundary layer, *J. Fluid Mech.*, **3**, No. 4, 344–356; (1962) Corrélations spatio-temporelles en écoulements turbulents, *Mécanique de la turbulence* (Coll. Intern. du CNRS à Marseille), Paris, Ed. CNRS, 419–445.

Feller, W., (1966) *An Introduction to Probability Theory and its Applications*, Vol. 2, Wiley, New York.

Feynman, R. P., (1948) Space-time approach to non-relativistic quantum mechanics, *Rev. Mod. Phys.*, **20**, No. 2, 367–387; (1951) On operator calculus having applications in quantum electrodynamics, *Phys. Rev.*, **84**, No. 1, 108–128.

Feynman, R. P., and A. R. Hibbs, (1965) *Quantum Mechanics and Path Integrals*, McGraw-Hill, New York.

Ffowcs Williams, J. E., (1969) Hydrodynamic noise, *Ann. Rev. Fluid Mech.*, **1**, 197–222.

Fiedler, F., and H. A. Panofsky, (1970) Atmospheric scales and spectral gaps, *Bull. Am. Met. Soc.*, **51**, No. 12, 1114–1119.

Fisher, M. J., and P. O. A. L. Davies, (1964) Correlation measurements in a non-frozen pattern of turbulence, *J. Fluid Mech.*, **18**, No. 1, 97–116.

Foiaş, C., (1970) Ergodic problems in functional spaces related to Navier-Stokes equations, *Proc. Internat. Conf. on Functional Analysis and Related Topics* (Tokyo, April 1969), University of Tokyo Press, 290–304; (1971) Solutions statistiques des équations d'évolutions non lineaires. *Problems in Non-Linear Analysis*, (CIME, Varenna, August 1970) Edizioni Cremonese, Roma, 131–188.

Fox, R. L., (1971) Solution for turbulent correlations using multipoint distribution functions, *Phys. Fluids*, **14**, No. 8, 1806–1808.

Frenkiel, F. N., (1948) The decay of isotropic turbulence, *J. Appl. Mech.*, **15**, No. 4, 311–321.

Frenkiel, F. N., and I. Katz, (1956) Studies of small-scale turbulent diffusion in the atmosphere, *J. Meteor.*, **13**, No. 4, 388–394.

Frenkiel, F. N., and P. S. Klebanoff, (1965a) Les asymétries d'ordre supérieur dans un écoulement turbulent, *C. r. Acad. Sci.*, **260**, No. 23, 6026–6029; (1956b) Two-dimensional probability distribution in a turbulent field, *Phys. Fluids*, 8, No. 12, 2291–2293; (1966) Space-time correlations in turbulence. *Dynamics of Fluids and Plasmas*, (S. I. Pai, ed.) Academic Press, New York, 257–274; (1967) Higher-order correlations in a turbulent field, *Phys. Fluids*, **10**, No. 3, 507–520; (1971) Statistical properties of velocity derivatives in a turbulent field, *J. Fluid Mech.*, **48**, No. 1, 183–208.

Frenzen, P., (1965) Determination of turbulence dissipation by Eulerian variance analysis, *Quart. J. Roy. Meteor. Soc.*, **91**, No. 387, 28–34.

Fried, D. L., (1969) Remote probing of the optical strenth of atmospheric turbulence and of wind velocity, *Proc. IEEE*, **57**, No. 4, 415–420.

Friedrichs, K. O., and H. N. Shapiro et al., (1957) Integration of functionals, *Inst. of Math. Sci.* Mimeographed notes, New York University.

Friehe, C. A., C. W. Van Atta, and C. H. Gibson, (1971) Jet turbulence: Dissipation rate measurements and correlations, *Proc. AGARD Spec. Meeting on Turbulent Shear Flows*, London.

Frisch, U., (1968) Wave propagation in random media. *Probabilistic methods in applied mathematics*, (A. T. Bharucha-Reid, ed.), Vol. 1, Academic Press, New York, 76–198.

Furutsu, K., (1963) On the statistical theory of electromagnetic waves in a fluctuating medium, *J. Res. Nat. Bur. Standards*, **D67**, No. 3, 303–323.

Galeev, A. A., and V. I. Karpman, (1963) Turbulence theory of weakly nonequilibrium tenuous plasma and the structure of shock waves, *Zh. Exper. Teor. Fiz.*, **44**, No. 2, 592–602 (1963).

Garger, E. K., (1969) The perspective photography of smoke plumes in a horizontal plane, *Meteor. i Gidrolog.*, No. 4, 96–101.

Gel'fand, I. M., and R. A. Minlos, (1954) Solution of quantum field equations, *Dokl. Akad. Nauk SSSR*, **97**, No. 2, 209–212.

Gel'fand, I. M., R. A. Minlos, and A. M. Yaglom, (1958) Functional Integrals, Proc. Third All-Union Math. Cong., *Izv. Akad. Nauk SSSR*, Moscow, Vol. 3, 521–531.

Gel'fand, I. M., and G. E. Shilov, (1964) Generalized Functions: Properties and Operations (*Generalized Functions*, Vol. 1), Academic Press, New York-London.

Gel'fand, I. M., and N. Ya. Vilenkin, (1964) Applications of Harmonic Analysis. *Generalized Functions*, Vol. 4, Academic Press, New York-London.

Gel'fand, I. M., and A. M. Yaglom, (1960) Integration in functional spaces and its application to quantum physics, *J. Math. Phys.*, **1**, No. 1, 48–69.

Ghosh, K. M., (1954) A note on the Karman's spectrum function of isotropic turbulence, *Proc. Nat. Inst. Sci. India*, **20**, No. 3, 336–340; (1955) Numerical solution to find out the spectrum function of isotropic turbulence with fourth power law fitting at small wave numbers, *Bull. Calcutta Math. Soc.*, **47**, No. 2, 71–76.

Gibson, C. H., (1968) Fine structure of scalar fields mixed by turbulence. I.

Zero-gradient points and minimal gradient surfaces. II. Spectral theory, *Phys. Fluids,* **11**, No. 11, 2305–2315; 2316–2327.

Gibson, C. H., C. C. Chen, and S. C. Lin, (1968) Measurements of turbulent velocity and temperature fluctuations in the wake of a sphere, *AIAA J.*, **6**, No. 4, 642–649.

Gibson, C. H., C. A. Friehe, and S. O. McConnell, (1971) Measurements of sheared turbulent scalar fields. To be published.

Gibson, C. H., and P. J. Masiello, (1971) Observations of the variability of dissipation rates of the turbulent velocity and temperature fields. *Statistical Models and Turbulence. Proceedings of the Symposium at San Diego, July 15-21, 1971*, (M. Rosenblatt and C. Van Atta, eds.), Lecture Notes in Physics 12, Springer-Verlag, Berlin-Heidelberg-New York, 427–453.

Gibson, C. H., and W. H. Schwarz, (1963a) Detection of conductivity fluctuations in a turbulent flow field, *J. Fluid Mech.*, **16**, No. 3, 357–364; (1963b) The universal equilibrium spectra of turbulent velocity and scalar fields. *J. Fluid Mech.*, **16**, No. 3, 365–384.

Gibson, C. H., G. R. Stegen, and S. McConnell, (1970) Measurements of the universal constant in Kolmogoroff's third hypothesis for high Reynolds number turbulence, *Phys. Fluids,* **13**, No. 10, 2448–2451.

Gibson, C. H., G. R. Stegen, and R. B. Williams, (1970) Statistics of the fine structure of turbulent velocity and temperature fields measured at high Reynolds number, *J. Fluid Mech.*, **41**, No. 1, 153–167.

Gibson, C. H., and R. B. Williams, (1969) Turbulence structure in the atmospheric boundary layer over the open ocean, *Proc. AGARD Spec. Sympos. on Aerodynamics Atmosph. Shear Flow,* Munich, Sept. 15–17, Paper No. 5.

Gibson, M. M., (1962) Spectra of turbulence at high Reynolds number, *Nature,* **195**, No. 4848, 1281–1283; (1963) Spectra of turbulence in a round jet, *J. Fluid Mech.*, **15**, No. 2, 161–173.

Gifford, F. J., (1955) A simultaneous Lagrangian-Eulerian turbulence experiment, *Monthly Weath. Rev.*, **83**, No. 12, 293–301; (1956) The relation between space and time correlations in the atmosphere, *J. Meteor.*, **13**, No. 3, 289–294, (1957a) Relative atmospheric diffusion of smoke puffs, *J. Meteor.*, **14**, No. 5, 410–414; (1957b) Further data on relative atmospheric diffusion, *J. Meteor.*, **14**, No. 5, 475–476; (1959a) The interpretation of meteorological spectra and correlations, *J. Meteor.*, **16**, No. 3, 344–346; (1959b) Smoke plumes as quantitative air pollution indices, *Intern. J. Air Poll.*, **2**, No. 1, 42–50; (1959c) Statistical properties of a fluctuating plume dispersion model, *Adv. Geophys.*, **6** (Atmospheric diffusion and air pollution). 117–137; (1960) Peak to average concentration ratios according to a fluctuating plume dispersion model, *Intern. J. Air Poll.*, **3**, No. 4, 253–260; (1962) The vertical variation of atmospheric eddy energy dissipation, *J. Atmosph. Sci.*, **19**, No. 2, 205–206.

Gödecke, K., (1935) Messungen der atmosphärischen Turbulenz in Bodennähe mit einer Hitzdrahtmethode, *Ann. Hydrogr.*, No. 10, 400–410.

Goldstein, S., (1951) On the law of decay of homogeneous isotropic turbulence and the theories of the equilibrium and similarity spectra, *Proc. Camb. Phil. Soc.*, **47**, No. 3, 554–574.

Golitsyn, G. S., (1960) The small-scale structure of turbulence, *Prikl. Mat. Mekh.*, **24**, No. 6, 1124–1129; (1961) Possible heating of the upper atmosphere by long-wave acoustic radiation, *Izv. Akad. Nauk SSSR*, Ser. Geofiz., No. 7, 1092–1093; (1962) Dissipation fluctuations in a locally isotropic turbulent flow, *Dokl. Akad. Nauk SSSR*, **144**, No. 3, 520–523; (1963) Correlation calculations in locally isotropic turbulent flow, *Prikl. Mat. Mekh.*, **27**, No. 1, 61–74; (1964) Time spectrum of microfluctuations in atmospheric pressure, *Izv. Akad. Nauk SSSR*, Ser. Geofiz., No. 8, 1253–1258.

Golitsyn, G. S., A. S. Gurvich, and V. N. Tatarskii, (1960) Studies of frequency spectra of fluctuations in the amplitude and phase difference of sound waves in a turbulent atmosphere, *Akust. Zh. (Acoustical Journal)*, **6**, No. 2, 187–197.

Gorshkov, N. F., (1966) Turbulent energy spectrum in the range of large wave numbers, *Izv. Akad. Nauk SSSR*, Fiz. Atmosf. i Okeana, **2**, No. 9, 989–992; (1967) Measurements of the pressure microfluctuation spectra in the atmospheric surface layer, *Izv. Akad. Nauk SSSR*, Fiz. Atmosf. i Okeana, **3**, No. 4, 447–451.

Gossard, E. E., (1960a) Power spectra of temperature, humidity, and refractive index from aircraft and tethered balloon measurements, *Trans. Inst. Radio Engrs on Antennas and Propagation*, **AP-8**, 186–201; (1960b) Spectra of atmospheric scalars, *J. Geophys. Res.*, **65**, No. 10, 3339–3351.

Gracheva, M. E., and A. S. Gurvich, (1965) Strong light-intensity fluctuations in the atmospheric surface layer, *Izv. Vyssh. Uchebn. Zaved., Radiofizika*, **8**, No. 4, 717–724.

Gracheva, M. E., A. S. Gurvich and M. A. Kallistratova, (1970a) Measurements of the mean amplitude level of a light wave propagating in a turbulent atmosphere, *Izv. Vyssh. Uchebn. Zaved., Radiofizika*, **13**, No. 1, 50–55; (1970b) Measurements of the variance of "strong" intensity fluctuations for laser radiation in the atmosphere, *Izv. Vyssh. Uchebn. Zaved., Radiofizika*, **13**, No. 1, 56–60.

Granotstein, V. L., S. J. Buchsbaum and D. S. Bugnolo (1966) Fluctuation spectrum of a plasma additive in a turbulent gas, *Physical Review Letters*, **16**, No. 12, 504–507.

Grant, H. L., (1958) The large eddies of turbulent motion, *J. Fluid Mech.*, **4**, No. 2, 149–190.

Grant, H. L., B. A. Hughes, W. M. Vogel, and A. Moilliet, (1968) The spectrum of temperature fluctuations in turbulent flow, *J. Fluid Mech.*, **34**, No. 3, 423–442.

Grant, H. L., and A. Moilliet, (1962) The spectrum of a cross-stream component of a turbulence in a tidal stream, *J. Fluid Mech.*, **13**, No. 2, 237–240.

Grant, H. L., A. Moilliet and W. M. Vogel, (1968) Some observations of the occurrence of turbulence in and above the thermocline, *J. Fluid Mech.*, **34**, No. 3, 443–448.

Grant, H. L., and I. C. T. Nisbet, (1957) The inhomogeneity of grid turbulence, *J. Fluid Mech.*, 2, No. 3, 263–272.

Grant, H. L., R. W. Steward and A. Moilliet, (1962) Turbulence spectra from a tidal channel, *J. Fluid Mech.*, **12**, No. 2, 241–268.

Green, H. S., (1952) *Molecular Theory of Fluids*, North Holland Publ. Co., Amsterdam.

Greenhow, J. S., (1959) Eddy diffusion and its effect on meteor trails, *J. Geophys. Res.*, **64**, No. 12, 2208–2209.

Gunnerson, C. G., (1960) Discussion on "Eddy diffusion in homogeneous turbulence" by G. T. Orlob, *J. Hydraul. Div. Proc. Amer. Soc. Civil Engrs*, **86**, No. 4, Part 1, 101–109.

Gurvich, A. S., (1960a) Experimental investigation of frequency spectra of the vertical wind velocity in the atmospheric surface layer, *Dokl. Akad. Nauk SSSR*, **132**, No. 4, 806–809; (1960b) Experimental investigation of frequency spectra and probability distribution functions for the vertical wind velocity, *Izv. Akad. Nauk SSSR*, Ser. Geofiz., No. 7, 1042–1055; (1960c) Measurement of the velocity-difference skewness in the atmospheric surface layer, *Dokl. Akad. Nauk SSSR*, **134**, No. 5, 1073–1075; (1961) Spectral composition of turbulent momentum flux, *Izv. Akad. Nauk SSSR*, Ser. Geofiz., No. 10, 1578–1579; (1962) Spectra of the vertical wind-velocity fluctuations and their relation to micrometeorological conditions, *Atmospheric Turbulence (Proc. In-ta Fiziki Atmosf. Akad. Nauk SSSR*, No. 4), 101–136; (1965) Spectra of vertical turbulent fluxes in the atmospheric surface layer, *Izv. Akad. Nauk SSSR*, Fiz. Atmosf. i Okeana, **1**, No. 7, 764–766; (1966) Probability distribution for the square of the velocity difference in a turbulent flow, *Izv. Akad. Nauk SSSR*, Fiz. Atmosf. i Okeana, **2**, No. 10, 1095–1098; (1967) Probability distribution for the square of the temperature difference between two points in a turbulent flow, *Dokl. Akad. Nauk SSSR*, **172**, No. 3, 554–557; (1968) Determination of turbulence characteristics from light propagation experiments, *Izv. Akad. Nauk SSSR*, Fiz. Atmosf. i Okeana, **4**, No. 2, 160–169.

Gurvich, A. S., M. A. Kallistratova and N. S. Time, (1968) Fluctuations of the characteristics of a laser light wave produced by its propagation in the atmosphere, *Izv. Vyssh. Uchebn. Zaved., Radiofizika*, **11**, No. 9, 1360–1370.

Gurvich, A. S., and A. I. Kon, (1964) Dependence of scintillation on the size of light source, *Izv. Vyssh. Uchebn. Zaved., Radiofizika*, **7**, No. 4, 790–792.

Gurvich, A. S., B. M. Koprov, L. R. Tsvang, and A. M. Yaglom, (1967) Data on the Small-Scale Structure of Atmospheric Turbulence, *Atmospheric Turbulence and Radio Wave Propagation*, (A. M. Yaglom and V. I. Tatarskii, eds.), Nauka Press, Moscow, 30–52.

Gurvich, A. S., and T. K. Kravchenko, (1962) Frequency spectrum of temperature fluctuations in the small-scale range. *Atmospheric Turbulence (Trudy In-ta Fiz. Atmosf. AN SSSR*, No. 4), 144–146.

Gurvich, A. S., and B. N. Meleshkin, (1966) Determination of the internal turbulent length scale from light intensity fluctuations, *Izv. Akad. Nauk SSSR*, Fiz. Atmosf. i Okeana, **2**, No. 7, 688–694.

Gurvich, A. S., V. V. Pakhomov and A. M. Cheremukhin, (1971) The isotropy of small-scale turbulent fluctuations of refractive index, *Izv. Akad. Nauk SSSR*, Fiz. Atmosf. i Okeana, **7**, No. 1, 76–80.

Gurvich, A. S., V. I. Tatarskii and L. R. Tsvang, (1958) Experimental investigation of statistical characteristics of the scintillations of a terrestrial light source, *Dokl. Akad. Nauk SSSR*, **123**, No. 4, 655–658; (1959) Scintillation of terrestrial light sources. Proc. Second Intern. Conf. on Stellar Scintillation, *Akad. Nauk SSSR Press*, Moscow.

Gurvich, A. S., and L. R. Tsvang, (1960) Spectral composition of a turbulent heat flux, *Izv. Akad. Nauk SSSR*, Ser. Geofiz., No. 10, 1547–1548.

Gurvich, A. S., and A. M. Yaglom, (1967) Breakdown of eddies and probability distributions for small-scale turbulence, *Phys. Fluids*, **10**, Supplement (Proc. Kyoto Sympos. on Boundary Layers and Turbulence), S59–S65.

Gurvich, A. S., and S. L. Zubkovskii, (1963) Experimental estimate of fluctuations in the turbulent energy dissipation, *Izv. Akad. Nauk SSSR*, Ser. Geofiz., No. 12, 1856–1858; (1965) Measurement of the fourth- and sixth-order moments of the velocity gradient, *Izv. Akad. Nauk SSSR*, Fiz. Atmosf. i Okeana, **1**, No. 8, 797–802; (1966) Estimate of the structure function of temperature fluctuation in the atmosphere, *Izv. Akad. Nauk SSSR*, Fiz. Atmosf. i Okeana, **2**, No. 2, 202–204.

Hamilton, Y. F., and L. S. Schulman, (1971) Path integrals and product integrals, *J. Math. Phys.*, **12**, No. 1, 160–164.

Hanzawa, M., (1953) On the eddy diffusion of pumices ejected from Myojin Reef in the Southern Sea of Japan, *Oceanogr. Mag.*, **4**, No. 4, 143–148.

Hasselmann, K., (1958) Zur Deutung der dreifachen Geschwindigkeitskorrelationen der isotropen Turbulenz, *Dtsch. hydrogr. Zs.*, **11**, No. 5, 207–217; (1967) Nonlinear interactions treated by the methods of theoretical physics (with application to generation of waves by wind). *Proc. Roy. Soc.*, **A299**, No. 1456, 77–100.

Hay, D. R., (1967) Stratification in the layer of frictional influence, *Atmospheric Turbulence and Radio Wave Propagation*, (A. M. Yaglom and V. I. Tatarskii, eds.) Nauka Press, Moscow, 241–248.

Heisenberg, W., (1948a) Zur statistichen Theorie der Turbulenz. *Z. Physik*, **124**, No. 7–12, 628–657; (1948b) On the theory of statistical and isotropic turbulence, *Proc. Roy. Soc. (London)*, **A195**, No. 1042, 402–406.

Hela, I., and Voipio, (1960) Tracer dyes as a means of studying turbulent diffusion in the sea, *Ann. Acad. Sci. Finnicae*, ser A., **VI**, No. 69, 3–9.

Herbstreit, J. M., and M. C. Thompson, (1955) Measurements of the phase of radio waves received over transmission paths with electrical lengths varying as a result of atmospheric turbulence, *Proc. Inst. Radio Engrs,* **43**, No. 10, 1391–1404.

Herring, J. R., (1965) Self-consistent field approach to turbulence theory, *Phys. Fluids,* **8**, No. 12, 2219–2225; (1966) Self-consistent field approach to nonstationary turbulence, *Phys. Fluids,* **9**, No. 11, 2106–2110.

Herring, J. R., and R. H. Kraichnan, (1971) Comparison of some approximations for isotropic turbulence. *Statistical Models and Turbulence. Proc. of Symp. at San Diego, July 15–21, 1971,* (M. Rosenblatt and C. Van Atta, eds.) Lecture notes in Physics 12, Springer-Verlag, Berlin-Heidelberg-New York, 148–194.

Heskestad, G., (1965) A generalized Taylor hypothesis with application for high Reynolds number turbulent shear flows, *J. Appl. Mech., Trans. ASME, Ser. E.,* **32**, No. 4, 735–739.

Hill, J. C., and C. A. Sleicher, (1969) Equations for errors in turbulence measurements with inclined hot wires, *Phys. Fluids,* **12**, No. 5, Part 1, 1126–1127.

Hinze, J. O., (1959) *Turbulence. An Introduction to its Mechanism and Theory,* McGraw-Hill, New York.

Hoffman, W. C., (1964) Wave propagation in a general random continuous medium, *Proc. Symp. Appl. Math.,* **16**, (Stochastic Processes in Mathematical Physics and Engineering), 117–144.

Hopf, E., (1952) Statistical hydromechanics and functional calculus, *J. Rat. Mech. Anal.* **1**, No. 1, 87–123; (1957) On the application of functional calculus to the statistical theory of turbulence, *Proc. Symp. Appl. Math.,* 7, 41–50; (1962) Remarks on the functional-analytic approach to turbulence, *Proc. Symp. Appl. Math.,* **13**, Hydrodynamic instability, 157–163.

Hopf, E., and E. W. Titt, (1953) On certain special solutions of the Φ-equation of statistical hydrodynamics, *J. Rat. Mech. Anal.,* **2**, No. 3, 587–592.

Hosokawa, I., (1968a) A functional treatise on statistical hydromechanics with random force action, *J. Phys. Soc. Japan,* **25**, No. 1, 271–278; (1968b) Direct computation of the functional integral expression for the correlation function of the turbulent velocity field, *Phys. Fluids,* **11**, No. 9, 2052–2054.

Hosokawa, I., and K. Yamomoto, (1970) Numerical study of the Burgers' model of turbulence based on the characteristic functional formalism, *Phys. Fluids,* **13**, No. 7, 1683–1692.

Howells, I. D., (1960) An approximate equation for the spectrum of a conserved scalar quantity in a turbulent fluid, *J. Fluid Mech.,* **9**, No. 1, 104–106.

Hutchings, J. W., (1955) Turbulence theory applied to large scale atmosphere phenomena, *J. Meteor.,* **12**, No. 3, 263–271.

Ichiye, T., and F. C. W. Olson, (1960) Uber die "neighbour diffusivity" in Ozean, *Dtsch. hydrogr. Zs.,* **13**, No. 1, 13–23.

Ievlev, V. M., (1970) Approximate equations of turbulent motion of incompressible fluid, *Izv. Akad. Nauk SSSR,* Ser. Mekh. Zhidk. i Gaza, No. 1, 91–103.

Inoue, E., (1950a) On the temperature fluctuations in a heated turbulent fluid, *Geophys. Notes,* Tokyo Univ., **3**, No. 34; (1950b) The application of turbulence theory to oceanography, *J. Meteor. Soc., Japan,* **28**, No. 11, 424–430; (1950–1951) On the turbulent diffusion in the atmosphere, I, II. *J. Meteor. Soc. Japan,* **28**, No. 12, 441–456; **29**, No. 7, 246–253; (1951a) Some remarks on the dynamical and thermal structure of a heated turbulent flow, *J. Phys. Soc. Japan,* **6**, No. 5, 392–396; (1951b) On the Lagrangian correlation coefficient and its application to the atmospheric diffusion phenomena. Rep. Geophys. Inst. Univ. Tokyo; (1952a) On the Lagrangian correlation coefficient for turbulent diffusion and its application to atmospheric diffusion phenomena, Geophys. Res. Pap. No. 19, 397–412; (1952b) Turbulent fluctuations in temperature in the atmosphere and oceans, *J. Meteor. Soc. Japan,* **30**, No. 9, 289–295; (1963) On the horizontal diffusion over the sea surface, *Proc. 1st Australasian Conf. Hydraulics Fluid Mech.*, Pergamon Press, London, 385–392.

Inoue, E., and K. Imai, (1955) Eulerian correlation of the atmospheric pressure fluctuations of medium scales, *J. Meteor. Soc. Japan,* **33**, No. 4, 169–173.

Ivanov, V. N., (1962) Dissipation of turbulent energy in the atmosphere, *Izv. Akad. Nauk SSSR,* Ser. Geofiz., No. 9, 1261–1267; (1964) Turbulent energy and its dissipation in the lower atmosphere, *Izv. Akad. Nauk SSSR,* Ser. Geofiz., No. 9, 1405–1413.

Jahnke, E., F. Emde and F. Lösch, (1960) *Tables of Higher Functions,* McGraw-Hill, New York.

Jain, P. C., (1962) Isotropic temperature fluctuations in isotropic turbulence, *Proc. Nat. Inst. Sci. India,* **A28**, No. 3, 401–416.

Joseph, J., and H. Sendner, (1958) Uber die horizontale Diffusion im Meere. *Dtsch. hydrogr. Zs.,* **11**, No. 12, 49–77; (1962) On the spectrum of the mean diffusion velocities in the ocean, *J. Geophys. Res.,* **67**, No. 8, 3201–3205.

Kac, M., (1951) On some connections between probability theory and differential and integral equations, *Proc. Second Berkeley Sympos. Math. Stat. and Probab.*, University of California Press, 189–215; (1959) *Probability and Related Topics in Physical Sciences,* Interscience Publ., London and New York.

Kadomtsev, B. B., (1965) *Plasma Turbulence.* Academic Press, London-New York.

Kaimal, J. C., J. Borkowski, S. Panchev, D. T. Gjessing, and L. Hasse, (1969) Anisotropy of the fine structure, *Radio Science,* **4**, No. 12, 1369–1370.

Kaimal, J. C., and D. A. Haugen, (1967) Characteristics of vertical velocity fluctuations observed on a 430-m tower, *Quart. J. Roy. Meteor. Soc.,* **93**, No. 397, 305–317.

Kaimal, J. C., J. C. Wyngaard, Y. Izumi, and O. R. Coté, (1972) Spectral characteristics of surface layer turbulence, *Quart. J. Roy. Meteor. Soc.,* **98**, No. 417, 563–589.

Kallistratova, M. A., (1959) Experimental investigation of the sound scattering in a turbulent atmosphere, *Dokl. Akad. Nauk SSSR,* **125,** No. 1, 69–72; (1962) Experimental investigation of the scattering of sound waves in a turbulent atmosphere. *Atmospheric Turbulence (Trudy In-ta Fiziki Atmosf. Akad. Nauk SSSR,* No. 4) 203–256.

Kampé de Feriet, J., (1939) Les fonctions aléatoires stationnaires et la théorie statistique de la turbulence homogène, *Ann. Soc. Sci. Bruxelles,* **59,** 145–194; (1948) Le tenseur spectral de la turbulence homogène, non isotrope dans un fluide incompressible, *C. r. Acad. Sci.,* **227,** No. 16, 760–761; (1953) Fonctions aléatoires et théorie statistique de la turbulence homogène, Théorie des fonctions aléatoire (A. Blanc-Lapierre and R. Fortet, eds) Masson, Paris, 568–623.

Kármán, T., von (1937) The fundamentals of the statistical theory of turbulence, *J. Aeronaut. Sci.,* **4,** No. 4, 131–138; (1948a) Sur la théorie statistique de la turbulence, *C. r. Acad. Sci.,* **226,** No. 26, 2108–2114; (1948b) Progress in the statistical theory of turbulence, *Proc. Nat. Acad. Sci. USA,* **34,** No. 11, 530–539.

Kármán , T., von, and L. Howarth (1938) On the statistical theory of isotropic turbulence, *Proc. Roy. Soc.,* **A164,** No. 917, 192–215.

Kármán, T., von, and C. C. Lin, (1949) On the concept of similarity in the theory of isotropic turbulence, *Rev. Mod. Phys.,* **21,** No. 3, 516–519; (1951) On the statistical theory of isotropic turbulence, *Adv. Appl. Mech.,* **2,** 1–19.

Keller, B. S., and A. M. Yaglom, (1970) Effect of fluctuations in energy dissipation rate on the functional form of the energy spectrum in the far short-wave range, *Izv. Akad. Nauk SSSR,* Ser. Mekh. Zhidk. i Gaza, No. 3, 70–79.

Keller, G., (1955) The relation between the structure of stellar shadow patterns and stellar scintillations, *J. Opt. Soc. Amer.,* **45,** No. 10, 845–851.

Keller, J. B., (1964) Stochastic equations and wave propagation in random media, *Proc. Symp. Appl. Math.,* **16,** (Stochastic Processes in Mathematical Physics and Engineering), 145–170.

Kellogg, W. W., (1956) Diffusion of smoke in the stratosphere, *J. Meteor.,* **13,** No. 3, 241–250.

Kennedy, D. A., and S. Corrsin, (1961) Spectral flatness factor and "intermittence" in turbulence and in nonlinear noise, *J. Fluid Mech.,* **10,** No. 2, 366–370.

Khang, W. H., and A. Siegel, (1970) The Cameron-Martin-Wiener method in turbulence and in Burgers' model: General formulae, and application to late decay, *J. Fluid Mech.,* **41,** No. 3, 593–618.

Khazen, E. M., (1963) Nonlinear theory of transition to turbulence, *Dokl. Akad. Nauk SSSR,* **153,** No. 6, 1284–1287.

Khinchin, A. Ya., (1934) Korrelationstheorie der stationären stochastischen Prozesse, *Math. Ann.,* **109,** No. 4, 604–615.

Kholmyanskii, M. Z., (1970) Investigation of the fluctuations of the wind velocity derivative in an atmospheric surface layer, *Izv. Akad. Nauk SSSR,*

Fiz. Atmosf. i Okeana, **6**, No. 4. 423–430; (1972) Measurements of the microturbulent fluctuations of the wind velocity derivative in an atmospheric surface layer, *Izv. Akad. Nauk SSSR,* Fiz. Atmosf. i Okeana, **8**, No. 8, 818–828.

Kilezhenko, V. P., (1964) Experimental investigation of horizontal turbulent diffusion of an impurity spot in the sea, *Materialy Rybokhoz. issled. Sev. basseina (Materials of Fishery Res. of Northern Basin),* No. 4, 101–105.

Kistler, A. L., and T. Vrebalovich, (1966) Grid turbulence at large Reynolds numbers, *J. Fluid Mech.*, **26**, No. 1, 37–47.

Klebanoff, P. S., (1955) Characteristics of turbulence in a boundary layer with zero pressure gradient, *Nat. Adv. Com. Aeronaut.*, Rep. No. 1247.

Klyatskin, V. I., (1966a) Sound emission by a set of vortices, *Izv. Akad. Nauk SSSR, Mekh. Zhidk. i Gaza,* **1**, No. 6, 995–1001; (1966b) Homogeneous and isotropic turbulence in a weakly compressible medium, *Izv. Akad. Nauk SSSR,* Fiz. Atmosf. i Okeana, **2**, No. 5, 474–485; (1969) Conditions of validity of the Markov random process approximation in problems connected with light propagation in a medium with random inhomogeneities of refractive index, *Zh. Exper. Teor. Fiz.*, **57**, No. 3(9), 952–958; (1971a) Asymptotic behaviour of the intensity fluctuations of a plane light wave propagating through a turbulent medium, *Zh. Exper. i Tech. Fiz.*, **60**, No. 4, 1300–1305; (1971b) Space-time description of a stationary and homogeneous turbulence, *Izv. Akad. Nauk SSSR, Ser. Mech. Zhidk. i Gaza,* No. 4, 120–127.

Klyatskin, V. I., and V. I. Tatarskii, (1970a) On parabolic equation approximation for problems of wave propagation in a medium with random inhomogeneities, *Zh. Exper. Teor. Fiz.*, **58**, No. 2, 624–634; (1970b) Theory of light beam propagation in a medium with random inhomogeneities, *Izv. Vyssh. Uchebn. Zaved., Radiofizika,* **13**, No. 7, 1061–1068; (1971) New method of successive approximations in the problem of wave propagation in a medium with random large-scale inhomogeneities, *Izv. Vyssh. Uchebn. Zaved., Radiofizika,* **14**, No. 7, 1400–1415.

Kolchinskii, I. G., (1957) Some results of observations of stellar scintillation at the Field Station Goloseevo of Main Astronomical Observatory, Acad. of Sciences of the Ukrainian SSR, *Astron. Zh.*, **34**, No. 4, 638–651.

Kolesnikova, V. N., and A. S. Monin, (1965) Spectra of oscillations in meteorological fields, *Izv. Akad. Nauk SSSR,* Fiz. Atmosf. i Okeana, **1**, No. 7, 653–669.

Kolmogorov, A. N., (1935) Zufällige Bewegungen, *Ann. Math.* **35**, No. 1, 116–117; (1940a) Curves in Hilbert space which are invariant under the one-parameter group of motions, *Dokl. Akad. Nauk SSSR,* **26**, No. 1, 6–9; (1940b) Wiener's spiral and some other interesting curves in Hilbert space, *Dokl. Akad. Nauk SSSR,* **26**, No. 2, 115–118; (1941a) Local structure of turbulence in an incompressible fluid at very high Reynolds numbers, *Dokl. Akad. Nauk SSSR,* **30**, No. 4, 299–303; (1941b) Logarithmically normal distribution of fragmentary particle sizes, *Dokl.*

*Akad. Nauk SSSR*, **31**, No. 2, 99–101; (1941c) Decay of isotropic turbulence in incompressible viscous fluids, *Dokl. Akad. Nauk SSSR*, **31**, No. 6, 538–541; (1941d) Energy dissipation in locally isotropic turbulence, *Dokl. Akad. Nauk SSSR*, **32**, No. 1, 19–21; (1941e) Stationary sequences in Hilbert space, *Byull. Mosk. Gos. Un-ta, Matematika (Bull. Moscow Univ., Ser. Math.)*, **2**, No. 6, 1–40; (1962a) Refined Description on the Local Structure of Turbulence in an Incompressible Viscous Fluid at High Reynolds Numbers (in French and Russian). *Mècanique de la Turbulence* (Coll. Intern. du CNRS à Marseille). Ed. CNRS, Paris, 447–458; A refinement of previous hypotheses concerning the local structure of turbulence in a viscous incompressible fluid at high Reynolds number, *J. Fluid Mech*, **13**, No. 1, 82–85.

Kon, A. I., (1969a) *The Effect of Geometrical Factors on Statistical Characteristics of Waves Propagating in a Turbulent Atmosphere*, Candidate Dissertation, Inst. Atmosph. Phys., Acad. Sci. USSR, Moscow; (1969b) The effect of finite size of a source and a receiver on light intensity fluctuations, *Izv. Vyssh. Uchebn., Radiofizika*, **12**, No. 5, 686–693.

Kon, A. I., and V. I. Tatarskii, (1964) Scintillation of sources having finite angular dimensions, *Izv. Vyssh. Uchebn. Zaved., Radiofizika*, **7**, No. 2, 306–312.

Koprov, B. M., (1965) Turbulent spectra of vertical wind-velocity fluctuations in the surface layer of the atmosphere developed in convection. *Izv. Akad. Nauk SSSR*, Fiz. Atmosf. i Okeana, **1**, No. 11, 1141–1149.

Koprov, B. M., and L. R. Tsvang, (1965) Direct airborne measurements of turbulent heat flux, *Izv. Akad. Nauk SSSR*, Fiz. Atmosf. i Okeana, **1**, No. 6, 643–647; (1966) Characteristics of small-scale turbulence in a stratified boundary layer, *Izv. Akad. Nauk SSSR*, Fiz. Atmosf. i Okeana, **2**, No. 11, 1142–1150.

Kovasznay, L. S. G., (1948) Spectrum of locally isotropic turbulence, *J. Aeronaut. Sci.*, **15**, No. 12, 745–753; (1953) Turbulence in supersonic flow, *J. Aeronaut. Sci.*, **20**, No. 10, 657–674, 682.

Kraichnan, R. H., (1953) The scattering of sound in a turbulent medium, *J. Acoust. Soc. Amer.*, **25**, No. 6, 1096–1104; (1957) Relation of fourth-order to second-order moments in stationary isotropic turbulence, *Phys. Rev.*, **107**, No. 6, 1485–1490; (1959) The structure of isotropic turbulence at very high Reynolds numbers, *J. Fluid Mech.*, **5**, No. 4, 497–543; (1961) Dynamics of nonlinear stochastic systems, *J. Math. Phys.*, **2**, No. 1, 124–148; (1962) The closure problem of turbulence theory, *Proc. Symp. Appl. Math.*, **13**, (Hydrodynamic instability), 199–225; (1964a) Decay of isotropic turbulence in the direct-interaction approximation, *Phys. Fluids*, **7**, No. 7, 1030–1048; (1964b) Approximations for steady-state isotropic turbulence, *Phys. Fluids*, **7**, No. 8, 1163–1168; (1964c) Kolmogorov's hypothesis and Eulerian turbulence theory, *Phys. Fluids*, **7**, No. 11, 1723–1734; (1965a) Lagrangian-history closure approximation for turbulence, *Phys. Fluids*, **8**, No. 4, 575–598 (errata, 1966, **9**, No. 9); (1965b)

Preliminary calculation of the Kolmogorov turbulence spectrum, *Phys. Fluids,* **8,** No. 5, 995–997; (1966a) Isotropic turbulence and inertial-range structure, *Phys. Fluids,* **9,** No. 9, 1728–1752; (1966b) Dispersion of particle pairs in homogeneous turbulence, *Phys. Fluids,* **9,** No. 10, 1937–1943; (1966c) Invariant principles and approximation in turbulence dynamics, *Dynamics of Fluids and Plasmas,* (S. I. Pai, ed.). Academic Press, New York, 239–255; (1967a) Inertial ranges in two-dimensional turbulence, *Phys. Fluids,* **10,** No. 7, 1417–1423; (1967b) Intermittency in the very small scales of turbulence, *Phys. Fluids,* **10,** No. 9, 2080–2082; (1968a) Lagrangian-history statistical theory for Burgers' equation, *Phys. Fluids,* **11,** No. 2, 265–277; (1968b) Small-scale structure of a scalar field convected by turbulence, *Phys. Fluids,* **11,** No. 5, 945–953; (1970a) Convergents to turbulence functions, *J. Fluid Mech.,* **41,** No. 1, 189–217; (1970b) Diffusion by a random velocity field, *Phys. Fluids,* **13,** No. 1, 22–31; (1971a) An almost-Markovian Galilean-invariant turbulence model, *J. Fluid Mech.,* **47,** No. 3, 513–524; (1971b) Inertial-range transfer in two- and three-dimensional turbulence, *J. Fluid Mech.,* **47,** No. 3, 525–536.

Kraichnan, R. H., and E. A. Spiegel, (1962) Model for energy transfer in isotropic turbulence, *Phys. Fluids,* **5,** No. 5, 583–588.

Krasil'nikov, V. A., (1945) Sound propagation in a turbulent atmosphere, *Dokl. Akad. Nauk SSSR,* **47,** No. 7, 486–489; (1947) Amplitude fluctuations during sound propagation through a turbulent atmosphere, *Dokl. Akad. Nauk SSSR,* **58,** No. 7, 1353–1356; (1949a) Effect of fluctuations in the refractive index of the atmosphere on the propagation of ultra-short radiowaves, *Izv. Akad. Nauk SSSR,* Ser. Geogr. i Geofiz., **13,** No. 1, 33–57; (1949b) Fluctuations in the angle of arrival in stellar scintillation phenomenon, *Dokl. Akad. Nauk SSSR,* **65,** No. 3, 291–294; (1953) Fluctuations in the phases of ultrasonic waves during their propagation through the atmospheric surface layer, *Dokl. Akad. Nauk SSSR,* **88,** No. 4, 657–660.

Krasil'nikov, V. A., and K. M. Ivanov-Shits, (1949) Some new experiments on sound propagation in the atmosphere, *Dokl. Akad. Nauk SSSR,* **67,** No. 4, 639–642.

Krasil'nikov, V. A., and V. I. Tatarskii, (1953) The dependence of the mean square phase and amplitude fluctuations on objective size during observations of stellar scintillation, *Dokl. Akad. Nauk SSSR,* **88,** No. 3, 435–438.

Krechmer, S. I., (1952) Experimental investigation of microfluctuations in the atmospheric temperature field, *Dokl. Akad Nauk SSSR,* **84,** No. 1, 55–58.

Krechmer, S. I., A. M. Obukhov, and N. Z. Pinus, (1952) Results of experimental investigation of microturbulence in the free atmosphere, *Trudy Tsentr. Aerolog. Obs.,* No. 6, 174–183.

Krzywoblocki, M. Z., (1952) On the generalized fundamental equations of isotropic turbulence in compressible fluids and in hypersonics, Proc. 1st US Nat. Congr. Appl. Mech., Chicago, 1951, New York, 827–835.

Kucharets, V. P., and L. R. Tsvang, (1969) Spectra of the turbulent heat flux in the boundary layer of the atmosphere, *Izv. Akad. Nauk SSSR,* Fiz. Atmosf. i Okeana, 5, No. 11, 1132–1142.

Kullenberg, G., (1969) Measurements of horizontal and vertical diffusion in coastal waters, *Kungl. Vetenskaps-och Vitterhets-Samhället, Göteborg,* ser. Geophysica, 2, 51; (1972) Apparent horizontal diffusion in stratified vertical shear flow, *Tellus,* 24, No. 1, 17–28.

Kumar, P., and R. Dash, (1970) Multi-point, multi-time velocity correlations and decay of homogeneous turbulence, *Phys. Fluids,* **13**, No. 3, 576–584.

Kuo, A. Y-S., and S. Corrsin, (1971) Experiments on internal intermittency and fine-structure distribution function in fully turbulent fluid, *J. Fluid Mech.,* **50**, No. 2, 285–320; (1972) Experiment on the geometry of the fine structure regions in fully turbulent fluid, *J. Fluid Mech.* **56**, No. 3, 447–479.

Landau, L. D., and E. M. Lifshits, (1963) *Fluid Mechanics,* Pergamon Press, London.

Lane, J. A., (1969) Some aspects of the fine structure of elevated layers in the troposphere, *Radio Science,* **4**, No. 12, 1111–1114.

Lappe, U. O., and B. Davidson, (1963) On the range of validity of Taylor's hypothesis and the Kolmogorov spectral law. *J. Atmosph. Sci.,* **20**, No. 6, 569–576.

Laufer, J., (1950) Some recent measurements in a two-dimensional turbulent channel, *J. Aeronaut. Sci.,* **17**, No. 5, 277–287; (1951) Investigation of turbulent flow in a two-dimensional channel, *Nat. Adv. Com. Aeronaut.,* Rep. No. 1033; (1954) The structure of turbulence in fully developed pipe flow, *Nat. Adv. Com. Aeronaut.,* Rep. No. 1174; (1962) Sound radiation from a turbulent boundary layer, *Mécanique de la turbulence* (Coll. Intern. du CNRS à Marseille), Paris, Ed. CNRS, 381–392.

Lee, D. A., (1965) Spectrum of homogeneous turbulence in the final stage of decay, *Phys. Fluids,* **8**, No. 10, 1911–1913.

Lee, D. A., and H. S. Tan, (1967) Study of inhomogeneous turbulence, *Phys. Fluids,* **10**, No. 6, 1124–1230.

Lee, J., (1965) Decay of scalar quantity fluctuations in a stationary isotropic turbulent velocity field, *Phys. Fluids,* **8**, No. 9, 1647–1658; (1966) Comparison of closure approximation theories of turbulent mixing, *Phys. Fluids,* **9**, No. 2, 363–372; (1971) On Phythian's perturbation theory for stationary homogeneous turbulence, *J. Phys. A (Gen. Phys.),* **4**, No. 1, 73–76.

Lee, L. L., (1965) A formulation of the theory of isotropic hydromagnetic turbulence in an incompressible fluid, *Ann. Phys.,* **32**, No. 2, 292–321.

Lee, T. D., (1950) Note on the coefficient of eddy viscosity in isotropic turbulence, *Phys. Rev.,* **77**, No. 6, 842–843.

Leith, C. E., (1967) Diffusion approximation to inertial energy transfer in isotropic turbulence, *Phys. Fluids,* **10**, No. 7, 1409–1416.

Leontovich, M. A., and V. A. Fock, (1948) Solution of the problem of electromagnetic wave propagation along the earth's surface by the parabolic equation method. *Investigations of Radio Wave Propagation,* No. II, Akad. Nauk SSSR Press, 13–39.

Lewis, R. M., and R. H. Kraichnan, (1962) A space-time functional formalism for turbulence, *Comm. Pure Appl. Math.,* **15,** No. 4, 397–411.

Liepmann, H. W., J. Laufer and K. Leipmann, (1951) On the spectrum of isotropic turbulence, *Nat. Adv. Com. Aeronaut.,* Tech. Note No. 2473.

Lighthill, M. J., (1952) On sound generated aerodynamically. I. General theory, *Proc. Roy. Soc.,* **A211,** No. 1107, 564–587; (1954) On sound generated aerodynamically. II. Turbulence as a source of sound, *Proc. Roy. Soc.,* **A222,** No. 1148, 1–32; (1962) Sound generated aerodynamically, *Proc. Roy. Soc.,* **A267,** No. 1329, 147–182.

Limber, D. N., (1951) Numerical results for pressure velocity correlations in homogeneous isotropic turbulence, *Proc. Nat. Acad. Sci., USA,* **37,** No. 4, 230–233.

Lin, C. C., (1948) Note on the law of decay of isotropic turbulence, *Proc. Nat. Acad. Sci., USA,* **34,** No. 4, 230–233; (1949) Remarks on the spectrum of turbulence, *Proc. 1st Symp. Appl. Math.* ("Non-linear problems in mechanics of continua"), 81–86; (1953a) On Taylor's hypothesis in wind tunnel turbulence, *Quart. Appl. Math.,* **10,** No. 4, 295–306; (1953b) A critical discussion of similarity concepts in isotropic turbulence, *Proc. Symp. Appl. Math.,* **4** (Fluid dynamics), 19–27; (1960a) On a theory of dispersion by continuous movements, I. *Proc. Nat. Acad. Sci., USA,* **46,** No. 4, 566–570; (1960b) On a theory of dispersion by continuous movements. II. Stationary anisotropic processes, *Proc. Nat. Acad. Sci., USA,* **46,** No. 8, 1147–1150.

Lin, C. C., and W. H. Reid, (1963) Turbulent flow. Theoretical aspects, Handbuch der Physik, Bd. VIII/2, Berlin-Göttingen-Heidelberg, Springer, 438–523.

Lin, J-T., (1972) Velocity spectrum of locally isotropic turbulence in the inertial and dissipation ranges, *Phys. Fluids,* **15,** No. 1, 205–207.

Ling, S. S., and T. T. Huang, (1970) Decay of weak turbulence, *Phys. Fluids,* **13,** No. 12, 2912–2924.

Livingston, P. M., P. M. Deitz and E. C. Alcaraz, (1970) Light propagation through a turbulent atmosphere: Measurements of the optical filter function, *J. Opt. Soc. Amer.,* **60,** No. 7, 925–935.

Loeffler, A. L., and R. G. Deissler, (1961) Decay of temperature fluctuations in homogeneous turbulence before the final period, *Intern. J. Heat and Mass Transfer,* **1,** No. 4, 312–324.

Loitsyanskii, L. G., (1939) Some basic laws for isotropic turbulent flow, *Trudy Tsentr. Aero.-Giedrodin. Inst.,* No. 440, 3–23.

Lumley, J. L., (1957) *Some Problems Connected with the Motion of Small Particles in a Turbulent Fluid,* Ph.D. Dissertation, Dept. of Mechanics, Johns Hopkins University, Baltimore, Maryland; (1964) The spectrum of nearly inertial turbulence in a stably stratified fluid, *J. Atmosph. Sci.,* **21,** No. 1, 99–102; (1965) Interpretation of time spectra measured in high-intensity shear flows, *Phys. Fluids,* **8**, No. 6, 1056–1062; (1966) Invariants in turbulent flow, *Phys. Fluids,* **9**, No. 11, 2111–2113; (1967a) The structure of inhomogeneous turbulent flows, *Atmospheric Turbulence and Radio Wave Propagation* (A. M. Yaglom and V. I. Tatarskii, eds) Nauka Press, Moscow, 166–188; (1967b) Similarity and the turbulent energy spectrum, *Phys. Fluids,* **10**, No. 4, 855–858; (1970) *Stochastic Tools in Turbulence,* Academic Press, New York-London; (1972) On the solution of equations describing small scale deformation. *Symposia Mathematica,* Vol. IX, Academic Press, New York, 315–334.

Lumley, J. L., and H. A. Panofsky, (1964) *The Structure of Atmospheric Turbulence,* Interscience Publ., New York-London-Sydney.

Lundgren, T. S., (1967) Distribution functions in the statistical theory of turbulence, *Phys. Fluids,* **10**, No. 5, 969–975; (1969) Model equation for nonhomogeneous turbulence, *Phys. Fluids,* **12**, No. 3, 485–497; (1971) A closure hypothesis for the hierarchy of equations for turbulent probability distribution functions. *Statistical Models and Turbulence. Proc. of the Symp. at San Diego, July 15–21, 1971,* (M. Rosenblatt and C. Van Atta, eds.) Lecture notes in Physics 12, Springer-Verlag, Berlin-Heidelberg-New York, 70–100.

Lyubimov, B. Ya., (1969) Lagrange's description of turbulence dynamics, *Dokl. Akad. Nauk SSSR,* **184**, No. 5, 1069–1071.

Lyubimov, B. Ya., and F. R. Ulinich, (1970a) Statistical equations of turbulent motion in the Lagrange's variables, *Prikl. Mat. Mek.,* **34**, No. 1, 24–31; (1970b) On the closure problem of turbulence theory, *Dokl, Akad, Nauk SSSR,* **191**, No. 3, 551–552.

MacCready, P. B., (1953) Atmospheric turbulence measurements and analysis, *J. Meteor.,* **10**, No. 4, 325–337; (1962a) Turbulence measurements by sailplane, *J. Geophys. Res.,* **67**, No. 3, 1041–1050; (1962b) The inertial subrange of atmospheric turbulence, *J. Geophys. Res.,* **67**, No. 3, 1051–1059.

MacEwen, G. F., (1950) A statistical model of instantaneous point and disc sources with application to oceanographic observations, *Trans. Amer. Geophys. Union,* **31**, No. 1, 35–46.

MacPhail, D. C., (1940) An experimental verification of the isotropy of turbulence produced by a grid, *J. Aeronaut. Sci.,* **8**, No. 2, 73–75.

Mandelbrot, B., (1965) Self-similar error clusters in communications systems and the concept of conditional stationarity, *IEEE Trans. Communication Technol.,* COM-13, No. 1, 71–90; (1967) Sporadic random functions and conditional spectral analysis; self-similar examples and limits, *Proc. 5th Berkeley Symp. Mathem. Statistics and Probability* Vol. 3, University of California Press, Berkeley and Los Angeles, 155–179; (1967b) Sporadic turbulence, *Phys. Fluids,* **10**, No. 9, Part II (Supplement), S302–S303: (1972) Possible refinement of the lognormal hypothesis concerning the distribution of energy dissipation in intermittent turbulence. *Statistical*

*Models and Turbulence, Proceedings of the Symposium at San Diego, July 15–21, 1971.* (M. Rosenblatt and C. Van Atta, eds.) Lecture Notes in Physics 12, Springer-Verlag, Berlin-Heidelberg-New York, 333–351.

Martin, H. C., (1966) Microstructure of temperature and humidity near the ground, Ph.D. thesis, Univ. of Western Ontario, London, Canada.

Mattioli, E., (1951) Le relazioni tra le funzioni di correlazione della velocità nella turbolenza omogenea e isotropica, *Atti Accad. Naz. Lincei,* **9**, No. 11, 260–264.

McBean, G. A., (1968) An investigation of the turbulence within the forest, *J. Appl. Meteor.*, **7**, No. 3, 410–416.

McShane, E. J., (1963) Integrals devised for special purposes, *Bull. Amer. Math. Soc.*, **69**, No. 5, 597–627.

Meecham, W. C., (1965) Turbulence energy principles for quasi-normal and Wiener-Hermite expansions, *Phys. Fluids,* **8**, No. 9, 1738–1739; (1970) Equilibrium characteristics of nearly normal turbulence, *J. Fluid Mech.*, **41**, No. 1, 179–188.

Meecham, W. C., and W. C. Clever, (1971) Use of Cameron-Martin-Wiener representations for nonlinear random process applications. *Statistical Models and Turbulence, Proc. of the Symp. at San Diego, July 15-21, 1971,* (M. Rosenblatt and C. Van Atta, eds.) Lecture Notes in Physics 12, Springer-Verlag, Berlin-Heidelberg-New York, 205–229.

Meecham, W. C., and D. -T. Jeng, (1968) Use of the Wiener-Hermite expansion for nearly normal turbulence, *J. Fluid Mech.*, **32**, No. 2, 225–249.

Meecham, W. C., and A. Siegel, (1964) Wiener-Hermite expansion in model turbulence at large Reynolds numbers, *Phys. Fluids,* **7**, No. 8, 1178–1190.

Meetz, K., (1956a) Das zeitliche Abklingen der Energiespektren in der homogenen isotropen Turbulenz als Anfangwertproblem, *Zs. Naturforsch.*, **11a**, No. 10, 832–847; (1956b) Das zeitliche Abklingen der Geschwindigkeits- und Drückkorelationen in der homogenen isotropen Turbulenz als Anfangswertproblem, *Zs. Naturforsch.*, **11a**, No. 10, 848–857.

Megaw, E. C. S., (1950) Scattering of electromagnetic waves by atmospheric turbulence, *Nature,* **166**, No. 4235, 1100–1104.

Meksyn, D., (1963) Differential integral equation for the spectrum of isotropic homogeneous turbulence, *Zs. Phys.*, **174**, No. 3, 301–313.

Millionshchikov, M. D., (1939a) Decay of homogeneous isotropic turbulence in a viscous incompressible fluid, *Dokl. Akad. Nauk SSSR,* **22**, No. 5, 236–240; (1939b) Decay of velocity fluctuations in wind tunnels, *Dokl. Akad. Nauk SSSR,* **22**, No. 5, 241–242; (1941a) Theory of homogeneous isotropic turbulence, *Dokl. Akad. Nauk SSSR,* **32**, No. 9, 611–614; (1941b) Theory of homogeneous isotropic turbulence, *Izv. Akad. Nauk SSSR,* Ser. Georgr. i Geofiz., **5**, No. 4–5, 433–446.

Mills, R. R., and S. Corrsin, (1959) Effect of contraction on turbulence and temperature fluctuations generated by a warm grid, *Nat. Aeronaut. Space Adm.* Memorandum 5-5-59W, 1–58.

Mills, R. R., Jr., A. L. Kistler, V. O'Brien, and S. Corrsin, (1958) Turbulence and temperature fluctuations behind a heated grid, *Nat. Adv. Com. Aeronaut.* Tech. Note 4288, 1–67.

Millsaps, K., (1955) The Obukhoff spectrum of homogeneous isotropic turbulence, *J. Aeronaut. Sci.*, **22**, No. 7, 511.

Mintzer, D. J., (1953–1954) Wave propagation in a randomly inhomogeneous medium. I–III, *J. Acoust. Soc. Amer.*, **25**, No. 5, 922–927; No. 6, 1107–1111; **26**, No. 2, 186–190.

Mirabel, A. P., (1969) An example of the application of the Millionshchikov hypothesis to the problem of isotropic turbulence decay, *Izv. Akad. Nauk SSSR*, Ser. Mech. Zhidk. i Gaza, No. 5, 171–175.

Miyake, M., M. Donelan and Y. Mitsuta, (1970) Airborne measurement of turbulent fluxes, *J. Geophys. Res.*, **75**, No. 24, 4506–4518.

Miyake, M., and G. McBean, (1970) On the measurement of vertical humidity transport over land, *Boundary-Layer Meteor.*, **1**, No. 1, 88–101.

Miyake, M., R. W. Stewart and R. W. Burling, (1970) Spectra and cospectra of turbulence over water, *Quart. J. Roy. Meteor. Soc.*, **96**, No. 407, 138–143.

Monin, A. S., (1955) Equation of turbulent diffusion, *Dokl. Akad. Nauk SSSR*, **105**, No. 2, 256–259; (1956) Horizontal mixing in the atmosphere, *Izv. Akad. Nauk SSSR*, Ser. Geofiz., No. 3, 327–345; (1958) Structure of atmospheric turbulence, *Teor. Veroyatn. i ee Primen.*, **3**, No. 3, 285–317; (1959a) Theory of locally isotropic turbulence, *Dokl. Akad. Nauk SSSR*, **125**, No. 3, 515–518; (1959b) On the similarity of turbulence in the presence of a mean vertical temperature gradient, *J. Geophys. Res.* **64**, No. 12, 2196–2197; (1959c) Turbulence in shear flow with stability, *J. Geophys. Res.*, **64**, No. 12, 2224–2225; (1960) Lagrangian turbulence characteristics, *Dokl. Akad. Nauk SSSR*, **134**, No. 2, 304–307; (1961) Some properties of sound scattering in a turbulent atmosphere, *Akust. Zh. (Acoustical Journal)*, **7**, No. 4, 457–461; (1962a) Empirical data on turbulence in the surface layer of atmosphere, *J. Geophys. Res.* **67**, No. 8, 3103–3109; (1962b) Structure of wind-velocity and temperature fields in the atmospheric surface layer. *Atmospheric Turbulence (Trudy In-ta Fiziki Atmosf. Akad. Nauk SSSR*, No. 4), 5–20; (1962c) Turbulent spectra in a temperature-inhomogeneous atmosphere, *Izv. Akad. Nauk SSSR*, Ser. Geofiz., No. 3, 397–407; (1962d) Lagrangian hydrodynamic equations for incompressible viscous fluids, *Prikl. Mat. Mekh.*, **26**, No. 2, 320–327; (1964) Solution of the turbulence problem by the method of perturbation theory, *Prikl. Mat. Mekh.*, **28**, No. 2, 319–325; (1967a) Equation for finite-dimensional probability distributions of a turbulent field, *Dokl. Akad. Nauk SSSR*, **177**, No. 5, 1036–1038; (1967b) Equations of turbulent motion, *Prikl. Mat. Mekh.*, **31**, No. 6, 1057–1068; (1969) On the interaction between the vertical and horizontal diffusion of admixtures in the sea, *Okeanologiya (Oceanology)*, **9**, No. 1, 76–81.

Monin, A. S., and A. M. Obukhov, (1958) Small oscillations of the atmosphere and adaptation of meteorological fields, *Izv. Akad. Nauk SSSR*, Ser. Geofiz., No. 11, 1360–1373.

Mordukhovich, M. I., (1970) Measurement of intensity fluctuation variance and of mean amplitude level for a laser light beam propagating along a strongly inhomogeneous terrain, *Izv. Vyssh. Uchebn. Zaved., Radiofizika*, **13**, No. 2, 275–280.

Moyal, J. E., (1952) The spectra of turbulence in a compressible fluid; eddy turbulence and random noise, *Proc. Camb. Phil. Soc.*, **48**, No. 2, 329–344.

Müller, E. A., and K. Matschat, (1958) Zur Lärmerzeugung durch abklingende homogene isotrope Turbulenz, *Zs. Flügwiss.*, **6**, No. 6, 161–170.

Münch, G., and A. D. Wheelon, (1958) Space-time correlations in stationary isotropic turbulence, *Phys. Fluids*, **1**, No. 6, 462–468.

Myrup, L. O., (1968) Atmospheric measurements of the buoyant subrange of turbulence, *J. Atmosph. Sci.*, **25**, No. 6, 1160–1164; (1969) Turbulence spectra in stable and convective layers in the free atmosphere, *Tellus*, **21**, No. 3, 341–354.

Nasmyth, P. W., (1970) *Oceanic Turbulence*, Ph.D. Dissertation, Inst. of Oceanography, Univ. of British Columbia, Vancouver, Canada.

Neumann, J., (1969) Equilibrium range spectra of turbulence, *J. Atmos. Sci.*, **26**, No. 5, 1155–1156.

Neumann, J., von, I. J. Schoenberg, (1941) Fourier integrals and metric geometry, *Trans. Amer. Math. Soc.*, **50**, No. 2, 226–251.

Norton, K. A., (1959) Recent experimental evidence favouring the $\rho K_1(\rho)$ correlation function for describing the turbulence of refractivity in the troposphere and stratosphere, *J. Atmosph. Terr. Phys.*, **15**, Nos. 3/4, 206–227.

Novikov, E. A., (1961a) Energy spectrum of turbulent flow of an incompressible fluid, *Dokl. Akad. Nauk SSSR*, **139**, No. 2, 331–334; (1961b) Electron-density fluctuations in the ionosphere, *Dokl. Akad. Nauk SSSR*, **139**, No. 3, 587–589; (1961c) *Hydromagnetic Turbulence in the Ionosphere*, Candidate Dissertation, *In-t Fiziki Atmosfery Akad. Nauk SSSR*, Moscow; (1961d) Solution of some equations with functional derivatives, *Usp. Mat. Nauk*, **16**, No. 2(98), 135–141; (1963a) Method of random forces in turbulence theory, *Zh. Exper. Teor. Fiz.*, **44**, No. 6, 2159–2168; (1963b) Variability of energy dissipation rate in a turbulent flow and the energy distribution over the spectrum, *Prikl. Mat. Mekh*, **27**, No. 5, 944–946; (1964) Functionals and the method of random forces in turbulence theory, *Zh. Exper. Teor. Fiz.*, **47**, No. 5(11), 1919–1926; (1965) High-order correlations in a turbulent flow, *Izv. Akad. Nauk SSSR*, Fiz. Atmosf. i Okeana, **1**, No. 8, 788–796; (1966) Mathematical model of intermittency in turbulent flow, *Dokl. Akad. Nauk SSSR*, **168**, No. 6, 1279–1282; (1967) Kinetic equations for vorticity field, *Dokl. Akad. Nauk SSSR*, **177**, No. 2, 299–301; (1969) Scale similarity for random fields, *Dokl. Akad. Nauk SSSR*, **184**, No. 5, 1072–1075; (1970) Intermittency and scale similarity of the structure of turbulent flow, *Prikl. Mat. Mekh.*, **35**, No. 2, 266–277.

Novikov, E. A., and R. W. Stewart, (1964) Intermittency of turbulence and spectrum of fluctuations in energy-dissipation, *Izv. Akad. Nauk SSSR*, Ser. Geofiz., No. 3, 408–413.

Nye, J. O., and R. S. Brodkey, (1967) The scalar spectrum in the viscous-convective subrange, *J. Fluid Mech.*, **29**, No. 1, 151–163.

O'Brien, E. E., (1963) Initial spectra of a scalar field transported by turbulence, *Phys. Fluids*, **6**, No. 7, 1016–1020; (1968) Qualitative test of the direct interaction hypothesis. *Phys. Fluids*, **11**, No. 10, 2087–2088.

O'Brien, E. E., and G. C. Francis, (1962) A consequence of the zero fourth cumulant approximation, *J. Fluid Mech.*, **13**, No. 3, 369–382.

O'Brien, E. E., and S. Pergament, (1964) Note on the unphysical spectral predictions of the joint normal distribution hypothesis, *Phys. Fluids*, **7**, No. 4, 609.

Obukhov, A. M., (1941a) Spectral energy distribution in a turbulent flow, *Dokl. Akad. Nauk SSSR*, **32**, No. 1, 22–24; (1941b) Spectral energy distribution in a turbulent flow, *Izv. Akad. Nauk SSSR*, Ser. Georgr. i Geofiz., **5**, No. 4–5, 453–466; (1941c) Sound scattering in a turbulent flow, *Dokl. Akad. Nauk SSSR*, **30**, No. 7, 611–614; (1942) On the theory of atmospheric turbulence, *Izv. Akad. Nauk SSSR*, Ser. Fiz., **6**, No. 1–2, 59–63; (1949a) Structure of the temperature field in a turbulent flow, *Izv. Akad. Nauk SSSR*, Ser. Geogr. i Geofiz., **13**, No. 1, 58–69; (1949b) Pressure fluctuations in a turbulent flow, *Dokl. Akad. Nauk SSSR*, **66**, No. 1, 17–20; (1949c) Local structure of atmospheric turbulence, *Dokl. Akad. Nauk SSSR*, **67**, No. 4, 643–646; (1951) Characteristics of wind microstructure in the atmospheric surface layer, *Izv. Akad. Nauk SSSR*, Ser. Geofiz., No. 3, 49–68; (1953) Effect of weak inhomogeneities in the atmosphere on sound and light propagation, *Izv. Akad. Nauk SSSR*, Ser. Geofiz., No. 2, 155–165; (1954) Statistical description of continuous fields, *Trudy Geofiz. In-ta Akad. Nauk SSSR*, No. 24(151), 3–42; (1959a) Effect of Archimedean forces on the structure of the temperature field in a turbulent flow, *Dokl. Akad. Nauk SSSR*, **125**, No. 6, 1246–1248; (1959b) Description of turbulence in terms of Lagrangian Variables, *Adv. in Geophysics*, **6** (Atmospheric Diffusion and Air Pollution), 113-115; (1962a) Some specific features of atmospheric turbulence *J. Geophys. Res.* **67**, No. 8, 3011–3014; (1962b) Some specific features of atmospheric turbulence, *J. Fluid Mech.* **13**, No. 1, 77–81.

Obukhov, A. N., and A. M. Yaglom, (1951) Microstructure of a turbulent flow, *Prikl. Mat. Mekh.*, **15**, No. 1, 3–26, Trans. as NACA TM 1350, June 1953; (1958) Microstructure of developed turbulence, *Proc. Third All-Union Mathematical Congress, An SSSR Press, Moscow*, 3, 542–557; (1959) On the microstructure of atmospheric turbulence, *Quart. J. Roy Meteor. Soc.* **85**, No. 364, 81–90.

Ogura, Y., (1953) The relation between the space- and time-correlation functions in a turbulent flow, *J. Meteor. Soc. Japan*, **31**, No. 11–12,

355–369; (1955) A supplementary note on the relation between the space- and time-correlation functions in a turbulent flow, *J. Meteor. Soc. Japan*, **33**, No. 1, 31–37; (1957) The influence of finite observation intervals on the measurement of turbulent diffusion parameters, *J. Meteor.*, **14**, No. 2, 176–181; (1958) On the isotropy of large-scale disturbances in the upper troposphere, *J. Meteor.*, **15**, No. 4, 375–382; (1959) Diffusion from a continuous source in relation to a finite observation interval, *Adv. Geophys.*, **6** (Atmospheric diffusion and air pollution), 149–159; (1962a) Energy transfer in a normally distributed and isotropic turbulent velocity field in two dimensions, *Phys. Fluids*, **5**, No. 4, 395–401; (1962b) Energy transfer in an isotropic turbulent flow, *J. Geophys. Res.*, **67**, No. 8, 3143–3149; (1963) A consequence of the zero-fourth-cumulant approximation in the decay of isotropic turbulence, *J. Fluid Mech.*, **16**, No. 1, 38–40.

Ogura, Y., and K. Miyakoda, (1953) Some remarks on the "turbulent element model" of the isotropic turbulence, *J. Meteor. Soc. Japan*, **31**, No. 6, 206–218.

Ogura, Y., Y. Sekiguchi and K. Miyakoda, (1953) Classification of turbulent diffusion in the atmosphere, *J. Meteorol. Soc. Japan*, **31**, No. 8, 271–285.

Ohji, M., (1961) A note on the hypothesis of zero fourth-order cumulants in homogeneous turbulence, *Repts Res. Inst. Appl. Mech. Kyushu Univ.*, **9**, No. 36, 163–166; (1962) Some considerations on the four-point dynamical equations of homogeneous turbulence, *Repts Res. Inst. Appl. Mech. Kyushu Univ.*, **10**, No. 38, 33–43.

Okubo, A., (1962a) A review of theoretical models for turbulent diffusion in the sea, *J. Oceanogr. Soc. Japan*, 20th Anniver., Volume, 286–320; (1962b) Horizontal diffusion from an instantaneous point source due to oceanic turbulence, *Techn. Rep. No. 32*, Chesapeake Bay Inst., Johns Hopkins Univ.; (1967) The effect of shear in an oscillatory current on horizontal diffusion from an instantaneous source, *Int. J. Oceanol. Limnol.*, **1**, No. 3, 194–204; (1968a) Some remarks on the importance of the "shear effect" on horizontal diffusion, *J. Oceanogr, Soc. Japan*, **24**, No. 2, 60–69; (1968b) A new set of oceanic diffusion diagrams, *Techn. Rep. No. 38*, Chesapeake Bay Inst., Johns Hopkins Univ.

Okubo, A., and R. V. Ozmidov, (1970) Empirical dependence of the horizontal eddy diffusivity in the ocean on the length scale of the cloud, *Izv. Akad. Nauk SSSR*, Fiz. Atmosf. i Okeana, **6**, No. 5, 534–536.

Olson, F. C. W., and T. Ichiye, (1959) Horisontal diffusion, *Science*, **140**, No. 3384, 1255.

Onsager, L., (1945) The distribution of energy in trubulence (abstr.), *Phys. Rev.*, **68**, No. 11–12, 286; (1949) Statistical hydrodynamics, *Nuovo cimento, (9)*, **6**, Suppl., No. 2, 279–287.

Oort, A. H., and A. Taylor, (1969) On the kinetic energy spectrum near the ground, *Monthly Weather Rev.*, **97**, No. 9, 623–636.

Orlob, G. T., (1959) Eddy diffusion in homogeneous turbulence, *J. Hydraul. Div. Proc. Amer. Soc. Civil Engrs*, **86**, No. 9, 75–101.

Orszag, S. A., (1967) Approximate calculation of the Kolmogorov-Obukhov constant, *Phys. Fluids,* **10,** No. 2, 454–455; (1970a) Analytical theories of turbulence, *J. Fluid Mech.,* **41,** No. 2, 363–386; (1970b) Comments on "Turbulent hydrodynamic line stretching: Consequences of isotropy," *Phys. Fluids,* **13,** No. 8, 2203–2204; (1970c) Indeterminacy of the moment problem for intermittent turbulence, *Phys. Fluids,* **13,** No. 9, 2211–2212; (1975) *The statistical theory of turbulence,* University Press, Cambridge.

Orszag, S. A., and M. D. Kruskal, (1966) Theory of turbulence, *Phys. Rev. Letters,* **16,** No. 11, 141–144; (1968) Formulation of the theory of turbulence, *Phys. Fluids,* **11,** No. 1, 43–60.

Ozmidov, R. V., (1957) Experimental investigation of horizontal turbulent diffusion in seawater and in an artificial reservoir of moderate depth, *Izv. Akad. Nauk SSSR,* Ser. Geofiz., No. 6, 756–764; (1958) Calculation of horizontal turbulent diffusion of impurity spots in seawater, *Dokl. Akad. Nauk SSSR,* **120,** No. 4, 761–763; (1959a) Investigation of meso-scale turbulent exchange in an ocean, using radar observations of floating buoys, *Dokl. Akad. Nauk SSSR,* **126,** No. 1, 63–65; (1959b) Dependence of the horizontal turbulent eddy diffusivity on the relative depth of a reservoir, *Izv. Akad. Nauk SSSR,* Ser. Geofiz., No. 8, 1242–1246; (1960a) Diffusion of an impurity in a homogeneous isotropic turbulent field, *Izv. Akad. Nauk SSSR,* Ser. Geofiz., No. 1, 174–175; (1960b) Rate of turbulent energy dissipation in seawater, and the universal dimensionless constant in the 4/3 law, *Izv. Akad. Nauk SSSR,* Ser. Geofiz., No. 8, 1234–1237; (1965) Energy distribution among motions of different scale in the ocean, *Izv. Akad. Nauk SSSR,* Fiz. Atmosf. i Okeana, **1,** No. 4, 439–444; (1968) *Horizontal Turbulence and Turbulent Exchange in the Ocean,* Nauka Press, Moscow.

Ozmidov, R. V., A. N. Gezentsvei and G. S. Karabashev, (1969) New data on admixture diffusion in the sea, *Izv. Akad. Nauk SSSR,* Fiz. Atmosf. i Okeana, **5,** No. 11, 1191–1204.

Palm, E., (1957) On diffusion of a cluster, *Inst. Weather and Climate Res.,* Rep. No. 2, Oslo.

Panchev, S., (1960) Sur la théory statistique de la turbulence, *C. r. Acad. Sci.,* **250,** No. 4, 661–662; (1969) Kovasznay's spectral theory of turbulence, *Phys. Fluids,* **12,** No. 4, 935–936.

Panchev, S., and D. Kesich, (1969) Energy spectrum of isotropic turbulence at large wavenumbers, *Dokl. Bolg. Akad. Nauk (Proc. Acad. Sci. Bulgaria),* **22,** No. 6, 627–630.

Panofsky, H. A., (1968) The structure constant for the index of refraction in relation to the gradient of index of refraction in the surface layer, *J. Geophys. Res.,* **73,** No. 18, 6047–6049; (1969a) Spectra of atmospheric variables in the boundary layer, *Radio Science,* **4,** No. 12, 1101–1109; (1969b) The spectrum of temperature, *Radio Science,* **4,** No. 12, 1143–1146.

Panofsky, H. A., H. E. Cramer and V. R. K. Rao, (1958) The relation between

Eulerian time and spece spectra, *Quart. J. Roy. Meteor. Soc.*, **84**, No. 361, 270–273.

Panofsky, H., and E. Mares, (1968) Recent measurements of cospectra for heat-flux and stress, *Quart. J. Roy. Meteor. Soc.*, **94**, No. 402, 581–585.

Panofsky, H. A., and R. A. McCormick, (1960) The spectrum of vertical velocity near the surface, *Quart. J. Roy. Meteor. Soc.*, **86**, No. 370, 495–503.

Panofsky, H. A., and F. Pasquill, (1963) The constant of the Kolmogorov law, *Quart. J. Roy. Meteor. Soc.*, **89**, No. 382, 550–555.

Panofsky, H. A., and J. Van der Hoven, (1955) Spectra and cross-spectra of velocity components in the mesometeorological range, *Quart. J. Roy. Meteor. Soc.*, **81**, No. 350, 603–606; (1956) Structure of small scale and middle scale turbulence at Brookhaven. *Sci. Rep. No. 1*, Pennsylvania State University.

Pao, Yih-Ho, (1964) Statistical behavior of a turbulent multicomponent mixture with first-order reactions, *AIAA Journal*, **2**, No. 9, 1550–1559; (1965) Structure of turbulent velocity and scalar fields at large wavenumbers, *Phys. Fluids*, **8**, No. 6, 1063–1075; (1968) Transfer of turbulent energy and scalar quantities at large wavenumbers, *Phys. Fluids*, **11**, No. 6, 1371–1372.

Paquin, J. E., and S. Pond, (1971) The determination of the Kolmogorov constants for velocity, temperature and humidity fluctuations from second- and third-order sturcture functions, *J. Fluid Mech.*, **50**, No. 2, 257–269.

Payne, F. R., and J. L. Lumley, (1966) One-dimensional spectra derived from an airborne hot-wire anemometer, *Quart. J. Roy. Meteor, Soc.*, **92**, No. 393, 397–401.

Pearson, E. A., (1958) Discussion on "The measurement and calculation of stream reaeration ratio" by D. J. O'Connor, *Proc. Seminar Oxygen Relat. in Streams*, Taft San. Engrg Center USPHS, Cincinnati, Ohio, 43–45.

Pearson, J. R. A., (1959) The effect of uniform distortion on weak homogeneous turbulence, *J. Fluid Mech.*, **5**, No. 2, 274–288.

Peierls, R. E., (1955) *Quantum Theory of Solids.* Clarendon Press, Oxford.

Pekeris, C. L., (1947) Note on the scattering of radiation in an inhomogeneous medium, *Phys. Rev.*, **71**, No. 4, 268–269; No. 7, 457.

Perntner, J. M., and F. M. Exner, (1910) Meteorologische Optik, Wien-Leipzig.

Phillips, O. M., (1960) On the generation of sound by supersonic turbulent shear layers, *J. Fluid Mech.*, **9**, No. 1, 1–28; (1967) On the Bolgiano and Lumley–Shur theories of the buoyancy subrange. *Atmospheric Turbulence and Radio Wave Propagation* (A. M. Yaglom and V. I. Tatarskii, eds) Nauka Press, Moscow, 121–129.

Phythian, R., (1969) Self-consistent perturbation series for stationary homogeneous turbulence, *J. Phys., A (Gen. Phys.)*, **2**, No. 2, 181–192; (1972) Comments on the paper "On Phythian's perturbation theory for stationary homogeneous turbulence," *J. Phys. A (Gen. Phys.)*, **5**, No. 2, 246–255.

Pinus, N. Z., (1966) Energy spectra of wind-velocity fluctuations in the free atmosphere, *Meteor. i Gidrol.*, No. 4, 3–11; (1969) On the energy sources of the meso- and microscale eddy motions in the free atmosphere, *Meteor. i Gidrolog.*, No. 1, 3–13.

Pinus, N. Z., and L. V. Shcherbakova, (1966) On the structure of the wind velocity field in the thermally stratified atmosphere, *Izv. Akad. Nauk SSSR*, Fiz., Atmosf. Okeana, **2**, No. 11, 1126–1134.

Pisareva, V. V., (1960) Range of validity of the method of smooth perturbations in the problem of radiation propagation through a medium containing inhomogeneities, *Akust. Zh. (Acoustical Journal)*, **6**, No. 1, 87–91.

Plate, E. J., and S. P. Arya, (1969) Turbulence spectra in a stably stratified boundary layer, *Radio Science*, **4**, No. 12, 1163–1168.

Polowchak, Van M., and H. A. Panofsky, (1968) The spectrum of daily temperatures as a climatic indicator, *Monthly Weather Rev.*, **96**, No. 9, 596–600.

Pond, S., (1965) Turbulence spectra in the atmospheric boundary layer over the sea, Ph.D. thesis, Inst. of Oceanography, Univ. of British Columbia, Vancouver, Canada.

Pond, S., S. D. Smith, P. F. Hamblin, and R. W. Burling (1966) Spectra of velocity and temperature fluctuations in the atmospheric boundary layer over sea, *J. Atmosph. Sci.*, **23**, No. 4, 376–386.

Pond, S., and R. W. Stewart, (1965) Measurement of statistical characteristics of small-scale turbulence, *Izv. Akad. Nauk SSSR*, Fiz. Atmosf. i Okeana, **1**, No. 9, 914–919.

Pond, S., R. W. Stewart and R. W. Burling, (1963) Turbulence spectra in the wind over waves, *J. Atmosph. Sci.*, **20**, No. 3, 319–324.

Portfors, E. A., and J. F. Keffer, (1969) Isotropy in initial period grid turbulence, *Phys. Fluids*, **12**, No. 7, 1519–1520.

Powell, A., (1962) Three-sound-pressures theorem, and its application in aerodynamically generated sound *J. Acoust. Soc. Am.*, **34**, No. 7, 902–906.

Priestley, C. H. B., (1959a) *Turbulent transfer in the lower atmosphere*, Chicago Univ. Press, Chicago; (1959b) The isotropic limit and the microscale of turbulence, *Adv. Geophys.*, **6**, (Atmospheric diffusion and air pollution) 97–100.

Protheroe, W. M., (1954) Preliminary report on stellar scintillation, *Contr. Perkins Observ.*, ser. 2, No. 4.

Proudman, I., (1951) A comparison of Heisenberg's spectrum of turbulence with experiment, *Proc. Cambr. Phil. Soc.*, **47**, No. 1, 158–176; (1952) The generation of noise by isotropic turbulence, *Proc. Roy. Soc.*, **A214**, No. 1116, 119–132; (1962) On Kraichnan's theory of turbulence, *Mécanique de la turbulence* (Coll. Intern. du CNRS à Marseille), Paris, Ed. CNRS, 107–112.

Proudman, I., and W. H. Reid, (1954) On the decay of a normally distributed and homogeneous turbulent velocity field, *Phil. Trans. Roy. Soc.*, **A247**, No. 926, 163–189.

Readings, C. J., and D. R. Rayment, (1969) The high-frequency fluctuation of the wind in the first kilometer of the atmosphere, *Radio Science,* **4,** No. 12, 1127–1131.

Record, F. A., and H. E. Cramer, (1966) Turbulent energy dissipation rates and exchange processes above a non-homogeneous surface, *Quart. J. Roy. Meteor. Soc.,* **92,** No. 394, 519–532.

Reid, W. H., (1955) On the stretching of material lines and surfaces in isotropic turbulence with zero fourth cumulants, *Proc. Cambr. Phil. Soc.,* **51,** No. 2, 350–362; (1956a) The skewness factor according to Obukhoff's transfer theory, *J. Aeronaut. Sci.,* **23,** No. 4, 379–380; (1956b) Two remarks on Heisenberg's theory of isotropic turbulence, *Quart. Appl. Math.,* **14,** No. 2, 201–205; (1956c) On the approach to the final period of decay in isotropic turbulence according to Heisenberg's transfer theory, *Proc. Nat. Acad. Sci. USA,* **42,** No. 6, 559–563; (1960) One-dimensional equilibrium spectra in isotropic turbulence, *Phys. Fluids,* **3,** No. 1, 72–77.

Reid, W. H., and D. L. Harris, (1959) Similarity spectra in isotropic turbulence, *Phys. Fluids,* **2,** No. 2, 139–146.

Reiter, E. R., and A. Burns, (1967) Atmospheric Structure and Clear Air Turbulence, *Atmospheric Turbulence and Radio Wave Propagation,* (A. M. Yaglom and V. I. Tatarskii, eds.) Nauka Press, Moscow, 53–64.

Repnikov, M. N., (1967) On the von Kármán-Howarth equation, *Prikl. Mat. Mekh.,* **31,** No. 2, 327.

Richardson, L. F., (1926) Atmospheric diffusion shown on a distance-neighbour graph, *Proc. Roy. Soc.,* **A110,** No. 756, 709–737; (1929) A search for the law of atmospheric diffusion, *Beitr. Phys. freien Atmosph.,* **15,** No. 1, 24–29; (1952) Transforms for the eddy-diffusion of clusters, *Proc. Roy. Soc.,* **A214,** No. 1116, 1–20.

Richardson, L. F., and H. Stommel, (1948) Note on eddy-diffusion in the sea, *J. Meteor.,* **5,** No. 5, 238–240.

Roberts, P. H., (1957) On the application of a statistical approximation to the theory of turbulent diffusion, *J. Math. Mech.,* **6,** No. 6, 781–799; (1961) Analytical theory of turbulent diffusion, *J. Fluid Mech.,* **11,** No. 2, 257–283.

Robertson, H. P., (1940) The invariant theory of isotropic turbulence, *Proc. Cambr. Phil. Soc.,* **36,** No. 2, 209–223.

Rosen, G., (1960) Turbulence theory and functional integration. I, II, *Phys. Fluids,* **3,** No. 4, 519–524, 525–528; (1967) Functional integration theory for incompressible fluid turbulence, *Phys. Fluids,* **10,** No. 12, 2614–2619; (1970) Evaluation of the characteristic functional for incompressible fluid turbulence, *Phys. Letters,* **A31,** No. 3, 142–143; (1971) Functional calculus theory for incompressible fluid turbulence, *J. Math. Phys.,* **12,** No. 5, 812–820.

Rosen, G., J. A. Okolowski and G. Eckstut, (1969) Functional integration theory for incompressible fluid turbulence. II, *J. Math. Phys.,* **10,** No. 3, 415–420.

Rotta, J. C., (1950) Das Spektrum isotroper Turbulenz im statistischen Gleichgewicht, *Ing.-Arch.*, **18**, No. 1, 60–76; (1953) Similarity theory of isotropic turbulence, *J. Aeronaut. Sci.*, **20**, No. 11, 769–778, 800.

Rozanov, Yu. A., (1967) *Stationary Random Processes.* Holden-Day, San Francisco.

Rytov, S. M., (1937) Diffraction of light by ultrasonic waves, *Izv. Akad. Nauk SSSR*, Ser. Fiz., No. 2, 223–259.

Saffman, P. G., (1963) On the fine-scale structure of vector fields convected by a turbulent fluid, *J. Fluid Mech.*, **16**, No. 4, 545–572; (1967a) The large scale structure of homogeneous turbulence, *J. Fluid Mech.*, **27**, No. 3, 581–593; (1967b) Note on decay of homogeneous turbulence, *Phys. Fluids*, **10**, No. 6, 1349; (1968) Lectures on homogeneous turbulence. *Topics in Nonlinear Physics*, (N. J. Zabusky, ed.) Springer-Verlag, Berlin-Heidelberg-New York, 485–614; (1969) Application of the Wiener-Hermite expansion to the diffusion of a passive scalar in a homogeneous turbulent flow, *Phys. Fluids*, **12**, No. 9, 1786–1789; (1970) Dependence on Reynolds number of high-order moments of velocity derivatives in isotropic turbulence, *Phys. Fluids*, **13**, No. 8, 2193–2194.

Sahashi, K., (1967) Estimation of evaporation rate by use of a sonic anemometer, *Special Contributions*, Geophys. Inst. Kyoto Univ., No. 7, 95–109.

Sandborn, V. A., (1959) Measurements of intermittency of turbulent motion in a boundary layer, *J. Fluid Mech.*, **6**, No. 2, 221–240.

Sandborn, V. A., and R. D. Marshall, (1965) Local isotropy in wind tunnel turbulence, Techn. Rep., Fluid dynamics and diffusion laboratory, Colorado State Univ., USA.

Sato, H., (1951) On the turbulence behind a row of parallel rods, *Proc. 1st Japan Nat. Congr. Appl. Mech.*, 469–473.

Schönfeld, J. C., (1962) Integral diffusivity, *J. Geophys. Res.*, **67**, No. 8, 3187–3199.

Schuert, E. A., (1970) Turbulent diffusion in the intermediate waters of the North Pacific Ocean, *J. Geophys. Res.*, **75**, No. 3, 673–682.

Schweber, S. S., H. A. Bethe and F. Hoffmann, (1955) *Mesons and Fields*, Vol. 1, Row, Peterson and Co., New York.

Sedov, L. I., (1944) Decay of isotropic turbulence in an incompressible fluid, *Dokl. Akad. Nauk SSSR*, **42**, No. 3, 121–124; (1959) *Similarity and Dimensional Methods in Mechanics.* Trans. by M. Friedman, (M. Holt, ed.) Academic Press, New York.

Segal, I., (1965) Algebraic integration theory, *Bull. Amer. Math. Soc.*, **71**, No. 3, 419–489.

Sen, N. R., (1951) On Heisenberg's spectrum of turbulence, *Bull. Calcutta Math. Soc.*, **43**, No. 1, 1–7; (1957) On decay of energy spectrum of isotropic turbulence, *Proc. Nat. Inst. Sci. India*, **A23**, No. 6, 530–533.

Seneca, J., (1955) Mesures de diffusivité turbulente sur des flocons de fumée, *J. Sci. Météor.*, **7**, No. 26, 221–225.

Sheih, C. M., H. Tennekes and J. L. Lumley, (1971) Airborne hot-wire measurements of the small-scale structure of atmospheric turbulence, *Phys. Fluids,* **14**, No. 2, 201–215.

Shilov, G. E., (1963) Integration in infinite-dimensional spaces and the Wiener integral, *Usp. Mat. Nauk,* **18**, No. 2(110), 99–120.

Shimanuki, A., (1961) Diffusion from the continuous source with finite release time, *Sci. Rep. Tohoku Univ., ser. V,* Geophys., **12**, No. 3, 184–190.

Shiotani, M., (1955) On the fluctuation of the temperature and turbulent structure near the ground, *J. Meteor. Soc. Japan,* **33**, No. 3, 117–123; (1957) On the statistics of fluctuations of wind velocity in the lowest atmosphere, *Proc. 6th Japan Nat. Congr. Appl. Mech.*, 311–314; (1963) Some notes on the structures of wind and temperature fields in the lowest air layers, *J. Meteor. Soc. Japan,* **41**, No. 5, 261–269.

Shirokova, T. A., (1959) Second approximation in the method of smooth perturbations, *Akust. Zh. (Acoustical Journal),* **5**, No. 4, 485–489.

Shishov, V. I., (1968) The theory of wave propagation in random media, *Izv. Vyssh. Uchebn. Zaved., Radiofizika,* **11**, No. 6, 866–875.

Shur, G. N., (1962) Experimental investigation of the energy spectra of atmospheric turbulence, *Trudy Tsentr. Aerolog. Obs.*, No. 43, 79–90; (1964) Spectral structure of turbulence in the free atmosphere on the basis of aeroplane measurements, *Trudy Tsentr. Aerolog. Obs.*, No. 53, 43–53.

Shut'ko, A. V., (1964) Statistical theory of turbulence, *Dokl. Akad. Nauk SSSR,* **158**, No. 5, 1058–1060.

Siegel, A., and W. -H. Kahng, (1969) Symmetry properties of Cameron-Martin-Wiener kernels, *Phys. Fluids,* **12**, No. 9, 1778–1785.

Silverman, R. A., (1956) Turbulent mixing theory applied to radio scattering, *J. Appl. Phys.*, **27**, No. 7, 699–705; (1957) Fading of radio waves scattered by dielectric turbulence, *J. Appl. Phys.* **28**, No. 4, 506–511.

Simmons, L. F. G., and C. Salter, (1938) An experimental determination of the spectrum of turbulence, *Proc. Roy. Soc.*, **A165**, No. 920, 73–89.

Sitnikov, K. A., (1958) Invariants for homogeneous and isotropic turbulence in a compressible viscous fluid, *Dokl. Akad. Nauk SSSR,* **122**, No. 1, 29–32.

Smith, F. B., and J. S. Hay, (1961) The expansion of clusters of particles in the atmosphere, *Quart. J. Roy. Meteor. Soc.*, **87**, No. 371, 82–101.

Smith, S. D., (1967) Thrust-anemometer measurements of wind-velocity spectra and of Reynolds stress over a coastal inlet, *J. Marine Res.*, **25**, No. 3, 239–262.

Staras, H., (1955) Forward scattering of radio waves by anisotropic turbulence, *Proc. Inst. Radio Engrs,* **43**, No. 10, 1374–1380.

Stewart, R. W., (1951) Triple velocity correlations in isotropic turbulence, *Proc. Camb. Phil. Soc.*, **47**, No. 1, 146–147; (1963) Regarding the agreement between measured values of the spectra and skeroness of locally isotropic turbulence, *Dokl. Akad. Nauk SSSR,* **152**, No. 3, 324–326. (1969) Turbulence and waves in a stratified atmosphere, *Radio Science,* **4**, No. 12, 1269–1278.

Stewart, R. W., and H. L. Grant, (1962) Determination of the rate of dissipation of turbulent energy near the sea surface in the presence of waves, *J. Geophys. Res.*, **67**, No. 8, 3177–3180.

Stewart, R. W., and A. A. Townsend, (1951) Similarity and self-preservation in isotropic turbulence, *Phil. Trans. Roy. Soc.*, **A243**, No. 867, 359–386.

Stewart, R. W., J. R. Wilson and R. W. Burling, (1970) Some statistical properties of small scale turbulence in an atmospheric boundary layer, *J. Fluid Mech.*, **41**, No. 1, 141–152.

Stilke, G., (1964) On methods and results of our refractive-index measurements carried out with kytoonborne radiosondes and airborne refractometers. *1964 World Conference on Radio Meteorology, Boulder, Colorado,* Ber. Inst. Radiometeorol., No. 9, Hamburg, Germany.

Stommel, H., (1949) Horizontal diffusion due to oceanic turbulence, *J. Marine Res.*, **8**, No. 3, 199–225.

Streit, L., (1965) An introduction to theories of integration over function spaces, *Acta Phys. Austrica,* Suppl. II, 2–21.

Suchkov, B. A., (1958) Sound-amplitude fluctuations in a turbulent medium, *Akust. Zh. (Acoustical Journal),* **4**, No. 1, 85–91.

Sullivan, P. J., (1971) Some data on the distance-neighbour function for relative diffusion, *J. Fluid Mech.*, **47**, No. 3, 601–607.

Sutton, G. W., (1968) Decay of passive scalar fluctuations in isotropic homogeneous turbulence, *Phys. Fluids,* **11**, No. 3, 671.

Swinbank, W. C., (1955) An experimental study of eddy transports in the lower atmosphere, *CSIRO, Div. Meteor. Phys.*, Tech. Pap. No. 2, Melbourne.

Synge, J. L., and C. C. Lin, (1943) On a statistical model of isotropic turbulence, *Trans. Roy. Soc. Can.*, **37**, Sec. 3, 45–79.

Syono, S., and K. Gambo, (1952) On numerical prediction (II), *J. Meteor. Soc. Japan,* **30**, No. 8, 264–271.

Takeuchi, K., (1962) On the nondimensional rate of dissipation of turbulent energy in the surface boundary layer, *J. Meteor. Soc. Japan,* ser. 2, **40**, No. 3, 127–135.

Tan, H. S., and S. C. Ling, (1963) Final stage decay of grid-produced turbulence, *Phys. Fluids,* **6**, No. 12, 1693–1699.

Tanaka, M., (1969) 0–5th cumulant approximation of inviscid Burgers turbulence, *J. Meteor. Soc. Japan,* **47**, No. 5, 373–383.

Tanenbaum, B. S., and D. Mintzer, (1960) Energy transfer in a turbulent fluid, *Phys. Fluids,* **3**, No. 4, 529–538.

Tani, I., and Y. Kobayashi, (1952) Experimental studies on compound jets, *Proc. 1st Japan Nat. Congr. Appl. Mech.*, 465–468.

Tank, W., (1957) The use of large-scale parameters in small-scale diffusion studies, *Bull. Amer. Meteor. Soc.*, **38**, No. 1, 6–12.

Tatarskii, V. I., (1965a) Microstructure of the temperature field in the atmospheric surface layer, *Izv. Akad. Nauk SSSR,* Ser Geofiz., No. 6, 689–699; (1956b) Amplitude and phase fluctuations for waves propagating in a weakly inhomogeneous atmosphere, *Dokl. Akad. Nauk SSSR*, **107**, No. 2, 245–248; (1958) Propagation of waves in locally isotropic turbulent medium with smoothly varying characteristics, *Dokl. Akad. Nauk SSSR*, **120**, No. 2, 289–292; (1959) Interpretation of observations of the scintillations of stars and distant terrestrial light sources, *Proc. Second All-Union Conf. on Stellar Scintillations, Izd. An SSSR*, Moscow; (1960) Radiophysical methods of investigating atmospheric turbulence, *Izv. Vyssh. Uchebn. Zaved.*, **3**, *Radiofizika* No. 4, 551–583; (1961a) *Wave Propagation in a Turbulent Medium*, McGraw-Hill, New York; (1961b) On the primitive functional and its application to the integration of certain equations involving functional derivatives, *Usp. Mat. Nauk*, **16**, No. 4(100), 179–186; (1962a) Second approximation in the theory of propagation of waves in a medium with random inhomogeneities, *Izv. Vyssh. Uchebn. Zaved., Radiofizika*, **5**, No. 3, 490–507; (1962b) Application of quantum field theory to the problem of the decay of homogeneous turbulence, *Zh. Exper. Teor. Fiz.*, **42**, No. 5, 1386–1396; (1967a) Estimation of light depolarization by turbulent atmospheric inhomogeneities, *Izv. Vyssh. Uchebn. Zaved., Radiofizika*, **10**, No. 12, 1762–1765; (1967b) Line-of-sight propagation fluctuations. *Atmospheric Turbulence and Radio Wave Propagation*, (A. M. Yaglom and V. I. Tatarskii, eds.), Nauka Press, Moscow, 314–329; (1969) Light propagation in a medium with random inhomogeneities of refractive index in the Markov random process approximation, *Zh. Exper. Teor. Fiz. (J. Exper. and Theor. Phys.)*, **56**, No. 6, 2106–2117; (1970) An approximation of Markov random process for propagation of short waves in a medium with random inhomogeneities, *Akad. Nauk SSSR, Otdel. Okeanologii, Fiz. Atmosf. Georgr. (Dept. Oceanology, Atmosph. Phys. and Geogr., Acad. Sci. USSR)*, 121 pp., preprint (full translation included in Tatarskii, 1971); (1971) The effects of the turbulent atmosphere on wave propagation, Israel Program for Sci. Translations, Jerusalem, and US Dept. of Commerce, Nat. Techn. Inform. Service, Springfield, Va.

Tatarskii, V. I., and G. S. Golitsyn, (1962) Scattering of electromagnetic waves by turbulent inhomogeneities in the troposphere. *Atmospheric Turbulence (Trudy In-ta Fiziki Atmosf. Akad. Nauk SSSR*, No. 4), 147–202.

Tatarskii, V. I., and L. N. Zhukova, (1959) Chromatic scintillation of stars, *Dokl. Akad. Nauk SSSR*, **124**, No. 3, 567–570.

Tatsumi, T., (1955) Theory of isotropic turbulence with the normal joint-probability distribution of velocity, *Proc. 4th Japan. Nat. Congr. Appl. Mech., Tokyo*, 307–311; (1957a) The theory of decay process of incompressible isotropic turbulence, *Proc. Roy. Soc.*, **A239**, No. 1216, 16–45; (1957b) The energy spectrum of incompressible isotropic turbulence, *Actes IX Congr. Internat. Mécan. Appl.*, **3**, Bruxelles, Univer. Bruxelles, 396–404.

Taylor, G. I., (1935) Statistical theory of turbulence. I–IV, *Proc. Roy. Soc.*, **A151**, No. 874, 421–478; (1937) The statistical theory of isotropic turbulence, *J. Aeronaut. Sci.*, **4**, No. 8, 311–315; (1938a) Production and dissipation of vorticity in a turbulent fluid, *Proc. Roy. Soc.*, **A164**, No. 918, 15–23; (1938b) The spectrum of turbulence, *Proc. Roy. Soc.*, **A164**, No. 919, 476–490.

Taylor, R. J., (1955) Some observations of wind velocity autocorrelations in the lowest layers of the atmosphere, *Austr. J. Phys.*, **8**, No. 4, 535–544; (1958) Thermal structures in the lowest layers of the atmosphere, *Austr. J. Phys.*, **11**, No. 2, 168–176; (1961) A new approach to the measurement of turbulent fluxes in the lower atmosphere, *J. Fluid Mech.*, **10**, No. 3, 449–458.

Tchen, C. M., (1954) Transport processes as foundation of the Heisenberg and Obukhoff theories of turbulence, *Phys. Rev.*, **93**, No. 1, 4–14; (1959) Diffusion of particles in turbulent flow, *Adv. Geophys.*, **6** (Atmospheric diffusion and air pollution), 165–173.

Tennekes, H., (1968a) Simple approximations to turbulent energy transfer in the universal equilibrium range, *Phys. Fluids,* **11**, No. 1, 246–247; (1968b) Simple model for the small-scale structure of turbulence, *Phys. Fluids,* **11**, No. 3, 669–670.

Teptin, G. M., (1965) Structure functions for turbulence in a stably stratified atmosphere, *Izv. Akad. Nauk SSSR,* Fiz. Atmosf. i Okeana, **1**, No. 10, 1091–1094.

Thompson, M. C., J. H. B. Janes and A. W. Kirkpatrick, (1960) An analysis of time variations in tropospheric refractive index and apparent radio path length, *J. Geophys. Res.*, **65**, No. 1, 193–201.

Tollmien, W., (1952–1953) Abnahme der Windkanalturbulenz nach dem Heisenbergschen Austauschsatz als Anfangswertproblem, *Wissensch, Zs. Techn. Hochschule Dresden,* **2**, No. 3, 443–448.

Townsend, A. A., (1947) The measurement of double and triple correlation derivatives in isotropic turbulence, *Proc. Cambr. Phil. Soc.*, **43**, No. 4, 560–570; (1948a) Local isotropy in the turbulent wake of a cylinder, *Austr. J. Sci. Res.*, **1**, No. 2, 161–174; (1948b) Experimental evidence for the theory of local isotropy, *Proc. Cambr. Phil. Soc.*, **44**, No 4, 560–565; (1951a) On the fine-scale structure of turbulence, *Proc. Roy. Soc.*, **A208**, No. 1095, 534–542; (1951b) The diffusion of heat spots in isotropic turbulence, *Proc. Roy. Soc.* **A209**, No. 1098, 418–430; (1951c) The structure of the turbulent boundary layer, *Proc. Cambr. Phil. Soc.*, **47**, No. 2, 375–395; (1954) The diffusion behind a line source in homogeneous turbulence, *Proc. Roy. Soc.*, **A224**, No. 1159, 487–512; (1965) The interpretation of stellar shadow-bands as a consequence of turbulent mixing, *Quart. J. Roy. Meteor. Soc.*, **91**, No. 387, 3–9.

Trout, D., and H. A. Panofsky, (1969) Energy dissipation near the tropapause, *Tellus,* **21**, No. 3, 355–358.

Tsvang, L. R., (1960a) Measurements of the frequency spectra of temperature fluctuations in the surface layer of the atmosphere, *Izv. Akad. Nauk SSSR,* Ser. Geofiz., No. 8, 1252–1262; (1960b) Measurements of the spectra of temperature fluctuations in the free atmosphere, *Izv. Akad. Nauk SSSR*, Ser. Geofiz., No. 11, 1674–1678; (1962) Measurements of turbulent heat fluxes and the spectra of temperature fluctuations. *Atmospheric Turbulence (Trudy In-ta Fiz. Atmosf. Akad. Nauk SSSR*, No. 4), 137–143; (1963) Some characteristics of temperature spectra in the atmospheric boundary layer, *Izv. Akad. Nauk SSSR*, Ser. Geofiz., No. 10, 1594–1600; (1969) Microstructure of temperature fields in the free atmosphere, *Radio Science*, **4**, No. 12, 1175–1177.

Tsvang, L. R., S. L. Zubkovskii, V. N. Ivanov, F. Ya. Klinov, and T. K. Kravchenko, (1963) Measurements of some turbulence properties in the lowest 300-meters of the atmosphere, *Izv. Akad. Nauk SSSR*, Ser. Geofiz., No. 5, 769–782.

Tsuji, H., (1955) Experimental studies on the characteristics of isotropic turbulence behind two grids, *J. Phys. Soc. Japan*, **10**, No. 7, 578–586; (1956) Experimental studies on the spectrum of isotropic turbulence behind two grids, *J. Phys. Soc. Japan*, **11**, No. 12, 1096–1104.

Tsuji, H., and F. R. Hama, (1953) Experiment on the decay of turbulence behind two grids, *J. Aeronaut. Sci.*, **20**, No. 12, 848–849.

Uberoi, M. S., (1953) Quadruple velocity correlations and pressure fluctuations in isotropic turbulence, *J. Aeronaut. Sci.*, **20**, No. 3, 197–204; **21**, No. 2, 142, 1954 (corrections); (1954) Correlations involving pressure fluctuations in homogeneous turbulence, *Nat. Adv. Com. Aeronaut.*, Tech. Note 3116; (1957) Equipartition of energy and local isotropy in turbulent flows, *J. Appl. Phys.*, **28**, No. 10, 1165–1170; (1963) Energy transfer in isotropic turbulence, *Phys. Fluids*, **6**, No. 8, 1048–1056.

Uberoi, M. S., and S. Corrsin, (1953) Diffusion of heat from a line source in isotropic turbulence, *Nat. Adv. Com. Aeronaut.*, Rep. No. 1142.

Uberoi, M. S., and P. Freymuth, (1969) Spectra of turbulence in wakes behind circular cylinders, *Phys. Fluids*, **12**, No. 7, 1359–1363; (1970) Turbulent energy balance and spectra of the axisymmetric wake, *Phys. Fluids*, **13**, No. 9, 2205–2210.

Uberoi, M. S., and S. Wallis, (1966) Small axisymmetric contraction of grid turbulence, *J. Fluid Mech.*, **24**, No. 3, 539–543; (1967) Effect of grid geometry on turbulence decay, *Phys. Fluids*, **10**, No. 6, 1216–1224; (1969) Spectra of grid turbulence, *Phys. Fluids*, **12**, No. 7, 1355–1358.

Ulinich, F. R., and B. Ya. Lyubimov, (1968) Statistical theory of turbulence of an incompressible fluid at large Reynolds numbers, *Zh. Exper. Teor. Fiz.*, **55**, No. 3(9), 951–965.

Van Atta, C. W., and W. Y. Chen, (1968) Correlation measurements in grid turbulence using digital harmonic analysis, *J. Fluid Mech.*, **34**, No. 3, 497–515; (1969a) Measurements of spectral energy transfer in grid turbulence, *J. Fluid Mech.*, **38**, No. 4, 743–763; (1969b) Correlation measurements in turbulence using digital Fourier analysis, *Phys. Fluids*, **12**, Supplement II, II-264, II-269; (1970) Structure functions of turbulence in the atmospheric boundary layer over the ocean, *J. Fluid Mech.*, **44**, No. 1, 145–159.

Van Atta, C. W., and J. Park, (1971) Statistical self-similarity and inertial subrange turbulence. *Statistical Models and Turbulence. Proceedings of the Symposium at San Diego, July 15–21, 1971,* (M. Rosenblatt and C. Van Atta, eds.) Lecture Notes in Physics 12, Springer-Verlag, Berlin-Heidelberg-New York, 402–426.

Van Atta, C. W., and T. T. Yeh, (1970) Some measurements of multi-point time correlations in grid turbulence, *J. Fluid Mech.*, **41**, No. 1, 169–178.

Van der Hoven, J., (1957) Power spectrum of horizontal wind speed in the frequency range from 0.0007 to 900 cycles per hour, *J. Meteor.*, **14**, No. 2, 160–164.

Varadhan, S. R. S., (1966) Asymptotic probabilities and differential equations, *Comm. Pure Appl. Math.*, **19**, No. 3, 261–286.

Villars, F., and V. F. Weisskopf, (1954) The scattering of electromagnetic waves by turbulent atmospheric fluctuations, *Phys. Rev.*, **94**, No. 2, 232–240; (1955) On the scattering of radio waves by turbulent fluctuations of the atmosphere, *Proc. Inst. Radio Engrs*, **43**, No. 10, 1232–1239.

Vinnichenko, N. K., and J. A. Dutton, (1969) Empirical studies of atmospheric structure and spectra in the free atmosphere, *Radio Science,* **4**, No. 12, 1115–1126.

Vinnichenko, N. K., N. Z. Pinus, and G. N. Shur, (1967) Some results of the experimental turbulence investigations in the troposphere. *Atmospheric Turbulence and Radio Wave Propagation*, (A. M. Yaglom and V. I. Tatarskii, eds.) Nauka Press, Moscow, 65–75.

Vlasova, T. G., F. A. Markus and A. M. Cheremukhin, (1971) Measurements of the coherence function of a light beam propagating in an atmosphere, *Izv. Vyssh. Uchebn. Zaved., Radiofisika,* **14**, No. 6, 876–879.

Volkov, Yu. A., (1969) Spectra of velocity and temperature fluctuations in the air flow over the wavy sea surface, *Izv. Akad. Nauk SSSR*, Fiz. Atmosf. Okeana, **5**, No. 12, 1251–1265.

Volkov, Yu. A., V. P. Kukharets and L. R. Tsvang, (1968) Turbulence in the boundary layer of the atmosphere above the steppe and the sea, *Izv. Akad. Nauk SSSR,* Fiz. Atmosf. Okeana, **4**, No. 10, 1026–1041.

Volkovitskaya, Z. I., and V. N. Ivanov, (1970) Dissipation rate of turbulent energy in an atmospheric boundary layer, *Izv. Akad. Nauk SSSR,* Fiz. Atmosf. i Okeana, **6**, No. 5, 435–444.

Walker, E. R., (1964) Atmospheric turbulence characteristics measured at Suffield experimental station, *Beitr. Phys. Atmosph.*, **37**, No. 1, 38–52.

Walton, J. J., (1968) Turbulent spectra from the Kraichnan-Spiegel approximation, *Phys. Fluids,* **11**, No. 2, 435–437.

Webb, E. K., (1964) Ratio of spectrum and structure-function constants in the inertial subrange, *Quart. J. Roy. Meteor. Soc.*, **90**, No. 385, 344–346.

Weiler, H. S., and R. W. Burling, (1967) Direct measurements of stress and spectra of turbulence in the boundary layer over the sea, *J. Atmosph. Sci.*, **24**, No. 6, 653–664.

Wiener, N., (1958) *Nonlinear Problems in Random Theory*, M.I.T. Press, Cambridge, Mass.

Weizsäcker, C. F. von, (1948) Das Spektrum der Turbulenz bei grossen Reynolds'schen Zahlen, *Z. Physik.*, **124**, No. 7–12, 614–627.

Welander, P., (1955) Studies on the general development of motion in a two-dimensional ideal fluid, *Tellus*, **7**, No. 2, 141–156.

Wentzel, D. G., (1958) On the spectrum of turbulence, *Phys. Fluids*, **1**, No. 3, 213–214.

Weyl, H., (1939) The classical groups, their invariants and representations, Princeton Univ. Press, Princeton, N.J.

Wheelon, A. D., (1957) Spectrum of turbulent fluctuations produced by turbulent mixing of gradients, *Phys. Rev.*, **105**, No. 6, 1706–1710; (1958) On the spectrum of a passive scalar mixed by turbulence, *J. Geophys. Res.*, **63**, No. 4, 849–850.

Wieghardt, K., (1941) Zusammenfassender Bericht über Arbeiten zur statistischen Turbulenztheorie, *Luftfahrforsch*, **18**, No. 1, 1–7.

Wilkins, E. M., (1958) Observations on the separation of pairs of neutral balloons and applications to atmospheric diffusion theory, *J. Meteor.*, **15**, No. 3, 324–327; (1960) Dissipation of energy by atmospheric turbulence, *J. Meteor.*, **17**, No. 1, 91–92; (1963) Decay rates for turbulent energy throughout the atmosphere, *J. Atmosph. Sci.*, **20**, No. 5, 473–476.

Wills, J. A. B., (1964) On convection velocities in turbulent shear flows, *J. Fluid Mech.*, **20**, No. 3, 417–432.

Wyld, H. W., (1961) Formulation of the theory of turbulence in an incompressible fluid, *Ann. Phys.*, **14**, No. 2, 143–165.

Wyngaard, J. C., (1971) The effect of velocity sensitivity on temperature derivative statistics in isotropic turbulence, *J. Fluid Mech.*, **48**, No. 4, 763–769.

Wyngaard, J. C., and O. R. Coté, (1971) The budgets of turbulent kinetic energy and temperature variance in the atmospheric surface layer, *J. Atmosph. Sci.*, **28**, No. 2, 190–201.

Wyngaard, J. C., Y. Izumi and S. A. Collins, Jr., (1971) Behavior of the refractive index structure parameter near the ground, *J. Opt. Soc. Amer.*, **61**, No. 12, 1646–1650.

Wyngaard, J. C., and Y. H. Pao, (1971) Some measurements of the fine structure of large Reynolds number turbulence. *Statistical Models and Turbulence. Proceedings of the Symposium at San Diego, July 15–21, 1971* (M. Rosenblatt and C. Van Atta, eds.), Lecture Notes in Physics, Springer-Verlag, Berlin-Heidelberg-New York.

Wyngaard, J. C., and H. Tennekes, (1970) Measurements of the small-scale structure of turbulence at moderate Reynolds numbers, *Phys. Fluids*, **13**, No. 8, 1962–1969.

Yaglom, A. M., (1948) Homogeneous and isotropic turbulence in a viscous compressible fluid, *Izv. Akad. Nauk SSSR*, Ser. Geogr. i Geofiz., **12**, No. 6,

501–522; (1949a) Acceleration field in a turbulent flow, *Dokl. Akad. Nauk SSSR,* **67**, No. 5, 795–798; (1949b) Local structure of the temperature field in a turbulent flow, *Dokl. Akad. Nauk SSSR,* **69**, No. 6, 743–746; (1955) Correlation theory of processes with stationary random increment of order *n*, *Matem. Sb.,* **37**, No. 1, 141–196; (1957) Some classes of random fields in *n*-dimensional space, related to stationary random processes, *Teor. Veroyatn. i ee Primenen.,* **2**, No. 3, 292–337; (1962a) *An Introduction to the Theory of Stationary Random Functions,* Prentice-Hall, Englewood Cliffs, N.J.; (1962b) Some mathematical models generalizing the model of homogeneous and isotropic turbulence, *J. Geophys. Res.,* **67**, No. 8, 3081–3087. (1963) Spectral representations for various classes of random functions, *Proc. Fourth All-Union Mathematical Congr., Izd. An SSSR,* Leningrad, Vol. 1, 250–273; (1966) Effect of fluctuations in energy dissipation rate on the form of turbulence characteristics in the inertial subrange, *Dokl. Akad. Nauk SSSR,* **166**, No. 1, 49–52; (1967a) Inequality for the joint moment of velocity and temperature derivatives in locally isotropic turbulence, *Izv. Akad. Nauk SSSR,* Fiz. Atmosf. i Okeana, **3**, No. 9; (1969) Asymptotic behavior of the turbulence spectrum in various models of spectral energy transfer. *Problems of Hydrodynamics and Continuum Mechanics* (L. I. Sedov 60th Birthday memorial volume), Philadelphia, Soc. Ind. Appl. Math., 781–802.

Yaglom, A. M., and V. I. Tatarskii (editors), (1967) *Atmospheric Turbulence and Radio Wave Propagation* (*Proc. Intern. Colloquium in Moscow*) Nauka Press, Moscow.

Yaglom, I. M., (1947) Homogeneous and isotropic turbulence in a compressible fluid, *Trudy Nauchno-Issled. Uchrezhd. Gidrometsluzhby SSSR,* Ser. 1, No. 30, 10–28.

Zakharov, V. E., (1965) Soluable model of weak turbulence, *Zh. Prikl. Mekh. Tekhn. Fiz.,* No. 1, 14–20.

Zhukova, L. N., (1958) Observation of stellar scintillations by the photoelectric method, *Izv. Glabn. Astron. Obs. Akad. Nauk SSSR,* **21**, No. 162, No. 3, 72–82.

Zimmerman, S. P., (1965) Turbulent atmospheric parameters by contaminant deposition, *J. Appl. Meteor.,* **4**, No. 2, 279–288.

Zubkovskii, S. L., (1962) Frequency spectra of the horizontal wind-velocity fluctuations in the atmospheric surface layer, *Izv. Akad. Nauk SSSR,* Ser. Geofiz., No. 10, 1425–1433; (1963) Experimental investigation of the vertical wind-velocity spectra in the free atmosphere, *Izv. Akad. Nauk SSSR,* Ser. Geofiz., No. 8, 1285–1288.

Zubkovskii, S. L., and B. M. Koprov, (1969) Experimental study of spectra of turbulent heat and momentum fluxes in the atmospheric surface layer, *Izv. Akad. Nauk SSSR,* Fiz. Atmosf. Okeana, **5**, No. 4, 323–331.

# SUPPLEMENTARY REMARKS TO VOLUME 1

1. In the presentation of the result of Arnold (1965) on pp. 158–160 of Vol. 1 the evaluation of the first variation $\delta G[\psi]|_{\psi=\Psi}$ was in fact carried out under the additional condition that the disturbance $\psi'$ does not change the velocity circulation of the flow along both the boundaries (i.e., that $\int_T \frac{\partial \psi'}{\partial z}\, dx = \int_B \frac{\partial \psi'}{\partial z}\, dx = 0$ where T and B mean top and bottom, and the integral is taken over a single period). However, this condition was not explicitly stated in the text. It is clear that it is sufficient for the condition to be valid at the initial instant $t = 0$ only, since the inviscid equations of motion imply the conservation of the velocity circulation. It is also easy to see that this condition does not imply any loss of generality. The reason is that if $\psi'$ is any small disturbance of a steady plane-parallel flow with the velocity profile $U(z)$ which does change the velocity circulation, then $\Psi + \psi'$ can be rewritten as $\Psi_1 + \psi_1'$ where $\Psi_1$ corresponds to a slightly different steady flow (with the velocity profile $U_1(z) = cU(z) + c_1$ where $c$ is close to one and $c_1$ is close to zero) and $\psi'$ is a small disturbance of this new steady flow which does not change the velocity circulation. The arguments presented on pp. 158-160 of Vol. 1 then show that the disturbed flow will be sufficiently close to the new steady flow for all $\tau > 0$; hence it will also be close to the original steady flow for all $t > 0$.

2. In Sect. 8 of Vol. 1 an attempt was made to interpret the approximate agreement of some of the observations of meteorological fields in the atmospheric surface layer with the predictions of the asymptotic theory of the free convection state above values of $-\zeta = -z/L$ of the order of 0.05 as an indication that the asymptotic theory derived for

$-\zeta \to \infty$ begins to be valid at such small values of $-\zeta$. It is easy to see that such an interpretation cannot be correct. In fact, the rate of production of turbulent energy by the wind shear is equal to $u_*^2 \frac{\partial \bar{u}}{\partial z}$ and the rate of production of the same energy by buoyancy is equal to $\frac{g}{T_0} \frac{q}{c_p \rho_0}$ (see, e.g., Sect. 6.5 in Vol. 1). By using any method of estimation of $\overline{wu}\, \frac{\partial \bar{u}}{\partial z}$ it is easy to show that buoyant production cannot exceed the mechanical production of turbulent energy at a height lower than $z \gtrsim -L$. Hence the free convection state with negligible mechanical production can be established only at heights considerably greater than $-L$. Nevertheless definite arguments can be given in favor of the suggestion that there exists a range of much smaller heights of the order of several hundredths or a few tenths of $-L$ where many of the free convection laws (e.g., (7.35)–(7.41) and Eqs. (7.87) for $f_5$ and $f_6$ but not those for $f_3$, $f_4$, and $f_7$) are approximately valid. These arguments are formulated in the recent papers by S. S. Zilitinkevich (1971) and R. Betchov and A. M. Yaglom (1971). However at heights of order $-L$ all these laws must begin to be violated and the second range of their applicability (which is the true free convection range in the sense used in Chapt. 4) will be apparently established only at much greater heights. The numerical coefficients in all these laws can be, of course, different for the two ranges of their validity; moreover the second range may even be nonexistent in the earth's atmosphere, since it may relate to heights too great for validity of the general similarity theory formulated in Chapt. 4 of Vol. 1.

3. The results announced in the preliminary note by Salwen and Grosch (1968) which were referred to at the end of Sect. 2.8 are now published in detail as a paper: Salwen, H., and C. E. Grosch (1972) The stability of Poiseuille flow in a pipe of circular cross-section, *J. Fluid Mech.*, **54**, No. 1, 93–112.

The paper by Izakson (1937) referred in Sect. 5.5 was simultaneously published in English. See Izakson, A., (1937) On the formula for the velocity distribution near walls, *Techn. Phys. USSR*, **4**, No. 2, 155–162.

The paper by Obukhov (1946) several times referred in Sect. 7 was recently translated into English. See Obukhov, A. M., (1971) Turbulence in an atmosphere with a non-uniform temperature, *Boundary-Layer Meteor.*, **2**, No. 1, 7–29.

The main material of the Russian paper by Spalding and Jayatilleke (1965) referred in Sect. 5.7 can also be found in English in Jayalilleke, C. L. V., (1969) The influence of Prandtl number and surface roughness on the resistance of the laminar sub-layer to momentum and heat transfer, *Progr. Heat Mass Transfer*, **1**, 193–329.

# REFERENCES

1. Betchov, R., A. M. Yaglom (1971) "*Remarks on the Similarity Laws for Turbulence in an Unstably Stratified Fluid,*" Izv. Akad. Nauk SSSR, Ser. Fiz. Atmosf. i Okeana, 7, No. 12, pp. 1270–1279.
2. Graebel, W. P., (1970) "*The Stability of Pipe Flow,*" Part 1, "*Asymptotic Analysis for Small Wave-numbers,*" J. Fluid Mech., *43*, No. 2, pp. 279–290.
3. Zilitinkevich, S. S., (1971) "*On Turbulence and Diffusion in Free Convection,*" Izv. Akad. Nauk SSSR, Ser. Fiz. Atmosf. i Okeana, 7, No. 12, pp. 1263–1269.

# ERRATA TO VOLUME 1

In the list line $\overline{13}$ denotes the 13th line from above; line $\underline{24}$ is the 24th line from below.

| Page No. | Line | Reads | Should Read |
|---|---|---|---|
| 17 | $\underline{2}$ | Kraychnan | Kraichnan |
| 18 | $\overline{13}$ | Kraichnana's | Kraichnan's |
| 64 | $\underline{5}$ | Yaglom (1949) | Yaglom (1948) |
| 65 | $\overline{2}$ | velocity, $u$ | velocity $u$ |
| 65 | Eq. (1.86) | $\nu\nabla_i^2$ | $\nu\nabla^2 u_i$ |
| 102 | $\overline{16}$ | Reed | Reid |
| 102 | $\underline{20}$ | Sun Dah-Dhen | Sun Dah-Chen |
| 110 | $\overline{12}$ | Jeffreys (1962) | (1926) |
| 112 | $\overline{8}$ | without any | (without any |
| 118 | $\overline{5}$ | $\sim \frac{U''(Z_0)\varphi(Z_0)}{U'(z_0)} n(z - z_0)$ | $\sim \frac{U''(z_0)\varphi(z_0)}{U'(z_0)} \ln(z - z_0)$ |
| 124 | $\underline{18}$ | Michalke (1968a) | (1969) |
| 125 | $\underline{8}$ | Schlichting (1960; 1959). | Schlichting (1960; 1959)]. |

| Page No. | Line | Reads | Should Read |
|---|---|---|---|
| 126 | $\underline{15}$ | Howard (1963) | (1963a) |
| 130 | $\overline{8}$ | Krilov | Krylov |
| 130 | $\underline{4}$ | $Re_{cr}$ 5767 | $Re_{cr} \approx 5767$ |
| 130 | $\underline{3}$ | 1.025/H. | 1.025/H). |
| 145 | $\overline{6}$ | Gill (1962) | Batchelor and Gill (1962) |
| 145 | $\overline{14}$ | Sexl and Spielberg (1959) | (1958) |
| 147 | $\overline{11}$ | $U(y, z) = (r, \varphi)$ | $U(y, z) = U(r, \varphi)$ |
| 155 | $\underline{7}$ | $Ra_{cr\ min}$ | $Ra_{cr\ min})$ |
| 156 | $\overline{8}$ | Howard (1963) | (1963b) |
| 158 | $\underline{18}$ | $\Delta\Psi = \frac{d^2\psi}{dz^2}$ | $\Delta\Psi = \frac{d^2\Psi}{dz^2}$ |
| 159 | $\overline{17}$ | $\Delta x'$ | $\Delta\psi'$ |
| 159 | $\underline{5}$ | $\delta G[\psi]\|_{\psi=\psi}$ | $\delta G[\psi]\|_{\psi=\Psi}$ |
| 167 | $\overline{15}$ | Howard (1963) | (1963b) |
| 168 | $\underline{21}$ | Yudovich (1967) | (1967c) |
| 168 | $\underline{4}$ | Yalovlev | Yakovlev |
| 171 | $\overline{7}$ | Watson (1960) | (1960a) |
| 181 | $\underline{13}$ | Busse (1966b) | (1967b) |
| 183 | $\underline{10}$ | $\delta_2 > 0, \delta_2 > 0$ | $\delta_1 > 0, \delta_2 > 0$ |
| 186 | $\underline{1}$ | Busse (1967) | (1967b) |
| 187 | $\underline{8}$ | Davis (1969) | (1968) |
| 189 | $\underline{14} - \underline{13}$ | Busse (1967) | (1967c) |
| 190 | $\overline{3}$ | Ponomarenko (1968b) | Ponomarenko (1968b)) |
| 190 | $\overline{8} - \overline{9}$ | $Ra_{cr} < Ra_2 > Ra_1$ | $Ra_{cr} < Ra_2 < Ra_1$ |
| 190 | $\underline{14}$ | or temperature | on temperature |
| 193 | $\underline{8}$ | in the case $\delta > 0$ | in the case $\delta < 0$ |
| 193 | $\underline{6}$ | $\delta < 0$, and $Re > Re_{cr}$ | $\delta < 0$ and $Re > Re_{cr}$ |
| 194 | Fig. 23d | $\delta > 0$ $\delta < 0$ | $\delta > 0$ $\gamma < 0$ |
| 196 | $\overline{12}$ | Sato (1956; 1960) | Sato (1959; 1960) |
| 208 | $\underline{8}$ | Rotta | Rota |
| 208 | $\underline{6}$ | Rotta | Rota |
| 221 | $\overline{7}$ | (3.14) | (3.24) |
| 223 | $\underline{9}$ | $\sqrt{B_2}$ | $\sqrt{b_2}$ |
| 223 | $\underline{8}$ | ... of $u$. | ... of $u$). |
| 224 | $\overline{6}$ | (or semiinvariants | (or semiinvariants) |
| 236 | Eq. (4.34) | $\overline{\}u[\theta(x)]\{}^2$ | $\}\overline{u[\theta(x)]}\{^2$ |
| 242 | $\overline{11}$ | $\Phi[\theta(x)]$ | $\Phi[\theta(x)]$ |
| 243 | $\overline{6}$ | $-\frac{i}{\varphi}$ | $-\frac{i}{6}$ |
| 245 | $\underline{12}$ | Loève (1955) | (1960) |
| 253 | Eq. (4.77) | $\overline{u(t_1)\dots u(t_N)u(t_1+\tau)}\dots u(t_N+\tau)$ | $\overline{u(t_1)\dots u(t_N)u(t_1+\tau)\dots u(t_N+\tau)}$ |

| Page No. | Line | Reads | Should Read |
|---|---|---|---|
| 268 | Eq. (5.16') | $\overline{u_0'^2}$ | $\overline{u_\phi'^2}$ |
| 269 | $\overline{4}$ | $z$ | $r$ |
| 271 | $\overline{2}$ | $\frac{\partial \bar{u}}{\partial z}$ | $\frac{d\bar{u}}{dz}$ |
| 272 | $\overline{14}$ | $\overline{u'w'}$ | $-\overline{u'w'}$ |
| 280 | $\overline{5}$ | Coantic (1966; 1967) | (1966; 1967b) |
| 280 | $\overline{6}-\overline{9}$ | The sentence "Other examples of . . . but must be considered satisfactory." is printed here erroneously; it must be placed on p. 281, line $\underline{20}$ (at the end of the paragraph). | |
| 282 | $\underline{9}$ | Hinze (1961) | (1959) |
| 282 | $\underline{22}$ | Levich (1959) | Levich (1962) |
| 283 | $\overline{3}$ | Ta'ai | Ts'ai |
| 284 | $\overline{2}$ | $\nu$ | $v'$ |
| 284 | $\overline{9}$ | $z_* \frac{\nu}{u_*}$ | $z_* = \frac{\nu}{u_*}$ |
| 285 | $\overline{5}$ | $\sigma_\nu/u_* = A_1$ | $\sigma_u/u_* = A_1$ |
| 286 | $\overline{11}$ | $f_1(h_0u_*/\nu, a\beta, \ldots)$ | $f_1(h_0u_*/\nu, a, \beta, \ldots)$ |
| 287 | $\overline{3}$ | $z_1 e^{\frac{1}{A\sqrt{\frac{1}{2}c_f(z_1)}}} = z_2 e^{\frac{1}{A\sqrt{\frac{1}{2}c_f(z_2)}}}$ | $z_1 e^{-\frac{1}{A\sqrt{\frac{1}{2}c_f(z_1)}}} = z_2 e^{-\frac{1}{A\sqrt{\frac{1}{2}c_f(z_2)}}}$ |
| 287 | $\overline{5}$ | $z_0 = ze^{\frac{1}{A\sqrt{\frac{1}{2}c_f(z)}}}$ | $z_0 = ze^{-\frac{1}{A\sqrt{\frac{1}{2}c_f(z)}}}$ |
| 288 | $\underline{9}$ | $\ln \frac{h_0 u_*}{\nu}$ | $\log \frac{h_0 u_*}{\nu}$ |
| 292 | $\overline{13}$ | $u_* = \sqrt{\frac{\tau}{\rho}}$ | $u_* = \sqrt{\frac{\tau_0}{\rho}}$ |
| 294 | $\overline{10}$ and $\underline{12}$ | Lettau (1967) | (1967a) |
| 297 | $\underline{3}, \underline{2}$ | $u_*f(zu_*/\nu)(2H_1 - z)u_*/\nu)$ | $u_*f(zu_*/\nu, (2H_1 - z)u_*/\nu)$ |
| 299 | caption to Fig. 31 | and (+) | and of F. Dönch (+) |
| 299 | $\underline{3}$ | Millikan (1938) | Millikan (1939) |
| 302 | Eq. (5.44') | $\frac{1}{\kappa}\frac{H_1}{h_0}$ | $\frac{1}{\kappa} \ln \frac{H_1}{h_0}$ |
| 303 | $\underline{10}$ | Millikan (1938) | Millikan (1939) |
| 304 | $\overline{3}$ | 1938 | 1939 |
| 308 | $\underline{12}$ | $C_0$ | $C$ |
| 319 | $\overline{3}$ | $\cos \pi_\eta$ | $\cos \pi\eta$ |
| 325 | Eqs. (5.63), (5.64), (5.68) | ln | log |

| Page No. | Line | Reads | Should Read |
|---|---|---|---|
| 326 | Eqs. (5.69), (5.70) | ln | log |
| 327 | $\underline{11}$ | Jayatillaka | Jayatilleke |
| 328 | Eq. (5.74) | $= \frac{j_0}{\rho \kappa u_*} \varphi(\ldots$ | $= - \frac{j_0}{\rho \kappa u_*} \varphi(\ldots$ |
| 330 | Eq. (5.76) | $= \frac{q_0}{\alpha c_p \rho \kappa u_* z} =$ | $= - \frac{q_0}{\alpha c_p \rho \kappa u_* z} = \cdots$ |
| 332 | $\overline{2}$ | $\bar{T}_+(z) = c_p \rho u_* \ldots$ | $\bar{T}_+(z) = - c_p \rho u_* \ldots$ |
| 332 | Fig. 43 | $T_+$ must be replaced by $\bar{T}_+$; $\ln z_+$ must be replaced by $\log z_+$ (decimal logarithm!). In the figure caption the equation must be printed as: $\bar{T}_+(z) = c_p \rho u_* [\bar{T}(0) - \bar{T}(z)]/q_0$ | |
| 334 | $\underline{2}$ | buld | bulk |
| 334 | $\underline{14}$ | $C(0.7$ | $C(0.7)$ |
| 335 | $\underline{16}$ | $\nu_m$ | $U_m$ |
| 335 | $\underline{15}$ | or | on |
| 338 | Eq. (5.81) | $\rho\chi$ | $\rho x$ |
| 338 | Eq. (5.81') | $\rho\chi$ | $\rho x$ |
| 339 | $\overline{2}$ | The minus sign is missing in the right-hand side. | |
| 339 | $\underline{11}$ | $\vartheta_1 - \vartheta_0$ | $\bar{\vartheta}_1 - \bar{\vartheta}_0$ |
| 339 | $\underline{12}$ | $\vartheta_1 - \vartheta_0$ | $\bar{\vartheta}_1 - \bar{\vartheta}_0$ |
| 340 | $\overline{1}$ | $\vartheta_0$ | $\bar{\vartheta}_0$ |
| 340 | $\overline{2}$ | $\vartheta_1$ | $\bar{\vartheta}_1$ |
| 340 | $\overline{6}$ | $c_p$ | $c_h$ |
| 340 | $\overline{7}$ | $\vartheta_0$ and $\vartheta_1$ | $\bar{\vartheta}_0$ and $\bar{\vartheta}_1$ |
| 340 | Eq. (5.77'') | $\vartheta_0$ | $\bar{\vartheta}_0$ |
| 340 | $\overline{15}$ | $\vartheta_1 - \vartheta_0$ | $\bar{\vartheta}_1 - \bar{\vartheta}_0$ |
| 340 | Eq. (5.82) | $\vartheta_2 - \vartheta_1$ | $\bar{\vartheta}_2 - \bar{\vartheta}_1$ |
| 340 | Eq. (5.82) | The minus sign is missing in the left-hand side | |
| 340 | $\underline{16}$ | $\vartheta_2 - \vartheta_1$ | $\bar{\vartheta}_2 - \bar{\vartheta}_1$ |
| 340 | $\underline{8}$ | The minus sign is missing in the left-hand side | |
| 340 | $\underline{5}$ | $\vartheta_m$ | $\bar{\vartheta}_m$ |
| 341 | $\underline{1}$ | $\frac{q_0}{c_p \rho (\chi + 0.2 u_* z)}$ | $-\frac{q_0}{c_p \rho (\chi + 0.2 u_* z)}$ |
| 342 | $\overline{15}$ | Jayatillaka | Jayatilleke |
| 342 | $\overline{13}$ | Ribaud (1961) | Ribaud (1941) |
| 343 | $\overline{4}$ | $\alpha \ln \frac{\beta_v}{\mathrm{Pr}}$ | $\alpha^{-1} \ln \frac{\beta_v}{\mathrm{Pr}}$ |
| 343 | $\underline{9}$ | $\delta_v$ | $\delta'_v$ |
| 343 | $\underline{8}$ | $-cz^{-m+1}$ | $-cz^{-m+1})$ |
| 344 | $\underline{14}$ | Lin, Moulton and Putman. . . (1933) | Lin, Moulton and Putman. . . (1953) |

| Page No. | Line | Reads | Should Read |
|---|---|---|---|
| 345 | $\underline{24} - \underline{25}$ | [see also Harriot and Hamilton (1965); Hubbard (1964); Hubbard and Lightfoot (1966); Kader (1969); and Gukhman and Kader (1969)]. | see also Harriot and Hamilton (1965)], Hubbard (1964) [see also Hubbard and Lightfoot (1966)], and Kader (1969) [see also Gukhman and Kader (1969)]. |
| 347 | | The first and the second paragraphs must be reversed in order (i.e., the second paragraph must be the first). | |
| 347 | $\underline{20}$ | Thompson | Thomson |
| 347 | $\underline{11}$ | Thompson | Thomson |
| 347 | $\underline{24}$ | case of the heat. . . | case of a rough wall the determination of the heat. . . |
| 347 | $\underline{23}$ | $\vartheta_1 - \vartheta_0$ | $\bar{\vartheta}_1 - \bar{\vartheta}_0$ |
| 347 | $\underline{15}$ | $\vartheta_0$ | $\bar{\vartheta}_0$ |
| 347 | $\underline{13}$ | $\kappa B$ | $\kappa \theta_* B$ |
| 347 | $\underline{9}$ | $B = c^{-1}(h_0 u_*/\nu)^{-k}(\mathrm{Pr})^{-n}$ | $B = c(h_0 u_*/\nu)^k \, \mathrm{Pr}^n$ |
| 354 | $\overline{8}$ | Shih-i Pai (1954) | Pai (1954) |
| 364 | $\underline{12}$ | Boussinesq (1877, 1899) | (1877, 1897) |
| 373 | $\overline{1}$ | $\tau = -\kappa \frac{\bar{u}'}{\bar{u}''}$ | $l = -\kappa \frac{d\bar{u}/dz}{d^2\bar{u}/dz^2}$ |
| 373 | $\overline{10}$ | Lettau (1967) | (1967b) |
| 378 | $\overline{2}$ | $\overline{\vartheta'} u_i$ (twice!) | $\overline{\vartheta' u_i'}$ (twice!) |
| 380 | $\underline{4}$ | equations; | equations); |
| 385 | $\underline{7}$ | $-(\overline{u_1'^2} - \overline{u_2'^2}) \sim (\overline{u_1'^2} - \overline{u_3'^2})$ | $-(\overline{u_1'^2} - \overline{u_2'^2}) - (\overline{u_1'^2} - \overline{u_3'^2})$ |
| 387 | $\underline{12}$ | $\overline{(\vartheta - \bar{\vartheta})^2 = \vartheta'^2}$ | $\overline{(\vartheta - \bar{\vartheta})^2} = \overline{\vartheta'^2}$ |
| 387 | Eq. (6.15′) | $u_\alpha' \overline{\vartheta'^2}$ | $\overline{u_\alpha' \vartheta'^2}$ |
| 388 | $\overline{2}$ | $\bar{u}_a \overline{\frac{\partial \vartheta'^2}{\partial x_a}}$ | $\bar{u}_a \frac{\partial \overline{\vartheta'^2}}{\partial x_a}$ |
| 390 | $\underline{10}$ | $\overline{u_1'^2} = \overline{u_2'^2} = \overline{u_3'^2} = 0$ | is printed erroneously and must be deleted |
| 393 | $\overline{13}$ | (1961a; | (1967a; |
| 393 | $\underline{12}$ | $\overline{\rho u_i} + \overline{\rho u_i'}$ | $\bar{\rho}\bar{u}_i + \overline{\rho u_i'}$ |
| 393 | Eq. (6.28) | $= \frac{1}{2} \overline{\rho u_\alpha u_\alpha} + \overline{\rho u_\alpha' u_\alpha} + \cdots$ | $= \frac{1}{2} \bar{\rho}\bar{u}_\alpha \bar{u}_\alpha + \overline{\rho u_\alpha'} \bar{u}_\alpha + \cdots$ |
| 393 | $\underline{6}$ | $= \frac{1}{2}(\overline{\rho u_\alpha' u_\alpha'} + \overline{\rho u_\alpha' u_\alpha})$ | $= \frac{1}{2}(\bar{\rho}\overline{u_\alpha' u_\alpha'} + \overline{\rho' u_\alpha' u_\alpha'})$ |
| 394 | Eq. (6.31) | $\overline{\rho u_i}, \overline{\rho u_i u_\alpha}, \overline{\rho u_i' u_\alpha}, \overline{\rho u_\alpha' u_i}, \overline{\rho X_i}$ | $\bar{\rho}\bar{u}_i, \bar{\rho}\bar{u}_i\bar{u}_\alpha, \overline{\rho u_i'}\bar{u}_\alpha, \overline{\rho u_\alpha'}\bar{u}_i, \bar{\rho}\bar{X}_i$ |
| 394 | Eq. (6.32) | $\overline{\rho u_i}, \overline{\rho u_i u_\alpha}, \overline{\rho u_i' u_\alpha'}$ | $\bar{\rho}\bar{u}_i, \bar{\rho}\bar{u}_i\bar{u}_\alpha, \bar{\rho}\overline{u_i' u_\alpha'}$ |
| 394 | $\underline{4}$ | $\frac{\partial \overline{\rho u_i}}{\partial t}$ | $\frac{\partial \bar{\rho}\bar{u}_i}{\partial t}$ |
| 395 | Eq. (6.34) | $\overline{\rho u_i' u_\alpha}, \overline{\rho u_\alpha' u_i}$ | $\overline{\rho u_i'}\bar{u}_\alpha, \overline{\rho u_\alpha'}\bar{u}_i$ |

| Page No. | Line | Reads | Should Read |
|---|---|---|---|
| 397 | $\overline{12}$ | The density fluctuations and the vertical velocity | The density fluctuations and the vertical velocity fluctuations |
| 399 | $\underline{3}$ | $\frac{\partial \bar{u}_3}{\partial x_3} = 0$, i.e. $\bar{u}_3 = 0$ | $\frac{\partial \bar{u}_3}{\partial x_3} = 0$, i.e. $\bar{u}_3 = 0$ (since its value at the wall is equal to zero) |
| 401 | Eq. (6.46) | $= -\frac{1}{2}\frac{\partial \overline{u'_a u'_a w'}}{\partial z} - \overline{u'w'} \cdots$ | $= -\overline{u'w'} \cdots$ |
| 401 | $\underline{12}$ | Klug, (1963) | Klug (1963) |
| 402 | $\underline{6}$ | Eq. (6.40). | Eq. (6.40)]. |
| 408 | $\underline{8}$ | $i\zeta \log \zeta$ | $i\zeta \ln \zeta$ |
| 408 | $\underline{7}$ | $i \log \kappa\zeta_0$ | $i \ln \kappa\zeta_0$ |
| 408 | $\underline{6}$ | $\kappa H/u_*$ | $\kappa G/u_*$ |
| 408 | Eq. (6.72) | log (twice!) | ln (twice!) |
| 408 | $\underline{3}$ | $\zeta = u_*/G$ | $\xi = u_*/G$ |
| 410 | Fig. 46(b) | $-d \cdot \text{sign} f$ | $-\alpha \cdot \text{sign} f$ |
| 410 | caption | $u_*/U$ | $u_*/G$ |
| 411 | $\underline{3}$ | log (twice!) | ln (twice!) |
| 411 | $\overline{7}$ | section | chapter |
| 412 | $\overline{2}$ | log | ln |
| 440 | $\underline{8}$ | difficult | not difficult |
| 464 | $\overline{2}$ | movements | moments |
| 488 | Fig. 62 | numbers 0.55 and 0.50 on $y$ axis must be replaced in order | |
| 508 | $\overline{6}$ | Bradley (1968), | Bradley (1968)], |
| 508 | $\overline{7}$ | Bradley (1968)]. | Bradley (1968). |
| 522 | $\overline{7} - \overline{8}$ | $C_2$ in their notation | ($C_2$ in their notation) |
| 545 | $\overline{13}$ | (1926b) | (1962b) |
| 563 | $\underline{5}$ | $c \approx 0.66$ | $c \approx 0.56b$ |
| 577 | $\overline{13}$ | $\overline{U}$ | $\overline{U}$ |
| 652 | Eq. (10.104′) | $\int_0^\infty Z\theta_{00}(Z,t)dZ \int_0^{\infty \theta_{00}(Z,t)dZ} =$ | $\int_0^\infty Z\theta_{00}(Z,t)dZ \int_0^\infty \theta_{00}(Z,t)dZ =$ |
| 667 | $\underline{15}$ | $\overline{u\vartheta}$ | $\bar{u}\bar{\vartheta}$ |
| 685 | $\underline{5}$ | $\mu = \lambda^2/\kappa$ | $\mu = \lambda^2/2\kappa$ |
| 685 | $\underline{9}$ | $\alpha =$ | $a =$ |
| 685 | $\underline{6}$ | $\alpha \to \infty, \ldots, W^2/\alpha \to 2K$ | $a \to \infty, \ldots, W^2/a \to 2K$ |
| 687 | Eq. (10.147) | 0 for $\xi > 1$ or $\xi < 0$ | 1 for $\xi > 0$ and 0 for $\xi < 0$ |
| 687 | $\underline{14}$ | $\frac{\partial \Psi_0}{\partial Z}$ | $Q\frac{\partial \Psi_0}{\partial Z}$ |
| 688 | $\overline{17}$ | $n = \lambda/2K$ | $n = \lambda/2\kappa$ |

| Page No. | Line | Reads | Should Read |
|---|---|---|---|
| 716 | $\underline{17}$ | Fridman | Friedmann |
| 718 | $\overline{13}$ | The paper by Kolmogorov (1935) is erroneously extracted from the general list of Kolmogorov's works. | |
| 721 | $\overline{21} - \overline{22}$ | The last reference in the list of Lindgren's papers must be deleted (it is the title of Lindzen's paper repeated on lines $\overline{23} - \overline{24}$). | |
| 727 | $\overline{8}$ | Thompson | Thomson |
| 732 | $\underline{20} - \underline{21}$ | Donnelly and Schwarz, 1968); | Donnelly and Schwarz, 1965); |
| 737 | $\overline{14}$ | Jayatillaka, (1968) | Jayatilleke, (1965) |
| 745 | $\overline{16}$ | Sun Dah-Dhen | Sun Dah-Chen |
| 745 | The last two lines on the page must be deleted | | |
| 746 | $\overline{12}$ | Tšveng | Tsvang |

# AUTHOR INDEX

Alcaraz, E. C., 705
Alekseev, V. G., 357, 358
Artem'ev, A. V., 719
Arya, S. P., 482, 501, 515

Backus, G., 294
Ball, F. K., 478, 566
Barabanenkov, Yu. N., 654
Bass, J., 229, 757, 763
Batchelor, G. K., 40, 50, 117, 128, 132, 143, 145, 148, 150, 163, 164, 165, 167, 168, 169, 170, 181, 193, 198, 205, 209, 210, 211, 233, 244, 253, 255, 256, 257, 258, 259, 271, 292, 375, 388, 408, 409, 422, 423, 433, 436, 437, 439, 445, 453, 462, 528, 537, 541, 543, 544, 545, 552, 555, 567, 574, 575, 579, 580, 582, 583, 595, 596, 599, 658
Bean, B. R., 510
Becker, H. A., 511, 512
Benton, G. S., 525
Beran, M. J., 297, 304, 706
Bergman, P. G., 667
Berman, S., 470
Betchov, R., 58, 307, 480
Bethe, H. A., 272, 295, 757
Blanch, G., 238, 239
Blokhintsev, D. I., 668
Bodner, S. E., 285
Bolgiano, R., 386, 388, 389, 391, 509
Booker, H. G., 565, 658
Borkowski, J., 457
Boston, N. E. J., 458, 481, 483, 490, 493, 501, 504, 515, 516, 598
Bovsheverov, V. M., 467
Bowden, K. F., 558, 560
Bowne, N. E., 565
Bradbury, L. J. S., 481, 484
Bradshaw, P., 206
Brier, G. W., 555
Brodkey, R. S., 514
Buchshaum, S. J., 515
Bugnolo, D. S., 515
Bull, G., 510
Burling, R. W., 457, 469, 471, 472, 474, 481, 483, 488, 490, 499, 518, 598, 606, 608, 612, 625, 629, 630, 632, 633, 634, 635
Burns, A. B., 469, 470
Busch, N. E., 456, 470, 476, 522
Businger, J., 469
Butler, M. E., 732
Byzova, N. L., 485, 522, 524, 565, 566, 567

Camac, M., 801
Canavan, G. H., 285
Cermak, J. E., 482
Chamberlain, J. W., 294
Champagne, F. H., 116, 368, 454, 460
Chandrasekhar, S., 40, 229, 238, 239, 293, 294, 320, 388, 408
Charnock, H., 561, 566
Chen, H. S., 116, 119, 120, 121, 127, 143, 195, 208, 209, 218, 236, 237, 243, 246, 247, 250, 453, 465, 466, 471, 472, 481, 484, 488, 489, 511, 519, 598, 599, 606, 608, 625, 629, 630, 642, 643, 644
Cheremukhin, A. M., 457, 718, 719
Chernov, L. A., 654, 706
Chisholm, J. H., 680
Chou, P. Y., 114, 243
Chuang, H., 482
Clay, J. P., 515
Clever, W. C., 285
Cocke, W. Y., 579, 580
Cohen, R., 565
Collins, S. A., Jr., 508
Collis, D. C., 258
Comte-Bellot, G., 11, 116, 129, 185, 194, 365, 366, 453, 482, 484

Corrsin, S., 11, 116, 120, 123, 129, 137, 138, 141, 143, 145, 146, 149, 163, 185, 194, 212, 245, 258, 340, 341, 365, 366, 368, 385, 394, 406, 416, 417, 426, 447, 448, 453, 454, 460, 547, 580, 596, 598, 599, 600, 601, 629, 630
Cote, O. R., 456, 457, 470, 476, 477, 478, 481, 482, 483, 501, 502, 503, 506, 518, 519, 522
Cramer, H., 5, 15, 368
Cramer, H. E., 464, 478, 485, 495, 775
Crow, S. C., 285
Csanady, G. T., 558, 560, 565

Dagkesamanskaya, I. M., 721
Daletskii, Yu. L., 803
Dash, R., 271
Davenport, A. G., 524
Davidson, B., 368, 451
Davies, P. O. A. L., 368
Deacon, E. L., 517
deBettencourt, J. T., 680
Defant, A., 558
Deissler, R. G., 270, 271, 530
Deitz, P. M., 705
Dikii, L. A., 523
Dolin, L. S., 706, 716, 717
Donean, M., 511
Donelan, M., 515, 518, 519
Donsker, M. D., 770, 804
Doob, J. L., 5, 12, 88
Dryden, H. L., 181, 462
Dugstad, I., 407
Dumas, R., 11, 206, 366, 368
Dunn, D. J., 456, 465, 700
Durst, C. S., 561
Dutton, J. A., 478, 501, 507, 525, 600

Eckstut, G., 292
Edmonds, F. N., Jr., 509
Edwards, S. F., 798, 802, 804
Elagina, L. G., 510, 519
Ellison, T. H., 216, 240, 647, 668
Ellsaesser, H. W., 479, 525
Emde, F., 279
Emmanuel, C. G., 510
Eschenroeder, A. Q., 221
Exner, F. M., 738

Favre, A., 11, 206, 366, 368
Feller, W., 614, 617, 624, 639
Ferguson, H., 238, 239
Feynman, R. P., 803, 810
Ffowes-Williams, J. E., 335
Fiedler, F., 522
Fisher, M. J., 368
Fock, V. A., 704
Foias, C., 750, 773
Fox, R. L., 315, 316
Francis, G. C., 289
Frenkiel, F. N., 11, 163, 243, 244, 246, 247, 249, 250, 366, 563, 564, 599
Frenzen, P., 478
Freymuth, P., 459, 460, 481, 484, 488, 490
Fried, D. L., 726
Friedrichs, K. O., 803
Friehe, C. A., 458, 481, 484, 501, 504, 515, 598, 603, 606, 608, 625, 626, 629, 630, 631
Frisch, U., 654
Furutsu, K., 770

Galeev, A. A., 801
Gambo, K., 525
Garger, E. K., 565, 566, 567
Gariglio, J., 11, 206, 366, 368
Gel'fand, I. M., 82, 94, 752, 803
Gezentsvei, A. N., 577
Ghosh, K. M., 241
Gibson, C. H., 206, 208, 209, 432, 445, 458, 462, 478, 489, 491, 501, 511, 513, 515, 598, 603, 606, 608, 625, 626, 627, 629, 630, 631, 632
Gibson, M. M., 458, 481, 484, 487, 490, 504
Gifford, F. J., 357, 366, 368, 451, 562, 563, 564, 565, 566, 567
Gjessing, D. T., 457
Gödecke, K., 463
Goldstein, S., 188, 194, 211, 221, 229
Golitsyn, G. S., 334, 375, 404, 405, 408, 409, 523, 524, 595, 601, 658, 701, 702
Gordon, W. E., 658
Gorshkov, N. F., 494, 517
Gossard, E. E., 368, 451, 509, 516, 523, 524
Gracheva, M. E., 705, 706
Granatstein, V. L., 515
Grant, H. L., 116, 481, 482, 483, 485, 487, 490, 501, 504, 507, 514, 588, 600, 634, 647
Green, H. S., 272, 316
Greenhow, J. S., 565, 566
Gunnerson, C. G., 558

Gurvich, A. S., 452, 457, 467, 470, 471, 473, 485, 494, 499, 500, 501, 503, 504, 517, 518, 596, 598, 604, 605, 615, 620, 625, 626, 628, 629, 650, 690, 694, 701, 702, 705, 706, 718, 719, 733

Hama, F. R., 194
Hamblin, P. F., 481, 483, 490, 499
Hamilton, Y. F., 803
Hanzawa, M., 558
Harris, V. G., 233, 240, 241, 368, 454, 460
Hasse, L., 457
Hasselmann, K., 123, 224, 801
Haugen, D. A., 478, 479, 485
Hay, D. R., 511, 565
Heisenberg, 210, 217, 229, 237, 240, 242, 243, 253, 260, 341, 355, 375, 408, 446, 556
Hela, I., 578
Herbstreit, J. M., 703
Herring, J. R., 306, 415, 801, 802
Heskestad, G., 367
Hibbs, A. R., 803
Hill, J. C., 116
Hinze, J. O., 221, 255
Ho, T. L., 706
Hoffmann, W. C., 272, 295, 654, 757
Hopf, E., 223, 282, 410, 745, 756, 775, 782
Hosokawa, I., 772, 804
Hottel, H. C., 511, 512
Howarth, L., 58, 70, 122, 142, 161, 177, 185, 189
Howells, I. D., 219, 445, 446
Huang, T. T., 168, 189
Hughes, B. A., 481, 483, 488, 501, 514
Hughes, B. A., 504
Hutchings, J. W., 525

Ichiye, T., 558, 559
Ievlev, V. M., 314
Imai, K., 525
Inoue, E., 360, 361, 385, 525, 558, 577
Ivanov, V. N., 467, 468, 478, 479, 485, 522, 524, 566
Ivanov-Shits, K. M., 699, 700
Izumi, Y., 456, 457, 476, 477, 478, 482, 483, 501, 503, 506, 508, 518, 519, 522

Jahnke, E., 279
Jain, P. C., 295
Janes, J. H. B., 509
Jeng, D. T., 285
Joseph, J., 577

Kac, M., 804
Kadomtsev, B. B., 307, 308, 409, 410, 801
von Kármán, T., 50, 58, 70, 122, 142, 161, 177, 185, 189, 190, 198, 206, 211, 220, 462
Kahn, A. B., 525
Kahng, W. H., 285
Kaimal, J. C., 456, 457, 476, 477, 478, 479, 482, 483, 485, 501, 503, 506, 518, 519, 522
Kallistratova, M. A., 683, 685, 705, 706
Kampé, de Fériet, J., 18, 20, 27, 527
Kantrowitz, A. R., 801
Karabashev, G. S., 577
Karpman, V. I., 801
Katz, I., 563, 564
Keffer, J. F., 116, 129, 195, 455
Keller, B. S., 589, 634, 647, 648, 654, 733
Kellogg, W. W., 563, 564
Kennedy, D. A., 596
Kesich, D., 406
Khazen, E. M., 273, 284
Khinchin, A. Ya., 8, 10
Kholmyanskii, M. Z., 598, 606, 609, 625, 626, 629, 630, 631, 646
Kilezhenko, V. P., 558, 578
Kirkpatrick, A. W., 509
Kistler, A. L., 120, 123, 137, 143, 146, 206, 453, 457
Klebanoff, P. S., 11, 243, 244, 246, 247, 249, 250, 366, 454, 493, 599
Klinov, F. Ya., 467, 468
Klyatskin, V. I., 335, 704, 706, 714, 715, 717, 721, 773
Kobayashi, Y., 454
Kolchinskii, I. G., 727, 728
Kolesnikova, V. N., 521, 522, 523, 524, 526
Kolmogorov, A. N., 5, 15, 80, 88, 91, 114, 163, 185, 341, 342, 347, 350, 353, 354, 396, 398, 462, 572, 573, 590, 611, 612, 618, 642, 650
Kon, A. I., 727, 728, 729, 731, 733
Koprov, B. M., 368, 451, 452, 455, 467, 470, 478, 479, 485, 494, 508, 518, 519, 520
Kovasznay, L. S. G., 214, 226, 321, 416
Kraichnan, R. H., 222, 224, 261, 268, 269, 270, 272, 284, 290, 295, 299, 306, 308, 309, 409, 411, 415, 530, 531, 568, 569, 570, 590, 601, 668, 682, 757, 762, 763, 765, 769, 781, 786, 788, 802

Krasil'nikov, V. A., 667, 668, 699, 700, 727, 729
Kravchenko, T. K., 467, 468, 499, 500
Kravtsov, Yu. A., 654
Krechmer, S. I., 495, 600
Krinks, R. W., 510
Kruskal, M. D., 267, 410
Krzywoblocki, M. Z., 321
Kucharets, V. P., 478, 479, 485, 509, 518, 519
Kullenberg, G., 558, 560, 561
Kumar, P., 271
Kuo, A. Y-S., 245, 426, 596, 598, 599, 600, 629, 630

Landau, L. D., 150, 331, 349, 360, 545, 584
Lane, J. A., 510
Lappe, U. O., 368, 451
Laufer, J., 206, 255, 335, 454, 462, 492
Lee, D. A., 117, 161, 164
Lee, J., 269, 271, 284, 290, 308, 801
Lee, L., 304
Lee, T. D., 233, 240
Leith, C. E., 215, 230, 234, 240, 285
Leontovich, M. A., 704
Lewis, R. M., 757, 762, 763, 765, 769, 786, 788
Liepmann, H. W., 206, 255, 462
Liepmann, K., 206, 255, 462
Lifshitz, E. M., 150, 331, 349, 360, 545, 584
Lighthill, M. J., 330, 335
Limber, D. N., 251
Lin, C. C., 114, 117, 123, 148, 182, 190, 191, 194, 198, 211, 234, 239, 240, 241, 258, 360, 365, 366, 481, 547, 549, 573
Lin, J. T., 406
Lin, S. C., 484, 489
Ling, S. S., 164, 168, 189
Lions, J. L., 804
Litrak, M. M., 801
Livingston, P. M., 705
Loeffler, A. L., 271
Loitsyanskii, L. G., 148, 156, 163
Losch, F., 279
Lumley, J. L., 2, 150, 168, 234, 367, 392, 420, 425, 456, 469, 470, 481, 484, 493, 517, 580, 598, 603, 606, 608, 625
Lundgren, T. S., 311, 315, 316, 416
Lyubimov, B. Ya., 313, 315, 317

MacCready, P. B., 464, 469, 470, 475, 476, 478, 493
MacEwen, G. F., 577
MacPhail, D. C., 119
Mandelbrot, B., 611, 637
Mares, E., 455, 518, 519
Markus, F. A., 719
Marshall, R. D., 482, 484, 488
Martin, H. C., 495, 497, 511
Masiello, P. J., 627, 629, 630, 631
Matschat, K., 334
Mattioli, E., 123
McBean, G. A., 511, 518, 519
McComb, W. D., 802
McConnell, S. O., 458, 481, 484, 501, 504, 515, 598, 606, 608, 625, 626, 627, 629, 630, 631
McCormick, R. A., 469, 475
McShane, E. J., 803
Meecham, W. C., 284, 285
Meetz, K., 235, 236, 239, 240, 255
Megaw, E. C. S., 658
Meksyn, D., 221
Meleshkin, B. N., 504
Millionshchikov, M. D., 68, 156, 158, 163, 242, 272, 292
Mills, R. R., 120, 123, 129, 137, 143, 146
Millsaps, K., 227
Minlos, R. A. 804
Mintzer, D. J., 236, 668
Mirabel, A. P., 284
Mitsuta, Y., 511, 515, 518, 519
Miyake, M., 457, 511, 515, 518, 519
Miyakoda, K., 219, 545
Mock, W. C., 462
Moilliet, A., 481, 482, 483, 485, 487, 488, 490, 501, 504, 507, 514, 588, 600, 647
Monin, A., 219, 310, 311, 403, 418, 463, 472, 473, 518, 521, 522, 523, 524, 526, 537, 560, 575, 576, 673, 682, 763, 791
Mordukhovich, M. I., 705
Morozov, S. A., 485, 522, 524
Moyal, J. E., 321
Muller, E. A., 334
Munch, G., 259
Myrup, L. O., 522

Nasmyth, P. W., 458, 481, 483, 490
Neumann, J., 88, 494
Nisbet, I. C. T., 116
Norton, K. A., 703

Novikov, E. A., 311, 424, 427, 431, 432, 439, 441, 443, 547, 551, 588, 595, 596, 609, 611, 615, 617, 620, 627, 637, 770, 772, 773, 796, 806
Nye, J. O., 514

O'Brien, E. E., 120, 123, 289, 290, 307
O'Brien, V., 137, 143, 146
Obukhov, A. N., 50, 56, 132, 201, 213, 216, 217, 227, 253, 256, 341, 350, 355, 369, 371, 376, 378, 381, 385, 389, 390, 393, 403, 407, 408, 409, 420, 463, 464, 523, 545, 547, 556, 567, 571, 590, 600, 611, 663, 668, 676
Ogura, Y., 219, 283, 284, 285, 289, 365, 525, 536, 545
Ohji, M., 271, 284
Okolowski, J. A., 292
Okubo, A., 560, 561, 565, 577, 578
Olson, F. C. W., 558, 559
Onsager, L., 340, 341, 355, 416
Oort, A. H., 522, 524
Orlob, G. T., 558, 560
Orszag, S. A., 261, 267, 272, 284, 307, 409, 410, 411, 579, 603, 637, 639, 640
Ozmidov, R. V., 526, 558, 560, 561, 566, 567, 574, 577, 578

Pakhomov, V. V., 457, 718
Palm, E., 575
Palowchak, 523
Panchev, S., 295, 406, 457
Panofsky, H., 368, 455, 456, 457, 469, 470, 474, 475, 476, 478, 485, 501, 502, 506, 508, 517, 518, 519, 521, 522, 523, 524
Pao, Y.-H., 141, 394, 416, 482, 483, 488, 493, 494, 598, 606, 608, 625, 646
Paquin, J. E., 467, 471, 472, 495, 503, 511, 515
Park, J., 245, 597, 598, 643, 644
Pasquill, F., 485
Patrick, R. M., 801
Payne, F. R., 469, 470
Pearson, J. R. A., 426, 558
Peierls, R. E., 801, 804
Pekeris, C. L., 668
Pergament, S., 290
Perntner, J. M., 738
Petschek, H. E., 801
Phillips, O. M., 330, 420
Phythian, R., 307, 801
Pinus, N. Z., 469, 522, 525, 527, 600
Pisareva, V. V., 705
Plate, E. J., 482, 501, 515
Pond, S., 457, 467, 471, 472, 481, 483, 488, 490, 495, 499, 500, 503, 511, 515, 596, 597, 598, 606, 607, 612, 625
Portfors, E. A., 116, 129, 195, 453, 455
Portman, P. A., 680
Priestley, C. H. B., 475, 493
Protheroe, W. M., 732, 736
Proudman, I., 77, 148, 167, 169, 170, 189, 205, 235, 239, 240, 264, 274, 279, 281, 282, 283, 307, 308, 333, 334, 462

Rao, V. E. K., 368
Raymont, D. R., 478, 479, 485
Readings, C. J., 478, 479, 485
Record, F. A., 464, 478, 485, 495
Reid, W. H., 77, 148, 167, 169, 227, 233, 234, 239, 240, 241, 264, 274, 279, 281, 282, 283, 288, 547, 549, 573, 582, 583
Reiter, E. R., 469, 470
Repnikov, M. N., 291
Richardson, L. F., 525, 537, 552, 553, 555, 556, 558, 574
Roberts, P. H., 290, 294, 530, 568
Robertson, H. P., 39
Roche, J. E., 680
Rosen, G., 292, 804, 808, 811
Rotta, J. C., 211, 239, 240, 241
Rozanov, Yu. A., 5
Rytov, S. M., 654, 663, 676

Saffman, P. G., 174, 175, 185, 261, 285, 425, 426, 431, 530, 531, 579, 603
Sahashi, K., 511
Salter, C., 10, 206
Sandborn, V. A., 482, 484, 488, 596, 599
Sato, H., 206
Schoenberg, I. J., 88
Schonfeld, J. C., 577
Schubauer, G. B., 462
Schuert, E. A., 560, 577, 578
Schulman, L. S., 803
Schwarz, W. H., 205, 206, 208, 209, 210, 462, 481, 484, 487, 503, 511, 513
Schweber, S. S., 295, 757
Scramstad, H. K., 462
Sedov, L. I., 159, 161, 177, 179, 181, 185, 186
Segal, I., 803
Sekiguchi, Y., 545
Sen, N. R., 241

Sendner, H., 577
Seneca, J., 563, 564
Shapiro, H. N., 803
Shcherbakova, L. V., 527
Sheih, C. M., 456, 469, 470, 481, 484, 493, 598, 603, 606, 608, 625
Shilov, G. E., 803
Shimanuki, A., 536
Shiotani, M., 464, 470, 495
Shirokova, T. A., 705
Shishov, V. I., 706, 721
Shmeter, 469, 522, 525, 527
Shur, G. N., 392, 420, 469, 522, 525, 527
Shut'ko, A. V., 309, 410
Siegel, A., 285
Silverman, R. A., 658, 678
Simmons, L. F. G., 10, 206
Sitnikov, K. A., 320, 328
Sleicher, C. A., 116
Smith, S. D., 457, 481, 483, 490, 499, 518, 565
Spiegel, E. A., 222
Staras, H., 658
Stegen, G. R., 458, 481, 484, 491, 501, 504, 515, 598, 606, 608, 625, 627, 629, 630, 631, 632
Stewart, R. W., 117, 120, 123, 165, 167, 189, 190, 191, 192, 193, 205, 206, 207, 218, 235, 236, 238, 239, 241, 246, 247, 457, 458, 462, 469, 470, 471, 472, 481, 482, 483, 485, 487, 488, 490, 493, 501, 518, 522, 588, 596, 597, 598, 600, 606, 607, 608, 609, 611, 612, 617, 625, 629, 630, 632, 633, 634, 635, 647
Stilke, G., 509
Stommel, H., 558, 559
Streit, L., 803
Suchkov, B. A., 700, 701
Sullivan, P. J., 576, 578
Suomi, V. E., 469
Sutton, G. W., 212
Swinbank, W. C., 478, 497
Synge, J. L., 114
Syono, S., 525

Takeuchi, K., 478, 497
Tan, H. S., 117, 164
Tanaka, M., 273
Tanenbaum, B. S., 236
Tani, I., 454
Tank, W., 563
Tatarskii, V. I., 467, 494, 496, 502, 510, 566 567, 654, 656, 658, 662, 666, 676, 690, 695, 697, 701, 702, 704, 705, 706, 714, 715, 717, 720, 725, 726, 727, 729, 731, 733, 735, 736, 737, 740, 741, 757, 759, 770, 773, 804, 808
Tatsumi, T., 264, 275, 276, 282, 283
Taylor, A., 522, 524
Taylor, G. I., 10, 11, 114, 116, 119, 143, 144, 258, 362, 464
Taylor, R. J., 478, 495, 497
Tchen, C. M., 216, 576
Tennekes, H., 406, 456, 469, 470, 481, 484, 493, 598, 602, 603, 606, 608, 625, 646
Teptin, G. M., 392
Thompson, M. C., 509, 703
Time, N. S., 705
Titt, E. W., 223, 282, 410, 782
Tollmien, W., 235, 236, 238
Townsend, A. A., 120, 129, 143, 145, 163, 164, 168, 189, 190, 191, 192, 193, 198, 205, 206, 207, 218, 225, 233, 235, 236, 238, 239, 241, 243, 244, 257, 258, 259, 271, 292, 424, 425, 427, 445, 454, 462, 487, 493, 537, 544, 582, 595, 596, 599, 601, 692
Trout, D., 478, 485
Tsuji, H., 193, 194
Tsvang, L. R., 368, 451, 452, 467, 468, 470, 478, 479, 485, 494, 497, 501, 505, 508, 509, 517, 518, 519, 666

Uberoi, M. S., 116, 129, 132, 143, 191, 194, 195, 196, 197, 206, 207, 208, 209, 236, 237, 238, 240, 244, 247, 248, 255, 257, 258, 341, 365, 453, 455, 459, 460, 481, 484, 488, 490
Ulinich, F. R., 313, 315, 317

van Atta, C. W., 116 , 119, 120, 121, 127, 143, 195, 208, 209, 236, 237, 243, 245, 246, 247, 250, 453, 465, 466, 471, 472, 481, 484, 488, 597, 598, 599, 603, 606, 608, 625, 629, 630, 642, 643, 644
van der Hoven, J., 469, 521
Varadhan, S. R. S., 804
Vilenkin, N. Ya., 82, 94, 752, 803
Villars, F., 386, 658
Vinnichenko, N. K., 478, 501, 507, 522, 525, 527, 600
Vlasova, T. G., 719

Vogel, W. M., 481, 483, 485, 488, 501 504, 507, 514, 600
Voipio, A., 578
Volkov, Yu. A., 469, 478, 479, 485, 499, 509
Volkovitskaya, Z. I., 478, 485
Vrebalovich, T., 206, 453, 457

Walker, E. R., 524
Wallis, S., 116, 129, 143, 194, 453
Walton, J. J., 234, 240
Webb, E. K., 356
Weiler, H. S., 457, 469, 474, 518
Weisskopf, V. F., 386, 658
Weizsacker, C. F. von, 217, 341, 355, 556
Welander, P., 424
Wentzel, D. G., 294
Wheelon, A. D., 259, 386
Wieghardt, K., 123
Wiener, N., 284, 285
Wilkins, E. M., 478, 561, 566
Williams, G. C., 511, 512
Williams, R. B., 458, 469, 478, 481, 484, 491, 501, 504, 515, 598, 629, 630, 631, 632
Wills, J. A. B., 368
Wilson, J. R., 471, 472, 598, 606, 608, 625, 629, 630, 632, 633, 635
Wyld, H. W., 294, 297, 298, 303, 705, 764, 765
Wyngaard, J. C., 456, 457, 458, 470, 476, 477, 478, 481, 482, 483, 488, 493, 501, 502, 503, 506, 507, 508, 518, 519, 522, 598, 606, 608, 625, 646

Yaglom, A. M., 2, 5, 18, 20, 42, 48, 50, 53, 63, 82, 94, 98, 117, 132, 136, 148, 160, 161, 222, 230, 231, 232, 253, 256, 321, 357, 358, 369, 371, 375, 386, 400, 403, 405, 406, 407, 408, 409, 452, 463, 470, 494, 510, 589, 611, 613, 615, 620, 626, 628, 629, 634, 642, 647, 648, 650, 804
Yamomoto, K., 804
Yeh, T. T., 247, 250

Zakharov, V. E., 801
Zhukova, L. N., 732, 737, 740, 741
Zimmerman, S. P., 478, 563, 566, 567
Zubkovskii, S. L., 455, 467, 468, 471, 476, 478, 479, 485, 501, 503, 518, 520, 596 604, 605, 625

# SUBJECT INDEX


Acceleration field
  in locally isotropic turbulence, 368
Asymptotic behavior
  of correlations, 169
  of spectra, 171

Balance equations
  for isotropic turbulence, 141
Buoyancy, 387

Chemical reactions
  asymptotic spectral behavior, 447
  closure hypothesis for, 416
Chromatic scintillation, 737
Closure
  by power series in Re, 295
Closure hypothesis
  constant skewness of differences, 403
  chemical reaction, 416
  Kraichnan LHDI, 409
  Millionshchikov (zero fourth cumulant), 407
  on spectral transfer, 405
  for stratification, 417
Compressible turbulence, 317
  final period of decay, 321
Correlation
  as Fourier transform of spectrum, 8, 10
  lateral, 39
    derivatives of, 44
  longitudinal, 39
    derivatives of, 44
  pressure, 130
  space-time, 290
  temperature, 136
  tensor, 38
  velocity, 117
Cumulants, 21

Decomposition
  of random field into potential and solenoidal, 56

Diagram technique, 297
Diffusion, 527
  closure hypothesis for moment equations, 567
  hypothesis on probability density, 573
  hypothesis on relative concentration distribution, 577
  Markovian model, 571
  of point pairs, 536
  relative, 551
    measurements, 556
  single particles, 527
Dissipation fluctuations, 584
  effect on small-scale properties, 640
  experimental, 625
  refined similarity hypothesis, 650
  spectrum of, 604
  statistical characteristics of, 594
  Yaglom model, 613

Equilibrium range similarity
  experimental verification, 486

Fields
  axisymmetric, 40
  helical, 40
  homogeneous increments, multi-dimensional, 95
  homogeneous increments, scalar, 93
  isotropic, mixed, 47
    potential, 50
    scalar, 30
    solenoidal, 49
    vector, 35
  locally homogeneous, 93
  locally isotropic, mixed, 103
    multi-dimensional, 101
    scalar, 98
    vector, 102
      potential, 111
      solenoidal, 112
Filters, effect of, 6

Final period (of decay)
of compressible turbulence, 321
experimental, 162
influence of spectrum singularity on, 174
isotropic theory, 152
Five-thirds law, 355
Fluxes, spectra of measurements, 517
Fourier-Stieltjes integral, 5
Frozen turbulence hypothesis, 361
Functional Characteristic, 743
with external forces, 763
space-time, 760
spectral equation for, 751
Functional equation solutions
Edwards scheme, 798
by expanding in Reynolds number, 791
by functional power series, 773
other expansions, 798
self-consistent approaches, 801
zero-order in Reynolds number, 783
Functional integrals, 802

Hopf equation, 750
spectral form, 756

Inertial subrange forms
experimental verifications, 461
Integral
Corrsin, 148
Loitsyanskii, 146
Intermittancy
of dissipation, 584
of small scale components, 600
Invariants
in isotropic compressible turbulence, 317

Kolmogorov similarity hypothesis, on dissipation, 590

Lagrangian statistics
in locally isotropic turbulence, 358
Local isotropy
defined, 341
experimental verification, 453
Log normality
of dissipations fluctuations, 625

Macrocomponent, of a field, 18, 20
Microcompoent, of a field, 18, 20
Millionschchikov (zero-fourth cumulant-quasi-Gaussian) hypothesis
applied to pressure correlations and spectra, 250
compared with data, 241
for estimating acceleration fluctuations, 256
for temperature fluctuations, 286
Moments,
higher order, 21
of isotropic fields, three-point, 75
two-point fourth order, 67
two-point third order, 64

Perturbation theory, 295
Pressure field, in locally isotropic turbulence, 368
Probability distributions, equations for, 310
Process, stationary increments
multi-dimensional, 91
scalar, 80

Quasi-equilibrium hypothesis (meso-scale), 210
Quasi-equilibrium range, defined, 351

Richardson's four-thirds law, 551

Scalar fields (other than temperature), measurements, 509
Scale
differential, 35
lateral, 46
longitudinal, 46
integral, 35
lateral, 46
longitudinal, 46
isotropic potential field, 56
isotropic solenoidal field, 56
Scattering, 674
Scintillation, 729
Self-preservation hypothesis
experimental verifications, 189
Kolmogorov (small scale), 197
conditions for existence of, in grid turbulence, 204
spectral formulations, 185
temperature fluctuations, 212
von Kármán, 177
weakened, 181
Similarity hypothesis
Kolmogorov, 345
for temperature field, 382
Skewness
of derivation, for isotropic fields, 58
Sound generation by turbulence, 328

Spectra
  cumulant, 21
  equations for, in local isotropy, 395
  fine-scale measurements, 479
  of fluxes (experimental), 517
  inertial sub-range velocity measurements, 467
  isotropic potential field, 51
  isotropic solenoidal field, 51
  Kolmogorov, 357
  large-scale measurements, 517
  pressure, isotropic equations for, 135
  temperature, equations for, 137
  of three-point moments for isotropic fields, 77
  of two-point fourth-order moments for isotropic fields, 69
  of two-point third-order moments for isotropic fields, 66
  velocity, isotropic equations for, 123
Spectral constants
  experimental determination, 489
Spectral transfer hypothesis, 212
  applied to decaying grid turbulence, 235
  Goldstein, 221
  Hasselmann (for correlation functions), 224
  Heisenberg, 217
    modified, 218
  Kovasznay, 214
  Leith, 215
  Obukhov, 215
    modified, 216
  self-preserving solutions, 237
  spectral form, asymptotic, 230
    experimental verifications, 232
    quasi-equilibrium, 225
  von Kármán, 219
    modified, 222
Spectral representations
  curl, 25
  derivatives of homogeneous fields, 23
  divergence, 25
  of homogeneous scalar fields, 16
  inversion formula for, 5
  locally homogeneous fields, multi-dimensional, 96
    scalar, 94
  solenoidal homogeneous vector field, 26
  stationary increment (processes), 86
  stationary processes, derivatives of, 12
    scalar, 1
    vector, 14
  of a vector field, 19
Spectrum
  cross, 19
  of dissipation fluctuations, 604
  as Fourier transform of correlation, 8, 10
  lateral, derivatives of, 44
    one-dimensional, 43
    three-dimensional, 41
  longitudinal, derivations of, 44
    one-dimensional, 43
    three-dimensional, 41
  nonnegative, 9
  one-dimensional, 32
  of a scalar field, 18
  of a stationary process, 7
  three dimensional, 17
Spectrum behavior, for dissipation range with chemical reactions, 447
  for temperature, large Prandtl number, 433
    small Prandtl number, 439
  for velocity, 421
Stratifications
  closure hypothesis for, in local isotropy, 417
  in locally isotropic turbulence, 387
Stretching, of material lines and surfaces, 578
Structure function, 83
  equations for, derived without large-scale isotorpy, 401
  measurements (velocity), 462
  of multi-dimensional process, 92
Structure, small scale, 337

Taylor's hypothesis, 11, 361
  applications to measurement, 449
Temperature field
  in locally isotropic turbulence, 377
Temperature field measurements
  dissipation range, 511
  inertial-convective subrange, 494
Third-order moment equations, 260
Third-order moment equations closure
  by cumulant discard, 271
  by moment discard, 267
Turbulence, isotropic
  experimental realization, 113
Two-thirds law, 354

Vorticity field
  in locally isotropic turbulences, 368

Wave propagation
  amplitude and phase fluctuations, 685
  electromagnetic, 653
  scattering, 674
  source, 668
  stellar images, chromatic scintillations, 737
    jitter, 721
    scintillations, 729
  telescope averaging, 729
  time spectra of intensity fluctuations, 733
  strong amplitude fluctuations, 704

A CATALOG OF SELECTED

# DOVER BOOKS

IN SCIENCE AND MATHEMATICS

# Astronomy

CHARIOTS FOR APOLLO: The NASA History of Manned Lunar Spacecraft to 1969, Courtney G. Brooks, James M. Grimwood, and Loyd S. Swenson, Jr. This illustrated history by a trio of experts is the definitive reference on the Apollo spacecraft and lunar modules. It traces the vehicles' design, development, and operation in space. More than 100 photographs and illustrations. 576pp. 6 3/4 x 9 1/4. 0-486-46756-2

EXPLORING THE MOON THROUGH BINOCULARS AND SMALL TELESCOPES, Ernest H. Cherrington, Jr. Informative, profusely illustrated guide to locating and identifying craters, rills, seas, mountains, other lunar features. Newly revised and updated with special section of new photos. Over 100 photos and diagrams. 240pp. 8 1/4 x 11. 0-486-24491-1

WHERE NO MAN HAS GONE BEFORE: A History of NASA's Apollo Lunar Expeditions, William David Compton. Introduction by Paul Dickson. This official NASA history traces behind-the-scenes conflicts and cooperation between scientists and engineers. The first half concerns preparations for the Moon landings, and the second half documents the flights that followed Apollo 11. 1989 edition. 432pp. 7 x 10. 0-486-47888-2

APOLLO EXPEDITIONS TO THE MOON: The NASA History, Edited by Edgar M. Cortright. Official NASA publication marks the 40th anniversary of the first lunar landing and features essays by project participants recalling engineering and administrative challenges. Accessible, jargon-free accounts, highlighted by numerous illustrations. 336pp. 8 3/8 x 10 7/8. 0-486-47175-6

ON MARS: Exploration of the Red Planet, 1958-1978--The NASA History, Edward Clinton Ezell and Linda Neuman Ezell. NASA's official history chronicles the start of our explorations of our planetary neighbor. It recounts cooperation among government, industry, and academia, and it features dozens of photos from Viking cameras. 560pp. 6 3/4 x 9 1/4. 0-486-46757-0

ARISTARCHUS OF SAMOS: The Ancient Copernicus, Sir Thomas Heath. Heath's history of astronomy ranges from Homer and Hesiod to Aristarchus and includes quotes from numerous thinkers, compilers, and scholasticists from Thales and Anaximander through Pythagoras, Plato, Aristotle, and Heraclides. 34 figures. 448pp. 5 3/8 x 8 1/2. 0-486-43886-4

AN INTRODUCTION TO CELESTIAL MECHANICS, Forest Ray Moulton. Classic text still unsurpassed in presentation of fundamental principles. Covers rectilinear motion, central forces, problems of two and three bodies, much more. Includes over 200 problems, some with answers. 437pp. 5 3/8 x 8 1/2. 0-486-64687-4

BEYOND THE ATMOSPHERE: Early Years of Space Science, Homer E. Newell. This exciting survey is the work of a top NASA administrator who chronicles technological advances, the relationship of space science to general science, and the space program's social, political, and economic contexts. 528pp. 6 3/4 x 9 1/4. 0-486-47464-X

STAR LORE: Myths, Legends, and Facts, William Tyler Olcott. Captivating retellings of the origins and histories of ancient star groups include Pegasus, Ursa Major, Pleiades, signs of the zodiac, and other constellations. "Classic." – *Sky & Telescope.* 58 illustrations. 544pp. 5 3/8 x 8 1/2. 0-486-43581-4

A COMPLETE MANUAL OF AMATEUR ASTRONOMY: Tools and Techniques for Astronomical Observations, P. Clay Sherrod with Thomas L. Koed. Concise, highly readable book discusses the selection, set-up, and maintenance of a telescope; amateur studies of the sun; lunar topography and occultations; and more. 124 figures. 26 halftones. 37 tables. 335pp. 6 1/2 x 9 1/4. 0-486-42820-6

# Chemistry

MOLECULAR COLLISION THEORY, M. S. Child. This high-level monograph offers an analytical treatment of classical scattering by a central force, quantum scattering by a central force, elastic scattering phase shifts, and semi-classical elastic scattering. 1974 edition. 310pp. 5 3/8 x 8 1/2. 0-486-69437-2

HANDBOOK OF COMPUTATIONAL QUANTUM CHEMISTRY, David B. Cook. This comprehensive text provides upper-level undergraduates and graduate students with an accessible introduction to the implementation of quantum ideas in molecular modeling, exploring practical applications alongside theoretical explanations. 1998 edition. 832pp. 5 3/8 x 8 1/2. 0-486-44307-8

RADIOACTIVE SUBSTANCES, Marie Curie. The celebrated scientist's thesis, which directly preceded her 1903 Nobel Prize, discusses establishing atomic character of radioactivity; extraction from pitchblende of polonium and radium; isolation of pure radium chloride; more. 96pp. 5 3/8 x 8 1/2. 0-486-42550-9

CHEMICAL MAGIC, Leonard A. Ford. Classic guide provides intriguing entertainment while elucidating sound scientific principles, with more than 100 unusual stunts: cold fire, dust explosions, a nylon rope trick, a disappearing beaker, much more. 128pp. 5 3/8 x 8 1/2. 0-486-67628-5

ALCHEMY, E. J. Holmyard. Classic study by noted authority covers 2,000 years of alchemical history: religious, mystical overtones; apparatus; signs, symbols, and secret terms; advent of scientific method, much more. Illustrated. 320pp. 5 3/8 x 8 1/2. 0-486-26298-7

CHEMICAL KINETICS AND REACTION DYNAMICS, Paul L. Houston. This text teaches the principles underlying modern chemical kinetics in a clear, direct fashion, using several examples to enhance basic understanding. Solutions to selected problems. 2001 edition. 352pp. 8 3/8 x 11. 0-486-45334-0

PROBLEMS AND SOLUTIONS IN QUANTUM CHEMISTRY AND PHYSICS, Charles S. Johnson and Lee G. Pedersen. Unusually varied problems, with detailed solutions, cover of quantum mechanics, wave mechanics, angular momentum, molecular spectroscopy, scattering theory, more. 280 problems, plus 139 supplementary exercises. 430pp. 6 1/2 x 9 1/4. 0-486-65236-X

ELEMENTS OF CHEMISTRY, Antoine Lavoisier. Monumental classic by the founder of modern chemistry features first explicit statement of law of conservation of matter in chemical change, and more. Facsimile reprint of original (1790) Kerr translation. 539pp. 5 3/8 x 8 1/2. 0-486-64624-6

MAGNETISM AND TRANSITION METAL COMPLEXES, F. E. Mabbs and D. J. Machin. A detailed view of the calculation methods involved in the magnetic properties of transition metal complexes, this volume offers sufficient background for original work in the field. 1973 edition. 240pp. 5 3/8 x 8 1/2. 0-486-46284-6

GENERAL CHEMISTRY, Linus Pauling. Revised third edition of classic first-year text by Nobel laureate. Atomic and molecular structure, quantum mechanics, statistical mechanics, thermodynamics correlated with descriptive chemistry. Problems. 992pp. 5 3/8 x 8 1/2. 0-486-65622-5

ELECTROLYTE SOLUTIONS: Second Revised Edition, R. A. Robinson and R. H. Stokes. Classic text deals primarily with measurement, interpretation of conductance, chemical potential, and diffusion in electrolyte solutions. Detailed theoretical interpretations, plus extensive tables of thermodynamic and transport properties. 1970 edition. 590pp. 5 3/8 x 8 1/2. 0-486-42225-9

# Physics

OPTICAL RESONANCE AND TWO-LEVEL ATOMS, L. Allen and J. H. Eberly. Clear, comprehensive introduction to basic principles behind all quantum optical resonance phenomena. 53 illustrations. Preface. Index. 256pp. 5⅜ x 8½. 0-486-65533-4

QUANTUM THEORY, David Bohm. This advanced undergraduate-level text presents the quantum theory in terms of qualitative and imaginative concepts, followed by specific applications worked out in mathematical detail. Preface. Index. 655pp. 5⅜ x 8½. 0-486-65969-0

ATOMIC PHYSICS (8th EDITION), Max Born. Nobel laureate's lucid treatment of kinetic theory of gases, elementary particles, nuclear atom, wave-corpuscles, atomic structure and spectral lines, much more. Over 40 appendices, bibliography. 495pp. 5⅜ x 8½. 0-486-65984-4

A SOPHISTICATE'S PRIMER OF RELATIVITY, P. W. Bridgman. Geared toward readers already acquainted with special relativity, this book transcends the view of theory as a working tool to answer natural questions: What is a frame of reference? What is a "law of nature"? What is the role of the "observer"? Extensive treatment, written in terms accessible to those without a scientific background. 1983 ed. xlviii+172pp. 5⅜ x 8½. 0-486-42549-5

AN INTRODUCTION TO HAMILTONIAN OPTICS, H. A. Buchdahl. Detailed account of the Hamiltonian treatment of aberration theory in geometrical optics. Many classes of optical systems defined in terms of the symmetries they possess. Problems with detailed solutions. 1970 edition. xv + 360pp. 5⅜ x 8½. 0-486-67597-1

PRIMER OF QUANTUM MECHANICS, Marvin Chester. Introductory text examines the classical quantum bead on a track: its state and representations; operator eigenvalues; harmonic oscillator and bound bead in a symmetric force field; and bead in a spherical shell. Other topics include spin, matrices, and the structure of quantum mechanics; the simplest atom; indistinguishable particles; and stationary-state perturbation theory. 1992 ed. xiv+314pp. 6⅛ x 9¼. 0-486-42878-8

LECTURES ON QUANTUM MECHANICS, Paul A. M. Dirac. Four concise, brilliant lectures on mathematical methods in quantum mechanics from Nobel Prize-winning quantum pioneer build on idea of visualizing quantum theory through the use of classical mechanics. 96pp. 5⅜ x 8½. 0-486-41713-1

THIRTY YEARS THAT SHOOK PHYSICS: THE STORY OF QUANTUM THEORY, George Gamow. Lucid, accessible introduction to influential theory of energy and matter. Careful explanations of Dirac's anti-particles, Bohr's model of the atom, much more. 12 plates. Numerous drawings. 240pp. 5⅜ x 8½. 0-486-24895-X

ELECTRONIC STRUCTURE AND THE PROPERTIES OF SOLIDS: THE PHYSICS OF THE CHEMICAL BOND, Walter A. Harrison. Innovative text offers basic understanding of the electronic structure of covalent and ionic solids, simple metals, transition metals and their compounds. Problems. 1980 edition. 582pp. 6⅛ x 9¼. 0-486-66021-4

A TREATISE ON ELECTRICITY AND MAGNETISM, James Clerk Maxwell. Important foundation work of modern physics. Brings to final form Maxwell's theory of electromagnetism and rigorously derives his general equations of field theory. 1,084pp. 5⅜ x 8½. Two-vol. set. Vol. I: 0-486-60636-8 Vol. II: 0-486-60637-6

QUANTUM MECHANICS: PRINCIPLES AND FORMALISM, Roy McWeeny. Graduate student-oriented volume develops subject as fundamental discipline, opening with review of origins of Schrödinger's equations and vector spaces. Focusing on main principles of quantum mechanics and their immediate consequences, it concludes with final generalizations covering alternative "languages" or representations. 1972 ed. 15 figures. xi+155pp. 5⅜ x 8½. 0-486-42829-X

INTRODUCTION TO QUANTUM MECHANICS With Applications to Chemistry, Linus Pauling & E. Bright Wilson, Jr. Classic undergraduate text by Nobel Prize winner applies quantum mechanics to chemical and physical problems. Numerous tables and figures enhance the text. Chapter bibliographies. Appendices. Index. 468pp. 5⅜ x 8½. 0-486-64871-0

METHODS OF THERMODYNAMICS, Howard Reiss. Outstanding text focuses on physical technique of thermodynamics, typical problem areas of understanding, and significance and use of thermodynamic potential. 1965 edition. 238pp. 5⅜ x 8½. 0-486-69445-3

THE ELECTROMAGNETIC FIELD, Albert Shadowitz. Comprehensive undergraduate text covers basics of electric and magnetic fields, builds up to electromagnetic theory. Also related topics, including relativity. Over 900 problems. 768pp. 5⅝ x 8¼. 0-486-65660-8

GREAT EXPERIMENTS IN PHYSICS: FIRSTHAND ACCOUNTS FROM GALILEO TO EINSTEIN, Morris H. Shamos (ed.). 25 crucial discoveries: Newton's laws of motion, Chadwick's study of the neutron, Hertz on electromagnetic waves, more. Original accounts clearly annotated. 370pp. 5⅜ x 8½. 0-486-25346-5

EINSTEIN'S LEGACY, Julian Schwinger. A Nobel Laureate relates fascinating story of Einstein and development of relativity theory in well-illustrated, nontechnical volume. Subjects include meaning of time, paradoxes of space travel, gravity and its effect on light, non-Euclidean geometry and curving of space-time, impact of radio astronomy and space-age discoveries, and more. 189 b/w illustrations. xiv+250pp. 8⅜ x 9¼. 0-486-41974-6

STATISTICAL PHYSICS, Gregory H. Wannier. Classic text combines thermodynamics, statistical mechanics and kinetic theory in one unified presentation of thermal physics. Problems with solutions. Bibliography. 532pp. 5⅜ x 8½. 0-486-65401-X

Paperbound unless otherwise indicated. Available at your book dealer, online at **www.doverpublications.com**, or by writing to Dept. GI, Dover Publications, Inc., 31 East 2nd Street, Mineola, NY 11501. For current price information or for free catalogues (please indicate field of interest), write to Dover Publications or log on to **www.doverpublications.com** and see every Dover book in print. Dover publishes more than 500 books each year on science, elementary and advanced mathematics, biology, music, art, literary history, social sciences, and other areas.